Amino Acids, Peptides and Proteins in Organic Chemistry

Edited by
Andrew B. Hughes

Further Reading

Fessner, W.-D., Anthonsen, T.

Modern Biocatalysis

Stereoselective and Environmentally Friendly Reactions

2009

ISBN: 978-3-527-32071-4

Sewald, N., Jakubke, H.-D.

Peptides: Chemistry and Biology

2009

ISBN: 978-3-527-31867-4

Lutz, S., Bornscheuer, U. T. (eds.)

Protein Engineering Handbook

2 Volume Set

2009

ISBN: 978-3-527-31850-6

Aehle, W. (ed.)

Enzymes in Industry

Production and Applications

2007

ISBN: 978-3-527-31689-2

Wiley-VCH (ed.)

Ullmann's Biotechnology and Biochemical Engineering

2 Volume Set

2007

ISBN: 978-3-527-31603-8

Budisa, N.

Engineering the Genetic Code

Expanding the Amino Acid Repertoire for the Design of Novel Proteins

2006

ISBN: 978-3-527-31243-6

Demchenko, A. V. (ed.)

Handbook of Chemical Glycosylation

Advances in Stereoselectivity and Therapeutic Relevance

2008

ISBN: 978-3-527-31780-6

Lindhorst, T. K.

Essentials of Carbohydrate Chemistry and Biochemistry

2007

ISBN: 978-3-527-31528-4

Amino Acids, Peptides and Proteins in Organic Chemistry

Volume 1 - Origins and Synthesis of Amino Acids

Edited by
Andrew B. Hughes

WILEY-VCH Verlag GmbH & Co. KGaA

The Editor

Dr. Andrew B. Hughes
La Trobe University
Department of Chemistry
Victoria 3086
Australia

Library of Congress Card No.: applied for

British Library Cataloguing-in-Publication Data
A catalogue record for this book is available from the British Library.

Bibliographic information published by the Deutsche Nationalbibliothek
The Deutsche Nationalbibliothek lists this publication in the Deutsche Nationalbibliografie; detailed bibliographic data are available in the Internet at http://dnb.d-nb.de.

Composition Thomson Digital, Noida, India
Printing Strauss GmbH, Mörlenbach
Bookbinding Litges & Dopf GmbH, Heppenheim
Cover Design Schulz Grafik Design, Fußgönheim

Printed in the Federal Republic of Germany
Printed on acid-free paper

ISBN: 978-3-527-32096-7

Contents

Amino Acids, Peptides and Proteins in Organic Chemistry. Vol.1 – Origins and Synthesis of Amino Acids.
Edited by Andrew B. Hughes

ISBN: 978-3-527-32096-7

List of Contributors

David J. Ager
DSM Pharma Chemicals
PMB 150
9650 Strickland Road, Suite 103
Raleigh, NC 27615
USA

Luigi Aurelio
Monash University
Department of Medicinal Chemistry
Victorian College of Pharmacy
381 Royal Parade
Parkville, Victoria 3052
Australia

Yamir Bandala
Centro de Investigación y de Estudios
Avanzados del Instituto Politécnico
Nacional
Departamento de Química
Apartado Postal 14-740
07000 México DF
México

A. Richard Chamberlin
University of California, Irvine
Departments of Chemistry
Pharmaceutical Sciences, and
Pharmacology
Irvine, CA 92697
USA

Roberto Fernández de la Pradilla
CSIC
Instituto de Química Orgánica
Juan de la Cierva 3
28006 Madrid
Spain

Enikő Forró
University of Szeged
Institute of Pharmaceutical Chemistry
Eötvös u. 6
6720 Szeged
Hungary

Stephen Freeland
University of Maryland
Baltimore County
Department of Biological Sciences
1000 Hilltop Circle,
Baltimore, MD 21250
USA

Ferenc Fülöp
University of Szeged
Institute of Pharmaceutical Chemistry
Eötvös u. 6
6720 Szeged
Hungary

Amino Acids, Peptides and Proteins in Organic Chemistry. Vol.1 – Origins and Synthesis of Amino Acids.
Edited by Andrew B. Hughes

ISBN: 978-3-527-32096-7

Antonio Guarna
Università degli Studi di Firenze
Polo Scientifico e Tecnologico
Dipartimento di Chimica Organica
"Ugo Schiff"
Via della Lastruccia 13
50019 Sesto Fiorentino
Firenze
Italy

Stephen A. Habay
University of California, Irvine
Department of Chemistry
Irvine, CA 92697
USA

Jane R. Hanrahan
University of Sydney
Faculty of Pharmacy
Sydney, NSW 2006
Australia

Graham A.R. Johnston
University of Sydney
Adrien Albert Laboratory of Medicinal Chemistry
Department of Pharmacology
Faculty of Medicine
Sydney, NSW 2006
Australia

Andrew B. Hughes
La Trobe University
Department of Chemistry
Victoria 3086
Australia

Eusebio Juaristi
Centro de Investigación y de Estudios Avanzados del Instituto Politécnico Nacional
Departamento de Química
Apartado Postal 14-740
07000 México DF
México

Bernard Kaptein
DSM Pharmaceutical Products
Innovative Synthesis & Catalysis
PO Box 18
6160 MD Geleen
The Netherlands

Steven M. Kennedy
University of California, Irvine
Department of Chemistry
Irvine, CA 92697
USA

Loránd Kiss
University of Szeged
Institute of Pharmaceutical Chemistry
Eötvös u. 6
6720 Szeged
Hungary

Z. Martins
Imperial College London
Department of Earth Science and Engineering
Exhibition Road
London SW7 2AZ
UK

Gloria Menchi
Università degli Studi di Firenze
Polo Scientifico e Tecnologico
Dipartimento di Chimica Organica
"Ugo Schiff"
Via della Lastruccia 13
50019 Sesto Fiorentino
Firenze
Italy

Steve S. Park
University of California, Irvine
Department of Chemistry
Irvine, CA 92697
USA

Caroline M. Reid
University of Glasgow, WestChem
Department of Chemistry
The Joseph Black Building
University Avenue
Glasgow G12 8QQ
UK

M.A. Sephton
Imperial College London
Department of Earth Science and
Engineering
Exhibition Road
London SW7 2AZ
UK

Hans E. Schoemaker
DSM Pharmaceutical Products
Innovative Synthesis & Catalysis
PO Box 18
6160 MD Geleen
The Netherlands

Theo Sonke
DSM Pharmaceutical Products
Innovative Synthesis & Catalysis
PO Box 18
6160 MD Geleen
The Netherlands

Peter Spiteller
Technische Universität München
Institut für Organische Chemie und
Biochemie II
Lichtenbergstraße 4
85747 Garching bei München
Germany

Madeleine Strickland
University of Bristol
School of Chemistry
Cantock's Close
Bristol BS8 1TS
UK

Andrew Sutherland
University of Glasgow, WestChem
Department of Chemistry
The Joseph Black Building
University Avenue
Glasgow G12 8QQ
UK

Andrea Trabocchi
Università degli Studi di Firenze
Polo Scientifico e Tecnologico
Dipartimento di Chimica Organica
"Ugo Schiff"
Via della Lastruccia 13
50019 Sesto Fiorentino
Firenze
Italy

Alma Viso
CSIC
Instituto de Química Orgánica
Juan de la Cierva 3
28006 Madrid
Spain

Christine L. Willis
University of Bristol
School of Chemistry
Cantock's Close
Bristol BS8 1TS
UK

Part One
Origins of Amino Acids

Amino Acids, Peptides and Proteins in Organic Chemistry. Vol.1 – Origins and Synthesis of Amino Acids.
Edited by Andrew B. Hughes

ISBN: 978-3-527-32096-7

1
Extraterrestrial Amino Acids

Z. Martins and M.A. Sephton

1.1
Introduction

The space between the stars, the interstellar medium (ISM), is composed of gas-phase species (mainly hydrogen and helium atoms) and submicron dust grains (silicates, carbon-rich particles, and ices). The ISM has many different environments based on its different temperatures (T_k), hydrogen density (n_H), and ionization state of hydrogen (for reviews, see [1–3]); it includes the diffuse ISM ($T_k \sim 100$ K, $n_H \sim 10$–$300\,cm^{-3}$), molecular clouds ($T_k \sim 10$–100 K, $n_H \sim 10^3$–$10^4\,cm^{-3}$; e.g., [4]) [molecular clouds are not uniform but instead have substructures [5] – they contain high-density clumps (also called dense cores; $n_H \sim 10^3$–$10^5\,cm^{-3}$), which have higher densities than the surrounding molecular cloud; even higher densities are found in small regions, commonly known as "hot molecular cores", which will be the future birth place of stars], and hot molecular clouds ($T_k \sim 100$–300 K, $n_H \sim 10^6$–$10^8\,cm^{-3}$; e.g., [6]). Observations at radio, millimeter, submillimeter, and infrared frequencies have led to the discovery of numerous molecules (currently more than 151) in the interstellar space, some of which are organic in nature (Table 1.1; an up-to-date list can be found at www.astrochemistry.net). The collapse of a dense cloud of interstellar gas and dust leads to the formation of a so-called solar nebula. Atoms and molecules formed in the ISM, together with dust grains are incorporated in this solar nebula, serving as building blocks from which future planets, comets, asteroids, and other celestial bodies may originate. Solar system bodies, such as comets (e.g., [7] and references therein; [8]), meteorites (e.g., [9, 10]), and interplanetary dust particles (IDPs [11, 12]) are known to contain extraterrestrial molecules, which might have a heritage from interstellar, nebular, and/or parent body processing. Delivery of these molecules to the early Earth and Mars during the late heavy bombardment (4.5–3.8 billion years ago) may have been important for the origin of life [13, 14]. Among the molecules delivered to the early Earth, amino acids may have had a crucial role as they are the building blocks of proteins and enzymes, therefore having implications for the origin of life. In this chapter we describe different extraterrestrial environments where amino acids may be present and

Amino Acids, Peptides and Proteins in Organic Chemistry. Vol.1 – Origins and Synthesis of Amino Acids.
Edited by Andrew B. Hughes

ISBN: 978-3-527-32096-7

Table 1.1 Summary of molecules identified in the interstellar and circumstellar medium[a].

Number of atoms											
2	**3**	**4**	**5**	**6**	**7**	**8**	**9**	**10**	**11**	**12**	**13**
H_2	H_3^+	CH_3	CH_4	C_2H_4	CH_3NH_2	CH_3CH_3	C_2H_5OH	CH_3CH_2CHO	CH_3C_6H	C_6H_6	$HC_{11}N$
CH	CH_2	NH_3	NH_4^+	CH_3OH	CH_3CCH	C_3H_4O	CH_3OCH_3	CH_3COCH_3	HC_9N		
CH^+	NH_2	H_3O^+	CH_2NH	CH_3CN	c-C_2H_4O	CH_2OHCHO	CH_3CH_2CN	$HOCH_2CH_2OH$			
NH	H_2O	C_2H_2	H_3CO^+	CH_3NC	CH_3CHO	CH_3COOH	CH_3CONH_2	NH_2CH_2COOH			
OH	H_2O^+	H_2CN	SiH_4	NH_2CHO	CH_2CHCN	$HCOOCH_3$	CH_3C_4H				
OH^+	C_2H	$HCNH^+$	c-C_3H_2	CH_3SH	C_6H	CH_2CCHCN	C_8H				
HF	HCN	H_2CO	H_2CCC	H_2CCCC	C_6H^-	CH_3C_3N	C_8H^-				
C_2	HNC	c-C_3H	CH_2CN	HCCCCH	HC_5N	C_6H_2	HC_7N				
CN	HCO	l-C_3H	H_2CCO	$H_2C_3N^+$		HCCCCCCH					
CN^+	HCO^+	HCCN	NH_2CN	c-H_2C_3O		C_7H					
CO	HN_2^+	HNCO	HCOOH	HC_2CHO							
CO^+	HOC^+	$HNCO^-$	C_4H	C_5H							
N_2^+	HNO	$HOCO^+$	C_4H^-	HC_4N							
SiH	H_2S	H_2CS	HC_3N	C_5N							
NO	H_2S^+	C_3N	HCCNC	SiC_4							
CF^+	C_3	C_3O	HNCCC								
HS	C_2O	HNCS	C_5								
HS^+	CO_2	SiC_3									
HCl	CO_2^+	C_3S									
SiC	N_2O										
SiN	HCS^+										
CP	NaCN										
CS	MgCN										

SiO	MgNC
PN	c-SiC_2
AlF	AlNC
NS	SiCN
SO	SiNC
SO^+	C_2S
NaCl	OCS
SiS	SO_2
AlCl	
S_2	
FeO	
KCl	

[a]Observations suggest the presence of polycyclic aromatic hydrocarbons in the interstellar gas (e.g., [244]).

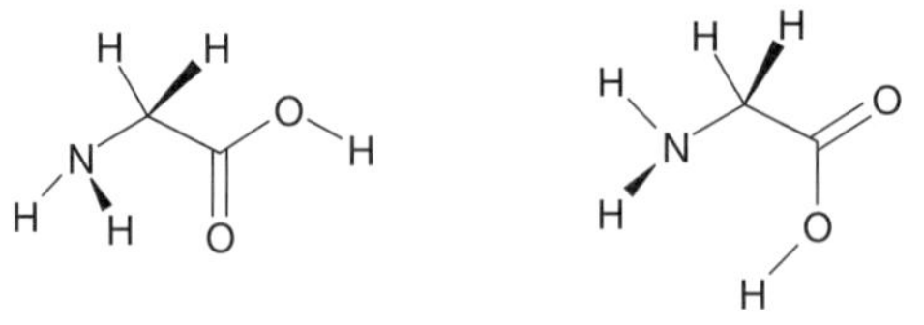

I II

Figure 1.1 Molecular structures of conformers I and II of glycine in the gas phase (adapted from [21, 29]).

detected, their proposed formation mechanisms, and possible contribution to the origin of life on Earth.

1.2 ISM

The search for amino acids, in particular for the simplest amino acid glycine (NH_2CH_2COOH), in the ISM has been carried on for almost 30 years [15–28]. While in theory glycine may have several conformers in the gas phase [29], astronomical searches have only focused on two (Figure 1.1, adapted from [21, 29]). Conformer I is the lowest energy form, while conformer II has a higher energy, larger dipole moment, and therefore stronger spectral lines [30]. Only upper limits of both conformers were found in the ISM until Kuan *et al.* [25] reported the detection of glycine in the hot molecular cores Sgr B2(N-LMH), Orion-KL, and W51 e1/e2. This detection has been disputed by Snyder *et al.* [26], who concluded that the spectral lines necessary for the identification of interstellar glycine have not yet been found. In addition, they argued that some of the spectral lines identified as glycine by Kuan *et al.* [25] could be assigned to other molecular species. Further negative results include the astronomical searches of Cunningham *et al.* and Jones *et al.* [27, 28], who claim that their observations rule out the detection of both conformers I and II of glycine in the hot molecular core Sgr B2(N-LMH). They conclude that it is unlikely that Kuan *et al.* [25] detected glycine in either Sgr B2(N-LMH) or Orion-KL. No other amino acid has been detected in the ISM. Despite these results, amino acids were proposed to be formed in the ISM by energetic processing on dust grain surfaces, which will then be evaporated, releasing the amino acids into the gas phase (solid-phase reactions), or synthesized in the gas phase via ion–molecule reactions (gas-phase reactions). These two processes will now be described in more detail (for a review, see, e.g., [31]).

1.2.1 Formation of Amino Acids in the ISM via Solid-Phase Reactions

Several mechanisms have been proposed for amino acid formation in the ISM. These include solid-phase reactions on interstellar ice grains by energetic processing,

Table 1.2 Abundances of interstellar ices (normalized to H_2O) in the high-mass protostellar objects W33A and NGC758:IRS9, in the low-mass protostellar object Elias 29, and the field star Elias 16.

Ice specie	W33A high-mass protostar	NGC758:IRS9 high-mass protostar	Elias 29 low-mass protostar	Elias 16 field star
H_2O	100	100	100	100
CO	9	16	5.6	25
CO_2	14	20	22	15
CH_4	2	2	<1.6	—
CH_3OH	22	5	<4	<3.4
H_2CO	1.7–7	5	—	—
NH_3	3–15	13	<9.2	<6
OCS	0.3	0.05	<0.08	—
C_2H_6	<0.4	<0.4	—	—
HCOOH	0.4–2	3	—	—
O_2	<20	—	—	—
OCN^-	3	1	<0.24	<0.4

Adapted from [31, 32, 52].

which may occur in cold molecular clouds (e.g., [32] and references therein). In these regions of the ISM, in which temperatures are very low (<50 K), atoms and molecules in the gas phase will be accreted onto the surface of dust grains leading to the formation of ice mantles [33, 34]. Diffusion of accreted atoms leads to surface reactions, forming additional species in the ice mantles. These interstellar ices are mainly composed of H_2O, CO, CO_2, CH_4, CH_3OH, and NH_3, with traces of other species (Table 1.2; [32, 34–38]). Once these ice grains are formed, energetic processes [e.g., cosmic rays and ultraviolet (UV) irradiation] may change the ice mantle composition.

A range of interstellar ice analogs have been irradiated at low temperatures (~10 K) to produce a variety of amino acids. Holtom *et al.* [39] used galactic cosmic ray particles to irradiate an ice mixture containing carbon dioxide (CO_2) and methylamine (CH_3NH_2), which produced hydroxycarbonyl (HOCO) and aminomethyl (CH_2NH_2) radicals. The recombination of these radicals would then form glycine and its isomer ($CH_3NHCOOH$). Briggs *et al.* [40] UV-irradiated a mixture of CO : H_2O : NH_3 (5 : 5 : 1) at 12 K for 24 h. This resulted in the formation of an organic residue, which included among other organic compounds, 0.27% of glycine. Ice mixtures containing H_2O : CH_3OH : NH_3 : CO : CO_2 (2 : 1 : 1 : 1 : 1 molar composition; [41]), and H_2O with NH_3, CH_3OH, and HCN (0.5–5% NH_3, 5–10% CH_3OH, and 0.5–5% HCN, relative to H_2O [42]) were UV-irradiated in high vacuum at below 15 K. While Bernstein *et al.* [42] obtained glycine and racemic mixtures (D/L ~ 1) of alanine and serine, a large variety of amino acids were found by Munõz Caro *et al.* [41]. These results were confirmed by Nuevo *et al.* [43, 44].

The exact formation pathway of amino acids in interstellar ices is unknown, but the Strecker synthesis ([42]; for more details, see Section 1.4), reactions on the surface of polycyclic aromatic hydrocarbon flakes [45], and radical–radical reactions [46, 47] have been proposed. As radical–radical reactions can occur with almost no activation energy [47], theoretical modeling suggests that glycine may be formed in interstellar ice mantles via the following radical–radical reaction sequence:

$$CO + OH \rightarrow COOH \quad (1.1)$$

$$CH_3 + COOH \rightarrow CH_3COOH \quad (1.2)$$

$$NH_2 + CH_3COOH \rightarrow NH_2CH_2COOH + H \quad (1.3)$$

Quantum chemical calculations indicate that amino acids may also be formed by recombination of the radicals COOH and CH_2NH_2, which are produced by dehydrogenation of H_2O and CH_3OH, and hydrogenation of HCN, respectively [47]. However, all radical–radical reactions described above require the radicals to diffuse into and/or onto the ice mantle which, as noted by Woon [47], may only occur at temperatures of 100 K or higher, much higher than the temperature in molecular clouds. Furthermore, Elsila *et al.* [48] used isotopic labeling techniques to test whether Strecker synthesis or radical–radical reactions were responsible for amino acid formation in interstellar ice analogs. Their results show that amino acid formation occurs via multiple routes, not matching the previously proposed Strecker synthesis or radical–radical mechanisms. Ultimately, the need for high UV flux to produce amino acids in ice mantles contrasts with the low expected efficiency of UV photolysis in dark molecular clouds [4]. This, together with the fact that amino acids have low resistance to UV photolysis [49], raises concerns about the amino acid formation in interstellar ices by UV photolysis.

1.2.2
Formation of Amino Acids in the ISM via Gas-Phase Reactions

Potential mechanisms alternative to solid-phase reactions include gas-phase formation of interstellar amino acids via ion–molecule reactions. Amino acids, once formed, could potentially survive in the gas phase in hot molecular cores, because the UV flux is sufficiently low (i.e., 300 mag of visual extinction) [31]. Alcohols, aminoalcohols, and formic acid evaporated from interstellar ice grains (Table 1.2; [50–53]) may produce amino acids in hot molecular cores through exothermic alkyl and aminoalkyl cation transfer reactions [52]. Aminoalkyl cation transfer from aminomethanol and aminoethanol to HCOOH can produce protonated glycine and β-alanine, respectively via the following reactions [52]:

$$NH_2CH_2OH_2^+ + HCOOH \rightarrow NH_2CH_2COOH_2^+ + H_2O \quad (1.4)$$

$$NH_2(CH_2)_2OH_2^+ + HCOOH \rightarrow NH_2(CH_2)_2COOH_2^+ + H_2O \quad (1.5)$$

An electron recombination will then produce the neutral amino acids. Further alkylation may produce a large variety of amino acids through elimination of a water molecule [31, 52].

Alternatively, Blagojevic *et al.* [54] have experimentally proven the gas-phase formation of protonated glycine and β-alanine, by reacting protonated hydroxylamine with acetic and propanoic acid, respectively:

$$NH_2OH^+ + CH_3COOH \rightarrow NH_2CH_2COOH^+ + H_2O \quad (1.6)$$

$$NH_2OH^+ + CH_3CH_2COOH \rightarrow NH_2(CH_2)_2COOH^+ + H_2O \quad (1.7)$$

Neutral amino acids could then be produced by dissociative recombination reactions [54].

Independently of the mechanism of synthesis (solid-phase or gas-phase reactions), once formed, amino acids would need to be resistant and survive exposure to cosmic rays and UV radiation in the ISM. The stability of amino acids in interstellar gas and on interstellar grains has been simulated [49]. Different amino acids [i.e., glycine, L-alanine, α-aminoisobutyric acid (α-AIB), and β-alanine] were irradiated in frozen argon, nitrogen, or water matrices to test their stability against space radiation. It was shown that these amino acids have very low stability against UV photolysis. Therefore, amino acids will not survive in environments subject to high UV flux, such as the diffuse ISM. This does not eliminate formation of amino acids in the ISM, but instead requires that amino acids are incorporated into UV-shielded environments such as hot molecular cores, in the interior of comets, asteroids, meteorites, and IDPs.

1.3 Comets

Comets are agglomerates of ice, organic compounds, and silicate dust, and are some of the most primitive bodies in the solar system (for reviews about comets, see, e.g., [55–57]). Comets were first proposed to have delivered prebiotic molecules to the early Earth by Chamberlin and Chamberlin [58]. Since then, space telescopes (such as the Hubble Space Telescope, Infrared Space Observatory, and Spitzer Space Telescope; e.g., [59–66]), ground-based observations (e.g., [67–69]), cometary fly-bys (Deep Space 1 mission, and Vega1, Vega2, Suisei, Sakigake, ICE, and Giotto spacecraft missions; e.g., [70–76]), impacts (Deep Impact mission, which impacted into the 9P/Tempel comet's nucleus; e.g., [77–79]), collection of dust from the coma of a comet (Stardust mission to comet Wild-2; e.g., [80–84]), and rendezvous missions (such as the Rosetta mission, which will encounter the comet 67P/Churyumov–Gerasimenko in 2014) advanced our knowledge about these dirty snowballs.

Several organic compounds have been detected in comets (Table 1.3; for reviews, see, e.g., [7, 8, 85]). Fly-by missions have suggested the presence of amino acids on comet Halley [71], but their presence could not be confirmed due to the limited resolution of the mass spectrometers on board the Giotto and Vega spacecrafts. In addition, only an upper limit of less than 0.15 of glycine relative to water has been determined in the coma of Hale–Bopp using radio telescopes (Table 1.3; [69]). Although several amino acid precursors (see Section 1.4), including ammonia,

Table 1.3 Molecular abundances of ices for comets Halley, Hyakutake, and Hale–Bopp.

Molecule	Halley	Hyakutake	Hale–Bopp
H_2O	100	100	100
H_2O_2		<0.04	<0.03
CO	15	6–30[a]	20[a]
CO_2	3	<7[b]	6[b]
CH_4	0.2–1.2	0.7	0.6
C_2H_2	~0.3	0.5	0.1
C_2H_6	~0.4	0.4	0.3
CH_3C_2H			<0.045
CH_3OH	1.3–1.7	2	2.4
H_2CO	0–5	0.2–1[a]	1.1[a]
HCOOH			0.08
CH3COOH			<0.06
$HCOOCH_3$			0.08
CH_3CHO			0.025
H_2CCO			<0.032
C_2H_5OH			<0.05
CH_3OCH_3			<0.45
CH_2OHCHO			<0.04
NH_3	0.1–2	0.5	0.7
HCN	~0.2	0.1	0.25
HNCO		0.07	0.10
HNC		0.01[a]	0.04[a]
CH_3CN		0.01	0.02
HC_3N			0.02
NH_2CHO			0.01
NH_2CH_2COOH			<0.15
C_2H_5CN			<0.01
CH_2NH			<0.032
HC_5N			<0.003
N_2O			<0.23
NH_2OH			<0.25
H_2S	0.04	0.8	1.5
OCS		0.1[a]	0.3[a]
SO			0.2–0.8[a]
CS_2		0.1c	0.2[c]
SO_2			0.23
H_2CS			0.02
S_2		0.005	
NaCl			<0.0008
NaOH			<0.0003

Abundances are normalized to H_2O and were measured at around 1 AU from the Sun. Adapted from [8, 32, 69, 86, 245].
[a]Extended sources (the abundance is model dependent).
[b]Measured at 2.9 AU from the Sun.
[c]Abundance deduced from CS.

HCN, formaldehyde, and cyanoacetylene, have been observed in the Hyakutake and Hale–Bopp comets [7], only a very limited number of carbonyl compounds necessary for the synthesis of amino acids were detected in comets [7, 32, 86]. The ultimate proof for the presence of amino acids in comets is a sample return mission such as Stardust, which collected dust from the coma of the Wild-2 comet using a lightweight material called aerogel [80]. Analyses of comet-exposed aerogel samples show a relative molar abundance of glycine that slightly exceeds that found in control samples, suggesting a cometary origin for this amino acid [83]. Compound-specific isotopic analyses of glycine present in comet-exposed aerogel samples have not yet been performed and therefore it has not been possible to ultimately constrain its origin. Other amino acids present in the comet-exposed aerogel samples included ε-amino-*n*-caproic acid, β-alanine, and γ-amino-*n*-butyric acid (γ-ABA). The similarity in the distribution of these amino acids in the comet-exposed sample, the witness tile (which witnessed all the terrestrial and space environments as the comet-exposed samples, but did not "see" comet Wild-2), and the Stardust impact location soil indicates a terrestrial origin (contamination) for these amino acids [83].

1.4 Meteorites

Meteorites are extraterrestrial objects that survived the passage through the Earth's atmosphere and the impact with the Earth's surface. Excepting the lunar and Martian meteorites [87–91], all meteorites are thought to have originated from extraterrestrial bodies located in the asteroid belt (e.g., [92–98]). Although unproven, it was also suggested that they could have originated from comets ([99–102] and references therein). Meteorites can be divided into iron, stony-iron, and stony meteorites. They can be further divided into classes according to their chemical, mineralogical, and isotopic composition (for reviews, see, e.g., [103–105]). A very primitive class of stony meteorites, named carbonaceous chondrites, has not been melted since their formation early in the history of the solar system, around 4.6 billion years ago (for reviews, see, e.g., [9, 10]). Within the class of carbonaceous chondrites, there are the CI-, CM-, CK-, CO-, CR-, CV-, CH-, and CB-type chondrites. Chondrites are also classified and grouped into petrographic types. This refers to the intensity of thermal metamorphism or aqueous alteration that has occurred on the meteorite parent body, ranging from types 1 to 6. A petrologic type from 3 to 1 indicates increasing aqueous alteration. A petrologic type from 3 to 6 indicates increasing thermal metamorphism.

Carbonaceous chondrites have a relatively high carbon content and can contain up to 3 wt% of organic carbon. More than 70% of it is composed of a solvent-insoluble macromolecular material, while less than 30% is a mixture of solvent-soluble organic compounds. Carbonaceous chondrites, as revealed by extensive analyses of the Murchison meteorite, have a rich organic inventory that includes organic compounds important in terrestrial biochemistry (Table 1.4). These include amino acids (e.g., [106–108]), carboxylic acids (e.g., [109, 110]), purines and pyrimidines

Table 1.4 Abundances (in ppm) of the soluble organic matter found in the Murchison meteorite.

Compounds	Concentration
Carboxylic acids (monocarboxylic)	332
Sulfonic acids	67
Amino acids	60
Dicarboximides	>50
Dicarboxylic acids	>30
Polyols	24
Ketones	17
Hydrocarbons (aromatic)	15–28
Hydroxycarboxylic acids	15
Hydrocarbons (aliphatic)	12–35
Alcohols	11
Aldehydes	11
Amines	8
Pyridine carboxylic acid	>7
Phosphonic acid	1.5
Purines	1.2
Diamino acids	0.4
Benzothiophenes	0.3
Pyrimidines	0.06
Basic *N*-heterocycles	0.05–0.5

Adapted from [9, 10, 161, 246].

(e.g., [111–113]), polyols [114], diamino acids [115], dicarboxylic acids (e.g., [116–119]), sulfonic acids [120], hydrocarbons (e.g., [121, 122]), alcohols (e.g., [123]), amines and amides (e.g., [124, 125]), and aldehydes and ketones [123].

The first evidence of extraterrestrial amino acids in a meteorite was obtained by Kvenvolden *et al.* [121], after analyzing a sample of the Murchison meteorite which had recently fallen in Australia in 1969. These authors detected several amino acids in this meteorite, including the nonprotein amino acids α-AIB and isovaline, which suggested an abiotic and extraterrestrial origin for these compounds. Since then, Murchison has been the most analyzed carbonaceous chondrite for amino acids, with more than 80 different amino acids identified, the majority of which are rare (or nonexistent) in the terrestrial biosphere (for reviews, see, e.g., [107, 108]). These amino acids have carbon numbers from C_2 through C_8, and show complete structural diversity (i.e., all isomers of a certain amino acid are present). They can be divided into two structural types, monoamino alkanoic acids and monoamino dialkanoic acids, which can occur as *N*-alkyl derivatives or cyclic amino acids, with structural preference in abundance order $\alpha > \gamma > \beta$. Branched-chain amino acid isomers predominate over straight ones and there is an exponential decline in concentration with increasing carbon number within homologous series.

Amino acids have also been reported in several other carbonaceous chondrites besides Murchison (Table 1.5). Within the CM2 group the total amino acid abundances and distributions are highly variable; Murray [126], Yamato (Y-) 74 662 [127, 128], and Lewis Cliff (LEW) 90 500 [129, 130] show an amino acid distribution and

Table 1.5 Summary of the average blank corrected amino acid concentration (in ppb) in the 6 M HCl acid-hydrolyzed hot-water extracts of carbonaceous chondrites.

	CM2									
Amino acid	**Murchison**	**Murray**	**Y-74 662**	**LEW 90 500**	**Y-791 198**	**Essebi**	**Nogoya**	**Mighei**	**ALHA 77 306**	**ALH 83 100**
D-Aspartic acid	100 ± 15	51 ± 31	160[a]	127 ± 24	226	72 ± 33	163 ± 135	105 ± 62	186[a]	29 ± 4
L-Aspartic acid	342 ± 103	65 ± 16	[a]	151 ± 73	280	134 ± 12	418 ± 106	145 ± 22	[a]	43 ± 6
D-Glutamic acid	537 ± 117	135 ± 50	532[a]	317 ± 55	530	49 ± 21	211 ± 74	111 ± 48	177[a]	21 ± 7
L-Glutamic acid	801 ± 200	261 ± 15	[a]	316 ± 55	588	130 ± 83	1003 ± 110	279 ± 24	[a]	23 ± 6
D-Serine	436 ± 227[a]	92 ± 32[a]	21[a]	<219	84	<62[a]	327 ± 84[a]	68 ± 23[a]	42[a]	<4
L-Serine	[a]	[a]	[a]	<235	126	[a]	—	—	[a]	<5
Glycine	2919 ± 433	2110 ± 144	2553	1448 ± 682	14 130	495 ± 6	1118 ± 729	788 ± 66	548	300 ± 75
β-Alanine	1269 ± 202	1063 ± 268	1247	442 ± 238	1425	1396 ± 157	796 ± 97	897 ± 72	160	338 ± 31
γ-ABA	1331 ± 472	717 ± 192	1753[e,f]	164 ± 21[d]	495	6425 ± 645	1548 ± 187	1136 ± 351	309	308 ± 68[d]
D,L-β-AIB[c]	343 ± 102	147 ± 88	ND	[d]	1836	ND	<124	ND	124	[d]
D-Alanine	720 ± 95	617 ± 79	1158[a]	343 ± 171	2895	128 ± 13	125 ± 92	240 ± 18	160[a]	110 ± 44
L-Alanine	956 ± 171	647 ± 58	[a]	352 ± 161	2913	113 ± 10	332 ± 120	347 ± 57	[a]	134 ± 19
D-β-ABA	708 ± 171[a]	424 ± 18[a]	[e,f]	155 ± 16	<1753	389 ± 180[a]	283 ± 144[a]	487 ± 50[a]	134[a]	33 ± 8
L-β-ABA	[a]	[a]	[e,f]	172 ± 40	1495	[a]	[a]	[a]	[a]	33 ± 12
α-AIB	2901 ± 328	1968 ± 350	381	2706 ± 377	22 630	208 ± 26	458 ± 346	740 ± 219	144	250 ± 40
D-α-ABA	295 ± 111	463 ± 68[a]	691[a]	431 ± 159[a]	907	<11	<151[a]	100 ± 70	93[a]	19 ± 5[a]
L-α-ABA	347 ± 98	[a]	[a]	[a]	907	<18	[a]	101 ± 78	[a]	[a]
D-Isovaline	350 ± 183	2834 ± 780[a]	ND	1306 ± 83[a]	4075[a]	<11	<72[a]	136 ± 49	ND	<10[a]
L-Isovaline	458 ± 209	[a]	ND	[a]	[a]	<18	[a]	159 ± 136	ND	[a]
Total	14 800	12 600	8500	8400	55 600	9500	6800	5800	2100	1600

(Continued)

Table 1.5 (*Continued*)

	CM2		CM1			CI1		CV3	C2
Amino acid	**Y-79 331**	**B-7904**	**ALH 88 045**	**MET 01 070**	**LAP 0227**	**Orgueil**	**Ivuna**	**Allende**	**Tagish Lake**
D-Aspartic acid	10[a]	20[a]	68 ± 18	106 ± 27	10 ± 1	28 ± 16	30 ± 2	<7	11 ± 1
L-Aspartic acid	[a]	[a]	99 ± 39	201 ± 61	29 ± 13	54 ± 18	146 ± 8	100 ± 42	83 ± 8
D-Glutamic acid	ND	ND	33 ± 9	54 ± 14	24 ± 3	15 ± 6	8 ± 1	<7	16 ± 2
L-Glutamic acid	ND	ND	174 ± 50	274 ± 74	64 ± 10	61 ± 31	372 ± 11	329 ± 41	306 ± 48
D-Serine	11[a]	25[a]	<174[a]	147 ± 52[a]	101 ± 13[a]	51 ± 26[a]	217 ± 12[a]	241 ± 149[a]	ND
L-Serine	[a]	[a]	[a]	[a]	[a]	[a]	[a]	[a]	ND
Glycine	7	19	369 ± 117	575 ± 159	55 ± 17	707 ± 80	617 ± 83	457 ± 121	147 ± 17
β-Alanine	4	1	197 ± 21	169 ± 39	16 ± 6	2052 ± 311	1401 ± 146	317 ± 28	64 ± 10
γ-ABA	ND	ND	168 ± 35	259 ± 61	95 ± 10	628 ± 294	~600	307 ± 129	77 ± 10
D,L-β-AIB[c]	ND	ND	ND	ND	ND	148 ± 70	84 ± 12	<6	ND
D-Alanine	[g]	2[a]	389 ± 87	630 ± 135	201 ± 14	69 ± 9	82 ± 22	<4	20 ± 5
L-Alanine	[g]	[a]	605 ± 149	1028 ± 239	228 ± 45	69 ± 9	157 ± 14	127 ± 92	75 ± 18
D-β-ABA	ND	ND	ND	ND	ND	332 ± 99[a]	438 ± 142[a]	<6[a]	<26[a]
L-β-ABA	ND	ND	ND	ND	ND	[a]	[a]	[a]	[a]
α-AIB	ND	ND	380 ± 97	881 ± 280	28 ± 10	39 ± 37	46 ± 33	<10	<27
D-α-ABA	ND	ND	ND	ND	ND	13 ± 11[a]	12 ± 7[a]	<4[a]	84 ± 40[a]
L-α-ABA	ND	ND	ND	ND	ND	[a]	[a]	[a]	[a]
D-Isovaline	ND	ND	ND	ND	ND	<194[a]	<163[a]	<35[a]	<56 ± [a]
L-Isovaline	ND	ND	ND	ND	ND	[a]	[a]	[a]	[a]
Total	30	70	2500	4300	850	4300	4200	1900	890

Amino acid	CR2				CR1
	EET 92 042	GRA 95 229	Renazzo	Shişr 033	GRO 95 577
D-Aspartic acid	409 ± 41	551 ± 75	49 ± 12	57 ± 5	13 ± 2
L-Aspartic acid	465 ± 24	576 ± 51	173 ± 80	189 ± 23	19 ± 4
D-Glutamic acid	3090 ± 422	3489 ± 389	<26	124 ± 7	16 ± 6
L-Glutamic acid	4468 ± 503	4209 ± 415	856 ± 188	489 ± 34	40 ± 3
D-Serine	742 ± 42[a]	1807 ± 84[a]	195 ± 112[a]	32 ± 11	50 ± 11[a]
L-Serine	[a]	[a]	[a]	140 ± 34	[a]
Glycine	24 975 ± 608	40 496 ± 1028	875 ± 188	417 ± 55	136 ± 14
β-Alanine	3046 ± 50	3143 ± 495	223 ± 55	62 ± 10	122 ± 6
γ-ABA	1512 ± 66	1914 ± 398	1092 ± 243	49 ± 23	54 ± 6
D,L-β-AIB[c]	1429 ± 333	2091 ± 405	<40	<2	30 ± 2
D-Alanine	21 664 ± 1009	52 465 ± 6860	<38	274 ± 14	74 ± 22
L-Alanine	22 297 ± 1583	51 141 ± 6272	<132	330 ± 59	96 ± 20
D-β-ABA	1327 ± 33	3903 ± 377	534 ± 24[a]	<6	49 ± 5[a]
L-β-ABA	1458 ± 99	4239 ± 494	[a]	<4	[a]
α-AIB	50 210 ± 870	30 257 ± 1226	<73	34 ± 5	48 ± 3
D-α-ABA	1123 ± 54	2956 ± 125	<89[a]	<5[a]	ND
L-α-ABA	1244 ± 28	2955 ± 120	[a]	[a]	ND
D-Isovaline	22 806 ± 459[a]	29 245 ± 2229[a]	349 ± 33[a]	<3	<131[a]
L-Isovaline	[a]	[a]	[a]	<5	[a]
Total	162 000	235 000	4300	2200	750

The associated errors are based on the standard deviation of the average value. Adapted from [126–128, 130–133, 136–138, 140–142].
[a]Enantiomers could not be detected individually, so the total (D + L) abundance of the amino acids is reported in the row for the D-enantiomer.
[b]Tentative identification.
[c]Optically pure standard not available for enantiomeric identification.
[d]γ-ABA and D,L-β-AIB could not be separated under the chromatographic conditions used, so the total abundance is reported in the cell corresponding to γ-ABA.
[e]γ-ABA and D,L-β-ABA could not be separated under the chromatographic conditions used, so the total abundance is reported in the cell corresponding to γ-ABA.
[f]The value corresponds mostly to γ-ABA.
[g]Detected but too small to estimate.
ND = not determined.

abundance similar to the CM2 Murchison. While the CM2 Y-791 198 has an extremely high total amino acid concentration (71 ppm [131, 132]), which is about 5 times as high as Murchison (15 ppb), the CM2s Essebi, Nogoya, Mighei [133], Allan Hills (ALHA) 77 306 [134–136], ALH 83 100 [130], Y-79 331, and Belgica (B-) 7904 [137] have much lower amino acid abundances, some being depleted in amino acids (Table 1.5). For Essebi, Botta *et al.* [133] consider that the high abundances (relative to glycine) of γ-ABA and β-alanine are derived from terrestrial contamination at the fall site.

CM1s chondrites were analyzed for the first time for amino acids by Botta *et al.* [138]. ALH 88 045, MET (Meteorite Hills) 01 070, and LAP (La Paz) 0227 have total amino acid concentration much lower than the average of the CM2s. According to Botta *et al.* [138], these results and the similar relative amino acid abundances between the CM1 class meteorites and the CM2 Murchison are explained by decomposition of a CM2-like amino acid distribution during extensive aqueous alteration in the CM1s meteorite parent body.

The CI1 chondrites Orgueil and Ivuna have total amino acid abundances of about 4.2 ppm, with β-alanine, glycine, and γ-ABA as the most abundant amino acids, while glycine and α-AIB are the most abundant amino acids in the CM2 chondrites Murchison and Murray [136]. The CV3 Allende [133, 139] and the ungrouped C2 Tagish Lake meteorites [133, 140] are essentially free of amino acids (total amino acid abundances of 2 and 1 ppm, respectively), with most of the amino acids probably being terrestrial contaminants.

The highest amino acid abundances ever measured in a meteorite were found on the CR2s EET 92 042 and GRA 95 229, with total amino acid concentrations of 180 and 249 ppm, respectively [141]. The most abundant amino acids present in these meteorites are the α-amino acids glycine, isovaline, α-AIB and alanine. The high $\delta^{13}C$ results together with the racemic enantiomeric ratios determined [141] for most amino acids indicate an extraterrestrial origin for these compounds (see Section 1.4.1). In addition, these authors analyzed the CR1 GRO 95 577, which was found to be depleted in amino acids (1 ppb). Other CRs analyzed include the CR2 chondrites Renazzo [133] and Shişr 033 [142]. Renazzo has a total amino acid abundance of only 4.8 ppm, which is similar to the CI chondrites Orgueil and Ivuna. This meteorite has a distinct amino acid distribution, with γ-ABA, glycine, and L-glutamic acid as the most abundant amino acids. Only upper limits for alanine and α-AIB were reported for Renazzo, while isovaline was tentatively identified [133]. The most abundant amino acids in the Shişr 033 meteorite are glycine, L-glutamic acid, L-alanine, and L-aspartic acid. In addition to this, Shişr 033 D/L protein amino acid ratios are smaller than 0.4 and in agreement with the D/L amino acid ratios of Shişr 033 fall-site soil. These results suggest extensive amino acid contamination of the meteorite (see Section 1.4.1). However, Shişr 033 contains a small fraction of extraterrestrial amino acids, as indicated by the presence of α-AIB [142].

Apart from carbonaceous chondrites, amino acid analyses have also been carried out on Martian meteorites. As our present knowledge of amino acids potentially present in Mars may be accessed from these meteorites (see also Section 1.6), amino acid analyses have been performed in the Martian meteorites EET 79 001 [143], ALH 84 001 [144], and Miller Range (MIL) 03 346 [145]. In all three samples, the meteoritic amino acid distribution was similar to the one in the Allan Hills ice, which suggested that the ice meltwater was the source of the amino acids in these meteorites. In addition, analysis of the Nakhla meteorite, which fell in Egypt, shows that the amino acid distribution (including the D/L ratios) is similar to the one in the sea-floor sediment from the Nile Delta [146].

1.4.1 Sources of Meteoritic Amino Acids (Extraterrestrial versus Terrestrial Contamination)

In order to determine if the amino acids present in carbonaceous chondrites are indigenous to the meteorites, four approaches are generally applied: (i) detection of amino acids that are unusual in the terrestrial environment, (ii) comparison of the absolute abundances of amino acids in the meteorites to the levels found in the fall-site environment (soil or ice), (iii) determination of enantiomeric ratios (D/L ratios), and (iv) determination of compound specific stable isotope ratios of hydrogen, carbon, and nitrogen.

1.4.1.1 Detection of Amino Acids that are Unusual in the Terrestrial Environment

The majority of the more than 80 different amino acids identified in carbonaceous meteorites are nonexistent (or rare) in terrestrial proteins (for a review, see, e.g., [108]). Extraterrestrial meteoritic nonprotein amino acids such as α-AIB, isovaline, β-ABA, and β-AIB have concentrations usually in the order of a few hundred parts per billion maximum (Table 1.5). However, Murchison [126] has a higher abundance of α-AIB (2901 ppb), while Murray and LEW90 500 [126, 129, 130] contain higher abundances of both α-AIB (1968 and 2706 ppb, respectively) and isovaline (2834 and 1306 ppb, respectively). The highest abundances of α-AIB, isovaline, β-ABA, and β-AIB were detected in the CR2 chondrites EET92 042 and GRA95 229 [141]. The CM2 Y791 198 ([131, 132]) contained similar abundances of α-AIB, β-ABA, and β-AIB as EET92 042 and GRA95 229, but lower abundance of isovaline (Table 1.5).

1.4.1.2 Determination of the Amino Acid Content of the Meteorite Fall Environment

Samples collected from meteorite fall sites have been analyzed for amino acids and their distribution compared to the one from the carbonaceous chondrites. Ice from the Antarctic regions of Allan Hills [143, 144] and La Paz [130, 147] contained only trace levels of aspartic acid, serine, glycine, alanine, and γ-ABA (less than 1 ppb of total amino acid concentration). No isovaline or β-ABA was detected above detection limits. Only an upper limit of α-AIB (<2 ppt) was detected in the Allan Hills ice [144], while a relatively high abundance (ranging between 25 and 46 ppt) of α-AIB was detected in the La Paz Antarctic ice [130, 147].

Soil samples from the Shişr 033 fall site show that the most abundant amino acids are L-glutamic acid, L-aspartic acid, glycine, and L-alanine, with nonprotein amino acids absent from the soil [142]. In addition, comparison of the protein amino acid enantiomeric ratios of Shişr 033 to those of the soil (D/L $<$ 0.4) shows agreement, indicating that most of the amino acids in this meteorite are terrestrial in origin. On the other hand, a soil sample collected close to the fall site of the Murchison meteorite showed much smaller amino acid relative concentrations (glycine = 1) when compared to the Murchison meteorite, indicating that the majority of the amino acids present in this meteorite are extraterrestrial in origin [133].

1.4.1.3 Determination of Enantiomeric Ratios

Chirality is a useful tool for determining the origin (biotic versus abiotic) of amino acids in meteorites. On Earth most proteins and enzymes are made of only the L-enantiomer of chiral amino acids; however, abiotic synthesis of amino acids yields racemic mixtures (D/L ~ 1). If we assume that meteoritic protein amino acids were racemic (D/L ~ 1) prior to the meteorite fall to Earth, then their D/L ratios can be used as a diagnostic signature to determine the degree of terrestrial L-amino acid contamination they have experienced. In fact, racemic amino acid ratios for protein (and nonprotein) amino acids in carbonaceous chondrites indicate an abiotic synthetic origin. Although racemic mixtures have been observed for most nonprotein chiral amino acids (Table 1.5), small L-enantiomeric excess for some nonprotein amino acids has been reported in the Murchison and Murray meteorites [148–150]. Six α-methyl-α-amino acids unknown or rare in the terrestrial biosphere (both diastereomers of α-amino-α,β-dimethyl-pentanoic acid, isovaline, α-methylnorvaline, α-methylnorleucine, and α-methylvaline) had L-enantiomeric excesses ranging from 2.8 to 9.2% in Murchison and from 1.0 to 6.0% in Murray [148, 149]. More specifically, Murchison has shown to have an L-enantiomeric excess of isovaline ranging from 0 to 15.2% with significant variation both between meteorite stones and even within the same meteorite stone [150]. The meteoritic enantiomeric excess of α-methyl-α-amino acids and the absence for the α-H-α-amino acids may be explained by the resistance to racemization of α-methyl-α-amino acids during aqueous alteration in the meteorite parent body, due to their lack of an α-hydrogen [149, 151, 152]. Another explanation could be a different amino acid formation process, namely pre-solar formation for the α-methyl-α-amino acids and subsequent incorporation into the parent body, followed by parent body formation of the α-H-α-amino acids [149, 152].

1.4.1.4 Determination of Compound-Specific Stable Isotope Ratios of Hydrogen, Carbon, and Nitrogen

For meteoritic nonchiral amino acids, such as glycine, α-AIB, β-ABA, and β-alanine, compound-specific stable isotope measurements are the only means to establish their origin. The abundances of stable isotopes are expressed in δ values. These indicate the difference in per mil (‰) between the ratio in the sample and the same ratio in the standard, as shown by:

$$\delta(‰) = \frac{(R_{\text{sample}} - R_{\text{standard}})}{R_{\text{standard}}} \times 1000 \tag{1.8}$$

where R represents $D/^1H$ for hydrogen, $^{13}C/^{12}C$ for carbon, and $^{15}N/^{14}N$ for nitrogen. The standards usually used are standard mean ocean water for hydrogen, Pee Dee Belemnite for carbon, and air for nitrogen.

Stable isotope analyses of the total amino acid fractions of the Murchison meteorite showed $\delta D = +1370‰$, $\delta^{15}N = +90‰$, and $\delta^{13}C = +23.1‰$ [153], which were later confirmed by Pizzarello *et al.* [154, 155], who obtained $\delta D = +1751‰$, $\delta^{15}N = +94‰$, and $\delta^{13}C = +26‰$. Stable isotope analyses were obtained for individual amino acids in different meteorites (Table 1.6; [126, 141, 150, 156–159]). These values (with a few exceptions, in which there is terrestrial contribution) are clearly outside the amino acid terrestrial range (from -70.5 to $+11.25‰$; [160]) and fall within the range of those measured for other indigenous polar organic compounds present in meteorites [161]. The highly enriched δD, $\delta^{15}N$, and $\delta^{13}C$ values determined for the meteoritic amino acids indicate primitive extraterrestrial organic matter.

The deuterium enrichment of amino acids is thought to be the result of interstellar chemical reactions (e.g., gas-phase ion–molecule reaction and reactions on interstellar grain surfaces) which formed the amino acid precursors. These reactions occur in the low temperatures of dense clouds ($T < 50$ K) in which deuterium fractionation is efficient (e.g., [162–164]). Meteoritic amino acids would have then been formed from their deuterium-enriched interstellar precursors and deuterium-depleted water ([165] and references therein) by synthesis (aqueous alteration) in the meteorite parent body (see Section 1.4.2). However, α-amino acids are more deuterium (and ^{13}C)-enriched than α-hydroxy acids [117], which is inconsistent with a Strecker-cyanohydrin-type synthesis from a common precursor [116, 166]. Differences may be explained by different reaction paths leading to different isotopic distributions [164]. The ^{15}N enrichment of amino acids is also thought to be due to chemical fractionation in interstellar ion–molecule exchange reactions [167, 168].

The hydrogen isotope composition of meteoritic amino acids follows a relatively simple pattern, in which δD varies more with the structure of their carbon chains (δD is higher for amino acids having a branched alkyl chain) than with the chain length [159]. On the other hand, $\delta^{13}C$ of α-amino acids (α-methyl-α- and α-H-α-amino acids) decreases with increasing carbon chain length (with the α-methyl-α-amino acids more ^{13}C-enriched than the corresponding α-H-α-amino acids), while $\delta^{13}C$ for non-α-amino acids remains unchanged or increases with increasing carbon chain length [158]. This suggests diverse synthetic processes for meteoritic amino acids, in particular that the amino acid carbon chain elongation followed at least two synthetic pathways [158, 159].

1.4.2 Synthesis of Meteoritic Amino Acids

Meteoritic amino acids are thought to be formed by a variety of synthetic pathways. Namely, it is suggested that α-amino acids form by a two-step process, in which the α-amino acid precursors (carbonyl compounds, ammonia, and HCN) were present (or formed) in a proto-solar nebula and were later incorporated into an asteroidal parent

Table 1.6 Compound-specific stable isotope composition (in ‰) of individual amino acids in carbonaceous chondrites.

Amino acid	Murchison			Murray		GRA 95 229	EET 92 042	Orgueil
	δD	$\delta^{13}C$	$\delta^{15}N$	δD	$\delta^{13}C$	$\delta^{13}C$	$\delta^{13}C$	$\delta^{13}C$
D,L-Aspartic acid[a]			+61					
D-Aspartic acid		+25				+35	+34	
L-Aspartic acid		−6				+33	+23	
D-Glutamic acid	+473	+29	+60			+47	+46	
L-Glutamic acid	+292	+7	+58			−18	−19	
Glycine		+22; +41	+37	+399		+34	+32	+22
β-Alanine	+461	+5	+61	+1063				+18
γ-ABA	+599	+19		+763				
D,L-β-AIB[a]	+967							
D-Alanine	+429	+22; +52	+60	+614		+42	+45	
L-Alanine	+360	+27; +39	+57	+510		+41	+50	
D,L-β-ABA[a]	+20.7							
α-AIB	+3058	+5; +43	+184	+3097		+32		
D-α-ABA	+1338	+29		+1633				
L-α-ABA	+1225	+28						
D,L-Isovaline[a]	+3419	+16; +22	+66	+3181		+51		
D-Isovaline		+11 to +20			+19			
L-Isovaline		+17 to +22		+3283	+20			
D-Norvaline		+15		+1505				
D-Valine				+2432				
L-Valine				+2266				
L-Leucine			+60					
D,L-α-Methylnorvaline[a]	+2686	+7		+3021				

D,L-α-Amino-α,β-dimethylbutyric acid[a]	+3318			+3604
L-α-Amino-α,β-dimethylbutyric acid[a]		+23		
D-*allo*-isoleucine				+2251
L-*allo*-isoleucine				+2465
L-isoleucine				+1819
D,L-γ-Aminovaleric acid[a]	+367	+29		+965
δ-Aminovaleric acid	+362	+24		+552
β-Amino-α,α-dimethylpropionic acid		+25		+2590
β-Amino-β-methylbutyric acid				+1141
D,L-γ-Amino-α-methylbutyric acid[a] + D,L-γ-amino-β-methylbutyric acid[a]	+1164			+1164
D,L-α-Aminoadipic acid[a]	+181			+619
L-α-Aminoadipic acid		+35		
D,L-β-Aminoadipic acid[a]		+37		
D,L-α-Methylglutamic acid[a]	+2049	+32		+1563
D,L-*threo*-β-Methylglutamic acid		+32		
D- or L-*threo*-β-Methylglutamic acid[b]				+1255
D- or L-*threo*-β-Methylglutamic acid[b]	+1352			+1408
D,L-*threo*-γ-Methylglutamic acid[a]		+36		
D,L-Proline[a]		+4	+50	+478
Cycloleucine	+425	+24		+850
Sarcosine	+962	+53	+129	+1337
D,L-*N*-Methylalanine[a]				+1267
N-Methyl-α-AIB				+3446

Adapted from [126, 141, 142, 150, 157–159].
[a]Isotopic value corresponds to the combined enantiomeric peaks.
[b]The chromatographic order of elution for the two enantiomers could not be established for lack of standards.

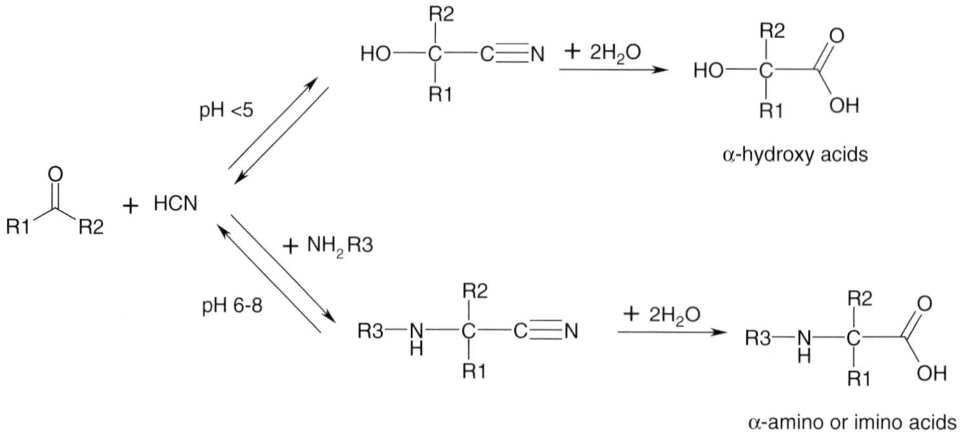

Figure 1.2 The Strecker-cyanohydrin synthetic pathway for the formation of α-amino α-hydroxy and imino acids (adapted from [116, 133, 171, 247]). R_1 and R_2 correspond to H or C_nH_{2n+1}. If R_3 corresponds to H then α-amino acids are produced; if R_3 is an amino acid then imino acids are produced.

body [169]. During aqueous alteration on the asteroidal parent body, Strecker-cyanohydrin synthesis would have taken place to form α-amino acids (Figure 1.2; [108, 116, 163, 166]). Since the carbonyl precursors (aldehydes and ketones) are thought to be synthesized by the addition of a single-carbon donor to the growing alkane chain, a decrease of the α-amino acid abundances with increasing chain length is expected. Also, synthesis of branched carbon chain analogs is expected to be favored over straight-carbon chain analogs (e.g., [108]), and this trend is observed in the EET 92 042 and GRA 95 229 meteorites [141]. Additional support for this hypothesis is the finding of α-amino acid, α-hydroxy acids [116, 117, 166], and imino acids [170] in carbonaceous meteorites. However, non-α-amino acids cannot be produced by the Strecker-cyanohydrin synthesis. Alternatively, meteoritic β-amino acids are thought to be synthesized by Michael addition of ammonia to α,β-unsaturated nitriles, followed by reduction/hydrolysis (Figure 1.3; e.g., [108] and references therein). These precursor molecules have been detected in the ISM (Table 1.1 and reference therein) and also in comets (Table 1.3 and references therein). A chemical reaction such as a Michael addition could occur on the parent body of meteorites. For example, the extensively aqueous altered CI chondrites Orgueil and Ivuna are rich in β-alanine, but depleted in α-amino acids. As suggested [138], this might indicate that the CI parent body was depleted in carbonyl compounds (aldehydes and ketones) necessary for the Strecker-cyanohydrin synthesis to occur. Additional synthetic pathways in the meteorite parent body have been proposed for non-α-amino acids (for a review, see, e.g., [108]). For example, hydrolysis of lactams and carboxy lactams, which have been detected in carbonaceous meteorites [125], gives the corresponding β-, γ-, and δ-amino acids, and dicarboxylic amino acids, respectively.

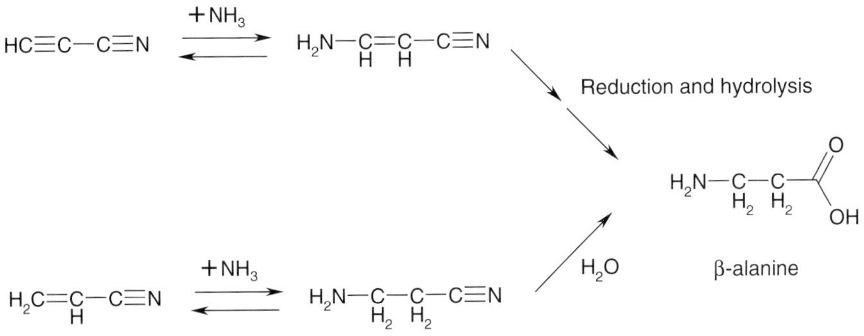

Figure 1.3 Michael-addition of ammonia to a α,β-unsaturated nitrile to form a β-amino alkylnitrile that is then hydrolyzed to form a β-amino acid (β-alanine) (adapted from [133]).

1.5 Micrometeorites and IDPs

Micrometeorites (MMs) and IDPs are thought to be the remains of comets and asteroids (for reviews, see, e.g., [171–176]). MMs are small extraterrestrial dust particles, typically in the range 50 μm to 2 mm [177], that have survived atmospheric entry. They are collected in deep-sea sediments [171, 178], Antarctic ice [179–181], and Greenland lake deposits [182–184]. On the other hand, IDPs are extraterrestrial particles (usually measuring less than 30 μm in size), which are collected from the Earth's stratosphere (at an altitude of ~20 km) by NASA aircraft (e.g., [171, 185, 186]).

Several organic molecules have been found in IDPs and MMs, such as ketone and aliphatic hydrocarbons [12, 187–189], and polycyclic aromatic hydrocarbons [11, 190]. Antarctic micrometeorites (AMMs) have also been analyzed for amino acids [191–193]. Most AMMs analyzed by Brinton *et al.* [191] had very low abundances of amino acids and high L-enantiomeric excess, with a distribution similar to that found in the Antarctic ice. One set of samples containing around 30 MMs was found to contain α-AIB at high levels (~280 ppm). However, the identification of α-AIB was tentative and needs further confirmation [191]. In addition, Glavin *et al.* [192] studied 455 AMMs and none of them contained α-AIB. A third study by Matrajt *et al.* [193] analyzed 300 AMMs and found α-AIB in around 100 MMs. These authors calculated that only around 14% of AMMs analyzed so far contained α-AIB.

1.6 Mars

The possibility of Mars harboring alien life (presently and/or in the past) is the focus of future space missions planned to the Red Planet. Their target compounds indicative of life include, among others, amino acids [194–196]. These are key biomolecules on Earth, but are also produced by abiotic reactions such as those occurring in the parent body of meteorites (see Section 1.4.2). Amino acids synthesized elsewhere in the solar system may therefore be delivered intact into the surface of Mars. In fact, large amounts of carbonaceous material are thought to be delivered to the surface of Mars by IDPs and meteorites every year [14, 197, 198]. However, in 1976 the Viking landers found no organic molecules above the parts per billion to parts per million levels on the surface of Mars (e.g., [199, 200]), even though they should have been able to detect amino acids on the order of tens of parts per million [201, 202]. This might be explained by chemical reactions occurring in the surface of Mars, leading to the destruction of organic molecules. For example, oxidant molecules may react with any potential organic compound present in the surface of the Martian soil, leading to their destruction. These oxidizing molecules may be formed by UV photolysis of the Martian atmosphere [203–207], interaction of Martian minerals with atmospheric H_2O_2 [208] or UV radiation [209], or by chemical weathering of silicates by low-temperature frost and adsorbed water in the Martian soil [210, 211]. In addition, amino acids directly exposed to Mars-like UV radiation are rapidly degraded [212–214]. For example, thin films of glycine and D-alanine have half-lives in the order of 10^4–10^5 s when irradiated under simulated noon-time Mars equatorial surface conditions [213].

If present, amino acids should therefore be in the subsurface of the Red Planet, shielded from exterior radiation. As shown by Kminek and Bada [215], amino acids can survive up to 3 billion years at a depth of more than 2 m. In particular, Aubrey *et al.* [216] found that the amino acids glycine and alanine have a half-life of up to 1.1 billion years, if buried under simulated Martian conditions. Similar results had previously been obtained by Kanavarioti and Mancinelli [217], based on amino acid decomposition rates in aqueous solutions. They found that a fraction of the amino acids phenylalanine, alanine, and pyroglutamic acid would have been preserved buried beneath the surface of Mars up to 3.5 billion years.

The mineralogical composition of Martian soils may also have an influence on the amino acid stability [216, 218, 219]. Peeters *et al.* [218, 219] have shown that different mineralogical compositions of Mars soil analogs lead to differences in amino acid stability (for a review about Mars soil analogs, see [220]); in particular, clay mineral matrices seem to have a shielding effect, protecting amino acids against destruction. In addition, Aubrey *et al.* [216] determined that amino acids can be preserved for geologically long periods (billions of years) in sulfate mineral matrices. These results suggest that locations on Mars containing clay and/or sulfate minerals (for a review, see, e.g., [221]) should be the prime targets for future missions with the goal to search for life, in particular for its amino acid constituents.

1.7
Delivery of Extraterrestrial Amino Acid to the Earth and its Importance to the Origin of Life

Independently of the environment where extraterrestrial amino acids were formed, these molecules were exogenously delivered to the early Earth during the period of late heavy bombardment 4.5–3.8 billion years ago. In fact, comets, meteorites, MMs, and IDPs are thought to have delivered tons of organic carbon per year to our planet during this period of time [13, 14, 222], just before life emerged (e.g., [223–225]). Although we do not know what the production rates of different organic compounds were on the early Earth (i.e., impact-shock synthesis, endogenous production by UV light or electrical discharge, and synthesis in submarine hydrothermal vents), the higher the content of key molecules in primitive extraterrestrial materials, the more likely it is that exogenous material played a role in the origin of life. As noted by Chyba and Sagan [14], the heavy bombardment may have delivered or produced organic molecules in the early Earth in quantities comparable to other sources, possibly playing an important role for the origin of life.

The survival rate of amino acids in extraterrestrial bodies under simulated Earth atmospheric entry has been studied by several authors. Amino acids present in the interior of meteorites bigger than 1 mm survive atmospheric deceleration ([226]; see also Section 1.4), as the meteorite only experiences pyrolytic temperatures (>600 °C) and melting on the surface (<1 mm depth [227]). This does not happen with smaller particles such as MMs and IDPs, which are uniformly heated. In fact, it is estimated that most MMs and IDPs are heated for a few seconds to peak temperatures of up to 1700 °C during atmospheric entry [228–230]. However, a small percentage of dust (MMs and IDPs) enters the atmosphere at temperatures below 700 °C [173, 228, 231–233]. Glavin and Bada [234] investigated the sublimation of amino acids from sub-100 μm Murchison meteorite grains to test the survival of amino acids in MMs during atmospheric entry. They found that under vacuum (800 mT) and at 550 °C, only glycine survived. All other amino acids (including α-AIB and isovaline) were completely destroyed, which is not surprising if we consider that, with a few exceptions, pure amino acids suffer thermal decomposition in the range of 200–600 °C [235]. Glavin and Bada [234] also found that methylamine and ethylamine, which are the α-decarboxylation products of glycine and alanine, respectively, were not detected. This indicates that α-decarboxylation did not occur or that these amines were also decomposed during the experiments. Although amino acids are expected to have a higher atmospheric entry survival in smaller cosmic dust particles (<50 μm) [228, 231], Matrajt *et al.* [236] found similar results to Glavin and Bada [234] using activated alumina with grain sizes ranging from 5 to 9 μm, which suggests that the amino acid survival is not greatly dependent on the grain size. On the other hand, Matrajt *et al.* [236] demonstrated that the combination of porosity and heatshield effect (i.e., ablative cooling) of extraterrestrial dust particles during atmospheric heating results in poor heat transfer to the particle's interior, providing thermal protection and allowing organic compounds to survive. This reinforces the idea that IDPs, which are around 10% organic carbon by mass, can decelerate in the atmosphere and deliver organic compounds intact [222], and might have been the major source of exogenous organics in the early Earth as suggested by Chyba and Sagan [14].

Nonprotein amino acids, α-AIB and racemic isovaline, have been detected within around 1 m above and below the sediments of the iridium-rich Cretaceous/Tertiary boundary at Stevns Klint, Denmark by Zhao and Bada [237]. These authors suggested that the sediments represented components of a large bolide (i.e., a comet), which collided with the Earth 65 million years ago. These results are not in agreement with xenon measurements at the Cretaceous/Tertiary boundary [238, 239] or with two-dimensional smoothed particle hydrodynamics simulations of cometary organic pyrolysis by impacts [13]. However, Chyba *et al.* [13] noted the lack of relevant kinetic data for high-temperature pyrolysis of organic molecules. Further work on amino acid survival in simulated large asteroidal and cometary shock impacts has been performed [240–243]. Peterson *et al.* [240] conducted a series of shock impact experiments over a pressure range of 3.5–32 GPa using powdered Murchison and Allende meteorite samples. These were previously extracted to eliminate their original amino acid content and subsequently doped with known amino acids. The results show that amino acids diminished substantially with increasing high pressures and new "daughter" amino acids were formed, in particular β-alanine, glycine, alanine, γ-ABA, and β-AIB. At 30 GPa, the abundances of the daughter compounds exceeded those of the remaining initial amino acids. However, as noted by Blank *et al.* [242], these authors did not refer to the porosity of the starting material or the temperature conditions of the experiments. Further shock experiments in the range 5–21 GPa and 139–597 °C and using aqueous amino acid solutions were performed by Blank *et al.* [242]. In all the experiments a large fraction of amino acids survived, supporting the hypothesis that organic compounds could survive impact processes. In addition, high-resolution hydrocode simulations of comet impacts, over different projectile radii, impact velocities, and angles, show that significant amounts of amino acids can be delivered intact to the Earth via kilometer-sized comet impacts [241, 243].

1.8 Conclusions

To date, extraterrestrial amino acids have only been unequivocally identified in meteorites and a few AMMs. However, the potential precursors of these prebiotic molecules are abundant in a variety of extraterrestrial environments, including the ISM and comets. This suggests that meteoritic amino acids may have a contribution from interstellar, nebular and/or parent body processing. The delivery of these prebiotic molecules to the early Earth during the period of heavy bombardment (4.5–3.8 billion years ago) may have provided the necessary feedstock for the evolution of emergent life systems.

Laboratory analyses of meteoritic material together with future missions (including sample return missions) to solar system planets (e.g., Mars Science Laboratory, ExoMars), satellites, asteroids, and comets may expand our inventory of amino acids in extraterrestrial environments.

Acknowledgments

Z.M. and M.A.S. acknowledge financial support from the Science and Technology Facilities Council.

References

1 Spitzer, L. Jr. (1985) Clouds between the stars. *Physica Scripta*, **T11**, 5–13.

2 van Dishoeck, E.F. (1998) The chemistry of diffuse and dark interstellar clouds, in *The Molecular Astrophysics of Stars and Galaxies* (eds T.W. Hartquist and D.A. Williams), Clarendon Press, Oxford, p. 53.

3 Wooden, D.H., Charnley, S.B., and Ehrenfreund, P. (2004) Composition and evolution of interstellar clouds, in *Comets II* (eds M.C. Festou, H.U. Keller, and H.A. Weaver), University of Arizona Press, Tucson, AZ, pp. 33–66.

4 Prasad, S.S. and Tarafdar, S.P. (1983) UV radiation field inside dense clouds – its possible existence and chemical implications. *Astrophysical Journal*, **267**, 603–609.

5 Evans, N.J. II (1999) Physical conditions in regions of star formation. *Annual Review of Astronomy and Astrophysics*, **37**, 311–362.

6 Millar, T.J. and Hatchell, J. (1998) Chemical models of hot molecular cores. *Faraday Discussions*, **109**, 15–30.

7 Crovisier, J. and Bockelée-Morvan, D. (1999) Remote observations of the composition of cometary volatiles. *Space Science Reviews*, **90**, 19–32.

8 Ehrenfreund, P., Irvine, W., Becker, L., Blank, J., Brucato, J.R., Colangeli, L., Derenne, S., Despois, D., Dutrey, A., Fraaije, H., Lazcano, A., owen, T., and Robert, F. (2002) Astrophysical and astrochemical insights into the origin of life. *Reports on Progress in Physics*, **65**, 1427–1487.

9 Botta, O. and Bada, J.L. (2002) Extraterrestrial organic compounds in meteorites. *Surveys in Geophysics*, **23**, 411–467.

10 Sephton, M.A. (2002) Organic compounds in carbonaceous meteorites. *Natural Product Reports*, **19**, 292–311.

11 Clemett, S.J., Maechling, C.R., Zare, R.N., Swan, P.D., and Walker, R.M. (1993) Identification of complex aromatic molecules in individual interplanetary dust particles. *Science*, **262**, 721–725.

12 Matrajt, G., Muñoz Caro, G.M., Dartois, E., D'Hendecourt, L., Deboffle, D., and Borg, J. (2005) FTIR analysis of the organics in IDPs: comparison with the IR spectra of the diffuse interstellar medium. *Astronomy and Astrophysics*, **433**, 979–995.

13 Chyba, C.F., Thomas, P.J., Brookshaw, L., and Sagan, C. (1990) Cometary delivery of organic molecules to the early Earth. *Science*, **249**, 366–373.

14 Chyba, C.F. and Sagan, C. (1992) Endogenous production, exogenous delivery and impact-shock synthesis of organic molecules: an inventory for the origins of life. *Nature*, **355**, 125–132.

15 Brown, R.D., Godfrey, P.D., Storey, J.W.V., Bassez, M.P., Robinson, B.J., Batchelor, R.A., McCulloch, M.G., Rydbeck, O.E.H., and Hjalmarson, A.G. (1979) A search for interstellar glycine. *Monthly Notices of the Royal Astronomical Society*, **186**, 5P–8P.

16 Hollis, J.M., Snyder, L.E., Suenram, R.D., and Lovas, F.J. (1980) A search for the lowest-energy conformer of interstellar glycine. *Astrophysical Journal*, **241**, 1001–1006.

17 Snyder, L.E., Hollis, J.M., Suenram, R.D., Lovas, F.J., Buhl, D., and Brown, L.W. (1983) An extensive galactic search for conformer II glycine. *Astrophysical Journal*, **268**, 123–128.

18 (a) Berulis, I.I., Winnewisser, G., Krasnov, V.V., and Sorochenko, R.L. (1985) A search for interstellar glycine. *Soviet Astronomy Letters*, **11**, 251–253; (b) Berulis, I.I., Winnewisser, G., Krasnov, V.V., and Sorochenko, R.L. (1985) *Chemical Abstracts*, **103**, 150320.

19 Guelin, M. and Cernicharo, J. (1989) Molecular abundances in the dense interstellar and circumstellar clouds,

in *Physics and Chemistry of Interstellar Molecular Clouds – mm and Sub-mm Observations in Astrophysics* (eds G. Winnewisser and T. Armstrong), Springer, Berlin, pp. 337–343.

20 Combes, F. and Nguyen-Q-Rieu, Wlodarczak, G. (1996) Search for interstellar glycine. *Astronomy and Astrophysics*, **308**, 618–622.

21 Snyder, L.E. (1997) The search for interstellar glycine. *Origins of Life and Evolution of the Biosphere*, **27**, 115–133.

22 Ceccarelli, C., Loinard, L., Castets, A., Faure, A., and Lefloch, B. (2000) Search for glycine in the solar type protostar IRAS 16293-2422. *Astronomy and Astrophysics*, **362**, 1122–1126.

23 Hollis, J.M., Pedelty, J.A., Snyder, L.E., Jewell, P.R., Lovas, F.J., Palmer, P., and Liu, S.-Y. (2003) A sensitive very large array search for small-scale glycine emission toward OMC-1. *Astrophysical Journal*, **588**, 353–359.

24 Hollis, J.M., Pedelty, J.A., Boboltz, D.A., Liu, S.-Y., Snyder, L.E., Palmer, P., Lovas, F.J., and Jewell, P.R. (2003) Kinematics of the Sagittarius B2(N-LMH) molecular core. *Astrophysical Journal*, **596**, L235–L238.

25 Kuan, Y.-J., Charnley, S.B., Huang, H.-C., Tseng, W.-L., and Kisiel, Z. (2003) Interstellar glycine. *Astrophysical Journal*, **593**, 848–867.

26 Snyder, L.E., Lovas, F.J., Hollis, J.M., Friedel, D.N., Jewell, P.R., Remijan, A., Ilyushin, V.V., Alekseev, E.A., and Dyubko, S.F. (2005) A rigorous attempt to verify interstellar glycine. *Astrophysical Journal*, **619**, 914–930.

27 Cunningham, M.R., Jones, P.A., Godfrey, P.D., Cragg, D.M., Bains, I., Burton, M.G., Calisse, P., Crighton, N.H.M., Curran, S.J., Davis, T.M., Dempsey, J.T., Fulton, B., Hidas, M.G., Hill, T., Kedziora-Chudczer, L., Minier, V., Pracy, M.B., Purcell, C., Shobbrook, J., and Travouillon, T. (2007) *Monthly Notices of the Royal Astronomical Society*, **376**, 1201–1210.

28 Jones, P.A., Cunningham, M.R., Godfrey, P.D., and Cragg, D.M. (2007) A search for biomolecules in Sagittarius B2 (LMH) with the Australia Telescope Compact Array. *Monthly Notices of the Royal Astronomical Society*, **374**, 579–589.

29 Császár, A.G. (1992) Conformers of gaseous glycine. *Journal of the American Chemical Society*, **114**, 9568–9575.

30 Lovas, F.J., Kawashima, Y., Grabow, J.-U., Suenram, R.D., Fraser, G.T., and Hirota, E. (1995) Microwave spectra, hyperfine structure, and electric dipole moments for conformers I and II of glycine. *Astrophysical Journal*, **455**, L201–L204.

31 Ehrenfreund, P., Charnley, S.B., and Botta, O. (2005) Voyage from dark clouds to the early Earth, in *Astrophysics of Life* (eds M. Livio, N. Reid and W.B. Sparks), Telescope Science Institute Symposium Series, vol. 16, Cambridge University Press, Cambridge, pp. 1–20.

32 Ehrenfreund, P. and Charnley, S.B. (2000) Organic molecules in the interstellar medium, comets, and meteorites: a voyage from dark clouds to the early Earth. *Annual Review of Astronomy and Astrophysics*, **38**, 427–483.

33 Sandford, S.A. and Allamandola, L.J. (1993) Condensation and vaporization studies of CH_3OH and NH_3 ices: major implications for astrochemistry. *Astrophysical Journal*, **417**, 815–825.

34 Sandford, S.A. (1996) The inventory of interstellar materials available for the formation of the solar system. *Meteoritics & Planetary Science*, **31**, 449–476.

35 Whittet, D.C.B., Schutte, W.A., Tielens, A.G.G.M., Boogert, A.C.A., De Graauw, T., Ehrenfreund, P., Gerakines, P.A., Helmich, F.P., Prusti, T., and Van Dishoeck, E.F. (1996) An ISO SWS view of interstellar ices: first results. *Astronomy and Astrophysics*, **315**, L357–L360.

36 Whittet, D.C.B. (1997) Interstellar ices studied with the Infrared Space Observatory. *Origins of Life and Evolution of the Biosphere*, **27**, 101–113.

37 Gibb, E.L., Whittet, D.C.B., Schutte, W.A., Boogert, A.C.A., Chiar, J.E., Ehrenfreund, P., Gerakines, P.A., Keane, J.V., Tielens, A.G.G.M., van Dishoeck, E.F., and Kerkhof, O. (2000) An inventory of interstellar ices toward the embedded protostar W33A. *Astrophysical Journal*, **536**, 347–356.

38 Gibb, E.L., Whittet, D.C.B., Boogert, A.C.A., and Tielens, A.G.G.M. (2004) Interstellar ice: the infrared space observatory legacy. *Astrophysical Journal Supplement Series*, **151**, 35–73.

39 Holtom, P.D., Bennett, C.J., Osamura, Y., Mason, N.J., and Kaiser, R.I. (2005) A combined experimental and theoretical study on the formation of the amino acid glycine (NH_2CH_2COOH) and its isomer ($CH_3NHCOOH$) in extraterrestrial ices. *Astrophysical Journal*, **626**, 940–952.

40 Briggs, R., Ertem, G., Ferris, J.P., Greenberg, J.M., McCain, P.J., Mendoza-Gomez, C.X., and Schutte, W. (1992) Comet Halley as an aggregate of interstellar dust and further evidence for the photochemical formation of organics in the interstellar medium. *Origins of Life and Evolution of the Biosphere*, **22**, 287–307.

41 Muñoz Caro, G.M., Meierhenrich, U.J., Schutte, W.A., Barbier, B., Arcones Segovia, A., Rosenbauer, H., Thiemann, W.H.-P., Brack, A., and Greenberg, J.M. (2002) Amino acids from ultraviolet irradiation of interstellar ice analogues. *Nature*, **416**, 403–406.

42 Bernstein, M.P., Dworkin, J.P., Sandford, S.A., Cooper, G.W., and Allamandola, L.J. (2002) Racemic amino acids from the ultraviolet photolysis of interstellar ice analogues. *Nature*, **416**, 401–403.

43 Nuevo, M., Meierhenrich, U.J., Munoz Caro, G.M., Dartois, E., d'Hendecourt, L., Deboffle, D., Auger, G., Blanot, D., Bredehoeft, J.-H., and Nahon, L. (2006) The effects of circularly polarized light on amino acid enantiomers produced by the UV irradiation of interstellar ice analogs. *Astronomy and Astrophysics*, **457**, 741–751.

44 Nuevo, M., Meierhenrich, U.J., d'Hendecourt, L., Munoz Caro, G.M., Dartois, E., Deboffle, D., Thiemann, W.H.-P., Bredehoeft, J.-H., and Nahon, L. (2007) Enantiomeric separation of complex organic molecules produced from irradiation of interstellar/ circumstellar ice analogs. *Advances in Space Research*, **39**, 400–404.

45 Mendoza, C., Ruette, F., Martorell, G., and Rodríguez, L.S. (2004) Quantum-chemical modeling of interstellar grain prebiotic chemistry: catalytic synthesis of glycine and alanine on the surface of a polycyclic aromatic hydrocarbon flake. *Astrophysical Journal*, **601**, L59–L62.

46 Sorrell, W.H. (2001) Origin of amino acids and organic sugars in interstellar clouds. *Astrophysical Journal*, **555**, L129–L132.

47 Woon, D.E. (2002) Pathways to glycine and other amino acids in ultraviolet-irradiated astrophysical ices determined via quantum chemical modelling. *Astrophysical Journal*, **571**, L177–L180.

48 Elsila, J.E., Dworkin, J.P., Bernstein, M.P., Martin, M.P., and Sandford, S.A. (2007) Mechanisms of amino acid formation in interstellar ice analogs. *Astrophysical Journal*, **660**, 911–918.

49 Ehrenfreund, P., Bernstein, M.P., Dworkin, J.P., Sandford, S.A., and Allamandola, L.J. (2001) The photostability of amino acids in space. *Astrophysical Journal*, **550**, L95–L99.

50 Charnley, S.B. (1997) On the nature of interstellar organic chemistry, in *Astronomical and Biochemical Origins and the Search for Life in the Universe* (eds C.B. Cosmovici, S. Bowyer and D. Werthimer), IAU Colloquium, vol. 161, Editrice Compositori, Bologna, p. 89.

51 Schutte, W.A., Boogert, A.C.A., Tielens, A.G.G.M., Whittet, D.C.B., Gerakines, P.A., Chiar, J.E., Ehrenfreund, P., Greenberg, J.M., Van Dishoeck, E.F., and De Graauw, T. (1999) Weak ice absorption features at 7.24 and

7.41 MU M in the spectrum of the obscured young stellar object W 33A. *Astronomy and Astrophysics*, **343**, 966–976.

52 Charnley, S.B., Ehrenfreund, P., and Kuan, Y.-J. (2001) Spectroscopic diagnostics of organic chemistry in the protostellar environment. *Spectrochimica Acta Part A – Molecular and Biomolecular Spectroscopy*, **57**, 685–704.

53 Liu, S.-Y., Mehringer, D.M., and Snyder, L.E. (2001) Observations of formic acid in hot molecular cores. *Astrophysical Journal*, **552**, 654–663.

54 Blagojevic, V., Petrie, S., and Bohme, D.K. (2003) Gas-phase syntheses for interstellar carboxylic and amino acids. *Monthly Notices of the Royal Astronomical Society*, **339**, L7–L11.

55 Greenberg, J.M. and Hage, J.I. (1990) From interstellar dust to comets – a unification of observational constraints. *Astrophysical Journal*, **361**, 260–274.

56 Mumma, M.J., Weissman, P.R., and Stern, S.A. (1993) Comets and origin of the solar system – reading the Rosetta stone, in *Protostars and Planets HI* (eds E.H. Levy and J.I. Lunine), University of Arizona Press, Tucson, AZ, pp. 1177–1252.

57 Bockelée-Morvan, D., Crovisier, J., Mumma, M.J., and Weaver, H.A. (2005) The composition of cometary volatiles, in *Comets II* (eds M. Festou, H.U. Keller and H.A. Weaver), University of Arizona Press, Tucson, AZ, pp. 391–423.

58 Chamberlin, T.C. and Chamberlin, R.T. (1908) Early terrestrial conditions that may have favored organic synthesis. *Science*, **28**, 897–910.

59 Weaver, H.A., A'Hearn, M.F., Arpigny, C., Boice, D.C., Feldman, P.D., Larson, S.M., Lamy, P., Levy, D.H., Marsden, B.G., Meech, K.J., Noll, S., Scotti, J.V., Sekanina, Z., Shoemaker, C.S., Shoemaker, E.M., Smith, T.E., Stern, S.A., Storrs, A.D., Trauger, J.T., Yeomans, D.K., and Zellner, B. (1995) The Hubble Space Telescope (HST) observing campaign on comet Shoemaker–Levy 9. *Science*, **267**, 1282–1288.

60 Crovisier, J. (1997) Infrared observations of volatile molecules in comet Hale–Bopp. *Earth Moon Planets*, **79**, 125–143.

61 Crovisier, J., Leech, K., Bockelée-Morvan, D., Brooke, T.Y., Hanner, M.S., Altieri, B., Keller, H.U., and Lellouch, E. (1997) The spectrum of comet Hale–Bopp (C/1995 01) observed with the Infrared Space Observatory at 2.9 AU from the Sun. *Science*, **275**, 1904–1907.

62 Lamy, P.L., Toth, I., and Weaver, H.A. (1998) Hubble Space Telescope observations of the nucleus and inner coma of comet 19P/1904 Y2 (Borrelly). *Astronomy and Astrophysics*, **337**, 945–954.

63 Lisse, C.M., A'Hearn, M.F., Groussin, O., Fernandez, Y.R., Belton, M.J.S., van Cleve, J.E., Charmandaris, V., Meech, K.J., and McGleam, C. (2005) Rotationally resolved 8–35 micron Spitzer Space Telescope observations of the nucleus of comet 9P/Tempel 1. *Astrophysical Journal*, **625**, L139–L142.

64 Gehrz, R.D., Reach, W.T., Woodward, C.E., and Kelley, M.S. (2006) Infrared observations of comets with the Spitzer Space Telescope. *Advances in Space Research*, **38**, 2031–2038.

65 Kelley, M.S., Woodward, C.E., Harker, D.E., Wooden, D.H., Gehrz, R.D., Campins, H., Hanner, M.S., Lederer, S.M., Osip, D.J., Pittichova, J., and Polomski, E. (2006) A Spitzer Study of Comets 2P/ Encke, 67P/Churyumov– Gerasimenko, and C/2001 HT50 (LINEAR-NEAT). *Astrophysical Journal*, **651**, 1256–1271.

66 Reach, W.T., Kelley, M.S., and Sykes, M.V. (2007) A survey of debris trails from short-period comets. *Icarus*, **191**, 298–322.

67 Irvine, W.M., Senay, M., Lovell, A.J., Matthews, H.E., McGonagle, D., and Meier, R. (2000) Detection of nitrogen sulfide in comet Hale–Bopp. *Icarus*, **143**, 412–414.

68 Bockelée-Morvan, D., Biver, N., Colom, P., Crovisier, J., Henry, F., Lecacheux, A., Davies, J.K., Dent, W.R.F., and Weaver, H.A. (2004) The outgassing and composition of comet 19P/Borrelly from radio observations. *Icarus*, **167**, 113–128.

69 Crovisier, J., Bockelee-Morvan, D., Colom, P., Biver, N., Despois, D., and Lis, D.C. (2004) The composition of ices in comet C/1995 O1 (Hale–Bopp) from radio spectroscopy. Further results and upper limits on undetected species. *Astronomy and Astrophysics*, **418**, 1141–1157.

70 Hirao, K. and Itoh, T. (1987) The Sakigake/Suisei encounter with comet P/Halley. *Astronomy and Astrophysics*, **187**, 39–46.

71 Kissel, J. and Krueger, F.R. (1987) The organic component in dust from comet Halley as measured by the PUMA mass spectrometer on Board VEGA 1. *Nature*, **326**, 755–760.

72 Jessberger, E.K., Christoforidis, A., and Kissel, J. (1988) Aspects of the major element composition of Halley's dust. *Nature*, **332**, 691–695.

73 Fomenkova, M.N. (1999) On the organic refractory component of cometary dust. *Space Science Reviews*, **90**, 109–114.

74 Soderblom, L.A., Becker, T.L., Bennett, G., Boice, D.C., Britt, D.T., Brown, R.H., Buratti, B.J., Isbell, C., Giese, B., Hare, T., Hicks, M.D., Howington-Kraus, E., Kirk, R.L., Lee, M., Nelson, R.M., Oberst, J., Owen, T.C., Rayman, M.D., Sandel, B.R., Stern, S.A., Thomas, N., and Yelle, R.V. (2002) Observations of comet 19P/Borrelly by the miniature integrated camera and spectrometer aboard Deep Space 1. *Science*, **296**, 1087–1091.

75 Nordholt, J.E., Reisenfeld, D.B., Wiens, R.C., Gary, S.P., Crary, F., Delapp, D.M., Elphic, R.C., Funsten, H.O., Hanley, J.J., Lawrence, D.J., McComas, D.J., Shappirio, M., Steinberg, J.T., Wang, J., and Young, D.T. (2003) Deep Space 1 encounter with comet 19P/Borrelly: ion composition measurements by the PEPE mass spectrometer. *Geophysical Research Letters*, **30**, 18/1–18/4.

76 Buratti, B.J., Hicks, M.D., Soderblom, L.A., Britt, D., Oberst, J., and Hillier, J.K. (2004) Deep Space 1 photometry of the nucleus of comet 19P/Borrelly. *Icarus*, **167**, 16–29.

77 A'Hearn, M.F., Belton, M.J.S., Delamere, W.A., Kissel, J., Klaasen, K.P., McFadden, L.A., Meech, K.J., Melosh, H.J., Schultz, P.H., Sunshine, J.M., Thomas, P.C., Veverka, J., Yeomans, D.K., Baca, M.W., Busko, I., Crockett, C.J., Collins, S.M., Desnoyer, M., Eberhardy, C.A., Ernst, C.M., Farnham, T.L., Feaga, L., Groussin, O., Hampton, D., Ipatov, S.I., Li, J.-Y., Lindler, D., Lisse, C.M., Mastrodemos, N., Owen, W.M., Richardson, J.E., Wellnitz, D.D., and White, R.L. (2005) Deep Impact: excavating comet Tempel 1. *Science*, **310**, 258–264.

78 Keller, H.U., Jorda, L., Kueppers, M., Gutierrez, P.J., Hviid, S.F., Knollenberg, J., Lara, L.-M., Sierks, H., Barbieri, C., Lamy, P., Rickman, H., and Rodrigo, R. (2005) Deep Impact observations by OSIRIS onboard the Rosetta spacecraft. *Science*, **310**, 281–283.

79 Thomas, P.C., Veverka, J., Belton, M.J.S., Hidy, A., A'Hearn, M.F., Farnham, T.L., Groussin, O., Li, J.-Y., McFadden, L.A., Sunshine, J., Wellnitz, D., Lisse, C., Schultz, P., Meech, K.J., and Delamere, W.A. (2007) The shape, topography, and geology of Tempel 1 from Deep Impact observations. *Icarus*, **187**, 4–15.

80 Brownlee, D., Tsou, P., Aleon, J., Alexander, C.M.O'D., Araki, T., Bajt, S., Baratta, G.A., Bastien, R., Bland, P., Bleuet, P., Borg, J., Bradley, J.P., Brearley, A., Brenker, F., Brennan, S., Bridges, J.C., Browning, N.D., Brucato, J.R., Bullock, E., Burchell, M.J., Busemann, H., Butterworth, A., Chaussidon, M., Cheuvront, A., Chi, M., Cintala, M.J., Clark, B.C., Clemett, S.J., Cody, G., Colangeli, L., Cooper, G., Cordier, P., Daghlian, C., Dai, Z., D'Hendecourt, L., Djouadi, Z., Dominguez, G., Duxbury, T., Dworkin, J.P., Ebel, D.S., Economou, T.E., Fakra, S., Fairey, S.A.J., Fallon, S.,

Ferrini, G., Ferroir, T., Fleckenstein, H., Floss, C., Flynn, G., Franchi, I.A., Fries, M., Gainsforth, Z., Gallien, J.-P., Genge, M., Gilles, M.K., Gillet, P., Gilmour, J., Glavin, D.P., Gounelle, M., Grady, M.M., Graham, G.A., Grant, P.G., Green, S.F., Grossemy, F., Grossman, L., Grossman, J.N., Guan, Y., Hagiya, K., Harvey, R., Heck, P., Herzog, G.F., Hoppe, P., Hoerz, F., Huth, J., Hutcheon, I.D., Ignatyev, K., Ishii, H., Ito, M., Jacob, D., Jacobsen, C., Jacobsen, S., Jones, S., Joswiak, D., Jurewicz, A., Kearsley, A.T., Keller, L.P., Khodja, H., Kilcoyne, A.L.D., Kissel, J., Krot, A., Langenhorst, F., Lanzirotti, A., Le, L., Leshin, L.A., Leitner, J., Lemelle, L., Leroux, H., Liu, M.-C., Luening, K., Lyon, I., MacPherson, G., Marcus, M.A., Marhas, K., Marty, B., Matrajt, G., McKeegan, K., Meibom, A., Mennella, V., Messenger, K., Messenger, S., Mikouchi, T., Mostefaoui, S., Nakamura, T., Nakano, T., Newville, M., Nittler, L.R., Ohnishi, I., Ohsumi, K., Okudaira, K., Papanastassiou, D.A., Palma, R., Palumbo, M.E., Pepin, R.O., Perkins, D., Perronnet, M., Pianetta, P., Rao, W., Rietmeijer, F.J.M., Robert, F., Rost, D., Rotundi, A., Ryan, R., Sandford, S.A., Schwandt, C.S., See, T.H., Schlutter, D., Sheffield-Parker, J., Simionovici, A., Simon, S., Sitnitsky, I., Snead, C.J., Spencer, M.K., Stadermann, F.J., Steele, A., Stephan, T., Stroud, R., Susini, J., Sutton, S.R., Suzuki, Y., Taheri, M., Taylor, S., Teslich, N., Tomeoka, K., Tomioka, N., Toppani, A., Trigo-Rodriguez, J.M., Troadec, D., Tsuchiyama, A., Tuzzolino, A.J., Tyliszczak, T., Uesugi, K., Velbel, M., Vellenga, J., Vicenzi, E., Vincze, L., Warren, J., Weber, I., Weisberg, M., Westphal, A.J. Wirick, S. Wooden, D. Wopenka, B. Wozniakiewicz, P. Wright, I. Yabuta, H. Yano, H. Young, E.D. Zare, R.N. Zega, T. Ziegler, K. Zimmerman, L. Zinner, E. and Zolensky, M. (2006) Comet 81P/Wild 2 under a microscope. *Science*, **314**, 1711–1716.

81 Flynn, G.J., Bleuet, P., Borg, J., Bradley, J.P., Brenker, F.E., Brennan, S., Bridges, J., Brownlee, D.E., Bullock, E.S., Burghammer, M., Clark, B.C., Dai, Z.R., Daghlian, C.P., Djouadi, Z., Fakra, S., Ferroir, T., Floss, C., Franchi, I.A., Gainsforth, Z., Gallien, J.-P., Gillet, P., Grant, P.G., Graham, G.A., Green, S.F., Grossemy, F., Heck, P.R., Herzog, G.F., Hoppe, P., Hoerz, F., Huth, J., Ignatyev, K., Ishii, H.A., Janssens, K., Joswiak, D., Kearsley, A.T., Khodja, H., Lanzirotti, A., Leitner, J., Lemelle, L., Leroux, H., Luening, K., MacPherson, G.J., Marhas, K.K., Marcus, M.A., Matrajt, G., Nakamura, T., Nakamura-Messenger, K., Nakano, T., Newville, M., Papanastassiou, D.A., Pianetta, P., Rao, W., Riekel, C., Rietmeijer, F.J.M., Rost, D., Schwandt, C.S., See, T.H., Sheffield-Parker, J., Simionovici, A., Sitnitsky, I., Snead, C.J., Stadermann, F.J., Stephan, T., Stroud, R.M., Susini, J., Suzuki, Y., Sutton, S.R., Taylor, S., Teslich, N., Troadec, D., Tsou, P., Tsuchiyama, A., Uesugi, K., Vekemans, B., Vicenzi, E.P., Vincze, L., Westphal, A.J., Wozniakiewicz, P., Zinner, E., and Zolensky, M.E. (2006) Elemental compositions of comet 81P/Wild 2 samples collected by stardust. *Science*, **314**, 1731–1735.

82 McKeegan, K.D., Aleon, J., Bradley, J., Brownlee, D., Busemann, H., Butterworth, A., Chaussidon, M., Fallon, S., Floss, C., Gilmour, J., Gounelle, M., Graham, G., Guan, Y., Heck, P.R., Hoppe, P., Hutcheon, I.D., Huth, J., Ishii, H., Ito, M., Jacobsen, S.B., Kearsley, A., Leshin, L.A., Liu, M.-C., Lyon, I., Marhas, K., Marty, B., Matrajt, G., Meibom, A., Messenger, S., Mostefaoui, S., Mukhopadhyay, S., Nakamura-Messenger, K., Nittler, L., Palma, R., Pepin, R.O., Papanastassiou, D.A., Robert, F., Schlutter, D., Snead, C.J., Stadermann, F.J., Stroud, R., Tsou, P.,

Westphal, A., Young, E.D., Ziegler, K., Zimmermann, L., and Zinner, E. (2006) Isotopic compositions of cometary matter returned by Stardust. *Science*, **314**, 1724–1728.

83 Sandford, S.A., Aleon, J., Alexander, C.M.O'D., Araki, T., Bajt, S., Baratta, G.A., Borg, J., Bradley, J.P., Brownlee, D.E., Brucato, J.R., Burchell, M.J., Busemann, H., Butterworth, A., Clemett, S.J., Cody, G., Colangeli, L., Cooper, G., D'Hendecourt, L., Djouadi, Z., Dworkin, J.P., Ferrini, G., Fleckenstein, H., Flynn, G.J., Franchi, I.A., Fries, M., Gilles, M.K., Glavin, D.P., Gounelle, M., Grossemy, F., Jacobsen, C., Keller, L.P., Kilcoyne, A.L.D., Leitner, J., Matrajt, G., Meibom, A., Mennella, V., Mostefaoui, S., Nittler, L.R., Palumbo, M.E., Papanastassiou, D.A., Robert, F., Rotundi, A., Snead, C.J., Spencer, M.K., Stadermann, F.J., Steele, A., Stephan, T., Tsou, P., Tyliszczak, T., Westphal, A.J., Wirick, S., Wopenka, B., Yabuta, H., Zare, R.N., and Zolensky, M.E. (2006) Organics captured from comet 81P/Wild 2 by the Stardust spacecraft. *Science*, **314**, 1720–1724.

84 Zolensky, M.E., Zega, T.J., Yano, H., Wirick, S., Westphal, A.J., Weisberg, M.K., Weber, I., Warren, J.L., Velbel, M.A., Tsuchiyama, A., Tsou, P., Toppani, A., Tomioka, N., Tomeoka, K., Teslich, N., Taheri, M., Susini, J., Stroud, R., Stephan, T., Stadermann, F.J., Snead, C.J., Simon, S.B., Simionovici, A., See, T.H., Robert, F., Rietmeijer, F.J.M., Rao, W., Perronnet, M.C., Simon, D.A., Okudaira, K., Ohsumi, K., Ohnishi, I., Nakamura-Messenger, K., Nakamura, T., Mostefaoui, S., Mikouchi, T., Meibom, A., Matrajt, G., Marcus, M.A., Leroux, H., Lemelle, L., Le, L., Lanzirotti, A., Langenhorst, F., Krot, A.N., Keller, L.P., Kearsley, A.T., Joswiak, D., Jacob, D., Ishii, H., Harvey, R., Hagiya, K., Grossman, L., Grossman, J.N., Graham, G.A., Gounelle, M., Gillet, P., Genge, M.J., Flynn, G., Ferroir, T., Fallon, S., Ebel, D.S., Dai, Z.R., Cordier, P., Clark, B., Chi, M., Butterworth, A.L., Brownlee, D.E., Bridges, J.C., Brennan, S., Brearley, A., Bradley, J.P., Bleuet, P., Bland, P.A., and Bastien, R. (2006) Mineralogy and petrology of comet 81P/Wild 2 nucleus samples. *Science*, **314**, 1735–1739.

85 Mumma, M.J., DiSanti, M.A., Dello Russo, N., Magee-Sauer, K., Gibb, E., and Novak, R. (2003) Remote infrared observations of parent volatiles in comets: a window on the early solar system. *Advances in Space Research*, **31**, 2563–2575.

86 Bockelée-Morvan, D., Lis, D.C., Wink, J.E., Despois, D., Crovisier, J., Bachiller, R., Benford, D.J., Biver, N., Colom, P., Davies, J.K., Gerard, E., Germain, B., Houde, M., Mehringer, D., Moreno, R., Paubert, G., Phillips, T.G., and Rauer, H. (2000) New molecules found in comet C/1995 O1 (Hale–Bopp). Investigating the link between cometary and interstellar material. *Astronomy and Astrophysics*, **353**, 1101–1114.

87 Bogard, D.D. and Johnson, P. (1983) Trapped noble gases indicate lunar origin for Antarctic meteorite. *Geophysical Research Letters*, **10**, 801–803.

88 Bogard, D.D. and Johnson, P. (1983) Martian gases in an Antarctic meteorite? *Science*, **221**, 651–654.

89 Marvin, U.B. (1983) The discovery and initial characterization of Allan Hills 81005: the first lunar meteorite. *Geophysical Research Letters*, **10**, 775–778.

90 Mayeda, T.K., Clayton, R.N., and Molini-Velsko, C.A. (1983) Oxygen and silicon isotopes in ALHA 81005. *Geophysical Research Letters*, **10**, 799–800.

91 Becker, R.H. and Pepin, R.O. (1984) The case for a Martian origin of the shergottites: nitrogen and noble gases in EETA 79001. *Earth and Planetary Science Letters*, **69**, 225–242.

92 Hiroi, T., Pieters, C.M., Zolensky, M.E., and Lipschutz, M.E. (1993) Evidence of thermal metamorphism on the C, G, B, and F asteroids. *Science*, **261**, 1016–1018.

93 Luu, J., Jewitt, D., and Cloutis, E. (1994) Near-infrared spectroscopy of primitive solar system objects. *Icarus*, **109**, 133–144.

94 Hiroi, T., Zolensky, M.E., Pieters, C.M., and Lipschutz, M.E. (1996) Thermal metamorphism of the C, G, B, and F asteroids seen from the 0.7 micron, 3 micron and UV absorption strengths in comparison with carbonaceous chondrites. *Meteoritics & Planetary Science*, **31**, 321–327.

95 Burbine, T.H., Binzel, R.P., Bus, S.J., and Clark, B.E. (2001) K asteroids and CO3/CV3 chondrites. *Meteoritics & Planetary Science*, **36**, 245–253.

96 Burbine, T.H., Buchanan, P.C., Binzel, R.P., Bus, S.J., Hiroi, T., Hinrichs, J.L., Meibom, A., and McCoy, T.J. (2001) Vesta, vestoids, and the howardite, eucrite, diogenite group: relationships and the origin of spectral differences. *Meteoritics & Planetary Science*, **36**, 761–781.

97 Hiroi, T., Zolensky, M.E., and Pieters, C.M. (2001) The Tagish Lake Meteorite: a possible sample from a D-type asteroid. *Science*, **293**, 2234–2236.

98 Hiroi, T. and Hasegawa, S. (2003) Revisiting the search for the parent body of the Tagish Lake meteorite – case of a T/D asteroid 308 Polyxo. *Antarctartic Meteorite Research*, **16**, 176–184.

99 Hartmann, W.K., Tholen, D.J., and Cruikshank, D.P. (1987) The relationship of active comets, extinct comets, and dark asteroids. *Icarus*, **69**, 33–50.

100 Campins, H. and Swindle, T.D. (1998) Expected characteristics of cometary meteorites. *Meteoritics & Planetary Science*, **33**, 1201–1211.

101 Lodders, K. and Osborne, R. (1999) Perspectives on the comet–asteroid–meteorite link. *Space Science Reviews*, **90**, 289–297.

102 Gounelle, M., Morbidelli, A., Bland, P.A., Spurny, P., Young, E.D., and Sephton, M.A. (2008) Meteorites from the outer solar system?, in *The Solar System Beyond Neptune* (eds M.A. Barucci, H. Boehnhardt, D.P. Cruikshank and A. Morbidelli), University of Arizona Press, Tucson, AZ, pp. 525–541.

103 Norton, O.R. (1998) What is a meteorite, in *Rocks from Space: Meteorites and Meteorite Hunters*, Mountain Press, Missoula, MT, pp. 175–240.

104 McSween H.Y. Jr. (1999) *Meteorites and their Parent Planets*, Cambridge University Press, Cambridge.

105 Weisberg, M.K., McCoy, T.J., and Krot, A.N. (2006) Systematics and evaluation of meteorite classification, in *Meteorites and the Early Solar System II* (eds D.S. Lauretta and H.Y. McSween Jr.), University of Arizona Press, Tucson, AZ, pp. 19–52.

106 Cronin, J.R. and Pizzarello, S. (1983) Amino acids in meteorites. *Advances in Space Research*, **3**, 5–18.

107 Cronin, J.R., Pizzarello, S., and Cruikshank, D.P. (1988) Organic matter in carbonaceous chondrites, planetary satellites, asteroids and comets, in *Meteorites and the Early Solar System* (eds J.F. Kerridhe and M.S. Matthews), University of Arizona Press, Tucson, AZ, pp. 819–857.

108 Cronin, J.R. and Chang, S. (1993) Organic matter in meteorites: Molecular and isotopic analyses of the Murchison meteorites, in *The Chemistry of Life's Origin* (eds J.M. Greenberg, C.X. Mendoza-Gomez and V. Pirronello), Kluwer, Dordrecht, pp. 209–258.

109 Lawless, J.G. and Yuen, G.U. (1979) Quantification of monocarboxylic acids in the Murchison carbonaceous meteorite. *Nature*, **282**, 396–398.

110 Yuen, G., Blair, N., Des Marais, D.J., and Chang, S. (1984) Carbon isotope composition of low molecular weight hydrocarbons and monocarboxylic acids from Murchison meteorite. *Nature*, **307**, 252–254.

111 Stoks, P.G. and Schwartz, A.W. (1979) Uracil in carbonaceous meteorites. *Nature*, **282**, 709–710.

112 Stoks, P.G. and Schwartz, A.W. (1981) Nitrogen-heterocyclic compounds in meteorites: significance and mechanisms

of formation. *Geochimica et Cosmochimica Acta*, **45**, 563–569.

113 Martins, Z., Botta, O., Fogel, M.L., Sephton, M.A., Glavin, D.P., Watson, J.S., Dworkin, J.P., Schwartz, A.W., and Ehrenfreund, P. (2008) Extraterrestrial nucleobases in the Murchison meteorite. *Earth and Planetary Science Letters*, **270**, 130–136.

114 Cooper, G., Kimmich, N., Belisle, W., Sarinana, J., Brabham, K., and Garrel, L. (2001) Carbonaceous meteorites as a source of sugar-related organic compounds for the early Earth. *Nature*, **414**, 879–883.

115 Meierhenrich, U.J., Muñoz Caro, G.M., Bredehöft, J.H., Jessberger, E.K., and Thiemann, W.H.-P. (2004) Identification of diamino acids in the Murchison meteorite. *Proceedings of the National Academy of Sciences of the United States of America*, **101**, 9182–9186.

116 Peltzer, E.T., Bada, J.L., Schlesinger, G., and Miller, S.L. (1984) The chemical conditions on the parent body of the Murchison meteorite: some conclusions based on amino, hydroxy, and dicarboxylic acids. *Advances in Space Research*, **4**, 69–74.

117 Cronin, J.R., Pizzarello, S., Epstein, S., and Krishnamurthy, R.V. (1993) Molecular and isotopic analyses of the hydroxy acids, dicarboxylic acids, and hydroxydicarboxylic acids of the Murchison meteorite. *Geochimica et Cosmochimica Acta*, **57**, 4745–4752.

118 Shimoyama, A. and Shigematsu, R. (1994) Dicarboxylic acids in the Murchison and Yamato-791198 carbonaceous chondrites. *Chemistry Letters*, 523–526.

119 Martins, Z., Watson, J.S., Sephton, M.A., Botta, O., Ehrenfreund, P., and Gilmour, I. (2006) Free dicarboxylic and aromatic acids in the carbonaceous chondrites Murchison and Orgueil. *Meteoritics & Planetary Science*, **41**, 1073–1080.

120 Cooper, G.W., Thiemens, M.H., Jackson, T.L., and Chang, S. (1997) Sulfur and hydrogen isotope anomalies in meteorite sulfonic acids. *Science*, **277**, 1072–1074.

121 Kvenvolden, K.A., Lawless, J., Pering, K., Peterson, E., Flores, J., Ponnamperuma, C., Kaplan, I.R., and Moore, C. (1970) Evidence for extraterrestrial amino-acids and hydrocarbons in the Murchison meteorite. *Nature*, **228**, 923–926.

122 Pering, K.L. and Ponnamperuma, C. (1971) Aromatic hydrocarbons in the Murchison meteorite. *Science*, **173**, 237–239.

123 Jungclaus, G.A., Yuen, G.U., Moore, C.B., and Lawless, J.G. (1976) Evidence for the presence of low-molecular-weight alcohols and carbonyl compounds in the Murchison meteorite. *Meteoritics*, **11**, 231–237.

124 Jungclaus, G., Cronin, J.R., Moore, C.B., and Yuen, G.U. (1976) Aliphatic amines in the Murchison meteorite. *Nature*, **261**, 126–128.

125 Cooper, G.W. and Cronin, J.R. (1995) Linear and cyclic aliphatic carboxamides of the Murchison meteorite: hydrolyzable derivatives of amino acids and other carboxylic acids. *Geochimica et Cosmochimica Acta*, **59**, 1003–1015.

126 Ehrenfreund, P., Glavin, D.P., Botta, O., Cooper, G., and Bada, J.L. (2001) Extraterrestrial amino acids in Orgueil and Ivuna: tracing the parent body of CI type carbonaceous chondrites. *Proceedings of the National Academy of Sciences of the United States of America*, **98**, 2138–2141.

127 Shimoyama, A., Ponnamperuma, C., and Yanai, K. (1979) Amino Acids in the Yamato carbonaceous chondrite from Antarctica. *Nature*, **282**, 394–396.

128 Shimoyama, A., Ponnamperuma, C., and Yanai, K. (1979) Amino acids in the Yamato-74662 meteorite, an Antarctic carbonaceous chondrite. *Memoirs National Institute of Polar Research*, **15**, 196–205.

129 Botta, O. and Bada, J.L. (2002) Amino acids in the Antarctic CM meteorite LEW90500 (abstract #1391). 33rd Lunar and Planetary Science Conference, Houston, TX.

130 Glavin, D.P., Dworkin, J.P., Aubrey, A., Botta, O., Doty, J.H. III, Martins, Z. and Bada, J.L. (2006) Amino acid analyses of Antarctic CM2 meteorites using liquid chromatography–time of flight–mass spectrometry. *Meteoritics & Planetary Science*, **41**, 889–902.

131 Shimoyama, A., Harada, K., and Yanai, K. (1985) Amino acids from the Yamato-791198 carbonaceous chondrite from Antarctica. *Chemistry Letters*, 1183–1186.

132 Shimoyama, A. and Ogasawara, R. (2002) Dipeptides and diketopiperazines in the Yamato-791198 and Murchison carbonaceous chondrites. *Origins of Life and Evolution of the Biosphere*, **32**, 165–179.

133 Botta, O., Glavin, D.P., Kminek, G., and Bada, J.L. (2002) Relative amino acid concentrations as a signature for parent body processes of carbonaceous chondrites. *Origins of Life and Evolution of the Biosphere*, **32**, 143–163.

134 Cronin, J.R., Pizzarello, S., and Moore, C.B. (1979) Amino acids in an Antarctic carbonaceous chondrite. *Science*, **206**, 335–337.

135 Holzer, G. and Oró, J. (1979) The organic composition of the Allan Hills carbonaceous chondrite (77306) as determined by pyrolysis–gas chromatography–mass spectrometry and other methods. *Journal of Molecular Evolution*, **13**, 265–270.

136 Kotra, R.K., Shimoyama, A., Ponnamperuma, C., and Hare, P.E. (1979) Amino acids in a carbonaceous chondrite from Antarctica. *Journal of Molecular Evolution*, **13**, 179–183.

137 Shimoyama, A. and Harada, K. (1984) Amino acid depleted carbonaceous chondrites (C2) from Antarctica. *Geochemical Journal*, **18**, 281–286.

138 Botta, O., Martins, Z., and Ehrenfreund, P. (2007) Amino acids in Antarctic CM1 meteorites and their relationship to other carbonaceous chondrites. *Meteoritics & Planetary Science*, **42**, 81–92.

139 Cronin, J.R. and Moore, C.B. (1971) Amino acid analyses of the Murchison, Murray, and Allende carbonaceous chondrites. *Science*, **172**, 1327–1329.

140 Kminek, G., Botta, O., Glavin, D.P., and Bada, J.L. (2002) Amino acids in the Tagish Lake Meteorite. *Meteoritics & Planetary Science*, **37**, 697–701.

141 Martins, Z., Alexander, C.M.O'D., Orzechowska, G.E., Fogel, M.L., and Ehrenfreund, P. (2007) Indigenous amino acids in primitive CR meteorites. *Meteoritics & Planetary Science*, **42**, 2125–2136.

142 Martins, Z., Hofmann, B.A., Gnos, E., Greenwood, R.C., Verchovsky, A., Franchi, I.A., Jull, A.J.T., Botta, O., Glavin, D.P., Dworkin, J.P., and Ehrenfreund, P. (2007) Amino acid composition, petrology, geochemistry, ^{14}C terrestrial age and oxygen isotopes of the Shisr 033 CR chondrite. *Meteoritics & Planetary Science*, **42**, 1581–1595.

143 McDonald, G.D. and Bada, J.L. (1995) A search for endogenous amino acids in the Martian meteorite EETA 79001. *Geochimica et Cosmochimica Acta*, **59**, 1179–1184.

144 Bada, J.L., Glavin, D.P., McDonald, G.D., and Becker, L. (1998) A search for endogenous amino acids in Martian meteorite ALH84001. *Science*, **279**, 362–365.

145 Glavin, D.P., Aubrey, A., Dworkin, J.P., Botta, O., and Bada, J.L. (2005) Amino acids in the Antarctic Martian meteorite MIL 03346 (abstract #1920). 36th Lunar and Planetary Science Conference, Houston, TX.

146 Glavin, D.P., Bada, J.L., Brinton, K.L.F., and McDonald, G.D. (1999) Amino acids in the Martian meteorite Nakhla. *Proceedings of the National Academy of Sciences of the United States of America*, **96**, 8835–8838.

147 Botta, O., Martins, Z., Emmenegger, C., Dworkin, J.P., Glavin, D.P., Harvey, R.P., Zenobi, R., Bada, J.L., and Ehrenfreund, P. (2008) Polycyclic aromatic hydrocarbons and amino acids in meteorites and ice samples from LaPaz icefield, Antarctica. *Meteoritics & Planetary Science*, **43**, 1465–1480.

148 Cronin, J.R. and Pizzarello, S. (1997) Enantiomeric excesses in meteoritic amino acids. *Science*, **275**, 951–955.

149 Pizzarello, S. and Cronin, J.R. (2000) Non-racemic amino acids in the Murray and Murchison meteorites. *Geochimica et Cosmochimica Acta*, **64**, 329–338.

150 Pizzarello, S., Zolensky, M., and Turk, K.A. (2003) Nonracemic isovaline in the Murchison meteorite: chiral distribution and mineral association. *Geochimica et Cosmochimica Acta*, **67**, 1589–1595.

151 Pollock, G.E., Cheng, C.-N., Cronin, S.E., and Kvenvolden, K.A. (1975) Stereoisomers of isovaline in the Murchison meteorite. *Geochimica et Cosmochimica Acta*, **39**, 1571–1573.

152 Cronin, J.R. and Pizzarello, S. (1999) Amino acid enantiomer excesses in meteorites: origin and significance. *Advances in Space Research*, **23**, 293–299.

153 Epstein, S., Krishnamurthy, R.V., Cronin, J.R., Pizzarello, S., and Yuen, G.U. (1987) Unusual stable isotope ratios in amino acid and carboxylic acid extracts from the Murchison meteorite. *Nature*, **326**, 477–479.

154 Pizzarello, S., Krishnamurthy, R.V., Epstein, S., and Cronin, J.R. (1991) Isotopic analyses of amino acids from the Murchison meteorite. *Geochimica et Cosmochimica Acta*, **55**, 905–910.

155 Pizzarello, S., Feng, X., Epstein, S., and Cronin, J.R. (1994) Isotopic analyses of nitrogenous compounds from the Murchison meteorite: ammonia, amines, amino acids, and polar hydrocarbons. *Geochimica et Cosmochimica Acta*, **58**, 5579–5587.

156 Engel, M.H., Macko, S.A., and Silfer, J.A. (1990) Carbon isotope composition of individual amino acids in the Murchison meteorite. *Nature*, **348**, 47–49.

157 Engel, M.H. and Macko, S.A. (1997) Isotopic evidence for extraterrestrial non-racemic amino acids in the Murchison meteorite. *Nature*, **389**, 265–268.

158 Pizzarello, S., Huang, Y., and Fuller, M. (2004) The carbon isotopic distribution of Murchison amino acids. *Geochimica et Cosmochimica Acta*, **68**, 4963–4969.

159 Pizzarello, S. and Huang, Y. (2005) The deuterium enrichment of individual amino acids in carbonaceous meteorites: a case for the presolar distribution of biomolecule precursors. *Geochimica et Cosmochimica Acta*, **69**, 599–605.

160 Scott, J.H., O'Brien, D.M., Emerson, D., Sun, H., McDonald, G.D., Salgado, A., and Fogel, M.L. (2006) An examination of the carbon isotope effects associated with amino acid biosynthesis. *Astrobiology*, **6**, 867–880.

161 Sephton, M.A. and Botta, O. (2005) Recognizing life in the Solar System: guidance from meteoritic organic matter. *International Journal of Astrobiology*, **4**, 269–276.

162 Tielens, A.G.G.M. (1983) Surface chemistry of deuterated molecules. *Astronomy and Astrophysics*, **119**, 177–184.

163 Millar, T.J., Bennett, A., and Herbst, E. (1989) Deuterium fractionation in dense interstellar clouds. *Astrophysical Journal*, **340**, 906–920.

164 Sandford, S.A., Bernstein, M.P., and Dworkin, J.P. (2001) Assessment of the interstellar processes leading to deuterium enrichment in meteoritic organics. *Meteoritics & Planetary Science*, **36**, 1117–1133.

165 Lerner, N.R., Peterson, E., and Chang, S. (1993) The Strecker synthesis as a source of amino acids in carbonaceous chondrites – deuterium retention during synthesis. *Geochimica et Cosmochimica Acta*, **57**, 4713–4723.

166 Peltzer, E.T. and Bada, J.L. (1978) α-Hydroxycarboxylic acids in the Murchison meteorite. *Nature*, **272**, 443–444.

167 Terzieva, R. and Herbst, E. (2000) The possibility of nitrogen isotopic fractionation in interstellar clouds. *Monthly Notices of the Royal Astronomical Society*, **317**, 563–568.

168 Aléon, J. and Robert, F. (2004) Interstellar chemistry recorded by nitrogen isotopes in Solar System organic matter. *Icarus*, **167**, 424–430.

169 Cronin, J.R., Cooper, G.W., and Pizzarello, S. (1995) Characteristics and formation of amino acids and hydroxy acids of the Murchison meteorite. *Advances in Space Research*, **15**, 91–97.

170 Lerner, N.R. and Cooper, G.W. (2005) Iminodicarboxylic acids in the Murchison meteorite: evidence of Strecker reactions. *Geochimica et Cosmochimica Acta*, **69**, 2901–2906.

171 Brownlee, D.E. (1985) Cosmic dust – collection and research. *Annual Review of Earth and Planetary Sciences*, **13**, 147–173.

172 Bradley, J.P. (1988) Analysis of chondritic interplanetary dust thin-sections. *Geochimica et Cosmochimica Acta*, **52**, 889–900.

173 Sandford, S.A. and Bradley, J.P. (1989) Interplanetary dust particles collected in the stratosphere: observations of atmospheric heating and constraints on their interrelationships and sources. *Icarus*, **82**, 146–166.

174 Taylor, S. and Brownlee, D.E. (1991) Cosmic spherules in the geologic record. *Meteoritics*, **26**, 203–211.

175 Rietmeijer, F.J.M. (1998) Interplanetary dust particles, *Reviews in Mineralogy and Geochemistry*, **36**, 2/1–2/95.

176 Genge, M.J. (2007) Micrometeorites and their implications for meteors. *Earth Moon Planets*, **102**, 525–535.

177 Love, S.G. and Brownlee, D.E. (1993) A direct measurement of the terrestrial mass accretion rate of cosmic dust. *Science*, **262**, 550–553.

178 Blanchard, M.B., Brownlee, D.E., Bunch, T.E., Hodge, P.W., and Kyte, F.T. (1980) Meteoroid ablation spheres from deep-sea sediments. *Earth and Planetary Science Letters*, **46**, 178–190.

179 Maurette, M., Olinger, C., Michel-Levy, M.C., Kurat, G., Pourchet, M., Brandstatter, F., and Bourot-Denise, M. (1991) A collection of diverse micrometeorites recovered from 100 tonnes of Antarctic blue ice. *Nature*, **351**, 44–47.

180 Taylor, A.D., Baggaley, W.J., and Steel, D.I. (1996) Discovery of interstellar dust entering the Earth's atmosphere. *Nature*, **380**, 323–325.

181 Taylor, S., Lever, J.H., and Harvey, R.P. (1998) Accretion rate of cosmic spherules measured at the South Pole. *Nature*, **392**, 899–903.

182 Maurette, M., Hammer, C., Brownlee, D.E., Reeh, N., and Thomsen, H.H. (1986) Placers of cosmic dust in the blue ice lakes of Greenland. *Science*, **233**, 869–872.

183 Maurette, M., Jehanno, C., Robin, E., and Hammer, C. (1987) Characteristics and mass distribution of extraterrestrial dust from the Greenland ice cap. *Nature*, **328**, 699–702.

184 Robin, E., Christophe Michel-Levy, N., Bourot-Denise, M., and Jéhanno, C. (1990) Crystalline micrometeorites from Greenland blue lakes: their chemical composition, mineralogy and possible origin. *Earth and Planetary Science Letters*, **97**, 162–176.

185 Sandford, S.A. (1987) The collection and analysis of extraterrestrial dust particles. *Fundamental Cosmic Physics*, **12**, 1–73.

186 Messenger, S. (2002) Opportunities for the stratospheric collection of dust from short-period comets. *Meteoritics & Planetary Science*, **37**, 1491–1505.

187 Flynn, G.J., Keller, L.P., Feser, M., Wirick, S., and Jacobsen, C. (2003) The origin of organic matter in the solar system: evidence from the interplanetary dust particles. *Geochimica et Cosmochimica Acta*, **67**, 4791–4806.

188 Flynn, G.J., Keller, L.P., Jacobsen, C., and Wirick, S. (2004) An assessment of the amount and types of organic matter contributed to the Earth by interplanetary dust. *Advances in Space Research*, **33**, 57–66.

189 Keller, L.P., Messenger, S., Flynn, G.J., Clemett, S., Wirick, S., and Jacobsen, C. (2004) The nature of molecular cloud material in interplanetary dust. *Geochimica et Cosmochimica Acta*, **68**, 2577–2589.

190 Clemett, S.J., Chillier, X.D.F., Gillette, S., Zare, R.N., Maurette, M., Engrand, C., and Kurat, G. (1998) Observation of indigenous polycyclic aromatic hydrocarbons in "giant" carbonaceous Antarctic micrometeorites. *Origins of Life and Evolution of the Biosphere*, **28**, 425–448.

191 Brinton, K.L.F., Engrand, C., Glavin, D.P., Bada, J.L., and Maurette, M. (1998) A search for extraterrestrial amino acids in carbonaceous Antarctic micrometeorites. *Origins of Life and Evolution of the Biosphere*, **28**, 413–424.

192 Glavin, D.P., Matrajt, G., and Bada, J.L. (2004) Re-examination of amino acids in Antarctic micrometeorites. *Advances in Space Research*, **33**, 106–113.

193 Matrajt, G., Pizzarello, S., Taylor, S., and Brownlee, D. (2004) Concentration and variability of the AIB amino acid in polar micrometeorites: implications for the exogenous delivery of amino acids to the primitive Earth. *Meteoritics & Planetary Science*, **39**, 1849–1858.

194 Cabane, M., Coll, P., Szopa, C., Israeel, G., Raulin, F., Sternberg, R., Mahaffy, P., Person, A., Rodier, C., Navarro-Gonzalez, R., Niemann, H., Harpold, D., and Brinckerhoff, W. (2004) Did life exist on Mars? Search for organic and inorganic signatures, one of the goals for "SAM" (sample analysis at Mars). *Advances in Space Research*, **33**, 2240–2245.

195 Bada, J.L., Sephton, M.A., Ehrenfreund, P., Mathies, R.A., Skelley, A.M., Grunthaner, F.J., Zent, A.P., Quinn, R.C., Josset, J.-L., Robert, F., Botta, O., and Glavin, D.P. (2005) Life on Mars: new strategies to detect life on Mars. *Astronomy & Geophysics*, **46**, 626–627.

196 Bada, J.L., Ehrenfreund, P., Grunthaner, F., Blaney, D., Coleman, M., Farrington, A., Yen, A., Mathies, R., Amudson, R., Quinn, R., Zent, A., Ride, S., Barron, L., Botta, O., Clark, B., Glavin, D., Hofmann, B., Josset, J.L., Rettberg, P., Robert, F., and Sephton, M. (2008) Urey: Mars Organic and Oxidant Detector. *Space Science Reviews*, **135**, 269–279.

197 Flynn, G.J. (1996) The delivery of organic matter from asteroids and comets to the early surface of Mars. *Earth Moon and Planets*, **72**, 469–474.

198 Bland, P.A. and Smith, T.B. (2000) Meteorite accumulations on Mars. *Icarus*, **144**, 21–26.

199 Biemann, K., Oro, J., Toulmin, P., Orgel, L.E., Nier, A.O., Anderson, D.M., Simmonds, P.G., Flory, D., Diaz, A.V., Rushneck, D.R., and Biller, J.A. (1976) Search for organic and volatile inorganic compounds in two surface samples from the Chryse Planitia region of Mars. *Science*, **194**, 72–76.

200 Biemann, K., Oro, J., Toulmin, P., Orgel, L.E., Nier, A.O., Anderson, D.M., Simmonds, P.G., Flory, D., Diaz, A.V., Rushneck, D.R., Biller, J.E., and Lafleur, A.L. (1977) The search for organic substances and inorganic volatile compounds in the surface of Mars. *Journal of Geophysical Research*, **82**, 4641–4658.

201 Benner, S.A., Devine, K.G., Matveeva, L.N., and Powell, D.H. (2000) The missing organic molecules on Mars. *Proceedings of the National Academy of Sciences of the United States of America*, **97**, 2425–2430.

202 Glavin, D.P., Schubert, M., Botta, O., Kminek, G., and Bada, J.L. (2001) Detecting pyrolysis products from bacteria on Mars. *Earth and Planetary Science Letters*, **185**, 1–5.

203 Kong, T.Y. and McElroy, M.B. (1977) Photochemistry of the Martian atmosphere. *Icarus*, **32**, 168–189.

204 Hunten, D.M. (1979) Possible oxidant sources in the atmosphere and surface of

Mars. *Journal of Molecular Evolution*, **14**, 71–78.

205 Bullock, M.A., Stoker, C.R., McKay, C.P., and Zent, A.P. (1994) A coupled soil-atmosphere model of H_2O_2 on Mars. *Icarus*, **107**, 142–154.

206 Nair, H., Allen, M., Anbar, A.D., Yung, Y.L., and Clancy, R.T. (1994) A photochemical model of the Martian atmosphere. *Icarus*, **111**, 124–150.

207 Encrenaz, Th., Bezard, B., Greathouse, T.K., Richter, M.J., Lacy, J.H., Atreya, S.K., Wong, A.S., Lebonnois, S., Lefevre, F., and Forget, F. (2004) Hydrogen peroxide on Mars: evidence for spatial and seasonal variations. *Icarus*, **170**, 424–429.

208 Quinn, R.C. and Zent, A.P. (1999) Peroxide-modified titanium dioxide: a chemical analog of putative Martian soil oxidants. *Origins of Life and Evolution of the Biosphere*, **29**, 59–72.

209 Yen, A.S., Kim, S.S., Hecht, M.H., Frant, M.S., and Murray, B. (2000) Evidence that the reactivity of the Martian soil is due to superoxide ions. *Science*, **289**, 1909–1912.

210 Huguenin, R.L., Miller, K.J., and Harwood, W.S. (1979) Frost-weathering on Mars – experimental evidence for peroxide formation. *Journal of Molecular Evolution*, **14**, 103–132.

211 Huguenin, R.L. (1982) Chemical weathering and the Viking biology experiments on Mars. *Journal of Geophysical Research*, **87**, 10069–10082.

212 Stoker, C.R. and Bullock, M.A. (1997) Organic degradation under simulated Martian conditions. *Journal of Geophysical Research*, **102**, 10881–10888.

213 Ten Kate, I.L., Garry, J.R.C., Peeters, Z., Quinn, R., Foing, B., and Ehrenfreund, P. (2005) Amino acid photostability on the Martian surface. *Meteoritics & Planetary Science*, **40**, 1185–1193.

214 Ten Kate, I.L., Garry, J.R.C., Peeters, Z., Foing, B., and Ehrenfreund, P. (2006) The effects of Martian near surface conditions on the photochemistry of amino acids. *Planetary and Space Science*, **54**, 296–302.

215 Kminek, G. and Bada, J.L. (2006) The effect of ionizing radiation on the preservation of amino acids on Mars. *Earth and Planetary Science Letters*, **245**, 1–5.

216 Aubrey, A.D., Cleaves, H.J., Chalmers, J.H., Skelley, A.M., Mathies, R.A., Grunthaner, F.J., Ehrenfreund, P., and Bada, J.L. (2006) Sulfate minerals and organic compounds on Mars. *Geology*, **34**, 357–360.

217 Kanavarioti, A. and Mancinelli, R.L. (1990) Could organic matter have been preserved on Mars for 3.5 billion years? *Icarus*, **84**, 196–202.

218 Peeters, Z., Quinn, R., Martins, Z., Becker, L., Brucato, J.R., Willis, P., Grunthaner, F., and Ehrenfreund, P. (2008) Mars regolith analogues – Interactions between mineralogical and organic compounds (abstract #1391). 39th Lunar and Planetary Science Conference, Houston, TX.

219 Peeters, Z., Quinn, R., Martins, Z., Sephton, M.A., Becker, L., Brucato, J.R., Willis, P., Grunthaner, F., and Ehrenfreund, P. (2009) Habitability on planetary surfaces: interdisciplinary preparation phase for future Mars missions. *International Journal of Astrobiology*, accepted.

220 Marlow, J.J., Martins, Z., and Sephton, M.A. (2008) Mars on Earth: soil analogues for future Mars missions. *Astronomy & Geophysics*, **49**, 2.20–2.23.

221 Chevrier, V. and Mathé, P.E. (2007) Mineralogy and evolution of the surface of Mars: a review. *Planetary and Space Science*, **55**, 289–314.

222 Anders, E. (1989) Pre-biotic organic matter from comets and asteroids. *Nature*, **342**, 255–257.

223 Schidlowski, M. (1988) A 3,800-million-year isotopic record of life from carbon in sedimentary rocks. *Nature*, **333**, 313–318.

224 Schopf, J.W. (1993) Microfossils of the early Archean apex chert: new evidence of the antiquity of life. *Science*, **260**, 640–646.

225 Moorbath, S. (2005) Palaeobiology: dating earliest life. *Nature*, **434**, 155.

226 Basiuk, V.A. and Douda, J. (1999) Pyrolysis of simple amino acids and nucleobases: survivability limits and implications for extraterrestrial delivery. *Planetary and Space Science*, **47**, 577–584.

227 Sears, D.W. (1975) Temperature gradients in meteorites produced by heating during atmospheric passage. *Modern Geology*, **5**, 155–164.

228 Love, S.G. and Brownlee, D.E. (1991) Heating and thermal transformation of micrometeoroids entering the earth's atmosphere. *Icarus*, **89**, 26–43.

229 Rietmeijer, F.J.M. (1996) The ultrafine mineralogy of molten interplanetary dust particle as an example of the quench regime of atmospheric entry heating. *Meteoritics & Planetary Science*, **31**, 237–242.

230 Greshake, A., Kloeck, W., Arndt, P., Maetz, M., Flynn, G.J., Bajt, S., and Bischoff, A. (1998) Heating experiments simulating atmospheric entry heating of micrometeorites: clues to their parent body sources. *Meteoritics & Planetary Science*, **33**, 267–290.

231 Flynn, G.J. (1989) Atmospheric entry heating – a criterion to distinguish between asteroidal and cometary sources of interplanetary dust. *Icarus*, **77**, 287–310.

232 Farley, K.A., Love, S.G., and Patterson, D.B. (1997) Atmospheric entry heating and helium retentivity of interplanetary dust particles. *Geochimica et Cosmochimica Acta*, **61**, 2309–2316.

233 Genge, M.J., Grady, M.M., and Hutchison, R. (1997) The textures and compositions of fine-grained Antarctic micrometeorites: implications for comparisons with meteorites. *Geochimica et Cosmochimica Acta*, **61**, 5149–5162.

234 Glavin, D.P. and Bada, J.L. (2001) Survival of amino acids in micrometeorites during atmospheric entry. *Astrobiology*, **1**, 259–269.

235 Rodante, F. (1992) Thermodynamics and kinetics of decomposition processes for standard α-amino acids and some of their dipeptides in the solid state. *Thermochimica Acta*, **200**, 47–61.

236 Matrajt, G., Brownlee, D., Sadilek, M., and Kruse, L. (2006) Survival of organic phases in porous IDPs during atmospheric entry: a pulse-heating study. *Meteoritics & Planetary Science*, **41**, 903–911.

237 Zhao, M. and Bada, J.L. (1989) Extraterrestrial amino acids in Cretaceous/Tertiary boundary sediments at Stevns Klint, Denmark. *Nature*, **339**, 463–465.

238 Wolbach, W.S., Lewis, R.S., and Anders, E. (1985) Cretaceous extinctions: evidence for wildfires and search for meteoritic material. *Science*, **230**, 167–170.

239 Lewis, R.S. and Wolbach, W.S. (1986) Search for noble gases at the K–T boundary. *Meteoritics*, **21**, 434–435.

240 Peterson, E., Horz, F., and Chang, S. (1997) Modification of amino acids at shock pressures of 3.5 to 32 GPa. *Geochimica et Cosmochimica Acta*, **61**, 3937–3950.

241 Pierazzo, E. and Chyba, C.F. (1999) Amino acid survival in large cometary impacts. *Meteoritics & Planetary Science*, **34**, 909–918.

242 Blank, J.G., Miller, G.H., Ahrens, M.J., and Winans, R.E. (2001) Experimental shock chemistry of aqueous amino acid solutions and the cometary delivery of prebiotic compounds. *Origins of Life and Evolution of the Biosphere*, **31**, 15–51.

243 Pierazzo, E. and Chyba, C.F. (2006) Impact delivery of pre-biotic organic matter to planetary surfaces, in *Comets and the Origins and Evolution of Life II* (eds P.J. Thomas, R. Hicks, C.F. Chyba and C.P. McKay), Springer, Heidelberg, pp. 137–168.

244 Tielens, A.G.G.M., Hony, S., van Kerckhoven, C., and Peeters, E. (1999) Interstellar and circumstellar PAHs. *Chemical Abstracts*, **131**, 329422.

245 Mumma, M.J., DiSanti, M.A., Magee-Sauer, K., Bonev, B.P., Villanueva, G.L., Kawakita, H., Dello Russo, N., Gibb, E.L.,

Blake, G.A., Lyke, J.E., Campbell, R.D., Aycock, J., Conrad, A., and Hill, G.M. (2005) Parent volatiles in comet 9P/Tempel 1: before and after impact. *Science*, **310**, 270–274.

246 Pizzarello, S., Huang, Y., Becker, L., Poreda, R.J., Nieman, R.A., Cooper, G., and Williams, M. (2001) The organic content of the Tagish Lake meteorite. *Science*, **293**, 2236–2239.

247 Sephton, M.A. (2004) Meteorite composition: organic matter in ancient meteorites. *Astronomy & Geophysics*, **45**, 8–14.

2
"Terrestrial" Amino Acids and their Evolution

Stephen Freeland

2.1
Introduction

During the twentieth century, it came as something of a surprise to learn that beneath the surface of impressively diverse phenotypes, all life on Earth shares a remarkably unvarying fundamental biochemistry. In particular, the discovery of DNA paved the way for Crick's [1] declaration of a "central dogma" to life's molecular biology: that in all organisms, information stored in the form of DNA is first transcribed into the chemical sister language of RNA, and then this RNA is quite literally decoded into proteins which form the structures and catalyze the reactions necessary for metabolism [1]. (The technical definition of a "code" is "a system of symbols ... used to represent assigned ... meanings." In the case of the genetic code, the symbols are nucleotides and the assigned meanings are amino acids. In recent years, there has been an increasingly tendency for the science-reporting media to refer to genomes or genomic material in general as "the genetic code." This shift in meaning not only creates ambiguity, but is an incorrect use of the word.)

Now, four decades later, one of the first facts that students of molecular biology learn is that life builds with a set of 20 amino acids: phenylalanine (Phe), leucine (Leu), isoleucine (Ile), methionine (Met), valine (Val), serine (Ser), proline (Pro), threonine (Thr), alanine (Ala), tyrosine (Tyr), histidine (His), glutamine (Gln), lysine (Lys), asparagine (Asn), aspartic acid (Asp), glutamic acid (Glu), cysteine (Cys), tryptophan (Trp), arginine (Arg), and glycine (Gly). However, any discussion as to how and why this particular set of building blocks came to play their central biological role must start by pointing out that (like most foundational generalizations of biology) the simple statement that life builds with 20 amino acids is quite simply untrue. It is true to say that *most naturally occurring* genomes (organisms) *directly genetically encode* this *standard set* of 20 amino acids, but each aspect of this statement deserves careful consideration: it is the qualifiers and their exceptions that provide an appropriate background to the clues that current science has amassed regarding how and why the "terrestrial amino acids" evolved.

Amino Acids, Peptides and Proteins in Organic Chemistry. Vol.1 – Origins and Synthesis of Amino Acids.
Edited by Andrew B. Hughes

ISBN: 978-3-527-32096-7

2.2
What are the 20 "Terrestrial" Amino Acids?

First and foremost, organisms utilize not 20 but hundreds if not thousands of amino acids within their metabolism. These range from the tens of intermediates (such as citrulline and ornithine) via which shared, fundamental biochemistry biosynthesizes and catabolizes the 20 standard amino acids, through to the hundreds, perhaps thousands of post-translational modifications that are made to amino acid residues after their initial incorporation into a protein chain (e.g., [2–4]), including some of the more arcane examples that are used in specialized roles such as toxins and antibiotics (e.g., [5, 6]). In this sense, our discussion begins by considering what precisely defines the "terrestrial" amino acids? To state that we are interested in the origin of the *genetically encoded* amino acids is a deceptively simple answer: what exactly does it mean to be "genetically encoded?" If we mean that genetic instructions specify the occurrence of a specific amino acid within molecular biology then we fail to clearly distinguish any subset from this cast of thousands. Functioning genomes comprise, by definition, the genetic material necessary for self replication. In other words, they are (to a first approximation) independent units of life. As such, all the information necessary to make metabolic intermediates, post-translational modifications, and other naturally biosynthesized amino acids is genetically encoded by the organism in which they occur. Meanwhile, many genomes lack the metabolic pathways necessary to biosynthesize subsets of the standard amino acid alphabet (in *Homo sapiens,* these "essential" amino acids comprise eight members – 40% of the standard alphabet [7]), such that the presence of these amino acids within the metabolism of the associated organism is not, in this sense, "genetically encoded." Moving beyond the genetically programmed existence of amino acids within living cells and refocusing on their direct incorporation into a growing peptide chain clarifies things a little. This excludes all post-translational modifications, and all amino acids whose function is restricted to that of metabolic intermediates and intracellular signals (e.g., [8]) (though of course there is increasing recognition that the "standard 20" are utilized in many important roles other than as residues within a protein sequence, e.g., [9]). However, even this concept of genetic coding still fails to distinguish the standard alphabet from amino acids that are used as building blocks in secondary versions of protein synthesis. For example, "nonribosomal peptide synthesis" occurs in at least two of the three domains of life – bacteria and eukaryotes [10, 11]. Here, large and complex enzymes polymerize short runs of amino acids that include such oddities as halogen-bearing side-chains and even D-amino acids. The results are short peptides of notably strong bioactivity, including antimicrobial peptides [12] and the deadly toxins of such colorfully named fungi as the Destroying Angel, *Amanita bisporigera* [13]. The full scope of amino acids used in this way remains unknown, but coupled to other forms of "noncanonical" protein synthesis (such as that found in bacterial cell walls [14]) these phenomena illustrate that not even the concept of "amino acids directly incorporated during protein synthesis" is enough to define the standard alphabet of 20. Rather, their uniqueness is inextricably tied to other, specific molecules that perform standard decoding of genes into protein products. In particular, they must be defined as being those that are found to naturally occur at one end of an appropriate transfer RNA (tRNA) molecule which in turn is ready, in the physical context of a ribosome, to recognize a specific triplet codon of messenger RNA (mRNA) nucleotides at the other end so as to effect translation. In recognizing that the definition of "genetically encoded" amino acids requires specific mention of associated molecular machinery, it might seem tempting to include additional elements of translation – for example, the appropriate aminoacyl-tRNA synthetase (aaRS) enzyme that ensures that each amino acid is attached onto the appropriate tRNA. However, at this point the definition starts to break down once more, as species occur within at least two of the three domains of life which prepare some of their pretranslational aminoacyl-tRNA complexes by modifying the amino acid residue *in situ*. Archaea, for example, can prepare for asparagine translation by first charging the $tRNA^{Asn}$ with an aspartic acid residue and then enzymatically converting this to an asparagine residue while it resides on the end of the tRNA. A similar process can occur in the formation of $tRNA^{Gln}$ [15]. Thus, the one and only defining characteristic shared by the 20 constituents of the standard amino acid alphabet is they may be found attached to a tRNA such that they can reliably decode one or more specific codon's worth of mRNA information.

2.2.1
The 21st and 22nd Genetically Encoded Amino Acids

Even here we run into complications (this is, after all, biology!) Specifically, within the diversity of life that has evolved on this planet, genomic research has now uncovered two additional amino acids that are directly genetically encoded by natural genomes in the sense that they occur on tRNA's in such a manner as to be directly incorporated into nascent peptides during ribosomal protein synthesis. The first of these to be discovered, selenocysteine (Sec), is used by representatives from all three domains of life (Bacteria, Archaea and Eukaryota) [16]. Here, the amino acid may be thought of as a cysteine in which the sulfur group has been replaced by the more reactive counterpart, selenium. Where it occurs, the amino acid has captured a "stop" codon, UGA (i.e., one that is normally a signal to "terminate protein translation"). Typical textbook accounts distinguish selenocysteine from the standard 20 amino acids by noting its translation is abnormal in three main aspects: (i) the amino acid is not directly charged onto its appropriate $tRNA^{Sec}$, but rather is made *in situ* by an enzyme that produces it from serine bound to the $tRNA^{Sec}$; (ii) the tRNA in question is unusual in structure and requires the presence of another protein (the SelB or mSelB) in order to compete its way into the ribosome for translation; and (iii) in order for UGA to be translated as selenocysteine, the mRNA in which its codon appears must also possess a second motif, the so-called SECIS (selenocysteine insertion sequence) element, that induces "misreading" of UGA from its usual meaning of "stop."

In fact, none of these objections makes selenocysteine translation qualitatively different from "normal" translation of the standard 20 amino acids. As mentioned above, translation of glutamine and asparagine in some archaeal lineages requires *in situ* enzyme modification of another amino acid bound to the appropriate tRNA; all aminoacyl-tRNA's use "elongation factor" proteins to complete translation and recent literature has seen a growing number of reports that the effective translation of "normal" amino acids can depend on a broader mRNA context than the existence of three nucleotides listed as an appropriate codon in textbook illustrations of the standard genetic code. (While this has been particularly noted for the codons that initiate and terminate translation, the concept of "codon context" now seems more widely applicable [17].) This serves as a general reminder that gene-to-protein translation is not equivalent to the "digital" process that converts streams of 1s and 0s into English sentences on my word-processor: rather it is performed by molecular machines which function with a given level of speed and accuracy that can be influenced by all sorts of contextual factors [18]).

More recently, researchers have found a 22nd amino acid, pyrrolysine (Pyl), encoded by the genomes of certain lineages of methanogenic bacteria within the domain Archaea [19]. (Pyrrolysine has also been detected within the genetic code of a single bacterium: *Desulfitobacterium hafniense*. This occurrence is generally interpreted as evidence for "lateral gene transfer", that is, that *D. hafniense* possesses the pyrrolysine-encoding machinery not by virtue of sharing a common ancestor with the Archaea, but rather by having "stolen" the genes by some means during more recent evolutionary history.) Once again this added amino acid has taken over a stop codon, this time UAG. Furthermore, it is associated with a *cis*-acting motif, the PYLIS (pyrrolysine insertion sequence), within the mRNA for effective translation. In fact, its biggest difference from selenocysteine translation is that pyrrolysine is charged directly onto its corresponding tRNA by a fairly normal-looking class II aminoacyl synthetase enzyme [20]: pyrrolysine is even less clearly distinguished than selenocysteine from the standard 20 amino acids!

As such, the 20 members of the standard amino acid alphabet can only be fully distinguished in terms of evolutionary history. Overwhelming evidence points to the idea that all living species (Bacteria, Archaea and Eukaryota) share a common evolutionary ancestor. Within the origins-of-life research community, the point at which these lineages meet is referred to as the Last Universal Common Ancestor (LUCA), and much time and effort has been spent in elucidating the genome and metabolism of this putative organism (e.g., [21–23]). At its simplest, the idea is that if a majority of representatives from all three domains contain a specific genomic trait, then the most parsimonious explanation is that it was present in LUCA. As such, the narrow occurrence of pyrrolysine makes it a clear candidate for post-LUCA emergence and even though selenocysteine is found in all three domains of life, its distribution is scattered and infrequent enough (e.g., it appears to be completely absent from plants and fungi) that it is widely regarded as a later addition, propagated by lateral transfer of genetic material between separate, coexisting species (e.g., [24]). Indeed, this view of selenocysteine as a late arrival is supported by the unusual molecular details of selenocysteine translation.

2.2.2 Do other Genetically Encoded Amino Acids Await Discovery?

Given that one of these two additional amino acids was noticed only recently, the question naturally arises as to whether any more await discovery? Several analyses have looked for tell-tale signs of new amino acids within the repertoire of genomic and proteomic sequence data already deposited into public databases. The first of these [25] built on standard methods for detecting gene homologs using BLAST (http://www.ncbi.nlm.nih.gov/BLAST/, i.e., using a computer algorithm to find known proteins that are statistically similar to a query sequence of interest). Noting that selenocysteine and pyrrolysine each use a "stop codon" of the standard genetic code, the researchers looked for putative protein-coding genes, "open reading frames" (ORF's), that match known proteins if decoding were to continue through the ORF's first apparent stop codon. The technique was able to successfully identify many known selenocysteine- and pyrrolysine-containing proteins, and to suggest previously uncharacterized genes as likely members of this group. It further correctly identified other genomes in which one or more stop codons are reassigned to one of the 20 amino acids of the standard alphabet (e.g., UGA, reassigned from "stop" to tryptophan in *Mycoplasma genitalium*; see [26] for a review of this interesting phenomenon). However, these authors did not report any suggestion of additional amino acids.

A different group of researchers continued to refocus the search explicitly on the possible existence of new amino acids and moved to the more sophisticated strategy of sifting through all known sequence data for tRNA genes, searching for those that appear suitable for decoding stop codons [27]. This time the analysis directly concluded that "the occurrence of additional amino acids that are widely distributed and genetically encoded is unlikely." Finally, a third group of researchers returned to the idea of searching for potential ORFs with internal stop codons [28]. Once again, the conclusion was that "no candidate for the 23rd amino acid was discovered." Thus, the current consensus is that life genetically encodes just 22 amino acids. However, while this work is fully competent as bioinformatics research, its conclusions are circumscribed by the general limitations of all such work: it is only as good as the assumptions on which it builds. Aside from the inherent restrictions of each technique (e.g., that known homologs exist for novel-amino-acid-containing proteins, that we know how to recognize a highly unusual tRNA, or that additional amino acids will have taken over a stop codon), a more general limitation applies. Recent advances in genome-sequencing technology have started to unveil the enormous scope of genomes that remain entirely unsequenced. For example, samples of surface-water biodiversity [29] join those of deeper-dwelling oceanic microbial communities (e.g., [30] and references therein) and soil [31] to show that current sequence data barely scratches the surface of what the biosphere has to offer. The human intestine alone may contain of the order of 40 000 largely uncharacterized microbial species [32]. It would thus seem premature to announce with any confidence that only 22 amino acids are used in ribosomal protein synthesis.

2.2.3 Genetic Engineering can Enlarge the Amino Acid Alphabet

The idea that further amino acids could one day be found gains support from another frontier, this time within applied molecular biology, where research has been steadily revealing a lack of physical or mechanical restrictions as to which amino acids can be incorporated into ribosomal peptide synthesis. Here, researchers are actively manipulating the molecular machinery of the genetic code – with enough success that it is now relatively formulaic (if not quite routine) to incorporate human-chosen amino acids into the otherwise natural protein translation machinery [33]. The earliest report of such an experiment was "top-down" in the sense of using artificial selection to breed strains of *Bacillus subtilis* that would incorporate 4-fluorotryptophan instead of tryptophan [34]. For a long time after this announcement, no follow-up studies were attempted, and when this approach was tried again more recently using *Escherichia coli*, it met with limited success [35]. The latter research team concluded that "the incorporation of unnatural amino acids into organismal proteomes may be possible but . . . extensive evolution may be required to reoptimize proteins and metabolism . . . complete replacement of an amino acid through selective breeding is far harder than originally thought" [36]. However, these same researchers then went on to breed, with relative ease, a bacteriophage that could grow successfully in media containing either tryptophan or 6-fluorotryptophan [37]. They noted that the evolved genome had undergone a relatively small series of mutations that allowed it to tolerate this amino acid "ambiguity" and concluded that "these results support the 'ambiguous intermediate' hypothesis for the emergence of divergent genetic codes, in which the adoption of a new genetic code is preceded by the evolution of proteins that can simultaneously accommodate more than one amino acid at a given codon." In other words, while the full replacement of one amino acid by another seems to be a major evolutionary undertaking, the addition of a new amino acid (particularly one that is in some way similar to an existing amino acid) can be surprisingly easy.

Meanwhile, a "bottom-up" engineering approach has used rational design of artificial aaRS enzymes and tRNAs to introduce more than 30 unnatural amino acids into bacterial, yeast, and mammalian cells [38]. Given that new reports appear at regular intervals (e.g., [39–41]), it would seem that this is a frontier whose limits lie far ahead of current achievements. As artificially constructed terrestrial amino acid alphabets start to dwarf that which emerged over the course of natural evolution, it becomes increasingly clear that the standard translational alphabet of just 20 (22) amino acids represents a small subset of what was possible.

2.2.4
Significance of Understanding the Origins of the Standard Alphabet

An introduction that whittles a precise definition for the 20 members of the standard amino acid alphabet may seem pedantic and time spent discussing the potential for confusion here somewhat contrived. In fact, the amino acids that lie outside (and at the blurred edges) of this definition not only have provided important insights into where the standard alphabet came from (Sections 1.1.2.3 and 1.1.2.4), but also serve to illustrate the significance of the search for its evolutionary causes. Two frontiers of life-science research stand to benefit significantly from greater understanding here: one ("astrobiology") is such pure science that it is sometimes mistaken for philosophical speculation, the other ("synthetic biology") is so applied and commercially relevant that it is sometimes mistaken for engineering.

Astrobiology is concerned with the prevalence, distribution, and characteristics of life within our universe. Although at present the only known example of life is that which has descended from a common ancestor on Earth, this seems increasingly likely to change. New insights from the physical and geophysical sciences are starting to suggest that the universe may be predisposed to the sorts of planetary conditions from which we emerged [42]. A natural next question is what (if anything) can science infer about the likely nature of life that originates elsewhere? Early attempts to extrapolate from Earth's biochemistry as a whole have reached very different conclusions about whether we should anticipate a more or less universal biochemistry for all life, whatever its origin (e.g., contrast [43] with [44]). Although the amino acids form a natural subset of biomolecules on which to focus the debate, even here ideas oscillate between extremes. Some have argued that the variation and distribution of proteins found in nature suggest that the particular protein families we find populating our biosphere were as inevitable to evolution as inorganic crystal structures are to physics [45]; others disagree entirely [46]. However, the debate has so far failed to consider the role of the amino acid alphabet from which protein folds are constructed: we simply do not know if an enlarged or different amino acid alphabet would unlock the gates to whole new areas of structure and function.

Meanwhile, applied molecular biology has made such great progress in the past generation that a basic understanding of life's central dogma has moved via recombinant DNA technology into genetic engineering of ever broader scope. Researchers now talk increasingly of "synthetic biology" [47], referring to biological systems that are no longer manipulations of naturally occurring counterparts but instead have been designed *de novo* by human imagination. While those approaching the topic from biology focus on designing and then synthesizing genes, chromosomes, and entire genomes, those coming from chemistry are manipulating the fundamental building blocks – nucleotides and amino acids. Thus, when the biologists and chemists are able to combine their science, the potential will exist to directly and empirically explore the potential for new alphabets to make new structures and functions.

Thus, these superficially dissimilar research directions meet to share a common question: what would be the consequence of altering the standard amino acid alphabet? A theoretical framework here would be a powerful addition to current knowledge and a logical place to start is by asking if there is any reason why evolution "chose" a particular set of 20 amino acids for direct genetic encoding? For one community answers would inform our expectations (and search) for extraterrestrial life, for the other they would guide the exploration of a new materials' science.

2.3 What do We Know about the Evolution of the Standard Amino Acid Alphabet?

Against this background, current knowledge regarding the origin and evolution of the standard amino acid alphabet has come from three directions: (i) the role of prebiotic chemistry in producing amino acids before life emerged on Earth, (ii) the role of metabolism in "inventing" new amino acids during early evolution and (iii) the role of natural selection in discriminating between different amino acid options.

2.3.1 Nonbiological, Natural Synthesis of Amino Acids

One of the simplest questions that can be asked regarding the standard alphabet is where the first amino acids came from. In the absence of Intelligent Design (for which no scientific evidence exists), either life got started without amino acids and only later "invented" them through metabolism or they result from nonbiological processes of chemical synthesis. Broadly speaking, the latter option may be viewed as part of the larger "heterotrophic" theory for the origin of life, which claims that Earth's biology originated from an accumulation of organic compounds (e.g., [48]). Thus, a legitimate and tractable question for chemists has been what conditions and reactants are necessary or conducive to the nonbiological synthesis of amino acids?

The earliest work here was speculative: both Oparin [49] and Haldane [50] proposed that an appropriate energy source applied to a "primordial soup" of carbon-containing molecules would be able to produce amino acids. It took three decades before Miller and Urey demonstrated this possibility empirically [51]: the first results of the now-famous spark tube experiments were reported in 1953, and concluded that methane, ammonia, water, and hydrogen could react in the presence of a suitable energy source to produce key members of the standard amino acid alphabet. Subsequent extensions to these experiments have seen researchers move beyond the original suggestion of an oceanic origin for life to simulate many specific alternatives (e.g., tidal pools, oceans covered by glaciers, hydrothermal vents, or even the atmosphere itself; see [52, 53] for reviews) and many different energy sources (e.g., electricity [54], heat [55], and radiation [56]). However, all results cluster around the finding that reliable and unambiguously plausible abiotic synthesis of amino acids requires chemically reducing conditions, and this remains a highly contentious point for the prebiotic atmosphere (e.g., [57–60]).

In light of this, some researchers have gone on to point out that if the Earth's atmosphere were not itself sufficiently reducing then organic compounds could have been brought to our planet by "dust particles, comets, and meteorites" [48]. In other words, extraterrestrial synthesis may have been a primary source for the organic compounds that led to the origin of life. Although this idea is an old one [61], recent estimates that "the Earth acquires 100 000–1 000 000 kg of such material each day" [62] have done much to raise its profile within the origins-of-life research community.

Attempts to understand the relevant chemistry here include direct simulations and remote-sensing observations of extraterrestrial chemistry (e.g., [63]); however, the older and more thoroughly developed information in this field comes from analysis of carbonaceous chondrite meteorites – a class of stony, extraterrestrial body that formed from accretion of dust in the primitive solar system and subsequently did not experience the high temperatures or other traumatic events that would have disrupted their geochemical composition (e.g., [64]). Compounds that have been found in such meteorites include "aliphatic hydrocarbons, aromatic hydrocarbons, amino acids, carboxylic acids, sulfonic acids, phosphonic acids, alcohols, aldehydes, ketones, sugars, amines, amides, nitrogen heterocycles, sulfur heterocycles and a relatively abundant high molecular weight macromolecular material" and current thinking suggests that they may have brought significant quantities of a rich variety of organics to prebiotic Earth [62]. The Murchison meteorite, which fell to Earth in 1969, is worthy of particular mention in that it has offered an excellent sample with which to study the typical abiotic chemistry because of its relatively good state of preservation and large mass (i.e., 100 kg). Since its discovery, research into the organic composition of the meteorite has been continuous (e.g., [65–77]), and has produced a highly detailed picture of the types and relative abundances of amino acids present. As a result, this particular meteorite has become "generally used as the standard reference for organic compounds in extraterrestrial material" [72] and a widespread view is that the meteorite "offer[s] an invaluable sample for the direct analysis of abiotic chemical evolution prior to the onset of life" [77].

Meanwhile, a smaller body of research has shifted focus away from "black box" observations of what various prebiotic mixtures can produce, concentrating instead on the thermodynamics of amino acid synthesis. Amend and Shock [78], for example, calculated the free energy of formation of the amino acids from CO_2, NH_4^+, and H_2 for conditions of low pressure and low temperature (consonant with "surface" synthesis), and for an alternative of high pressure and high temperature (that would better represent the situation in a deep-sea "hydrothermal vent"). Their primary finding was that this difference in conditions could completely alter the landscape of prebiotic plausibility: many of the large, chemically complex amino acids that are never observed in Miller–Urey-type experiments switch from high to low energy of formation when the conditions change.

The combined output of these experiments, analyses, observations, and critiques has built into a sizeable body of literature on abiotic synthesis of amino acids (e.g., [61–63, 79–81] and references therein). An authoritative review would not only require an entire book chapter to itself, but would distract from the bigger landscape of thinking about the origin of the standard amino acid alphabet. For present purposes, a more enlightening view comes from extracting the generalizations that can be made about this work. A first major step here was made by Wong and Bronskill [82], who argued that beneath all the varying details is an underlying consistency that the standard amino acids can be divided into two groups: prebiotically plausible and prebiotically implausible (or evolutionary "early" versus "late"). To give one specific observation, eight of the 20 standard amino acids have been identified within the Murchison meteorite (Gly, Ala, Asp, Glu, Val, Ile, Leu, and Pro). Spark-tube experiments typically produce exactly the same eight, in broadly similar proportions, as well as small amounts of a ninth (Thr) [83]. A much more recent, more thorough meta-analysis that compared the contents of meteorites with prebiotic simulations produced a very similar picture (Figure 2.1), and went on to calculate a ranking of prebiotic plausibility (from "most confidently early" to "most confidently late" arrivals in the genetic code) [84]. By cross-referencing this ranking

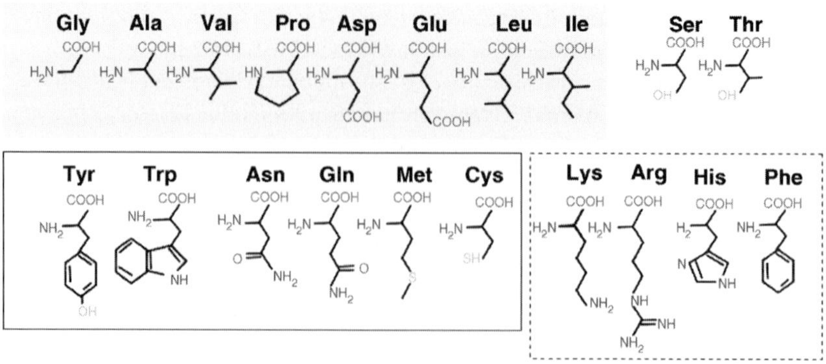

Figure 2.1 The 20 members of the standard amino acid alphabet. Those shown against a gray background (Gly, Ala, Val, Pro, Asp, Glu, Leu, and Ile) have been found in the Murchison meteorite [69]; those enclosed by a solid black box (Tyr, Trp, Asn, Gln, Met, and Cys) have not been observed as products of any major prebiotic synthesis or meteorite study. Those shown in a dashed gray box have the weakest support from prebiotic syntheses: Arg and His have only been produced in the presence of catalysts, whereas Lys and Phe have also been produced in experiments using high-temperature/pressure conditions simulating a hydrothermal vent. Further details may be found in Higgs and Pudritz [84].

system with the thermodynamic calculations of Amend and Shock [78], the authors were able to show a surprisingly strong correlation ($R = 0.96$) between how often an amino acid showed up in prebiotic simulations and meteorites, and its energy of formation at "surface conditions." (The authors note that this correlation does not exist for energy of formation at putative hydrothermal conditions.) In short, the typical amino acid contents of meteorites match the typical products of prebiotic simulations, which in turn match the free energy of formation predicted for low temperature and low pressure. Taken together, this adds up to a strong inference that before life got started, nonbiological processes would have produced around half of the amino acids that we see in today's standard amino acid alphabet.

This potentially important finding comes with one caveat. Scientific consensus should always be examined carefully: it is strong and useful only when it builds from genuinely independent results. In the case of amino acid abiotic synthesis, this is only partly the case. The match between prebiotic simulations and extraterrestrial chemistry is gratifying in that these are two relatively independent sources of data: when Miller and Urey began the spark tube experiments, they had no idea that meteorite analysis would one day be possible (indeed the controversy over reducing conditions on early Earth highlights the possibility that the two sources of data could easily have emerged in serious disagreement). Within the prebiotic simulations themselves, however, the choice of initial conditions is not as varied as one might perhaps hope – few studies have focused on high temperatures and high pressures where, as Amend and Shock observe, the products of abiotic synthesis are likely to be very different [78]. Since the underlying point of their thermodynamic calculations was to highlight this difference, it is not so very surprising that a consensus of prebiotic simulations (which are in effect swamped by examples of low temperature/pressure assumptions) is in agreement with energy of formation for surface synthesis rather than that associated with high temperature and high pressure. Indeed, one way of reading Higgs and Pudritz meta-analysis [84] is to say that generally speaking, for any variation of "surface" conditions (from room temperature and 1 atm pressure down to the near-zero temperature and pressure of space) amino acids tend to be produced in direct proportion to a fairly consistent picture thermodynamic stability – and this means that small, chemically simple amino acids appear first and in greatest quantity. Nothing here argues against the possibility that amino acids were prebiotically synthesized under very different conditions.

Hydrothermal vents are worthy of special consideration because they have recently been gathering more and more attention within the origins-of-life research community as different strands of evidence point to the idea that early life (and perhaps the origin of life) began at high temperature and pressure (e.g., [85–88]). However, research surrounding the abiotic chemistry of hydrothermal vents is relatively young and so far its proponents have tended towards the alternative, more recently developed "autotrophic" theory for the origin of life [89] according to which the importance of prebiotic chemistry lies not in the stockpiling of useful organic molecules, but rather in the idea that the synthetic pathways themselves eventually became living metabolism. (For example, one contemporary version of this idea is the "pioneer organism theory" which envisages that life began in "a momentary, mechanistically definite origin by autocatalytic carbon fixation within a hot, volcanic flow in contact with transition metal catalysts" [89]; in this particular case, the abiotic chemistry that occurred in this environment is argued to have become the central metabolism of an iron–sulfur metabolism.) As a result, researchers have not sketched a clear picture of the prebiotic amino acid composition or abundances that were likely associated with this environment [88, 90]. The larger field of abiotic synthesis urgently needs these sorts of facts and figures if it is to start formulating direct tests that discriminate, based on these two different chemistries, which model fits best with other clues for the origin of the amino acid alphabet.

2.3.2
Biosynthetic Theories for the Evolutionary Expansion of the Standard Amino Acid Alphabet

Accepting for the moment that the constituent members of the standard amino acid alphabet may be divided into those that are prebiotically plausible and those that are prebiotically implausible, the next question to be tackled is where did the latter class of molecules come from? A general answer here does in fact come from the broader considerations of the autotrophic theory which suggests that from its inception, life synthesized its own organic molecules. Taking this view at face value, the origin of amino acids lies in the evolution of metabolism. However, there is no need to suppose that such metabolism initially involved all 20 standard amino acids: it makes sense to suppose that a pool of abiotically plausible pathways allowed life to get started and then subsequent evolution added increasingly sophisticated molecules into the mix [91]. One of the earliest attempts to develop these ideas with reference to the amino acids came when Dillon [92] considered genus *Thioploca*'s growth solely on

H_2O, H_2S, NH_3, and CO_2, and hypothesized that its metabolic pathways manipulated such simple molecules into the amino acids of the standard alphabet. While his assumed metabolic pathways were largely guesses that turned out to be wrong in many details, it is of course true that modern organisms can and do synthesize amino acids from simple molecules. It thus remains tempting to picture a network of prebiotic reaction pathways that interconverts simple, abiotically plausible molecules and becomes coopted into metabolism. Along these lines, Morowitz *et al.* [93] have focused on some of the most central metabolic pathways that are conserved throughout all living systems, namely those of the Krebs cycle (also known as the citric acid cycle or the tricarboxylic acid cycle). In particular, they examined the reverse-Krebs cycle (a simple inversion of the "normal" pathways) which is used by some bacteria to produce the "starting organic compounds for the synthesis of all major classes of biomolecules." Calculating the energetics of abiotic reaction pathways, they found that the reverse Krebs cycle was "statistically favored among competing redox relaxation pathways under early-earth conditions" and went on to conclude that "this feature drove its emergence and also accounts for its evolutionary robustness and universality" [94]. In other words, some of the most central and highly conserved metabolism of modern life may directly reflect an autocatalytic cycle of abiotic reactions that occurred on a prebiotic Earth.

For our purposes, the most useful aspect of such autotrophic thinking is a logically independent idea that it has absorbed along the way, namely that ancient metabolic processes may be those we still see in operation today. In fact, this is a truly old idea that was originally applied to the heterotrophic theory [95]: the underlying rationale is that as organisms evolve, characteristics that emerged early on may become "locked in" by increasing numbers of new evolutionary innovations that make adaptive sense only in relation to them. Thus, each fundamental biochemical pathway may have arisen just once and then been evolutionarily frozen into place. If this is true, then modern metabolism contains a footprint of the evolutionary history by which it emerged, and much can be learned about the evolution of the amino acid alphabet simply by examining the biosynthetic pathways by which they are produced.

In this spirit, Wong rendered Dillon's earlier focus on amino acid biosynthesis rather more sophisticated by identifying true, conserved metabolic pathways, and coined the term "coevolution theory" to explain the metabolic sources of some amino acids and their evolutionary relationship with genetic code [96, 97]. According to coevolution theory, amino acids are grouped into two categories, precursors and products. The former includes the amino acids available prebiotically and the latter are the biosynthetic derivatives of their corresponding precursor amino acids. During the expansion of the amino acid alphabet the structure of genetic code evolved in step with this series of biosynthetic "inventions": product amino acids took over (by tRNA duplication and modification) a subset of the codons that formerly belonged to precursors. Present-day charging of glutamine, asparagine, and selenocysteine tRNAs (Section 1.1) may be viewed as "molecular fossils" of this very process, and the implied tRNA evolutionary relationships have recently gained support from the observation that phenylalanine and tyrosine tRNAs, both of which belong to the same metabolic pathway, display high conservation [98, 99]. Coupled to other observations, Wong has gone so far as to declare his coevolution hypothesis "proven" [100]. However, things are not quite as clear-cut as this suggests. The precise details of the coevolution model (the details of biosynthetic pathways and the assignment of products and precursors) remain contentious [101–103]. This is not to suggest that the central concept of an amino acid alphabet expanding through innovative biosynthesis is wrong, but rather that the specific steps and pathways of this expansion require careful scrutiny [103]. Indeed, Wong's specific model of code/amino acid/metabolism coevolution is best seen as an example of the bigger, broader concept that a small, prebiotically plausible amino acid alphabet expanded by recruiting new members from biosynthetic pathways during early evolution.

2.3.3
Evidence for a Smaller Initial Amino Acid Alphabet

Various clues regarding biosynthetic expansion of the standard amino acid alphabet have come from four different directions: analysis of metabolic pathways (e.g., [104, 105]), analysis of molecules directly involved in ribosomal protein synthesis (both tRNAs, e.g., [106–108], and aaRSs, e.g., [109–112]), ideas surrounding the prebiotic and biosynthetic emergence of the nucleic acid alphabet (e.g., [113–117]), and analysis of amino acid composition of present-day genomes (e.g., [118, 119]).

Of particular note here are studies that focus explicitly on protein structure and function. It was first proposed by Orgel [120] and later demonstrated by others [121, 122] that a native-like protein structure can be formed by simply arranging hydrophobic and hydrophilic amino acids in simple, repeating patterns. Subsequent bioinformatics analysis of modern protein sequences has suggested that as few as six amino acids would be a minimum necessary for a simple fold structure in primitive proteins [123]. Although other estimates place this figure slightly higher [124–127], a conservative consensus is that an amino acid alphabet of about half the size we see today would be sufficient for a simple proteome. It is interesting to note that this is in broad agreement with the number of amino acids that prebiotic simulation experiments can agree upon (e.g., [84]).

An interesting tweak of this thinking is to suggest that the amino acid alphabet could even have been larger during some points in primordial evolution, and subsequently "evolved simplicity" (lost amino acids). This particular concept of a standard amino acid alphabet that *lost* amino acids was first described in detail by Jukes [128] as he considered an explanation for the discrepancy between the number of codons assigned to arginine ($n = 6$) and its surprisingly low usage in protein sequences. Specifically, he proposed that arginine was an "intruder," which replaced its metabolic precursor, ornithine, by forming a stronger bond to ornithine's cognate tRNA. Some supporting evidence for this hypothesis has come from the phylogeny of aaRSs: "the presence of arginine in class Ia [which includes mainly hydrophobic amino acids] might reflect an early use in proteins of its precursor ornithine, whose amino-propyl side-chain is more hydrophobic than the guanidine-propyl side-chain of arginine" [117].

Taken as a whole, many of the specific claims deduced from these different studies contradict one another. Just as with prebiotic syntheses, one approach to see through the competing details is to perform a meta-analysis. Trifonov [129] has done just this, seeking a "consensus order" by which amino acids entered the genetic code by giving an independent vote to each of 40 proposed models for genetic code expansion taken from the published scientific literature. According to slight variations in precise assumptions, he derived several slightly different ranking systems that list the 20 amino acids in order of evolutionary arrival, from first to last. Amongst other findings, he noted that "Nine amino acids of Miller's imitation of primordial environment are all ranked as topmost" (i.e., Gly, Ala, Val, Asp, Glu, Pro, Ser, Leu, and Thr). Although this concept of meta-analysis is much needed in the field, and the results here are interesting, any attempt to generate a consensus order is subject to the same caveat as that described for the meta-analysis of prebiotic plausibility: it must build from independent data sources. Within the tens of different proposals for amino acid alphabet expansion, many have simply borrowed ideas from one another as building blocks in different overarching pictures, leading to a situation in which the most popular ideas are most likely to get further "tweaked" into new scientific publications and thus get extra "votes" during the derivation of a consensus. Scientific truth is not democratic and it would be interesting to see further meta-analyses that explicitly seek to disentangle this phenomenon in order to compare genuinely independent lines of evidence.

2.3.4
Proteins as Emergent Products of an RNA World

Meanwhile, the more general concept that metabolism-reveals-evolutionary history has been developed in a different direction that calls into question the very need for any prebiotic synthesis of amino acids. Building from Crick's [130] observation that "tRNA looks like natures attempt to make RNA do the work of a protein" (and his speculation that the original ribosome might have contained no proteins at all), White [131] went on to formally propose that "A metabolic system composed of nucleic acid enzymes … existed prior to the evolution of ribosomal protein synthesis" (noting that this "rationalizes the fact that many coenzymes are nucleotides or heterocyclic bases which could be derived from nucleotides"). Subsequently this idea has grown into the RNA world hypothesis [132] according to which living systems used RNA both for genetic information and for catalytically active molecules before the evolutionary emergence of Crick's central dogma (e.g., [133]). In strong versions of this hypothesis, all proteins were later evolutionary arrivals such that a fully functioning and sophisticated metabolism was already in place before any amino acids were needed for protein synthesis (e.g., [134]).

Although the broad concept of an RNA world has gained enormous momentum and a correspondingly large scientific literature, it is not without its problems. First and foremost, while amino acids are (relatively) easy to make under plausible prebiotic conditions, nucleic acids are not (e.g., [135]). Though some details have changed a little since it was written, Shapiro's ruthless account [136] remains an important read for anyone interested in forming a balanced opinion here. Coupled to this, a second and less direct criticism, at least of the "strong" RNA world hypothesis, is that the combined weight of data that adds up to support the idea of "early" (prebiotically plausible) and "late" (biosynthetically derived) amino acids seems like an awful lot of coincidence if complex ribozymes were in place to efficiently synthesize whatever was needed by an RNA world, independent of prebiotic synthesis.

2.3.5 Stereochemical Rationale for Amino Acid Selection

These cautionary notes aside, the RNA world hypothesis has produced another direct line of thinking as to how and why the standard amino acid alphabet arose. "Stereochemical theories" for the origin of the genetic code propose that amino acids were selected based on their direct chemical affinities to short RNA motifs.

From a historical perspective, these ideas date from early speculations as to how the genetic code would be found to operate: once it was known that nucleic acids encoded proteins, thinkers such as Gamow proposed various models of "direct templating" according to which the physical surface of DNA or RNA would produce a series of shapes into which the amino acids would naturally bind, ready for polymerization into a protein (see [137] for an overview). The discovery of tRNA's roles as physical "adaptors" that recognize genetic information (i.e., a codon) at one end and carry an amino acid at the physically distant other end put paid to such early ideas. However, parallel discoveries that the genetic code appeared universal across life led some to surmise that only one genetic code was possible and further models were developed which posited suitably shaped pockets in various parts of the codon, anticodon, or tRNA (reviewed in [138]). When placed alongside earlier explanations, the problem that emerged was the sheer diversity of often contradictory results that could apparently be obtained by appropriately motivated model building. Although subsequent efforts sought empirical evidence for interactions between amino acids and individual bases or nucleotides (see [139, 140] for reviews) no clear, testable rationale has ever been offered for the evolutionary significance of such associations to triplet decoding beyond a general assertion that this is how genetic coding started.

Meanwhile, the rise of the RNA world hypothesis led to a resurgence of interest in the possibility for RNA/amino acid interactions, especially when the emergence of associated technologies for selecting RNA molecules with specific bioactivities provided opportunities to gather further empirical data. In particular, "SELEX" experiments (Systematic Evolution of Ligands by EXponential enrichment [141]) enable researchers to find nucleic acid oligomers, known as aptamers, that exhibit specific binding affinities. The process is one of molecular artificial selection, isolating aptamers that bind a particular target by selective amplification over several generations. Of all amino acids, arginine has received the most attention to date: several different laboratories have selected and characterized arginine aptamers, and subsequent meta-analysis of all the published aptamers found a strong association between all six arginine codons and arginine-binding sites [140]. Subsequently, a total of eight amino acids have been shown to associate with aptamers in which codons or anticodons are especially prevalent [142] and only two negative results have been published (for valine, no aptamers have been found, and for glutamine, those that have been found show no codon or anticodon over-abundance). Thus, empirical evidence for direct stereochemical affinities between amino acids and their corresponding nucleotide encodings is stronger now than ever.

Taken at face value, these observed associations have the potential to explain life's choice of encoded amino acids as those that interacted favorably with associated RNA motifs. However, we remain far from being able to make that inference with any degree of confidence. The argument for molecular fossils is essentially "if it ain't broke, then evolution won't fix it," but in the case of genetic coding it is clear that the emergence of tRNA's and protein synthetases did introduce major disruptions to any primordial system of direct RNA/amino acid templating, and synthetic biology is making it clear that once this sophisticated adaptor system is in place, there are very few restrictions as to which amino acid(s) can be associated with which codons. It is thus not clear why these direct affinities should have survived the evolutionary upheavals that physically separated amino acids from their corresponding codons/anticodons. In this light it becomes notable that aptamer results show no consistency with other directions of thinking about the origin of the standard amino acid alphabet. Propositions of "early" and "late" amino acids are fairly uniform in the view that arginine was not prebiotically plausible (see [79, 82, 129]); tryptophan is likewise regarded as the primary example of an amino acid that required a biosynthetic origin and entered the code late, whereas valine shows up as one of the strongest candidates for an "early" amino acid (see [129] and references therein). One possible explanation is that the origin of protein translation within an RNA world occurred much later (and in a qualitatively different environment) from the origin of life [142]. This makes sense, but (as noted above) offers no explanation as for the wealth of data that combines to support the idea that prebiotically plausible amino acids arrived first and were later augmented by biosynthetic growth (a point that has not been lost on some proponents of the RNA world [143]).

2.4
Amino Acids that Life Passed Over: A Role for Natural Selection?

All of the work described so far makes a subtle inference that the major problem of explaining the emergence of the standard amino acid alphabet lies in understanding how its 20 constituents could have been made and/or incorporated into genetic coding. These insights are necessary but not sufficient for a comprehensive explanation of the standard amino acid alphabet's emergence: while an extensive scientific literature concerns possible origins of the standard 20, a dearth of information surrounds the amino acids that were likely available to early life but not incorporated into genetic coding [144].

For example, there has been much discussion of the fact that eight members of the standard alphabet are found within the Murchison meteorite, but very little

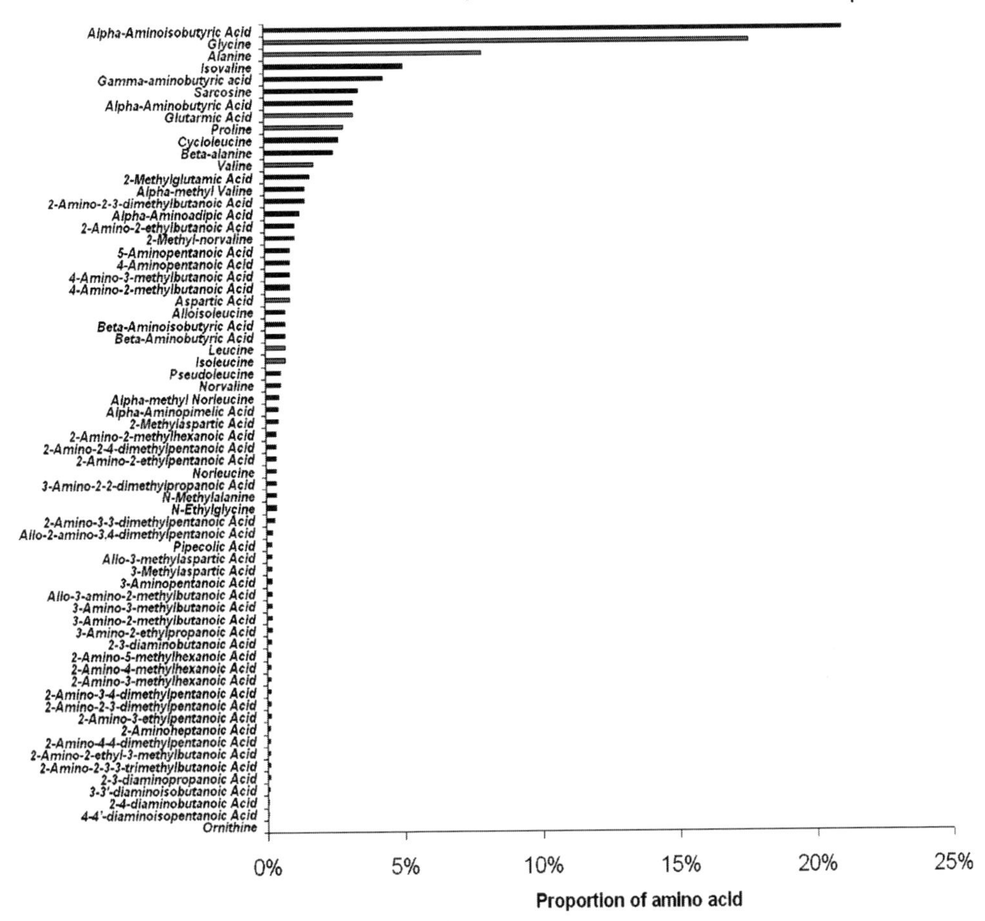

Figure 2.2 The 66 amino acids that have been identified within the Murchison meteorite to date, listed by abundance (as a percentage of total amino acid content). It is notable that the single most abundant amino acid, α-isobutyric acid, is not one used by biology and that those eight which overlap with the standard amino acid alphabet (shown in red) are by no means the most abundant.

concerning the fact that 58 other amino acids are also present (Figure 2.2). Similarly, while there has been a lot of discussion as to how biosynthetic pathways can illustrate the evolutionary invention of prebiotically implausible amino acids, excepting Jukes commentary on ornithine, virtually none of it addresses the fact that tens of other amino acids occur within these conserved biosynthetic pathways but were never picked up by the genetic code. Indeed, it would be naïve to assume that these biochemical pathways were the only ones that life could have evolved when faced with the profusion of unusual amino acids that are biosynthesized by modern organisms for use in noncoding capacities. So why did nature "choose" these particular 20 (22) amino acids for genetic coding?

Until recently, the only direct consideration of this question came from a single review article by two leaders of the prebiotic chemistry community [145]. In considering amino acids "... on the basis of the availability in the primitive ocean, function in proteins, the stability of the amino acid and its peptides, stability to racemization, and stability on the transfer RNA" the authors reached the conclusion that "If life were to arise on another planet, we would expect that the catalysts would be poly-α-amino acids and that about 75% of the amino acids would be the same as on the earth." Some of the general reasoning was strong and straightforward. For example, the authors suggest that α-amino acids were selected over alternatives with longer carbon backbones, as the latter would lack the rigidity to support stable protein secondary or tertiary structures (though even here, recent developments in biochemistry suggest a careful reevaluation may be required [146]). However, the specific explanations offered for individual members of the standard alphabet were offered as verbal arguments without any clear, testable predictions. Indeed, the authors themselves describe the choices of the sulfur and aromatic amino acids as "not compelling" and express frank puzzlement at the absence of α-amino-*n*-butyric acid, norvaline and norleucine [145].

Of course many other adaptive criteria can be proposed, but the challenge is to render such ideas as testable science: unless this is achieved then at best we have some good ideas for the themes involved in amino acid alphabet selection; at worst, we have untestable explanations that critics could dismiss as "adaptive storytelling" [147]. In principle, the solution is straightforward: to place the members of the standard alphabet within some sort of bigger, quantified "chemical space" populated by plausible alternatives and ask whether nature's choice distinguishes itself in any statistically significant fashion. Certainly the work concerning prebiotic chemistry and biosynthetic growth of the code infers various scenarios for plausible alternatives. The problem, however, is that not much is known about the physical and chemical properties of other amino acids. It is quite simply costly and time-consuming to carefully investigate the physical and chemical properties of different molecules. Thus, while the biomedical research community has had plenty of motivation to do so for the members of the standard amino acid alphabet (indeed, a public database is dedicated to the hundreds of different measurements that have been published here [148]), no equivalent motivation has driven corresponding measurements for the much larger pool of noncoding amino acids. In this context, just two direct studies have developed this strategy to date [149, 150].

2.4.1
Were the Standard Amino Acids Chosen for High Biochemical Diversity?

One of the most common adaptive explanations for nature's choice of amino acids (usually stated as an assumption within work of a different primary focus) is that the standard amino acid alphabet was selected for its biochemical diversity. For example, a typical view is that "The natural repertoire of 20 amino acids presumably reflects the combined requirements of providing a diversity of chemical functionalities, and providing enough structural diversity that sequences are likely to define unique three-dimensional shapes" [151]. This idea of an evolutionary drive towards increasing biochemical diversity is especially prevalent among those who argue for the RNA world, where the relative sizes of the nucleic acid and amino acid alphabets provide an intuitive idea that encoded amino acids emerged because "proteins provided a greater catalytic versatility than nucleic acids (20 versus 4 building blocks)" [152].

However, there is actually surprisingly little evidence for this idea at present. Anecdotally, we may note that the presence within the standard alphabet of valine, leucine, and isoleucine (all similar-sized, hydrophobic amino acids that interchange freely in molecular evolution) suggests that diversity alone is unlikely to form a good explanation. This is especially noteworthy when coupled to the observation that these three amino acids are widely accepted as being present at the earliest stages of code evolution, when the entire alphabet was much smaller. Recently, one study has attempted to initiate a more comprehensive and compelling analysis of the idea [149]. Starting with a demonstration that existing chemoinformatics software can predict fundamental biochemical properties of amino acids with surprisingly good accuracy [153] it went on to test the members of the standard alphabet against three different definitions of "plausible alternatives" [149]. The method was to quantify biochemical diversity of an amino acid alphabet as simple statistical variance in size, charge, and hydrophobicity, and thus to ask what percentage of randomly selected alphabets exhibit a higher diversity than that observed in nature? The three tests performed were (i) a comparison of the eight members of the standard alphabet found in the Murchison meteorite with random alphabets of equivalent size drawn from the full contents of the Murchison meteorite, (ii) a comparison of the full standard alphabet with this same pool of prebiotically plausible alternatives, and (iii) a comparison of the full standard alphabet with a larger pool of alternatives in which the contents of the Murchison meteorite were augmented by amino acids found as intermediates in the conserved metabolic pathways of central metabolism.

The results obtained were complex, but may be summarized as follows. First, the full set of 20 standard amino acids is statistically more diverse than random samples of equivalent size drawn from a pool of prebiotically plausible alternatives (as defined by the contents of the Murchison meteorite). This is true for size and charge, but not for hydrophobicity unless the probability of entry into the standard alphabet was proportional to prebiotic abundance (again as defined by abundance within the Murchison meteorite). This in itself is somewhat disconcerting given that all signs are that side-chain hydrophobicity is the single most important factor in determining how protein sequences fold into a specific structure with a specific function and things were even less intuitive for the other tests. When the pool of prebiotically plausible amino acids is enlarged to include amino acid intermediates of conserved metabolism, the standard amino acid alphabet does not appear especially diverse in any of the three fundamental properties tested. Finally, when samples of eight amino acids are drawn from the contents of the Murchison meteorite (to reflect the fact that eight of these amino acids did end up in the standard alphabet), life's choice again appeared unremarkable. In short, the first attempt to test the widespread idea that the standard amino acid alphabet might represent an especially diverse set of amino acids found no such evidence.

Several interpretations are possible: that the standard amino acids are unusually diverse, but the Murchison meteorite offers a poor model for the appropriate pool of plausible alternatives, that the definition of "biochemically diverse" is poorly represented in these tests, or that the standard alphabet really is not especially diverse. All of these possibilities are worthy of further consideration, and in the course of this work the authors produced a public database of amino acids and associated properties to encourage other researchers to test alternative hypotheses [154]. However, the take-home message is that carefree claims that natural selection steered amino acid alphabet evolution towards major biochemical diversity need careful examination.

2.4.2 Were the Standard Amino Acids Chosen for "Cheap" Biosynthesis?

One possible factor that could favor evolution of a lower biochemical diversity can be derived from the observation that bulky amino acids, such as phenylalanine and tyrosine, are used much less within "natural" proteins than simple and smaller alternatives [155, 156]. Coupled to the observation that many naturally occurring amino acids not found in the alphabet (e.g., those used in post-translational modification) tend to be large and chemically complex, it seems plausible that simple, economic principles operated during alphabet formation: perhaps entry into the standard alphabet was restricted to the smallest and cheapest amino acids capable of forming a functional protein library.

Although this idea has yet to be tested directly, a related study suggests that such an approach would be worthwhile. Expanding the work of Shock and Amend, Ji *et al.* [150] used a "semi-empirical quantum method" to determine free energies of formation not just for the 20 amino acids of the standard alphabet, but for all of their isomers. (The assumption here is that if either prebiotic chemistry or metabolism is capable of producing a side-chain of any particular size and complexity, then rearrangements of these atoms were equally plausible.) In comparing members of the standard alphabet against this pool of alternatives, it was found that the specific isomers that nature consistently chose were those associated with the lowest (or near-lowest) free energy. Aside from the obvious interpretation – that there may be some adaptive value to using stable amino acids – the authors point out that this effect is strongest for small, simple, prebiotically plausible amino acids where (according to thermodynamics) low-free-energy isomers would likely have been most abundant. In a later commentary on this finding [157], one of the authors explicitly connects this notion of abundance to cost-efficiency, suggesting that "employing the standard amino acids (especially small ones) to build primitive life is, at least in part, a measure of economic selection." This is entirely consistent with the parallel analysis by Higgs and Pudritz [79].

2.4.3 Were the Standard Amino Acids Chosen to Speed Up Evolution?

A final, less well-developed idea for the advantages of a less diverse amino acid alphabet comes from another observation regarding the standard genetic code: that the pattern of codon assignments within the standard genetic code appears to be optimized so that wherever a mutation or misreading of any single nucleotide changes the amino acid "meaning" of a codon, the two amino acids in question share similar biochemical properties (see [158] for a full review of the extensive scientific literature associated with this idea). By and large, this has been interpreted as evidence that natural selection shaped a pattern of codon/amino acid relationships that would minimize the deleterious effects of genetic errors (i.e., taking the standard alphabet as a given, how should they be encoded relative to one another?). However, there is no reason why such a process should not have gone further, guiding which particular amino acids were incorporated into the standard alphabet while biosynthetic expansion took place [159] (or, indeed, which particular stereochemical affinities were selected over others [160]). Moreover, a few contributions to the scientific literature regarding an "error minimizing" genetic code have pointed out that this type of pattern for codon/amino acid assignments could have another, more subtle effect – on the speed and efficiency of evolution itself [161, 162]; to understand this last point, we need to reconsider the role coded amino acids play.

Throughout this chapter I have referred to the "amino acid alphabet" to emphasize the role of the 20 standard amino acids as combinatorial building blocks of coded proteins. Viewed in this way, they are analogous to Lego building blocks and the assumption that evolution drove the development of biochemical diversity here amounts to the suggestion that a more diverse set of building blocks allows a greater range of structures to be built. However, there is another side to encoding amino acids that this analogy overlooks. Amino acids must substitute for one another over the course of time as protein sequences evolve and diversify. If, from the pool of possible amino acids, an alphabet is selected in which each member is as different from the others as possible, then this creates a very rugged "fitness landscape" for evolution: each and every mutation will lead to the substitution of a chemically different amino acid, and a likely disruption to the corresponding protein structure and function. Evolutionary theory predicts that mutations with these properties will be removed by "purifying selection" and thus not contribute to long-term evolution. By ensuring that amino acids are assigned to codons in such a manner as to minimize the size of these disruptions, a genetic code can potentially change the landscape upon which evolution walks, allowing a series of successful, functional intermediates between proteins of different structure and function. This in turn can lead to a faster, more efficient process of protein evolution by smoothing the pathways of change. At present, a single study has tested this idea [163], simulating molecular evolution under different genetic codes in which the amino acid alphabet is held constant, but the pattern of codon assignments is changed. Results supported the prediction, and there is no reason why this logic should not apply to the broader question of *which* amino acids enter the code as well as *where* they enter. This offers one potential interpretation of the finding that the standard alphabet does not appear especially diverse: it could be that an ideal amino acid alphabet is a "goldilocks" one in which we have just enough diversity and not too much!

2.5
Why Does Life Genetically Encode L-Amino Acids?

A final word should go to the issue of amino acid chirality. Almost all amino acids can exist as either of two enantiomers, meaning that the four chemical groups attached to the central, α-carbon atom can be organized in one of two, mirror-image, arrangements (L- and D-chirality). All the members of the standard amino acid alphabet are L-enantiomers. Why this should be so remains largely a mystery. Clearly, homochirality of amino acids (or nucleotides) is distinctly advantageous. Enantiomers are, by definition, different shapes that cannot be rotated or superimposed onto one another, thus molecular machinery that recognizes and "handles" one version typically cannot work with the mirror image (indeed natural living systems primarily use D-amino acids in chemical warfare as highly effective toxins or as structural defenses that inhibit biological degradation). This phenomenon of "cross-enantiomeric inhibition" offers a clear reason as to why life should have come to focus on one particular chirality, but does not suggest which, if either, would be preferable. Since the problem was first identified in the 1950s the challenge has been to find some ubiquitous, nonbiological process that skews chemistry towards one particular chirality. An enormous number of ingenious and creative ideas have been put forward to answer this question (e.g., see [164, 165] for recent reviews). Unfortunately many (probably most) have lacked the scientific credentials of associated predictions and/or testability. Among the most popular ideas, the one with the longest history is that of circularly polarized light that occurs during supernovae explosions (e.g., [166] and references therein). However, the idea that this connects meaningfully with prebiotic synthesis has been roundly criticized elsewhere (e.g., [167]). More recently, another idea that has gathered some interest comes from the observation that the weak nuclear force can promote the formation of nonracemic molecular mixtures, but again the effect is weak and leaves many unconvinced (see [168] for an overview). As such, the general consensus at present is that "mirror-image" life would be entirely possible, and that our own evolutionary history could equally well have gone either way. Some, in fact, have gone so far as to suggest that a separate origin of life on Earth that used the opposite chirality could be living under our noses [169].

An interesting footnote here is that at the time of writing this chapter, a new analysis of pristine meteorite contents has just reported a heavy chiral content for key organic molecules [170]. The explanation remains far from clear – it could involve circular dichroism, quantum effects or something entirely different – but it seems that the topic of chirality has plenty of surprises left to offer.

2.6
Summary, Synthesis, and Conclusions

The idea that there are 20 "terrestrial" amino acids obscures a far richer and more complex reality (Figure 2.3). Naturally occurring genomes directly, genetically encode up to 22 amino acids that cannot be clearly distinguished on chemical, physical, or

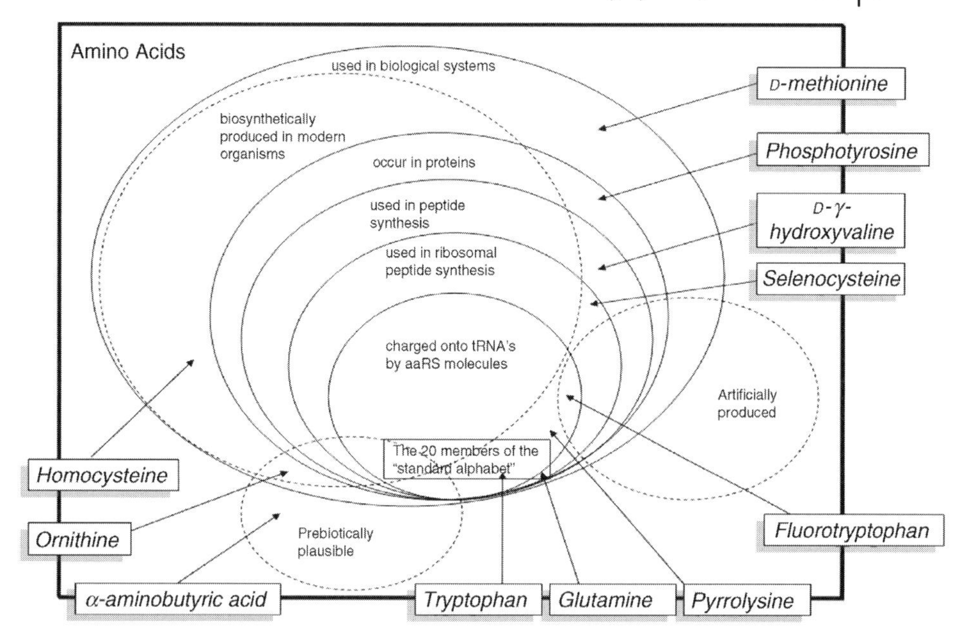

Figure 2.3 The complex reality of biological amino acids. The 20 members of the standard amino acid only form a small subset of those used in biological systems. They only partially overlap with what are thought to be prebiotically plausible molecules, and they fit no simple definition that does not involve mention of specific translational machinery and a concept of a common evolutionary ancestor for all modern life on Earth! Some examples of amino acids that lie around the edge of the standard 20 are also illustrated.

biological grounds from a much larger pool of alternatives. Aside from indirect genetic coding (such as simple biosynthesis, post-translational modification, non-ribosomal peptide synthesis) various representatives from this larger pool perhaps once were, perhaps still are, and certainly could be incorporated into standard genetic coding. Meanwhile, it seems entirely likely that the amino acid alphabet continues to expand in some lineages when decoding of selenocysteine and pyrrolysine is carefully compared with equivalent procedures for the "standard" amino acids and with theories for the mechanism by which the standard code emerged.

This throws wide open the question of why life on Earth evolved a particular "standard alphabet" of 20. At present, the best clues we have as to how and why life chose these amino acids comes from three branches of research: investigations of the natural abiotic processes that produce amino acids, analysis of evolutionary footprints hidden in modern metabolism (including theories as to how the standard genetic code developed), and analyses of putative adaptive properties for the standard alphabet.

Combining the insights harvested here, it seems highly likely that the "standard amino acid alphabet" started out much smaller than the 20 we see today (perhaps around half this number) and subsequently grew through the biosynthetic invention (and incorporation) of additional members. However, while the general idea of biosynthetic growth of the alphabet is strongly supported by a wide diversity of evidence (including genetic engineering experiments that have quickly and easily selected for genomes that add fluorotryptophan to the alphabet, and natural genetic codes that have added pyrrolysine and selenocysteine), the precise details by which the standard alphabet emerged remain far less clearly understood than some reports would suggest. Moreover, theories of alphabet growth do nothing to explain why certain biosynthetic inventions were selected over others: indeed it remains unclear what factors may have guided natural selection in "choosing" the members of the standard alphabet. At one extreme, it is possible that the standard alphabet developed in the only way it could due to the necessity for direct, stereochemical interactions between amino acids and their nucleic acid code words. Problems for this "strong" stereochemical view center around its lack of agreement with most other evidence for the origin and growth of the amino acid alphabet. Meanwhile, the widespread assumption that the amino acid alphabet was selected to build diverse protein structures and functions is currently associated with a surprising lack of direct evidence. Though more work here is urgently needed, two possible checks on the adaptive value of biochemical diversity concern cost and evolutionary potential. The first of these suggests that biochemical diversity may not be necessary to build diverse structures and functions, such that a choice of small, simple amino acids make better economic sense. The second suggests that an alphabet of diverse meanings could actually impede the efficiency with which proteins evolve such that we might expect selection for amino acids that, far from adding to biochemical diversity of the alphabet actually "cushion" the effects of mutation and/or mistranslation by resembling pre-existing members of the alphabet.

A better understanding of how and why life "chose" a particular amino acid alphabet during primordial evolution would offer insight to astrobiologists as they consider what we might expect from life that originates elsewhere in the universe. It would also offer guiding insight to the pioneers of synthetic biology who are set to engineer genes and proteins that have never been seen in nature (including those that contain "unnatural" amino acids). As we look to the future, continued growth in computing power and accessibility joins with exponential growth in databases of molecular biology and corresponding advances in our understanding of protein folding and chemical biology to offer great potential for theoreticians to derive testable hypotheses regarding the formation of the amino acid standard alphabet. Meanwhile, the potential for biomedical and other commercial applications of synthetic proteins that contain unnatural amino acids seems likely to drive the empirical exploration of this theory, feeding new ideas and insights back into our understanding of how the standard alphabet evolved.

References

1 Crick, F.H.C. (1958) On protein synthesis. *Symposia of the Society for Experimental Biology*, **12**, 139–163.

2 Uy, R. and Wold, F. (1977) Posttranslational covalent modification of proteins. *Science*, **198**, 890–896.

3 Seo, J. and Lee, K.-J. (2004) Post-translational modifications and their biological functions: proteomic analysis and systematic approaches. *Journal of Biochemistry and Molecular Biology*, **37**, 35–44.

4 Mann, M. and Jensen, O.N. (2003) Proteomic analysis of post-translational modifications. *Nature Biotechnology*, **21**, 255–261.

5 Sahl, H.-G., Jack, R.W., and Bierbaum, G. (1995) Biosynthesis and biological activities of lantibiotics with unique post-translational modifications. *European Journal of Biochemistry*, **230**, 827–853.

6 Mor, A., Amiche, M., and Nicolas, P. (1992) Enter a new post-translational modification: D-amino acids in gene-encoded peptides. *Trends in Biochemical Sciences*, **17**, 481–485.

7 Young, V.R. (1994) Adult amino acid requirements: the case for a major revision in current recommendations. *The Journal of Nutrition*, **124** (8 Suppl.), 1517S–1523.

8 van Sluijters, D.A., Dubbelhuis, P.F., Blommaart, E.F.C., and Meijer, A.J. (2000) Amino-acid-dependent signal transduction. *The Biochemical Journal*, **351**, 545–550.

9 Meijer, A.J. (2003) Amino acids as regulators and components of nonproteinogenic pathways. *The Journal of Nutrition*, **133** (6 Suppl. 1), 2057S–2062S.

10 Zuber, P. (1991) Non-ribosomal peptide synthesis. *Current Opinion in Cell Biology*, **3**, 1046–1050.

11 Schwarzer, D., Finking, R., and Marahiel, M.A. (2003) Nonribosomal peptides: from genes to products. *Natural Product Reports*, **20**, 275–287.

12 Volkmann, R.A. and Heck, S.D. (1998) Biosynthesis of D-amino acid-containing peptides: exploring the role of peptide isomerases. *EXS*, **85**, 87–105.

13 Hallen, H.E., Luo, H., Scott-Craig, J.S., and Walton, J.D. (2007) Gene family encoding the major toxins of lethal *Amanita* mushrooms. *Proceedings of the National Academy of Sciences of the United States of America*, **104**, 19097–19101.

14 Grutters, M., van Raaphorst, W., Epping, E., Helder, W., de Leeuw, J.W., Glavin, D.P., and Bada, J. (2002) Preservation of amino acids from *in situ*-produced bacterial cell wall peptidoglycans in northeastern Atlantic continental margin sediments. *Limnology and Oceanography*, **47**, 1521–1524.

15 Francklyn, C. (2003) tRNA synthetase paralogs: evolutionary links in the transition from tRNA-dependent amino acid biosynthesis to *de novo* biosynthesis. *Proceedings of the National Academy of Sciences of the United States of America*, **100**, 9650–9652.

16 Hatfield, D.L. and Gladyshev, V.N. (2002) How selenium has altered our understanding of the genetic code. *Molecular and Cellular Biology*, **22**, 3565–3576.

17 Moura, G., Pinheiro, M., Silva, R., Miranda, I., Afreixo, V., Dias, G., Freitas, A., Oliveira, J.L., and Santos, M.A.S. (2005) Comparative context analysis of codon pairs on an ORFeome scale. *Genome Biology*, **6**, R28.

18 Nie, L., Wu, G., and Zhang, W. (2006) Correlation of mRNA expression and protein abundance affected by multiple sequence features related to translational efficiency in *Desulfovibrio vulgaris*: a quantitative analysis. *Genetics*, **174**, 2229–2243.

19 Srinivasan, G., James, C.M., and Krzycki, J.A. (2002) Pyrrolysine encoded by UAG in Archaea: charging of a UAG-decoding specialized tRNA. *Science*, **296**, 1459–1462.

20 Zhang, Y., Baranov, P.V., Atkins, J.F., and Gladyshev, V.N. (2005) Pyrrolysine and selenocysteine use dissimilar decoding strategies. *The Journal of Biological Chemistry*, **280**, 20740–20751.

21 Mat, W.-K., Xue, H., and Wong, J.T.-F. (2008) The genomics of LUCA. *Frontiers*

in *Bioscience: A Journal and Virtual Library*, **13**, 5605–5613.

22 Mushegian, A. (2008) Gene content of LUCA, the last universal common ancestor. *Frontiers in Bioscience: A Journal and Virtual Library*, **13**, 4657–4666.

23 Sobolevsky, Y. and Trifonov, E.N. (2006) Protein modules conserved since LUCA. *Journal of Molecular Evolution*, **63**, 622–634.

24 Copeland, P.R. (2005) Making sense of nonsense: the evolution of selenocysteine usage in proteins. *Genome Biology*, **6**, 221.

25 Chaudhuri, B.N. and Yeates, T.O. (2005) A computational method to predict genetically encoded rare amino acids in proteins. *Genome Biology*, **6**, R79.

26 Knight, R.D., Freeland, S.J., and Landweber, L.F. (2001) Rewiring the keyboard: evolvability of the genetic code. *Nature Reviews. Genetics*, **2**, 49–58.

27 Lobanov, A.V., Kryukov, G.V., Hatfield, D.L., and Gladyshev, V.N. (2006) Is there a twenty third amino acid in the genetic code? *Trends in Genetics*, **22**, 357–360.

28 Fujita, M., Mihara, H., Goto, S., Esaki, N., and Kanehisa, M. (2007) Mining prokaryotic genomes for unknown amino acids: a stop-codon-based approach. *BMC Bioinformatics*, **8**, 225.

29 Williamson, S.J., Rusch, D.B., Yooseph, S., Halpern, A.L., Heidelberg, K.B., Glass, J.I., Andrews-Pfannkoch, C., Fadrosh, D., Miller, C.S., Sutton, G., Frazier, M., and Venter, J.C. (2008) The sorcerer II global ocean sampling expedition: metagenomic characterization of viruses within aquatic microbial samples. *PLoS ONE*, **3**, e1456.

30 Santelli, C.M., Orcutt, B.N., Banning, E., Bach, W., Moyer, C.L., Sogin, M.L., Staudigel, H., and Edwards, K.J. (2008) Abundance and diversity of microbial life in ocean crust. *Nature*, **453**, 653–656.

31 Fierer, N., Breitbart, M., Nulton, J., Salamon, P., Lozupone, C., Jones, R., Robeson, M., Edwards, R.A., Felts, B., Rayhawk, S., Knight, R., Rohwer, F., and Jackson, R.B. (2007) Metagenomic and small-subunit rRNA analyses reveal the genetic diversity of bacteria, Archaea, fungi, and viruses in soil. *Applied and Environmental Microbiology*, **73**, 7059–7066.

32 Gill, S.R., Pop, M., Deboy, R.T., Eckburg, P.B., Turnbaugh, P.J., Samuel, B.S., Gordon, J.I., Relman, D.A., Fraser-Liggett, C.M., and Nelson, K.E. (2006) Metagenomic analysis of the human distal gut microbiome. *Science*, **312**, 1355–1359.

33 Bacher, J.M., Hughes, R.A., Wong, J.T.-F., and Ellington, A.D. (2004) Evolving new genetic codes. *Trends in Ecology & Evolution*, **19**, 69–75.

34 Wong, J.T.F. (1983) Membership mutation of the genetic code: loss of fitness by tryptophan. *Proceedings of the National Academy of Sciences of the United States of America*, **80**, 6303–6306.

35 Bacher, J.M. and Ellington, A.D. (2001) Selection and characterization of*Escherichia coli* variants capable of growth on an otherwise toxic tryptophan analogue. *Journal of Bacteriology*, **183**, 5414–5425.

36 Bacher, J.M. and Ellington, A.D. (2007) Global incorporation of unnatural amino acids in *Escherichia coli*. *Methods in Molecular Biology*, **352**, 23–34.

37 Bacher, J.M., Bull, J.J., and Ellington, A.D. (2003) Evolution of phage with chemically ambiguous proteomes. *BMC Evolutionary Biology*, **3**, 24.

38 Xie, J. and Schultz, P.G. (2006) A chemical toolkit for proteins – an expanded genetic code. *Nature Reviews. Molecular Cell Biology*, **7**, 775–782.

39 Liu, C.C. and Schultz, P.G. (2006) Recombinant expression of selectively sulfated proteins in *Escherichia coli*. *Nature Biotechnology*, **24**, 1436–1440.

40 Ohta, A., Murakami, H., Higashimura, E., and Suga, H. (2007) Synthesis of polyester by means of genetic code reprogramming. *Chemistry & Biology*, **14**, 1315–1322.

41 Guo, J., Wang, J., Anderson, J.C., and Schultz, P.G. (2008) Addition of an alpha-hydroxy acid to the genetic code of bacteria. *Angewandte Chemie (International Edition in English)*, **47**, 722–725.

42 Barrow, J.D., Morris, S.C., Freeland, S.J., and Harper, C.L. (eds) (2007) *Fitness of the Cosmos for Life: Biochemistry and Fine-tuning*, Cambridge University Press, Cambridge.

43 Pace, N.R. (2001) The universal nature of biochemistry. *Proceedings of the National Academy of Sciences of the United States of America*, **98**, 805–808.

44 Benner, S.A., Ricardo, A., and Carrigan, M.A. (2004) Is there a common chemical model for life in the universe? *Current Opinion in Chemical Biology*, **8**, 672–689.

45 Denton, M.J., Marshall, C.J., and Legge, M. (2002) The protein folds as platonic forms: new support for the pre-Darwinian conception of evolution by natural law. *Journal of Theoretical Biology*, **219**, 325–342.

46 Qian, J., Luscombe, N.M., and Gerstein, M. (2001) Protein family and fold occurrence in genomes: power-law behaviour and evolutionary model. *Journal of Molecular Biology*, **313**, 673–681.

47 Benner, S.A. and Sismour, A.M. (2005) Synthetic biology. *Nature Reviews. Genetics*, **6**, 533–543.

48 Lazcano, A. and Miller, S.L. (1999) On the origin of metabolic pathways. *Journal of Molecular Evolution*, **49**, 424–431.

49 (a) Oparin, A.I. (1924) *The Origin of Life*, Moscow Worker, Moscow; (b) English translation of this Russian work first appears as: Oparin, A.I. (1968) *The Origin and Development of Life (NASA TTF-488)*, GPO, Washington, DC.

50 (a) Haldane, J.B.S. (1929) The origin of life. *Rationalist Annual*, **3**, 148–153; (b) Reprinted as an Appendix in Bernal, J.D. (1967) *The Origin of Life*, World, New York.

51 Miller, S.L. (1953) A production of amino acids under possible primitive earth conditions. *Science*, **117**, 528–529.

52 Lazcano, A. and Miller, S.L. (1996) The origin and early evolution of life: prebiotic chemistry, the pre-RNA world, and time. *Cell*, **85**, 793–798.

53 Cleaves, H.J., Chalmers, J.H., Lazcano, A., Miller, S.L., and Bada, J.L. (2008) A reassessment of prebiotic organic synthesis in neutral planetary atmospheres. *Origins of Life and Evolution of the Biosphere*, **38**, 105–115.

54 Hill, R.D. (1992) An efficient lightning energy source on the early Earth. *Origins of Life and Evolution of the Biosphere*, **22**, 277–285.

55 Ogata, Y., Imai, E.-I., Honda, H., Hatori, K., and Matsuno, K. (2000) Hydrothermal circulation of seawater through hot vents and contribution of interface chemistry to prebiotic synthesis. *Origins of Life and Evolution of the Biosphere*, **30**, 527–537.

56 Garzón, L. and Garzón, L. (2001) Radioactivity as a significant energy source in prebiotic synthesis. *Origins of Life and Evolution of the Biosphere*, **31**, 3–13.

57 Tian, F., Toon, O.B., Pavlov, A.A., and De Sterck, H. (2005) A hydrogen-rich early Earth atmosphere. *Science*, **308**, 1014–1017.

58 Chyba, C.F. (2005) Atmospheric science. Rethinking earth's early atmosphere. *Science*, **308**, 962–963.

59 Goldblatt, C., Lenton, T.M., and Watson, A.J. (2006) Bistability of atmospheric oxygen and the great oxidation. *Nature*, **443**, 683–686.

60 Kasting, J.F. (2006) Earth sciences: ups and downs of ancient oxygen. *Nature*, **443**, 643–645.

61 Oró, J. (1961) Comets and the formation of biochemical compounds on the primitive earth. *Nature*, **190**, 389–390.

62 Sephton, M.A. (2002) Organic compounds in carbonaceous meteorites. *Natural Product Reports*, **19**, 292–311.

63 Brack, A. (2007) From interstellar amino acids to prebiotic catalytic peptides: a review. *Chemistry & Biodiversity*, **4**, 665–679.

64 Brown, P.G., Hildebrand, A.R., Zolensky, M.E., Grady, M., Clayton, R.N., Mayeda,

T.K., Tagliaferri, E., Spalding, R., MacRae, N.D., Hoffman, E.L., Mittlefehldt, D.W., Wacker, J.F., Bird, J.A., Campbell, M.D., Carpenter, R., Gingerich, H., Glatiotis, M., Greiner, E., Mazur, M.J., McCausland, P.J., Plotkin, H., and Mazur, T.R. (2000) The fall, recovery, orbit, and composition of the Tagish Lake meteorite: a new type of carbonaceous chondrite. *Science*, **290**, 320–325.

65 Kvenvolden, K., Lawless, J., Pering, K., Peterson, E., Flores, J., Ponnamperuma, C., Kaplan, I.R., and Moore, C. (1970) Evidence for extraterrestrial amino-acids and hydrocarbons in the Murchison meteorite. *Nature*, **228**, 923–926.

66 Oró, J., Gibert, J., Lichtenstein, H., Wikstrom, S., and Flory, D.A. (1971) Amino-acids, aliphatic and aromatic hydrocarbons in the Murchison Meteorite. *Nature*, **230**, 105–106.

67 Wolman, Y., Haverland, W.J., and Miller, S.L. (1972) Nonprotein amino acids from spark discharges and their comparison with the Murchison meteorite amino acids. *Proceedings of the National Academy of Sciences of the United States of America*, **69**, 809–811.

68 Lawless, J.G. (1973) Amino acids in the Murchison meteorite. *Geochimica et Cosmochimica Acta*, **37**, 2207–2212.

69 Cronin, J.R. and Pizzarello, S. (1983) Amino acids in meteorites. *Advances in Space Research*, **3**, 5–18.

70 Engel, M.H., Macko, S.A., and Silfer, J.A. (1990) Carbon isotope composition of individual amino acids in the Murchison meteorite. *Nature*, **348**, 47–49.

71 Cooper, G.W., Onwo, W.M., and Cronin, J.R. (1992) Alkyl phosphonic acids and sulfonic acids in the Murchison meteorite. *Geochimica et Cosmochimica Acta*, **56**, 4109–4115.

72 Cooper, G., Kimmich, N., Belisle, W., Sarinana, J., Brabham, K., and Garrel, L. (2001) Carbonaceous meteorites as a source of sugar-related organic compounds for the early earth. *Nature*, **414**, 879–883.

73 Meierhenrich, U.J., Munoz Caro, G.M., Bredehoft, J.H., Jessberger, E.K., and Thiemann, W.H.-P. (2004) Identification of diamino acids in the Murchison meteorite. *Proceedings of the National Academy of Sciences of the United States of America*, **101**, 9182–9186.

74 Strasdeit, H. (2005) New studies on the Murchison meteorite shed light on the pre-RNA world. *ChemBioChem*, **6**, 801–803.

75 Gorrell, I.B., Wang, L., Marks, A.J., Bryant, D.E., Bouillot, F., Goddard, A., Heard, D.E., and Kee, T.P. (2006) On the origin of the Murchison meteorite phosphonates. Implications for pre-biotic chemistry. *Journal of the Chemical Society, Chemical Communications*, 1643–1645.

76 Levin, E.M., Bud'ko, S.L., Mao, J.D., Huang, Y., and Schmidt-Rohr, K. (2007) Effect of magnetic particles on NMR spectra of Murchison meteorite organic matter and a polymer-based model system. *Solid State Nuclear Magnetic Resonance*, **31**, 63–71.

77 Pizzarello, S. (2007) The chemistry that preceded life's origin: a study guide from meteorites. *Chemistry & Biodiversity*, **4**, 680–693.

78 Amend, J.P. and Shock, E.L. (1998) Energetics of amino acid synthesis in hydrothermal ecosystems. *Science*, **281**, 1659–1662.

79 Lazcano, A. and Bada, J.L. (2003) The 1953 Stanley L. Miller experiment: Fifty years of prebiotic organic chemistry. *Origins of Life and Evolution of the Biosphere*, **33**, 235–242.

80 Bernstein, M. (2006) Prebiotic materials from on and off the early earth. *Philosophical Transactions of the Royal Society of London. Series B, Biological Sciences*, **361**, 1689–7000 (and subsequent discussion 1700–1702).

81 Shock, E.L. (2002) Astrobiology: seeds of life? *Nature*, **416**, 380–381.

82 Wong, J.T.-F. and Bronskill, P.M. (1979) Inadequacy of prebiotic synthesis as origin of proteinous amino acids. *Journal of Molecular Evolution*, **13**, 115–125.

83 Miller, S.L. and Orgel, L.E. (1974) *The Origins of Life on the Earth*, Prentice-Hall Inc., New Jersey.

84 Higgs, P.G. and Pudritz, R.E. (2007) From protoplanetary disks to prebiotic amino acids and the origin of the genetic code, in *Planetary Systems and the Origins of Life*, vol. 3 (eds R.E. Pudritz, P.G. Higgs, and J. Stone), Cambridge Series in Astrobiology, Cambridge University Press, Cambridge.

85 Martin, W. and Russell, M.J. (2007) On the origin of biochemistry at an alkaline hydrothermal vent. *Philosophical Transactions of the Royal Society of London. Series B, Biological Sciences*, **362**, 1887–1925.

86 Walter, M.R. (1996) Ancient hydrothermal ecosystems on earth: a new palaeobiological frontier. *Ciba Foundation Symposium*, **202**, 112–127, discussion 127–130.

87 Simoneit, B.R.T. (1995) Evidence for organic synthesis in high temperature aqueous media-facts and prognosis. *Origins of Life and Evolution of the Biosphere*, **25**, 119–140.

88 Bada, J.L., Fegley, B. Jr., Miller, S.L., Lazcano, A., Cleaves, H.J., Hazen, R.M., and Chalmers, J. (2007) Debating evidence for the origin of life on earth. *Science*, **315**, 937–938.

89 Wächtershäuser, G. (1990) Evolution of the first metabolic cycles. *Proceedings of the National Academy of Sciences of the United States of America*, **87**, 200–204.

90 Wächtershäuser, G. and Huber, C. (2007) Response to 'debating evidence for the origin of life on earth'. *Science*, **315**, 938–939.

91 Cody, G.D. and Scott, J.H. (2007) The roots of metabolism, in *Planets and Life: The Emerging Science of Astrobiology* (eds W.T. Sullivan and J.A. Barross), Cambridge University Press, Cambridge.

92 Dillon, L.S. (1973) The origins of the genetic code. *The Botanical Review*, **39**, 301–345.

93 Morowitz, H.J., Kostelnik, J.D., Yang, J., and Cody, G.D. (2000) The origin of intermediary metabolism. *Proceedings of the National Academy of Sciences of the United States of America*, **97**, 7704–7708.

94 Smith, E. and Morowitz, H.J. (2004) Universality in intermediary metabolism. *Proceedings of the National Academy of Sciences of the United States of America*, **101**, 13168–13173.

95 Horowitz, N.H. (1945) The evolution of biochemical syntheses. *Proceedings of the National Academy of Sciences of the United States of America*, **31**, 153–157.

96 Wong, J.T.-F. (1975) A co-evolution theory of the genetic code. *Proceedings of the National Academy of Sciences of the United States of America*, **72**, 1909–1912.

97 Wong, J.T.-F. (1976) The evolution of a universal genetic code. *Proceedings of the National Academy of Sciences of the United States of America*, **73**, 2336–2340.

98 Chaley, M.B., Korotkov, E.V., and Phoenix, D.A. (1999) Relationships among isoacceptor tRNAs seems to support the coevolution theory of the origin of the genetic code. *Journal of Molecular Evolution*, **48**, 168–177.

99 Xue, H., Tong, K.-L., Marck, C., Grosjean, H., and Wong, J.T.-F. (2003) Transfer RNA paralogs, evidence for genetic code-amino acid biosynthesis coevolution and an archaeal root of life. *Gene*, **310**, 59–66.

100 Wong, J.T.-F. (2007) Question 6: coevolution theory of the genetic code: a proven theory. *Origins of Life and Evolution of the Biosphere*, **37**, 403–408.

101 Amirnovin, R. (1997) An analysis of the metabolic theory of the origin of the genetic code. *Journal of Molecular Evolution*, **44**, 473–476.

102 Ronneberg, T.A., Landweber, L.F., and Freeland, S.J. (2000) Testing a biosynthetic theory of the genetic code, fact or artifact? *Proceedings of the National Academy of Sciences of the United States of America*, **97**, 13690–13695.

103 Di Giulio, M. (2001) A blind empiricism against the coevolution theory of the

origin of the genetic code. *Journal of Molecular Evolution*, **53**, 724–732.

104 Taylor, F.J.R. and Coates, D. (1989) The code within the codons. *Bio Systems*, **22**, 177–187.

105 Cunchillos, C. and Lecointre, G. (2005) Integrating the universal metabolism into a phylogenetic analysis. *Molecular Biology and Evolution*, **22**, 1–11.

106 Moller, W. and Janssen, G.M. (1990) Transfer RNAs for primordial amino acids contain remnants of a primitive code at position 3 to 5. *Biochimie*, **72**, 361–368.

107 Rodin, S., Rodin, A., and Ohno, S. (1996) The presence of codon–anticodon pairs in the acceptor stem of tRNAs. *Proceedings of the National Academy of Sciences of the United States of America*, **93**, 4537–4542.

108 Schimmel, P. (1996) Origin of genetic code: a needle in the haystack of tRNA sequences. *Proceedings of the National Academy of Sciences of the United States of America*, **93**, 4521–4522.

109 Cavalcanti, A.R.O., Leite, E.S., Neto, B.B., and Ferreira, R. (2004) On the classes of aminoacyl-tRNA synthetases, amino acids and the genetic code. *Origins of Life and Evolution of the Biosphere*, **34**, 407–420.

110 Klipcan, L. and Safro, M. (2004) Amino acid biogenesis, evolution of the genetic code and aminoacyl-tRNA synthetases. *Journal of Theoretical Biology*, **228**, 389–396.

111 Ribas de Pouplana, L. and Schimmel, P. (2001) Aminoacyl-tRNA synthetases: potential markers of genetic code development. *Trends in Biochemical Sciences*, **26**, 591–596.

112 Wetzel, R. (1995) Evolution of the aminoacyl-tRNA synthetases and the origin of the genetic code. *Journal of Molecular Evolution*, **40**, 545–550.

113 Crick, F.H.C., Brenner, S., Klug, A., and Pieczenik, G. (1976) A speculation on the origin of protein synthesis. *Origins of Life*, **7**, 389–397.

114 Eigen, M. and Schuster, P. (1978) The hypercycle. A principle of natural self-organization. Part C, the realistic hypercycle. *Die Naturwissenschaften*, **65**, 341–369.

115 Hartman, H. (1995) Speculations on the origin of the genetic code. *Journal of Molecular Evolution*, **40**, 541–544.

116 Jimenez-Sanchez, A. (1995) On the origin and evolution of the genetic code. *Journal of Molecular Evolution*, **41**, 712–716.

117 Lehmann, J., Riedo, B., and Dietler, G. (2004) Folding of small RNAs displaying the GNC base-pattern, implications for the self-organization of the genetic system. *Journal of Theoretical Biology*, **227**, 381–395.

118 Brooks, D.J., Fresco, J.R., and Singh, M. (2004) A novel method for estimating ancestral amino acid composition and its application to proteins of the last universal ancestor. *Bioinformatics*, **20**, 2251–2257.

119 Jordan, I.K., Kondrashov, F.A., Adzhubei, I.A., Wolf, Y.I., Koonin, E.V., Kondrashov, A.S., and Sunyaev, S. (2005) A universal trend of amino acid gain and loss in protein evolution. *Nature*, **433**, 633–638.

120 Orgel, L.E. (1968) Evolution of the genetic apparatus. *Journal of Molecular Biology*, **38**, 381–393.

121 DeGrado, W.F., Wasserman, Z.R., and Lear, J.D. (1989) Protein design, a minimalist approach. *Science*, **243**, 622–628.

122 Kamtekar, S., Schiffer, J.M., Xiong, H., Babik, J.M., and Hecht, M.H. (1993) Protein design by binary patterning of polar and nonpolar amino acids. *Science*, **262**, 1680–1685.

123 Kolaskar, A.S. and Ramabrahmam, V. (1982) Obligatory amino acids in primitive proteins. *Bio Systems*, **15**, 105–109.

124 Fan, K. and Wang, W. (2003) What is the minimum number of letters required to fold a protein? *Journal of Molecular Biology*, **328**, 921–926.

125 Fernandez, A. (2004) Lower limit to the size of the primeval amino acid alphabet. *Zeitschrift Fur Naturforschung C – A Journal of Biosciences*, **59**, 151–152.

126 Murphy, L.R., Wallqvist, A., and Levy, R.M. (2000) Simplified amino acid alphabets for protein fold recognition and implications for folding. *Protein Engineering*, **13**, 149–152.

127 Romero, P., Obradovic, Z., and Dunker, A.K. (1999) Folding minimal sequences: the lower bound for sequence complexity of globular proteins. *FEBS Letters*, **462**, 363–367.

128 Jukes, T.H. (1973) Arginine as an evolutionary intruder into protein synthesis. *Biochemical and Biophysical Research Communications*, **53**, 709–714.

129 Trifonov, E.N. (2000) Consensus temporal order of amino acids and evolution of the triplet code. *Gene*, **261**, 139–151.

130 Crick, F.H. (1968) Origin of the genetic code. *Journal of Molecular Biology*, **38**, 367–379.

131 White, H.B. (1976) Coenzymes as fossils of an earlier metabolic state. *Journal of Molecular Evolution*, **7**, 101–104.

132 Gilbert, W. (1986) Origin of life: the RNA world. *Nature*, **319**, 618.

133 Chen, X., Li, N., and Ellington, A.D. (2007) Ribozyme catalysis of metabolism in the RNA world. *Chemistry & Biodiversity*, **4**, 633–655.

134 Benner, S.A., Ellington, A.D., and Tauer, A. (1989) Modern metabolism as a palimpsest of the RNA world. *Proceedings of the National Academy of Sciences of the United States of America*, **86**, 7054–7058.

135 Müller, U.F. (2006) Re-creating an RNA world. *Cellular and Molecular Life Sciences*, **63**, 1278–1293.

136 Shapiro, R. (1986) *A Skeptics Guide to the Creation of Life on Earth*, Bantam Press, New York.

137 Hayes, B. (1998) The invention of the genetic code. *American Scientist*, **86**, 8–14.

138 Knight, R.D., Freeland, S.J., and Landweber, L.F. (1999) Selection, history and chemistry: the three faces of the genetic code. *Trends in Biochemical Sciences*, **24**, 241–247.

139 Yarus, M. (1998) Amino acids as RNA ligands: a direct-RNA-template theory for the code's origin. *Journal of Molecular Evolution*, **47**, 109–117.

140 Knight, R.D. and Landweber, L.F. (1998) Rhyme or reason: RNA–arginine interactions and the genetic code. *Chemistry & Biology*, **5**, R215–R220.

141 Djordjevic, M. (2007) SELEX experiments: new prospects, applications and data analysis in inferring regulatory pathways. *Biomolecular Engineering*, **24**, 179–189.

142 Yarus, M., Caporaso, J.G., and Knight, R. (2005) Origins of the genetic code: the escaped triplet theory. *Annual Review of Biochemistry*, **74**, 179–198.

143 Hughes, R.A., Robertson, M.P., Ellington, A.D., and Levy, M. (2004) The importance of prebiotic chemistry in the RNA world. *Current Opinion in Chemical Biology*, **8**, 629–633.

144 Lu, Y. and Freeland, S.J. (2006) On the evolution of the standard amino-acid alphabet. *Genome Biology*, **7**, 102.

145 Weber, A.L. and Miller, S.L. (1981) Reasons for the occurrence of the twenty coded protein amino acids. *Journal of Molecular Evolution*, **17**, 273–284.

146 Guthöhrlein, E.W., Malesević, M., Majer, Z., and Sewald, N. (2007) Secondary structure inducing potential of beta-amino acids: torsion angle clustering facilitates comparison and analysis of the conformation during MD trajectories. *Biopolymers*, **88**, 829–839.

147 Gould, S.J. and Lewontin, R.C. (1979) The spandrels of San Marco and the Panglossian paradigm: a critique of the adaptationist programme. *Proceedings of the Royal Society of London. Series B, Biological Sciences*, **205**, 581–598.

148 Kawashima, S., Pokarowski, P., Pokarowska, M., Kolinski, A., Katayama, T., and Kanehisa, M. (2008) AAindex: amino acid index database, progress report 2008. *Nucleic Acids Research*, **36**, D202–D205.

149 Lu, Y. and Freeland, S.J. (2007) A quantitative investigation of the chemical space surrounding amino acid alphabet formation. *Journal of Theoretical Biology*, **250**, 349–361.

150 Ji, H.-F., Shen, J.L., and Zhang, H.-Y. (2005) Low-lying free energy levels of amino acids and its implications for origins of life. *Journal of Molecular Structure*, **756**, 109–112.

151 Hinds, D.A. and Levitt, M. (1996) From structure to sequence and back again. *Journal of Molecular Biology*, **258**, 201–209.

152 Szathmary, E. (1999) The origin of the genetic code: amino acids as cofactors in an RNA world. *Trends in Genetics*, **15**, 223–229.

153 Lu, Y. and Freeland, S.J. (2006) Testing the potential for computational chemistry to quantify physiochemical properties of the non-proteinaceous amino acids. *Astrobiology*, **6**, 606–624.

154 Lu, Y., Bulka, B., des Jardins, M., and Freeland, S.J. (2007) Amino acid quantitative structure property relationship database: a web-based platform for quantitative investigations of amino acids. *Protein Engineering Design and Selection*, **20**, 347–351.

155 Akashi, H. and Gojobori, T. (2002) Metabolic efficiency and amino acid composition in the proteomes of *Escherichia coli* and *Bacillus subtilis*. *Proceedings of the National Academy of Sciences of the United States of America*, **99**, 3695–3700.

156 Dufton, M.J. (1997) Genetic code synonym quotas and amino acid complexity: cutting the cost of proteins? *Journal of Theoretical Biology*, **187**, 165–173.

157 Zhang, H.-Y. (2007) Exploring the evolution of standard amino-acid alphabet: when genomics meets thermodynamics. *Biochemical and Biophysical Research Communications*, **359**, 403–405.

158 Freeland, S.J., Wu, T., and Keulmann, N. (2003) The case for an error minimizing standard genetic code. *Origins of Life and Evolution of the Biosphere*, **33**, 457–477.

159 Stoltzfus, A. and Yampolsky, L.Y. (2007) Amino acid exchangeability and the adaptive code hypothesis. *Journal of Molecular Evolution*, **65**, 456–462.

160 Caporaso, J.G., Yarus, M., and Knight, R. (2005) Error minimization and coding triplet/binding site associations are independent features of the canonical genetic code. *Journal of Molecular Evolution*, **61**, 597–607.

161 Aita, T., Urata, S., and Husimi, Y. (2000) From amino acid landscape to protein landscape: analysis of genetic codes in terms of fitness landscape. *Journal of Molecular Evolution*, **50**, 313–323.

162 Freeland, S.J. (2002) The Darwinian genetic code: an adaptation for adapting? *Genetic Programming and Evolvable Machines*, **3**, 113–127.

163 Zhu, W. and Freeland, S.J. (2006) The standard genetic code enhances adaptive evolution of proteins. *Journal of Theoretical Biology*, **239**, 63–70.

164 Deamer, D.W., Dick, R., Thiemann, W., and Shinitzky, M. (2007) Intrinsic asymmetries of amino acid enantiomers and their peptides: a possible role in the origin of biochirality. *Chirality*, **19**, 751–763.

165 Caglioti, L., Holczknecht, O., Fujii, N., Zucchi, C., and Palyi, G. (2006) Astrobiology and biological chirality. *Origins of Life and Evolution of the Biosphere*, **36**, 459–466.

166 Bailey, J., Chrysostomou, A., Hough, J.H., Gledhill, T.M., McCall, A., Clark, S., Menard, F., and Tamura, M. (1998) Circular polarization in star-formation regions: implications for biomolecular homochirality. *Science*, **281**, 672–674.

167 Hecht, J. (1998) Inner circles: a strange light from space may account for life's love of the left. *New Scientist*, **159**, 11.

168 Service, R.F. (1999) Does life's handedness come from within? *Science*, **286**, 1282–1283.

169 Davies, P. (2007) Are aliens among us? *Scientific American*, **297**, 36–43.

170 Pizzarello, S., Huang, Y., and Alexandre, M.R. (2008) Molecular asymmetry in extraterrestrial chemistry: Insights from a pristine meteorite. *Proceedings of the National Academy of Sciences of the United States of America*, **105**, 3700–3704.

Part Two
Production/Synthesis of Amino Acids

Amino Acids, Peptides and Proteins in Organic Chemistry. Vol.1 – Origins and Synthesis of Amino Acids.
Edited by Andrew B. Hughes

ISBN: 978-3-527-32096-7

3
Use of Enzymes in the Synthesis of Amino Acids

Theo Sonke, Bernard Kaptein, and Hans E. Schoemaker

3.1
Introduction

Amino acids, natural as well as synthetic ones, are extensively used in the food, feed, agrochemical, and pharmaceutical industries. The total amino acid market has been estimated at approximately US$4.5 billion in 2004 [1]. Many proteinogenic amino acids are used as infusion solutions, whereas the essential amino acids (e.g., lysine, threonine, and methionine) are used as feed additives. Monosodium L-glutamate (more commonly known as MSG) is widespread as a taste enhancer/seasoning agent, the commercial production of which dates back to 1908 [2]. L-Aspartic acid and L-phenylalanine methyl ester together form the low-calorie sweetener aspartame (**1**) [3]. However, not only proteinogenic amino acids are applied; synthetic amino acids are also intermediates in pharmaceuticals and agrochemicals production. D-Phenylglycine and D-*p*-hydroxyphenylglycine (D-HPG) are produced in quantities of several thousands of tons per year for the synthesis of the semisynthetic broad-spectrum antibiotics ampicillin (**2a**), amoxicillin (**2b**), and others [4, 5]. Other pharmaceuticals containing unnatural amino acids are, for example, the antihypertensive drugs Aldomet [L-α-methyl- 3,4-dihydroxyphenylalanine (L-α-methyl-DOPA)] (**4**) [6], the angiotensin-converting enzyme inhibitor ramipril (**3**) [7], and many of the HIV-protease inhibitors of which atazanavir (**5**) is an example [8]. D-Valine is used as a building block of the pyrethroid insecticide fluvalinate (**6**) [9], whereas the α-methyl-substituted amino acid α-methylvaline is a structural element in the herbicide arsenal (**7**) (currently marketed as its isopropylamine salt by BASF) and related herbicides [10, 11], and L-α-methylphenylglycine is a part of the fungicide fenamidone (**8**) [12, 13].

Although many amino acids appear in living organisms, only a few of them are actually isolated from nature for commercial purposes. L-Cysteine and L-4-hydroxyproline are some of the examples in which extraction of a natural protein-rich feedstock like hair, keratin, and feather is still a commercially viable process [2]. For most proteinogenic amino acids fermentation is nowadays the preferred route for their preparation [1]. Examples include L-lysine and L-glutamic acid,

Amino Acids, Peptides and Proteins in Organic Chemistry. Vol.1 – Origins and Synthesis of Amino Acids.
Edited by Andrew B. Hughes

ISBN: 978-3-527-32096-7

which are produced in amounts of multihundreds of kilotons per year. However, fermentation is also the production method of choice for proteinogenic amino acids with a much smaller market volume, such as L-phenylalanine, L-aspartic acid, and L-isoleucine [2].

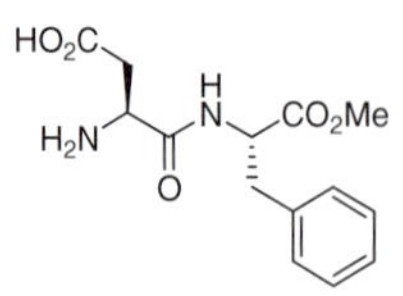

Aspartame **1**

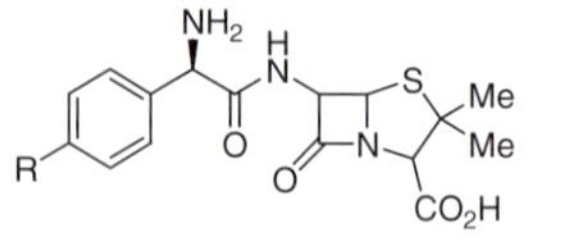

Ampicillin (R = H) **2a**
Amoxycillin (R = OH) **2b**

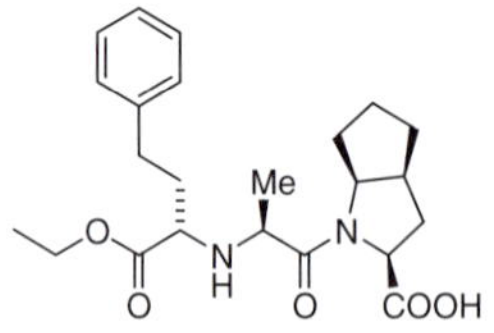

Ramipril **3**

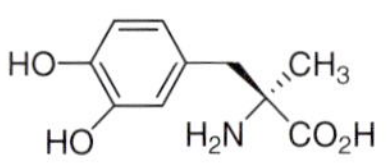

L-α-MethylDOPA **4**

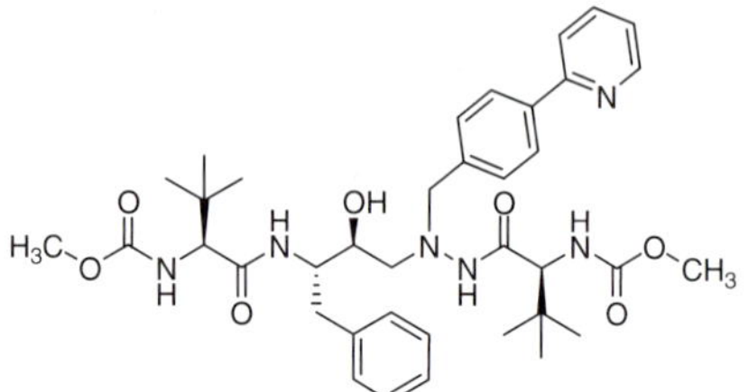

Atazanavir **5**

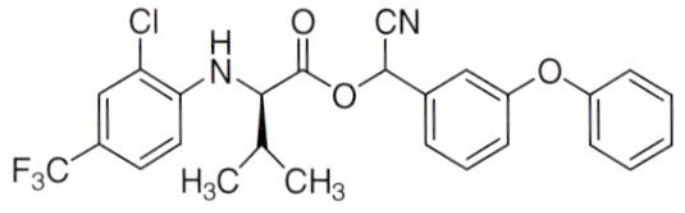

Fluvalinate **6**

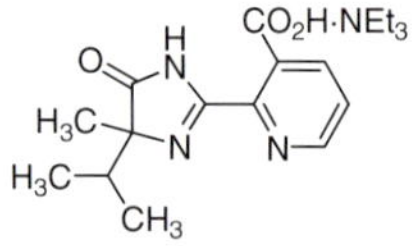

Arsenal **7**

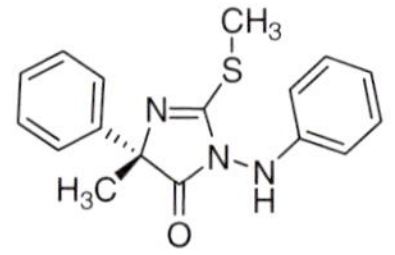

Fenamidone **8**

Of the proteinogenic amino acids, D,L-methionine and glycine are the only ones manufactured synthetically. Both L- and D,L-methionine have a similar effect as a feed additive, because the animal organism is converting the D-enantiomer into its nutritive optical antipode. As a consequence, methionine has been fed as a racemic mixture for more than 50 years [14].

Only a few examples are known in which amino acids are produced by chemical (catalytic) asymmetric synthesis. The asymmetric hydrogenation developed by Monsanto for L-phenylalanine is today only used for the production of L-DOPA [15]. Many other asymmetric routes to enantiopure α-H- and α,α-disubstituted α-amino acids have been developed on a laboratory scale, but to our knowledge none of them has been scaled-up beyond pilot-plant scale [16, 17].

3.2
Chemo-Enzymatic Processes to Enantiomerically Pure Amino Acids

Commercialization of the first biocatalytic method for enantiomerically pure amino acids goes back to 1954, when Tanabe Seiyaku implemented an aminoacylase-based resolution process for the production of several L-amino acids (Section 3.2.1) [18].

Since then many different biocatalytic procedures for the production of L- as well as D-amino acids have been described, but only a handful have been developed beyond the laboratory bench and implemented at full industrial scale.

Enzymatic resolution steps have an inherent maximum yield per cycle of 50% unless the unwanted enantiomer can be racemized. A second category of chemo-enzymatic processes rely on an enzyme-catalyzed asymmetric synthesis as a key step. Consequently, these processes have a maximum theoretical yield of 100% per cycle.

In the last decades a number of chemo-enzymatic processes for the production of enantiomerically pure amino acids have been commercialized by different companies [19]. In the following sections an overview is given of their most important features, with the main focus on typical process characteristics like scope, limitations, and different modes of operation through the years.

3.3 Acylase Process

One of the most widespread methods for the biocatalytic synthesis of L-amino acids is the resolution of *N*-acetyl-D,L-amino acids (**9**) by the enzyme *N*-acyl-L-amino acid amidohydrolase (aminoacylase, acylase I; EC 3.5.1.14) [18, 20]. In this process, the enzyme selectively hydrolyses the *N*-acetyl-L-amino acid to the corresponding L-amino acid without touching the D-substrate (Scheme 3.1). Equimolar amounts of acetic acid are formed as a side-product. After the reaction has gone to completion, the amino acid product can be separated from the nonreacted substrate through ion exchange chromatography or crystallization. The racemic *N*-acetylated amino acid substrates are fairly easily accessible through acetylation of D,L-amino acids with acetyl chloride or acetic acid anhydride in alkali in a Schotten–Baumann reaction [21] or via amidocarbonylation [22].

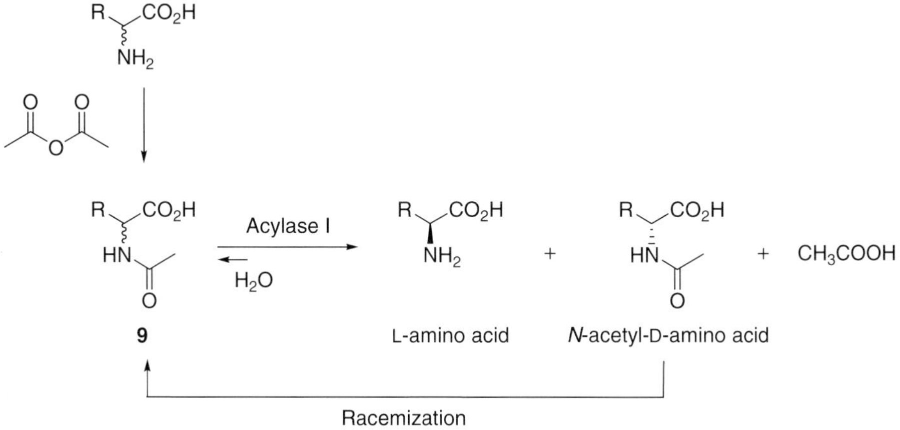

Acylase I

H_2O

+ CH_3COOH

9

L-amino acid

N-acetyl-D-amino acid

Racemization

Scheme 3.1 Enantiospecific hydrolysis of *N*-acetyl-D,L-amino acids (**9**) by *A. oryzae* acylase I.

A drawback of the use of acylases is the fact that the hydrolysis reaction of *N*-acetyl amino acids is equilibrium-limited, but this equilibrium is well on the side of hydrolysis [23]. Typical equilibrium constants range from 3.7 M for methionine to 12.5 M for norleucine (pH 7.5, 25 °C), which translates to equilibrium conversions of 89–96% based on 0.5 M of *N*-acetyl amino acid [24]. This incomplete conversion results in a lower enantiomeric excess of the remaining *N*-acetyl-D-amino acid, which certainly complicates the cost-efficient production of D-amino acids by this process.

After introducing the acylase process in 1954 using the acylase I from *Aspergillus oryzae*, the Japanese company Tanabe Seiyake further improved its competitiveness by switching from a batch reaction with the native enzyme to a continuous mode of operation in a fixed-bed reactor in 1969. Successful immobilization of the *A. oryzae* acylase I onto diethylaminoethyl cellulose DEAE-Sephadex at an industrial scale was essential for the switch to this continuous enzymatic process. This first immobilized enzyme reactor system led to 40% lower production costs compared to the batch process [18].

Degussa implemented the acylase I process from *A. oryzae* in 1982 [19]. Currently, this process is operated in an enzyme membrane reactor using a hollow-fiber membrane with a cut-off value of 10 kDa. This results in over 99.9% retention of the 73 kDa acylase. Using this setup Degussa is producing several hundred tons per year of enantiomerically pure L-amino acids, mostly L-methionine and L-valine [25].

Since the acylase process is a typical example of an enzymatic kinetic resolution many attempts have been undertaken to racemize the remaining *N*-acetyl-D-amino acid. The high pK_a of approximately 15 [19] of the *N*-acetyl amino acids necessitates rather extreme conditions to effect this racemization [25, 26], precluding simultaneous resolution and racemization steps. This problem was solved by the discovery of an *N*-acylamino acid racemase, in several actinomycetes strains, first by scientists at Takeda [27–29] and later also by scientists at Degussa [30]. More recently, an *N*-acylamino acid racemase was also identified in the bacterium *Deinococcus radiodurans* [31]. Although the reaction conditions for these acylamino acid racemases seem to be compatible with those for acylase I (e.g., metal ion dependency, pH, temperature) and proof-of-principle for a dynamic kinetic resolution (DKR) acylase process has been demonstrated on the laboratory scale [31–33], there are no clear indications that such a combined process has already been implemented at the commercial scale [19].

Over the years several microbial amidohydrolases have been isolated that readily and enantioselectively convert, for example, *N*-acetyl-L-proline and its derivatives [34, 35], a class of substrates not accepted by the acylase I. Interestingly, the enzyme from *Comamonas testosteroni* can also resolve racemic *N*-acylated *N*-alkyl-amino acids, thereby opening a novel route to enantiomerically pure *N*-alkyl-amino acids [36]. Several Japanese groups have invested in the isolation of suitable D-aminoacylases. Although such D-specific enzymes have been found in, for example, *Streptomyces* sp. [37], *Pseudomonas* sp. [38], and *Alcaligenes* sp. [39], for obtaining enantiopure products purification from the L-aminoacylase background activity present in the microorganisms is necessary. Such tedious and expensive purification complicates their commercial application [20].

3.4 Amidase Process

Another process for the production of enantiomerically pure α-H-α-amino acids was developed at DSM in the mid-1970s. This process, which is based on the enzymatic kinetic resolution of racemic α-H-α-amino acid amides (**10**) with L-selective amide hydrolases, was commercialized by DSM in the mid-1980s for the production of several L- and D-amino acids [4, 40, 41]. As a rule, the amide substrates are directly accessible by Strecker synthesis on their corresponding aldehydes followed by hydrolysis under mild basic conditions in the presence of catalytic amounts of an aldehyde or ketone [42]. Different from the acylase process, the amide substrate is thus a precursor of the amino acid to be prepared. If needed, the racemic amino acid amides can also be prepared on a laboratory scale by the alkylation of *N*-acetamidomalonate esters or of imino esters and amides derived from glyoxylic acid [43]. Another advantage over the acylase process is the fact that the amide hydrolysis is not thermodynamically limited, which implies that the conversion can, in principle, be quantitative at every substrate concentration applied [19]. Consequently, D-amino acids with near absolute enantiomeric excess are also accessible through this process applying L-selective amide hydrolases. Alternatively, D-selective amino acid amide hydrolases can be used for the production of D-amino acids. Although suitable D-selective hydrolases have among others been identified in *Ochrobactrum anthropi* strains SV3 [44] and C1-38 [45], their large-scale application is complicated by the fact that they need to be separated from the L-amide hydrolases that are present in almost all microorganisms.

The amidase process at DSM has long been operated with permeabilized whole cells of *Pseudomonas putida* ATCC 12 633. Because this biocatalyst has a nearly 100% enantioselectivity for the hydrolysis of most L-amides (enantiomeric ratio $E > 200$), both the L-amino acid and the D-amino acid amide are obtained in almost 100% e.e. at 50% conversion. The hydrolysis reaction further furnishes 1 equiv. of ammonia (compared to the L-amino acid amide) as side-product. After completion of the hydrolysis reaction 1 equiv. of benzaldehyde is added to form the Schiff base of the unreacted D-amino acid amide, which precipitates from the solution. In this way the L-acid and D-amide can easily be separated (Scheme 3.2). Alternatively, separation of acid and amide is possible with a basic ion-exchange resin.

Apart from its exquisite enantioselectivity, the *P. putida* whole-cell biocatalyst is characterized by its broad substrate specificity. So far over 100 different α-H-α-amino acid amides have been successfully resolved. The size of the side-chain may range from the small methyl group in alanine to the very bulky group of lupinic acid [46]. Furthermore, the alkyl or aryl side-chains may contain heteroatoms like sulfur, nitrogen, and oxygen. Also cyclic amino acid amides like proline amide and piperidine-2-carboxyamide, and amino acid amides with alkenyl and alkynyl substituents are substrates for this biocatalyst [47, 48]. Protein purification experiments identified an L-aminopeptidase that contributes to a considerable extent to the broad substrate specificity of *P. putida* ATCC 12 633 [49]. This enzyme performs optimally at pH 9.0–9.5 and 40 °C in the presence of 0.2–20 mM Mn^{2+} ions as an activating

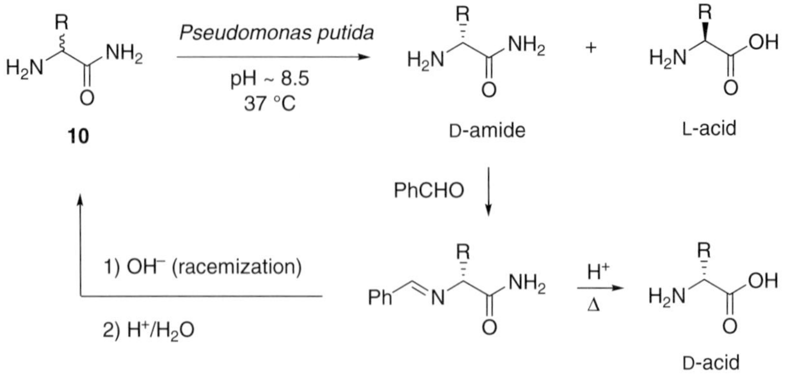

Scheme 3.2 Enzymatic resolution of α-H-α-amino acid amides (**10**) by *P. putida*.

divalent metal ion. Cloning of the gene encoding this aminopeptidase by reversed genetics and subsequent overexpression in *E. coli* led to a highly efficient whole-cell biocatalyst [50]. Protein database searches revealed that this L-aminopeptidase belongs to the leucine aminopeptidase family of proteins, which catalyze the hydrolysis of amino acids from the N-terminus of (poly)peptide chains. Their, in general, broad substrate specificity correlates well with the fact that the purified *P. putida* L-aminopeptidase displays activity towards a broad range of α-H-α-amino acid amides and dipeptides.

The added value of this highly active recombinant *E. coli* based whole-cell aminopeptidase biocatalyst became manifest in the preparation of a set of unsaturated α-H-α-amino acids (Scheme 3.3) [51]. In general, the resolution of these unsaturated amino acid amides with the recombinant *E. coli*-based system as well as

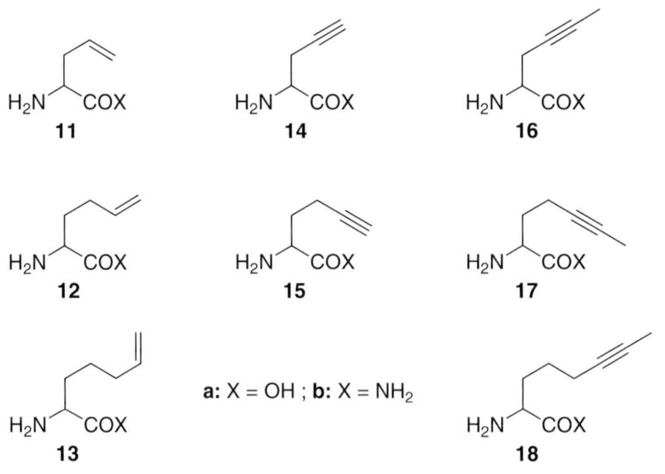

Scheme 3.3 The unsaturated amino acids resolved by Wolf *et al.* [51] using a recombinant *E. coli*-based whole-cell biocatalyst expressing the *P. putida* L-aminopeptidase.

with the wild-type *P. putida* cells proceeded smoothly yielding the L-acid and D-amide in high enantiomeric excess (>95%) at 50% conversion. Owing to the increased expression of the L-aminopeptidase in the *E. coli* cells they were applied in a cell : substrate ratio of 1 : 500 only, whereas this ratio was 1 : 10 with the *P. putida* cells. More interestingly, however, L-3-butynylglycine **15a** and especially its methylated homolog **17a** were obtained in a moderate enantiomeric excess only (91 and 70%, respectively) using *P. putida* cells, whereas the resolution reactions with the recombinant *E. coli* biocatalyst resulted in a superior enantiomeric excess of both L-acids (97 and 99%, respectively). Further experiments made clear that this unsatisfactory low enantiomeric excess of both L-acids with *P. putida* cells as biocatalyst originates from the presence of an amino acid racemase with narrow substrate specificity that is absent in the recombinant *E. coli* whole-cell biocatalyst [51].

This latter biocatalyst was subsequently used for the preparation of the L- and D-enantiomers of amino acids **11a–18a** on the multigram scale using the procedure depicted in Scheme 3.4. After standard work-up of the unreacted D-amides, these were hydrolyzed under very mild conditions by the nonselective amidase present in *Rhodococcus erythropolis* NCIMB 11 540 cells. Chemical hydrolysis was not an option for this type of unsaturated amide, since the harsh acidic conditions needed would lead to decomposition of some of the side-chains. All L-acids were obtained in above 98% e.e. except for **18a**, which was obtained in an enantiomeric excess of 96%. The enantiomeric excesses of the D-acids appeared to be excellent without exception.

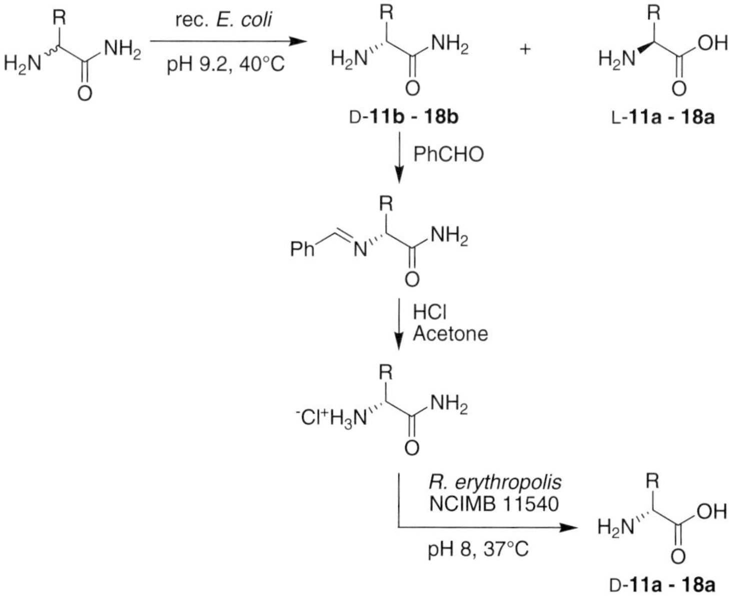

R
H_2N NH_2
O
rec. *E. coli*
pH 9.2, 40°C
R
H_2N NH_2
O
D-**11b** - **18b**
+
R
H_2N OH
O
L-**11a** - **18a**
PhCHO
R
Ph N NH_2
O
HCl
Acetone
R
$^-Cl^+H_3N$ NH_2
O
R. erythropolis
NCIMB 11540
pH 8, 37°C
R
H_2N OH
O
D-**11a** - **18a**

Scheme 3.4 Optimized amidase based process for the multi-gram synthesis of enantiomerically pure L- and D-unsaturated amino acids.

Like the acylase process, the amidase process is a resolution and thus it is hampered by a maximum yield of 50%. Racemization of the remaining D-amino acid amide, which can be easily done via formation of the benzaldehyde Schiff base under basic conditions, can lead to 100% yield (Scheme 3.2) [52]. As formation of this Schiff base is also the basis for the separation of the L-acid and D-amide, racemization can be performed without any additional step. Nevertheless, it was envisioned that a fully enzymatic DKR process combining the action of an enantioselective amidase and an α-H-α-amino acid amide racemase in one vessel would be a more cost-efficient process. Scientists at DSM recently identified the required α-H-α-amino acid amide racemase in a novel *O. anthropi* strain [53]. This amino acid amide racemase (AmaR), which appeared to be pyridoxal 5′-phosphate (PLP)-dependent, combines good thermostability with a broad pH optimum (6–10), which are two prerequisites for large-scale application. Apart from lactams, AmaR also racemizes a range of linear α-H-α-amino acid amides, although with lower activity, especially in the case of amino acid amides with a Cβ branched side-chain. Furthermore, the racemase is strongly inhibited by a couple of divalent metal ions, including Mn^{2+}, that is essential for the optimal performance of the *P. putida* L-aminopeptidase. This conflict was solved by careful tuning of the Mn^{2+} concentration, which showed that 0.6 mM of this divalent metal ion is the best compromise [54]. The applicability of AmaR for a DKR amidase process was demonstrated for the production of L-aminobutyric acid using a cell-free extract of an *E. coli* strain expressing AmaR and the endogenous L-selective amino acid amide hydrolase. Whereas the blank reaction with solely the L-amide hydrolase stopped near 50% conversion, the presence of AmaR led to formation of L-aminobutyric acid in 84% conversion and 96.3% e.e. [54].

AmaR has the highest homology (52% identity) with the α-amino-ε-caprolactam (ACL) racemase from *Achromobacter obae*. This ACL racemase has been developed by scientists at Toray Industries for the industrial production of L-lysine from D,L-ACL [26, 55]. In contrast to earlier reports [56], Asano and Yamaguchi recently reinvestigated this ACL racemase and found that it is also able to racemize linear α-H-α-amino acid amides, although with at least 35-fold lower specific activity than for ACL [57]. Furthermore, they showed that by combining this enzyme with the D-aminopeptidase from *O. anthropi* C1-38 [45], synthesis of near stoichiometric amounts of D-amino acids from L-amino acid amides is possible [58].

3.4.1 Amidase Process for α,α-Disubstituted α-Amino Acids

Inspired by the advantages of the amidase process, DSM scientists developed a similar enzymatic kinetic resolution process for the production of enantiopure α,α-disubstituted amino acids. Although alternative routes are available [59, 60], in this case the Strecker synthesis also is the most direct way to prepare the racemic disubstituted amino acid amides, but the hydrolysis of the aminonitrile intermediate needs more harsh conditions (e.g., benzaldehyde/pH 14, concentrated H_2SO_4, or HCl-saturated formic acid) because of the increased steric hindrance [61]. Since the *P. putida* biocatalysts require an α-hydrogen atom for activity, novel amidase biocatalysts needed to be identified. Screening led to the identification of *Mycobacterium neoaurum* ATCC 25 795, a biocatalyst that affords the (*S*)-α,α-disubstituted amino acids and the corresponding (*R*)-amides in almost 100% e.e. at 50% conversion for most α-methyl-substituted compounds ($E > 200$) [62, 63]. Only for glycine amides with two small substituents at the chiral center is the enantioselectivity decreased. Generally, α-methyl-substituted amino acid amides are hydrolyzed with high activity, but increasing the size of the smallest substituent to ethyl, propyl, or allyl dramatically reduces the activity, especially if the largest substituent does not contain a -CH_2- spacer at the chiral carbon atom [62]. In line with this, α-H-α-amino acid amides are also good substrates and are hydrolyzed enantioselectively. The enzyme that was responsible for the enantioselective hydrolysis of D,L-α-methylvaline amide has been purified and characterized [64]. This enzyme, which was classified as an amino amidase, is active toward a rather broad range of both α-H- and α-alkyl-substituted amino acid amides, which are hydrolyzed with moderate to high L-selectivity – the lowest enantioselectivity was obtained toward alanine amide ($E \geq 25$). However, this biocatalyst is inactive toward dialkyl amino acid amides with very bulky substituents and α-hydroxy acid amides.

To close this gap in the range of compounds that can be produced by the amidase technology, a novel amidase biocatalyst was identified in a classical screening program [65]. This novel biocatalyst, *O. anthropi* NCIMB 40 321, is characterized by its extremely broad substrate specificity including α-H- and (bulky) α,α-dialkyl-substituted amino acid amides, α-hydroxy acid amides, and *N*-hydroxyamino acid amides, its very good temperature stability and especially its relaxed pH profile. Although the amidase displays its highest activity at pH 8.5, 55% of this activity is still retained at pH 5.0. This enables the hydrolysis of substrates, which are only very poorly soluble at weakly alkaline conditions, just by performing the hydrolysis reaction at slightly acidic conditions. This feature of the *O. anthropi* amidase biocatalyst appeared to be crucial in the resolution of the α-H-α-amino β-hydroxy acid amides **19**, which are intermediates in novel routes to the antibiotics thiamphenicol (**20a**) and florfenicol (**20b**) [66]. In this case, the enzymatic resolution reaction was performed at pH 5.6–6.0 to ensure a fair solubility of the otherwise insoluble amides. Also in the preparation of (*S*)-α-methyl-3,4-dichlorophenylalanine **21**, an intermediate for cericlamine HCl, a potent and selective synaptosomal 5-hydroxytryptamine (serotonin) uptake inhibitor under development [67, 68], the activity of the *O. anthropi* amidase at low pH (in this case pH 5.3 was used) was a decisive factor [69].

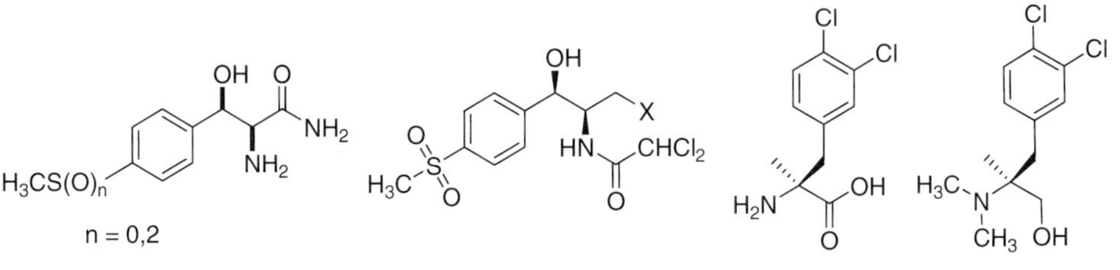

$H_3CS(O)_n$, n = 0,2 — **19**

20a (X=OH)
20b (X=F)

21

Cericlamine

Recently, the purification, cloning, and heterologous expression in *E. coli* of the most important amidase (LamA) from *O. anthropi* NCIMB 40 321 has been reported [70]. LamA displays activity toward a broad range of substrates consisting of α-hydrogen- and (bulky) α,α-disubstituted α-amino acid amides, α-hydroxy acid amides, and α-*N*-hydroxyamino acid amides, and is thus responsible on its own for the extremely broad substrate specificity of the *O. anthropi* whole cells. For all substrates investigated, only the L-enantiomer was hydrolyzed ($E > 150$). LamA appeared to be a metallo-enzyme, as it was strongly inhibited by the metal-chelating compounds ethylenediaminetetraacetic acid (EDTA) and 1,10-phenanthroline. The activity of the EDTA-treated enzyme could be restored by the addition of Zn^{2+} (to 80%), Mn^{2+} (to 400%), and Mg^{2+} (to 560%). The L-amidase gene encodes a polypeptide of 314 amino acids with clear homology to the acetamidase/formamidase family of proteins including the stereoselective amidases from *E. cloacae* [71], *Thermus* sp. [72], and *Klebsiella oxytoca* [73]. The *Enterobacter* and *Thermus* amidases (67 and 53% identical to LamA) have been developed by Mitsubishi researchers. Both amidases are highly L-selective toward D,L-*tert*-leucine amide and are also active toward lactate amide, implying that these amidases can also convert α-hydroxy acid amides. The amidase from *K. oxytoca* (28% identity to LamA) has been developed by Lonza for the (*R*)-selective hydrolysis of racemic 3,3,3-trifluoro-2-hydroxy-2-methylpropanamide [74].

Altogether, the *O. anthropi*L-amidase has a unique set of properties for application in the fine-chemicals industry.

3.5 Hydantoinase Process

A third commercialized chemo-enzymatic production method for enantiomerically pure α-H-α-amino acids is based on the enantioselective hydrolysis of racemic 5-monosubstituted hydantoins by D- or L-selective hydantoinases, often combined with D- or L-selective *N*-carbamoylases [25, 75, 76]. In 1979, Kanega implemented a first generation of such processes for the production of D-HPG (**24**) [75] – the side-chain for the semisynthetic β-lactam antibiotic amoxicillin and nowadays produced in several thousands of tons annually using the hydantoinase process. This first-generation process involved one enzymatic and two chemical conversions (Scheme 3.5).

In this process the D-hydantoin is selectively hydrolyzed to *N*-carbamoyl-D-HPG (**23**) by a D-specific hydantoinase in immobilized whole resting cells of the bacterium *Bacillus brevis* [77]. The slightly alkaline conditions (pH 8.0) lead to spontaneous racemization of the remaining L-hydantoin (see below), resulting in a quantitative conversion into *N*-carbamoyl-D-HPG. In the last step the *N*-carbamoyl-D-HPG is chemically hydrolyzed to the desired D-HPG using an equimolar amount of HNO_2.

Alternatively, Recordati commercialized a similar process employing resting cells of an *Agrobacterium radiobacter* strain, which contained both a D-hydantoinase and a strictly D-selective *N*-carbamoylamino acid amido hydrolase (D-*N*-carbamoylase)

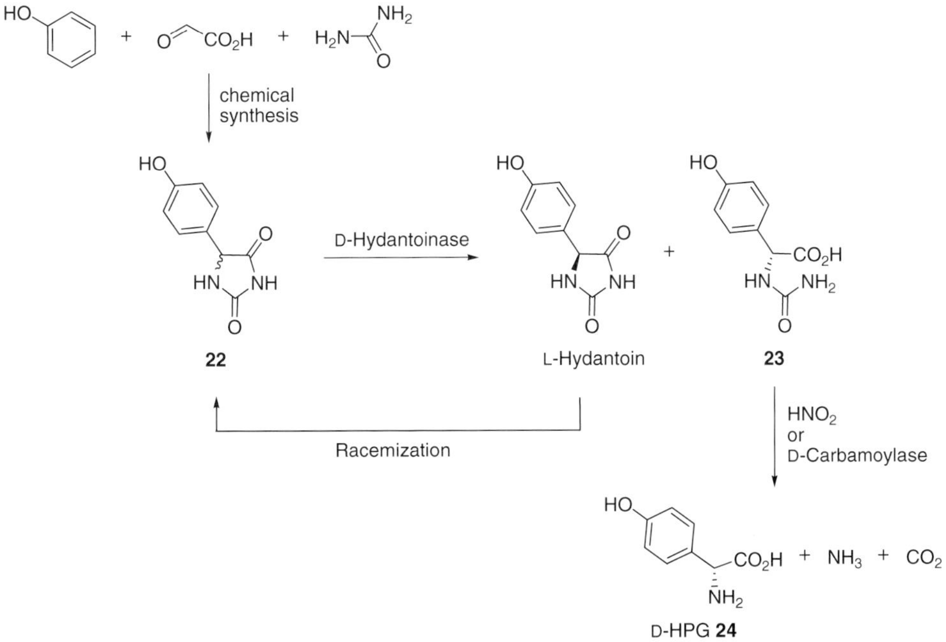

Scheme 3.5 Industrial production of D-HPG (**24**) by the D-hydantoinase process.

activity [78]. Consequently, the initially formed *N*-carbamoyl-D-HPG is irreversibly converted *in situ* to D-HPG in nearly 100% yield [79]. The carbamoylase also broadens the scope of the D-hydantoinase process to α-amino acids that cannot withstand the HNO_2 treatment, like D-tryptophan, D-citrulline, or D-pyridylalanine [76]. Since the *A. radiobacter* cells have a very broad substrate specificity, Degussa in collaboration with Recordati used the *A. radiobacter* biomass for the production in bulk amounts of a broad array of D-amino acids [80].

Industrial application of D-*N*-carbamoylases has long been complicated by their limited stability and their inhibition by ammonium ions. The inhibition by ammonium ions has been tackled by *in situ* removal via absorption to a silicate complex [81, 82] or via formation of a poorly soluble ammonium salt, for example, $MgNH_4PO_4$, by performing the reaction in the presence of $MgHPO_4$ [83], making industrial application more viable. More stable D-*N*-carbamoylases could, for instance, be obtained by a combination of random mutagenesis and screening for increased thermal [84, 85] or oxidative stability [86], or by immobilization [87]. Since 1995, Kanega has been applying a more thermostable variant of the D-*N*-carbamoylase of *Agrobacterium* sp. KNK712 as an immobilized biocatalyst in combination with a D-hydantoinase in a second-generation D-HPG process [75].

Chemical racemization of the hydantoins at slightly alkaline conditions (pH 8–9) is strongly dependent on the stabilizing effect of the side-chain in the 5-position. At pH 8.5 and 40 °C, racemization half-lives of the hydantoins range from 0.21 h in the case of the *p*-hydroxyphenyl side-chain to 120 h in the case of the *tert*-butyl side-chain [76].

Therefore, for 5-monoalkyl-substituted hydantoins the rate of racemization can become the rate-limiting step in the overall process.

Based on the observation that racemic 5-isopropylhydantoin was converted into D-valine in around 75–85% conversion by resting *A. radiobacter* cells, Battilotti and Barberini suggested that an enzyme might be responsible for the racemization of the remaining L-hydantoin [88]. A few years later hydantoin racemases could indeed be purified from *Arthrobacter aurescens* DSM 3747 [89] and *Pseudomonas* sp. NS671 [90], which were subsequently cloned and heterologously expressed in *E. coli.* [91, 92]. Recently, more hydantoin racemases were characterized and cloned, for instance, from *Agrobacterium tumefaciens* [93], *Sinorhizobium meliloti* [94], and *Microbacterium liquefaciens* [95]. In particular, the new hydantoin racemase from an *A. radiobacter* strain has industrial potential because it does not suffer from substrate inhibition – a drawback of many other hydantoin racemases [96].

Next to the enzymes acting on D-hydantoins and *N*-carbamoyl-D-amino acids, a number of strains have been identified that convert racemic 5-substituted hydantoins into L-amino acids. An example is the process for L-methionine using *Pseudomonas* sp. NS671 [97], *Bacillus stearothermophilus* NS1122A [98], and *Arthrobacter* sp. DSM 7330 [99]. L-Hydantoinases and L-carbamoylases from different microorganisms have been purified and characterized, and genes encoding these enzymes have been cloned and heterologously expressed [76]. Interestingly, the hydantoinases were not always fully L-specific [100], but a strictly L-specific carbamoylase in these microorganisms without exception resulted in the L-amino acids.

The recombinant expression of the three required enzymes in an easy to ferment microorganism not only leads to higher expression levels than in the wild-type microorganisms [101], but also enables the fine-tuning of the amount of each of these three enzymes securing an optimal flux through this mini-pathway without accumulation of intermediates [102]. This has also been achieved by reversing the enantioselectivity and improving the activity of the hydantoinase from an *Arthrobacter* sp. by directed evolution [103]. These developments will lead to more productive biocatalysts and consequently lower costs. It is expected that this will drive the replacement of the wild-type strains by more active and tuned recombinant production strains as well as the implementation of new processes. The availability of production hosts that are optimized for the production of D-amino acids [104, 105], the straightforward synthesis of the racemic hydantoins from cheap raw materials via a number of different routes [76] as well as the low side-product formation will certainly stimulate this.

3.6 Ammonia Lyase Processes

The second category of chemo-enzymatic processes to form α-H-α-amino acids rely on an enzyme-catalyzed asymmetric synthesis, which implies that they can theoretically utilize up to 100% of the substrate per cycle. One such process platform makes use of ammonia lyases, a class of enzymes that catalyze the reversible addition of ammonia to carbon–carbon double bonds [106].

HO_2C–CH=CH–CO_2H + NH_3 ⇌ (Aspartase) HO_2C–CH$_2$–CH(NH_2)–CO_2H

fumaric acid — L-aspartic acid **25**

Scheme 3.6 Aspartase-catalyzed synthesis route to L-aspartic acid (**25**) from fumaric acid.

3.6.1
Aspartase-Catalyzed Production of L-Aspartic Acid

L-Aspartate ammonia lyase (aspartase; EC 4.3.1.1) is one of the few ammonia lyases that is applied in industry, namely for the large-scale conversion of inexpensive fumaric acid into L-aspartic acid (**25**) via addition of ammonia (Scheme 3.6). This amino acid is widely used, for example, for food flavoring (mainly in Japan), in parenteral and enteral nutrition, as an acidulant, and as a precursor for the artificial sweetener aspartame (**1**) [26, 107].

A number of companies are operating such an aspartase-based process, but each in a somewhat different format. Although started in 1953 with a batchwise process [26], Tanabe Seiyaku switched to a continuous mode of operation in 1973 applying whole *E. coli* cells immobilized in a polyacrylamide gel lattice [108, 109]. The high immobilization yield (>70%) and excellent biocatalyst stability (half-life of 120 days at 37 °C) reduced the production costs to about 60% compared to the conventional batch process [18, 110]. The process was further improved by the introduction of κ-carrageenan as immobilization matrix in 1978 [18, 111] and of a hyper-aspartase-producing *E. coli* variant in 1983 [112].

Kyowa Hakko Kogyo, Mitsubishi Petrochemical, and W. R. Grace & Co. have implemented different process formats after similar optimization trajectories [26, 107, 113–117].

All these processes have in common that they are carried out with a 2- to 3-fold excess of ammonia to drive the reaction equilibrium from fumaric acid in the direction of L-aspartic acid ($K_{eq} = [\text{L-Asp}]/([\text{fumaric acid}]\cdot[NH_3]) = 4.2 \times 10^2\ M^{-1}$) [118]. This leads to almost stoichiometric conversion and excellent enantiomeric purity of greater than 99.9%.

More active aspartase biocatalysts have been obtained by cloning and overexpression of the aspartase-encoding gene from *Brevibacterium flavum* MJ-233 [119] and *E. coli* [120, 121], and by protein engineering via site-directed mutagenesis and directed evolution [122, 123]. A highly thermostable aspartase has been identified in *Bacillus* sp. YM55-1 [124]. On-going engineering projects will certainly benefit from the recently elucidated X-ray crystal structures of the *E. coli* aspartase [125] and of the thermostable aspartase from *Bacillus* sp. YM55-1 [126].

Production costs can be further reduced by combining aspartase with maleate isomerase (EC 5.2.1.1) and using maleic acid, derived from cheap maleic anhydride, as a substrate [127, 128].

The continuous optimization programs of the different players in this field have made the aspartase-based processes some of the most efficient enzyme processes known today [19] [space-time yield: 60 000 g/(l day); initial substrate concentration: 1.5–2 M [129]] and economically more attractive than fermentation.

3.6.2
Production of L-Alanine from Fumaric Acid by an Aspartase–Decarboxylase Cascade

Although L-alanine (**26**) is applied in enteral and parenteral nutrition, and as a food additive because of its sweet taste and bacteriostatic properties, the annual world production of this amino acid is approximately 500 tons only [130]. Although studies have proved that L-alanine can be fermented in high yield and optical purity [131], it is still produced enzymatically from L-aspartic acid by irreversible decarboxylation with L-aspartate β-decarboxylase (EC 4.1.1.12) (Scheme 3.7) [19]. L-Aspartate β-decarboxylase needs PLP as a prosthetic group and is allosterically activated by α-keto acids [132].

In 1965, Tanabe Seiyaku started production of L-alanine from L-aspartic acid as a batch process applying whole cells of the strain *Pseudomonas dacunhae*, a strain with high L-aspartate β-decarboxylase activity [133]. To improve the productivity and competitiveness, a switch was made to a continuous process in 1982 [134], which employed sonicated *P. dacunhae* cells immobilized in κ-carrageenan in a pressurized fixed-bed reactor [133, 134].

In the L-alanine manufacturing process the decarboxylase reaction can be combined with the aspartase-catalyzed synthesis of L-aspartic acid from fumaric acid as depicted in Scheme 3.7 [135, 136]; however, because of the different pH optima of the two enzymes a two-step production process is more efficient [118]. Such a two-step process has been operated by Tanabe Seiyaku since 1982 [137].

Since the aspartate β-decarboxylase is highly L-enantioselective it can also be used for the production of D-aspartic acid (**27**), together with L-alanine (**26**), from racemic aspartic acid via a kinetic resolution (Scheme 3.8). Racemic aspartic acid is easily prepared by the chemical addition of ammonia to the α,β-double bond in fumaric acid.

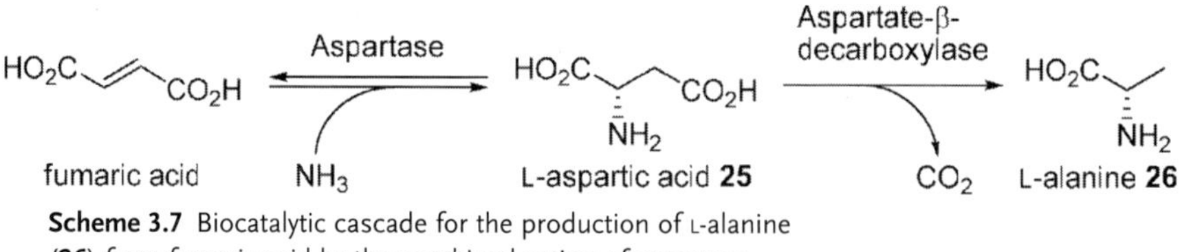

fumaric acid ⇄ (Aspartase, NH_3) L-aspartic acid **25** → (Aspartate-β-decarboxylase, −CO_2) L-alanine **26**

Scheme 3.7 Biocatalytic cascade for the production of L-alanine (**26**) from fumaric acid by the combined action of aspartase and L-aspartate β-decarboxylase.

DL-aspartic acid → (Aspartate-β-decarboxylase, −CO_2) L-alanine **26** + D-aspartic acid **27** + CO_2

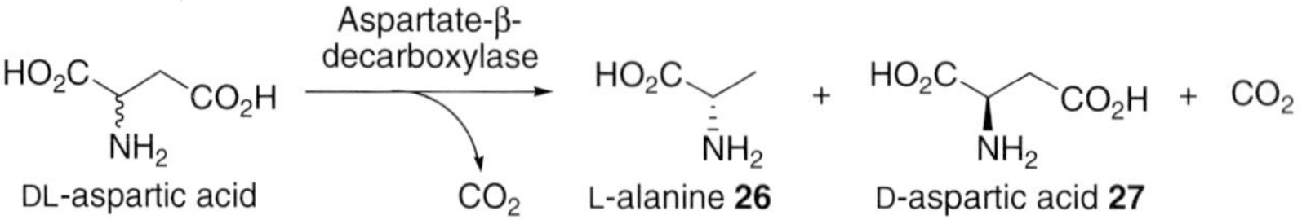

Scheme 3.8 Aspartate β-decarboxylase-catalyzed kinetic resolution process to D-aspartic acid (**27**) and L-alanine (**26**).

Both products can subsequently be separated via selective crystallization at different pH [26]. The continuous version of this process to D-aspartic acid and L-alanine using the pressurized plug-flow reactor and *P. dacunhae* cells in immobilized form is operated by Tanabe Seiyaku since 1989 [18].

3.6.3 Phenylalanine Ammonia Lyase-Catalyzed Production of L-Phenylalanine and Derivatives

Apart from aspartase, L-phenylalanine ammonia lyase (PAL, EC 4.3.1.5) is a second lyase of commercial importance. This enzyme is found ubiquitously in plants and also in specific microorganisms, especially yeasts and fungi. It catalyzes the non-oxidative deamination of L-phenylalanine (**28**) into ammonia and *trans*-cinnamic acid (**29**), which is the committed step in the phenylpropanoid pathways, which lead to a great variety of lignins, flavonoids, and coumarins [138].

The use of PAL-containing yeast cells (mostly *Rhodotorula rubra, Rhodotorula glutinis,* and *Rhodosporidium toruloides*) in the reverse reaction [139, 140] has been considered for the production of L-phenylalanine (Scheme 3.9). More than 12 000 metric tonnes [141] of this is applied as a building block for the sweetener aspartame annually.

The conversion of *trans*-cinnamic acid and ammonia into L-phenylalanine is thermodynamically quite unfavorable ($K_{eq} = 4.7$ M) [118], necessitating high concentrations of ammonia and an elevated pH to drive the reaction into the direction of L-phenylalanine formation [142]. Further drawbacks of the PAL process are low enzyme-specific activity, low enzyme stability, and strong substrate inhibition [141]. The specific activity of the PALs from the yeasts *R. toruloides* [143, 144] and *R. glutinis* [145, 146] at 30 °C is 2.5–5 U/mg only. In addition, the low PAL expression level of below 1% led to a relatively low cellular activity (e.g., for *R. glutinis* 35 U/g cell dry weight) [147] and consequently high cell loading in the bioconversion step. Owing to the profound lack of stability, the reuse of PAL is not a feasible option [147].

Fed-batch operation has been used to circumvent inhibition of PAL by *trans*-cinnamic acid above 50 mM [147, 148] and has actually been commercialized by Genex to L-phenylalanine titers of 43 g/l [space-time yield: 8 g L-phenylalanine/(l·day)] [26, 149].

The PAL-mediated bioconversion to L-phenylalanine has been improved over the last decades using both anaerobic (by N_2-sparging) and static conditions [149, 150], or by applying reducing conditions, for example, by the addition of 2-mercaptoethanol or thioglycolic acid [151, 152], as well as by addition of other types of compounds [142, 150, 151] and the omission of chlorine ions [150, 153]. All these stabilizing factors

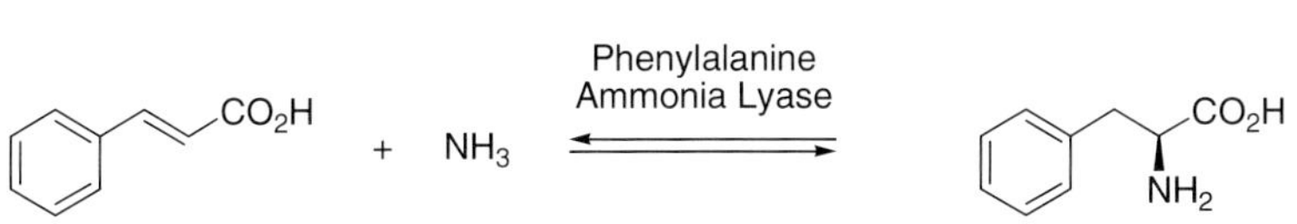

Phenylalanine Ammonia Lyase

CO_2H + NH_3 ⇌ CO_2H, NH_2

cinnamic acid **29** L-phenylalanine **28**

Scheme 3.9 Phenylalanine ammonia lyase-catalyzed production of L-phenylalanine (**28**).

could successfully be combined, enabling the reuse of the yeast cells in six to eight consecutive L-phenylalanine synthesis runs [150, 151].

In a different approach to improve the PAL-catalyzed L-phenylalanine synthesis, mutants of *Rhodotorula graminis* with increased cellular activity and an improved stability during fermentation and bioconversion have been obtained by classical strain improvement [154, 155]. Alternatively, a superior *R. rubra* strain was obtained by enrichment [156].

More recently, PALs from plants [157] and prokaryotic origin [145, 158, 159] have become more readily available for bioconversion reactions through their overexpression in efficient enzyme production systems. Diversa recently isolated a set of 18 new PALs by a sequence-based approach [160], 16 of which were from bacterial species whose genome sequence was already publicly available and erroneously had been annotated as genes encoding a histidine ammonia lyase.

Owing to its relaxed substrate specificity nowadays PAL is only used for the synthesis of non-natural L-phenylalanine analogs, characterized by small volumes and high prices, by Mitsui [161], Great Lakes [162], and others [162–165].

3.7 Aminotransferase Process

A second enzyme-catalyzed asymmetric process relies on the action of aminotransferases (EC 2.6.1.X), also frequently named transaminases, which catalyze the reversible transfer of an amino group from a (preferably cheap) amino acid donor to an α-keto acid acceptor yielding a new amino acid along with an α-keto acid side-product (Scheme 3.10). Aminotransferases belong to the large and diverse group of PLP-dependent enzymes. The PLP prosthetic group serves as an amine acceptor in the first of two distinct half-reactions and subsequently transfers the amino group to the α-keto acid in the second half-reaction.

Aminotransferases are ubiquitous enzymes in nature, playing an essential role in the biosynthesis of most proteinogenic amino acids, and in the supply of specific D-amino acids for peptidoglycan and secondary metabolite biosynthesis [166–168].

Aminotransferases are generally quite active with typical specific activities up to 400 U/mg of protein [169], and they are highly enantioselective and therefore very suitable as biocatalysts for the production of enantiomerically pure α-H-α-amino acids [170–172]. Both highly selective D- and L-aminotransferases are known. Furthermore, the substrate specificity of these enzymes is rather broad, which also enables their application in the synthesis of enantiopure non-natural amino acids. Finally, in contrast to amino acid dehydrogenases (AADHs) (Section 3.8), they do not require external cofactor regeneration.

R^1 CO_2H / NH_2 + R CO_2H / O ⇄ (L-Aminotransferase) R^1 CO_2H / O + R CO_2H / NH_2

L-amino acid + α-keto acid ⇄ α-keto acid + L-amino acid

Scheme 3.10 General reaction catalyzed by an L-selective aminotransferase.

The application of aminotransferases for synthetic purposes, however, suffers from one major disadvantage – the equilibrium constant of the reaction is generally near unity. For example, the apparent equilibrium constants K'_{eq} for the synthesis of L-alanine, L-valine, L-leucine, and L-*tert*-leucine using L-glutamate as the amino donor appear to be 1.86, 0.53, 0.37, and 0.16, respectively [170, 173], resulting in a reduced yield of the desired product and a complex product mixture with severe complications for downstream processing. Over the years different approaches have been developed to overcome this incomplete conversion, of which the use of an excess of a cheap amino acid donor is maybe the most obvious one. Schulz *et al.*, for instance, used the highly stable immobilized *E. coli* 4-aminobutyrate : 2-ketoglutarate transaminase (EC 2.6.1.19) and a 4-fold molar excess of L-glutamate to convert the 2-keto acid precursor 2-oxo-4-[(hydroxy)(methyl)phosphinoyl]butyric acid (**31**) to L-phosphinothricin (**30**), the active ingredient of the herbicide Basta (Bayer CropScience), in over 90% yield [174]. Space-time yields of above 50 g/(l h) have been reached in a continuous production process.

Precipitation of the amino acid product is a second option to drive the equilibrium towards product formation, for instance, as applied in the synthesis of L-homophenylalanine (pH 2–9: solubility 2 mM) in greater than 99% e.e. and 94% conversion from 840 mM 2-oxo-4-phenylbutyric acid using 900 mM L-aspartic acid as the amine donor and the aromatic amino acid aminotransferase from *Enterobacter* sp. BK2K-1 overexpressed in *E. coli* as the catalyst [175]. A similar precipitation-driven approach has been reported by Tosoh for the synthesis of 3-(2-naphthyl)-L-alanine in 93% yield and greater than 99% e.e. from its α-keto acid (180 mM) applying L-glutamic acid (360 mM) as amino donor and the hyper-thermostable aminotransferase from *Thermococcus profundus* [176].

A more generally applicable and economically attractive approach to shift the equilibrium is the use of aspartic acid as the amine donor. The formation of oxaloacetate and its essentially irreversible decarboxylation to pyruvate and carbon dioxide shifts the equilibrium of the aminotransferase reaction almost completely in the direction of the desired amino acid (Scheme 3.11). Although oxaloacetate

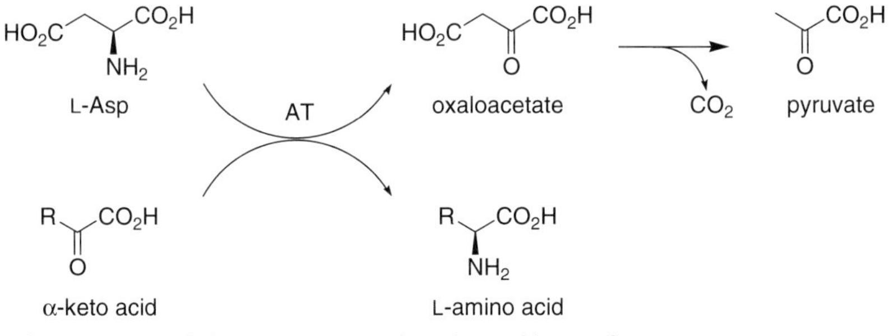

Scheme 3.11 Coupled reaction system to drive the equilibrium of L-aspartic acid-dependent aminotransferase (AT) reactions to completion. The decarboxylation of oxaloacetate can be accelerated by multivalent metal ions as well as by the enzyme oxaloacetate decarboxylase.

decarboxylates spontaneously under physiological conditions [177], a real efficient aminotransferase process requires a further rate enhancement by the addition of multivalent metal ions [178] or by applying the enzyme oxaloacetate decarboxylase (EC 4.1.1.3) [179]. Use of the *E. coli* aspartate aminotransferase AspC (EC 2.6.1.1) and the oxaloacetate decarboxylase from *P. putida* ATCC 950 for the production of L-phenylalanine from phenylpyruvate and L-aspartic acid resulted in a conversion of no less than 97%, whereas the identical reaction without the oxaloacetate decarboxylase furnished L-phenylalanine in 42% conversion only [170, 172].

Not all amino acid aminotransferases can utilize L-aspartic acid as the amino donor. In those cases the aminotransferase can be coupled with AspC to regenerate the intermediate amine donor, such as L-glutamic acid in the case of the branched-chain aminotransferase IlvE (EC 2.6.1.42) (Scheme 3.12) [170, 180]. This principle of coupled aminotransferases was, amongst others, applied by Rozzell *et al.* for the branched-chain α-H-α-amino acids L-valine, L-leucine, and L-isoleucine [180], and by Bartsch *et al.* for an improved process to L-phosphinothricin (**30**) [181].

Although this enzyme-coupled process concept was a major improvement, its application is still complicated by the formation of significant amounts of L-alanine from pyruvate, catalyzed by a side-activity of many of the broad-spectrum aminotransferases. This inherent drawback has been elegantly solved by scientists of NSC Technologies by engineering of novel biosynthetic pathways. These pathways contain the enzyme acetolactate synthase, which dimerizes the pyruvate side-product to acetolactate (**33**), that undergoes spontaneous decarboxylation resulting in the overall formation of acetoin (**34**) as the final volatile side-product (see Scheme 3.13

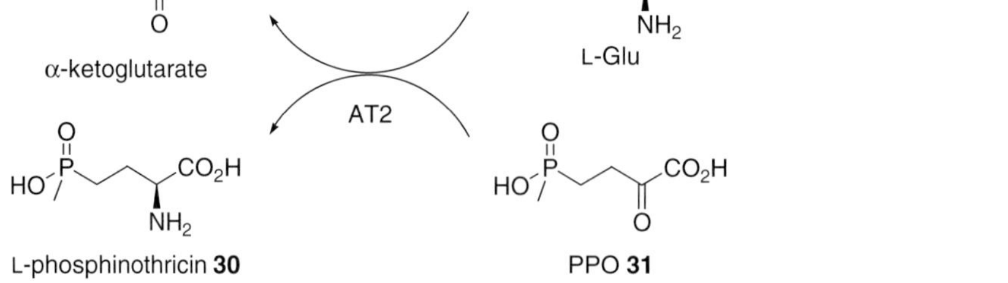

L-Asp — AT1 — oxaloacetate — CO_2 — pyruvate

α-ketoglutarate — L-Glu

L-phosphinothricin **30** — AT2 — PPO **31**

Scheme 3.12 AgrEvo process for the production of the herbicide ingredient L-phosphinothricin as example of a coupled aminotransferase process to drive the equilibrium of an L-glutamate-dependent aminotransferase reaction (AT2) in the synthesis direction by decarboxylation of oxaloacetate. AT1: AspC (glutamate : oxaloacetate aminotransferase, EC 2.6.1.1); AT2: for example, 4-aminobutyrate : 2-ketoglutarate aminotransferase (EC 2.6.1.19).

for similar reaction artificial pathway) [182]. The strength of this approach was demonstrated by Fotheringham *et al.* for the efficient ton-scale production of L-2-aminobutyric acid from cheap L-threonine applying a single recombinant *E. coli* strain expressing the genes encoding the *E. coli* aromatic aminotransferase (TyrB), *Bacillus subtilis* acetolactate synthase (AlsS), and *E. coli* threonine deaminase (IlvA) [183, 184].

In an attempt to completely eliminate alanine formation, novel aspartic acid-independent reaction pathways have recently been developed. These pathways rely on ω-aminotransferases for the regeneration of L-glutamate from α-ketoglutarate by transfer of the side-chain amino group from the donors L-lysine or L-ornithine. The subsequent spontaneous cyclization of the aldehydes drives the formation of the α-H-α-amino acid product to near completion. This system was applied for the preparation of L-aminobutyric acid from 2-ketobutyric acid in 92% yield using L-ornithine as the amino group donor [185, 186]. A similar approach was successful in the preparation of L-*tert*-leucine (**35**) from trimethylpyruvic acid and L-ornithine by combining the ornithine δ-aminotransferase with the *E. coli* branched-chain aminotransferase (IlvE).

3.7.1
Aminotransferase-Catalyzed Production of D-α-H-α-Amino Acids

As already briefly referred to above, bacteria can also contain D-amino acid aminotransferases (EC 2.6.1.21) for the biosynthesis of the peptidoglycan amino acids D-alanine and D-glutamic acid and of a broad range of D-amino acids used in the synthesis of secondary metabolites, such as peptide-based antibiotics. However, the number of known D-amino acid aminotransferases is limited. These enzymes also contain PLP as a coenzyme [187] and show a similar broad substrate specificity as their L-selective counterparts [168, 188, 189].

The D-amino acid aminotransferase from *Bacillus* sp. YM-1 has been studied in most detail; its gene has been cloned and overexpressed in *E. coli* [190], and its crystal structure has been solved [191, 192]. These studies showed that this enzyme has a unique fold not seen for any other PLP-dependent enzyme, except for the structure of the L-branched-chain amino acid aminotransferase from *E. coli* [193], suggesting a common ancestral gene [190].

For D-aminotransferase reactions D-amino acids are required as amine donors. As these D-amino acids are expensive, the process is combined with highly specific racemases, such as for aspartic acid (EC 5.1.1.13), glutamic acid (EC 5.1.1.3), and alanine (EC 5.1.1.1), to generate these amine donors *in situ* from L-amino acids. As an example, the D-amino acid aminotransferase from *Bacillus sphaericus* [194] and the aspartic acid racemase from *Streptococcus thermophilus* [195] have been combined with the threonine deaminase and acetolactate synthase, already mentioned above, in a recombinant microorganism for the production of D-2-aminobutyric acid (**32**) (Scheme 3.13) [171, 183, 196]. This process to make D-amino acids can also be operated with living whole cells under fermentative conditions, as was exemplified by the production of 4.2 g/l D-phenylalanine of greater than 99% enantiomeric purity

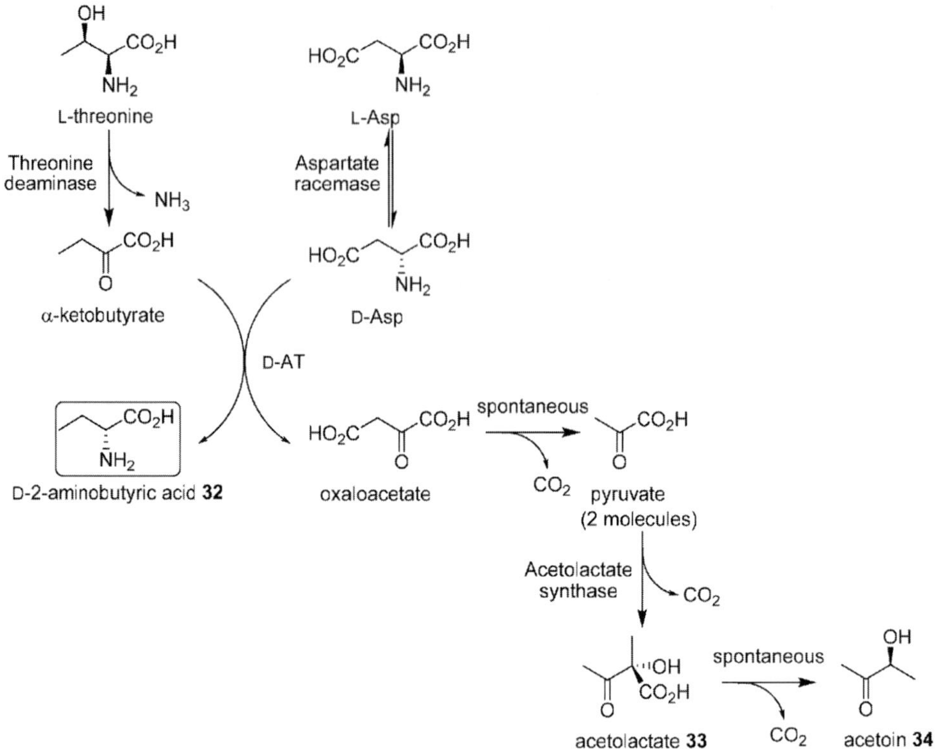

Scheme 3.13 Synthetic biochemical pathway for the production of D-aminobutyric acid (**32**) employing a D-amino acid aminotransferase (D-AT). Both substrates are synthesized *in situ* (i.e., α-ketobutyrate from L-threonine by the action of a threonine deaminase and D-aspartic acid from L-aspartic acid by the action of the enzyme aspartate racemase).

in a fed-batch fermentation with feeding of glucose and D,L-alanine as the amino group donor [182, 197]. Other types of multistep D-aminotransferase reaction pathway using D-alanine as the amino group donor have also been described [196, 198, 199].

The strict stereo-conservation of the D-amino acid aminotransferases mentioned above urges for D-amino acids as the amino group donor [191, 192, 200], whereas in the 1980s van den Tweel *et al.* identified an unusual aminotransferase from *P. putida* strain LW-4 [201], able to catalyze the formation of D-HPG (**24**) from *p*-hydroxyphenylglyoxylate with L-glutamic acid as the amino donor [202, 203]. This novel type of stereo-inverting aminotransferase (EC 2.6.1.72), of which the gene was only recently cloned and sequenced [204], can also be applied for the formation of D-phenylglycine, but has no activity for the other D-α-H-α-amino acids [205]. Due to the low cost of these two aromatic D-amino acids and the expensive α-keto acid precursors, this novel biocatalyst is still of limited commercial relevance [196]. However, this may change once the scope of this aminotransferase is broadened to other classes of D-amino acids by protein engineering – a process which will certainly benefit from the structural information that recently became available [206].

3.8 AADH Process

AADHs (EC 1.4.1.X) form a second class of enzymes that catalyze the reductive amination of α-keto-acids to α-amino acids, in this case with the concomitant oxidation of the cofactor NAD(P)H (Scheme 3.14). These enzymes are found in prokaryotes and eukaryotes, catalyzing the oxidative deamination of amino acids, which is the first step in their degradation for use as carbon and nitrogen sources as well as for energy [207, 208].

In contrast to aminotransferase-catalyzed reactions, the thermodynamics of the AADH reactions dictates that the equilibrium is usually far on the side of the aminated products with a typical K_{eq} of $9 \times 10^{12}\,M^{-2}$ (leucine) and $2.2 \times 10^{13}\,M^{-2}$ (phenylalanine) [208, 209].

Only a few of the many AADHs identified are of synthetic interest, amongst which are alanine dehydrogenase (EC 1.4.1.1), glutamate dehydrogenase (EC 1.4.1.2-4), and particularly phenylalanine dehydrogenase (EC 1.4.1.20) and leucine dehydrogenase (EC 1.4.1.9) [207, 210–212]. Except for the alanine dehydrogenase from *Phormidium lapideum* [210], all AADHs are L-selective.

Most of the synthetic work with leucine dehydrogenases has been performed with the enzymes from *Thermoactinomyces intermedius* [213] and from several *Bacillus* sp. [214, 215], which appeared to be NADH-specific. The *Bacillus* sp. leucine dehydrogenases accept α-keto acids with hydrophobic, aliphatic, branched, and unbranched carbon chains from four to six carbon atoms and some alicyclic α-keto acids, but not aromatic substrates, like L-phenylalanine [212, 214, 215]. On the other hand, the NADH-specific phenylalanine dehydrogenase enzymes found in a *Brevibacterium* strain [216], a *Rhodococcus* sp. [208, 217], and from *T. intermedius* [218] show a very broad substrate specificity, accepting α-keto acids with aromatic as well as aliphatic side-chains. The substrate ranges of phenylalanine dehydrogenase and leucine dehydrogenase are thus complementary.

Implementation of AADHs as a competing L-amino acid production technology requires an efficient system to regenerate the oxidized cofactor, NAD^+. This is achieved, for example, by the simultaneous oxidation of glucose to gluconic acid using the glucose dehydrogenase from *Bacillus megaterium* [219]. Another,

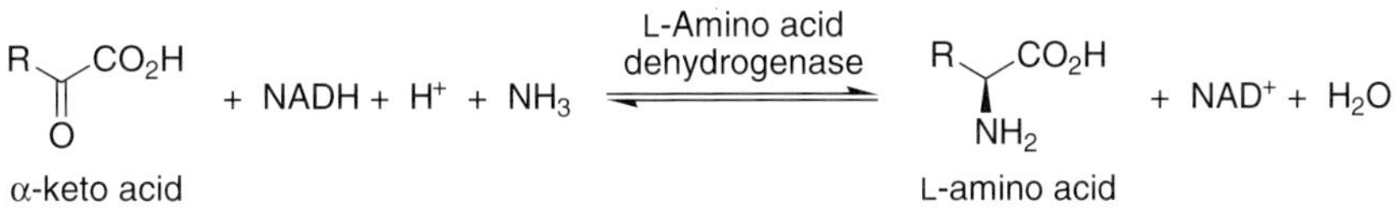

$$\underset{\alpha\text{-keto acid}}{R\text{–}C(=O)\text{–}CO_2H} + NADH + H^+ + NH_3 \xrightleftharpoons[]{\text{L-Amino acid dehydrogenase}} \underset{\text{L-amino acid}}{R\text{–}CH(NH_2)\text{–}CO_2H} + NAD^+ + H_2O$$

Scheme 3.14 General reaction catalyzed by L-selective amino acid dehydrogenases.

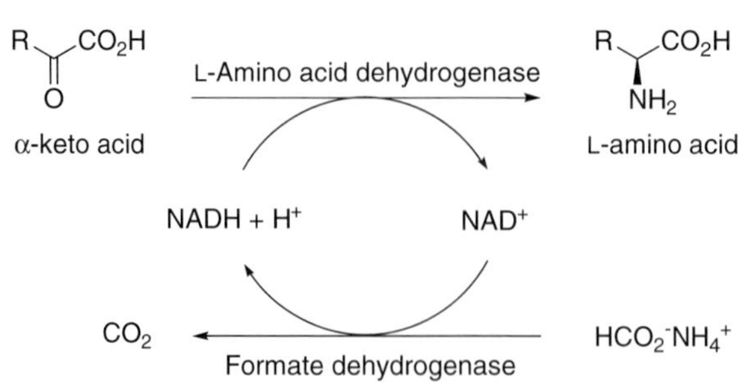

Scheme 3.15 Reductive amination process to enantiomerically pure L-amino acids with formate dehydrogenase-based cofactor regeneration.

more frequently used regeneration system is based on the formate dehydrogenase from the yeast *Candida boidinii* (EC 1.2.1.2) [220] introduced by Whitesides *et al.* (Scheme 3.15) [221, 222]. Formate dehydrogenase catalyzes the NAD^+-dependent oxidation of formate to carbon dioxide. The low-cost availability of formate dehydrogenase and formate [223, 224], the irreversibility of the reaction, and the scalable downstream processing add to the distinct advantages for industrial application of this system [222], despite low specific activity of 6.5 U/mg and instability of formate dehydrogenase under process conditions [223].

By using an enzyme-membrane reactor (EMR) concept in which both enzymes are completely retained [225], Kragl *et al.* developed a cost-efficient process, either in a continuous or in a repetitive batch mode [226, 227]. Further improvements were made by the use of polyethylene glycol (PEG) enlarged NAD^+ to prevent leakage of the expensive cofactor from the EMR. In this way the efficiency of use of this cofactor (total turnover number) could be maximized [221, 225, 228]. The EMR has been used in a continuous mode in the synthesis of the unnatural amino acid L-*tert*-leucine (**35**) from trimethylpyruvic acid by the action of the *Bacillus cereus* leucine dehydrogenase and *C. boidinii* formate dehydrogenase and applying PEG-enlarged NAD^+ – a process that has been commercialized by Degussa [225].

It has to be noted here that because of the considerable price reduction in recent years native NAD^+ can now be used instead of the PEG-enlarged cofactor in this type of continuous process without additional costs [226]. This led Kragl *et al.* to develop a new continuous L-*tert*-leucine (**35**) process in the EMR with native NAD^+ as cofactor, reaching an average conversion of 93% and a space-time yield of 366 g/(l·day) [227].

Krix *et al.*, in collaboration with Degussa, developed a repetitive batch EMR process using native NAD^+ and partially purified leucine dehydrogenase from *B. cereus* or *B. stearothermophilus* and formate dehydrogenase from *C. boidinii* for the gram-scale synthesis of several unnatural aliphatic α-H-α-amino acids with bulky side-chains [215]. The average enzyme recovery was above 95% (often >99%) for the leucine dehydrogenases and 70% for the less stable formate dehydrogenase. In all cases the L-amino acids were obtained in greater than 99.9% e.e. This process was scaled to 400 l for the synthesis of 30 kg of L-neopentylglycine (**36**) to prove its commercial-scale feasibility [215, 229].

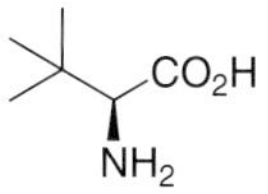

L-*tert*-leucine **35** 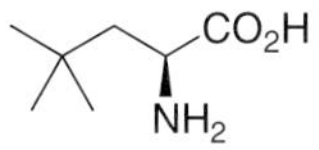 L-neopentylglycine **36** 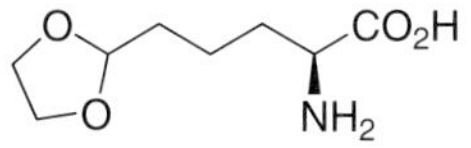 (*S*)-allysine ethylene acetal **37**

Apart from processes relying on the use of isolated enzymes, whole-cell processes are increasingly used. Asano *et al.*, for instance, reported on the use of permeabilized *B. sphaericus* and *C. boidinii* whole cells as a source of phenylalanine dehydrogenase and formate dehydrogenase, respectively, for the synthesis of L-phenylalanine from phenylpyruvate [230]. At Bristol-Myers Squibb a reductive amination route to L-allysine ethylene acetal (**37**) has been successfully run on semi-production scale using phenylalanine dehydrogenase from *T. intermedius* ATCC 33 205, in combination with the formate dehydrogenase from *C. boidinii* SC13 822 [231]. Since the *T. intermedius* cells quickly lysed at the end of the large-scale fermentation process, its phenylalanine dehydrogenase gene was expressed in *E. coli*. Using this recombinant *E. coli* strain in combination with the heat-dried *C. boidinii* cells, almost 200 kg of product **37** was prepared with an average yield of 91% and greater than 98% e.e. A further decrease in the cost price of this process appeared to be possible by development of a recombinant *Pichia pastoris* strain coexpressing the *T. intermedius* phenylalanine dehydrogenase gene and the endogenous formate dehydrogenase gene. Using this strain in heat-dried form, 15 kg of **37** was produced in 97% yield and greater than 98% e.e. [231].

Two single, *E. coli* based whole-cell biocatalysts for the asymmetric reductive amination of α-keto acids have been constructed by Galkin *et al.* by expression of the formate dehydrogenase from *Mycobacterium vaccae* [232] in combination with either the leucine dehydrogenase [213] or the phenylalanine dehydrogenase [233] from *T. intermedius*. With these cells L-leucine, L-valine, L-norvaline, L-methionine (combination leucine dehydrogenase–formate dehydrogenase), and L-phenylalanine and L-tyrosine (combination phenylalanine dehydrogenase–formate dehydrogenase) were produced in high chemical yields (>88%) and excellent enantioselectivity (>99.9%) [198]. Recently, Degussa also reported on the construction of an *E. coli* whole-cell biocatalyst that is based on the coexpression of the gene encoding a stabilized formate dehydrogenase mutant [223] and the *B. cereus* leucine dehydrogenase gene [234, 235]. Application of this new whole-cell biocatalyst for the synthesis of L-*tert*-leucine (**35**) and L-neopentylglycine (**36**) showed that this reaction proceeded even without the addition of an external cofactor [235, 236]. Only at substrate concentrations exceeding 0.5 M did addition of a low amount of additional cofactor (1–10 μM) appear to be necessary to get full conversion. Alternatively, near quantitative conversion of the substrate trimethylpyruvic acid could be obtained with fed-batch operation of the process.

Until recently, synthesis of D-α-H-α-amino acids via enzymatic reductive amination was impossible because a suitable D-AADH had not been identified. In 2006, however, engineering of the enzyme meso-2,6-diaminopimelic acid D-hydrogenase from *Corynebacterium glutamicum* led to the first known highly stereoselective D-AADH [237]. Application of this mutant D-AADH in the reductive amination direction in combination with glucose dehydrogenase to regenerate the $NADP^+$ cofactor led to the synthesis of D-amino acids in an enantiomeric excess of 95% or higher. The sole exception was D-alanine, which was obtained in 77% e.e. only because of the presence of an alanine racemase in the partially purified enzyme preparation used.

3.9 Conclusions

In this chapter an overview is given of the main chemo-enzymatic platforms for the production of enantiomerically pure amino acids that have been commercialized by different companies. Depending on the critical step, these platforms can be divided into resolution-based processes (acylase, amidase, and hydantoinase processes) and processes that rely on asymmetric synthesis (ammonia lyase-, aminotransferase-, and AADH-based processes). Many factors determine the commercial attractiveness of a biocatalytic process, such as substrate costs and availability, biocatalyst productivity, space-time yield, and costs for downstream processing. As a consequence, it is very difficult to agree on the most preferred process platform. Resolution-based processes have the intrinsic drawback that the maximum yield is 50% only and the remaining substrate is wasted, but this is of less importance in the case of cheaply available raw materials or simple external recycling loops. Furthermore, future developments of DKR processes of the acylase- and amidase-based processes will, in principle, lead to a near-quantitative conversion of the racemic substrate. Also, the hydantoinase-based processes, although based on a kinetic resolution, have a maximum yield of 100% due to spontaneous or enzymatic racemization of the nonhydrolyzed hydantoin.

Processes relying on an asymmetric synthesis, on the other hand, intrinsically have a theoretical yield of 100% per cycle. However, this advantage may be counterbalanced by much higher substrate costs or, even worse, limited substrate availability at large-scale. The AADH route is certainly one of the most attractive methods to produce L-α-H-α-amino acids today, but it is not suitable for the production of α,α-disubstituted amino acids. Ammonia lyase-based methods, on the other hand, seem to be especially suited for niche applications because of the generally limited substrate spectrum of these enzymes and thermodynamic limitations. Nevertheless, the L-aspartate ammonia lyase-based synthesis of L-aspartic acid is one of the most efficient biocatalytic processes currently known. Thus, each of these process platforms has its specific advantages and disadvantages. The attractiveness of the different platforms concepts will further depend on company-specific knowledge, intellectual property rights, and equipment. It is therefore expected that each of these processes will continue to be in operation, at least for the next decade.

References

1 Leuchtenberger, W., Huthmacher, K., and Drauz, K. (2005) Biotechnological production of amino acids and derivatives: current status and prospects. *Applied Microbiology and Biotechnology*, **69**, 1–8.

2 Ikeda, M. (2003) Amino acid production processes. *Advances in Biochemical Engineering/Biotechnology*, **79**, 1–35.

3 Oyama, K. (1992) The industrial production of aspartame, in *Chirality in Industry* (eds A.N. Collins, G.N. Sheldrake, and J. Crosby), John Wiley & Sons, Ltd, Chichester, pp. 237–247.

4 Kamphuis, J., Boesten, W.H.J., Kaptein, B., Hermes, H.F.M., Sonke, T., Broxterman, Q.B., Van den Tweel, W.J.J., and Schoemaker, H.E. (1992) The production and uses of optically pure natural and unnatural amino acids, in *Chirality in Industry* (eds A.N. Collins, G.N. Sheldrake, and J. Crosby), John Wiley & Sons, Ltd, Chichester, pp. 187–208.

5 Wegman, M.A., Janssen, M.H.A., van Rantwijk, F., and Sheldon, R.A. (2001) Towards biocatalytic synthesis of β-lactam antibiotics. *Advanced Synthesis and Catalysis*, **343**, 559–576.

6 Kleemann, A., Engel, J., Reichert, D., and Kutscher, B. (1999) *Pharmaceutical Substances: Syntheses, Patents, Applications*, Thieme, Stuttgart, pp. 1213–1215.

7 Teetz, V., Geiger, R., Henning, R., and Urbach, H. (1984) Synthesis of a highly active angiotensin converting enzyme inhibitor: 2-[*N*-[(S)-1-ethoxycarbonyl-3-phenylpropyl]-L-alanyl]-(1*S*,3*S*,5*S*)-2-azabicyclo[3.3.0]octane-3-carboxylic acid (Hoe 498). *Arzneimittelforschung*, **34**, 1399–1401.

8 Bold, G., Fässler, A., Capraro, H.-G., Cozens, R., Klimkait, T., Lazdins, J., Mestan, J., Poncioni, B., Rösel, J., Stover, D., Tintelnot-Blomley, M., Acemoglu, F., Beck, W., Boss, E., Eschbach, M., Hürlimann, T., Masso, E., Roussel, S., Ucci-Stoll, K., Wyss, D., and Lang, M. (1998) New aza-dipeptide analogues as potent and orally absorbed HIV-1 protease inhibitors: candidates for clinical development. *Journal of Medicinal Chemistry*, **41**, 3387–3401.

9 Henrick, C.A. and Garcia, B.A. (1973) Esters and thiolesters of amino acids, processes for their production, and compositions including them, GB 1,588,111 to Zoecon Corporation. *Chemical Abstracts*, **78**, 123297.

10 Stepek, W.J. and Nigro, M.M. (1986) Novel process for the preparation of aminonitriles useful for the preparation of herbicides, EP 0,123,830 to American Cyanamid Company. *Chemical Abstracts*, **105**, 148198.

11 Tomlin, C.D.S. (2003) *The Pesticide Manual*, British Crop Protection Council, Alton, pp. 555–556.

12 Genix, P., Guesnet, J.-L., and Lacroix, G. (2003) Chemistry and stereo-chemistry of fenamidone. *Pflanzenschutz – Nachrichten Bayer*, **56**, 421–434.

13 Lacombe, J.-P., Patty, L., and Steiger, D. (2001) Fenamidone. An antimildew compound for grapevine and potatoes. *Phytocoenologia*, **535**, 42–44.

14 Mueller, U. and Huebner, S. (2003) Economic aspects of amino acids production. *Advances in Biochemical Engineering/Biotechnology*, **79**, 137–170.

15 Knowles, W.S. (2004) Asymmetric hydrogenations – the Monsanto L-DOPA process, in *Asymmetric Catalysis on Industrial Scale: Challenges, Approaches and Solutions* (eds H.-U. Blaser and E. Schmidt), Wiley-VCH Verlag GmbH, Weinheim, pp. 23–38.

16 Nájera, C. and Sansano, J.M. (2007) Catalytic asymmetric synthesis of α-amino acids. *Chemical Reviews*, **107**, 4584–4671.

17 Vogt, H. and Bräse, S. (2007) Recent approaches towards the asymmetric synthesis of α,α-disubstituted α-amino acids. *Organic and Biomolecular Chemistry*, 5, 406–430.

18 Chibata, I., Tosa, T., and Shibatani, T. (1992) The industrial production of optically active compounds by immobilized biocatalysts, in *Chirality in Industry* (eds A.N. Collins, G.N. Sheldrake, and J. Crosby), John Wiley & Sons, Chichester, Ltd, pp. 351–370.

19 Bommarius, A.S., Schwarm, M., and Drauz, K. (2001) Comparison of different chemoenzymatic process routes to enantiomerically pure amino acids. *Chimia*, **55**, 50–59.

20 Bommarius, A.S. (2002) Hydrolysis of *N*-acylamino acids, in *Enzyme Catalysis in Organic Synthesis* (eds K. Drauz and H. Waldman), Wiley-VCH Verlag GmbH, Weinheim, pp. 741–760.

21 Sonntag, N.O.V. (1953) The reactions of aliphatic acid chlorides. *Chemical Reviews*, **52**, 237–416.

22 Beller, M., Eckert, M., and Moradi, W.A. (1999) First amidocarbonylation with nitriles for the synthesis of *N* -acyl amino acids. *Synlett*, 108–110.

23 Wandrey, C. and Flaschel, E. (1979) Process development and economic aspects in enzyme engineering. Acylase L-methionine system. *Advances in Biochemical Engineering*, **12**, 147–218.

24 Galaev, I.Y. and Švedas, V.K. (1982) A kinetic study of hog kidney aminoacylase. *Biochimica et Biophysica Acta*, **701**, 389–394.

25 Bommarius, A.S., Drauz, K., Groeger, U., and Wandrey, C. (1992) Membrane bioreactors for the production of enantiomerically pure α-amino acids, in *Chirality in Industry* (eds A.N. Collins, G.N. Sheldrake, and J. Crosby), John Wiley & Sons, Ltd, Chichester, pp. 371–397.

26 Liese, A., Seelbach, K., and Wandrey, C. (2000) *Industrial Biotransformations*, Wiley-VCH Verlag GmbH, Weinheim.

27 Tokuyama, S. and Hatano, K. (1995) Purification and properties of thermostable *N*-acylamino acid racemase from *Amycolatopsis* sp. TS-1-60. *Applied Microbiology and Biotechnology*, **42**, 853–859.

28 Tokuyama, S., Hatano, K., and Takahashi, T. (1994) A novel enzyme, *N*-acylamino acid racemase, in actinomycetes. Part 1. Discovery of a novel enzyme, *N*-acylamino acid racemase in an Actinomycete: screening, isolation and identification. *Bioscience, Biotechnology, and Biochemistry*, **58**, 24–27.

29 Tokuyama, S., Miya, H., Hatano, K., and Takahashi, T. (1994) Purification and properties of a novel enzyme, *N*-acylamino acid racemase, from *Streptomyces atratus* Y-53. *Applied Microbiology and Biotechnology*, **40**, 835–840.

30 May, O., Verseck, S., Bommarius, A., and Drauz, K. (2002) Development of dynamic kinetic resolution processes for biocatalytic production of natural and nonnatural L-amino acids. *Organic Process Research & Development*, **6**, 452–457.

31 Hsu, S.-K., Lo, H.-H., Kao, C.-H., Lee, D.-S., and Hsu, W.-H. (2006) Enantioselective synthesis of L-homophenylalanine by whole cells of recombinant*Escherichia coli* expressing L-aminoacylase and *N*-acylamino acid racemase genes from *Deinococcus radiodurans* BCRC12827. *Biotechnology Progress*, **22**, 1578–1584.

32 Tokuyama, S. (2001) Discovery and application of a new enzyme *N*-acylamino acid racemase. *Journal of Molecular Catalysis B – Enzymatic*, **12**, 3–14.

33 Tokuyama, S. and Hatano, K. (1996) Overexpression of the gene for *N*-acylamino acid racemase from *Amycolatopsis* sp. TS-1-60 in *Escherichia coli* and continuous produciton of optically active methionine by a bioreactor. *Applied Microbiology and Biotechnology*, **44**, 774–777.

34 Groeger, U., Leuchtenberger, W., and Drauz, K. (1991) Substantially purified *N*-acyl-L-proline acylase from *Comamonas testosteroni* DSM 5416 and Alcaligenes denitrificans DSM 5417, US 5,120,652 to Degussa AG. *Chemical Abstracts*, **115**, 130669.

35 Kikuchi, M., Koshiyama, I., and Fukushima, D. (1983) A new enzyme, proline acylase (*N*-acyl-L-proline amidohydrolase) from *Pseudomonas* species. *Biochimica et Biophysica Acta*, **744**, 180–188.

36 Groeger, U., Drauz, K., and Klenk, H. (1992) Enzymatic preparation of enantiomerically pure *N*-alkyl amino acids. *Angewandte Chemie (International Edition in English)*, **31**, 195–197.

37 Sugie, M. and Suzuki, H. (1980) Optical resolution of DL-amino acids with D-aminoacylase of *Streptomyces*. *Agricultural and Biological Chemistry*, **44**, 1089–1095.

38 Sakai, K., Oshima, K., and Moriguchi, M. (1991) Production and characterization of *N*-acyl-D-glutamate amidohydrolase from *Pseudomonas* sp. strain 5f-1. *Applied and Environmental Microbiology*, **57**, 2540–2543.

39 Yang, Y.-B., Lin, C.-S., Tseng, C.-P., Wang, Y.-J., and Tsai, Y.-C. (1991) Purification and characterization of D-aminoacylase from *Alcaligenes faecalis* DA1. *Applied and Environmental Microbiology*, **57**, 1259–1260.

40 Kamphuis, J., Boesten, W.H.J., Broxterman, Q.B., Hermes, H.F.M., van Balken, J.A.M., Meijer, E.M., and Schoemaker, H.E. (1990) New developments in the chemo-enzymatic production of amino acids. *Advances in Biochemical Engineering/Biotechnology*, **42**, 133–186.

41 Kamphuis, J., Meijer, E.M., Boesten, W.H.J., Broxterman, Q.B., Kaptein, B., Hermes, H.F.M., and Schoemaker, H.E. (1992) Production of natural and synthetic L- and D-amino acids by aminopeptidases and amino amidases, in *Biocatalytic Production of Amino Acids and Derivatives* (eds J.D. Rozzell and F. Wagner), Hanser, Munich, pp. 177–206.

42 Boesten, W.H.J. (1977) Process for preparing α-amino-acid amides, GB 1,548,032 to DSM/Stamicarbon BV. *Chemical Abstracts*, **87**, 39839.

43 Hyett, D.J., Didonè, M., Milcent, T.J.A., Broxterman, Q.B., and Kaptein, B. (2006) A new method for the preparation of functionalized unnatural α-H-α-amino acid derivatives. *Tetrahedron Letters*, **47**, 7771–7774.

44 Asano, Y., Mori, T., Hanamoto, S., Kato, Y., and Nakazawa, A. (1989) A new D-stereospecific amino acid amidase from *Ochrobactrum anthropi*. *Biochemical and Biophysical Research Communications*, **162**, 470–474.

45 Asano, Y., Nakazawa, A., Kato, Y., and Kondo, K. (1989) Properties of a novel D-stereospecific aminopeptidase from *Ochrobactrum anthropi*. *The Journal of Biological Chemistry*, **264**, 14233–14239.

46 Shadid, B., van der Plas, H.C., Boesten, W.H.J., Kamphuis, J., Meijer, E.M., and Schoemaker, H.E. (1990) The synthesis of L-(−)- and D-(+)-lupinic acid. *Tetrahedron*, **46**, 913–920.

47 Rutjes, F.P.J.T. and Schoemaker, H.E. (1997) Ruthenium-catalyzed ring closing olefin metathesis of non-natural α-amino acids. *Tetrahedron Letters*, **38**, 677–680.

48 Wolf, L.B., Tjen, K.C.M.F., Rutjes, F.P.J.T., Hiemstra, H., and Schoemaker, H.E. (1998) Pd-catalyzed cyclization reactions of acetylene-containing α-amino acids. *Tetrahedron Letters*, **39**, 5081–5084.

49 Hermes, H.F.M., Sonke, T., Peters, P.J.H., van Balken, J.A.M., Kamphuis, J., Dijkhuizen, L., and Meijer, E.M. (1993) Purification and characterization of an L-aminopeptidase from *Pseudomonas putida* ATCC 12633. *Applied and Environmental Microbiology*, **59**, 4330–4334.

50 Sonke, T., Kaptein, B., Boesten, W.H.J., Broxterman, Q.B., Kamphuis, J.,

Formaggio, F., Toniolo, C., Rutjes, F.P.J.T., and Schoemaker, H.E. (2000) Aminoamidase-catalyzed preparation and further transformations of enantiopure α-hydrogen- and α,α-disubstituted α-amino acids, in *Stereoselective Biocatalysis* (ed. R.N. Patel), Marcel Dekker, New York, pp. 23–58.

51 Wolf, L.B., Sonke, T., Tjen, K.C.M.F., Kaptein, B., Broxterman, Q.B., Schoemaker, H.E., and Rutjes, F.P.J.T. (2001) A biocatalytic route to enantiomerically pure unsaturated α-H-α-amino acids. *Advanced Synthesis and Catalysis*, **343**, 662–674.

52 Boesten, W.H.J., Schoemaker, H.E., and Dassen, B.H.N. (1987) Process for racemizing an optically active *N*-benzylidene amino-acid amide, EP 0,199,407 to Stamicarbon BV. *Chemical Abstracts*, **107**, 40327.

53 Boesten, W.H.J., Raemakers-Franken, P.C., Sonke, T., Euverink, G.J.W., and Grijpstra, P. (2003) Polypeptides having α-H-α-amino acid amide racemase activity and nucleic acids encoding the same, WO 2003/106691 to DSM IP Assets BV. *Chemical Abstracts*, **140**, 55597.

54 Sonke, T. (2008) Novel developments in the chemo-enzymatic synthesis of enantiopure α-hydrogen- and α,α-disubstituted α-amino acids and derivatives, PhD Thesis, University of Amsterdam.

55 Fukumura, T. (1977) Conversion of D- and DL-α-amino-ε-caprolactam into L-lysine using both yeast cells and bacterial cells. *Agricultural and Biological Chemistry*, **41**, 1327–1330.

56 Ahmed, S.A., Esaki, N., Tanaka, H., and Soda, K. (1983) Properties of α-amino-ε-caprolactam racemase from *Achromobacter obae*. *Agricultural and Biological Chemistry*, **47**, 1887–1893.

57 Asano, Y. and Yamaguchi, S. (2005) Discovery of amino acid amides as new substrates for α-amino-ε-caprolactam racemase from *Achromobacter obae*. *Journal of Molecular Catalysis B – Enzymatic*, **36**, 22–29.

58 Asano, Y. and Yamaguchi, S. (2005) Dynamic kinetic resolution of amino acid amide catalyzed by D-aminopeptidase and α-amino-ε-caprolactam racemase. *Journal of the American Chemical Society*, **127**, 7696–7697.

59 Kaptein, B., Boesten, W.H.J., Broxterman, Q.B., Schoemaker, H.E., and Kamphuis, J. (1992) Synthesis of α,α-disubstituted α-amino acid amides by phase-transfer catalyzed alkylation. *Tetrahedron Letters*, **33**, 6007–6010.

60 Roos, E.C., López, M.C., Brook, M.A., Hiemstra, H., Speckamp, W.N., Kaptein, B., Kamphuis, J., and Schoemaker, H.E. (1993) Synthesis of α-substituted α-amino acids via cationic intermediates. *The Journal of Organic Chemistry*, **58**, 3259–3268.

61 Becke, F., Fleig, H., and Pässler, P. (1971) General method for the preparation of amides from their corresponding nitriles. II. *Annalen Der Chemie – Justus Liebig*, **749**, 198–201.

62 Kaptein, B., Boesten, W.H.J., Broxterman, Q.B., Peters, P.J.H., Schoemaker, H.E., and Kamphuis, J. (1993) Enzymatic resolution of α,α-disubstituted α-amino acid esters and amides. *Tetrahedron: Asymmetry*, **4**, 1113–1116.

63 Kruizinga, W.H., Bolster, J., Kellogg, R.M., Kamphuis, J., Boesten, W.H.J., Meijer, E.M., and Schoemaker, H.E. (1988) Synthesis of optically pure α-alkylated α-amino acids and a single-step method for enantiomeric excess determination. *The Journal of Organic Chemistry*, **53**, 1826–1827.

64 Hermes, H.F.M., Tandler, R.F., Sonke, T., Dijkhuizen, L., and Meijer, E.M. (1994) Purification and characterization of an L-amino amidase from *Mycobacterium neoaurum* ATCC 25795. *Applied and Environmental Microbiology*, **60**, 153–159.

65 Van den Tweel, W.J.J., van Dooren, T.J.G.M., de Jonge, P.H., Kaptein, B.,

Duchateau, A.L.L., and Kamphuis, J. (1993) *Ochrobactrum anthropi* NCIMB 40321: a new biocatalyst with broad-spectrum L-specific amidase activity. *Applied Microbiology and Biotechnology*, **39**, 296–300.

66 Kaptein, B., van Dooren, T.J.G.M., Boesten, W.H.J., Sonke, T., Duchateau, A.L.L., Broxterman, Q.B., and Kamphuis, J. (1998) Synthesis of 4-sulfur-substituted (2*S*,3*R*)-3-phenylserines by enzymatic resolution. Enantiopure precursors for thiamphenicol and florfenicol. *Organic Process Research & Development*, **2**, 10–17.

67 Gouret, C.J., Porsolt, R., Wettstein, J.G., Puech, A., Soulard, C., Pascaud, X., and Junien, J.L. (1990) Biochemical and pharmacological evaluation of the novel antidepressant and serotonin uptake inhibitor (2-(3,4-dichlorobenzyl)-2-dimethylamino-1-propanol hydrochloride. *Arzneimittelforschung*, **40**, 633–640.

68 Gouret, C.J., Wettstein, J.G., Porsolt, R.D., Puech, A., and Junien, J.L. (1990) Neuropsychopharmacologial profile of JO 1017, a new antidepressant and selective semtonin uptake inhibitor. *European Journal of Pharmacology*, **183**, 1478.

69 Kaptein, B., Moody, H.M., Broxterman, Q.B., and Kamphuis, J. (1994) Chemo-enzymatic synthesis of (*S*)-(+)-cericlamine and related enantiomerically pure 2,2-disubstituted-2-aminoethanols. *Journal of the Chemical Society, Perkin Transactions 1*, 1495–1498.

70 Sonke, T., Ernste, S., Tandler, R.F., Kaptein, B., Peeters, W.P.H., van Assema, F.B.J., Wubbolts, M.G., and Schoemaker, H.E. (2005) L-Selective amidase with extremely broad substrate specificity from *Ochrobactrum anthropi* NCIMB 40321. *Applied and Environmental Microbiology*, **71**, 7961–7973.

71 Nakamura, T. and Yu, F. (2000) Amidase gene, US 6,617,139 to Mitsubishi Rayon Co. Ltd. *Chemical Abstracts*, **133**, 319048.

72 Katoh, O., Akiyama, T., and Nakamura, T. (2003) Novel amide hydrolase gene, EP 1,428,876 to Mitsubishi Rayon Co. Ltd. *Chemical Abstracts*, **138**, 233991.

73 Shaw, N.M., Naughton, A., Robins, K., Tinschert, A., Schmid, E., Hischier, M.-L., Venetz, V., Werlen, J., Zimmermann, T., Brieden, W., de Riedmatten, P., Roduit, J.-P., Zimmermann, B., and Neumüller, R. (2002) Selection, purification, characterisation, and cloning of a novel heat-stable stereo-specific amidase from *Klebsiella oxytoca*, and its application in the synthesis of enantiomerically pure (*R*)- and (*S*)-3,3,3-trifluoro-2-hydroxy-2-methylpropionic acids and (*S*)-3,3,3-trifluoro-2-hydroxy-2-methylpropionamide. *Organic Process Research & Development*, **6**, 497–504.

74 Brieden, W., Naughton, A., Robins, K., Shaw, N.M., Tinschert, A., and Zimmermann, T. (1998) Method of preparing (*S*)- or (*R*)-3,3,3-trifluoro-2-hydroxy-2-methylpropionic acid, US 6,773,910 to Lonza AG. *Chemical Abstracts*, **128**, 153206.

75 Ogawa, J. and Shimizu, S. (2000) Stereoselective synthesis using hydantoinases and carbamoylases, in *Stereoselective Biocatalysis* (ed. R.N. Patel), Marcel Dekker, New York, pp. 1–21.

76 Pietzsch, M. and Syldatk, C. (2002) Hydrolysis and formation of hydantoins, in *Enzyme Catalysis in Organic Synthesis* (eds K. Drauz and H. Waldman), Wiley-VCH Verlag GmbH, Weinheim, pp. 761–799.

77 Baldaro, E.M. (1993) Chemo-enzymatic production of D-amino acids. *Pharmaceutical Manufacturing International*, 137–139.

78 Olivieri, R., Fascetti, E., Angelini, L., and Degen, L. (1979) Enzymatic conversion of *N*-carbamoyl-D-amino acids to D-amino acids. *Enzyme and Microbial Technology*, **1**, 201–204.

79 Olivieri, R., Fascetti, E., Angelini, L., and Degen, L. (1981) Microbial transformation of racemic hydantoins to

D-amino acids. *Biotechnology and Bioengineering*, **23**, 2173–2183.

80 Drauz, K., Kottenhahn, M., Makryaleas, K., Klenk, H., and Bernd, M. (1991) Chemoenzymatic syntheses of ω-ureido D-amino acids. *Angewandte Chemie (International Edition in English)*, **30**, 712–714.

81 Kim, G.J. and Kim, H.S. (1994) Adsorptive removal of inhibitory byproduct in the enzymatic production of optically active D-*p*-hydroxyphenylglycine from 5-substituted hydantoin. *Biotechnology Letters*, **16**, 17–22.

82 Kim, G.-J. and Kim, H.-S. (1995) Optimization of the enzymatic synthesis of D-*p*-hydroxyphenylglycine from DL-5-substituted hydantoin using D-hydantoinase and *N*-carbamoylase. *Enzyme and Microbial Technology*, **17**, 63–67.

83 Boesten, W.H.J. and Kierkels, J.G.T. (2002) Process for the preparation of enantiomer-enriched amino acids, WO 2002/061107 to DSM N.V. *Chemical Abstracts*, **137**, 139489.

84 Ikenaka, Y., Nanba, H., Yajima, K., Yamada, Y., Takano, M., and Takahashi, S. (1998) Increase in thermostability of *N*-carbamyl-D-amino acid amidohydrolase on amino acid substitutions. *Bioscience, Biotechnology, and Biochemistry*, **62**, 1668–1671.

85 Ikenaka, Y., Nanba, H., Yajima, K., Yamada, Y., Takano, M., and Takahashi, S. (1998) Relationship between an increase in thermostability and amino acid substitutions in *N*-carbamyl-D-amino acid amidohydrolase. *Bioscience, Biotechnology, and Biochemistry*, **62**, 1672–1675.

86 Oh, K.-H., Nam, S.-H., and Kim, H.-S. (2002) Improvement of oxidative and thermostability of *N*-carbamyl-D-amino acid amidohydrolase by directed evolution. *Protein Engineering*, **15**, 689–695.

87 Nanba, H., Ikenaka, Y., Yamada, Y., Yajima, K., Takano, M., Ohkubo, K., Hiraishi, Y., Yamada, K., and Takahashi, S. (1998) Immobilization of *N*-carbamyl-D-amino acid amidohydrolase. *Bioscience, Biotechnology, and Biochemistry*, **62**, 1839–1844.

88 Battilotti, M. and Barberini, U. (1988) Preparation of D-valine from D,L-5-isopropylhydantoin by stereoselective biocatalysis. *Journal of Molecular Catalysis*, **43**, 343–352.

89 Syldatk, C., Müller, R., Pietzsch, M., and Wagner, F. (1992) Microbial and enzymatic production of L-amino acids from DL-5-monosubstituted hydantoins, in *Biocatalytic Production of Amino Acids & Derivatives* (eds D. Rozzell and F. Wagner), Hanser, Munich, pp. 129–176.

90 Watabe, K., Ishikawa, T., Mukohara, Y., and Nakamura, H. (1992) Purification and characterization of the hydantoin racemase of *Pseudomonas* sp. strain NS671 expressed in *Escherichia coli*. *Journal of Bacteriology*, **174**, 7989–7995.

91 Watabe, K., Ishikawa, T., Mukohara, Y., and Nakamura, H. (1992) Identification and sequencing of a gene encoding a hydantoin racemase from the native plasmid of *Pseudomonas* sp. strain NS671. *Journal of Bacteriology*, **174**, 3461–3466.

92 Wiese, A., Pietzsch, M., Syldatk, C., Mattes, R., and Altenbuchner, J. (2000) Hydantoin racemase from *Arthrobacter aurescens* DSM 3747: heterologous expression, purification and characterization. *Journal of Biotechnology*, **80**, 217–230.

93 Martínez-Rodríguez, S., Las Heras-Vázquez, F.J., Clemente-Jiménez, J.M., and Rodríguez-Vico, F. (2004) Biochemical characterization of a novel hydantoin racemase from *Agrobacterium tumefaciens* C58. *Biochimie*, **86**, 77–81.

94 Martínez-Rodríguez, S., Las Heras-Vázquez, F.J., Mingorance-Cazorla, L., Clemente-Jiménez, J.M., and Rodríguez-Vico, F. (2004) Molecular cloning, purification, and biochemical characterization of hydantoin racemase from the legume symbiont *Sinorhizobium* *meliloti* CECT 4114. *Applied and Environmental Microbiology*, **70**, 625–630.

95 Suzuki, S., Onishi, N., and Yokozeki, K. (2005) Purification and characterization of hydantoin racemase from *Microbacterium liquefaciens* AJ 3912. *Bioscience, Biotechnology, and Biochemistry*, **69**, 530–536.

96 Boesten, W.H.J., Kierkels, J.G.T., van Assema, F.B.J., Ruiz Perez, L.M., Gonzales Pacanowska, D., Gonzales Lopez, J., and De La Escalera Huesco, S. (2003) Hydantoin racemase, WO 2003/100050 to DSM N.V. *Chemical Abstracts*, **140**, 14387.

97 Watabe, K., Ishikawa, T., Mukohara, Y., and Nakamura, H. (1992) Cloning and sequencing of the genes involved in the conversion of 5-substituted hydantoins to the corresponding L-amino acids from the native plasmid of *Pseudomonas* sp. strain NS671. *Journal of Bacteriology*, **174**, 962–969.

98 Ishikawa, T., Mukohara, Y., Watabe, K., Kobayashi, S., and Nakamura, H. (1994) Microbial conversion of DL-5-substituted hydantoins to the corresponding L-amino acids by *Bacillus stearothermophilus* NS1122A. *Bioscience, Biotechnology, and Biochemistry*, **58**, 265–270.

99 Wagner, T., Hantke, B., and Wagner, F. (1996) Production of L-methionine from D,L-5-(2-methylthioethyl) hydantoin by resting cells of a new mutant strain of *Arthrobacter* species DSM 7330. *Journal of Biotechnology*, **46**, 63–68.

100 May, O., Siemann, M., Pietzsch, M., Kiess, M., Mattes, R., and Syldatk, C. (1998) Substrate-dependent enantioselectivity of a novel hydantoinase from *Arthrobacter aurescens* DSM 3745: purification and characterization as new member of cyclic amidases. *Journal of Biotechnology*, **61**, 1–13.

101 Nozaki, H., Takenaka, Y., Kira, I., Watanabe, K., and Yokozeki, K. (2005) D-Amino acid production by *E. coli* co-expressed three genes encoding hydantoin racemase, D-hydantoinase and *N*-carbamoyl-D-amino acid amidohydrolase. *Journal of Molecular Catalysis B – Enzymatic*, **32**, 213–218.

102 Wilms, B., Wiese, A., Syldatk, C., Mattes, R., and Altenbuchner, J. (2001) Development of an *Escherichia coli* whole cell biocatalyst for the production of L-amino acids. *Journal of Biotechnology*, **86**, 19–30.

103 May, O., Nguyen, P.T., and Arnold, F.H. (2000) Inverting enantioselectivity by directed evolution of hydantoinase for the improved production of L-methionine. *Nature Biotechnology*, **18**, 317–320.

104 May, O., Buchholz, S., Schwarm, M., Drauz, K., Turner, R.J., and Fotheringham, I. (2003) Mutants for the preparation of D-amino acids, WO 2004/042047 to Degussa AG. *Chemical Abstracts*, **140**, 405577.

105 Turner, R.J., Aikens, J., Royer, S., DeFilippi, L., Yap, A., Holzle, D., Somers, N., and Fotheringham, I.G. (2004) D-Amino acid tolerant hosts for D-hydantoinase whole cell biocatalysts. *Engineering in Life Sciences*, **4**, 517–520.

106 Wubbolts, M.G. (2002) Addition of amines to C= C bonds, in *Enzyme Catalysis in Organic Synthesis* (eds K. Drauz and H. Waldman), Wiley-VCH Verlag GmbH, Weinheim, pp. 866–872.

107 Calton, G.J. (1992) The enzymatic production of L-aspartic acid, in *Biocatalytic Production of Amino Acids and Derivatives* (eds J.D. Rozzell and F. Wagner), Hanser, Munich, pp. 3–21.

108 Chibata, I., Tosa, T., and Sato, T. (1973) Process for the production of L-aspartic acid, US 3,791,926 to Tanabe Seiyaku Co. *Chemical Abstracts*, **79**, 30499.

109 Tosa, T., Sato, T., Mori, T., and Chibata, I. (1974) Basic studies for continuous production of L-aspartic acid by immobilized *Escherichia coli* cells. *Applied Microbiology*, **27**, 886–889.

110 Sato, T., Mori, T., Tosa, T., Chibata, I., Furui, M., Yamashita, K., and Sumi, A.

(1975) Engineering analysis of continuous production of L-aspartic acid by immobilized *Escherichia coli* cells in fixed beds. *Biotechnology and Bioengineering*, **17**, 1797–1804.

111 Sato, T., Nishida, Y., Tosa, T., and Chibata, I. (1979) Immobilization of *Escherichia coli* cells containing aspartase activity with κ-carrageenan. Enzymic properties and application for L-aspartic acid production. *Biochimica et Biophysica Acta*, **570**, 179–186.

112 Umemura, I., Takamatsu, S., Sato, T., Tosa, T., and Chibata, I. (1984) Improvement of production of L-aspartic acid using immobilized microbial cells. *Applied Microbiology and Biotechnology*, **20**, 291–295.

113 Sato, T. and Tosa, T. (1993) Production of L-aspartic acid, in *Industrial Application of Immobilized Biocatalysts* (eds A. Tanaka, T. Tosa, and T. Kobayashi), Marcel Dekker, New York, pp. 15–24.

114 Terasawa, M., Yukawa, H., and Takayama, Y. (1985) Production of L-aspartic acid from *Brevibacterium* by the cell re-using process. *Process Biochemistry*, **20**, 124–128.

115 Wood, L.L. and Calton, G.J. (1984) A novel method of immobilization and its use in aspartic acid production. *Nature Biotechnology*, **2**, 1081–1084.

116 Wood, L.L. and Calton, G.J. (1983) Immobilization of cells with a polyazatidine prepolymer, US 4,732,851 to Purification Engineering Inc. *Chemical Abstracts*, **99**, 211194.

117 Yamagata, H., Terasawa, M., and Yukawa, H. (1994) A novel industrial process for L-aspartic acid production using an ultrafiltrationmembrane. *Catalysis Today*, **22**, 621–628.

118 Jandel, A.-S., Hustedt, H., and Wandrey, C. (1982) Continuous production of L-alanine from fumarate in a two-stage membrane reactor. *European Journal of Applied Microbiology and Biotechnology*, **15**, 59–63.

119 Asai, Y., Inui, M., Vertes, A., Kobayashi, M., and Yukawa, H. (1995) Cloning and sequence determination of the aspartase-encoding gene from *Brevibacterium flavum* MJ-233. *Gene*, **158**, 87–90.

120 Kisumi, M., Komatsubara, S., and Taniguchi, T. (1985) Method for producing L-aspartic acid, US 4,692,409 to Tanabe Seiyaku Co. *Chemical Abstracts*, **102**, 111473.

121 Nishimura, N., Taniguchi, T., and Komatsubara, S. (1989) Hyperproduction of aspartase by a catabolite repression-resistant mutant of *Escherichia coli* B harboring multicopy *asp*A and *par* recombinant plasmids. *Journal of Fermentation and Bioengineering*, **67**, 107–110.

122 Murase, S., Takagi, J.S., Higashi, Y., Imaishi, H., Yumoto, N., and Tokushige, M. (1991) Activation of aspartase by site-directed mutagenesis. *Biochemical and Biophysical Research Communications*, **177**, 414–419.

123 Wang, L.-j., Kong, X.-d., Zhang, H.-y., Wang, X.-p., and Zhang, J. (2000) Enhancement of the activity of L-aspartase from *Escherichia coli* W by directed evolution. *Biochemical and Biophysical Research Communications*, **276**, 346–349.

124 Kawata, Y., Tamura, K., Yano, S., Mizobata, T., Nagai, J., Esaki, N., Soda, K., Tokushige, M., and Yumoto, N. (1999) Purification and characterization of thermostable aspartase from *Bacillus* sp. YM55-1. *Archives of Biochemistry and Biophysics*, **366**, 40–46.

125 Shi, W., Dunbar, J., Jayasekera, M.M.K., Viola, R.E., and Farber, G.K. (1997) The structure of L-aspartate ammonia-lyase from *Escherichia coli*. *Biochemistry*, **36**, 9136–9144.

126 Fujii, T., Sakai, H., Kawata, Y., and Hata, Y. (2003) Crystal structure of thermostable aspartase from *Bacillus* sp. YM55-1: structure-based exploration of functional sites in the aspartase family. *Journal of Molecular Biology*, **328**, 635–654.

127 Goto, M., Nara, T., Tokumaru, I., Fugono, N., Uchida, Y., Terasawa, M., and Yukawa,

H. (1996) Method of producing fumaric acid, EP 0,693,557 to Mitsubishi Chemical Corporation. *Chemical Abstracts*, **124**, 143771.

128 Kobayashi, M., Terasawa, M., and Yukawa, H. (1999) L-Aspartic acid, in *Encyclopedia of Bioprocess Technology – Fermentation, Biocatalysis, and Bioseparation* (eds M.C. Flickinger and S.W. Drew), John Wiley & Sons, Inc., New York, pp. 210–213.

129 Rozzell, J.D. (1999) Biocatalysis at commercial scale. Myths and realities. *Chimica Oggi – Chemistry Today*, **17**, 42–47.

130 Kumagai, H. (2006) Amino acid production, in *The Prokaryotes*, 3rd edn, Vol. 1 (eds M. Dwerkin, S. Falkow, E. Rosenberg, K-.H. Schleifer, and E. Stackebrandt), Springer, Berlin, pp. 756–765.

131 Hashimoto, S.-i. and Katsumata, R. (1998) L-Alanine fermentation by an alanine racemase-deficient mutant of the DL-alanine hyperproducing bacterium *Arthrobacter oxydans* HAP-1. *Journal of Fermentation and Bioengineering*, **86**, 385–390.

132 Tate, S.S. and Meister, A. (1969) Regulation of the activity of L-aspartate β-decarboxylase by a novel allosteric mechanism. *Biochemistry*, **8**, 1660–1668.

133 Calton, G.J. (1992) The enzymatic preparation of L-alanine, in *Biocatalytic Production of Amino Acids and Derivatives* (eds J.D. Rozzell and F. Wagner), Hanser, Munich, pp. 59–74.

134 Furui, M. and Yamashita, K. (1983) Pressurized reaction method for continuous production of L-alanine by immobilized *Pseudomonas dacunhae* cells. *Journal of Fermentation Technology*, **61**, 587–591.

135 Goto, M., Nara, T., Terasawa, M., and Yukawa, H. (1991) Process for producing l-alanine, EP 0,386,476 to Mitsubishi Petrochemical Co., Ltd. *Chemical Abstracts*, **114**, 80071.

136 Takamatsu, S., Umemura, I., Yamamoto, K., Sato, T., Tosa, T., and Chibata, I. (1982) Production of L-alanine from ammonium fumarate using two immobilized microorganisms. Elimination of side reactions. *European Journal of Applied Microbiology and Biotechnology*, **15**, 147–152.

137 Tosa, T., Takamatsu, S., Furui, M., and Chibata, I. (1984) Continuous production of L-alanine: successive enzyme reactions with two immobilized cells. *Annals of the New York Academy of Sciences*, **434**, 450–453.

138 Hanson, K.R. and Havir, E.A. (1978) An introduction to the enzymology of phenylpropanoid biosynthesis. *Recent Advances in Phytochemistry*, **12**, 91–137.

139 Nelson, R.P. (1976) Immobilized microbial cells, US 3,957,580 to Pfizer Inc. *Chemical Abstracts*, **84**, 2022.

140 Robers, F.F. Jr., Hamsher, J.J., and Nelson, R.P. (1976) Production of l-phenylalanine, GB 1,489,468 to Pfizer Inc. *Chemical Abstracts*, **84**, 178203.

141 Fotheringham, I.G. (1999) Phenylalanine, in *Encyclopedia of Bioprocess Technology – Fermentation, Biocatalysis, and Bioseparation* (eds M.C. Flickinger and S.W. Drew), John Wiley & Sons, Inc., New York, pp. 1943–1954.

142 Kishore, G.M. (1985) Stabilization of L-phenylalanine ammonia-lyase enzyme, EP 0,136,996 to Monsanto Company. *Chemical Abstracts*, **103**, 21245.

143 Adachi, O., Matsushita, K., Shinagawa, E., and Ameyama, M. (1990) Crystallization and properties of L-phenylalanine ammonia-lyase from *Rhodosporidium toruloides*. *Agricultural and Biological Chemistry*, **54**, 2839–2843.

144 Rees, D.G. and Jones, D.H. (1996) Stability of L-phenylalanine ammonia-lyase in aqueous solution and as the solid state in air and organic solvents. *Enzyme and Microbial Technology*, **19**, 282–288.

145 Abell, C.W. and Shen, R.S. (1987) Phenylalanine ammonia-lyase from the yeast *Rhodotorula glutinis. Methods in Enzymology*, **142**, 242–248.

146 Fritz, R.R., Hodgins, D.S., and Abell, C.W. (1976) Phenylalanine ammonia-lyase. Induction and purification from yeast and clearance in mammals. *The Journal of Biological Chemistry*, **251**, 4646–4650.

147 Yamada, S., Nabe, K., Izuo, N., Nakamichi, K., and Chibata, I. (1981) Production of L-phenylalanine from *trans*-cinnamic acid with *Rhodotorula glutinis* containing L-phenylalanine ammonia-lyase activity. *Applied and Environmental Microbiology*, **42**, 773–778.

148 Takaç, S., Akay, B., and Özdamar, T.H. (1995) Bioconversion of *trans*-cinnamic acid to L-phenylalanine by L-phenylalanine ammonia-lyase of *Rhodotorula glutinis*: parameters and kinetics. *Enzyme and Microbial Technology*, **17**, 445–452.

149 Vollmer, P.J., Montgomery, J.P., Schruber, J.J., and Yang, H.-H. (1985) Method for stabilizing the enzymic activity of phenylalanine ammonia lyase during l-phenylalanine production, EP 0,143,560 to Genex Corporation. *Chemical Abstracts*, **103**, 69802.

150 Evans, C.T., Conrad, D., Hanna, K., Peterson, W., Choma, C., and Misawa, M. (1987) Novel stabilization of phenylalanine ammonia-lyase catalyst during bioconversion of *trans*-cinnamic acid to L-phenylalanine. *Applied Microbiology and Biotechnology*, **25**, 399–405.

151 El-Batal, A.I. (2002) Optimization of reaction conditions and stabilization of phenylalanine ammonia lyase-containing *Rhodotorula glutinis* cells during bioconversion of *trans*-cinnamic acid to L-phenylalanine. *Acta Microbiologica Polonica*, **51**, 139–152.

152 Vollmer, P.J. and Schruben, J.J. (1986) Stabilization of phenylalanine ammonia-lyase in a bioreactor using reducing agents, US 4,574,117 to Genex Corporation. *Chemical Abstracts*, **104**, 166923.

153 Evans, C.T., Hanna, K., Payne, C., Conrad, D., and Misawa, M. (1987) Biotransformation of *trans*-cinnamic acid to L-phenylalanine: optimization of reaction conditions using whole yeast cells. *Enzyme and Microbial Technology*, **9**, 417–421.

154 Orndorff, S.A., Costantino, N., Stewart, D., and Durham, D.R. (1988) Strain improvement of *Rhodotorula graminis* for production of a novel L-phenylalanine ammonia-lyase. *Applied and Environmental Microbiology*, **54**, 996–1002.

155 Orndorff, S.A. and Durham, D.R. (1989) Phenylalanine ammonia lyase-producing strains, US, 4,757,015 to Genex Corporation. *Chemical Abstracts*, **110**, 73881.

156 Evans, C.T., Hanna, K., Conrad, D., Peterson, W., and Misawa, M. (1987) Production of phenylalanine ammonia-lyase (PAL): isolation and evaluation of yeast strains suitable for commercial production of L-phenylalanine. *Applied Microbiology and Biotechnology*, **25**, 406–414.

157 Baedeker, M. and Schulz, G.E. (1999) Overexpression of a designed 2.2 kb gene of eukaryotic phenylalanine ammonia-lyase in *Escherichia coli. FEBS Letters*, **457**, 57–60.

158 Xiang, L. and Moore, B.S. (2005) Biochemical characterization of a prokaryotic phenylalanine ammonia lyase. *Journal of Bacteriology*, **187**, 4286–4289.

159 Xiang, L. and Moore, B.S. (2006) Biochemical characterization of a prokaryotic phenylalanine ammonia lyase [Correction]. *Journal of Bacteriology*, **188**, 5331.

160 Weiner, D., Varvak, A., Richardson, T., Podar, M., Burke, E., and Healey, S. (2006) Lyase enzymes, nucleic acids encoding them and methods for making and using them, WO 2006/099207 to Diversa Corporation. *Chemical Abstracts*, **145**, 309308.

161 Yanaka, M., Ura, D., Takahashi, A., and Fukuhara, N. (1994) Production of beta-substituted alanine derivative,

JP 6,113,870 to Mitsui Toatsu Chemicals Inc. *Chemical Abstracts*, **121**, 155941.

162 Liu, W. (1991) Synthesis of optically active phenylalanine analogs using *Rhodotorula graminis*, US 5,981,239 to Great Lakes Chemical Corp. *Chemical Abstracts*, **131**, 321632.

163 de Vries, J.G., de Lange, B., de Vries, A.H.M., Mink, D., van Assema, F.B.J., Maas, P.J.D., and Hyett, D.J. (2006) Process for the preparation of enantiomerically enriched indoline-2-carboxylic acid, EP 1,676,838 to DSM IP Assets BV. *Chemical Abstracts*, **145**, 124453.

164 Gloge, A., Zoñ, J., Kövári, Á., Poppe, L., and Rétey, J. (2000) Phenylalanine ammonia-lyase: the use of its broad substrate specificity for mechanistic investigations and biocatalysis – synthesis of L-arylalanines. *Chemistry – A European Journal*, **6**, 3386–3390.

165 Renard, G., Guilleux, J.C., Bore, C., Malta-Valette, V., and Lerner, D.A. (1992) Synthesis of L-phenylalanine analogs by *Rhodotorula glutinis*. Bioconversion of cinnamic acids derivatives. *Biotechnology Letters*, **14**, 673–678.

166 Thorne, C.B., Gómez, C.G., and Housewright, R.D. (1955) Transamination of D-amino acids by *Bacillus subtilis*. *Journal of Bacteriology*, **69**, 357–362.

167 Thorne, C.B. and Molnar, D.M. (1955) D-amino acid transamination in *Bacillus anthracis*. *Journal of Bacteriology*, **70**, 420–426.

168 Yonaha, K., Misono, H., Yamamoto, T., and Soda, K. (1975) D-amino acid aminotransferase of *Bacillus sphaericus*. Enzymologic and spectrometric properties. *The Journal of Biological Chemistry*, **250**, 6983–6989.

169 Rozzell, J.D. (1987) Immobilized aminotransferases for amino acid production. *Methods in Enzymology*, **136**, 479–497.

170 Crump, S.P. and Rozzell, J.D. (1992) Biocatalytic production of amino acids by transamination, in *Biocatalytic Production of Amino Acids and Derivatives* (eds J.D. Rozzell and F. Wagner), Hanser, Munich, pp. 43–58.

171 Fotheringham, I.G., Pantaleone, D.P., and Taylor, P.P. (1997) Biocatalytic production of unnatural amino acids, mono esters, and *N*-protected derivatives. *Chimica Oggi – Chemistry Today*, **15**, 33–37.

172 Rozzell, J.D. and Bommarius, A.S. (2002) Transaminations, in *Enzyme Catalysis in Organic Synthesis* (eds K. Drauz and H. Waldman), Wiley-VCH Verlag GmbH, Weinheim, pp. 873–893.

173 Tewari, Y.B., Goldberg, R.N., and Rozzell, J.D. (2000) Thermodynamics of reactions catalysed by branched-chain-amino-acid transaminase. *Journal of Chemical Thermodynamics*, **32**, 1381–1398.

174 Schulz, A., Taggeselle, P., Tripier, D., and Bartsch, K. (1990) Stereospecific production of the herbicide phosphinothricin (glufosinate) by transamination: isolation and characterization of a phosphinothricin-specific transaminase from *Escherichia coli*. *Applied and Environmental Microbiology*, **56**, 1–6.

175 Cho, B.-K., Seo, J.-H., Kang, T.-W., and Kim, B.-G. (2003) Asymmetric synthesis of L-homophenylalanine by equilibrium-shift using recombinant aromatic L-amino acid transaminase. *Biotechnology and Bioengineering*, **83**, 226–234.

176 Hanzawa, S., Oe, S., Tokuhisa, K., Kawano, K., Kobayashi, T., Kudo, T., and Kakidani, H. (2001) Chemo-enzymatic synthesis of 3-(2-naphthyl)-L-alanine by an aminotransferase from the extreme thermophile, *Thermococcus profundus*. *Biotechnology Letters*, **23**, 589–591.

177 Bessman, S.P. and Layne, E.C. Jr. (1950) Stimulation of the non-enzymatic decarboxylation of oxalacetic acid by amino acids. *Archives of Biochemistry*, **26**, 25–32.

178 Walter, J.F. and Sherwin, M.B. (1986) Improved transamination process for producing L-amino acids, GB 2,161,159 to W R Grace & Co. *Chemical Abstracts*, **104**, 205538.

179 Rozzell, J.D. (1985) Production of L-amino acids by transamination, US 4,518,692 to Genetics Institute Inc. *Chemical Abstracts*, **102**, 219635.

180 Rozzell, J.D. (1987) Production of amino acids using coupled aminotransferases, US 4,826,766 to Genetics Institute Inc. *Chemical Abstracts*, **107**, 57455.

181 Bartsch, K., Schneider, R., and Schulz, A. (1996) Stereospecific production of the herbicide phosphinothricin (glufosinate): purification of aspartate transaminase from *Bacillus stearothermophilus*, cloning of the corresponding gene, *aspC*, and application in a coupled transaminase process. *Applied and Environmental Microbiology*, **62**, 3794–3799.

182 Fotheringham, I. (2000) Engineering biosynthetic pathways: new routes to chiral amino acids. *Current Opinion in Chemical Biology*, **4**, 120–124.

183 Ager, D.J., Fotheringham, I.G., Li, T., Pantaleone, D.P., and Senkpeil, R.F. (2000) The large scale synthesis of "unnatural" amino acids. *Enantiomer*, **5**, 235–243.

184 Fotheringham, I. and Taylor, P.P. (2006) Microbial pathway engineering for amino acid manufacture, in *Handbook of Chiral Chemicals*, 2nd edn (ed. D.J. Ager), CRC Press, Boca Raton, FL, pp. 31–45.

185 Fotheringham, I.G., Li, T., Senkpeil, R.F., and Ager, D. (2000) Transaminase biotransformation process employing glutamic acid, WO 2000/23609 to NSC Technologies LLC. *Chemical Abstracts*, **132**, 292811.

186 Li, T., Kootstra, A.B., and Fotheringham, I.G. (2002) Nonproteinogenic α-amino acid preparation using equilibrium shifted transamination. *Organic Process Research & Development*, **6**, 533–538.

187 Soda, K., Yonaha, K., Misono, H., and Osugi, M. (1974) Purification and crystallization of D-amino acid aminotransferase of *Bacillus sphaericus*. *FEBS Letters*, **46**, 359–363.

188 Lee, S.-G., Hong, S.-P., Song, J.J., Kim, S.-J., Kwak, M.-S., and Sung, M.-H. (2006) Functional and structural characterization of thermostable D-amino acid aminotransferases from *Geobacillus* spp. *Applied and Environmental Microbiology*, **72**, 1588–1594.

189 Tanizawa, K., Masu, Y., Asano, S., Tanaka, H., and Soda, K. (1989) Thermostable D-amino acid aminotransferase from a thermophilic *Bacillus* species. Purification, characterization, and active site sequence determination. *The Journal of Biological Chemistry*, **264**, 2445–2449.

190 Tanizawa, K., Asano, S., Masu, Y., Kuramitsu, S., Kagamiyama, H., Tanaka, H., and Soda, K. (1989) The primary structure of thermostable D-amino acid aminotransferase from a thermophilic *Bacillus* species and its correlation with L-amino acid aminotransferases. *The Journal of Biological Chemistry*, **264**, 2450–2454.

191 Peisach, D., Chipman, D.M., Van Ophem, P.W., Manning, J.M., and Ringe, D. (1998) Crystallographic study of steps along the reaction pathway of D-amino acid aminotransferase. *Biochemistry*, **37**, 4958–4967.

192 Sugio, S., Petsko, G.A., Manning, J.M., Soda, K., and Ringe, D. (1995) Crystal structure of a D-amino acid aminotransferase: how the protein controls stereoselectivity. *Biochemistry*, **34**, 9661–9669.

193 Okada, K., Hirotsu, K., Sato, M., Hayashi, H., and Kagamiyama, H. (1997) Three-dimensional structure of *Escherichia coli* branched-chain amino acid aminotransferase at 2.5 Å resolution. *Journal of Biochemistry*, **121**, 637–641.

194 Fotheringham, I.G., Bledig, S.A., and Taylor, P.P. (1998) Characterization of the genes encoding D-amino acid transaminase and glutamate racemase, two D-glutamate biosynthetic enzymes of *Bacillus sphaericus* ATCC 10208. *Journal of Bacteriology*, **180**, 4319–4323.

195 Yohda, M., Okada, H., and Kumagai, H. (1991) Molecular cloning and nucleotide sequencing of the aspartate racemase gene from lactic acid bacteria *Streptococcus thermophilus*. *Biochimica et Biophysica Acta*, **1089**, 234–240.

196 Taylor, P.P., Pantaleone, D.P., Senkpeil, R.F., and Fotheringham, I.G. (1998) Novel biosynthetic approaches to the production of unnatural amino acids using transaminases. *Trends in Biotechnology*, **16**, 412–418.

197 Fotheringham, I.G., Taylor, P.P., and Ton, J.L. (1998) Preparation of D-amino acids by direct fermentative means, US 5,728,555 to Monsanto Company. *Chemical Abstracts*, **128**, 216437.

198 Galkin, A., Kulakova, L., Yoshimura, T., Soda, K., and Esaki, N. (1997) Synthesis of optically active amino acids from α-keto acids with *Escherichia coli* cells expressing heterologous genes. *Applied and Environmental Microbiology*, **63**, 4651–4656.

199 Galkin, A., Kulakova, L., Yamamoto, H., Tanizawa, K., Tanaka, H., Esaki, N., and Soda, K. (1997) Conversion of α-keto acids to D-amino acids by coupling of four enzyme reactions. *Journal of Fermentation and Bioengineering*, **83**, 299–300.

200 Martínez del Pozo, A., Merola, M., Ueno, H., Manning, J.M., Tanizawa, K., Nishimura, K., Soda, K., and Ringe, D. (1989) Stereospecificity of reactions catalyzed by bacterial D-amino acid transaminase. *The Journal of Biological Chemistry*, **264**, 17784–17789.

201 Van den Tweel, W.J.J., Smits, J.P., and de Bont, J.A. (1986) Microbial metabolism of D- and L-phenylglycine by *Pseudomonas putida* LW-4. *Archives of Microbiology*, **144**, 169–174.

202 Van den Tweel, W.J.J., Ogg, R.L.H.P., and de Bont, J.A.M. (1988) Process for the preparation of a D-α-amino acid from the corresponding α-keto acid, EP 0,315,786 to Stamicarbon BV. *Chemical Abstracts*, **108**, 166114.

203 Van den Tweel, W.J.J., Smits, J.P., Ogg, R.L.H.P., and de Bont, J.A.M. (1988) The involvement of an enantioselective transaminase in the metabolism of D-3- and D-4-hydroxyphenylglycine in *Pseudomonas putida* LW-4. *Applied Microbiology and Biotechnology*, **29**, 224–230.

204 Müller, U., van Assema, F., Gunsior, M., Orf, S., Kremer, S., Schipper, D., Wagemans, A., Townsend, C.A., Sonke, T., Bovenberg, R., and Wubbolts, M.G. (2006) Metabolic engineering of the *E. coli* L-phenylalanine pathway for the production of D-phenylglycine (D-Phg). *Metabolic Engineering*, **8**, 196–208.

205 Wiyakrutta, S. and Meevootisom, V. (1997) A stereo-inverting D-phenylglycine aminotransferase from *Pseudomonas stutzeri* ST-201: purification, characterization and application for D-phenylglycine synthesis. *Journal of Biotechnology*, **55**, 193–203.

206 Kongsaeree, P., Samanchart, C., Laowanapiban, P., Wiyakrutta, S., and Meevootisom, V. (2003) Crystallization and preliminary X-ray crystallographic analysis of D-phenylglycine aminotransferase from *Pseudomonas stutzeri* ST201. *Acta Crystallographica. Section D, Biological Crystallography*, **59**, 953–954.

207 Brunhuber, N.M.W. and Blanchard, J.S. (1994) The biochemistry and enzymology of amino acid dehydrogenases. *Critical Reviews in Biochemistry and Molecular Biology*, **29**, 415–467.

208 Brunhuber, N.M.W., Thoden, J.B., Blanchard, J.S., and Vanhooke, J.L. (2000) *Rhodococcus*L-phenylalanine dehydrogenase: kinetics, mechanism, and structural basis for catalytic specifity. *Biochemistry*, **39**, 9174–9187.

209 Sanwal, B.D. and Zink, M.W. (1961) L-Leucine dehydrogenase of *Bacillus* *cereus. Archives of Biochemistry and Biophysics*, **94**, 430–435.

210 Bommarius, A.S. (2002) Reduction of C= N bonds, in *Enzyme Catalysis in Organic Synthesis* (eds K. Drauz and H. Waldman), Wiley-VCH Verlag GmbH, Weinheim, pp. 1047–1063.

211 Hummel, W. and Kula, M.R. (1989) Dehydrogenases for the synthesis of chiral compounds. *European Journal of Biochemistry*, **184**, 1–13.

212 Ohshima, T. and Soda, K. (2000) Stereoselective biocatalysis: amino acid dehydrogenases and their applications, in *Stereoselective Biocatalysis* (ed. R.N. Patel), Marcel Dekker, New York, pp. 877–902.

213 Ohshima, T., Nishida, N., Bakthavatsalam, S., Kataoka, K., Takada, H., Yoshimura, T., Esaki, N., and Soda, K. (1994) The purification, characterization, cloning and sequencing of the gene for a halostable and thermostable leucine dehydrogenase from *Thermoactinomyces intermedius*. *FEBS Journal*, **222**, 305–312.

214 Bommarius, A.S., Drauz, K., Hummel, W., Kula, M.-R., and Wandrey, C. (1994) Some new developments in reductive amination with cofactor regeneration. *Biocatalysis*, **10**, 37–47.

215 Krix, G., Bommarius, A.S., Drauz, K., Kottenhahn, M., Schwarm, M., and Kula, M.-R. (1997) Enzymatic reduction of α-keto acids leading to L-amino acids, D- or L-hydroxy acids. *Journal of Biotechnology*, **53**, 29–39.

216 Hummel, W., Weiß, N., and Kula, M.R. (1984) Isolation and characterization of a bacterium possessing L-phenylalanine dehydrogenase activity. *Archives of Microbiology*, **137**, 47–52.

217 Hummel, W., Schütte, H., Schmidt, E., Wandrey, C., and Kula, M.R. (1987) Isolation of L-phenylalanine dehydrogenase from *Rhodococcus* sp. M4 and its application for the production of L-phenylalanine. *Applied Microbiology and Biotechnology*, **26**, 409–416.

218 Ohshima, T., Takada, H., Yoshimura, T., Esaki, N., and Soda, K. (1991) Distribution, purification, and characterization of thermostable phenylalanine dehydrogenase from thermophilic actinomycetes. *Journal of Bacteriology*, **173**, 3943–3948.

219 Hanson, R.L., Schwinden, M.D., Banerjee, A., Brzozowski, D.B., Chen, B.-C., Patel, B.P., McNamee, C.G., Kodersha, G.A., Kronenthal, D.R., Patel, R.N., and Szarka, L.J. (1999) Enzymatic synthesis of L-6-hydroxynorleucine. *Bioorganic and Medicinal Chemistry*, **7**, 2247–2252.

220 Schütte, H., Flossdorf, J., Sahm, H., and Kula, M.R. (1976) Purification and properties of formaldehyde dehydrogenase and formate dehydrogenase from *Candida boidinii. European Journal of Biochemistry*, **62**, 151–160.

221 Kula, M.R. and Wandrey, C. (1987) Continuous enzymatic transformation in an enzyme-membrane reactor with simultaneous NADH regeneration. *Methods in Enzymology*, **136**, 9–21.

222 Shaked, Z. and Whitesides, G.M. (1980) Enzyme-catalyzed organic synthesis: NADH regeneration by using formate dehydrogenase. *Journal of the American Chemical Society*, **102**, 7104–7105.

223 Slusarczyk, H., Felber, S., Kula, M.-R., and Pohl, M. (2000) Stabilization of NAD-dependent formate dehydrogenase from *Candida boidinii* by site-directed mutagenesis of cysteine residues. *European Journal of Biochemistry*, **267**, 1280–1289.

224 Weuster-Botz, D., Paschold, H., Striegel, B., Gieren, H., Kula, M.-R., and Wandrey, C. (1994) Continuous computer controlled production of formate dehydrogenase (FDH) and isolation on a pilot scale. *Chemical Engineering & Technology*, **17**, 131–137.

225 Wöltinger, J., Drauz, K., and Bommarius, A.S. (2001) The membrane reactor in the fine chemicals industry. *Applied Catalysis A – General*, **221**, 171–185.

226 Kragl, U., Vasic-Racki, D., and Wandrey, C. (1992) Continuous processes with soluble enzymes. *Chemie Ingenieur Technik*, **64**, 499–509.

227 Kragl, U., Kruse, W., Hummel, W., and Wandrey, C. (1996) Enzyme engineering aspects of biocatalysis: cofactor regeneration as example. *Biotechnology and Bioengineering*, **52**, 309–319.

228 Wichmann, R., Wandrey, C., Bückmann, A.F., and Kula, M.R. (1981) Continuous enzymatic transformation in an enzyme membrane reactor with simultaneous NAD(H) regeneration. *Biotechnology and Bioengineering*, **23**, 2789–2802.

229 Bommarius, A.S., Schwarm, M., and Drauz, K. (1998) Biocatalysis to amino acid-based chiral pharmaceuticals – examples and perspectives. *Journal of Molecular Catalysis B – Enzymatic*, **5**, 1–11.

230 Asano, Y., Yamada, A., Kato, Y., Yamaguchi, K., Hibino, Y., Hirai, K., and Kondo, K. (1990) Enantioselective synthesis of (*S*)-amino acids by phenylalanine dehydrogenase from *Bacillus sphaericus*: use of natural and recombinant enzymes. *The Journal of Organic Chemistry*, **55**, 5567–5571.

231 Hanson, R.L., Howell, J.M., LaPorte, T.L., Donovan, M.J., Cazzulino, D.L., Zannella, V., Montana, M.A., Nanduri, V.B., Schwarz, S.R., Eiring, R.F., Durand, S.C., Wasylyk, J.M., Parker, W.L., Liu, M.S., Okuniewicz, F.J., Chen, B.-C., Harris, J.C., Natalie, K.J., Ramig, K., Swaminathan, S., Rosso, V.W., Pack, S.K., Lotz, B.T., Bernot, P.J., Rusowicz, A., Lust, D.A., Tse, K.S., Venit, J.J., Szarka, L.J., and Patel, R.N. (2000) Synthesis of allysine ethylene acetal using phenylalanine dehydrogenase from *Thermoactinomyces intermedius*. *Enzyme and Microbial Technology*, **26**, 348–358.

232 Galkin, A., Kulakova, L., Tishkov, V., Esaki, N., and Soda, K. (1995) Cloning of formate dehydrogenase gene from a methanol-utilizing bacterium *Mycobacterium vaccae* N10. *Applied Microbiology and Biotechnology*, **44**, 479–483.

233 Takada, H., Yoshimura, T., Ohshima, T., Esaki, N., and Soda, K. (1991) Thermostable phenylalanine dehydrogenase of *Thermoactinomyces intermedius*: cloning, expression, and sequencing of its gene. *Journal of Biochemistry*, **109**, 371–376.

234 Gröger, H., Werner, H., Altenbuchner, J., Menzel, A., and Hummel, W. (2005) Process for preparing optically active amino acids using a whole-cell catalyst, WO 2005/093081 to Degussa AG. *Chemical Abstracts*, **143**, 365753.

235 Menzel, A., Werner, H., Altenbuchner, J., and Gröger, H. (2004) From enzymes to "designer bugs" in reductive amination: a new process for the synthesis of L-*tert*-leucine using a whole cell-catalyst. *Engineering in Life Sciences*, **4**, 573–576.

236 Gröger, H., May, O., Werner, H., Menzel, A., and Altenbuchner, J. (2006) A "second-generation process" for the synthesis of L-neopentylglycine: asymmetric reductive amination using a recombinant whole cell catalyst. *Organic Process Research & Development*, **10**, 666–669.

237 Vedha-Peters, K., Gunawardana, M., Rozzell, J.D., and Novick, S.J. (2006) Creation of a broad-range and highly stereoselective D-amino acid dehydrogenase for the one-step synthesis of D-amino acids. *Journal of the American Chemical Society*, **128**, 10923–10929.

4
β-Amino Acid Biosynthesis

Peter Spiteller

4.1
Introduction

4.1.1
Importance of β-Amino Acids and their Biosynthesis

In contrast to α-amino acids, the essential constituents of proteins in all living organisms, β-amino acids – except β-alanine (**1**) and β-aminoisobutyric acid (**2**) – only occur in secondary metabolites [1]. Nevertheless, β-amino acids are an important class of compounds, which are often part of bioactive secondary metabolites in all five kingdoms of living organisms, namely protista, bacteria, fungi, plants, and animals [1]. For instance, β-amino acids occur in peptides, cyclopeptides, depsipeptides, glycopeptides, alkaloids, and terpenes, and often contribute decisively to their bioactivity [1]. The biological importance of β-amino acids in peptides is often based on their capability to improve the stability of peptides towards degradation by peptidases [2]. Thus, β-amino acids play an important role in the development of drugs [2]. Accordingly, synthetic chemists have developed a huge variety of enantioselective syntheses for different β-amino acids [3].

However, organic syntheses of β-amino acids usually differ fundamentally from biosynthetic pathways nature has invented [4]. Insights in the biosynthesis of β-amino acids enable us to check whether a structure suggested for a new natural product is in accordance with a known biosynthetic pathway. In addition, a detailed understanding of the biosynthesis of β-amino acids opens a variety of new perspectives that are discussed in Section 4.3.

4.1.2
Scope of this Chapter

While previous reviews have dealt either preferentially with the synthesis of different types of β-amino acids [3, 5] or their occurrence in nature [1, 6], only a few reviews have discussed aspects of their biosynthesis [4]. In addition, these reviews usually

Amino Acids, Peptides and Proteins in Organic Chemistry. Vol.1 – Origins and Synthesis of Amino Acids.
Edited by Andrew B. Hughes

ISBN: 978-3-527-32096-7

focused on a certain class of enzymes (e.g., aminomutases [4]). Moreover, in recent years our knowledge of the biosynthesis of β-amino amino acids has expanded rapidly, since new methods, such as the identification, sequencing, and analysis of biosynthetic gene clusters, were applied to the elucidation of the biosynthesis of β-amino amino acids. As a consequence, the most important types of 2,3-aminomutases are now known in detail [7]. Similarly, at least some biosynthetic pathways of the large group of polyketide-type β-amino acids have been elucidated. Consequently, it is now possible for the first time to draw a comprehensive picture of the biosynthesis of most β-amino acids and to predict the general biosynthesis of most β-amino acids from their structural characteristics even if their biosynthesis has not yet been investigated.

4.2
Biosynthesis of β-Amino Acids

β-Alanine (**1**) and β-aminoisobutyric acid (**2**) occur in a multitude of organisms, while other β-amino acids are usually only present in a limited number of organisms [1]. Moreover, at least three different biosynthetic pathways are known that lead to the generation of β-alanine or β-aminoisobutyric acid [4]. Therefore, these special biosynthetic pathways are discussed first in a separate section (Section 4.2.1).

Apart from β-alanine (**1**) and β-aminoisobutyric acid (**2**), β-amino acids generally either originate from proteinogenic α-amino acids or from a *de novo* synthesis. β-Amino acids that are biosynthetically derived from proteinogenic α-amino acids can be divided into:

- β-amino acids analogous to proteinogenic α-amino acids (Section 4.2.2).
- α,β-amino acids (Section 4.2.3) and
- α-keto-β-amino acids (Section 4.2.4).

In contrast, the largest group of β-amino acids, the

- polyketide-type β-amino acids (Section 4.2.5),

is synthesized *de novo* from acetate units by polyketide synthases (PKSs).

4.2.1
Biosynthesis of β-Alanine and β-Aminoisobutyric Acid

4.2.1.1 β-Alanine

β-Alanine (**1**) is present in all living organisms, since some primary metabolites, such as coenzyme A (CoA), contain a β-alanine moiety [8]. In mammals, β-alanine has been found in free form [8]. In some plants, β-alanine betaine serves as an osmoprotectant [9, 10]. β-Alanine has been found also in fungi (e.g., as a constituent of some cryptophycins [11, 12] such as **3**, the destruxins [13] such as **4**, leualacin [14, 15], and the leucinostatins [16]) as well as in marine sponges and cyanobacteria, where it occurs as a constituent of the theonellamides C–F [17, 18], the theonellapeptolides [19], and the yanucamides [20, 21].

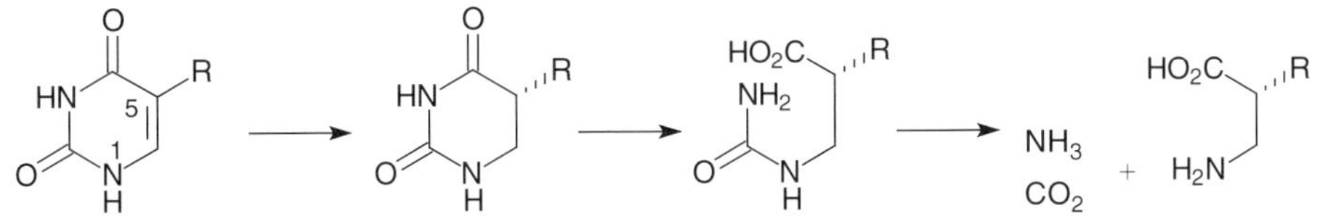

R = H: uracil (**5**) — R = H: **7** — R = H: **9** — R = H: β-alanine (**1**)

R = CH_3: thymine (**6**) — R = CH_3: **8** — R = CH_3: **10** — R = CH_3: (*R*)-β-AiB [(*R*)-**2**]

Scheme 4.1 Biosynthesis of β-alanine and (*R*)-β-aminoisobutyric acid from pyrimidines.

Three different pathways are known that lead to β-alanine (**1**) [4]. For instance, uracil (**5**) is degraded to β-alanine [8]. Similarly, thymine (**6**) is transformed to (*R*)-β-aminoisobutyric acid [(*R*)-**2**], since the enzymes involved in the degradation reactions are not very specific (Scheme 4.1) [8]. In the first step, the pyrimidine **5** (or **6**) is hydrogenated by dihydrouracil dehydrogenase to yield the corresponding dihydropyrimidine **7** (or **8**), which is hydrolyzed in the second step by β-dihydropyrimidinase to *N*-carbamoyl-β-alanine (**9**) and *N*-carbamoyl-β-aminoisobutyric acid (**10**), respectively. These compounds are finally hydrolyzed by *N*-carbamoyl-β-alanine amidohydrolase to β-alanine (**1**) and (*R*)-β-aminoisobutyric acid [(*R*)-**2**], respectively [8] (Scheme 4.1).

In yeasts, β-alanine originates from spermine (**11**), which is oxidized via 3-aminopropanal (**12**) to β-alanine (Scheme 4.2) [22].

In contrast to mammals that generate β-alanine mainly from uracil (**5**), in *Escherichia coli* β-alanine (**1**) is produced mainly from L-aspartic acid (**13**) by L-aspartate-α-decarboxylase (Scheme 4.3) (Section 4.2.1.2) [23, 24].

Recently, β-alanine (**1**) has also been generated directly from α-alanine via an alanine 2,3-aminomutase, which was obtained by genetic engineering from lysine 2,3-aminomutase (Section 4.2.2.5) [25].

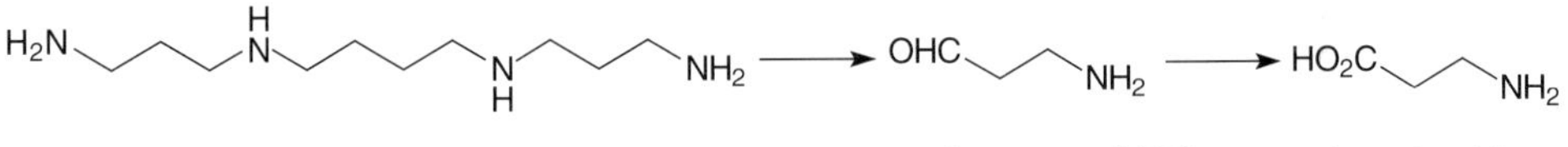

spermine (**11**) → 3-aminopropanal (**12**) → β-alanine (**1**)

Scheme 4.2 Biosynthesis of β-alanine from spermine (**11**).

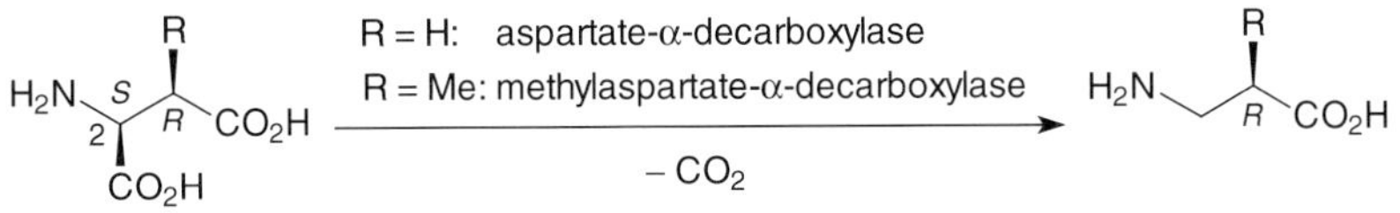

R = H: aspartate-α-decarboxylase
R = Me: methylaspartate-α-decarboxylase
$- CO_2$

R = H: L-aspartic acid (**13**) — R = H: β-alanine (**1**)

R = Me: (2*S*,3*R*)-methylaspartic acid (**14**) — R = Me: β-aminoisobutyric acid [(R)-**2**]

Scheme 4.3 Biosynthesis of β-alanine (**1**) in *E. coli* and of (*R*)-β-aminoisobutyric acid [(*R*)-**2**] in *Nostoc* sp. ATCC 53789.

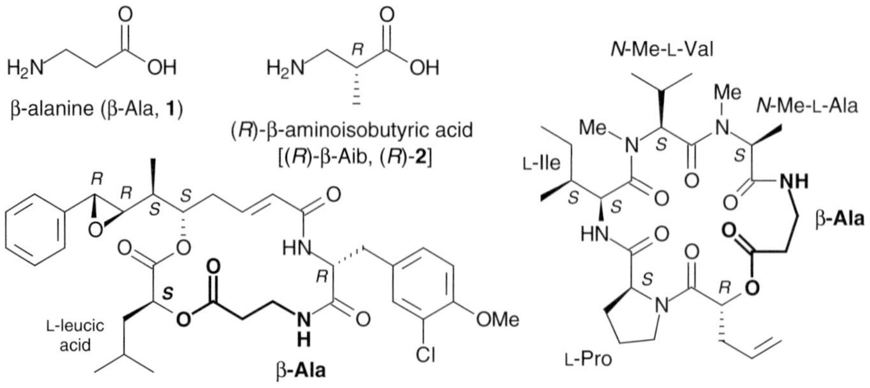

Figure 4.1 β-Alanine, β-aminoisobutyric acid, and selected natural products that contain β-alanine.

4.2.1.2 β-Aminoisobutyric Acid

(*R*)-β-Aminoisobutyric acid [(*R*)-**2**] is a degradation product in the catabolism of thymine (**6**) and therefore widespread in nature (Scheme 4.1) [8]. (*R*)-β-Aminoisobutyric acid has been found in free form in plants (e.g., in *Lunaria annua* [26]). It occurs as a constituent in phascolosomine isolated from the worm *Phascolon strombi* [27] and in some cryptophycins [12, 28] [e.g., in cryptophycin 1, which only differs from **3** (Figure 4.1) by the presence of an (*R*)-β-Aib residue instead of a β-Ala residue]. The cryptophycins are antitumor-active depsipeptides, isolated from terrestrial cyanobacteria. Recently, identification, cloning, and sequencing of the biosynthetic gene cluster [29] responsible for the cryptophycin biosynthesis of the lichen cyanobacterial symbiont *Nostoc* sp. ATCC 53789 and feeding experiments with labeled precursors revealed that the (*R*)-β-aminoisobutryric acid (**2**) moiety of cryptophycin 1 originates from (2*S*,3*R*)-3-methylaspartatic acid (**14**), which is decarboxylated by the β-methylaspartate-α-decarboxylase CrpG analogously to the biosynthesis of β-alanine (**1**) from L-aspartic acid (**13**) by aspartate-α-decarboxylase (Scheme 4.3) [30].

In mammals, free (*S*)-β-aminoisobutyric acid [(*S*)-**2**] is present, which originates from L-valine (**15**) via the intermediates **16–20** (Scheme 4.4) [8].

4.2.2 Biosynthesis of β-Amino Acids by 2,3-Aminomutases from α-Amino Acids

β-Amino acids that are structurally closely related to corresponding proteinogenic α-amino acids (Figure 4.2) are usually generated by various types of 2,3-aminomutases which shift the amino group from the α- to the β-position. So far, it has been shown that β-lysine (**21**), 3,5-diaminohexanoic acid (**22**), β-arginine (**23**), β-phenylalanine (**24**), 3-phenylisoserine (**25**), β-tyrosine (**26**), β-3,4-dihydroxyphenylalanine

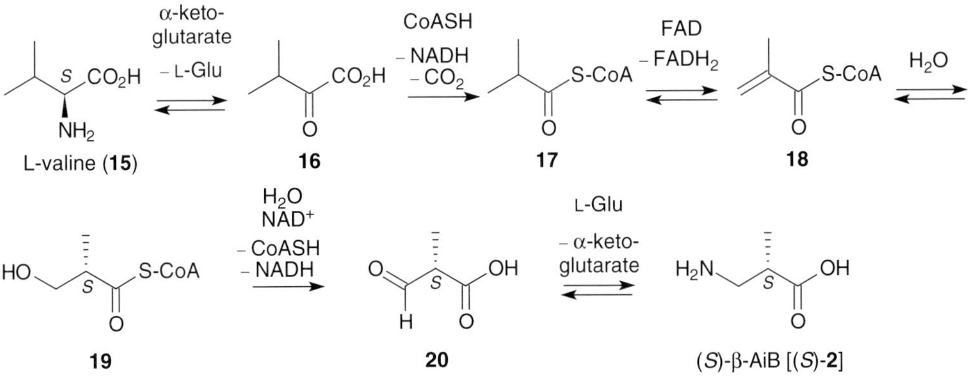

Scheme 4.4 Biosynthesis of (S)-β-aminoisobutyric acid [(S)-2] from L-valine (15).

(β-DOPA) (27), 3-chloro-β-DOPA (28), β-glutamate (29), β-glutamine (30), β-leucine (31), and β-alanine (1) can be generated by the action of aminomutases (Figure 4.2).

The occurrence of aminomutases was first recognized during investigations of the metabolism of lysine in *Clostridium* species by Stadtman [31] and Barker [32, 33] in the 1960s. By the action of two aminomutases α-lysine is first converted to β-lysine (21) and then to 3,5-diaminohexanoic acid (22) [31, 34].

So far, three types of 2,3-aminomutases are known [7, 35]. The first type requires *S*-adenosylmethionine (SAM), pyridoxal 5′-phosphate (PLP), and iron–sulfur clusters as cofactors. In this case, the amino group is shifted via a radical mechanism.

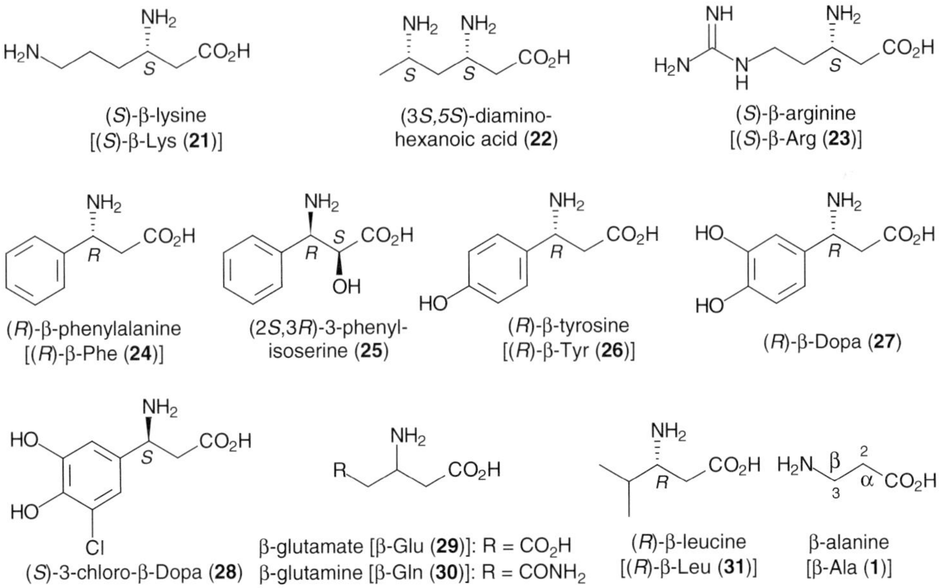

Figure 4.2 β-Amino acids derived from α-amino acids by the action of aminomutases.

The second group of aminomutases is ATP-dependent, while the third requires a 4-methylideneimidazol-5-one (MIO) moiety generated autocatalytically in the active site of the enzyme [7]. The 5,6-aminomutase from *Clostridium sticklandii* that converts β-lysine (**21**) to 3,5-diaminohexanoic acid (**22**) is closely related to the SAM-dependent aminomutases, but requires cobalamine and PLP as cofactors [36].

2,3-Aminomutases are known mainly from bacteria, but do also occur sometimes in fungi and plants, whereas there are no reports on the presence of 2,3-aminomutases or β-amino acids in animals except β-alanine (**1**) or β-aminoisobutyric acid (**2**) [1].

4.2.2.1 **β-Lysine, β-Arginine, and Related β-Amino Acids**

(*S*)-β-Lysine (**21**) occurs in free form in several *Clostridium* species (e.g., in *Clostridium subterminale* SB4 [32]). In Archaea, such as *Methanosarcina thermophila*, N^{ε}-acetyl-β-lysine serves as an osmolyte [37]. β-Lysine is present as a constituent of several secondary metabolites from bacteria, such as in albothricin (**32**) [38] (Figure 4.3), streptothricin F (**33**) [39], viomycin (**34**) [40] (Figure 4.4), tuberactionomycin O (**35**) [41], capreomycins IA (**36**) and IB (**37**) [42], lysinomycin [43], myomycin [44], and tallysomycin A [45].

β-Arginine (**23**) is known as a constituent of LL-BM547β (**38**) (Figure 4.4) isolated from the *Nocardia* sp. Lederle culture BM547 [46] and of blasticidin S (**39**) (Figure 4.5) isolated from *Streptomyces griseochromogenes* [47].

The biosynthesis of (*S*)-β-lysine (**21**) in *C. subterminale* SB4 has been studied in detail [48, 49], both by feeding experiments with labeled precursors, and by cloning, sequencing and heterologous overexpression [50] of the lysine 2,3-aminomutase in *E. coli*. The hexameric enzyme with a molecular weight of 285 kDa requires SAM and PLP as cofactors and the presence of iron ions [51, 52]. Incubation experiments of cell free extracts of *C. subterminale* SB4 with (2*RS*)-[3-^{13}C,2-^{15}N]lysine revealed that the amino group is shifted in an intramolecular reaction from the (2*S*)-position to the (3*S*)-position in (*S*)-β-lysine [53]. Similar feeding experiments with (2*RS*,3*R*)-[3-^{2}H] lysine and (2*RS*,3*S*)-[3-^{2}H]lysine demonstrated that the *pro*-(3*R*) hydrogen of L-lysine (**40**) migrates to the *pro*-(2*R*) position in (*S*)-β-lysine (**21**), while the *pro*-(3*S*) hydrogen remains at C-3 (Scheme 4.5) [53].

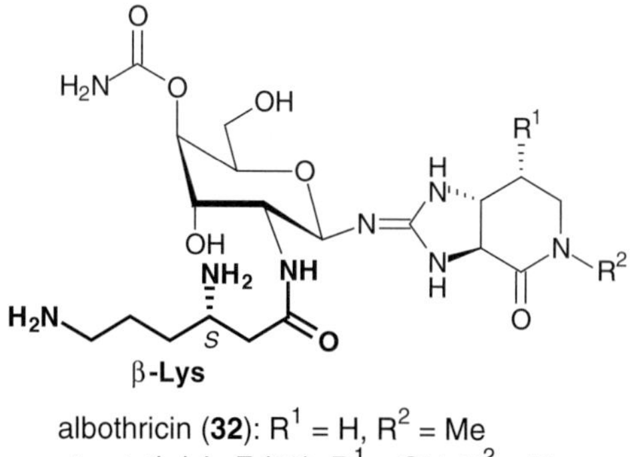

albothricin (**32**): R^1 = H, R^2 = Me
streptothricin F (**33**): R^1 = OH, R^2 = H

Figure 4.3 Albothricin and streptothricin F.

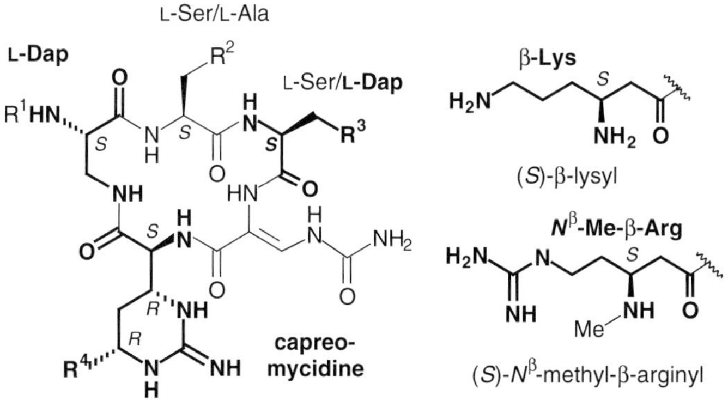

tuberactinomycin B (viomycin, **34**): R^1 = (*S*)-β-lysyl, R^2 = OH, R^3 = OH, R^4 = OH
tuberactinomycin O (**35**): R^1 = (*S*)-β-lysyl, R^2 = OH, R^3 = OH, R^4 = H
capreomycin IA (**36**): R^1 = H, R^2 = OH, R^3 = NH-[(*S*)-β-lysyl], R^4 = H
capreomycin IB (**37**): R^1 = H, R^2 = H, R^3 = NH-[(*S*)-β-lysyl], R^4 = H
LL-BM547β (**38**): R^1 = (*S*)-N^{β}-methyl-β-arginyl, R^2 = OH, R^3 = OH, R^4 = OH

Figure 4.4 Viomycin and related natural products.

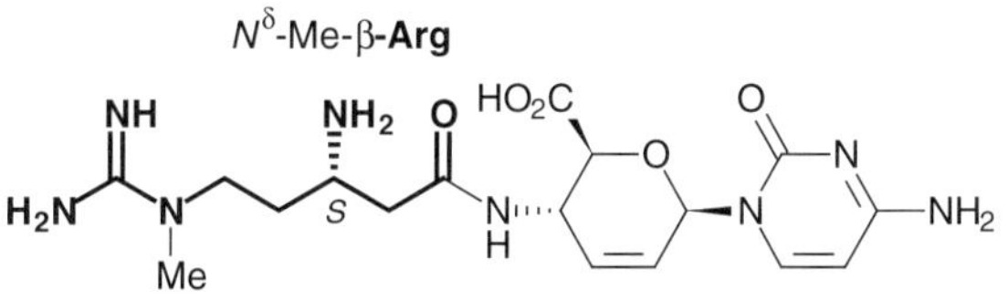

Figure 4.5 Blasticidin S (**39**).

Experiments by Frey *et al.* revealed that SAM is essential for the generation of radicals in the course of the reaction similarly to adenosylcobalamin which serves as a radical-generating cofactor in methylmalonyl-CoA mutase [54]. The latter enzyme catalyses the rearrangement of methylmalonyl-CoA to succinyl-CoA [54]. Since SAM is a much smaller molecule compared to adenosylcobalamin, Frey termed it "a poor man's adenosylcobalamin" [55, 56]. However, when it became evident that SAM initiates a greater diversity of radical reactions than cobalamine itself, Frey then

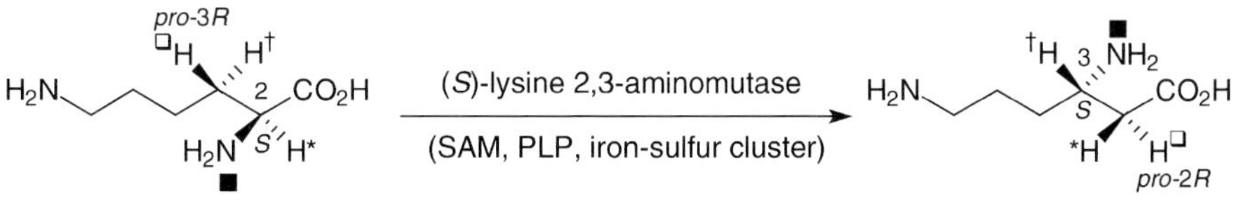

Scheme 4.5 Biosynthesis of (*S*)-β-lysine (**21**) from L-α-lysine (**40**) in *C. subterminale* SB4.

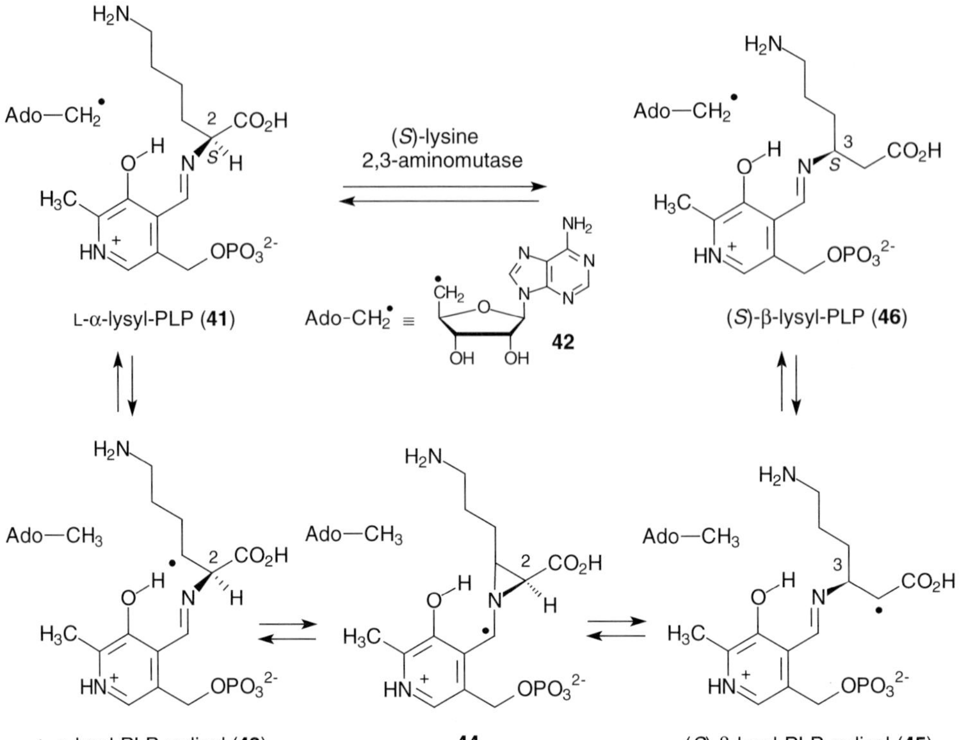

Scheme 4.6 Reaction mechanism for the generation of β-lysyl-PLP (**46**) from α-lysyl-PLP (**41**) by the SAM-dependent lysine 2,3-aminomutase from *C. subterminale* SB4.

referred to SAM as "a rich man's adenosylcobalamin" or as "a wolf in sheep's clothing" [57]. The aminomutase reaction starts with the formation of the α-lysyl aldimine (**41**) from α-lysine (**40**) and PLP (Scheme 4.6) [57, 58]. In the second step a 5′-deoxyadenosyl radical (**42**) generated from SAM removes the *pro*-(3*R*) hydrogen from the α-lysyl aldimine **41**. In the next step, the α-lysyl radical **43** is isomerized via **44** to the corresponding β-lysyl radical **45**. Transfer of a hydrogen from the methyl group of 5′-deoxadenosine to the β-lysyl radical leads to the β-lysyl aldimine **46** and regeneration of the 5′-deoxyadenosyl radical. Finally, the β-lysyl aldimine is hydrolyzed to β-lysine (**21**) [57].

The course of the reaction was investigated in detail (e.g., the occurrence of the β-lysyl-PLP radical **46** was confirmed by electron spin resonance experiments [59, 60]).

The biosynthesis of β-arginine (**23**) has been investigated first by feeding experiments with stable isotopes in the blasticidin S (**39**)-producing bacterium *Streptomyces griseochromogenes* [61] (Figure 4.5). In the course of the biosynthesis of β-arginine (**23**) the nitrogen migrates in an intramolecular reaction from the α-position to the β-position analogously to the β-lysine biosynthesis in *C. subterminale* SB4 [61]. Moreover, sequencing of the gene cluster responsible for the biosynthesis of

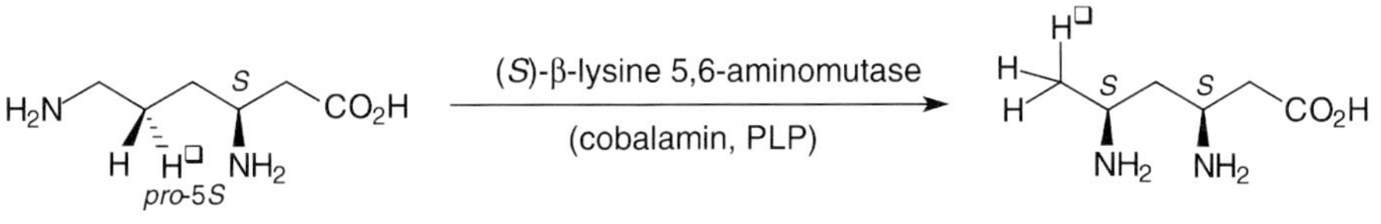

L-β-lysine (**21**) — (*S*)-β-lysine 5,6-aminomutase (cobalamin, PLP) → (3*S*,5*S*)-diaminohexanoic acid (**22**)

Scheme 4.7 Biosynthesis of (3*S*,5*S*)-diaminohexanoic acid (**22**) from β-lysine (**21**) in *C. sticklandii*.

blasticidin S revealed that *blsG*, the gene encoding the β-arginine 2,3-aminomutase, is 48% identical and 65% similar to the corresponding gene of the lysine 2,3-aminomutase in *C. subterminale* SB4, suggesting the same reaction mechanism and cofactor requirement for both enzymes [62].

Clostridium sticklandii contains a β-lysine 5,6-aminomutase which converts β-lysine (**21**) to 3,5-diaminohexanoic acid (**22**) [36] (Scheme 4.7). Feeding experiments [63], elucidation of the cofactor requirement of this aminomutase [36], and cloning, sequencing, and analysis of the corresponding gene [64] suggest that it follows, in general, the same reaction mechanism as the SAM-dependent lysine 2,3-aminomutase from *C. subterminale* SB4, but uses cobalamine instead of SAM as cofactor (Scheme 4.7).

4.2.2.2 β-Phenylalanine, β-Tyrosine, and Related β-Amino Acids

β-Phenylalanine (**24**) occurs in natural products as a constituent of some peptides; for instance, it is present in andrimid (**47**) (Figure 4.6), isolated from cultures of an *Enterobacter* species [65, 66], in the moiramides A, B (**48**), and C (**49**), isolated from the bacterium *Pseudomonas fluorescens* [66], in the astins, such as astin A (**50**), isolated from the roots of the plant *Aster tataricus* [67], in cyclochlorotine (**51**), a toxin from the rice mold *Penicillium islandicum* [68], in the edeines D_1 [69] and F_1 [70], isolated from cultures of *Bacillus brevis* Vm4, and in the pyloricidins from cultures of the *Bacillus* sp. HC-70 [71].

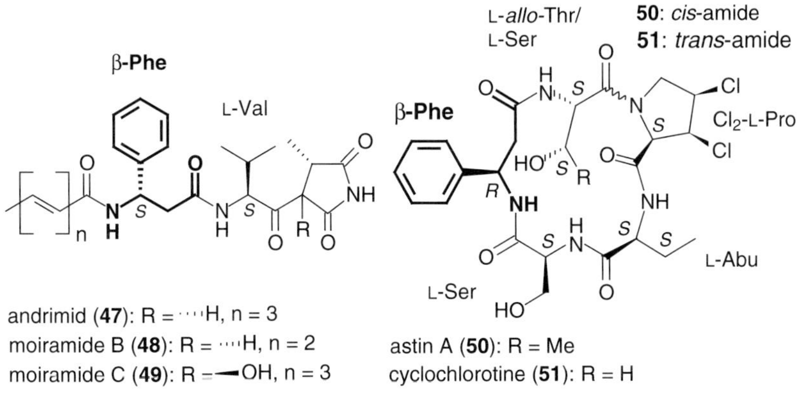

Figure 4.6 Selected natural products containing a β-phenylalanine residue.

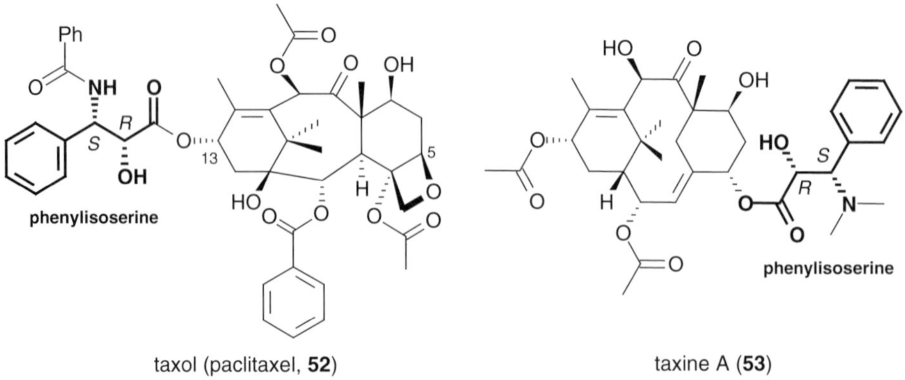

phenylisoserine

taxol (paclitaxel, **52**)

phenylisoserine

taxine A (**53**)

Figure 4.7 Selected natural products containing a phenylisoserine residue.

Moreover, β-phenylalanine serves as an intermediate in the biosynthesis of the 3-phenylisoserine (**25**) moiety of the antitumor-active terpenoid antibiotic taxol (**52**) isolated from the stem bark of the western yew, *Taxus brevifolia* [72] (Figure 4.7). 3-Phenylisoserine (**25**) is also present in taxine A (**53**) and taxine B isolated from the English yew, *Taxus baccata* [73].

β-Tyrosine (**26**) is a constituent of peptides such as the chondramides C (**54**) [74] and D (**55**), occurring in the myxomycete *Chondromyces crocatus* [75, 76], the edeines A_1 (**56**) and B_1 (**57**) [77], the geodiamolides H (**58**) and I (**59**) isolated from a marine sponge of the genus *Cymbastela* [78], and jaspamide (**60**), which has been found in different marine sponge genera [79] (e.g., in *Jaspis* species) (Figure 4.8).

In addition, β-tyrosine (**26**) is an intermediate in the biosynthesis of the β-amino acid (*R*)-β-DOPA (**27**), occurring in free form in fruiting bodies of the mushroom *Cortinarius violaceus* [80]. Moreover, it is an intermediate in the biosynthesis of the (*S*)-3-chloro-4,5-dihydroxy-β-phenylalanine (**28**) moiety of the enediyne antitumor antibiotic C-1027 isolated from *Streptomyces globisporus* [81, 82]. C-1027 consists of an acidic protein with a molecular weight of 15 000 Da and the labile enedyine chromophore **61** [82] (Figure 4.9).

So far, the biosynthesis of β-phenylalanine (**24**) has only been investigated in detail in the case of the important antitumor antibiotic taxol (**52**) [7, 83]. The first step in the biosynthesis of the phenylisoserine (**25**) side-chain of taxol is the conversion of L-phenylalanine (**62**) to (3*R*)-β-phenylalanine (**24**) by a phenylalanine 2,3-aminomutase whose reaction mechanism has been studied in detail [83]. Feeding experiments with labeled precursors revealed that the nitrogen is shifted in an intramolecular reaction from C-2 to C-3, while the *pro*-(3*S*) hydrogen of L-phenylalanine migrates to C-2 of β-phenylalanine and the *pro*-(3*R*) hydrogen remains at C-3 (Scheme 4.8) [83].

The biochemical analysis of the phenylalanine 2,3-aminomutase from *T. brevifolia* revealed that it does not require external cofactors [83], suggesting the same reaction mechanism for the phenylalanine 2,3-aminomutase as deduced for the tyrosine 2,3-aminomutase SgcC4 which relies on the electronegativity of an autocatalytically formed MIO (**63**) moiety at the active site of the enzyme [84] (Scheme 4.9). This assumption was confirmed by cloning the corresponding gene using a strategy

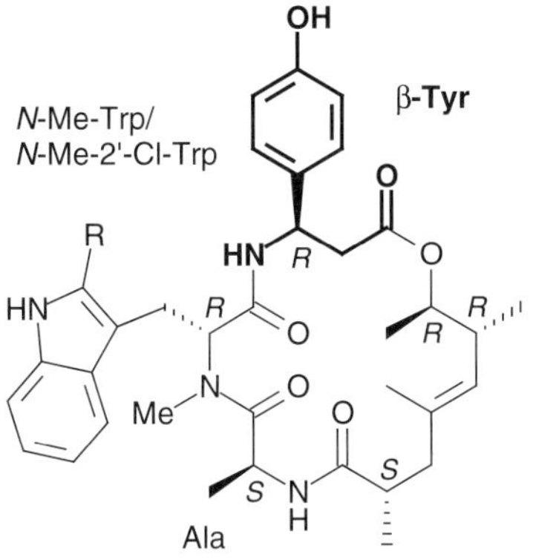

chondramide C (**54**): R = H
chondramide D (**55**): R = Cl

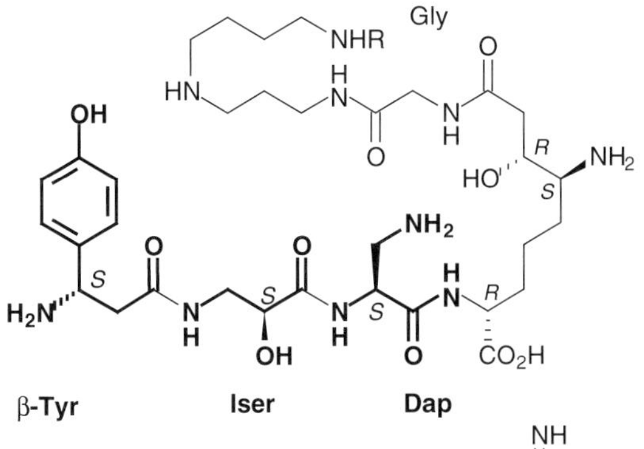

edeine A_1 (**56**): R = H edeine B_1 (**57**): R = $C(=NH)NH_2$

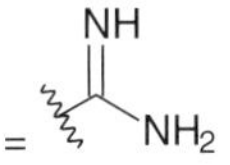

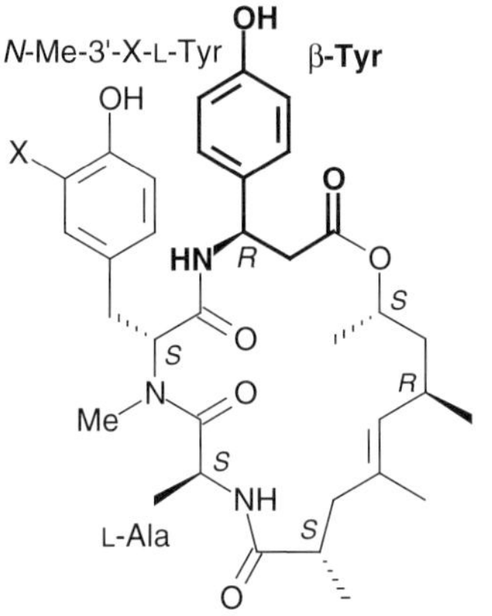

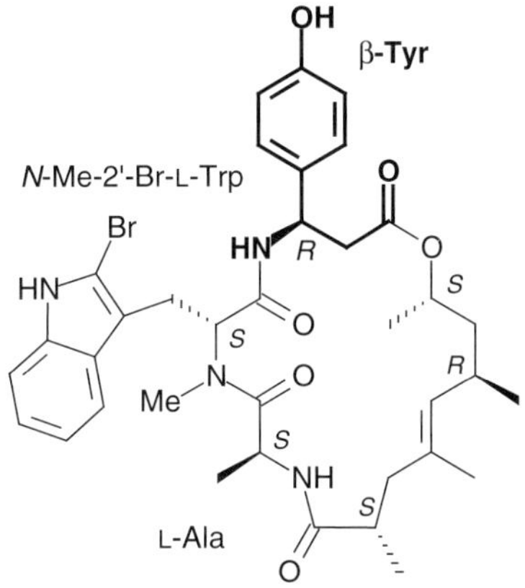

geodiamolide H (**58**): X = I
geodiamolide I (**59**): X = Br

jaspamide (jasplakinolide, **60**)

Figure 4.8 Selected natural products containing a β-tyrosine residue.

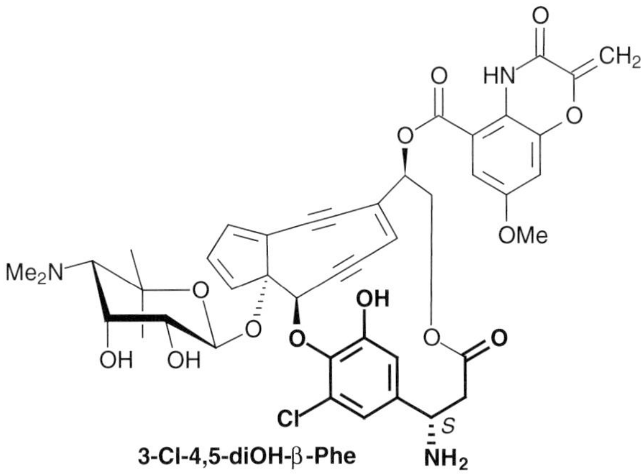

3-Cl-4,5-diOH-β-Phe

Figure 4.9 Chromophore **61** of C-1027.

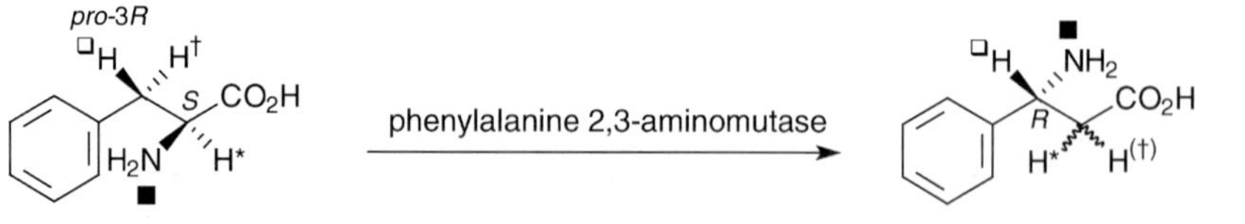

Scheme 4.8 Biosynthesis of (R)-β-phenylalanine (**24**) from L-α-phenylalanine (**62**) in *T. brevifolia*.

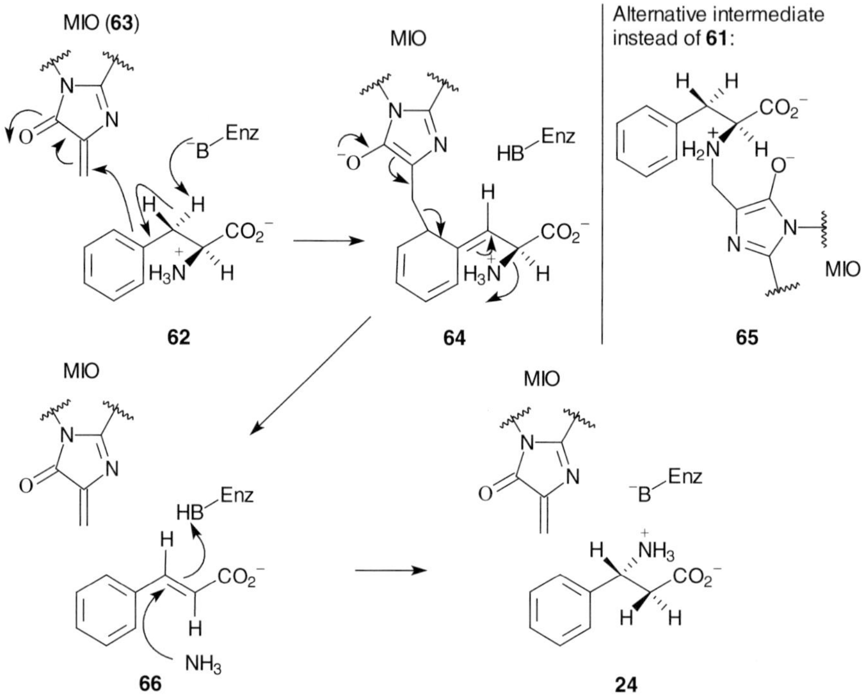

Scheme 4.9 Hypothetical reaction mechanism for the generation of (3R)-β-phenylalanine (**24**) from L-phenylalanine (**62**) by the ammonia lyase-type phenylalanine 2,3-aminomutase from *Taxus* species.

based on sequence homology to ammonia lyase-type aminomutases [7]. The phenylalanine 2,3-aminomutase consisting of 698 amino acid residues shares a common Ala–Ser–Gly motif with other ammonia lyases that autocatalytically forms the active MIO moiety [7, 85]. Experiments with the overexpressed enzyme also demonstrate, that it catalyses the enantiospecific reverse reaction to L-phenylalanine (**62**) both with (3R)- and (3S)-β-phenylalanine. Since (3S)-β-phenylalanine is converted 20 times faster to L-phenylalanine than the corresponding (3R) enantiomer by a combination of reverse and forward reactions, the mutase is able to epimerize (3S)-β-phenylalanine effectively to (3R)-β-phenylalanine (**24**). In the first step of the reaction the electrophilic MIO (**63**) moiety attacks either the *ortho* carbon of the aromatic ring resulting in the intermediate **64** or the amino residue of L-phenylalanine (**62**) resulting in the intermediate **65**, thus enabling the elimination of H^+ and NH_3 from L-phenylalanine and leading to the generation of cinnamic acid (**66**). During the course of the reaction *trans*-cinnamic acid (**66**) was found to be generated as a minor compound, thus confirming an ammonia lyase-type reaction mechanism. Subsequent addition of H^+ and NH_3 finally yields (*R*)-β-phenylalanine (**24**) (Scheme 4.9) [7].

The biosynthesis both of (*S*)-β-tyrosine [(*S*)-**26**] in *Streptomyces globisporus* [84] – producing the antibiotic C-1027 (Figure 4.9) – and of (*R*)-β-tyrosine [(*R*)-**24**] in *C. crocatus* [86] – the producer of the chondramides (Figure 4.8) – is also catalyzed by a 2,3-aminomutase. The identification of the tyrosine 2,3-aminomutase genes *sgcC4* [84] and *cmdF* [86] from the corresponding gene cluster, followed by sequencing, cloning, and heterologous expression of the aminomutase genes in *E. coli* BL-21 revealed that these tyrosine 2,3-aminomutases exhibit significant sequence homologies to a family of ammonia lyases which use MIO (**63**) as cofactor. Consequently, the tyrosine 2,3-aminomutases SgcC4 [84] and CmdF [86] belong to the ammonia lyase-type of aminomutases which also include the phenylalanine 2,3-aminomutase from *Taxus cuspidata* [7]. The mechanism of the mutase reaction of both tyrosine 2,3-aminomutases is likely to be the same as that already discussed for the phenylalanine 2,3-aminomutase from *Taxus* species (Scheme 4.9). Nevertheless, there is an essential difference between both bacterial tyrosine 2,3-aminomutases regarding the facial selectivity [86]. While CmdF generates (*R*)-β-tyrosine [(*R*)-**26**], SgcC4 produces (*S*)-β-tyrosine [(*S*)-**26**] from the common substrate L-tyrosine. As well as the phenylalanine 2,3-aminomutase from *Taxus* species, both tyrosine 2,3-aminomutases seem to possess an additional racemase activity, thus after prolonged incubation times (*S*)-β-tyrosine is converted to (*R*)-β-tyrosine and vice versa [84].

Much less is known about the course of the aminomutase reaction generating the β-tyrosine (**26**) moiety of the edeine antibiotics occurring in *B. brevis* Vm4 [87]. Incubation experiments with the purified tyrosine aminomutase revealed that the reaction requires ATP, but not PLP [87]. Hence, this tyrosine aminomutase differs from all other aminomutases known so far. Feeding experiments with ^{14}C- and ^{3}H-labeled compounds showed that the *pro*-(3*S*) hydrogen in L-tyrosine is lost during the reaction, while the *pro*-(3*R*) hydrogen remains in the molecule [88]. In addition, most of the hydrogen at C-2 and of the nitrogen present in L-tyrosine is lost [88]. However, feeding experiments with ^{14}C- and ^{3}H-labeled compounds are not as reliable as those with ^{13}C- and ^{2}H-labeled compounds, because an incorporation can be analyzed only in the latter case by nuclear magnetic resonance and mass spectrometry. Therefore, further experiments are required to obtain detailed insights into the reaction mechanism of this tyrosine 2,3-aminomutase.

So far, the biosynthesis of (*R*)-β-DOPA (**27**) in the mushroom *Cortinarius violaceus* has only been investigated by feeding experiments [89]. According to these experiments, (*R*)-β-DOPA (**27**) is generated via (*R*)-β-tyrosine from L-tyrosine by a 2,3-aminomutase. Feeding experiments with labeled precursors demonstrate that the amino group is shifted intramolecularly from the (2*S*)- to the (3*S*)-position in the course of the reaction [89]. Nevertheless, it is still unknown which type of aminomutase is present in the fungus, since the cofactor requirement of this fungal tyrosine 2,3-aminomutase has not been investigated so far.

4.2.2.3 β-Glutamate and β-Glutamine

β-Glutamate (**29**) and β-glutamine (**30**) (Figure 4.2) occur as osmolytes in some Archaea; for instance, β-glutamate has been found in *Methanococcus thermolithotrophicus* [90]. Recently, a gene has been identified in the bacterium *Clostridium difficile* that encodes a protein that is analogous to lysine 2,3-aminomutase, but converts L-glutamate to β-glutamate [91]. Therefore, at least this glutamate aminomutase is likely to belong to the SAM-dependent aminomutases.

β-Glutamine (**30**) is present in the halophilic methanogen *Methanococcus portucalensis* as an osmolyte when grown in media containing more than 2 M NaCl [92]. The glutamine synthase of *M. portucalensis* is also capable of accepting β-glutamate as substrate, suggesting that β-glutamine is derived from β-glutamate [92].

4.2.2.4 β-Leucine

A non-cobalamine-dependent 2,3-aminomutase activity converting leucine to (*R*)-β-leucine (**31**) has been detected in the plant *Andrographis paniculata* [93], but no further reliable reports on the occurrence of β-leucine are known so far [94].

4.2.2.5 β-Alanine

The SAM-dependent lysine 2,3-aminomutase from *C. subterminale* SB4 (Section 4.2.2.1) exhibits some cross-reactivity with L-alanine and is therefore able to convert it to β-alanine (**1**). In order to generate an enzyme with an improved alanine 2,3-aminomutase activity, the lysine 2,3-aminomutase from *Porphyromonas gingivalis* was recently genetically engineered by substitution of seven amino acids [25].

4.2.3 Biosynthesis of α,β-Diamino Acids from α-Amino Acids

4.2.3.1 General Biosynthesis of α,β-Diamino Acids

The biosynthesis of 2,3-diamino acids usually starts either from a proteinogenic α-amino-β-hydroxy acid, such as threonine or serine [95], or a proteinogenic α-amino acid is first hydroxylated in the β-position by an oxygenase to an α-amino-β-hydroxy acid [96]. Elimination of water then leads to the corresponding 2,3-didehydro amino acid which suffers a nucleophilic attack by ammonia or the amino group moiety of an amino acid resulting in the generation of a 2,3-diamino acid or a 2,3-diamino acid moiety.

4.2.3.2 Structures and Occurrence of α,β-Diamino Acids in Nature

Examples for 2,3-diamino acids which occur either in their free form or as constituents in natural products are diaminopropanoic acid (Dap, **67**), 2,3-diaminobutanoic acid (2,3-Dab, **68**), capreomycidine (**69**), and streptolidine (**70**) (Figure 4.10).

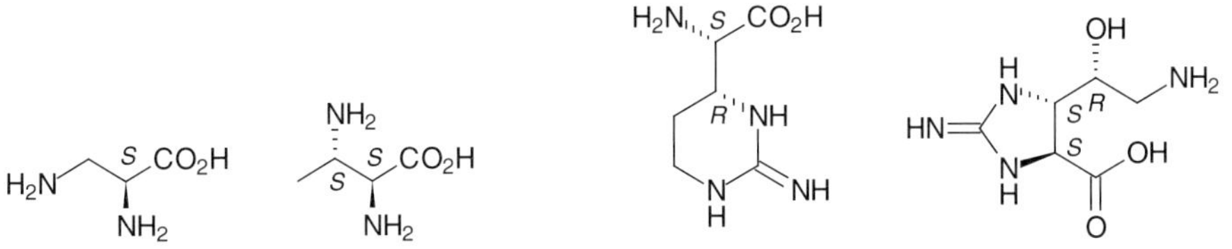

(S)-Dap (**67**) (2S,3S)-2,3-Dab (**68**) (2S,3R)-capreomycidine (**69**) streptolidine (**70**)

Figure 4.10 Examples for α,β-diamino acid residues occurring in nature.

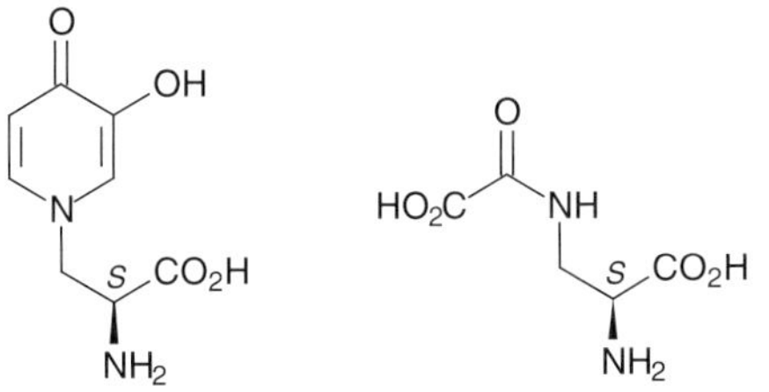

mimosine (**71**) L-β-ODAP (**72**)

Figure 4.11 Selected natural products isolated from plants containing a (S)-Dap (**67**) residue.

Dap (**67**) is the most common 2,3-diamino acid in nature (Figure 4.10). It occurs in its free form and as derivatives [97] such as (*S*)-mimosine (**71**) [98] and β-*N*-oxalyl-L-α,β-diaminopropionic acid (β-ODAP, **72**) [99] in several plants (Figure 4.11). β-ODAP is a neurotoxin [100] present in the leaves and seeds of several *Lathyrus*, *Lens*, and *Pisum* species (Figure 4.11).

In addition, (*S*)-Dap (**67**) is a constituent of the bleomycins (Figure 4.14), which are glycopeptide antibiotics isolated from *Streptomyces verticillus* [101]. It is also present in viomycin (**34**) [40], tuberactinomycin O (**35**) [41], and the capreomycins IA (**36**) and IB (**37**) [42], tuberculostatic cyclic peptide antibiotics from *Streptomyces* species that are structurally closely related to each other (Figure 4.4). (*S*)-Dap (**67**) can also be found in the cyclotheonamides (Figure 4.16), cyclic pentapeptides isolated from a marine sponge of the genus *Theonella* [102], in the edeines A_1 (**56**), B_1 (**57**) (Figure 4.8), D_1, and F_1, antibiotic oligopeptides isolated from *B. brevis* Vm4 [69, 70, 77], and in some keramamides (Figure 4.16), bioactive cyclic peptides isolated from a *Theonella* species [103]. (*S*)-Dap (**67**) also occurs formally as a constituent of the cephalosporins (e.g., **73**) and the penicillins (e.g., **74**) originally isolated from cultures of *Penicillium* species [104] (Figure 4.12).

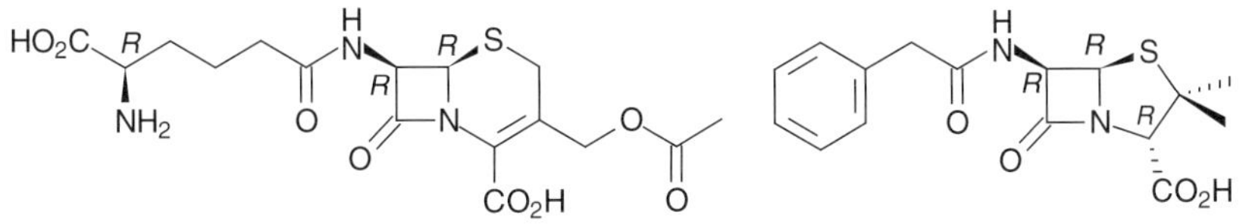

cephalosporin C (**73**) penicillin G (**74**)

Figure 4.12 Selected β-lactam antibiotics.

Like (*S*)-Dap (**67**), 2,3-Dab (**68**) can also be found in nature in its free form. For instance, both the (2*R*,3*S*) isomer and the (2*S*,3*S*) isomer of 2,3-Dab are present in the roots of *Lotus tenuis* [105]. In addition, 2,3-Dab (**68**) serves as a constituent in the aciculitins A (**75**), B (**76**), and C (**77**) (Figure 4.13), bicyclic glycopeptides isolated from the marine sponge *Aciculites orientalis* [106], in the papuamides A (**78**) and B (**79**) (Figure 4.13), cyclic depsipeptides isolated from sponges of the genus *Theonella* [107], and in quinaldopeptin, a cyclopeptide from *Streptoverticillium album* Q132-6 [108].

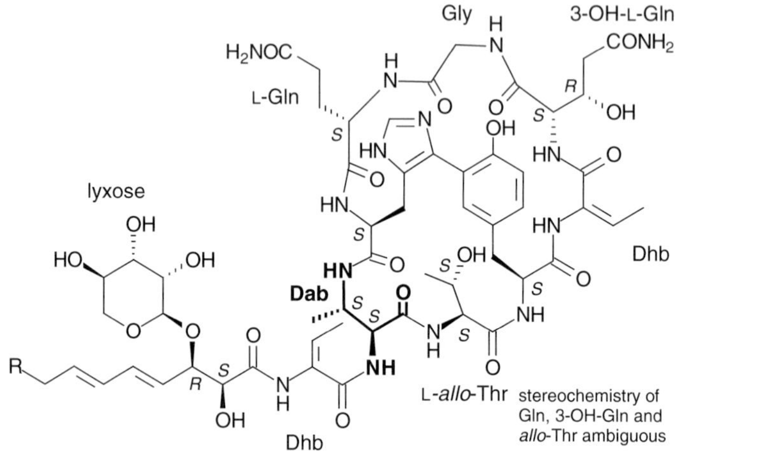

aciculitin A (**75**): R = C_5H_{11} aciculitin B (**76**): R = C_6H_{13} aciculitin C (**77**): R = C_7H_{15}

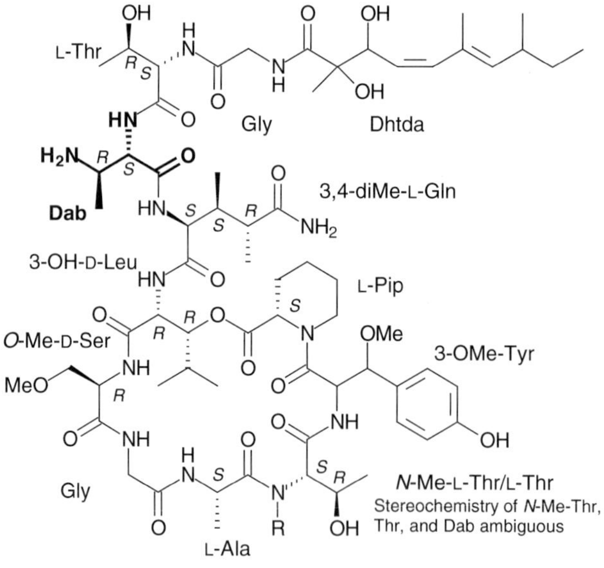

papuamide A (**78**): R = Me papuamide B (**79**): R = H

Figure 4.13 Selected natural products containing a 2,3-Dab (**68**) residue.

Capreomycidine (**69**) is a cyclic guanidino amino acid occurring as a constituent of the tuberculostatic capreomycins [42] and the structurally closely related compounds LL-BM547α, LL-BM547β [46], the tuberactinomycins [41], and viomycin (**34**) [40] (Figure 4.4), while streptolidine (**70**) is present in the streptothricins, for instance in streptothricin F (**33**) [39] (Figure 4.3), broad-spectrum peptidyl nucleoside antibiotics that were isolated from *Streptomyces lavendulae*.

4.2.3.3 Biosynthesis of Selected α,β-Diamino Acids

The biosynthesis of compounds containing a 2,3-Dap (**67**) moiety has been investigated to some extent in the case of β-ODAP (**72**) [95] (Figure 4.11) and of the bleomycins [109] (Figure 4.14). Also, the biosynthesis of the penicillins (Figure 4.12) is well known [104]. In contrast, the biosynthesis of 2,3-Dab (**68**) has not been investigated so far, while the biosynthesis of the (2*S*,3*R*)-capreomycidine (**69**) moiety [110] of viomycin (**34**) and the streptolidine (**70**) moiety [111] have been studied in some detail.

4.2.3.3.1 Biosynthesis of β-ODAP The grass pea (*Lathyrus sativus*) is a drought-resistant legume which is consumed as food in Northern India. However, it contains β-ODAP (**72**), a neurotoxin that can cause the disease neurolathyrism in humans, especially if the seeds are consumed in large quantities [99]. Therefore, the toxin β-ODAP has raised considerable interest and also its biosynthesis has been studied by feeding experiments, for instance with isoxazolin-5-one (**80**) and β-(isoxazolin-5-one-2-yl)-L-alanine (BIA, **81**). The incorporation of *O*-acetyl-[3-^{14}C]serine [112] and ^{14}C-labeled BIA (**81**) [95] revealed that serine is a precursor of the Dap (**67**) moiety in

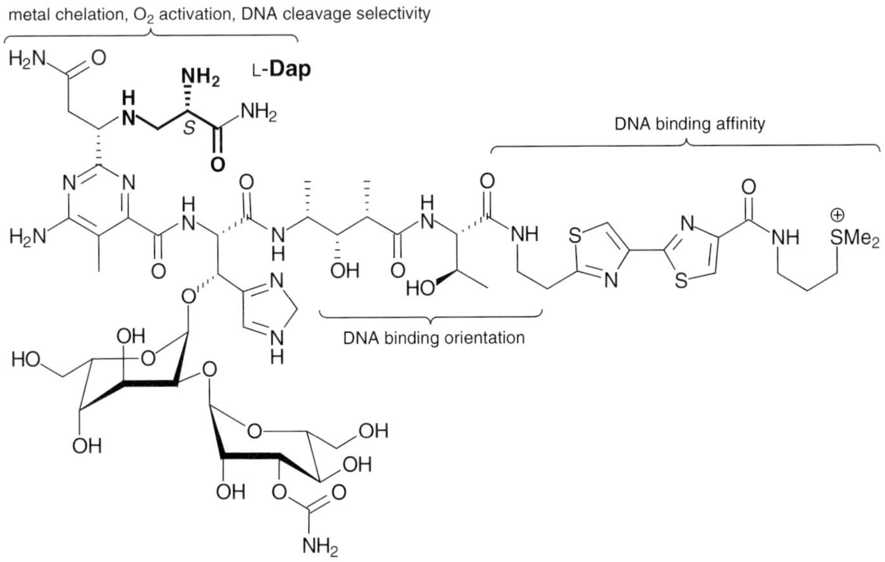

Figure 4.14 Bleomycin A$_2$ (**82**).

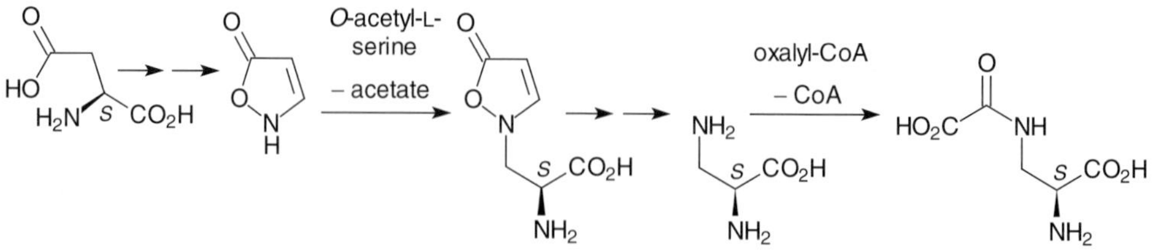

aspartic acid (**13**) **80** L-BIA (**81**) L-Dap (**67**) L-β-ODAP (**72**)

Scheme 4.10 Proposed biosynthesis of L-Dap (**67**) and β-ODAP (**72**) in *Lathyrus sativus*.

β-ODAP and in BIA (**81**), which is a precursor of β-ODAP (**72**). Incubation experiments with *O*-acetylserine and isoxazolin-5-one (**80**) with purified cysteine synthase from *L. sativus* yielded BIA and demonstrated that the Dap moiety in BIA (**81**) and in β-ODAP originates from serine and isoxazolin-5-one (**80**) [112] which might be generated from aspartic acid (**13**) [113] (Scheme 4.10).

4.2.3.3.2 **Biosynthesis of the α,β-Diaminopropanoic Acid Moiety in the Bleomycins**
The bleomycins are glycopeptide antibiotics that have gained widespread interest due to the outstanding antitumor activities of some representatives, such as bleomycin A_2 (**82**) [101] (Figure 4.14). Its biological activity is based on a sequence-specific binding and subsequent cleavage of DNA. The Dap (**67**) moiety is essential for the bioactivity, because it is part of a metal-binding domain, capable of forming a complex with Fe^{2+} ions that activate molecular oxygen and generate hydroperoxyl radicals which initiate the cleavage of the DNA [101].

Some years ago, the gene cluster responsible for the biosynthesis of the bleomycins in *Streptomyces verticillus* ATCC 15 003 was identified, cloned and analyzed [109]. The bleomycin gene cluster turned out to encode both type I PKSs and nonribosomal peptide synthases (NRPSs) (Section 4.2.5.2). The genes *blmVI* and *blmV* of the bleomycin gene cluster encode NRPS that are responsible for the biosynthesis of the Dap (**67**) moiety [109]. In the course of the reaction serine is first transferred to a peptidyl carrier protein (PCP) (Scheme 4.11). Then seryl-PCP (**83**) is dehydrated to yield 2,3-dehydroseryl-PCP (**84**). Nucleophilic attack from the amino group of asparagine at C-3 of the 2,3-dehydroserine derivative **84** yields the dipeptide **85** (Scheme 4.11) [109].

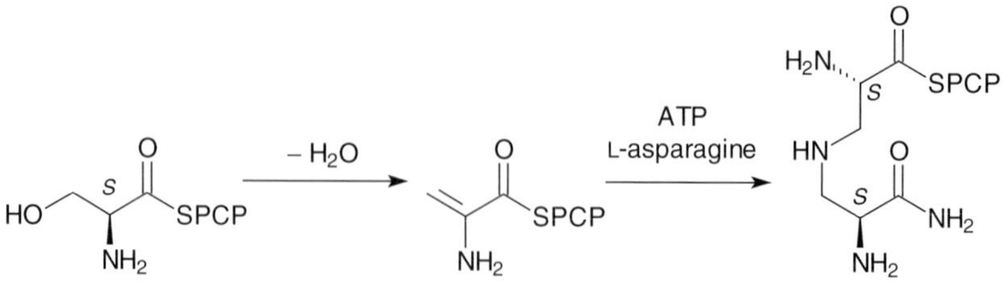

83 **84** **85**

Scheme 4.11 Biosynthesis of the diaminopropanoic acid moiety in bleomycin.

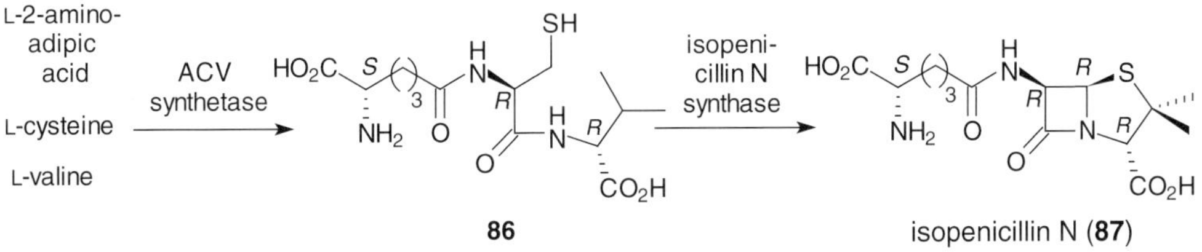

L-2-amino-adipic acid, L-cysteine, L-valine → ACV synthetase → 86 → isopenicillin N synthase → isopenicillin N (87)

Scheme 4.12 Biosynthesis of isopenicillin N (**87**).

4.2.3.3.3 **Biosynthesis of the Penicillins** Formally the penicillins, probably the most prominent of all antibiotics, contain an α,β-diaminopropanoic acid moiety. However, free α,β-diaminopropanoic acid or a derivative thereof is not an intermediate in the penicillin biosynthesis [104]. Instead, the tripeptide **86** – generated from L-2-aminoadipic acid, L-cysteine, and L-valine (**15**) – undergoes an oxidative cyclization reaction catalyzed by the enzyme isopenicillin N synthase yielding **87** which contains a derivatized α,β-diaminopropanoic acid moiety [104] (Scheme 4.12).

4.2.3.3.4 **Biosynthesis of the Capreomycidine Moiety in Viomycine** Viomycin (**34**) [40] (Figure 4.4), the capreomycins [42] and the tuberactinomycins [41] belong to the tuberactinomycin family, a class of cyclic pentapeptide antibiotics which are used for the treatment of multidrug-resistant tuberculosis infections. Viomycin (**34**) and all other related representatives of the tuberactinomycin family contain both a 2,3-Dap (**67**) and either a capreomycidine (**69**) or a hydroxylated capreomycidine moiety (Figure 4.4).

Feeding experiments revealed that arginine (**88**) serves as the precursor of capreomycidine (**69**) and a feeding experiment with [2,3,3,5,5-$^2H_5$]arginine suggested that the generation of capreomycidine (**69**) proceeds via a 2,3-dehydroarginyl intermediate [114]. Some years ago, the gene cluster for the viomycin biosynthesis in *Streptomyces vinaceus* was identified, cloned, and analyzed [115]. The gene cluster contains two genes encoding the enzymes VioC and VioD, which have been overexpressed in *E. coli* and which are responsible for the biosynthesis of capreomycidine (**69**) from L-arginine [96, 110]. VioC is an α-ketoglutarate-dependent nonheme iron dioxygenase that catalyzes the hydroxylation of L-arginine at C-3 leading to the intermediate **89** (Scheme 4.13) [96]. VioD then catalyzes the replacement of the 3-hydroxy group at C-3 with its own guanidino group yielding capreomycidine (**69**). The intermediates **90**, **91**, and **92** are stabilized by the cofactor PLP (Scheme 4.13) [110].

4.2.3.3.5 **Biosynthesis of the Streptolidine Moiety in Streptothricin F** Streptothricin F (**33**) (Figure 4.3) is a broad-spectrum antibiotic isolated originally from *S. lavendulae*. It contains a streptolidine (**70**) and a β-lysine (**21**) residue [44]. According to feeding experiments, the biosynthesis of the steptolidine (**70**) moiety starts from L-arginine (**88**) [116, 117]. Originally, it was proposed that the streptolidine biosynthesis involves capreomycidine (**69**) as an intermediate which

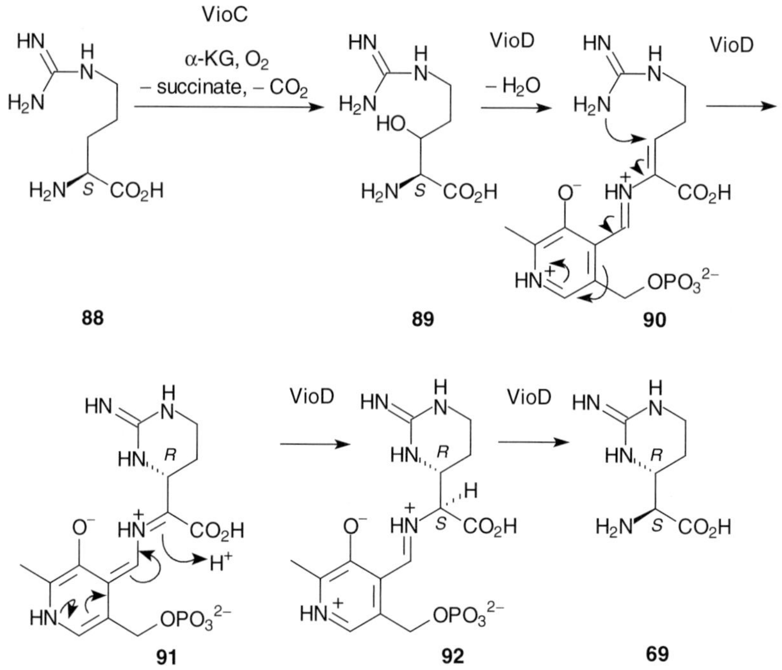

Scheme 4.13 Biosynthesis of capreomycidine (**69**) in *S. vinaceus*.

rearranges to the streptolidine (**70**). However, labeled capreomycidine was not incorporated into streptothricin F (**33**) [111]. Therefore, it remains an open question whether the streptolidine (**70**) moiety is generated via **69** or via **93** and **94** in nature (Scheme 4.14).

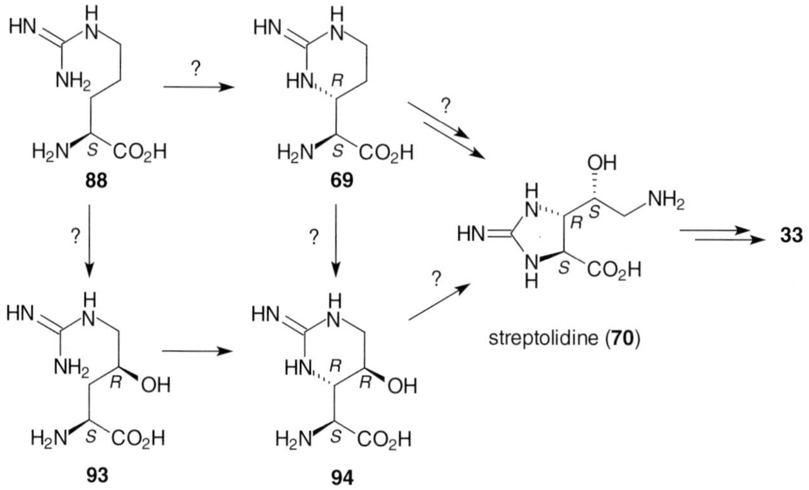

Scheme 4.14 Hypothetical pathways for the biosynthesis of the streptolidine (**70**) residue from L-arginine in *S. lavendulae*.

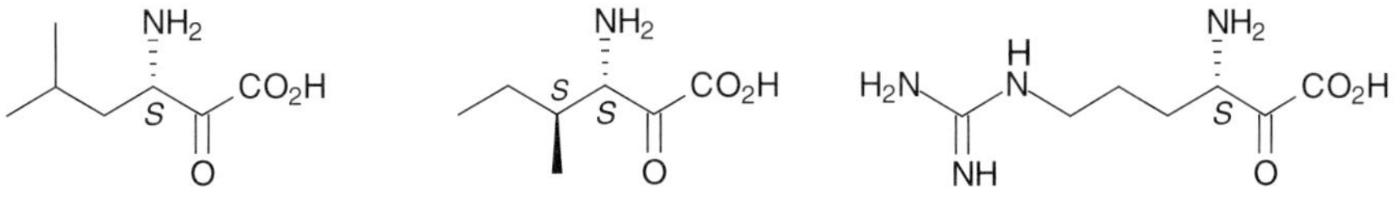

α-keto-homoleucine (**95**) α-keto-homoisoleucine (**96**) α-keto-homoarginine (**97**)

Figure 4.15 α-Keto-β-amino acids.

4.2.4
Biosynthesis of α-Keto-β-Amino Acids from α-Amino Acids

α-Keto-homoleucine (**95**), α-keto-homoisoleucine (**96**), and α-keto-homoarginine (**97**) are known as constituents of natural products (Figure 4.15).

α-Keto-homoleucine (**95**) and α-keto-homoisoleucine (**96**), respectively, are constituents of the keramamides B (**98**), C (**99**) and D (**100**) [103, 118] and of orbiculamide A (**101**) [119], **96** occurs in the keramamides F (**102**) and J (**103**), while α-keto-homoarginine (**97**) is present in the cyclotheonamides A (**104**) and B (**105**) [102] (Figure 4.16).

Although the biosynthesis of α-keto-homoleucine (**95**), α-keto-homoisoleucine (**96**), and α-keto-homoarginine (**97**) has not been investigated yet, it appears reasonable to speculate that all these β-amino acids originate from the corresponding α-amino acids or the corresponding α-keto acids, for instance, **95** might be produced from leucine or 4-methyl-2-oxopentanoic acid. However, the α-keto-β-amino acids are not generated directly by carboxylation with a carboxylase relying on biotin as cofactor, since the carboxyl group in biotin can only act as electrophile, but not as nucleophile. Therefore, a more complicated reaction sequence is required to explain the production of α-keto-β-amino acids from α-amino acids.

4.2.5
De Novo Biosynthesis of β-Amino Acids by PKSs

4.2.5.1 Introduction

Those β-amino acids that are generated by PKSs probably constitute the largest group among all β-amino acids [1]. Usually they are present as constituents of nonribosomal peptides which are produced by NRPSs [120, 121] that are closely related to PKSs [122]. These multienzyme complexes generate peptides, cyclopeptides, and cyclodepsipeptides from monomeric amino acids – such as proteinogenic α-amino acids, all kinds of β-amino acids, γ-amino acids, or D-amino acids – as building blocks in a multistep process similar to an assembly line [120, 123]. While β-amino acids derived from α-amino acids are usually produced by 2,3-aminomutases, the polyketide-type β-amino acids originate from β-oxo acids that are generated by PKSs from acetate and propionate units [122], and which are transaminated to polyketide-type β-amino acids (Scheme 4.15) [124].

Although we often do not know exactly why bacteria produce such a variety of natural products containing β-amino acids, it seems reasonable to assume that they

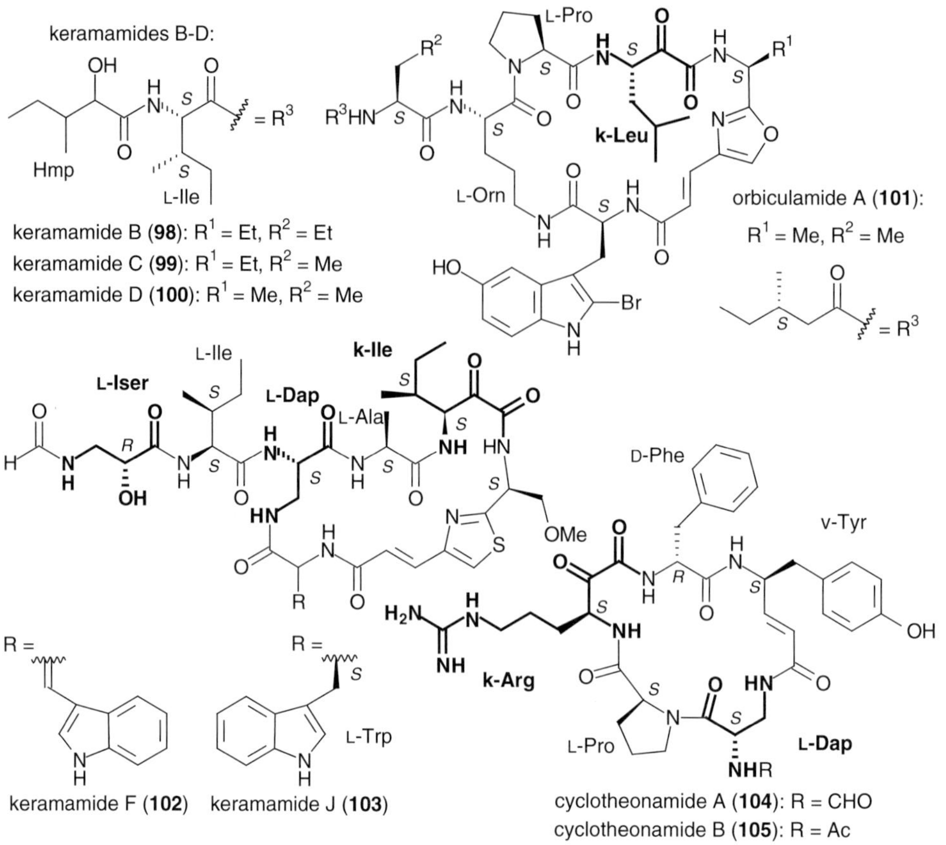

Figure 4.16 Selected natural products containing an α-keto-β-amino acid residue.

use these compounds as a chemical defense against predators and competitors [125,126].

Most of the polyketide-type β-amino acids have only been known for 25 years. This is no accident, because most polyketide-type β-amino acids occur in marine organisms, such as sponges, which were not investigated extensively before [127, 128]. However, the true producers of the polyketide-type β-amino acids are not the sponges themselves, but bacteria present as symbionts in the respective marine organism [129]. In general, both marine and terrestrial bacteria are known to be very skilful in the biosynthesis of bioactive polyketides. Unfortunately, endosymbiotic bacteria usually cannot be cultivated [130]. Therefore, the biosynthesis of most polyketide-type β-amino acids has not been confirmed by feeding experiments or by analysis of the corresponding genes. Instead, most of the β-amino acids can only be assigned to be polyketide-type β-amino acids due to structural characteristics, such as the position of methyl groups, the presence of oxo groups and the position of double bonds which are typical for compounds of polyketide biosynthetic origin [122].

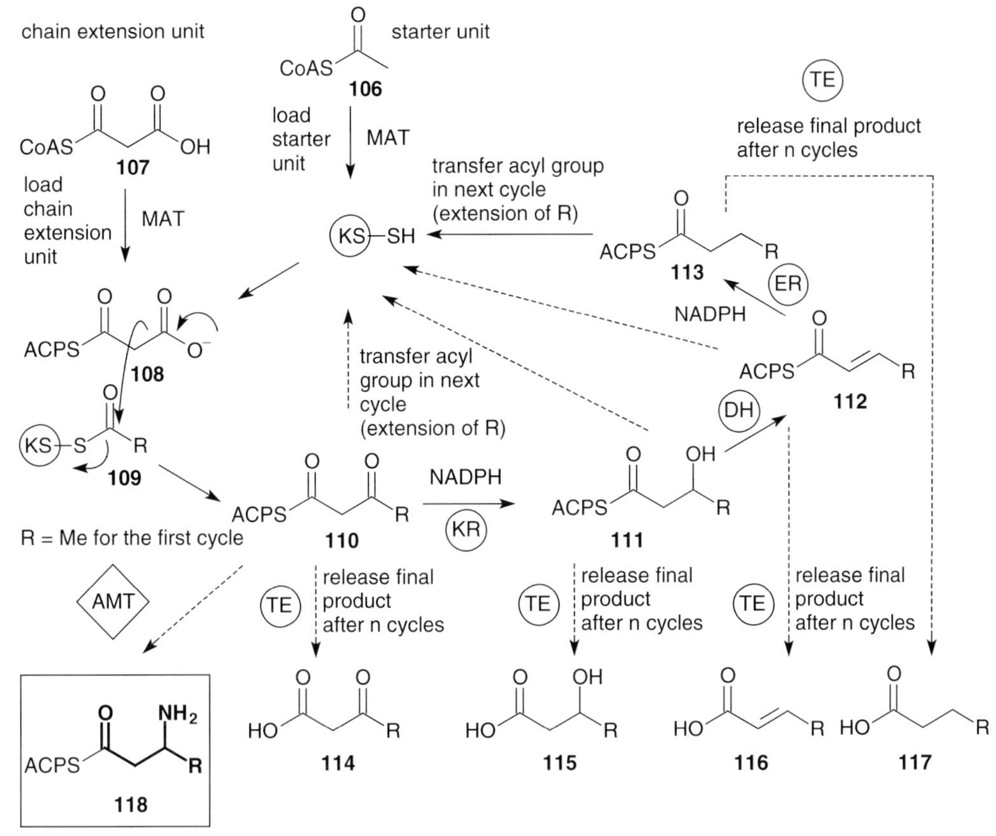

Scheme 4.15 General biosynthesis of polyketide-type β-amino acids **118** and of polyketides **114–117**. MAT: malonyl acetyl transferase; see text for other abbreviations.

4.2.5.2 **General Biosynthesis of Polyketide-Type β-Amino Acids**

The reason for the great structural variety of this class of compounds can be attributed to the way polyketides are synthesized in nature. First the starter unit – usually acetyl-CoA (**106**) is loaded to the ketoacyl synthase (KS) domain and acetate is transformed to malonyl-CoA (**107**) and then loaded to the acyl carrier protein (ACP) yielding malonyl-ACP (**108**). In general, a polyketide is generated by iterative condensation reactions of the polyketide chain with malonyl-ACP (**108**) and **109** leading to a chain elongation of two carbons in each iterative cycle (Scheme 4.15) [122]. One chain elongation cycle starts with the mandatory condensation step catalyzed by a KS leading to a β-keto acid **110**. The following three steps are facultative, since not every PKS domain contains a ketoacyl reductase (KR), a dehydratase (DH), and a enoyl reductase (ER) (Scheme 4.15, dotted arrows directed to KS-SH). The KR reduces the β-keto acid **109** to the corresponding β-hydroxy acid **111**, while the DH is able to convert **111** to the α,β-unsaturated acid **112** and the enoyl reductase hydrogenates the double bond to give **113**. Depending on the presence of a KR, DH, or ER domain at the respective elongation step, the final product can possess oxo groups or hydroxy groups at odd carbon atoms or a double bond between C-2 and C-3 or C-4 and C-5 (and so on).

After the last elongation cycle, the final product is usually released by a thioesterase (TE) from the ACP yielding the polyketide **114**, **115**, **116**, or **117** [122]. However, in the case of polyketide-type β-amino acids, an alternative aminotransferase (AMT) catalyzes the conversion of the β-keto acid **110** to the β-amino acid **118** [124].

Since the reactions which are performed by PKSs can vary with each chain elongation step, a huge variety of different compounds can be generated by PKSs [122]. Instead of acetate units, propionate units are sometimes used for chain elongation reactions leading to the presence of additional methyl groups at carbons with an even atom number [122]. Moreover, additional methyl groups can be introduced by SAM-dependent *C*-, *O*- and *N*-methyl transferases. In addition, a huge variety of starter units are known to induce polyketide synthesis [122]. For instance, apart from acetate and propionate, also aromatic carboxylic acids, such as benzoic acid or phenyl acetic acid, are used.

By these means, a polyketide-type β-amino acid **118** is generated. Usually the β-amino acid **118** is not released by a TE from the enzyme complex. Instead, it is processed further by NRPSs [124]. Thus, other amino acid moieties are condensed with the polyketide-type β-amino acid (Scheme 4.19). In the last step, the peptide is released from the enzyme complex by action of a TE [124]. The above-described biosynthesis of nonribosomal peptides with a β-amino acid moiety has two consequences. First, polyketide-type β-amino acids usually do not occur in free form in nature, since they are processed further and only the final product is released from the enzyme complex. Second, the polyketide-type β-amino acid **118** usually serves as a starter amino acid for a NRPS.

4.2.5.3 Structures and Occurrence of Polyketide-Type β-Amino Acids in Nature

So far, aliphatic polyketide-type β-amino acids, such as 3-amino-2-methylbutanoic acid (Amba, **119**), aminopentanoic acid (Apa, **120**), 3-amino-2-methylpentanoic acid (Map, **121**), 3-amino-2,4-dimethylpentanoic acid (Admpa, **122**), 3-amino-2-methylhexanoic acid (Amha, **123**), 3-amino-7-octynoic acid (Aoya, **124**), 3-amino-2-methyl-7-octynoic acid (Amoa, **125**), and (*R*)-3-aminopalmitic acid (**126**), are only known as constituents of peptides and cyclopeptides (Figure 4.17).

For instance, Amba (**119**) is present in the cyclic pentadepsipeptide dolastatin D (**127**) from the sea hare *Dolabella auricularia* [131] and in the cyclic depsipeptide guineamide B (**128**) from the marine cyanobacterium *Lyngbya majuscula* [132]. Apa (**120**) occurs in the cyclic pentadepsipeptide obyanamide (**129**) from *Lyngbya coniferoides* [133] (Figure 4.18).

Map (**121**) has been found as constituent of the dolastatins 11 (**130**) [134] and 12 (**131**) [134], of guineamide A [132] and majusculamide C (**132**) [135]. Admpa (**122**) is a constituent of dolastatin 16 (**133**) [136] and of homodolastatin 16 (**134**) [137] (Figure 4.19).

Amha (**123**) occurs in depsipeptides such as guineamide D (**135**) [132], kulokekahilide-1 [138], malevamide B [139], and the ulongamides A–F [140]. Aoya (**124**)

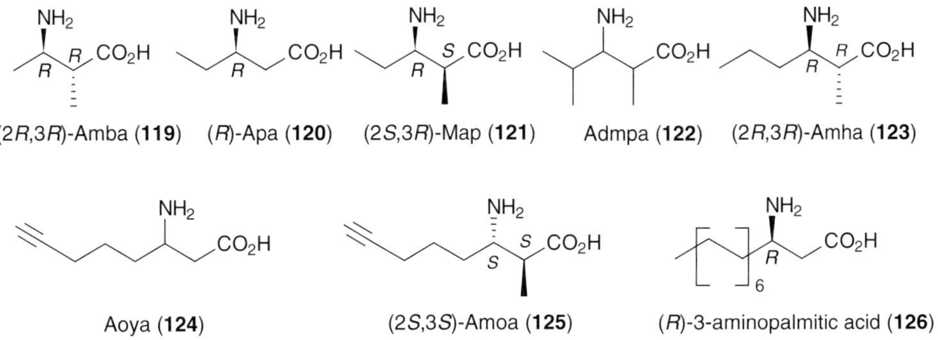
(2R,3R)-Amba (119)
(R)-Apa (120)
(2S,3R)-Map (121)
Admpa (122)
(2R,3R)-Amha (123)
Aoya (124)
(2S,3S)-Amoa (125)
(R)-3-aminopalmitic acid (126)

Figure 4.17 Aliphatic polyketide-type β-amino acids.

has been found as constituent of dolastatin 17 (**136**) [141], and Amoa (**125**) in guineamide C (**137**) [132], malevamide C [139], onchidin [142], and ulongapeptin [143] (Figure 4.20).

Long-chain β-amino acids, such as (*R*)-3-aminopalmitic acid (**126**), are constituents of cyclic octapeptides including bacillomycin F (**138**) [144], the iturins A (**139**) and C (**140**) [145] (Figure 4.21), and mycosubtilin (**141**) [146] isolated from cultures of *B. subtilis*. Each iturin occurs in nature as a mixture of congeners, distinguished in the long-chain β-amino acid moiety. Therefore, for example, several different iturins A (**139**), named as iturin A1, iturin A2, and so on, are known. In addition, long-chain β-amino acids are also constituents of the rhodopeptins, cyclic tetrapeptides from *Rhodococcus* sp. Mer-N1033 [147].

Several aliphatic polyketide-type β-amino acids which contain hydroxy groups, such as (2*S*,3*R*)-3-amino-2-hydroxy-5-methylhexanoic acid (Ahmh, **142**), 3-amino-2-hydroxy-6-methyloctanoic acid (Hamo, **143**), 3-amino-2,5,7,8-tetrahydroxy-10-methylundecanoic acid (Aound, **144**), 3-amino-2-hydroxydecanoic acid (Ahda, **145**),

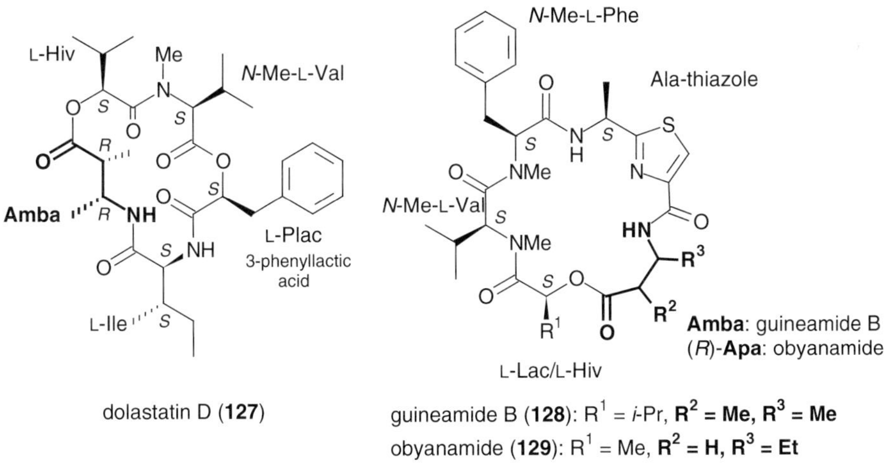
L-Hiv
Me
N-Me-L-Val
Amba
L-Plac
3-phenyllactic acid
L-Ile
dolastatin D (127)
N-Me-L-Phe
Ala-thiazole
N-Me-L-Val
L-Lac/L-Hiv
Amba: guineamide B
(R)-Apa: obyanamide
guineamide B (128): R1 = i-Pr, R2 = Me, R3 = Me
obyanamide (129): R1 = Me, R2 = H, R3 = Et

Figure 4.18 Natural products containing an Amba or an Apa residue.

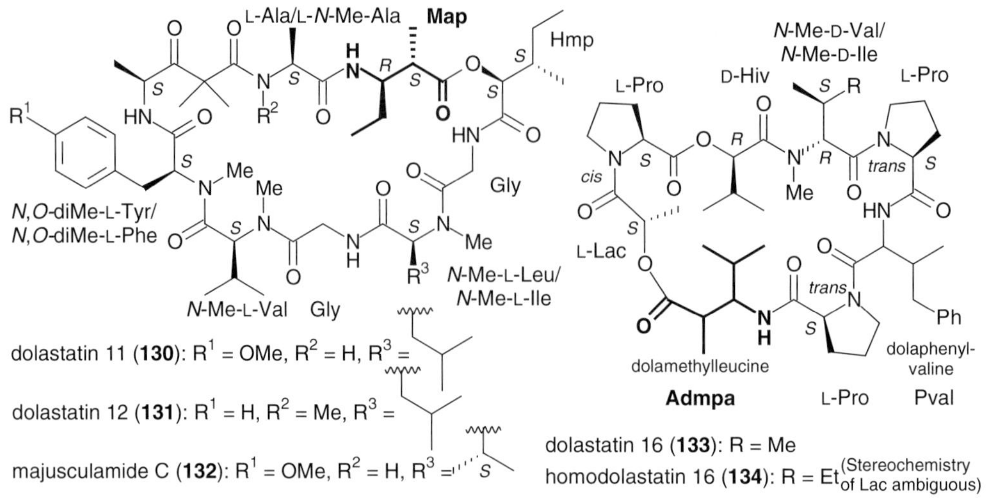

Figure 4.19 Selected natural products containing a Map or Admpa residue.

(2*R*,3*R*,4*S*)-3-amino-2-hydroxy-4-methylpalmitic acid (Hamp, **146**), (2*R*,3*R*,4*S*)-3-amino-14-chloro-2-hydroxy-4-methylpalmitic acid (Champ, **147**), and (2*R*,3*R*,4*S*)-3-amino-14-oxo-2-hydroxy-4-methylstearic acid (Ahmos, **148**), are known as constituents of a variety of peptides (Figure 4.22).

Ahmh (**142**) is present in amastatin (**149**), an aminopeptidase A inhibitor from *Streptomyces* sp. ME98-M3 [148, 149], while Hamo (**143**) occurs in the cyclic octapeptide perthamide B (**150**) isolated from a marine sponge of the genus *Theonella* [150]. Aound (**144**) is a constituent of the cyclic undecapeptide schizotrin A (**151**) [151] and Ahda (**145**) forms part of microginin (**152**) [152, 153] (Figure 4.23).

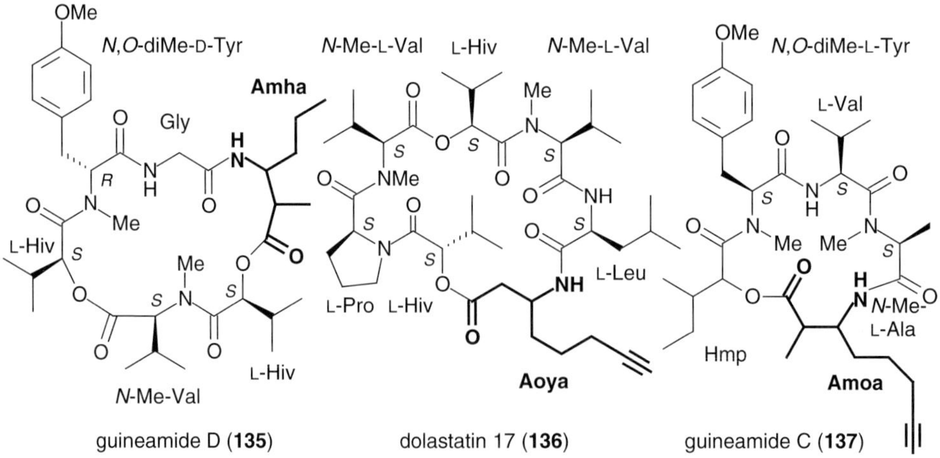

Figure 4.20 Selected natural products containing an Amha, Aoya, or Amoa residue.

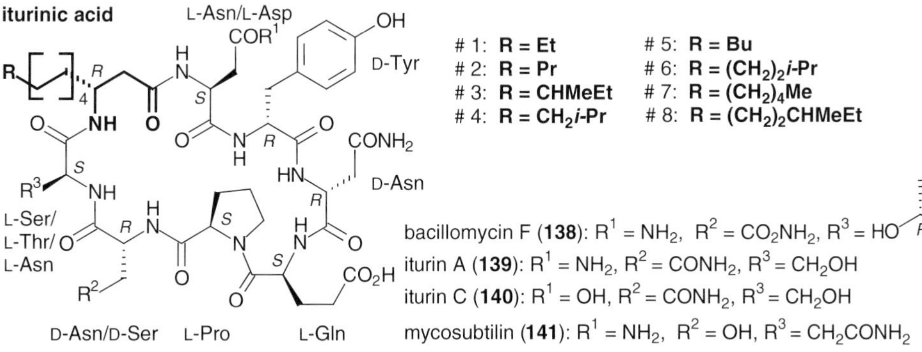

Figure 4.21 Natural products containing long-chain β-amino acids.

The puwainaphycins A (**153**) and B (**154**) contain an Ahmos (**148**) moiety [155], while in the puwainaphycins C (**155**) and D (**156**) the chlorine-containing Champ (**147**) [155] has been found. Hamp (**146**) is present in the cyclic decapeptides calophycin [154] and puwainaphycin E (**157**) [155] (Figure 4.24).

In some natural products polyketide-type aromatic β-amino acids are present, such as (2*S*,3*R*)-3-amino-2-hydroxyphenylbutanoic acid (Ahpa, **158**), (2*S*,3*R*,5*R*)-3-amino-2,5-dihydroxy-8-phenyloctanoic acid (Ahoa, **159**), (3*S*,4*S*,5*E*,7*E*)-3-amino-4-hydroxy-6-methyl-8-phenylocta-5,7-dienoic acid (Apoa/Ahmp, **160**), (3*S*,4*S*,5*E*,7*E*)-3-amino-4-hydroxy-6-methyl-8-(4-bromophenyl)octa-5,7-dienoic acid (Aboa, **161**), (2*S*,3*R*,5*S*)-3-amino-2,5,9-trihydroxy-10-phenyldecanoic acid (Ahpda, **162**), (2*S*,3*S*,8*S*,9*S*)-3-amino-9-methoxy-2,8-dimethyl-10-phenyldeca-4,6-dienoic acid (Adda, **163**), (2*S*,3*R*,4*S*,5*S*,6*S*,11*E*)-3-amino-6-methyl-12-(*p*-methoxyphenyl)-2,4,5-trihydroxydodec-11-enoic acid (Ammtd, **164**), and (2*S*,3*R*,4*S*,5*S*,7*E*)-3-amino-8-phenyl-2,4,5-trihydroxyoct-7-enoic acid (Apto, **165**) (Figure 4.25).

Ahpa (**158**) is a constituent of the dipeptide bestatin (**166**), an aminopeptidase inhibitor isolated from *Streptomyces olivoreticuli* [156]. It is also present in phebestin (**167**), an inhibitor of aminopeptidase N [157], and in probestin (**168**) [158], an inhibitor of aminopeptidase M [159] (Figure 4.26).

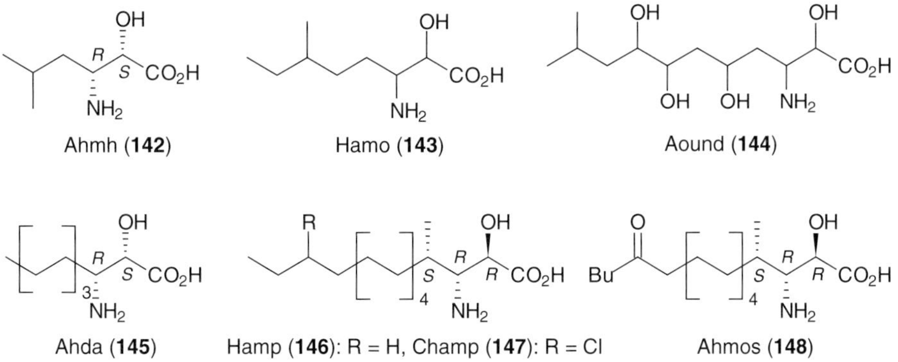

Figure 4.22 Aliphatic polyketide-type β-amino acids containing hydroxy groups.

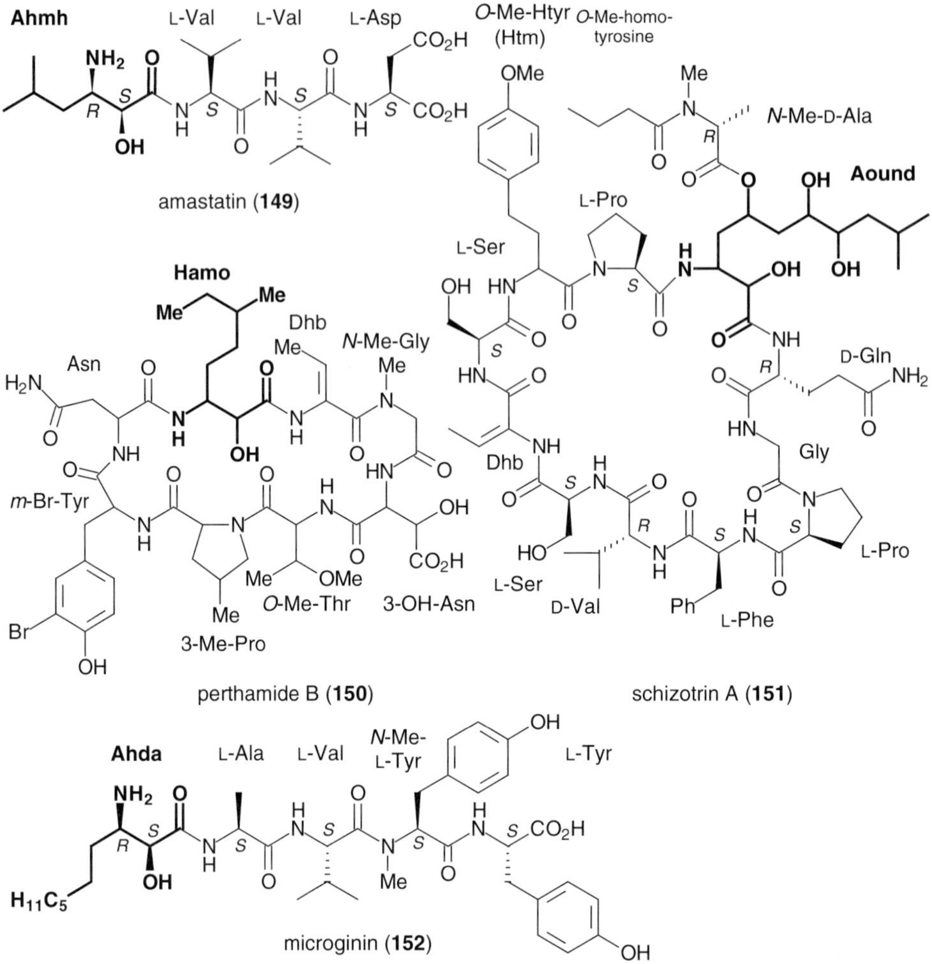

Figure 4.23 Selected natural products containing an Ahmh, Hamo, Aound, or Ahda residue.

Ahoa (**159**) is present in the cyclic heptapeptide nostophycin (**169**) isolated from a cyanobacterium of the genus *Nostoc* [160], while Apoa/Ahmp (**160**) and its *p*-bromophenyl derivative Aboa (**161**) occur in theonegramide [161], the theonellamides A (**170**) and D (**171**) [17, 18], and theopalauamide [162], which are bicyclic dodecapeptides isolated from marine sponges belonging to the genus *Theonella* (Figure 4.27).

Ahpda (**162**) was found in scytonemin A (**172**), a cyclopeptide isolated from the terrestrial cyanobacterium of the genus *Scytonema* [163] (Figure 4.28).

Adda (**163**) is a constituent of deleterious toxins from cyanobacteria such as the microcystins, for instance **173** and **174** [164, 165], motuporin (**175**) [165], and nodularin (**176**) [166] (Figure 4.29).

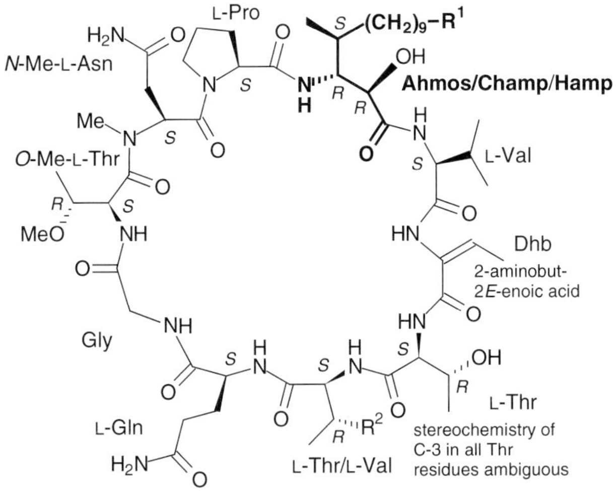

puwainaphycin A (**153**): **R^1 = COBu**, R^2 = OH
puwainaphycin B (**154**): **R^1 = COBu**, R^2 = H
puwainaphycin C (**155**): **R^1 = CHClEt**, R^2 = OH
puwainaphycin D (**156**): **R^1 = CHClEt**, R^2 = H
puwainaphycin E (**157**): **R^1 = Pr**, R^2 = OH

Figure 4.24 Selected natural products containing an Ahmos, Champ, or Hamp residue.

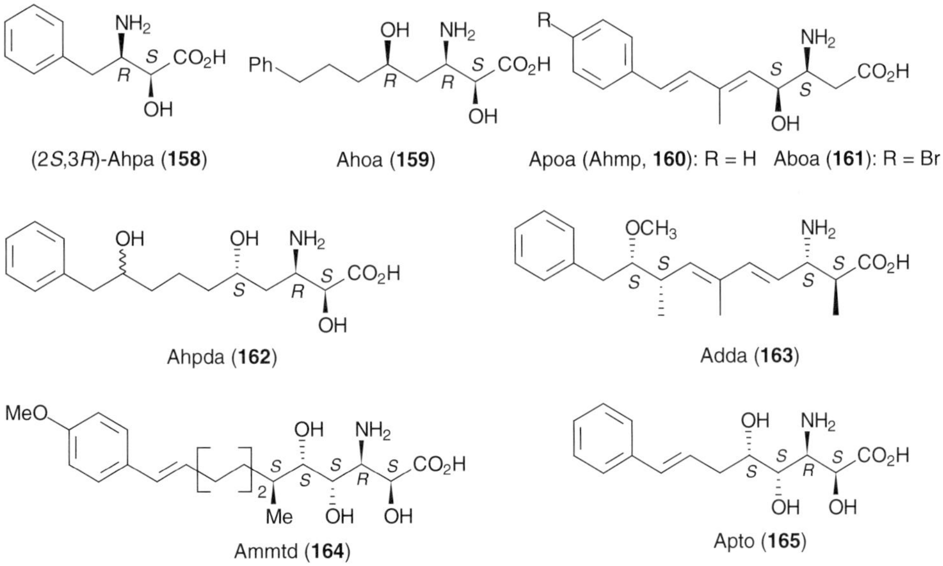

(2*S*,3*R*)-Ahpa (**158**) Ahoa (**159**) Apoa (Ahmp, **160**): R = H Aboa (**161**): R = Br

Ahpda (**162**) Adda (**163**)

Ammtd (**164**) Apto (**165**)

Figure 4.25 Aromatic polyketide-type β-amino acids.

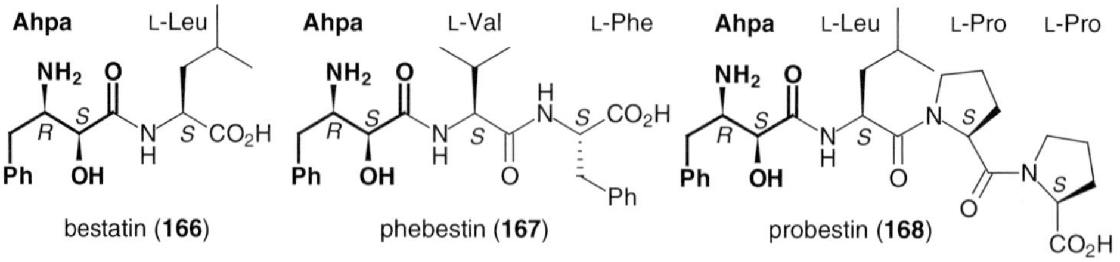

bestatin (**166**) phebestin (**167**) probestin (**168**)

Figure 4.26 Selected natural products containing an Ahpa residue.

A variety of aromatic 3-amino-2,4,5-trihydroxy acids, for instance Ammtd (**164**) and Apto (**165**) are known to be constituents of the microsclerodermins A (**177**), B (**178**), C (**179**), and D (**180**), hexapeptides that occur in marine sponges of the order Lithidista, a *Theonella* species, and a *Microscleroderma* species [167, 168] (Figure 4.30).

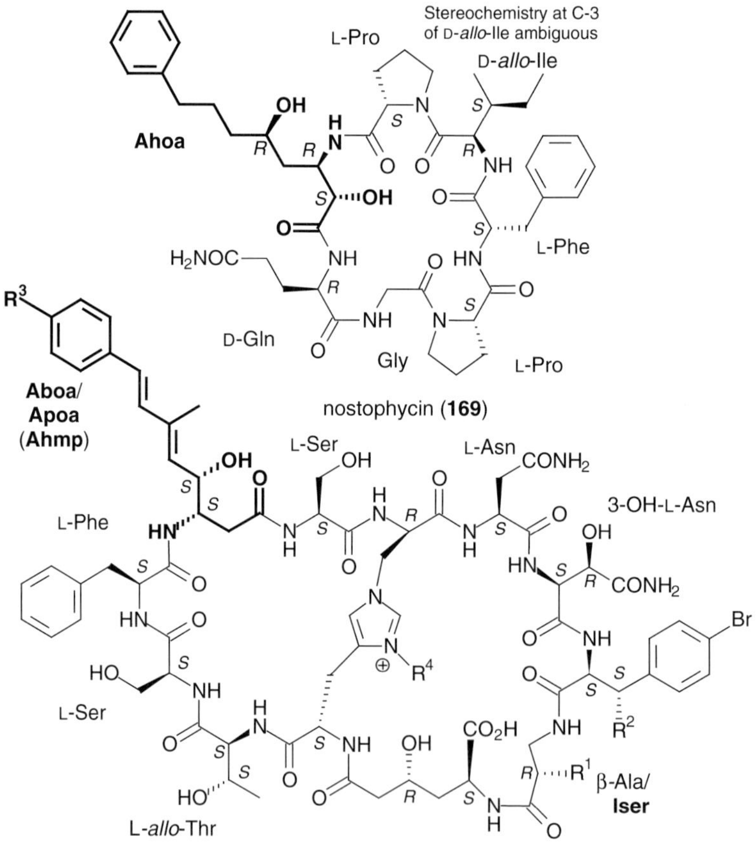

nostophycin (**169**)

theonellamide A (**170**): R^1 = OH, R^2 = Me, $\mathbf{R^3 = H}$, R^4 = β-D-Gal
theonellamide D (**171**): R^1 = H, R^2 = H, $\mathbf{R^3 = Br}$, R^4 = β-D-Ara

Figure 4.27 Selected natural products containing an Ahoa, Apoa, or Aboa residue.

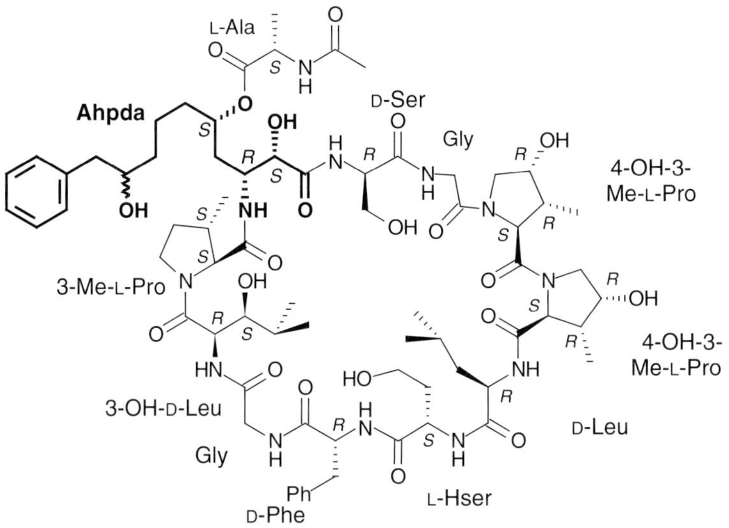

Figure 4.28 Scytonemin A (**172**), a natural product containing an Ahpda residue.

4.2.5.4 Biosynthesis of Selected Polyketide-Type β-Amino Acids

The biosynthesis of polyketide-type β-amino acids has only been studied in a few cases in detail including the biosynthesis of long-chain polyketide-type β-amino acids from the iturins A (**139**) and C (**140**) [169], bacillomycin F (**138**), and mycosubtilin (**141**) [170] (Figure 4.21), the biosynthesis of the Ahda (**162**) moiety in microginin (**152**) [171] (Figure 4.23), the Ahpa (**158**) moiety in bestatin (**166**) [172] (Figure 4.26), and the Adda (**163**) moiety in the microcystins [124] (Figure 4.29).

4.2.5.4.1 Long-Chain β-Amino Acids Occurring as Constituents of the Iturins

The iturins [145], for example, iturin A (**139**) and the closely related antibiotic mycosubtilin

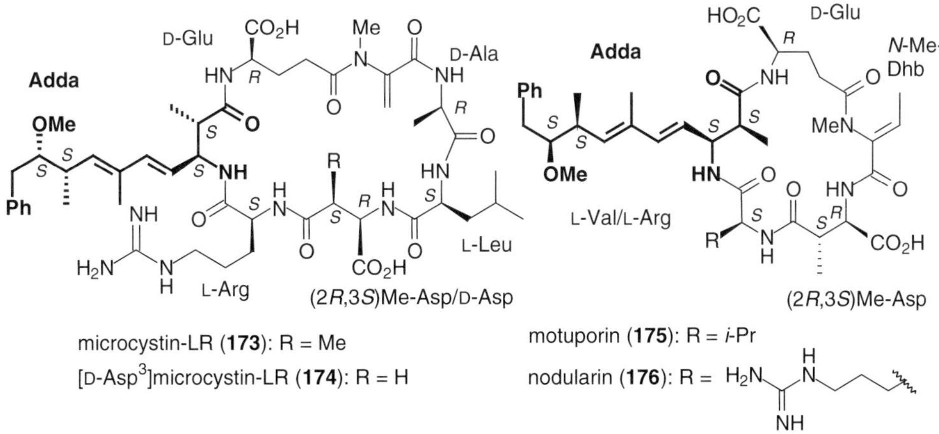

microcystin-LR (**173**): R = Me
[D-Asp³]microcystin-LR (**174**): R = H

motuporin (**175**): R = *i*-Pr
nodularin (**176**): R = H_2N–C(=NH)–NH–$(CH_2)_3$–

Figure 4.29 Selected natural products containing an Adda residue.

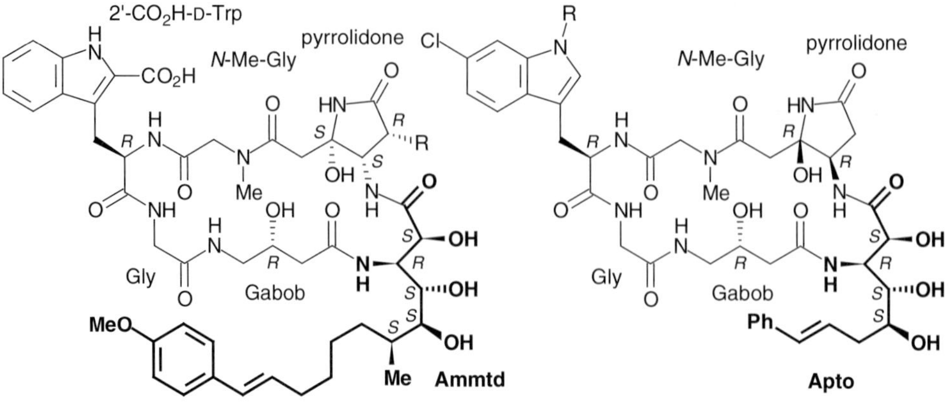

microsclerodermin A (**177**): R = OH
microsclerodermin B (**178**): R = H
microsclerodermin C (**179**): R = $CONH_2$
microsclerodermin D (**180**): R = H

Figure 4.30 Microsclerodermins, natural products containing 3-amino-2,4,5-trihydroxy acid residues.

(**141**) [146] and bacillomycin F (**138**) [144] are antifungal cyclooctapeptides produced by *B. subtilis*, which all contain a long-chain β-amino acid moiety (Figure 4.21).

The gene clusters that encode the biosynthesis of **141** [170] and iturin A (**139**) [169] have been identified, sequenced, and characterized. According to these results, the long-chain β-amino acid moiety is generated by condensation of an activated acid, such as **181**, with malonyl-CoA and transamination of the resulting β-keto thioester **182** by an AMT to give the β-amino thioester **183**, which subsequently undergoes condensation with asparagine to form the dipeptide **184** by an NRPS. In iturin, the biosynthesis of the long-chain β-amino acid and the dipeptide **184** is catalyzed by iturin A, a 449-kDa protein that consists of three modules homologous to fatty acid synthases, AMTs, and peptide synthases (Scheme 4.16) [169].

The proteins iturin B and iturin C are peptide synthases which are responsible for the generation of the final cyclooctapeptide **139** from the dipeptide **184** [169].

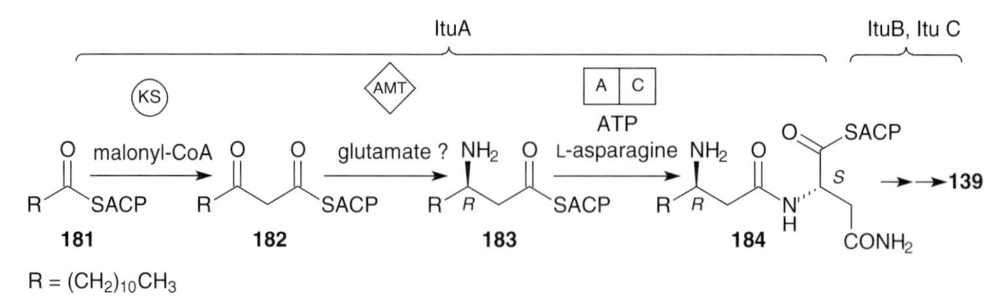

Scheme 4.16 Biosynthesis of the long-chain β-amino acid **183**, the dipeptide **184**, and iturin A (**139**) by ItuA, ItuB and ItuC.

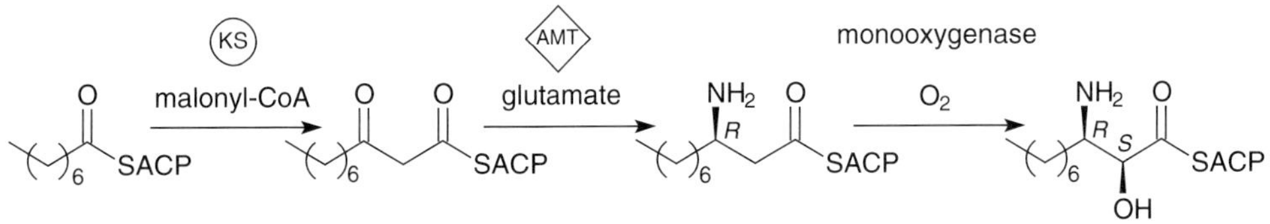

Scheme 4.17 Biosynthesis of the (2*S*,3*R*)-Ahda-ACP (**188**) residue of microginin (**152**).

4.2.5.4.2 **Biosynthesis of the Ahda Moiety in Microginin** The pentapeptide microginin (**152**) containing a (2*S*,3*R*)-Ahda (**145**) moiety has been isolated from the cyanobacterium *Microcystis aeruginosa* NIES-100 (Figure 4.23) [152]. It is an inhibitor of the angiotensin-converting enzyme [152].

Recently, the gene cluster responsible for the biosynthesis of microginin (**152**) has been identified, sequenced, and analyzed [171]. According to this analysis, the (2*S*,3*R*)-Ahda (**145**) residue is generated starting from octanoic acid. An adenylation domain activates octanoic acid as an acyl adenylate which is then transferred to an acyl carrier protein (Scheme 4.17). An elongation module of the PKS consisting of a KS, an acyl transferase (AT), and an ACP is responsible for the condensation of a malonyl-CoA residue with the activated octanoic acid **185** to the 3-oxodecanoic acid **186**. The 3-oxodecanoic acid **186** is then transaminated by an AMT yielding the β-amino acid **187**. A monooxygenase catalyses the hydroxylation of the β-amino thioester **187** at C-2 leading to (2*S*,3*R*)-Ahda-ACP (**188**). The alanine, valine, and the two tyrosine residues are added to the (2*S*,3*R*)-Ahda moiety by action of elongation modules of NRPSs [171].

4.2.5.4.3 **Biosynthesis of the Ahpa Residue in Bestatin** Bestatin (**166**) is a dipeptide consisting of (2*S*,3*R*)-Ahpa (**158**) and L-leucine isolated from *Streptomyces olivoreticuli*. It acts as a potent inhibitor of several peptidases [156]. The biosynthesis of bestatin (**166**) has been studied by feeding experiments [172]. These experiments revealed that the (2*S*,3*R*)-Ahpa (**158**) moiety is probably generated from the activated phenylacetic acid **189** as starter unit and from one malonyl-CoA moiety via **190** and **191**, since acetic acid and phenylalanine – which is likely to be transformed to phenylacetic acid – is incorporated into bestatin as well as L-leucine (Scheme 4.18).

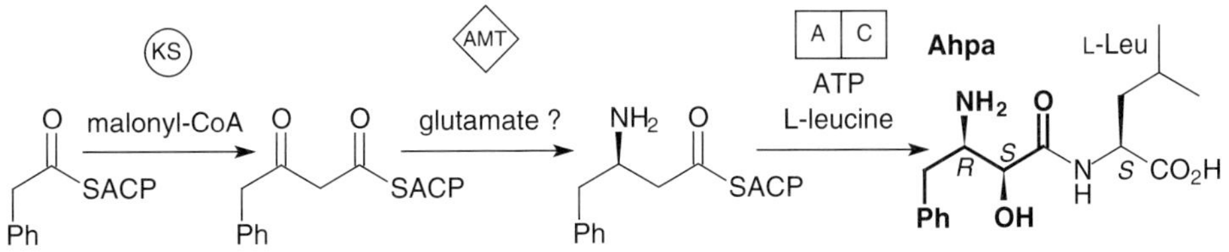

Scheme 4.18 Hypothetical biosynthesis of (2*S*,3*R*)-Ahpa-ACP (**191**) and bestatin (**166**).

4.2.5.4.4 **Biosynthesis of the Adda Residue in the Microcystins** Blooms of toxic cyanobacteria constitute a significant threat to human health. The most important toxins produced by these blooms are the microcystins [164, 165] (Figure 4.29), known from species such as *Microcystis aeruginosa*. Therefore, these compounds became objects of intensive investigations and the biosynthesis of these important compounds has been studied in detail both by feeding experiments [173] and by characterization of the biosynthetic gene cluster [124] from *M. aeruginosa* PCC7806. All microcystins, such as microcystin-LR (**173**) [174], are cyclic heptapeptides containing the polyketide-type β-amino acid Adda (**163**) (Figure 4.29). The Adda moiety is also present in motuporin (**175**) [165] and nodularin [166], cyclic pentapeptides closely related to the microcystins (Figure 4.29).

The analysis of the microcystin biosynthesis gene cluster revealed that the operon *mcyD–J* encodes PKS–NRPS modules which catalyze the formation of the β-amino acid Adda (**163**) and its linkage to D-glutamate, while a second smaller operon consisting of *mcyA–C* encodes the NRPS modules for the extension of the dipeptide to the heptapeptide including the cyclization of the heptapeptide to the final product [124]. As shown in Scheme 4.19, the biosynthesis of Adda (**163**) is initiated by McyG, a 294-kDa polypeptide of mixed NRPS and PKS function, consisting of a NRPS adenylation domain which activates phenyl acetate and a malonyl-specific KS, AT, C-methyl transferase (CM), DH, KR, and ACP domain. Thus, starting from **192** the biosynthesis leads via the diketide **193**, the triketide **194**, the tetraketide **195**, and the activated Adda residue **196** to the dipeptide **197** [124]. McyJ serves as *O*-methyl transferase, while McyD is a large 435-kDa polypeptide consisting of two type I PKS modules, each containing a KS, AT, KR, DH, and ACP domain, which generates the tetraketide **195**. In addition, the N-terminal module contains a putative CM domain. McyE is a 393-kDa polypeptide of mixed NRPS and PKS function. The N-terminal region of this protein is a PKS module consisting of KS, AT, CM, ACP, and AMT domains. This PKS module is linked to a NRPS module, consisting of two condensation, an adenylation, and a thiolation domain(s). The first condensation domain is apparently responsible for the formation of the peptide bond between Adda-ACP (**196**) and D-glutamate, which is probably generated by McyF, a putative racemase, from L-glutamate. The dipeptide **197** is then transformed to the final cycloheptapeptide **173** by the NRPS McyA–C [124].

4.2.6
β-Amino Acids Whose Biosynthesis is Still Unknown

Despite the fact that the biosynthesis of most β-amino acids has either been elucidated or can be deduced from their structures, there are some β-amino acids whose biosynthesis is still unknown (Figure 4.31). For instance, this applies to (*S*)-β-homolysine (**198**), cispentacin (**199**), and 2-pyrrolidinylacetic acid (**200**) occurring in plants such as *Nicotiana tabaccum* [175]. (*S*)-β-Homolysine is a constituent of resormycin, a herbicidal and antifungal tripeptide isolated from *Streptomyces platensis* MJ953-SF5 [176]. Cispentacin has been found in free form in *Bacillus cereus* L450-B2 and exhibits antifungal activity [177]. It is also a constituent of the

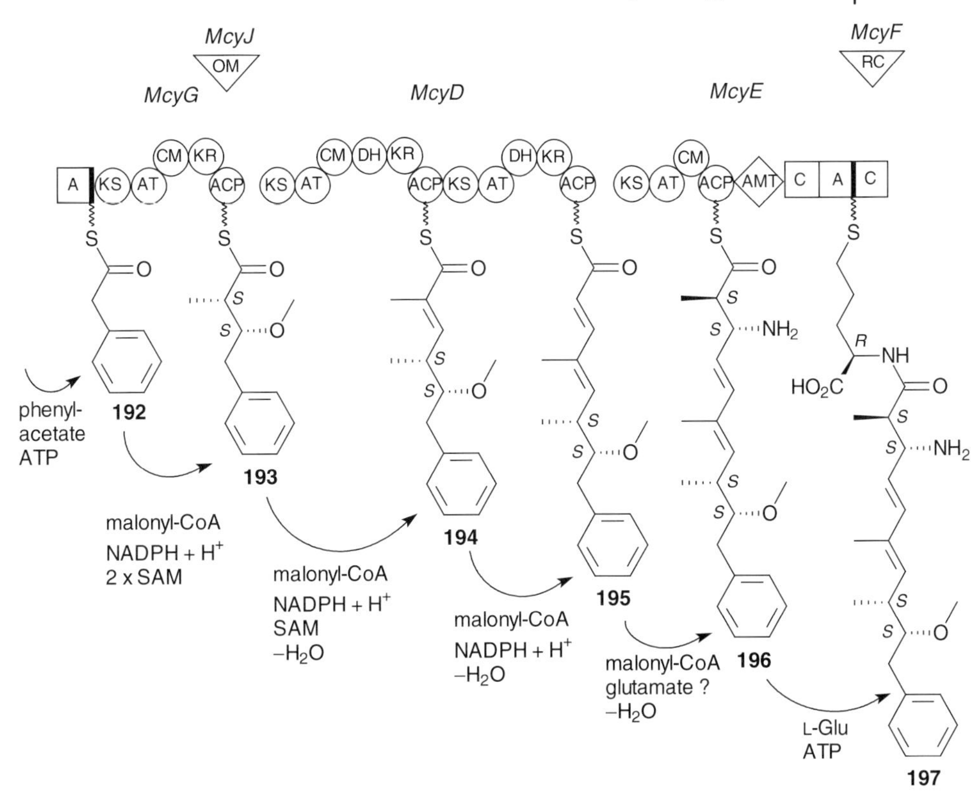

Scheme 4.19 Model for the biosynthesis of Adda-ACP (**196**) in the microcystins. Each circle represents a PKS and each rectangle a NRPS domain. A: aminoacyl adenylation; C: condensation; OM: *O*-methyl transferase; RC: racemase; see text for other abbreviations.

nucleoside antibiotic amipurimycin isolated from *Streptomyces novoguineensis* T-36496 [178].

It can be speculated that (*S*)-β-homolysine (**198**) either is generated by a *de novo* synthesis following the polyketide biosynthetic pathway or by a biosynthesis starting from L-lysine. 2-Pyrrolidinylacetic acid (**200**) might be generated from β-lysine (**21**) by formal loss of ammonia.

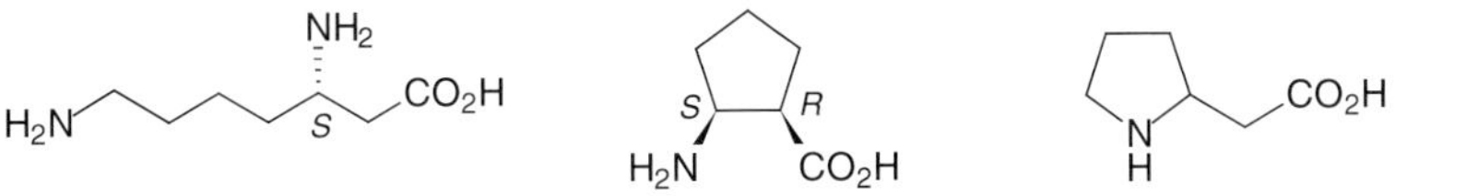

(*S*)-β-homolysine (**198**) cispentacin (**199**) 2-pyrrolidinylacetic acid (**200**)

Figure 4.31 Selected β-amino acids whose biosynthesis is unknown.

4.3 Conclusions and Future Prospects

The elucidation of many important biosynthetic pathways leading to β-amino acids, in combination with the knowledge on the occurrence of certain β-amino acids in nature, allows the conclusion that the capabilities to synthesize β-amino acids are not equally distributed among animals, plants, fungi, bacteria, and Protista. Animals are obviously only able to synthesize β-alanine (**1**) and β-aminoisobutyric acid (**2**) (Figure 4.1). Plants and fungi are capable of producing β-amino acids directly from proteinogenic α-amino acids by the action of aminomutases; for instance, a phenylalanine 2,3-aminomutase is involved in the biosynthesis of the phenylisoserine (**25**) moiety of taxol (**52**) [7] (Figure 4.7). In addition, plants are able to generate α,β-diamino acids, such as β-ODAP (**72**) [95] (Figure 4.11). Like animals, plants, and fungi, bacteria are able to produce β-amino acids from α-amino acids. However, bacteria also are able to generate polyketide-type β-amino acids by *de novo* biosynthesis from acetate and propionate units [1]. Consequently, bacteria are the most promising sources for the detection of new β-amino acids.

It can be expected that the biosynthesis of many more β-amino acids will be elucidated within the next decade based on an improved knowledge of the biosynthesis of peptides which contain polyketide-type β-amino acids. Even if many of the bacteria producing β-amino acids cannot be cultivated, since they rely on a symbiotic relationship with marine organisms, the biosynthesis of their β-amino acids can be elucidated by expression of their genes in a foreign host [130]. By these means, it will also be possible to detect genes that encode enzymes capable of producing free β-amino acids or compounds containing β-amino acids that have not been directly isolated from the corresponding source organism.

Since several of the most potent natural products known today, such as the bleomycins [101] (Figure 4.14) and the penicillins [104] (Figure 4.12), contain a β-amino acid moiety, the introduction of β-amino acid moieties will remain an important tool of medicinal chemistry in drug development. Moreover, β-amino acids play a decisive role in the optimization of the properties of peptidomimetics [2]. In addition to organic syntheses of β-amino acids, the rapidly growing knowledge of the biosynthesis of β-amino acids will allow production of many β-amino acids. For instance, if the genes responsible for the biosynthesis of a certain β-amino acid are known, it can be overexpressed in fast growing easy to handle organisms, such as *E. coli*. Moreover, by genetic engineering of genes responsible for the biosynthesis of β-amino acids and peptides containing β-amino acid moieties, new unnatural β-amino acids and peptides will be effectively generated in the future [179]. Thus, genetic engineering also could play an important role in the search for and the development of new lead structures. First experiments to modify the specificity of enzymes have already been reported. For instance, lysine 2,3-aminomutase was modified by genetic engineering to accept alanine as substrate, thus producing β-alanine (**1**) instead of β-lysine (**21**) [25].

References

1 Spiteller, P. and von Nussbaum, F. (2005) β-Amino acids in natural products, in *Enantioselective Synthesis of β-Amino Acids* (eds V. Soloshonok and E. Juaristi), Wiley-VCH Verlag GmbH, Wienheim, pp. 19–91.

2 Steer, D.L., Lew, R.A., Perlmutter, P., Smith, A.I., and Aguilar, M.-I. (2002) *Current Medicinal Chemistry*, **9**, 811–822.

3 Soloshonok, V. and Juaristi, E. (eds) (2005) *Enantioselective Synthesis of β-Amino Acids*, Wiley-VCH Verlag GmbH, Wienheim.

4 Spiteller, P. and von Nussbaum, F. (2004) Biosynthesis of β-amino acids, in *Highlights in Bioorganic Chemistry: Methods and Applications* (eds C. Schmuck and H. Wennemers), Wiley-VCH Verlag GmbH, Wienheim, pp. 90–106.

5 Cardillo, G. and Tomasini, C. (1996) *Chemical Society Reviews*, **25**, 117–128.

6 von Nussbaum, F. and Spiteller, P. (2004) β-Amino acids in nature, in *Highlights in Bioorganic Chemistry: Methods and Applications* (eds C. Schmuck and H. Wennemers), Wiley-VCH Verlag GmbH, Wienheim, pp. 63–89.

7 Walker, K.D., Klettke, K., Akiyama, T., and Croteau, R. (2004) *The Journal of Biological Chemistry*, **279**, 53947–53954.

8 Griffith, O.W. (1986) *Annual Review of Biochemistry*, **55**, 855–878.

9 Rathinasabapathi, B., Sigua, C., Ho, J., and Gage, D.A. (2000) *Physiologia Plantarum*, **109**, 225–231.

10 Hanson, A.D., Rathinasabapathi, B., Rivoal, J., Burnet, M., Dillon, M.O., and Gage, D.A. (1994) *Proceedings of the National Academy of Sciences of the United States of America*, **91**, 306–310.

11 Golakoti, T., Ogino, J., Heltzel, C.E., Le Husebo, T., Jensen, C.M., Larsen, L.K., Patterson, G.M.L., Moore, R.E., Mooberry, S.L., Corbett, T.H., and Valeriote, F.A. (1995) *Journal of the American Chemical Society*, **117**, 12030–12049.

12 Tius, M.A. (2002) *Tetrahedron*, **58**, 4343–4367.

13 Pedras, M.S.C., Zaharia, L.I., and Ward, D.E. (2002) *Phytochemistry*, **59**, 579–596.

14 Hamano, K., Kinoshita, M., Tanzawa, K., Yoda, K., Ohki, Y., Nakamura, T., and Kinoshita, T. (1992) *The Journal of Antibiotics*, **45**, 906–913.

15 Hamano, K., Kinoshita, M., Furuya, K., Miyamoto, M., Takamatsu, Y., Hemmi, A., and Tanzawa, K. (1992) *The Journal of Antibiotics*, **45**, 899–905.

16 Mori, Y., Tsuboi, M., Suzuki, M., Fukushima, K., and Arai, T. (1982) *The Journal of Antibiotics*, **35**, 543–544.

17 Matsunaga, S. and Fusetani, N. (1995) *The Journal of Organic Chemistry*, **60**, 1177–1181.

18 Matsunaga, S., Fusetani, N., Hashimoto, K., and Wälchli, M. (1989) *Journal of the American Chemical Society*, **111**, 2582–2588.

19 Kitagawa, I., Lee, N.K., Kobayashi, M., and Shibuya, H. (1991) *Tetrahedron*, **47**, 2169–2180.

20 Sitachitta, N., Williamson, R.T., and Gerwick, W.H. (2000) *Journal of Natural Products*, **63**, 197–200.

21 Xu, Z., Peng, Y., and Ye, T. (2003) *Organic Letters*, **5**, 2821–2824.

22 White, W.H., Gunyuzlu, P.L., and Toyn, J.H. (2001) *The Journal of Biological Chemistry*, **276**, 10794–10800.

23 Williamson, J.M. and Brown, G.M. (1979) *The Journal of Biological Chemistry*, **254**, 8074–8082.

24 Ramjee, M.K., Genschel, U., Abell, C., and Smith, A.G. (1997) *The Biochemical Journal*, **323**, 661–669.

25 Balatskaya, S., Borup, B., Chatterjee, R., Dhawan, I., Mitchell, K.W., Mundorff, E., Partridge, L., Tobin, M., and Fox, R.J. (2007) WO 2007047773.

26 Larsen, P.O. (1965) *Acta Chemica Scandinavica*, **19**, 1071–1078.

27 Guillou, Y. and Robin, Y. (1973) *The Journal of Biological Chemistry*, **248**, 5668–5672.

28 Eggen, M.J. and Georg, G.I. (2002) *Medicinal Research Reviews*, **22**, 85–101.

29 Magarvey, N.A., Beck, Z.Q., Golakoti, T., Ding, Y., Huber, U., Hemscheidt, T.K., Abelson, D., Moore, R.E., and Sherman, D.H. (2006) *ACS Chemical Biology*, **1**, 766–779.

30 Beck, Z.Q., Burr, D.A., and Sherman, D.H. (2007) *ChemBioChem*, **8**, 1373–1375.

31 Bray, R.C. and Stadtman, T.C. (1968) *The Journal of Biological Chemistry*, **243**, 381–385.

32 Costilow, R.N., Rochovansky, O.M., and Barker, H.A. (1966) *The Journal of Biological Chemistry*, **241**, 1573–1580.

33 Chirpich, T.P., Zappia, V., Costilow, R.N., and Barker, H.A. (1970) *The Journal of Biological Chemistry*, **245**, 1778–1789.

34 Tsai, L. and Stadtman, T.C. (1968) *Archives of Biochemistry and Biophysics*, **125**, 210–225.

35 Mutatu, W., Klettke, K.L., Foster, C., and Walker, K.D. (2007) *Biochemistry*, **46**, 9785–9794.

36 Baker, J.J., van der Drift, C., and Stadtman, T.C. (1973) *Biochemistry*, **12**, 1054–1063.

37 Sowers, K.R., Robertson, D.E., Noll, D., Gunsalus, R.P., and Roberts, M.F. (1990) *Proceedings of the National Academy of Sciences of the United States of America*, **87**, 9083–9087.

38 Ohba, K., Nakayama, H., Furihata, K., Furihata, K., Shimazu, A., Seto, H., Ōtake, N., Yang, Z.-Z., Xu, L.-S., and Xu, W.-S. (1986) *The Journal of Antibiotics*, **39**, 872–875.

39 Khokhlov, A.S. and Shutova, K.I. (1972) *The Journal of Antibiotics*, **25**, 501–508.

40 Finlay, A.C., Hobby, G.L., Hochstein, F., Lees, T.M., Lenert, T.F., Means, J.A., P'An, S.Y., Regna, P.P., Routien, J.B., Sobin, B.A., Tate, K.B., and Kane, J.H. (1951) *American Review of Tuberculosis*, **63**, 1.

41 Izumi, R., Noda, T., Ando, T., Take, T., and Nagata, A. (1972) *The Journal of Antibiotics*, **25**, 201–207.

42 Nomoto, S., Teshima, T., Wakamiya, T., and Shiba, T. (1977) *The Journal of Antibiotics*, **30**, 955–959.

43 Kurath, P., Rosenbrook, W. Jr., Dunnigan, D.A., McAlpine, J.B., Egan, R.S., Stanaszek, R.S., Cirovic, M., Mueller, S.L., and Washburn, W.H. (1984) *The Journal of Antibiotics*, **37**, 1130–1143.

44 French, J.C., Bartz, Q.R., and Dion, H.W. (1973) *The Journal of Antibiotics*, **26**, 272–283.

45 Konishi, M., Saito, K., Numata, K., Tsuno, T., Asama, K., Tsukiura, H., Naito, T., and Kawaguchi, H. (1977) *The Journal of Antibiotics*, **30**, 789–805.

46 McGahren, W.J., Morton, G.O., Kunstmann, M.P., and Ellestad, G.A. (1977) *The Journal of Organic Chemistry*, **42**, 1282–1286.

47 Takeuchi, S., Hirayama, K., Ueda, K., Sakai, H., and Yonehara, H. (1958) *The Journal of Antibiotics*, **11**, 1–5.

48 Frey, P.A. and Booker, S.J. (2001) *Advances in Protein Chemistry*, **58**, 1–45.

49 Frey, P.A. (2001) *Annual Review of Biochemistry*, **70**, 121–148.

50 Ruzicka, F.J., Lieder, K.W., and Frey, P.A. (2000) *Journal of Bacteriology*, **182**, 469–476.

51 Song, K.B. and Frey, P.A. (1991) *The Journal of Biological Chemistry*, **266**, 7651–7655.

52 Petrovich, R.M., Ruzicka, F.J., Reed, G.H., and Frey, P.A. (1992) *Biochemistry*, **31**, 10774–10781.

53 Aberhart, D.J., Gould, S.J., Lin, H.-J., Thiruvengadam, T.K., and Weiller, B.H. (1983) *Journal of the American Chemical Society*, **105**, 5461–5470.

54 Marsh, E.N.G. and Drennan, C.L. (2001) *Current Opinion in Chemical Biology*, **5**, 499–505.

55 Frey, P.A., Ballinger, M.D., and Reed, G.H. (1998) *Biochemical Society Transactions*, **26**, 304–310.

56 Frey, P.A. (1993) *The FASEB Journal*, **7**, 662–670.

57 Frey, P.A. and Magnusson, O.T. (2003) *Chemical Reviews*, **103**, 2129–2148.

58 Frey, P.A. and Reed, G.H. (2000) *Archives of Biochemistry and Biophysics*, **382**, 6–14.

59 Ballinger, M.D., Reed, G.H., and Frey, P.A. (1992) *Biochemistry*, **31**, 949–953.

60 Ballinger, M.D., Frey, P.A., and Reed, G.H. (1992) *Biochemistry*, **31**, 10782–10789.

61 Prabhakaran, P.C., Woo, N.-T., Yorgey, P.S., and Gould, S.J. (1988) *Journal of the American Chemical Society*, **110**, 5785–5791.

62 Cone, M.C., Yin, X., Grochowski, L.L., Parker, M.R., and Zabriskie, T.M. (2003) *ChemBioChem*, **4**, 821–828.

63 Rétey, J., Kunz, F., Arigoni, D., and Stadtman, T.C. (1978) *Helvetica Chimica Acta*, **61**, 2989–2998.

64 Chang, C.H. and Frey, P.A. (2000) *The Journal of Biological Chemistry*, **275**, 106–114.

65 Fredenhagen, A., Tamura, S.Y., Kenny, P.T.M., Komura, H., Naya, Y., Nakanishi, K., Nishiyama, K., Sugiura, M., and Kita, H. (1987) *Journal of the American Chemical Society*, **109**, 4409–4411.

66 Needham, J., Kelly, M.T., Ishige, M., and Andersen, R.J. (1994) *The Journal of Organic Chemistry*, **59**, 2058–2063.

67 Morita, H., Nagashima, S., Takeya, K., and Itokawa, H. (1993) *Chemical & Pharmaceutical Bulletin*, **41**, 992–993.

68 Yoshioka, H., Nakatsu, K., Sato, M., and Tatsuno, T. (1973) *Chemistry Letters*, 1319–1322.

69 Wojciechowska, H., Ciarkowski, J., Chmara, H., and Borowski, E. (1972) *Experientia*, **28**, 1423–1424.

70 Wojciechowska, H., Zgoda, W., Borowski, E., Dziegielewski, K., and Ulikowski, S. (1983) *The Journal of Antibiotics*, **36**, 793–798.

71 Nagano, Y., Ikedo, K., Fujishima, A., Izawa, M., Tsubotani, S., Nishimura, O., and Fujino, M. (2001) *The Journal of Antibiotics*, **54**, 934–947.

72 Wani, M.C., Taylor, H.L., Wall, M.E., Coggon, P., and McPhail, A.T. (1971) *Journal of the American Chemical Society*, **93**, 2325–2327.

73 Graf, E., Kirfel, A., Wolff, G.-J., and Breitmaier, E. (1982) *Liebigs Annalen der Chemie*, 376–381.

74 Waldmann, H., Hu, T.-S., Renner, S., Menninger, S., Tannert, R., Oda, T., and Arndt, H.-D. (2008) *Angewandte Chemie (International Edition in English)*, **47**, 6473–6477.

75 Kunze, B., Jansen, R., Sasse, F., Höfle, G., and Reichenbach, H. (1995) *The Journal of Antibiotics*, **48**, 1262–1266.

76 Jansen, R., Kunze, B., Reichenbach, H., and Höfle, G. (1996) *Liebigs Annalen der Chemie*, 285–290.

77 Hettinger, T.P. and Craig, L.C. (1970) *Biochemistry*, **9**, 1224–1232.

78 Tinto, W.F., Lough, A.J., McLean, S., Reynolds, W.F., Yu, M., and Chan, W.R. (1998) *Tetrahedron*, **54**, 4451–4458.

79 Zabriskie, T.M., Klocke, J.A., Ireland, C.M., Marcus, A.H., Molinski, T.F., Faulkner, D.J., Xu, C., and Clardy, J.C. (1986) *Journal of the American Chemical Society*, **108**, 3123–3124.

80 von Nussbaum, F., Spiteller, P., Rüth, M., Steglich, W., Wanner, G., Gamblin, B., Stievano, L., and Wagner, F.E. (1998) *Angewandte Chemie (International Edition in English)*, **37**, 3292–3295.

81 Otani, T., Minami, Y., Marunaka, T., Zhang, R., and Xie, M. (1988) *The Journal of Antibiotics*, **41**, 1580–1585.

82 Yoshida, K.-I., Azuma, R., Saeki, M., and Otani, T. (1993) *Tetrahedron Letters*, **34**, 2637–2640.

83 Walker, K.D. and Floss, H.G. (1998) *Journal of the American Chemical Society*, **120**, 5333–5334.

84 Christenson, S.D., Liu, W., Toney, M.D., and Shen, B. (2003) *Journal of the American Chemical Society*, **125**, 6062–6063.

85 Steele, C.L., Chen, Y., Dougherty, B.A., Li, W., Hofstead, S., Lam, K.S., Xing, Z., and Chiang, S.-J. (2005) *Archives of Biochemistry and Biophysics*, **438**, 1–10.

86 Rachid, S., Krug, D., Weissman, K.J., and Müller, R. (2007) *The Journal of Biological Chemistry*, **282**, 21810–21817.

87 Kurylo-Borowska, Z. and Abramsky, T. (1972) *Biochimica et Biophysica Acta*, **264**, 1–10.

88 Parry, R.J. and Kurylo-Borowska, Z. (1980) *Journal of the American Chemical Society*, **102**, 836–837.

89 Spiteller, P., Rüth, M., von Nussbaum, F., and Steglich, W. (2000) *Angewandte Chemie (International Edition in English)*, **39**, 2754–2756.

90 Robertson, D.E., Lesage, S., and Roberts, M.F. (1989) *Biochimica et Biophysica Acta*, **992**, 320–326.

91 Ruzicka, F.J. and Frey, P.A. (2007) *Biochimica et Biophysica Acta*, **1774**, 286–296.

92 Robinson, P., Neelon, K., Schreier, H.J., and Roberts, M.F. (2001) *Applied and Environmental Microbiology*, **67**, 4458–4463.

93 Freer, I., Pedrocchi-Fantoni, G., Picken, D.J., and Overton, K.H. (1981) *Journal of the Chemical Society, Chemical Communications*, 80–82.

94 Aberhart, D.J. (1988) *Analytical Biochemistry*, **169**, 350–355.

95 Kuo, Y.-H., Khan, J.K., and Lambein, F. (1994) *Phytochemistry*, **35**, 911–913.

96 Yin, X. and Zabriskie, T.M. (2004) *ChemBioChem*, **5**, 1274–1277.

97 Evans, C.S., Qureshi, M.Y., and Bell, E.A. (1977) *Phytochemistry*, **16**, 565–570.

98 Mostad, A., Roemming, C., and Rosenqvist, E. (1973) *Acta Chemica Scandinavica*, **27**, 164–176.

99 Yan, Z.-Y., Spencer, P.S., Li, Z.-X., Liang, Y.-M., Wang, Y.-F., Wang, C.-Y., and Li, F.-M. (2006) *Phytochemistry*, **67**, 107–121.

100 Spencer, P.S., Allen, C.N., Kisby, G.E., Ludolph, A.C., Ross, S.M., and Roy, D.N. (1991) *Advances in Neurology*, **56**, 287–299.

101 Boger, D.L. and Cai, H. (1999) *Angewandte Chemie (International Edition in English)*, **38**, 448–476.

102 Fusetani, N., Matsunaga, S., Matsumoto, H., and Takebayashi, Y. (1990) *Journal of the American Chemical Society*, **112**, 7053–7054.

103 Itagaki, F., Shigemori, H., Ishibashi, M., Nakamura, T., Sasaki, T., and Kobayashi, J. (1992) *The Journal of Organic Chemistry*, **57**, 5540–5542.

104 Schofield, C.J., Baldwin, J.E., Byford, M.F., Clifton, I., Hajdu, J., Hensgens, C., and Roach, P. (1997) *Current Opinion in Structural Biology*, **7**, 857–864.

105 Shaw, G.J., Ellingham, P.J., Bingham, A., and Wright, G.J. (1982) *Phytochemistry*, **21**, 1635–1637.

106 Bewley, C.A., He, H., Williams, D.H., and Faulkner, D.J. (1996) *Journal of the American Chemical Society*, **118**, 4314–4321.

107 Ford, P.W., Gustafson, K.R., McKee, T.C., Shigematsu, N., Maurizi, L.K., Pannell, L.K., Williams, D.E., de Silva, E.D., Lassota, P., Allen, T.M., Soest, R.V., Andersen, R.J., and Boyd, M.R. (1999) *Journal of the American Chemical Society*, **121**, 5899–5909.

108 Toda, S., Sugawara, K., Nishiyama, Y., Ohbayashi, M., Ohkusa, N., Yamamoto, H., Konishi, M., and Oki, T. (1990) *The Journal of Antibiotics*, **43**, 796–808.

109 Shen, B., Du, L., Sanchez, C., Edwards, D.J., Chen, M., and Murrell, J.M. (2002) *Journal of Natural Products*, **65**, 422–431.

110 Yin, X., McPhail, K.L., Kim, K.-J., and Zabriskie, T.M. (2004) *ChemBioChem*, **5**, 1278–1281.

111 Jackson, M.D., Gould, S.J., and Zabriskie, T.M. (2002) *The Journal of Organic Chemistry*, **67**, 2934–2941.

112 Ikegami, F., Ongena, G., Sakai, R., Itagaki, S., Kobori, M., Ishikawa, T., Kuo, Y.-H., Lambein, F., and Murakoshi, I. (1993) *Phytochemisty*, **33**, 93–98.

113 Randoux, T., Braekman, J.C., Daloze, D., and Pasteels, J.M. (1991) *Die Naturwissenschaften*, **78**, 313–314.

114 Gould, S.J. and Minott, D.A. (1992) *The Journal of Organic Chemistry*, **57**, 5214–5217.

115 Yin, X., O'Hare, T., Gould, S.J., and Zabriskie, T.M. (2003) *Gene*, **312**, 215–224.

116 Gould, S.J., Lee, J., and Wityak, J. (1991) *Bioorganic Chemistry*, **19**, 333–350.

117 Martinkus, K.J., Tann, C.-H., and Gould, S.J. (1983) *Tetrahedron*, **39**, 3493–3505.

118 Kobayashi, J., Itagaki, F., Shigemori, H., Ishibashi, M., Takahashi, K., Ogura, M., Nagasawa, S., Nakamura, T., Hirota, H., Ohta, T., and Nozoe, S. (1991) *Journal of the American Chemical Society*, **113**, 7812–7813.

119 Fusetani, N., Sugawara, T., Matsunaga, S., and Hirota, H. (1991) *Journal of the American Chemical Society*, **113**, 7811–7812.

120 Schwarzer, D., Finking, R., and Marahiel, M.A. (2003) *Natural Product Reports*, **20**, 275–287.

121 Finking, R. and Marahiel, M.A. (2004) *Annual Review of Microbiology*, **58**, 453–488.

122 Staunton, J. and Weissman, K.J. (2001) *Natural Product Reports*, **18**, 380–416.

123 Fischbach, M.A. and Walsh, C.T. (2006) *Chemical Reviews*, **106**, 3468–3496.

124 Tillett, D., Dittmann, E., Erhard, M., von Döhren, H., Börner, T., and Neilan, B.A. (2000) *Chemistry & Biology*, **7**, 753–764.

125 Engel, S., Jensen, P.R., and Fenical, W. (2002) *Journal of Chemical Ecology*, **28**, 1971–1985.

126 Yang, X., Strobel, G., Stierle, A., Hess, W.M., Lee, J., and Clardy, J. (1994) *Plant Science*, **102**, 1–9.

127 Blunt, J.W., Copp, B.R., Hu, W.-P., Munro, M.H.G., Northcote, P.T., and Prinsep, M.R. (2008) *Natural Product Reports*, **25**, 35–94.

128 Bewley, C.A. and Faulkner, D.J. (1998) *Angewandte Chemie (International Edition in English)*, **37**, 2162–2178.

129 Fusetani, N. and Matsunaga, S. (1993) *Chemical Reviews*, **93**, 1793–1806.

130 Piel, J. (2006) *Current Medicinal Chemistry*, **13**, 39–50.

131 Sone, H., Nemoto, T., Ishiwata, H., Ojika, M., and Yamada, K. (1993) *Tetrahedron Letters*, **34**, 8449–8452.

132 Tan, L.T., Sitachitta, N., and Gerwick, W.H. (2003) *Journal of Natural Products*, **66**, 764–771.

133 Williams, P.G., Yoshida, W.Y., Moore, R.E., and Paul, V.J. (2002) *Journal of Natural Products*, **65**, 29–31.

134 Pettit, G.R., Kamano, Y., Kizu, H., Dufresne, C., Herald, C.L., Bontems, R.J., Schmidt, J.M., Boettner, F.E., and Nieman, R.A. (1989) *Heterocycles*, **28**, 553–558.

135 Carter, D.C., Moore, R.E., Mynderse, J.S., Niemczura, W.P., and Todd, J.S. (1984) *The Journal of Organic Chemistry*, **49**, 236–241.

136 Pettit, G.R., Xu, J.-P., Hogan, F., Williams, M.D., Doubek, D.L., Schmidt, J.M., Cerny, R.L., and Boyd, M.R. (1997) *Journal of Natural Products*, **60**, 752–754.

137 Davies-Coleman, M.T., Dzeha, T.M., Gray, C.A., Hess, S., Pannell, L.K., Hendricks, D.T., and Arendse, C.E. (2003) *Journal of Natural Products*, **66**, 712–715.

138 Kimura, J., Takada, Y., Inayoshi, T., Nakao, Y., Goetz, G., Yoshida, W.Y., and Scheuer, P.J. (2002) *The Journal of Organic Chemistry*, **67**, 1760–1767.

139 Horgen, F.D., Yoshida, W.Y., and Scheuer, P.J. (2000) *Journal of Natural Products*, **63**, 461–467.

140 Luesch, H., Williams, P.G., Yoshida, W.Y., Moore, R.E., and Paul, V.J. (2002) *Journal of Natural Products*, **65**, 996–1000.

141 Pettit, G.R., Xu, J.-P., Hogan, F., and Cerny, R.L. (1998) *Heterocycles*, **47**, 491–496.

142 Rodríguez, J., Fernández, R., Quiñoá, E., Riguera, R., Debitus, C., and Bouchet, P. (1994) *Tetrahedron Letters*, **35**, 9239–9242.

143 Williams, P.G., Yoshida, W.Y., Quon, M.K., Moore, R.E., and Paul, V.J. (2003) *Journal of Natural Products*, **66**, 651–654.

144 Peypoux, F., Marion, D., Maget-Dana, R., Ptak, M., Das, B.C., and Michel, G. (1985) *European Journal of Biochemistry*, **153**, 335–340.

145 Bonmatin, J.-M., Laprévote, O., and Peypoux, F. (2003) *Combinatorial*

Chemistry & High Throughput Screening, **6**, 541–556.

146 Peypoux, F., Pommer, M.T., Marion, D., Ptak, M., Das, B.C., and Michel, G. (1986) *The Journal of Antibiotics*, **39**, 636–641.

147 Chiba, H., Agematu, H., Dobashi, K., and Yoshioka, T. (1999) *The Journal of Antibiotics*, **52**, 700–709.

148 Aoyagi, T., Tobe, H., Kojima, F., Hamada, M., Takeuchi, T., and Umezawa, H. (1978) *The Journal of Antibiotics*, **31**, 636–638.

149 Tobe, H., Morishima, H., Naganawa, H., Takita, T., Aoyagi, T., and Umezawa, H. (1979) *Agricultural and Biological Chemistry*, **43**, 591–596.

150 Gulavita, N.K., Pomponi, S.A., Wright, A.E., Yarwood, D., and Sills, M.A. (1994) *Tetrahedron Letters*, **35**, 6815–6818.

151 Pergament, I. and Carmeli, S. (1994) *Tetrahedron Letters*, **35**, 8473–8476.

152 Okino, T., Matsuda, H., Murakami, M., and Yamaguchi, K. (1993) *Tetrahedron Letters*, **34**, 501–504.

153 Bunnage, M.E., Burke, A.J., Davies, S.G., and Goodwin, C.J. (1995) *Tetrahedron: Asymmetry*, **6**, 165–176.

154 Moon, S.-S., Chen, J.L., Moore, R.E., and Patterson, G.M.L. (1992) *The Journal of Organic Chemistry*, **57**, 1097–1103.

155 Gregson, J.M., Chen, J.-L., Patterson, G.M.L., and Moore, R.E. (1992) *Tetrahedron*, **48**, 3727–3734.

156 Umezawa, H., Aoyagi, T., Suda, H., Hamada, M., and Takeuchi, T. (1976) *The Journal of Antibiotics*, **29**, 97–99.

157 Nagai, M., Kojima, F., Naganawa, H., Hamada, M., Aoyagi, T., and Takeuchi, T. (1997) *The Journal of Antibiotics*, **50**, 82–84.

158 Yoshida, S., Nakamura, Y., Naganawa, H., Aoyagi, T., and Takeuchi, T. (1990) *The Journal of Antibiotics*, **43**, 149–153.

159 Aoyagi, T., Yoshida, S., Nakamura, Y., Shigihara, Y., Hamada, M., and Takeuchi, T. (1990) *The Journal of Antibiotics*, **43**, 143–148.

160 Fujii, K., Sivonen, K., Kashiwagi, T., Hirayama, K., and Harada, K.-I. (1999) *The Journal of Organic Chemistry*, **64**, 5777–5782.

161 Bewley, C.A. and Faulkner, D.J. (1994) *The Journal of Organic Chemistry*, **59**, 4849–4852.

162 Schmidt, E.W., Bewley, C.A., and Faulkner, D.J. (1998) *The Journal of Organic Chemistry*, **63**, 1254–1258.

163 Helms, G.L., Moore, R.E., Niemczura, W.P., Patterson, G.M.L., Tomer, K.B., and Gross, M.L. (1988) *The Journal of Organic Chemistry*, **53**, 1298–1307.

164 Craig, M. and Holmes, C.F.B. (2000) Freshwater hepatotoxins: microcystin and nodularin, mechanisms of toxins and effects on health, in *Seafood and Freshwater Toxins: Pharmacology, Physiology, and Detection* (ed. L.M. Botana), Marcel Dekker, New York, pp. 643–671.

165 Rinehart, K.L., Namikoshi, M., and Choi, B.W. (1994) *Journal of Applied Phycology*, **6**, 159–176.

166 De Silva, E.D., Williams, D.E., Andersen, R.J., Klix, H., Holmes, C.F.B., and Allen, T.M. (1992) *Tetrahedron Letters*, **33**, 1561–1564.

167 Bewley, C.A., Debitus, C., and Faulkner, D.J. (1994) *Journal of the American Chemical Society*, **116**, 7631–7636.

168 Schmidt, E.W. and Faulkner, D.J. (1998) *Tetrahedron*, **54**, 3043–3056.

169 Tsuge, K., Akiyama, T., and Shoda, M. (2001) *Journal of Bacteriology*, **183**, 6265–6273.

170 Duitman, E.H., Hamoen, L.W., Rembold, M., Venema, G., Seitz, H., Saenger, W., Bernhard, F., Reinhardt, R., Schmidt, M., Ullrich, C., Stein, T., Leenders, F., and Vater, J. (1999) *Proceedings of the National Academy of Sciences of the United States of America*, **96**, 13294–13299.

171 Kramer, D. (2007) EP1792981.

172 Okuyama, A., Naganawa, H., Harada, S., Aoyagi, T., and Umezawa, H. (1986) *Biochemistry International*, **12**, 627–631.

173 Moore, R.E., Chen, J.L., Moore, B.S., Pattersen, G.M.L., and Carmichael, W.W. (1991) *Journal of the American Chemical Society*, **113**, 5083–5084.

174 Rinehart, K.L., Harada, K.-I., Namikoshi, M., Chen, C., Harvis, C.A., Munro, M.H.G., Blunt, J.W., Mulligan, P.E., Beasley, V.R., Dahlem, A.M., and Carmichael, W.W. (1988) *Journal of the American Chemical Society*, **110**, 8557–8558.

175 Tomita, H., Mizusaki, S., and Tamaki, E. (1964) *Agricultural and Biological Chemistry*, **28**, 451–455.

176 Igarashi, M., Kinoshita, N., Ikeda, T., Kameda, M., Hamada, M., and Takeuchi, T. (1997) *The Journal of Antibiotics*, **50**, 1020–1025.

177 Konishi, M., Nishio, M., Saitoh, K., Miyaki, T., Oki, T., and Kawagushi, H. (1989) *The Journal of Antibiotics*, **42**, 1749–1755.

178 Goto, T., Toya, Y., Ohgi, T., and Kondo, T. (1982) *Tetrahedron Letters*, **23**, 1271–1274.

179 Aguilar, M.-I., Purcell, A.W., Devi, R., Lew, R., Rossjohn, J., Smith, A.I., and Perlmutter, P. (2007) *Organic and Biomolecular Chemistry*, **5**, 2884–2890.

5
Methods for the Chemical Synthesis of Noncoded α-Amino Acids found in Natural Product Peptides

Stephen A. Habay, Steve S. Park, Steven M. Kennedy, and A. Richard Chamberlin

5.1
Introduction

Noncoded amino acids, also referred to as "nonribosomal" or "nonproteinogenic" amino acids [1], are those building blocks that are not incorporated normally into polypeptides synthesized by the ribosome. Thus, they do not include the 20 known coded amino acids found in nature. These noncoded residues are, however, sometimes found in ribosomally synthesized peptides as a consequence of some post-translational modification [2]. Additionally, these unusual amino acids can be found in natural product peptides that are synthesized by many diverse organisms, such as bacteria, fungi, plants, and marine sponges. These organisms prepare and link together the noncoded amino acids on assembly line-like enzyme complexes known as nonribosomal peptide synthetases [3]. The natural product peptides produced from these tiny chemical factories provide us with some of the most powerful molecular tools for biology and medicine known today [4]. Antibiotics, immunosuppresants, anticancer compounds, siderophores, antivirals, anti-inflammatories, and insecticides are a few of the diverse biological activities of natural product peptides [5]. Consequently, the ability to obtain these and other amino acid-containing compounds in sufficient quantities has driven the synthetic community to develop a great spectrum of chemical methods of synthesis [6–10].

The chemical synthesis of natural product peptides begins with the preparation of its constituent amino acids. Many elegant and efficient syntheses of *unnatural* noncoded α-amino acids have been described over the years, with a plethora of excellent reviews available on this topic [11–17]. However, no compilation of synthetic methods for *natural* noncoded α-amino acids exists currently. This chapter will focus on the preparation of nonproteinogenic α-amino acids that are building blocks of peptide natural products. Due to limitations of space, natural products consisting of single amino acids (e.g., furanomycin, dysiherbaine, kaitocephalin, kainic acid, etc.) and dipeptides (e.g., diketopiperazines) will not be discussed here; alternative compilations are available for these compounds [18–20]. We will instead focus on natural product peptides of three or more amino acids. Additionally, we make no

Amino Acids, Peptides and Proteins in Organic Chemistry. Vol.1 – Origins and Synthesis of Amino Acids.
Edited by Andrew B. Hughes

ISBN: 978-3-527-32096-7

attempt to examine the patent literature or enzymatic methods of production, focusing exclusively on the literature of chemical synthetic methods.

There are many categories of noncoded amino acids found in natural products [5]. For example, α,β-dehydroamino acids, halogenated derivatives, β-amino acids, *N*-alkyl amino acids, and α,α-dialkyl residues are some of the building blocks not utilized by the ribosome. To avoid overlap with other chapters of this volume, the above mentioned compounds will be excluded from our discussion here. We will instead highlight synthetic endeavors of three specific categories of noncoded α-amino acids. A section on the synthesis of noncoded cyclic amino acids (CAAs) will examine five-membered ring (proline-based) and six-membered ring (pipecolic acid-based) amino acids, and bicyclic residues (octahydroindoles and hydropyrroloindoles). A section on the chemical synthesis of noncoded amino acids that are modifications of coded amino acids represents those residues found in peptides that undergo post-translational alteration. These constituent amino acids are typically those that are modified via hydroxylation, methylation, glycosylation, and so on. Finally, those noncoded amino acids containing elaborate, and often times hydrophobic, side-chain residues are profiled.

5.2
Noncoded CAAs

Naturally occurring nonproteinogenic CAAs are excellent building blocks for the construction of rigid molecules because of the inherent conformational restraint exerted by the cyclic feature. CAAs are frequently used in biomimetic polymer design and they are also incorporated into peptidomimetic molecules to probe ligand–receptor interactions. In addition, a variety of CAAs is found in numerous bioactive natural products that serve as a novel scaffold for therapeutic development. Due to these unique properties, CAAs have attracted both synthetic and medicinal chemists over many years. This section provides an overview of recent endeavors in synthesizing naturally occurring nonproteinogenic CAAs, focusing mainly on the CAAs that are incorporated in natural products (Figure 5.1).

Both *trans* and *cis* diastereomers of 3-hydroxyprolines (3-OH Pro), (2*S*,3*S*)-hydroxylproline and (2*R*,3*S*)-proline, respectively, are two of the most frequently occurring unusual nonproteinogenic CAAs isolated from peptide natural products and complex alkaloids such as castanospermine [21], slaframine [22], detoxinine [23, 24], mucrorin D [25], telomycin [26], and polyhydroxylated alkaloids [27]. In particular, castanospermines, slaframine, and detoxinine are known as potent glycosidase inhibitors [28, 29], where their analogs serve as lead compounds in the development of novel antiviral agents. Due to the prevalence of 3-OH Pro in various bioactive natural products, many useful asymmetric synthetic methods have been reported to prepare both diastereomers of 3-OH Pro [30–39].

Larcheveque *et al.* introduced an exquisite synthesis of (2*S*,3*S*)-3-OH Pro isomer in a concise manner using a reductive cyanation strategy [34]. Their synthetic work began by transforming L-malic acid into a γ-iodide species in four steps followed by

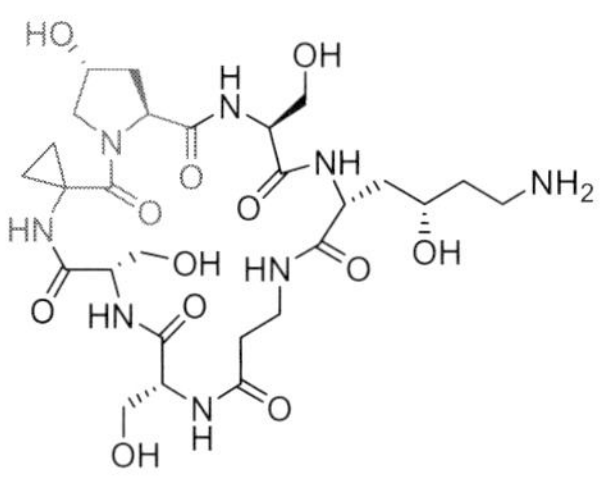

Serinocyclin A

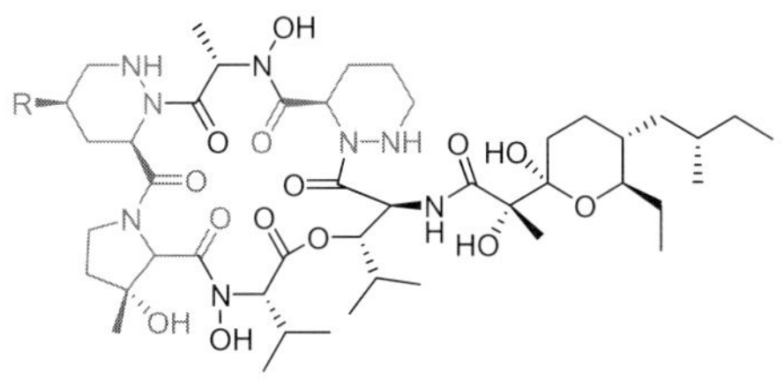

Polyoxypeptin A R = OH

Polyoxypeptin B R = H

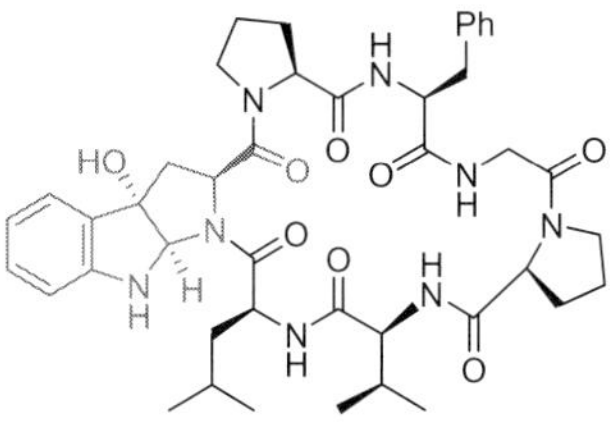

Phakellistatin 3

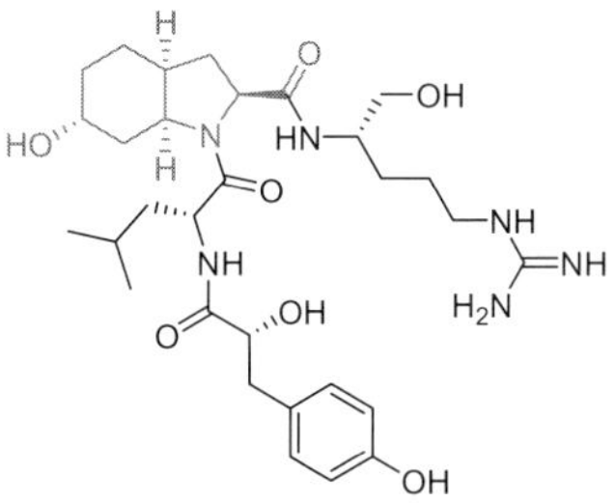

Aeruginosin 298A

Figure 5.1 Example of natural products containing nonproteinogenic cyclic amino acid.

alkylation of the γ-carbon with allylamine, which cyclized spontaneously into the *N*-allyl-γ-lactam derivative (Scheme 5.1). After silyl protection of the secondary alcohol, the protected intermediate **1** was treated with diisobutylaluminium hydride (DIBAL-H) to induce the formation of an iminium ion *in situ*, followed by nucleophilic attack by a cyanide ion to afford *N*-allylated pyrrolidine species **2** in 93% yield with excellent diastereoselectivity. The authors have proposed that the high diastereoselection comes from nucleophilic attack at the less hindered face of the iminium species opposite to the *tert*-butyldimethylsilyl (TBS) silyloxy ether. The cyanide functionality was then hydrolyzed and esterified to the methyl ester under acidic conditions while removing the silyl protecting group simultaneously. After a straightforward deprotection, methyl ester hydrolysis, and purification protocol, enantiopure (2*S*,3*S*)-3-OH Pro **3** was obtained in a 17% overall yield starting from L-malic acid.

Utilization of inexpensive sugars as starting materials is a well-known approach in synthetic organic chemistry for gaining facile entry into optically active advanced intermediates. Careful manipulations of the functional groups on the sugar molecule can eliminate the use of chiral catalysts or chiral auxiliaries, which can be expensive and can add more synthetic steps, although sometimes trivial, to completion of a target molecule. As an example, Park *et al.* have used relatively inexpensive sugar molecules, D-glucono-δ-lactone and L-gluconic acid γ-lactone, to prepare two diastereomers of 3-OH Pro [35].

D-Glucono-δ-lactone **4** was treated with *p*-toluenesulfonic acid and 1,1-dimethoxyethane in methanol to afford a diol-protected methyl ester species (Scheme 5.2) [40].

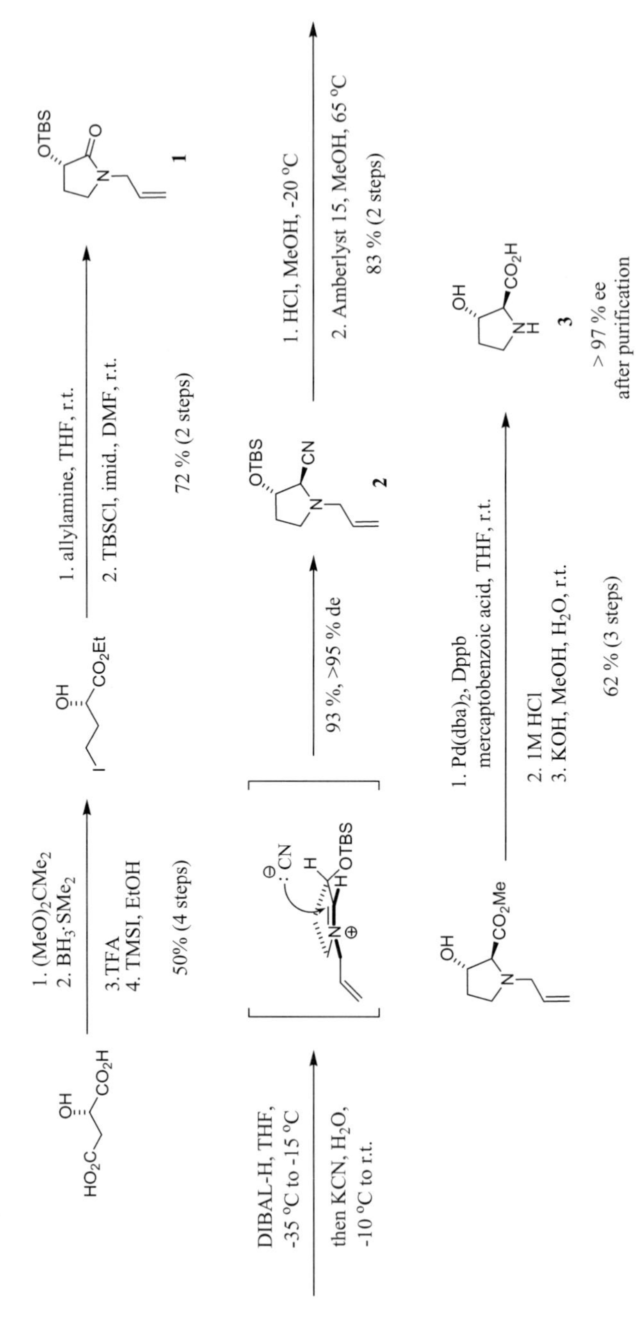

Scheme 5.1 Larcheveque synthesis of enantiopure (2*S*,3*S*)-3-OH Pro.

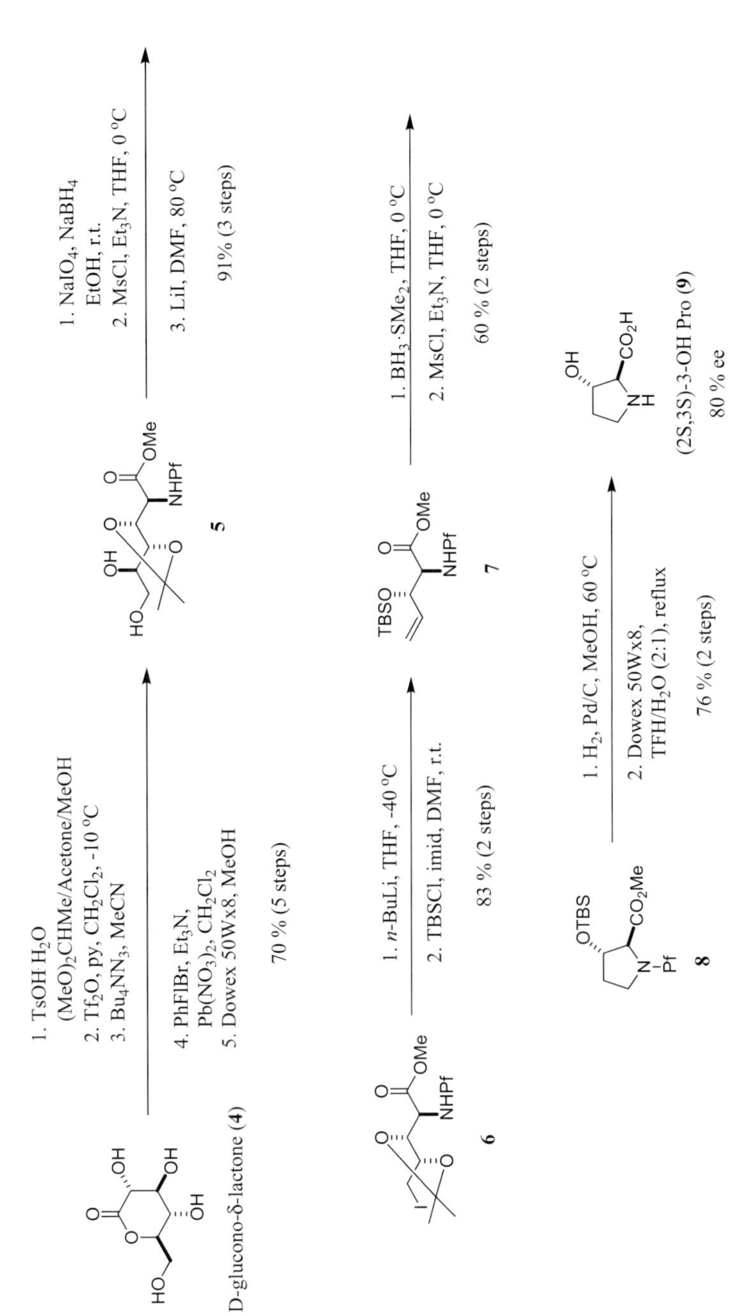

Scheme 5.2 Park synthesis of the (2*S*,3*S*)-3-OH Pro.

The unprotected secondary alcohol was transformed into a viable leaving group using triflic anhydride and the resulting intermediate underwent an S_N2 reaction with tetrabutylammonium azide to yield the manno-azide compound. The azide functionality was reduced and protected as the 9-phenylfluoren-9-yl (PhFl)-protected amine followed by selective cleavage of the terminal isopropylidene to afford **5** [41]. The terminal diol **5** underwent oxidative cleavage to the aldehyde, followed by $NaBH_4$ reduction, mesylation of the primary alcohol, and a substitution reaction with lithium iodide to give 2,3-isopropylidene iodide **6**. Removal of the isopropylidene group and dealkyloxyhalogenation occurred simultaneously under basic conditions. The free alcohol was protected with a TBS group to afford the pentenoic acid derivative **7**. The terminal alkene of the pentenoic acid derivative **7** was oxidized via hydroboration to a primary alcohol intermediate, followed by mesylation to induce an intramolecular amination to produce the *N*-protected proline ester **8**. Finally, all protecting groups were removed and the methyl ester was hydrolyzed to the carboxylic acid yielding (2*S*,3*S*)-3-OH Pro **9** in good yield with acceptable enantiomeric excess.

To complete the synthesis of the (2*R*,3*S*)-3-OH Pro isomer **11**, the same synthetic strategy was used (starting from L-gluconic acid γ-lactone **10**) (Scheme 5.3) [35, 42].

Jurczak *et al.* have synthesized the (2*R*,3*S*)-3-OH Pro isomer in a very efficient manner starting from L-serinal (Scheme 5.4) [30]. The L-serinals were prepared from an L-serine 2,2,6,6-tetramethylpiperidinooxy (TEMPO) oxidation protocol which is known for minimal epimerizatin of the α-stereocenter. Various L-serinal derivatives and Lewis acids were screened to find a diastereoselective 1,2-addition reaction with allyltrimethylsilane. The optimum result was obtained with *N*-Cbz-*O*-TBS-L-serinal **12** and 1 equiv. of $SnCl_4$, giving the *syn* product **13** as the major diastereomer (98 : 2 *syn* : *anti*). Silyl protection of the resulting *syn* adduct, followed by oxidation of the terminal alkene produced an aldehyde species **14** that spontaneously cyclized to the hemi-aminal **15**. Upon treatment with $NaBH_3CN$ the hemi-aminal **15** was reduced to a *N*-protected pyrrolidine derivative. Selective desilylation revealed primary alcohol **16** which was oxidized to the carboxylic acid with a catalytic amount of ruthenium

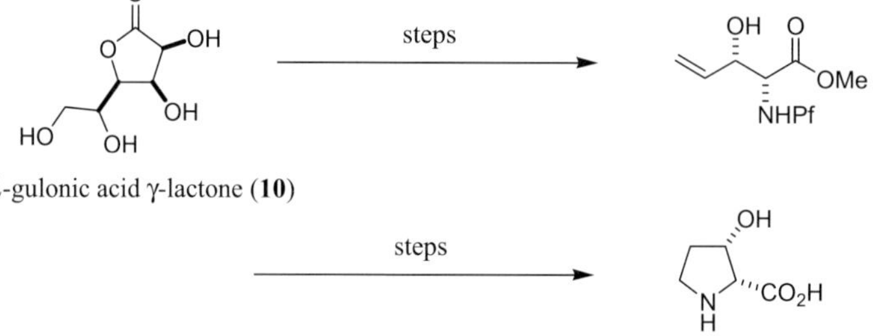

L-gulonic acid γ-lactone (**10**)

steps

steps

(2*R*,3*S*)-3-OH Pro (**11**)
29 % overall yield
92 % ee

Scheme 5.3 Park synthesis of the (2*R*,3*S*)-3-OH Pro.

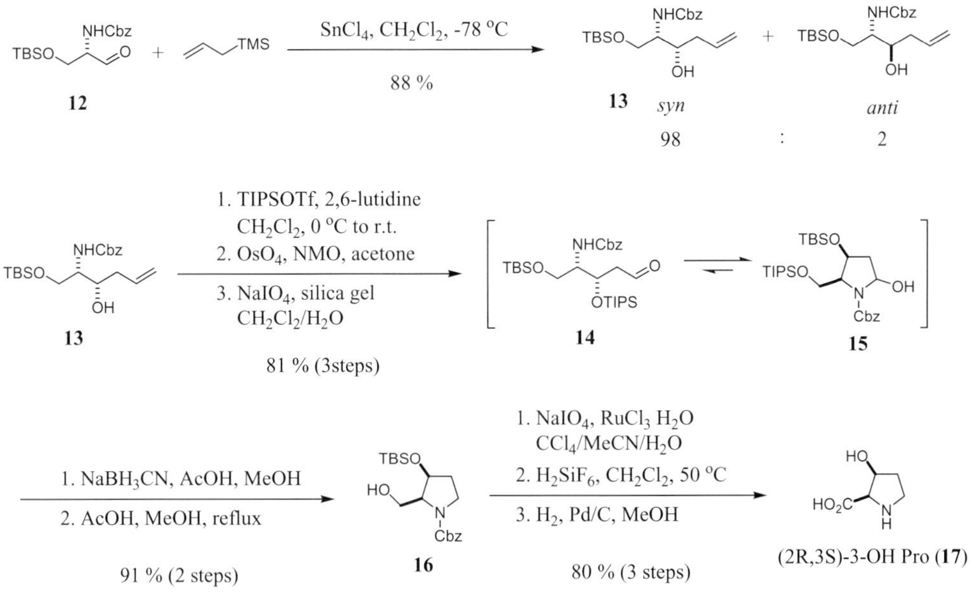

Scheme 5.4 Jurczak's synthesis of (2*R*,3*S*)-3-OH Pro from L-serinals.

trichloride in the presence of $NaIO_4$. The remaining protecting groups were removed under standard conditions to afford the (2*R*,3*S*)-3-OH Pro isomer **17** in enantiomerically pure form.

Discovery of natural products that can induce apoptosis of cancer cells has been a major research area in development of novel anticancer agents [43]. In 1998, Umezawa *et al.* isolated hexadepsipeptides polyoxytoxins A and B from *Streptomyces* culture broth. These compounds are believed to be potent inducers of apoptosis in human pancreatic carcinoma adenocarcinoma AsPC-1 cells [44–47]. Apart from their interesting biological activity, polyoxytoxins A and B contain the unusual CAA (2*S*,3*R*)-3-hydroxy-3-methylprolines (3-OH MePro) **20** (Scheme 5.5). Recently, enantioselective preparation of 3-OH MePro has been the focus of many research groups in response to the need for methodology to produce 3-OH MePro in large quantities for the total synthesis of the polyoxytoxins [48–57].

Hamada *et al.* have reported three synthetic methods to prepare 3-OH MePro to date, with the latest two versions deserving attention due to the efficiency of the synthesis [49, 56, 57]. In their 2002 report, 3-OH MePro was synthesized by a tandem Michael-aldol reaction using *N*-1-naphthylsulfonylglycine (*R*)-binaphthyl ester **18** and methyl vinyl ketone as the starting materials (Scheme 5.5) [56]. Addition of 1 equiv. of lithium chloride was necessary as an additive to control the diastereoselectivity of the reaction. Although the origin of this selectivity is ambiguous at present, the reaction afforded 77% yield of product **19** with an impressive 91 : 8 diastereomeric ratio favoring the desired (2*R*,3*S*)-*syn* adduct. The final product **20** was obtained as a *p*-toluenesulfonic acid salt in a total of five steps, starting from the chiral glycine

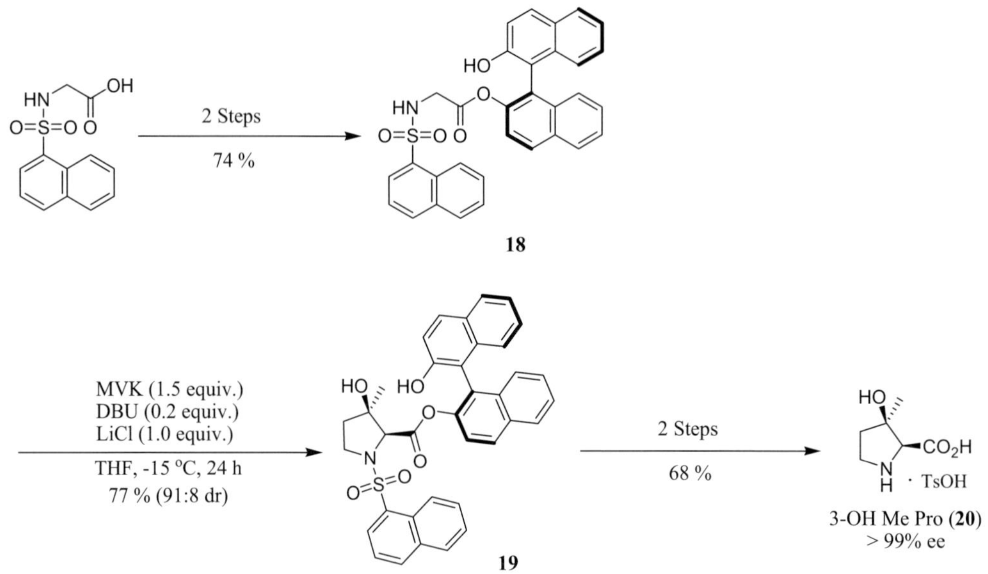

2 Steps
74 %
18

MVK (1.5 equiv.)
DBU (0.2 equiv.)
LiCl (1.0 equiv.)
THF, -15 °C, 24 h
77 % (91:8 dr)
19
2 Steps
68 %
· TsOH
3-OH Me Pro (**20**)
> 99% ee

Scheme 5.5 Hamada's synthesis of 3-OH MePro from naphthylsulfonylglycine.

derivative **18**, in 39% overall yield and in essentially enantiomerically pure form (>99% e.e. after recrystallization).

In the most recent publication by Hamada *et al.*, an intramolecular aldol reaction was featured as the key step [57]. The substrate for the intramolecular aldol reaction was prepared in three steps: tosylallylamide conjugate addition to methyl vinyl ketone, dihydroxylation of the terminal alkene, and oxidative cleavage to afford the aldehyde species **21** with a 77% overall yield. In the key step, 3-OH MePro was used in catalytic amount to produce **22** in good yield with high enantioselectivity (Scheme 5.6). The reason for the high selectivity of the intramolecular aldol reaction is not clear at this moment. However, Hamada *et al.* proposed that the reaction goes through a closed transition state **23** where the 3-hydroxyl group on the catalyst assists in ordering the transition state. Intermediate **22** was then transformed into the desired product **24**, with standard functional group manipulations in good yield and with excellent enantioselectivity.

Yao *et al.* prepared 3-OH MePro from the inexpensive starting material, 3-butyn-1-ol **25**, utilizing a Sharpless dihydroxylation reaction as the key step [50]. The alkyne starting material was converted into a (2*Z*)-α,β-unsaturated ester in three steps followed by the Sharpless dihydroxylation of the trisubstituted olefin, using AD-mix-α, to afford the dihydroxylated compound in 90% yield with 91% enantiomeric excess (Scheme 5.7). The dihydroxylated compound was transformed into cyclic sulfate **26** that was then selectively opened with sodium azide. After a series of reduction, protection and deprotection, diol **27** was obtained in good yield. Subsequent treatment of the diol species with methane sulfonyl chloride and triethylamine followed by ester hydrolysis and Boc deprotection gave 3-OH MePro **28** as the final product in 40% overall yield.

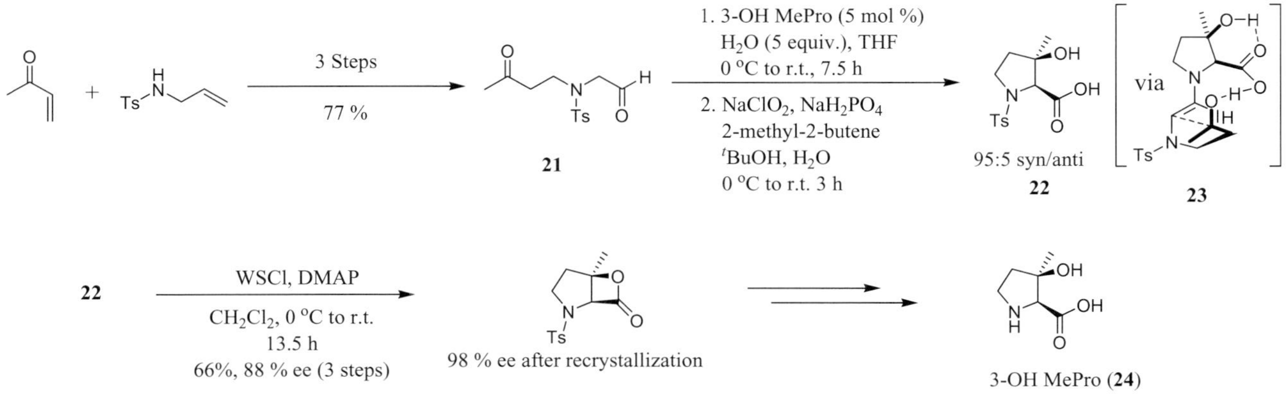

3 Steps

77 %

21

1. 3-OH MePro (5 mol %), THF
H_2O (5 equiv.),
0 °C to r.t., 7.5 h
2. $NaClO_2$, NaH_2PO_4
2-methyl-2-butene
^tBuOH, H_2O
0 °C to r.t. 3 h

95:5 syn/anti
22

via

23

22

WSCl, DMAP
CH_2Cl_2, 0 °C to r.t.
13.5 h
66%, 88 % ee (3 steps)

98 % ee after recrystallization

3-OH MePro (**24**)

Scheme 5.6 Hamada's intramolecular aldol synthesis of 3-OH MePro.

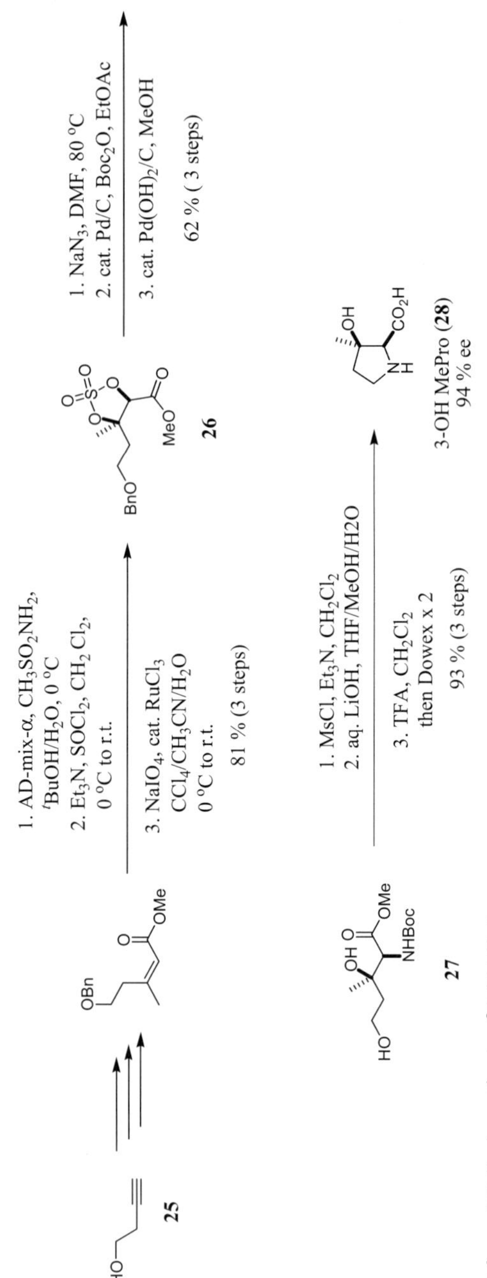

Scheme 5.7 Yao's synthesis of 3-OH MePro.

(2*S*,3*S*,4*S*)-3-Hydroxy-4-methylproline (Hmp) **32** (Scheme 5.8) is another proline derivative found in many biologically important natural products such as echinocandins [58, 59], mulundocanins [60], aculeacin A [61, 62], and nostopeptins [63]. Among these natural products, the echinocandins are potent antifungal agents where the analogs, LY303366-anidulafungin (Eraxis) and FK463/micafungin (Mycamine), are currently being used in the clinic. Also, nostopeptins are known to be elastase inhibitors [63]. Elastases are involved in many inflammatory diseases such as pulmonary emphysema, rheumatoid arthritis, and adult respiratory distress syndrome [64]. Due to the important bioactivity of natural products containing the Hmp moiety, many research groups have disclosed reports involving preparation of free Hmp and its protected derivatives [31, 65–73].

The first stereocontrolled synthesis of Hmp was completed by Ohfune *et al.* en route to a total synthesis of echinocandin D [65, 68]. The critical intermediate, *O*-benzyl-epoxy alcohol **29**, was prepared by known methods developed by Kishi *et al.* to establish the stereocenter at C-3 (Scheme 5.8) [74]. The epoxy alcohol **29** was oxidized to carboxylic acid **30** and the epoxide was opened selectively with aqueous ammonia. The free amine was Boc-protected, followed by esterification of the carboxylic acid with diazomethane. The benzyl group was then removed, followed by tosylation of the primary alcohol to provide the *O*-tosylated methyl ester **31**. The resulting *O*-tosylated methyl ester compound was treated with 1 equiv. of sodium hydride to close the ring into an Hmp precursor. After removal of all protecting groups, enantiopure Hmp **32** was obtained in 18% overall yield.

One year after Ohfune's synthesis of Hmp, Evans *et al.* reported a different approach to Hmp utilizing an oxazolidinone chiral auxiliary and hydroboration–cycloalkylation strategy [66]. To install the C-2 and C-3 stereocenters, an aldol reaction was carried out with bromoacetyl oxazolidinone **33a** and methacrolein mediated by dibutylboron triflate to afford the *syn* adduct **33b** diastereoselectively (Scheme 5.9). The bromide was displaced with inversion of stereochemistry by treatment with sodium azide. The chiral auxiliary was removed in the presence of bromomagnesium methoxide to give methyl ester **34** directly. Addition of dicyclohexylborane to methyl ester **34** followed by treatment with 1 N HCl induced cyclization into the pyrrolidine through a migratory insertion pathway and hydrolysis of the aminoborane, to establish the stereochemistry at C-4 while cleaving the N−B bond in one pot, yielding Hmp methyl ester **35** in 72% yield. The stereochemical outcome of the reaction was rationalized by a closed transition state in its lowest energy conformation where the migrating alkyl and departing diazonium are orbitally aligned in an *anti* orientation.

Huang *et al.* prepared Hmp by asymmetric induction from the innate chirality of the starting material [72]. Their synthesis began by stepwise treatment of chiral methylmalic acid **36** with acetyl chloride and *p*-methoxy benzyl amine, giving the protected malimide **37** as an inseparable mixture of diastereomers, favoring the *cis* product by a 9 : 1 ratio (Scheme 5.10). The secondary alcohol was protected with benzyl bromide, affording the fully protected malimide derivative **38**. To the resulting malimide, 2-lithiofuran, generated *in situ* from furan and *n*BuLi at −78 °C, was added, forming a racemic mixture of the adduct. This racemic mixture was treated

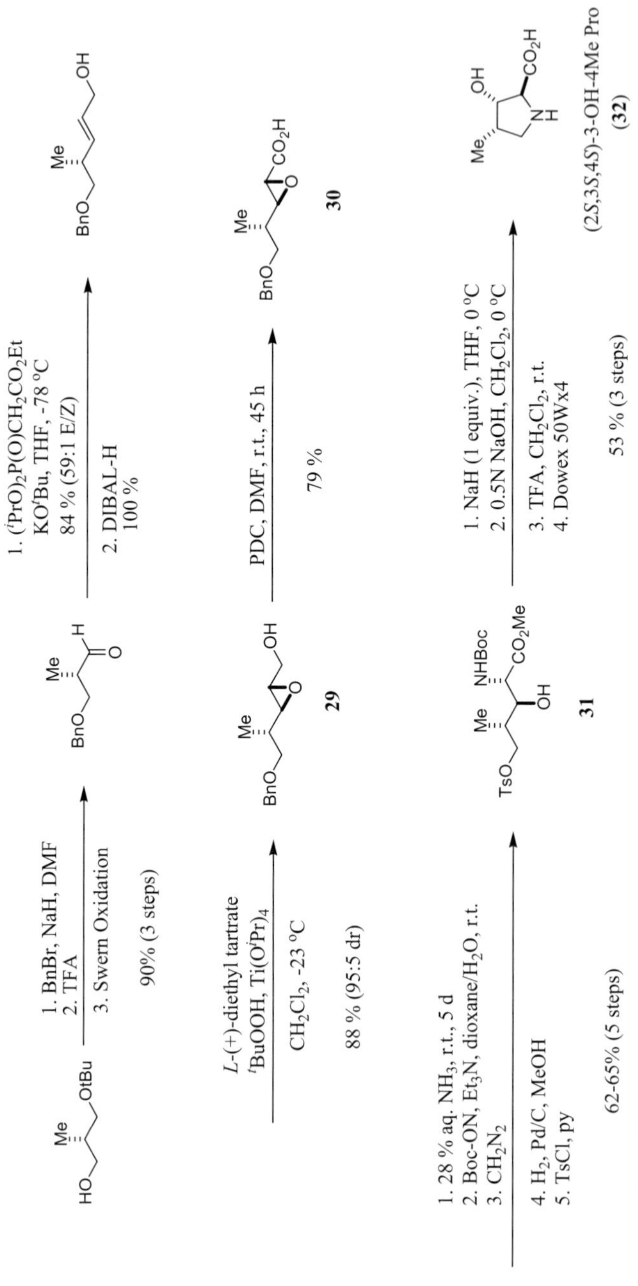

Scheme 5.8 Ohfune's synthesis of enantiopure Hmp.

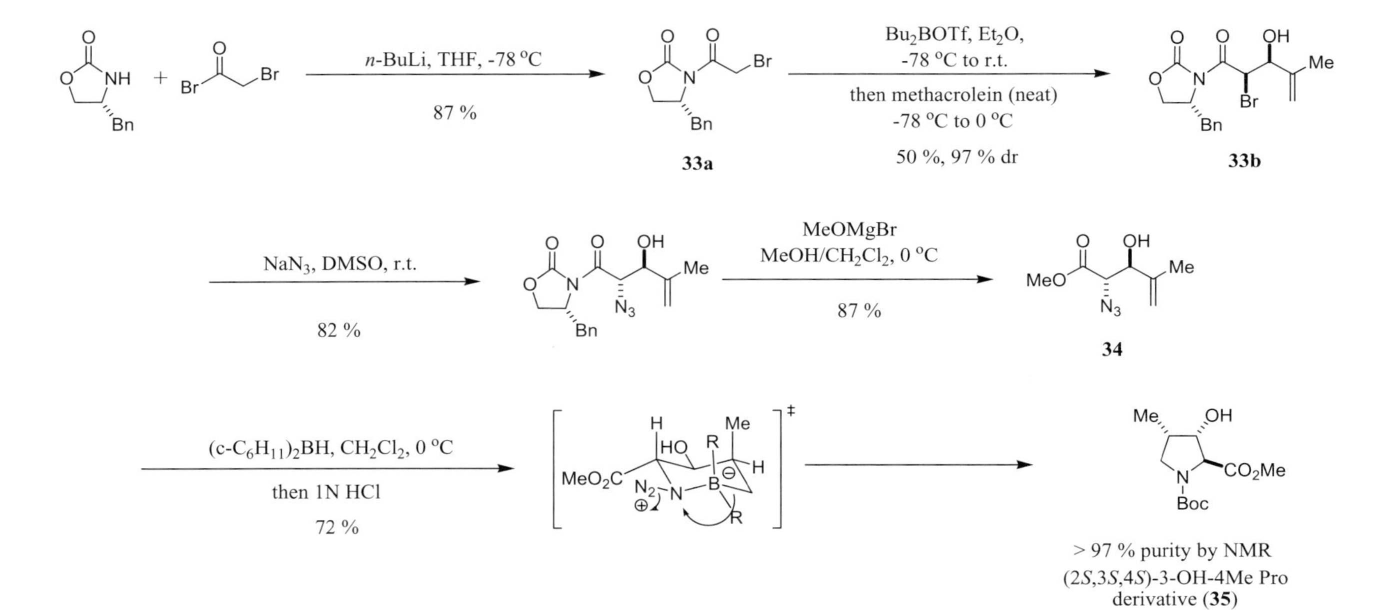

Scheme 5.9 Evan's synthesis of (2*S*,3*S*,4*S*)-3-OH-4-MePro derivative.

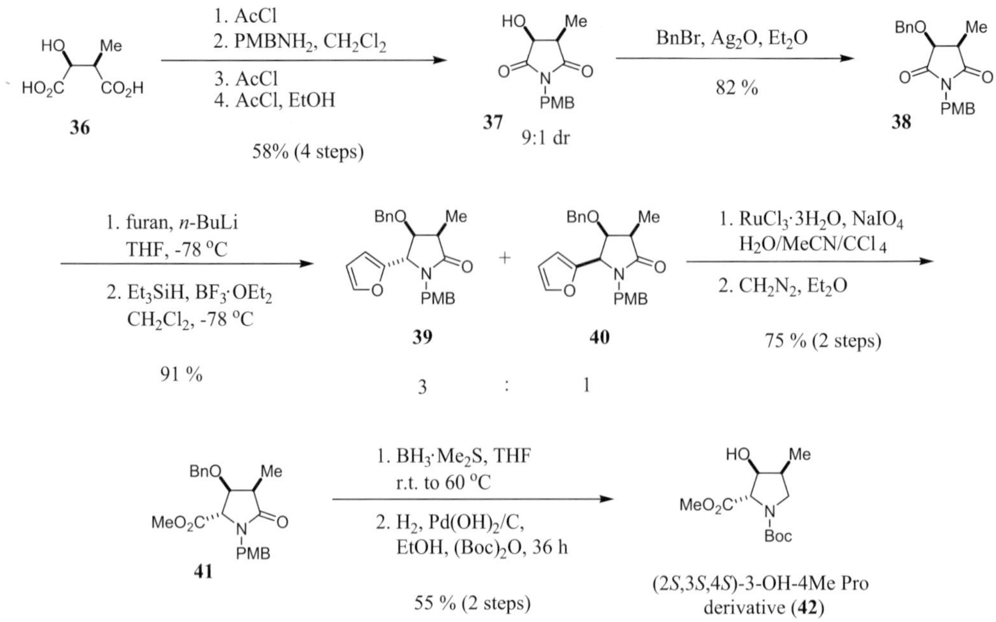

Scheme 5.10 Huang's synthesis of the (2S,3S,4S)-3-OH-4MePro derivative.

with boron trifluoride diethyl etherate and triethylsilane to give a separable mixture of the coupled products **39** and **40** in a ratio of 3 : 1. The reaction is believed to go through an *N*-acyliminium intermediate and the stereoselection comes from the addition of hydride from silane at the more hindered face. The 2-furyl group was transformed into a carboxylic acid chemoselectively under oxidative conditions using ruthenium trichloride and then it was esterified with diazomethane to afford the methyl ester γ-lactam **41**. Finally the lactam carbonyl was selectively reduced with borane dimethylsulfide, the benzyl and *p*-methoxy benzyl groups were removed simultaneously by treatment with Pearlman's catalyst, and the pyrrolidine nitrogen was protected with Boc_2O to afford the *N*-Boc-methyl ester Hmp derivative **42**.

Pipecolic acid, a noncoded CAA, is a six-membered ring derivative of proline. It is isolated as a secondary metabolite from some plants and fungi [75]. Also, it is found to be incorporated in pharmaceutically important natural products such as immunosuppressors FK506 [76] and rapamycin [77], oxytocin antagonist L-356209 [78], antiprotozal agent ampicidin [79], antifungal agent petriellin A [80], and antitumor antibiotic sandramycin [81]. Although pipecolic acid is available commercially, the enantiomerically pure form of it is quite expensive, which can be a problem when large quantities are needed. These reasons fostered interest in its large-scale enantioselective preparation. Currently, many synthetic methods (for preparing pipecolic acid) have been reported in the literature and have been compiled in review articles [82, 83]. In addition, microwave-assisted hydrogenation reactions have been used in preparation of pipecolic acid recently [84, 85]. In this section, a few interesting examples have been selected for discussion, including the most recent methodology.

Corey *et al.* elegantly assembled pipecolic acid using chiral phase transfer catalysis [86]. Their synthesis began with an asymmetric alkylation of the protected glycine **43** with 1-chloro-4-iodobutane in the presence of cinchonidine catalyst **45**, affording the alkylated product **44** in high yield with excellent enantiopurity (Scheme 5.11). The exceptionally high enantioselectivity (200 : 1) is believed to come from the selective facial screening exerted by the tight ion pair consisting of the cinchonidinium cation and the ester enolate [87]. The diphenyl imine was reduced subsequently with sodium borohydride and the resulting amine **46** cyclized under basic conditions. Finally, (*S*)-pipecolic acid *tert*-butyl ester **47** was obtained by removal of the nitrogen protecting group under typical hydrogenolysis conditions.

In Riera's preparation of pipecolic acid, the synthesis was accomplished by Sharpless asymmetric epoxidation of alcohol **48** to provide the epoxide **49** in high yield and high enantiomeric purity (Scheme 5.12) [88]. The epoxide was opened regioselectively by allylamine and treated subsequently with Boc_2O, under basic conditions, to give the *N*-protected diol **50**. The two terminal alkenes of the diol species underwent treatment with Grubbs catalyst, followed by reduction of the endocyclic double bond. The final product, *N*-Boc-(*S*)-pipecolic acid **51**, was acquired by oxidative cleavage of the diol. Recrystallization of each intermediate gives essentially enantiopure product.

Fadel *et al.* synthesized (*R*)-pipecolic acid by an asymmetric Strecker synthesis [89]. The dihydropyran **52** was opened in the presence of a chiral amine and sodium cyanide, under acidic conditions, to give the secondary amine **54** with high yield and excellent diastereoselectivity (Scheme 5.13). Fadel *et al.* rationalized the high diastereoselection using an Ahn–Eisentein model, depicted in **53**, where cyanide ion attacks the less hindered *si* face of the iminium species. The amino nitrile was converted to the amide under acidic conditions. The piperidine ring was closed by treatment with I_2/PPh_3/imidazole. Next, the nitrogen protecting group was removed with Pearlman's catalyst, followed by acidic hydrolysis of the amide to give (*R*)-pipecolic acid **55** with high enantiomeric purity.

In 1994, Murakami *et al.* isolated a new class of serine protease inhibitor called aeruginosin 298A from the blue-green alga *Microcystis aeruginosa* [90]. Since then many different aeruginosins have been discovered and isolated from natural sources [91]. Aeruginosins have grasped immediate attention in the scientific community due to possible uses as anticoagulants and anticancer agents [90, 92–102]. Aeruginosins are linear peptides containing a novel octahydroindole core, L-2-carboxy-6-hydroxyoctahydroindole (L-Choi), which is a noncoded CAA. The key to the total synthesis of all aeruginosins lies presumably in the efficiency of the method used to prepare L-Choi enantioselectively.

The first synthesis of the L-Choi motif was completed by Bonjoch *et al.* in 1996 and again in 2001 with revised absolute configuration of aeruginosin 298 A/B [103, 104]. Bonjoch's group started their synthesis of L-Choi by Birch reduction of L-Tyr(OMe)-OH **56** (Scheme 5.14). The resultant enol ether was cleaved by treatment with HCl to yield the intermediate cyclohexenone **57**, which underwent intramolecular 1,4-

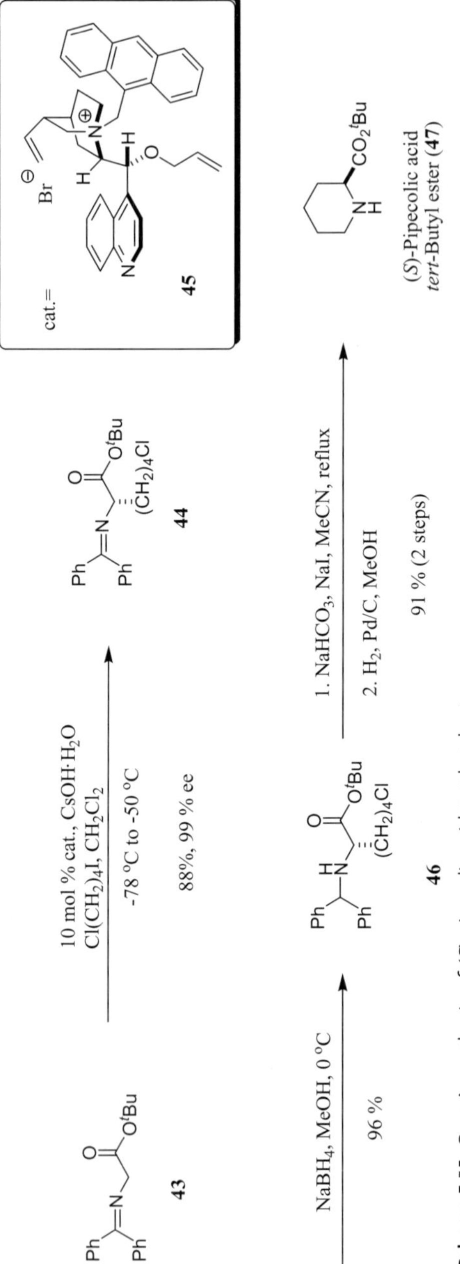

Scheme 5.11 Corey's synthesis of (*S*)-pipecolic acid *tert*-butyl ester.

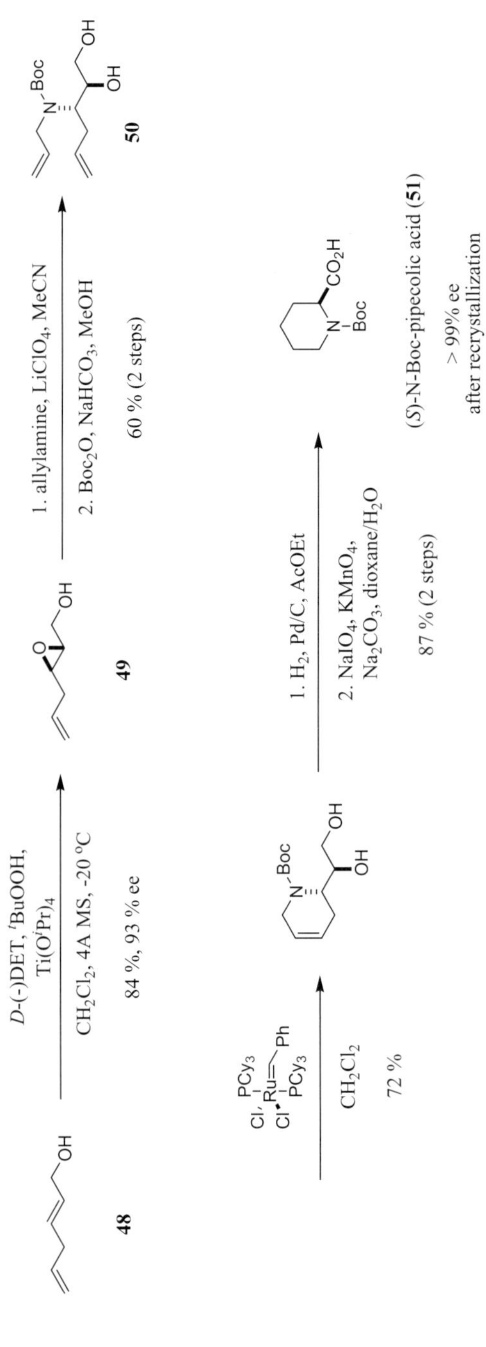

Scheme 5.12 Riera's preparation of (*S*)-*N*-Boc-pipecolic acid.

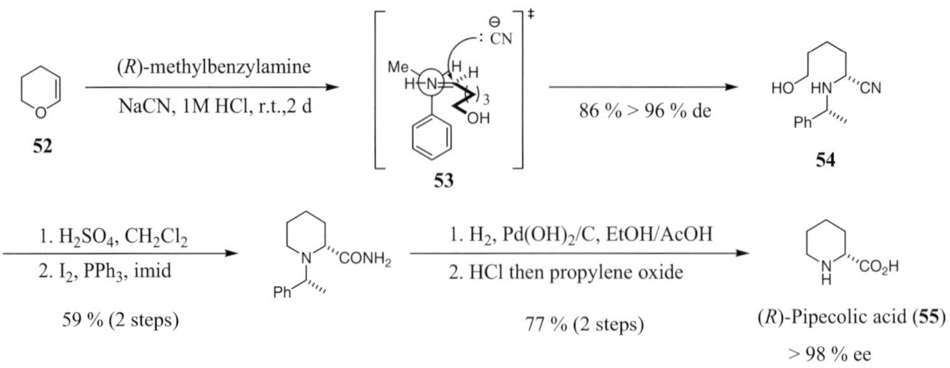

52 (R)-methylbenzylamine, NaCN, 1M HCl, r.t., 2 d → [53]‡ → 86 % > 96 % de → 54

1. H_2SO_4, CH_2Cl_2
2. I_2, PPh_3, imid
59 % (2 steps)

1. H_2, $Pd(OH)_2/C$, EtOH/AcOH
2. HCl then propylene oxide
77 % (2 steps)

(R)-Pipecolic acid (55) > 98 % ee

Scheme 5.13 Fadel's synthesis of (*R*)-pipecolic acid.

conjugate addition to afford a mixture of the desired *endo* **59** and *exo* **58** bicyclic products. The *exo* adduct **58** was converted to the *endo* **59** (the thermodynamically favored isomer) under acidic conditions. After switching the nitrogen protecting group, the bicycle was treated with LS-Selectride to reduce the ketone functionality to an alcohol with 8 : 1 selectivity favoring the desired stereochemistry. By this efficient route, Bonjoch's group was able to prepare the *N*-Boc-protected methyl ester derivative of L-Choi **60** in six steps with 18% overall yield. Wipf *et al.* have developed a route with similar efficiency to protected L-Choi where the key transformation was the oxidative ring-closing reaction of Cbz-L-Tyr-OH with iodobenzene diacetate, giving a high diastereomeric ratio favoring the desired product [105].

Analogous to Bonjoch's synthesis of L-Choi, Shibasaki *et al.* used a 1,4-conjugate addition reaction to prepare a protected L-Choi derivative [106, 107]. In Shibasaki *et al.*'s synthesis, the precursor for the 1,4-conjugate addition reaction was built using a chiral phase transfer-catalyzed glycine alkylation affording both high yield and high enantiomeric excess from relatively simple starting materials **61** and **62** (Scheme 5.15). The resulting intermediate **63** was treated with acid, to unmask the ketone, induce alkene migration, and 1,4-conjugate addition. Treatment with Bonjoch's conditions reported by Bonjoch then led to the desired product **65** in enantiomerically pure form.

Currently, the most efficient route to the L-Choi core (Scheme 5.16) is the method reported by Hanessian *et al.* [101, 102]. One interesting facet of their approach is that they avoid the step involving acid-mediated thermodynamic equilibration of the *endo*/*exo* mixture of the fused-ring compound, which can be problematic in some cases. Hanessian *et al.* began their synthesis by diastereoselective alkylation of the protected glutamate **66** at the γ-carbon using 3-butenol triflate as an electrophile [108]. The alkylated compound **67** underwent intramolecular cyclization at elevated temperature once the free secondary amine was exposed. Following conversion of the imide carbonyl **68** into acetate **69**, the substituted pyrrolidine compound was cyclized through a Lewis acid-mediated aza-Prins reaction. The cyclization reaction afforded the desired bicyclic bromide **71** as single isomer with good yield. Hanessian *et al.* have rationalized the outcome of the reaction using their transition state model **70** where

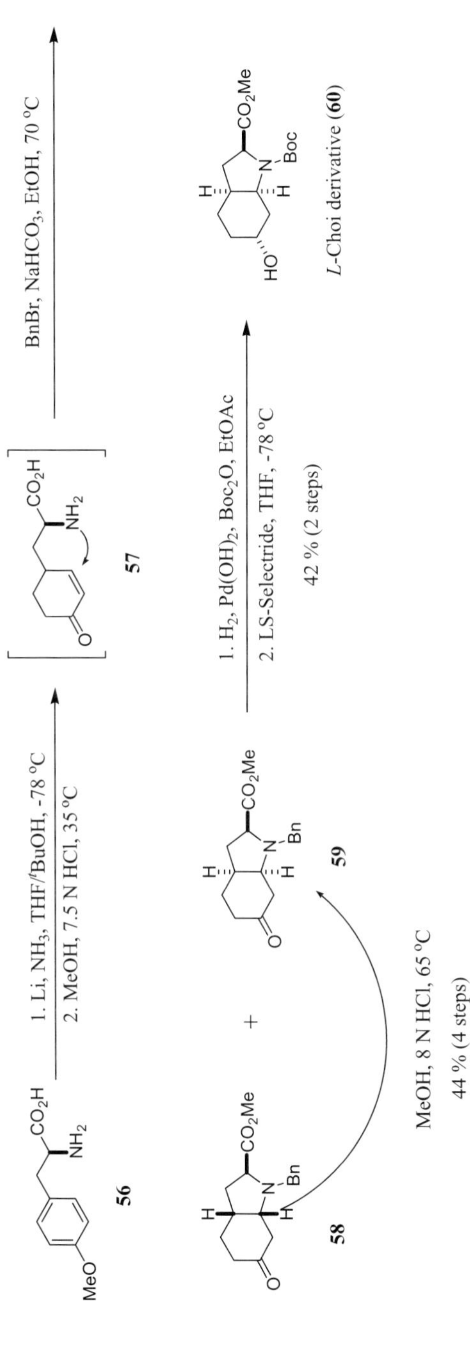

Scheme 5.14 Bonjoch's synthesis of *N*-Boc-protected methyl ester derivative of L-Choi.

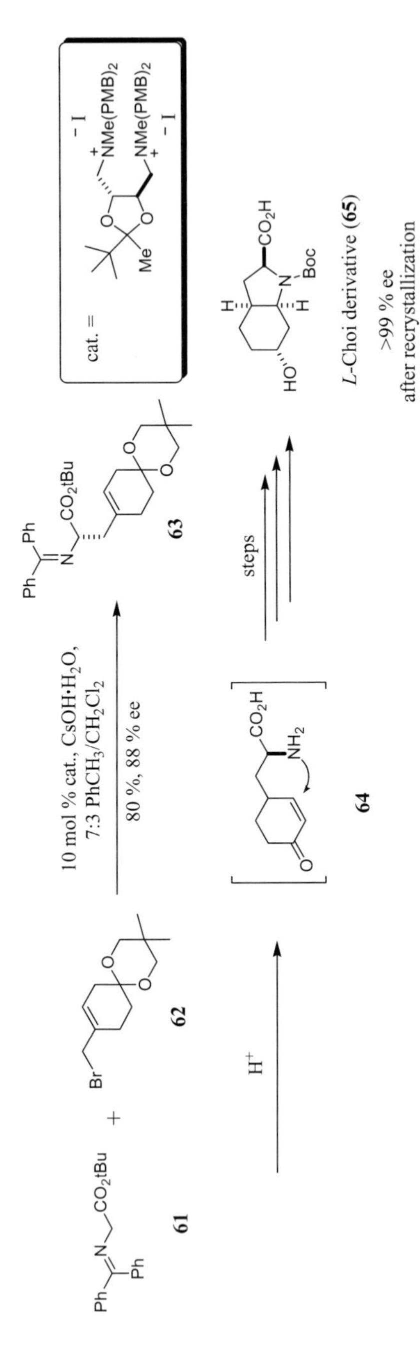

Scheme 5.15 Shibasaki's synthesis of L-Choi derivative.

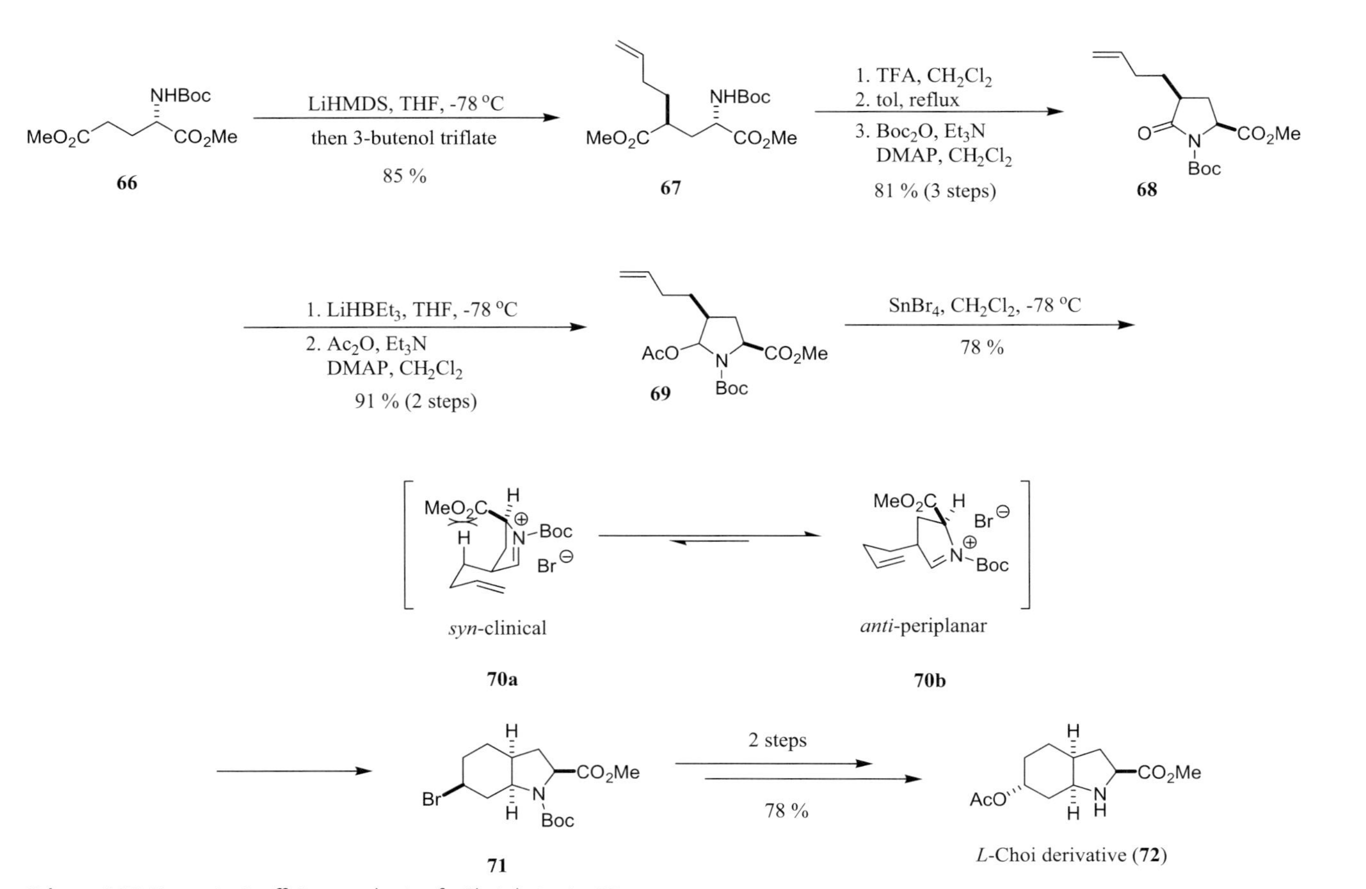

Scheme 5.16 Hanessian's efficient synthesis of L-Choi derivative **72**.

syn-clinal approach of the alkene tether creates an unfavorable steric clash between the axial hydrogen and the methyl ester substituent of the pyrrolidine ring. However, in the case of the *anti*-periplanar approach of the alkene tether, such an interaction is absent. The L-Choi derivative **72** was obtained in 36% yield, starting from L-*N*-Boc-Glu (OMe)-OMe, after displacement of bromine by acetate and Boc removal.

Hydropyrroloindole (Hpi) is another structural motif found in nonproteinogenic natural CAAs incorporated into many natural products, such as chloptosin [109], himastatin [110, 111], okaramines [112], phakellistatins [113], gypsetin [114], kapakahines [115], omphalotins [116], and sporidesmins [117]. These Hpi-containing natural products are known to exhibit impressive antibacterial and anticancer activities; consequently, those compounds and Hpi have become important synthetic targets. Shioiri *et al.* [118], Christophersen *et al.* [119], and Van Vranken *et al.* [120] used a photooxidative method to transform the tryptophan of structure **73** into the Hpi of phakellistatin 3 **74**, although at the expense of low product yield, low diastereoselectivity, and laborious purification of the desired product **74** (Scheme 5.17) [118–120].

On the other hand, Savige has prepared Hpi by epoxidation of tryptophan using *m*-chloroperoxybenzoic acid (*m*CPBA) followed by intramolecular ring closure where the nucleophile was the α-amine [121]. However, this approach was not

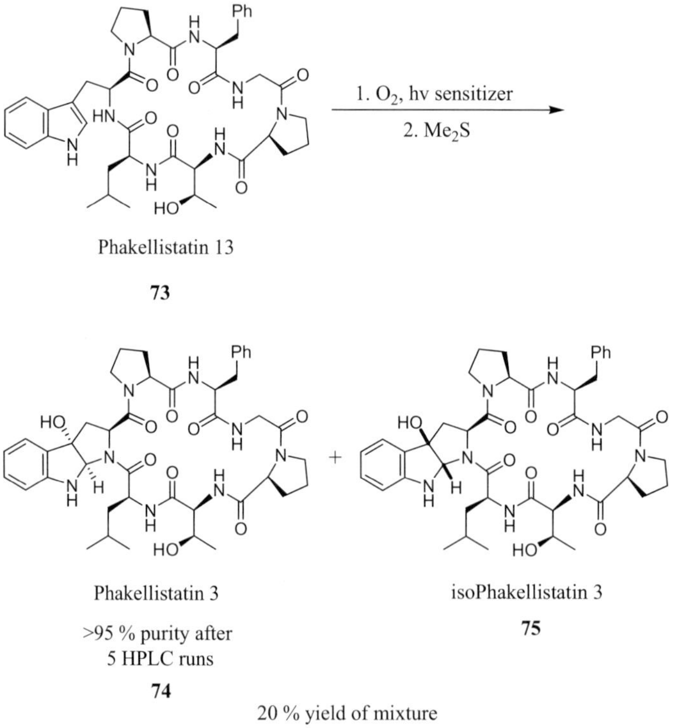

Phakellistatin 13

73

1. O_2, hv sensitizer

2. Me_2S

Phakellistatin 3

>95 % purity after
5 HPLC runs

74

isoPhakellistatin 3

75

20 % yield of mixture

Scheme 5.17 Photooxidative conversion of tryptophan to afford Hpi.

free from the problems seen in the photooxidation method mentioned above. Danishefsky *et al.* overcame all the difficulties related to the Hpi synthesis by using 3,3-dimethyldioxirane (DMDO) as the oxidant. In their synthesis, both *syn-cis* **77** and *anti-cis* **76** were prepared independently as single isomers (Scheme 5.18) [122]. It is interesting to note that, for preparation of the *anti-cis* isomer, intramolecular ring closure was mediated by the bromination of the C-2–C-3 π-bond of the indole core **78**. The dihydropyrroloindole intermediate **79** was treated with DMDO to afford *anti* product **80** with stereoselectivity greater than 15 : 1.

In contrast, the *syn-cis* isomer **83** was prepared by direct oxidation of the C-2–C-3 π-bond on the indole core **82** by DMDO followed by spontaneous intramolecular epoxide ring opening by the α-trityl amine to produce the desired isomer **83** stereospecifically (Scheme 5.19) [122].

Recently, Perrin *et al.* adapted Danishefsky's strategy to prepare the Hpi moiety in linear peptides [123]. Their strategy minimizes protecting group manipulation and increases the ease of incorporation of the Hpi moiety into peptide natural products. The tryptophanylated amino acid **84** was treated with DMDO to produce diastereomers **85** and **86** (Scheme 5.20). Perrin *et al.* reported that the mixture of products can be separated readily, giving access to both diastereomers, *syn-cis* **85** and *anti-cis* **86**, in pure form.

In 2003, Ley *et al.* reported the most efficient synthesis of the *syn-cis* isomer of the Hpi amino acid **91** while en route to the total synthesis of (+)-okaramine C [124, 125]. Ley *et al.*'s method started by protection of the indole nitrogen with Boc_2O and esterification of the carboxylic acid functionality on Cbz-L-Trp **87** (Scheme 5.21). In the key synthetic step, the fully protected tryptophan **88** was treated with *N*-phenylselenyl phthalimide (*N*-PSP) to induce intramolecular cyclization, followed by oxidative cleavage of the Se–C bond with excess *m*CPBA. Under optimized conditions, Ley *et al.* obtained the diastereomerically pure product **90** in near quantitative yield. After removal of the Boc and Cbz groups, the *syn-cis* isomer of Hpi methyl ester **91** was obtained in 74% overall yield, starting from L-*N*-Cbz-Trp-OH **87**, as a single product.

5.3
Noncoded Amino Acids by Chemical Modification of Coded Amino Acids

Numerous variants of β-hydroxy-α-amino acids (Figure 5.2) are found in many complex natural products, such as azinothricin [126], A83586C [127], papuamides [128], polyoxypeptines [44, 46], GE3 [129], vancomycin [130], and cyclosporins [131]. These natural products are very important in both clinical and therapeutic development due to the antibiotic, anticancer, and immunosuppressant activities they possess. Vancomycin, in particular, has drawn much attention in the clinic owing to the fact that it is the final line of defense against methicillin-resistant *Staphylococcus aureus* infection. In addition, β-hydroxy-α-amino acids are used as chiral building blocks in preparing precursors to important molecules [132–137]. Currently, various synthetic methods have been developed to prepare β-hydroxy-α-

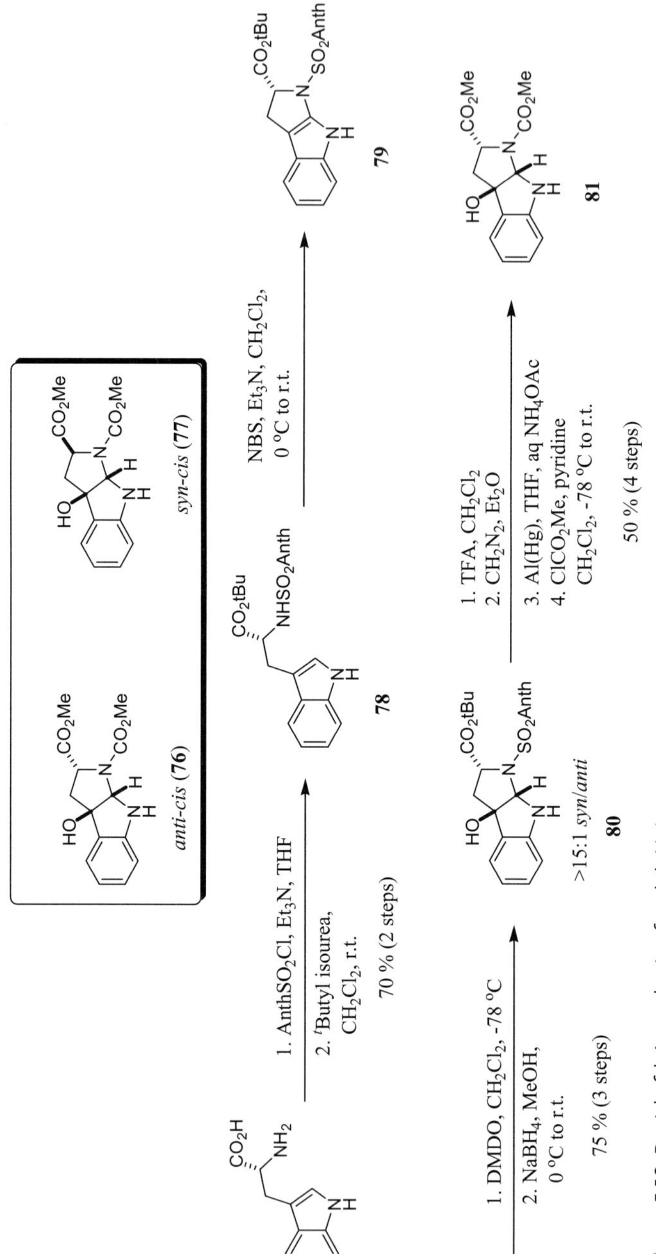

Scheme 5.18 Danishefsky's synthesis of *anti-cis* Hpi.

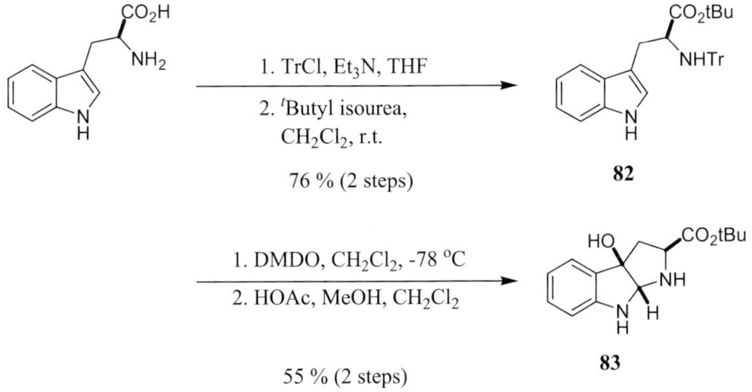

1. TrCl, Et_3N, THF
2. ^{t}Butyl isourea, CH_2Cl_2, r.t.
76 % (2 steps)

82

1. DMDO, CH_2Cl_2, -78 °C
2. HOAc, MeOH, CH_2Cl_2
55 % (2 steps)

83

Scheme 5.19 Danishefsky's synthesis of *syn-cis* Hpi.

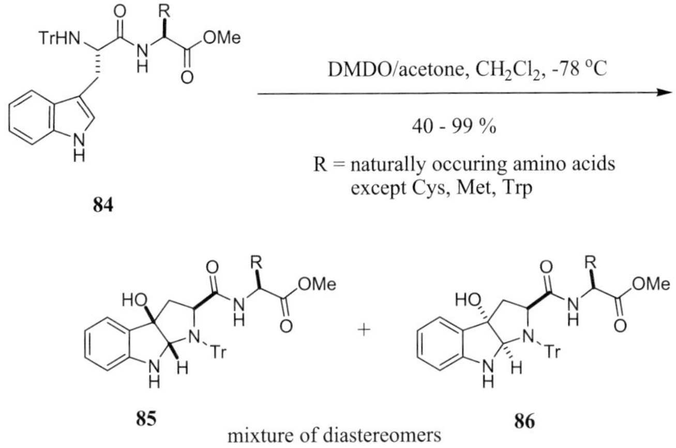

84

DMDO/acetone, CH_2Cl_2, -78 °C
40 - 99 %
R = naturally occuring amino acids except Cys, Met, Trp

85 + 86
mixture of diastereomers

Scheme 5.20 Perrin's synthesis of an Hpi dipeptide.

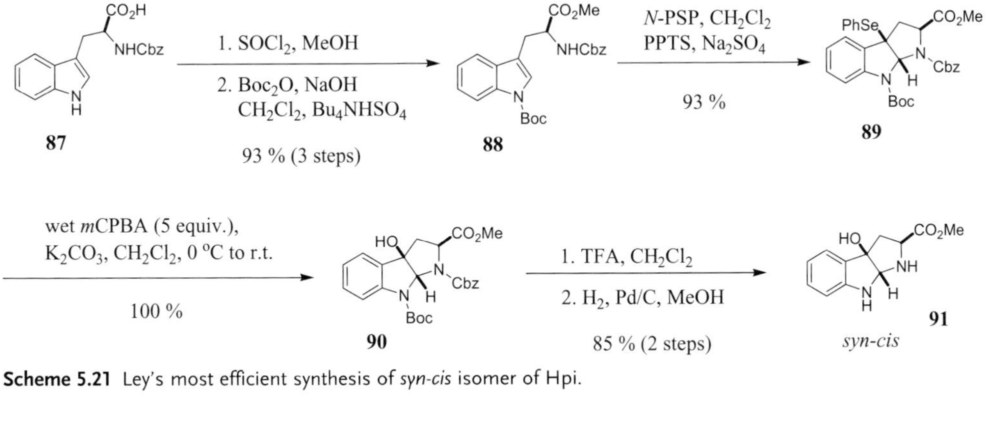

87

1. $SOCl_2$, MeOH
2. Boc_2O, NaOH CH_2Cl_2, Bu_4NHSO_4
93 % (3 steps)

88

N-PSP, CH_2Cl_2 PPTS, Na_2SO_4
93 %

89

wet *m*CPBA (5 equiv.), K_2CO_3, CH_2Cl_2, 0 °C to r.t.
100 %

90

1. TFA, CH_2Cl_2
2. H_2, Pd/C, MeOH
85 % (2 steps)

91 *syn-cis*

Scheme 5.21 Ley's most efficient synthesis of *syn-cis* isomer of Hpi.

Papuamide B

A83586C R_1 = Me, R_2 = Me, R_3 = Me
Azinothricin R_1 = Me, R_2 = CH_2OMe, R_3 = Et
GE3 R_1 = ^{i}Bu, R_2 = Me, R_3 = Me

Cyclosporin A

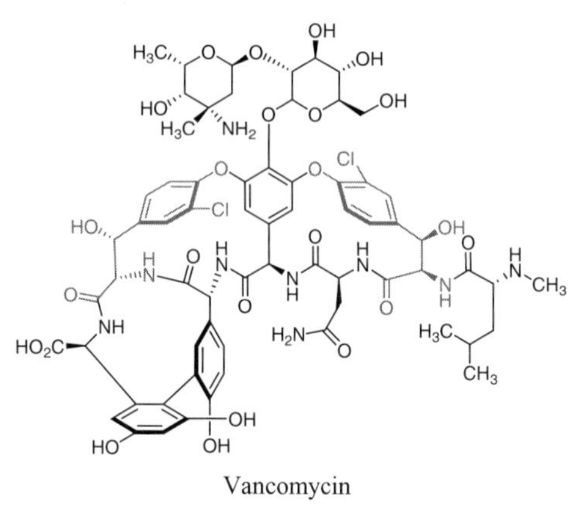

Vancomycin

Polyoxypeptin A R = OH
Polyoxypeptin B R = H

Figure 5.2 Representative natural products containing β-hydroxy-α-amino acids.

amino acids stereoselectively, which include asymmetric aldol reaction [138–156], alkylation [153–156], electrophilic amination and allylic amination [157, 158], conjugate addition [159], cyanation [160], enantioselective hydrogenation [161], selective hydrolysis [135], rearrangement [162–165], regioselective aziridine ring opening [135, 166–168], dynamic kinetic resolution (DKR) [169–175], Sharpless asymmetric aminohydroxylation [176–178], Sharpless asymmetric dihyroxylation [179–182], Sharpless asymmetric epoxidation [183–185], and Strecker synthesis [14, 135, 186]. In this section, some of these methodologies are showcased. The reader is also directed to a review by Hamada for complete discussion of asymmetric synthesis of β-hydroxy-α-amino acids [187, 279, 280].

To synthesize β-hydroxyleucine, an asymmetric aldol reaction has been the method of choice. Corey *et al.* reported a scalable synthesis of β-hydroxyleucine using a chiral boron reagent **92** to obtain two diastereomers of 3-hydroxyleucine [150]. The (*Z*)-enolate, produced by reacting *tert*-butyl bromoacetate **93** with triethylamine and chiral bromoborane **92**, was alkylated with isobutyraldehyde to afford the *anti* adduct **94** in excellent yield with high diastereoselection (Scheme 5.22). Under basic conditions, the bromide was displaced from the *anti* adduct **94** to form epoxide **95**. The epoxide ring **95** was opened with benzylamine, followed by benzyl group removal with Pearlman's catalyst to give (2*S*,3*S*)-3-hydroxyleucine **97** in 50% overall yield, starting from *tert*-butyl bromo acetate, with acceptable enantiomeric purity. To obtain the (2*R*,3*S*)-3-hydroxyleucine isomer **100**, the secondary alcohol of the *anti* aldol adduct **94** was protected with a silyl group, followed by stereocenter inversion at C-2 by displacement of the bromide with excess sodium azide. Azide **98** was reduced and removal of all protecting groups afforded the desired isomer **100** as a methanesulfonate salt, in high yield.

Maruoka *et al.* also investigated asymmetric aldol reactions to prepare β-hydroxyleucine and other β-hydroxy-α-amino acid derivatives [143]. In their synthesis, isobutyraldehyde **101** was reacted with a glycinate Schiff's base **102** under biphasic conditions in the presence of chiral ammonium salt **103**. Under optimized conditions, the *anti* aldol adduct **104**, leading to (2*S*,3*S*)-3-hydroxyleucine, was obtained in good yield with high diastereoselection and enantiomeric purity (Scheme 5.23). The high diastereoselection in this reaction originates presumably from formation of a sterically congested ammonium enolate salt. The large ring of the quaternary ammonium cation induces formation of (*E*)-enolate **105a** geometry selectively over the (*Z*)-enolate **105b** while blocking the *re* face of the enolate preferentially. In addition, the overwhelming steric hindrance between the R substituent of the approaching aldehyde, and the large ring structure of the ammonium cation (**105b**) directs the attack on the aldehyde to the conformation where the R substituent is gauche to both the 2-imino moiety and the *tert*-butoxy group of the enolate (**105a**).

Barbas *et al.* successively prepared β-hydroxyleucine and other β-hydroxy-α-amino acid derivatives using an organocatalyst for the aldol reaction [145]. The glycine aldehyde derivative **106** and isobutyraldehyde underwent an aldol reaction in the presence of a catalytic amount of L-proline. The resulting aldol adduct **107** was oxidized to the carboxylic acid under Pinnick oxidation conditions, followed by

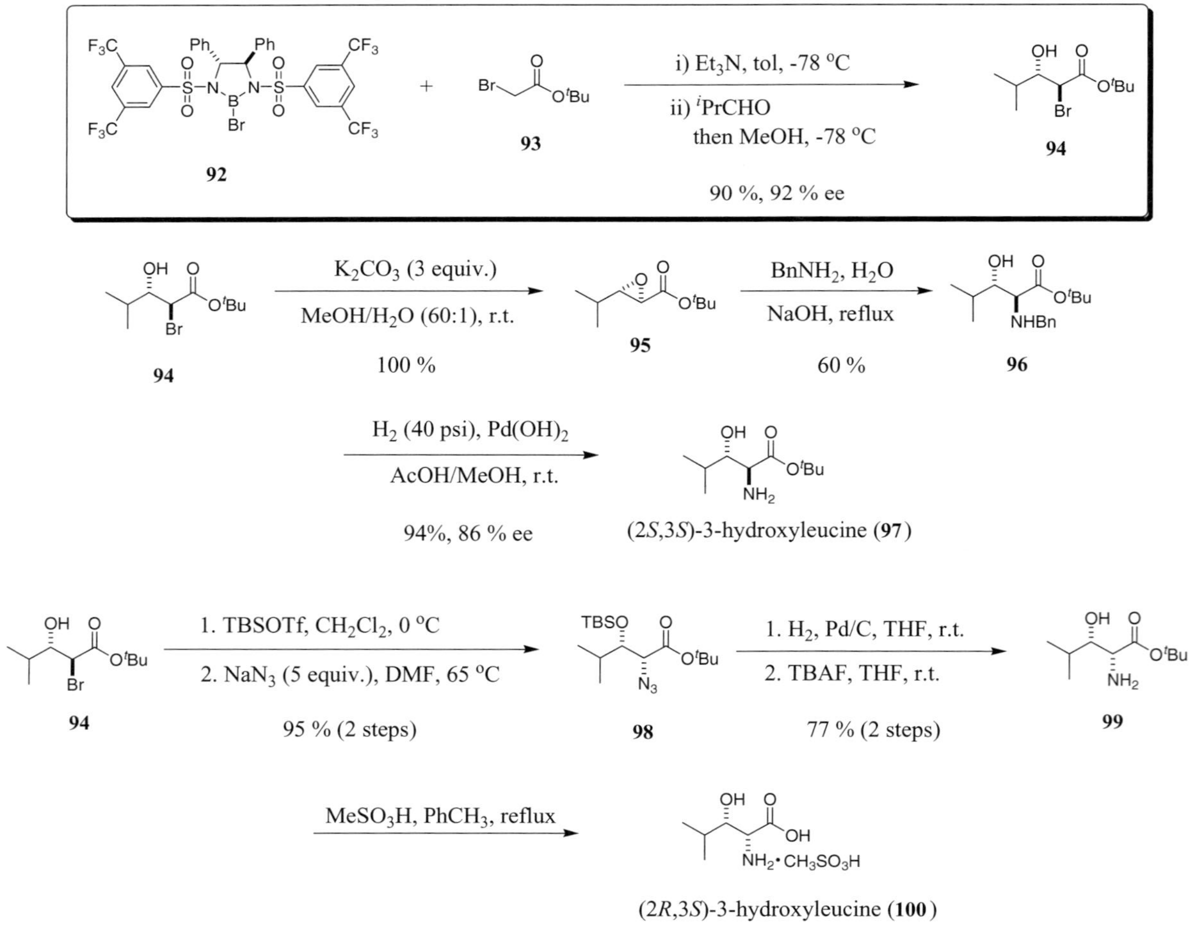

F_3C, CF_3, Ph, Ph, O=S=O, N, B, Br, CF_3, F_3C

92 + **93** (Br, O, O^tBu) → i) Et_3N, tol, -78 °C; ii) ^iPrCHO then MeOH, -78 °C; 90 %, 92 % ee → **94** (OH, O, O^tBu, Br)

94 → K_2CO_3 (3 equiv.), MeOH/H_2O (60:1), r.t.; 100 % → **95** (O, O, O^tBu) → $BnNH_2$, H_2O, NaOH, reflux; 60 % → **96** (OH, O, O^tBu, NHBn)

→ H_2 (40 psi), $Pd(OH)_2$, AcOH/MeOH, r.t.; 94%, 86 % ee → (OH, O, O^tBu, NH_2) (2*S*,3*S*)-3-hydroxyleucine (**97**)

94 → 1. TBSOTf, CH_2Cl_2, 0 °C; 2. NaN_3 (5 equiv.), DMF, 65 °C; 95 % (2 steps) → **98** (TBSO, O, O^tBu, N_3) → 1. H_2, Pd/C, THF, r.t.; 2. TBAF, THF, r.t.; 77 % (2 steps) → **99** (OH, O, O^tBu, NH_2)

→ $MeSO_3H$, $PhCH_3$, reflux → (OH, O, OH, $NH_2 \cdot CH_3SO_3H$) (2*R*,3*S*)-3-hydroxyleucine (**100**)

Scheme 5.22 Corey's synthesis of β-hydroxyleucine.

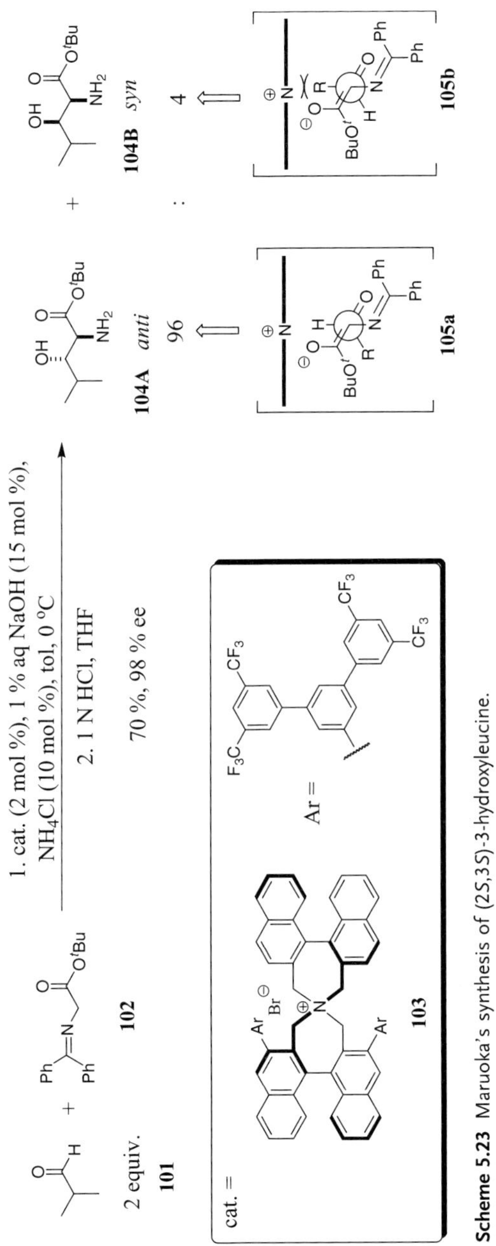

Scheme 5.23 Maruoka's synthesis of (2*S*,3*S*)-3-hydroxyleucine.

esterification with trimethylsilyldiazomethane to give the *N*-protected (2*S*,3*S*)-3-hydroxyleucine derivative **108** in 73% yield with great diastereoselectivity and enantioselectivity. Of key importance to this reaction was the use of the phthalimide as a nitrogen protecting group. The enamine intermediate can react potentially with the aldehyde in two different modes (A versus B in Scheme 5.24). The authors stated that the regioselection of the aldol reaction, via path A, was optimal only with the phthalimide protecting group.

An interesting aldol strategy was used to prepare two diastereomers of β-hydroxy-α-amino acids by Evans *et al.* The Evans group's work was based on earlier reports made by Ito and Hayashi about stereoselective synthesis of oxazoline derivatives [161]. In addition, Suga and Ibata have used similar methods to synthesize oxazoline derivatives in the presence of uncharacterized Lewis acid catalysts [188]. Evans *et al.* improved the methodology with the development of chiral (salen)Al catalyst **114**. 5-Methoxyoxazole **109** was treated with the aluminum catalyst **114** followed by the addition of the aryl aldehyde (table in Scheme 5.25), which yielded 2-oxazoline-4-carboxylates **111a** and **111b** in good yields with high diastereoselectivities and enantioselectivities. Moreover, *cis*-2-oxazoline-4-carboxylates **112** can be epimerized to the *trans* isomers **113** by treatment with catalytic 1,8-diazabicyclo[5.4.0]undec-7-ene (DBU). The oxazoline products **113** can be treated with acid to open the ring structure to afford β-hydroxyphenylalanine and β-hydroxytyrosine derivatives. A few advantages of Evans's methodology over other methods include the reduction of catalyst loading and decreased reaction time. However, one drawback of this reaction is that aliphatic aldehydes are unreactive under standard reaction conditions.

Stereoselective hydrogenation via DKR is another popular method used to synthesize a variety of β-hydroxy-α-amino acids (Scheme 5.26). This methodology was pioneered by Noyori *et al.* by utilizing the Ru-(*R*)-2,2′-bis(diphenylphosphanyl)-1,1′-binaphthyl (BINAP) under 100 atm of hydrogen to obtain *syn*-selective β-hydroxy-α-amino acids from α-amino-β-keto esters [171]. In contrast, Hamada *et al.* used a Ru-(*S*)-BINAP catalyst to prepare *anti*-selective β-hydroxy-α-amino acids [169]. In both cases, the yield of product was good, with high diastereo- and enantioselectivity. It was speculated, although it is not yet completely clear, that the difference in the diastereoselectivity between the two methods arises from alternative cyclic transition states. In Hamada *et al.*'s proposed mechanism, the substrate–catalyst complex forms a free amine-assisted five-membered ring transition state **115** for the hydrogenation reaction. On the other hand, for Noyori *et al.*'s method, the substrate is *N*-benzoyl-protected, thus the hydrogenation probably goes through a six-membered transition state **116**.

Recently, White *et al.* applied a Pd(II)-promoted allylic C—H amination method to build an oxazolidinone derivative diastereoselectively, which was used as a precursor in preparation of 3-hydroxyleucine [158]. Enantioenriched homoallylic *N*-tosyl carbamate **117** was treated with a catalytic amount of bis-sulfoxide palladium species **118** in the presence of phenyl benzoquinone to give the desired oxazolidinone **120** in moderate yield but good diastereomeric ratio. The basis of the methodology is utilization of a weak internal nucleophile, *N*-tosyl carbamate, to eliminate the

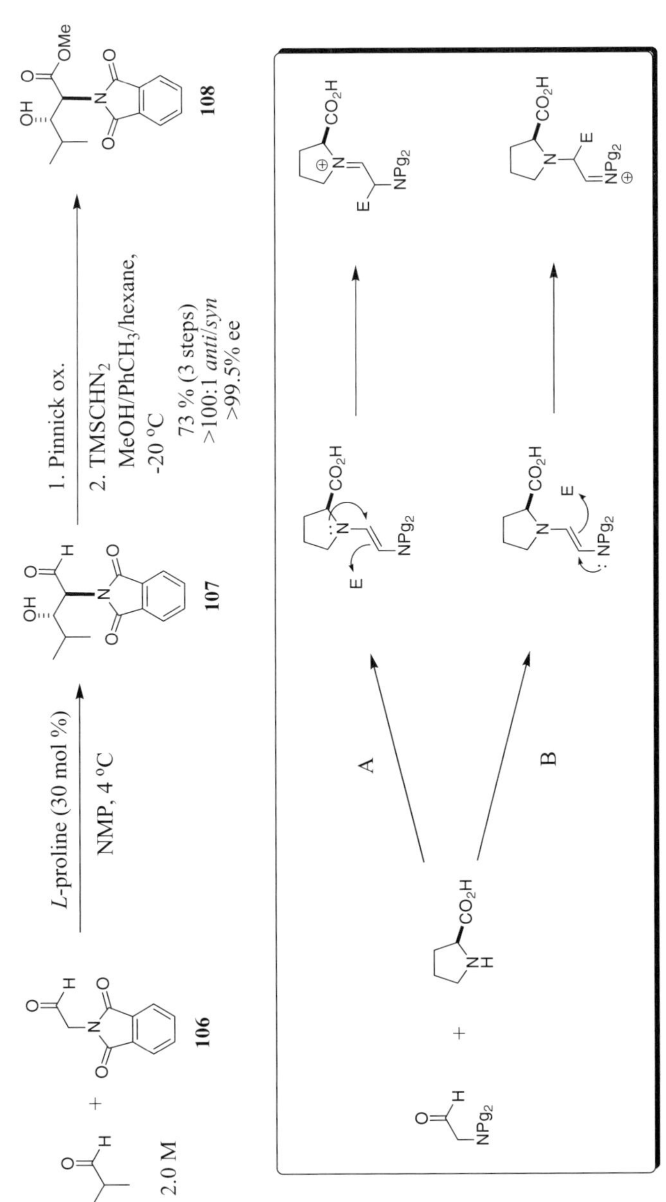

Scheme 5.24 Barbas' organocatalytic synthesis of the protected (2S,3S)-3-hydroxyleucine.

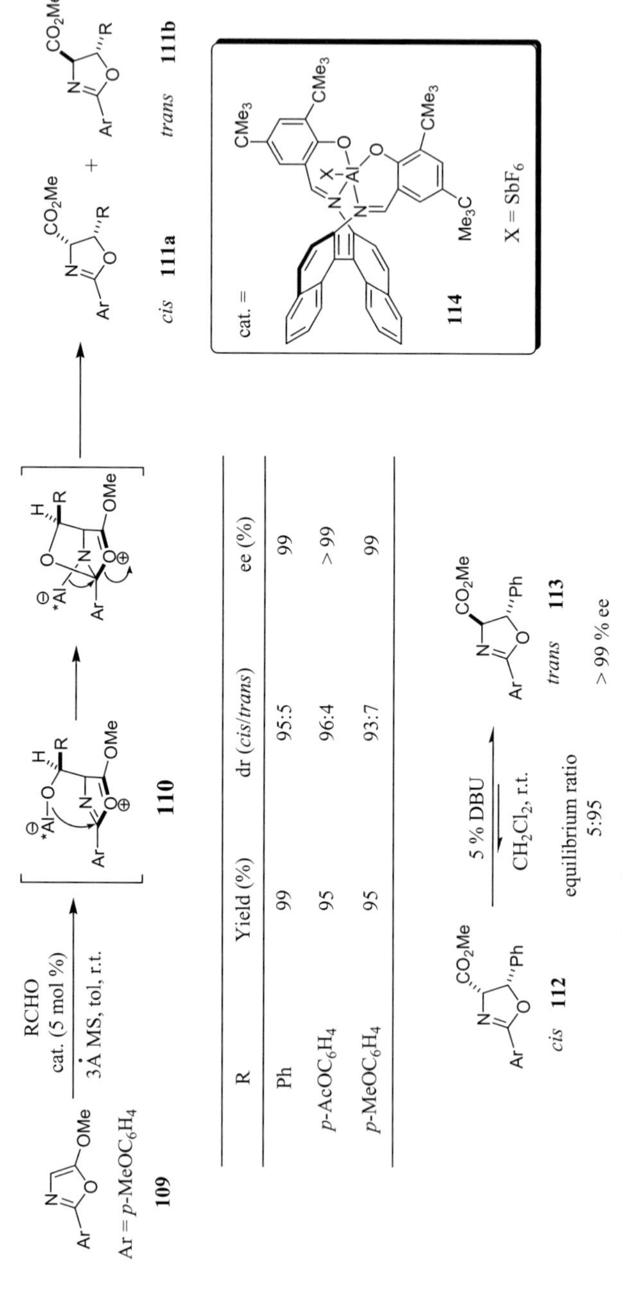

R	Yield (%)	dr (*cis*/*trans*)	ee (%)
Ph	99	95:5	99
p-AcOC$_6$H$_4$	95	96:4	> 99
p-MeOC$_6$H$_4$	95	93:7	99

Scheme 5.25 Evans' synthesis of β-hydroxyphenylalanines and β-hydroxytyrosines.

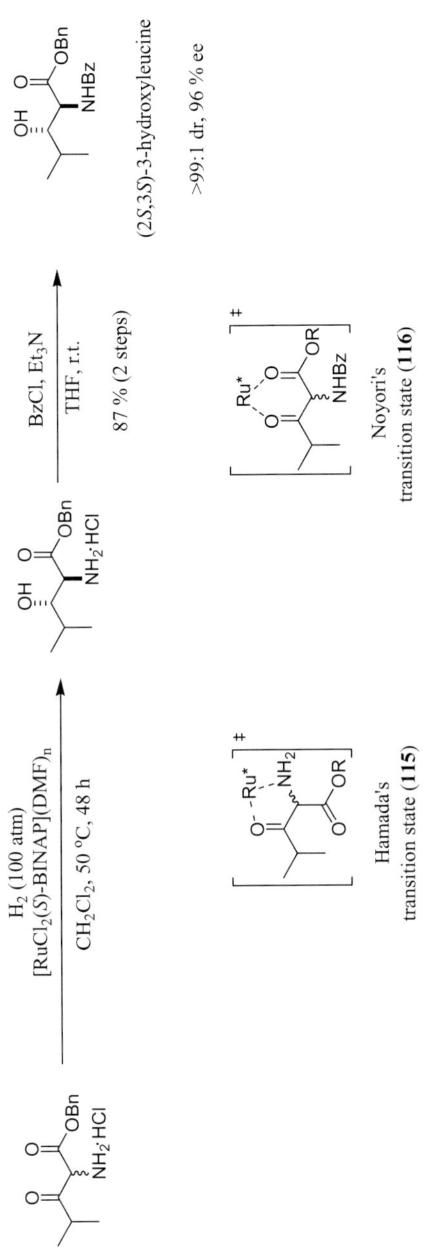

H_2 (100 atm)
$[RuCl_2(S)\text{-BINAP}](DMF)_n$
CH_2Cl_2, 50 °C, 48 h

BzCl, Et_3N
THF, r.t.
87 % (2 steps)

(2*S*,3*S*)-3-hydroxyleucine
>99:1 dr, 96 % ee

Hamada's transition state (**115**)

Noyori's transition state (**116**)

Scheme 5.26 Stereoselective hydrogenation via DKR.

possibility of electrophilic allylic α-C–H bond cleavage. The allylic C–H cleavage was mediated by the Pd(II) catalyst followed by formation of a π-allyl palladium species **119**. The acetate counter ion of palladium deprotonates N–H of the *N*-tosyl carbamate and then the ring was closed to give the oxazolidinone **120**. Another important factor in this reaction was the use of benzoquinone which reoxidizes Pd(0) and regenerates the acetate base for the next catalytic cycle. To finish the synthesis, the terminal olefin **120** was oxidized to the carboxylic acid **121** with a catalytic amount of ruthenium catalyst in the presence of sodium periodate. The tosyl group was removed under standard conditions followed by oxazolidinone hydrolysis to afford enantiopure (2*R*,3*S*)-3-hydroxyleucine **122** (Scheme 5.27).

Panek *et al.* investigated the Sharpless asymmetric aminohydroxylation to prepare enantiopure 3-hydroxyleucine [177]. The Panek group started their synthetic investigation of this reaction using an α,β-unsaturated alkyl ester with the $(DHQ)_2$-AQN ligand; however, it gave the product with the undesired regiochemistry. The regiochemistry was reversed by using *para*-substituted aryl ester substrates **123**. Under optimized conditions, the reaction gave the desired product **124** in moderate yield with high regio- and enantioselection (Scheme 5.28). The same reaction conditions were used on a 40-mmol scale and gave good yield of product without suffering a decrease in selectivity. The regio- and enantioselectivity was further improved by a single round of recrystallization. The reversal of the regiochemistry was accounted for by the difference in ester geometry of alkyl ester substrates versus aryl ester substrates. Based on a nuclear Overhauser enhancement experiment of the two substrates, the alkyl ester adopts an *s-cis* conformation, while the aryl ester is *s-trans*. The authors speculated that this difference alters the geometrical build up of the ligand-bound osmium species and the substrate complex between the alkyl ester and aryl ester substrates [189]. In addition, it was found that the rate of the reaction was increased by increasing the electron-donating ability of the substituent on the *para* position of the aryl ring. To rationalize this effect, the authors proposed that the electron donating ability of the aryl ring stabilizes the partial positive charge formed on the olefinic carbons in the transition state.

Ariza *et al.* prepared two diastereomers of 3-hydroxyleucine, (2*S*,3*S*) and (2*R*,3*S*), from a single starting material utilizing a Pd(0)-catalyzed allylic alkylation (Scheme 5.29) [190]. The isopropyl-2-yn-1,4-diol **125** was diverged into (*Z*) **127** and (*E*) **126** unsaturated diols via stereoselective reduction of the alkyne. Each isomer was treated with tosyl isocyanate to give the dicarbamate compounds followed by treatment with Pd(0) catalyst to provide the desired oxazolidinone derivatives **128** and **129**. For both isomers, the reaction gave a high yield of product with excellent diastereoselective ratio. The olefins were oxidized to the aldehydes by ozonolysis then further oxidized to carboxylic acids using sodium chlorite and hydrogen peroxide to give the oxazolidinone precursors that can be transformed into 3-hydroxyleucines **130** and **131** via tosyl group removal and oxazolidinone ring opening. It is interesting that the olefin geometry at the beginning of the synthesis determines the stereochemical outcome of the product. In the case of the (*E*)-olefin isomer **126**, the Pd(0) catalyst approaches the olefin from both faces without selectivity, which gives rise to two possible conformations of the π-allyl complex (Scheme 5.29). The π-allyl

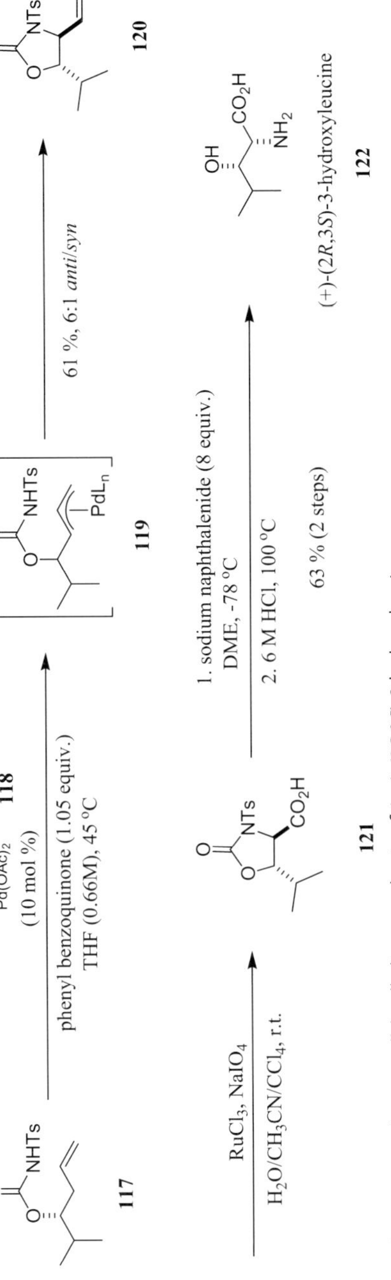

Scheme 5.27 White's π-allyl palladium synthesis of (+)-(2S,3S)-3-hydroxyleucine.

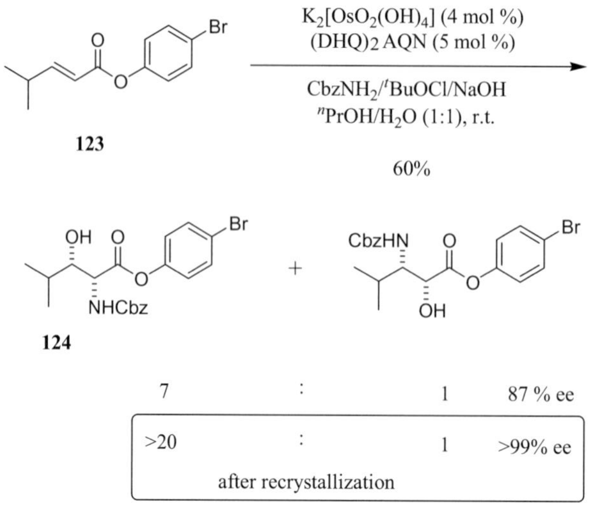

$K_2[OsO_2(OH)_4]$ (4 mol %)
(DHQ)2 AQN (5 mol %)
$CbzNH_2/^tBuOCl/NaOH$
$^nPrOH/H_2O$ (1:1), r.t.
60%

123 → 124 + regioisomer

7 : 1 87 % ee

>20 : 1 >99% ee after recrystallization

Scheme 5.28 Sharpless aminohydroxylation synthesis of enantiopure 3-hydroxyleucine.

complex, which leads to formation of undesired product, equilibrates into a conformation that minimizes steric hindrance between ligands on the palladium and the alkyl chain of the substrate. On the other hand, a different type of π-allyl complex (Scheme 5.29) was proposed to explain the stereoselection of the reaction with (*Z*)-olefin substrate **127**. Both faces of the (*Z*)-olefin substrate are homotopic giving rise to two possible π-allyl complex geometries. In this case, steric hindrance between the alkyl chain and the carbamate group of the substrate determines the stereochemistry of the product. The π-allyl complex with the alkyl chain within the van der Waals radii of the carbamate group equilibrates into a conformation in which the unfavorable interaction can be avoided.

3-Hydroxyaspartic acid is another type of β-hydroxy-α-amino acid that is valuable in pharmaceutical research. Hanessian *et al.* were able to obtain two diastereomers, (3*R*)- and (3*S*)-hydroxy-L-aspartic acid from L-aspartic acid dimethyl ester by treating it with either an oxaziridine reagent or a molybdenum reagent as an oxidizing agent under basic conditions [191]. Amouroux *et al.* prepared the two diastereomers of 3-hydroxy-D-aspartic acid using L- and D-diethyl tartrate as starting materials [192]. Another interesting synthesis was completed by Cardillo *et al.* where *N*-benzoyl L-aspartic acid dimethyl ester **132** was oxidized at C-3 in three synthetic steps (Scheme 5.30) [194]. The key to their reaction was the formation of a dianion in a regioselective manner, followed by iodination at C-3 *in situ* to induce intramolecular ring closure, to give the *trans* oxazoline compound **133** in high yield with good diastereomeric ratio. The final product, (3*S*)-hydroxy-L-aspartic acid **135**, was obtained by the hydrolysis of the oxazoline derivative **133** in an acidic medium. Although this methodology is efficient, one disadvantage is that the oxazoline **133** is always accompanied by the undesired elimination product **134**.

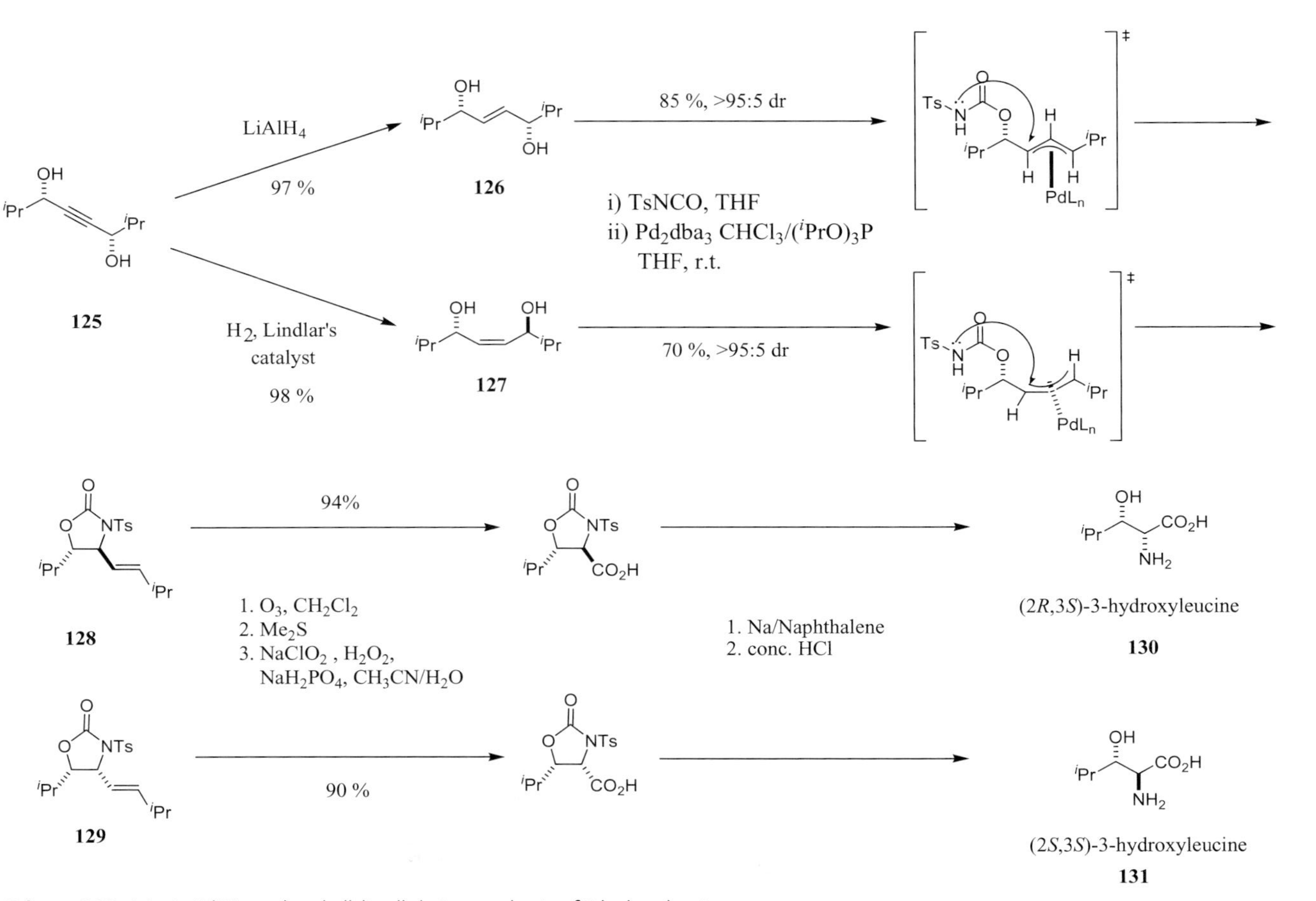

Scheme 5.29 Ariza's Pd(0)-catalyzed allylic alkylation synthesis of 3-hydroxyleucine.

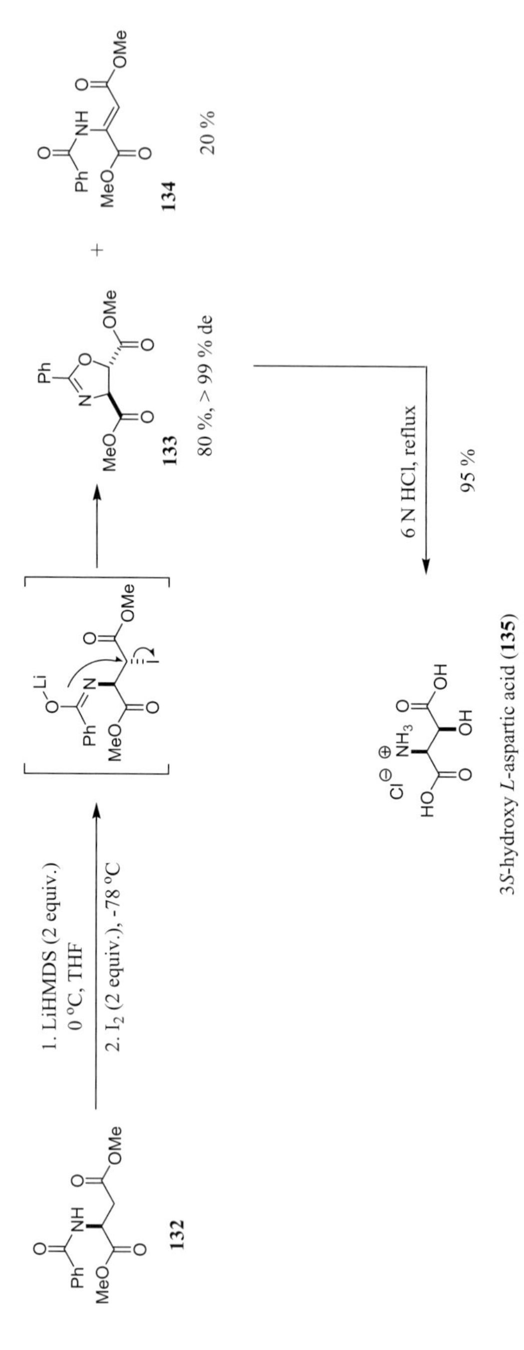

Scheme 5.30 Cardillo's synthesis of (3S)-hydroxy-L-aspartic acid.

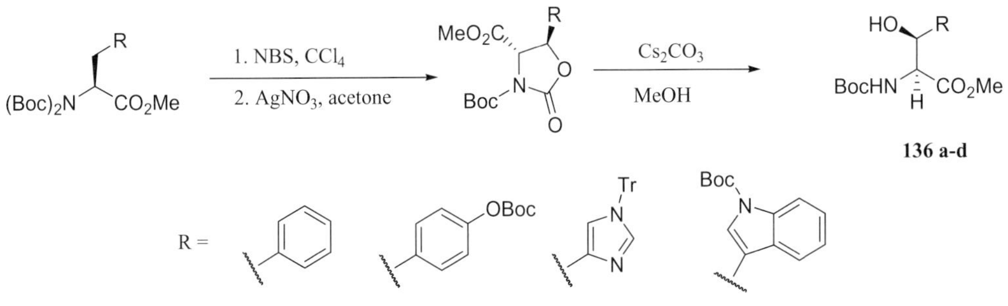

(Boc)$_2$N CO$_2$Me with R; 1. NBS, CCl$_4$; 2. AgNO$_3$, acetone; MeO$_2$C, Boc N, O; Cs$_2$CO$_3$, MeOH; HO, R, BocHN, H, CO$_2$Me

136 a-d

R = OBoc, Tr, N, N, Boc, N

136 a-d aromatic groups

Scheme 5.31 Crich's synthesis of β-hydroxy aromatic amino acids.

β-Hydroxyhistidine is an unusual amino acid constituent found in such peptide natural products as exochelin MN [133], PF244 [194], corrugatin [195], tallysomycin [196], and the important anticancer family of bleomycins [197]. Both the *threo* and *erythro* isomers have been synthesized previously through various aldol strategies [198, 199]. Most recently, Crich *et al.* have been able to make β-hydroxyhistidine **136** and other β-hydroxy aromatic amino acids **136a–d** through the use of a radical bromination/oxazolidinone cyclization strategy (Scheme 5.31) [200].

β-Methyl aspartate is an extremely important noncoded amino acid due to its attractive biological activities [135, 201], and its presence in protein phosphatase inhibitor peptides like motuporin [5] and the microcystin family [202]. As a result, many diverse synthetic routes to its structure have been developed. The classic laboratory method of producing β-methylaspartate is to methylate directly the β-carbon of an aspartic acid diester **137**. Alkylation with methyl iodide occurs in excellent yield, but with poor diastereoselectivity (2 : 3, *syn* : *anti* for **138**). To prevent epimerization of the α-carbon of aspartate during methylation, Wolf and Rapoport first developed the PhFl protecting group **139** (Scheme 5.32) [203]. In addition to its sheer steric shielding, the PhFl group acts as a stereoelectronic guard against enolization of the α-stereocenter. Bond rotation to the reactive conformation that

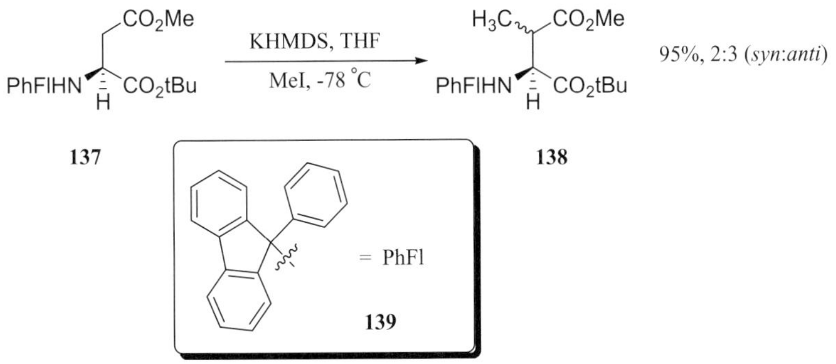

CO$_2$Me, PhFlHN, H, CO$_2$tBu; KHMDS, THF; MeI, -78 °C; H$_3$C, CO$_2$Me, PhFlHN, H, CO$_2$tBu; 95%, 2:3 (*syn:anti*)

137 **138**

= PhFl

139

Scheme 5.32 Classic β-methylaspartate synthesis by β-carbon alkylation.

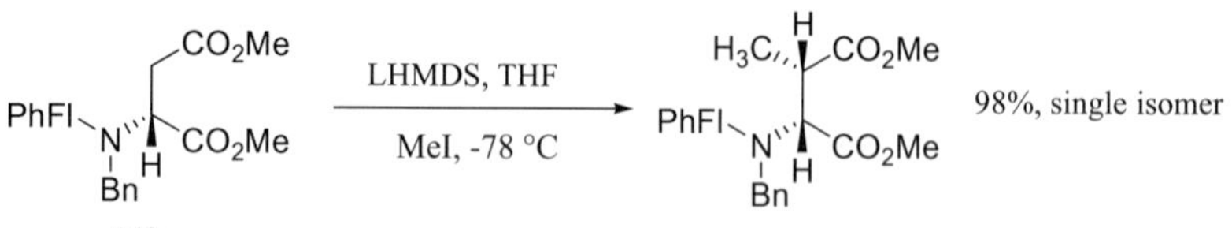

Scheme 5.33 Chamberlin's improved methylation towards synthesis of β-methylaspartate.

maximizes orbital overlap between the α-proton and the π-system of the carbonyl is prevented by severe developing allylic strain [5].

Vastly improved diastereoselectivity for methylation has been achieved through careful choice of reaction conditions and substrate protecting groups. For example, Chamberlin *et al.* [5] observed complete regioselectivity and stereoselectivity of methylation using lithium hexamethyldisilazide as the base with PhFl/benzyl-protected aspartate dimethyl ester **140**. Preparation of the D-*syn*-β-methyl diastereomer **141** has been done on scales of up to 30 g of material, with no chromatography, in almost quantitative yield (Scheme 5.33).

Alternatively, the L-*syn*-β-methylaspartate **143** can be accessed through methylation of the closed β-lactam form of L-aspartic acid **142**. Hanessian *et al.* demonstrated that diastereoselective alkylations of aspartate-derived β-lactams are useful for the synthesis of many optically pure 3-alkyl aspartates (Scheme 5.34) [191].

Alkylation of chiral glycine equivalents, as opposed to aspartate residues, has also provided β-alkyl aspartates in good yields and with excellent selectivity. Attachment of a chiral auxiliary to the glycine unit can provide enantiodiscrimination during the alkylation event with racemic electrophiles to give single isomer products. For example, Davies *et al.* have demonstrated alkylation of a chiral diketopiperazine moiety with racemic 2-bromopropanoate, leading to 3-methylaspartates in a diastereomeric ratio of (94.5 : 5.5, *trans*: *cis*). After hydrolysis of the diketopiperazine ring and distillation, both *syn*- and *anti*-β-methylaspartates can be obtained in good yield and greater than 98% e.e. [204, 205].

Alkylation of glycine equivalents has also been done under asymmetric phase-transfer conditions. An interesting example from Maruoka's laboratory allows access to the *syn*-diastereomers of β-methylphenylalanine in 88% yield with 95% e.e. [206]. Subsequent controlled epimerization yields the *anti* diastereomers, which lends the method to practical use for obtaining all four stereoisomers of β-methylphenylala-

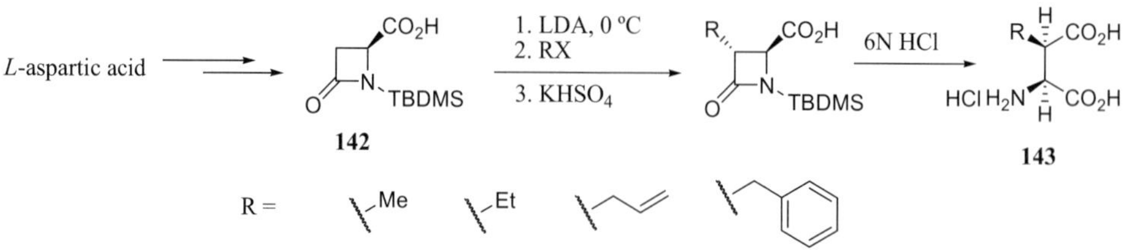

Scheme 5.34 Hanessian's synthesis of L-*syn*-β-alkyl aspartate derivatives.

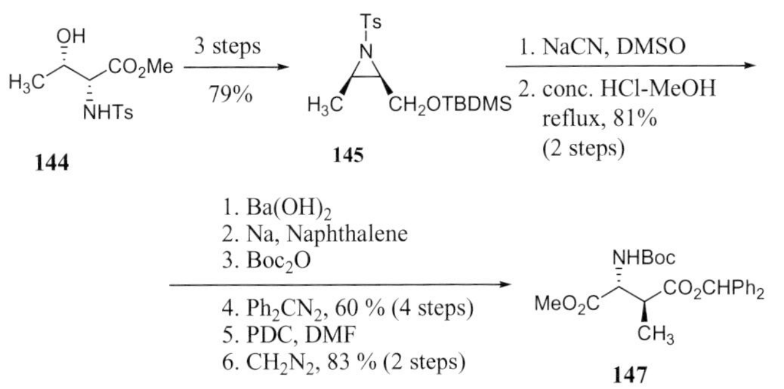

Scheme 5.35 Valentekovich and Schreiber's synthesis of D-*syn*-β-methylaspartate.

nine. The extension of this reaction to β-methylaspartate, however, led to decreased yield and stereoselectivity.

Valentekovich and Schreiber prepared a differentially protected D-*syn*-β-methylaspartate residue during the course of their total synthesis of motuporin (Scheme 5.35) [207]. Although lengthy, the synthesis allows access to an intermediate lactone that can be used to prepare two other fragment portions of motuporin. The preparation of D-*syn*-β-methylaspartate begins from *p*-toluenesufonyl-protected D-threonine methyl ester **144**. The common lactone **146** is synthesized in five steps through an intermediate tosyl aziridine **145**. Hydrolysis of the lactone **146**, an exchange of nitrogen protecting groups, protection of the carboxyl group, and oxidation/protection of the α-hydroxy moiety leads to an orthogonally protected aspartate **147**.

An interesting alternative strategy for the synthesis of β-methylaspartates utilizes an ester enolate Claisen rearrangement of a chiral amino acid ester [208] (Scheme 5.36). The clear benefit of this method is the ability to produce both the *syn* and *anti* diastereomers, depending upon which starting olefin geometry is chosen. Ohfune *et al.* prepared these compounds from chiral Boc-protected glycine esters **148*E*** and **148*Z*** [209]. Upon treatment with lithium diisopropylamide and zinc chloride, the intermediate ester enolate undergoes rearrangement, presumably through a six-membered ring transition state, to give either the *syn* or *anti* isomers

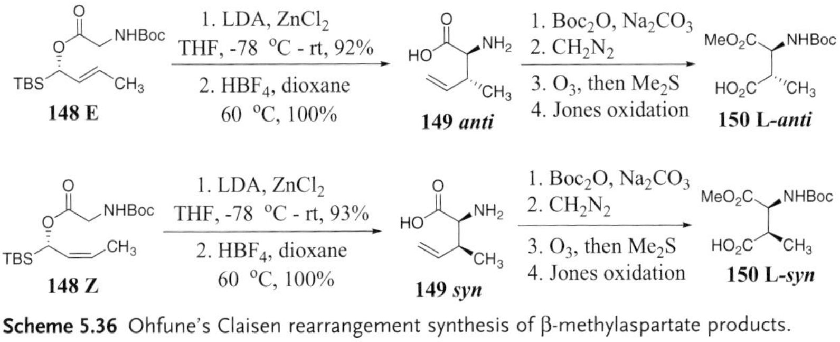

Scheme 5.36 Ohfune's Claisen rearrangement synthesis of β-methylaspartate products.

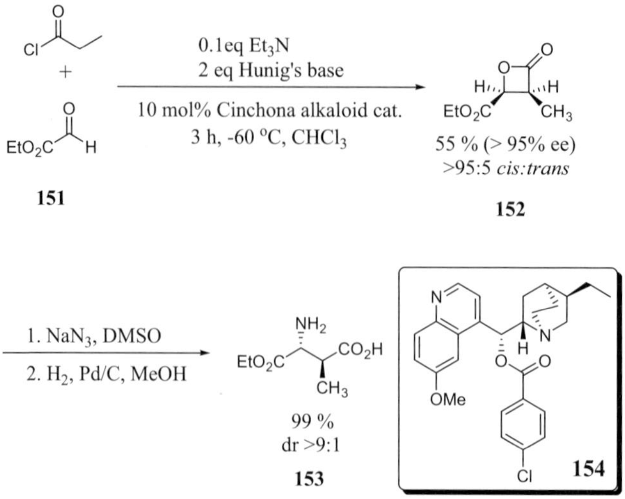

Scheme 5.37 Armstrong's organocatalytic synthesis of *anti*-β-methylaspartate ethyl ester.

149 *anti* and **149 *syn*** with greater than 20 : 1 selectivity. Subsequent amino acid protection and oxidative cleavage of the olefin yields the β-methylaspartate products **150 *anti*** and **150 *syn*** [201].

The latest strides in organocatalytic methods have also led to contributions in the field of amino acid synthesis. Armstrong's group has developed a synthesis of *anti*-β-methylaspartate ethyl ester **153** from the nucleophilic opening of a chiral β-lactone **152** with azide followed by catalytic hydrogenation (Scheme 5.37) [210]. The chiral β-lactone derives from the cinchona alkaloid (**154**)-catalyzed enantioselective coupling of ethyl glyoxylate **151** with an *in situ*-derived methyl ketene. The coupling reaction proceeds in moderate yield but with excellent enantioselectivity.

(2*S*,3*S*,4*R*)-3,4-Dimethylglutamine (DiMeGln) is one of the nonproteinogenic amino acids incorporated in cyclic depsipeptide natural products such as papuamides [128], callipeltins [211, 212], neamphamide [213], and mirabamides [214] (Figure 5.3). These cyclic depsipeptides exhibit both antitumor and anti-HIV activities, and therefore they are recognized as interesting synthetic targets for further study in discovery of novel therapeutic agents. One of the main dilemmas in total synthesis of the previously mentioned cyclic depsipeptide was developing an efficient method to prepare DiMeGln stereoselectively. Currently, several methods have been developed to synthesize DiMeGln. Hamada and Lipton used (*S*)-pyroglutamic acid as a chiral synthon followed by installing methyl groups at C-3 and C-4 with an organocuprate reagent [215–217]. The problematic step in Hamada's synthesis was the pyrrolidinone ring-opening step with ammonia, which gave a low yield of product. Lipton *et al.* solved the problem by the use of a catalytic amount of KCN to provide the *N*-Boc-DiMeGln derivative [216] and used a lanthanide catalyst for the *N*-Fmoc-DiMeGln derivative [217]. On the other hand, Ma *et al.* used δ-valerolactone as the crucial intermediate, which was opened under ammonia/methanol conditions without the use of any catalyst, to obtain the DiMeGln derivative [218].

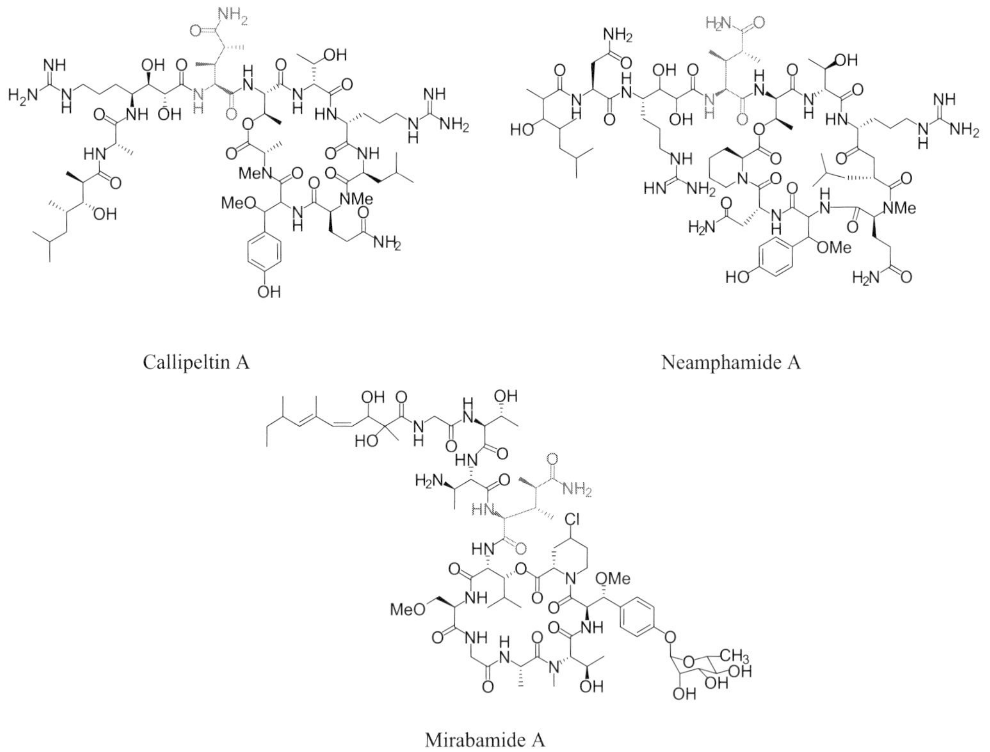

Callipeltin A

Neamphamide A

Mirabamide A

Figure 5.3 Representative natural products containing DiMeGln.

The most efficient synthesis of DiMeGln was reported by Joullie *et al.* en route to the total synthesis of callipeltin A [219]. The key step of Joullie's synthesis was an asymmetric Michael addition where a camphorsultam chiral auxiliary was used to control the stereochemistry of the product (Scheme 5.38). The (+)-camphorsultam **155** was treated with *trans*-crotonyl chloride under basic conditions to afford the Michael addition precursor. The resulting product **156** was added into the reaction medium containing *N,N*-dibenzylpropionyl amide **157** and lithium diisopropylamide giving the desired Michael adduct **158** with moderate diastereoselectivity. The diastereoselectivity of the Michael addition was further improved with the use of a bulkier protecting group for the propionyl amide. To complete the synthesis of the DiMeGln derivative **159**, the Michael adduct **158** was stereoselectively azidated at C-2, reduced to the amine, and Boc-protected, followed by removal of the chiral auxiliary by hydrolysis.

5.4 Noncoded Amino Acids with Elaborate Side-Chains

The peptide natural product cyclosporine A **160** (Scheme 5.39) has been of tremendous medical use as an effective immunosuppressive agent for organ transplantation

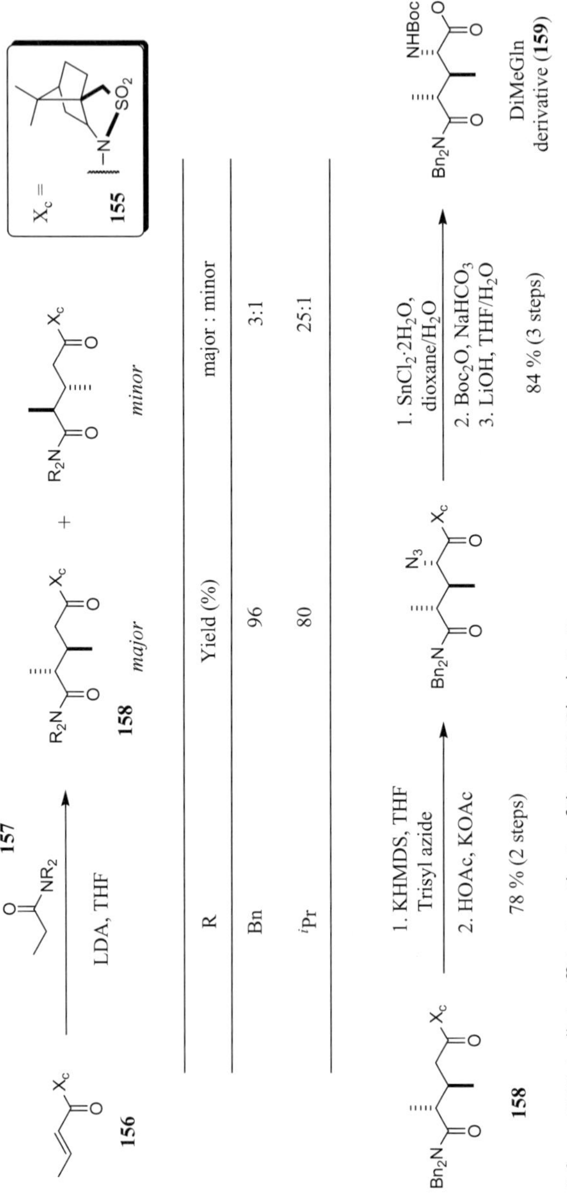

Scheme 5.38 Joullie's efficient synthesis of the DiMeGln derivative.

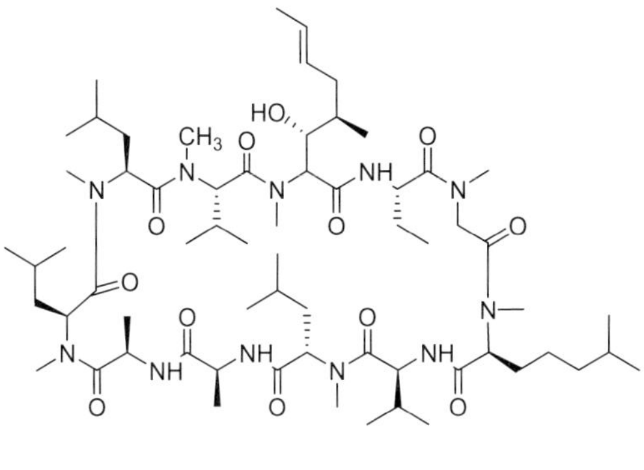

Cyclosporin A **160**

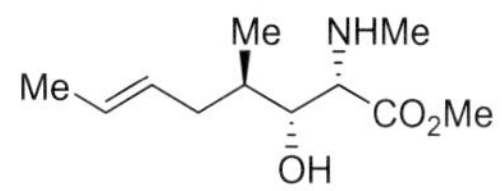

Scheme 5.39 Cyclosporin A and the noncoding amino acid MeBmt.

and for treatment of autoimmune diseases such as psoriasis and rheumatoid arthritis. In addition, further research has indicated the likelihood of its use in the treatment of HIV, hepatitis C, and malaria. Owing to such a broad palate of medical uses, chemical synthesis of cyclosporine A **160** and its constituent amino acids is of great importance. The unusual amino acid (2*S*,3*R*,4*R*,6*E*)-3-hydroxy-4-methyl-2-(methylamino)-6-octenoic acid (MeBmt) **161** of cyclosporine has been the focus of numerous synthetic endeavors over the years, two of which are profiled here.

The first synthesis of MeBmt was reported by Wenger in 1983 as a 24-step linear route from tartaric acid [220]. Evans and Weber used a diastereoselective aldol method of synthesizing MeBmt that provided the amino acid in high optical purity (>99% d.e.) after chromatography and recrystallization (Scheme 5.40) [43, 142]. While the Evan's synthesis provided MeBmt **162**, efforts continue in search of a more efficient synthesis that will allow for the production of multigram quantities of MeBmt. Many other syntheses of MeBmt have been reported in the last 20 years, including aldol reaction approaches by Rich *et al.* [221], Seebach *et al.* [222], Tung *et al.* [223], Togni *et al.* [224], and Rapoport *et al.* [225]. In all of the aldol addition

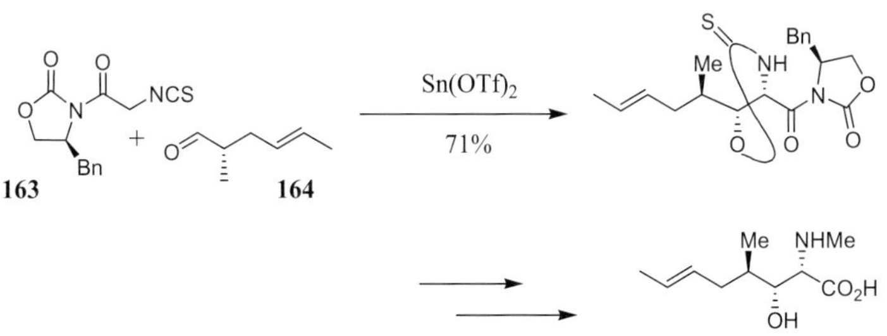

Scheme 5.40 Evan's synthesis of MeBmt.

methods for MeBmt synthesis, an α-alkyl branched aldehyde **164** and a chiral glycine synthon **163** are necessary.

Many methods of chiral epoxide opening to prepare MeBmt have also been explored. These include reports by Rich *et al.* [223] and Savignac *et al.* [226] in which Sharpless asymmetric epoxidation of allylic alcohols was employed to form the epoxide stereochemistry. A limitation to chiral epoxide methods is the necessity of introducing an *N*-methyl amine at a late stage in the synthesis. In their formal synthesis of MeBmt, Raghavan and Rasheed formed a chiral epoxide from an enantiopure bromohydrin as one of the key stereogenic steps (Scheme 5.41). The epoxide was opened stereospecifically in a carbamate formation step that introduced the *N*-methyl amine **167** [227].

Various other routes to MeBmt have also been explored. Organometallic catalysis has also contributed to the synthesis of MeBmt. Cook and Shanker reported a synthesis of MeBmt employing a palladium-catalyzed equilibrium to obtain a chiral oxazoline intermediate [228]. For a radical-based approach to MeBmt, see Matsunaga *et al.*'s work [229]. A novel synthesis of MeBmt starting from D-glucose was reported by Rao *et al.* [230]. Tuch *et al.* used D-isoascorbic acid as a starting material for their 1,4-conjugate addition synthesis of MeBmt [231].

A small collection of cyclic tetrapeptides contain the (2*S*,9*S*)-2-amino-8-oxo-9,10-epoxydecanoic acid (Aoe) residue. The Aoe amino acid in these peptides, which include chlamydocin [232], HC-toxin [233], trapoxin B [234], WF-3161 [235], and Cyl-2 [236], is thought to act as an irreversible histone deacetylase inhibitor, presumably by covalent attachment of the epoxide moiety within the enzyme active site [237]. This group of natural product peptides has garnered much attention as a result of this and other notable bioactivities [238]. Unfortunately, due to the epoxy ketone sensitivity to nucleophilic attack, a suitable precursor residue is first usually incorporated into the peptide structure and then derivatized at a late stage to the epoxy ketone in order to preserve the side-chain. The Aoe amino acid, however, has been synthesized, in protected form, in the course of exploring new synthetic methodology.

There are several syntheses of Aoe reported [239–241]. Among those, an interesting example from Katsuki *et al.* [242] showcases a stereoselective alkylation of a chiral glycine equivalent (Scheme 5.42). Alkylation of the chiral glycine amide **170** with the protected triflate, followed by hydrolysis, produced the free amino acid **171** in good yield and with 98% e.e. Protection of the amino acid and oxidation of the side-chain hydroxyl led to the intermediate aldehyde **172**. Addition of vinyl magnesium bromide, asymmetric epoxidation, and oxidation gave Cbz-Aoe methyl ester **173**.

Of the many late-stage modifications of the Aoe-containing cyclic tetrapeptides, there are several notable examples where the epoxy ketone of Aoe is installed or unmasked. Schmidt's synthesis of WF-3161 makes use of an α-chloro acetonide as the precursor to the epoxy ketone (Scheme 5.43). Hydrolysis of the acetonide **174**, intramolecular chloride displacement, and oxidation yields the natural product **175**.

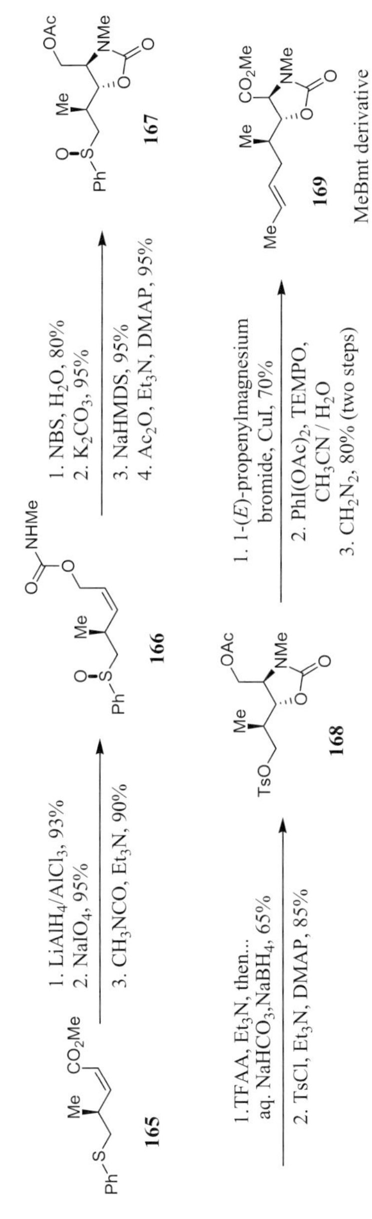

Scheme 5.41 Raghavan's formal synthesis of MeBmt.

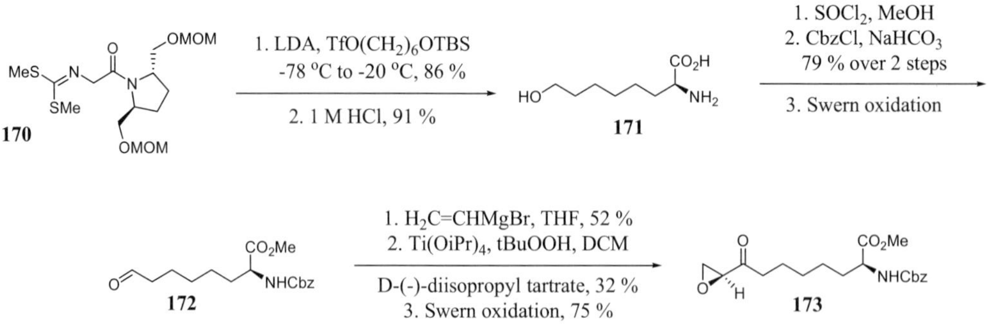

170 → **171**: 1. LDA, $TfO(CH_2)_6OTBS$, -78 °C to -20 °C, 86 %; 2. 1 M HCl, 91 %

171 → **172**: 1. $SOCl_2$, MeOH; 2. CbzCl, $NaHCO_3$, 79 % over 2 steps; 3. Swern oxidation

172 → **173**: 1. $H_2C{=}CHMgBr$, THF, 52 %; 2. $Ti(OiPr)_4$, tBuOOH, DCM, D-(-)-diisopropyl tartrate, 32 %; 3. Swern oxidation, 75 %

Scheme 5.42 Katsuki's chiral glycine alkylation method of Aoe synthesis.

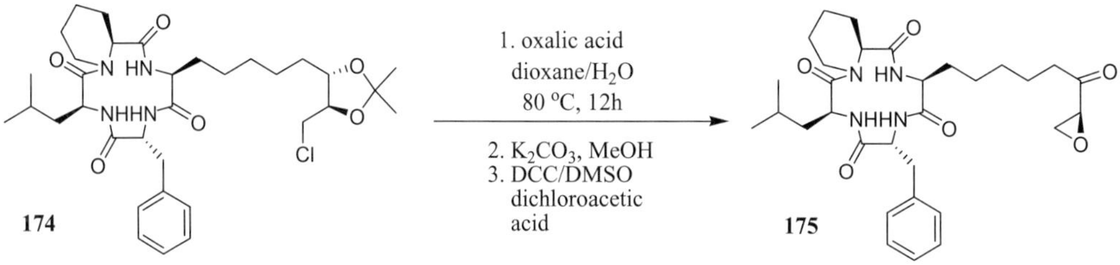

174 → **175**: 1. oxalic acid, dioxane/H_2O, 80 °C, 12h; 2. K_2CO_3, MeOH; 3. DCC/DMSO, dichloroacetic acid

Scheme 5.43 Schmidt's synthesis of Aoe with late-stage epoxide unmasking.

The tuberactinomycin and the muraymycin classes of natural product contain the guanidine-bearing peptide derivative capreomycidine. Most notably, the capreomycins **176**, tuberactinomycins **177**, and viomycins **177** have been the focus of many synthetic efforts (Scheme 5.44). Consequently, the synthesis of the noncoding capreomycidine amino acid has been successfully achieved in the research laboratories of synthetic chemists. Johnson *et al.* reported the first synthesis of racemic capreomycidines followed by fractional crystallization of the picrates to provided capreomycidine and epi-capreomycidine in 1968 [38, 243, 244]. Efforts toward an asymmetric synthesis were undertaken by Shiba *et al.* [245] in the 1970s. Shiba *et al.*'s route started from a β-hydroxy-L-ornithine derivative obtained through enzymatic resolution of the racemate. While plagued by low stereoselection and poor yielding steps, they reported the first total synthesis of capreomycin IA (45 total steps and 0.008% overall yield) in 1978 [246].

Demong and Williams reported the asymmetric synthesis of capreomycidine (six steps, 28% yield) en route to his total synthesis of capreomycin IB [247, 248]. By invoking an enolate-aldimine strategy, condensation of benzylimine **179** with the lithium enolate of chiral glycinate **180** resulted in an inseparable 3.3 : 1 mixture of diastereomers **181** epimeric at the β-carbon (Scheme 5.45). The mixture was carried through a sequence of guanidinylation to **182**, silyl-deprotection to **183**, and intra-

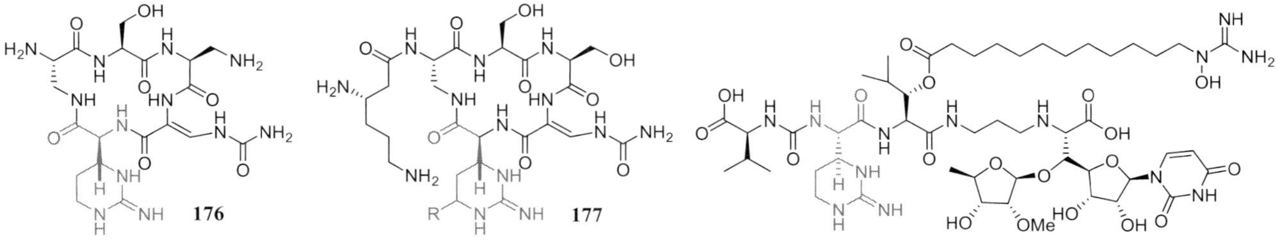

Capreomycin IIA **176**

Tuberactinomycin O; R=H
Tuberactinomycin B (Viomycin); R=OH **177**

Muraymycin A1 **178**

Scheme 5.44 Natural products containing the noncoding amino acid capreomycidine.

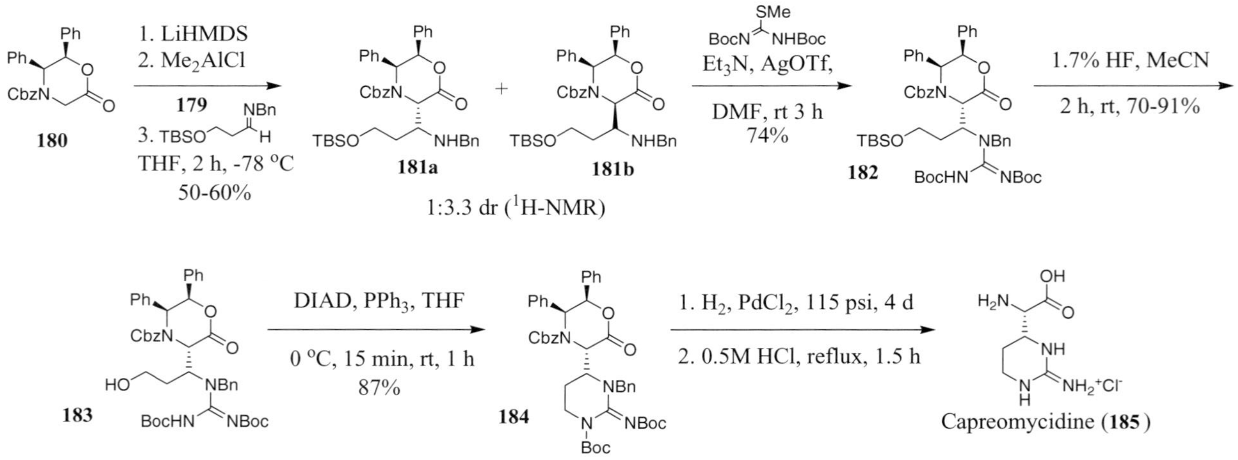

Scheme 5.45 William's concise synthesis of capreomycidine.

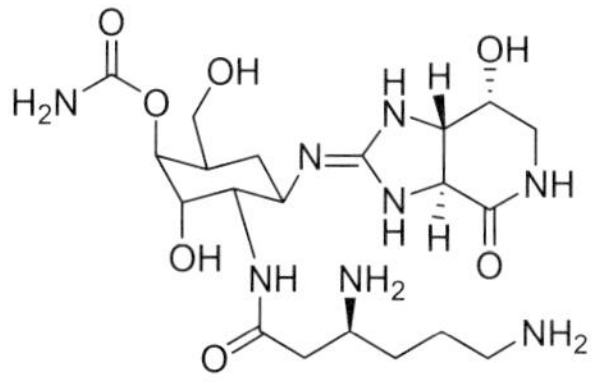

Streptothricin F (**186**)

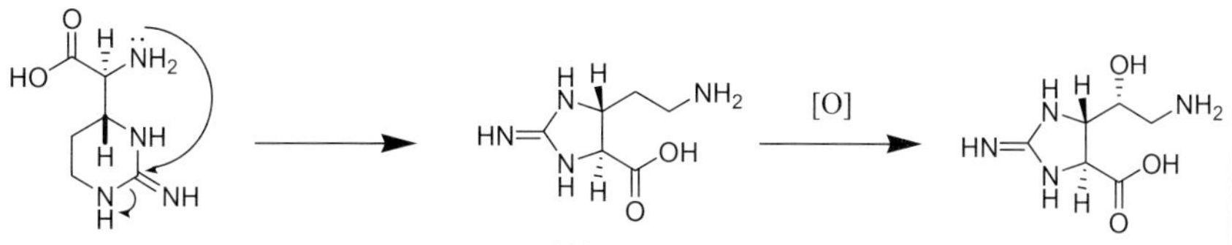

(2*S*,3*R*)-Capreomycidine (**187**) **188** [O] Streptolidine (**189**)

Scheme 5.46 Proposed biosynthetic rearrangement of capreomycidine to streptolidine.

molecular Mitsunobu substitution to provide the cyclic guanidine **184**. Global deprotection of **184** ended this concise synthesis of capreomycidine **185**, which was used to complete Demong and Williams' total synthesis of capreomycin IB (27 total steps and 2% overall yield).

During studies on the biosynthesis of the peptidyl nucleoside antibiotic streptothricin F **186** the conversion of arginine to streptolidine **189** has been proposed to include capreomycidine **187** as an intermediate (Scheme 5.46). Thomas *et al.* [249] helped build on previous studies to elucidate the biochemical mechanism behind this conversion [250, 251] in 2003.

To gain insight into this hypothesis Zabriskie *et al.* synthesized L-[*guanidine*-^{13}C] capreomycidine **195** (Scheme 5.47) [250]. Starting with Garner's aldehyde **190**, a sequence of Grignard addition of allyl magnesium chloride followed by silyl protection and ozonolysis with reductive workup produced a mixture of diastereomeric primary alcohols **191a** and **191b** separable by chromatography. The [^{13}C] guanidine was later installed through condensation of [^{13}C]cyanogen bromide in reaction with the diamine **193**. Deprotection and cyclization provided the ^{13}C-incorporated capreomycidine **195** in 10% yield, over four steps, from guanidine **194**.

The echinocandins (Scheme 5.48) are a hexapeptide-containing group of natural products that include multiple hydroxylated amino acids (**196–198**) and often exhibit potent fungicidal biological activity. The rare β-hydroxylated amino acid (2*S*,3*R*)-3-hydroxyhomotyrosine (Hht) **196** is noteworthy for its incorporation into the echinocandins, as well as, related natural products (pneumocandins, mulundocandins, lypopeptide L-688,786). Hht is also present in the antifungal drugs anidulafungin, caspofungin, and micafungin. Consequently, the Hht amino acid

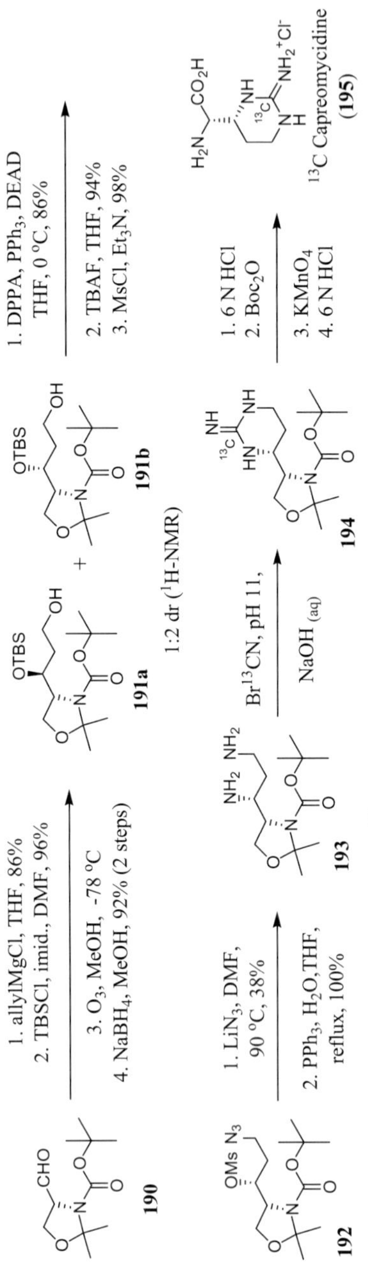

Scheme 5.47 Zabriskie's synthesis of L-[*guanidine*-^{13}C]capreomycidine.

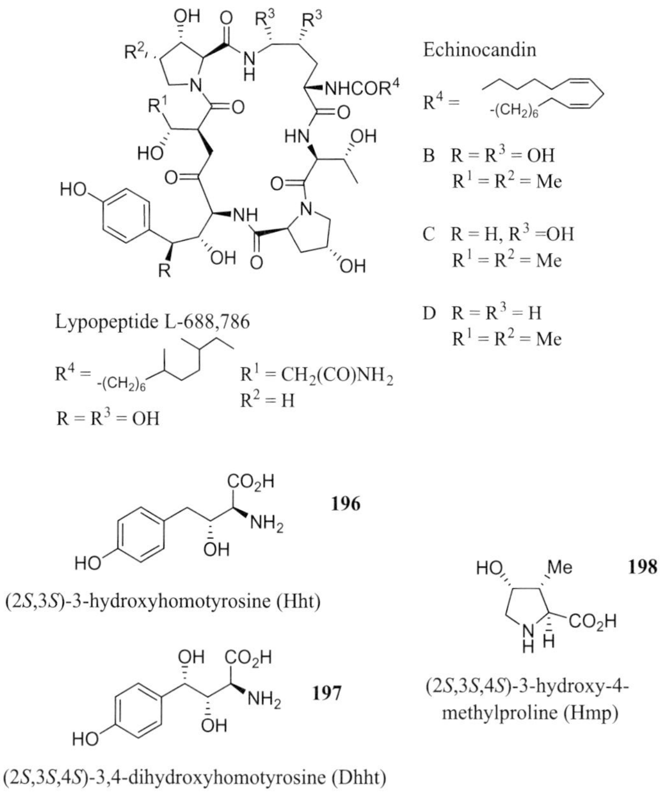

Echinocandin

R^4 = $-(CH_2)_6$–

B $R = R^3 = OH$
$R^1 = R^2 = Me$

C $R = H, R^3 = OH$
$R^1 = R^2 = Me$

D $R = R^3 = H$
$R^1 = R^2 = Me$

Lypopeptide L-688,786

R^4 = $-(CH_2)_6$–

$R^1 = CH_2(CO)NH_2$
$R^2 = H$

$R = R^3 = OH$

196
(2*S*,3*S*)-3-hydroxyhomotyrosine (Hht)

197
(2*S*,3*S*,4*S*)-3,4-dihydroxyhomotyrosine (Dhht)

198
(2*S*,3*S*,4*S*)-3-hydroxy-4-methylproline (Hmp)

Scheme 5.48 Echinocandin natural products and constituent noncoding amino acids.

has been synthesized through multiple routes. Ohfune *et al.* completed the first synthesis of Hht [65].

The Ohfune synthesis of Hht begins with chiral 2-amino-3-butenol derived from L- or D-methionine (Scheme 5.49). The allylic amine **199** undergoes regioselective and stereoselective epoxidation upon treatment with *m*CPBA. After acetate protection of

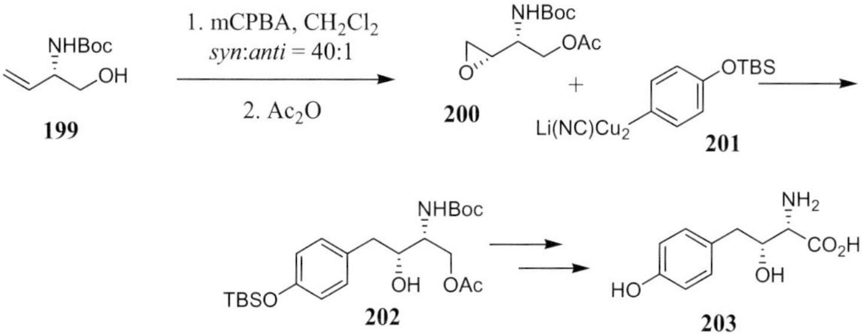

199 → 1. mCPBA, CH_2Cl_2, *syn*:*anti* = 40:1; 2. Ac_2O → **200** + **201** ($Li(NC)Cu_2$) → **202** → → **203**

Scheme 5.49 Ohfune's synthesis of Hht.

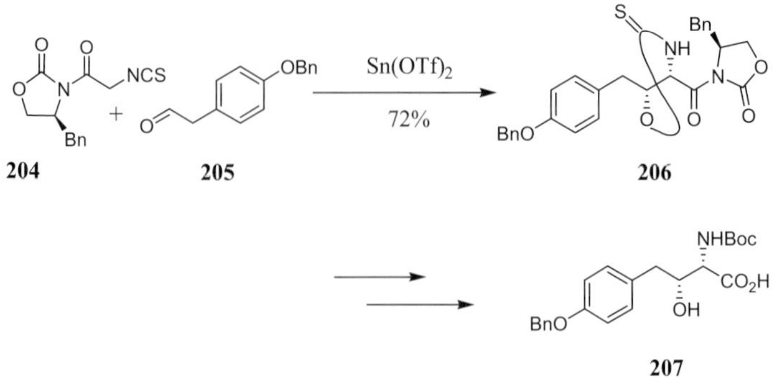

Scheme 5.50 Evan's aldol synthesis of *N*-Boc-protected Hht.

the primary hydroxyl group, the epoxide **200** was subjected to nucleophilic epoxide opening with the phenyl cuprate **201**. A sequence of protecting group manipulation and oxidation of the primary hydroxyl to the carboxylic acid provides Hht **203** in moderate to good yields (60–80% per step) and excellent stereoselectivity (>40 : 1 *syn* : *anti* from the epoxidation step). Ohfune *et al.* used this route to Hht in his total synthesis of echinocandin D [252].

Evans *et al.* reported a diastereoselective aldol approach to the synthesis of Hht **196** and related β-hydroxy-α-amino acids [142, 253] including Hmp **198**. The stereoselective aldol addition of (isothiocyanoacetyl)oxazolidinone **204** and *p*-(benzyloxy)phenylacetaldehyde **205** proceeded in 72% yield. Removal of the chiral auxiliary (95% yield) followed by Boc protection of the nitrogen and sulfur/oxygen exchange (95%) afforded the corresponding *N*-Boc-oxazolidinone **206**. Cleavage of the oxazolidinone with lithium hydroxide proceeded in 83% yield to provide the differentially protected Hht **207** (Scheme 5.50). This aldol method was used by Evans *et al.* to complete a total synthesis of echinocandin D [66].

In the lipopeptide series of natural products, as well as in echinocandin B, the Hht amino acid **196** is replaced by (2*S*,3*S*,4*S*)-3,4-dihydroxyhomotyrosine (Dhht) **197**. It has been accounted by Zambias *et al.* that Dhht is important for the biological activity of echinocandin B [254]. Palomo *et al.* have reported the synthesis of Dhht [255]. The synthesis from alkene **208** begins with a sequence of Sharpless asymmetric dihydroxylation, acetal protection of the 1,2-diol and DIBAL-H reduction of the methyl ester to provide aldehyde **209**. Subsequent imine formation followed by asymmetric carboxylation and cyclization allowed access to the β-lactam **210**. In the cyclization step the acetoxy functionality proved necessary to make the transformation stereospecific. TEMPO oxidation of the acetoxy-containing β-lactam **210** produced the *N*-carboxy anhydride **211** in excellent yield. The *N*-carboxy anhydride **211** was used to couple with the differentially protected L-threonine **212**. Reductive benzyl deprotection provided the masked Dhht-containing dipeptide **213** used to further the synthetic efforts towards echinocandin B. As shown in Scheme 5.51, the synthesis

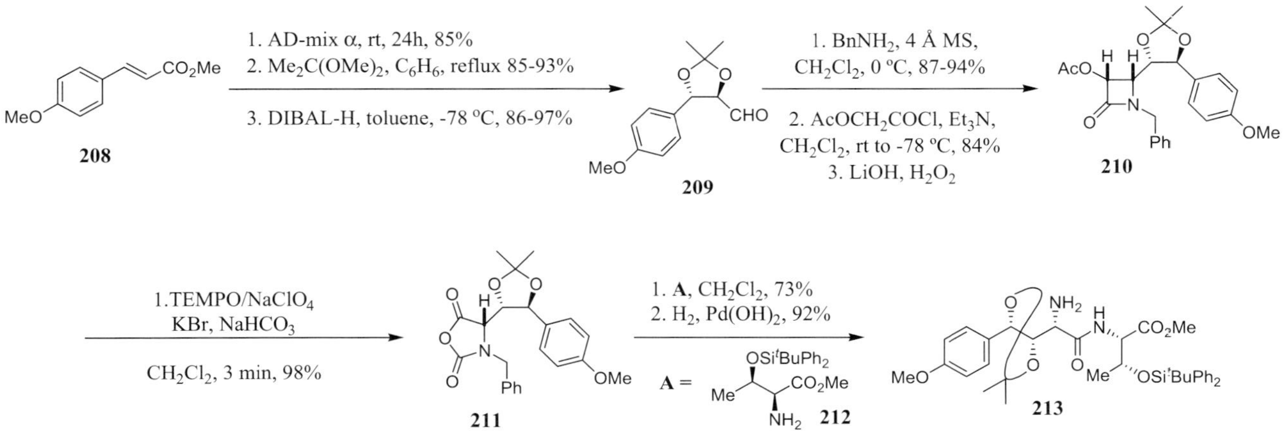

Scheme 5.51 Palomo's synthesis of Dhht.

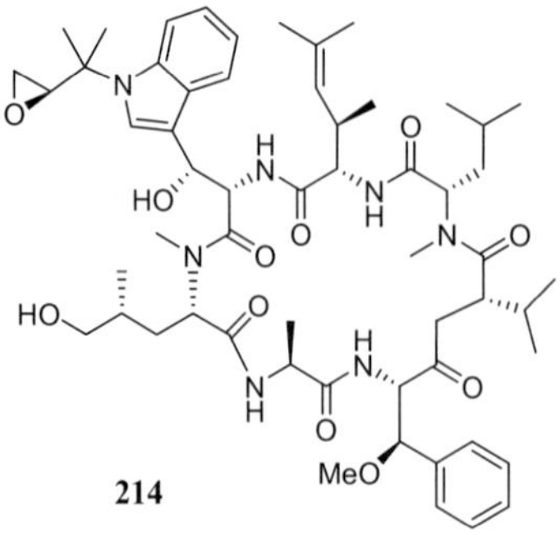

214

Cyclomarin A

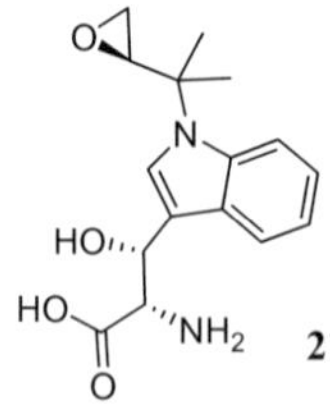

215

N-(1',1'-dimethyl-2',3'-epoxypropyl)-3-hydroxytryptophan

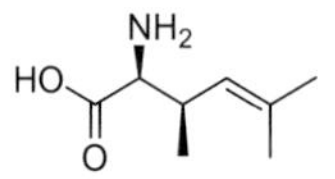

(2*S*,3*R*)-2-amino-3,5-dimethylhex-4-enoic acid
216

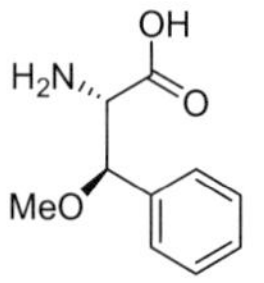

217

(2*S*,3*R*) β-methoxyphenylalanine

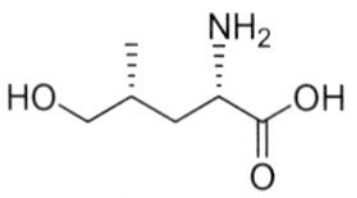

(2*S*,4*R*) δ-hydroxyleucine
218

Scheme 5.52 Cyclomarin A and constituent noncoding amino acids.

of **213** proceeded stereospecifically in good to excellent yields. It is interesting to note that echinocandin B can be easily converted into echinocandin C in one step using sodium cyanoborohydride [256].

Adding to the long list of marine cyclopeptides that contain interesting noncoded amino acids isolated in recent years [257–260], the cyclomarins A (**214**), B, and C were obtained from Mission Bay, California. Clardy *et al.* isolated and characterized these structures [261]. Synthesis of the novel amino acids present in these compounds has attracted the attention of multiple research groups [257–260, 262, 263]. Four of these amino acids have been synthesized by both the Joullie and Yao laboratories (Scheme 5.52).

Joullie *et al.* reported a concise synthesis of (2*S*,4*R*)-δ-hydroxyleucine **218** employing Evan's asymmetric alkylation to provide aldehyde **219** in greater than 96% d.e. (Scheme 5.53) [264]. The Davis asymmetric Strecker reaction that followed with a 91% d.e. led to the benzyl-protected (2*S*,4*R*)-δ-hydroxyleucine **220**. The amino acid (2*S*,3*R*)-β-methoxyphenylalanine **217** was obtained in 91% d.e. using the Grignard addition of phenyl magnesium bromide to Lajoie's serine aldehyde **221** [264].

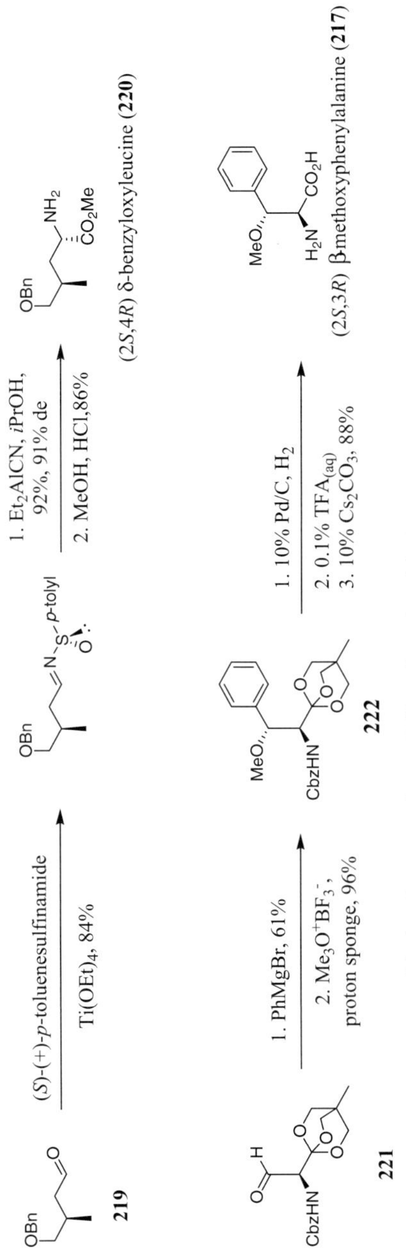

Scheme 5.53 Joullie's (2*S*,4*R*)-δ-hydroxyleucine and (2*S*,3*R*)-β-methoxyphenylalanine.

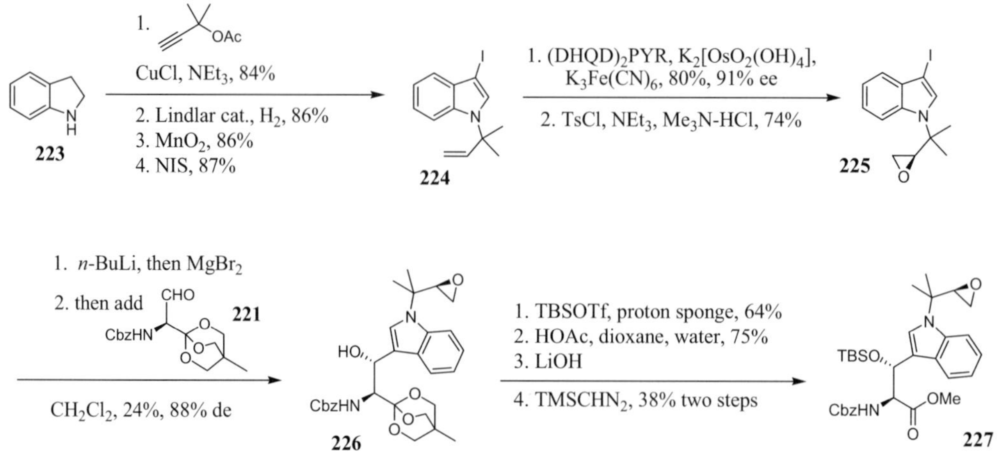

Scheme 5.54 *N*-(1′1′-Dimethyl-2′,3′-epoxypropyl)-3-hydroxytryptophan synthesis by Joullie.

Synthesis of the epoxide **215** was reported by Joullie *et al.* [265]. The key stereogenic steps were a stereoselective Sharpless dihydroxylation (91% e.e.) and addition of the Grignard derived from indole **225** to Lajoie's serine aldehyde **221** (88% d.e.). The Grignard addition unfortunately took place in a low yield of 24%. Protective group manipulation at the end of the synthesis provides the differentially protected hydroxytryptophan **227**, albeit in lower yield for the last two steps (Scheme 5.54).

By using an enantioselective [3,3]-Claisen rearrangement as the key stereogenic step Yao *et al.* obtained the *N*-TFA amino acid ester **229** in good yield and 88% e.e. [266]. Subsequent to troublesome phthaloyl protection to give the amine **230**, installation of the *gem*-dimethyl alkene proved difficult due to the low yielding Wittig olefination providing the alkene **231** in 40% yield. In any case, the (2*S*,3*R*)-2-Fmoc-amino-3,5-dimethylhex-4-enoic acid **232** was synthesized and used in Yao's total synthesis of cyclomarin C (Scheme 5.55).

As the last resort for treating antibiotic-resistant bacterial infections, vancomycin **233**, is well known as the glycopeptide antibiotic of choice. Still, bacterial resistance to vancomycin is increasing. Many common strains of Enterococci resistant to vancomycin possess a pathway in which the terminal dipeptide of the cell wall peptidoglycan precursor is modified. For a comprehensive review of glycopeptide and lipoglycopeptide antibiotics, see the review by Walsh *et al.* and references therein [267]. Teicoplanin, ristocetin **234**, and ereomycin are structurally related compounds of which teicoplanin and ristocetin contain not only the signature AB-fragment biaryl bis-amino acid **235**, but also have the FG-fragment aryl ether bis-amino acid **236** (Scheme 5.56). Recently, the need for new antibiotics has led to an increase in synthetic work towards the unique noncoded α-amino acids present in these natural products.

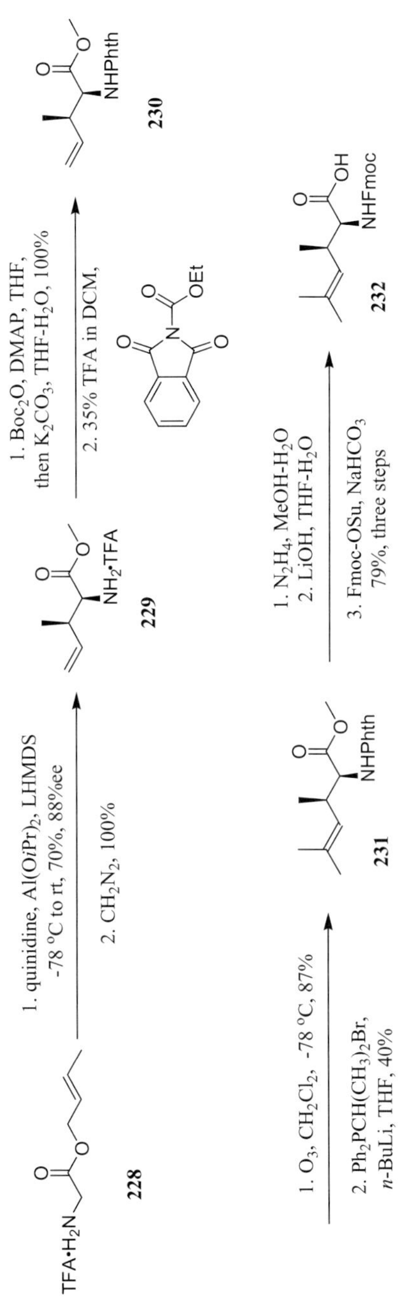

Scheme 5.55 Yao's synthesis of (2*S*,3*R*)-2-Fmoc-amino-3,5-dimethylhex-4-enoic acid.

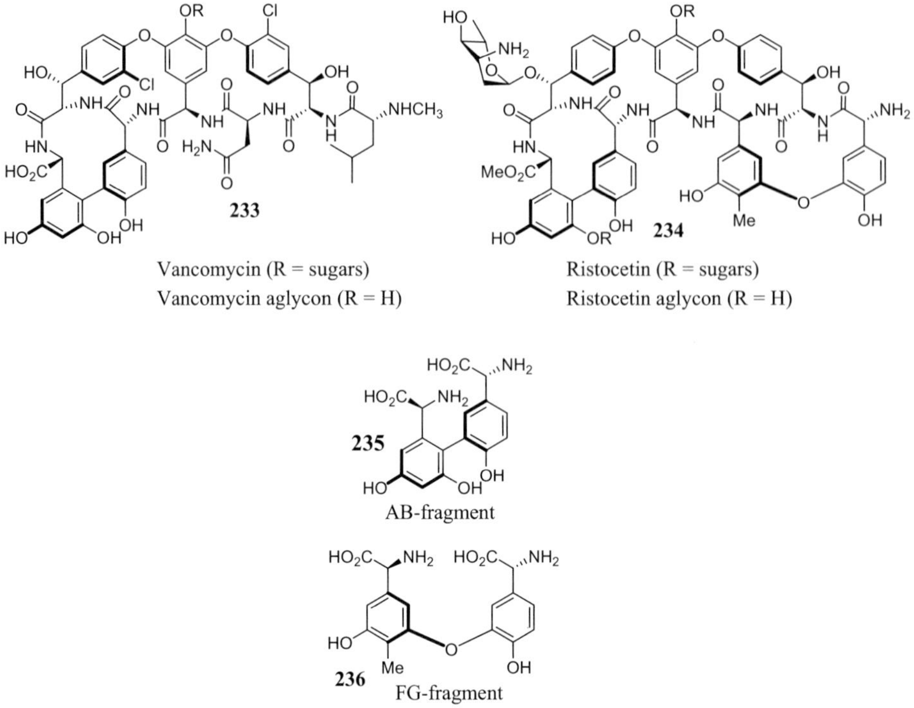

233

Vancomycin (R = sugars)
Vancomycin aglycon (R = H)

234

Ristocetin (R = sugars)
Ristocetin aglycon (R = H)

235
AB-fragment

236
FG-fragment

Scheme 5.56 Vancomycin, ristocetin, and aryl-containing amino acid fragments.

The total synthesis of both vancomycin and vancomycin aglycon were completed by the research groups of Evans and Nicolaou [268–272], and later by Boger [273–275]. During these synthetic efforts two distinct tactics were used to form the biaryl-containing amino acids of vancomycin. Evans *et al.* employed his well-known diastereoselective aldol reaction [268] to set the α-stereocenter of the amino acid, whereas Nicolaou *et al.* and Boger *et al.* chose a Sharpless amino hydroxylation approach [270, 274]. To couple the two aryl groups **239** and **240**, various transition metal catalyzed reactions have been used including Suzuki coupling, as employed in Nicolaou *et al.*'s synthesis (Scheme 5.57) [270, 274]. The 2 : 1 mixture of (*S* : *R*) atropisomers **241** and **242** were carried through the rest of the synthesis of vancomycin. Boger's total synthesis also employed a Suzuki coupling that provided a 2 : 1 (*S* : *R*) mixture, but thermal equilibration at 120 °C increased this ratio to 3 : 1 (*S* : *R*) [274].

Another approach was employed in Evans *et al.*'s synthesis of vancomycin [268]. They chose an oxidative cyclization strategy to connect the AB-fragments **243**. This worked well for C–C bond formation, but provided the unnatural atropisomer **244**

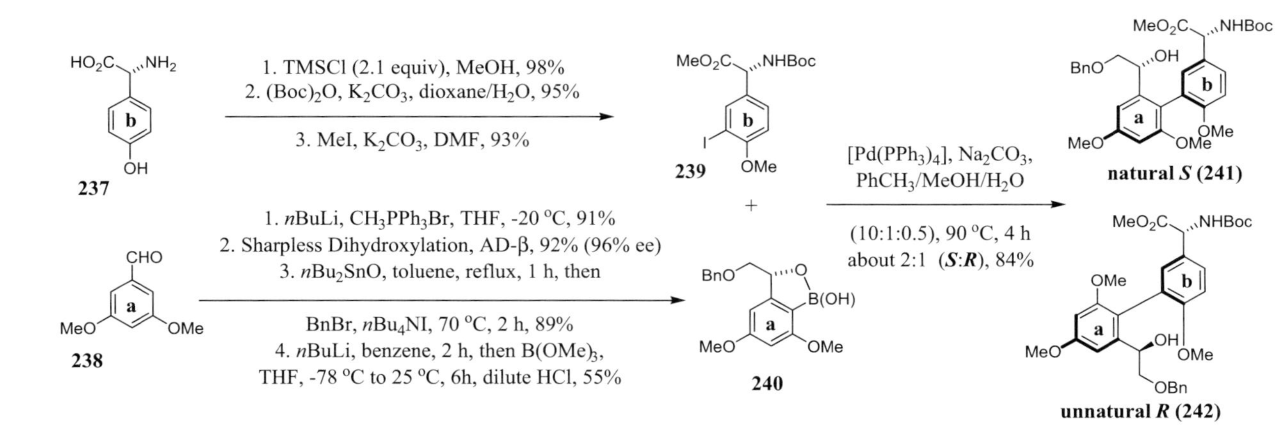

Scheme 5.57 Nicolaou's synthesis of the AB-fragment of vancomycin.

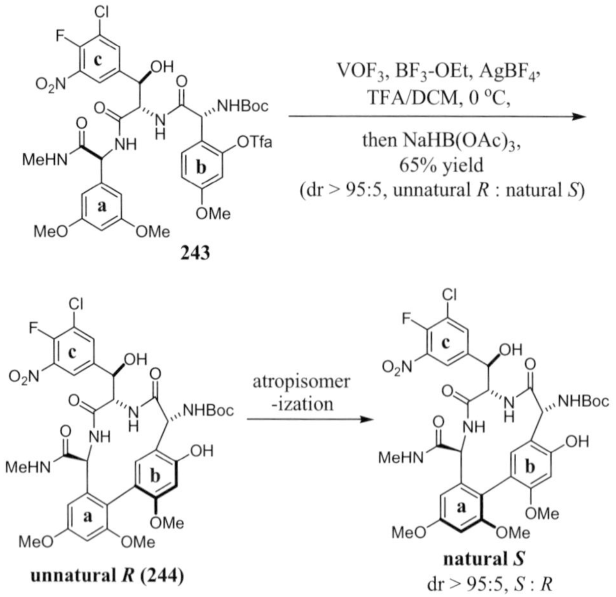

Scheme 5.58 Evan's synthesis of the AB-fragment of vancomycin.

as the major product. Later in the total synthesis an atropisomerization (MeOH, 55 °C, 24 h) allowed for conversion of the unnatural (*R*) isomer to the natural (*S*) isomer providing a diastereomeric ratio greater than 95 : 5 of the desired (*S*) natural configuration (Scheme 5.58).

The glycopeptides ristocetin **234** and teicoplanin, while structurally related to vancomycin, contain many differences in structure. One of these distinctions relevant to this discussion on noncoded α-amino acids with elaborate side-chains is the diaryl ether-containing side-chain that connects two of the amino acid residues in these cyclic heptapeptides (Scheme 5.56).

Of the synthesis efforts towards ristocetin and teicoplanin, the Boger total synthesis of ristocetin [276] and teicoplanin [277] illustrates one method for constructing the diaryl ether fragment. By employing Sharpless asymmetric aminohydroxylation, Boger's synthesis of both ristocetin and teicoplanin quickly built the requisite coupling fragments **245** and **246**. Next, intermolecular aromatic nucleophilic substitution (S_NAr) of the two fragments resulted in optimized yield of 70% for diaryl ether **247** formation (Scheme 5.59). In other work toward the diaryl ether fragment, Pearson *et al.* have explored the application of arene–ruthenium complexes as S_NAr substrates [278].

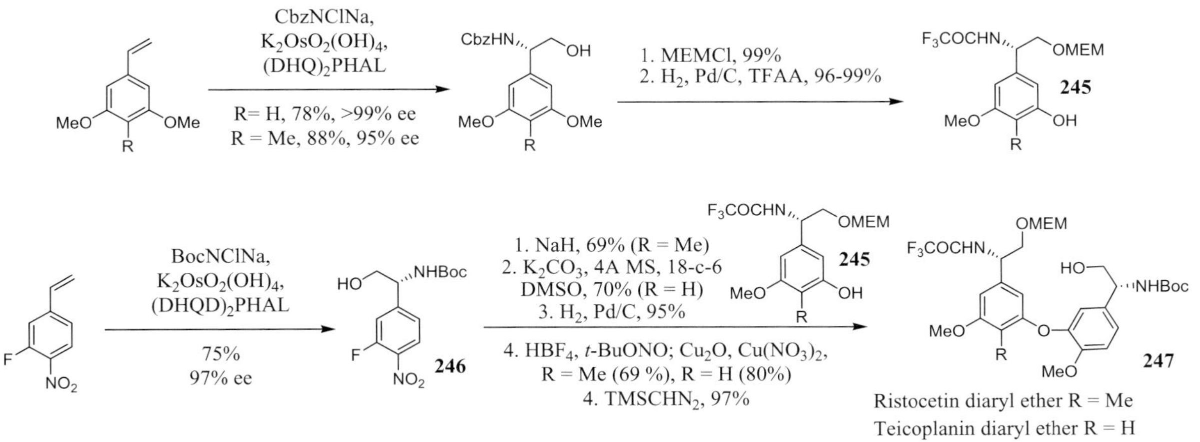

CbzNClNa, $K_2OsO_2(OH)_4$, $(DHQ)_2PHAL$

R= H, 78%, >99% ee
R = Me, 88%, 95% ee

1. MEMCl, 99%
2. H_2, Pd/C, TFAA, 96-99%

245

BocNClNa, $K_2OsO_2(OH)_4$, $(DHQD)_2PHAL$

75%
97% ee

246

1. NaH, 69% (R = Me)
2. K_2CO_3, 4A MS, 18-c-6 DMSO, 70% (R = H)
3. H_2, Pd/C, 95%

4. HBF_4, *t*-BuONO; Cu_2O, $Cu(NO_3)_2$, R = Me (69 %), R = H (80%)
4. $TMSCHN_2$, 97%

245

247

Ristocetin diaryl ether R = Me
Teicoplanin diaryl ether R = H

Scheme 5.59 Boger's synthesis of the biaryl ether fragments of ristocetin and teicoplanin.

5.5 Conclusions

In recent years, with the advent of new synthetic technologies such as catalytic asymmetric synthesis, chiral phase transfer catalysis, and organocatalysis, the diversity of complex amino acids that can be synthesized has increased greatly. We hope this chapter provides a substantial launching point for the reader to begin their own investigation into the chemical synthesis of unusual, yet natural amino acids. As more and more interesting and unusual natural products are discovered, the possibility for development of synthetic methods for these useful and biologically important amino acids grows as well.

References

1 Barrett, G.C. and Elmore, D.T. (1998) *Amino Acids and Peptides*, Cambridge University Press, Cambridge.

2 Brooks, D.A. (2007) Protein: cotranslational and posttranslational modification in organelles, in *Encyclopedia of Life Sciences*, vol. **1**, John Wiley & Sons, Ltd, Chichester, pp. 204–208.

3 Doekel, S. and Marahiel, M.A. (2001) Biosynthesis of natural products on modular peptide synthetases. *Metabolic Engineering*, **3**, 64–77.

4 Felnagle, E.A., Jackson, E.E., Chan, Y.A., Podevels, A.M., Berti, A.D., McMahon, M.D., and Thomas, M.G. (2008) Nonribosomal peptide synthetases involved in the production of medically relevant natural products. *Molecular Pharmacology*, **5**, 191–211.

5 Humphrey, J.M. and Chamberlin, A.R. (1997) Chemical synthesis of natural product peptides: coupling methods for the incorporation of noncoded amino acids into peptides. *Chemical Reviews*, **97**, 2243–2266.

6 Chan, W.C. and Higton, A. (2002) Amino acids. *Amino Acids, Peptides and Proteins*, **33**, 1–82.

7 Chan, W.C., Higton, A., and Davies, J.S. (2006) Amino acids. *Amino Acids, Peptides and Proteins*, **35**, 1–73.

8 Scott, W.L., O'Donnell, M.J., Delgado, F., and Alsina, J. (2002) A solid-phase synthetic route to unnatural amino acids with diverse side-chain substitutions. *The Journal of Organic Chemistry*, **67**, 2960–2969.

9 O'Donnell, M.J. (2004) The enantioselective synthesis of α-amino acids by phase-transfer catalysis with achiral Schiff base esters. *Accounts of Chemical Research*, **37**, 506–517.

10 Ma, J.-M. (2003) Recent developments in the catalytic asymmetric synthesis of α- and β-amino acids. *Angewandte Chemie (International Edition in English)*, **42**, 4290–4299.

11 Ager, D.J. and Fotheringham, I.G. (2001) Methods for the synthesis of unnatural amino acids. *Current Opinion in Drug Discovery & Development*, **4**, 800–807.

12 Ager, D.J. (2002) Asymmetric hydrogenation and other methods for the synthesis of unnatural amino acids and derivatives. *Current Opinion in Drug Discovery & Development*, **5**, 892–905.

13 Mazurkiewicz, R., Kuznik, A., Grymel, M., and Pazdzierniok-Holewa, A. (2007) α-Amino acid derivatives with a Cα−P bond in organic synthesis. *ARKIVOC*, 193–216.

14 Cativiela, C. and Diaz-de-Villegas, M.D. (2007) Recent progress on the

stereoselective synthesis of acyclic quaternary α-amino acids. *Tetrahedron: Asymmetry*, **18**, 569–623.

15 Perdih, A. and Dolenc, M.S. (2007) Recent advances in the synthesis of unnatural α-amino acids. *Current Organic Chemistry*, **11**, 801–832.

16 Ohfune, Y. and Shinada, T. (2003) Asymmetric Strecker route toward the synthesis of biologically active α,α-disubstituted α-amino acids. *Bulletin of the Chemical Society of Japan*, **76**, 1115–1129.

17 Bonauer, C., Walenzyk, T., and König, B. (2006) α,β-Dehydroamino acids. *Synthesis*, **1**, 1–20.

18 Hunt, S. (1985) Degradation of amino acids accompanying *in vitro* protein hydrolysis, in *Chemistry and Biochemistry of the Amino Acids* (ed. G.C. Barrett), Chapman & Hall, London, pp. 55–138.

19 Fischer, P.M. (2003) Diketopiperazines in peptide and combinatorial chemistry. *Journal of Peptide Science*, **9**, 9–35.

20 Martins, M.B. and Carvalho, I. (2007) Diketopiperazines: biological activity and synthesis. *Tetrahedron*, **63**, 9923–9932.

21 Hohenschutz, L.D., Bell, E.A., Jewess, P.J., Leworthy, D.P., Pryce, R.J., Arnold, E., and Clardy, J. (1981) Castanospermine, A 1,6,7,8-tetrahydroxyoctahydroindolizine alkaloid, from seeds of *Castanospermum australe*. *Phytochemistry*, **20**, 811–814.

22 Gardiner, R.A., Rinehart, K.L., Snyder, J.J., and Broquist, H.P. (1968) Slaframine. Absolute stereochemistry and a revised structure. *Journal of the American Chemical Society*, **90**, 5639–5640.

23 Kakinuma, K., Otake, N., and Yonehara, H. (1972) The structure of detoxin D1. Selective antagonist of blasticidin S. *Tetrahedron Letters*, **13**, 2509–2512.

24 Ohfune, Y. and Nishio, H. (1984) Acyclic stereocontrolled synthesis of (−)-detoxinine. *Tetrahedron Letters*, **25**, 4133–4136.

25 Tschesche, R., David, S.T., Uhlendorf, J., and Fehlhaber, H.W. (1972) Alkaloids from Rhamnaceae. XV. Mucronine D, a further peptide alkaloid from *Zizyphus mucronata*. *Chemische Berichte*, **105**, 3106–3114.

26 Sheehan, J.C., Mania, D., Nakamura, S., Stock, J.A., and Maeda, K. (1968) The structure of telomycin. *Journal of the American Chemical Society*, **90**, 462–470.

27 Sardina, F.J. and Rapoport, H. (1996) Enantiospecific synthesis of heterocycles from α-amino acids. *Chemical Reviews*, **96**, 1825–1872.

28 Hausler, H., Kawakami, R.P., Mlaker, E., Severn, W.B., Wrodnigg, T.M., and Stutz, A.E. (2000) Sugar analogues with basic nitrogen in the ring as anti-infectives. *Journal of Carbohydrate Chemistry*, **19**, 435–449.

29 Provencher, L., Steensma, D.H., and Wong, C.-H. (1994) Five-membered ring aza sugars as potent inhibitors of α-L-rhamnosidase (naringinase) from *Penicillium decumbens*. *Bioorganic and Medicinal Chemistry*, **2**, 1179–1188.

30 Jurczak, J., Prokopowicz, P., and Golebiowski, A. (1993) Highly stereoselective synthesis of *cis*-(2*R*,3*S*)-3-hydroxyproline. *Tetrahedron Letters*, **34**, 7107–7110.

31 Herdeis, C., Hubmann, H.P., and Lotter, H. (1994) Chiral pool synthesis of *trans*-(2*S*,3*S*)-3-hydroxyproline and castanodiol from *S*-pyroglutamic acid. *Tetrahedron: Asymmetry*, **5**, 119–128.

32 Dell'Uomo, N., Di Giovanni, M.C., Misiti, D., Zappia, G., and Delle Monache, G. (1996) A straightforward synthesis of (2*S*,3*R*)-3-hydroxyproline and *trans*-(2*R*,3*S*)-2-hydroxymethyl-3-hydroxypyrrolidine. *Tetrahedron: Asymmetry*, **7**, 181–188.

33 Poupardin, O., Greck, C., and Genet, J.-P. (1998) Rapid asymmetric synthesis of highly functionalized C5 chiral synthons. Practical preparation of *trans*-3-hydroxy-D-proline. *Synlett*, 1279–1281.

34 Durand, J.-O., Larcheveque, M., and Petit, Y. (1998) Reductive cyanation: a key step for a short synthesis of (−)-(2*S*,3*S*)-3-

hydroxyproline. *Tetrahedron Letters*, **39**, 5743–5746.

35 Lee, J.H., Kang, J.E., Yang, M.S., Kang, K.Y., and Park, K.H. (2001) Efficient synthesis of 3-hydroxyprolines and 3-hydroxyprolinols from sugars. *Tetrahedron*, **57**, 10071–10076.

36 Gryko, D., Prokopowicz, P., and Jurczak, J. (2002) The stereochemical course of addition of allyltrimethylsilane to protected L-alaninals and L-serinals in the presence of Lewis acids. Total synthesis of *cis*-(2*R*,3*S*)-3-hydroxyproline. *Tetrahedron: Asymmetry*, **13**, 1103–1113.

37 Sinha, S., Tilve, S., and Chandrasekaran, S. (2005) A convenient synthesis of *trans*-3-hydroxy-L-proline. *ARKIVOC*, 209–217.

38 Huang, P.-Q. and Huang, H.-Y. (2004) An improved asymmetric synthesis of unusual amino acid (2*S*,3*S*)-3-hydroxyproline. *Synthetic Communications*, **34**, 1377–1382.

39 Mulzer, J., Meier, A., Buschmann, J., and Luger, P. (1996) Total synthesis of *cis*- and *trans*-3-hydroxy-D-proline and (+)-detoxinine. *The Journal of Organic Chemistry*, **61**, 566–572.

40 Csuk, R., Hugener, M., and Vasella, A. (1988) A new synthesis of *N*-acetylneuraminic acid. *Helvetica Chimica Acta*, **71**, 609–618.

41 Kim, J.H., Yang, M.S., Lee, W.S., and Park, K.H. (1998) Chirospecific synthesis of 1,4-dideoxy-1,4-imino-D-arabinitol and 1,4-dideoxy-1,4-imino-L-xylitol via one-pot cyclisation. *Journal of the Chemical Society, Perkin Transactions 1*, 2877–2880.

42 Lee, B.W., Jeong, I.-Y., Yang, M.S., Choi, S.U., and Park, K.H. (2000) A short and efficient synthesis of 2*R*,3*R*,4*R*-3,4-dihydroxyproline, 1,4-dideoxy-1,4-imino-L-xylitol, 2*R*,3*R*,4*R*,5*S*-3,4,5-trihydroxypipecolic acid, and 1,5-dideoxy-1,5-imino-L-iditol. *Synthesis*, 1305–1309.

43 Daniel, P.T., Koert, U., and Schuppan, J. (2006) Apoptolid: induction of apoptosis by a natural product. *Angewandte Chemie (International Edition in English)*, **45**, 872–893.

44 Umezawa, K., Nakazawa, K., Uemura, T., Ikeda, Y., Kondo, S., Naganawa, H., Kinoshita, N., Hashizume, H., Hamada, M., Takeuchi, T., and Ohba, S. (1998) Polyoxypeptin isolated from streptomyces: a bioactive cyclic depsipeptide containing the novel amino acid 3-hydroxy-3-methylproline. *Tetrahedron Letters*, **39**, 1389–1392.

45 Umezawa, K., Nakazawa, K., Uchihata, Y., and Otsuka, M. (1999) Screening for inducers of apoptosis in apoptosis-resistant human carcinoma cells. *Advances in Enzyme Regulation*, **39**, 145–156.

46 Umezawa, K., Nakazawa, K., Ikeda, Y., Naganawa, H., and Kondo, S. (1999) Polyoxypeptins A and B produced by *Streptomyces*: apoptosis-inducing cyclic depsipeptides containing the novel amino acid (2*S*,3*R*)-3-hydroxy-3-methylproline. *The Journal of Organic Chemistry*, **64**, 3034–3038.

47 Chen, W.H., Horoszewicz, J.S., Leong, S.S., Shimano, T., Penetrante, R., Sanders, W.H., Berjian, R., Douglass, H.O., Martin, E.W., and Chu, T.M. (1982) Human pancreatic adenocarcinoma: *in vitro* and *in vivo* morphology of a new tumor line established from ascites. *In vitro*, **18**, 24–34.

48 Noguchi, Y., Uchiro, H., Yamada, T., and Kobayashi, S. (2001) Synthetic study of polyoxypeptin: stereoselective synthesis of (2*S*,3*R*)-3-hydroxy-3-methylproline. *Tetrahedron Letters*, **42**, 5253–5256.

49 Makino, K., Kondoh, A., and Hamada, Y. (2002) Synthetic studies on polyoxypeptins: stereoselective synthesis of (2*S*,3*R*)-3-hydroxy-3-methylproline using SmI_2-mediated cyclization. *Tetrahedron Letters*, **43**, 4695–4698.

50 Qin, D.-G., Zha, H.-Y., and Yao, Z.-J. (2002) Enantioselective synthesis of (2*S*,3*R*)-3-hydroxy-3-methylproline. A novel amino acid found in polyoxypeptins. *The Journal of Organic Chemistry*, **67**, 1038–1040.

51 Shen, J.-W., Qin, D.-G., Zhang, H.-W., and Yao, Z.-J. (2003) Studies on the

synthesis of (2*S*,3*R*)-3-hydroxy-3-methylproline via C2– N bond formation. *The Journal of Organic Chemistry*, **68**, 7479–7484.

52 Merino, P., Revuelta, J., Tejero, T., Cicchi, S., and Goti, A. (2004) Fully stereoselective nucleophilic addition to a novel chiral pyrroline *N*-oxide: total syntheses of (2*S*,3*R*)-3-hydroxy-3-methylproline and its (2*R*)-epimer. *European Journal of Organic Chemistry*, 776–782.

53 Davis, F.A., Ramachandar, T., and Liu, H. (2004) Asymmetric synthesis of α-amino 1,3-dithioketals from sulfinimines (*N*-sulfinyl imines). Synthesis of (2*S*,3*R*)-(−)-3-hydroxy-3-methylproline. *Organic Letters*, **6**, 3393–3395.

54 Chen, Z. and Ye, T. (2005) Diastereoselective synthesis of the acyl side-chain and amino acid (2*S*,3*R*)-3-hydroxy-3-methylproline fragments of polyoxypeptin A. *Synlett*, 2781–2785.

55 Haddad, M. and Larcheveque, M. (2005) Chemoenzymatic synthesis of *N*-Boc protected (2*S*,3*R*)-3-hydroxy-3-methylproline. *Tetrahedron: Asymmetry*, **16**, 2243–2247.

56 Makino, K., Nagata, E., and Hamada, Y. (2005) Practical synthesis of (2*S*,3*R*)-3-hydroxy-3-methylproline, a constituent of papuamides, using a diastereoselective tandem Michael-aldol reaction. *Tetrahedron Letters*, **46**, 8159–8162.

57 Yoshitomi, Y., Makino, K., and Hamada, Y. (2007) Organocatalytic synthesis of (2*S*,3*R*)-3-hydroxy-3-methyl-proline (OHMePro), a component of polyoxypeptins, and relatives using OHMePro itself as a catalyst. *Organic Letters*, **9**, 2457–2460.

58 Udodong, U.E., Turner, W.W., Astleford, B.A., Brown, F., Clayton, M.T., Dunlap, S.E., Frank, S.A., Grutsch, J.L., LaGrandeur, L.M., Verral, D.E., and Werner, J.A. (1998) Studies on the phosphorylation of LY303366. *Tetrahedron Letters*, **39**, 6115–6118.

59 Benz, F., Knusel, F., Nuesch, J., Treichler, H., Voser, W., Nyfeler, R., and Keller-Schierlein, W. (1974) Metabolic products of microoganisms. 143. Echinocandin B, a new polypeptide antibiotic from *Aspergillus nidulans* var *echinulatus*. Isolation and components. *Helvetica Chimica Acta*, **57**, 2459–2477.

60 Mukhopadhyay, T., Roy, K., Bhat, R.G., Sawant, S.N., Blumbach, J., Ganguli, B.N., Fehlhaver, H.W., and Kogler, H. (1992) Deoxymulundocandin – a new echinocandin type antifungal antibiotic. *Journal of Antibiotics*, **45**, 618–623.

61 Satoi, S., Yagi, A., Asano, K., Mizuno, K., and Watanabe, T. (1977) Studies on aculeacin. II. Isolation and characterization of aculeacins B, C, D, E, F and G. *Journal of Antibiotics*, **30**, 303–307.

62 Mizuno, K., Yagi, A., Satoi, S., Takada, M., Hayashi, M., Asano, K., and Matsuda, T. (1977) Studies on aculeacin. I. Isolation and characterization of aculeacin A. *Journal of Antibiotics*, **30**, 297–302.

63 Okino, T., Qi, S., Matsuda, H., Murakami, M., and Yamaguchi, K. (1997) Nostopeptins A and B, elastase inhibitors from the cyanobacterium *Nostoc minutum*. *Journal of Natural Products*, **60**, 158–161.

64 Groutas, W.C. (1987) Inhibitors of leukocyte elastase and leukocyte Cathepsin G. Agents for the treatment of emphysema and related ailments. *Medicinal Research Reviews*, **7**, 227–241.

65 Kurokawa, N. and Ohfune, Y. (1986) Total synthesis of echinocandins I. Stereocontrolled syntheses of the constituent amino acids. *Journal of the American Chemical Society*, **108**, 6041–6043.

66 Evans, D.A. and Weber, A.E. (1987) Synthesis of the cyclic hexapeptide echinocandin D. New approaches to the asymmetric synthesis of β-hydroxy α-amino acids. *Journal of the American Chemical Society*, **109**, 7151–7157.

67 Mulzer, J., Becker, R., and Brunner, E. (1989) Synthesis of hydroxylated 1-azabicyclo[3.1.0]hexane and prolinol derivatives by stereo- and regiocontrolled

Staudinger aminocyclization. Application to the nonproteinogenic amino acid (2*S*,3*S*,4*S*)-3-hydroxy-4-methylproline (HMP) and its enantiomer. *Journal of the American Chemical Society*, **111**, 7500–7504.

68 Kurokawa, N. and Ohfune, Y. (1993) Synthetic studies on antifungal cyclic peptides, echinocandins. Stereoselective total synthesis of echinocandin D via a novel peptide coupling. *Tetrahedron*, **49**, 6195–6222.

69 Langlois, N. (1998) Synthesis of the nonproteinogenic amino acid (2*S*,3*S*,4*S*)-3-hydroxy-4-methylproline, a constituent of echinocandins. *Tetrahedron: Asymmetry*, **9**, 1333–1336.

70 Langlois, N. and Rakotondradany, F. (2000) Diastereoselective synthesis of (2*S*,3*S*,4*S*)-3-hydroxy-4-methylproline, a common constituent of several antifungal cyclopeptides. *Tetrahedron*, **56**, 2437–2448.

71 Raghavan, S. and Ramakrishna Reddy, S. (2003) Practical, efficient, stereoselective, formal synthesis of (2*R*,3*R*,4*R*)-3-hydroxy-4-methylproline. *Tetrahedron Letters*, **44**, 7459–7462.

72 Meng, W.-H., Wu, T.-J., Zhang, H.-K., and Huang, P.-Q. (2004) Asymmetric syntheses of protected (2*S*,3*S*,4*S*)-3-hydroxy-4-methylproline and 4′-*tert*-butoxyamido-2′-deoxythymidine. *Tetrahedron: Asymmetry*, **15**, 3899–3910.

73 Mohapatra, D.K., Mondal, D., Chorghade, M.S., and Gurjar, M.K. (2006) General strategy for a short and efficient synthesis of 3-hydroxy-4-methylprolines (HMP). *Tetrahedron Letters*, **47**, 9215–9219.

74 Nagoka, H. and Kishi, Y. (1981) Further synthetic studies on rifamycin S. *Tetrahedron*, **37**, 3873–3888.

75 Zacharius, R.M., Thompson, J.F., and Steward, F.C. (1952) The detection, isolation and identification of (−)-pipecolic acid as a constituent of plants. *Journal of the American Chemical Society*, **74**, 2949.

76 Tanaka, H., Kuroda, A., Marusawa, H., Hatanaka, H., Kino, T., Goto, T., Hashimoto, M., and Taga, T. (1987) Structure of FK506, a novel immunosuppressant isolated from *Streptomyces*. *Journal of the American Chemical Society*, **109**, 5031–5033.

77 Vezina, C., Kudelski, A., and Sehgal, S.N. (1975) Rapamycin (AY 22,989), a new antifungal antibiotic. I. Taxonomy of the producing streptomycete and isolation of the active principle. *Journal of Antibiotics*, **28**, 721–726.

78 Pettibone, D.J., Clineschmidt, B.V., Anderson, P.S., Freidinger, R.M., Lundell, G.F., Koupal, L.R., Schwartz, C.D., Williamson, J.M., Goetz, M.A., Hensens, O.D., Liesch, J.M., and Springer, J.P. (1989) A structurally unique, potent, and selective oxytocin antagonist derived from *Streptomyces silvensis*. *Endocrinology*, **125**, 217–222.

79 Darkin-Rattray, S.J., Gurnett, A., Myers, R.W., Dulski, P.M., Crumley, T.M., Allocco, J.J., Cannova, C., Meinke, P.T., Colletti, S.L., Bednarek, M., Singh, S., Goetz, M., Dombrowski, A., Polishook, J., and Schmatz, D. (1996) Apicidin: a novel antiprotozoal agent that inhibits parasite histone deacetylase. *Proceedings of the National Academy of Sciences of the United States of America*, **93**, 13143–13147.

80 Lee, K.K., Gloer, J.B., Scott, J.A., and Malloch, D. (1995) Petriellin A: a novel antifungal depsipeptide from the coprophilous fungus *Petriella sordida*. *The Journal of Organic Chemistry*, **60**, 5384–5385.

81 Boger, D.L., Chen, J.-H., and Saionz, K.W. (1996) (−)-Sandramycin: total synthesis and characterization of DNA binding properties. *Journal of the American Chemical Society*, **118**, 1629–1644.

82 Couty, F. (1999) Asymmetric syntheses of pipecolic acid and derivatives. *Amino Acids*, **16**, 297–320.

83 Kadouri-Puchot, C. and Comesse, S. (2005) Recent advances in asymmetric synthesis of pipecolic acid and derivatives. *Amino Acids*, **29**, 101–130.

84 Heller, E., Lautenschlager, W., and Holzgrabe, U. (2005) Microwave-enhanced hydrogenations at medium pressure using a newly constructed reactor. *Tetrahedron Letters*, **46**, 1247–1249.

85 Vanier, G.S. (2007) Simple and efficient microwave-assisted hydrogenation reactions at moderate temperature and pressure. *Synlett*, 131–135.

86 Corey, E.J., Noe, M.C., and Xu, F. (1998) Highly enantioselective synthesis of cyclic and functionalized α-amino acids by means of a chiral phase transfer catalyst. *Tetrahedron Letters*, **39**, 5347–5350.

87 Corey, E.J., Xu, F., and Noe, M.C. (1997) A rational approach to catalytic enantioselective enolate alkylation using a structurally rigidified and defined chiral quaternary ammonium salt under phase transfer conditions. *Journal of the American Chemical Society*, **119**, 12414–12415.

88 Ginesta, X., Pericas, M.A., and Riera, A. (2002) Straightforward entry to the pipecolic acid nucleus. Enantioselective synthesis of baikiain. *Tetrahedron Letters*, **43**, 779–782.

89 Fadel, A. and Lahrache, N. (2007) An efficient synthesis of enantiomerically pure (*R*)-pipecolic acid, (*S*)-proline, and their *N*-alkylated derivatives. *The Journal of Organic Chemistry*, **72**, 1780–1784.

90 Murakami, M., Okita, Y., Matsuda, H., Okino, T., and Yamaguchi, K. (1994) Aeruginosin 298-A, a thrombin and trypsin inhibitor from the blue-green alga microcystis aeruginosa (NIES-298). *Tetrahedron Letters*, **35**, 3129–3132.

91 Ersmark, K., Del Valle, J.R., and Hanessian, S. (2008) Chemistry and biology of the aeruginosin family of serine protease inhibitors. *Angewandte Chemie (International Edition in English)*, **47**, 1202–1223.

92 Murakami, M., Ishida, K., Okino, T., Okita, Y., Matsuda, H., and Yamaguchi, K. (1995) Aeruginosins 98-A and B, trypsin inhibitors from the blue-green alga *Microcystis aeruginosa* (NIES-98). *Tetrahedron Letters*, **36**, 2785–2788.

93 Matsuda, H., Okino, T., Murakami, M., and Yamaguchi, K. (1996) Aeruginosins 102-A and B, new thrombin inhibitors from the cyanobacterium *Microcystis viridis* (NIES-102). *Tetrahedron*, **52**, 14501–14506.

94 Shin, H.J., Matsuda, H., Murakami, M., and Yamaguchi, K. (1997) Aeruginosins 205A and -B, serine protease inhibitory glycopeptides from the cyanobacterium *Oscillatoria agardhii* (NIES-205). *The Journal of Organic Chemistry*, **62**, 1810–1813.

95 Noel, A., Gilles, C., Bajou, K., Devy, L., Kebers, F., Lewalle, J.M., Maquoi, E., Munaut, C., Remacle, A., and Foidart, J.M. (1997) Emerging roles for proteinases in cancer. *Invasion & Metastasis*, **17**, 221–239.

96 Kodani, S., Ishida, K., and Murakami, M. (1998) Aeruginosin 103-A, a thrombin inhibitor from the cyanobacterium *Microcystis viridis*. *Journal of Natural Products*, **61**, 1046–1048.

97 Banker, R. and Carmeli, S. (1999) Inhibitors of serine proteases from a waterbloom of the cyanobacterium *Microcystis* sp. *Tetrahedron*, **55**, 10835–10844.

98 Ploutno, A., Shoshan, M., and Carmeli, S. (2002) Three novel protease inhibitors from a natural bloom of the cyanobacterium *Microcystis aeruginosa*. *Journal of Natural Products*, **65**, 973–978.

99 Carroll, A.R., Pierens, G.K., Fechner, G., de Leone, P., Ngo, A., Simpson, M., Hyde, E., Hooper, J.N.A., Bostrom, S.-L., Musil, D., and Quinn, R.J. (2002) Dysinosin A: A novel inhibitor of factor VIIa and thrombin from a new genus and species of Australian sponge of the family Dysideidae. *Journal of the American Chemical Society*, **124**, 13340–13341.

100 Carroll, A.R., Buchanan, M.S., Edser, A., Hyde, E., Simpson, M., and Quinn, R.J. (2004) Dysinosins B–D, inhibitors of factor VIIa and thrombin from the Australian sponge *Lamellodysidea chlorea*. *Journal of Natural Products*, **67**, 1291–1294.

101 Hanessian, S., Tremblay, M., and Petersen, J.F.W. (2004) The *N*-acyloxyiminium ion aza-Prins route to octahydroindoles: total synthesis and structural confirmation of the antithrombotic marine natural product oscillarin. *Journal of the American Chemical Society*, **126**, 6064–6071.

102 Hanessian, S., DelValle, J.R., Xue, Y., and Blomberg, N. (2006) Total synthesis and structural confirmation of chlorodysinosin A. *Journal of the American Chemical Society*, **128**, 10491–10495.

103 Bonjoch, J., Catena, J., Isabal, E., Lopez-Canet, M., and Valls, N. (1996) Synthesis of the octahydroindole core of aeruginosins: a new bicyclic α-amino acid. *Tetrahedron: Asymmetry*, **7**, 1899–1902.

104 Valls, N., Lopez-Canet, M., Vallribera, M., and Bonjoch, J. (2001) First total syntheses of aeruginosin 298-A and aeruginosin 298-B, based on a stereocontrolled route to the new amino acid 6-hydroxyoctahydroindole-2-carboxylic acid. *Chemistry – A European Journal*, **7**, 3446–3460.

105 Wipf, P. and Methot, J.-L. (2000) Total synthesis and stereochemical revision of (+)-aeruginosin 298-A. *Organic Letters*, **2**, 4213–4216.

106 Ohshima, T., Gnanadesikan, V., Shibuguchi, T., Fukuta, Y., Nemoto, T., and Shibasaki, M. (2003) Enantioselective syntheses of aeruginosin 298-A and its analogues using a catalytic asymmetric phase-transfer reaction and epoxidation. *Journal of the American Chemical Society*, **125**, 11206–11207.

107 Fukuta, Y., Ohshima, T., Gnanadesikan, V., Shibuguchi, T., Nemoto, T., Kisugi, T., Okino, T., and Shibasaki, M. (2004) Enantioselective syntheses and biological studies of aeruginosin 298-A and its analogs: application of catalytic asymmetric phase-transfer reaction. *Proceedings of the National Academy of Sciences of the United States of America*, **101**, 5433–5438.

108 Hanessian, S. and Margarita, R. (1998) 1,3-Asymmetric induction in dianionic allylation reactions of amino acid derivatives-synthesis of functionally useful enantiopure glutamates, pipecolates and pyroglutamates. *Tetrahedron Letters*, **39**, 5887–5890.

109 Umezawa, K., Ikeda, Y., Uchihata, Y., Naganawa, H., and Kondo, S. (2000) Chloptosin, an apoptosis-inducing dimeric cyclohexapeptide produced by streptomyces. *The Journal of Organic Chemistry*, **65**, 459–463.

110 Leet, J.E., Schroeder, D.R., Krishnan, B.S., and Matson, J.A. (1990) Himastatin, a new antitumor antibiotic from *Streptomyces hygroscopicus*. II. Isolation and characterisation. *Journal of Antibiotics*, **43**, 961–966.

111 Leet, J.E., Schroeder, D.R., Golik, J., Matson, J.A., Doyle, T.W., Lam, K.S., Hill, S.E., Lee, M.S., Whitney, J.L., and Krishnan, B.S. (1996) Himastatin, a new antitumor antibiotic from *Streptomyces hygroscopicus*. III. Structural elucidation. *Journal of Antibiotics*, **49**, 299–311.

112 Hayashi, H., Furutsuka, K., and Shiono, Y. (1999) Okaramines H and I, new okaramine congeners, from *Aspergillus aculeatus*. *Journal of Natural Products*, **62**, 315–317.

113 Pettit, G.R., Tan, R., Herald, D.L., Williams, M.D., and Cerny, R.L. (1994) Antineoplastic agents. 277. Isolation and structure of phakellistatin 3 and isophakellistatin 3 from a republic of comoros marine sponge. *The Journal of Organic Chemistry*, **59**, 1593–1595.

114 Schkeryantz, J.M., Woo, J.C.G., Siliphaivanh, P., Depew, K.M., and Danishefsky, S.J. (1999) Total synthesis of gypsetin, deoxybrevianamide E, brevianamide E, and tryprostatin B: novel constructions of 2,3-disubstituted indoles. *Journal of the American Chemical Society*, **121**, 11964–11975.

115 Nakao, Y., Yeung, B.K.S., Yoshida, W.Y., Scheuer, P.J., and Kelly-Borges, M. (1995)

Kapakahine B, a cyclic hexapeptide with an α-carboline ring system from the marine sponge *Cribrochalina olemda*. *Journal of the American Chemical Society*, **117**, 8271–8272.

116 Buchel, E., Martini, U., Mayer, A., Anke, H., and Sterner, O. (1998) Omphalotins B, C and D, nematicidal cyclopeptides from *Omphalotus olearius*. Absolute configuration of omphalotin A. *Tetrahedron*, **54**, 5345–5352.

117 Francis, E., Rahman, R., Safe, S., and Taylor, A. (1972) Sporidesmins. Part XII. Isolation and structure of sporidesmin G, a naturally-occurring 3,6-epitetrathio-2,5-piperazinedione. *Journal of the Chemical Society, Perkin Transactions 1*, 470–472.

118 Sakai, A., Tani, H., Aoyama, T., and Shioiri, T. (1998) Enantioselective photosensitized oxygenation. Its application to *N*β-(methoxycarbonyl) tryptamine and determination of absolute configuration of the product. *Synlett*, 257–258.

119 Anthoni, U., Christophersen, C., Nielsen, P.H., Christoffersen, M.W., and Sorensen, D. (1998) Cyclisations of tryptophans. VI. Cyclisation of L-tryptophan dipeptides by oxygenation with singlet oxygen. *Acta Chemica Scandinavica*, **52**, 958–960.

120 Greenman, K.L., Hach, D.M., and Van Vranken, D.L. (2004) Synthesis of phakellistatin 13 and oxidation to phakellistatin 3 and isophakellistatin 3. *Organic Letters*, **6**, 1713–1716.

121 Savige, W.E. (1975) New oxidation products of tryptophan. *Australian Journal of Chemistry*, **28**, 2275–2287.

122 Kamenecka, T.M. and Danishefsky, S.J. (2001) Discovery through total synthesis: a retrospective on the himastatin problem. *Chemistry – A European Journal*, **7**, 41–63.

123 May, J.P., Fournier, P., Pellicelli, J., Patrick, B.O., and Perrin, D.M. (2005) High yielding synthesis of 3-α-hydroxypyrrolo[2,3-β]indoline dipeptide methyl esters: synthons for expedient introduction of the hydroxypyrroloindoline moiety into larger peptide-based natural products and for the creation of tryptathionine bridges. *The Journal of Organic Chemistry*, **70**, 8424–8430.

124 Ley, S.V., Cleator, E., and Hewitt, P.R. (2003) A rapid stereocontrolled synthesis of the 3-α-hydroxy-pyrrolo[2,3-β]indole skeleton, a building block for 10-β-hydroxy-pyrazino[1′,2′:1,5]pyrrolo[2,3-β] indole-1,4-diones. *Organic and Biomolecular Chemistry*, **1**, 3492–3494.

125 Hewitt, P.R., Cleator, E., and Ley, S.V. (2004) A concise total synthesis of (+)-okaramine C. *Organic and Biomolecular Chemistry*, **2**, 2415–2417.

126 Maehr, H., Liu, C.M., Palleroni, N.J., Smallheer, J., Todaro, L., Williams, T.H., and Blount, J.F. (1986) Microbial products. VIII. Azinothricin, a novel hexadepsipeptide antibiotic. *Journal of Antibiotics*, **39**, 17–25.

127 Smitka, T.A., Deeter, J.B., Hunt, A.H., Mertz, F.P., Ellis, R.M., Boeck, L.D., and Yao, R.C. (1988) A83586C, a new depsipeptide antibiotic. *Journal of Antibiotics*, **41**, 726–733.

128 Ford, P.W., Gustafson, K.R., McKee, T.C., Shigematsu, N., Maurizi, L.K., Pannell, L.K., Williams, D.E., Dilip de Silva, E., Lassota, P., Allen, T.M., Van Soest, R., Andersen, R.J., and Boyd, M.R. (1999) Papuamides A–D – HIV-inhibitory and cytotoxic depsipeptides from the sponges *Theonella mirabilis* and *Theonella swinhoei* collected in Papua New Guinea. *Journal of the American Chemical Society*, **121**, 5899–5909.

129 Sakai, Y., Yoshida, T., Tsujita, T., Ochiai, K., Agatsuma, T., Saitoh, Y., Tanaka, F., Akiyama, T., Akinaga, S., and Mizukami, T. (1997) GE3, a novel hexadepsipeptide antitumor antibiotic, produced by *Streptomyces* sp. I. Taxonomy, production, isolation, physico-chemical properties, and biological activities. *Journal of Antibiotics*, **50**, 659–664.

130 Zhang, A.J. and Burgess, K. (1999) Totalsynthese von Vancomycin. *Angewandte Chemie*, **111**, 666–669.

131 Faulds, D., Goa, K.L., and Benfield, P. (1993) Cyclosporin: a review of its pharmacodynamic and pharmacokinetic properties, and therapeutic use in immunoregulatory disorders. *Drugs*, **45**, 953–1040.

132 Lotz, B.T. and Miller, M.J. (1993) Diastereoselective synthesis of the carbacephem framework. *The Journal of Organic Chemistry*, **58**, 618–625.

133 Dong, L. and Miller, M.J. (2002) Total synthesis of exochelin MN and analogues. *The Journal of Organic Chemistry*, **67**, 4759–4770.

134 Tanner, D. (1994) Chiral aziridines – their synthesis and use in stereoselective transformations. *Angewandte Chemie (International Edition in English)*, **33**, 599–619.

135 Esslinger, C.S., Agarwal, S., Gerdes, J., Wilson, P.A., Davis, E.S., Awes, A.N., O'Brien, E., Mavencamp, T., Koch, H.P., Poulsen, D.J., Rhoderick, J.F., Chamberlin, A.R., Kavanaugh, M.P., and Bridges, R.J. (2005) The substituted aspartate analogue L-β-*threo*-benzyl-aspartate preferentially inhibits the neuronal excitatory amino acid transporter EAAT3. *Neuropharmacology*, **49**, 850–861.

136 Pansare, S.V. and Vederas, J.C. (1987) Reaction of β-hydroxy α-amino acid derivatives with (diethylamino)sulfur trifluoride (DAST). Synthesis of β-fluoro α-amino acids. *The Journal of Organic Chemistry*, **52**, 4804–4810.

137 Badorrey, R., Cativiela, C., Díaz-de-Villegas, M.D., and Gálvez, J.A. (2002) Highly convergent stereoselective synthesis of chiral key intermediates in the synthesis of Palinavir from imines derived from L-glyceraldehyde. *Tetrahedron*, **58**, 341–354.

138 Bold, G., Duthaler, R.O., and Riediker, M. (1989) Enantioselective synthesis of D-*threo*-β-hydroxy-α-amino acids with titanium–carbohydrate complexes. *Angewandte Chemie (International Edition in English)*, **28**, 497–498.

139 Nakatsuka, T., Miwa, T., and Mukaiyama, T. (1981) A new method for enantioselective synthesis of β-hydroxy-α-amino acids. *Chemistry Letters*, 279–282.

140 Kobayashi, J., Nakamura, M., Mori, Y., Yamashita, Y., and Kobayashi, S. (2004) Catalytic enantio- and diastereoselective aldol reactions of glycine-derived silicon enolate with aldehydes: an efficient approach to the asymmetric synthesis of anti-β-hydroxy-α-amino acid derivatives. *Journal of the American Chemical Society*, **126**, 9192–9193.

141 Chevolot, L., Chevolot, A.M., Gajhede, M., Larsen, C., Anthoni, U., and Christophersen, C. (1985) Marine alkaloids. 10. Chartelline A: a pentahalogenated alkaloid from the marine bryozoan *Chartella papyracea*. *Journal of the American Chemical Society*, **107**, 4542–4543.

142 Evans, D.A. and Weber, A.E. (1986) Asymmetric glycine enolate aldol reactions: synthesis of cyclosporin's unusual amino acid, MeBmt. *Journal of the American Chemical Society*, **108**, 6757–6761.

143 Maruoka, K., Tayama, E., and Ooi, T. (2004) Stereoselective terminal functionalization of small peptides for catalytic asymmetric synthesis of unnatural peptides. *Proceedings of the National Academy of Sciences of the United States of America*, **101**, 5824–5829.

144 MacMillan, J.B. and Molinski, T.F. (2002) Lobocyclamide B from Lyngbya confervoides. Configuration and asymmetric synthesis of β-hydroxy-α-amino acids by (−)-sparteine-mediated aldol addition. *Organic Letters*, **4**, 1883–1886.

145 Thayumanavan, R., Tanaka, F., and Barbas, C.F. (2004) Direct organocatalytic asymmetric aldol reactions of α-amino aldehydes: expedient syntheses of highly enantiomerically enriched anti-β-hydroxy-α-amino acids. *Organic Letters*, **6**, 3541–3544.

146 Iwanowicz, E.J., Blomgren, P., Cheng, P.T.W., Smith, K., Lau, W.F., Pan, Y.Y., Gu,

H.H., Malley, M.F., and Gougoutas, J.Z. (1998) The enantioselective synthesis of anti-β-hydroxy α-amino acids via the reaction of lithium enolates of glycine bearing an oxazolidine chiral auxiliary with aldehydes. *Synlett*, 664–666.

147 Belokon, Y.N., Kochetkov, K.A., Ikonnikov, N.S., Strelkova, T.V., Harutyunyan, S.R., and Saghiyan, A.S. (2001) A new synthesis of enantiomerically pure *syn*-(*S*)-β-hydroxy-α-amino acids via asymmetric aldol reactions of aldehydes with a homochiral Ni(II)–glycine/(*S*)-BPB Schiff base complex. *Tetrahedron: Asymmetry*, **12**, 481–485.

148 Di Felice, P., Porzi, G., and Sandri, S. (1999) A new stereoselective approach to β-hydroxy-α-aminoacids and dipeptides. *Tetrahedron: Asymmetry*, **10**, 2191–2201.

149 Caddick, S., Parr, N.J., and Pritchard, M.C. (2000) *syn*-Selective boron mediated aldol condensations for the asymmetric synthesis of β-hydroxy-α-amino acids. *Tetrahedron Letters*, **41**, 5963–5966.

150 Corey, E.J., Lee, D.H., and Choi, S. (1992) An enantioselective synthesis of (2*S*,3*S*)- and (2*R*,3*S*)-3-hydroxyleucine. *Tetrahedron Letters*, **33**, 6735–6738.

151 Kanemasa, S., Mori, T., Wada, E., and Tatsukawa, A. (1993) *anti*-Selective aldol reactions of titanium enolates of *N*-alkylideneglycinates. Stereoselective synthesis of anti-isomers of β-hydroxy-α-amino esters. *Tetrahedron Letters*, **34**, 677–680.

152 Alker, D., Hamblett, G., Harwood, L.M., Robertson, S.M., Watkin, D.J., and Williams, C.E. (1998) Application of enantiopure templated azomethine ylids to β-hydroxy-α-amino acid synthesis. *Tetrahedron*, **54**, 6089–6098.

153 Blaskovich, M.A., Evindar, G., Rose, N.G.W., Wilkinson, S., Luo, Y., and Lajoie, G.A. (1998) Stereoselective synthesis of *threo* and *erythro* β-hydroxy and β-disubstituted-β-hydroxy α-amino acids. *The Journal of Organic Chemistry*, **63**, 3631–3646.

154 Okamoto, N., Hara, O., Makino, K., and Hamada, Y. (2002) Diastereoselective synthesis of all stereoisomers of β-methoxytyrosine, a component of papuamides. *The Journal of Organic Chemistry*, **67**, 9210–9215.

155 Avenoza, A., Cativiela, C., Corzana, F., Peregrina, J.M., and Zurbano, M.M. (2000) Asymmetric synthesis of all isomers of α-methyl-β-phenylserine. *Tetrahedron: Asymmetry*, **11**, 2195–2204.

156 Williams, L., Zhang, Z., Shao, F., Carroll, P.J., and Joullie, M.M. (1996) Grignard reactions to chiral oxazolidine aldehydes. *Tetrahedron*, **52**, 11673–11694.

157 Guanti, G., Banfi, L., and Narisano, E. (1988) Enantiospecific and diastereoselective synthesis of *anti* α-hydrazino- and α-amino-β-hydroxyacids through electrophilic amination of β-hydroxyesters. *Tetrahedron*, **44**, 5553–5562.

158 Fraunhoffer, K.J. and White, M.C. (2007) *syn*-1,2-Amino alcohols via diastereoselective allylic C–H amination. *Journal of the American Chemical Society*, **129**, 7274–7276.

159 Hirama, M., Hioki, H., and Ito, S. (1988) New entry to *syn*-β-hydroxy-α-amino acids. *Tetrahedron Letters*, **29**, 3125–3128.

160 Oba, M., Mita, A., Kondo, Y., and Nishiyama, K. (2005) Facile synthesis of 3-hydroxyglutamic acids via cyanation of chiral *N*-acyliminium cation derived from (*S*)-malic acid. *Synthetic Communications*, **35**, 2961–2966.

161 Kuwano, R., Okuda, S., and Ito, Y. (1998) Catalytic asymmetric synthesis of β-hydroxy-α-amino acids: highly enantioselective hydrogenation of β-oxy-α-acetamidoacrylates. *The Journal of Organic Chemistry*, **63**, 3499–3503.

162 Ruble, J.C. and Fu, G.C. (1998) Enantioselective construction of quaternary stereocenters: rearrangements of *O*-acylated azlactones catalyzed by a planar-chiral derivative of 4-(pyrrolidino) pyridine. *Journal of the American Chemical Society*, **120**, 11532–11533.

163 Tomasini, C. and Vecchione, A. (1999) Novel synthesis of 4-carboxymethyl 5-alkyl/aryl oxazolidin-2-ones by rearrangement of 2-carboxymethyl 3-alkyl/aryl *N-tert*-butoxycarbonyl aziridines. *Organic Letters*, **1**, 2153–2156.

164 Fanning, K.N., Jamieson, A.G., and Sutherland, A. (2005) Stereoselective β-hydroxy-α-amino acid synthesis via an ether-directed, palladium-catalyzed aza-Claisen rearrangement. *Organic and Biomolecular Chemistry*, **3**, 3749–3756.

165 Cardillo, G., Gentilucci, L., Gianotti, M., and Tolomelli, A. (2001) Asymmetric synthesis of 5-isopropyl-oxazoline-4-imide as *syn*-hydroxyleucine precursor. *Tetrahedron: Asymmetry*, **12**, 563–569.

166 Saito, S., Bunya, N., Inaba, M., Moriwake, T., and Torii, S. (1985) A facile cleavage of oxirane with hydrazoic acid in DMF. A new route to chiral β-hydroxy-α-amino acids. *Tetrahedron Letters*, **26**, 5309–5312.

167 Lee, K.-D., Suh, J.-M., Park, J.-H., Ha, H.-J., Choi, H.G., Park, C.S., Chang, J.W., Lee, W.K., Dong, Y., and Yun, H. (2001) New synthesis and ring opening of *cis*-3-alkylaziridine-2-carboxylates. *Tetrahedron*, **57**, 8267–8276.

168 Adams, Z.M., Jackson, R.F.W., Palmer, N.J., Rami, H.K., and Wythes, M.J. (1999) Stereoselective syntheses of protected β-hydroxy-α-amino acids using (arylthio) nitrooxiranes. *Journal of the Chemical Society-Perkin Transactions 1*, 937–948.

169 Makino, K., Goto, T., Hiroki, Y., and Hamada, Y. (2004) Stereoselective synthesis of *anti*-β-hydroxy-α-amino acids through dynamic kinetic resolution. *Angewandte Chemie (International Edition in English)*, **43**, 882–884.

170 Mordant, C., Dunkelmann, P., Ratovelomanana-Vidal, V., and Genet, J.-P. (2004) Dynamic kinetic resolution: an efficient route to *anti*-α-amino-β-hydroxy esters via Ru-SYNPHOS catalyzed hydrogenation. *Journal of the Chemical Society, Chemical Communications*, 1296–1297.

171 Noyori, R., Ikeda, T., Ohkuma, T., Widhalm, M., Kitamura, M., Takaya, H., Akutagawa, S., Sayo, N., Saito, T., Taketomi, T., and Kumobayashi, H. (1989) Stereoselective hydrogenation via dynamic kinetic resolution. *Journal of the American Chemical Society*, **111**, 9134–9135.

172 Coulon, E., Cristina, M., Cano de Andrade, C., Ratovelomanana-Vidal, V., and Genet, J.-P. (1998) An efficient synthesis of (2*S*,3*R*)-3-hydroxylysine via ruthenium catalyzed asymmetric hydrogenation. *Tetrahedron Letters*, **39**, 6467–6470.

173 Bourdon, L.H., Fairfax, D.J., Martin, G.S., Mathison, C.J., and Zhichkin, P. (2004) An enantioselective synthesis of nitrogen protected 3-arylserine esters. *Tetrahedron: Asymmetry*, **15**, 3485–3487.

174 Shiraiwa, T., Saijoh, R., Suzuki, M., Yoshida, K., Nishimura, S., and Nagasawa, H. (2003) Preparation of optically active *threo*-2-amino-3-hydroxy-3-phenylpropanoic acid (*threo*-β-phenylserine) via optical resolution. *Chemical & Pharmaceutical Bulletin*, **51**, 1363–1367.

175 Makino, K., Iwasaki, M., and Hamada, Y. (2006) Enantio- and diastereoselective hydrogenation via dynamic kinetic resolution by a cationic iridium complex in the synthesis of β-hydroxy-α-amino acid esters. *Organic Letters*, **8**, 4573–4576.

176 Park, H., Cao, B., and Joullie, M.M. (2001) Regioselective asymmetric aminohydroxylation approach to a β-hydroxyphenylalanine derivative for the synthesis of ustiloxin D. *The Journal of Organic Chemistry*, **66**, 7223–7226.

177 Morgan, A.J., Masse, C.E., and Panek, J.S. (1999) Reversal of regioselection in the Sharpless asymmetric aminohydroxylation of aryl ester substrates. *Organic Letters*, **1**, 1949–1952.

178 Tao, B., Schlingloff, G., and Sharpless, K.B. (1998) Reversal of regioselection in the asymmetric aminohydroxylation of cinnamates. *Tetrahedron Letters*, **39**, 2507–2510.

179 Shao, H., Rueter, J.K., and Goodman, M. (1998) Novel enantioselective synthesis of α-methylthreonines and α,β-dimethylcysteines. *The Journal of Organic Chemistry*, **63**, 5240–5244.

180 Rao, A.V.R., Chakraborty, T.K., Reddy, K.L., and Rao, A.S. (1994) An expeditious approach for the synthesis of β-hydroxy aryl α-amino acids present in vancomycin. *Tetrahedron Letters*, **35**, 5043–5046.

181 Hale, K.J., Manaviazar, S., and Delisser, V.M. (1994) A practical new asymmetric synthesis of (2*S*,3*S*)- and (2*R*,3*R*)-3-hydroxyleucine. *Tetrahedron*, **50**, 9181–9188.

182 Aurelio, L., Brownlee, R.T.C., and Hughes, A.B. (2004) Synthetic preparation of *N*-methyl-α-amino acids. *Chemical Reviews*, **104**, 5823–5846.

183 Caldwell, C.G. and Bondy, S.S. (1990) A convenient synthesis of enantiomerically pure (2*S*,3*S*)- or (2*R*,3*R*)-3-hydroxyleucine. *Synthesis*, 34–36.

184 Jungheim, L.N., Shepherd, T.A., Baxter, A.J., Burgess, J., Hatch, S.D., Lubbehusen, P., Wiskerchen, M., and Muesing, M.A. (1996) Potent human immunodeficiency virus type 1 protease inhibitors that utilize noncoded D-amino acids as P2/P3 ligands. *Journal of Medicinal Chemistry*, **39**, 96–108.

185 Righi, G., Rumboldt, G., and Bonini, C. (1995) A simple route to *syn* α-amino-β-hydroxy esters by C-2 regioselective opening of α,β-epoxy esters with metal halides. *Tetrahedron*, **51**, 13401–13408.

186 Zandbergen, P., Brussee, J., Van der Gen, A., and Kruse, C.G. (1992) Stereoselective synthesis of β-hydroxy-α-amino acids from chiral cyanohydrins. *Tetrahedron: Asymmetry*, **3**, 769–774.

187 Makino, K. and Hamada, Y. (2005) Asymmetric synthesis of β-hydroxy-α-amino acid. *Journal of Synthetic Organic Chemistry Japan*, **63**, 1198–1208.

188 Suga, H., Ikai, K., and Ibata, T. (1999) *Cis* and enantioselective synthesis of 2-oxazoline-4-carboxylates through Lewis acid-catalyzed formal [3 + 2] cycloaddition of 5-alkoxyoxazoles with aldehydes. *The Journal of Organic Chemistry*, **64**, 7040–7047.

189 Wallis, J.M. and Kochi, J.K. (1988) Direct osmylation of benzenoid hydrocarbons. Charge-transfer photochemistry of osmium tetraoxide. *The Journal of Organic Chemistry*, **53**, 1679–1686.

190 Amador, M., Ariza, X., Garcia, J., and Sevilla, S. (2002) Stereodivergent approach to β-hydroxy α-amino acids from C2-symmetrical alk-2-yne-1,4-diols. *Organic Letters*, **4**, 4511–4514.

191 Hanessian, S., Sumi, K., and Vanasse, B. (1992) The stereocontrolled synthesis of (2*S*,3*R*)-3-alkyl-L-aspartic acids using a 2-azetidinone framework as a chiral template. *Synlett*, 33–34.

192 Charvillon, F.B. and Amouroux, R. (1997) Synthesis of 3-hydroxylated analogues of D-aspartic acid β-hydroxamate. *Synthetic Communications*, **27**, 395–403.

193 Cardillo, G., Gentilucci, L., Tolomelli, A., and Tomasini, C. (1999) A practical method for the synthesis of β-amino α-hydroxy acids. Synthesis of enantiomerically pure hydroxyaspartic acid and isoserine. *Synlett*, 1727–1730.

194 Hancock, D.K. and Reeder, D.J. (1993) Analysis and configuration assignments of the amino acids in a pyoverdine-type siderophore by reversed-phase high-performance liquid chromatography. *Journal of Chromatography. A*, **646**, 335–343.

195 Risse, D., Beiderbeck, H., Taraz, K., Budzikiewicz, H., and Gustine, D. (1998) Corrugatin, a lipopeptide siderophore from *Pseudomonas corrugata*. *Zeitschrift fur Naturforschung. C, Journal of Biosciences*, **53**, 295–304.

196 Sznaidman, M.L. and Hecht, S.M. (2001) Studies on the total synthesis of tallysomycin. Synthesis of the threonylbithiazole moiety containing a structurally unique glycosylcarbinolamide. *Organic Letters*, **3**, 2811–2814.

197 Hecht, S.M. (2000) Bleomycin: new perspectives on the mechanism of action. *Journal of Natural Products*, **63**, 158–168.

198 Owa, T.O., Otsuka, M., and Ohno, M. (1988) Synthetic studies on an antitumor antibiotic, bleomycin. XXV. An efficient synthesis of *erythro*-β-hydroxy-L-histidine, the pivotal amino acid of bleomycin–iron (II)–oxygen complex. *Chemistry Letters*, 1873–1874.

199 Owa, T.O., Otsuka, M., and Ohno, M. (1988) Enantioselective synthesis of *erythro*-β-hydroxy-L-histidine, the pivotal amino acid of bleomycin–Fe(II)–O_2 complex. *Chemistry Letters*, 83–86.

200 Crich, D. and Banerjee, A. (2006) Expedient synthesis of *threo*-β-hydroxy-α-amino acid derivatives: phenylalanine, tyrosine, histidine, and tryptophan. *The Journal of Organic Chemistry*, **71**, 7106–7109.

201 Sakaguchi, K., Yamamoto, M., Kawamoto, T., Yamada, T., Shinada, T., Shimamoto, K., and Ohfune, Y. (2004) Synthesis of optically active β-alkyl aspartate via [3,3] sigmatropic rearrangement of α-acyloxytrialkylsilane. *Tetrahedron Letters*, **45**, 5869–5872.

202 Gulledge, B.M., Aggen, J.B., Huang, H.-B., Nairn, A.C., and Chamberlin, A.R. (2002) The microcystins and nodularins: cyclic polypeptide inhibitors of PP1 and PP2A. *Current Medicinal Chemistry*, **9**, 1991–2003.

203 Wolf, J.P. and Rapoport, H. (1989) Conformationally constrained peptides. Chirospecific synthesis of 4-alkyl-substituted γ-lactam-bridged dipeptides from L-aspartic acid. *The Journal of Organic Chemistry*, **54**, 3164–3173.

204 Bull, S.D., Davies, S.G., Garner, A.C., and Mujtaba, N. (2001) The asymmetric synthesis of (2*R*,3*R*)- and (2*R*,3*S*)-3-methyl-aspartates via an enantiodiscrimination strategy. *Synlett*, 781–784.

205 Bull, S.D., Davies, S.G., Epstein, S.W., Garner, A.C., Mujtaba, N., Roberts, P.M., Savory, E.D., Smith, A.D., Tamayo, J.A., and Watkin, D.J. (2006) Enantiodiscrimination of racemic electrophiles by diketopiperazine enolates: asymmetric synthesis of methyl 2-amino-3-aryl-butanoates and 3-methyl-aspartates. *Tetrahedron*, **62**, 7911–7925.

206 Ooi, T., Kato, D., Inamura, K., Ohmatsu, K., and Maruoka, K. (2007) Practical stereoselective synthesis of β-branched α-amino acids through efficient kinetic resolution in the phase-transfer-catalyzed asymmetric alkylations. *Organic Letters*, **9**, 3945–3948.

207 Valentekovich, R.J. and Schreiber, S.L. (1995) Enantiospecific total synthesis of the protein phosphatase inhibitor motuporin. *Journal of the American Chemical Society*, **117**, 9069–9070.

208 Sakaguchi, K., Fujita, M., Suzuki, H., Higashino, M., and Ohfune, Y. (2000) Reverse Brook rearrangement of 2-alkenyl trialkylsilyl ether. Synthesis of optically active (1-hydroxy-2-alkynyl) trialkylsilane. *Tetrahedron Letters*, **41**, 6589–6592.

209 Sakaguchi, K., Suzuki, H., and Ohfune, Y. (2001) Chirality transferring [3,3] sigmatropic rearrangement of (1-acyloxy-2-alkenyl)trialkylsilane. Synthesis of optically active vinylsilane-containing α-amino acid. *Chirality*, **13**, 357–365.

210 Armstrong, A., Geldart, S.P., Jenner, C.R., and Scutt, J.N. (2007) Organocatalytic synthesis of β-alkylaspartates via β-lactone ring opening. *The Journal of Organic Chemistry*, **72**, 8091–8094.

211 Zampella, A., D'Auria, M.V., Gomez-Paloma, L., Casapullo, A., Minale, L., Debitus, C., and Henin, Y. (1996) Callipeltin A, an anti-HIV cyclic depsipeptide from the New Caledonian Lithistida sponge *Callipelta* sp. *Journal of the American Chemical Society*, **118**, 6202–6209.

212 D'Auria, M.V., Zampella, A., Paloma, L.G., Minale, L., Debitus, C., Roussakis, C., and Le Bert, V. (1996) Callipeltins B and C; bioactive peptides from a marine Lithistida sponge *Callipelta* sp. *Tetrahedron*, **52**, 9589–9596.

213 Oku, N., Gustafson, K.R., Cartner, L.K., Wilson, J.A., Shigematsu, N., Hess, S., Pannell, L.K., Boyd, M.R., and McMahon, J.B. (2004) Neamphamide A, a new HIV-inhibitory depsipeptide from the Papua New Guinea marine sponge *Neamphius huxleyi*. *Journal of Natural Products*, **67**, 1407–1411.

214 Plaza, A., Gustchina, E., Baker, H.L., Kelly, M., and Bewley, C.A. (2007) Mirabamides A–D, depsipeptides from the sponge *Siliquariaspongia mirabilis* that inhibit HIV-1 fusion. *Journal of Natural Products*, **70**, 1753–1760.

215 Okamoto, N., Hara, O., Makino, K., and Hamada, Y. (2001) Stereoselective synthesis of (3*S*,4*R*)-3,4-dimethyl-(*S*)-glutamine and the absolute stereochemistry of the natural product from papuamides and callipeltin. *Tetrahedron: Asymmetry*, **12**, 1353–1358.

216 Acevedo, C.M., Kogut, E.F., and Lipton, M.A. (2001) Synthesis and analysis of the sterically constrained L-glutamine analogues (3*S*,4*R*)-3,4-dimethyl-L-glutamine and (3*S*,4*R*)-3,4-dimethyl-L-pyroglutamic acid. *Tetrahedron*, **57**, 6353–6359.

217 Calimsiz, S. and Lipton, M.A. (2005) Synthesis of *N*-Fmoc-(2*S*,3*S*,4*R*)-3,4-dimethylglutamine: an application of lanthanide-catalyzed transamidation. *The Journal of Organic Chemistry*, **70**, 6218–6221.

218 Xie, W., Ding, D., Li, G., and Ma, D. (2008) Total synthesis and structure assignment of papuamide B, a potent marine cyclodepsipeptide with anti-HIV properties. *Angewandte Chemie (International Edition in English)*, **47**, 2844–2848.

219 Liang, B., Carroll, P.J., and Joullie, M.M. (2000) Progress toward the total synthesis of callipeltin A (I): asymmetric synthesis of (3*S*,4*R*)-3,4-dimethylglutamine. *Organic Letters*, **2**, 4157–4160.

220 Wenger, R.M. (1983) Synthesis of cyclosporine I. Synthesis of enantiomerically pure (2*S*,3*R*,4*R*,6*E*)-3-hydroxy-4-methyl-2-methylamino-6-octenoic acid starting from tartaric acid. *Helvetica Chimica Acta*, **66**, 2308–2321.

221 Aebi, J.D., Dhaon, M.K., and Rich, D.H. (1987) A short synthesis of enantiomerically pure (2*S*,3*R*,4*R*,6*E*)-3-hydroxy-4-methyl-2-(methylamino)-6-octenoic acid, the unusual C9 amino acid found in the immunosuppressive peptide cyclosporine. *The Journal of Organic Chemistry*, **52**, 2881–2886.

222 Seebach, D., Juaristi, E., Miller, D.D., Schickli, C., and Weber, T. (1987) Addition of chiral glycine, methionine, and vinylglycine enolate derivatives to aldehydes and ketones in the preparation of enantiomerically pure α-amino-β-hydroxy acids. *Helvetica Chimica Acta*, **70**, 237–261.

223 Tung, R.D. and Rich, D.H. (1987) Total synthesis of the unusual cyclosporine amino acid MeBMT. *Tetrahedron Letters*, **28**, 1139–1142.

224 Togni, A., Pastor, S.D., and Rihs, G. (1989) Application of the gold(I)-catalyzed aldol reaction to a stereoselective synthesis of (2*S*,3*R*,4*R*,6*E*)-3-hydroxy-4-methyl-2-(methylamino)oct-6-enoic acid (MeBmt), cyclosporin's unusual amino acid. *Helvetica Chimica Acta*, **72**, 1471–1478.

225 Lubell, W.D., Jamison, T.F., and Rapoport, H. (1990) *N*-(9-Phenylfluoren-9-yl)-α-amino ketones and *N*-(9-phenylfluoren-9-yl)-α-amino aldehydes as chiral educts for the synthesis of optically pure 4-alkyl-3-hydroxy-2-amino acids. Synthesis of the C-9 amino acid MeBmt present in cyclosporin. *The Journal of Organic Chemistry*, **55**, 3511–3522.

226 Savignac, M., Durand, J.O., and Genêt, J.P. (1994) A short synthesis of the unusual amino acid of cyclosporine (4*R*)-4-[(*E*)-2-butenyl]-4,*N*-dimethyl-L-threonine (MeBmt). *Tetrahedron: Asymmetry*, **5**, 717–722.

227 Raghavan, S. and Rasheed, M.A. (2004) Modular and stereoselective formal synthesis of MeBmt, an unusual amino acid constituent of cyclosporin A. *Tetrahedron*, **60**, 3059–3065.

228 Cook, G.R. and Shanker, P.S. (2001) Stereoselective synthesis of MeBmt and methyl (4*R*,5*S*)-5-isopropyl-2-phenyloxazoline-4-carboxylate by a Pd-catalyzed equilibration. *The Journal of Organic Chemistry*, **66**, 6818–6822.

229 Matsunaga, H., Ishizuka, T., and Kunieda, T. (2005) Synthetic utility of five-membered heterocycles – chiral functionalization and applications. *Tetrahedron*, **61**, 8073–8094.

230 Rao, A.V.R., Yadav, J.S., Chandrasekhar, S., and Rao, C.S. (1989) Highly stereoselective approach for β-hydroxy-α-amino acids from D-glucose: the synthesis of MeBmt. *Tetrahedron Letters*, **30**, 6769–6772.

231 Tuch, A., Saniere, M., Le Merrer, Y., and Depezay, J.-C. (1997) Formal synthesis of an unusual amino acid component of cyclosporin, involving stereocontrolled nucleophilic 1,4-addition. *Tetrahedron: Asymmetry*, **8**, 1649–1659.

232 Schmidt, U., Lieberknecht, A., Griesser, H., and Bartkowiak, F. (1984) Stereoselective total synthesis of chlamydocin and dihydrochlamydocin. *Angewandte Chemie (International Edition in English)*, **23**, 318–320.

233 Walton, J.D. (2006) HC-toxin. *Phytochemistry*, **67**, 1406–1413.

234 Kijima, M., Yoshida, M., Sugita, K., Horinouchi, S., and Beppu, T. (1993) Trapoxin, an antitumor cyclic tetrapeptide, is an irreversible inhibitor of mammalian histone deacetylase. *The Journal of Biological Chemistry*, **268**, 22429–22435.

235 Schmidt, U., Beutler, U., and Lieberknecht, A. (1989) Total synthesis of the antitumor antibiotic WF-3161. *Angewandte Chemie (International Edition in English)*, **28**, 333–334.

236 Cavelier-Frontin, F., Pepe, G., Verducci, J., Siri, D., and Jacquier, R. (1992) Prediction of the best linear precursor in the synthesis of cyclotetrapeptides by molecular mechanic calculations. *Journal of the American Chemical Society*, **114**, 8885–8890.

237 Bruserud, O., Stapnes, C., Ersvaer, E., Gjertsen, B.T., and Ryningen, A. (2007) Histone deacetylase inhibitors in cancer treatment: a review of the clinical toxicity and the modulation of gene expression in cancer cells. *Current Pharmaceutical Biotechnology*, **8**, 388–400.

238 Nishino, H., Tomizaki, K.-Y., Kato, T., Nishino, N., Yoshida, M., and Komatsu, Y. (1999) Synthesis of cyclic tetrapeptides containing non-natural imino acids. *Peptide Science*, **35**, 189–192.

239 Jacquier, R., Lazaro, R., Raniriseheno, H., and Viallefont, P. (1984) Synthese stereospecifique de l'ester methylique de l'acide amino-2 oxo-8 epoxy-9,10 decanoique (2*S*,9*R*) *N*-protégé. *Tetrahedron Letters*, **25**, 5525–5528.

240 Ikegami, S., Hayama, T., Katsuki, T., and Yamaguchi, M. (1986) Asymmetric synthesis of α-amino acids by alkylation of a glycine amide derivative bearing chiral 2,5-disubstituted pyrrolidine as an amine component. *Tetrahedron Letters*, **27**, 3403–3406.

241 Baldwin, J.E., Adlington, R.M., Godfrey, C.R.A., and Patel, V.K. (1991) A new homolytic method for the stereospecific synthesis of (2*S*,9*S*)-2-amino-8-oxo-9,10-epoxydecanoic acid in protected form. *Journal of the Chemical Society, Chemical Communications*, 1277–1279.

242 Ikegami, S., Uchiyama, H., Hayama, T., Katsuki, T., and Yamaguchi, M. (1988) Asymmetric synthesis of α-amino acids by alkylation of *N*-[*N*-bis-(methylthio) methyleneglycyl]-2,5-bis (methoxymethoxymethyl)pyrrolidine and enantioselective synthesis of protected (2*S*,9*S*)-2-amino-8-oxo-9,10-epoxydecanoic acid. *Tetrahedron*, **44**, 5333–5342.

243 Bycroft, B.W., Cameron, D., Croft, L.R., and Johnson, A.W. (1968) Synthesis and stereochemistry of capreomycidine [α-(2-iminohexahydro-4-pyrimidinyl)glycine]. *Journal of the Chemical Society, Chemical Communications*, 1301–1302.

244 Johnson, A.W., Bycroft, B.W., and Cameron, D. (1971) Synthesis of capreomycidine and epicapreomycidine, the epimers of α-(2-iminohexahydro-4-pyrimidyl)glycine. *Journal of the Chemical Society, C: Organic*, 3040–3047.

245 Shiba, T., Ukita, T., Mizuno, K., Teshima, T., and Wakamiya, T. (1977) Total synthesis of L-capreomycidine. *Tetrahedron Letters*, **31**, 2681–2684.

246 Nomoto, S., Teshima, T., Wakamiya, T., and Shiba, T. (1978) Total synthesis of capreomycin. *Tetrahedron*, **34**, 921–927.

247 DeMong, D.E. and Williams, R.M. (2001) The asymmetric synthesis of (2*S*,3*R*)-capreomycidine. *Tetrahedron Letters*, **42**, 3529–3532.

248 DeMong, D.E. and Williams, R.M. (2003) Asymmetric synthesis of (2*S*,3*R*)-capreomycidine and the total synthesis of capreomycin IB. *Journal of the American Chemical Society*, **125**, 8561–8565.

249 Ju, J., Ozanick, S.G., Shen, B., and Thomas, M.G. (2004) Conversion of (2*S*)-arginine to (2*S*,3*R*)-capreomycidine by VioC and VioD from the viomycin biosynthetic pathway of *Streptomyces* sp. Strain ATCC11861. *ChemBioChem*, **5**, 1281–1285.

250 Yin, X., McPhail, K.L., Kim, K., and Zabriskie, T.M. (2004) Formation of the nonproteinogenic amino acid 2*S*,3*R*-capreomycidine by VioD from the viomycin biosynthesis pathway. *ChemBioChem*, **5**, 1278–1281.

251 Yin, X. and Zabriskie, T.M. (2004) VioC is a non-heme iron, α-ketoglutarate-dependent oxygenase that catalyzes the formation of 3*S*-hydroxy-L-arginine during viomycin biosynthesis. *ChemBioChem*, **5**, 1274–1277.

252 Kurokawa, N. and Ohfune, Y. (1986) Total synthesis of echinocandins. II. Total synthesis of echinocandin D via efficient peptide coupling reactions. *Journal of the American Chemical Society*, **108**, 6043–6045.

253 Evans, D.A., Sjogren, E.B., Weber, A.E., and Conn, R.E. (1987) Asymmetric synthesis of *anti*-β-hydroxy-α-amino acids. *Tetrahedron Letters*, **28**, 39–42.

254 Zambias, R.A., Hammond, M.L., Heck, J.V., Bartizal, K., Trainor, C., Abruzzo, G., Schmatz, D.M., and Nollstadt, K.M. (1992) Preparation and structure-activity relationships of simplified analogs of the antifungal agent cilofungin: a total synthesis approach. *Journal of Medicinal Chemistry*, **35**, 2843–2855.

255 Palomo, C., Oiarbide, M., and Landa, A. (2000) A strategy for the asymmetric aminohomologation of α,β-dihydroxy aldehydes: application to the synthesis of the southwest tripeptide segment of echinocandin B. *The Journal of Organic Chemistry*, **65**, 41–46.

256 Balkovec, J.M. and Black, R.M. (1992) Reduction studies of antifungal echinocandin lipopeptides. One step conversion of echinocandin B to echinocandin C. *Tetrahedron Letters*, **33**, 4529–4532.

257 Jimeno, J.M. (2002) A clinical armamentarium of marine-derived anti-cancer compounds. *Anti-Cancer Drugs*, **13** (Suppl. 1), S15–S19.

258 Vera, M.D. and Joullie, M.M. (2002) Natural products as probes of cell biology: 20 years of didemnin research. *Medicinal Research Reviews*, **22**, 102–145.

259 Taguchi, T. (2003) Development of marine-derived anti-cancer compounds. *Gan to Kagaku Ryoho*, **30**, 579.

260 Fusetani, N. and Matsunaga, S. (1993) Bioactive sponge peptides. *Chemical Reviews*, **93**, 1793–1805.

261 Renner, M.K., Shen, Y.-C., Cheng, X.-C., Jensen, P.R., Frankmoelle, W., Kauffman, C.A., Fenical, W., Lobkovsky, E., and Clardy, J. (1999) Cyclomarins A–C, new antiinflammatory cyclic peptides produced by a marine bacterium (*Streptomyces* sp). *Journal of the American Chemical Society*, **121**, 11273–11276.

262 Sugiyama, H., Yokokawa, F., Aoyama, T., and Shioiri, T. (2001) Synthetic studies of *N*-reverse prenylated indole. An efficient synthesis of antifungal indole alkaloids and *N*-reverse prenylated tryptophan. *Tetrahedron Letters*, **42**, 7277–7280.

263 Wen, S.-J., Zhang, H.-W., and Yao, Z.-J. (2002) Synthesis of a fully protected (2*S*,3*R*)-*N*-(1′,1′-dimethyl-2′-propenyl)-3-hydroxytryptophan from tryptophan. *Tetrahedron Letters*, **43**, 5291–5294.

264 Hansen, D.B. and Joullié, M.M. (2005) A stereoselective synthesis of (2*S*,3*R*)-β-methoxyphenylalanine: a component of cyclomarin A. *Tetrahedron: Asymmetry*, **16**, 3963–3969.

265 Hansen, D.B., Lewis, A.S., Gavalas, S.J., and Joullié, M.M. (2006) A stereoselective synthetic approach to (2*S*,3*R*)-*N*-(1′,1′-dimethyl-2,′3′-epoxypropyl)-3-hydroxytryptophan, a component of cyclomarin A. *Tetrahedron: Asymmetry*, **17**, 15–21.

266 Wen, S.-J. and Yao, Z.-J. (2004) Total synthesis of cyclomarin C. *Organic Letters*, **6**, 2721–2724.

267 Kahne, D., Leimkuhler, C., Lu, W., and Walsh, C. (2005) Glycopeptide and lipoglycopeptide antibiotics. *Chemical Reviews*, **105**, 425–448.

268 Evans, D.A., Dinsmore, C.J., Watson, P.S., Wood, M.R., Richardson, T.I., Trotter, B.W., and Katz, J.L. (1998) Nonconventional stereochemical issues in the design of the synthesis of the vancomycin antibiotics: challenges imposed by axial and nonplanar chiral elements in the heptapeptide aglycons. *Angewandte Chemie (International Edition in English)*, **37**, 2704–2708.

269 Evans, D.A., Wood, M.R., Trotter, B.W., Richardson, T.I., Barrow, J.C., and Katz, J.L. (1998) Total syntheses of vancomycin and eremomycin aglycons. *Angewandte Chemie (International Edition in English)*, **37**, 2700–2704.

270 Nicolaou, K.C., Takayanagi, M., Jain, N.F., Natarajan, S., Koumbis, A.E., Bando, T., and Ramanjulu, J.M. (1998) Total synthesis of vancomycin aglycon – part 3: final stages. *Angewandte Chemie (International Edition in English)*, **37**, 2717–2719.

271 Nicolaou, K.C., Jain, N.F., Natarajan, S., Hughes, R., Solomon, M.E., Li, H., Ramanjulu, J.M., Takayanagi, M., Koumbis, A.E., and Bando, T. (1998) Total synthesis of vancomycin aglycon – part 2: synthesis of amino acids 1–3 and construction of the AB-COD-DOE ring skeleton. *Angewandte Chemie (International Edition in English)*, **37**, 2714–2716.

272 Nicolaou, K.C., Natarajan, S., Li, H., Jain, N.F., Hughes, R., Solomon, M.E., Ramanjulu, J.M., Boddy, C.N.C., and Takayanagi, M. (1998) Total synthesis of vancomycin aglycon – part 1: synthesis of amino acids 4–7 and construction of the AB-COD ring skeleton. *Angewandte Chemie (International Edition in English)*, **37**, 2708–2714.

273 Boger, D.L., Miyazaki, S., Kim, S.H., Wu, J.H., Castle, S.L., Loiseleur, O., and Jin, Q. (1999) Total synthesis of the vancomycin aglycon. *Journal of the American Chemical Society*, **121**, 10004–10011.

274 Boger, D.L., Miyazaki, S., Kim, S.H., Wu, J.H., Loiseleur, O., and Castle, S.L. (1999) Diastereoselective total synthesis of the vancomycin aglycon with ordered atropisomer equilibrations. *Journal of the American Chemical Society*, **121**, 3226–3227.

275 Crowley, B.M. and Boger, D.L. (2006) Total synthesis and evaluation of [ψ[CH_2NH]Tpg4]vancomycin aglycon: reengineering vancomycin for dual D-Ala-D-Ala and D-Ala-D-Lac binding. *Journal of the American Chemical Society*, **128**, 2885–2892.

276 Crowley, B.M., Mori, Y., McComas, C.C., Tang, D., and Boger, D.L. (2004) Total synthesis of the ristocetin aglycon. *Journal of the American Chemical Society*, **126**, 4310–4317.

277 Boger, D.L., Kim, S.H., Mori, Y., Weng, J.H., Rogel, O., Castle, S.L., and McAtee, J.J. (2001) First and second generation total synthesis of the teicoplanin aglycon. *Journal of the American Chemical Society*, **123**, 1862–1871.

278 Pearson, A.J., Ciurea, D.V., and Velankar, A. (2008) Studies toward the total synthesis of ristocetin A aglycone using arene–ruthenium complexes as SNAr substrates: construction of an advanced tricyclic intermediate. *Tetrahedron Letters*, **49**, 1922–1926.

279 Juaristi, E. (ed.) (1997) *Enantioselective Synthesis of β-Amino Acids*, Wiley-VCH Verlag GmbH, Weinheim.

280 Juaristi, E. and López-Ruiz, H. (1999) Recent advances in the enantioselective synthesis of β-amino acids. *Current Medicinal Chemistry*, **6**, 983–1004.

6
Synthesis of *N*-Alkyl Amino Acids

Luigi Aurelio and Andrew B. Hughes

6.1
Introduction

Among the numerous reactions of nonribosomal peptide synthesis, *N*-methylation of amino acids is one of the common motifs. Consequently, the chemical research community interested in peptide synthesis and peptide modification has generated a sizeable body of literature focused on the synthesis of *N*-methyl amino acids (NMA). That literature is summarized herein.

Alkyl groups substituted on to nitrogen larger than methyl are exceedingly rare among natural products. However, medicinal chemistry programs and peptide drug development projects are not limited to *N*-methylation. While being a much smaller body of research, there is a range of methods for the *N*-alkylation of amino acids and those reports are also covered in this chapter.

The literature on *N*-alkyl, primarily *N*-methyl amino acids comes about due to the useful properties that the *N*-methyl group confers on peptides. *N*-Methylation increases lipophilicity, which has the effect of increasing solubility in nonaqueous solvents and improving membrane permeability. On balance this makes peptides more bioavailable and makes them better therapeutic candidates.

One potential disadvantage is the methyl group removes the possibility of hydrogen bonding and so binding events may be discouraged. It is notable though that the *N*-methyl group does not fundamentally alter the identity of the amino acid. Some medicinal chemists have taken advantage of this fact to deliberately discourage binding of certain peptides that can still participate in the general or partial chemistry of a peptide. A series of recent papers relating to Alzheimer's disease by Doig *et al.* [1–3] considers the use of small peptidic ligands bearing *N*-methyl amide bonds as a means of interrupting or reversing amyloid protein aggregation into toxic fibrils or lumps. Similar, related studies have been published by Gordon *et al.* [4] and Kapurniotu *et al.* [5].

Viewed from another point, the removal of the possibility of hydrogen bonding may improve the efficacy of a peptide by increasing its proteolytic resistance.

Amino Acids, Peptides and Proteins in Organic Chemistry. Vol.1 – Origins and Synthesis of Amino Acids.
Edited by Andrew B. Hughes

ISBN: 978-3-527-32096-7

Generally, the first event in an enzymic proteolytic event is recognition of the target amide bond by hydrogen bonding. Numerous examples of model or lead peptides acquiring increased proteolytic stability through site-specific *N*-methylation are known [6–12].

Thus, *N*-methylation and *N*-alkylation are accepted tools in peptide and peptidomimetic drug design. This leads to the requirement for methods to prepare the required monomers in forms suitable for solution and solid-phase peptide synthesis. Accordingly, in the synthetic literature summarized in this chapter attention is given, where possible, to the integrity of asymmetric centers that particular methods enjoy. A method that provides the *N*-methyl (*N*-alkyl) amino acid in high yield but as a racemate typically finds little use.

This chapter describes methods for the synthesis of *N*-methyl and larger *N*-alkyl amino acids. It addresses the synthetic challenges of *N*-methylation including regiospecific methylation, mono-*N*-methylation, and development of racemization-free chemistry. The synthetic methods reviewed reveal the difficulty that chemists have had in incorporating a single methyl group at the α-amino position and the problems encountered in applying these methods to the common 20 naturally occurring L-amino acids. Toward the end of the chapter, a specific section on *N*-alkylation where it differs from *N*-methylation is presented.

6.2
N-Methylation via Alkylation

6.2.1
S_N2 Substitution of α-Bromo Acids

The first published procedure for the *N*-methylation of α-amino acids dates back to 1915 pioneered by Emil Fischer *et al.* [13, 14]. This work provided a foundation for *N*-methyl analog synthesis utilizing *N*-tosyl amino acids and α-bromo acids as intermediates. Fischer *et al.* prepared *N*-methyl derivatives of alanine, leucine, and phenylalanine by nucleophilic displacement of optically active (*R*)-α-bromo acids (Scheme 6.1) [14]. Using this approach they made *N*-methyl derivatives of alanine, leucine, and phenylalanine with the L-configuration (Scheme 6.1).

The α-bromo acids are commonly obtained via diazotization of the parent amino acid (Figure 6.1) [15]. The reaction gives retention of configuration and this results in a "Walden inversion" [16], which forms an intermediate diazonium ion that is attacked intramolecularly, in S_N2 fashion, by the neighboring carboxylate group to form the highly reactive cyclic lactone **1** [17]. A second nucleophilic addition again in the S_N2 mode by a bromide ion provides the optically active α-bromo acids **2** with net retention of the original amino acid chirality. Consequently, substitution with excess methylamine at 0 °C provides NMAs with opposite configuration to the parent amino acids.

Izumiya and Nagamatsu extended this methodology to other amino acids such as tyrosine [15], methionine [18d], arginine [18b], and ornithine [18b]. A representative

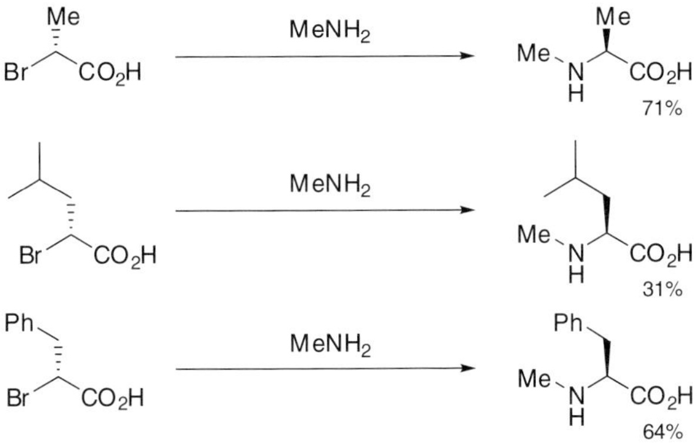

Scheme 6.1

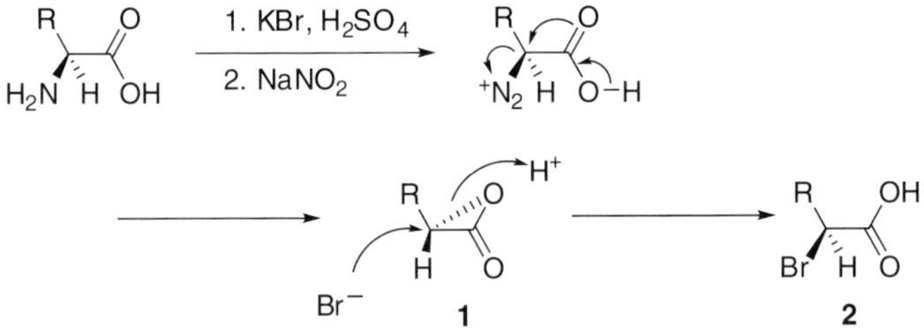

Figure 6.1 Mechanism of α-bromo acid formation via diazotization.

example is given in Scheme 6.2 in which *N*-methyl-D-tyrosine (D-surinamine) **3** is prepared by diazotization of *O*-methyl-L-tyrosine **5** to give the optically active α-bromo acid **4**. Displacement with methylamine at 100 °C in a sealed tube provided *N*-methyl-D-tyrosine **3**.

Izumiya combined both methods developed by Fischer to make NMAs [18a,c,e] of hydroxy-amino acids via α-bromo acids and *N*-tosyl amino acids. 3-Methoxy-2-bromoalkanoic acids were prepared from alkenoic acids as precursors (Scheme 6.3).

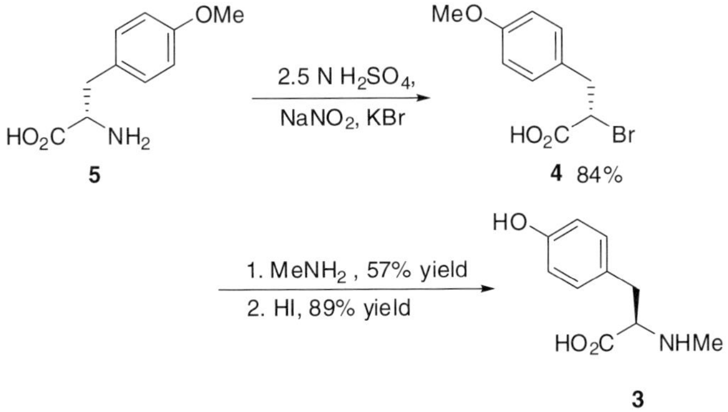

Scheme 6.2

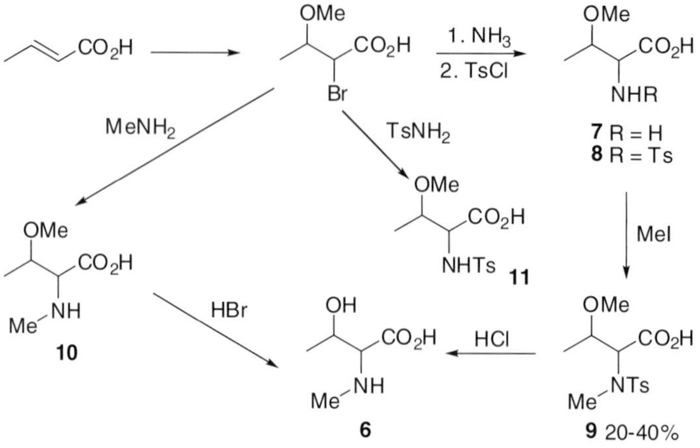

Scheme 6.3

Izumiya describes two paths to NMAs. This is shown by the preparation of *N*-methylthreonine **6**. The first pathway involves amination with ammonia to generate *O*-methylthreonine **7**. Tosylation provides **8** and *N*-methylation with methyl iodide under basic conditions gave the *N*-methylated fully protected threonine **9**. The tosyl and *O*-methyl groups were then removed under acidic conditions to give *N*-methylthreonine **6**. The second sequence used methylamine for the amination to make *N*,*O*-dimethylthreonine **10** and then *O*-demethylating with HBr to provide **6**. These sequences provided racemic serine, threonine and its diastereoisomers, and β-hydroxyvaline. In a variation, the tosyl path could be made more efficient by amination with *p*-toluenesulfonamide to give **11**.

α-Bromo acids can be replaced by triflates in S_N2 displacements. Effenberger *et al.* [19] synthesized *N*-methyl-D-alanine **12** (Scheme 6.4) in this way. Ethyl-L-lactate was converted to the triflate **13** and then treatment of the triflate with *N*-benzyl-*N*-methyl amine supplied fully protected ethyl-*N*-benzyl-*N*-methyl-D-alaninate **14**. The

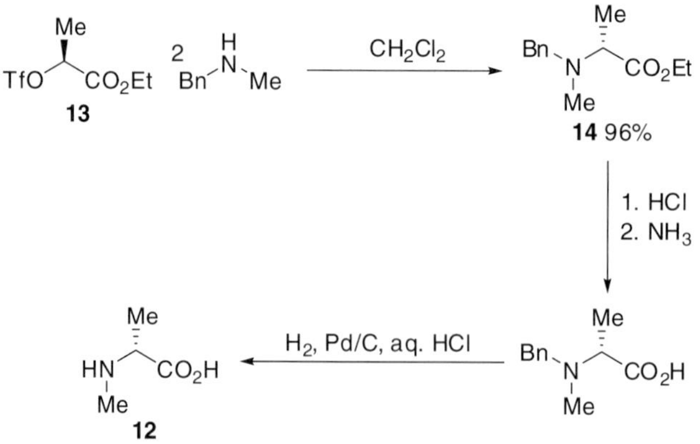

Scheme 6.4

excellent leaving group capability of the trifluoromethanesulfonate is the advantage of this technique even with weak amine nucleophiles at room temperature and below [19], and the fact that excess amine and high temperatures in sealed vessels are not required as in Izumiya's method (Scheme 6.2).

The synthesis of NMAs by S_N2 substitution of α-bromo acids is generally a short and simple sequence. However, it does come with limitations. The yields of product NMAs are low to moderate, the displacement using secondary amines is not reported, and epimerization is not entirely eradicated [20b]. Quitt *et al.* [20] established an epimerization-free reductive amination of a range of NMAs that revealed, by comparison of optical data, that some epimerization was occurring in the α-bromo acid substitution with the addition of methylamine at 0 °C. This approach to NMA synthesis was essentially abandoned, as Fischer and Izumiya are the sole contributors to the literature.

The alternative Effenberger *et al.* [19] approach involving triflate displacement is more mild, but suitable carboxyl protection is required. The increasing availability of lactates commercially and synthetically makes the triflate approach a far more viable procedure than the use of α-bromo acids as intermediates for NMAs as this technique was shown to provide optically pure derivatives and provides avenues to *N*-alkyl amino acids since secondary amines can also be utilized.

6.2.2 *N*-Methylation of Sulfonamides, Carbamates, and Amides

One of the common methods of *N*-methylation by alkylation is to use amide-like protection with various sulfonamides, carbamates, and (indeed) amides. Amide protection enhances NH acidity permitting deprotonation under basic conditions and in the presence of an alkylating reagent provides the NMAs, which will be discussed in three sections below. Alternatively, the Mitsunobu protocol can be employed in the synthesis of NMAs with various sulfonamide protecting groups due to the acidity of the sulfonamide nitrogen.

6.2.2.1 Base-Mediated Alkylation of *N*-Tosyl Sulfonamides

Fischer and Lipschitz [13] describe the preparation of *N*-tosyl α-amino acids (Scheme 6.5). They treated *N*-tosyl α-amino acids with sodium hydroxide at 65–70 °C and used methyl iodide as the alkylating agent. An advantage of *N*-tosyl protection is the high degree of crystallinity of the product NMAs, but a major drawback is the removal of the tosyl group, which can require vigorous conditions. The *N*-tosyl NMAs were subjected to acid hydrolysis with concentrated HCl for up to 8 h at 100 °C, to provide the free NMA. The other problem is that this method does proceed with epimerization in the methylation step in which sodium hydroxide was used at elevated temperatures. This was revealed by Quitt *et al.* [20] through comparison of optical rotation values.

The temperature is a major contributing factor to this epimerization process since the method of Hlavácek *et al.* [21, 22] revealed that *N*-tosyl amino acid isopropyl and *tert*-butyl esters of alanine and valine, when treated with sodium hydroxide and

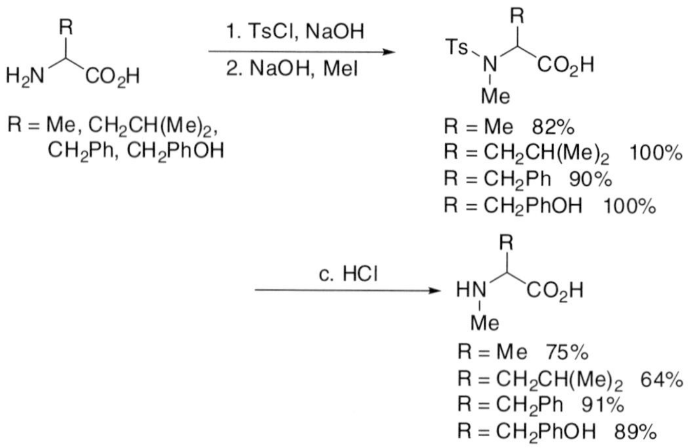

H_2N–CH(R)–CO_2H → (1. TsCl, NaOH; 2. NaOH, MeI) → Ts–N(Me)–CH(R)–CO_2H

R = Me, $CH_2CH(Me)_2$, CH_2Ph, CH_2PhOH

R = Me 82%
R = $CH_2CH(Me)_2$ 100%
R = CH_2Ph 90%
R = CH_2PhOH 100%

→ (c. HCl) → HN(Me)–CH(R)–CO_2H

R = Me 75%
R = $CH_2CH(Me)_2$ 64%
R = CH_2Ph 91%
R = CH_2PhOH 89%

Scheme 6.5

dimethylsulfate at 0 °C, showed no epimerization and remained optically active. Isolated NMAs were assessed by comparison of optical data with that of Quitt *et al.* [20]. This was a biphasic reaction, and detergent was included to improve phase mixing and also helped in removing traces of unreacted starting materials. Pure *N*-methyl amino acid derivatives of leucine, valine, phenylalanine, alanine, and ornithine were isolated in near quantitative yields from the methylation step [22]. By treating the *tert*-butyl esters with trifluoroacetic acid (TFA) and isopropyl esters with refluxing 4 M HCl, the free acids could be obtained. Subsequent tosyl group removal was accomplished with calcium metal in liquid ammonia or with HBr at reflux in the presence of phenol.

6.2.2.2 Base Mediated Alkylation of *N*-Nitrobenzenesulfonamides

Sulfonamide protection has been used for site-selective *N*-methylation on solid support [23]. Since the sulfonamide NH is far more acidic than amide NHs, selective deprotonation of sulfonamides was achieved in the presence of amides and as a result selective methylation of sulfonamides was possible. The N-terminal amino acid (resin bound) as the free amine was protected as the *o*-nitrobenzenesulfonamide (*o*-NBS), which can be removed selectively and with milder conditions when compared to *N*-tosyl protection, using a thiol and base, 1,8-diazabicyclo[5.4.0]undec-7-ene (DBU). The sulfonamide was treated with the guanidinium base, 7-methyl-1,5,7-triazabicyclo[4.4.0]dec-5-ene (MTBD), and alkylated with methyl *p*-nitrobenzenesulfonate (Scheme 6.6). This combination of sulfonamide protection, base deprotonation, and alkylation provided site-selective *N*-methylation, without methylation elsewhere in the growing peptide. It was found that the use of the guanidinium base MTBD was critical in achieving high yields and selectivity since weaker bases gave poor or no yields and stronger bases resulted in uncontrolled methylation of the amide backbone. The less vigorous conditions of methylation and deprotection with this method provide a useful alternative approach to *N*-tosyl protection.

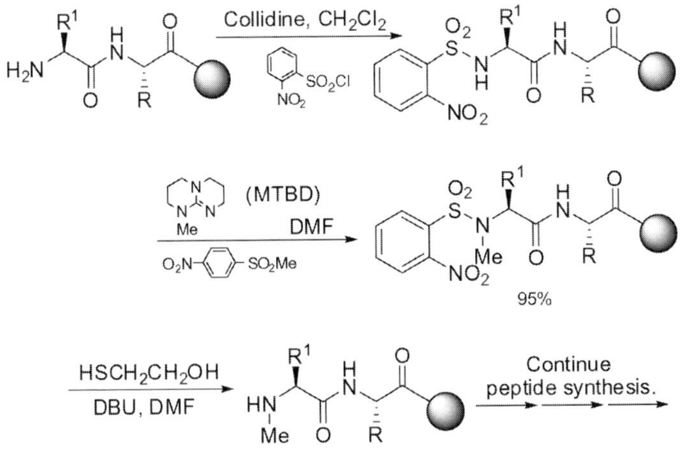

Scheme 6.6

o-NBS protection has also been used in solution phase synthesis of NMAs as their methyl esters. Albanese *et al.* [24] alkylated the intermediate *o*-NBS amides in solution phase by treating the sulfonamides with solid potassium carbonate, triethylbenzylammonium chloride (TEBA) as phase-transfer catalyst [24] and alkyl halides providing *N*-nitrobenzenesulfonamido-*N*-alkyl amino acid esters at 25 or 80 °C. These transformations were accomplished with valine, phenylalanine, and phenylglycine in 87, 86, and 91% yields, respectively for their *N*-methyl derivatives. The use of TEBA enabled the non-nucleophilic base potassium carbonate to be utilized, whereas *N*-alkylation was considerably reduced in the absence of TEBA. Removal of the NBS group is affected by thiophenol/potassium carbonate/acetonitrile at 80 °C or potassium thiophenoxide/dimethylformamide (DMF) at 25 °C leaving the methyl ester intact. Biron and Kessler [25] solved the problem of methyl ester cleavage in their synthesis of *N*-methylated amino acids with *o*-NBS protection. They converted *o*-NBS amino acid methyl esters to the *N*-methylated analog with dimethylsulfate/DBU and then cleaved the methyl ester under S_N2 dealkylation conditions with LiI in refluxing ethyl acetate. Nuclear magnetic resonance and high-performance liquid chromatography analysis of the de-esterified products revealed no epimerization had occurred.

An even milder approach to *N*-methylating amino acid sulfonamides is under the neutral diazomethylating conditions. Di Gioia *et al.* [26a] found that by treating *N*-nosyl amino acid methyl esters with a large excess of diazomethane, the corresponding NMA esters were obtained in quantitative yield for alanine, phenylalanine, valine, leucine, and isoleucine. The *N*-nosyl group was removed with 3 equiv. of mercaptoacetic acid in the presence of 8 equiv. of sodium methoxide at 50 °C, to provide the free amines in greater than 84% yields. Treating *N*-acetyl amino acid methyl esters gave almost no *N*-methylation. Di Gioia *et al.* [26b] extended this methodology to the synthesis of several *N*-methyl-*N*-nosyl-β-amino acids.

N-Methylation by alkylating sulfonamides is advantageous in that the increased acidity of the sulfonamide nitrogen can allow for selective methylation in a peptide [23] on a solid support or an orthogonally protected amino acid monomer. The Fischer method is undesirable since degrees of epimerization occur and the vigorous conditions for removing the tosyl group are undesirable for many sensitive amino acid residues. It is also an inappropriate protecting group in peptide synthesis since the conditions for removal also cleave peptide bonds by acid hydrolysis. The *N-o*-NBS or *N*-nosyl protections are significant improvements, having the advantage of mild deprotection conditions while still allowing *N*-alkylation and easy work-up in solution or solid phase. The method of Di Gioia *et al.* [26a] involving diazomethane, while elaborate, is performed under neutral conditions, but it is to be used with great caution due to the *explosive* and *toxic* nature of diazomethane! In the case of alkyl ester protection, it is not recommended to include such a protecting group that is usually removed by hydroxide or other strong bases, especially if there are no other ionizable sites in the amino acid other than the α-center where NMAs are concerned. However, the studies conducted by Biron and Kessler [25] have revealed that S_N2 dealkylation of methyl esters with LiI results in demethylation and the chiral integrity of the NMAs is retained.

6.2.2.3 *N*-Methylation via Silver Oxide/Methyl Iodide

N-Methylation of carbamate protected peptides and their peptide bonds was first described by Das *et al.* [27]. Permethylation of peptides improved their use in mass spectrometry studies. Their intentions were purely based on the fact that oligopeptides are less volatile due to hydrogen bonding and *N*-methylation of peptide bonds alleviates the volatility problem by removing the possibility of hydrogen bonding. Their procedure involved treatment of substrate *N*-acyl peptides with excess methyl iodide and silver oxide in DMF. The final methylated products showed higher volatility and allowed mass spectral analysis at lower temperatures in the ion source.

Olsen [28] expanded the methylation procedure of Das *et al.* [27] to include carbamate protected α-amino acids. The yields of mono-*N*-methyl amino acid methyl esters like alanine and valine were routinely in the range 93–98% (Scheme 6.7). However, *N*-methylation of residues such as cysteine, arginine, methionine, aspartic acid, serine, and threonine did not provide successful candidates using this procedure.

Okamoto *et al.* [29] extended Olsen's procedure to other amino acids, and NMAs of glutamic acid and serine were successfully synthesized (Scheme 6.7). Most of the *N*-methyl amino acids **15** were isolated in crystalline form as their dicyclohexylamine salts **16** following ester saponification. However, it was found that the optical rotation data for *N*-methyl derivatives of serine and glutamic acid were lower than reported values.

The silver oxide/methyl iodide method for *N*-methylation is a mild and racemization-free process. However, the final NMAs are obtained as their methyl esters if the free acid is employed. These derivatives are then subjected to saponification if the free acid is required. This has been shown to compromise the chiral integrity of the NMAs. In addition, this method is not always reproducible since the quality of silver oxide reflects upon the conversion of amino acid to its NMA analog and therefore fresh silver oxide is necessary for good conversions. Alternatively, *N*-carbamoyl amino acids with suitable ester protection that does not require saponification for

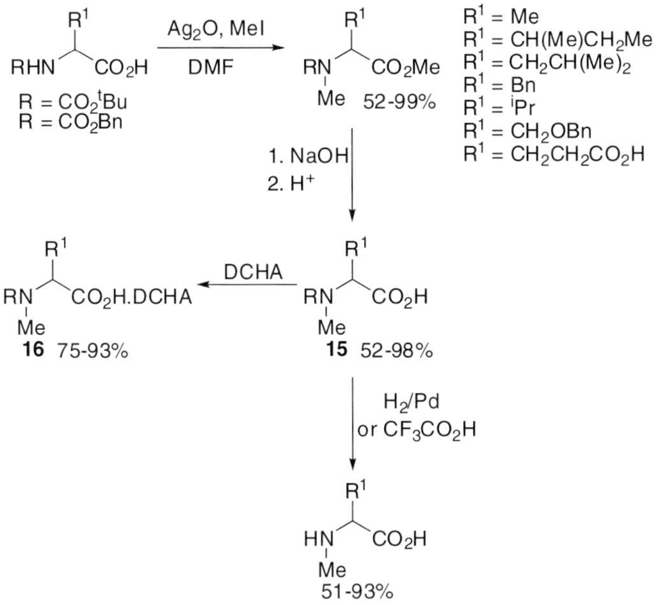

R^1 = Me
R^1 = $CH(Me)CH_2Me$
R^1 = $CH_2CH(Me)_2$
R^1 = Bn
R^1 = ^iPr
R^1 = CH_2OBn
R^1 = $CH_2CH_2CO_2H$

R = $CO_2{}^tBu$
R = CO_2Bn

Ag_2O, MeI, DMF: 52-99%
1. NaOH 2. H^+: **15** 52-98%
DCHA: **16** 75-93%
H_2/Pd or CF_3CO_2H: 51-93%

Scheme 6.7

removal should be employed in such a procedure to preclude saponification [28]. Tam *et al.* [30] synthesized *N*-methyl derivatives of α-*N*-Boc, side-chain *N*-phthaloyl-protected ornithine and lysine by protecting the carboxyl group as a benzyl ester. Silver oxide/methyl iodide-mediated *N*-methylation was achieved and the benzyl ester was removed under hydrogenolytic conditions to afford the free acids.

6.2.2.4 N-Methylation via Sodium Hydride/Methyl Iodide

The most broadly applied method for NMA synthesis is *N*-methylating *N*-acyl and *N*-carbamoyl amino acids with sodium hydride and methyl iodide developed by Benoiton *et al.* [31–34]. Benoiton *et al.* had synthesized a large range of NMAs using excess sodium hydride and methyl iodide. Many other contributors to the field have since utilized this method and variations thereof in producing NMAs. Benoiton *et al.* [31] initially attempted *N*-methylation employing *N*-acyl, *N*-tosyl, and *N*-carbamoyl α-amino acids **17**. Treating *N*-protected amino acids with sodium hydride and methyl iodide in tetrahydrofuran (THF)/DMF at 80 °C for 24 h produced *N*-methyl methyl esters **18**, which required a large excess of methyl iodide (8 equiv.) for optimal yields (Scheme 6.8). The methyl ester was saponified at 35 °C in methanol/THF to give the corresponding free acids **19**.

The use of alkaline conditions in the formation of the *N*-methyl group and removal of the methyl ester causes varying degrees of undesired epimerization [32–34]. Therefore, a direct route to *N*-methyl amino acids **19** was accomplished without esterification by lowering the reaction temperature to 0 °C. McDermott and Benoiton [35] found that reaction temperature was an important factor in avoiding

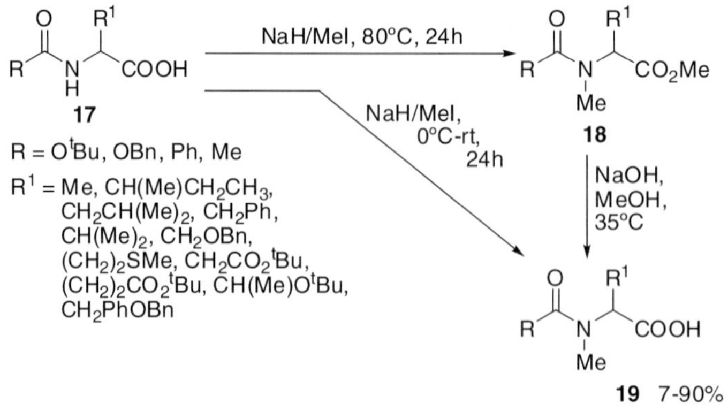

17

NaH/MeI, 80°C, 24h

18

NaH/MeI, 0°C-rt, 24h

NaOH, MeOH, 35°C

$R = O^tBu$, OBn, Ph, Me

R^1 = Me, $CH(Me)CH_2CH_3$, $CH_2CH(Me)_2$, CH_2Ph, $CH(Me)_2$, CH_2OBn, $(CH_2)_2SMe$, $CH_2CO_2{}^tBu$, $(CH_2)_2CO_2{}^tBu$, $CH(Me)O^tBu$, CH_2PhOBn

19 7-90%

Scheme 6.8

the formation of methyl esters (Scheme 6.8) and also identified acidic reaction conditions other than basic conditions that caused epimerization of *N*-methyl amino acid-containing dipeptides. It was found the anhydrous HBr/acetic acid used for *N*-Cbz removal caused epimerization and revealed the susceptibility of NMAs to epimerize in peptide synthesis during standard peptide synthesis [34].

McDermott and Benoiton [33, 34] undertook a systematic study of the extent of epimerization of NMA residues in peptides during hydrolysis and peptide synthesis. It was concluded that appreciable epimerization occurred with aqueous hydroxide due to the absence of ionizable groups other than the α-center. Analysis of the acid-catalyzed epimerization showed that anhydrous HBr/acetic acid caused epimerization depending on several factors such as acid strength, solvent polarity, and time. It was found that including water in the acidic mixtures suppressed epimerization completely as did HCl mixtures in place of HBr mixtures. The epimerization studies were extended to include coupling reactions between NMA peptides via the mixed anhydride activation approach, and they identified factors such as ionic strength and solvent polarity as controlling epimerization during peptide bond formation via the mixed anhydride activation/coupling procedure. Polar solvents and increased ionic strength of the solvent medium due to tertiary amine salts of hydrochlorides or *p*-toluenesulfonates promoted epimerization and in the absence of these factors less epimerization was observed. Only DCC/*N*-hydroxysuccinimide as an activating agent gave stereochemically pure coupled products. Furthermore, they found that an excess of base did not promote epimerization.

Another type of protecting group exploited yet rarely used are phosphoramides. Coulton *et al.* [36] synthesized a number of α-amino acid diphenylphosphinamides **20** (Scheme 6.9) that were methylated using the conditions of Benoiton *et al.* [31, 34, 35, 37]. The diphenylphosphinamide protecting group is acid labile and the product NMAs are highly crystalline, yet the downfall of this procedure was that the optical rotation data for the zwitterionic form did not agree well with those reported. This suggests that some epimerization may have occurred and the authors acknowledge this discrepancy, which reveals the need for further investigation of the stereochemical integrity of the product NMAs.

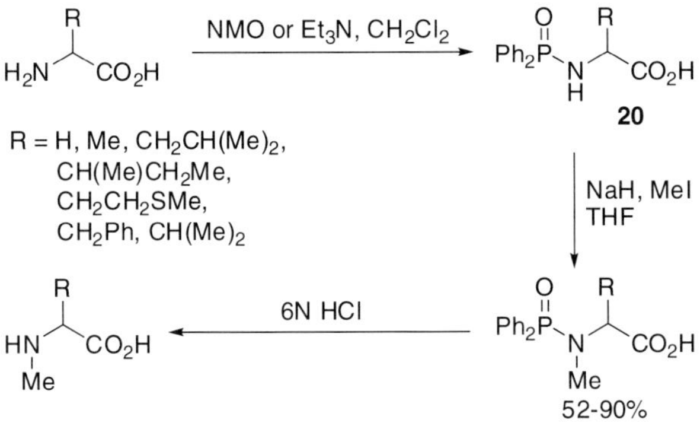

Scheme 6.9

Belagali *et al.* [38] utilized a similar approach to Benoiton with *N*-Boc-L-amino acids (Scheme 6.10), but took the *N*-Boc-L-amino acids **21** and treated them with sodium hydride/methyl iodide under the Benoiton conditions [31b]; however, they found that the yields of the *N*-methyl derivatives **22** were in the range 30–40%. Switching to more forceful conditions depicted in Scheme 6.10, by treating **21** with finely powdered potassium hydroxide, tetrabutylammonium hydrogen sulfate, and dimethylsulfate gave the *N*-methyl-*N*-Boc-L-amino acids **22** in low yields. However, utilizing sodium hexamethyldisilazide as base, the yields were greatly improved (68–72%) and the NMAs **23** were isolated after cleavage of the *N*-Boc group with TFA.

Burger and Hollweck [39], applied Benoiton's procedure to methylate 4-trifluoromethyl-1,3-oxazolidine-2,5-diones **24** known as Leuchs anhydrides (Scheme 6.11). The *N*-methylated Leuchs anhydrides **25** are activated towards nucleophilic attack and upon treatment with esterified amino acids peptide bond formation of α,α-dialkylated amino acids at the C-terminus could be achieved which is generally difficult [39].

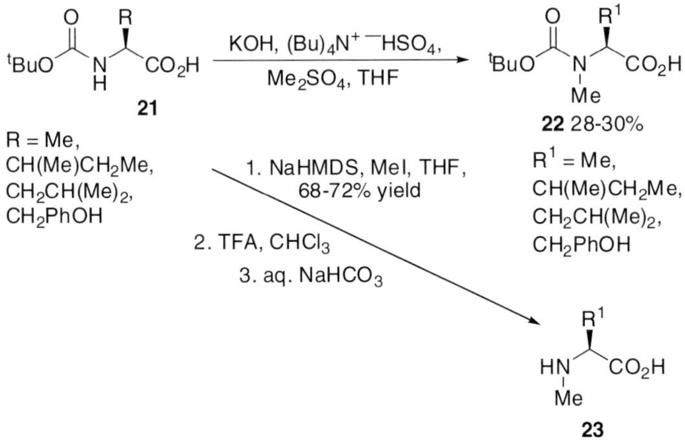

Scheme 6.10

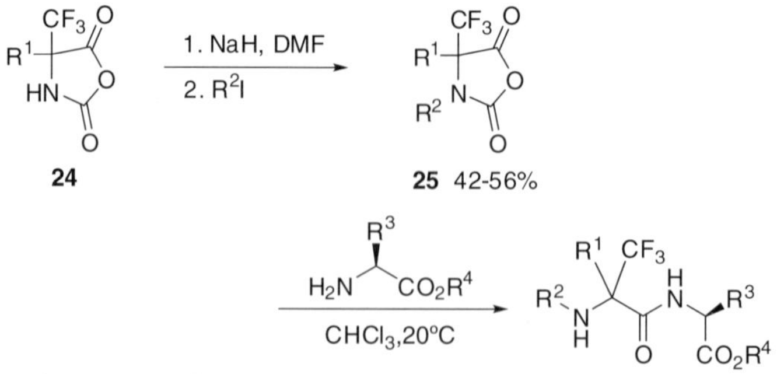

1. NaH, DMF; 2. R^2I

24 → **25** 42-56%

$H_2NCH(R^3)CO_2R^4$, $CHCl_3$, 20°C → 49-95% yield

	R^1	R^2	R^3	R^4
a	$CH_2CH(Me)_2$	Me	Bn	$C(Me)_3$
b	Bn	Me	Bn	$C(Me)_3$
c	Ph	Me	Me	$C(Me)_3$
d	Bn	Me	Me	$C(Me)_3$
e	Ph	Me	H	Me

Scheme 6.11

By slightly modifying the Benoiton method (Scheme 6.12), Prashad *et al.* [40] *N*-methylated dipeptides **26**, amino acid amides **27**, and amino acids **28**. They did this by treating the substrates with sodium hydride in THF and then methylated the resulting anion with dimethylsulfate, in the presence of catalytic amounts of water. The

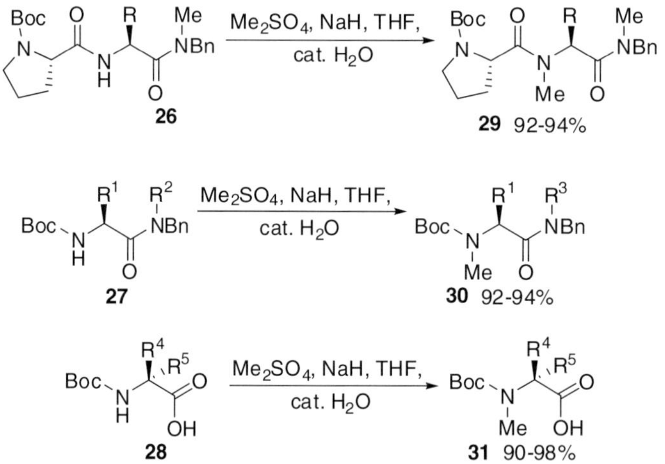

26 → Me_2SO_4, NaH, THF, cat. H_2O → **29** 92-94%

27 → Me_2SO_4, NaH, THF, cat. H_2O → **30** 92-94%

28 → Me_2SO_4, NaH, THF, cat. H_2O → **31** 90-98%

R	R^1	R^2	R^3	R^4	R^5
a 2MN*	**a** 2MN	Me	Me	**a** $CHMe_2$	H
b Bn	**b** 2MN	H	Me	**b** Bn	H
c Me	**c** Bn	Me	Me	**c** 2MN	H
	d Bn	H	Me	**d** CH_2CHMe_2	H
	e Me	Me	Me	**e** $-CH_2CH_2CH_2CH_2-$	
	f Me	H	Me		

*2MN=2-methylnaphthalene

Scheme 6.12

authors found that higher yields of *N*-methylation were achieved, since the addition of water produces dry sodium hydroxide that has better solubility in THF compared to sodium hydride and consequently makes for excellent yields of **29**, **30**, and **31**.

A number of NMA derivatives have been synthesized by the sodium hydride/methyl iodide method developed by Benoiton and NMAs manufactured by this method have been employed in a number of natural product syntheses. This method has generally been accepted as a mild and practical procedure that enables the *N*-methylation of a number of *N*-acyl and *N*-carbamoyl amino acids that are readily available. In the case of Fmoc-protected amino acids this method is not applicable due to the base lability of this protecting group. To avoid esterification, low temperatures are required in the methylation and epimerization is not entirely avoided [31–35]. As Prashad *et al.* [40] noted, sodium hydride does not have high solubility in THF and the sodium salt of the substrate amino acid formed by treatment with sodium hydride has low solubility, as is the case with Boc-Ala-OH. Twice the volume of organic solvent is required due to precipitation during the reaction otherwise the reaction is incomplete [31]. The addition of phase transfer catalysts and catalytic amounts of water to increase the solubility of reagents and intermediates has been a successful strategy to overcome some problems of this method.

6.2.2.5 N-Methylation of Trifluoroacetamides

The trifluoroacetamide is a protecting group that is scarcely used in amino acid protection. There are several advantages with this group in that it is easily introduced and very mild conditions are used to remove it (aqueous potassium carbonate). One other advantage is the increased acidity it confers on the NH proton. Liu *et al.* [41] exploited this property and synthesized *N*-methylphenylalanine analogs under mild conditions (Scheme 6.13). The *N*-methylation step proceeded in anhydrous acetone and potassium carbonate with methyl iodide as the alkylating agent. This produced the methyl ester. Both protecting groups were then removed with aqueous potassium carbonate.

6.2.2.6 N-Methylation via the Mitsunobu Reaction

Alkaline reagents can cause varying degrees of epimerization, particularly if the N- and C-termini are protected, making the α-center the most acidic site and prone to enolization. One variation on this approach was to exploit *N*-tosyl amino acids for use in the Mitsunobu reaction due to the acidity of the NH that the tosyl group bestows.

Papaioannou *et al.* [42] used the Mitsunobu protocol [43] to *N*-alkylate *N*-tosyl amino acid esters **32** and **33** (Scheme 6.14) without epimerizing the products for the

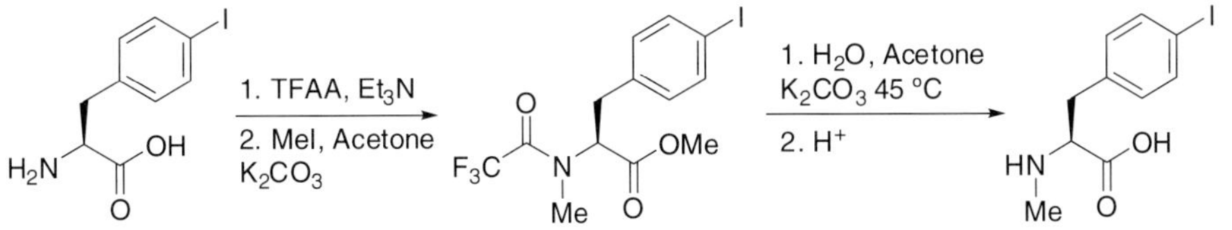

1. TFAA, Et_3N
2. MeI, Acetone, K_2CO_3

1. H_2O, Acetone, K_2CO_3 45 °C
2. H^+

Scheme 6.13

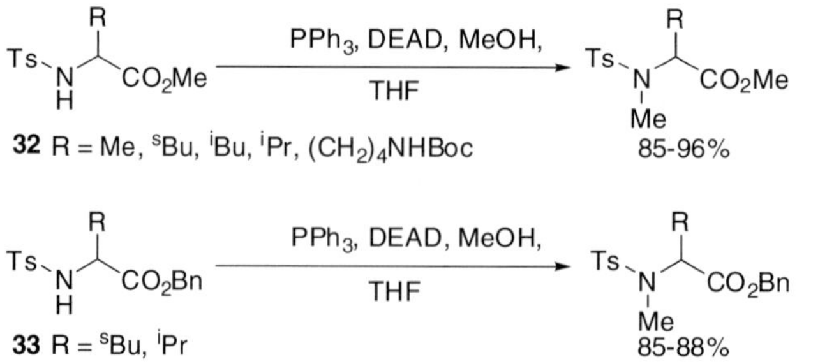

Ts-NH-CH(R)-CO_2Me → (PPh$_3$, DEAD, MeOH, THF) → Ts-N(Me)-CH(R)-CO_2Me 85-96%

32 R = Me, ^{s}Bu, ^{i}Bu, ^{i}Pr, $(CH_2)_4$NHBoc

Ts-NH-CH(R)-CO_2Bn → (PPh$_3$, DEAD, MeOH, THF) → Ts-N(Me)-CH(R)-CO_2Bn 85-88%

33 R = ^{s}Bu, ^{i}Pr

Scheme 6.14

methylation step. Papaioannou *et al.* saponified *N*-methyl-*N*-tosyl-L-valine methyl ester to evaluate the degree of epimerization. They found that saponifying with sodium hydroxide in methanol at room temperature produced up to 44% of the D-enantiomer. Alternatively, deprotection with iodotrimethylsilane effectively removed the methyl ester without epimerization. This reagent, however, is nonselective in that many other protecting groups are susceptible to cleavage with iodotrimethylsilane [44]. Alternatively, the S_N2 dealkylation method of Biron and Kessler [25] would be well suited. The benzyl esters were removed under hydrogenolytic conditions, which did not epimerize the NMAs and were the preferred choice for carboxyl protection in this case. The tosyl group was cleaved with sodium in liquid ammonia providing optically active NMAs.

Wisniewski and Kolodziejczyk [45] used the 2,2,5,7,8-pentamethylchroman-6-sulfonyl (Pmc) group, which has increased lability to acid conditions compared to *N*-toluenesulfonamides, to protect the amino acid nitrogen. The *N*-Pmc-protected amino acid *tert*-butyl and benzyl esters **34** (Scheme 6.15) were subjected to

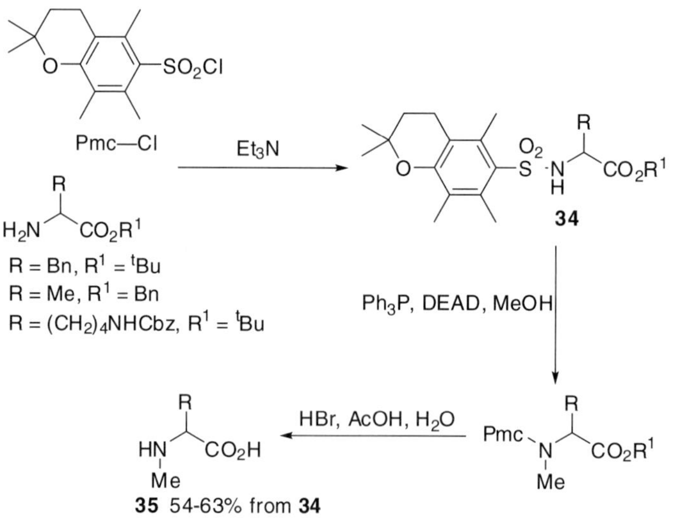

Pmc—Cl (SO_2Cl) + H_2N-CH(R)-CO_2R^1 → (Et_3N) → Pmc-SO_2-NH-CH(R)-CO_2R^1 **34**

R = Bn, R^1 = ^{t}Bu
R = Me, R^1 = Bn
R = $(CH_2)_4$NHCbz, R^1 = ^{t}Bu

34 → (Ph_3P, DEAD, MeOH) → Pmc-N(Me)-CH(R)-CO_2R^1 → (HBr, AcOH, H_2O) → HN(Me)-CH(R)-CO_2H

35 54-63% from **34**

Scheme 6.15

Mitsunobu conditions yielding, after deprotection of the Pmc group with HBr/ AcOH/H_2O (conditions reported by Benoiton *et al.* [32–34] to suppress epimerization), three NMAs **35** in 54–63% yield from **34**.

Yang and Chiu [46] applied a strategy similar to Miller and Scanlan to synthesize Fmoc-*N*-methyl amino acid forms of alanine, valine, phenylalanine, tryptophan, lysine, serine, and aspartic acid that were preloaded on 2-chloro-trityl resin with yields ranging from 86 to 100%. Yang and Chiu [46] *N*-methylated the corresponding 2-NBS under Mitsunobu conditions or with finely powdered potassium carbonate and methyl iodide, and noted that alcohols other than methanol could be used to provide the *N*-alkyl amino acids under Mitsunobu conditions [46]. The sulfonamide group was removed with sodium thiophenoxide and the free amine was carbamoylated with Fmoc-Cl/diisopropylethylamine and then cleaved from the resin with 0.5% TFA/ dichloromethane to provide the Fmoc-*N*-methyl amino acids, which were generally isolated in greater than 90% yield. The methylated amino acids thus isolated were found to be racemization free [46].

The Mitsunobu protocol for *N*-methylating *N*-sulfonyl amino acids is an effective racemization-free method for NMA synthesis. The use of *N*-nosyl protection over *N*-tosyl has provided a means for ready introduction and removal of sulfonamide type protection and the neutral conditions of the Mitsunobu reaction permit a variety of protecting groups that can be included in an orthogonal protection scheme. It would be preferable to limit this procedure to solid-phase synthetic schemes since the monomeric amino acid requires carboxyl protection as it will also be alkylated and excess reagents can be effectively washed away.

6.3 *N*-Methylation via Schiff's Base Reduction

6.3.1 Reduction of Schiff's Bases via Transition Metal-Mediated Reactions

An alternate method of alkylation for introducing methyl groups to the α-amino position is through reductive amination. This simple method is quite flexible in that groups other than methyl can be introduced by varying the carbonyl source. There are several methods developed for reducing the intermediate Schiff's bases that involve transition metal-catalyzed hydrogenation, borohydride reduction, and the Leuckart reaction. Borane reduction of formamides has also been included at the end of this section since it involves reduction. Schiff's base reduction is particularly attractive since the Schiff's base formation is a straightforward process performed by simply combining the aldehyde and the amine together in an appropriate solvent, and then reducing the intermediate imine that forms. *N*-Alkylation of amino acids by the Schiff's base approach works well for aldehydes other than formaldehyde [47–50], since steric hindrance conferred by the alkyl group and amino acid side-chains helps to minimize or prevent dialkylation. This steric limitation does not apply to formaldehyde. In reported attempts to mono *N*-methylate amino acids with formaldehyde, a combination of *N,N*-dimethylation, *N*-monomethylation, and starting material results [48, 51]. This can be rationalized by the fact that secondary amines are more nucleophilic than primary amines. When the Schiff's base intermediate is reduced to the *N*-methyl species, this species can form another Schiff's base or iminium ion with formaldehyde, since it is the smallest aldehyde. Therefore, equivalent amounts of formaldehyde will result in the mixtures observed, as was the case for Keller-Schierlein *et al.* [51] who synthesized N^{α}-methyl-N^{δ}-benzyloxycarbonyl-L-ornithine from N^{δ}-benzyloxycarbonyl-L-ornithine. When treating N^{δ}-benzyloxycarbonyl-L-ornithine with formalin and reducing the mixture with sodium borohydride, a crude mixture of di- and mono-*N*-methyl amino acids and starting material was recovered. After chromatographic purification, N^{α}-methyl-N^{δ}-benzyloxycarbonyl-L-ornithine was obtained in only 35% yield.

In a series of papers, Bowman *et al.* [52–54] describe the *N,N*-dimethylation of amino acids with formalin over palladium-on-charcoal catalyst under hydrogenolytic conditions. The work in this paper was concerned with dimethylation, and provided quantitative yields of the *N,N*-dimethyl amino acids of alanine, valine, leucine, phenylalanine, tyrosine, cysteine, aspartic acid, and glutamic acid. It was noted that the *N,N*-dimethyl derivative of aspartic acid was epimerized in aqueous solution at 100 °C [52]. Ikutani [55] applied the method of Bowman to synthesize *N,N*-dimethyl amino acids of glycine, alanine, leucine, phenylalanine, and tyrosine, which were then converted to *N*-oxides with peroxide. This was also the approach Poduska [56] used in dimethylating lysine derivatives.

The second paper [53] extends the methodology to the mono *N*-alkylation of valine, leucine, and phenylglycine with various aldehydes in ethanol or aqueous ethanol. In this case *N,N*-dialkylglycine can also be produced, whereas amino acids other than glycine were only mono-*N*-alkylated [53]. The last paper [54] describes the reductive alkylation of peptides for identifying the N-terminal amino acid in the chain, employing the same protocols as the two previous papers [54].

The *N,N*-dimethylation and mono-*N*-alkylation of amino acids performed via palladium catalysis is a cheap, effective, and epimerization-free route to these alkylated derivates. As mentioned, monomethylation is impractical with this method as dimethyl amino acids and starting material are byproducts that require tedious purification for their removal.

6.3.2
Reduction of Schiff's Bases via Formic Acid: The Leuckart Reaction

The Leuckart reaction is a method involving the reduction of imines in the presence of formic acid. The procedure developed for amino acid *N*-methylation heats *N*-benzyl amino acids in formic acid in the presence of formalin until CO_2 ceases to effervesce from the solution. This is the only type of reductive amination with formic acid/formalin to produce NMAs; no other variations have been described in the literature so far (Scheme 6.16). This method developed by Quitt *et al.* [20] reveals the variety of different functional groups that tolerate these conditions for *N*-methylating *N*-benzyl amino acids. The two amino acids, lysine and arginine, that

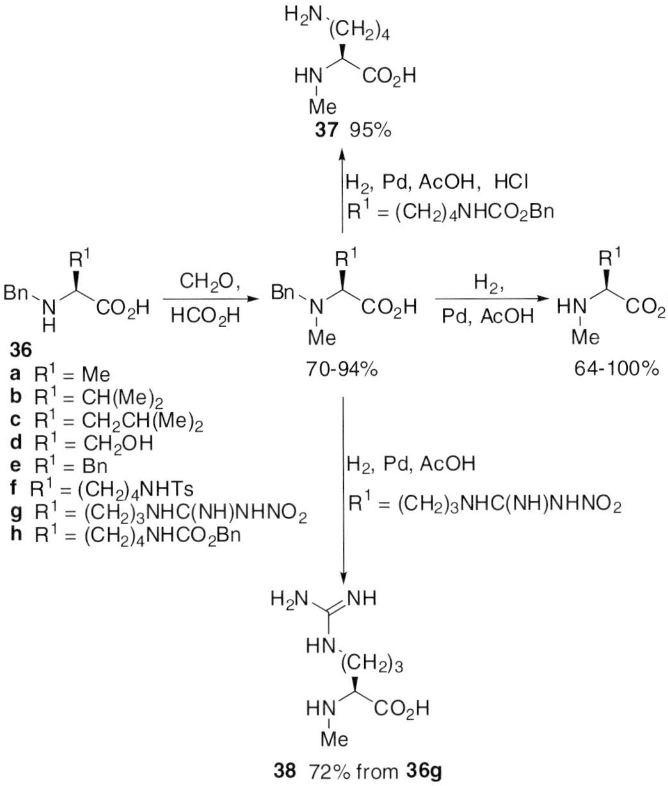

Scheme 6.16

present difficulties for some other methods were successfully *N*-methylated via reductive amination with formaldehyde and formic acid to give structures **37** and **38**, respectively. To date, the physical data obtained from these derivatives have provided a benchmark for the comparison of synthetic NMAs due to the mildness of this epimerization-free method. Ebata *et al.* [57] extended the methodology to other amino acids such as aspartic acid, isoleucine, threonine, and glycine with success, albeit the reactions were low yielding. Various other groups have also used this methodology to prepare *N*-methyl amino acid derivatives as part of synthesis [58] and other studies [59].

6.3.3
Quaternization of Imino Species

Another less-common method for NMA synthesis is forming quaternary iminium salts. This approach for amino acid monomethylation is appealing in that the imino group can only be alkylated once and this prohibits possible dialkylation. This procedure was applied by Eschenmoser *et al.* [60] in the formation of *N*-methyltryptophan (L-abrine) **39**, in which the *N*-chlorobutyroyl amide **40** was treated with silver

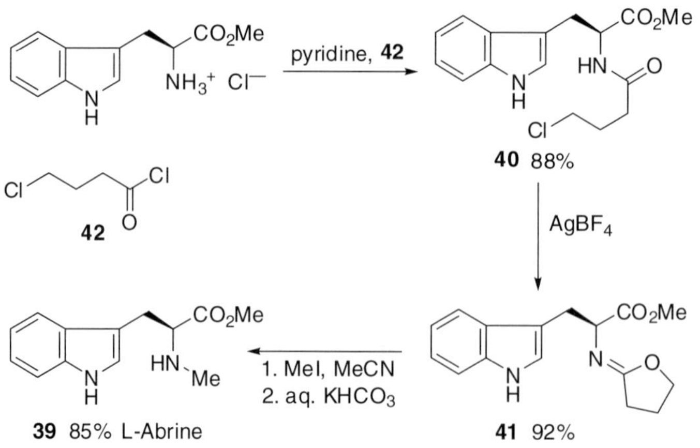

Scheme 6.17

tetrafluoroborate resulting in the iminolactone **41** (Scheme 6.17). Treating the imine with methyl iodide followed by hydrolysis with aqueous potassium carbonate provided the *N*-methyltryptophan **39**. The conversions of **40** through to **39** can be performed in one pot in 85% yield and, notably, the process was epimerization free.

Amidines of amino acid esters generated by reaction with DMF dimethyl acetal **43** have been utilized as intermediates in the formation of NMAs [61]. Methylsulfate or methyltriflate quaternization of the resulting amidine **44** gives an iminium salt **45**, which when hydrolyzed gives the *N*-methyl amino acid **46** (Scheme 6.18). It was found that the amidines were more reactive than simple alkyl Schiff's bases since amidines are more basic and the amidines that were prepared directly from the free amino acid and DMF dimethylacetal in refluxing toluene were epimerized. By simply utilizing amino acid esters enabled lower temperatures and reaction times for the formation of the amidines resulting in stereochemical integrity being intact. The amidine esters also enabled the alkylation with methyltriflate or dimethylsulfate

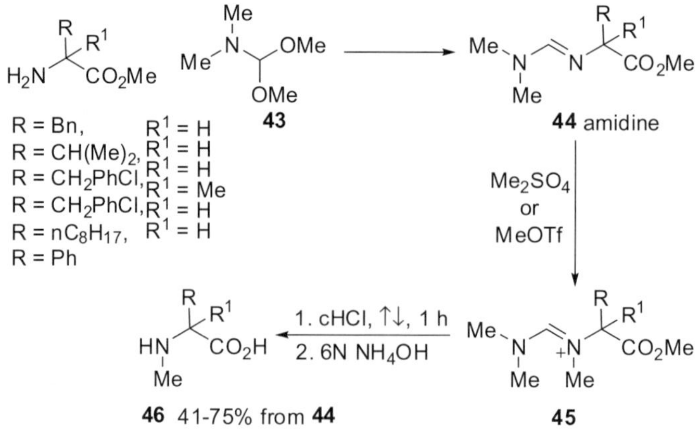

Scheme 6.18

under more mild conditions and this was tested with phenylglycine. The amidine of phenylglycine methyl ester was reacted with methyltriflate in dichloromethane at room temperature and after hydrolysis gave optically active *N*-methylphenylglycine. It was noted that these conditions are particularly mild as phenylglycine is prone to racemization.

6.3.4
Reduction of Schiff's Bases via Borohydrides

Borohydride reductions are alternative approaches to transition metal-catalyzed reduction of Schiff's base intermediates; however, borohydrides such as potassium, sodium, and lithium borohydride are seldom used to reduce Schiff's base intermediates since yields are compromised by competing side-products, particularly the direct reduction of the aldehyde [62]. Borohydrides such as sodium cyanoborohydride are more suited to this application especially in the *N*-alkylation of amino acid esters with aldehydes [48, 51, 63], and triacetoxyborohydride has been recommended as a replacement reducing agent to sodium cyanoborohydride since less-toxic side-products are formed and better yields and reproducibility of results can be obtained with this mild reducing agent [49, 50].

The reductive amination of proteins with formaldehyde in the presence of sodium cyanoborohydride to produce *N,N*-dimethylated proteins has been described [62]. The reaction was regiospecific, with methylation occurring only at the N-terminus and at lysyl side-chains and was a means of "labeling" the protein for further studies. Jentoft and Dearborn [62] discuss the superiority of sodium cyanoborohydride over sodium borohydride in its mildness and specificity for reductive amination.

Polt *et al.* [47] have utilized *N*-diphenylmethyl imine (ketimine) esters of amino acids and in a one-pot procedure reduced the intermediates with sodium cyanoborohydride to *N*-diphenylmethyl amino acid esters and then condensed these secondary amines with excess formaldehyde or other aldehydes in the presence of excess sodium cyanoborohydride providing *N*-diphenylmethyl-*N*-methyl amino esters. The fully protected NMAs were hydrogenolyzed over palladium catalyst to afford the *N*-methyl amino acids. In this way tryptophan was monoalkylated without competing Pictet–Spengler cyclization nor was there any mention of methylation occurring at the indole nitrogen [47]. This procedure was applied to alanine, serine, threonine, leucine, and tryptophan, and is closely related to the approach of Quitt *et al.* [20] One important note is that it was generally observed that 5–19% of unmethylated *N*-diphenylmethyl amino acid esters were recovered along with the starting ketimine [47].

Kaljuste and Undén [64] reported the mono-*N*-methylation of resin-bound terminal amino acid residues on solid phase. The authors make use of the acid-labile 4,4′-dimethoxydiphenylmethyl (4,4′-dimethoxydityl) group for nitrogen protecting terminal amino acid residues [65]. *N*-Methylation was performed with formaldehyde, acetic acid and sodium cyanoborohydride in DMF. This reaction proceeded in yields in the range 56–99% for most common amino acids. One of the problems associated with the procedure is that up to three methylation cycles were required for some amino acids in order to complete the methylation. It was noted that side-chain functionalized amino acids needed longer reaction times that then lead to undesirable side-products that could be avoided by decreasing the reaction time, but then incomplete methylation occurred.

6.3.5
Borane Reduction of Amides

Although the reduction of amino acid amides to *N*-alkyl amino acids [66–68], diverges from the parent topic title of Schiff's base reductions, its inclusion in this section is warranted due to its similarity with borohydride reductions of imine intermediates. Krishnamurthy [67] made use of the selective reduction of various formanilides and some alkyl formamides with excess borane/dimethylsulfide complex ($BH_3{\cdot}SMe_2$). The two-step process gave high purity *N*-methyl anilides in 80–100% yield. The method allows for mono-*N*-methylation without the problems associated with dimethylation and no methylation of imines. Chu *et al.* [69] exploited this strategy by reducing *N*-formyl-D-tryptophan methyl ester with $BH_3{\cdot}SMe_2$. The reduction gave, after work-up, *N*-methyl-D-tryptophan methyl ester in 56% yield.

Hall *et al.* [70] reduced amino acid amides in solution and on solid support with diborane in THF, and then treated the product with iodine to promote oxidative cleavage of the borane-amine adducts. In this fashion, amino acid formates coupled to Wang resin were reduced with diborane in greater than 72% yield and greater than 75% purity for the amino acids alanine, valine, serine, and phenylalanine.

The reduction of Schiff's base intermediates is a very mild and racemization-free process. Quitt's method [20] of reductive amination of *N*-benzyl amino acids is to date an efficient cheap and mild method for the synthesis of most NMAs, and has been used frequently for comparison of physicochemical data. A similar approach, employing sodium cyanoborohydride reduction of *N*-diphenylmethyl amino acid ester Schiff's base intermediates in solution and solid phase was described by Polt *et al.* [47] and Kaljuste and Undén [64], respectively. Their work revealed the efficacy of this approach as applied to a wide variety of amino acids, albeit on a small scale. Although in principal this technique is similar to Quitt's method, there are more manipulations involved and the C-terminus must be protected, and excessive amounts of formaldehyde and reducing agent are required to force complete methylation. However, it was shown in the work of Polt *et al.* [47] that small degrees of incomplete methylation were observed.

Reductive amination involving transition metal hydrogenolysis is somewhat limited to dimethylation with formaldehyde, but monoalkylation with aldehydes other than formaldehyde is possible [52–54].

One uncommon technique is the reduction of *N*-formyl amino acids. This approach is obvious since monoformylation of amino acids is readily achieved and therefore concerns of dialkylation and the need for multistep syntheses are eradicated. The problem is the carboxylic acid needs protection since the borane can reduce the acids to alcohols. The technique is further limited to amino acids without other amide groups (i.e., asparagine and glutamine) that may also be reduced.

6.4 *N*-Methylation by Novel Methods

The following section is a compilation of more elaborate methods for the synthesis of *N*-methyl amino acids. While some of the methods use techniques discussed previously for installing the *N*-methyl group, the methods in this section were devised to prepare especially unusual NMAs required typically for natural product syntheses. These techniques devised for unusual NMA syntheses in most cases were applicable for certain *N*-methylated derivatives and are often not appropriate for other NMAs.

6.4.1 1,3-Oxazolidin-5-ones

Ben-Ishai [71] reported the synthesis of oxazolidin-5-ones **47** (Scheme 6.19) by refluxing *N*-Cbz-protected amino acids with paraformaldehyde in the presence of an acid catalyst. The five-membered heterocyclic intermediates **47** resemble *N*-hydroxymethyl amides and display distinct carbonyl stretches in the infrared region between 1790 and 1810 cm^{-1}. The oxazolidin-5-one ring is susceptible to nucleophilic attack. Amines open them to form amides [71, 72] and alcohols to form esters [73]. Ben-Ishai established this nucleophilic susceptibility by treating the 1,3-oxazolidin-5-one **47c** with an equivalent amount of benzyl amine in alcohol to afford the *N*-hydroxymethyl amide **48c**. Hydrogenation of the *N*-hydroxymethyl intermediate provided *N*-methylglycine (sarcosine) **49c**. It was noted that treating the *N*-hydroxymethyl amide **48c** with an extra equivalent of benzyl amine affects the removal of the *N*-hydroxymethyl moiety to provide *N*-Cbz-glycine benzyl amide. The reductive cleavage of oxazolidin-5-ones to NMAs was not realized until the work of Freidinger *et al.* [74] (see below).

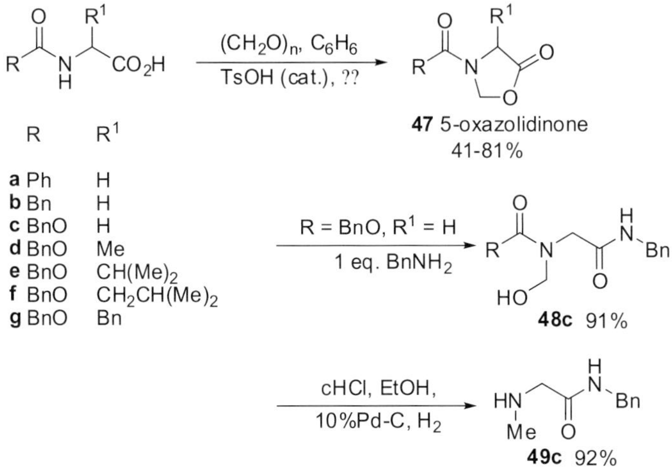

$(CH_2O)_n$, C_6H_6; TsOH (cat.), ?? → **47** 5-oxazolidinone 41-81%

	R	R^1
a	Ph	H
b	Bn	H
c	BnO	H
d	BnO	Me
e	BnO	$CH(Me)_2$
f	BnO	$CH_2CH(Me)_2$
g	BnO	Bn

R = BnO, R^1 = H; 1 eq. $BnNH_2$ → **48c** 91%

cHCl, EtOH, 10%Pd-C, H_2 → **49c** 92%

Scheme 6.19

Auerbach *et al.* [75] have shown that *N*-hydroxymethyl (or *N*-methylol) amides analogous to structure **48** could be reduced with triethylsilane/TFA in chloroform to the *N*-methyl amide. Their reduction proceeds by hydride transfer from the silane to an acyliminium ion derived from the *N*-hydroxymethyl amide under the acidic conditions and also showed that this reduction proceeds via a palladium-catalyzed hydrogenation in the presence of TFA [75].

Freidinger *et al.* [74] recognized the potential of oxazolidin-5-ones as stable lactones that are analogous to methylols and could be converted to *N*-methylated derivatives under the conditions described by Auerbach *et al.* [75]. By using Fmoc-protected amino acids, they extended the range of substrates that can be converted to 1,3-oxazolidin-5-ones with alkanals including paraformaldehyde. Treating these substrates with triethylsilane/TFA gave the expected *N*-Fmoc-*N*-methyl amino acids and also some *N*-alkyl derivatives. This sequence was applied to Fmoc-protected alanine, valine, methionine, phenylalanine, lysine, serine, and histidine. Freidinger *et al.* [74] also conducted epimerization studies on the technique using nuclear magnetic resonance analysis of the ^{13}C satellites of the methoxyl signal as internal reference peaks of D- and L-methyl-*N*-Fmoc-*N*-methyl-*O*-benzylserinate, and observed that no detectable epimerization occurred in the reductive cleavage reaction. This technique has also been applied to the TFA-stable *N*-Cbz-protected amino acids and a large range of *N*-Cbz-*N*-methyl amino acids have been synthesized [76, 77].

This technique was further extended to *N*-Boc-protected amino acids by Reddy *et al.* [78], who prepared 1,3-oxazolidin-5-ones **50** with *N*-Cbz and *N*-Boc protection (Scheme 6.20). They applied a different approach to the reduction of the oxazolidin-5-ones by hydrogenation over palladium catalyst under neutral conditions. The *N*-Cbz compounds were converted to NMAs with concomitant removal of the *N*-Cbz group **51** ($R^2 = H$) and the *N*-Boc derivatives were reduced to the corresponding *N*-Boc-*N*-methyl amino acids **51** ($R^2 = Boc$). This was the first report of success in the use of hydrogenation of *N*-Boc-protected oxazolidin-5-ones as a means of producing the *N*-methyl group directly. However, Itoh [72] reported that the 1,3-oxazolidin-5-one ring becomes reactive by removal of the *N*-protection and this was also the experience of Aurelio *et al.* [77], in the case of *N*-Cbz protected 1,3-oxazolidin-5-ones. Itoh [72] studied the reactions of *N*-Cbz 1,3-oxazolidin-5-ones and, in particular, the

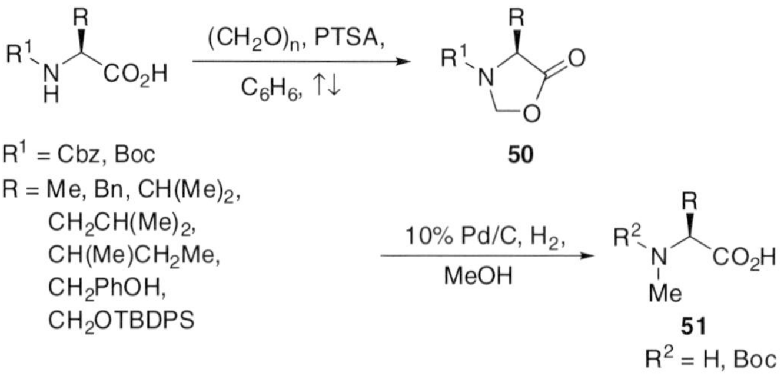

$(CH_2O)_n$, PTSA, C_6H_6, ↑↓

R^1 = Cbz, Boc
R = Me, Bn, $CH(Me)_2$, $CH_2CH(Me)_2$, $CH(Me)CH_2Me$, CH_2PhOH, $CH_2OTBDPS$

50

10% Pd/C, H_2, MeOH

51
R^2 = H, Boc

Scheme 6.20

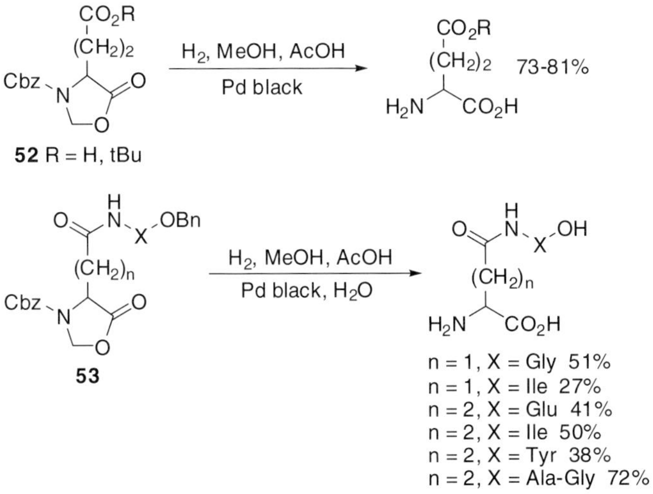

CO_2R, $(CH_2)_2$, Cbz, N, O, O

52 R = H, tBu

H_2, MeOH, AcOH

Pd black

CO_2R, $(CH_2)_2$, H_2N, CO_2H 73-81%

O, H, N, X, OBn, $(CH_2)_n$, Cbz, N, O, O

53

H_2, MeOH, AcOH

Pd black, H_2O

O, H, N, X, OH, $(CH_2)_n$, H_2N, CO_2H

n = 1, X = Gly 51%
n = 1, X = Ile 27%
n = 2, X = Glu 41%
n = 2, X = Ile 50%
n = 2, X = Tyr 38%
n = 2, X = Ala-Gly 72%

Scheme 6.21

hydrogenolyzes of the derivatives **52** and **53** (Scheme 6.21). These conversions did not produce *N*-methyl amino acids, but instead the methylene carbon was cleaved entirely from the lactone substrate, providing the parent amino acids. Williams and Yuan [79] also observed this result.

Aurelio *et al.* [77] prepared the *N*-Cbz-oxazolidin-5-ones **54** of a variety α-amino acids (Scheme 6.22). Substrates with reactive side-chains were included in the oxazolidin-5-one formation with varying degrees of success. Threonine and serine, in particular, were prone to oxazolidinine formation as was cysteine in forming *N*-Cbz-thiazolidines by reaction with the side-chain hydroxyl and thiol, respectively. Side-chain protection was thus necessary for oxazolidin-5-one formation of these amino acids as well as amino acids with basic side-chains [77c]. Amino acids like tyrosine, glutamic acid, and methionine were converted to the corresponding oxazolidin-5-one, and reduction of several of these substrates by catalytic hydrogenation gave varying amounts of the free α-amino acid **55** in accord with Itoh [72]. Resorting to the conditions applied by Freidinger *et al.* [74], triethylsilane/TFA provided the NMAs **56**. Use of the hydrogenolytic conditions of Reddy *et al.* [78], to reduce *N*-Boc oxazolidin-5-ones **57**, did not result in any of the expected NMA **58** [32]. Instead, two products, **59** and **60**, were recovered.

One reported successful reduction of *N*-Boc-oxazolidin-5-one of methionine in triethylsilane/TFA mixture was achieved by Willuhn *et al.* [80] in the synthesis of *N*-methylhomocysteine derivatives. The reduction under these conditions provided the NMA of methionine with concomitant removal of the Boc group.

A similar protocol to that of Freidinger was applied to Fmoc-protected *p*-aminomethylbenzoic acid substrates (Scheme 6.23) [81]. The methylol derivatives **61** and **62** were isolated after work-up with varying degrees of decomposition back to starting material when the starting Fmoc-*p*-aminomethylbenzoic acid was treated with formaldehyde in acetic acid. By treating the methylol derivatives with a triethylsilane/TFA

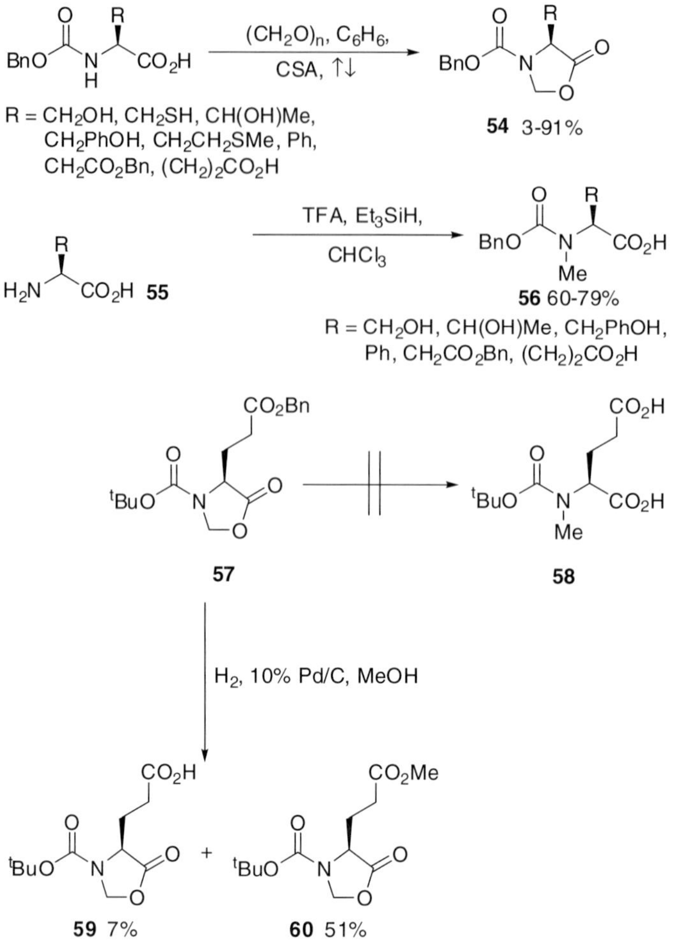

$(CH_2O)_n$, C_6H_6, CSA, ↑↓

R = CH_2OH, CH_2SH, CH(OH)Me, CH_2PhOH, CH_2CH_2SMe, Ph, CH_2CO_2Bn, $(CH_2)_2CO_2H$

54 3-91%

H_2N CO_2H **55**

TFA, Et_3SiH, $CHCl_3$

56 60-79%

R = CH_2OH, CH(OH)Me, CH_2PhOH, Ph, CH_2CO_2Bn, $(CH_2)_2CO_2H$

57 **58**

H_2, 10% Pd/C, MeOH

59 7% **60** 51%

Scheme 6.22

mixture, the corresponding *N*-Fmoc-*N*-methyl amino acids **63** were isolated. This was an unpredictable route due to degrees of reversion to starting materials and so a one-pot process was developed which did not involve the isolation of the often unstable methylol intermediate. The substrate was exposed to TFA and 40% formaldehyde solution for 30 min and then treating the intermediate methylol with triethylsilane providing good to excellent yields (up to 92% yield) of *N*-methylated product.

A variation on the theme of 1,3-oxazolidin-5-ones employed 2,2-bis(trifluoromethyl)-1,3-oxazolidin-5-ones **64** as cyclic aminals by condensing amino acids with hexafluoroacetone (Scheme 6.24) [82]. The 2,2-bis(trifluoromethyl)-1,3-oxazolidin-5-ones **64** were used as a means of protecting the carboxyl group and providing a single valence on the α-nitrogen for the desired reaction. The aminal **64** was chloromethylated with paraformaldehyde in the presence of thionyl chloride providing the

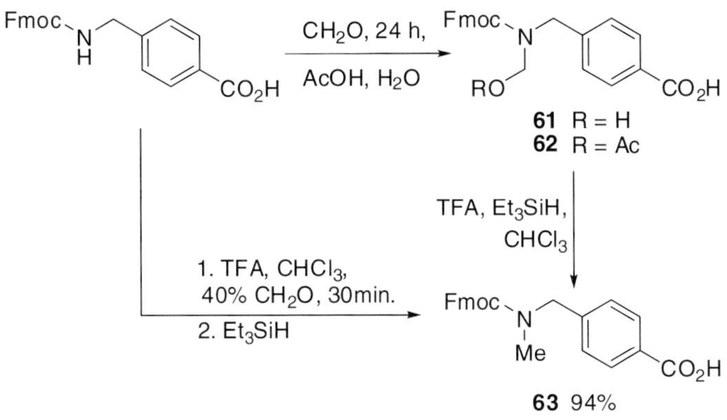

Fmoc-NH-CH_2-C_6H_4-CO_2H

CH_2O, 24 h, AcOH, H_2O

Fmoc-N(CH_2OR)-CH_2-C_6H_4-CO_2H

61 R = H
62 R = Ac

TFA, Et_3SiH, $CHCl_3$

1. TFA, $CHCl_3$, 40% CH_2O, 30min.
2. Et_3SiH

Fmoc-N(Me)-CH_2-C_6H_4-CO_2H

63 94%

Scheme 6.23

chloromethyl amine **65** which was converted to the *N*-methyl-oxazolidin-5-one **66** with triethylsilane and TFA. Acidolysis of the *N*-methylated derivatives **66** with isopropanol or methanol allows for the isolation of either the NMA **67** or the NMA methyl ester **68**, respectively.

Treating cysteine with formalin provides thiazolidine **69** (Scheme 6.25) which was employed as an intermediate for *N*-methylcysteine in the synthesis of

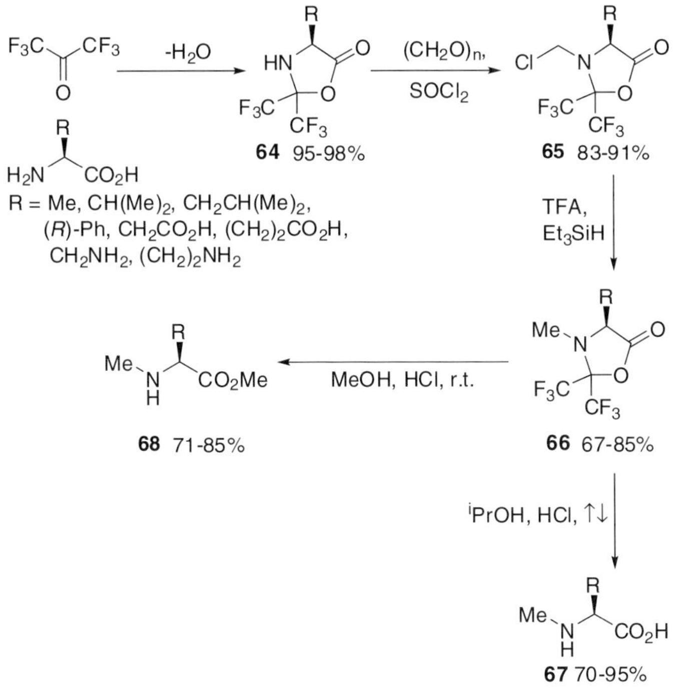

F_3C-CO-CF_3

H_2N-CH(R)-CO_2H

R = Me, $CH(Me)_2$, $CH_2CH(Me)_2$, (*R*)-Ph, CH_2CO_2H, $(CH_2)_2CO_2H$, CH_2NH_2, $(CH_2)_2NH_2$

$-H_2O$

64 95-98%

$(CH_2O)_n$, $SOCl_2$

65 83-91%

TFA, Et_3SiH

66 67-85%

MeOH, HCl, r.t.

Me-NH-CH(R)-CO_2Me

68 71-85%

^iPrOH, HCl, ↑↓

Me-NH-CH(R)-CO_2H

67 70-95%

Scheme 6.24

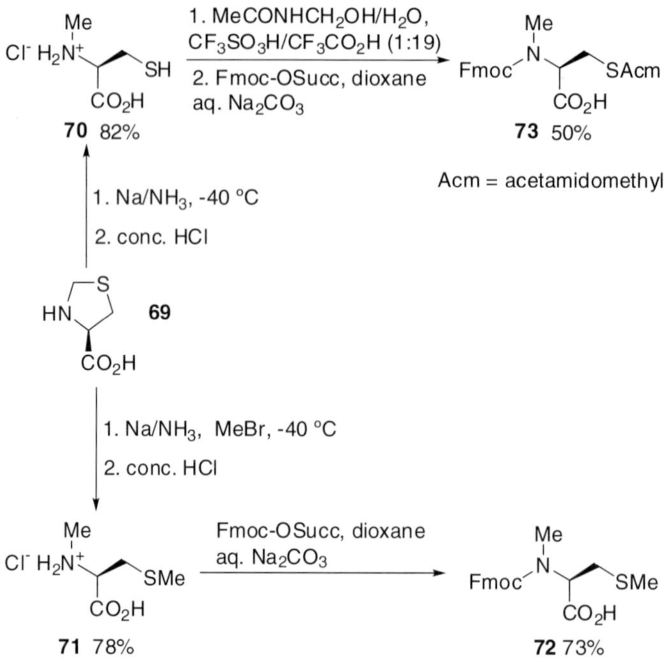

1. $MeCONHCH_2OH/H_2O$, CF_3SO_3H/CF_3CO_2H (1:19)
2. Fmoc-OSucc, dioxane aq. Na_2CO_3

70 82% → **73** 50%

Acm = acetamidomethyl

1. Na/NH_3, -40 °C
2. conc. HCl

69

1. Na/NH_3, MeBr, -40 °C
2. conc. HCl

Fmoc-OSucc, dioxane aq. Na_2CO_3

71 78% → **72** 73%

Scheme 6.25

[1-(*N*-methyl-hemi-L-cysteine)]-oxytocin [83]. Reduction of the thiazolidine intermediate with sodium in liquid ammonia gave *N*-methylcysteine that was treated in the same pot with an equivalent amount of benzyl chloride providing the *N*-methyl-*S*-benzyl-L-cysteine in 90% yield (the addition of 1 equiv. of water is crucial in suppressing dimerization) [83]. Final *N*-protection with CbzCl afforded *N*-Cbz-*N*-methyl *S*-benzyl-L-cysteine in 84% yield. Liu *et al.* [84] used the same protocol in making Fmoc derivatives of *N*-methyl-L-cysteine (Scheme 6.26). The thiazolidine **69** [85] was reduced to provide *N*-methyl-L-cysteine **70**. *In situ* treatment with methyl bromide provides the *S*-methyl derivative **71** that was treated with Fmoc-succinimide to give the NMA **72**. Alternatively, treatment of **70** with *N*-hydroxymethyl acetamide and a TFA/trifluoromethanesulfonic acid mixture provided an *S*-acetamidomethyl intermediate that was converted to the Fmoc derivative **73**.

The synthesis of *N*-methylcysteine is a challenging task and, in particular, an appropriate derivative for Fmoc solid-phase application available by a small number of manipulations is desirable. Ruggles *et al.* [86] synthesized *N*-Fmoc-*N*-methyl-Cys (*S*-*t*Bu)-OH **76g** (Scheme 6.26) from the commercially available *N*-Fmoc-Cys (*S*-*t*Bu)-OH **74g** via an oxazolidin-5-one intermediate that was reduced with a triethylsilane/TFA mixture. The authors conducted a study into various classical methods for installing the *N*-methyl moiety with discouraging results. By resorting to the 1,3-oxazolidin-5-one method, it was found that most conversions of a series of derivatives **74a–g** were accompanied by the thiazolidinine **77** formation, even in the reductive step. It was found that the acid-stable *tert*-butylthio protecting group

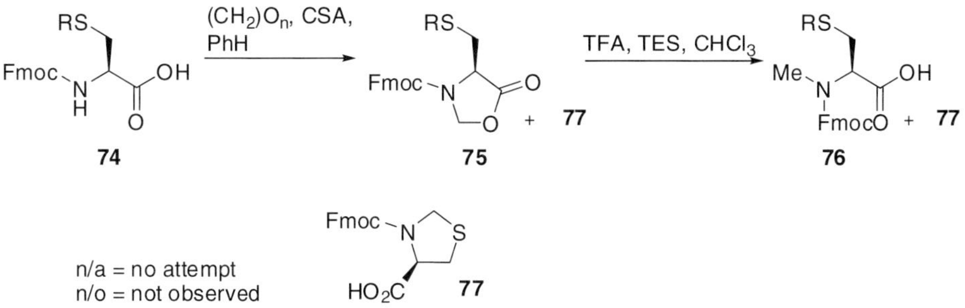

$(CH_2)O_n$, CSA, PhH; TFA, TES, $CHCl_3$

74 → 75 + 77 → 76 + 77

n/a = no attempt
n/o = not observed

74	75	% of 77	76	% of 77
74a R = Trt	**75a 76%**	**n/o**	**76a n/o**	**54**
74b R = Mob	**75b 55%**	**42**	**76b 64%**	**34**
74c R = Meb	**75c 85%**	**15**	**76c n/a**	**n/a**
74d R = Bn	**75d 98%**	**2**	**76d n/a**	**n/a**
74e R = tBu	**75e 41%**	**59**	**76e n/a**	**n/a**
74f R = Acm	**75f 20%**	**80**	**76f n/a**	**n/a**
74g R = StBu	**75g 99%**	**11**	**76g 89%**	**n/o**

Scheme 6.26

(*S*-*t*Bu) was the highest yielding in both steps and it was amenable to their synthetic protocol for construction of *N*-methylated small molecule mimics of cyclocystine [86].

Arvidsson *et al.* [87] have studied the reduction of Fmoc-protected 1,3-oxazolidin-5-ones and 1,3-oxazinan-6-ones with different Lewis acids in place of TFA. The authors found that 2 equiv. of aluminum chloride could replace TFA and the reaction time was reduced nearly to a sixth under standard conditions. It was also shown that lactonization and reductions could be performed under microwave irradiation. Several minutes were required for both manipulations, improving yields considerably in most cases.

The oxazolidin-5-one intermediate offers an advantage over the direct alkylation procedures in that the methylene bridge between the α-nitrogen and carboxyl groups offers simultaneous N- and C-terminal protection, and thus side-chain manipulations are possible. Furthermore, the methylene bridge can be smoothly converted to the NMA by reduction under acidic conditions. Although the triethylsilane/TFA combination is a versatile choice for reduction [77], the expense of these reagents and problems in removing trace amounts of TFA make the Lewis acid reduction an enticing practical improvement.

The synthesis of *N*-methylcysteine via reduction of the thiazolidine intermediate is a cost-effective and scalable procedure for making the *N*-methyl derivative. The manipulations involved are trivial and the added advantage is the fact that regioselective alkylation of the thiol group facilitates synthesis of a variety of cysteine derivatives. The recent approach by Ruggles *et al.* [86] utilizing commercially available *N*-Fmoc-Cys(*S*-*t*Bu)-OH for oxazolidinone formation and reduction to the NMA provides a derivative amenable to Fmoc solid-phase synthesis in only two steps.

6.4.2
Asymmetric Syntheses

Few contributors to the field have constructed NMAs by methods that require the α-center be created. The common reason for this is that the methodologies enable the synthesis of quite unusual NMAs with unnatural side-chains. The following section involves diverse methodologies that incorporate chiral auxiliaries that confer the required asymmetry on the α-carbon under construction.

A simple technique that utilizes L-proline as an auxiliary was reported by Poisel and Schmidt [88], in the synthesis of *N*-methylphenylalanine (Scheme 6.27). Azlactone **78** is readily prepared from *N*-acetyl-glycine and benzaldehyde under basic conditions. It is then treated with L-proline to form an arylidenedioxopiperazine **79a**. The chiral dioxopiperazine is methylated with classical sodium hydride/methyl iodide conditions providing **79b**. Subjecting **79b** to standard hydrogenation conditions with palladium metal catalyst gives *N*-methyl-L-phenylalanine-L-proline diketopiperazine in 90% e.e. The diketopiperazine was hydrolyzed under acidic conditions affording the free *N*-methylphenylalanine **80**.

Pandey *et al.* [89] condensed *N*-benzylsarcosine **82** and L-prolinol **83** and then through an intramolecular photosensitized electron transfer cyclization formed the chiral auxiliary **84** (93% d.e., Scheme 6.28). The intermediate ether **84** is formed through an iminium ion **81** and upon treatment with Grignard reagents yields *N*-methyl amino acid-L-prolinol dipeptides **85**. It was also shown that Lewis acid-mediated alkylation was possible and provided higher stereoselectivity than the Grignard approach. Hydrolysis of the dipeptides **85** with either aqueous HCl or methanolic HCl provided the corresponding *N*-benzyl-*N*-methyl amino acids or esters **86**, respectively, and L-prolinol **83**, which was recovered in 96%.

Agami *et al.* [90, 91] constructed various NMAs using the chiral morpholine **87** as a template (Scheme 6.29). By condensing *N*-methyl-D-phenylglycinol, glyoxal, and thiophenol, the morpholine **87** was obtained as a single stereoisomer [90]. Treating the morpholine **87** with organometallic reagents displaces the thiophenyl ether

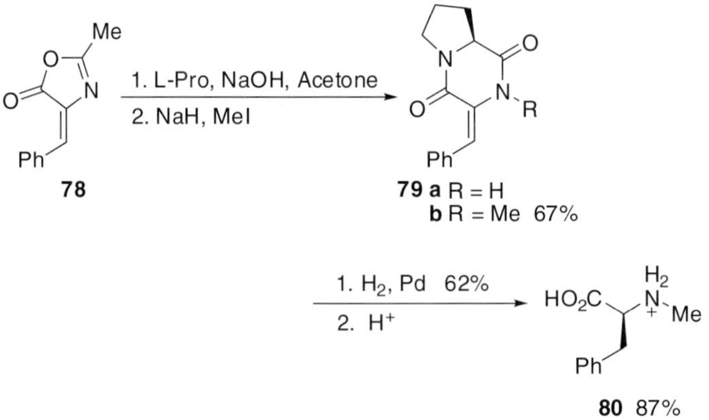

1. L-Pro, NaOH, Acetone
2. NaH, MeI
78
79 a R = H
b R = Me 67%
1. H_2, Pd 62%
2. H^+
80 87%

Scheme 6.27

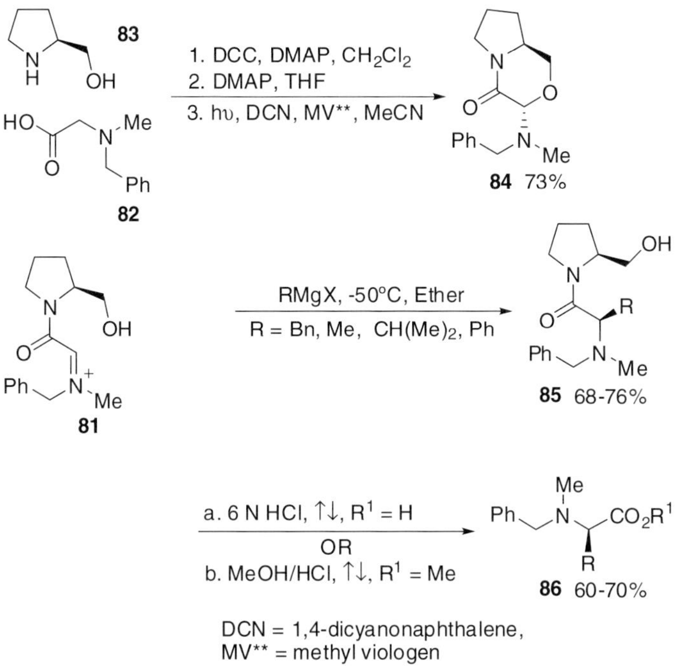

1. DCC, DMAP, CH_2Cl_2
2. DMAP, THF
3. hν, DCN, MV**, MeCN

84 73%

RMgX, -50°C, Ether
R = Bn, Me, $CH(Me)_2$, Ph

85 68-76%

a. 6 N HCl, ↑↓, R^1 = H
OR
b. MeOH/HCl, ↑↓, R^1 = Me

86 60-70%

DCN = 1,4-dicyanonaphthalene,
MV** = methyl viologen

Scheme 6.28

moiety in a stereocontrolled fashion with excellent control, giving in most cases above 98% e.e. Organozincates displaced with retention of configuration and organocuprates displaced with inversion of configuration. Oxidation of the hemiacetal under Swern conditions affords the lactone **88**, which can be completely epimerized with potassium *tert*-butoxide at 40 °C. The NMA was isolated by treating the lactone with vinyl chloroformate to give the acyclic carbamate **89**. Hydrolysis with methanolic HCl cleaves the carbamate and transesterifies the chlorophenethyl ester to the corresponding methyl ester **90**.

Oppolzer *et al.* [92] utilized acylated camphorsultams as auxiliaries in the production of NMAs. Selective hydroxamination of the enolate **91** enabled the monomethylation under reductive alkylation conditions with methanolic formaldehyde, providing the precursors **92** with enantiomeric excesses greater than 99% (Scheme 6.30). Reduction of **92** with zinc dust provided the (*N*-alkylamino)acylsultams **93** which were hydrolyzed under basic conditions to afford the (*S*)-configured NMAs **94** in high yields (90–100%). By applying the same synthetic sequence to the camphorsultam of opposite configuration, (*R*)-configured *N*-methyl α-amino acids can be synthesized with equal efficiency. One advantage of this versatile technique is that the diversity of side-chains in the final NMAs can be made by simply altering the acyl function attached to the camphorsultam auxiliary.

Pseudoephedrine has made its way as a chiral auxiliary into the asymmetric synthesis of NMAs as described by Myers *et al.* [93]. By amidating sarcosine with (*R,R*)-pseudoephedrine, the chiral auxiliary **96** was used in the synthesis of amino

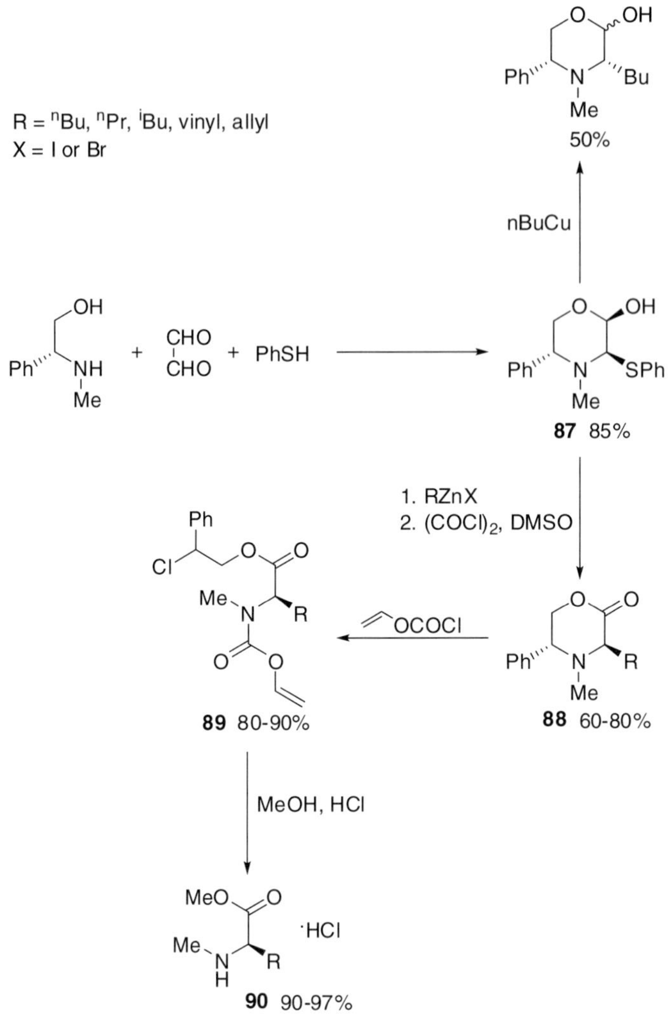

R = ^{n}Bu, ^{n}Pr, ^{i}Bu, vinyl, allyl
X = I or Br

Scheme 6.29

acids and *N*-methyl amino acids (Scheme 6.31). Lithiation of **96** forms the intermediate enolate **97** that was quenched with alkyl halides providing the NMA derivatives **98** with good stereocontrol. Where R = Bn, the alkylation product **98** (*N*-methylphenylalanine) was isolated in 93% yield and 88% d.e. in the crude state and the diastereomeric excess was improved to 99% by recrystallization. Where the alkyl substituent was an ethyl group, the product **98** was isolated in 77% yield with 94% d.e. after purification.

An imaginative approach to NMAs was applied by Grieco and Bahsas [94] in the synthesis of a variety of NMAs. Amino acid esters and amides were treated with formaldehyde and the intermediate iminium ion was trapped with excess cyclopentadiene via an aza-Diels–Alder pericyclic reaction as 2-azanorbornenes **100**

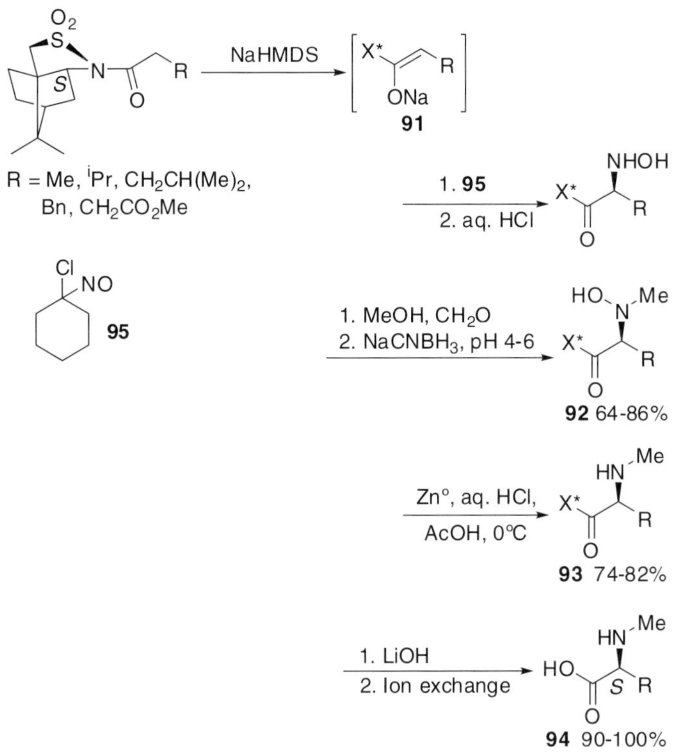

Scheme 6.30

(Scheme 6.32). In the presence of TFA, cycloreversion occurs and the intermediate iminium ion **101** is intercepted by silane affording the NMA esters and amides **102** in yields ranging from 67 to 92%.

The reductive alkylation of optically active scalemic azides **103** has found use providing intermediates in the synthesis of several NMAs (Scheme 6.33). Dorow and Gingrich [95] treated several azido acids, esters, and amides with dimethylbromoborane providing the NMAs. The synthetic sequence was subjected to an

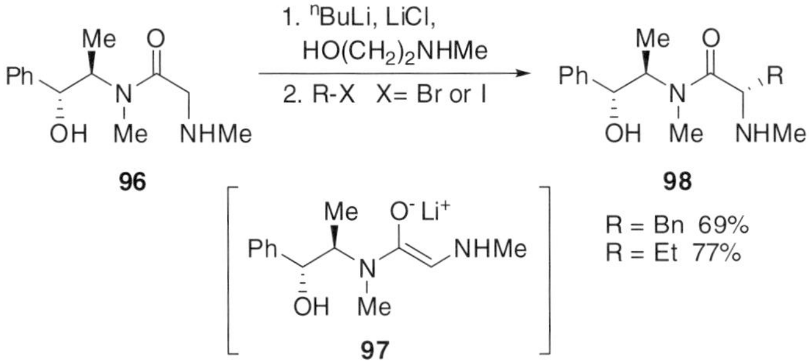

Scheme 6.31

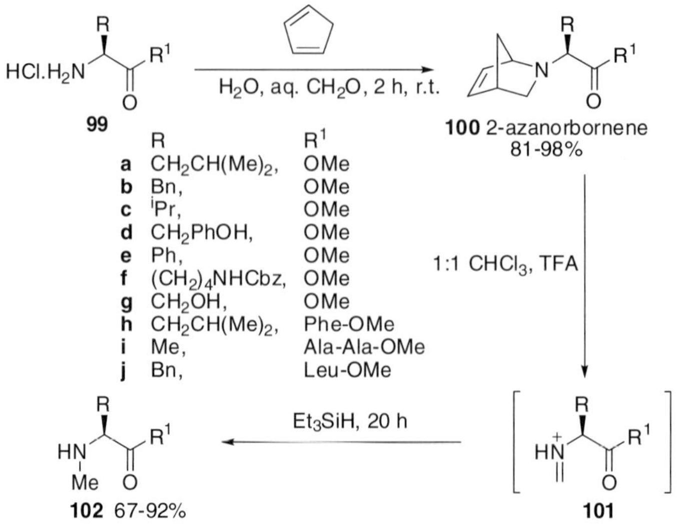

HCl.H_2N **99** → (H_2O, aq. CH_2O, 2 h, r.t.) → **100** 2-azanorbornene 81-98% → (1:1 $CHCl_3$, TFA) → [**101**] → (Et_3SiH, 20 h) → **102** 67-92%

	R	R^1
a	$CH_2CH(Me)_2$,	OMe
b	Bn,	OMe
c	^{i}Pr,	OMe
d	CH_2PhOH,	OMe
e	Ph,	OMe
f	$(CH_2)_4NHCbz$,	OMe
g	CH_2OH,	OMe
h	$CH_2CH(Me)_2$,	Phe-OMe
i	Me,	Ala-Ala-OMe
j	Bn,	Leu-OMe

Scheme 6.32

epimerization test with 2-azidophenylacetic acid **104** (99% e.e.), a precursor to *N*-methylphenylglycine, a standard used in epimerization studies of NMA syntheses. It was found that treatment of **104** with dimethylbromoborane at 40 °C gave the product *N*-methylphenylglycine in 68% yield with 38% e.e. When the same conditions were applied to **104** at 20 °C, (*S*)-*N*-methylphenylglycine was obtained in 99% yield, but the enantiomeric excess was not revealed and it was hypothesized that the increased temperature contributed to the possibility of enolization. The lower temperatures used suppressed enolization, providing optically active NMAs.

A clever method for *N*-methylating α-amino acids was devised by Laplante and Hall [96]. Amino acids bound on solid support were treated with pinacol chloromethylboronic ester under basic conditions with the highly hindered base, 1,2,2,6,6-pentamethylpiperidine (PMP) (Scheme 6.34). Peroxide treatment of the aminomethylboronic ester adduct provides the free NMA. To date this is the only procedure that utilizes amino acids in an unprotected form by which mono-*N*-methylation is achieved. The *N*-methylation is based on a 1,2-carbon-to-nitrogen migration of boron in α-aminoalkylboronic esters. The free amine **105** bound to either Wang or SASRIN resin (an acronym coined from super acid sensitive resin) was treated with an excess of the boronic ester (5 equiv.) that achieves dialkylation.

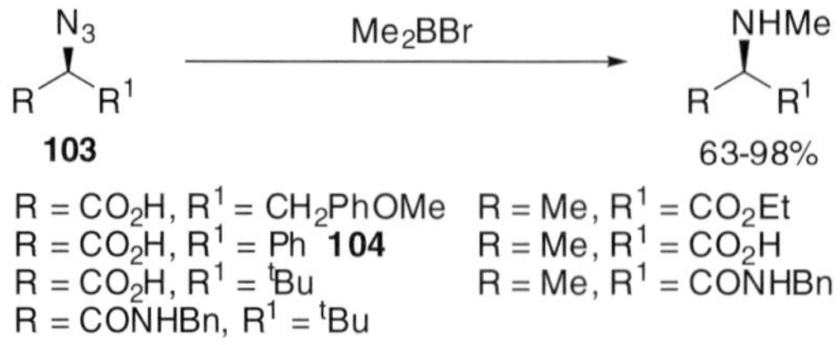

103 (N_3) → (Me_2BBr) → NHMe product, 63-98%

R = CO_2H, R^1 = CH_2PhOMe R = Me, R^1 = CO_2Et
R = CO_2H, R^1 = Ph **104** R = Me, R^1 = CO_2H
R = CO_2H, R^1 = ^{t}Bu R = Me, R^1 = CONHBn
R = CONHBn, R^1 = ^{t}Bu

Scheme 6.33

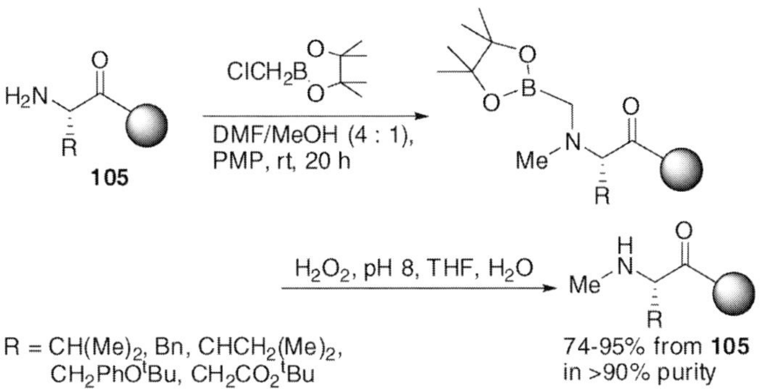

105

$ClCH_2B(pin)$, DMF/MeOH (4 : 1), PMP, rt, 20 h

H_2O_2, pH 8, THF, H_2O

R = $CH(Me)_2$, Bn, $CHCH_2(Me)_2$, CH_2PhO^tBu, $CH_2CO_2{}^tBu$

74-95% from **105** in >90% purity

Scheme 6.34

This is followed by hydrogen peroxide treatment in a pH 8 buffered solution and was designated as a "repair mechanism" that removes over alkylated sites. The dialkylation/peroxide process provided NMAs with greater than 90% purity. It was shown that using equivalent amounts of the boronic ester always resulted in varying degrees of alkylation. However, there are limitations to this procedure in that oxidizable candidates such as methionine were not suitable for these *N*-methylation conditions.

6.4.3 Racemic Syntheses

Up until now the processes reviewed for *N*-methylation of α-amino acids have focused on chiral methodologies whether they include optically active amino acids as starting materials or construction of optically active NMAs through asymmetric synthesis. One obvious facet of NMA chemistry is the propensity of NMAs to epimerize through certain reactions and the majority of the synthetic routes discussed thus far have been developed in order to eliminate this problem. Racemic amino acids are rarely employed in synthetic applications but have been evaluated as potential therapeutics [97]. One obvious disadvantage of racemic mixtures is that if single enantiomers are required, a resolution process must follow. However, one advantage of racemic substrates is that conditions which usually racemize amino acids and, in particular NMAs, are compatible with racemic syntheses.

The earliest account of NMA synthesis via azlactam intermediates was reported by Guerrero *et al.* [98] who employed intermediate **106** in the synthesis of *N*-methyl-3,4-dihydroxyphenylalanine (DOPA) **107** (Scheme 6.35) as opposed to the azlactone asymmetric synthesis developed by Poisel and Schmidt [88] several decades later. Creatinine **108** and vanillin **109** were condensed under classical azlactone conditions with acetic anhydride (dehydrating source) and fused sodium acetate (base) to provide the azlactam **106**. By utilizing creatinine **108** the *N*-methyl group is already in place. Reduction of the benzal group with sodium amalgam also results in concomitant removal of the acetate. Hydrolysis with barium hydroxide removes the formamidine moiety, and final reduction with red phosphorus and hydroiodic acid provided racemic *N*-methyl-DOPA **107**.

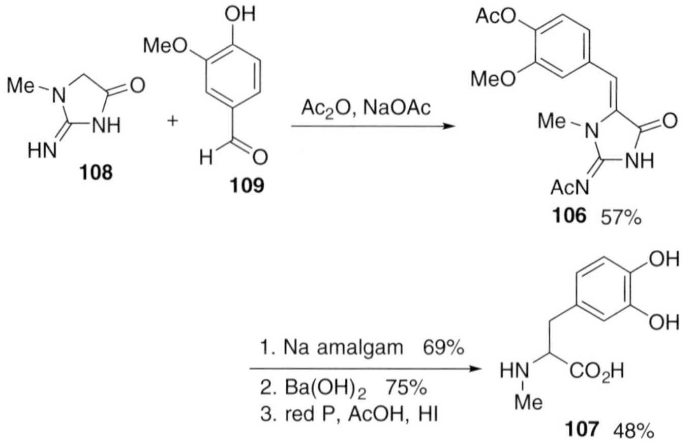

Scheme 6.35

Racemic unsaturated alkyl side-chain NMAs have been synthesized through a sulfone intermediate applied by Alonso *et al.* [99]. The *N*-Boc-*N*-methylsulfone **110** (Scheme 6.36) was lithiated and quenched with ethyl chloroformate providing the *N*-Boc-NMA ester **111**. The intermediate **111** was used to synthesize various unsaturated NMAs via two paths. One path involved the palladium catalyzed allylation with various allyl carbonates and the other involved epoxide ring opening of 2-vinyloxirane, affording α-tosyl-γ,δ-unsaturated-*N*-methyl amino acids **112** and **113**, respectively. The sulfone derivatives **112** and **113** were then treated with magnesium powder in methanol which caused desulfonylation at room temperature, as these intermediates were unstable. It was revealed that the nucleophilic substitutions were highly regioselective and completely stereoselective for compounds **114b**, **c**, **e**, and **f**, affording only the *E*-stereoisomers.

A novel guanidine-based NMA was synthesized by Larsen *et al.* [97] based on analogs of the antidiabetic/antiobesity agent, 3-guanidinopropionic acid **115** (Scheme 6.37). The racemic aminonitrile **117** was synthesized from the starting aldehyde **116**, and then oxidized to the carboxylic acid and hydrolyzed with HCl to the dihydrochloride salt **118**. The dihydrochloride salt **118** was derivatized with the guanilating agent 2-methyl-2-thiopseudourea sulfate and the *N*-benzyl group was removed under standard hydrogenation conditions to provide the NMA **119**.

As mentioned in the introduction to this section, if single enantiomers are required from a racemic synthesis then a resolution of racemate is to follow. Groeger *et al.* [100] synthesized a variety of racemic *N*-methylaminonitriles by condensing various aldehydes with methylamine and hydrogen cyanide in yields ranging from 79 to 89% (Scheme 6.38). Hydrolysis of the nitriles to the carboxylic acids and chloroacetylation provided the *N*-protected NMAs, that were resolved by the *N*-acyl-L-proline-acylase specific for (*S*)-configured NMAs. A complete loss of activity was observed if the substrate had α-substituents that were longer than two carbons or branched.

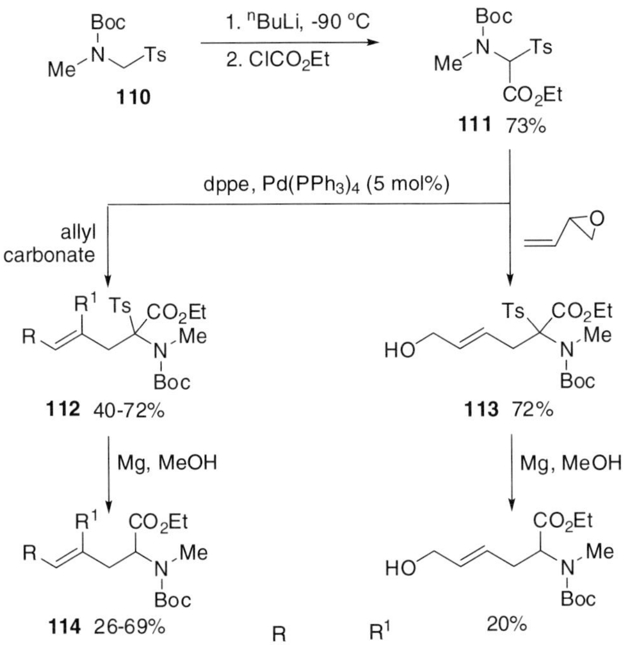

	R	R[1]
a	H	H
b	Me	H
c	Me	H
d	H	Me
e	Ph	H
f	CH_2=CH	H

dppe=1,2-bis (diphenylphosphine) ethane

Scheme 6.36

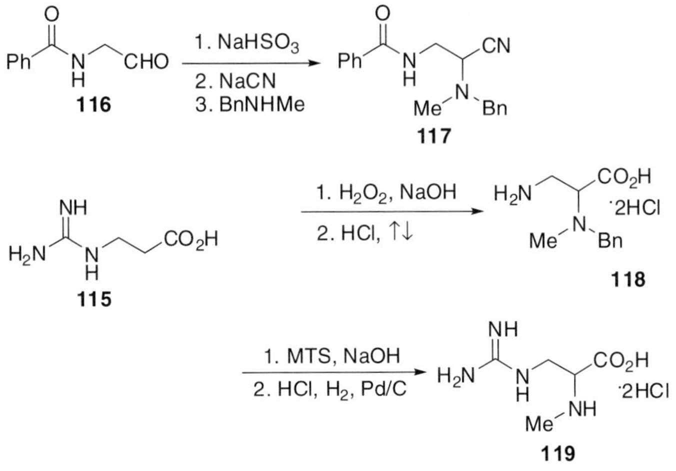

MTS=2-methyl-2-thiopseudourea sulfate

Scheme 6.37

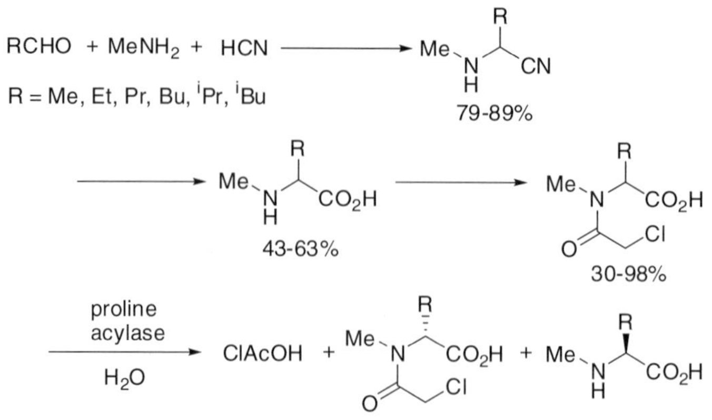

RCHO + $MeNH_2$ + HCN → Me-NH-CH(R)-CN (79-89%)

R = Me, Et, Pr, Bu, ^iPr, ^iBu

→ Me-NH-CH(R)-CO_2H (43-63%) → Me-N(COCH$_2$Cl)-CH(R)-CO_2H (30-98%)

proline acylase, H_2O → ClAcOH + Me-N(COCH$_2$Cl)-CH(R)-CO_2H + Me-NH-CH(R)-CO_2H

Scheme 6.38

6.5
N-Alkylation of Amino Acids

The following sections are devoted to the synthesis of *N*-alkyl α-amino acids with alkyl chain lengths longer than one carbon unit. One of the first methods for synthesizing various *N*-alkyl α-amino acids was reported in 1949 by Gal [101], who treated various racemic α-bromo acids with alkyl amines. Many of the methods described earlier for *N*-methylating α-amino acids are applicable for *N*-alkylation and in particular the techniques described in Section 6.3. *N*-Methylation via Schiff's base reduction is most pertinent to this topic. The work of Bowman [52–54] described in Section 6.3, has shown the ease with which mono *N*-alkylating α-amino acids with various aldehydes is compared to mono *N*-methylation under palladium catalyzed hydrogenation conditions. One should note that the area of *N*-alkylation has not been studied in as great detail as *N*-methylation since these substituents are rarely seen in natural products; however, it is an attractive functional group in materials and peptidomimetic chemistry. Some *N*-alkylated amino acids are used as *N*-protecting groups and in particular the benzyl-type groups (dityl and trityl also) are used in *N*-methylation procedures due to their ease of removal (see Section 6.3). The following sections are broken down by the type of technique employed to install the alkyl substituent rather than by *N*-alkyl α-amino acid since most authors employ a range of alkyl groups in a technique.

6.5.1
Borohydride Reduction of Schiff's Bases

This section comprises the majority of techniques that are used to synthesize *N*-alkyl α-amino acids. This simple procedure of adding an aldehyde to an amino acid in basic, acidic, or neutral media and reducing the resultant Schiff's base with a borohydride reducing agent offers the chemist a practical means to prepare *N*-alkyl α-amino acids by using simple zwitterionic amino acids without carboxyl protection.

Again, dialkylation is not problematic since steric hindrance does not allow for this to occur.

6.5.1.1 Sodium Borohydride Reductions

Preparation of *N*-benzyl α-amino acids is readily accomplished by the simple expedient of forming the Schiff's base of the amino acid and the appropriate benzaldehyde in aqueous base or alcoholic aqueous base mixtures [20, 102]. After stirring for a period of time to complete the Schiff's base formation, the intermediate is reduced with excess sodium borohydride in portions providing the *N*-benzyl α-amino acids in yields ranging from 40–90% and high purities. This simple procedure allows for excess benzaldehyde to be used without the problem of dialkylation occurring and there is no need for protecting the acid as an ester (Scheme 6.39).

6.5.1.2 Sodium Cyanoborohydride Reductions

In comparison to other reducing agents, sodium cyanoborohydride is the most frequently used borohydride in the reduction of Schiff's bases to form *N*-alkyl α-amino acids. Ohfune *et al.* [48a] used a variety of aldehydes and ketones to form Schiff's bases with five amino acids (methionine, phenylalanine, valine, serine, and glutamic acid) and these were reduced in one pot with sodium cyanoborohydride (Scheme 6.40). Methionine was used as a model compound for optimization of the reaction conditions. It was found methanol was the best solvent to use with 0.7 equiv. of sodium cyanoborohydride to affect complete reduction to the *N*-alkyl species. The products precipitated from the reaction media as the zwitterion and were simply washed with methanol providing essentially pure compound in yields ranging from 51 to 96%. Ando and Shioiri [103] applied the same protocol as Ohfune *et al.* [48a] in the synthesis of *N*-alkyl amino acids and methyl esters. In the case of methyl esters, acetic acid was added to the medium until pH 6 was reached.

Bitan *et al.* [104] synthesized a range of N^{α}-functionalized alkyl amino acids as building units for *N*-backbone cyclic peptides. Aldehydes of varying chain lengths containing heteroatoms are condensed under the conditions of Ohfune *et al.* [48a]. In

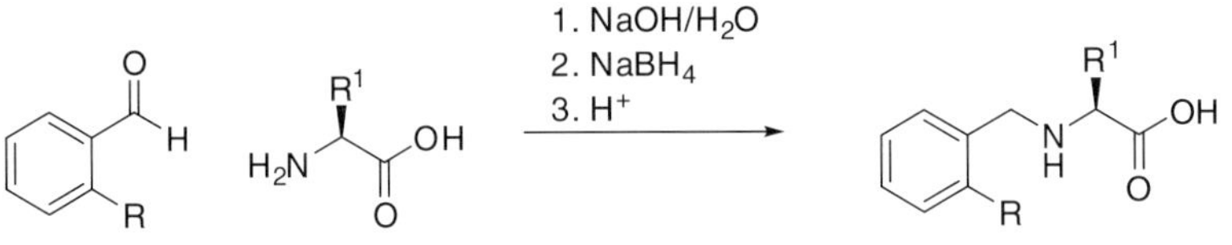

1. $NaOH/H_2O$
2. $NaBH_4$
3. H^+

R = OH, R^1 = gly	R = H, R^1 = ala
R = OH, R^1 = ala	R = H, R^1 = val
R = OH, R^1 = aib	R = H, R^1 = leu
R = OH, R^1 = val	R = H, R^1 = phe
R = OH, R^1 = his	R = H, R^1 = ser
R = OH, R^1 = tyr	R = H, R^1 = lys(Tos)
R = OH, R^1 = trp	R = H, R^1 = lys(Cbz)
	R = H, R^1 = arg(NO_2)

Scheme 6.39

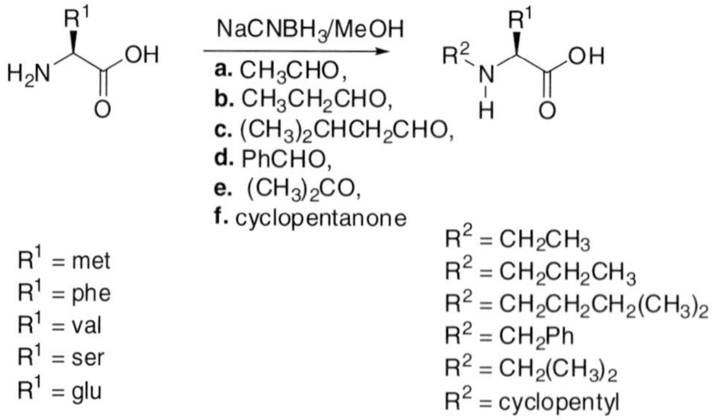

$NaCNBH_3/MeOH$

a. CH_3CHO,
b. CH_3CH_2CHO,
c. $(CH_3)_2CHCH_2CHO$,
d. $PhCHO$,
e. $(CH_3)_2CO$,
f. cyclopentanone

R^1 = met
R^1 = phe
R^1 = val
R^1 = ser
R^1 = glu

$R^2 = CH_2CH_3$
$R^2 = CH_2CH_2CH_3$
$R^2 = CH_2CH_2CH_2(CH_3)_2$
$R^2 = CH_2Ph$
$R^2 = CH_2(CH_3)_2$
R^2 = cyclopentyl

Scheme 6.40

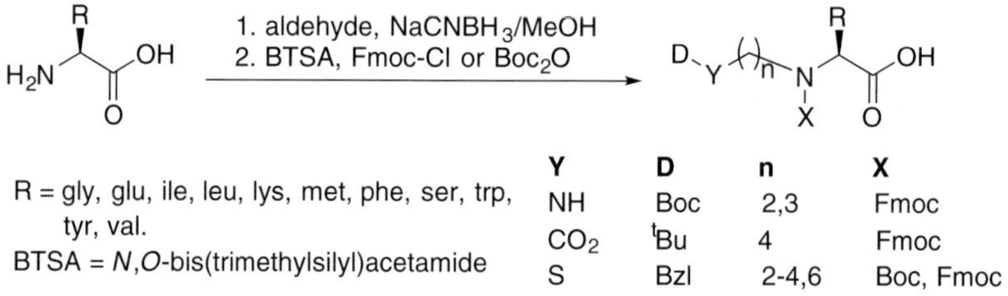

1. aldehyde, $NaCNBH_3/MeOH$
2. BTSA, Fmoc-Cl or Boc_2O

R = gly, glu, ile, leu, lys, met, phe, ser, trp, tyr, val.
BTSA = *N*,*O*-bis(trimethylsilyl)acetamide

Y	D	n	X
NH	Boc	2,3	Fmoc
CO_2	^{t}Bu	4	Fmoc
S	Bzl	2-4,6	Boc, Fmoc

Scheme 6.41

this fashion a number of orthogonally protected *N*-alkyl α-amino acids suitable for Fmoc solid-phase synthesis were produced (Scheme 6.41).

6.5.1.3 Sodium Triacetoxyborohydride Reductions

Ramanjulu and Joullié [49] synthesized various *N*-alkyl amino acid esters employing various aldehydes and reducing the imine thus formed with sodium triacetoxyborohydride. The authors state that better yields and reproducibility of results are obtained with this reducing agent compared to the cyanoborohydride. Rückle *et al.* [50] sought to improve on this technique by synthesizing various *N*-ethyl amino acids using excess acetaldehyde in the dehydrating solvent trimethyl orthoformate (Scheme 6.42). After 30 min imine formation was complete, and the excess acetaldehyde was removed by concentration and then the imine was reduced with excess triacetoxyborohydride in yields ranging from 57 to 85%.

6.5.2 *N*-Alkylation of Sulfonamides

The reader should consult Section 6.2.2 as the techniques described there are related to the alkylation protocol described here.

6.5.2.1 Base-Mediated Alkylation of Benzene Sulfonamides

In 1995, Fukuyama *et al.* [105] reported the *N*-alkylation of nosyl-protected amines via alkylation with alkyl halides and Mitsunobu reaction with alcohols. The nosyl

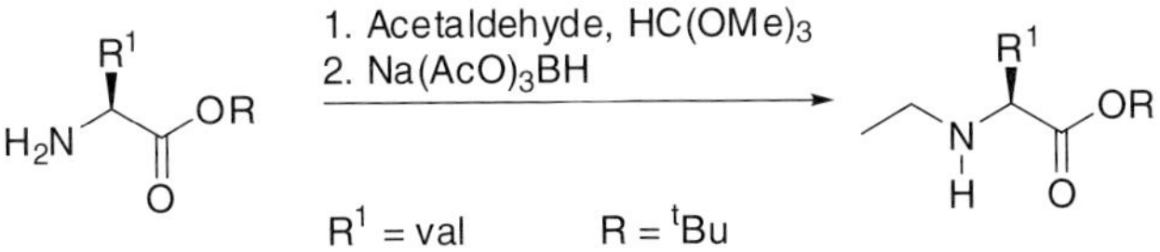

1. Acetaldehyde, $HC(OMe)_3$
2. $Na(AcO)_3BH$

R^1	R
R^1 = val	R = ^tBu
R^1 = val	R = Me
R^1 = leu	R = ^tBu
R^1 = phe	R = ^tBu
R^1 = lys(Cbz)	R = ^tBu
R^1 = thr	R = ^tBu
R^1 = thr	R = Bn
R^1 = thr(OBn)	R = Bn
R^1 = thr(O^{t}Bu)	R = ^tBu

Scheme 6.42

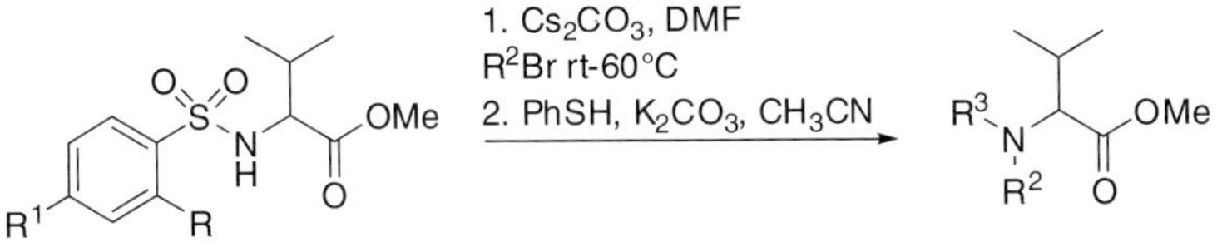

1. Cs_2CO_3, DMF
R^2Br rt-60 °C
2. PhSH, K_2CO_3, CH_3CN

R = NO_2	R^1 = H	R^2 = allyl	R^3 = H
R = NO_2	R^1 = H	R^2 = benzyl	R^3 = H
R = NO_2	R^1 = H	R^2 = butyl	R^3 = H
R = NO_2	R^1 = H	R^2 = 4-pentenyl	R^3 = H
R = H	R^1 = NO_2	R^2 = 4-pentenyl	R^3 = H
R = CN	R^1 = H	R^2 = allyl	R^3 = 4-CNPhSO_2
R = CN	R^1 = H	R^2 = butyl	R^3 = 4-CNPhSO_2

Scheme 6.43

protection protocol has many applications in amino acid synthesis, due to its ease of introduction and removal (see Section 6.2.2). Bhatt *et al.* [106] exploited this with their synthesis of oxopiperazines, and Bowman and Coghlan [107] revealed the ease with which unsaturated alkyl groups can be installed by using D- and L-valine methyl esters as models (Scheme 6.43). The authors describe two sets of conditions used to install the alkyl groups as "mild" in which the alkylation takes place at room temperature and "vigorous" when taking place at 60 °C. The 4-cyanobenzenesulfonyl group was also utilized in the protocol to overcome the sluggish allylation and butylation in the nosyl series. Although the alkylations were rapid and high yielding the conditions to remove the nosyl derivatives (PhSH, K_2CO_3, CH_3CN) did not remove the 4-cyano-benzenesulfonyl group and resulted in decomposition [107].

6.5.3
Reduction of *N*-Acyl Amino Acids

The reader should consult Section 6.3.5 as some of the methods described include examples of the *N*-alkyl amino acids from the acylated precursors; in particular,

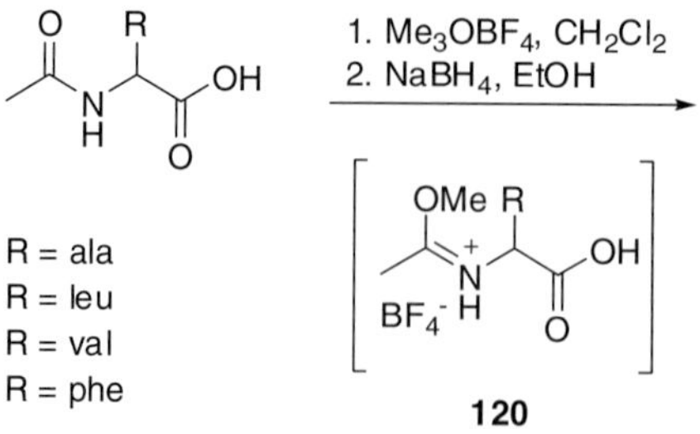

1. Me_3OBF_4, CH_2Cl_2
2. $NaBH_4$, EtOH

R = ala
R = leu
R = val
R = phe

BF_4^-

120

Scheme 6.44

the borane reduction of amides by Hall *et al.* [70] in solution and solid phase producing *N*-ethyl and *N*-propyl amino acid derivatives.

6.5.3.1 Reduction of Acetamides

The pioneering work of Benoiton *et al.* [31–34] in the *N*-methylation of carbamates and amides using sodium hydride and methyl iodide has been exploited in the synthesis of many natural products, and has been accepted as a mild procedure for *N*-methylating amino acid derivatives. Chen and Benoiton [66] have also devised a mild room temperature reduction of acetamides using Meerweins reagent [108] (trimethyloxonium tetrafluoroborate, Scheme 6.44). By treating the acetamides with trimethyloxonium tetrafluoroborate an imino ether fluoroborate intermediate **120** forms that is reduced to the alkyl substituent with sodium borohydride. In this fashion *N*-ethyl-D,L-amino acids were isolated in 41–55% yield. One chiral amino acid *N*-Ac-L-Leu-OH was submitted to the reaction conditions and the isolated *N*-ethyl-L-Leu-OH showed an optical rotation similar to a reported value, yet no other information on the chiral purity of this derivative was communicated.

6.5.4 Novel Methods for *N*-Alkylating α-Amino Acids

Only a few reports on the synthesis of *N*-alkyl amino acid synthesis by novel methods have been published and the reader is urged to refer to Section 6.4 as, again, some of the methods there are applicable to *N*-alkylation.

6.5.4.1 Asymmetric Synthesis of *N*-Alkyl α-Amino Acids

One report by Kadyrov *et al.* [109] describes the asymmetric synthesis of *N*-benzyl amino acids by condensing different α-keto acids with benzyl amines and reducing the imines enantioselectively with rhodium catalysts under pressure in a hydrogen atmosphere (Scheme 6.45). After optimization of the reaction conditions and utilization of suitable chiral ligands (Scheme 6.45) a range of chiral (*R*) amino acids **121** were produced in good yield and up to 98% e.e.

6.5.4.2 *N*-Alkylation of 1,3-Oxazolidin-5-ones

The 2,2-bis(trifluoromethyl)-1,3-oxazolidin-5-ones (Scheme 6.24) employed by Spengler and Burger [82] in the synthesis of *N*-methylamino acids were utilized by

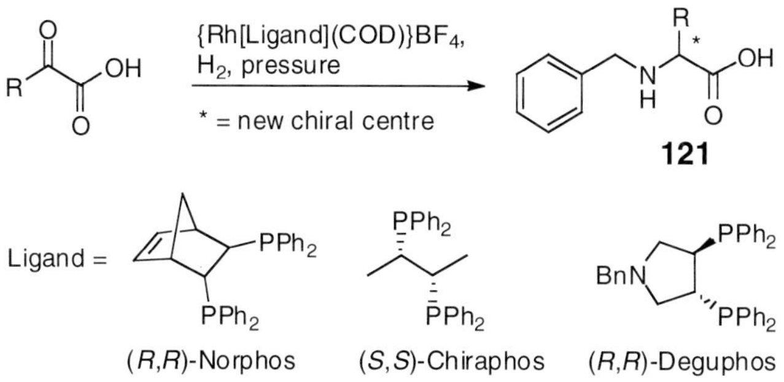

{Rh[Ligand](COD)}BF_4, H_2, pressure

* = new chiral centre

121

Ligand = (R,R)-Norphos (S,S)-Chiraphos (R,R)-Deguphos

R = $PhCH_2$, Me, Ph, $HOOCCH_2CH_2$, $HOOCCH_2$, $PhCH_2CH_2$, Me_2CHCH_2, Me_3CCH_2

Scheme 6.45

Schedel and Burger [110] in the synthesis of *N*-ethylamino acids (Scheme 6.46). By treating the 1,3-oxazolidin-5-ones with Cu(I) cyanide and 2 equiv. of methyl lithium the authors found that 20–50% of the NMA was formed along with the *N*-ethyl species. By reducing methyl lithium to one equivalent, high yields of the *N*-ethyl-2,2-bis(trifluoromethyl)-1,3-oxazolidin-5-ones were isolated and only trace amounts of the *N*-methyl derivative were detected. It was also shown that the free *N*-ethylamino acid could be isolated by treating the oxazolidin-5-one with dilute HCl or hydroxamates by treatment with hydroxylamine and dipeptide amides by exposure to amino acid amides.

It is well documented that the synthesis of *N*-methylamino acids is accompanied by difficulties that arise by installing a single methyl unit at the α-nitrogen. On the other hand, *N*-alkylation with more than one carbon unit has been shown to be relatively straightforward and less transformations are involved, particularly with reductive amination reactions as described by Ohfune *et al.* [48a]. The area of *N*-alkylation is not as broadly studied as *N*-methylation and this can be attributed to the fact that *N*-methylamino acids are highly prominent in natural products compared to *N*-alkylamino acids.

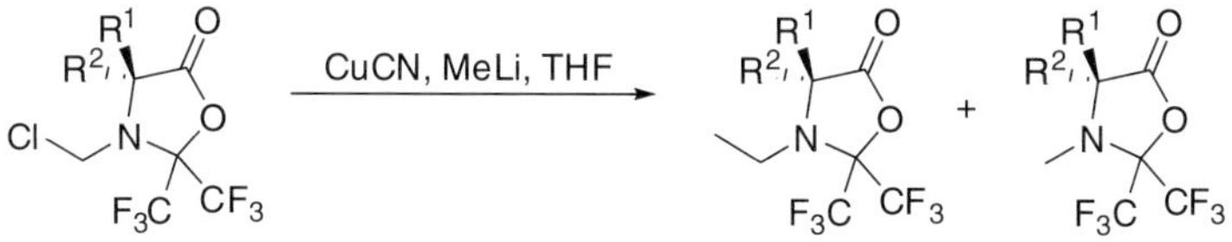

CuCN, MeLi, THF

$R^1 = CH_3$ $R^2 = H$
$R^1 = CH(CH_3)_2$ $R^2 = H$
$R^1 = CH_2CH(CH_3)_2$ $R^2 = H$
$R^1 = H$ $R^2 = Ph$
$R^1 = Ph$ $R^2 = H$

Scheme 6.46

References

1 Hughes, E., Burke, R.M., and Doig, A.J. (2000) *The Journal of Biological Chemistry*, **275**, 25109.

2 Doig, A.J., Hughes, E., Burke, R.M., Su, T.J., Heenan, R.K., and Lu, J. (2002) *Biochemical Society Transactions*, **30**, 537.

3 Mason, J.M., Kokkoni, N., Stott, K., and Doig, A.J. (2003) *Current Opinion in Structural Biology*, **13**, 526.

4 Gordon, D.J., Tappe, R., and Meredith, S.C. (2002) *The Journal of Peptide Research*, **60**, 37.

5 Kapurniotu, A., Schmauder, A., and Tenidis, K. (2002) *Journal of Molecular Biology*, **315**, 339.

6 Fairlie, D.P., Abbenante, G., and March, D.R. (1995) *Current Medicinal Chemistry*, **2**, 654.

7 Ostresh, J.M., Husar, G.M., Blondelle, S., Dorner, B., Weber, P.A., and Houghten, R.A. (1994) *Proceedings of the National Academy of Sciences of the United States of America*, **91**, 11138.

8 Miller, S.M., Simon, R.J., Ng, S., Zuckermann, R.N., Kerr, J.M., and Moos, W.H. (1995) *Drug Development Research*, **35**, 20.

9 Turker, R.K., Hall, M.M., Yamamoto, M., Sweet, C.S., and Bumpus, F.M. (1972) *Science*, **177**, 1203.

10 Haviv, F., Fitzpatrick, T.D., Swenson, R.E., Nichols, C.J., Mort, N.A., Bush, E.N., Diaz, G., Bammert, G., Nguyen, A., Rhutasel, N.S., Nellans, H.N., Hoffman, D.J., Johnson, E.S., and Greer, J. (1993) *Journal of Medicinal Chemistry*, **36**, 363.

11 Cody, W.L., He, J.X., Reily, M.D., Haleen, S.J., Walker, D.M., Reyner, E.L., Stewart, B.H., and Doherty, A.M. (1997) *Journal of Medicinal Chemistry*, **40**, 2228.

12 Payne, J.W. (1972) *Journal of General Microbiology*, **71**, 259.

13 Fischer, E. and Lipschitz, W. (1915) *Chemische Berichte*, **48**, 360.

14 Fischer, E. and Mechel, L.V. (1916) *Chemische Berichte*, **49**, 1355.

15 Izumiya, N. and Nagamatsu, A. (1952) *Bulletin of the Chemical Society of Japan*, **25**, 265.

16 (a) Winstein, S. (1939) *Journal of the American Chemical Society*, **61**, 1635; (b) Winstein, S. and Lucas, H.J. (1939) *Journal of the American Chemical Society*, **61**, 2845; (c) Brewster, P., Hiron, F., Hughes, E.D., Ingold, C.K., and Rao, P.A.D.S. (1950) *Nature*, **166**, 179.

17 Grossman, R.B. (1999) *The Art of Writing Reasonable Organic Reaction Mechanisms*, Springer, New York.

18 (a) Izumiya, N. (1952) *Kyushu Memoirs of Medical Science*, **3**, 1; (b) Izumiya, N. (1951) *Journal of the Chemical Society of Japan (Nippon Kagaku Zasshi)*, **72**, 550; (c) Izumiya, N. (1951) *Journal of the Chemical Society of Japan (Nippon Kagaku Zasshi)*, **72**, 784; (d) Izumiya, N. (1951) *Journal of the Chemical Society of Japan (Nippon Kagaku Zasshi)*, **72**, 26; (e) Izumiya, N. (1951) *Journal of the Chemical Society of Japan (Nippon Kagaku Zasshi)*, **72**, 700.

19 Effenberger, F., Burkard, U., and Willfahrt, J. (1986) *Liebigs Annalen der Chemie*, 314.

20 Quitt, P., Hellerbach, J., and Vogler, K. (1963) *Helvetica Chimica Acta*, **46**, 327; (b) Quitt, P. (1963) In *Peptides: Proceedings of the 5th European Symposium*, Pergamon, Oxford, p. 165.

21 Hlavácek, J., Poduska, K., Sorm, F., and Sláma, K. (1976) *Collection of Czechoslovak Chemical Communications*, **41**, 2079.

22 Hlavácek, J., Fric, I., Budesinsky, M., and Bláha, K. (1988) *Collection of Czechoslovak Chemical Communications*, **53**, 2473.

23 Miller, S.C. and Scanlan, T.S. (1997) *Journal of the American Chemical Society*, **119**, 2301.

24 Albanese, D., Landini, D., Lupi, V., and Penso, M. (2000) *European Journal of Organic Chemistry*, 1443.

25 Biron, E. and Kessler, H. (2005) *The Journal of Organic Chemistry*, **70**, 5183.
26 (a) Di Gioia, M.L., Leggio, A., Le Pera, A., Liguori, A., Napoli, A., Siciliano, C., and Sindona, G. (2003) *The Journal of Organic Chemistry*, **68**, 7416; (b) Belsito, E., Di Gioia, M.L., Greco, A., Leggio, A., Liguori, A., Perri, F., Siciliano, C., and Viscomi, M.C. (2007) *The Journal of Organic Chemistry*, **72**, 4798.
27 Das, B.C., Gero, S.D., and Lederer, E. (1967) *Biochemical and Biophysical Research Communications*, **29**, 211.
28 Olsen, R.K. (1970) *The Journal of Organic Chemistry*, **35**, 1912.
29 Okamoto, K., Abe, H., Kuromizu, K., and Izumiya, N. (1974) *Memoirs of the Faculty of Science, Kyushu University – Series C*, **9**, 131.
30 Tam, J.P., Spetzler, J.C., and Rao, C. (1993) In *Peptides: Biology and Chemistry: Proceedings of the Chinese Peptide Symposium*, ESCOM, Leiden, p. 285.
31 (a) Coggins, J.R. and Benoiton, N.L. (1971) *Canadian Journal of Chemistry*, **49**, 1968; (b) Cheung, S.T. and Benoiton, N.L. (1977) *Canadian Journal of Chemistry*, **55**, 906.
32 Benoiton, N.L., Kuroda, K., Cheung, S.T., and Chen, F.M.F. (1979) *Canadian Journal of Biochemistry*, **57**, 776.
33 McDermott, J.R. and Benoiton, N.L. (1973) *Canadian Journal of Chemistry*, **51**, 2555.
34 McDermott, J.R. and Benoiton, N.L. (1973) *Canadian Journal of Chemistry*, **51**, 2562.
35 McDermott, J.R. and Benoiton, N.L. (1973) *Canadian Journal of Chemistry*, **51**, 1915.
36 Coulton, S., Moore, G.A., and Ramage, R. (1976) *Tetrahedron Letters*, 4005.
37 Stoochnoff, B.A. and Benoiton, N.L. (1973) *Tetrahedron Letters*, 21.
38 Belagali, S.L., Mathew, T., and Himaja, M. (1995) *Indian Journal of Chemistry Section B – Organic Chemistry Including Medicinal Chemistry*, **34**, 45.
39 Burger, K. and Hollweck, W. (1994) *Synlett*, 751.
40 Prashad, M., Har, D., Hu, B., Kim, H.-Y., Repic, O., and Blacklock, T.J. (2003) *Organic Letters*, **5**, 125.
41 Liu, S., Gu, W., Lo, D., Ding, X.-Z., Ujiki, M., Adrian, T.E., Soff, G.A., and Silverman, R.B. (2005) *Journal of Medicinal Chemistry*, **48**, 3630.
42 Papaioannou, D., Athanassopoulos, C., Magafa, V., Karamanos, N., Stavropoulos, G., Napoli, A., Sindona, G., Aksnes, D.W., and Francis, G.W. (1994) *Acta Chemica Scandinavica*, **48**, 324.
43 Mitsunobu, O. (1981) *Synthesis*, 1.
44 Olah, G.A. and Narang, S.C. (1982) *Tetrahedron*, **38**, 2225.
45 Wisniewski, K. and Kolodziejczyk, A.S. (1997) *Tetrahedron Letters*, **38**, 483.
46 (a) Yang, L. and Chiu, K. (1997) In *Proceedings of the 15th American Peptide Symposium*, Kluwer, Boston, MA, p. 341; (b) Yang, L. and Chiu, K. (1997) *Tetrahedron Letters*, **38**, 7307.
47 Chruma, J.J., Sames, D., and Polt, R. (1997) *Tetrahedron Letters*, **38**, 5085.
48 (a) Ohfune, Y., Kurokawa, N., Higuchi, N., Saito, M., Hashimoto, M., and Tanaka, T. (1984) *Chemistry Letters*, 441; (b) Ohfune, Y., Higuchi, N., Saito, M., Hashimoto, M., and Tanaka, T. (1984) *Peptide Chemistry*, 89.
49 Ramanjulu, J.M. and Joullié, M.M. (1996) *Synthetic Communications*, **26**, 1379.
50 Rückle, T., Dubray, B., Hubler, F., and Mutter, M. (1999) *Journal of Peptide Science*, **5**, 56.
51 Keller-Schierlein, W., Hagmann, L., Zähner, H., and Huhn, W. (1988) *Helvetica Chimica Acta*, **71**, 1528.
52 Bowman, R.E. and Stroud, H.H. (1950) *Journal of the Chemical Society*, 1342.
53 Bowman, R.E. (1950) *Journal of the Chemical Society*, 1346.
54 Bowman, R.E. (1950) *Journal of the Chemical Society*, 1349.
55 Ikutani, Y. (1968) *Bulletin of the Chemical Society of Japan*, **41**, 1679.
56 Poduska, K. (1958) *Chemicke Listy*, **52**, 153.

57 Ebata, M., Takahashi, Y., and Otsuka, H. (1966) *Bulletin of the Chemical Society of Japan*, **39**, 2535.

58 Brockmann, H. and Lackner, H. (1967) *Chemische Berichte*, **100**, 353.

59 Eloff, J.N. (1980) *Zeitschrift fur Pflanzenphysiologie*, **98**, 411.

60 Peter, H., Brugger, M., Schreiber, J., and Eschenmoser, A. (1963) *Helvetica Chimica Acta*, **46**, 577.

61 O'Donnell, M.J., Bruder, W.A., Daugherty, B.W., Liu, D., and Wojciechowski, K. (1984) *Tetrahedron Letters*, **25**, 3651.

62 Jentoft, N. and Dearborn, D.G. (1983) *Methods in Enzymology*, **91**, 570.

63 Lane, C.F. (1975) *Synthesis*, 135.

64 Kaljuste, K. and Undén, A. (1993) *International Journal of Peptide and Protein Research*, **42**, 118.

65 Hanson, R.W. and Law, H.D. (1965) *Journal of the Chemical Society*, 7285.

66 Chen, F.M.F. and Benoiton, N.L. (1977) *Canadian Journal of Chemistry*, **55**, 1433.

67 Krishnamurthy, S. (1982) *Tetrahedron Letters*, **23**, 3315.

68 McKennon, M.J., Meyers, A.I., Drauz, K., and Schwarm, M. (1993) *The Journal of Organic Chemistry*, **58**, 3568.

69 Chu, K.S., Negrete, G.R., and Konopelski, J.P. (1991) *The Journal of Organic Chemistry*, **56**, 5196.

70 Hall, D.G., Laplante, C., Manku, S., and Nagendran, J. (1999) *The Journal of Organic Chemistry*, **64**, 698.

71 Ben-Ishai, D. (1957) *Journal of the American Chemical Society*, **79**, 5736.

72 Itoh, M. (1969) *Chemical & Pharmaceutical Bulletin*, **17**, 1679.

73 Allevi, P., Cighetti, G., and Anastasia, M. (2001) *Tetrahedron Letters*, **42**, 5319.

74 Freidinger, R.M., Hinkle, J.S., Perlow, D.S., and Arison, B.H. (1983) *The Journal of Organic Chemistry*, **48**, 77.

75 Auerbach, J., Zamore, M., and Weinreb, S.M. (1976) *The Journal of Organic Chemistry*, **41**, 725.

76 Cipens, G., Slavinskaya, V.A., Sile, D., Korchagova, E.K., Katkevich, M.Y., and Grigoreva, V.D. (1992) *Khimiya Geterotsiklicheskikh Soedinenii*, 681.

77 (a) Aurelio, L., Brownlee, R.T.C., Hughes, A.B., and Sleebs, B.E. (2000) *Australian Journal of Chemistry*, **53**, 425; (b) Aurelio, L., Brownlee, R.T.C., and Hughes, A.B. (2002) *Organic Letters*, **4**, 3767; (c) Aurelio, L., Box, J.S., Brownlee, R.T.C., Hughes, A.B., and Sleebs, M.M. (2003) *The Journal of Organic Chemistry*, **68**, 2652.

78 (a) Reddy, G.V., Rao, G.V., and Iyengar, D.S. (1998) *Tetrahedron Letters*, **39**, 1985; (b) Reddy, G.V. and Iyengar, D.S. (1999) *Chemistry Letters*, 299.

79 Williams, R.M. and Yuan, C. (1994) *The Journal of Organic Chemistry*, **59**, 6190.

80 Willuhn, M., Platzek, J., Ottow, E., Petrov, O., Borm, C., Hinz, D., Mann, G., Lister-James, J., and Wilson, D.M. (2003) PCT WO03042163.

81 Luke, R.W.A., Boyce, P.G.T., and Dorling, E.K. (1996) *Tetrahedron Letters*, **37**, 263.

82 (a) Spengler, J. and Burger, K. (1998) *Synthesis*, 67; (b) Burger, K. and Spengler, J. (2000) *European Journal of Organic Chemistry*, 199; (c) Burger, K., Spengler, J., Hennig, L., Herzschuh, R., and Essawy, S.A. (2000) *Monatshefte fur Chemie*, **131**, 463.

83 Yamashiro, D., Aanning, H.L., Branda, L.A., Cash, W.D., Murti, V.V.S., and Du Vigneaud, V. (1968) *Journal of the American Chemical Society*, **90**, 4141.

84 Liu, J.-F., Tang, X.-X., and Jiang, B. (2002) *Synthesis*, 1499.

85 Ratner, S. and Clarke, H.T. (1937) *Journal of the American Chemical Society*, **59**, 200.

86 Ruggles, E.L., Flemer, S. Jr., and Hondal, R.J. (2008) *Biopolymers*, **90**, 61.

87 (a) Zhang, S., Govender, T., Norström, T., and Arvidsson, P.I. (2005) *The Journal of Organic Chemistry*, **70**, 6918; (b) Govender, T. and Arvidsson, P.I. (2006) *Tetrahedron Letters*, **47**, 1691.

88 Poisel, H. and Schmidt, U. (1973) *Chemische Berichte*, **106**, 3408.

89 Pandey, G., Reddy, P.Y., and Das, P. (1996) *Tetrahedron Letters*, **37**, 3175.

90 Agami, C., Couty, F., Hamon, L., Prince, B., and Puchot, C. (1990) *Tetrahedron*, **46**, 7003.

91 Agami, C., Couty, F., Prince, B., and Puchot, C. (1991) *Tetrahedron*, **47**, 4343.

92 Oppolzer, W., Cintas-Moreno, P., Tamura, O., and Cardinaux, F. (1993) *Helvetica Chimica Acta*, **76**, 187.

93 Myers, A.G., Gleason, J.L., Yoon, T., and Kung, D.W. (1997) *Journal of the American Chemical Society*, **119**, 656.

94 Grieco, P.A. and Bahsas, A. (1987) *The Journal of Organic Chemistry*, **52**, 5746.

95 Dorow, R.L. and Gingrich, D.E. (1995) *The Journal of Organic Chemistry*, **60**, 4986.

96 Laplante, C. and Hall, D.G. (2001) *Organic Letters*, **3**, 1487.

97 Larsen, S.D., Connell, M.A., Cudahy, M.M., Evans, B.R., May, P.D., Meglasson, M.D., O'Sullivan, T.J., Schostarez, H.J., Sih, J.C., Stevens, F.C., Tanis, S.P., Tegley, C.M., Tucker, J.A., Vaillancourt, V.A., Vidmar, T.J., Watt, W., and Yu, J.H. (2001) *Journal of Medicinal Chemistry*, **44**, 1217.

98 Guerrero, T.H. and Deulofeu, V. (1937) *Chemische Berichte*, **70**, 947.

99 Alonso, D.A., Costa, A., and Nájera, C. (1997) *Tetrahedron Letters*, **38**, 7943.

100 Groeger, U., Drauz, K., and Klenk, H. (1992) *Angewandte Chemie (International Edition in English)*, **31**, 195.

101 Gal, E.M. (1949) *Journal of the American Chemical Society*, **71**, 2253.

102 Yang, C.-T., Vetrichelvan, M., Yang, X., Moubaraki, B., Murray, K.S., and Vittal, J.J. (2004) *Journal of The Chemical Society, Dalton Transactions*, 113.

103 Ando, A. and Shioiri, T. (1989) *Tetrahedron*, **45**, 4969.

104 Bitan, G., Muller, D., Kasher, R., Gluhov, E.V., and Gilon, C. (1997) *Journal of the Chemical Society, Perkin Transactions 1*, 1501.

105 Fukuyama, T., Jow, C.-K., and Cheung, M. (1995) *Tetrahedron Letters*, **36**, 6373.

106 Bhatt, U., Mohamed, N., Just, G., and Roberts, G. (1997) *Tetrahedron Letters*, **38**, 3679.

107 Bowman, W.R. and Coghlan, D.R. (1997) *Tetrahedron*, **53**, 15787.

108 Meerwein, H., Hinz, G., Hofmann, P., Kroning, E., and Pfeil, E. (1937) *Journal fur Praktische Chemie*, **147**, 257.

109 Kadyrov, R., Riermeier, T.H., Dingerdissen, U., Tararov, V., and Börner, A. (2003) *The Journal of Organic Chemistry*, **68**, 4067.

110 Schedel, H. and Burger, K. (2000) *Monatshefte fur Chemie*, **131**, 1011.

7
Recent Developments in the Synthesis of β-Amino Acids

Yamir Bandala and Eusebio Juaristi

7.1
Introduction

In the last two decades the synthesis of β-amino acids has been of increasing interest owing to their presence in biologically active compounds with outstanding pharmacological properties, either in free form or as constituents in various peptides. Indeed, β-amino acids are the main components of several medicinally useful molecules. For these reasons, numerous methodologies for the synthesis of racemic and enantiomerically pure compounds have emerged [1]. The present chapter provides an overview of the more significant advances achieved in the synthesis of β-amino acids – both racemic and enantioenriched – from 2000 to early 2008. These methodologies can be separated into 10 main categories that will be discussed in the subsequent sections (Sections 7.2–7.11).

7.2
Synthesis of β-Amino Acids by Homologation of α-Amino Acids

The Arndt–Eistert synthesis is a process in which a carboxylic acid is transformed into a higher homolog containing one additional carbon atom. The key step in the Arndt–Eistert procedure consists of the catalyzed Wolff rearrangement of a diazoketone to form a ketene derivative. Thus, acid chlorides react with diazomethane to give diazoketones, which in the presence of a nucleophile (e.g., water) and a suitable catalyst afford the carboxylic acid homolog. Light, heat, or transition metal catalysts such as silver oxide have been employed to catalyze the Wolff rearrangement to produce the desired carboxylic acid homolog (Scheme 7.1) [2].

Already in the 1940s it was shown that the Wolff rearrangement of diazoketones containing a center of chirality adjacent to the carbonyl function takes place with retention of configuration. This finding constitutes the basis for the application of the Arndt–Eistert reaction for the synthesis of enantiopure β-amino acids from their α-amino acid counterparts. Taking advantage of the ready availability, low cost,

Amino Acids, Peptides and Proteins in Organic Chemistry. Vol.1 – Origins and Synthesis of Amino Acids.
Edited by Andrew B. Hughes

ISBN: 978-3-527-32096-7

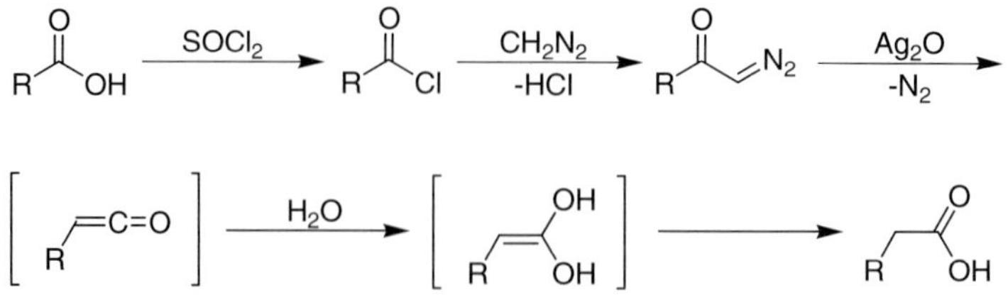

Scheme 7.1 Arndt–Eistert homologation of carboxylic acids via Wolff rearrangement of a diazoketone intermediate.

and high enantiomeric purity of natural α-amino acids, their direct elongation to prepare β-amino acids following the Arndt–Eistert procedure has found many applications [2].

Illustrative examples of this methodology were described by Linder *et al.* [3a] and Achmatowicz *et al.* [3b]. The suitably protected α-amino acid used by Linder *et al.* was activated as the mixed anhydride derivative and treated with diazomethane to produce the corresponding diazoketone; rearrangement in the presence of water furnished the desired β-amino acid [3a]. On the other hand, using D-alanine, D-phenylalanine, and D-valine as starting materials, Achmatowicz *et al.* synthesized several chiral dioxocyclam derivatives. In the first step of this procedure, the amino group of the starting amino acids was protected by treatment with phthalic anhydride to afford the *N*-phthaloyl derivatives, which were converted into the corresponding acid chlorides and then reacted with an excess of diazomethane to give the respective diazoketones. The key transformation (i.e., the Wolff rearrangement) was carried out by treating the methanolic solution of the diazoketone with a catalytic amount of silver benzoate, which induced the formation of the rearranged methyl esters **1** in good yields. β^3-Amino esters **1** were then converted into the desired chiral dioxocyclams (Scheme 7.2) [3b].

In this context, Suresh Babu *et al.* have demonstrated the effectiveness of alternative activating agents for the generation of the diazoketone intermediates [4].

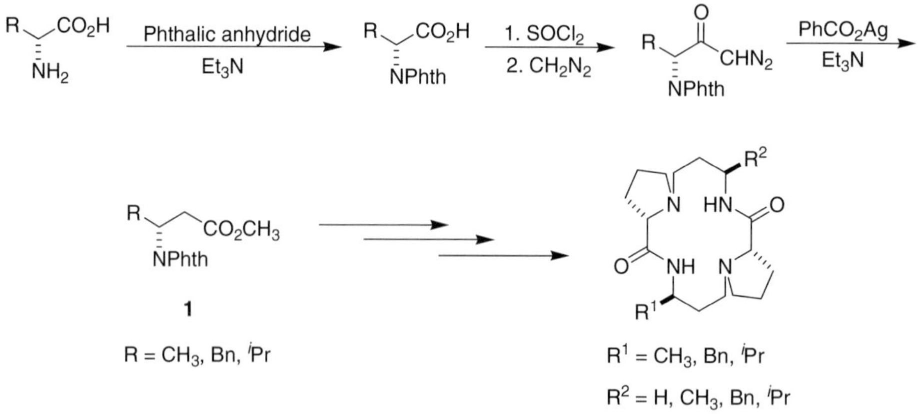

Scheme 7.2

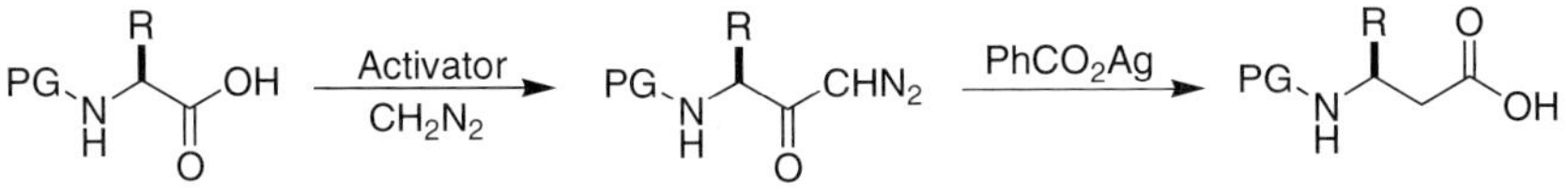

PG = Fmoc, Boc or Cbz

Activator = TBTU, BOP, PyBOP, HBTU/DIEA, p-TsCl, DAST, Boc_2O, FmocCl

Scheme 7.3 TBTU: *O*-benzothiazol-1-yl-*N,N,N′,N′*-tetramethyluronium tetrafluoroborate; BOP: benzotriazolyl-*N*-oxy-tris(dimethylamino)-phosphonium hexafluorophosphate, PyBOP: benzotriazol-1-yl-oxytripyrrolidinephosphonium hexafluorophosphate; HBTU/DIEA: *O*-benzothiazol-1-yl-*N,N,N′,N′*-tetramethyluronium hexafluorophosphate/*N,N*-diisopropylethylamine; p-TsCl: *p*-toluenesulfonyl chloride; DAST: (diethylamino)sulfur trifluoride; FmocCl: 9-fluorenylmethyl chloroformate.

In particular, the use of *O*-benzothiazol-1-yl-*N,N,N′,N′*-tetramethyluronium tetrafluoroborate, benzotriazolyl-*N*-oxy-tris(dimethylamino)phosphonium hexafluorophosphate, benzotriazol-1-yl-oxytripyrrolidinephosphonium hexafluorophosphate, *O*-benzothiazol-1-yl-*N,N,N′,N′*-tetramethyluronium hexafluorophosphate/*N,N*-diisopropylethylamine, *p*-toluenesulfonyl chloride, (diethylamino)sulfur trifluoride, Boc_2O, or 9-fluorenylmethyl chloroformate proved to be effective in the activation of the carboxyl group of *N*-protected α-amino acids to give the Fmoc-, Boc-, or Cbz-protected α-amino diazoketone derivatives that in the presence of $PhCO_2Ag$ in dioxane/H_2O were converted into the corresponding β^3-amino acids (Scheme 7.3) [4].

An interesting procedure for the preparation of *N*-Cbz-*N*-methyl β-amino acids was described for Hughes and Sleebs [5]. In the original approach, *N*-Cbz-α-amino acids were converted into the diazoketone and then treated with CF_3CO_2Ag to give the corresponding β^3-amino acid via Wolff rearrangement. Reaction of the β^3-amino acid with paraformaldehyde under acid catalysis (camphorsulfonic acid and acetic acid) gave the key 1,3-oxazinan-6-ones **2**. Subsequent reductive cleavage, effected with triethylsilane and trifluoroacetic acid (TFA), afforded the expected *N*-methyl β^3-amino acids in moderate yield (Scheme 7.4) [5]. In an alternative strategy, the

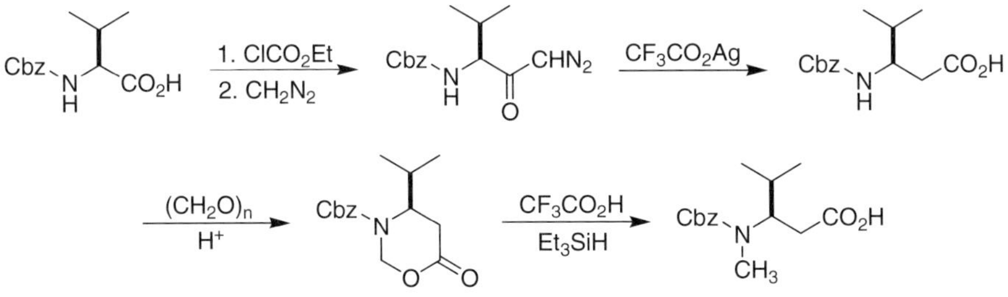

Scheme 7.4

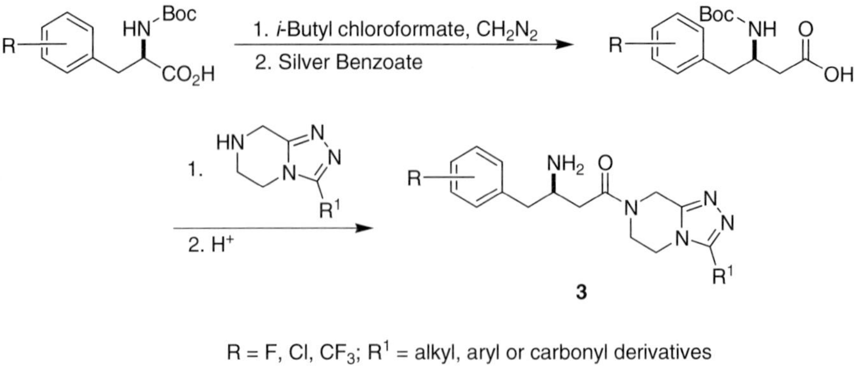

1. *i*-Butyl chloroformate, CH_2N_2
2. Silver Benzoate

1. [triazolopiperazine]
2. H^+

3

R = F, Cl, CF_3; R^1 = alkyl, aryl or carbonyl derivatives

Scheme 7.5

N-methyl α-amino acid was homologated by the Arndt–Eistert procedure to afford the corresponding *N*-methyl β-amino acid [6].

Recently, Govender and Arvidsson developed a microwave-based methodology for the conversion of Fmoc-*N*-protected β^3-amino acids into the corresponding *N*-methyl β^3-amino acid. Indeed, heating the protected β^3-amino acid with paraformaldehyde and *p*-toluene sulfonic acid in acetonitrile under microwave irradiation for 3 min at 130 °C gave the intermediate oxazinanone, which was reduced with $AlCl_3$ and triethylsilane in dry CH_2Cl_2 to afford the Fmoc-*N*-methylated amino acid [7].

A recent application of the homologation strategy in medicinal chemistry was described by Kim *et al.*, who synthesized a series of chiral β^3-amino acids, which by subsequent coupling with 3-substituted triazolopiperazines provided a series of dipeptidyl peptidase inhibitors **3** (Scheme 7.5) [8].

On the other hand, the Arndt–Eistert protocol has been subjected to several modifications. For example, Suresh Babu *et al.* have described the use of microwave irradiation in the Wolff rearrangement step obtaining several Fmoc-, Boc-, or Cbz-β-amino acids with retention of configuration [9]. In the same context, Koch and Podlech described that silica gel and a catalytic amount of silver trifluoroacetate added to the *N*-protected α-amino diazoketone in a suitable solvent produces, in only 15 min, the β-amino acid in excellent yields [10].

Another interesting homologation method for the synthesis of β-amino acids was described by Moumne *et al.* [11] by means of the Reformatsky reaction of configurationally stable α-bromo carboxylic acid esters with dibenzylidene iminium trifluoroacetate (a Mannich-type electrophile). Subsequent debenzylation with ammonium formate over palladium-charcoal was followed by saponification and *N*-Boc protection leading to racemic *N*-protected β^2-amino acids **4** (Scheme 7.6). Enantioselective versions of this reaction have been explored by the authors [11].

Recently, Gray *et al.* [12] obtained several highly enantioenriched β^3-amino esters through application of the Kowalsky ester homologation, which involves an ynolate **5** as a key intermediate (Scheme 7.7).

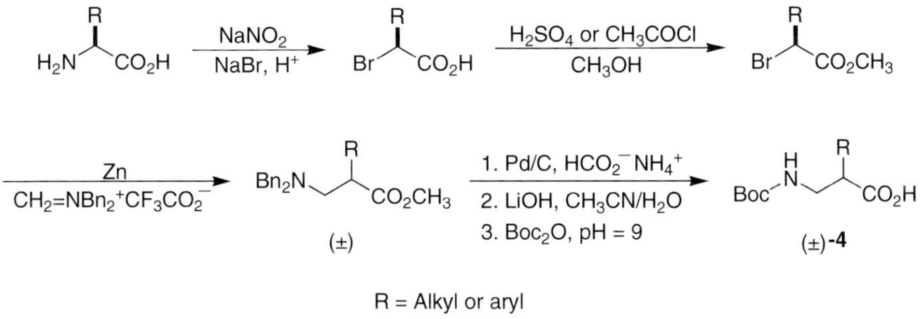

R = Alkyl or aryl

Scheme 7.6

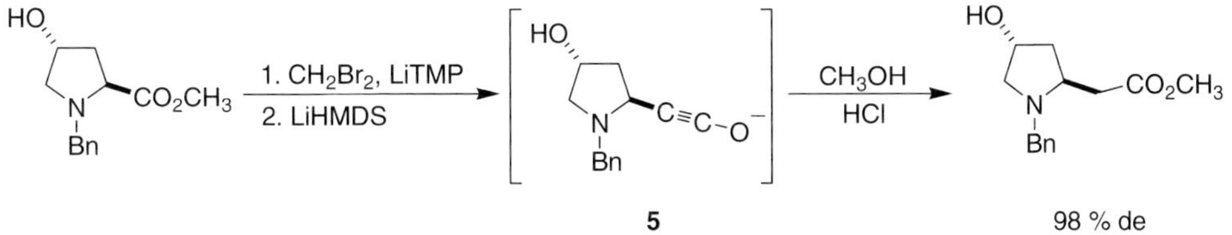

Scheme 7.7

In a recent report, Caputo and Longobardo [13] described the preparation of C-2-deuterated β³-amino acids **6** from *N*-Boc-protected 2-amino alcohols-1,1-d_2, which were obtained by sodium borodeuteride reduction of the parent *N*-Boc-protected α-amino acid mixed anhydrides. The deuterated *N*-Boc-2-aminoalcohols were converted into the corresponding 2-aminoiodides by reaction with triphenylphosphine/iodine complex, in the presence of imidazole. Finally, the aminoiodides were treated with potassium cyanide in anhydrous dimethylsulfoxide at 50 °C, to afford the corresponding β³-amino nitriles, which were then hydrolyzed to give the expected C-2 fully deuterated methyl β³-amino ester hydrochlorides **6**·HCl (Scheme 7.8).

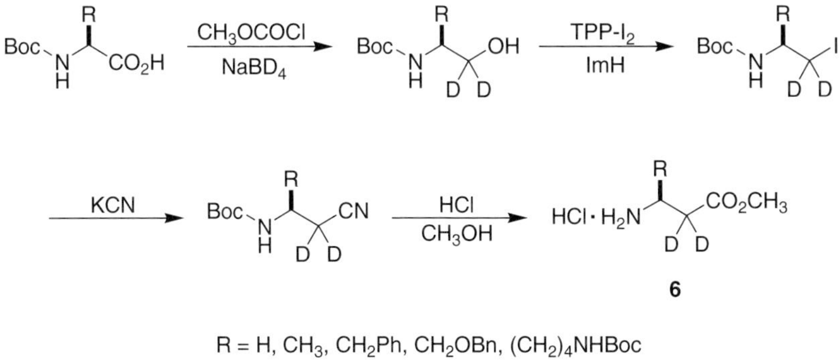

R = H, CH_3, CH_2Ph, CH_2OBn, $(CH_2)_4NHBoc$

Scheme 7.8 TPP-I_2: triphenylphosphine-iodine complex.

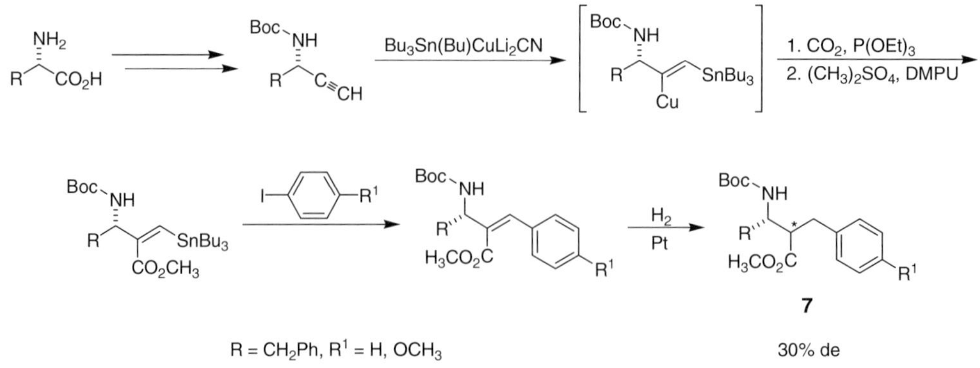

R = CH_2Ph, R^1 = H, OCH_3 30% de

Scheme 7.9 DMPU: 1,3-dimethyl-3,4,5,6-tetrahydro-2(1*H*)-pyrimidinone.

An interesting methodology for the preparation of unsaturated β-amino acids derivatives was described by Reginato *et al.* [14], who converted α-amino acids into the corresponding aldehydes followed by homologation to the alkyne derivative. Addition of stannylcuprate to this alkynyl derivative gave an intermediate which could be trapped with a suitable electrophile to introduce the carboxylic functionality, obtaining in this way the β-amino esters **7** (Scheme 7.9).

In an interesting development, Seo *et al.* [15] obtained α-hydroxy-β-amino acids **8** from the corresponding α-amino acids. This synthetic procedure begins with the chemoselective reduction of *N*-Boc-*N*-Bn-protected α-amino esters with lithium aluminum hydride. Swern oxidation of the resulting (2-hydroxyethyl) carbamates furnished the corresponding chiral aldehydes, which underwent an *E*-selective Horner–Wadsworth–Emmons reaction to afford allylic carbamates. Chemoselective reduction of the ester moiety with diisobutylaluminum hydride generated the expected allylic alcohols, which were converted into the corresponding oxazolidinones with trifluoromethanesulfonic anhydride in pyridine at −78 °C. The final conversion of the oxazolidinone products into the corresponding α-hydroxy-β-amino acids was achieved by dihydroxylation of the vinyl oxazolidinones with osmium tetroxide followed by cleavage in the presence of periodate, and subsequent oxidation with potassium permanganate. The oxazolidinone ring in these products was easily cleaved without epimerization at the C-5 position, by reflux in aqueous potassium hydroxide, to give *N*-benzyl-protected α-hydroxy-β-amino acids in good yield (75%). Finally, the amino group was deprotected by hydrogenolysis to give the desired free α-hydroxy-β-amino acids **8** (Scheme 7.10).

Sánchez-Obregón *et al.* [16] synthesized enantiomerically pure (2*R*,3*S*)-3-amino-2-hydroxybutanoic acid **9** from available (*S*)-alanine sulfoxide derivatives. The synthetic sequence involves five steps: (i) stereoselective reduction of the starting β-keto sulfoxide, (ii) protection of the resulting hydroxy sulfoxides, (iii) nonoxidative Pummerer reaction, (iv) oxidation of the resulting primary alcohols into acids, and (v) final deprotection (Scheme 7.11).

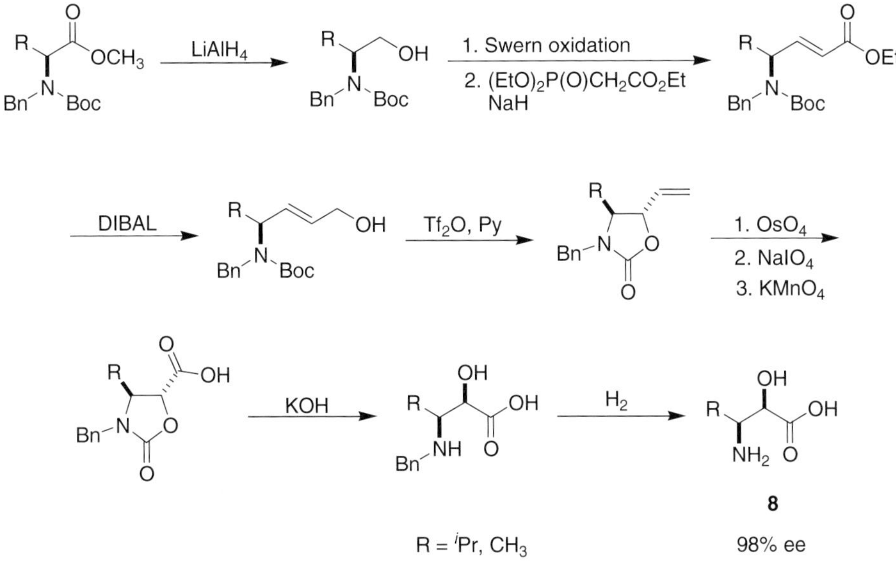

8

$R = {}^{i}Pr, CH_3$ 98% ee

Scheme 7.10 DIBAL: diisobutylaluminum hydride; TBMDSCl: *tert*-butyldimethylsilyl chloride.

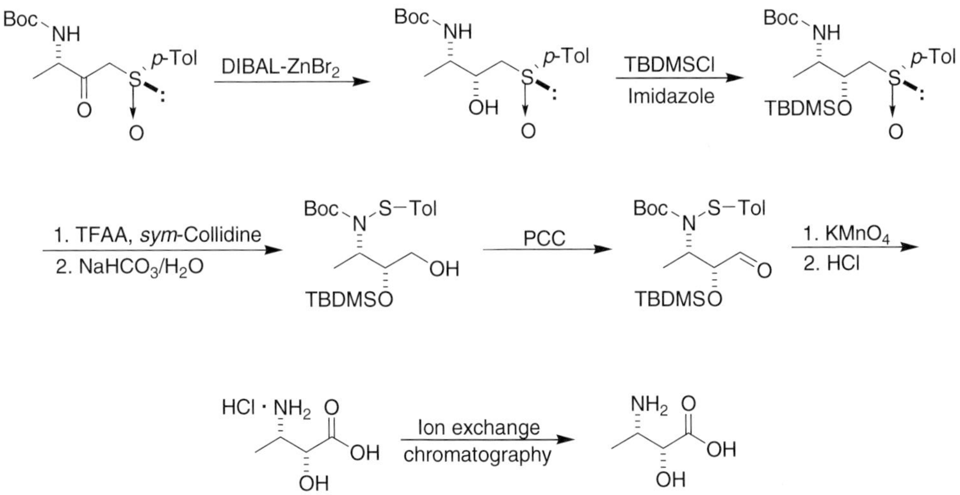

9

≥ 99% de

Scheme 7.11

7.3
Chiral Pool: Enantioselective Synthesis of β-Amino Acids from Aspartic Acid, Asparagine, and Derivatives

The chiral pool refers to the utilization of inexpensive, readily available, and enantiomerically pure natural products as substrates to be converted into chiral derivatives via conventional organic synthesis. In this strategy, the goal is that the built-in chirality in the natural product is preserved or altered in a controlled fashion along the remainder of the reaction sequence. Common chiral starting materials include monosaccharides and amino acids. In this regard, aspartic acid, asparagine, and their derivatives are ideal starting materials for the synthesis of enantiomerically pure β-amino acids, since they already contain the desired β-amino acid unit. Furthermore, both enantiomers of aspartic acid and asparagine are readily available.

In a salient application of this concept, Park *et al.* [17] described the synthesis of all four stereoisomers of aziridinyl acetic acid 2-substituted derivative **10**, a novel class of β-amino acids with an aziridine ring, starting with aspartic acid. The synthetic strategy involves the stereoselective incorporation of an alkyl group at the β-position of fully protected enantiopure aspartic acid followed by the construction of an aziridine ring by manipulation of the α-carboxylate and α-amino groups. Removal of the protecting groups yielded the target compounds (Scheme 7.12).

In this context, Seki *et al.* [18] reported a stereoselective synthesis of β-benzyl-α-alkyl-β-amino acids **12** from L-aspartic acid, which was subjected to a series of reactions to obtain (4*S*,5*R*)-5-phenyloxazolidin-2-one-4-acetate **11**. Alkylation of this phenyloxazolidinone with alkyl halides and subsequent hydrogenation afforded *anti*-disubstituted β-amino acids **11** with high stereoselectivities (Scheme 7.13).

Starting from (*S*)-asparagine, Ávila-Ortiz *et al.* [19] developed a synthesis of enantioenriched β^2-homo-3,4-dihydroxyphenylalanine (DOPA), (*R*)-**13**, a novel β-amino acid homolog of DOPA, the successful therapeutic agent in the treatment of

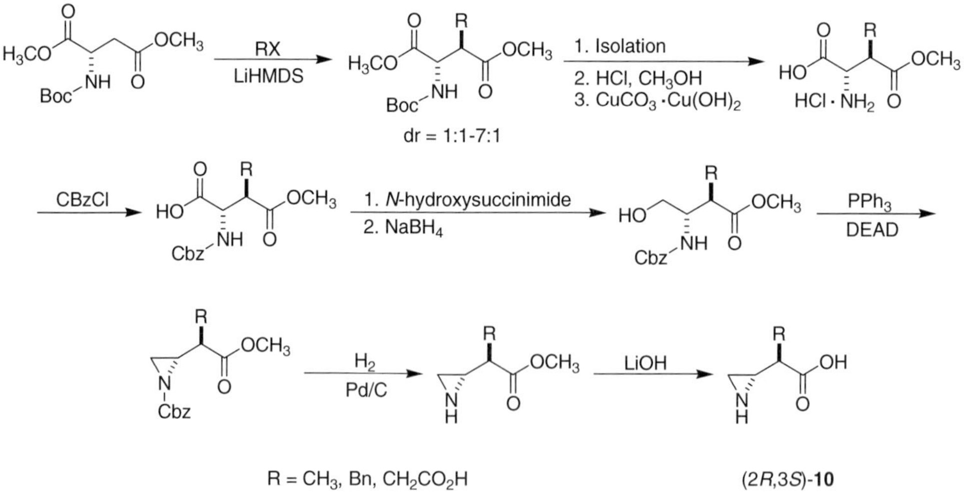

R = CH_3, Bn, CH_2CO_2H (2*R*,3*S*)-**10**

Scheme 7.12

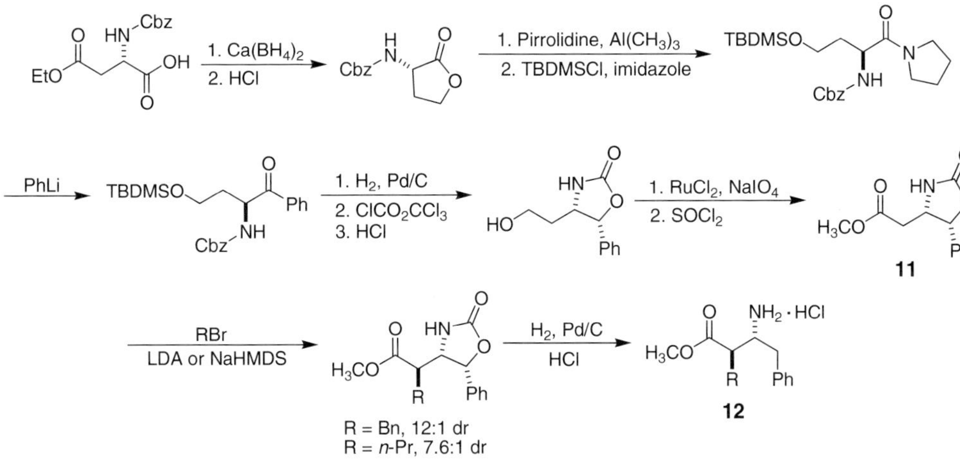

Scheme 7.13

Parkinson's disease. β^2-homo-DOPA was obtained via the diastereoselective alkylation of enantiopure pyrimidinone with veratryl iodide. Once separated, the diastereomeric derivatives were hydrolyzed with 57% HBr and the desired β-amino acids were purified by silica gel chromatography (Scheme 7.14). Alternatively [19], nearly enantiopure (*R*)- and (*S*)-β^2-homo-DOPA were obtained by resolution of racemic *N*-benzyloxycarbonyl-2-(3,4-dibenzyloxybenzyl)-3-aminopropionic acid with (*R*)- or (*S*)-α-phenylethylamine, followed by catalytic hydrogenolysis (Scheme 7.15).

An interesting method which employs (*R*)- or (*S*)-asparagine as starting material for the enantioselective preparation of α-substituted α,β-diamino acids was described by Castellanos *et al.* [20]. Enantiopure pyrimidinone (*R*)-**14** [21] was lithiated with lithium diisopropylamide (LDA) and the derived enolate was aminated with di-*tert*-butyl azodicarboxylate with a diastereoselectivity greater than 98%. Subsequent hydrolysis/hydrogenolysis/hydrolysis gave the expected diamino acids (*S*)-**15** in enantiopure form (Scheme 7.16).

In this context, starting from (*S*)-asparagine, Iglesias-Arteaga *et al.* [22] synthesized enantiopure 1-benzoyl-2(*S*)-*tert*-butyl-3-methylperhydropyrimidin-4-one (*S*)-**16**, a useful starting material for the enantioselective synthesis of α-substituted β-amino acids, by a modification of the original procedure [21] with several advantages: (i) the replacement of toxic lead tetraacetate by safer reagents [diacetoxyiodobenzene (DIB)/iodine] in the oxidative decarboxylation, (ii) *N*-methylation can be conveniently carried out with lithium diisopropylamide and methyl iodide (avoiding the employment of toxic dimethyl sulfate), and (iii) the use of milder reaction conditions and shorter reaction times (Scheme 7.17).

Based on this precedent, Díaz-Sánchez *et al.* [23] developed a one-pot procedure for the synthesis of 1-benzoyl-2(*S*)-substituted-5-iodo-dihyhydropyrimidin-4-one, for instance (*S*)-**17**, by tandem decarboxylation/β-iodination of the corresponding 6-carboxy-perhydropyrimidin-4-ones and subsequent Sonogashira coupling of the halogenated heterocyclic enones with various terminal alkynes to produce 1-benzoyl-2(*S*)-isopropyl-5-alkynyl-2,3-dihyhydropyrimidin-4-ones (*S*)-**18**, in good

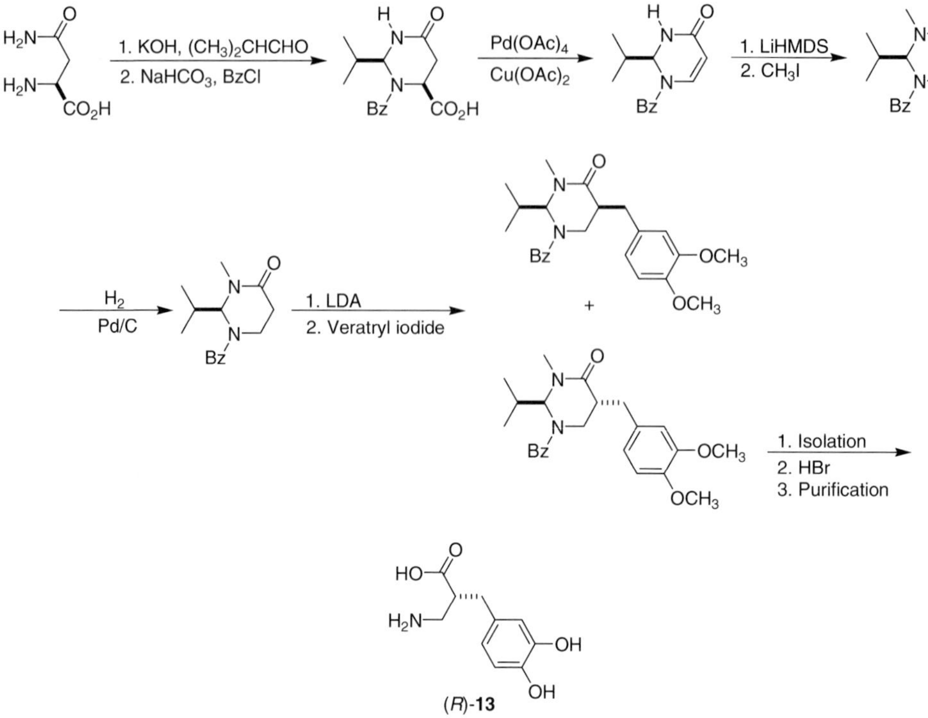

Scheme 7.14

yields (59–88%). Hydrogenation of the unsaturated alkynyl moieties in the Sonogashira product with Raney nickel followed by acid hydrolysis under microwave irradiation afforded the α-substituted β-amino acid (*S*)-2-(aminomethyl)-4-phenylbutanoic acid, (*S*)-**19**, in high enantiomeric excess (Scheme 7.18).

7.4
Synthesis of β-Amino Acids by Conjugate Addition of Nitrogen Nucleophiles to Enones

Conjugate addition of a nitrogen nucleophile to α,β-unsaturated carboxylic acid derivatives is one of the most useful and simplest methods for the formation of N–C bonds and, in particular, for the synthesis of β-amino acids.

7.4.1
Achiral β-Amino Acids

Many catalysts have been developed for the conjugate addition of aliphatic amines to α,β-unsaturated compounds to produce the β-amino carbonyl moiety. For example,

Scheme 7.15

Scheme 7.16

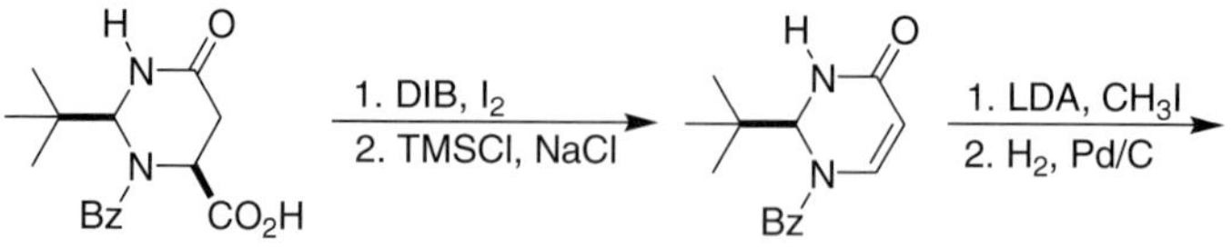

1. DIB, I_2
2. TMSCl, NaCl

1. LDA, CH_3I
2. H_2, Pd/C

(*S*)-**16**

Scheme 7.17

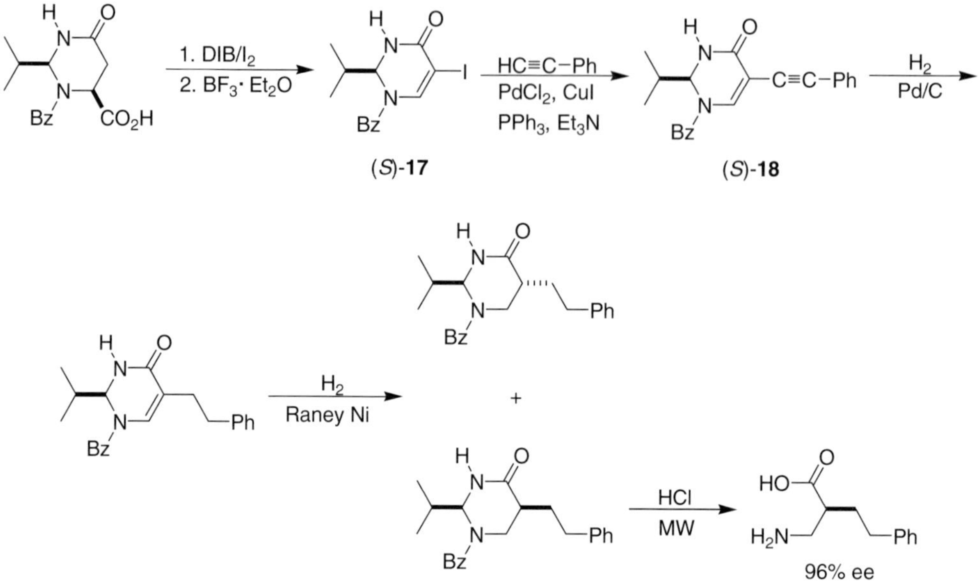

1. DIB/I_2
2. $BF_3 \cdot Et_2O$

(*S*)-**17**

$HC{\equiv}C{-}Ph$
$PdCl_2$, CuI
PPh_3, Et_3N

(*S*)-**18**

H_2
Pd/C

H_2
Raney Ni

HCl
MW

96% ee

(*S*)-**19**

Scheme 7.18

Iovel *et al.* [24] described in 2001 the addition reaction of several acetamides to ethyl acrylate in the presence of an equimolar amount of the CsF–$Si(OEt)_4$ system in benzene to give the corresponding ethyl esters of *N*-substituted β-amino acids (Scheme 7.19).

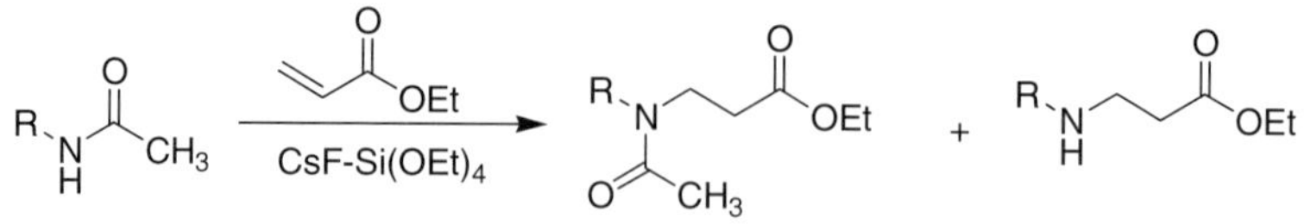

CsF-$Si(OEt)_4$

R = 3-$CF_3C_6H_4$, 4-$CF_3C_6H_4$, 2-pyridyl, 4-methyl-2-pyridyl, 3-methyl-2-pyridyl, Ph, 1-naphtyl, 5-quinolyl, benzyl, picolyl

Scheme 7.19

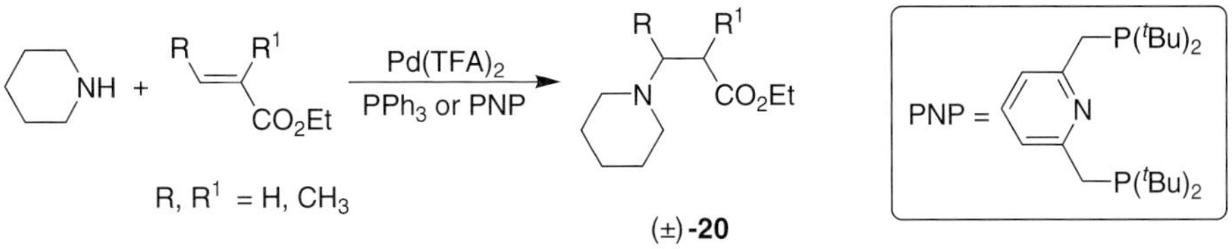

Scheme 7.20

An interesting development in the conjugate addition of amines to the C=C double bond of acrylic acid derivatives was reported by Kawatsura and Hartwig [25], who aided by a high-throughput colorimetric assay examined the effectiveness of several transition metal catalysts in this reaction. The colorimetric assay of reference led to the discovery of suitable catalysts for the addition of piperidine to methacrylonitrile, crotononitrile, ethyl crotonate, and ethyl methacrylate, as well as for the addition reactions of butylamine and aniline with methacrylonitrile. Furthermore, a novel catalytic protocol for the addition of piperidine to activated olefins at 100 °C was also discovered. Thus, treating the acrylate derivative with piperidine in presence of Pd $(TFA)_2$ and PPh_3 or PNP gave the corresponding β-amino carbonyl derivatives **20** (Scheme 7.20).

On the other hand, Amore *et al.* [26] reported a rapid and simple, microwave-promoted synthesis of *N*-aryl functionalized β-amino esters by means of Michael addition reactions. The reactions are performed neat, under microwave irradiation at 200 °C for 20 min in the presence of a catalytic amount of acetic acid. The esters can be easily hydrolyzed to the corresponding *N*-aryl functionalized β-amino acids [e.g., β-amino acid **21** (Scheme 7.21)].

Using Ce(IV) ammonium nitrate (CAN; 3 mol%) in aqueous media, Varala *et al.* [27] efficiently catalyzed the aza-Michael reaction of amines with α,β-unsaturated carbonyl compounds to produce the corresponding β-amino carbonyl compounds in good to excellent yields (55–99%) under mild conditions.

In an innovative contribution, Xu *et al.* [28] reported the use of the basic ionic liquid 3-butyl-1-methylimidazolium hydroxide [bmim]OH as an efficient catalyst and recoverable solvating medium for Michael additions between amines and α,β-unsaturated compounds. The ionic liquid could be reused at least 8 times with consistent catalytic activity.

Additional catalysts have been developed for the conjugate addition of nitrogen nucleophiles to enones, among them boric acid [29], $CuAlCO_3$ hydrotalcite [30], Cu(II) acetylacetonate immobilized in ionic liquids [31], bromodimethylsulfonium bromide [32], potassium pyrrolate in coordinating solvents or by lithium pyrrolate in

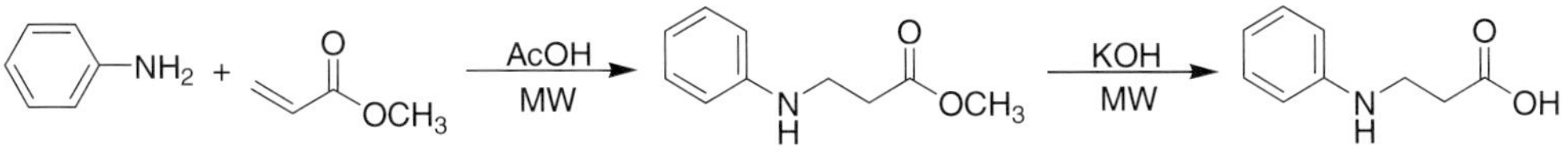

Scheme 7.21

the presence of coordinating ligands [33], solvent-free silica support [34], and $PdCl_2$ on supercritical CO_2 atmosphere [35].

7.4.2
Enantioselective Approaches

There are three main procedures to achieve asymmetric induction in the conjugate addition on nitrogen nucleophiles to enones: (i) addition of 'chiral ammonia' to the acceptor, (ii) addition of an achiral nitrogen nucleophile to a chiral acceptor, and (iii) asymmetric catalysis [36].

(i) *Addition of "chiral ammonia" to an acceptor.* The general strategy consists in the use of chiral ammonia equivalents, usually a benzylic asymmetric secondary amine, to a Michael acceptor, usually an unsaturated ester. The reaction produces a new stereogenic center in the position β to the ester group. The diastereoselectivity of the reaction depends on the starting ester, on the nature of the chiral ammonia equivalent that is employed, as well as on the particular reaction conditions.

(ii) *Addition of an achiral nitrogen nucleophile to a chiral acceptor.* The efficiency of this methodology depends on the degree of stereoselectivity induced by the chiral auxiliary present in the chiral acceptor.

(iii) *Asymmetric catalysis.* In this approach, the control of the stereoselectivity in the formation of the new center of chirality is achieved by the presence of a chiral environment generated principally by chiral ligands.

7.4.2.1 Addition of "Chiral Ammonia" Equivalents to Conjugated Prochiral Acceptors

Davies, the recognized pioneer and leader of this methodology, has shown the versatility of this approach in the synthesis of a large number of chiral β-amino acids [37]. For example, the addition of *N*-benzyl (*R*)- or (*S*)-phenylethylamine to a variety of alkyl or aryl unsaturated substrates proceeded with excellent diastereoselectivity and allowed the preparation of a great number of enantiopure β^3-amino acids (Scheme 7.22).

Based in this methodology, Coleman *et al.* [38] prepared a series of dihydrobenzofuran β-amino acids such as **23**, starting from bromobenzofuran derivatives that provided the dihydrobenzofuran propanoic acid ethyl ester **22** with excellent diastereoselectivity. Hydrogenolysis of this ester provided the corresponding dihydrobenzofuran β-amino acid (*S*)-**23** (Scheme 7.23).

Using Davies' chiral auxiliary, Bull *et al.* [39] prepared a variety of β-amino acids or β-lactams substituted at the β position with alkyl, aryl, or allyl groups. Furthermore, Podlech *et al.* [40] have described the use of the *p*-methoxyphenethylamine auxiliary in the formation of enantiomerically pure β-amino acids; this auxiliary can be cleaved using oxidative conditions (e.g., CAN) (Scheme 7.24).

In this context, Herrera *et al.* [41] reported the preparation of chiral α-hydroxy-β-amino acid derivatives based on the diastereoselective addition of (*R*)-methylbenzylamine to representative captodative olefins (Scheme 7.25).

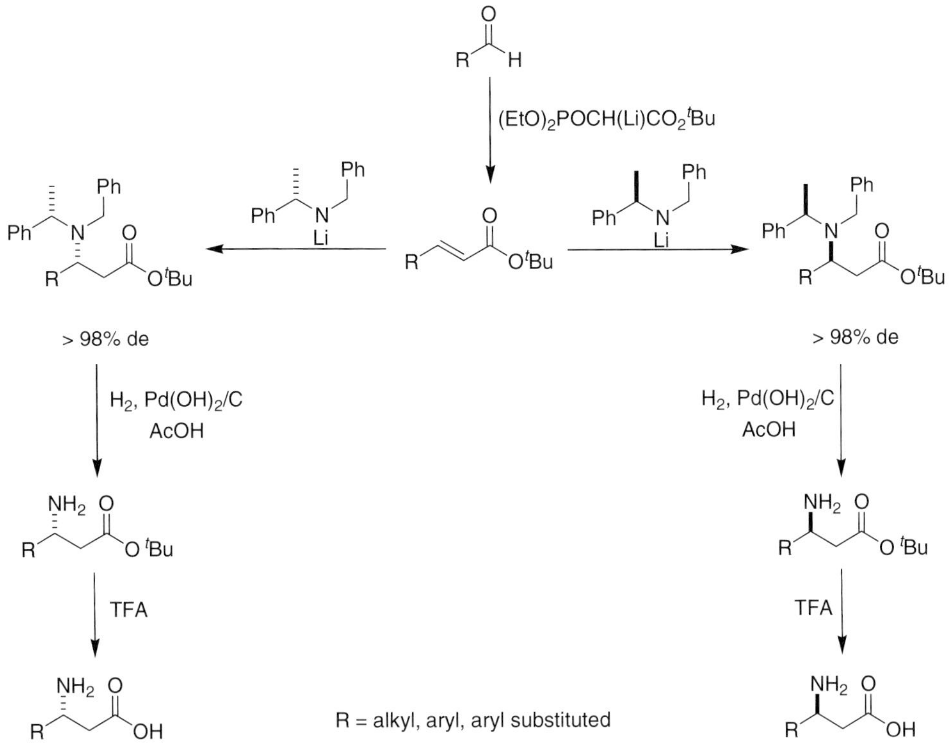

$(EtO)_2POCH(Li)CO_2{}^tBu$

> 98% de

H_2, $Pd(OH)_2/C$, AcOH

TFA

R = alkyl, aryl, aryl substituted

Scheme 7.22

Lee *et al.* [42] described the Michael addition of chiral hydroxylamine (*S*)-**24** derived from (*S*)-α-methylbenzylamine to α-alkylacrylates followed by cyclization giving a diastereomeric mixture of α-substituted isoxazolidinones. These diastereomeric heterocycles were separated by column chromatography and the subsequent hydrogenation of the purified α-substituted isoxazolidinones followed by Fmoc protection afforded enantiomerically pure, protected α-substituted β^2-amino acids **25** (Scheme 7.26).

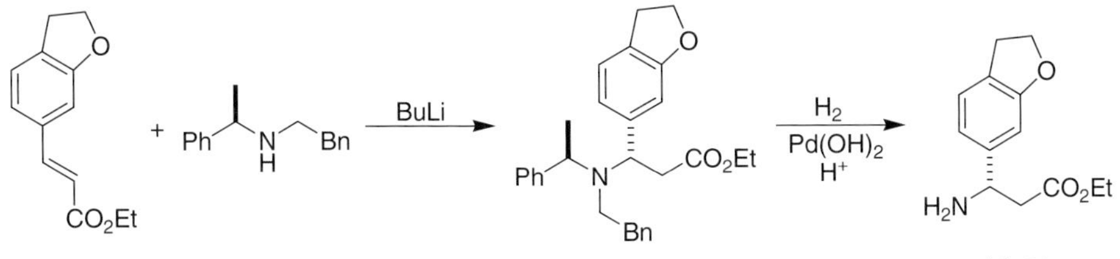

22 + Ph(Me)CH–NH–CH₂Bn —BuLi→ dr > 95:5 —H_2, $Pd(OH)_2$, H^+→ (*S*)-**23**

Scheme 7.23

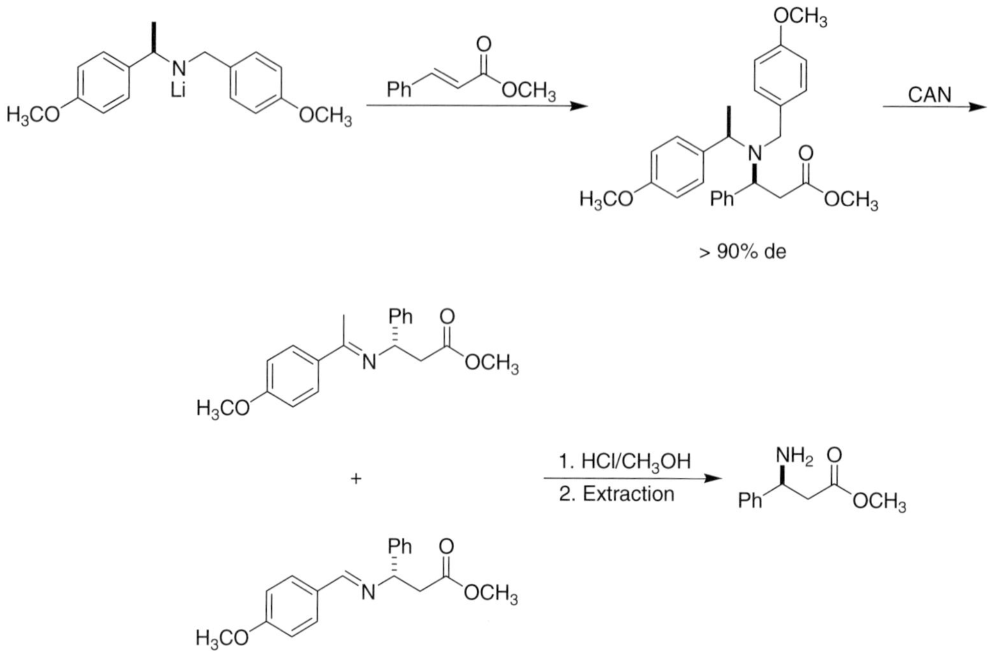

Scheme 7.24

7.4.2.2 Addition of a Nitrogen Nucleophile to a Chiral Acceptor

Sharma *et al.* [43] reported the synthesis of relevant glycosyl β-amino esters **26** through diastereoselective aza-Michael addition to the requisite sugar-based Michael acceptors. Indeed, treatment of glycosyl α,β-unsaturated ester precursors with benzylamine in the presence of tetra-*n*-butyl ammonium fluoride at ambient temperature afforded the desired β-amino glycosyl ester with good diastereomeric excess (Scheme 7.27).

Other interesting examples using unsaturated sugar substrates were described by Fernández *et al.* [44], who report the synthesis of sugar β-amino acids by the

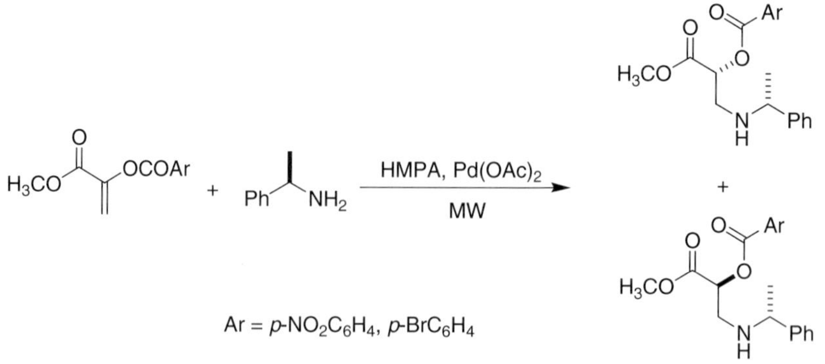

Ar = *p*-$NO_2C_6H_4$, *p*-BrC_6H_4

Scheme 7.25 HMPA: hexamethylphosphoramide.

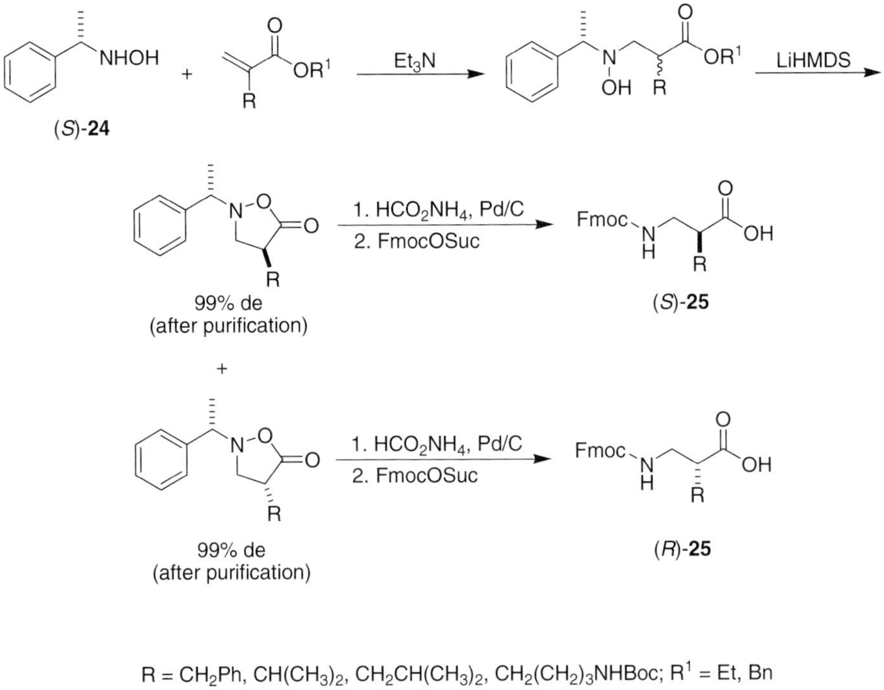

(S)-24 NHOH + OR[1] → Et$_3$N → LiHMDS →

99% de (after purification) + 99% de (after purification)

1. HCO_2NH_4, Pd/C
2. FmocOSuc

(S)-25 (R)-25

R = CH_2Ph, $CH(CH_3)_2$, $CH_2CH(CH_3)_2$, $CH_2(CH_2)_3NHBoc$; R^1 = Et, Bn

Scheme 7.26 TBAF: tetrabutylammonium fluoride.

diastereoselective transformation of D-glyceraldehyde into the corresponding β-amino acid ester β^2-sugar.

In this context, Etxebarria *et al.* [45] have prepared nonracemic β-amino esters in good yields and enantioselectivities using the diastereoselective conjugate addition of nitrogen nucleophiles to α,β-unsaturated amides containing the (*S,S*)-(+)-pseudo-ephedrine, (*S,S*)-**27**, chiral auxiliary (Scheme 7.28).

In similar fashion, the synthesis of fused furanosyl β-amino esters **28** from unsaturated sugar esters was described by Taillefumier *et al.* [46], by means of a

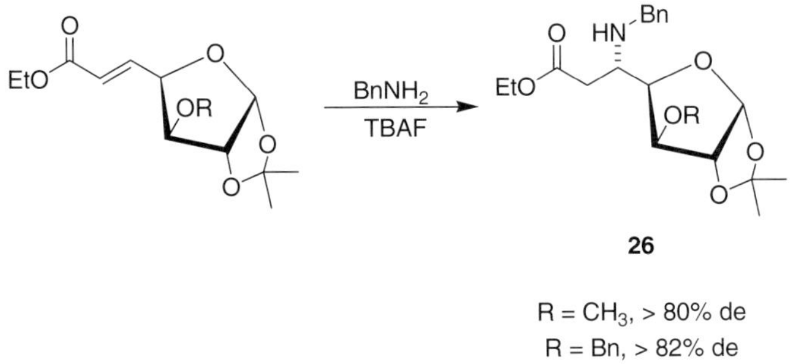

$BnNH_2$, TBAF

26

R = CH_3, > 80% de
R = Bn, > 82% de

Scheme 7.27

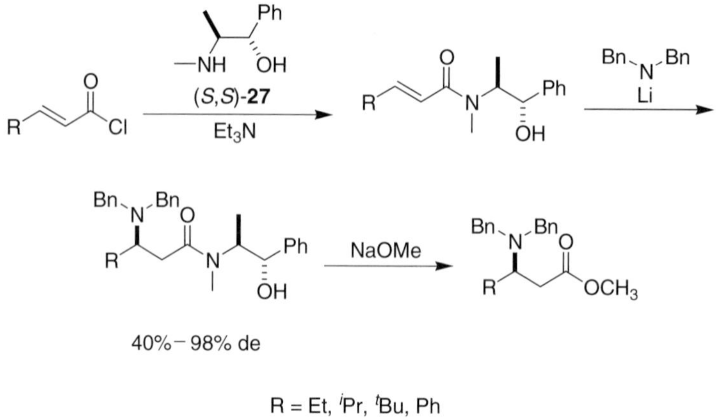

Scheme 7.28

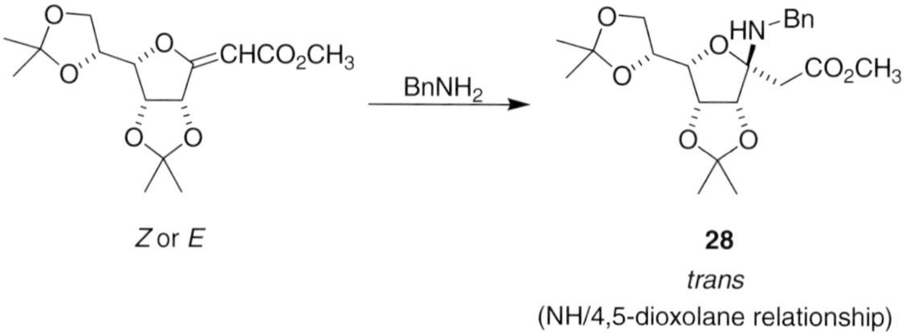

Scheme 7.29

diastereoselective 1,4-addition of benzylamine on the substrate glycosylidenes (Scheme 7.29).

7.4.2.3 Asymmetric Catalysis

Using a complex formed between 1,1′-bi-2-naphthol (BINOL) and rare earth-alkali metals such as yttrium and dysprosium, Yamagiwa *et al.* [47] obtained a series of β^3-amino esters and β^3-amino carbonyl derivatives, for example **29**, with excellent enantiomeric excess. In this system, the rare earth-alkali metal heterobimetallic complex functions as a Lewis acid–Lewis acid cooperative catalyst that activates the conjugated amide for nucleophilic addition (Scheme 7.30).

Rimkus and Sewald [48] have described an enantioselective addition of diethylzinc or mixed diorganozinc compounds to methyl 3-nitropropenoate catalyzed by Cu(I) complexes with BINOL-phosphoramidite ligands, giving rise to the formation of 2-alkyl-3-nitropropanoates, for example **30**, with enantioselectivities up to 92% and chemical yields up to 94%. β-Nitroesters **30** can be easily reduced by catalytic hydrogenation and subsequently hydrolyzed to give the desired β-amino acids (Scheme 7.31).

MacMillan's group [49] has development an efficient catalyst formed by chiral imidazolidinone (*R*,*R*)-**31** and *p*-toluene sulfonic acid. Exposure of 2-hexenal to

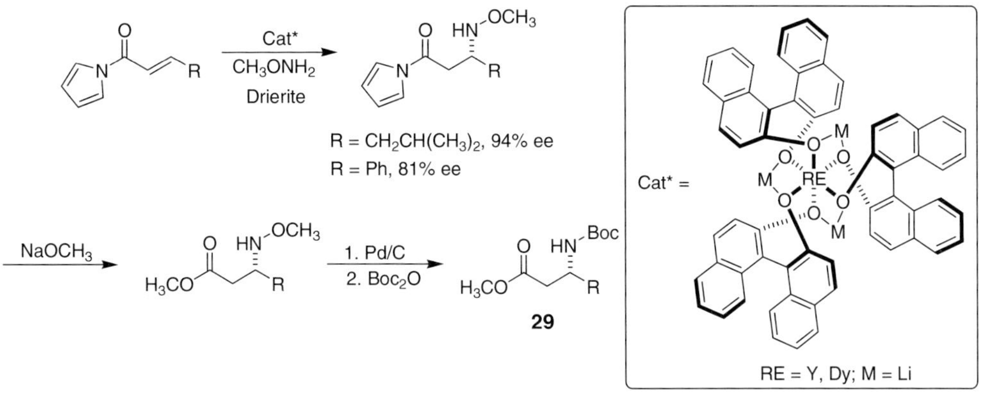

Scheme 7.30

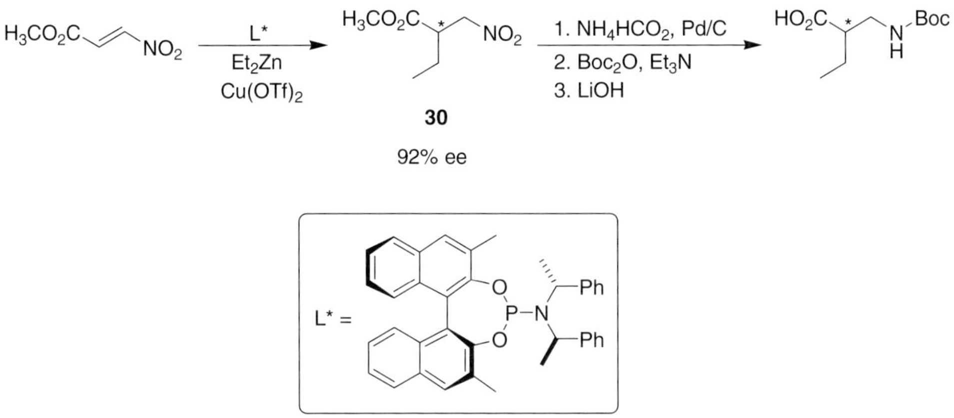

Scheme 7.31

benzyl *tert*-butyldimethylsilyloxycarbamate in the presence of the imidazolidinone catalyst followed by *in situ* Pinnick oxidation provided the corresponding β-amino acid with excellent enantioselectivity. Removal of the N–O bond could be accomplished under mild reducing conditions (Zn/AcOH) (Scheme 7.32).

Using an organocatalyst derived from L-proline, Córdova *et al.* [50] carried out the highly enantioselective conjugate addition of Cbz- or Boc-protected methoxyamines

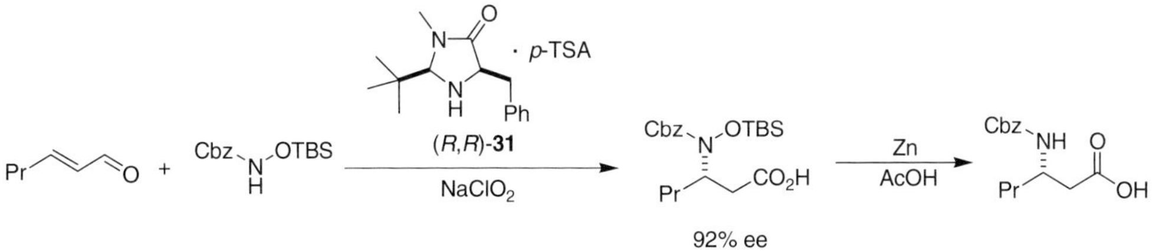

Scheme 7.32

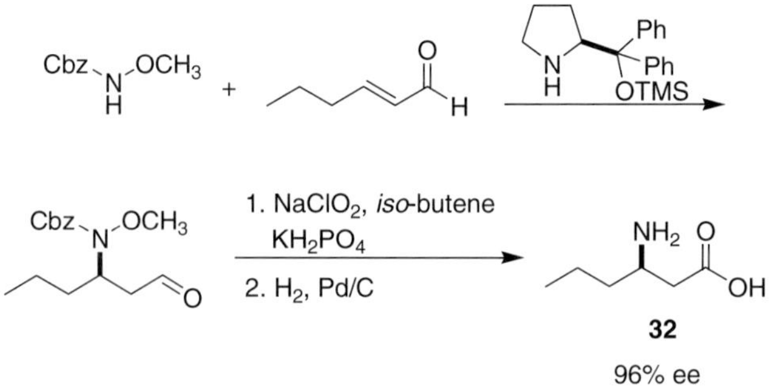

Scheme 7.33

to α,β-unsaturated aldehydes. The reaction gave access to the corresponding β-amino aldehydes that were then converted to β-amino acids, for example **32**, in good yields and high enantiomeric excess (Scheme 7.33).

In a related development, Sibi and Itoh [51] reported the conjugate addition of *O*-protected hydroxylamines to pyrazole-derived enoates with high enantioselectivity when aromatic chiral thioureas were used as activators. A wide variety of substrates undergo conjugate amine addition providing access to the desired enantioenriched β-amino acid derivatives. For example, the reaction of a pyrazole-derived enoate with $BnONH_2$ gave the addition product with 87% e.e. in the presence of chiral thiourea **33** (Scheme 7.34). The structural attributes present in optimal thiourea **33** suggest that it operates as a bifunctional organocatalyst [51].

Sammis and Jacobsen [52] described the application of readily available (salen)Al (III) catalysts in the highly enantioselective conjugate addition of hydrogen cyanide to α,β-unsaturated imides. The cyanide adducts can be converted readily into a variety of useful chiral building blocks, including α-substituted-β-amino acids (Scheme 7.35).

Using the same principle, Palomo *et al.* [53] examined a series of reactions carried out with α′-hydroxy enone and benzyl or *tert*-butyl carbamate in the presence of 10 mol% of chiral Lewis acid. This study revealed that chiral bis(oxazoline)copper complexes afforded the desired β-amino-protected carbonyl adducts **34** in good yield and high enantiomeric excess. Treatment of adducts **34** with sodium metaperiodate

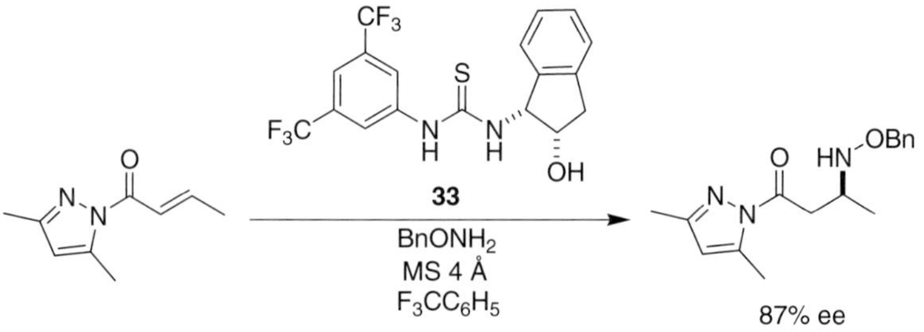

Scheme 7.34

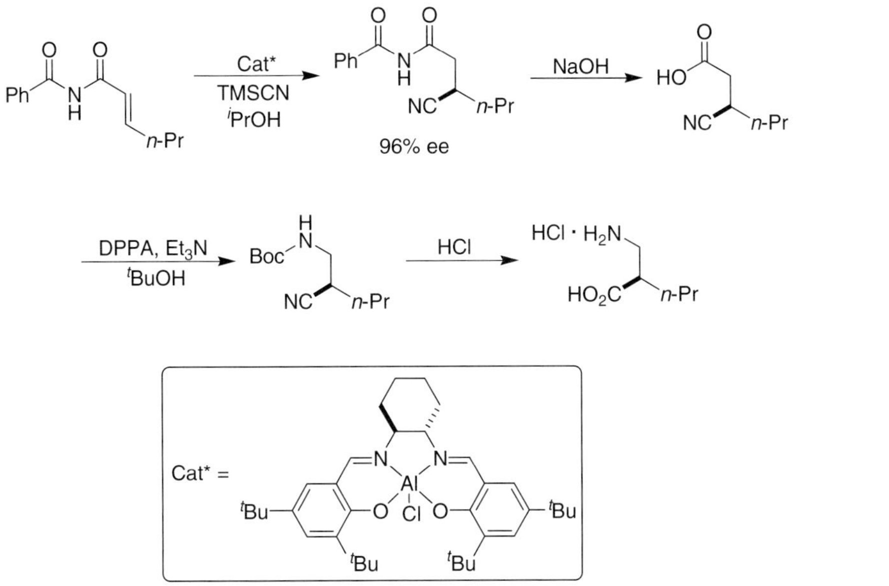

Scheme 7.35 DPPA: diphenylphosphoryl azide.

in a methanol/water mixture afforded *N*-protected β^2-substituted β-amino acid (Scheme 7.36).

In this context, Li *et al.* [54] found that the conjugate addition of primary aromatic amines to *N*-alkenoylcarbamates catalyzed by 2–5 mol% of cationic palladium complex $[[(R)\text{-BINAP}]Pd(MeCN)_2][TfO]_2$ [BINAP = 2,2′-bis(diphenylphosphino) 1,1′-binaphthyl] furnishes *N*-Boc-β-amino amides – suitable precursors to β-amino acids – with enantiomeric excesses greater than 99%.

In another interesting contribution from Sibi's group [55], 1,4-addition of *N*-benzylhydroxylamine to imides in the presence of $Mg(NTf_2)_2$ and a chiral-cyclopropylidene oxazole derivative gave α,β-*trans*-disubstituted *N*-benzylisoxazolidinones **35** in high diastereo- and enantioselectivities. The product isoxazolidinones were

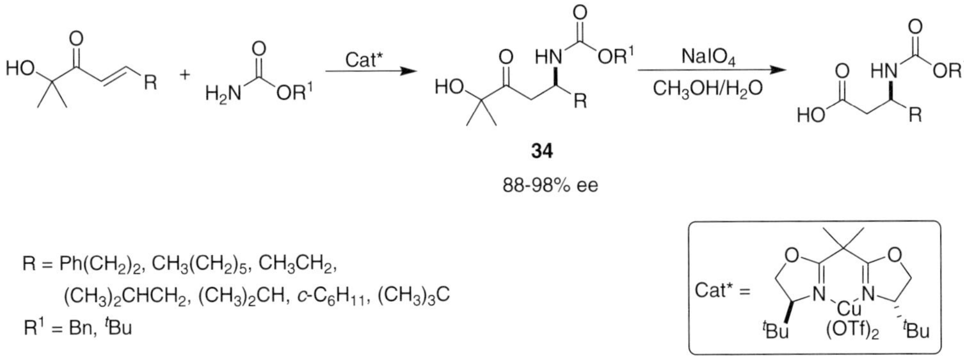

Scheme 7.36

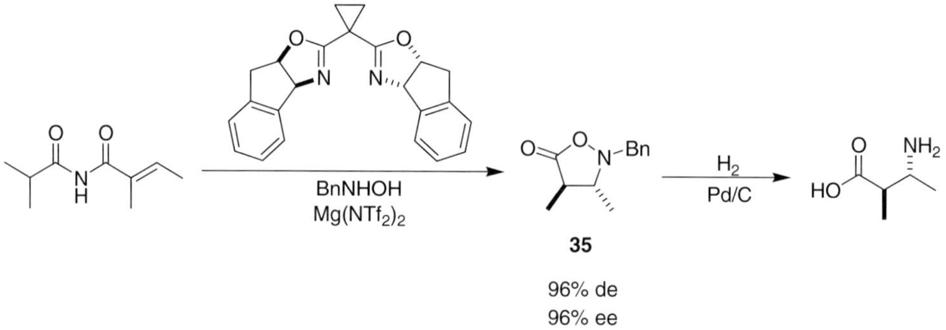

BnNHOH
$Mg(NTf_2)_2$
35
96% de
96% ee
H_2
Pd/C

Scheme 7.37

hydrogenolyzed to provide enantiomerically pure α,β-disubstituted β-amino acids (Scheme 7.37).

In related work, Kikuchi *et al.* [56] described the asymmetric conjugate addition of *O*-benzylhydroxylamine to α,β-unsaturated 3-acyloxazolidin-2-ones by the $RE(OTf)_3$ (RE = Sc, Y, La, Pr, Nd, Sm, Yb)/2,6-bis[(*S*)-4-isopropyloxazolin-2-yl]pyridine (*i*Pr-PYBOX) complex in the synthesis of several β-amino carbonyl compounds in good enantiomeric excess.

7.5
Synthesis of β-Amino Acids via 1,3-Dipolar Cycloaddition

The 1,3-dipolar cycloaddition, also known as the Huisgen cycloaddition, is the reaction between a 1,3-dipole and a dipolarophile, usually substituted alkenes, that leads to five-membered (hetero)cycles (Scheme 7.38) [57].

1,3-Dipolar compounds contain one or more heteroatoms and can be described as having at least one mesomeric structure that represents a charged dipole (Scheme 7.39).

The regioselectivity of the reaction depends on electronic and steric effects, and is somewhat predictable. Many reactions can be performed with high regioselectivity, and even enantioselective transformations of prochiral substrates have been described [58]. Some interesting examples are described herein.

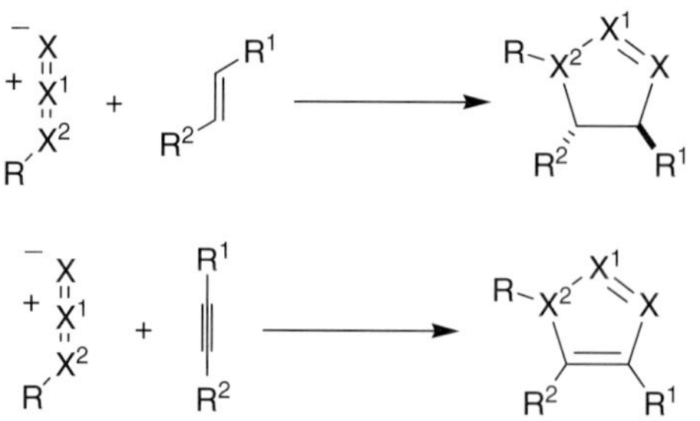

Scheme 7.38

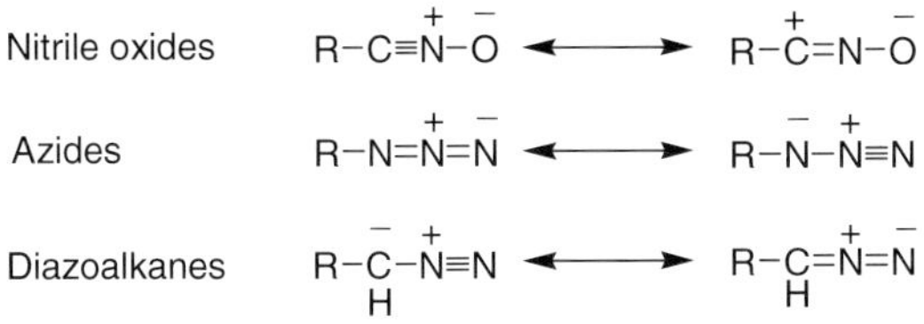

Nitrile oxides R–C≡N⁺–O⁻ ⟷ R–C⁺=N–O⁻

Azides R–N=N⁺=N⁻ ⟷ R–N⁻–N⁺≡N

Diazoalkanes R–CH⁻–N⁺≡N ⟷ R–CH=N⁺=N⁻

Scheme 7.39

Torssell and Somfai [59] have developed a three-component approach for the enantioselective synthesis of *syn*-α-hydroxy-β-amino esters based on the $Rh_2(OAc)_4$-catalyzed 1,3-dipolar cycloaddition of carbonyl ylides to imines. By using chiral α-methylbenzylimines as dipolarophiles an asymmetric version of the cycloaddition was realized. For example, addition of ethyl diazoacetate to α-methylbenzylimines, benzaldehyde, and $Rh_2(OAc)_4$ catalyst over 1 h at room temperature afforded the desired cycloadduct, which upon hydrolysis with *p*-toluene sulfonic acid yielded *syn*-α-hydroxy-β-amino esters **36** in modest enantioselectivity (Scheme 7.40).

On the other hand, using allyl aldehydes Hanselmann *et al.* [60] synthesized a series of isoxazolidines **37** via a $MgBr_2$-induced, chelation-controlled diastereoselective 1,3-dipolar cycloaddition reaction with *N*-hydroxyphenylglycinol as a chiral auxiliary. The diastereomerically pure isoxazolidines were further transformed into cyclic and acyclic β-amino acid derivatives via hydrogenation and subsequent oxidation (Scheme 7.41).

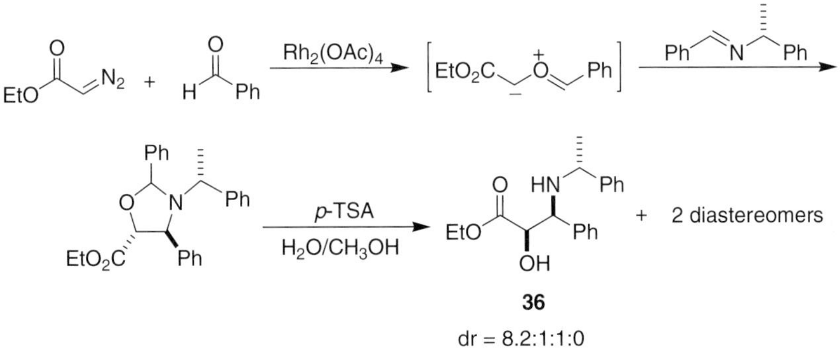

EtO, N2, H, Ph, Rh2(OAc)4, EtO2C, O, Ph, Ph, N, Ph

Ph, O, N, Ph, EtO2C, Ph, p-TSA, H2O/CH3OH, EtO, O, HN, Ph, Ph, OH, + 2 diastereomers

36

dr = 8.2:1:1:0

Scheme 7.40

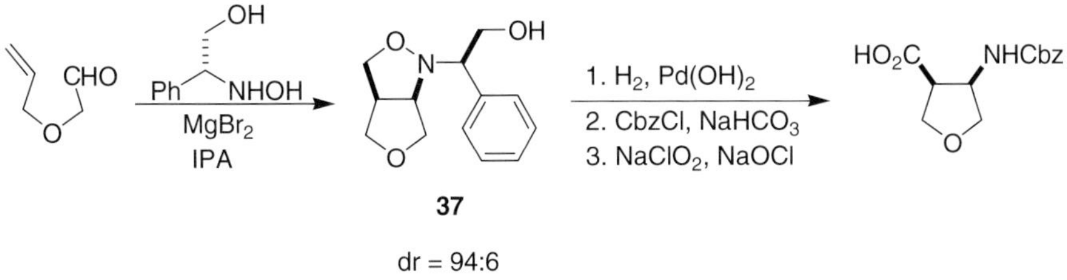

CHO, O, OH, Ph, NHOH, MgBr2, IPA, O, N, OH, O, 1. H2, Pd(OH)2, 2. CbzCl, NaHCO3, 3. NaClO2, NaOCl, HO2C, NHCbz, O

37

dr = 94:6

Scheme 7.41

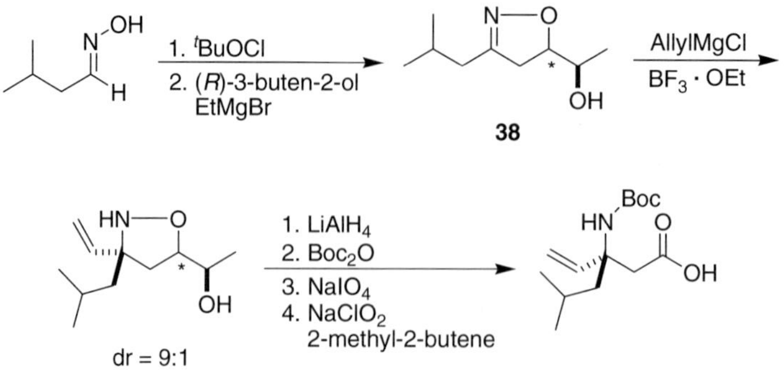

Scheme 7.42

In this context, Minter *et al.* [61] made use of a 1,3-dipolar cycloaddition reaction between oximes and (*R*)-3-buten-2-ol in the presence of EtMgBr to provide chiral isoxazolines, for example **38**, which were readily transformed into the corresponding $\beta^{3,3}$- and $\beta^{2,3,3}$-amino acids by N–O bond reduction with $LiAlH_4$, followed by protection of the free amine with Boc_2O and oxidative cleavage of the diol with $NaIO_4$ (Scheme 7.42).

In an extensive contribution, Kawakami *et al.* [62] reported the asymmetric synthesis of β-amino acids by addition of chiral enolates, such as (*S*)-**39**, to *N*-acyloxyiminium ions generated by the reaction of nitrones with acyl halides (Scheme 7.43). Reversal of diastereoselectivity shown in Scheme 7.43 was observed by the reactions of the Ti(IV) enolates. Indeed, using these reactions all four stereoisomers of α-methyl β-phenylalanines could be prepared in highly diastereoselective fashion [62].

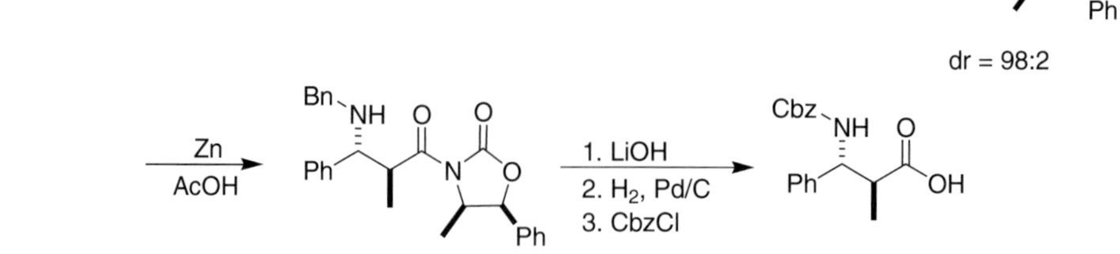

Scheme 7.43

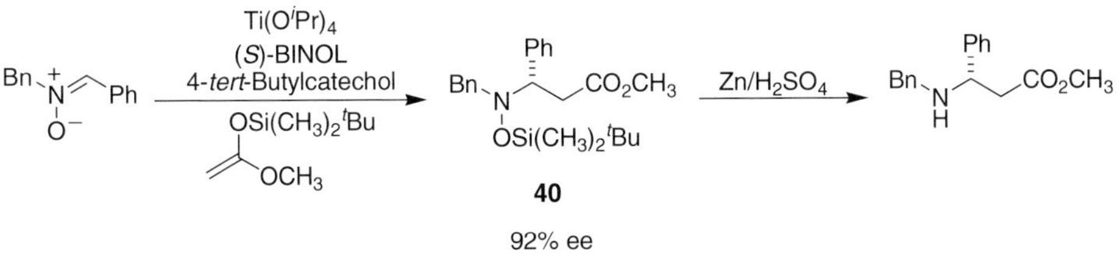

Scheme 7.44

Another interesting example where use is made of nitrones was described by Murahashi *et al.* [63]. These researchers report a catalytic and enantioselective synthesis of enantioenriched β-amino acid derivatives. For example, chiral *N*-(*tert*-butyldimethylsilyloxy)-β-amino esters such as **40**, were prepared by addition of a *tert*-butyldimethylsilyl ketene acetal to nitrones in the presence of titanium tetraisopropoxide, (*S*)-BINOL, and phenols or catechols (Scheme 7.44).

The synthesis of novel α-epoxy-β-amino acids was recently described by Luisi *et al.* [64] via lithiation of 2-oxazolinyloxirane **41** followed by reaction with (*Z*)-*N*-*tert*-butyl-α-phenylnitrone affording a good yield of the trioxadiazadispiro derivative **42** in a completely diastereoselective manner. The dispirocyclic compound **42** was treated with aqueous oxalic acid to give the epoxy-5-isoxazolidinone **43** which was reduced (H_2, Pd/C, MeOH) to the α-epoxy-β-amino acid **44** (Scheme 7.45).

By contrast, Shindo *et al.* [65] described the first anionic inverse electron-demand 1,3-dipolar cycloaddition of ynolates with nitrones to provide 5-isoxazolidinones **45** *en route* to 2,3-disubstituted β-amino acids (Scheme 7.46).

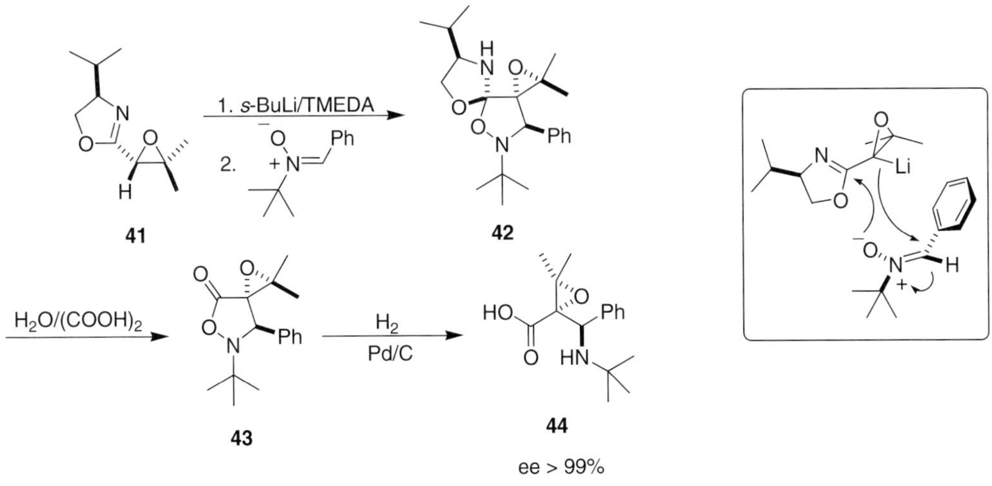

Scheme 7.45 TMEDA: *N*,*N*,*N*′,*N*′-tetramethylethylenediamine.

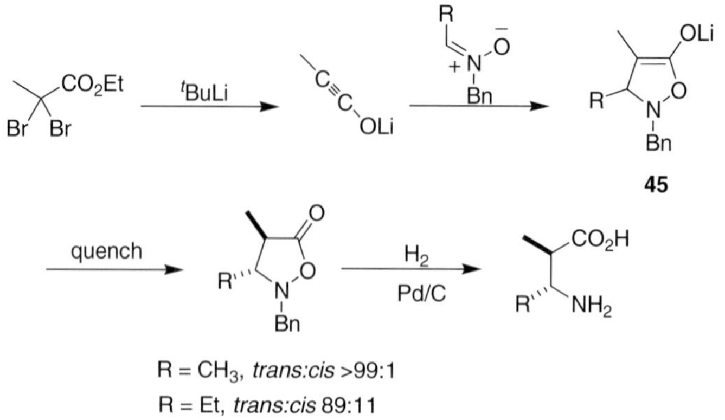

R = CH_3, *trans:cis* >99:1
R = Et, *trans:cis* 89:11

Scheme 7.46

7.6
Synthesis of β-Amino Acids by Nucleophilic Additions

This section presents salient synthetic reports describing the stereoselective preparation of substituted β-amino acids by means of nucleophilic 1,2-additions of enolates or organometallic reagents to imines, nitrones, or oximes [66].

7.6.1
Aldol- and Mannich-Type Reactions

In a pioneering approach, Roers and Verdine [67] carried out the enantio- and diastereoselective synthesis of *N*-protected β^2- and β^3-substituted amino acids. The key steps in this procedure are a catalytic asymmetric aldol reaction and a modified Curtius rearrangement to form chiral oxazolidinones **46** as intermediates, which following *N*-Boc protection and chemoselective ring opening, produce the desired β-amino acids (Scheme 7.47).

On the other hand, Takahashi *et al.* [68] described the LiOAc-catalyzed reaction between trimethylsilyl enolates and aldimines to afford the racemic β-amino carbonyl derivatives in good to high yields under mild conditions. For example, TsN=CHPh reacts with $Me_2C{=}C(OMe)OSiMe_3$ in aqueous *N,N*-dimethylformamide (DMF) at room temperature in the presence of $PhCO_2Li$ catalyst to give quantitative yields of $TsNHCHPhCMe_2CO_2Me$ **47** (Scheme 7.48).

Based on the same principle, Muraoka *et al.* [69] developed a rhodium-catalyzed method to produce β-amino ketones from the reaction of an aldimine with an α,β-unsaturated esters and a hydrosilane.

Making use of chiral *N*-acyl thiazolidinethione enolates, Ambhaikar *et al.* [70] developed an interesting synthesis of *N*-acyl-α,β-disubstituted β-amino acid derivatives **49**. As shown in Scheme 7.49, *O*-methyl benzaldoxime afforded the corresponding azacyclobutane **48**, which was successfully converted to the corresponding *N*-acyl β-amino carbonyl compound by simple exposure to benzoyl chloride. Final hydroly-

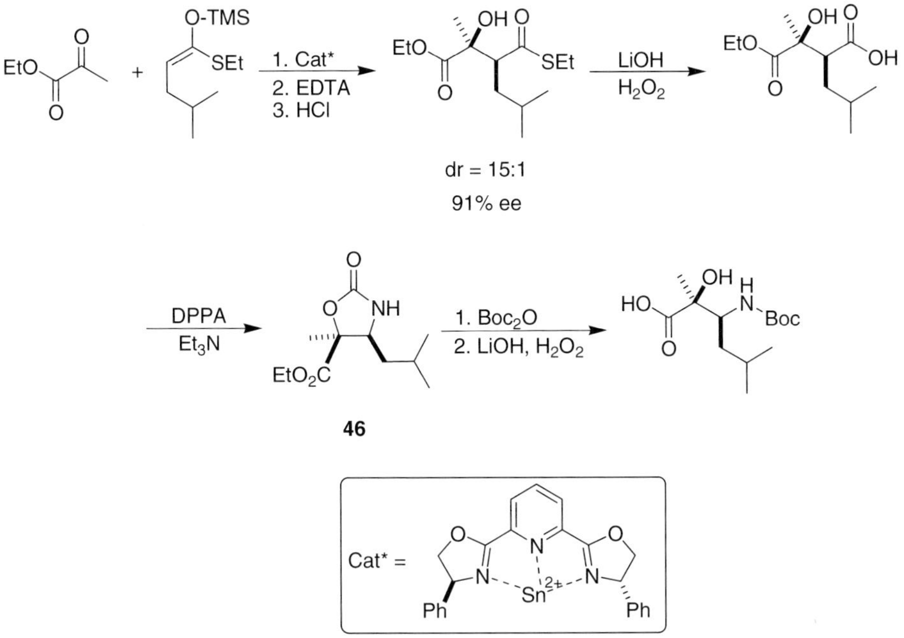

EtO, O-TMS, SEt, 1. Cat*, 2. EDTA, 3. HCl, OH, dr = 15:1, 91% ee, LiOH, H_2O_2, DPPA, Et_3N, NH, EtO_2C, **46**, 1. Boc_2O, 2. LiOH, H_2O_2, HO, Boc

Cat* = N, O, Sn2+, Ph

Scheme 7.47 EDTA: ethylenediaminetetraacetic acid.

sis of the acyl iminium intermediate afforded the desired *N*-acyl-α,β-disubstituted β-amino carbonyl compound **49** (Scheme 7.49).

Koriyama *et al.* [71] carried out the diastereoselective addition reaction of ester enolates and Grignard reagents to enantiomerically pure *N*-sulfinimines, observing contrasting diastereofacial selectivity depending on the metal, solvent, and additive employed.

In an extensive and systematic work, Tang and Ellman [72] have demonstrated the versatility of *tert*-butanesulfinyl aldimines as electrophilic substrates for the addition of enolates to produce β-substituted, α,β- and β,β-disubstituted, α,β,β- and α,α,β-trisubstituted, and α,α,β,β-tetrasubstituted β-amino acid derivatives. For example, the reaction of *tert*-butanesulfinylimine **50** and $MeCO_2Me$ in the presence of Ti$(OiPr)_3Cl$ and LDA in THF at −78 °C gave the β-(*tert*-butanesulfinylamino) acid ester **51** in yields greater than 70% and with diastereoselectivities higher than 98%. The *N*-sulfinyl-β-amino ester products were saponified with LiOH to afford the *N*-sulfinyl-β-amino acid or treated with acid to remove the sulfinyl group and afford the corresponding β-amino ester (Scheme 7.50).

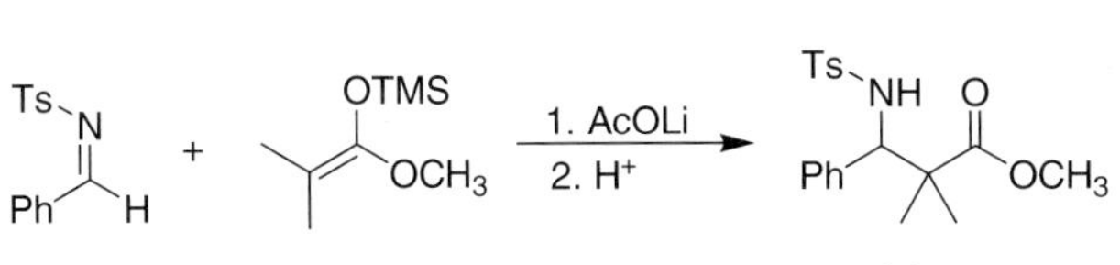

Ts, N, Ph, H, OTMS, OCH_3, 1. AcOLi, 2. H^+, NH, O, (±)-**47**

Scheme 7.48

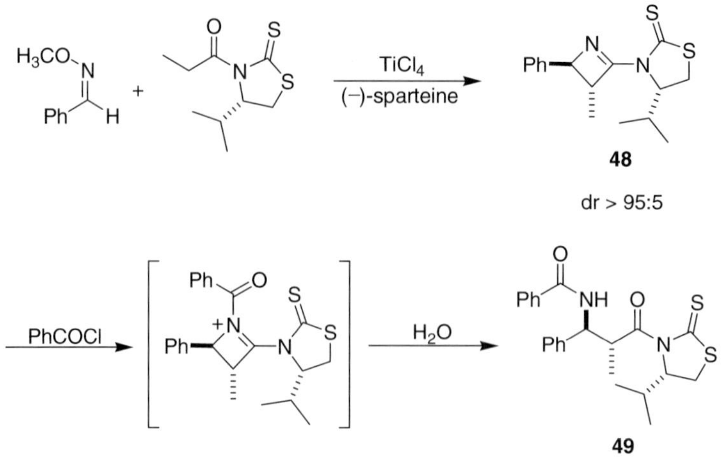

Scheme 7.49

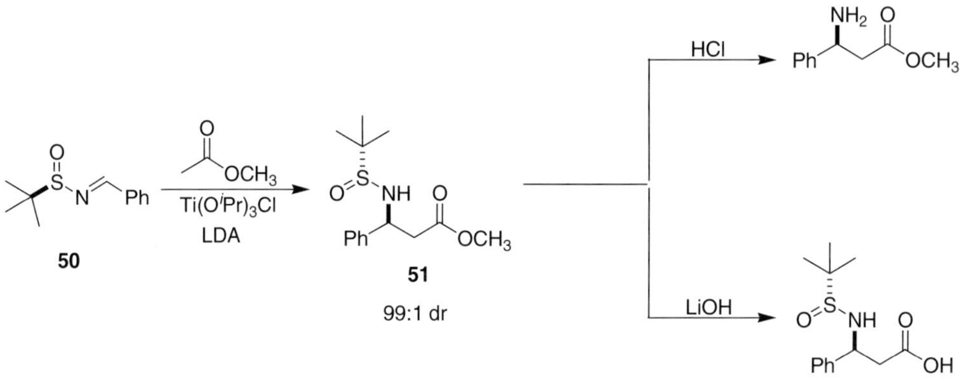

Scheme 7.50

Equally relevant, Davis *et al.* [73] prepared *N*-sulfinyl β-amino carbonyl compounds by use of chiral sulfinimines as auxiliaries in the addition of the sodium enolate of methyl acetate to the corresponding chiral imine. This addition reaction affords the expected chiral methyl *N*-(*p*-toluenesulfinyl)-3-amino-3-phenyl propanoate **52** (Scheme 7.51).

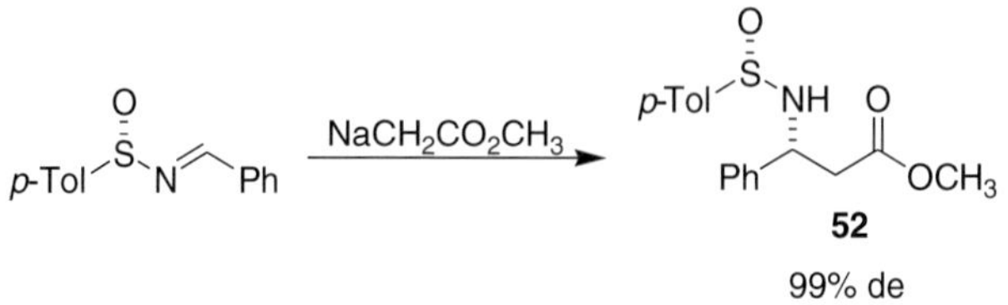

Scheme 7.51

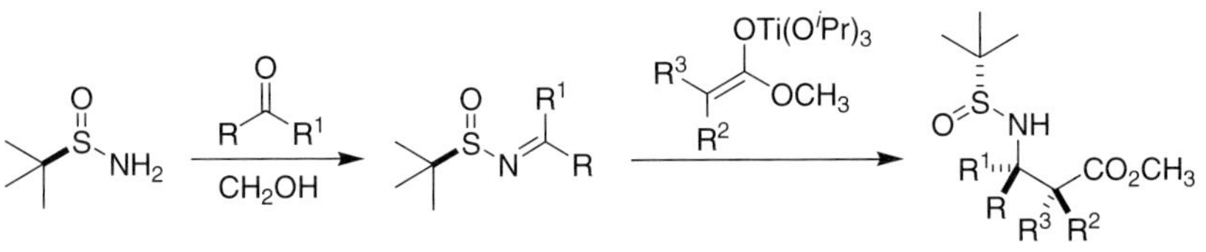

53

dr = 90:10 - 99:1

R = CH_3, Ph, ^{i}Pr, ^{i}Bu, Et; R^1 = H, CH_3; R^2 = CH_3, Bn; R^3 = H, CH_3

Scheme 7.52

By the same token, Ellman *et al.* [74] reported that the use of *tert*-butanesulfinamide for condensation with aldehydes and ketones provides *tert*-butanesulfinyl imines in high yields. Indeed, the *tert*-butanesulfinyl group activates the imines for nucleophilic addition to afford highly enantioenriched β-amino acids **53** (Scheme 7.52).

Soloshonok *et al.* [75] described the synthesis of β-substituted α,α-difluoro-β-amino acids via Reformatsky reaction between $BrZnCF_2CO_2Et$ and Davis' *N*-sulfinylimines, obtaining high stereochemical selectivities (diastereomeric ratio > 9 : 1) on the target β-amino acids. Furthermore, Kanai *et al.* [76] made use of a Reformatsky-type reaction too, with aldimines, ethyl bromoacetate, and a rhodium catalyst to obtain both β-amino esters and β-lactams. In this regard, Vidal *et al.* [77] described a similar approach using a polymeric support.

Interestingly, Adrian and Snapper [78] have developed a nickel-catalyzed Reformatsky-type three-component condensation reaction that affords β-amino carbonyl compounds at both gram and microscale settings. This reaction uses an inexpensive nickel catalyst ($NiCl_2[PPh_3]_3$), an α-halo ester (activated with zinc metal), an aldehyde, and aniline to afford the desired compounds. These researchers have further demonstrated the utility of this approach in the generation in parallel of a 64 β-amino carbonyl compound library.

On the other hand, Fustero's group [79] has developed a novel and effective method for the diastereoselective synthesis of fluorinated β-amino acid derivatives **54** starting from fluorinated imidoyl chlorides and ester enolates, by means of chemical reduction with $ZnI_2/NaBH_4$ and using (−)-8-phenylmenthol as a chiral auxiliary. The process allows for the preparation of enantiopure α-substituted β-fluoroalkyl β-amino esters (Scheme 7.53).

In a related development, a series of unsaturated β-amino acids were synthesized by Bellassoued *et al.* [80] by means of the addition of butenoate lithium enolates to aromatic aldimines promoted by zinc bromide. In all the cases, β-amino acids resulting from exclusive α-attack were obtained as the *anti* diastereoisomer. A tentative explanation in terms of six-membered cyclic transition states (Zimmerman–Traxler model) of the stereoselectivity in favor of the *anti* isomer was advanced by the authors and included in Scheme 7.54.

Della Rosa *et al.* [81] described a novel methodology for the addition of enediolates of carboxylic acids to isocyanates yielding, in one step, α-monosubstituted β-amido

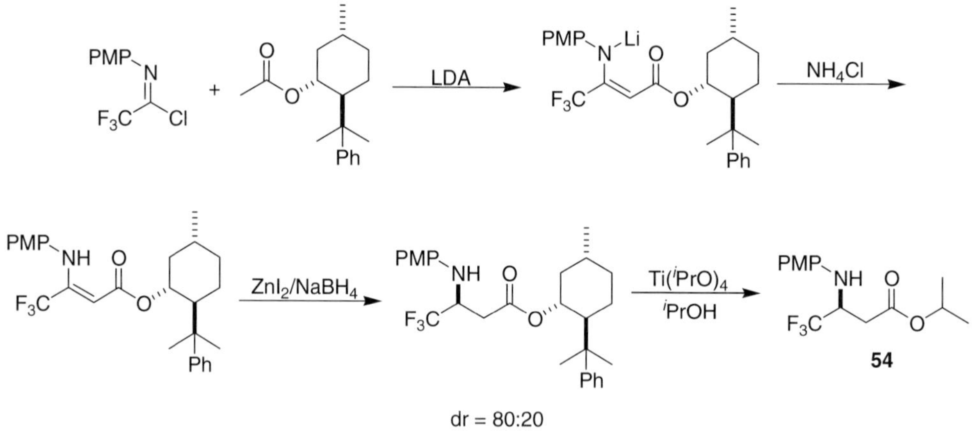
PMP
N
F3C
Cl
+
O
O
Ph
LDA
PMP
N
Li
O
F3C
O
Ph
NH4Cl
PMP
NH
O
F3C
O
Ph
ZnI2/NaBH4
PMP
NH
O
F3C
O
Ph
dr = 80:20
Ti(iPrO)4
iPrOH
PMP
NH
O
F3C
O
54

Scheme 7.53

acids. The reaction of enediolates, generated using lithium cyclohexylisopropylamide as the base, with isocyanates is complete in 1 h at −78 °C. The amido acid **55** is isolated and subsequent chemoselective reduction leads to several unsaturated β-amino acids (Scheme 7.55).

Lurain and Walsh [82] reported a general and enantioselective catalytic method for the synthesis of β-unsaturated β-amino acids. This methodology utilizes the highly enantioselective addition of a vinylzinc species to an aldehyde. The allylic alcohol product **56** is then subjected to Overman imidate rearrangement conditions ([3,3]-sigmatropic trichloroacetimidate rearrangement) to install the nitrogen, and the chirality established in the vinylation step is transferred to the resultant allylic amine. One-pot deprotection and oxidation of a pendant oxygen leads to the γ-unsaturated β-amino acid derivatives (Scheme 7.56). It should be noted that both the D and L configurations of the amino acids are accessible via this methodology.

In this context, Joffe *et al.* [83] developed an efficient and stereoselective titanium enolate based three-component condensation reaction (aldehyde, aniline, and chlor-

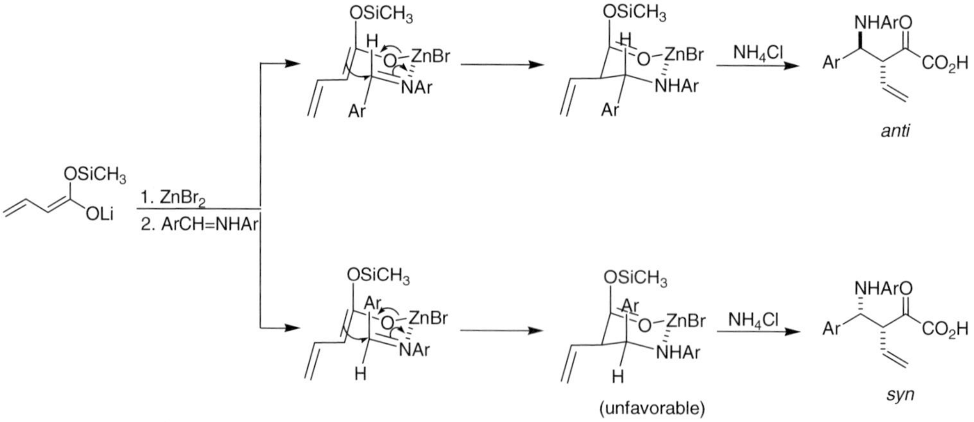
OSiCH3
OLi
1. ZnBr2
2. ArCH=NHAr
OSiCH3
H
O
ZnBr
NAr
Ar
OSiCH3
H
O
ZnBr
NHAr
Ar
NH4Cl
NHAr
O
Ar
CO2H
anti
OSiCH3
Ar
O
ZnBr
NAr
H
OSiCH3
Ar
O
ZnBr
NHAr
H
(unfavorable)
NH4Cl
NHAr
O
Ar
CO2H
syn

Scheme 7.54

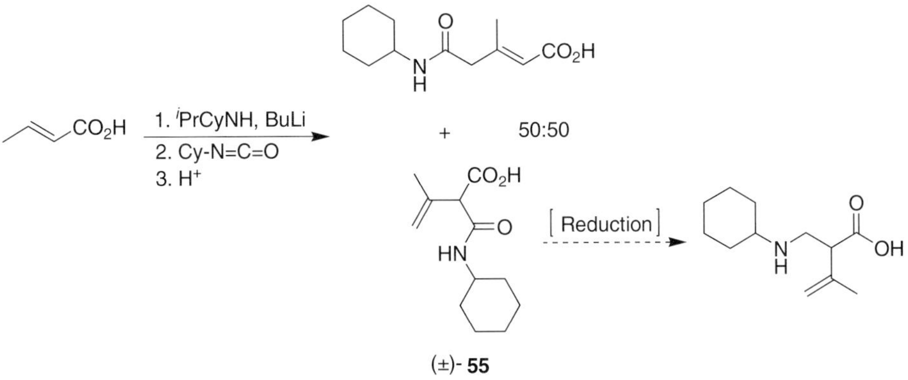

Scheme 7.55

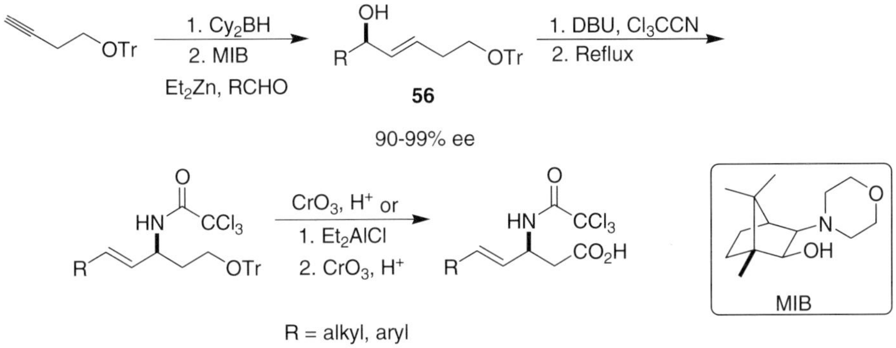

Scheme 7.56 DBU: 1,8-diazabicyclo[5.4.0]undec-7-ene.

otitanium enolates) that affords α,β-disubstituted β-amino carbonyl compounds. There are four sites for substitution with this multiple-component reaction and so it has significant potential for the preparation of α,β-disubstituted β-amino carbonyl libraries.

Making use of an addition reaction on racemic substrates and a subsequent enzymatic resolution, Rossen *et al.* [84] described the enantioselective synthesis of several β^2-amino acids. This protocol involves the addition of ketene diethyl acetal to an *in situ* prepared acyl imine. Novozym 525 L (a *Candida antarctica* lipase B) conduced then to the β-amino acids with good enantiomeric excess (Scheme 7.57).

In the area of organocatalysis, Ricci *et al.* [85] used cinchona-derived organocatalysts to achieve the reaction of malonic acid half thioester **57** with aryl *N*-tosyl imines. This protocol allowed for the formation of the desired protected β-amino thioester with moderate enantioselectivity (Scheme 7.58).

7.6.2
Morita–Baylis–Hillman-Type Reactions

A useful procedure for the formation of β-amino carbonyl compounds is the aza-Morita–Baylis–Hillman (aza-MBH) reaction, where an electron-deficient alkene

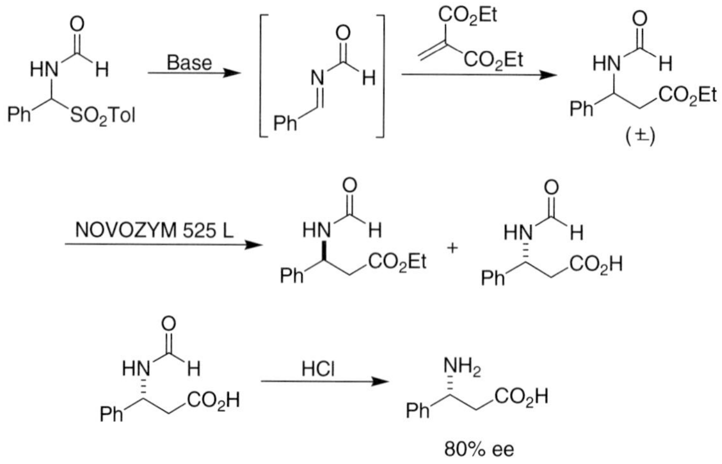

Scheme 7.57

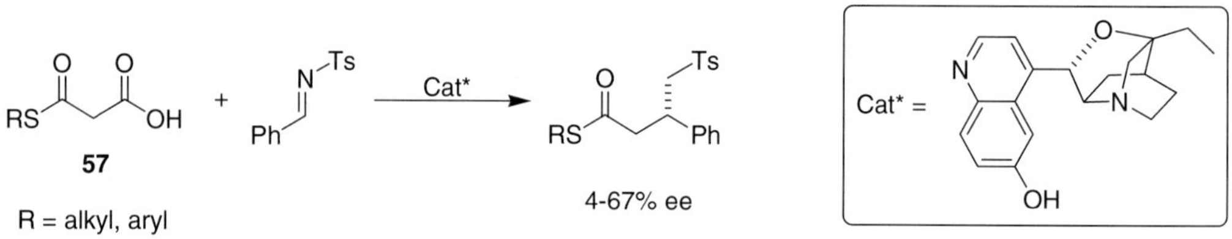

Scheme 7.58

such as an α,β-unsaturated carbonyl compound reacts with an imine in the presence of a nucleophile. Aza-MBH reactions are known in asymmetric synthesis by making use of chiral ligands.

Utzumi *et al.* [86] propose an access route to aza-MBH-type products from β-substituted α,β-unsaturated aldehydes and α-imino esters protected with a *p*-methoxyphenyl (PMP) group in the presence of (*S*)-proline and imidazole under mild conditions. For example, (*E*)-**58** was isolated in 99% e.e. from a 16 : 1 *E*/*Z* product mixture (Scheme 7.59).

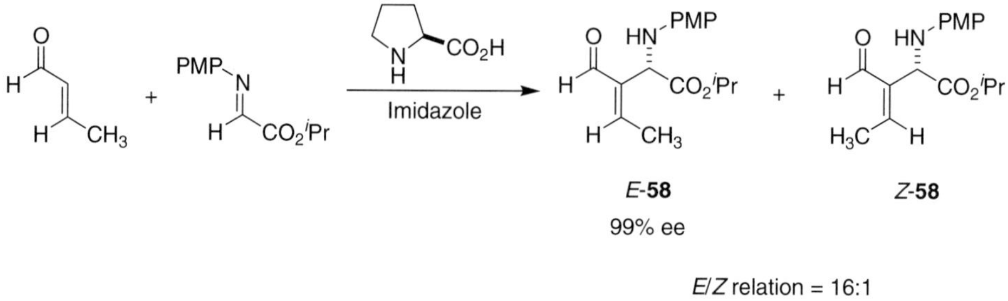

Scheme 7.59

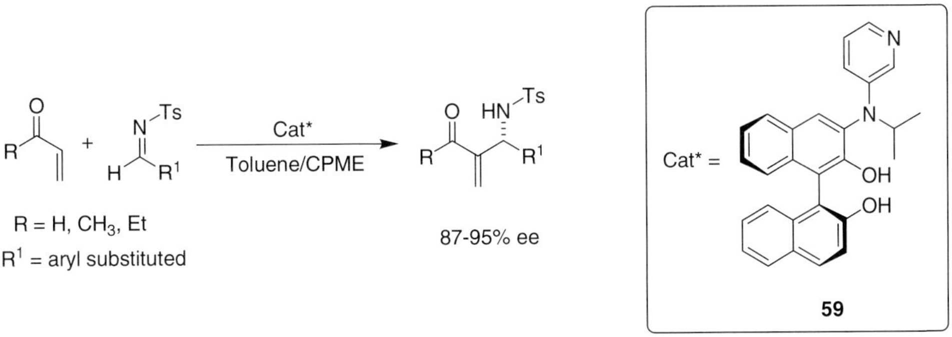

Scheme 7.60 CPME: cyclopentylmethyl ether.

Based on this type of reaction, Matsui *et al.* [87] demonstrated the efficiency of the bifunctional organocatalyst (*S*)-3-(*N*-isopropyl-*N*-3-pyridinylaminomethyl)-BINOL, **59**. The reaction proved to be significantly influenced by the position of the Lewis base attached to BINOL. A proper selection of the acidic and basic functionalities responsible for the activation of the substrate and for fixing the conformation of the intermediate complex is required to promote the reaction with high enantiocontrol (Scheme 7.60).

In this context, Xu and Shi [88] have developed highly efficient aza-MBH reactions by the appropriate combination of Michael acceptors and Lewis base promoters. Most of the reported aza-MBH reactions of *N*-tosylated imines with α,β-unsaturated enones can be accomplished within 1 h under mild reaction conditions.

On the other hand, Sergeeva *et al.* [89] have described a convenient synthesis of fluorinated β-amino acids via the aza-MBH reaction. Beginning with [2,2,2-trifluoro-1-(trifluoromethyl)ethylidene]carbamic acid, 1,1-dimethylethyl ester, *N*-[2,2,2-trifluoro-1-(trifluoromethyl)ethylidene]benzamide and acrylic acid esters, these researchers obtained the expected vinylic product **60**, which by cuprate conjugate addition provides ready access to α-substituted β,β-bis(trifluoromethyl) β-amino acids (Scheme 7.61).

Nemoto *et al.* [90] described a general and highly enantioselective allylic amination of racemic MBH adduct derivatives using several *P*-chirogenic diaminophosphine oxides and the presence of palladium catalyst ($[\eta^3\text{-}C_3H_5PdCl]_2$). The cyclic reaction products could be converted into chiral cyclic β-amino acids; for example, the β-amino ester **61** was obtained in 99% yield and 99% e.e. (Scheme 7.62).

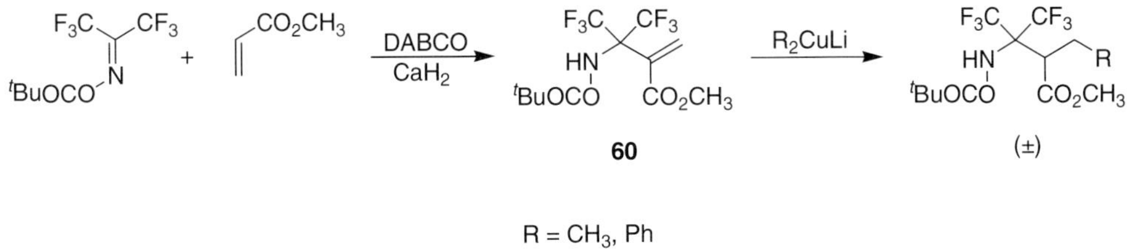

Scheme 7.61

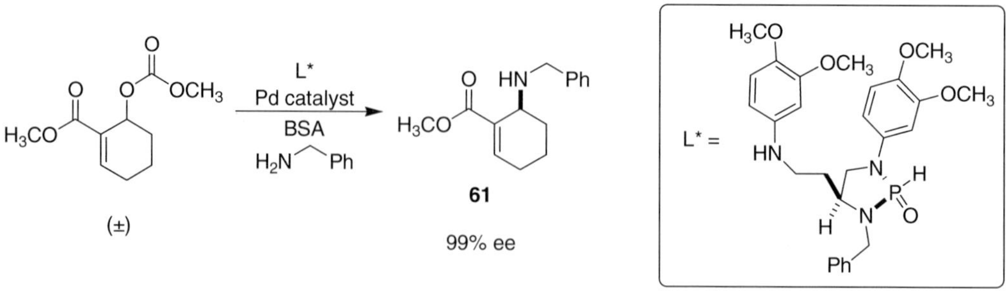

Scheme 7.62

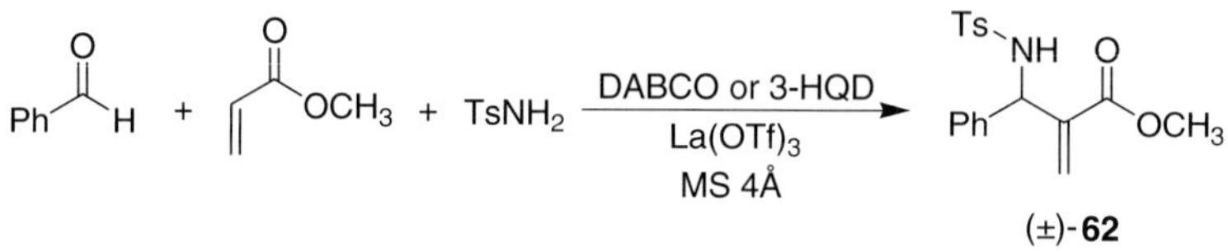

Scheme 7.63

Balan and Adolfsson [91] have developed a protocol based on a three-component reaction mixture of arylaldehydes, sulfonamides, and a Michael acceptor to afford moderate to high yields of α-methylene-β-sulfonylamido carbonyl derivatives **62**. The reactions are catalyzed by base (1,4-diazabicyclo[2,2,2]octane or 3-3-hydroxyquinuclidine) and Lewis acid [$La(OTf)_3$], in the presence of molecular sieves (4 Å). The products are easily isolated by simple extraction (Scheme 7.63).

7.6.3 Mannich-Type Reactions

A most useful reaction for the formation of β-amino carbonyl compounds is the one-step Mannich reaction. This reaction is a multicomponent condensation of a nonenolizable aldehyde, a primary or secondary amine, and an enolizable carbonyl compound to afford aminomethylated products (Scheme 7.64). This reaction has been widely explored by a great number of research groups. Some representative publications will be described in this work.

Among salient developments, Das *et al.* [92] report the one-pot, three-component condensation of aldehydes, ketones or keto esters, and benzyl carbamate in acetonitrile, with $Sc(OTf)_3$ or $Yb(OTf)_3$ as catalysts to furnish the corresponding Cbz-protected β-amino ketones. On the other hand, Khan's group [93] has described

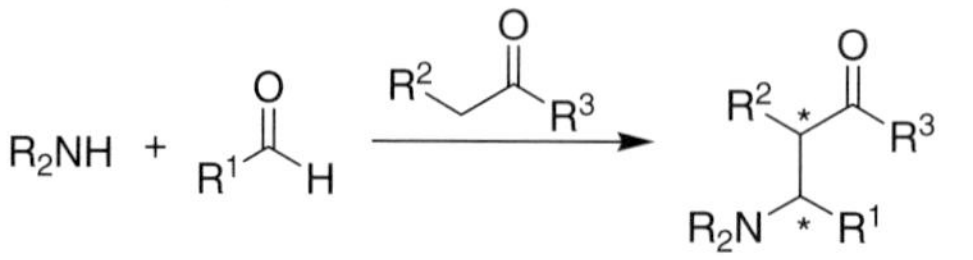

Scheme 7.64

the use of $CeCl_3{\cdot}H_2O$ as a Lewis acid catalyst for this purpose, achieving good diastereoselectivities in the Mannich products. Furthermore, Shou *et al.* [94] described the three-component synthesis of β-amino carbonyl compounds via Zn$(OTf)_2$-catalyzed cascade reaction of anilines with aromatic aldehydes and carbonyl compounds. This reaction was also applied to the synthesis of an α,β-amino carbonyl compound library on polyethylene glycol. Wu *et al.* [95] have synthesized a series of novel β-amino carbonyl derivatives via a three-component Mannich reaction of aromatic amines with aromatic aldehydes and active methylene compounds catalyzed by a (salen)Zn complex under mild reaction conditions. Exploring the use of rare earth elements, Wang *et al.* [96] examined several rare earth perfluorooctanoates $[RE(PFO)_3]$ for the synthesis of β-amino carbonyl compounds. On the other hand, Loh *et al.* [97] used $InCl_3$ as catalyst in a one-pot Mannich reaction in water to give high yields of the formation of β-amino ketones, esters, or acids with the recovery of the catalyst when the reaction is complete. Alternatively, Dondoni *et al.* [98] have carried out the preparation of *C*-galactosyl and *C*-ribosyl β-amino acids by reaction of formyl *C*-glycoside, *p*-methoxybenzyl amine, and a ketene silyl acetal in the presence of catalytic $InCl_3$ in MeOH at room temperature for 12 h. In this sense, Phunkan *et al.* [99] described that iodine is a very effective catalyst for a Mannich reaction between an aryl aldehyde, an aryl ketone, and benzyl carbamate, to produce Cbz-protected β-aryl β-amino carbonyl compounds in high yields. Furthermore, Azizi *et al.* [100] showed that heteropoly acids (such as $H_3PW_{12}O_{40}$ or $H_3PMo_{12}O_{40}$) efficiently catalyzed the one-pot, three-component Mannich reaction of ketones with aromatic aldehydes and different amines in water at ambient temperature to afford the corresponding β-amino carbonyl compounds **63** in good yields and with moderate diastereoselectivity (Scheme 7.65).

In an interesting contribution, Takaya *et al.* [101] reported the synthesis of racemic β-amino-β-trifluoromethyl carbonyl compounds by treatment of trifluoroacetaldehyde ethyl hemiacetal with *p*-anisidine in diethyl ether in the presence of molecular sieves of 4 Å at room temperature for 1.5 h. This protocol afforded the hemiaminal **64**, which reacted with the selected silyl enol ether in the presence of $GaCl_3$ to give the *N*-aryl-substituted β-amino carbonyl compound. The *N*-aryl substituent was readily

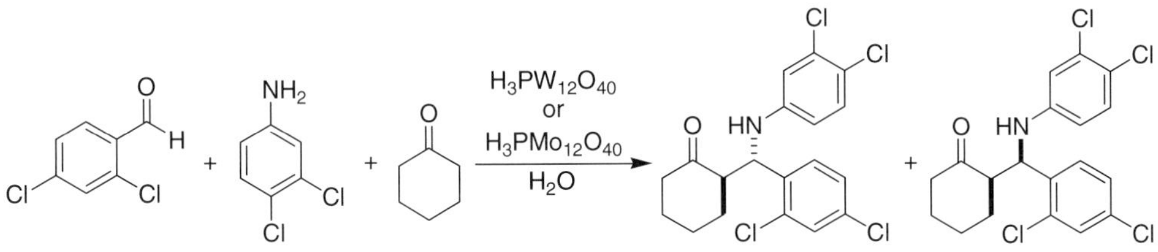

$H_3PW_{12}O_{40}$ or $H_3PMo_{12}O_{40}$, H_2O

anti-**63** *syn*-**63**

Relation *anti/syn*: $H_3PW_{12}O_{40}$, 41:59
$H_3PMo_{12}O_{40}$, 39:61

Scheme 7.65

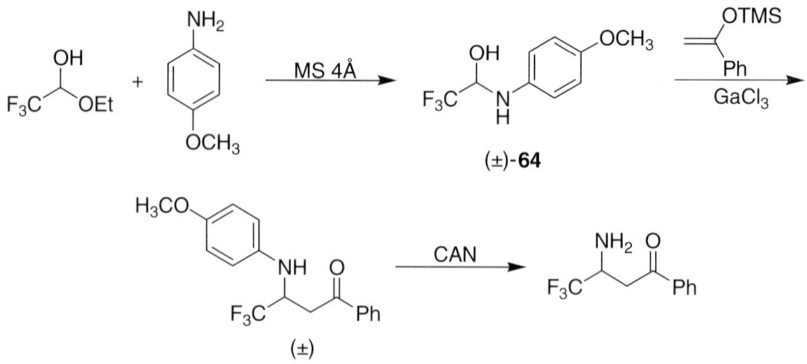

Scheme 7.66

removed with CAN in acetonitrile/water (9 : 1) at 0 °C for 1 h to give the expected β-amino-β-trifluoromethyl ketone in a good yield (Scheme 7.66).

Ollevier and Nadeau [102] disclosed the development of a bismuth-catalyzed Mannich-type three-component reaction that combines an aldehyde, aniline, and silyl ketene acetal to afford compounds with a β-amino ester core structure. This protocol works more efficiently in the presence of 2 mol% of $Bi(OTf)_3 \cdot nH_2O$ (with $1 < n < 4$).

Using a Brønsted acidic ionic liquid containing nucleophilic 1-methylimidazole and triphenylphosphine with 1,4-butane sultone, inorganic anions, *p*-toluenesulfonic acid, and TFA, Sahoo *et al.* [103] succeeded in developing a catalytic Mannich reaction that affords β-amino carbonyl compounds in excellent yield and short reaction times. The ionic liquid was easily separated from the reaction mixture by water extraction and was recycled four times without any loss in activity. Other ionic liquids that have been used in this type of approach include 1-butyl-3-methylimidazolium hydroxide [bmim]OH [104], 4-(trimethylammonium) butanesulfonic acid bisulfate [TMBSA]HSO_4 [105], ethyl methylimidazolium triflate [emim]OTf [106], and surfactant (sodium dodecylsulfate) in aqueous media [107].

In a pioneering organocatalytic approach, Wenzel and Jacobsen [108] reported the enantioselective synthesis of β-amino esters using a chiral urea or thiourea as catalyst. For example, a variety of β-amino esters were prepared in 87–99% yields and 86–98% e.e. from the *N*-Boc aldimines and silyl ketene acetals in the presence of nonracemic urea catalyst **65** (Scheme 7.67).

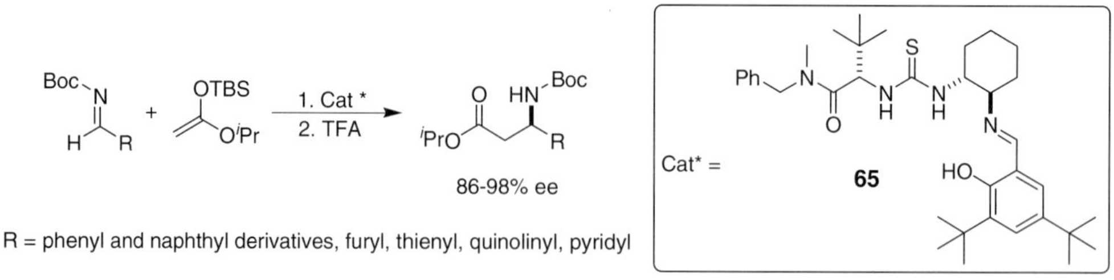

R = phenyl and naphthyl derivatives, furyl, thienyl, quinolinyl, pyridyl

Scheme 7.67

In this regard, Song *et al.* [109] have described the application of cooperative hydrogen-bonding catalysis with a readily accessible bifunctional cinchona alkaloid catalyst bearing either a 6′- or 9-thiourea functionality. Soon after, Yamaoka *et al.* [110] reported the use of a related chiral thiourea that catalyzes the formation of β-amino carbonyl compounds with high enantio- and diastereoselectivities.

Córdova *et al.* [111] described a proline-catalyzed reaction of *N*-PMP-protected α-imino ethyl glyoxylate with unmodified aliphatic aldehydes, which provided a general and very mild entry to either enantiomer of β-amino acids and derivatives in high yield and stereoselectivity. For example, $OHCCH_2R$ and $PMPN{=}CHCO_2Et$ reacted in the presence of L-proline in dioxane for 2–24 h at room temperature to give protected β-amino acids with enantiomeric excess values near 99%. The authors found that the diastereoselectivity of the reaction increased with the bulkiness of the substituents of the aldehyde. This approach provides facile access to substituted β-lactams [e.g., **66** (Scheme 7.68)].

Related work reporting the use of proline in the preparation of diastereomerically and enantiomerically pure β-substituted-β-amino derivatives has been described by Chowdari *et al.* [112], Yang *et al.* [113], Cobb *et al.* [114], Notz *et al.* [115], Vesely *et al.* [116], and most recently by Chi *et al.* [117].

An interesting modification of the proline catalyst was realized by Davies *et al.* [118], who discovered that rhodium prolinate catalyst $Rh_2[(S)\text{-DOSP}]_4$ [DOSP (4-dodecylphenylsulfonyl)prolinate] induces the asymmetric C–H activation reactions of methyl aryldiazoacetates in the synthesis of β-amino esters derivatives **67** (Scheme 7.69).

In an excellent contribution, Moody [119] discusses the application of oxime ethers prepared from (*R*)- or (*S*)-*O*-(1-phenylbutyl)hydroxylamine **68** as versatile intermediates for the asymmetric synthesis of β-amino acids. For example, β^3-substituted-β-amino acid derivatives were prepared by addition of an allyllithium to a ketoxime followed by functional group transformation (Scheme 7.70).

Another interesting method was described by Has-Becker *et al.* [120] who applied a new protocol for the alkenylation of carbonylic phosphonates. Thus, aldehydes react

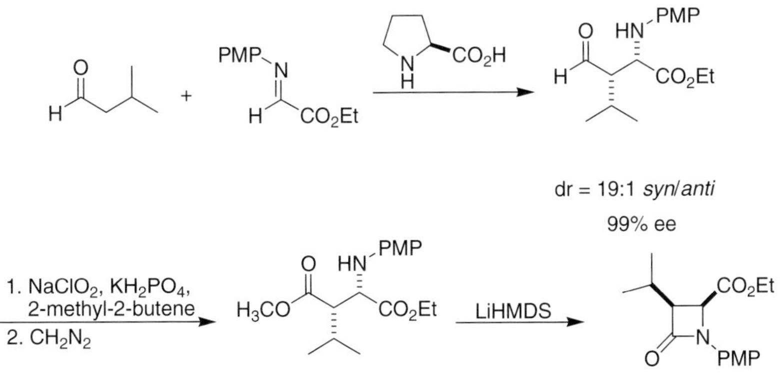

PMP, N, H, CO_2Et, CO_2H, O, HN, PMP, CO_2Et

dr = 19:1 *syn*/*anti*

99% ee

1. $NaClO_2$, KH_2PO_4, 2-methyl-2-butene
2. CH_2N_2

H_3CO, O, HN, PMP, CO_2Et, LiHMDS, CO_2Et, N, PMP, O

66

Scheme 7.68

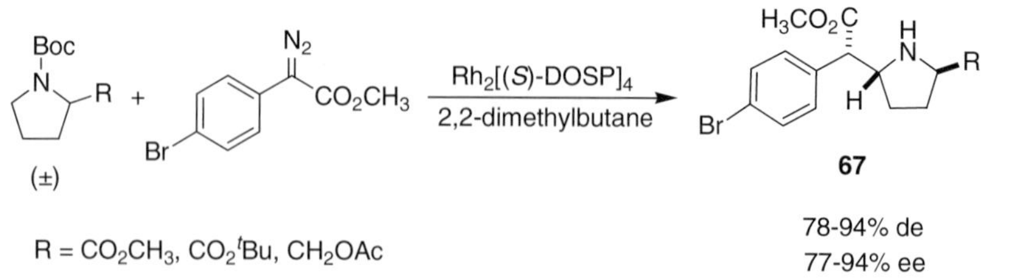

Scheme 7.69

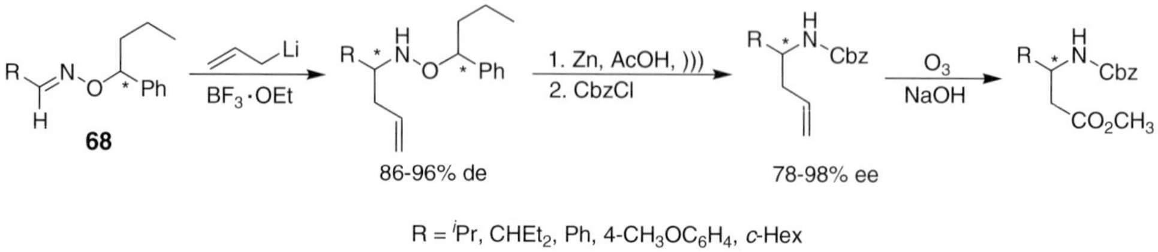

Scheme 7.70

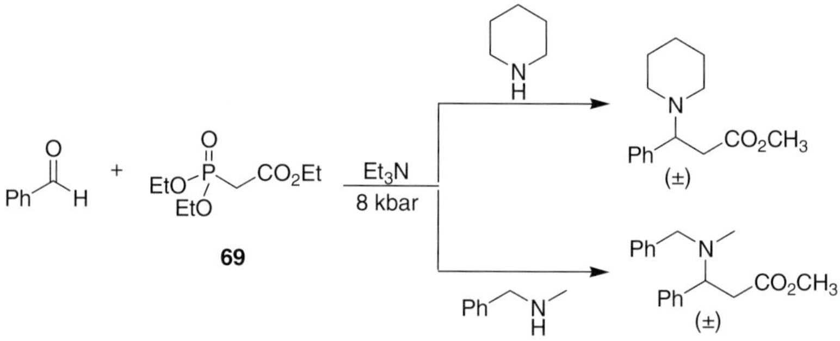

Scheme 7.71

with phosphonates such as **69**, and a domino process combining the Horner–Wadsworth–Emmons reaction with a Michael reaction results in the one-pot synthesis of β-amino esters (Scheme 7.71).

7.7
Synthesis of β-Amino Acids by Diverse Addition or Substitution Reactions

There exist several recent reports where the synthesis of various β-amino carbonylic compounds is achieved by appropriate substitution methodologies. For example, using a (*R*)-(+)-camphor derivative, Palomo *et al.* [121] described the synthesis of several β-amino esters and β-amino acids with excellent enantiomeric excess values. In particular, the lithium enolate of the methyl ketone **70** obtained from (1*R*)-(+)-

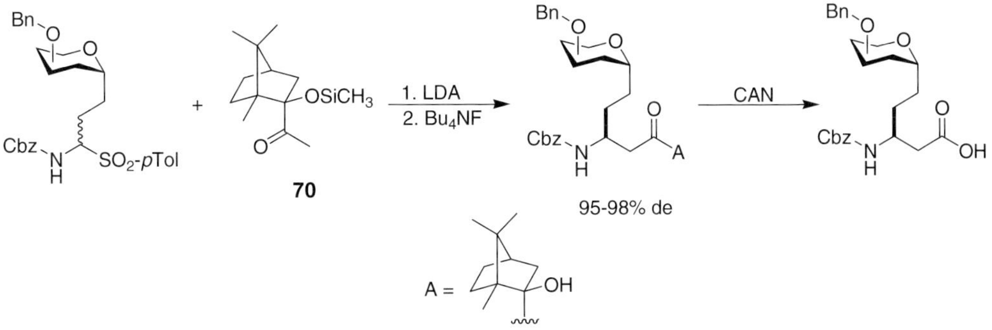

Scheme 7.72

camphor, reacted with sulfones of the general type $4\text{-MeC}_6H_4SO_2CHRNHR^1$ to give adducts that were converted to β-amino acids (e.g., glucopyranose-amino acids) (Scheme 7.72).

On the other hand, Fadini and Togni [122] have used complexes of the type [Ni(PPP)(solvent)]X_2 as catalysts in the asymmetric hydroamination of activated olefins containing ferrocenyl tridentate ligands. Fadini and Togni carried out the preparation of several products, for example **71**, which were then readily converted to the corresponding β-amino acid derivatives (Scheme 7.73).

Shin *et al.* [123] reported a novel asymmetric synthetic route to β-amino acids from a chiral 4-phenyloxazolidinone, which acts both as a chiral auxiliary and as an amine source. Thus, propionyl-phenyloxazolidinone **72** was stereoselectively reduced and the hydroxyl group protected as a trimethylsilyl ether, which was treated with $H_2C{=}C(OEt)OSiMe_2tBu$ to give the corresponding ethyl ester **73** as precursor to highly enantioenriched β-amino acids. Indeed, a two-step deprotection procedure of ethyl ester **73** provides the desired β-amino acid **74** (Scheme 7.74).

A related methodology was employed by Huguenot and Brigaud [124] in the synthesis of β-trifluoromethyl β-amino acid, (*R*)-$F_3CCH(NH_2)CH_2CO_2H$, from nonracemic trifluoromethyloxazolidines. Reaction of trifluoromethyloxazolidine

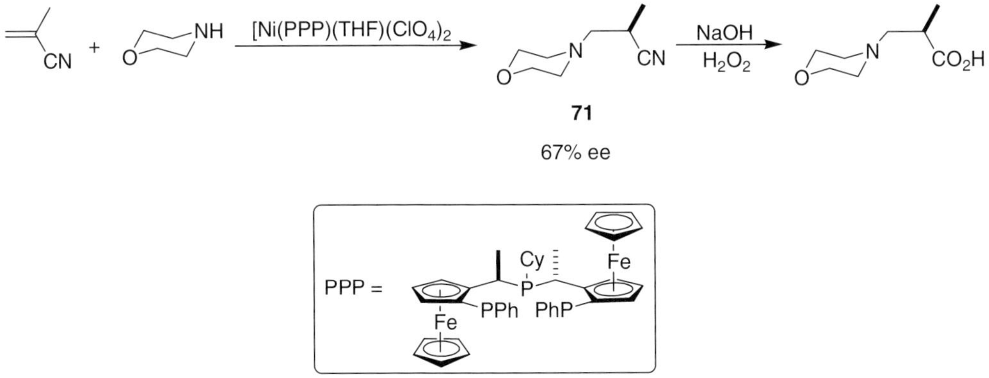

Scheme 7.73

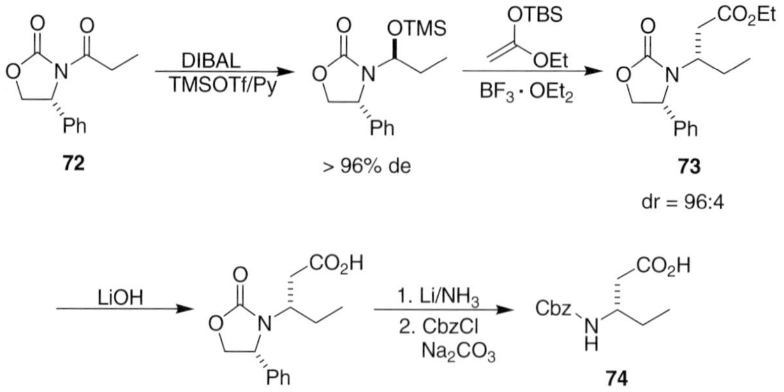

Scheme 7.74

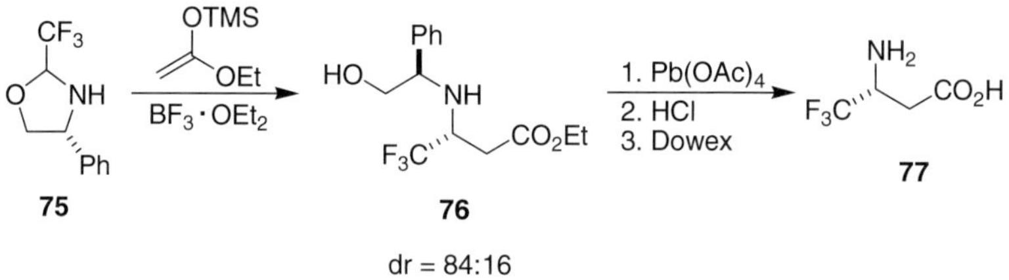

Scheme 7.75

75 with trimethylsilyl ketene acetals or trimethylsilyl enol ethers in the presence of $BF_3 \cdot OEt_2$ yields nonracemic β-amino esters **76** or β-amino ketones stereoselectively. Cleavage of the (*R*)-α-(hydroxymethyl)benzyl groups with $Pb(OAc)_4$ followed by acid hydrolysis provided the enantioenriched chiral β-amino acid **77** (Scheme 7.75).

In a detailed study, Hegedus *et al.* [125] used a lithiation/stannylation protocol on optically active *N*-propargyloxazolidinones to produce enantioenriched α-oxazolidinonyl-allenylstannanes **78**. Reaction of these with aldehydes in the presence of $BF_3 \cdot OEt_2$ produced β-hydroxypropargylamines, with high *syn* diastereo- and enantioselectivity. These compounds were converted to γ-hydroxy-β-amino acids, by oxidative cleavage of the alkyne (Scheme 7.76).

7.8 Synthesis of β-Amino Acids by Stereoselective Hydrogenation of Prochiral 3-Aminoacrylates and Derivatives

An alternative method for the synthesis of enantiopure β-amino acid derivatives involves the asymmetric hydrogenation of 3-amino acrylate substrates, usually by application of chiral rhodium or ruthenium organometallic complexes as catalysts. The configuration of the new center of chirality can be controlled by use of either chiral ligands or a chiral functionality on the acrylate substrate [126].

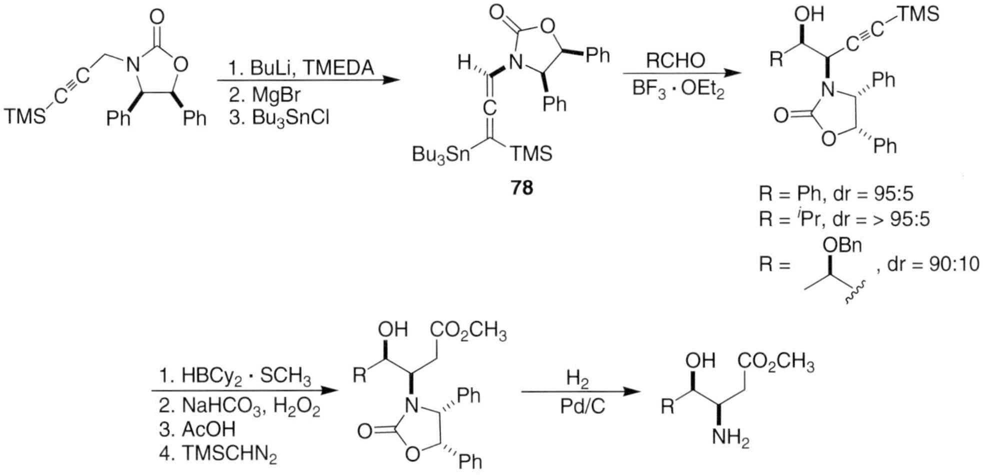

Scheme 7.76

7.8.1
Reductions Involving Phosphorus-Metal Complexes

In 2001, Saylik *et al.* [127] reported the enantioselective hydrogenation of several α,β-unsaturated methyl esters **79** bearing a phthalimidomethyl substituent at the α-carbon using a (*S*)-BINAP-Ru(II) catalyst. These researchers obtained β-amino acid precursors with enantiomeric excesses in the range of 84% (Scheme 7.77).

Examining a series of representative rhodium catalysts, Feringa *et al.* [128] discovered that chiral monodentate phosphoramidite ligands **80** or **81** induce high enantiomeric excess and full conversions in the hydrogenation of (*Z*)- and (*E*)-β-dehydroamino acid derivates containing both aliphatic and aromatic side-chains (Scheme 7.78).

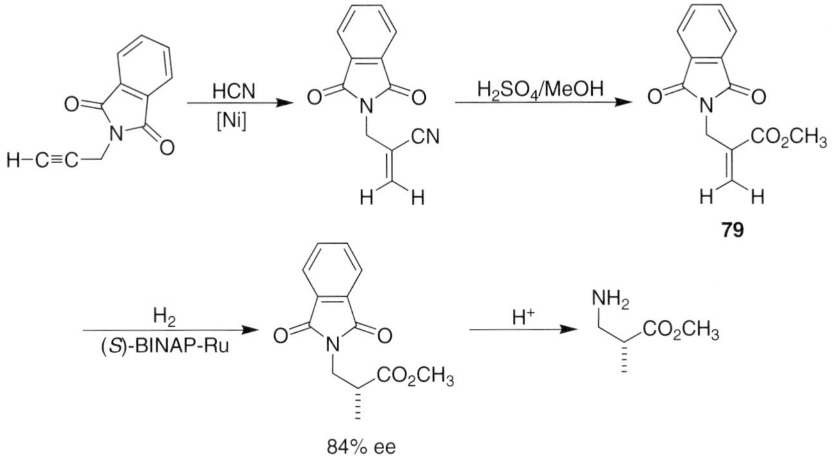

Scheme 7.77

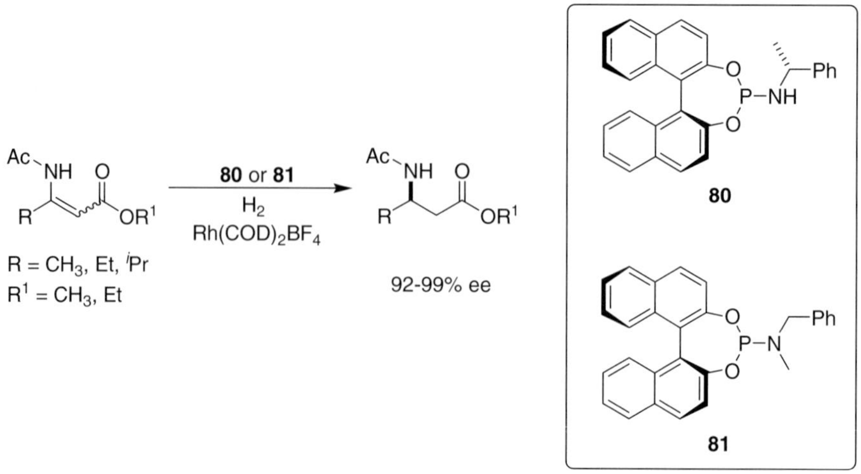
Ac
NH O
R
OR1
R = CH3, Et, iPr
R1 = CH3, Et
80 or 81
H2
Rh(COD)2BF4
Ac
NH O
R
OR1
92-99% ee
O
P–NH
O
Ph
80
O
P–N
O
Ph
81

Scheme 7.78

In this context, Ohashi *et al.* [129] have developed a rhodium norbornadiene (Rh [nbd]$^{+}BF_4^{-}$) complex containing unsymmetrical bis(phosphanes), **82**, which achieve the enantioselective hydrogenation of (*Z*)-dehydro-β-amino acid and enamide derivatives as substrates (Scheme 7.79).

In a salient development, Tang *et al.* [130] reported an asymmetric hydrogenation of (*Z*)-β-(acylamino)acrylic acid derivates with chiral bisphosphepin ligand **83**, which combines both C_2-chirality and stereogenic phosphorus centers. For example, excellent enantioselectivities and reactivities were observed in the rhodium-catalyzed asymmetric hydrogenation of (*Z*)-Ar(NHAc)C=CHCO$_2$Me substrates giving (*S*)-Ar (NHAc)CHCH$_2$CO$_2$Me β-amino acids (Scheme 7.80).

In another pioneering development, Holz *et al.* [131] described the applicability of chiral bisphospholane ligand **84** bearing a maleic anhydride backbone. Rhodium MalPHOS catalyst was particularly effective in the enantioselective hydrogenation of (*Z*)-configured β-acylamido acrylates derivatives, in particular when polar solvents were used (Scheme 7.81).

In this context, Reetz and Li [132] have obtained β-amino acid derivatives in high yields and enantiomeric excess by the use of mixtures of two different chiral

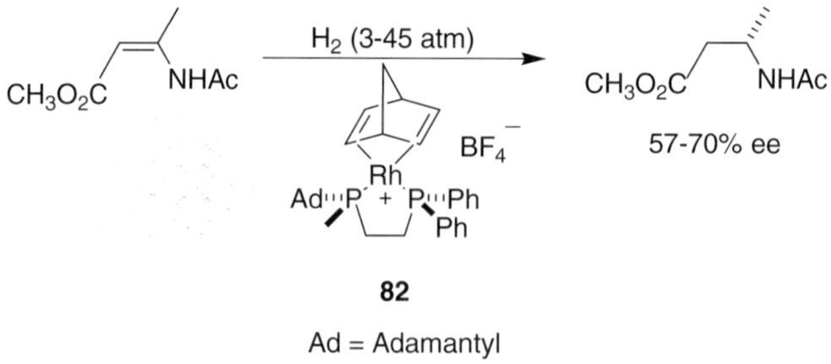
CH3O2C
NHAc
H2 (3-45 atm)
BF4
Rh
Ad P + P Ph
Ph
CH3O2C
NHAc
57-70% ee
82
Ad = Adamantyl

Scheme 7.79

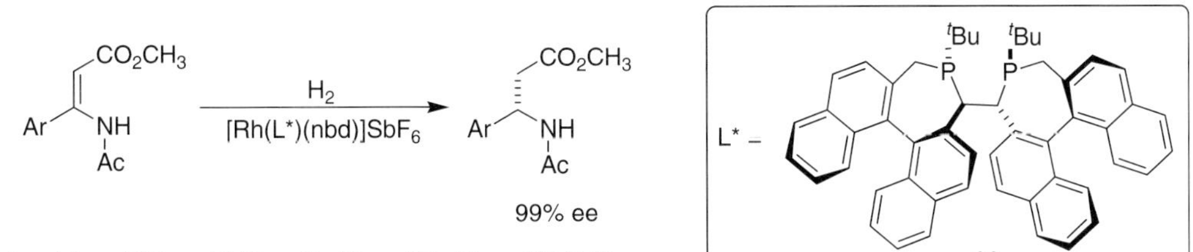

H_2, $[Rh(L^*)(nbd)]SbF_6$; 99% ee

Ar = Ph, p-F-Ph, p-Cl-Ph, p-Br-Ph, p-CH_3-Ph, p-CH_3O-Ph, p-BnO-Ph; o-CH_3-Ph, o-CH_3O-Ph

L* = **83**

Scheme 7.80

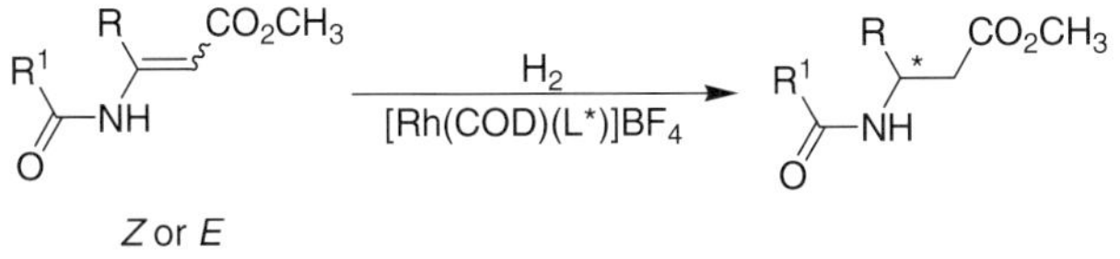

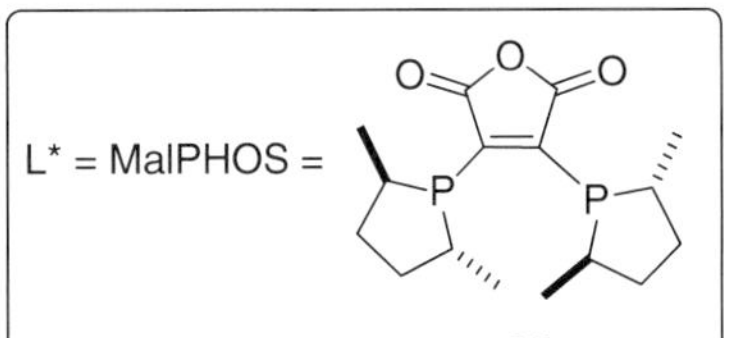

H_2, $[Rh(COD)(L^*)]BF_4$

Z or E

L* = MalPHOS = **84**

R = CH_3, R^1 = CH_3; (Z) 83% ee (R), (E) 97% ee (R)
R = CH_3, R^1 = Ph; (Z) 83% ee (R), (E) 98% ee (R)
R = Et, R^1 = CH_3; (Z) 81% ee (R), (E) 98% ee (R)
R = ^{i}Pr, R^1 = CH_3; (Z) 80% ee (S), (E) 98% ee (S)
R = Ph, R^1 = CH_3; (Z) 85% ee (S), (E) 84% ee (S)

Scheme 7.81

monodentate *P*-ligands in the rhodium-catalyzed hydrogenation of β-acylamino acrylates. Mixtures of BINOL-derived phosphites and phosphonites turned out to be particularly effective (94–99% e.e.). In similar fashion, Monti *et al.* [133] reported that enantiomeric excesses up to 98% were obtained for the dehydro-α-amino acids, by using the best combination of a phosphite and a phosphoramidite in the rhodium-catalyzed asymmetric hydrogenation of methyl (*Z*)-3-acetamidocrotonate.

In an extensive work, Enthaler *et al.* [134] have shown that monodentate chiral 4,5-dihydro-3*H*-dinaphthophosphepine ligands **85** can be used in various rhodium- and ruthenium-catalyzed hydrogenations allowing the synthesis of β-amino acid derivatives with good to high enantioselectivities (Scheme 7.82).

In similar fashion, Hsiao *et al.* [135] employed an interesting catalyst in the direct asymmetric hydrogenation of unprotected enamino esters and amides. Catalyzed by Rh complexes with Josiphos type chiral ligands **86** and **87**, this method gives β-amino esters and amides in high yield and high enantiomeric excess (82–97% e.e.), without the need of acyl protection/deprotection steps (Scheme 7.83).

7.8.2
Reductions Involving Catalytic Hydrogenations

An interesting reduction procedure was presented by Cohen *et al.* [136] in the enantioselective synthesis of esters of β-aryl-β-amino acids via reduction of

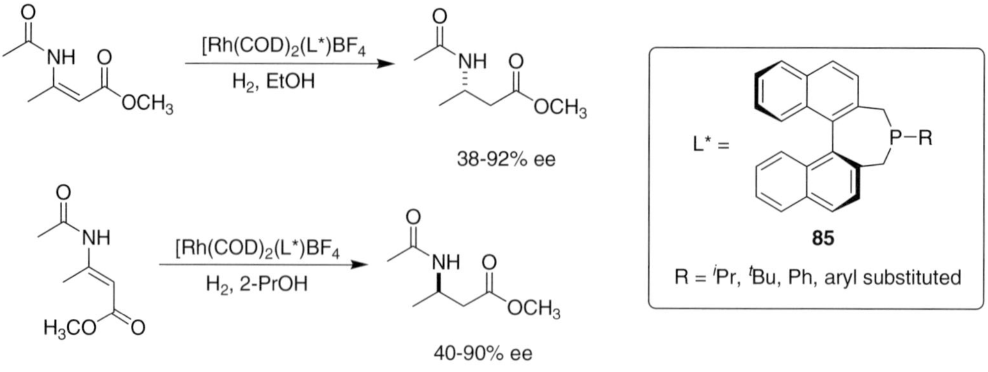

[Rh(COD)$_2$(L*)BF$_4$ / H$_2$, EtOH → 38-92% ee

[Rh(COD)$_2$(L*)BF$_4$ / H$_2$, 2-PrOH → 40-90% ee

L* = 85; R = ^{i}Pr, ^{t}Bu, Ph, aryl substituted

Scheme 7.82

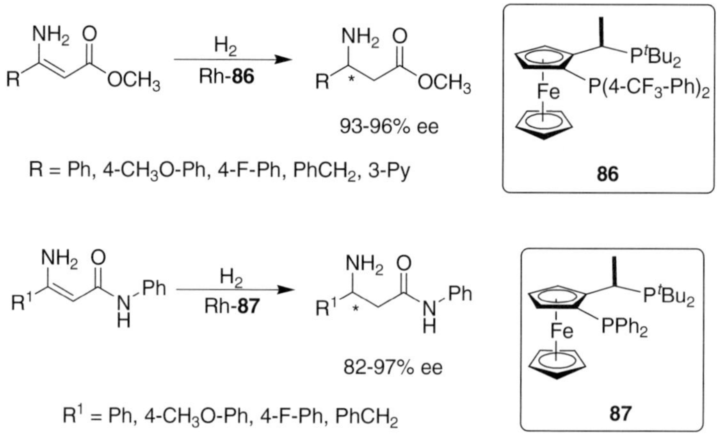

H$_2$ / Rh-**86** → 93-96% ee

R = Ph, 4-CH$_3$O-Ph, 4-F-Ph, PhCH$_2$, 3-Py

86: P^{t}Bu$_2$, P(4-CF$_3$-Ph)$_2$, Fe

H$_2$ / Rh-**87** → 82-97% ee

R^1 = Ph, 4-CH$_3$O-Ph, 4-F-Ph, PhCH$_2$

87: P^{t}Bu$_2$, PPh$_2$, Fe

Scheme 7.83

enantioenriched esters of *N*-(*p*-methoxy-α-methylbenzyl)enamines by catalytic hydrogenation with Pd(OH)$_2$/C followed by debenzylation with Et$_3$SiH and HCO$_2$H (Scheme 7.84).

Cimarelli *et al.* [137] carried out the stereoselective reduction of β-enamino esters **88** with sodium triacetoxyborohydride in acetic acid to give the corresponding benzyl β-amino esters, which were hydrogenolyzed with Pd(OH)$_2$/C, obtaining several open chain and cyclic β-amino acids (Scheme 7.85).

7.9
Synthesis of β-Amino Acids by use of Chiral Auxiliaries: Stereoselective Alkylation

A chiral auxiliary is an optically active compound that is temporarily incorporated into a substrate molecule to induce the selective formation of one of more new stereogenic centers. Following the creation of the new stereocenter(s), the auxiliary is generally removed (and sometimes reused).

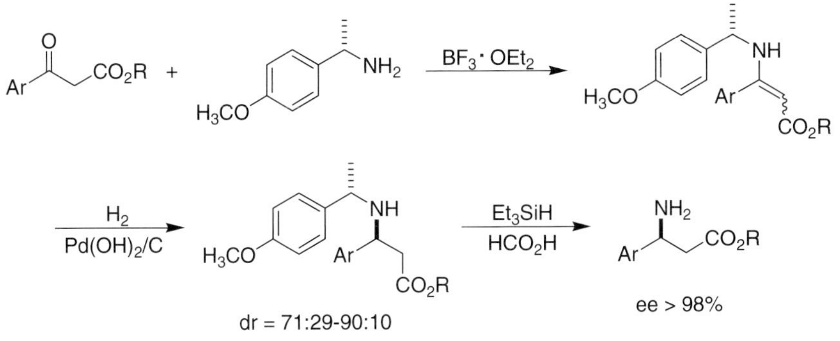

$BF_3 \cdot OEt_2$; H_2, $Pd(OH)_2/C$; Et_3SiH, HCO_2H

dr = 71:29-90:10

ee > 98%

Ar = pyridyl- or phenyl-substituted

R = CH_3, CH_2CH_3

Scheme 7.84

For instance, Sibi and Deshpande [138] reported the synthesis of β-amino acids via functionalization of linear dicarboxylic acid derivatives in a regio- and stereoselective manner. The two carboxy groups are differentiated by forming an ester at one end and by attachment of a chiral auxiliary (oxazolidinone **89**) to the other carboxy group. Having differentiated the two ends, the first step consists of a regio- and stereoselective functionalization at the carbon α to the imide. The second step involves the selective removal of either the imide or the ester functionality followed by a Curtius rearrangement of the free carboxy group with retention of stereochemistry (Scheme 7.86).

Based on the use of pseudoephedrine as chiral auxiliary, Nagula *et al.* [139] showed that the stereoselective alkylation of β-alanine derivatives provides a convenient,

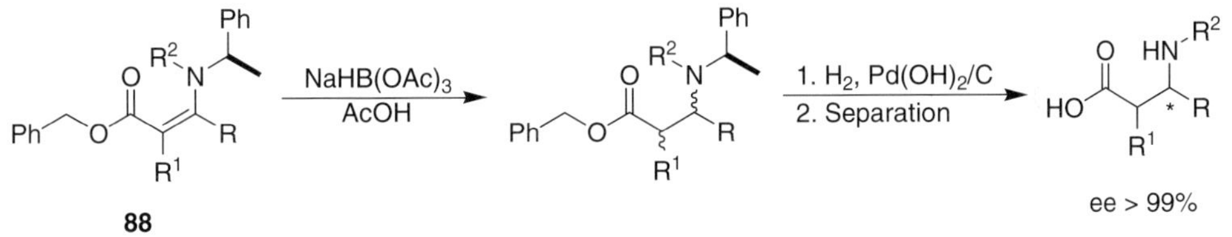

88

$NaHB(OAc)_3$, AcOH; 1. H_2, $Pd(OH)_2/C$ 2. Separation

ee > 99%

R	R^1	R^2	Configuration
CH_3	H	H	(*R*) or (*S*)
CH_3	—$(CH_2)_2$—		(*R*,*R*)
^{i}Pr	H	H	(*S*)
Ph	H	H	(*R*)
—$(CH_2)_3$—		H	(1*S*,2*R*)
—$(CH_2)_4$—		H	(1*S*,2*R*)

Scheme 7.85

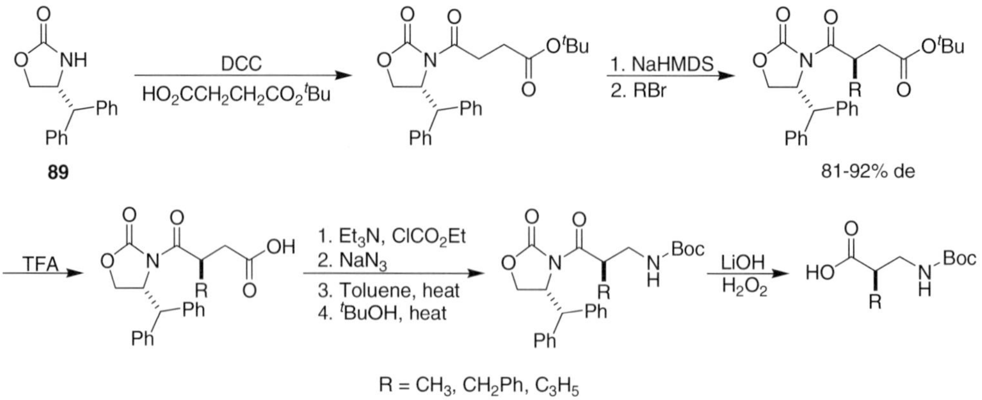

Scheme 7.86 DCC: dicyclohexylcarbodiimide

enantioselective route to α-alkyl β-amino acids. Indeed, Boc-β-alanine **90** was coupled to (*R*,*R*)-pseudoephedrine using a mixed anhydride method. Lithiation of the pseudoephedrine amide was accomplished with hexamethyldisilazane in the presence of excess LiCl. The deprotection step generally gave the desired product in good yield and with a high degree of stereoselectivity (Scheme 7.87).

Along similar lines, Iza *et al.* [140] described the use of (*S*,*S*)-pseudoephedrine as chiral auxiliary in the synthesis of β-substituted α-alkyl-β-amino amides, acids and lactams with full stereocontrol (Scheme 7.88).

Gutiérrez-García *et al.* [141] and Guzmán-Mejía *et al.* [142] describe the use of (*R*,*R*)-bis(α-phenylethyl)amine in the preparation of (*S*)-α-phenylethyl-β-alanine derivative **91** which, was alkylated with high diastereoselectivity. Hydrogenolysis of the alkylated product and final hydrolysis (4 N HCl) under microwave irradiation resulted in the formation of (*S*)-α-benzyl-β-alanine (Scheme 7.89).

On the other hand, Agami *et al.* [143] reported a synthetic method that involves the use of (*S*)-phenylglycinol-derived dihydrooxazolo pyrimidinone **92**, which was used in diastereoselective Michael additions with different organocuprate reagents. Following hydrogenolysis, the dihydropyrimidine dione was formed, and subse-

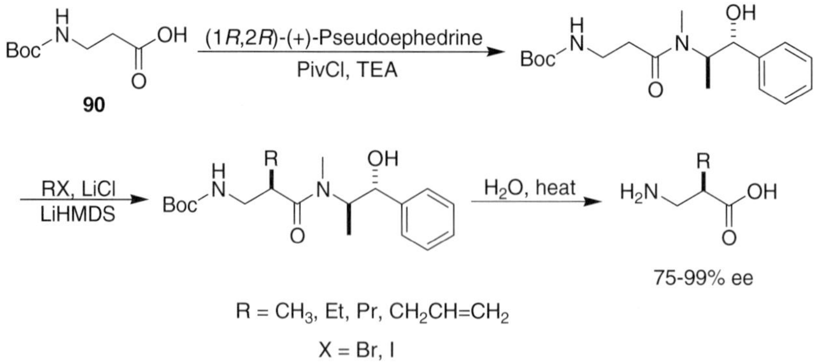

Scheme 7.87 PivCl: pivaloyl chloride.

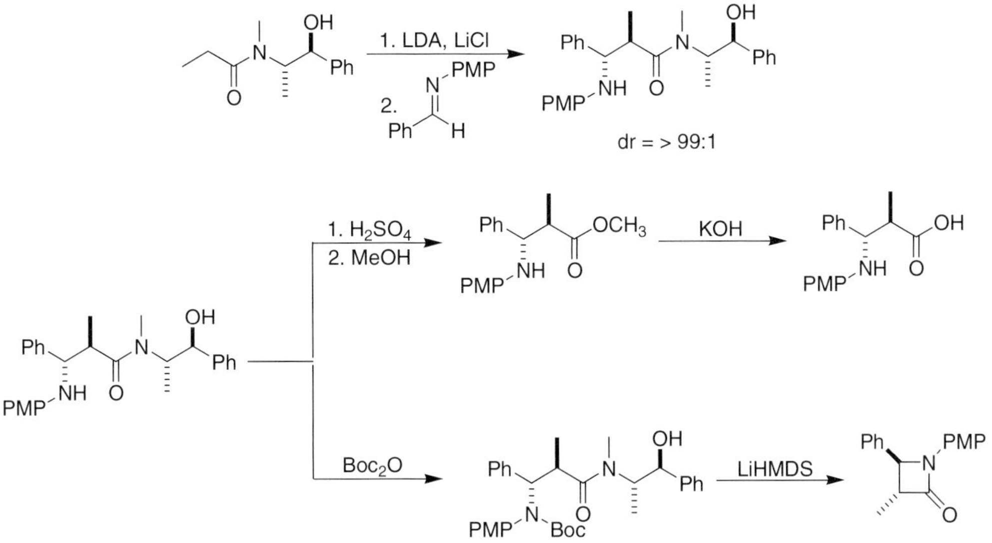

1. LDA, LiCl
2. PMP-N=CH-Ph
dr = > 99:1

1. H_2SO_4
2. MeOH
KOH
Boc_2O
LiHMDS

Scheme 7.88

quent hydrolysis and hydrogenolysis afforded the desired substituted β-amino acid optically active (Scheme 7.90).

A successful application of a chiral oxazolidinone auxiliary in the synthesis of *N*-Fmoc-protected β^2-homo isoleucine, tyrosine, and methionine was described by Sebesta and Seebach [144]. The diastereoselective amidomethylation of the titanium-enolates of 3-acyl-4-isopropyl-5,5-diphenyloxazolidin-2-ones **93** with $CbzNHCH_2OMe/TiCl_4$ proceeds in good yields and diastereoselectivities. Removal of the chiral auxiliary with LiOH or NaOH gave the *N*-Cbz-protected β-amino acids, which were

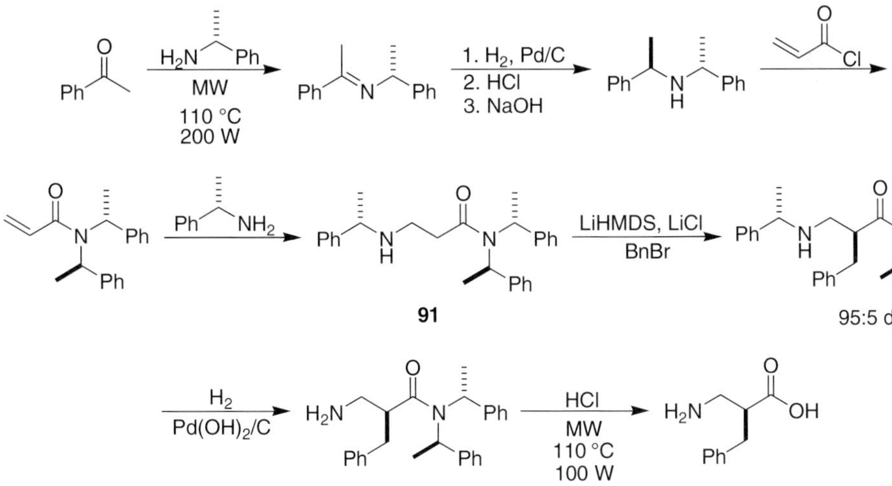

MW
110 °C
200 W
1. H_2, Pd/C
2. HCl
3. NaOH
LiHMDS, LiCl
BnBr
91
95:5 dr
H_2
$Pd(OH)_2/C$
HCl
MW
110 °C
100 W

Scheme 7.89

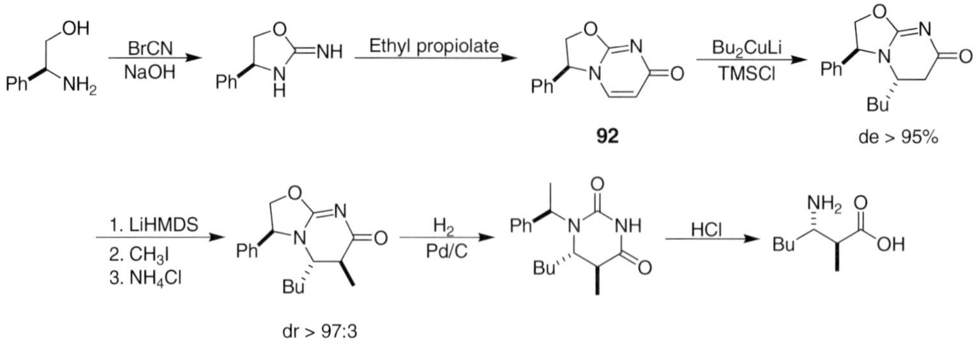

Scheme 7.90

subjected to an *N*-Cbz removal by hydrogenolysis and subsequent *N*-Fmoc protection (Scheme 7.91).

Aoyagi *et al.* [145] described the stereocontrolled asymmetric synthesis of α-hydroxy-β-amino acids via the Lewis acid-promoted cyanation of chiral 1,4-oxazines **94** (prepared from the enantiopure aminoalcohol) with trimethylsilyl cyanide. The removal of the Cbz group with H_2 and Pd/C, and subsequent base-catalyzed hydrolysis of the resulting cyano compounds with KOH proceeded with excellent stereoselectivity, providing access to diastereomerically pure oxazine-2-carboxylic acids which were readily converted to each enantiomer of the α-hydroxy-β-amino acids (Scheme 7.92).

7.10
Synthesis of β-Amino Acids via Radical Reactions

Among the three most relevant chemical reaction species (i.e., cations, anions, and radicals), the last have recently drawn much attention due to their potential utility in the synthesis of biologically active acyclic and cyclic compounds.

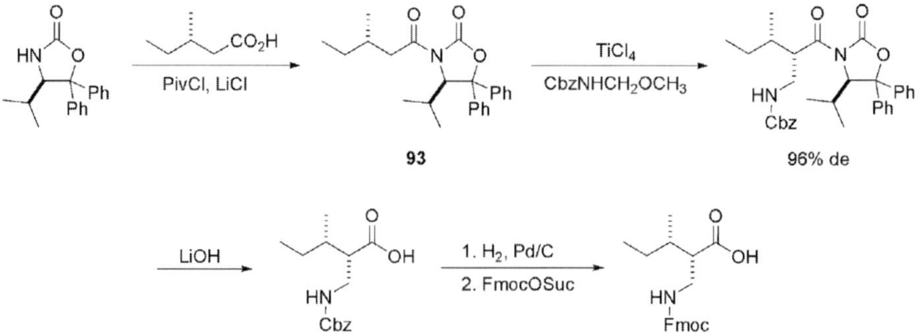

Scheme 7.91 Suc: succinimide.

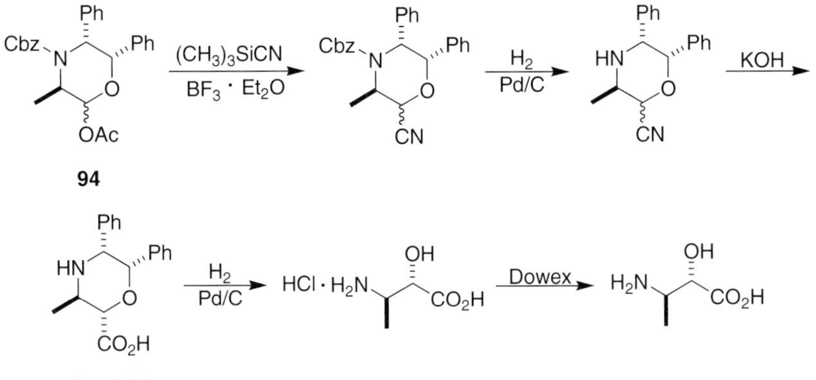

Scheme 7.92

For instance, Miyabe *et al.* [146] have described the synthesis of enantiomerically pure β-amino acid derivatives by means of diastereoselective radical addition to oxime esters **95** bearing Oppolzer's camphorsultam auxiliary. Alkyl addition to the oxime substrate was realized with triethylborane in the presence of $BF_3{\cdot}OEt_2$. Hydrogenolysis of the benzyloxy group and final removal of the sultam auxiliary by hydrolysis afforded the enantiomerically pure α,β-dialkyl-β-amino acids (Scheme 7.93).

In 2001, Huck *et al.* [147] described the synthesis of structurally diversified α-substituted β-amino acid derivatives via conjugate radical addition or Sonogashira coupling to unsaturated *N*-Boc-protected β-amino ester **96** (Scheme 7.94).

In this context, Miyata *et al.* [148] described the sulfanyl radical addition–cyclization of unsaturated oxime ethers for the synthesis of racemic cyclic β-amino acids. Thus, upon treatment with thiophenol in the presence of 2,2′-azobisisobutyronitrile, the starting oxime ethers underwent stereoselective sulfanyl radical addition–cyclization to give 2-(phenylsulfanylmethyl)-cycloalkylamines **97**. This method was successfully applied to the synthesis of 2-aminocyclopentanecarboxylic acid and 4-amino-3-pyrrolidinecarboxylic acid **98** (Scheme 7.95).

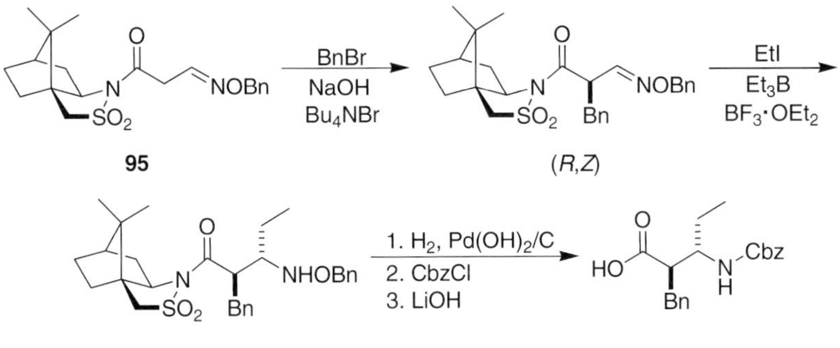

Scheme 7.93

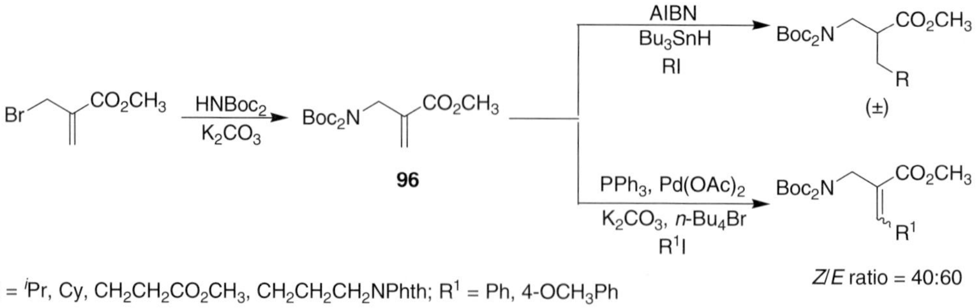

R = ^{i}Pr, Cy, $CH_2CH_2CO_2CH_3$, $CH_2CH_2CH_2NPhth$; R^1 = Ph, 4-OCH_3Ph

Scheme 7.94

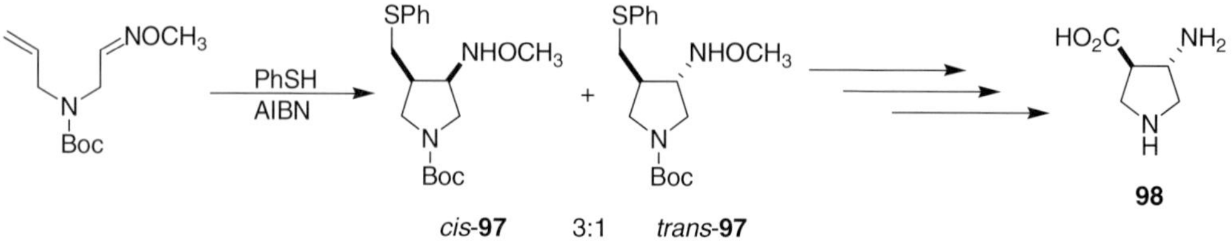

Scheme 7.95

7.11 Miscellaneous Methods for the Synthesis of β-Amino Acids

Ramachandran and Burghardt [149] developed an interesting methodology consisting of the allylation reaction of *N*-silyl-imines with *B*-allyldiisopinocampheylborane followed by oxidative workup to furnish the desired homoallylic amines in good yields and high enantiomeric excess. Protection of the homoallylic amines with Boc_2O, followed by oxidation of the olefinic bond provided the desired *N*-Boc-protected β-amino acids, which were isolated as hydrochloride salts (Scheme 7.96).

In another interesting development, Papa and Tomasini [150] described the synthesis of 2,2,3-trisubstituted *N*-benzoylaziridines **99**, which were transformed into polysubstituted α- or β-amino acids. Indeed, the *N*-benzoylaziridine ring can undergo ring opening at either C-2 or C-3 with concomitant expansion. Following this protocol, α-substituted α-hydroxy β-amino acids were obtained in racemic form (Scheme 7.97).

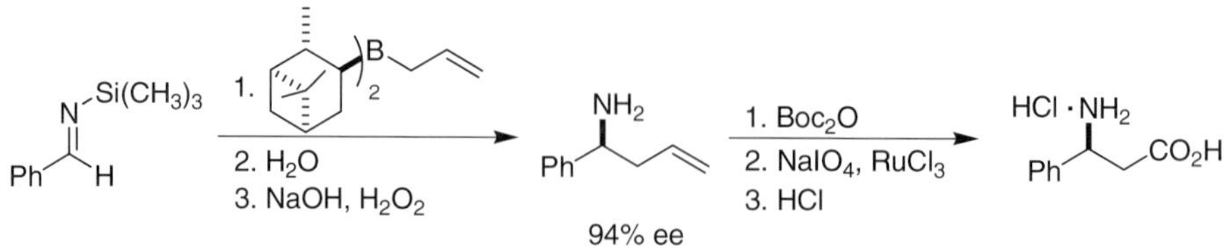

Scheme 7.96

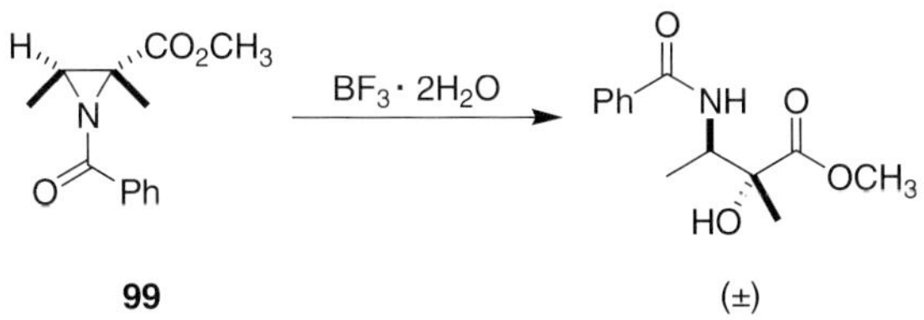

BF3 · 2H2O
99
(±)

Scheme 7.97

Other research groups that have reported aziridine opening in the synthesis of β-amino acids are those of Beresford [151], Mordini [152], Davis [153], Enders [154], Yamauchi [155], Wu [156], Matthews [157] and Crousse [158].

MacNevin *et al.* [159] described the stereoselective synthesis of unsaturated azetines **100** in the preparation of substituted β-amino acid derivatives employing an imidazolidinone chiral auxiliary in the presence of $TiCl_4$ and (−)-sparteine. Subsequent ring opening with benzoyl chloride and removal of the auxiliary provided the corresponding β-amino carbonyl derivatives in good yields (Scheme 7.98).

By contrast, Felpin *et al.* [160] reported a method for the regioselective preparation of deuterated β-amino acids based on the rearrangement of α-aminocyclopropanone hydrates. In this context, Seebach *et al.* [161] have described the synthesis of several β-amino acids labeled with ^{13}C and ^{15}N and for incorporation in peptidic sequences.

On the other hand, Schwarz *et al.* [162] reported the formation of α-cyclopropane-substituted β-amino acids by reaction of cyclopentanone with methyl cyanoacetate under azeotropic dehydration conditions. Nitromethane addition and subsequent cyclization under basic conditions provided spirocyclopropane **101**, which underwent chemoselective reduction with a sodium borohydride/Co(II) reagent system.

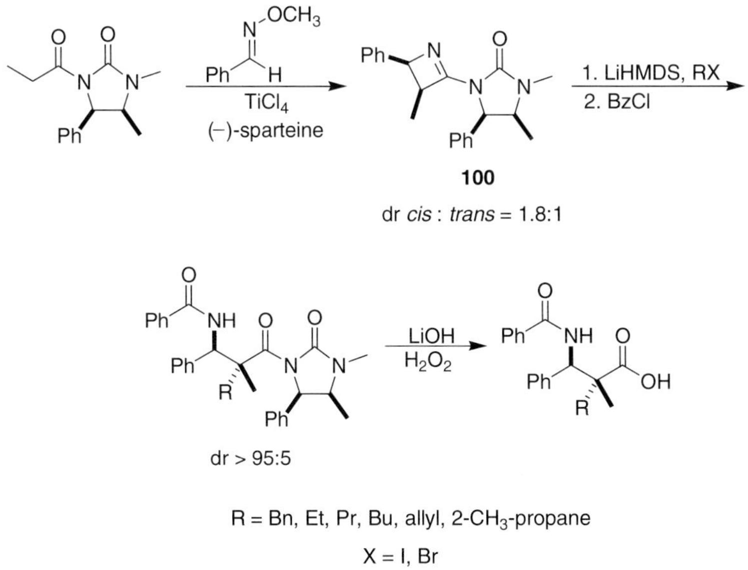

TiCl4
(−)-sparteine
1. LiHMDS, RX
2. BzCl
100
dr cis : trans = 1.8:1
dr > 95:5
LiOH
H2O2
R = Bn, Et, Pr, Bu, allyl, 2-CH3-propane
X = I, Br

Scheme 7.98

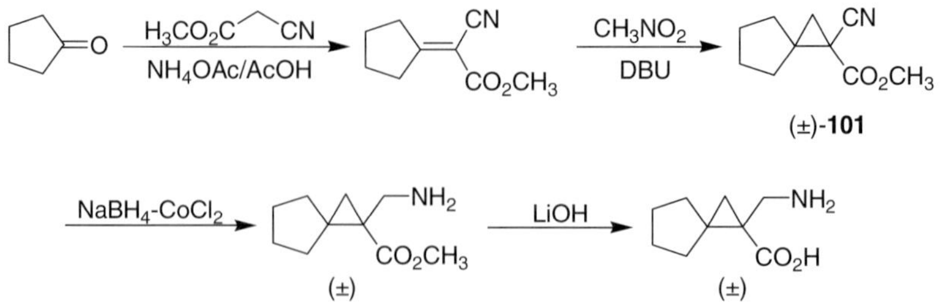

Scheme 7.99

The resulting aminoester was subsequently hydrolyzed under basic conditions to give the expected racemic β-amino acid (Scheme 7.99).

On the other hand, Moutevelis *et al.* [163] developed an efficient method for the synthesis of enantiopure β^3-amino acids based on the ruthenium-catalyzed oxidation of a phenyl group to a carboxylic acid (Scheme 7.100).

Another interesting method was described by Espino *et al.* [164] based on the cyclization of sulfamate esters via rhodium-catalyzed C–H bond oxidation/insertion reaction. Thus, oxathiazinanes **102** were prepared from the stereoselective intramolecular oxidative cyclization of sulfamate esters, using $Rh_2(OAc)_4$, $PhI(OAc)_2$, and MgO in CH_2Cl_2. Nucleophilic ring opening of oxathiazinanes with water followed by oxidation afforded the enantioenriched β-amino acids (Scheme 7.101).

In a related procedure, Freitag and Metz [165] developed a route for the preparation of racemic *N*-sulfonyl β-amino acids from *N*-Boc protected methanesulfonamide **103** employing a zinc-mediated allylation of cyclic *N*-sulfonyl imines as key step (Scheme 7.102).

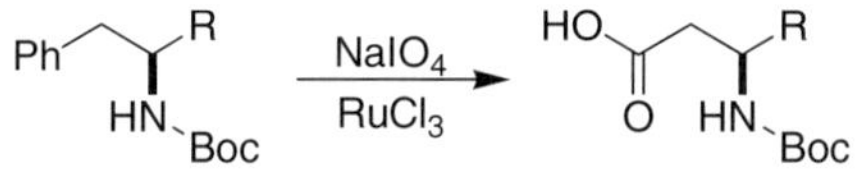

R = $CH_2CH_2CH_2CH_3$, $CH_2CH_2CO_2CH_3$, CH_2CO_2H, $CH_2CH_2CO_2H$

Scheme 7.100 TEMPO: 2,2,6,6-tetramethylpiperidinooxy.

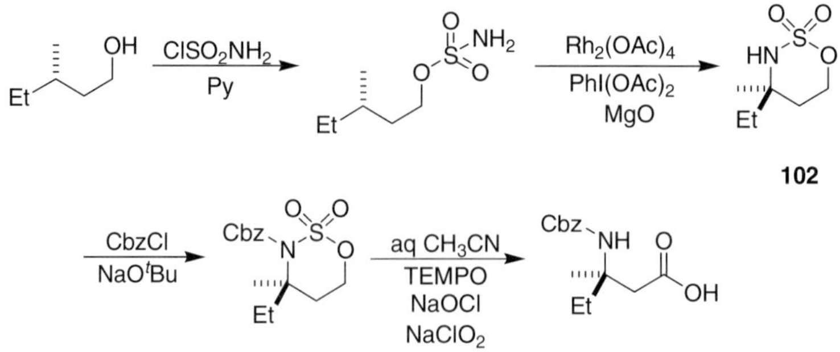

Scheme 7.101

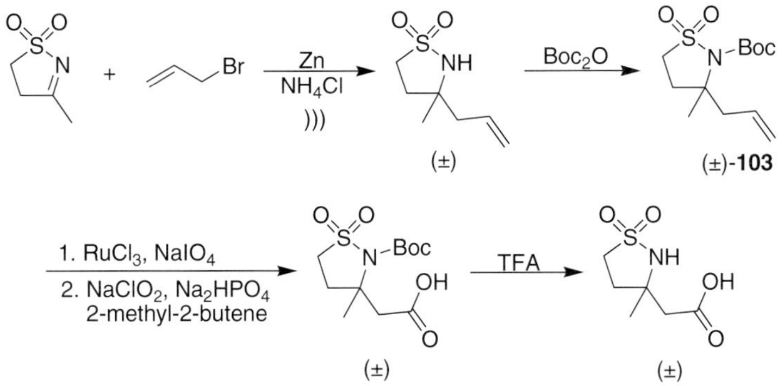

Scheme 7.102

The synthesis of β-amino acids presenting sugar rings in their structure has been described by several research groups. For example, Jayakanthan and Vankar [166] employed sugar-derived dienes **104** in Diels–Alder reactions with methyl α-nitro acrylate and ethyl β-nitro acrylate to form the corresponding cycloadducts which were then converted into conformationally constrained *C*-glycosyl β-amino acids (Scheme 7.103).

In this context, Inaba *et al.* [167] described a convenient preparation of both stereoisomers of a glycosylated β-amino acid. The β-*C*-glycoside was formed by the S_N2 reaction of α-acetobromoglucose **105** with the sodium salt of ethyl cyanoacetate. Subsequent reduction, *N*-protection, and final hydrolysis provide the desired β-amino acids (Scheme 7.104).

On the other hand, Nelson and Spencer [168] developed an interesting enantioselective synthesis of β-amino acids based on the asymmetric Al(III)-catalyzed acyl halide–aldehyde cyclocondensation reaction using representative aldehydes with acetyl bromide. This procedure afforded enantiomerically enriched β-(*R*)-substituted β-lactones **106**, which underwent ring opening with NaN_3. Subsequent hydrogenolysis gave the desired chiral β-amino acids (Scheme 7.105).

Fringuelli *et al.* [169], employed $Cu(NO_3)_2$ in the stereospecific opening of α,β-epoxycarboxylic acids such as **107** with NaN_3. The resulting azide was then reduced to furnish the final α-hydroxy-β-amino acid (Scheme 7.106).

Singh *et al.* [170] developed an interesting procedure for the synthesis of ω-heterocyclic-β-amino acids from (1*R*,3*R*)- and (1*R*,3*S*)-5-oxo-1-(1-phenyl-ethyl)-pyrrolidine-3-carboxylic acid methyl ester **108**, which were treated with Lawesson's reagent to give the corresponding thiolactam in quantitative yield. Sulfur extrusion

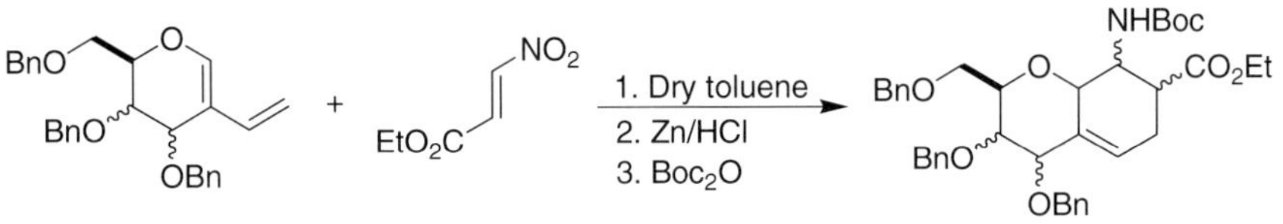

Scheme 7.103

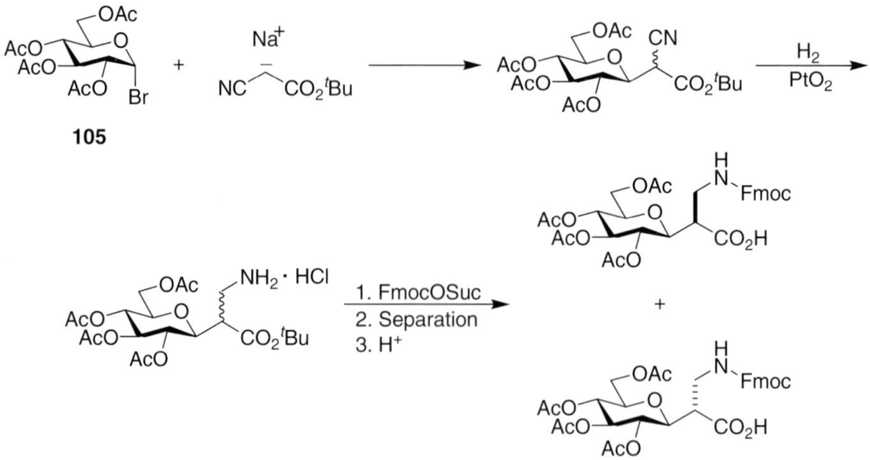

105

Scheme 7.104

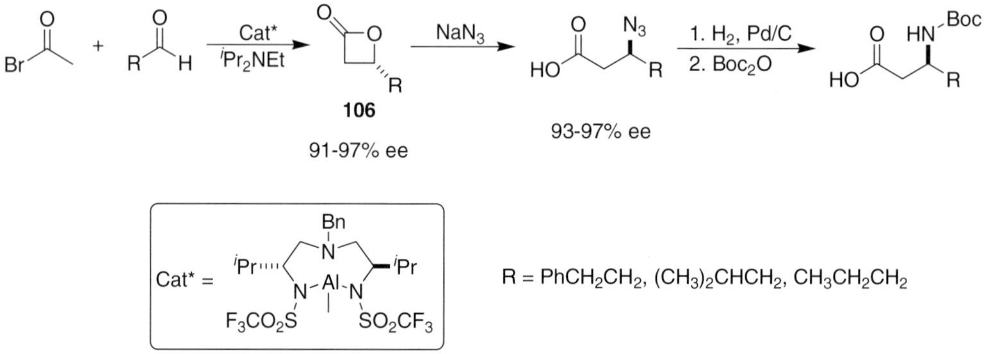

106

91-97% ee

93-97% ee

R = $PhCH_2CH_2$, $(CH_3)_2CHCH_2$, $CH_3CH_2CH_2$

Scheme 7.105

with various phenacyl bromides afforded 1-methyl-5-(2-aryl-2-oxo-ethylidene)-pyrrolidine 3-carboxylic acid methyl esters in good yields. These enaminones underwent ring-chain-transformation reaction to give pure ω-heterocyclic-β-amino acids in good yields (Scheme 7.107).

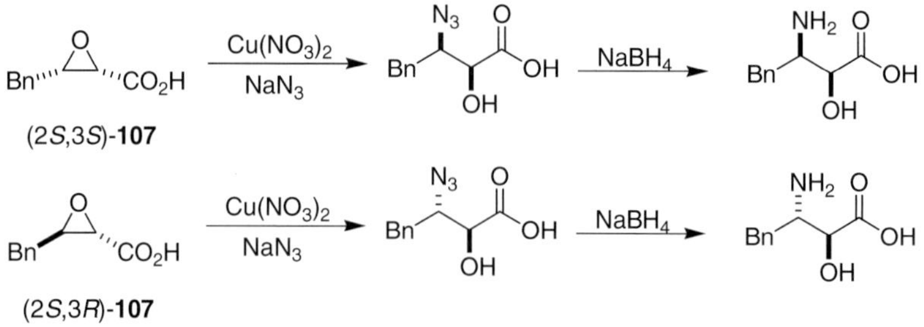

(2*S*,3*S*)-**107**

(2*S*,3*R*)-**107**

Scheme 7.106

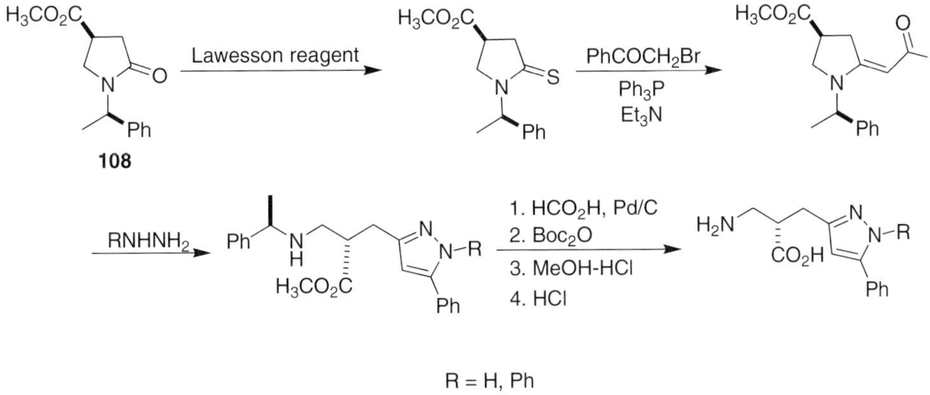

Scheme 7.107

Busnel and Baudy-Floc'h [171] described the preparation of novel N^{β}-Fmoc-aza-β^3-amino acids **109** by successive nucleophilic substitution of *N*-protected hydrazines with methyl bromoacetate (Scheme 7.108).

Given the medicinal and biological significance of cyclic β-amino acids, Fülöp *et al.* [172] have presented two excellent reviews on the developments in the field (see also Chapter 8). Miller and Nguyen [173] have also summarized salient methods that have been developed to prepare chiral cyclic β-amino acids in high enantiomeric purity, while Kuhl *et al.* [174] have discussed the application of five-, six-, and seven-membered alicyclic β-amino acids in medicinal chemistry.

Particularly interesting is the report of Perlmutter *et al.* [175], who describe the diastereoselective condensation under Lewis acid conditions ($SnCl_4$) of pyridyl thioester and an enantiopure imine to yield a diastereomeric mixture of β-lactam derivatives **110**, which were separated by chromatography. Treatment of the β-lactams with chlorotrimethylsilane in the presence of benzyl alcohol gave the desired metathesis substrates, which were subjected to Grubbs' catalyst under high-dilution conditions to afford the desired cyclic β-amino esters (Scheme 7.109).

Other successful methodology for the synthesis of several cyclic β-amino acids was described by Gauzy *et al.* [176] based on the [2 + 2]-photochemical cycloaddition reaction of uracils with ethylene followed by controlled degradation of the heterocyclic ring. For example, uracil derivatives **111** underwent [2 + 2]-photochemical

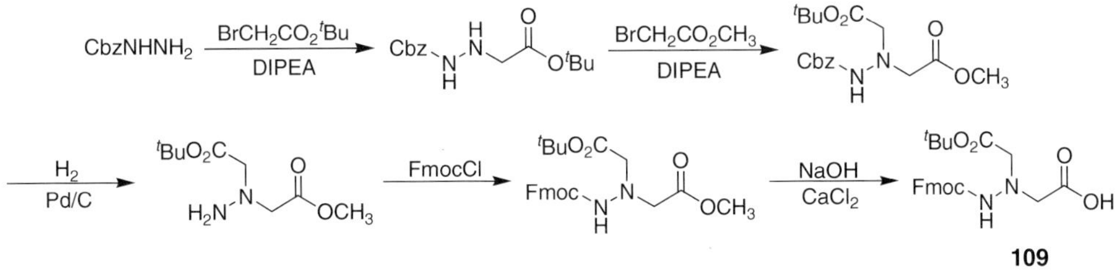

Scheme 7.108

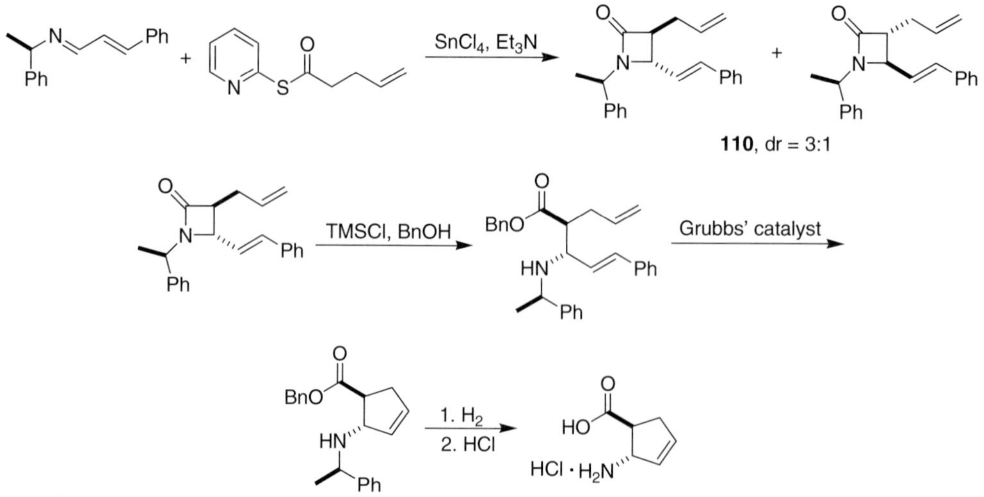

$SnCl_4$, Et_3N

TMSCl, BnOH

Grubbs' catalyst

1. H_2
2. HCl

110, dr = 3:1

Scheme 7.109

reaction with ethylene to give a cyclobutane adduct in good yield. The dihydropyrimidine ring in the cyclobutane adduct was then opened using aqueous NaOH to give the urea derivative, which was treated with $NaNO_2$ to give *cis*-cyclobutane β-amino acids in good yield (Scheme 7.110).

In other relevant work, Wang *et al.* [177] described the synthesis of five-membered cyclic β-amino acids by chemoselective reduction of the carbonyl group in cyclic β-keto ester **112** followed by Michael addition of (*R*)-α-phenylethylamine and subsequent hydrogenolysis (Scheme 7.111).

As part of their systematic work in this area, Izquierdo *et al.* [178] have described the synthesis of the enantiomerically pure cyclobutane β-amino acid derivatives from the hemi ester **113** prepared through chemoenzymatic hydrolysis of the corresponding *meso*-dimethyl ester (Scheme 7.112).

On the other hand, the enantioselective opening of β-lactams by means of enzymatic agents has been achieved by Fülöp *et al.* [179]. Indeed, 1,2-dipolar

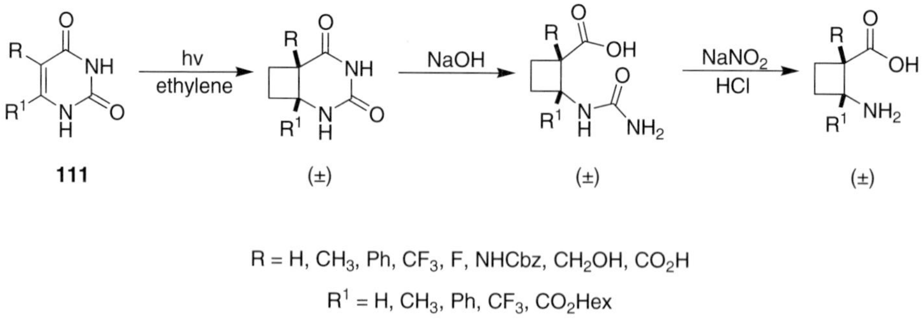

hν, ethylene; NaOH; $NaNO_2$, HCl

111 (±) (±) (±)

R = H, CH_3, Ph, CF_3, F, NHCbz, CH_2OH, CO_2H

R^1 = H, CH_3, Ph, CF_3, CO_2Hex

Scheme 7.110

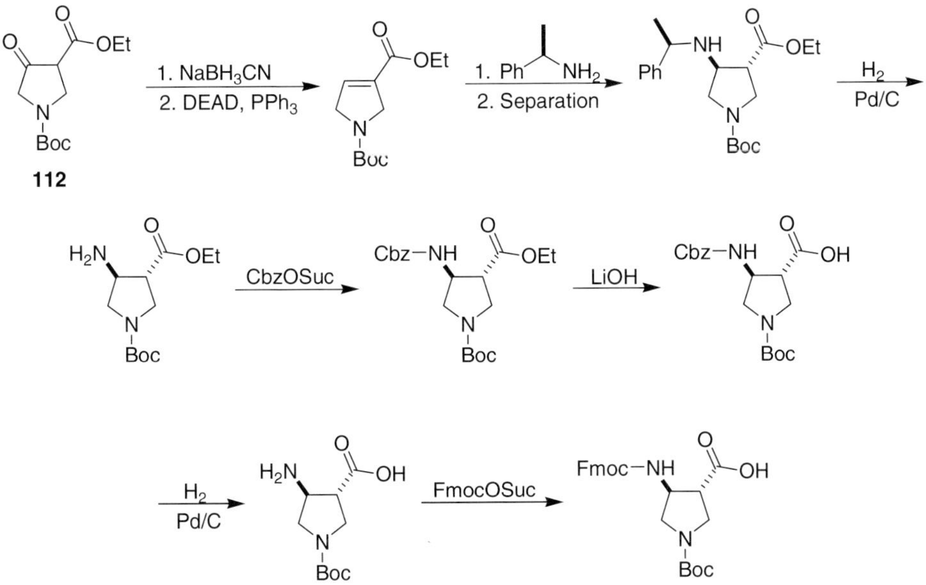

Scheme 7.111 DEAD: diethyl azodicarboxylate.

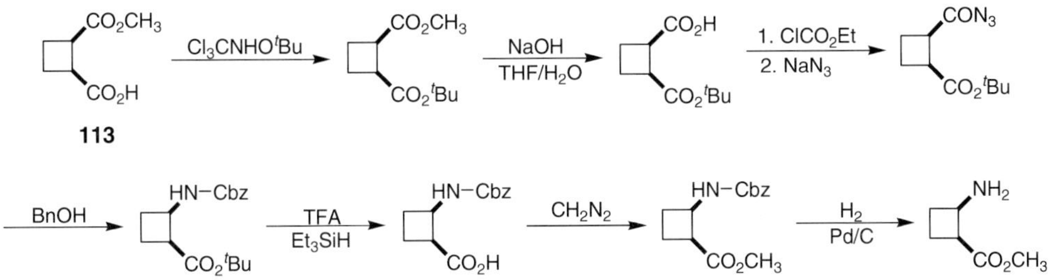

Scheme 7.112

cycloaddition of chlorosulfonyl isocyanate to several dienes afforded the racemic azetidinones **114**, which were resolved via opening of the β-lactam ring employing lipase B from *Candida antarctica* (CAL-B) [179] (Scheme 7.113).

7.12
Conclusions

The great relevance of β-amino acids and derivatives in various areas of biological chemistry is evidenced by the great number of methodologies that are available for the preparation of β-amino acids with various substitution patterns in both chiral and

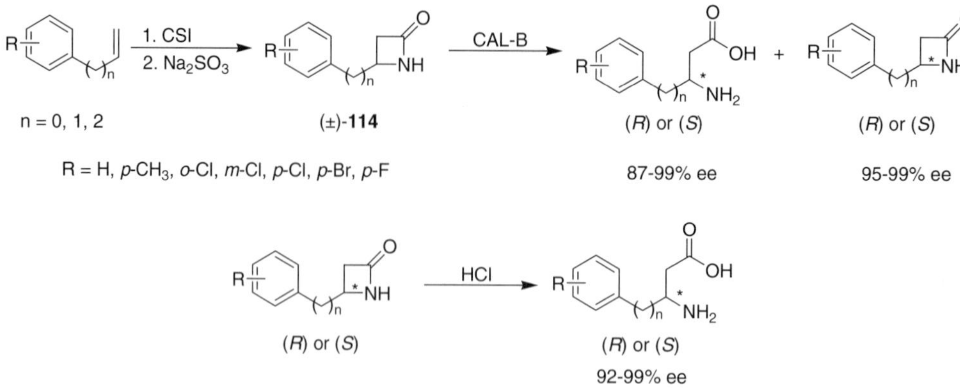

Scheme 7.113 CAL-B: lipase B from *Candida antarctica*.

racemic form. Each method has its own advantages and limitations and for this reason the development of additional novel and efficient processes for their synthesis will continue to attract the attention of synthetic chemists.

7.13 Experimental Procedures

7.13.1 Representative Experimental Procedure: Synthesis of (*S*)-3-(*tert*-Butyloxycarbonylamino)-4-phenylbutanoic Acid [3a]

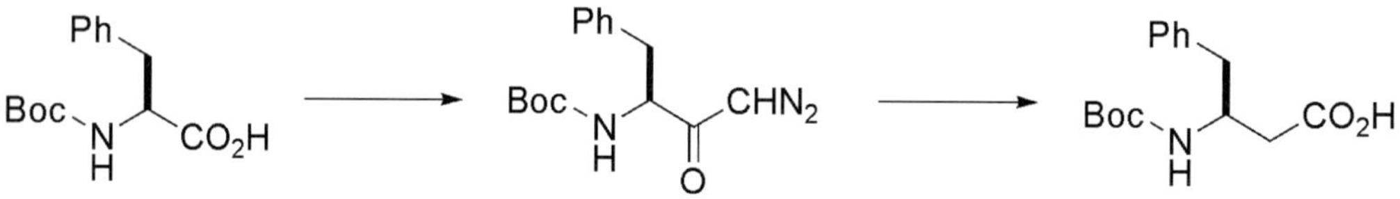

First Step: (*S*)-3-(*tert*-Butyloxycarbonylamino)-1-diazo-4-phenylbutan-2-one A 1-l, three-necked, round-bottomed flask was equipped with a magnetic stirring bar, nitrogen gas inlet, bubble counter, and a rubber septum on the center neck. The apparatus was dried under a rapid stream of nitrogen with a heat gun. After the flask cooled to room temperature, the rate of nitrogen flow was reduced, and Boc-phenylalanine (25.0 g, 94.2 mmol) and anhydrous THF (250 ml) were added. The flask was immersed in an ice-water bath before the addition of Et_3N (13.1 ml, 94.0 mmol). After 15 min, ethyl chloroformate (9.45 ml, 94.0 mmol) was added and the reaction mixture was stirred for another 15 min, and a white precipitate of Et_3NHCl appeared; the stirring was then stopped. The septum was replaced by a funnel, from which an ethereal solution of diazomethane (about 25 ml; *caution*!) was added. Stirring was resumed for about 5 s and the nitrogen stream was stopped. After 45 min, the remainder of the diazomethane solution (about 85 ml) was added. The cooling bath was removed and the solution was allowed to react for 3 h without stirring. With stirring, 0.5 N acetic acid (5 ml) was added carefully to destroy unreacted diazomethane, and then saturated aqueous $NaHCO_3$ solution (75 ml) was added carefully. The aqueous layer was separated in a separatory funnel and the organic layer was washed with saturated aqueous sodium chloride (75 ml). The organic layer was dried with $MgSO_4$, filtered, and the solvents were removed under vacuum on a rotary evaporator. The crude product was further dried under high vacuum for 3 h. The crude material was used directly in the next step.

Second Step: (*S*)-3-(*tert*-Butyloxycarbonylamino)-4-phenylbutanoic Acid A 500-ml, three-necked flask was equipped with a nitrogen gas inlet, bubble counter, septum, and a magnetic stirring bar. The flask was carefully wrapped in aluminum foil (to exclude light during the reaction). The crude diazoketone from the preceding step was dissolved in THF (380 ml) and added to the flask under an atmosphere of nitrogen. Deionized water (38 ml) was added, the flask was immersed in a dry ice/acetone bath and the solution was cooled to $-25\,^{\circ}C$ (temperature of the acetone cooling bath) for 30 min. Silver trifluoroacetate (2.72 g, 12.3 mmol) was placed in a 50-ml Erlenmeyer flask and quickly dissolved in Et_3N (39 ml, 279 mmol). The resulting solution was added to the diazoketone solution in one portion (via a syringe). The solution was allowed to warm to room temperature overnight. Evolution of nitrogen started at a bath temperature of about $-15\,^{\circ}C$. The solution was transferred to a 1-l, round-bottomed flask and the reaction vessel was rinsed with EtOAc (2×10 ml). The solution was evaporated to dryness with a rotary evaporator and the residue was stirred for 1 h with saturated aqueous $NaHCO_3$ solution (100 ml). The black mixture was transferred into a 1-l separatory funnel with water (150 ml) and EtOAc (200 ml) and the resulting mixture was shaken well. The clear aqueous layer was separated and put aside, leaving an organic phase containing a suspension of black solid. Brine (30 ml) was added to the organic phase and the resulting mixture was shaken vigorously. Saturated, aqueous $NaHCO_3$ solution (30 ml) was added, the medium was shaken again, and the layers were separated. The black solid was carried away with the aqueous phase, which was now combined with the first-separated aqueous phase. The organic layer was washed with three additional portions of saturated aqueous $NaHCO_3$ solution (30 ml each) and all the aqueous layers were combined. The first organic layer was put aside and not used further. The combined aqueous layers containing a black suspension were extracted with EtOAc (50 ml) and the EtOAc layer was then back-extracted with two portions of saturated aqueous $NaHCO_3$ solution (25 ml each), which were combined with the original aqueous layers. The EtOAc phase was put aside and not used further. All the combined aqueous layers were extracted again with EtOAc (50 ml), which was washed with saturated aqueous $NaHCO_3$ solution (2×20 ml). The organic layer was put aside and not used further. All the combined aqueous layers were then transferred to a 2-l, round-bottomed flask equipped with a magnetic stirring bar and about 10 drops of Congo red indicator and ethyl acetate (100 ml) were added. The flask was immersed in an ice-water bath, the solution was stirred and 5.0 N (17.5 wt%) HCl was added dropwise through an addition funnel until the color of the indicator changes from red to blue. The solution was placed in a 1-l separatory funnel and the organic layer was separated. The aqueous layer was additionally extracted with three portions of EtOAc (100 ml each). The combined organic layers were dried over $MgSO_4$ and evaporated on a rotary evaporator. Residual EtOAc was azeotropically removed by adding CH_2Cl_2 (10 ml) 3 times and evaporating on the rotary evaporator. TFA and traces of solvent were removed under high vacuum. The product crystallized slowly to give essentially pure material (16.9–17.1 g, 57.6–61.2 mmol, 61–65%) and could be recrystallized (Et_2O/light petroleum 1 : 1; about 100 ml) to yield 12.1 g product (43.3 mmol, 46%).

7.13.2
Representative Experimental Procedure: Synthesis of (*S*)-2-(Aminomethyl)-4-phenylbutanoic Acid, (*S*)-19 [20–23]

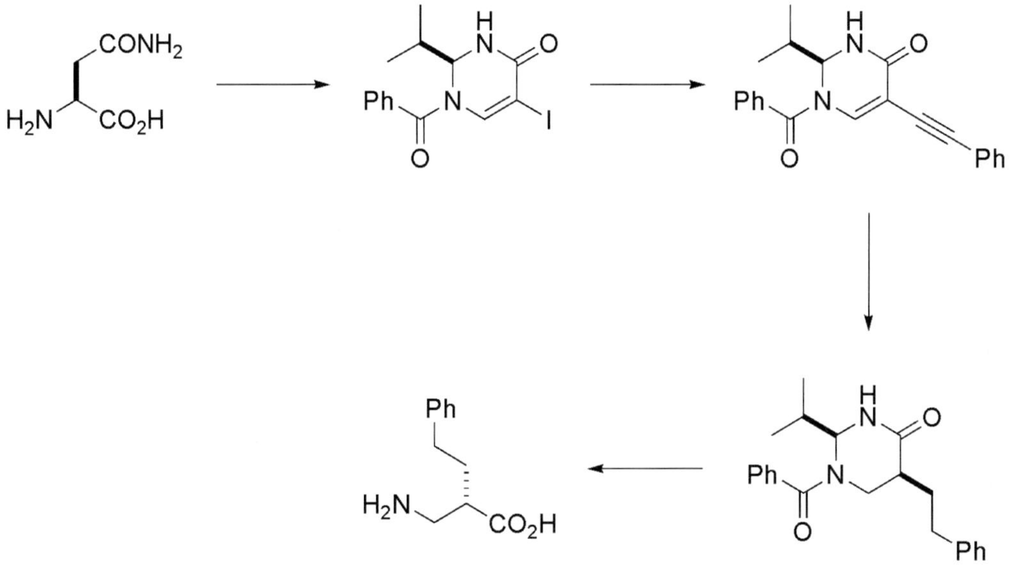

First Step: (*S*)-1-Benzoyl-5-iodo-2-isopropyl-2,3-dihydropyrimidin-4(1*H*)-one (*S*)-17 In an Erlenmeyer flask (1 l), equipped with a magnetic stirring bar, was placed KOH (13.2 g, 0.2 mol) and was dissolved in distilled H_2O (300 ml). Then, under stirring, (*S*)-asparagine monohydrate (30 g, 0.2 mol) and isobutyraldehyde (36.3 ml, 0.4 mmol) were added, and the mixture was vigorously stirred for 6 h at room temperature. Then, the mixture was cooled to 0 °C in an ice bath, and $NaHCO_3$ (8.4 g, 0.1 mol) and benzoyl chloride (11.6 ml, 0.1 mol) were added. After 30 min of stirring, additional portions of $NaHCO_3$ and benzoyl chloride (8.4 g and 11.6 ml, respectively) were added, and the resulting mixture was stirred at 0 °C for 1 h and then at room temperature for 1 h. Unreacted isobutyraldehyde and benzoyl chloride were removed by extraction with CH_2Cl_2 (200 ml). The aqueous phase was adjusted to pH 2.0 with 10% aqueous HCl solution, and the desired product was extracted with CH_2Cl_2 (3 × 500 ml). The organic layers were combined, dried (Na_2SO_4), and concentrated in a rotary evaporator at reduced pressure. The resulting crude solid was recrystallized from hexane/acetone 8:2 to afford 47.0 g (yield = 81%) of pure material. After that, a suspension of the obtained carboxylic acid (1.0 mmol), DIB (644 mg, 2.0 mmol), and iodine (253 mg, 1.0 mmol) in CH_2Cl_2 (20 ml) was stirred at ambient temperature until thin-layer chromatography (TLC) showed disappearance of the starting material (4–4.5 h). $BF_3 \cdot OEt_2$ (0.25 ml, 2 equiv.) was added and the resulting mixture was stirred for 1 h. Following addition of CH_2Cl_2 (20 ml), the reaction mixture was washed with four 15-ml portions of 5% aqueous $Na_2S_2O_3$, two 10-ml portions of 3% aqueous $NaHCO_3$, and two 10-ml portions of brine, dried over anhydrous Na_2SO_4, and evaporated to afford the 5-iodoenone (*S*)-**17** as a yellowish syrup that was purified by flash chromatography (yield = 78%).

Second Step: Synthesis of 5-Alkynyl Enones (*S*)-18 To an oven-dried round-bottom flask with a magnetic stirrer bar were added iodoenone (*S*)-**17** (0.54 mmol), $PdCl_2$ (8 mol% based on iodoenone), CuI (8 mol% based on iodoenone), and PPh_3 (0.16 equiv.). The flask was evacuated and backfilled with nitrogen gas. Acetonitrile (6 ml) was added followed by Et_3N (3 equiv.). The terminal alkyne was then added and the reaction mixture was stirred at room temperature until the reaction was completed (30 min to 1 h). The reaction mixture was evaporated to afford a dark syrup and the crude product was purified by flash chromatography (yield = 88%).

Third Step: (2*S*,5*S*)-1-Benzoyl-2-isopropyl-5-phenethyltetrahydro-pyrimidin-4(1*H*)-one In a hydrogenation flash was placed 263 mg (0.581 mmol) of acetylenic product (*S*)-**18** and 26 mg of 10% w/w Pd/C catalyst under a nitrogen atmosphere before the slow addition of MeOH (80 ml). The resulting mixture was pressurized to 60 psi of hydrogen and mechanically stirred at ambient temperature for 22 h, filtered through Celite and the filtrate was concentrated in a rotary evaporator. The reaction product was purified by flash chromatography (eluent EtOAc/hexane 4 : 6) to afford 214 mg (80% yield) of the alkynyl reduction product, as a white solid. Later, in a hydrogenation flask was placed the acetylenic derivative (1.25 g, 3.63 mmol), obtained in the previous step, and dry MeOH (60 ml) under a nitrogen atmosphere before the addition of Raney nickel (4.6 g) suspended in MeOH (60 ml) and acetic acid (1.2 ml). The resulting mixture was pressurized to 1750 psi of hydrogen and mechanically stirred at 80 °C for 24 h. The reaction mixture was filtered and the filtrate was concentrated in a rotary evaporator to afford the crude product consisting of a mixture of diastereomers. The major product was purified by flash chromatography (eluent EtOAc/hexane 6 : 4) to give 0.51 g of product (40% yield) that corresponded to the compound with configuration (2*S*,5*S*). Finally, this product (150 mg, 0.43 mmol) was placed in a 25-ml round-bottom flask and of 4.0 N HCl solution (20 ml) added. The resulting mixture was set on a microwave apparatus, adapted with a condenser and subjected to microwave irradiation (200 W) during 6 h at a temperature of 98 °C (cooled with a flow of air). The reaction mixture was allowed to cool to room temperature and was then extracted with three portions of CH_2Cl_2 (20 ml). The aqueous phase was concentrated at reduced pressure and the residue was purified by silica gel chromatography, eluent isopropanol/MeOH/NH_4OH (5 : 1 : 1) to give (*S*)-**19** (55 mg, 66%) as a white powder.

7.13.3
Representative Experimental Procedure: Synthesis of β^3-Amino Acids by Conjugate Addition of Homochiral Lithium *N*-Benzyl-*N*-(α-methylbenzyl)amide [37a]

NH2
(*R* or *S*)
R CO2R1
R N H CO2R1
R H2N CO2R1

To a stirred solution of (*R*)- or (*S*)-*N*-benzyl-*N*-α-methyl benzylamine (1.86 g, 8.8 mmol) in THF (10 ml) was slowly added *n*-BuLi (3.5 ml, 2.5 M in hexanes, 8.8 mmol) with cooling to −78 °C. After 30 min, a solution of α,β-unsaturated ester (5.5 mmol) also at −78 °C was transferred via a cannula. The resulting solution was stirred at −78 °C for 3 h before quenching with saturated aqueous NH_4Cl (5 ml). Upon warming to room temperature, the product was extracted with CH_2Cl_2. The organic fractions were combined, dried (Na_2SO_4), and the solvent was removed in vacuum. The residue was taken up in CH_2Cl_2 and the excess auxiliary extracted with 10% aqueous citric acid. The organic layer was washed with saturated aqueous $NaHCO_3$, dried (Na_2SO_4), and the solvent removed under vacuum to yield the crude product, which was used without further purification. After that, a stirred solution of this lithium amide adduct (4.5 mmol) in methanol (20 ml), distilled water (2 ml) and acetic acid (0.5 ml) was degassed before $Pd(OH)_2$ on carbon (0.5 g) addition. A hydrogen balloon was attached and the resulting suspension was stirred for 24 h under 1 atm of hydrogen. The reaction mixture was filtered through Celite and the solvent was removed under vacuum. The residue was partitioned between CH_2Cl_2 (20 ml) and saturated aqueous $NaHCO_3$ (10 ml). The organic phase was separated and the aqueous phase was extracted with CH_2Cl_2 (2 × 20 ml). The combined organic layers were dried (Na_2SO_4) and the solvent was removed under vacuum. The residue was purified by passage through a short plug of silica (doped 1% Et_3N; wash with 1 : 1 Et_2O/pentane; elute 9 : 1 Et_2O/MeOH) to yield the β-amino ester product which was used without further purification. Finally, TFA (1 ml) was added dropwise to a stirred solution of the β-amino ester (200 μmol) in CH_2Cl_2 (1 ml) under an atmosphere of argon, at room temperature. The solution was stirred for 15 h at room temperature. The solvents were removed under vacuum and coevaporated with ethereal HCl. The residue was subjected to ion-exchange chromatography (Dowex 80W-X8, elute with 1.0 M NH_4OH) to yield the free β^3-substituted β-amino acid.

7.13.4 Representative Experimental Procedure: Synthesis of Cyclic and Acyclic β-Amino Acid Derivatives by 1,3-Dipolar Cycloaddition [60]

General Procedure for 1,3-Dipolar Cycloaddition To a suspension of 2-(hydroxyamino)-2-phenylethanol (1.87 g, 12.2 mmol, 1.2 equiv.) in anhydrous THF or CH_2Cl_2 (55 ml) was added anhydrous $MgBr_2$ (2.25 g, 12.2 mmol, 1.2 equiv.). The solution was stirred at room temperature for 10 min. 2-Propanol (0.94 ml, 12.23 mmol, 1.2 equiv.) was added and after an additional 10 min of stirring, a solution of the corresponding aldehyde-allyl derivative (10.2 mmol, 1 equiv.) in anhydrous THF or CH_2Cl_2 (6 ml) was added. The mixture was stirred at 40–65 °C for 1–72 h (follow by TLC), poured into 5% aqueous $NaHCO_3$/ice and extracted with EtOAc. The combined organic extracts were washed with water, brine, dried over Na_2SO_4, and evaporated at reduced pressure. The residue was purified by flash column chromatography on silica gel, eluting with EtOAc/hexanes, to yield **37** derivatives in good yields (89–95%).

General Procedure for the Synthesis of *N*-Boc-Protected Cyclic and Acyclic β-Amino Acids To a solution of **37** derivatives (3.0 mmol) in MeOH (48 ml) was added 20 wt% $Pd(OH)_2$ on carbon catalyst. The mixture was hydrogenated at 1 atm overnight, filtered through Hyflo and evaporated at reduced pressure. The residue was dissolved in THF/water 2 : 1 (30 ml). $NaHCO_3$ (7.5 mmol, 2.5 equiv.) and di-*tert*-butyl dicarbonate (3.3 mmol, 1.1 equiv.) were added at 0 °C. The mixture was stirred at room temperature overnight, poured into 1.0 M HCl/ice, and extracted with EtOAc. The combined organic extracts were washed with water, brine, dried over Na_2SO_4, and evaporated *in vacuo*. The residue was purified by flash column chromatography on silica gel, eluting with EtOAc/hexanes 1 : 2, to yield the corresponding alcohol

(68–70%). To a solution of the alcohol (0.58 mmol, 1 equiv.) in MeCN/water 1 : 1 (8 ml) was added $RuCl_3 \cdot H_2O$ (0.03 mmol, 0.05 equiv.) and $NaIO_4$ (1.8 mmol, 3 equiv.) at 0 °C. The mixture was stirred at room temperature for 2 h, poured into 5% aqueous $NaHCO_3$/ice, and extracted with EtOAc. The combined organic extracts were washed with water and brine, dried over Na_2SO_4 and decolorized with activated carbon (5 mg), filtered through Hyflo, and evaporated at reduced pressure. The residue was purified by flash column chromatography on silica gel to yield the corresponding cyclic or acyclic β-amino acid in good yield (82–92%).

7.13.5
Representative Experimental Procedure: Synthesis of (*R*)-3-*tert*-Butoxycarbonylamino-3-phenylpropionic Acid Isopropyl Ester using a Mannich-Type Reaction [108]

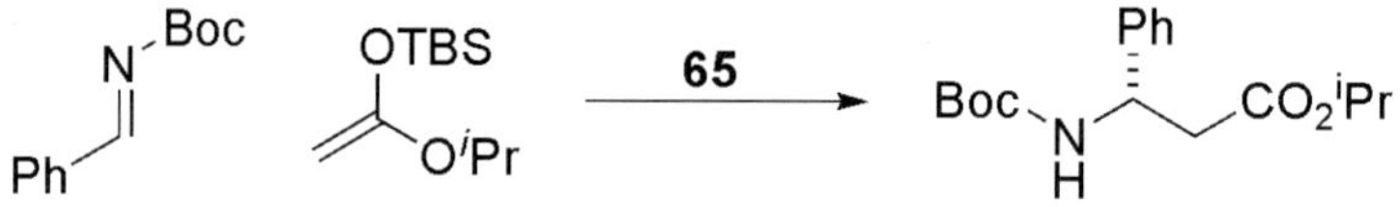

A 5-ml flask was charged sequentially with the urea catalyst **65** (0.025 mmol, 0.05 equiv.) and anhydrous toluene (250 μl). Benzaldehyde *N*-Boc imine (0.5 mmol, 1.0 equiv.) was then added in one portion with stirring. Once the solution was homogeneous, the flask was immersed in a dry ice/acetone bath and cooled to −40 °C. Then, the corresponding silyl ketene acetal (1.0 mmol, 2.0 equiv.) was slowly added along the flask wall over a 10-min period. The flask was sealed under an atmosphere of nitrogen and stirred at −40 °C. After 48 h, excess silyl ketene acetal was quenched at −40 °C via the rapid addition of a 3.0 M solution of TFA in toluene (500 μl; cooled to −20 °C prior to addition). The reaction was allowed to warm to 5 °C, and then partitioned between saturated aqueous $NaHCO_3$ solution and CH_2Cl_2 (1 : 1, 2 ml). The aqueous layer was extracted with CH_2Cl_2 (3 × 2 ml), and the combined organic extracts were dried over anhydrous Na_2SO_4, filtered, and concentrated under vacuum. The resulting residue was purified via flash chromatography on silica gel with EtOAc/hexane (90 : 10), to yield the (*R*)-3-*tert*-butoxycarbonylamino-3-phenylpropionic acid isopropyl ester as a white solid (95%, 97% e.e.).

7.13.6
Representative Experimental Procedure: General Procedure for the Hydrogenation of (*Z*)- and (*E*)-β-(Acylamino) acrylates by Chiral Monodentate Phosphoramidite Ligands [128a]

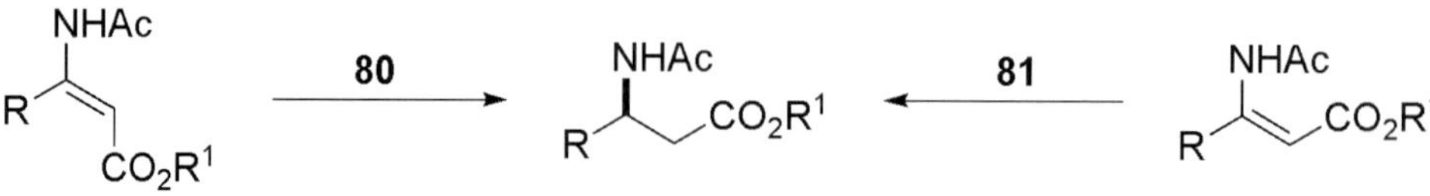

In a Schlenk tube equipped with a septum and a magnetic stirring bar, a mixture of $Rh(COD)_2BF_4$ (COD = cyclooctadiene) (5.1 mg, 12.5 μmol) and ligand **80** for (*Z*)-β-(acylamino)acrylates or **81** for (*E*)-β-(acylamino)acrylates (25 μmol) was dissolved in CH_2Cl_2 (1.25 ml). In a glass tube, an aliquot of this solution (1 ml) was added to a mixture of the (*Z*)-β-dehydroamino acid derivative (0.5 mmol) in isopropanol (4 ml) or the (*E*)-β-dehydroamino acid derivative (0.5 mmol) in CH_2Cl_2 (4 ml). This small glass tube was placed in an autoclave with eight reactors (Endeavor), and purged twice with nitrogen and once with hydrogen. Then, the autoclave was pressurized with hydrogen to 10 bar and the reaction was stirred at room temperature. The resulting mixture was filtered through a short silica column and conversion was determined by reference to the ^{1}H-nuclear magnetic resonance spectrum and the enantiomeric excess was determined by capillary gas chromatography, where the configuration of products were assigned as (*R*) or (*S*) by comparing the sign of optical rotations with that of the reported ones.

The racemic products were prepared by hydrogenation of the β-dehydroamino acid derivative using 10% Pd/C (10%) in MeOH under 1 atm of hydrogen for 16 h.

7.13.7 Representative Experimental Procedure: Synthesis of Chiral α-Substituted β-Alanine [141]

H_2N–CH_2CH_2–CO_2H ⟶ Cbz–NH–CH_2CH_2–CO_2H ⟶ Cbz–NH–CH_2CH_2–C(O)–N(CH(CH$_3$)Ph)$_2$ ⟶ Cbz–NH–CH_2–CH(R)–C(O)–N(CH(CH$_3$)Ph)$_2$ ⟶ Cbz–NH–CH_2–CH(R)–CO_2H

First Step: Benzyl 3-(bis((*R*)-1-phenylethyl) amino)-3-oxopropylcarbamate A solution containing β-aminopropionic acid (10.0 g, 112 mmol) in 1.0 N NaOH (112 ml) was cooled in an ice bath and treated with benzylchloroformate (17.6 ml, 21.0 g, 0.12 mol), followed by slow addition of 1.0 N NaOH (112 ml). The resulting mixture was allowed to warm to ambient temperature and stirred overnight. The crude product was extracted with portions of Et_2O (3 × 100 ml) and the aqueous phase was acidified with 6.0 N HCl to pH < 4.0. The precipitate that developed was dissolved in CH_2Cl_2, dried over anhydrous Na_2SO_4, and concentrated in the rotary evaporator. Recrystallization of the resulting residue from EtOAc/EtOH (10 : 1) afforded *N*-benzyloxycarbonyl-3-aminopropionic acid (21.5 g, 85%); this crystalline material (0.5 g, 2.2 mmol) was dissolved in 6.0 ml of CH_2Cl_2 and three drops of DMF. The resulting solution was treated with thionyl chloride (0.65 ml, 9.0 mmol), and heated to reflux for 4 h. The solvent was removed in the rotary evaporator and the residue was washed 3 times with toluene, and then it was redissolved in CH_2Cl_2 (4.0 ml) and added to a solution of (*R,R*)-*N,N*-bis-(α-phenylethyl)amine (1.0 g, 4.4 mmol) in toluene (4.0 ml) under nitrogen at 0 °C. The reaction mixture was allowed to warm and stirred at ambient temperature for 14 h. The solvent was removed in the rotary evaporator and the product was redissolved in Et_2O, dried over anhydrous Na_2SO_4, filtered, and concentrated at reduced pressure to give an oil that was purified by flash chromatography (hexane/EtOAc, 9 : 1–7 : 3) to afford benzyl 3-(bis(*R*)-1-phenylethyl)amino)-3-oxopropylcarbamate (0.60 g, 62%).

Second Step: Benzyl (*S*)-3-(bis((*R*)-1-phenylethyl) amino)-2-alkyl-3-oxopropylcarbamate To a solution of 2.6 N *n*-BuLi (0.43 ml, 1.02 mmol) in THF (5 ml) at −78 °C and under nitrogen atmosphere was added dropwise a solution of benzyl 3-(bis(*R*)-1-phenylethyl)amino)-3-oxopropylcarbamate (0.20 g, 0.46 mmol) in THF (10 ml). The resulting solution was stirred for 1 h at −78 °C and then the alkylating agent (1.1 equiv.) was added with continuous stirring. The reaction mixture was stirred at −78 °C for 3–4 h before quenching with aqueous NH_4Cl. The product was extracted with portions of EtOAc (3 × 5 ml); the combined organic extracts were dried over anhydrous Na_2SO_4, filtered and concentrated at reduced pressure. Final purification of the diastereomeric mixture of products was accomplished by flash chromatography (hexane/EtOAc 80 : 20).

Third Step: (*S*)-2-Alkyl-3-aminopropionic Acid In a hydrogenation flask was placed the diastereomeric mixture of (*R,R*)-*N′,N′*-bis(α-phenylethyl)-*N*-benzyloxycarbonyl-(2*S*)-2-alkyl propionamide (0.34 g, 0.77 mmol) (diastereomeric ratio = 75 : 25), 10% Pd/C catalyst (0.035 g), and MeOH (25 ml). The flask was pressurized to 1 atm of hydrogen and stirred at room temperature for 6 h. The reaction mixture was filtered through Celite and the filtrate was concentrated in a rotary evaporator to afford the deprotected amine (0.23 g, 98%). This product was transferred to a glass ampoule and dissolved in 4.0 N HCl (8.0 ml) and heated to 90 °C for 14 h. The crude product was washed with portions of CH_2Cl_2 (3 × 20 ml), the aqueous phase was concentrated *in vacuo*, and the residue was adsorbed to acidic ion exchange resin Dowex 50W-X4. The residue was washed with distilled water until the washings came out neutral and then the free amino acid was recovered with 0.1 N aqueous NH_4OH. Evaporation afforded the chiral α-substituted β-amino acid (0.07 g, 93%).

7.13.8
Representative Experimental Procedure: Synthesis of Chiral β-Amino Acids by Diastereoselective Radical Addition to Oxime Esters [146]

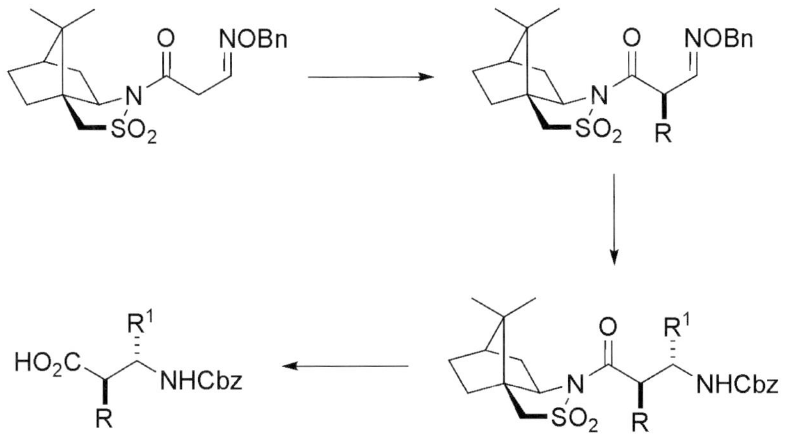

General Procedure for Alkylation of 95 To a solution of **95** (7.7 mmol) in CH_2Cl_2 (190 ml) were added an alkyl bromide (8.5 mmol), tetra-*n*-butylammonium bromide (0.77 mmol), and 5.0 N NaOH (7.5 ml) under a nitrogen atmosphere at 20 °C. After being stirred at the same temperature for 1 h, the reaction mixture was diluted with saturated aqueous NH_4Cl and then it was extracted with CH_2Cl_2. The organic phase was dried over $MgSO_4$ and concentrated at reduced pressure. Purification by flash column chromatography (EtOAc/hexane 1 : 4) afforded the alkylated **95**. The diastereomerically pure products (*R,Z*) were obtained by recrystallization from hexane/EtOAc.

General Procedure for Alkyl Radical Addition To a solution of alkylated **95** (2.08 mmol) in toluene (15 ml) were added an alkyl iodide (62.5 mmol), $BF_3{\cdot}OEt_2$ (6.25 mmol), and Et_3B (1.0 M in hexane, 6.25 mmol) at 20 °C. After being stirred at the same temperature for 3 min, $BF_3{\cdot}OEt_2$ (6.25 mmol) and Et_3B (6.25 mmol) were added twice. After being stirred at the same temperature for 3 min, the reaction mixture was diluted with saturated aqueous $NaHCO_3$ and then it was extracted with CH_2Cl_2. The organic phase was dried over $MgSO_4$ and concentrated at reduced pressure. Purification by preparative TLC (EtOAc/hexane 1 : 4) afforded the alkylated products. The subsequent reduction with H_2 and $Pd(OH)_2$ as catalyst in MeOH at 20 °C, *N*-protection with CbzCl, and final removal of the chiral auxiliary with LiOH in H_2O/THF gives the *N*-protected α,β-disubstituted β-amino acid.

References

1 (a) Juaristi, E., Quintana, D., and Escalante, J. (1994) *Aldrichimica Acta*, **27**, 3–11; (b) Cole, D.C. (1994) *Tetrahedron*, **50**, 9517–9582; (c) Cardillo, G. and Tomasini, C. (1996) *Chemical Society Reviews*, **25**, 117–128; (d) Juaristi, E. (ed.) (1997) *Enantioselective Synthesis of β-Amino Acids*, Wiley-VCH Verlag GmbH, Weinheim; (e) Juaristi, E. and López-Ruiz, H. (1999) *Current Medicinal Chemistry*, **6**, 983–1004; (f) Abele, S. and Seebach, D. (2000) *European Journal of Organic Chemistry*, 1–15; (g) Fülöp, F. (2001) *Chemical Reviews*, **101**, 2181–2204; (h) Liu, M. and Sibi, M.P. (2002) *Tetrahedron*, **58**, 7991–8035; (i) Ma, J.-A. (2003) *Angewandte Chemie (International Edition in English)*, **42**, 4290–4299; (j) Córdova, A. (2004) *Accounts of Chemical Research*, **37**, 102–112; (k) Juaristi, E. and Soloshonok, V.A. (eds) (2005) *Enantioselective Synthesis of β-Amino Acids*, 2nd edn, Wiley-VCH Verlag GmbH, Weinheim; (l) Dondoni, A. and Massi, A. (2006) *Accounts of Chemical Research*, **39**, 451–463; (m) Ager, D.J. (ed.) (2006) Synthesis of non-natural amino acids, in *Handbook of Chiral Chemicals*, 2nd edn, DSM Pharma Chemicals, Raleigh, NC.

2 Kirmse, W. (2002) *European Journal of Organic Chemistry*, 2193–2256.

3 (a) Linder, M.R., Steurer, S., and Podlech, J. (2002) *Organic Syntheses*, **79**, 154–164; (b) Achmatowicz, M., Szumna, A., Zielinski, T., and Jurczak, J. (2005) *Tetrahedron*, **61**, 9031–9041.

4 (a) Patil, B.S., Vasanthakumar, G.-R., and Suresh Babu, V.V. (2003) *Synthetic Communications*, **33**, 3089–3096; (b) Vasanthakumar, G.R. and Suresh Babu, V.V. (2003) *Indian Journal of Chemistry, Section B*, **42**, 1691–1695; (c) Vasanthakumar, G.-R. and Suresh Babu, V.V. (2003) *The Journal of Peptide Research*, **61**, 230–236; (d) Vasanthakumar, G.-R. and Suresh Babu, V.V. (2002) *Synthetic Communications*, **32**, 651–657; (e) Ananda, K., Gopi, H.N., and Suresh Babu, V.V. (2000) *The Journal of Peptide Research*, **55**, 289–294; (f) Vasanthakumar, G.-R., Patil, B.S., and Suresh Babu, V.V. (2002) *Journal of the Chemical Society, Perkin Transactions 1*, 2087–2089; (g) Kantharaju and Suresh Babu, V.V. (2004) *Indian Journal of Chemistry, Section B*, **43B**, 2152–2158.

5 Hughes, A.B. and Sleebs, B.E. (2005) *Australian Journal of Chemistry*, **58**, 778–784.

6 (a) Hughes, A.B. and Sleebs, B.E. (2006) *Helvetica Chimica Acta*, **89**, 2611–2637; (b) Sleebs, B.E. and Hughes, A.B. (2007) *The Journal of Organic Chemistry*, **72**, 3340–3352.

7 Govender, T. and Arvidsson, P.I. (2006) *Tetrahedron Letters*, **47**, 1691–1694.

8 Kim, D., Kowalchick, J.E., Edmondson, S.D., Mastracchio, A., Xu, J., Eiermann, G.J., Leiting, B., Wu, J.K., Pryor, K.D., Patel, R.A., He, H., Lyons, K.A., Thornberry, N.A., and Weber, A.E. (2007) *Bioorganic & Medicinal Chemistry Letters*, **17**, 3373–3377.

9 (a) Patil, B.S., Vasanthakumar, G.-R., and Suresh Babu, V.V. (2003) *Letters in Peptide Science*, **9**, 231–233; (b) Patil, B.S. and Suresh Babu, V.V. (2005) *Indian Journal of Chemistry, Section B*, **44**, 2611–2613.

10 Koch, K. and Podlech, J. (2005) *Synthetic Communications*, **35**, 2789–2794.

11 (a) Moumne, R., Lavielle, S., and Karoyan, P. (2006) *The Journal of Organic Chemistry*, **71**, 3332–3334; (b) Moumne, R., Denise, B., Parlier, A., Lavielle, S., Rudler, H., and Karoyan, P. (2007) *Tetrahedron Letters*, **48**, 8277–8280; (c) Moumne, R., Denise, B., Guitot, K., Rudler, H., Lavielle, S., and Karoyan, P. (2007) *European Journal of Organic Chemistry*, 1912–1920.

12 Gray, D., Concellon, C., and Gallagher, T. (2004) *The Journal of Organic Chemistry*, **69**, 4849–4851.

13 Caputo, R. and Longobardo, L. (2007) *Amino Acids*, **32**, 401–404.

14 Reginato, G., Mordini, A., Valacchi, M., and Piccardi, R. (2002) *Tetrahedron: Asymmetry*, **13**, 595–600.

15 (a) Seo, W.D., Curtis-Long, M.J., Jeong, S.H., Jun, T.H., Yang, M.S., and Park, K.H. (2007) *Synthesis*, 209–214; (b) see also Jung, D.Y., Kang, S., Chang, S., and Kim, Y.H., (2006) *Synlett*, 86–90.

16 Sánchez-Obregón, R., Salgado, F., Ortiz, B., Díaz, E., Yuste, F., Walls, F., and García Ruano, J.L. (2007) *Tetrahedron*, **63**, 10521–10527.

17 Park, J.-I., Tian, G.R., and Kim, D.H. (2001) *The Journal of Organic Chemistry*, **66**, 3696–3703.

18 Seki, M., Shimizu, T., and Matsumoto, K. (2000) *The Journal of Organic Chemistry*, **65**, 1298–1304.

19 Ávila-Ortiz, C.G., Reyes-Rangel, G., and Juaristi, E. (2005) *Tetrahedron*, **61**, 8372–8381.

20 Castellanos, E., Reyes-Rangel, G., and Juaristi, E. (2004) *Helvetica Chimica Acta*, **87**, 1016–1024.

21 Juaristi, E. (2003) 1-Benzoyl-2(*S*)-*tert*-butyl-3-methyl-perhydropyrimidin-4-one, in *Handbook of Reagents for Organic Synthesis. Chiral Reagents for Asymmetric Synthesis* (ed. L.A. Paquette), John Wiley & Sons, Ltd, Chichester, pp. 53–56.

22 Iglesias-Arteaga, M.A., Castellanos, E., and Juaristi, E. (2003) *Tetrahedron: Asymmetry*, **14**, 577–580.

23 Díaz-Sánchez, B.R., Iglesias-Arteaga, M.A., Melgar-Fernández, R., and Juaristi, E. (2007) *The Journal of Organic Chemistry*, **72**, 4822–4825.

24 Iovel, I., Golomba, L., Popelis, J., Gaukhman, A., and Lukevics, E. (2001) *Applied Organometallic Chemistry*, **15**, 67–74.

25 Kawatsura, M. and Hartwig, J.F. (2001) *Organometallics*, **20**, 1960–1964.

26 Amore, K.M., Leadbeater, N.E., Miller, T.A., and Schmink, J.R. (2006) *Tetrahedron Letters*, **47**, 8583–8586.

27 Varala, R., Sreelatha, N., and Adapa, S.R. (2006) *Synlett*, 1549–1553.

28 Xu, J.-M., Wu, Q., Zhang, Q.-Y., Zhang, F., and Lin, X.-F. (2007) *European Journal of Organic Chemistry*, 1798–1802.

29 Chaudhuri, M.K., Hussain, S., Kantam, M.L., and Neelima, B. (2005) *Tetrahedron Letters*, **46**, 8329–8331.

30 Kantam, M.L., Neelima, B., and Reddy, C.V. (2005) *Journal of Molecular Catalysis A – Chemical*, **241**, 147–150.

31 Kantam, M.L., Neeraja, V., Kavita, B., Neelima, B., Chaudhuri, M.K., and Hussain, S. (2005) *Advanced Synthesis and Catalysis*, **347**, 763–766.

32 Khan, A.T., Parvin, T., Gazi, S., and Choudhury, L.H. (2007) *Tetrahedron Letters*, **48**, 3805–3808.

33 Richardson, R.D., Hernandez-Juan, F.A., and Dixon, D.J. (2006) *Synlett*, 77–80.

34 Basu, B., Das, P., and Hossain, I. (2004) *Synlett*, 2630–2632.

35 Zou, B., Jiang, H.-F., and Wang, Z.-Y. (2007) *European Journal of Organic Chemistry*, 4600–4604.

36 Xu, L.-W. and Xia, C.-G. (2005) *European Journal of Organic Chemistry*, 633–639.

37 (a) Davies, S.G., Mulvaney, A.W., Russell, A.J., and Smith, A.D. (2007) *Tetrahedron: Asymmetry*, **18**, 1554–1566; (b) Chippindale, A.M., Davies, S.G., Iwamoto, K., Parkin, R.M., Smethurst, C.A.P., Smith, A.D., and Rodriguez-Solla, H. (2003) *Tetrahedron*, **59**, 3253–3265; (c) Davies, S.G., Iwamoto, K., Smethurst, C.A.P., Smith, A.D., and Rodriguez-Solla, H. (2002) *Synlett*, 1146–1148; (d) Davies, S.G., Garrido, N.M., Kruchinin, D., Ichihara, O., Kotchie, L.J., Price, P.D., Mortimer, A.J.P., Russell, A.J., and Smith, A.D. (2006) *Tetrahedron: Asymmetry*, **17**, 1793–1811; (e) Davies, S.G., Diez, D., Dominguez, S.H., Garrido, N.M., Kruchinin, D., Price, P.D., and Smith, A.D. (2005) *Organic and Biomolecular Chemistry*, **3**, 1284–1301.

38 Coleman, P.J., Hutchinson, J.H., Hunt, C.A., Lu, P., Delaporte, E., and Rushmore,

T. (2000) *Tetrahedron Letters*, **41**, 5803–5806.

39 Bull, S.D., Davies, S.G., Kelly, P.M., Gianotti, M., and Smith, A.D. (2001) *Journal of the Chemical Society, Perkin Transactions 1*, 3106–3111.

40 Podlech, J. (2000) *Synthetic Communications*, **30**, 1779–1786.

41 Herrera, R., Jiménez-Vázquez, H.A., and Tamariz, J. (2005) *ARKIVOC*, 233–249.

42 Lee, H.-S., Park, J.-S., Kim, B.M., and Gellman, S.H. (2003) *The Journal of Organic Chemistry*, **68**, 1575–1578.

43 Sharma, G.V.M., Reddy, V.G., Chander, A.S., and Reddy, K.R. (2002) *Tetrahedron: Asymmetry*, **13**, 21–24.

44 (a) Fernández, F., Otero, J.M., Estévez, J.C., and Estévez, R.J. (2006) *Tetrahedron: Asymmetry*, **17**, 3063–3066; (b) See also: See also: Moglioni, A.G., Muray, E., Castillo, J.A., Álvarez-Larena, A., Moltrasio, G.Y., Branchadell, V., and Ortuno, R.M. (2002) *The Journal of Organic Chemistry*, **67**, 2402–2410.

45 Etxebarria, J., Vicario, J.L., Badia, D., and Carrillo, L. (2004) *The Journal of Organic Chemistry*, **69**, 2588–2590.

46 Taillefumier, C., Lakhrissi, Y., Lakhrissi, M., and Chapleur, Y. (2002) *Tetrahedron: Asymmetry*, **13**, 1707–1711.

47 Yamagiwa, N., Qin, H., Matsunaga, S., and Shibasaki, M. (2005) *Journal of the American Chemical Society*, **127**, 13419–13427.

48 (a) Rimkus, A. and Sewald, N. (2003) *Organic Letters*, **5**, 79–80; (b) Rimkus, A. and Sewald, N. (2002) *Organic Letters*, **4**, 3289–3291; see, also: (c) Eilitz, U., Lessmann, F., Seidelmann, O., and Wendisch, V. (2003) *Tetrahedron: Asymmetry*, **14**, 3095–3097; (d) Lewandowska, E. (2007) *Tetrahedron*, **63**, 2107–2122.

49 Chen, Y.K., Yoshida, M., and MacMillan, D.W.C. (2006) *Journal of the American Chemical Society*, **128**, 9328–9329.

50 (a) Vesely, J., Ibrahem, I., Rios, R., Zhao, G.-L., Xu, Y., and Córdova, A. (2007) *Tetrahedron Letters*, **48**, 2193–2198; (b) Ibrahem, I., Rios, R., Vesely, J., Zhao, G.-L., and Córdova, A. (2007) *Journal of the Chemical Society, Chemical Communications*, 849–851.

51 Sibi, M.P. and Itoh, K. (2007) *Journal of the American Chemical Society*, **129**, 8064–8065.

52 Sammis, G.M. and Jacobsen, E.N. (2003) *Journal of the American Chemical Society*, **125**, 4442–4443.

53 Palomo, C., Oiarbide, M., Halder, R., Kelso, M., Gomez-Bengoa, E., and García, J.M. (2004) *Journal of the American Chemical Society*, **126**, 9188–9189.

54 Li, K., Cheng, X., and Hii, K.K. (2004) *European Journal of Organic Chemistry*, 959–964.

55 Sibi, M.P., Prabagaran, N., Ghorpade, S.G., and Jasperse, C.P. (2003) *Journal of the American Chemical Society*, **125**, 11796–11797.

56 Kikuchi, S., Sato, H., and Fukuzawa, S.-I. (2006) *Synlett*, 1023–1026.

57 (a) Huisgen, R. (1963) *Angewandte Chemie (International Edition in English)*, **2**, 633–645; (b) Huisgen, R. (1963) *Angewandte Chemie (International Edition in English)*, **2**, 565–598.

58 See for example: Sewald, N. (2003) *Angewandte Chemie (International Edition in English)*, **42**, 5794–5795.

59 (a) Torssell, S. and Somfai, P. (2006) *Advanced Synthesis and Catalysis*, **348**, 2421–2430; (b) Torssell, S., Kienle, M., and Somfai, P. (2005) *Angewandte Chemie (International Edition in English)*, **44**, 3096–3099.

60 Hanselmann, R., Zhou, J., Ma, P., and Confalone, P.N. (2003) *The Journal of Organic Chemistry*, **68**, 8739–8741.

61 (a) Minter, A.R., Fuller, A.A., and Mapp, A.K. (2003) *Journal of the American Chemical Society*, **125**, 6846–6847; (b) Fuller, A.A., Chen, B., Minter, A.R., and Mapp, A.K. (2005) *Journal of the American Chemical Society*, **127**, 5376–5383; (c) Fuller, A.A., Chen, B., Minter, A.R., and Mapp, A.K. (2004) *Synlett*, 1409–1413.

62 Kawakami, T., Ohtake, H., Arakawa, H., Okachi, T., Imada, Y., and Murahashi, S.-I. (2000) *Bulletin of the Chemical Society of Japan*, **73**, 2423–2444.

63 Murahashi, S., Imada, Y., Kawakami, T., Harada, K., Yonemushi, Y., and Tomita, N. (2002) *Journal of the American Chemical Society*, **124**, 2888–2889.

64 (a) Luisi, R., Capriati, V., Degennaro, L., and Florio, S. (2003) *Organic Letters*, **5**, 2723–2726; (b) Luisi, R., Capriati, V., Florio, S., and Vista, T. (2003) *The Journal of Organic Chemistry*, **68**, 9861–9864.

65 (a) Shindo, M., Itoh, K., Tsuchiya, C., and Shishido, K. (2002) *Organic Letters*, **4**, 3119–3121; (b) Shindo, M., Ohtsuki, K., and Shishido, K. (2005) *Tetrahedron: Asymmetry*, **16**, 2821–2831.

66 Taggi, A.E., Hafez, A.M., and Lectka, T. (2003) *Accounts of Chemical Research*, **36**, 10–19.

67 Roers, R. and Verdine, G.L. (2001) *Tetrahedron Letters*, **42**, 3563–3565.

68 (a) Takahashi, E., Fujisawa, H., and Mukaiyama, T. (2004) *Chemistry Letters*, **33**, 936–937; (b) Takahashi, E., Fujisawa, H., Yanai, T., and Mukaiyama, T. (2005) *Chemistry Letters*, **34**, 468–469.

69 Muraoka, T., Kamiya, S.-I., Matsuda, I., and Itoh, K. (2002) *Journal of the Chemical Society, Chemical Communications*, 1284–1285.

70 Ambhaikar, N.B., Snyder, J.P., and Liotta, D.C. (2003) *Journal of the American Chemical Society*, **125**, 3690–3691.

71 Koriyama, Y., Nozawa, A., Hayakawa, R., and Shimizu, M. (2002) *Tetrahedron*, **58**, 9621–9628.

72 Tang, T.P. and Ellman, J.A. (2002) *The Journal of Organic Chemistry*, **67**, 7819–7832.

73 (a) Davis, F.A., Prasad, K.R., Nolt, M.B., and Wu, Y. (2003) *Organic Letters*, **5**, 925–927; (b) Davis, F.A. and Song, M. (2007) *Organic Letters*, **9**, 2413–2416; (c) Davis, F.A., Zhou, P., and Chen, B.-C. (1998) *Chemical Society Reviews*, **27**, 13–18.

74 (a) Ellman, J.A. (2003) *Pure and Applied Chemistry*, **75**, 39–46; (b) Ellman, J.A., Owens, T.D., and Tang, T.P. (2002) *Accounts of Chemical Research*, **35**, 984–995.

75 (a) Sorochinsky, A., Voloshin, N., Markovsky, A., Belik, M., Yasuda, N., Uekusa, H., Ono, T., Berbasov, D.O., and Soloshonok, V.A. (2003) *The Journal of Organic Chemistry*, **68**, 7448–7454; (b) Soloshonok, V.A., Ohkura, H., Sorochinsky, A., Voloshin, N., Markovsky, A., Belik, M., and Yamazaki, T. (2002) *Tetrahedron Letters*, **43**, 5445–5448.

76 Kanai, K., Wakabayashi, H., and Honda, T. (2002) *Heterocycles*, **58**, 47–51.

77 (a) Vidal, A., Nefzi, A., and Houghten, R.A. (2001) *The Journal of Organic Chemistry*, **66**, 8268–8272; (b) Gouge, V., Jubault, P., and Quirion, J.-C. (2004) *Tetrahedron Letters*, **45**, 773–776.

78 Arian, J.C. Jr., and Snapper, M.L. (2003) *The Journal of Organic Chemistry*, **68**, 2143–2150.

79 (a) Fustero, S., Pina, B., Salavert, E., Navarro, A., Ramírez de Arellano, M.C., and Simón Fuentes, A. (2002) *The Journal of Organic Chemistry*, **67**, 4667–4679; (b) Fustero, S., Salavert, E., Pina, B., Ramírez de Arellano, C., and Asensio, A. (2001) *Tetrahedron*, **57**, 6475–6486; (c) Fustero, S., Díaz, M.D., Navarro, A., Salavert, E., and Aguilar, E. (2001) *Tetrahedron*, **57**, 703–712.

80 Bellassoued, M., Grugier, J., and Lensen, N. (2002) *Journal of Organometallic Chemistry*, **662**, 172–177.

81 Della Rosa, C., Gil, S., Rodríguez, P., and Parra, M. (2006) *Synthesis*, 3092–3098.

82 Lurain, A.E. and Walsh, P.J. (2003) *Journal of the American Chemical Society*, **125**, 10677–10683.

83 Joffe, A.L., Thomas, T.M., and Adrian, J.C. (2004) *Tetrahedron Letters*, **45**, 5087–5090.

84 Rossen, K., Jakubec, P., Kiesel, M., and Janik, M. (2005) *Tetrahedron Letters*, **46**, 1819–1821.

85 (a) Ricci, A., Pettersen, D., Bernardi, L., Fini, F., Fochi, M., Perez, H.R., and

Sgarzani, V. (2007) *Advanced Synthesis and Catalysis*, **349**, 1037–1040; (b) Marianacci, O., Micheletti, G., Bernardi, L., Fini, F., Fochi, M., Pettersen, D., Sgarzani, V., and Ricci, A. (2007) *Chemistry – A European Journal*, **13**, 8338–8351; (c) Fini, F., Bernardi, L., Herrera, R.P., Pettersen, D., Ricci, A., and Sgarzani, V. (2006) *Advanced Synthesis and Catalysis*, **348**, 2043–2046.

86 Utsumi, N., Zhang, H., Tanaka, F., and Barbas, C.F. III (2007) *Angewandte Chemie (International Edition in English)*, **46**, 1878–1880.

87 Matsui, K., Takizawa, S., and Sasai, H. (2005) *Journal of the American Chemical Society*, **127**, 3680–3681.

88 Xu, Y.-M. and Shi, M. (2004) *The Journal of Organic Chemistry*, **69**, 417–425.

89 (a) Sergeeva, N.N., Golubev, A.S., and Burger, K. (2001) *Synthesis*, 281–285; (b) Sergeeva, N.N., Golubev, A.S., Hennig, L., Findeisen, M., Paetzold, E., Oehme, G., and Burger, K. (2001) *Journal of Fluorine Chemistry*, **111**, 41–44.

90 Nemoto, T., Fukuyama, T., Yamamoto, E., Tamura, S., Fukuda, T., Matsumoto, T., Akimoto, Y., and Hamada, Y. (2007) *Organic Letters*, **9**, 927–930.

91 Balan, D. and Adolfsson, H. (2001) *The Journal of Organic Chemistry*, **66**, 6498–6501.

92 Das, B., Majhi, A., Reddy, K.R., and Suneel, K. (2007) *Journal of Molecular Catalysis A – Chemical*, **274**, 83–86.

93 Khan, A.T., Choudhury, L.H., Parvin, T., and Ali, M.A. (2006) *Tetrahedron Letters*, **47**, 8137–8141.

94 (a) Yang, Y.-Y., Shou, W.-G., and Wang, Y.-G. (2006) *Tetrahedron*, **62**, 10079–10086; (b) Shou, W.-G., Yang, Y.-Y., and Wang, Y.-G. (2006) *Tetrahedron Letters*, **47**, 1845–1847.

95 Wu, M., Jing, H., and Chang, T. (2007) *Catalysis Communications*, **8**, 2217–2221.

96 Wang, L., Han, J., Sheng, J., Tian, H., and Fan, Z. (2005) *Catalysis Communications*, **6**, 201–204.

97 Loh, T.-P., Liung, S.B.K.W., Tan, K.-L., and Wei, L.-L. (2000) *Tetrahedron*, **56**, 3227–3237.

98 (a) Dondoni, A., Massi, A., Sabbatini, S., and Bertolasi, V. (2004) *Tetrahedron Letters*, **45**, 2381–2384; (b) Dondoni, A., Massi, A., and Sabbatini, S. (2005) *Chemistry – A European Journal*, **11**, 7110–7125.

99 Phukan, P., Kataki, D., and Chakraborty, P. (2006) *Tetrahedron Letters*, **47**, 5523–5525.

100 Azizi, N., Torkiyan, L., and Saidi, M.R. (2006) *Organic Letters*, **8**, 2079–2082.

101 Takaya, J., Kagoshima, H., and Akiyama, T. (2000) *Organic Letters*, **2**, 1577–1579.

102 Ollevier, T. and Nadeau, E. (2006) *Synlett*, 219–222.

103 Sahoo, S., Joseph, T., and Halligudi, S.B. (2006) *Journal of Molecular Catalysis A – Chemical*, **244**, 179–182.

104 Gong, K., Fang, D., Wang, H.-L., and Liu, Z.-L. (2007) *Monatshefte fur Chemie*, **138**, 1195–1198.

105 Fang, D., Luo, J., Zhou, X.-L., and Liu, Z.-L. (2007) *Catalysis Letters*, **116**, 76–80.

106 Akiyama, T., Suzuki, A., and Fuchibe, K. (2005) *Synlett*, 1024–1026.

107 Itoh, J., Fuchibe, K., and Akiyama, T. (2006) *Synthesis*, 4075–4080.

108 Wenzel, A.G. and Jacobsen, E.N. (2002) *Journal of the American Chemical Society*, **124**, 12964–12965.

109 (a) Song, J., Wang, Y., and Deng, L. (2006) *Journal of the American Chemical Society*, **128**, 6048–6049; (b) Song, J., Shih, H.-W., and Deng, L. (2007) *Organic Letters*, **9**, 603–606.

110 Yamaoka, Y., Miyabe, H., Yasui, Y., and Takemoto, Y. (2007) *Synthesis*, 2571–2575.

111 Córdova, A., Watanabe, S.-I., Tanaka, F., Notz, W., and Barbas, C.F. III, (2002) *Journal of the American Chemical Society*, **124**, 1866–1867.

112 Chowdari, N.S., Suri, J.T., and Barbas, C.F. III (2004) *Organic Letters*, **6**, 2507–2510.

113 (a) Yang, J.W., Stadler, M., and List, B. (2007) *Angewandte Chemie (International*

Edition in English), **46**, 609–611; (b) Yang, J.W., Stadler, M., and List, B. (2007) *Nature Protocols*, **2**, 1937–1942.

114 Cobb, A.J.A., Shaw, D.M., and Ley, S.V. (2004) *Synlett*, 558–560.

115 Notz, W., Tanaka, F., Watanabe, S., Chowdari, N.S., Turner, J.M., Thayumanavan, R., and Barbas, C.F. (2003) *The Journal of Organic Chemistry*, **68**, 9624–9634.

116 Vesely, J., Rios, R., Ibrahem, I., and Córdova, A. (2006) *Tetrahedron Letters*, **48**, 421–425.

117 Chi, Y., English, E.P., Pomerantz, W.C., Horne, W.S., Joyce, L.A., Alexander, L.R., Fleming, W.S., Hopkins, E.A., and Gellman, S.H. (2007) *Journal of the American Chemical Society*, **129**, 6050–6055.

118 (a) Davies, H.M.L., Venkataramani, C., Hansen, T., and Hopper, D.W. (2003) *Journal of the American Chemical Society*, **125**, 6462–6468; (b) Davies, H.M.L. and Ni, A. (2006) *Journal of the Chemical Society, Chemical Communications*, 3110–3112.

119 Moody, C.J. (2004) *Journal of the Chemical Society, Chemical Communications*, 1341–1351.

120 Has-Becker, S., Bodmann, K., Kreuder, R., Santoni, G., Rein, T., and Reiser, O. (2001) *Synlett*, 1395–1398.

121 (a) Palomo, C., Oiarbide, M., González-Rego, M.C., Sharma, A.K., García, J.M., González, A., Landa, C., and Linden, A. (2000) *Angewandte Chemie (International Edition in English)*, **39**, 1063–1065; (b) Palomo, C., Oiarbide, M., Landa, A., González-Rego, M.C., García, J.M., González, A., Odriozola, J.M., Martín-Pastor, M., and Linden, A. (2002) *Journal of the American Chemical Society*, **124**, 8637–8643.

122 Fadini, L. and Togni, A. (2004) *Chimia*, **58**, 208–211.

123 Shin, D.-Y., Jung, J.-K., Seo, S.-Y., Lee, Y.-S., Paek, S.-M., Chung, Y.K., Shin, D.M., and Suh, Y.-G. (2003) *Organic Letters*, **5**, 3635–3638.

124 (a) Huguenot, F. and Brigaud, T. (2006) *The Journal of Organic Chemistry*, **71**, 2159–2162; (b) Lebouvier, N., Laroche, C., Huguenot, F., and Brigaud, T. (2002) *Tetrahedron Letters*, **43**, 2827–2830.

125 (a) Ranslow, P.B.D., Hegedus, L.S., and de los Rios, C. (2004) *The Journal of Organic Chemistry*, **69**, 105–111; (b) Hegedus, L.S., Ranslow, P., Achmatowicz, M., de los Ríos, C., Hyland, C., García-Frutos, E.M., and Salman, S. (2006) *Pure and Applied Chemistry*, **78**, 333–339.

126 Gennari, C., Monti, C., and Piarulli, U. (2006) *Pure and Applied Chemistry*, **78**, 303–310.

127 Saylik, D., Campi, E.M., Donohue, A.C., Jackson, W.R., and Robinson, A.J. (2001) *Tetrahedron: Asymmetry*, **12**, 657–667.

128 (a) Peña, D., Minnaard, A.J., de Vries, J.G., and Feringa, B.L. (2002) *Journal of the American Chemical Society*, **124**, 14552–14553; (b) de Vries, A.H.M., Boogers, J.A.F., van den Berg, M., Peña, D., Minnaard, A.J., Feringa, B.L., and de Vries, J.G. (2003) *PharmaChem*, **2**, 33–36; (c) Hoen, R., Tiemersma-Wegman, T., Procuranti, B., Lefort, L., de Vries, J.G., Minnaard, A.J., and Feringa, B.L. (2007) *Organic and Biomolecular Chemistry*, **5**, 267–275.

129 Ohashi, A., Kikuchi, S.-I., Yasutake, M., and Imamoto, T. (2002) *European Journal of Organic Chemistry*, 2535–2546.

130 Tang, W., Wang, W., Chi, Y., and Zhang, X. (2003) *Angewandte Chemie (International Edition in English)*, **42**, 3509–3511.

131 Holz, J., Monsees, A., Jiao, H., You, J., Komarov, I.V., Fischer, C., Drauz, K., and Borner, A. (2003) *The Journal of Organic Chemistry*, **68**, 1701–1707.

132 Reetz, M.T. and Li, X. (2004) *Tetrahedron*, **60**, 9709–9714.

133 Monti, C., Gennari, C., Piarulli, U., de Vries, J.G., de Vries, A.H.M., and Lefort, L. (2005) *Chemistry – A European Journal*, **11**, 6701–6717.

134 Enthaler, S., Erre, G., Junge, K., Holz, J., Boerner, A., Alberico, E., Nieddu, I., Gladiali, S., and Beller, M. (2007) *Organic*

Process Research & Development, **11**, 568–577.

135 Hsiao, Y., Rivera, N.R., Rosner, T., Krska, S.W., Njolito, E., Wang, F., Sun, Y., Armstrong, J.D., Grabowski, E.J.J., Tillyer, R.D., Spindler, F., and Malan, C. (2004) *Journal of the American Chemical Society*, **126**, 9918–9919.

136 Cohen, J.H., Abdel-Magid, A.F., Almond, H.R. Jr., and Maryanoff, C.A. (2002) *Tetrahedron Letters*, **43**, 1977–1981.

137 (a) Cimarelli, C., Palmieri, G., and Volpini, E. (2001) *Synthetic Communications*, **31**, 2943–2953; (b) Palmieri, G. and Cimarelli, C. (2006) *ARKIVOC*, 104–126.

138 Sibi, M.P. and Deshpande, P.K. (2000) *Journal of the Chemical Society, Perkin Transactions 1*, 1461–1466.

139 Nagula, G., Huber, V.J., Lum, C., and Goodman, B.A. (2000) *Organic Letters*, **2**, 3527–3529.

140 Iza, A., Vicario, J.L., Carrillo, L., and Badia, D. (2006) *Synthesis*, 4065–4074.

141 (a) Gutiérrez-García, V.M., López-Ruiz, H., Reyes-Rangel, G., and Juaristi, E. (2001) *Tetrahedron*, **57**, 6487–6496; (b) Gutiérrez-García, V.M., Reyes-Rangel, G., Muñoz-Muñiz, O., and Juaristi, E. (2001) *Journal of the Brazilian Chemical Society*, **12**, 652–660; (c) Gutiérrez-García, V.M., Reyes-Rangel, G., Muñoz-Muñiz, O., and Juaristi, E. (2002) *Helvetica Chimica Acta*, **85**, 4189–4199.

142 Guzmán-Mejía, R., Reyes-Rangel, G., and Juaristi, E. (2007) *Nature Protocols*, **2**, 2759–2766.

143 Agami, C., Cheramy, S., Dechoux, L., and Melaimi, M. (2001) *Tetrahedron*, **57**, 195–200.

144 Sebesta, R. and Seebach, D. (2003) *Helvetica Chimica Acta*, **86**, 4061–4072.

145 Aoyagi, Y., Jain, R.P., and Williams, R.M. (2001) *Journal of the American Chemical Society*, **123**, 3472–3477.

146 (a) Miyabe, H., Ueda, M., and Naito, T. (2004) *Synlett*, 1140–1157; (b) Miyabe, H., Fujii, K., and Naito, T. (1999) *Organic Letters*, **1**, 569–572.

147 Huck, J., Receveur, J.-M., Roumestant, M.-L., and Martinez, J. (2001) *Synlett*, 1467–1469.

148 Miyata, O., Muroya, K., Kobayashi, T., Yamanaka, R., Kajisa, S., Koide, J., and Naito, T. (2002) *Tetrahedron*, **58**, 4459–4479.

149 Ramachandran, P.V. and Burghardt, T.E. (2005) *Chemistry – A European Journal*, **11**, 4387–4395.

150 Papa, C. and Tomasini, C. (2000) *European Journal of Organic Chemistry*, 1569–1576.

151 Beresford, K.J.M., Church, N.J., and Young, D.W. (2006) *Organic and Biomolecular Chemistry*, **4**, 2888–2897.

152 (a) Mordini, A., Russo, F., Valacchi, M., Zani, L., Degl'Innocenti, A. and Reginato, G. (2002) *Tetrahedron*, **58**, 7153–7163; (b) Mordini, A., Sbaragli, L., Valacchi, M., Russo, F., and Reginato, G. (2002) *Journal of the Chemical Society, Chemical Communications*, 778–779.

153 Davis, F.A., Deng, J., Zhang, Y., and Haltiwanger, R.C. (2002) *Tetrahedron*, **58**, 7135–7143.

154 Enders, D. and Gries, J. (2005) *Synthesis*, 3508–3516.

155 Yamauchi, Y., Kawate, T., Itahashi, H., Katagiri, T., and Uneyama, K. (2003) *Tetrahedron Letters*, **44**, 6319–6322.

156 Wu, J., Hou, X.-L., and Dai, L.-X. (2000) *The Journal of Organic Chemistry*, **65**, 1344–1348.

157 Matthews, J.L., McArthur, D.R., and Muir, K.W. (2002) *Tetrahedron Letters*, **43**, 5401–5404.

158 Crousse, B., Narizuka, S., Bonnet-Delpon, D., and Begue, J.-P. (2001) *Synlett*, 679–681.

159 MacNevin, C.J., Moore, R.L., and Liotta, D.C. (2008) *The Journal of Organic Chemistry*, **73**, 1264–1269.

160 Felpin, F.-X., Doris, E., Wagner, A., Valleix, A., Rousseau, B., and Mioskowski, C. (2001) *The Journal of Organic Chemistry*, **66**, 305–308.

161 Seebach, D., Sifferlen, T., Bierbaum, D.J., Rueping, M., Jaun, B., Schweizer, B., Schaefer, J., Mehta, A.K., O'Connor, R.D.,

Meier, B.H., Ernst, M., and Glattli, A. (2002) *Helvetica Chimica Acta*, **85**, 2877–2917.

162 Schwarz, J.B., Gibbons, S.E., Graham, S.R., Colbry, N.L., Guzzo, P.R., Le, V.-D., Vartanian, M.G., Kinsora, J.J., Lotarski, S.M., Li, Z., Dickerson, M.R., Su, T.-Z., Weber, M.L., El-Kattan, A., Thorpe, A.J., Donevan, S.D., Taylor, C.P., and Wustrow, D.J. (2005) *Journal of Medicinal Chemistry*, **48**, 3026–3035.

163 Moutevelis-Minakakis, P., Sinanoglou, C., Loukas, V., and Kokotos, G. (2005) *Synthesis*, 933–938.

164 Espino, C.G., Wehn, P.M., Chow, J., and Du Bois, J. (2001) *Journal of the American Chemical Society*, **123**, 6935–6936.

165 Freitag, D. and Metz, P. (2006) *Tetrahedron*, **62**, 1799–1805.

166 Jayakanthan, K. and Vankar, Y.D. (2005) *Organic Letters*, **7**, 5441–5444.

167 Inaba, Y., Yano, S., and Mikata, Y. (2007) *Tetrahedron Letters*, **48**, 993–997.

168 Nelson, S.G. and Spencer, K.L. (2000) *Angewandte Chemie (International Edition in English)*, **39**, 1323–1325.

169 Fringuelli, F., Pizzo, F., Rucci, M., and Vaccaro, L. (2003) *The Journal of Organic Chemistry*, **68**, 7041–7045.

170 Singh, R.K., Sinha, N., Jain, S., Salman, M., Naqvi, F., and Anand, N. (2005) *Tetrahedron*, **61**, 8868–8874.

171 Busnel, O. and Baudy-Floc'h, M. (2007) *Tetrahedron Letters*, **48**, 5767–5770.

172 (a) Palko, M., Kiss, L., and Fülöp, F. (2005) *Current Medicinal Chemistry*, **12**, 3063–3083; (b) Fülöp, F., Martinek, T.A., and Toth, G.K. (2006) *Chemical Society Reviews*, **35**, 323–334.

173 Miller, J.A. and Nguyen, S.T. (2005) *Mini-Reviews in Organic Chemistry*, **2**, 39–45.

174 Kuhl, A., Hahn, M.G., Dumic, M., and Mittendorf, J. (2005) *Amino Acids*, **29**, 89–100.

175 Perlmutter, P., Rose, M., and Vounatsos, F. (2003) *European Journal of Organic Chemistry*, 756–760.

176 Gauzy, C., Saby, B., Pereira, E., Faure, S., and Aitken, D.J. (2006) *Synlett*, 1394–1398.

177 Wang, X., Espinosa, J.F., and Gellman, S.H. (2000) *Journal of the American Chemical Society*, **122**, 4821–4822.

178 Izquierdo, S., Martín-Vilà, M., Moglioni, A.G., Branchadell, V., and Ortuño, R.M. (2002) *Tetrahedron: Asymmetry*, **13**, 2403–2405.

179 (a) Forró, E., Paál, T., Tasnádi, G., and Fülöp, F. (2006) *Advanced Synthesis and Catalysis*, **348**, 917–923; (b) Tasnádi, G., Forró, E., and Fülöp, F. (2007) *Tetrahedron: Asymmetry*, **18**, 2841–2844; (c) Park, S., Forró, E., Grewal, H., Fülöp, F., and Kazlauskas, R.J. (2003) *Advanced Synthesis and Catalysis*, **345**, 986–995; (d) Forró, E. and Fülöp, F. (2001) *Tetrahedron: Asymmetry*, **12**, 2351–2358.

8
Synthesis of Carbocyclic β-Amino Acids

Loránd Kiss, Enikő Forró, and Ferenc Fülöp

8.1
Introduction

During the last 20 years the synthesis of carbocyclic β-amino acids has received considerable attention due to their significant biological potential [1–7]. This class of compounds is present in natural products, and also in antibiotics such as the antifungal cispentacin (**1**) and icofungipen (**2**). Cispentacin is, additionally, a component of the antibiotic amipurimycin. Hydroxy-functionalized derivatives such as oryzoxymycin (**3**) are likewise active antibacterial agents.

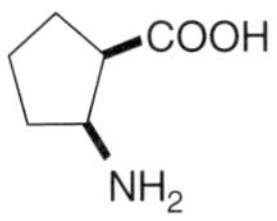

cispentacin (**1**) 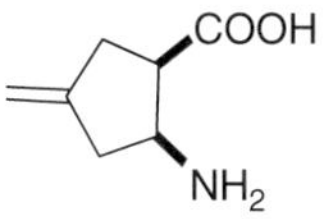 icofungipen (**2**) 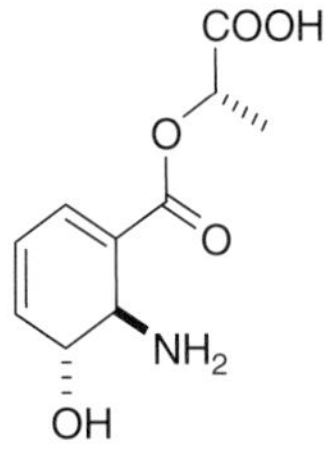 oryzoxymycin (**3**)

Conformationally constrained cyclic β-amino acids have been a topic of great interest in peptide chemistry in recent years [1–7]. Cyclic β-amino acids have a large range of uses as building blocks for the preparation of modified biologically active peptide analogs. Possessing an extra carbon atom between the amino and carboxylic groups, these β-amino acids have greater potential for structural diversity than their α-analogs. As a consequence of the availability of a vast number of stereo- and regioisomers, together with the possibility for further functionalizations on the ring, the structural diversity of these β-amino acids is significantly enhanced. Moreover, these β-amino acids are in general more stable to hydrolysis or enzymatic degradation than their α-analogs. This leads to the enhanced stability of peptides in which they are incorporated. Carbocyclic β-amino acids are used as starting substances for the synthesis of heterocyclic compounds, potential pharmacons, and natural product analogs. Their enantiomerically pure forms can serve as chiral auxiliaries in asymmetric transformations. The chemistry of carbocyclic β-amino

Amino Acids, Peptides and Proteins in Organic Chemistry. Vol.1 – Origins and Synthesis of Amino Acids.
Edited by Andrew B. Hughes

ISBN: 978-3-527-32096-7

acids was reviewed by our group in 2001 [1]; the present chapter highlights the chemistry of these compounds published since 2000, with the aim of offering the reader an insight into the most recent developments in this field.

8.2
Synthesis of Carbocyclic β-Amino Acids

Owing to their significance, a number of methods have been developed for the synthesis of these conformationally restricted molecules in racemic and enantiomerically pure form. The largest groups of carbocyclic β-amino acids and the most intensively investigated are the five- and six-membered derivatives. One main method of preparation consists of the amidation of 1,2-dicarboxylic anhydrides **4**, followed by Hoffmann rearrangement of the resulting amide **5** to amino acid **7**. The same starting material, the *meso* anhydride **4**, undergoes partial hydrolysis, followed by the Curtius rearrangement, to give 2-aminocyclohexanecarboxylic acid **7** (Scheme 8.1) ([1] and references cited therein).

A widely used method for the synthesis of carbocyclic β-amino acids is the transformation of cycloalkenes via the corresponding bicyclic β-lactams. The 1,2-dipolar cycloaddition of chlorosulfonyl isocyanate (CSI) to cycloalkenes **8** led regioselectively to the cycloalkane-fused β-lactams **10**, whose acidic hydrolysis resulted in lactam ring opening and formation of the carbocyclic β-amino acid hydrochlorides **11** (Scheme 8.2) ([1] and references cited therein).

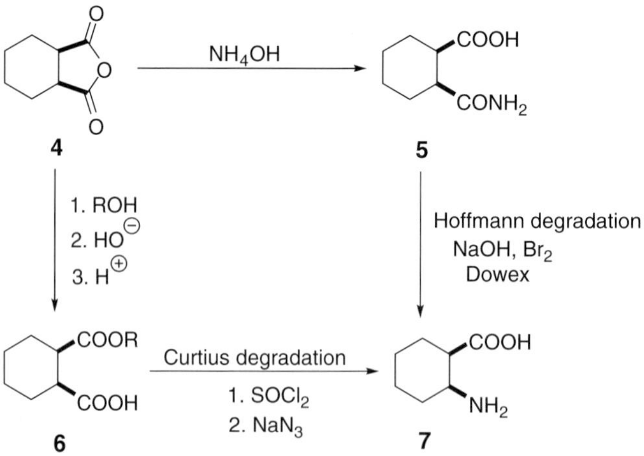

Scheme 8.1 Synthesis of carbocyclic β-amino acids by Curtius and Hoffmann rearrangements.

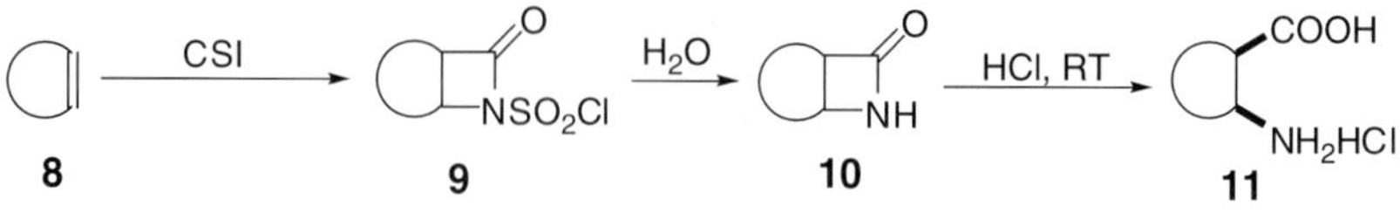

Scheme 8.2 Synthesis of carbocyclic β-amino acids from β-lactams.

8.2.1 Synthesis of Carbocyclic β-Amino Acids via Lithium Amide-Promoted Conjugate Addition

Among the various strategic methods for the synthesis of carbocyclic β-amino acids, the conjugate addition of an amine nucleophile to an α,β-unsaturated carboxylic acid derivative (e.g., **12**) is one of the most attractive procedures for stereoselective synthesis. Several lithium amides derived from chiral amines (e.g., **13**) have been used as ammonia equivalents in these transformations. Davies *et al.* made extensive use of this stereoselective addition method, which, after removal of the chiral auxiliary from **14** and subsequent hydrolysis, provided (1*R*,2*S*)-2-aminocyclopentanecarboxylic acid **1** in high enantiomeric purity (>98% e.e.) (Scheme 8.3) [8].

This methodology has been extended to the synthesis of substituted carbocyclic β-amino acid derivatives. *tert*-Butyl 3-methylcyclopentenecarboxylate **15** was applied for the synthesis of 3-methyl-substituted cispentacin and 3-methyltranspentacin enantiomers [9]. The conjugate addition of lithium (*S*)-*N*-benzyl-*N*-α-methylbenzylamide to the unsaturated 3-substituted ester **15** generated three β-amino ester diastereomers (**16–18**) in a ratio of 95.5 : 1.7 : 2.8 (Scheme 8.4); it proved possible to separate these diastereomers by column chromatography. Removal of the chiral auxiliary by hydrogenolysis of **16** in the presence of $Pd(OH)_2/C$ and subsequent ester hydrolysis furnished the 3-methylcispentacin enantiomer in greater than 98% e.e.

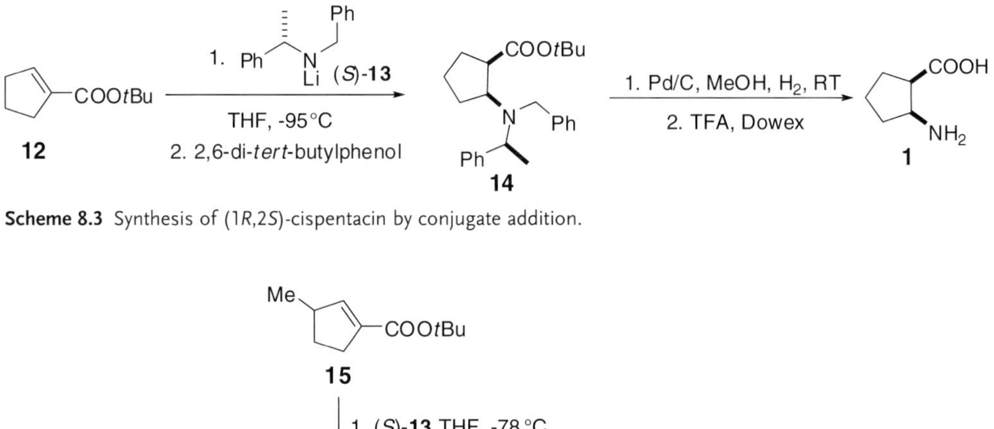

Scheme 8.3 Synthesis of (1*R*,2*S*)-cispentacin by conjugate addition.

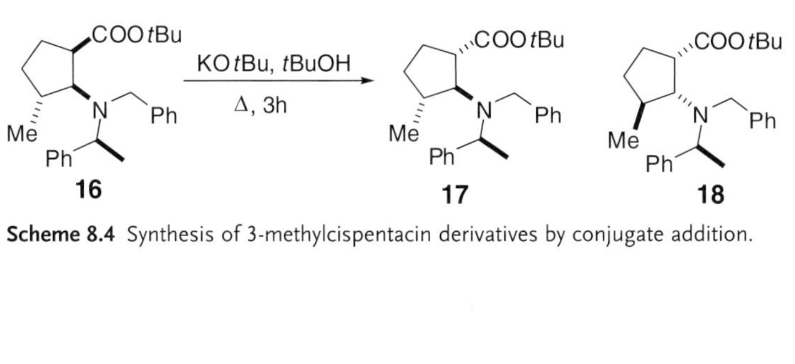

Scheme 8.4 Synthesis of 3-methylcispentacin derivatives by conjugate addition.

To prepare 3-methyltranspentacin, aminocarboxylate **17** was first obtained in a larger quantity from diastereoisomer **16** by epimerization on treatment with KO*t*Bu in *t*BuOH; this afforded **17** in quantitative yield and a diastereomeric excess of 99%. *N*-Deprotection of **17**, followed by ester hydrolysis, gave the desired 3-methyltranspentacin in 97% e.e. [9].

By using this conjugate addition methodology, Davies *et al.* also prepared 5-substituted derivatives of cispentacin. The addition of lithium amide (*S*)-**13** to 5-isopropyl-, 5-phenyl-, and 5-*tert*-butylcyclopentenecarboxylates proceeded with a high degree of selectivity, the major derivatives formed having the amine *anti* to the 5-alkyl group. *N*-Deprotection by hydrogenolysis, subsequent acid hydrolysis of the ester, and ion-exchange chromatography furnished 5-substituted cispentacin derivatives in 98% e.e. [10].

The asymmetric synthesis of a series of functionalized derivatives was achieved by using amine conjugate addition processes. Four stereoisomers of 2-amino-5-carboxymethylcyclopentanecarboxylate were prepared from diene **19** in reactions with (*R*)- and (*S*)-*N*-benzyl-*N*-α-methylbenzamide. Stereoisomers **20** and **21**, resulting from the addition of (*R*)-**13** to **19**, could be separated by chromatography. Removal of the chiral auxiliary and ester hydrolysis furnished enantiomers **22** and **23** with in greater than 99% e.e. (Scheme 8.5) [11].

The above strategy was extended for the preparation of analogous cyclohexane β-amino acids and other functionalized derivatives with a cyclopentane or cyclohexane skeleton. Starting from a diene-dicarboxylate, the above conjugate addition procedure with the chiral amide (*S*)-**13** resulted in, with high diastereoselectivity, the corresponding cyclohexanecarboxylate, which was readily converted to the cyclohexane β-amino diacid [12].

An excellent application of the conjugate addition/ring-closure method was presented for the synthesis of hydroxy-functionalized cyclopentane or cyclohexane

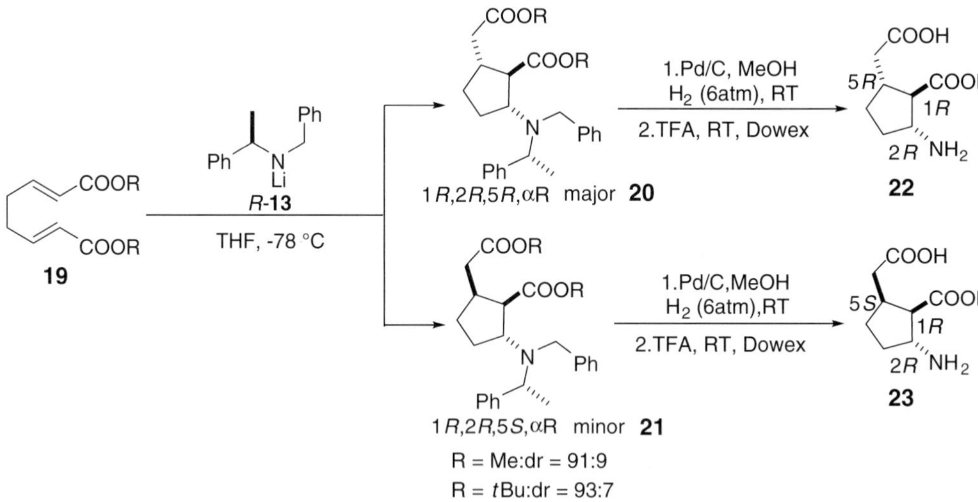

Scheme 8.5 Synthesis of 2-amino-5-carboxymethylcyclopentanecarboxylate stereoisomers.

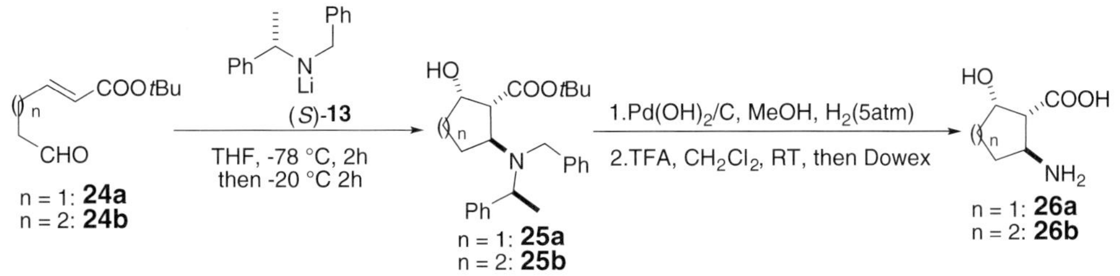

Scheme 8.6 Synthesis of hydroxylated 2-aminocycloalkanecarboxylic acids.

β-amino acids. Suitable starting materials for this purpose were formyl carboxylates **24a** and **24b**, whose transformation according to the known procedures resulted in (1*R*,2*S*,5*S*)-2-amino-5-hydroxycyclopentanecarboxylic acid **26a** and (1*R*,2*S*,5*S*)-2-amino-6-hydroxycyclohexanecarboxylic acid **26b** (Scheme 8.6) [12].

8.2.2 Synthesis of Carbocyclic β-Amino Acids by Ring-Closing Metathesis

Carbocyclic β-amino acids can be formed in ring-closing metathesis reactions. An advantage of this procedure is that it furnishes unsaturated carbocyclic β-amino acids, whose olefinic bond may be further functionalized. Chippindale *et al.* combined diastereoselective conjugate addition and ring-closing metathesis for the synthesis of carbocyclic β-amino acid derivatives [13].

The first step, stereoselective addition of the chiral lithium amide, generating the two C stereocenters, was followed by synthesis of dieno amino esters **29a**,**b** and **31** in several steps. These dienes underwent ring-closing metathesis upon treatment with a ruthenium alkylidene catalyst to afford the desired unsaturated carbocyclic β-amino ester derivatives **30a**,**b** and **32** (Scheme 8.7) [13].

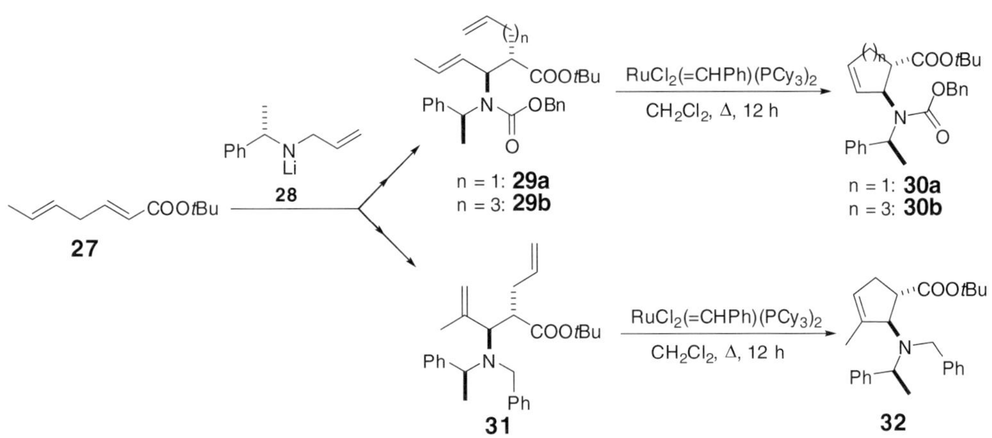

Scheme 8.7 Formation of 2-aminocycloalkanecarboxylic esters by ring-closing metathesis.

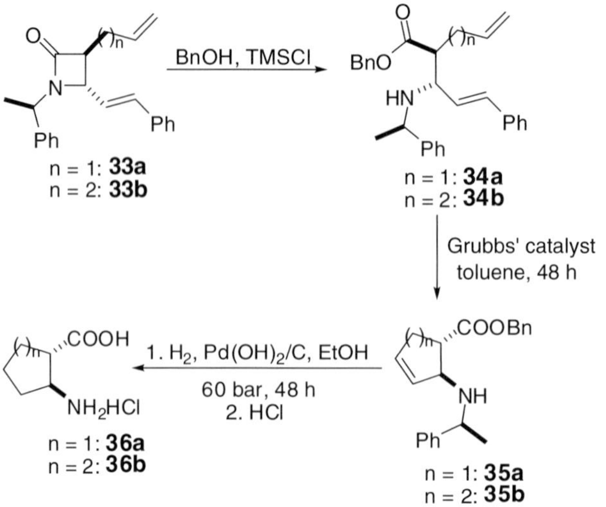

Scheme 8.8 Synthesis of *trans*-2-aminocycloalkanecarboxylic acids by ring-closing metathesis.

β-Lactam dienes **33a,b** are other useful starting materials, which can be converted by lactam ring opening, then ring-closing metathesis, to the unsaturated *trans* β-amino ester derivatives **35a,b** containing a cyclopentene or cyclohexene ring (Scheme 8.8). The stereocenters of the starting lactam remain unaffected during the process, resulting stereoselectively in *trans*-β-aminocyclopentanecarboxylic and *trans*-β-aminocyclohexanecarboxylic acid hydrochlorides **36a,b** [14].

With the application of ring-closing metathesis, a general method for the synthesis of fluorinated five-, six- and seven-membered cyclic β-amino esters was developed by Fustero *et al.* Fluorinated analogs of 2-aminocyclopentane-, 2-cyclohexane-, and 2-cycloheptanecarboxylates were synthesized through the cross-metathesis of fluorinated imidoyl chlorides **37a–c** and ethyl acrylate, followed by ring closure of the resulting imino esters **39a–c** in the presence of LDA. Catalytic hydrogenation of cyclic β-imino esters **41a–c** furnished the five-, six-, and seven-membered difluorinated cyclic *cis*-β-aminocarboxylates **42a–c** (Scheme 8.9) [15].

The asymmetric version of this procedure was achieved by using (−)-8-phenylmenthol as chiral auxiliary. Starting from the phenylmenthyl ester, the base-mediated ring-closure reaction afforded the optically active fluorinated cyclic β-amino ester **44** (Scheme 8.10) [15].

8.2.3
Syntheses from Cyclic β-Keto Esters

A widely used method for the synthesis of five- and six-membered cyclic β-amino acids is the reduction of oximes or enamines, which can be readily prepared from the corresponding carbocyclic β-keto esters. The reductive amination of β-keto esters offers an excellent opportunity for the synthesis of the carbocyclic β-amino acids

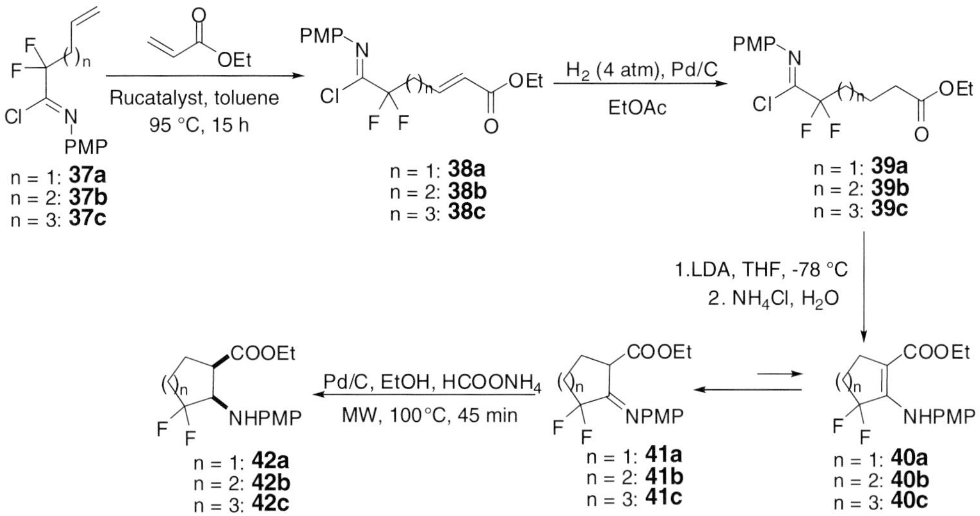

Scheme 8.9 Synthesis of fluorinated 2-aminocycloalkanecarboxylic acids.

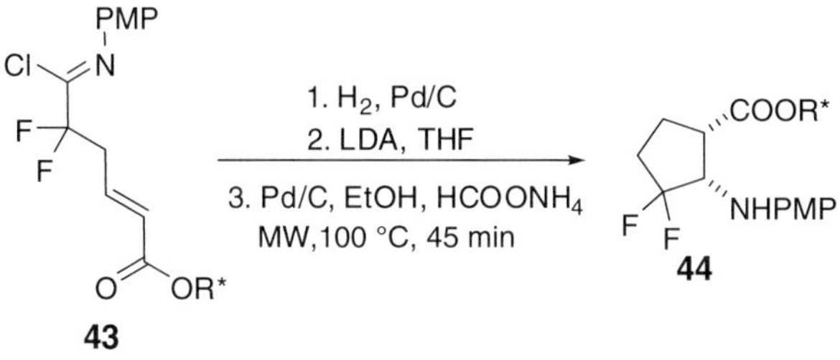

R* = 1*R*,2*S*,5*R*-8-phenylmenthyl

Scheme 8.10 Synthesis of a chiral fluorinated 2-aminocyclopentanecarboxylic acid derivative.

in enantiomerically pure form. LePlae *et al.* have prepared the Fmoc-protected β-aminocyclopentanecarboxylic acid enantiomer from keto ester **45** in four steps [16]. The chiral auxiliary (*S*)-α-methylbenzylamine was reacted with **45** to result in enamine **46**. The next step proved critical in this synthetic procedure, since not only did the reduction of **46** lead to side-products, but the yield of the desired product (**47**) was rather low (20%). Removal of the chiral auxiliary, protection of the amino group and oxidation of the alcohol functionality afforded amino acid enantiomer **49** (Scheme 8.11).

An alternative route was found by the same research group, in which the enamine derived from keto ester **45** and (*S*)-α-methylbenzylamine was reduced with $NaCNBH_3$ [16]. Hydrolysis of ester **50**, removal of the chiral auxiliary and Fmoc protection provided the enantiomerically pure amino acid **49** in 25% overall yield (Scheme 8.11).

Peelen *et al.* synthesized 4,4-disubstituted derivatives of 2-aminocyclopentane-carboxylic acid enantiomers from 4,4-disubstituted β-keto esters (R = Me, Ph,

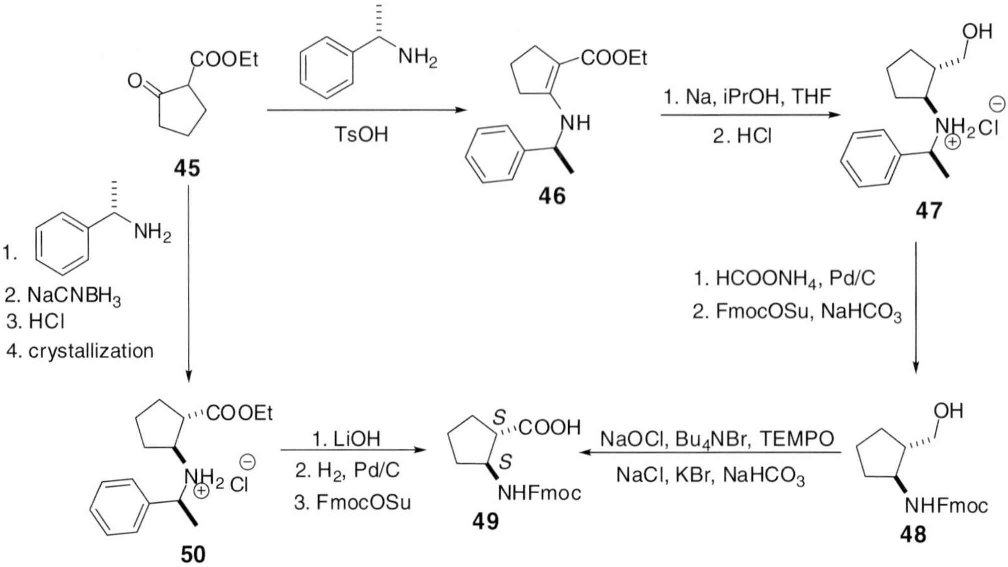

Scheme 8.11 Synthesis of a Fmoc-protected 2-aminocyclopentanecarboxylic acid enantiomer from a 2-keto ester.

COO*t*Bu, CH_2O*t*Bu) [17]. The keto ester was reacted with (*R*)-α-methylbenzylamine, followed by reduction with $NaCNBH_3$, to give a diastereomeric mixture of the corresponding *trans* amino esters. These isomers could be separated by chromatography, and then converted to both enantiomers of 4,4-disubstituted 2-aminocyclopentanecarboxylates.

Tang *et al.* reported a generally applicable catalytic method for the asymmetric synthesis of five-, six-, seven-, and eight-membered carbocyclic β-amino acid derivatives via asymmetric hydrogenation [18]. A ruthenium catalyst and several chiral phosphine ligands were used for this purpose. The *cis*-selective reduction resulted in the enantiomerically pure β-amino ester with *cis*-relative stereochemistry in 99% e.e. (Scheme 8.12).

Although the asymmetric hydrogenation furnished the *cis* β-amino ester derivative, the *trans* isomer could easily be prepared by the known epimerization reaction

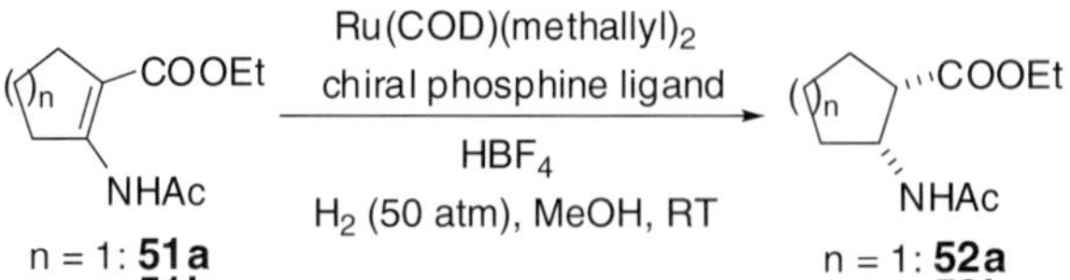

n = 1: **51a**
n = 2: **51b**
n = 3: **51c**
n = 4: **51d**

n = 1: **52a**
n = 2: **52b**
n = 3: **52c**
n = 4: **52d**

Scheme 8.12 Synthesis of 2-aminocycloalkanecarboxylic acid derivatives by asymmetric reduction.

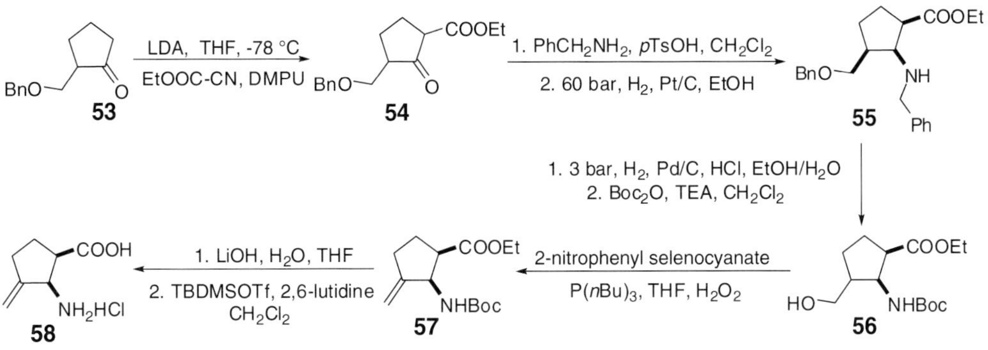

Scheme 8.13 Synthesis of icofungipen analog **58** from a 2-keto ester derivative.

at C-1 in the presence of NaOMe [18]. Asymmetric hydrogenation of dehydroaminocarboxylates with 85% e.e. was reported by Wu and Hoge [19]. New three-hindered quadrant chiral ligands in rhodium complexes were used in the reaction. There have recently been new developments as regards the asymmetric hydrogenation of dehydroamino acids. New chiral monodentate phosphine ligands were used on a multi-10-g scale in these transformations and up to 94% e.e. was achieved [20].

The reductive amination of β-keto esters is also a suitable method for the preparation of functionalized carbocyclic β-amino acid derivatives. Mittendorf *et al.* synthesized an icofungipen analog (**58**) in racemic form in seven steps involving reductive amination of keto ester **54** [21] (Scheme 8.13).

8.2.4
Cycloaddition Reactions: Application in the Synthesis of Carbocyclic β-Amino Acids

Diels–Alder cycloaddition is a widely used and efficient strategy for the formation of cyclic compounds with the generation of new stereocenters. A carbocyclic β-amino acid can be formed by this method. A mixture of *cis*- and *trans*-2-amino-3-cyclohexenecarboxylates (**61** and **62**) was synthesized in a ratio of 4 : 1 by the Diels–Alder reaction of 1,3-butadienecarbamate **59** with acrylate **60** [22].

The two diastereoisomers were separated by chromatography, ester and *N*-deprotected with iodotrimethylsilane (TMSI), and purified by ion-exchange chromatography to give the *cis* and *trans* cyclohexene β-amino acids **63** and **64** (Scheme 8.14).

Through this cycloaddition, not only the formation of a ring system, but also the introduction of functional groups may be achieved. Both the carboxylic and the amino group of a β-amino acid can be introduced on to the cyclohexane moiety via the 1,3-dipolar cycloaddition of a nitrone. Thus, a chiral nitrone was generated from the reaction of aldehyde **65** and (*R*)-*N*-hydroxyphenylglycinol **66**. Isoxazolidine **67** was formed in an intramolecular cycloaddition, the presence of the chelating metal (Mg) and the additive (isopropanol) proving very important (Scheme 8.15). The *cis*-fused derivative that was formed highly diastereoselectively (diastereomeric ratio 96 : 4) was easily converted to the Boc-protected cispentacin **68** [23].

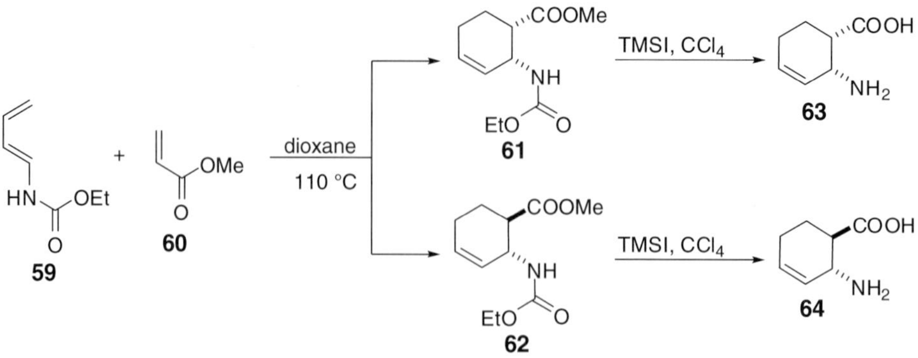

Scheme 8.14 Synthesis of 2-aminocyclohexenecarboxylic acid stereoisomers by Diels–Alder cycloaddition.

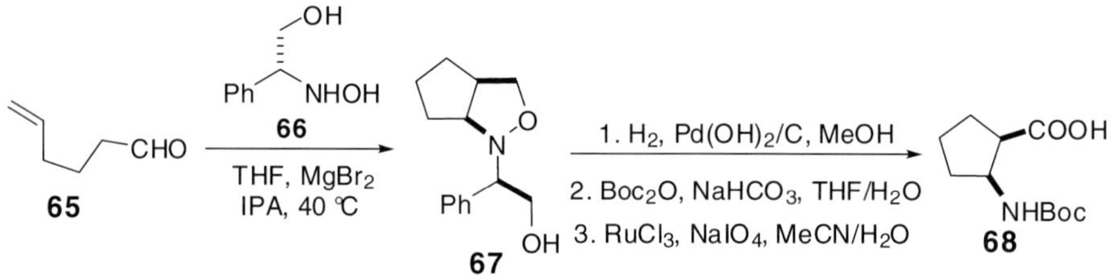

Scheme 8.15 Synthesis of Boc-protected cispentacin via dipolar cycloaddition of a chiral nitrone.

Racemic nitrones were also used effectively for the synthesis of enantiomerically pure cispentacin. In this case, a nitrone was added to the chiral, enantiomerically pure dithioacetal **69**, yielding the isoxazolidine **70** as a single isomer. Cleavage of the N–O bond, removal of the sulfuric moiety, and debenzylation furnished (−)-cispentacin **1** (Scheme 8.16) [24].

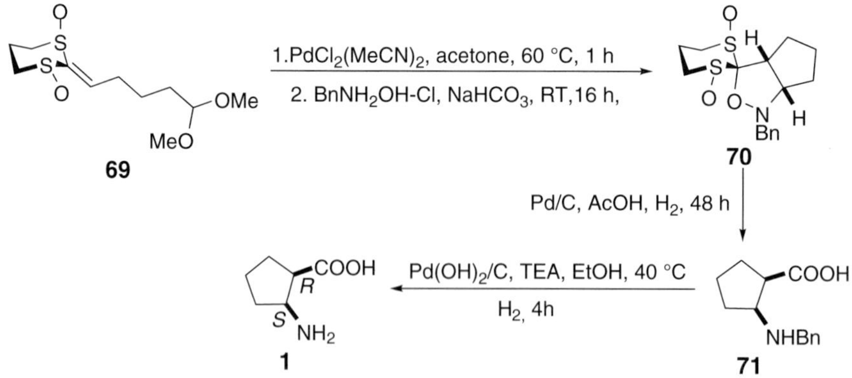

Scheme 8.16 Synthesis of a cispentacin enantiomer by cycloaddition of a nitrone.

8.2.5
Synthesis of Carbocyclic β-Amino Acids from Chiral Monoterpene β-Lactams

A convenient approach for the synthesis of enantiomerically pure carbocyclic β-amino acids is the transformation of readily available chiral monoterpenes. The reactivity of CSI towards the double bond of the monoterpene and then the opening of the ring of the resulting β-lactam are the key steps in these transformations. Thus α-pinene **72** underwent CSI cycloaddition stereo- and regioselectively to furnish β-lactam **73**. Opening of the Boc-protected lactam **74** afforded enantiomerically pure *N*-Boc-protected acid **75** or ester **76**. Finally, these derivatives were converted facilely into the α-pinene β-amino acid derivatives **77** and **79** (Scheme 8.17) [25, 26].

Via a similar strategy, enantiomerically pure β-amino acids **83** and **86** were simply synthesized from the readily available (+)-3-carene (**80**) (Scheme 8.18). The addition of CSI to the double bond led stereo- and regioselectively to the corresponding β-lactam **81**. Lactam ring opening yielded not only β-amino acids or esters, but also the corresponding enantiomerically pure β-amino alcohols by $LiAlH_4$ reduction [27].

In the recently published synthesis of δ-pinene-based β-amino acids [28], δ-pinene **87** was submitted to CSI cycloaddition to afford regio- and stereoselectively the corresponding lactam **88** (Scheme 8.19).

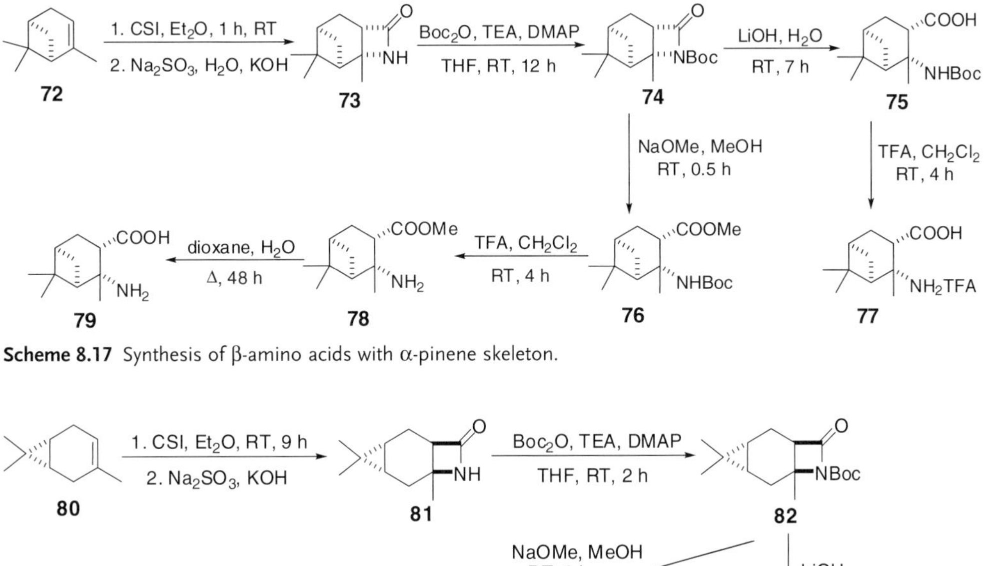

1. CSI, Et_2O, 1 h, RT; 2. Na_2SO_3, H_2O, KOH; Boc_2O, TEA, DMAP, THF, RT, 12 h; LiOH, H_2O, RT, 7 h; NaOMe, MeOH, RT, 0.5 h; TFA, CH_2Cl_2, RT, 4 h; dioxane, H_2O, Δ, 48 h; COOH, NH, NBoc, NHBoc, COOMe, NH_2, NH_2TFA; 72, 73, 74, 75, 76, 77, 78, 79

Scheme 8.17 Synthesis of β-amino acids with α-pinene skeleton.

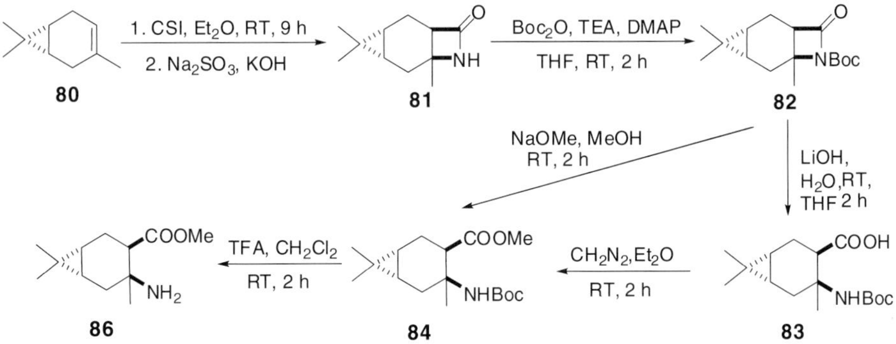

1. CSI, Et_2O, RT, 9 h; 2. Na_2SO_3, KOH; Boc_2O, TEA, DMAP, THF, RT, 2 h; NaOMe, MeOH, RT, 2 h; LiOH, H_2O,RT, THF 2 h; CH_2N_2,Et_2O, RT, 2 h; TFA, CH_2Cl_2, RT, 2 h; NH, NBoc, COOH, COOMe, NHBoc, NH_2; 80, 81, 82, 83, 84, 86

Scheme 8.18 Synthesis of β-amino acids from (+)-3-carene.

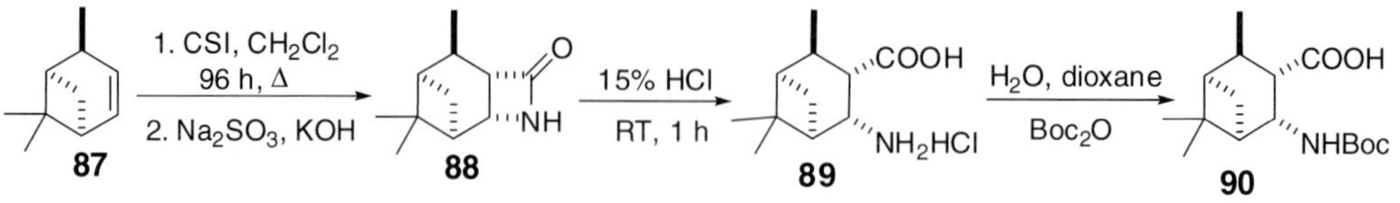
1. CSI, CH_2Cl_2
96 h, Δ
2. Na_2SO_3, KOH
87
88
NH
O
15% HCl
RT, 1 h
89
COOH
NH_2HCl
H_2O, dioxane
Boc_2O
90
COOH
NHBoc

Scheme 8.19 Synthesis of Boc-protected δ-pinene β-amino acid.

By known procedures, the lactam ring opening of **88** furnished the new carbocyclic β-amino acids **89** and **90**. Similar transformations were carried out on the (+)-enantiomer of isopinocamphenol [28].

8.2.6
Synthesis of Carbocyclic β-Amino Acids by Enantioselective Desymmetrization of *meso* Anhydrides

In general, asymmetric synthetic methods suffer from limitations regarding the possibility of the large-scale preparation of the required enantiomers. Desymmetrization of the readily available *meso* anhydrides allows large-scale work. Enantioselective alcoholysis of the anhydride in the presence of the chiral quinine provides the corresponding half-ester, which is followed by a Curtius rearrangement to give the amino ester enantiomer. Mittendorf *et al.* successfully applied the desymmetrization procedure for the preparation of amino acid **2** in high enantiomeric excess (99.5% e.e.) [29]. Alcoholysis of anhydride **92** with cinnamyl alcohol in the presence of a stoichiometric amount of quinine afforded **93**. Curtius rearrangement and deprotection resulted in icofungipen enantiomer **2** (Scheme 8.20).

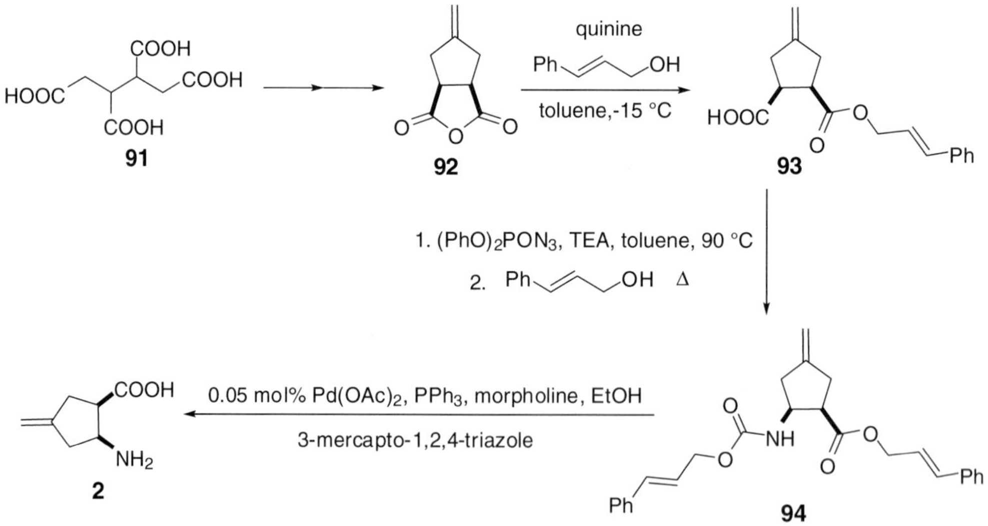
COOH
HOOC
COOH
COOH
91
92
quinine
Ph OH
toluene,-15 °C
HOOC
93
Ph
1. $(PhO)_2PON_3$, TEA, toluene, 90 °C
2. Ph OH Δ
COOH
NH_2
2
0.05 mol% $Pd(OAc)_2$, PPh_3, morpholine, EtOH
3-mercapto-1,2,4-triazole
NH
Ph
94
Ph

Scheme 8.20 Synthesis of icofungipen by desymmetrization of a *meso* anhydride.

Bolm *et al.* have demonstrated the general applicability of the desymmetrization of easily accessible *meso* anhydrides by cinchona-mediated ring opening with benzyl alcohol, leading to different hemiesters with high enantioselectivity (99% e.e.). These hemiesters were transformed to optically active carbocyclic β-amino acids containing a cyclobutane, cyclopentane, norbornane, or norbornene skeleton [30].

Hamersak *et al.* recently described the preparation of four *endo* isomers of icofungipen (**2**) in optically pure form by quinine-mediated desymmetrization of the corresponding racemic anhydride and then a Curtius rearrangement [31]. Anhydride **92** could be isomerized in the presence of *p*-toluenesulfonic acid to **95**. Its alcoholysis in the presence of quinine resulted in monoesters **96** and **97** in a ratio of 3 : 7. All attempts aimed at their separation failed, but the Curtius rearrangement products **98** and **99** could be separated by crystallization from EtOH.

Deprotection of **98** and **99** furnished β-amino acid enantiomers **100** and **101** in 99 and 96% e.e., respectively. When **99** was subjected first submitted to base-mediated epimerization and then to deprotection the third diastereoisomer **102**, was formed in 96% e.e. (Scheme 8.21).

The fourth isomer, **104**, was obtained by *exo–endo* isomerization of enantiomerically pure icofungipen in the presence of chlorotrimethylsilane and NaI (Scheme 8.22).

8.2.7 Miscellaneous

The Strecker synthesis is a well-known general method for the synthesis of α-amino acids. Fondekar *et al.* applied its asymmetric version for the synthesis of carbocyclic amino acids, by using (*R*)-α-methylbenzylamine as chiral auxiliary [32]. The carbonyl group of the starting material, **105**, was reacted with (*R*)-α-methylbenzylamine to furnish imine **106**. The imine functionality offered the opportunity for introduction of the carboxylic group. Nitrile addition afforded a mixture of diastereoisomers **107**, which could be separated by chromatography after hydrolysis of the nitrile group to isomers **108** and **109**. Removal of the chiral auxiliary and deprotection gave α,β-diaminocyclohexanecarboxylic acids **110** and **111** with excellent enantiomeric excess values of 99% (Scheme 8.23).

The synthesis of all four enantiomers of *cis*- and *trans*-2-aminocyclohexanecarboxylic acids was reported by Priego *et al.* [33]. Enantiomerically pure quinazolinone **112** was reduced diastereoselectively by hydrogenation with PtO_2, resulting in octahydroquinazolinone diastereomers **113**, **114**, and **115** in a ratio of 6 : 3 : 1.

Both *cis*-annelated derivatives **113** and **114** could be epimerized in the presence of KO*t*Bu, giving the corresponding *trans*-fused derivatives **115** and **116**, respectively, in good yields (Scheme 8.24). Subsequently, the four octahydroquinazolinones were converted by ring opening with HCl, followed by ion-exchange chromatography, to the two *cis*- and the two *trans*-2-aminocarboxylic acid enantiomers [33].

Carbocycles can be obtained from acyclic compounds by the radical addition-cyclization of sulfanylic compounds. The method also permits the synthesis of carbocyclic β-amino acids [34]. Cyclization was achieved from oximes or hydrazones

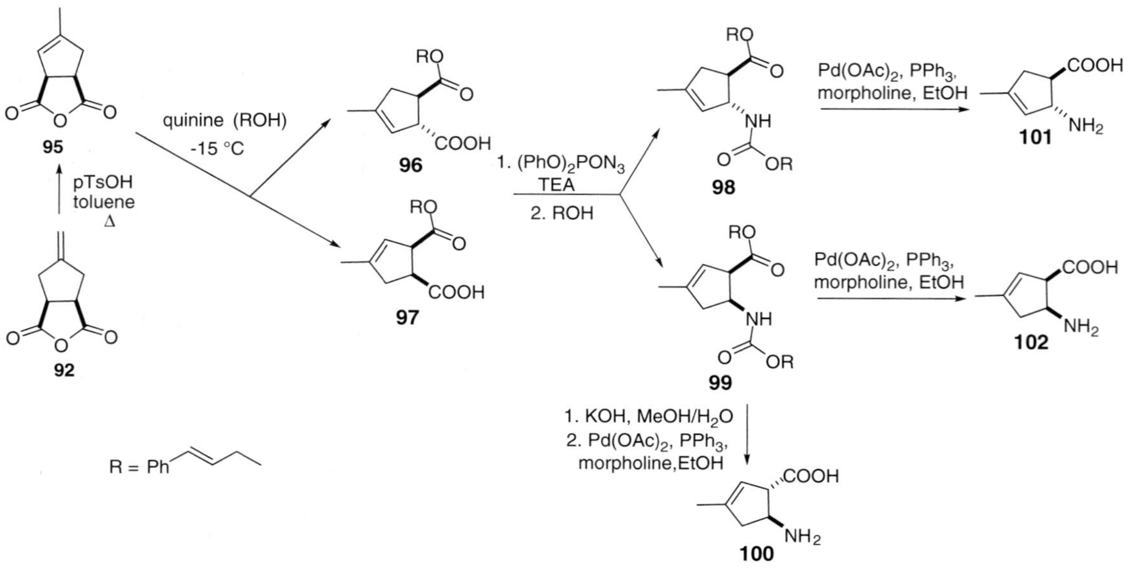

Scheme 8.21 Synthesis of icofungipen analogs by desymmetrization.

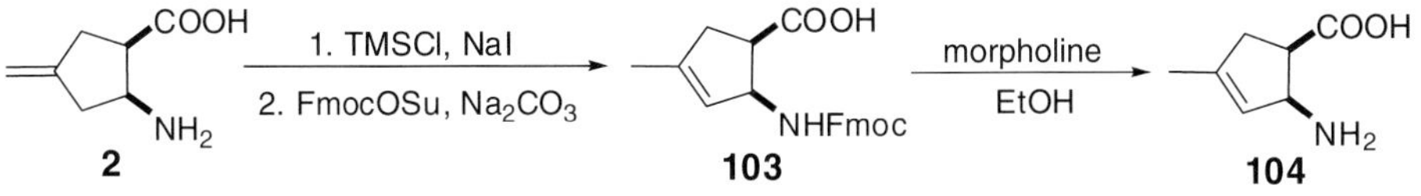

Scheme 8.22 Synthesis of icofungipen analog **104**.

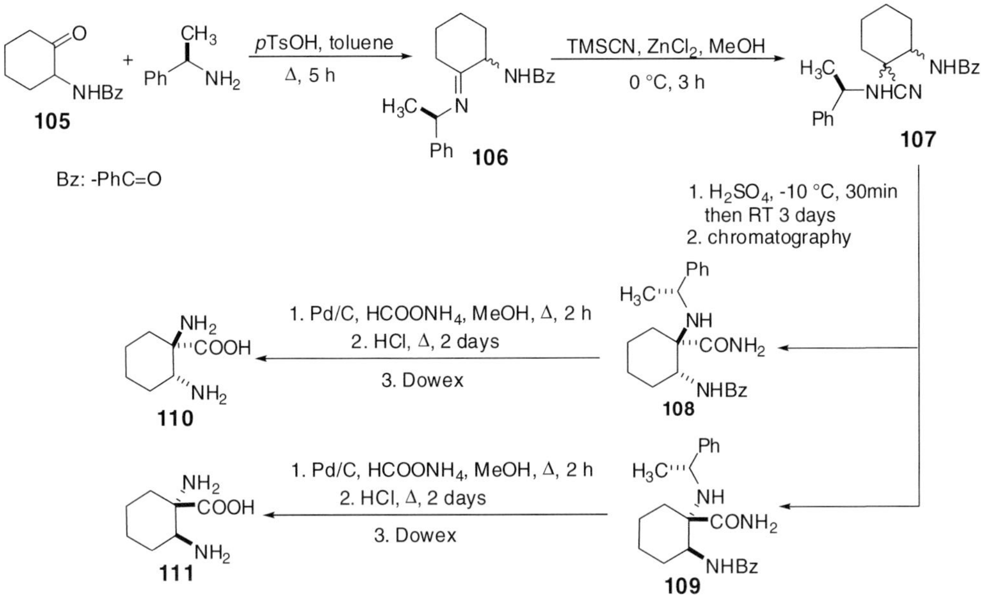

Scheme 8.23 Synthesis of α,β-diaminocyclohexanecarboxylic acid enantiomers.

with thiophenol and 2,2′-azobisisobutyronitrile (AIBN) as radical initiator, furnishing diastereomeric mixtures **117** + **118** and **119** + **120** in ratios of 3.3 : 1 and 2 : 1, respectively. The diastereomeric mixtures **117** + **118** and **119** + **120** were transformed to the Boc-protected amine diastereomers **121**, which were converted through sulfoxide **122** by pyrolysis to the *exo* methylene derivative **123**. Hydroboration of

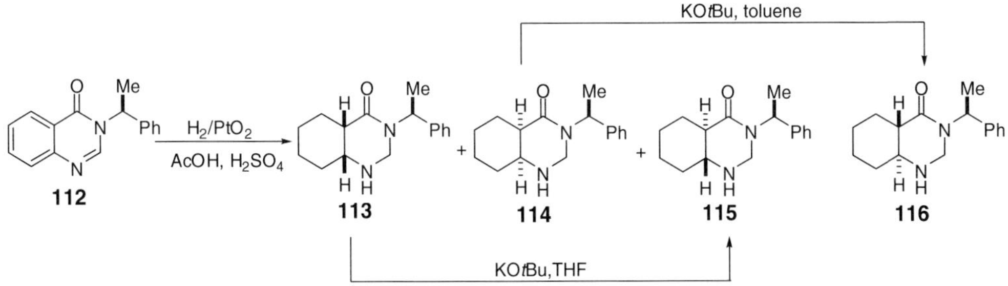

Scheme 8.24 Synthesis of octahydroquinazolinone diastereomers.

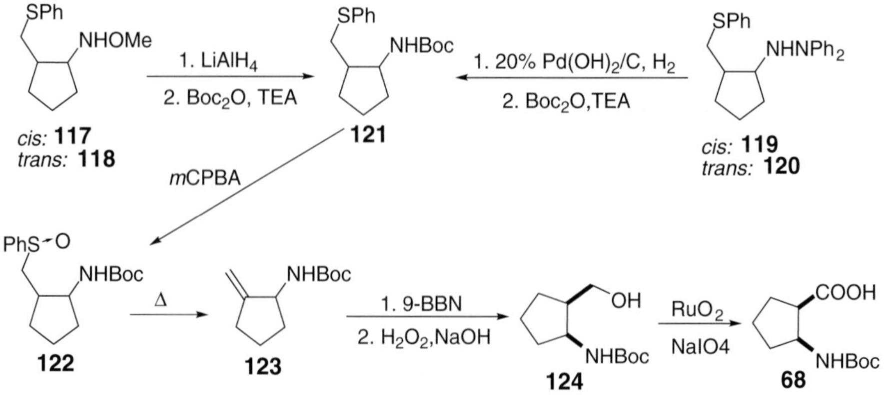

Scheme 8.25 Synthesis of Boc-protected cispentacin by sulfanyl radical cyclization (9-BBN: 9-Borabicyclo[3.3.1] nonane).

the olefinic bond gave alcohol **124** with high regio- and diastereoselectivity. Finally, oxidation of **124** resulted in the Boc-protected cispentacin **68** (Scheme 8.25). Via this procedure, both carboxylic β-amino acids and heterocyclic analogs of cispentacin could be synthesized, revealing the great advantage of this method.

Although numerous good methods are available for the synthesis of five- and six-membered cyclic *trans* β-amino acids, there are very few examples of the synthesis of analogs with larger ring systems, such as cycloheptane or cyclooctane derivatives. A simple and convenient synthesis of these starts from the readily available *cis*-2-hydroxycycloalkanecarboxylates (**125** or **126**), involving substitution of the hydroxy group with azide by inversion via the tosylates **127** or **128**, resulting in *trans*-2-azidocarboxylates **129** and **130** [35]. Reduction of the azido group followed by ester hydrolysis furnished the corresponding cycloheptane and cyclooctane *trans* β-amino acids **133** and **134** (Scheme 8.26). The method was also extended to the synthesis of the enantiomerically pure materials.

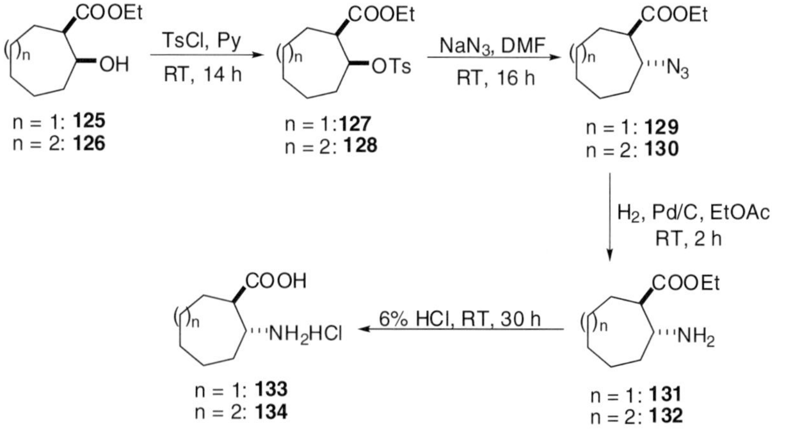

Scheme 8.26 Synthesis of seven- and eight-membered carbocyclic *trans* β-amino acids.

8.2.8
Synthesis of Small-Ring Carbocyclic β-Amino Acid Derivatives

The three-membered 2-aminocyclopropanecarboxylic acids appeared very attractive as building elements for the synthesis of peptides. However, their usage has been limited, probably because of their instability. In general, only the protected derivatives demonstrate stability [36]. The most common routes for the synthesis of 2-amino-cyclopropanecarboxylic acid derivatives consist of the addition of carbene analogs such as diazomethane or diazocarboxylates to an olefinic bond ([36] and references therein). Carboxylate-substituted cyclopropane could be constructed by the reaction of α-bromo carboxylate **135** with methyl acrylate [37]. The reaction resulted in the diastereomers **136** and **137** in a 4 : 1ratio. After hydrolysis and treatment with AcCl, diester **136** yielded anhydride **139**, which was then converted by esterification and Curtius rearrangement to Boc-protected aminocyclopropanecarboxylate **141** (Scheme 8.27).

Despite the building block potential in the peptide synthesis of β-aminocyclobutanecarboxylic acids as conformationally restricted derivatives, very little access to these compounds is known [30, 38]. Gauzy *et al.* developed an enantioselective procedure for the synthesis of (1*S*,2*R*)- and (1*R*,2*S*)-2-aminocyclobutanecarboxylic acids **150** and **151** via chiral uracil mimic **149** [39].

The photocycloaddition of ethylene **142** to **143** gave a mixture of *cis*-annelated diastereomers **144** and **145**. This mixture could be separated by chromatography; subsequent transformations, finally involving heterocyclic ring hydrolysis led to enantiomers **150** and **151** in overall yields of 33 and 20%, with greater than 97% e.e. (Scheme 8.28).

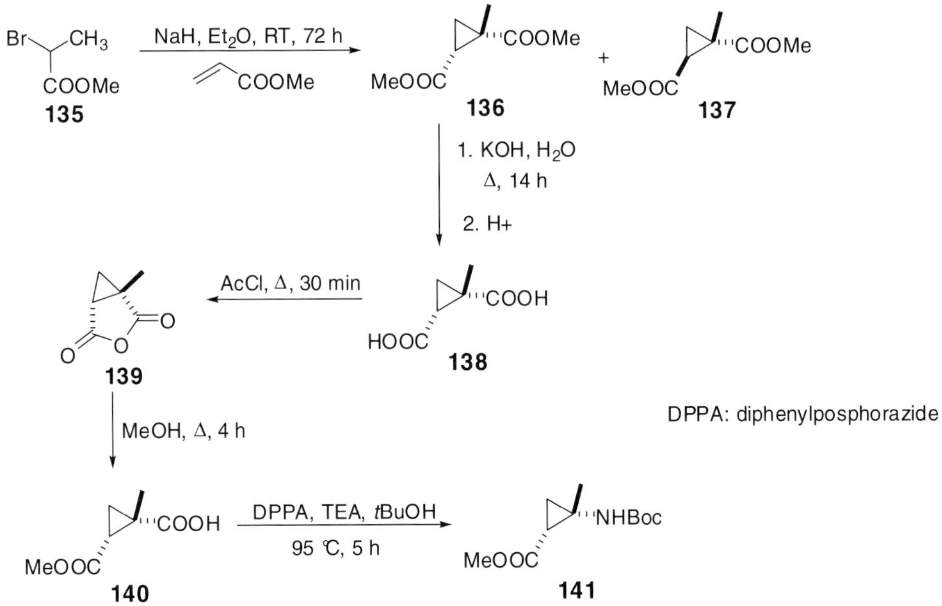

DPPA: diphenylposphorazide

Scheme 8.27 Synthesis of three membered β-amino acid derivative.

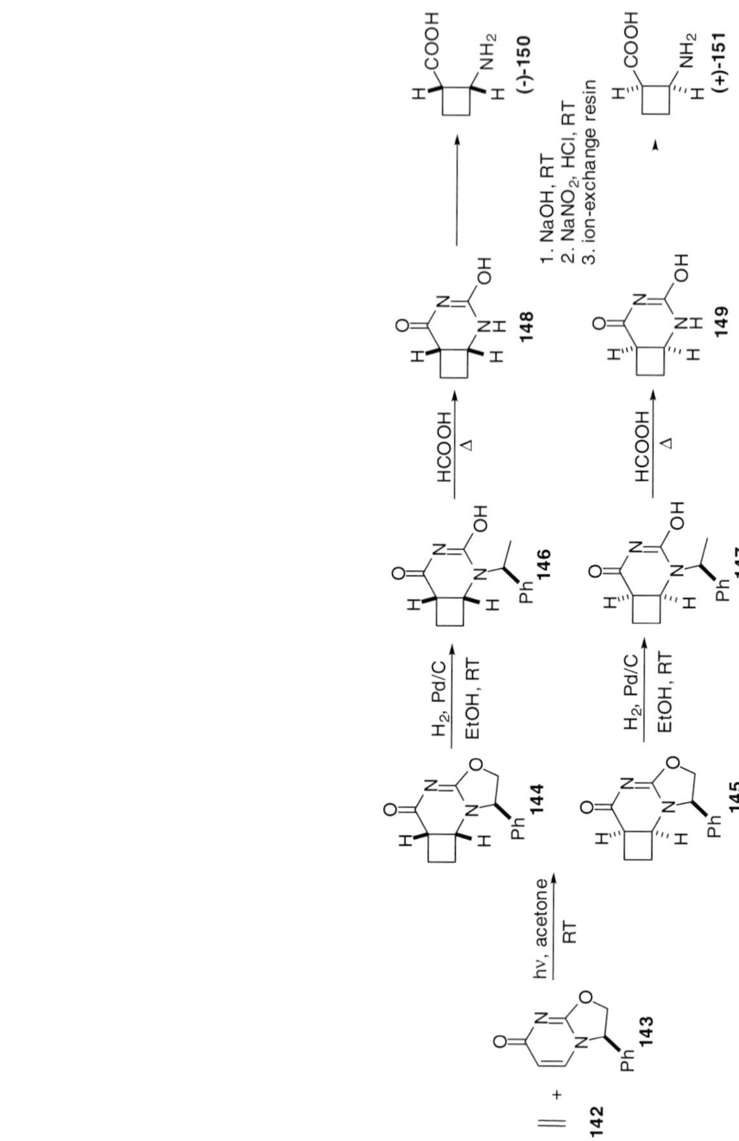

Scheme 8.28 Synthesis of 2-aminocyclobutanecarboxylic acid enantiomers.

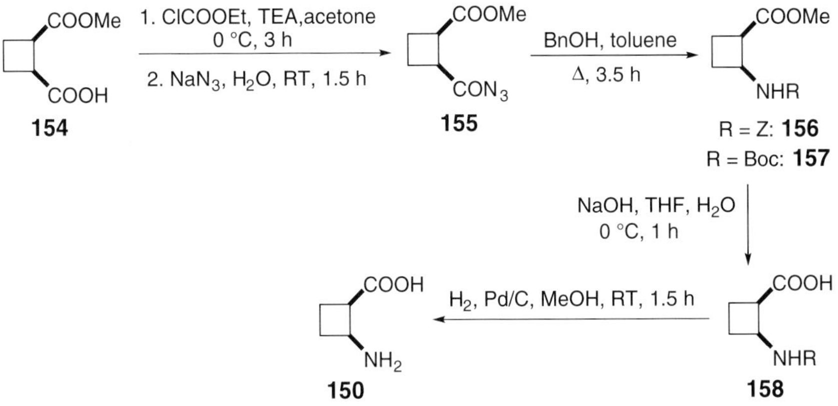

1. ClCOOEt, TEA, acetone, 0 °C, 3 h; 2. NaN_3, H_2O, RT, 1.5 h

BnOH, toluene, Δ, 3.5 h

R = Z: **156**
R = Boc: **157**

NaOH, THF, H_2O, 0 °C, 1 h

H_2, Pd/C, MeOH, RT, 1.5 h

Scheme 8.29 Synthesis of 2-aminocyclobutanecarboxylic acid enantiomers.

Starting from the chiral uracil derivative, Fernandes *et al.* synthesized all four stereoisomers of 2-aminocyclobutanecarboxylic acid in enantiomerically pure form, using a similar synthetic sequence involving photochemical [2 + 2] cycloaddition and isomerization of the *cis* to the *trans* β-amino acid [40]. They also reported the incorporation of cyclobutane *cis* β-amino acids into peptides [41]. This strategy of cycloaddition of an olefinic bond to an uracil derivative has been successfully applied for the synthesis of hydroxymethylated 2-aminocyclobutanecarboxylic acid stereoisomers **152** and **153** [42].

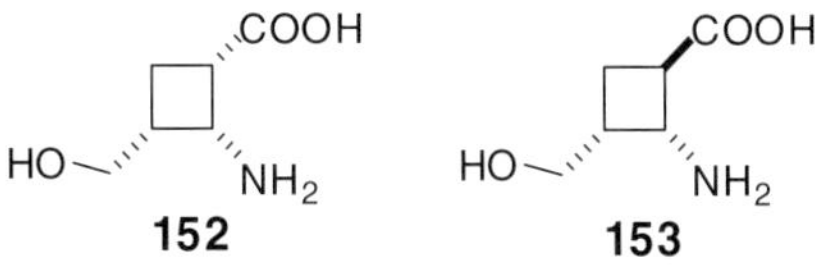

152 **153**

Izquierdo *et al.* synthesized both enantiomers of 2-aminocyclobutanecarboxylic acids from the enantiomerically pure half ester **154**, which was prepared by desymmetrization of *meso*-cyclobutane-1,2-dicarboxylic acid methyl ester [43].

The synthetic route involved the Curtius rearrangements of azidocarbonyl esters **155** and **161** from the same chiral monoester **154** (Schemes 8.29 and 8.30).

8.3
Synthesis of Functionalized Carbocyclic β-Amino Acid Derivatives

Functionalization of the carbocycle of a β-amino acid has generated great interest in recent years. Introduction of a functional group (e.g., hydroxy, amino, etc.) onto the carbocycle of an amino acid may have considerable influence on the structure and biological activity of subsequent peptides.

A series of novel β-amino acids, including *cis*-2-amino-4-hydroxycyclopentanecarboxylic acid diastereomers, were synthesized and tested for *in vitro* antifungal activity against *Candida albicans* by Mittendorf *et al.* [21]. Compounds **164** were prepared as

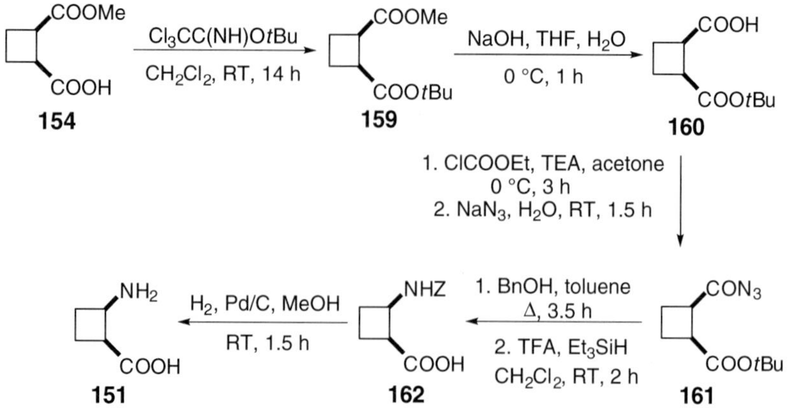

Scheme 8.30 Synthesis of 2-aminocyclobutanecarboxylic acid enantiomers.

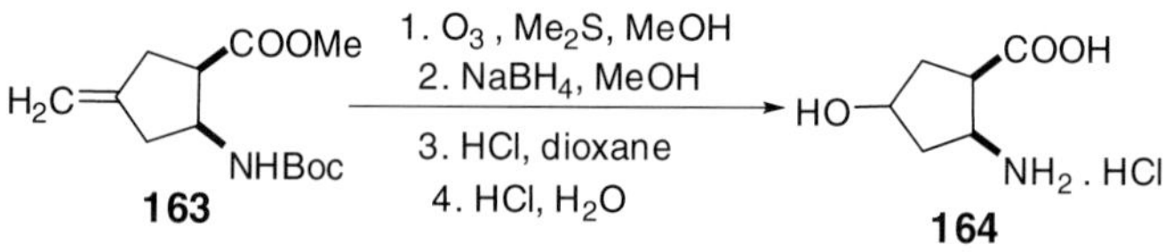

Scheme 8.31 Synthesis of hydroxylated 2-aminocyclopentanecarboxylic acid.

depicted in Scheme 8.31 from the enantiomerically pure icofungipen **2**. Ozonolysis of the olefinic bond of *N*-Boc-protected icofungipen **163**, reduction of the carbonyl group, hydrolysis, and deprotection resulted in a diastereomeric (3 : 1) mixture of hydroxylated compounds **164** (Scheme 8.31).

In 2003, Bunnage *et al.* reported the asymmetric synthesis of (−)-oryzoxymycin (**173**) [44]. The Diels–Alder reaction of furan with nitroacrylate **165** gave a mixture of cycloadducts favoring the *endo* nitro isomer **167** (4 : 1). Subsequent conversion of the bicyclic protected amino ester **167** via ester **169** afforded the lactate-coupled product **170** as a mixture of two diastereomers. Separation of the diastereomers proved impossible; hence, the authors devised an enantioselective preparation of ester **169**. Pig liver esterase (PLE)-catalyzed selective hydrolysis of bicyclic ester **168** gave, after ring opening and hydrolysis, enantiomers **169** and **171**. With the enantiopure ester, a sequence similar to that for the racemic compound was followed, giving optically pure oryzoxymycin **173** as its trifluoroacetic acid salt. Reduction of (+)-**169** afforded the perhydro amino acid derivative (1*R*,2*S*,3*S*)-**172** (Scheme 8.32).

Masesane and Steel described a simple synthesis of 3,4-dihydroxy derivatives of 2-aminocyclohexanecarboxylic acid, based on the reductive opening of epoxide intermediates derived from cyclohexadienes **174** and **175** [45]. Epoxidation of **174** or **175** selectively afforded the corresponding epoxides **176** and **177** as the only products. Reduction of **177** furnished the all-*cis* isomer **178** (Scheme 8.33).

Treatment of **169** with *m*-chloroperbenzoic acid (*m*CPBA) in CH_2Cl_2 gave a separable mixture of epoxides **179** and **180** in a ratio of 9 : 1 (method A). Interestingly, the epoxidation of **169** in MeCN instead of CH_2Cl_2 changed the stereoselectivity

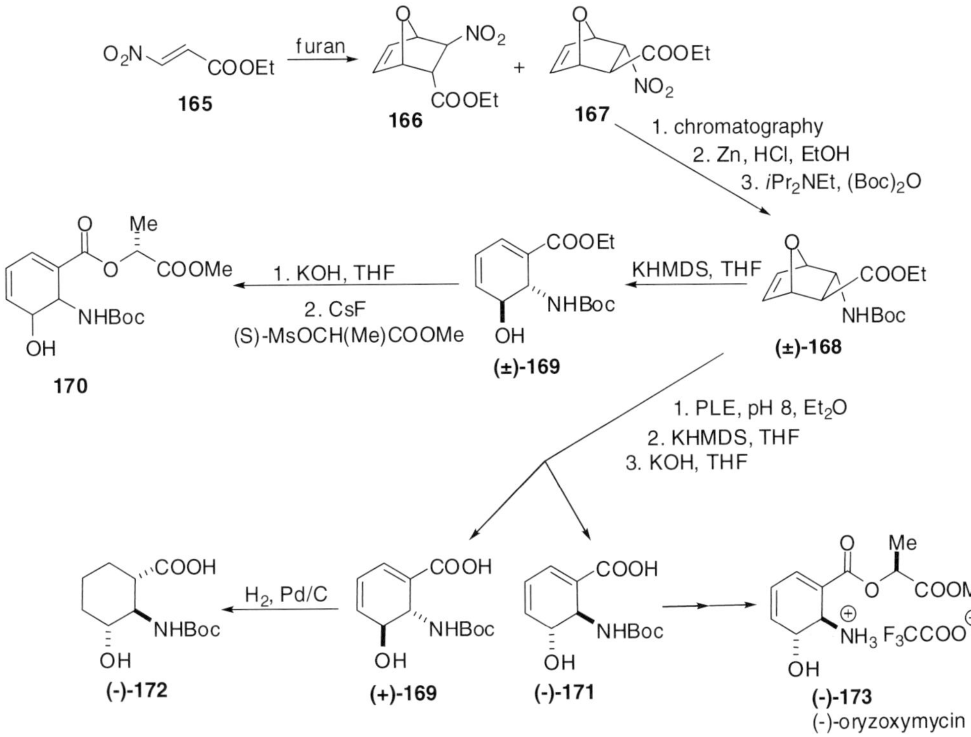

Scheme 8.32 Synthesis of (−)-oryzoxymycin.

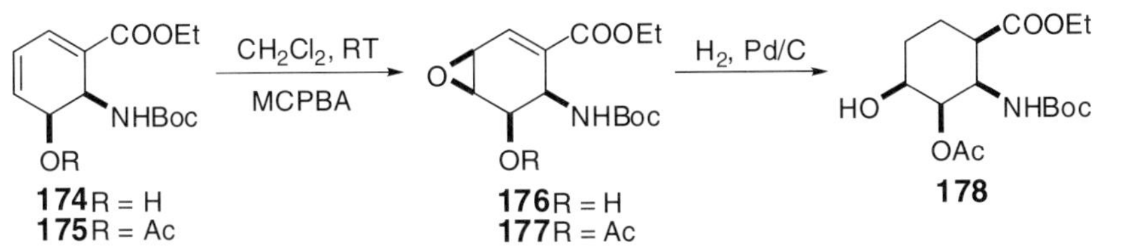

Scheme 8.33 Synthesis of 3,4-dihydroxylated ethyl β-aminocyclohexanecarboxylate by epoxidation.

and resulted in a 2 : 1 mixture of these isomers, favoring the *cis*-epoxyalcohol **180** (method B). Reductive opening of the separated epoxides **179** and **180** in the presence of Pd/C in a hydrogen atmosphere afforded the diacetyl derivatives **181** and **182**, respectively, as the only isomers (Scheme 8.34).

Stereoselective iodooxazine formation was the key step in the synthesis of all-*cis*-2-amino-4-hydroxycyclohexanecarboxylic acid (**190**), when *cis*-2-amino-4-cyclohex-ene-carboxylic acid derivatives **183a–c** were used as starting materials. The reactions of **183a–c** with *N*-iodosuccinimide or *N*-bromosuccinimide proceeded stereoselectively, furnishing bicyclic 1,3-oxazinone (**184**) and 1,3-oxazine derivatives (**187a**,**b** and **188a**,**b**). These derivatives were dehalogenated with Bu_3SnH, resulting in **185** and **189a**,**b**, which were easily converted to the all-*cis* amino acid **190** and its bicyclic oxazinone derivative **186** (Scheme 8.35) [46].

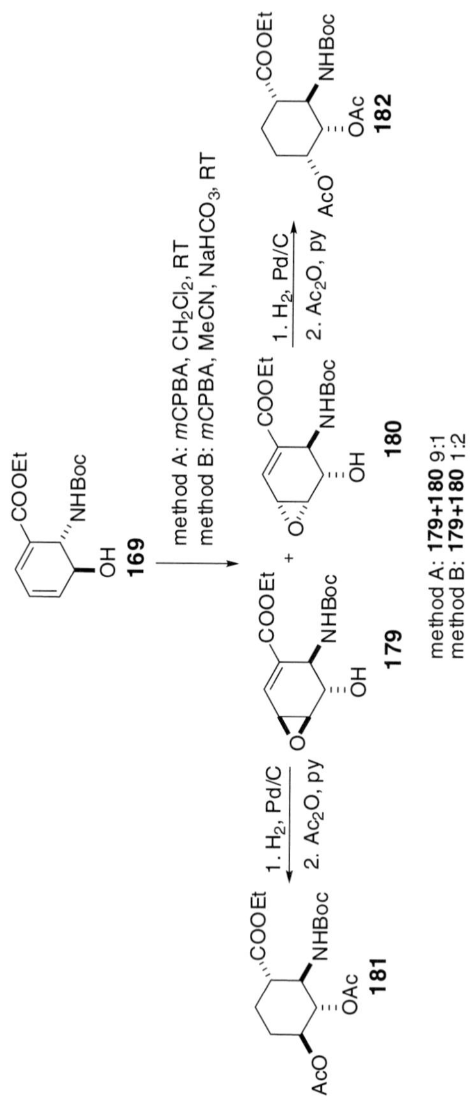

Scheme 8.34 Synthesis of 3,4-dihydroxylated ethyl β-aminocyclohexanecarboxylate diastereomers by epoxidation.

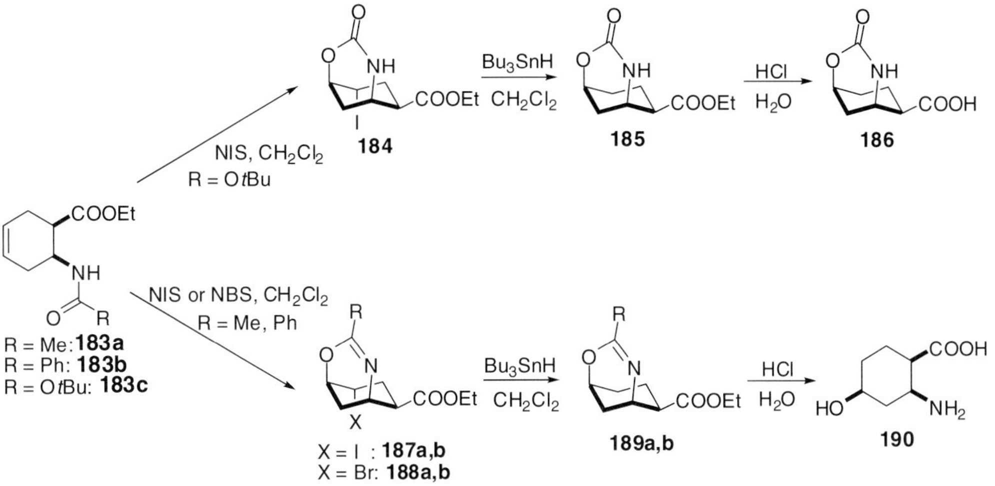

NIS, CH_2Cl_2
R = O*t*Bu
184
Bu_3SnH
CH_2Cl_2
185
HCl
H_2O
186
R = Me: 183a
R = Ph: 183b
R = O*t*Bu: 183c
NIS or NBS, CH_2Cl_2
R = Me, Ph
X = I : 187a,b
X = Br: 188a,b
189a,b
190

Scheme 8.35 Synthesis of all-*cis* 4-hydroxy-2-aminocyclohexanecarboxylic acid by iodooxazine formation.

Similarly, ethyl *trans*-2-acetylaminocyclohex-4-ene-carboxylate (**191**) was converted to the corresponding 2-*trans*-amino-4-*trans*-hydroxycyclohexanecarboxylic acid (**193**) (Scheme 8.36).

Another approach to hydroxylated carbocyclic β-amino acids is iodolactonization. The reactions of *cis*-*N*-acyl amino acids (**194a,b**) with I_2/KI in slightly alkaline medium produced iodolactones **195a,b**. Removal of the iodine was effected by reduction with Bu_3SnH, giving lactones **196a,b**. The *N*-Boc lactone was finally converted to the all-*cis* isomer of 5-hydroxylated amino acid **197** by acid hydrolysis and ion-exchange chromatography (Scheme 8.37) [46].

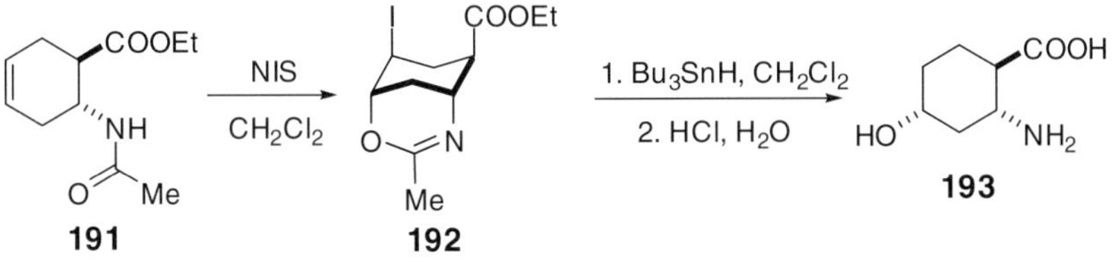

191
NIS
CH_2Cl_2
192
1. Bu_3SnH, CH_2Cl_2
2. HCl, H_2O
193

Scheme 8.36 Synthesis of 4-hydroxy-2-aminocyclohexanecarboxylic acid by iodooxazine formation.

194a,b
I_2, KI, $NaHCO_3$
CH_2Cl_2, H_2O
195a,b
Bu_3SnH
CH_2Cl_2
196a,b
HCl
H_2O
R = O*t*Bu
197
R = Ph : a; R = O*t*Bu : b

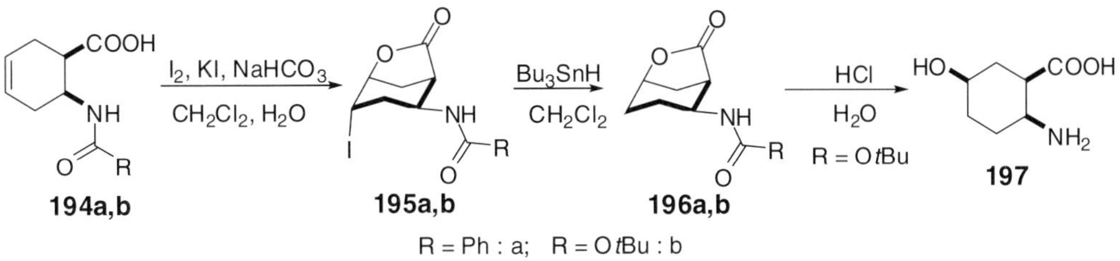

Scheme 8.37 Synthesis of all-*cis*-5-hydroxy-2-aminocyclohexanecarboxylic acid by iodolactonization.

In an analogous way, from *trans-tert*-butoxycarbonylamino-4-cyclohexenecarboxylic acid the corresponding 2-*trans*-amino-5-*cis*-hydroxycyclohexane-carboxylic acid (**198**) was obtained [44]. The syntheses of 4-hydroxy- and 5-hydroxy-2-aminocyclohexane-carboxylic acid enantiomers were carried out similarly as for the racemic compounds resulting in the products with 98% e.e. or grater [46].

When subjected to the iodolactone procedure, the bicyclic β-lactam derived from 1,3-cyclohexadiene furnished *all-cis*-2-amino-3-hydroxycyclohexanecarboxylic acid (**199**), while application of the iodooxazine procedure led not only to the all-*cis*-2-amino-3-hydroxycyclohexanecarboxylic acid, but also to all-*cis*-2-amino-4-hydroxy-cyclohexanecarboxylic acid **190** [47].

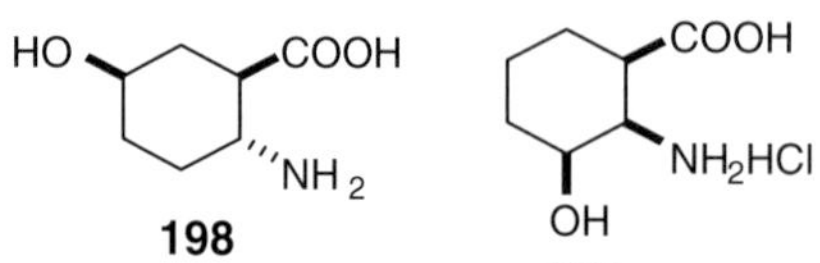

HO COOH NH2 **198**

COOH NH2HCl OH **199**

Another successful route to hydroxylated β-aminocyclohexanecarboxylic acid derivatives is based on stereoselective epoxidation and regioselective oxirane ring opening [48]. In this procedure, the unsaturated *trans* amino ester **200** underwent epoxidation diastereoselectively to epoxy derivative **201**, in which the oxirane ring and the C-2 carbamate were situated on the same side of the cyclohexane skeleton. On treatment with $NaBH_4$, epoxide **201** afforded regioselectively 5-hydroxy amino ester **202**, which was readily transformed to 5-hydroxy-2-aminocyclohexanecarboxylic acid **203** (Scheme 8.38).

This *cis*-selective epoxidation method was successfully applied for the synthesis of other diastereomers of hydroxylated 2-aminocyclohexanecarboxylic acids [49]. Epoxidation of *cis*-2-amino ester **204***cis*-diastereoselectively gave epoxy derivative **205**, whose oxirane opening and subsequent deprotection furnished regioselectively 4-hydroxy amino acids **190** or **193** depending on the reaction conditions (Scheme 8.39).

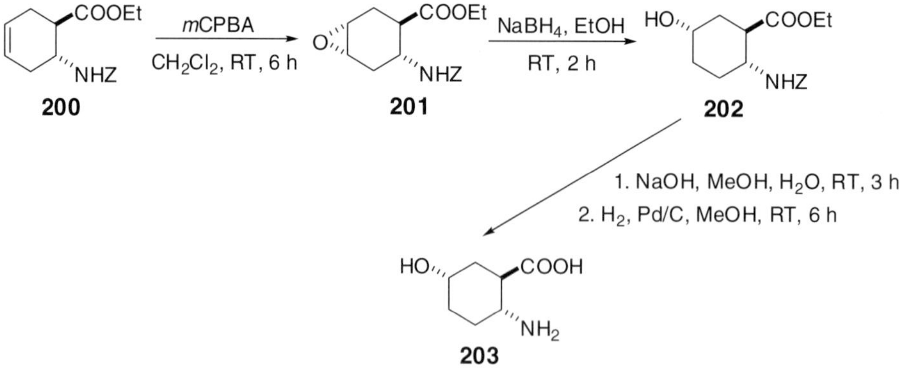

COOEt NHZ **200** → mCPBA, CH_2Cl_2, RT, 6 h → O COOEt NHZ **201** → $NaBH_4$, EtOH, RT, 2 h → HO COOEt NHZ **202**

1. NaOH, MeOH, H_2O, RT, 3 h
2. H_2, Pd/C, MeOH, RT, 6 h

HO COOH NH2 **203**

Scheme 8.38 Synthesis of 5-hydroxylated-2-aminocyclohexanecarboxylic acid by stereoselective epoxidation.

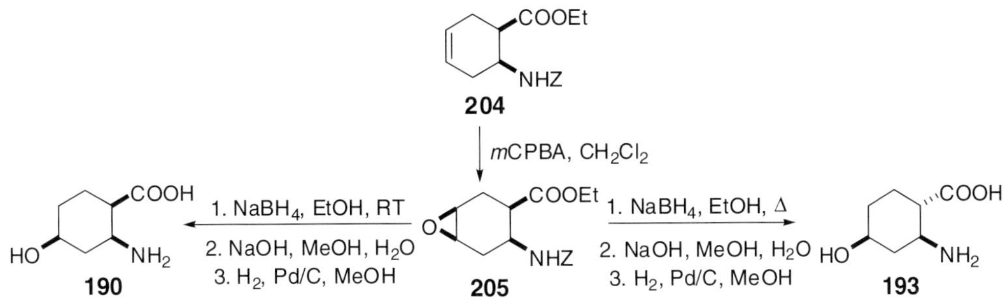

Scheme 8.39 Synthesis of 4-hydroxylated-2-amino-cyclohexanecarboxylic acid diastereomers by stereoselective epoxidation.

Both *cis*- and *trans*-selective epoxidation can be efficiently used for hydroxy group introduction onto a cyclohexane β-amino acid skeleton [49]. Owing to the opposite selectivity, this strategy allows the preparation of other new hydroxylated derivatives. *trans*-Selective epoxidation on the Boc-protected β-lactam derived from cyclohexadiene **206** gave the epoxy lactam **207**. Opening of lactam **207** with NaOEt at 0 °C yielded epoxy ester **209**. The same reaction at room temperature resulted in isomerization on C-1, giving **208**. Regioselective oxirane ring opening in **208** or **209** with $NaBH_4$ furnished the 4-hydroxylated and the 5-hydroxylated ester, respectively (Scheme 8.40). The 4-hydroxy-2-aminocyclohexanecarboxylic ester was subsequently hydrolyzed and deprotected, affording amino acid **211**.

Polyhydroxylated β-amino acid derivatives in enantiomerically pure form were synthesized by Soengas *et al.* from a nitrohexofuranose enantiomer [50, 51] prepared in nine steps from idofuranose **212**. The synthesized compound **216** may be regarded as a trihydroxy derivative of cispentacin (Scheme 8.41).

An extra amino group was introduced onto the skeleton of a 2-aminocyclohexanecarboxylic acid by further transformations of the hydroxylated derivatives **217** and **220**. The synthetic routes involved substitution of the hydroxy group via the

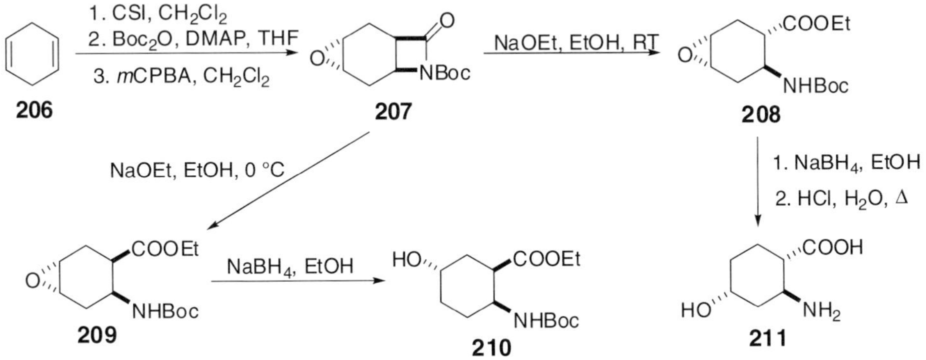

Scheme 8.40 Synthesis of hydroxylated cyclohexanecarboxylic acid derivatives by epoxidation of β-lactam.

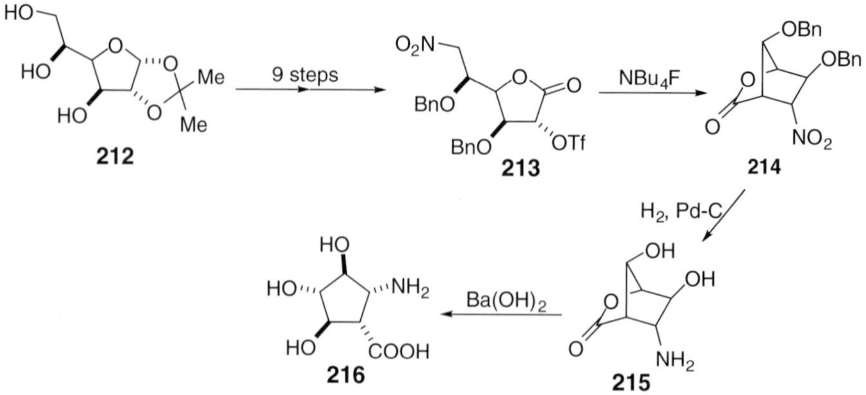

Scheme 8.41 Synthesis of polyhydroxylated 2-aminocyclopentanecarboxylic acid.

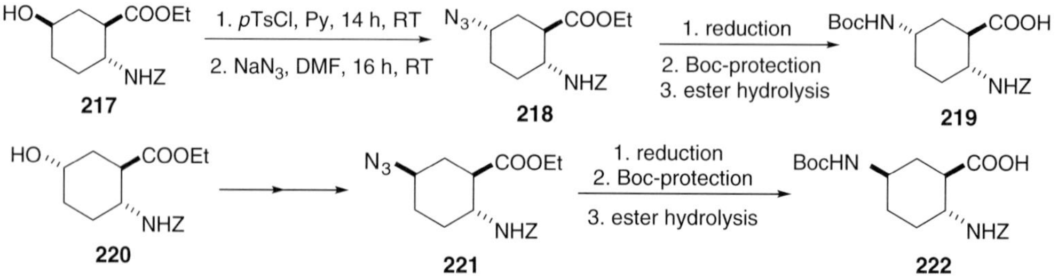

Scheme 8.42 Synthesis of orthogonally protected diaminocyclohexanecarboxylic acids.

tosylate with azide, followed by reduction and then protection of the formed amino groups [52] (Scheme 8.42). Two orthogonally protected β,γ-diamino acid diastereomers (**219** and **222**) were prepared by this simple method.

Orthogonally protected cyclopentane β,γ-diamino acid derivatives were synthesized with opposite selectivities by the epoxidation method (Schemes 8.39 and 8.40). Depending on the substrate, either the *cis* (**225**) or *trans* epoxides (**227** and **228**) were prepared [53] (Scheme 8.43).

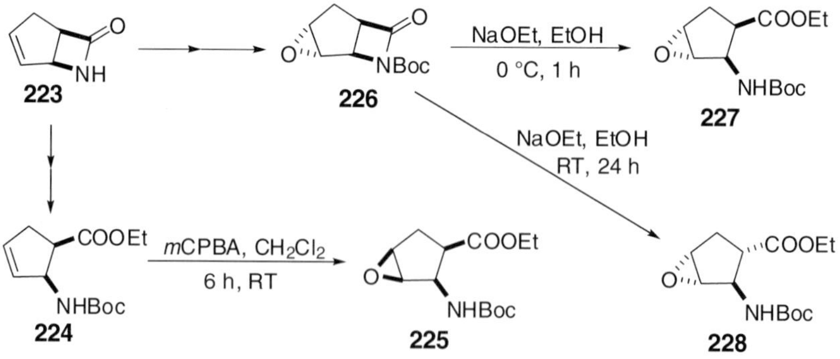

Scheme 8.43 Synthesis of epoxy 2-aminocyclopentanecarboxylate diastereomers.

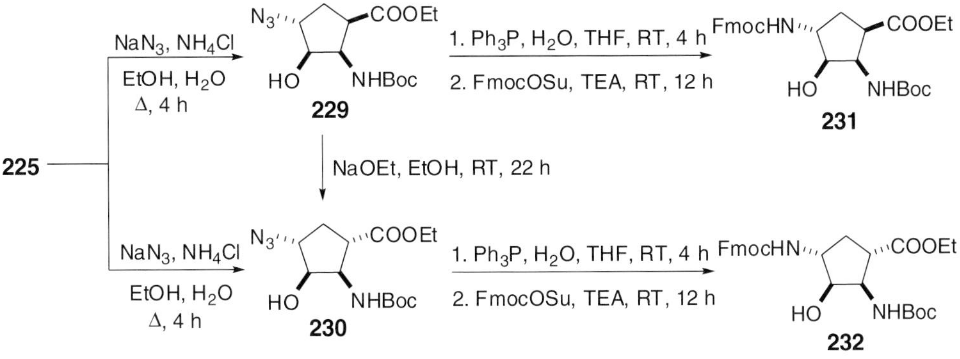

Scheme 8.44 Synthesis of orthogonally protected β,γ-diaminocarboxylates.

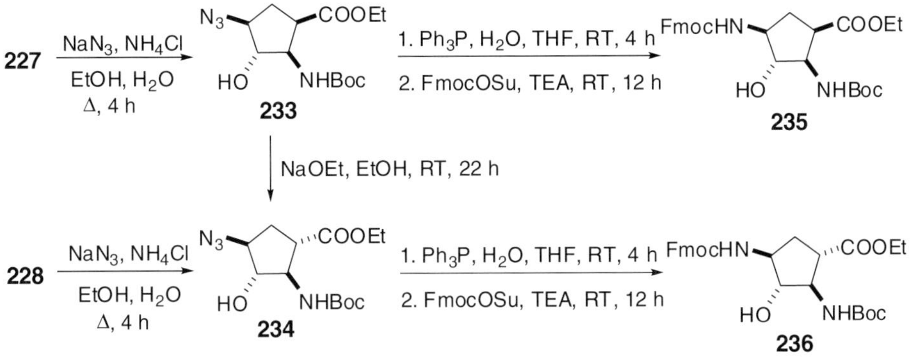

Scheme 8.45 Synthesis of orthogonally protected β,γ-diaminocarboxylates.

An amino moiety was introduced by opening of the oxirane ring of these epoxides with NaN_3, which in all cases regioselectively furnished the corresponding 4-azido esters **229**, **230**, **233**, and **234** (Schemes 8.44 and 8.45).

Subsequent reduction of the azido group, followed by protection of the amines, afforded four diastereomers of the orthogonally protected β,γ-diamino esters **231**, **232**, **235**, and **236** (Schemes 8.44 and 8.45). The syntheses were also performed for the enantiomers.

These orthogonally protected cyclopentane diamino esters are of interest not only as conformationally constrained potential building blocks in peptide chemistry, but also as precursors for the synthesis of carbocyclic nucleosides.

8.4 Enzymatic Routes to Carbocyclic β-Amino Acids

In the past few years, new enantioselective enzymatic routes to enantiopure carbocyclic β-amino acids have been developed. For example, nitrile-hydrolyzing strains of *Rhodococcus* sp. have been introduced as new catalysts for the preparation of β-amino acid enantiomers. A new direct enzymatic method to enantiopure β-amino acids through the lipase-catalyzed ring cleavage of carbocyclic β-lactams in an organic solvent was recently reported [61]. A very efficient enzymatic technique that affords enantiopure carbocyclic *cis* and *trans* β-amino acids through the lipase-catalyzed hydrolysis of the corresponding β-amino esters has also been described [68].

8.4.1
Enantioselective *N*-Acylations of β-Amino Esters

The earlier-designed enzymatic *N*-acylation of small ring-fused β-amino esters [54] has been successfully optimized for new, medium and large alicyclic substrates. Excellent enantioselectivities ($E > 200$) were observed for the *Candida antarctica* lipase CAL-A-catalyzed (*R*)-selective *N*-acylation of methyl β-aminocycloalkanecarboxylates **237–239** when 2,2,2-trifluoroethyl butanoate was used as acyl donor in iPr_2O at room temperature (Scheme 8.46) [55]. The same method was used for the kinetic resolution of a homoadamantane derivative: methyl *cis*-5-aminotricyclo [4.3.1.13.8]undecane-4-carboxylate ($E > 200$) [55].

8.4.2
Enantioselective *O*-Acylations of *N*-Hydroxymethylated β-Lactams

A lipase-catalyzed asymmetric acylation of the primary OH group of *N*-hydroxymethylated β-lactams with smaller-fused rings [56, 57] has subsequently been extended to the kinetic resolution of larger ring-fused *N*-hydroxymethylated β-lactams (**246–249**) (Scheme 8.47). Good to excellent enantioselectivities were observed for the acylations of the OH group of racemic **246**, **248**, and **249** in the

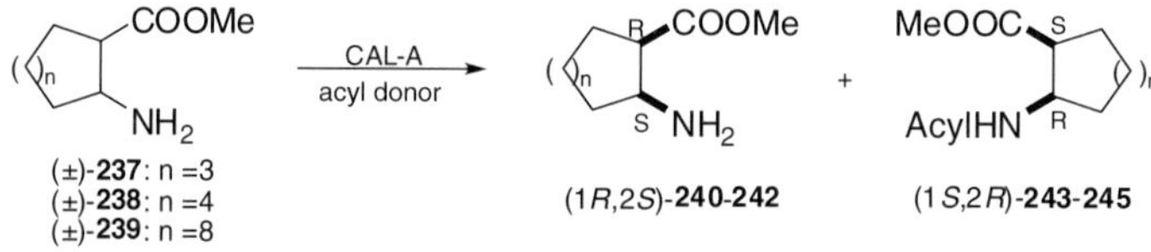

Scheme 8.46 Enzymatic *N*-acylation of β-amino esters.

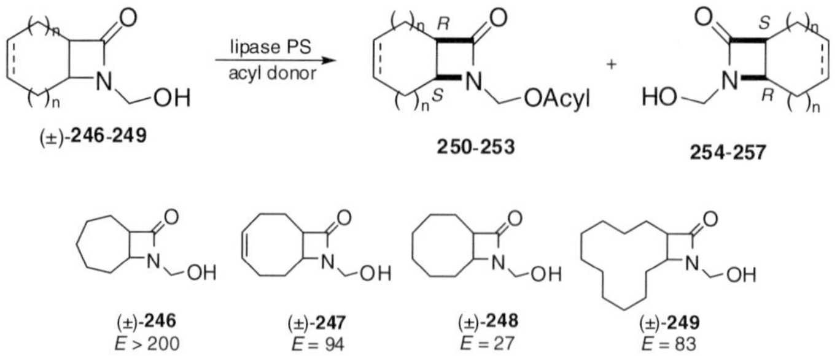

Scheme 8.47 Enantioselective *O*-acylation of *N*-hydroxymethylated β-lactams.

presence of *Pseudomonas cepacia* lipase PS with trifluoroethyl butanoate in Me_2CO at room temperature [55]. The method was also applied for the resolution of *cis*-10-hydroxymethyl-10-azatetracyclo[7.2.0.1$^{2.6}$.1$^{4.8}$]tridecan-11-one, a homoadamantane derivative ($E > 200$) [58]. Good enantioselectivity ($E = 94$) was achieved for the (*S*)-selective acylations of 9-hydroxymethyl-9-azabicyclo[6.2.0]dec-4-en-10-one [**247**] when lipase PS and vinyl butyrate were used in iPr_2O at $-15\,°C$ [59]. *trans*-13-Hydroxymethyl-13-azabicyclo[10.2.0]tetradecan-14-one was resolved with vinyl butanoate and *C. antarctica* lipase CAL-B in Me_2CO ($E = 26$) [55]. The ring opening of lactam enantiomers with 36% HCl or 22% HCl/EtOH afforded the corresponding enantiomeric β-amino acid or ester hydrochloride. The method was applied for the synthesis of all four enantiomers of 1-aminoindane-2-carboxylic acid, a new cispentacin benzolog ($E > 200$) [60], where formally reverse (*R*) selectivity was observed for the acylation. The same stereochemical demands as earlier were fulfilled around the asymmetric centers, only the substituent priority on the substrates differing.

8.4.3
Enantioselective Ring Cleavage of β-Lactams

Forró and Fülöp [61] developed a direct enzymatic method for the enantioselective ring cleavage of unactivated β-lactams (Scheme 8.48). Independently of the ring size, high enantioselectivities ($E > 200$) were observed when the ring-opening reactions of **258–261** were performed with H_2O in iPr_2O at $60\,°C$. The products were easily separated through an organic solvent/H_2O extraction. The method was scaled up and used for the preparation, for example, (1*R*,2*S*)-2-aminocyclopentanecarboxylic acid [cispentacin, **262** ($n = 1$)] (>99% e.e.). 1,4-Ethyl and 1,4-ethylene-bridged cispentacin enantiomers (>99% e.e.) were prepared with H_2O and Lipolase in toluene at $65\,°C$ [62]. Other extensions of this method included the ring cleavage of the unsaturated β-lactams 6-azabicyclo[3.2.0]hept-3-en-7-one, 7-azabicyclo[4.2.0]oct-4-en-8-one, 7-azabicyclo[4.2.0]oct-3-en-8-one, and 9-azabicyclo[6.2.0]dec-4-en-10-one with H_2O and Lipolase in iPr_2O at $70\,°C$ ($E > 200$) [63]. It was noteworthy that

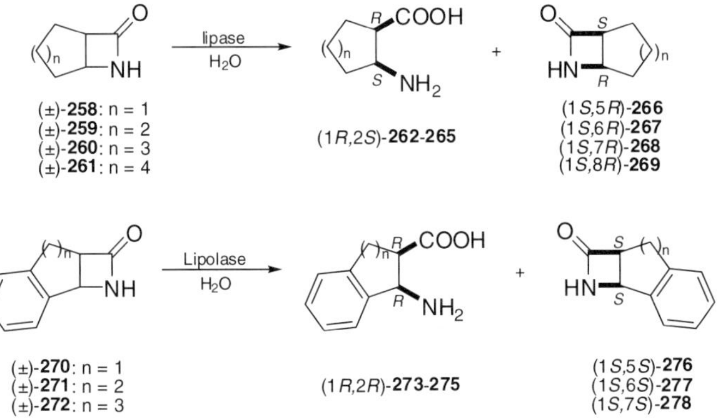

lipase, H_2O

(±)-**258**: n = 1
(±)-**259**: n = 2
(±)-**260**: n = 3
(±)-**261**: n = 4

(1*R*,2*S*)-**262-265**

(1*S*,5*R*)-**266**
(1*S*,6*R*)-**267**
(1*S*,7*R*)-**268**
(1*S*,8*R*)-**269**

Lipolase, H_2O

(±)-**270**: n = 1
(±)-**271**: n = 2
(±)-**272**: n = 3

(1*R*,2*R*)-**273-275**

(1*S*,5*S*)-**276**
(1*S*,6*S*)-**277**
(1*S*,7*S*)-**278**

Scheme 8.48 Enantioselective ring cleavage of β-lactams.

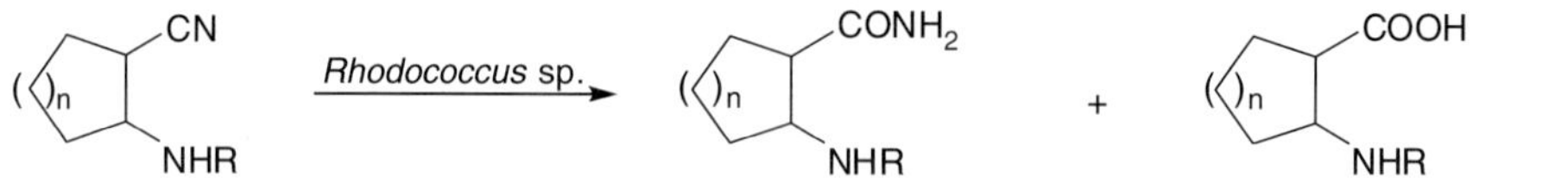

CN / NHR — *Rhodococcus* sp. → CONH$_2$ / NHR + COOH / NHR

cis- or *trans*-(±)-**279**: n = 1
cis- or *trans*-(±)-**280**: n = 2
R = Ts or Bz

cis- or *trans*-**281**: n = 1
cis- or *trans*-**282**: n = 2
R = Ts or Bz

cis- or *trans*-**283**: n = 1
cis- or *trans*-**284**: n = 2
R = Ts or Bz

Scheme 8.49 Nitrile hydrolysis by whole cells from *Rhodococcus* sp.

Lipolase in an organic solvent favored the opposite enantiomer from that in the case of lactamase present in the whole cells of *Rhodococcus globerulus* (NCIMB 41 042) or *R. equi* in phosphate buffer at pH 7 [64]. The enantioselective ring opening reactions of tricyclic β-lactams (**270–272**), resulting in benzocispentacin and its new six- and seven-membered homologs, were also carried out with the methodology presented above [65]. The enantioselective ring cleavage of *trans*-13-azabicyclo[10.2.0]tetradecan-14-one proceeded with high enantioselectivity ($E > 200$) [66].

8.4.4 Biotransformation of Carbocyclic Nitriles

Klempier *et al.* [67] described the preparation of β-amino amides and carbocyclic acids by whole cells of *R. equi* A4, *Rhodococcus* sp. R312 and *R. erythropolis* NCIMB 11 540 (Scheme 8.49). Racemic *trans*-**279** gave the amides *trans*-**281** with high enantiomeric excess (≥94% for R = Bz and >99% for R = Ts) and acids *trans*-**283** with low to moderate enantiomeric excess (≤75% for R = Bz and ≤14% for R = Ts), while *trans*-**280** yielded the acid enantiomers *trans*-**284** with high enantiomeric excess (>95% for R = Bz and ≥87% for R = Ts) and amides *trans*-**282** with low to moderate enantiomeric excess (≤67% for R = Bz and ≤77% for R = Ts). The biotransformations of *cis* five- and six-membered 2-amino nitriles with the above-mentioned catalysts have also been described; the enantiomeric excess values of the resulting *cis* compounds were found to be much lower. The absolute configurations of the products were not determined.

8.4.5 Enantioselective Hydrolysis of β-Amino Esters

The lipase-catalyzed hydrolysis of carbocyclic β-amino esters was equally applicable for the synthesis of *cis* and *trans* β-amino acids [68, 69]. Good to excellent enantioselectivities (usually $E > 100$) were observed when the Lipolase-catalyzed (*S*)-selective hydrolysis of *cis*-**285–288** and (*R*)-selective hydrolysis of *trans*-**298** and **299** (Scheme 8.50) were performed with 0.5 equiv. H_2O in $i$$Pr_2O$ at 65 °C. No correlation was found between the size of the cycloalkane ring or *cis–trans* isomers and the reaction rates. The resulting β-amino acid and the unreacted β-amino ester enantiomers were easily separated through organic solvent/H_2O extraction.

Steel *et al.* [44] reported the preparation of enantiopure ethyl 3-endo-*tert*-butoxycarbonylamino-7-oxabicyclo[2.2.1]hept-5-ene-2-exo-carboxylate **306** through the

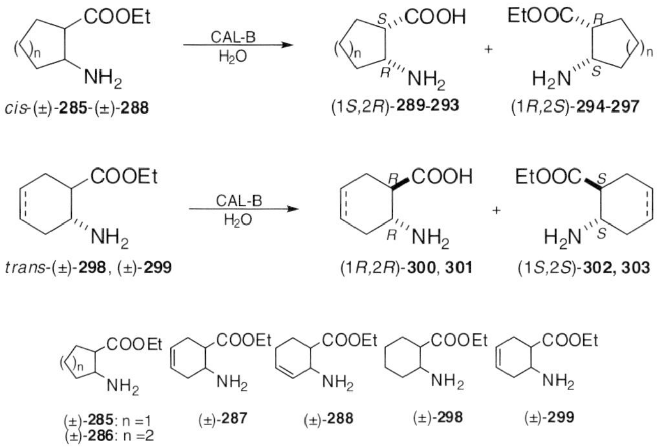

Scheme 8.50 Enantioselective hydrolysis of β-amino esters.

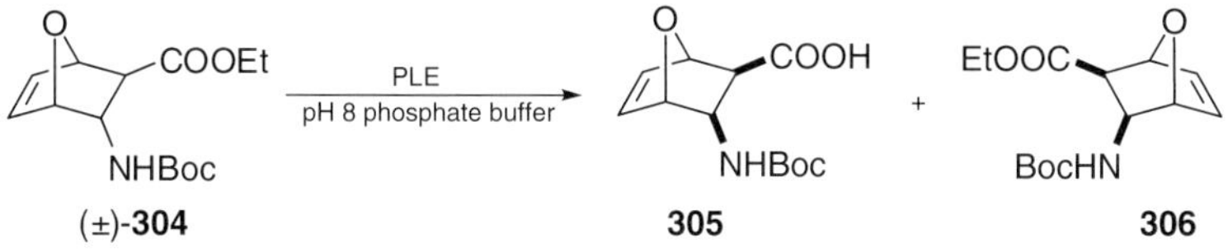

Scheme 8.51 Enantioselective hydrolysis of β-amino esters.

PLE-catalyzed enantioselective hydrolysis of racemic β-amino ester **304** in pH 8 phosphate buffer/Et_2O at room temperature (Scheme 8.51). The products (**305** and **306**) were separated by a preparative chiral high-performance liquid chromatographic (HPLC) method (Chiralpak AD, *n*-heptane/EtOH 95 : 5).

8.4.6
Analytical Methods for the Enantiomeric Separation of Carbocyclic β-Amino Acids

Péter *et al.* used HPLC direct and indirect methods for the enantioseparation of five- and six-membered cyclic *cis* and *trans* β-amino acids. The direct separations [70, 71] were based on chiral stationary phases (CSPs) [Crownpak CR(+)], while the indirect separations [72–74] were based on achiral stationary phases and separation of the diastereomers formed by precolumn derivatization with chiral derivatizing reagents 1-fluoro-2,4-dinitrophenyl-5-L-alanine amide (FDAA, Marfey's reagent), 2,3,4,6-tetra-*O*-acetyl-β-D-glucopyranosyl isothiocyanate (GITC), and (1*S*,2*S*)-1,3-diacetoxy-1-(4-nitrophenyl)-2-propyl-isothiocyanate [75]. The isomers of β-amino acids with a bicyclo[2.2.1]heptane or heptene skeleton were separated in a similar way, a Crownpak CR(+) column being used for the direct resolution, and GITC and FDAA as chiral derivatizing reagents for the indirect separation [76]. Gasparrini *et al.* described the application of a CSP containing a macrocyclic glycopeptide antibiotic, A-40,926, in the direct chromatographic resolution of five- and six-membered cyclic *cis* and *trans* β-amino acids [77]. Two macrocyclic antibiotic CSPs based on native teicoplanin and teicoplanin aglycone, ChirobioticT and Chirobiotic TAG were successfully used for the HPLC separation of carbocyclic *cis* and *trans* β-amino acid enantiomers by using reversed-phase or polar organic mobile-mode systems [78]. Direct reversed-phase HPLC methods were developed for the baseline resolutions of bicyclic β-amino acids, the six- and seven-membered homologs of *cis*-1-amino-4,5-benzocyclopentane-2-carboxylic acid (benzocispentacin). The separations were performed on CSP columns containing Chirobiotic T, Chirobiotic TAG, vancomycin (Chirobiotic V), vancomycin aglycone (Chirobiotic VAG), ristocetin (Chirobiotic R), or a new dimethylphenyl β-cyclodextrin-based carbamate-derivatized Cyclobond DMP [79].

An easy-to-operate indirect gas chromatography (GC) method has been developed for the enantio-separation of carbocyclic β-amino acids (saturated and unsaturated, mono- and bicyclic, *cis* and *trans*) [62, 63, 65, 68]. The enantioseparation was performed for the double derivatized samples, first with CH_2N_2 (under a well-working hood), then with $(MeCO)_2O$ in the presence of 4-dimethylaminopyridine and pyridine. Both derivatizations were carried out in a test-tube, and the samples were then analyzed by using GC on a chiral column (Chirasil-Dex CB or Chirasil-L-Val). The operating conditions: injection temperature: 250 °C; detector temperature: 270 °C; column temperature up to an optimized program, for example, 120°C (7 min)–190 °C (20 °C/min) in the case of benzocispentacin [65]; carrier gas: nitrogen, flow rate 1.0 ml/min, split 20 : 1; make-up gas: 20 ml/min.

8.5 Conclusions and Outlook

It may be concluded that there has been tremendous development in this field in recent years and a great deal is known concerning the synthesis of cyclic β-amino acids, but more research is clearly still necessary at a basic level. This may result in access to new, highly substituted new cyclic β-amino acids, and to the development of simple and cheaper routes to enantiomers.

The olefinic moiety of cycloalkene β-amino acids offers an excellent opportunity for the regio- and stereoselective introduction of polar substituents (hydroxy, amino, azido, etc.) on the ring. The availability of highly substituted cyclic β-amino acids has aroused broad interest in peptide chemistry and may result in new directions in connection with functionalized carbocyclic nucleosides.

A large number of synthetic techniques are currently available for the synthesis of cyclic β-amino acid enantiomers, but they continue to suffer from limitations such as difficulty in large-scale synthesis, practicability, and cost. The achievement of successful desymmetrization methodology, a significant challenge, could furnish a useful approach to the preparation of these cyclic β-amino acids in enantiomerically pure form. Direct enzymatic routes, such as those involving lipase-catalyzed kinetic resolution (maximum yield 50%) appear to be excellent methods. Apart from the catalytic methods, dynamic kinetic resolution techniques (which theoretically allow 100% conversion) will undoubtedly acquire enormous significance in the future.

8.6 Experimental Procedures

8.6.1 Synthesis of Hydroxy Amino Ester Ethyl (1*R**,2*S**,4*S**)-2-(Benzyloxycarbonylamino)-4-hydroxycyclohexanecarboxylate (205a) by Oxirane Ring Opening of 205 with Sodium Borohydride [49]

205 → 205a ($NaBH_4$, EtOH, RT, 2 h)

To a solution of epoxide **205** (1.6 g, 5.0 mmol) in EtOH (45 ml), $NaBH_4$ (480 mg, 12.6 mmol) was added in portions. The mixture was stirred at room temperature for 2 h and the solvent was then evaporated at reduced pressure. The residue was taken up in EtOAc (120 ml) and the solution was washed with H_2O (3×80 ml), dried (Na_2SO_4), and concentrated *in vacuo*. The crude oily product was chromatographed over silica gel (*n*-hexane/EtOAc 1 : 2) to give a white solid; yield: 70%; melting point (m.p.) 50–53 °C. ^{1}H-nuclear magnetic resonance (NMR) [400 MHz, dimethylsulfoxide (DMSO), 25 °C, trimethylsilane (TMS)]: $\delta = 1.14$ (t, $J = 7.2$ Hz, 3H, CH_3), 1.41–1.63 (m, 3H, H3, H6), 1.69–1.80 (m, 2H, H5), 1.85–1.95 (m, 1H, H6), 2.68–2.75 (m, 1H, H1), 3.63–3.72 (m, 1H, H4), 3.77–3.89 (m, 1H, H2), 4.00 (q, $J = 7.2$ Hz, 1H, CH_2), 4.97–5.20 (m, 2H, OCH_2), 6.89 (d, $J = 8.4$ Hz, 1H, NH), 7.27–7.39 (m, 5H, Ar). ^{13}C-NMR (100 MHz, DMSO): $\delta = 14.8$, 21.8, 31.5, 37.3, 44.1, 48.3, 60.5, 66.1, 67.0, 128.5, 128.6, 129.1, 138.0, 155.9, 173.5.

8.6.2 Synthesis of Bicyclic β-Lactam (1*R**,5*S**)-6-azabicyclo[3.2.0]hept-3-en-7-one (223) by the Addition of Chlorosulfonyl Isocyanate to Cyclopentadiene [53]

Cyclopentadiene → 223 (1. CSI, Et_2O, 0 °C, 1 h; 2. Na_2SO_3, H_2O)

A solution of cyclopentadiene (22 ml, 316 mmol) in dry Et_2O (80 ml) was added dropwise under stirring at −10 °C to a solution of chlorosulfonyl isocyanate (14.2 ml, 163.2 mmol) in dry Et_2O (60 ml) and the mixture was stirred at this temperature for 1 h. Then an aqueous solution of Na_2SO_3 (240 ml) was added to the mixture together with a 10% aqueous solution of KOH giving pH 8. After stirring the mixture for 30 min at 0 °C, the layers were separated and the aqueous phase was extracted with Et_2O (3 × 100 ml). The organic layers were dried (Na_2SO_4) and the solvent was evaporated and crystallized from $i$$Pr_2O$ giving lactam **223**.

^{1}H-NMR (400 MHz, D_2O, TMS): δ = 2.16–2.48 (m, 1H, CH_2), 2.66–2.74 (m, 1H, CH_2), 3.78–3.83 (m, 1H, H-1), 4.48–4.51 (m, 1H, H-5), 5.92–6.02 (m, 2H, H-3 and H-4), 6.79–6.82 (brs, 1H, NH). Analysis: calculated for C_6H_7NO: C, 66.04; H, 6.47; N, 12.84; found: C, 66.01; H, 6.43; N, 12.80.

8.6.3
Synthesis of β-Amino Ester Ethyl *cis*-2-aminocyclopent-3-enecarboxylate Hydrochloride (223a) by Lactam Ring-Opening Reaction of Azetidinone 223 [53]

O
NH
223
HCl/EtOH, 0 °C
1 h
COOEt
NH_2-HCl
223a

Lactam **223** (18 g, 165.2 mmol) was dissolved in anhydrous EtOH (100 ml) and 30% HCl/EtOH (4020 ml) was added dropwise under stirring at 0 °C. After 30 min, the solution was treated with Et_2O (200 ml) and the precipitate was filtered off giving ester **223a**.

A yellowish-white solid; yield: 89%. m.p. 198–200 °C. ^{1}H-NMR (400 MHz, D_2O, TMS): δ = 1.29 (t, J = 7.10 Hz, 3H), 2.77–2.83 (m, 2H), 3.52–3.61 (m, 1H), 4.20–4.29 (m, 2H), 4.47–4.51 (m, 1H), 5.82–5.86 (m, 1H), 6.27–6.31 (m, 1H).

8.6.4
Synthesis of Epoxy Amino Ester Ethyl (1*R**,2*R**,3*R**,5*S**)-2-(*tert*-butoxycarbonylamino)-6-oxabicyclo[3.1.0]hexane-3-carboxylate (225) by Epoxidation of Amino Ester 224 [53]

COOEt
NHBoc
224
MCPBA, CH_2Cl_2
0 °C, 6 h
O
COOEt
NHBoc
225

To a solution of amino ester **224** (9 mmol) in CH_2Cl_2 (80 ml) was added *m*CPBA (11 mmol) at 0 °C. After the mixture was stirred for 6 h, CH_2Cl_2 (70 ml) was added and the resulting mixture was washed with saturated $NaHCO_3/H_2O$ (3 × 100 ml). The organic layer was then dried (Na_2SO_4) and concentrated under reduced pressure. The crude product was chromatographed on silica gel (*n*-hexane/EtOAc 3 : 1) yielding epoxide **225**.

A white solid; yield: 64%. m.p. 50–55 °C. ^{1}H-NMR (400 MHz, $CDCl_3$, 25 °C, TMS): δ = 1.27 (t, J = 7.15 Hz), 1.46 (s, 9H), 1.92–1.99 (m, 1H), 2.65–2.70 (m, 1H), 2.74–2.81 (m, 1H), 3.44–3.45 (m, 1H), 3.53–3.54 (m, 1H), 4.10–4.20 (m, 2H),

4.38–4.45 (m, 1H), 6.40 (brs, 1H). Analysis: calculated for $C_{13}H_{21}NO_5$: C, 57.55; H, 7.80; N, 5.16; found: C, 57.81; H, 7.56; N, 5.39.

8.6.5
Synthesis of Azido Ester Ethyl (1*R**,2*R**,3*R**,4*R**)-4-Azido-2-(*tert*-butoxycarbonylamino)-3-hydroxycyclopentanecarboxylate (229) by Oxirane Ring Opening of 225 with Sodium Azide [53]

COOEt NHBoc O **225** — NaN_3, NH_4Cl, EtOH/H_2O, Δ, 4 h → N_3 COOEt HO NHBoc **229**

To a solution of amino ester **225** (4.5 mmol) in EtOH (20 ml) and water (2 ml) were added NaN_3 (0.6 g, 9 mmol, 2 equiv.) and NH_4Cl (50 mg, 0.9 mmol, 20 mol%), and the mixture was stirred under reflux for 4 h. The mixture was then concentrated under reduced pressure, and the residue was taken up in EtOAc (80 ml), washed with water (3 × 70 ml), dried (Na_2SO_4), and concentrated in vacuum. The residue was purified by crystallization (*n*-hexane/EtOAc).

White crystals; yield: 73%. m.p. 100–102 °C. ^{1}H-NMR (400 MHz, $CDCl_3$, 25 °C, TMS): $\delta = 1.28$ (t, $J = 7.15$ Hz, 3H), 1.44 (s, 9H), 1.91–1.99 (m, 1H), 2.43–2.50 (m, 1H), 3.33–3.38 (m, 1H), 3.74 (brs, 1H), 3.94–4.00 (m, 1H), 4.01–4.04 (m, 1H), 4.11–4.24 (m, 2H), 4.32–4.42 (m, 1H), 5.25 (brs, 1H). ^{13}C-NMR (400 MHz, $CDCl_3$, TMS): $\delta = 14.8$, 29.0, 32.7, 45.2, 54.8, 62.5, 66.9, 78.0, 80.6, 156.1, 177.1. Mass spectroscopy (MS): [electrospray (ES), +ve] $m/z = 337$ (M + Na).

8.6.6
Isomerization of Azido Amino Ester 229 to Ethyl (1*S**,2*R**,3*R**,4*R**)-4-Azido-2-(*tert*-butoxycarbonylamino)-3-hydroxycyclopentanecarboxylate (230) [53]

N_3 COOEt HO NHBoc **229** — NaOEt, EtOH, RT, 22 h → N_3 COOEt HO NHBoc **230**

To a solution of azido amino ester **229** (2.8 g, 9 mmol) in anhydrous EtOH (30 ml), NaOEt (0.73 g, 1.2 equiv.) was added and the mixture was stirred for 22 h. It was then diluted with EtOAc (100 ml), washed with H_2O (3 × 80 ml), dried (Na_2SO_4), and concentrated. The crude oily product was chromatographed on silica gel (*n*-hexane/EtOAc 2 : 1) giving azido ester **230**.

White crystals; yield: 28%. m.p. 40–45 °C. ^{1}H-NMR (400 MHz, $CDCl_3$, 25 °C, TMS): $\delta = 1.27$ (t, $J = 7.15$ Hz, 3H), 1.45 (s, 9H), 1.86–1.96 (m, 1H), 2.41–2.50 (m, 1H), 2.84–2.91 (m, 1H), 3.16 (brs, 1H), 3.84–3.90 (m, 1H), 4.10–4.22 (m, 4H), 5.09 (brs, 1H). ^{13}C-NMR (400 MHz, $CDCl_3$, TMS): $\delta = 14.8$, 29.0, 31.3, 46.7, 57.1, 61.8, 66.1, 77.0, 78.0, 156.6, 174.2. MS: (ES, +ve) $m/z = 337$ (M + Na).

8.6.7
Lipase-Catalyzed Enantioselective Ring Cleavage of 4,5-Benzo-7-azabicyclo[4.2.0]octan-8-one (271), Synthesis of (1*R*,2*R*)- and (1*S*,2*S*)-1-Amino-1,2,3,4-tetrahydronaphthalene-2-carboxylic Acid Hydrochlorides (307 and 308) [65]

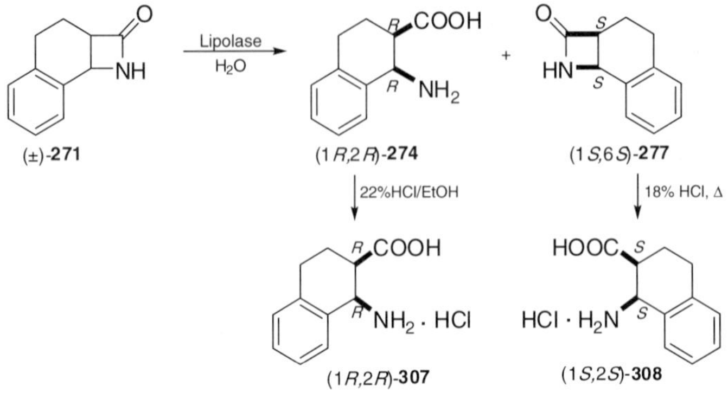

Racemic **271** (4.00 g, 23.1 mmol) was dissolved in iPr_2O (80 ml). Lipolase (4 g, 50 mg/ml) and H_2O (206 μl, 11.6 mmol) were added and the mixture was shaken in an incubator shaker at 70 °C for 7 h. The reaction was stopped by filtering off the enzyme at 50% conversion. The solvent was evaporated *in vacuo* and the residue (1*S*,6*S*)-**277** crystallized out [1.89 g, 47%; was recrystallized from iPr_2O[$[\alpha]_D^{25} = +313$ ($c = 0.3$ in $CHCl_3$); m.p. 140–141 °C; e.e. = 99%]. The filtered enzyme was washed with distilled H_2O (3×25 ml) and the H_2O was evaporated at reduced pressure, yielding the crystalline β-amino acid (1*R*,2*R*)-**274** [1.87 g, 42%; $[\alpha]_D^{25} = +25$ ($c = 0.3$ in H_2O); m.p. 269–270 °C with sublimation (recrystallized from H_2O and Me_2CO), e.e. = 99%]. When **274** (500 mg) was treated with 22% HCl/EtOH (10 ml), (1*R*,2*R*)-**307** was obtained [483 mg, 81%; $[\alpha]_D^{25} = +28$ ($c = 0.3$ in H_2O); m.p. 238–240 °C with sublimation (recrystallized from EtOH and Et_2O), e.e. = 99%].

(1*S*,6*S*)-**277** (500 mg, 2.875 mmol) was dissolved in 18% HCl (20 ml) and refluxed for 3 h. The solvent was evaporated *in vacuo*, and the product was recrystallized from EtOH and Et_2O, which afforded white crystals of (1*S*,2*S*)-**308** [570 mg, 86%; $[\alpha]_D^{25} = -27.9$ ($c = 0.3$ in H_2O); m.p. 260–261 °C; e.e. = 99%].

^{1}H-NMR (400 MHz, $CDCl_3$, 25 °C, TMS) data for **277**: $\delta = 1.59$–2.86 (m, 4H, $2 \times CH_2$), 3.71–3.73 (m, 1H, C*H*CO), 4.68 (d, $J = 4.9$ Hz, 1H, C*H*N), 5.96 (bs, 1H, N*H*), 7.19–7.3 (m, 4H, C_6H_4). ^{13}C-NMR (100.62 MHz, $CDCl_3$) $\delta = 23.6$, 27.5, 50.8, 52.2, 127.3, 129.2, 129.7, 130.2, 134.7, 140.0, 170.4. Analysis: calculated for $C_{11}H_{11}NO$: C, 76.28; H, 6.40; N, 8.09; found: C, 76.32; H, 6.44; N, 8.11.

^{1}H-NMR (400 MHz, D_2O, 25 °C, TMS) data for **274**: $\delta = 1.98$ (dd, $J = 9.9$, 4.8 Hz, 1H, CH_2), 2.22 (dd, $J = 11.1$, 1.9 Hz, 1H, CH_2), 2.84–3.01 (m, 3H, CH_2 and C*H*CO), 4.67 (bs, 1H, C*H*N), 7.29–7.39 (m, 4H, C_6H_4). ^{13}C-NMR (100 MHz, D_2O) $\delta = 21.7$, 28.1, 43.4, 50.3, 127.0, 129.6, 129.8, 130.1, 131.3, 137.6, 181.4. Analysis: calculated for $C_{11}H_{13}NO_2$: C, 69.09; H, 6.85; N, 7.32; found: C, 69.11; H, 6.85; N, 7.33.

^{1}H-NMR (400 MHz, D_2O, 25 °C, TMS) data are similar for **307** and **308**: $\delta = 2.06$–2.31 (m, 2H, CH_2), 2.91–2.99 (m, 2H, CH_2), 3.13–3.18 (m, 1H, C*H*CO),

4.83 (d, $J = 3.4$ Hz, 1H, C*H*N), 7.24–7.37 (m, 4H, C_6H_4). ^{13}C-NMR (100.62 MHz, D_2O) $\delta = 20.7$, 27.5, 41.7, 50.0, 127.2, 129.2, 130.0, 130.2, 130.5, 137.1, 177.0. Analysis: calculated for $C_{11}H_{13}NO_2{\cdot}HCl$: C, 58.03; H, 6.20; N, 6.15; found: for **307**: C, 58.17; H, 6.31; N, 6.10; found: for **308**: C, 57.88; H, 6.39; N, 6.12.

The progress of the reaction was followed by taking samples from the reaction mixture at intervals and analyzing them by GC. The enantiomeric excess value for the unreacted β-lactam enantiomer was determined by GC on a Chromopak Chiralsil-Dex CB column (25 m) [120 °C for 7 min → 190 °C (temperature rise 20 °C/min; 140 kPa; retention times (min), **277**: 16.57 (antipode: 15.82)), while the e.e. value for the β-amino acid produced was determined by using a gas chromatograph equipped with a chiral column after double derivatization with (i) diazomethane [*caution*: the derivatization with diazomethane should be performed in a fume hood!], and (ii) acetic anhydride in the presence of 4-dimethylaminopyridine and pyridine [(Chromopak Chirasil-Dex CB column, 120 °C for 7 min → 190 °C (temperature rise 20 °C/min; 140 kPa; retention times (min), **274**: 15.71 (antipode: 15.52)] (GC chromatograms).

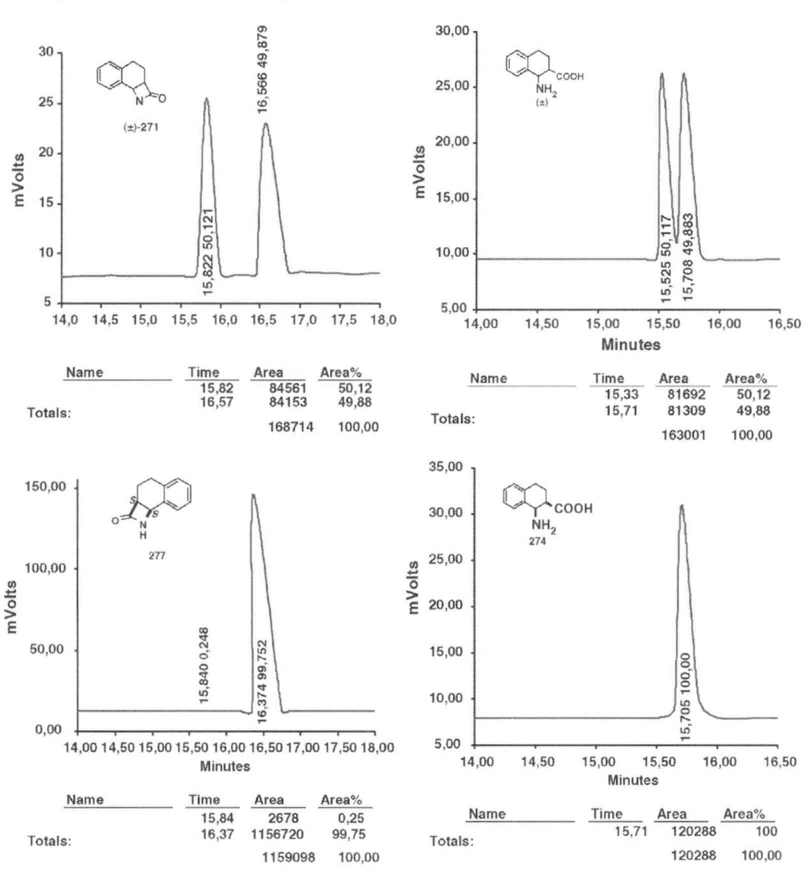

Name	Time	Area	Area%
	15,82	84561	50,12
	16,57	84153	49,88
Totals:		168714	100,00

Name	Time	Area	Area%
	15,33	81692	50,12
	15,71	81309	49,88
Totals:		163001	100,00

Name	Time	Area	Area%
	15,84	2678	0,25
	16,37	1156720	99,75
Totals:		1159098	100,00

Name	Time	Area	Area%
	15,71	120288	100
Totals:		120288	100,00

GC chromatograms for (±)-**271** and **277**

GC chromatograms for racemic 1-amino-1,2,3,4-tetrahydronaphtalene-2-carboxylic acid and **274**

8.6.8
Lipase-Catalyzed Enantioselective Hydrolysis of Ethyl *trans*-2-aminocyclohexane-1-carboxylate (298), Synthesis of (1*R*,2*R*)- and (1*S*,2*S*)-2-Aminocyclohexane-1-carboxylic Acid Hydrochlorides (309 and 310) [68]

(±)-**298** → CAL-B, H_2O → (1*R*,2*R*)-**300** + (1*S*,2*S*)-**302**

(1*R*,2*R*)-**300** → 22% HCl/EtOH → (1*R*,2*R*)-**309**

(1*S*,2*S*)-**302** → 18% HCl, Δ → (1*S*,2*S*)-**310**

Lipolase (1.8 g, 30 mg/ml) and H_2O (105 μl, 5.84 mmol) were added to the racemic **298** (2 g, 11.7 mmol) in *i*Pr_2O (60 ml), and the mixture was shaken in an incubator shaker at 70 °C for 72 h. The reaction was stopped by filtering off the enzyme at 50% conversion. The solvent was evaporated at reduced pressure and the residue was immediately hydrolyzed by refluxing (4 h) with 18% aqueous HCl solution (20 ml) to give the β-amino acid hydrochloride (1*S*,2*S*)-**310** (964 mg, 48%; $[\alpha]_D^{25} = +51$ ($c = 0.2$ in H_2O); m.p. 198–202 °C (recrystallized from EtOH and Et_2O); e.e. = 99%]. The filtered enzyme was washed with distilled H_2O (3 × 25 ml), and the H_2O was evaporated *in vacuo*, yielding the crystalline β-amino acid (1*R*,2*R*)-**300** [808 mg, 48%; $[\alpha]_D^{25} = -64$ ($c = 0.3$ in H_2O); m.p. 263–264 °C (recrystallized from H_2O and $(CH_3)_2CO$); e.e. = 99%]. When (1*R*,2*R*)-**300** (300 mg) was treated with 22% HCl/EtOH (25 ml), (1*R*,2*R*)-**309** was obtained [322 mg, 89%; $[\alpha]_D^{25} = -51$ ($c = 0.2$ in H_2O); m.p. 195–197 °C (recrystallized from EtOH and Et_2O), e.e. = 99%].

^{1}H-NMR (400 MHz, D_2O, 25 °C, TMS) data for **300**: δ = 1.29–2.11 (m, 8H, 4 × CH_2) 2.14-2.23 (m, 1H, H-1) 3.20-3.27 (m, 1H, H-2). Analysis: calculated for $C_7H_{13}NO_2$: C 58.72, H 9.15, N 9.78; found: C 58.84, H 9.16, N 9.60.

^{1}H-NMR (400 MHz, D_2O, 25 °C, TMS) data are similar for **309** and **310**: δ = 1.27–2.18 (m, 8H, 4 × CH_2), 2.49–2.54 (m, 1H, H-1), 3.35–3.42 (m, 1H, H-2). Analysis: calculated for $C_7H_{13}NO_2{\cdot}HCl$: C 46.80, H 7.86, N 7.80; found for **309**: C 46.57, H 7.69, N 7.75; found for **310**: C 46.77, H 7.91, N 7.82.

The e.e. values for the unreacted β-amino ester **302** and the β-amino acid enantiomer produced **300** were determined by GC on a Chromopak Chiralsil-Dex CB column (25 m) after double derivatization, presented earlier [Chromopak Chirasil-Dex CB column, 120 °C for 2 min → 190 °C, rate of temperature rise 20 °C/min, 140 kPa, retention times (min): **302**: 10.81 (antipode 10.93), **300**: 8.86 (antipode 8.78)].

References

1 Fülöp, F. (2001) The chemistry of 2-aminocycloalkanecarboxylic acids. *Chemical Reviews*, **101**, 2181–2204.

2 Park, K.H. and Kurth, M.J. (2002) Cyclic amino acid derivatives. *Tetrahedron*, **58**, 8629–8659.

3 Liu, M. and Sibi, M.P. (2002) Recent advances in the stereoselective synthesis of β-amino acids. *Tetrahedron*, **58**, 7991–8035.

4 Miller, J.A. and Nguyen, S.T. (2005) The enantioselective synthesis of conformationally constrained cyclic β-amino acids. *Mini-Reviews in Organic Chemistry*, **2**, 39–45.

5 Fülöp, F., Martinek, T.A., and Tóth, G.K. (2006) Application of alicyclic β-amino acids in peptide chemistry. *Chemical Society Reviews*, **35**, 323–334.

6 Liljeblad, A. and Kanerva, L.T. (2006) Biocatalysis, a profound tool in the preparation of highly enantiopure β-amino acids. *Tetrahedron*, **62**, 5831–5854.

7 Forró, E. and Fülöp, F. (2004) Direct and indirect enzymatic methods for the preparation of enantiopure β-amino acids and derivatives from β-lactams. *Mini-Reviews in Organic Chemistry*, **1**, 93–102.

8 Davies, S.G., Garrido, N.M., Kruchinin, D., Ichihara, O., Kotchie, L.J., Price, P.D., Mortimer, A.J.P., Russell, A.J., and Smith, A.D. (2006) Homochiral lithium amides for the asymmetric synthesis of β-amino acids. *Tetrahedron: Asymmetry*, **17**, 1793–1811.

9 Bunnage, M.E., Chippindale, A.M., Davies, S.G., Parkin, R.M., Smith, A.S., and Withey, J.M. (2003) Asymmetric synthesis of (1*R*,2*S*,3*R*)-3-methylcispentacin and (1*S*,2*S*,3*R*)-3-methyltranspentacin by kinetic resolution of *tert*-butyl (±)-3-methylcyclopentene-1-carboxylate. *Organic and Biomolecular Chemistry*, **1**, 3698–3707.

10 Davies, S.G., Garner, A.C., Long, M.J.C., Morrison, R.M., Roberts, P.M., Savory, E.D., Smith, A.D., Sweet, M.J., and Withey, J.M. (2005) Kinetic resolution and parallel kinetic resolution of methyl (±)-5-alkyl-cyclopentene-1-carboxylates for the asymmetric synthesis of 5-alkyl-cispentacin derivatives. *Organic and Biomolecular Chemistry*, **3**, 2762–2775.

11 Urones, J.G., Garrido, N.M., Diez, D., Hammoumi, M.M.E., Dominguez, S.H., Casaseca, J.A., Davies, S.G., and Smith, A.D. (2004) Asymmetric synthesis of the stereoisomers of 2-amino-5-carboxymethyl-cyclopentane-1-carboxylate. *Organic and Biomolecular Chemistry*, **2**, 364–372.

12 Davies, S.G., Diez, D., Dominguez, S.H., Garrido, N.M., Kruchinin, D., Price, P.D., and Smith, A.D. (2005) Cyclic β-amino acid derivatives: synthesis via lithium amide promoted tandem asymmetric conjugate addition–cyclisation reactions. *Organic and Biomolecular Chemistry*, **3**, 1284–1301.

13 Chippindale, A.M., Davies, S.G., Iwamoto, K., Parkin, R.M., Smethurst, C.A.P., Smith, A.D., and Rodriguez-Solla, H. (2003) Asymmetric synthesis of cyclic β-amino acids and cyclic amines via sequential diastereoselective conjugate addition and ring closing metathesis. *Tetrahedron*, **59**, 3253–3265.

14 Perlmutter, P., Rose, M., and Vounatsos, F. (2003) A stereoselective synthesis of five- and six-membered cyclic β-amino acids. *European Journal of Organic Chemistry*, 756–760.

15 Fustero, S., Sanchez-Rosello, M., Sanz-Cervera, J.F., Acena, J.L., del Pozo, C., Fernandez, B., Bartolome, A., and Asensio, A. (2006) Asymmetric synthesis of fluorinated cyclic β-amino acid derivatives through cross metathesis. *Organic Letters*, **8**, 4633–4636.

16 LePlae, P.R., Umezawa, N., and Lee, H.-S. and Gellman, S.H. (2001) An efficient route to either enantiomer

of *trans*-2-aminocyclopentanecarboxylic acid. *The Journal of Organic Chemistry*, **66**, 5629–5632.

17 Peelen, T.J., Chi, Y., English, E.P., and Gellman, S.H. (2004) Synthesis of 4,4-disubstituted 2-aminocyclopentanecarboxylic acid derivatives and their incorporation into 12-helical β-peptides. *Organic Letters*, **6**, 4411–4414.

18 Tang, W., Wu, S., and Zhang, X. (2003) Enantioselective hydrogenation of tetrasubstituted olefins of cyclic β-(acylamino)acrylates. *Journal of the American Chemical Society*, **125**, 9570–9571.

19 Wu, H.P. and Hoge, G. (2004) Highly enantioselective asymmetric hydrogenation of β-acetamido dehydroamino acid derivatives using a three-hindered quadrant rhodium catalyst. *Organic Letters*, **6**, 3645–3647.

20 Enthaler, S., Erre, G., Junge, K., Holz, J., Börner, A., Alberico, E., Nieddu, I., Gladiali, S., and Beller, M. (2007) Development of practical rhodium phosphine catalysts for the hydrogenation of β-dehydroamino acid derivatives. *Organic Process Research & Development*, **11**, 568–577.

21 Mittendorf, J., Kunisch, F., Matzke, M., Militzer, H.-C., Schmidt, A., and Schönfeld, W. (2003) Novel antifungal β-amino acids: synthesis and activity against *Candida albicans*. *Bioorganic & Medicinal Chemistry Letters*, **13**, 433–436.

22 Choi, S. and Silverman, R.B. (2002) Inactivation and inhibition of γ-aminobutyric acid aminotransferase by conformationally restricted vigabatrin analogues. *Journal of Medicinal Chemistry*, **45**, 4531–4539.

23 Hanselmann, R., Zhou, J., Ma, P., and Confalone, P.N. (2003) Synthesis of cyclic and acyclic β-amino acids via chelation-controlled 1,3-dipolar cycloaddition. *The Journal of Organic Chemistry*, **68**, 8739–8741.

24 Aggarwal, V.K., Roseblade, S., and Alexander, R. (2003) The use of enantiomerically pure ketene dithioacetal bis(sulfoxides) in highly diastereoselective intramolecular nitrone cycloadditions. Application in the total synthesis of the β-amino acid (−)-cispentacin and the first asymmetric synthesis of *cis*-(3*R*,4*R*)-4-amino-pyrrolidine-3-carboxylic acid. *Organic and Biomolecular Chemistry*, **1**, 684–691.

25 Szakonyi, Z. and Fülöp, F. (2003) Mild and efficient ring opening of monoterpene-fused β-lactam enantiomers. Synthesis of novel β-amino acid derivatives. *ARKIVOC*, 225–232.

26 Szakonyi, Z., Balázs, Á., Martinek, T.A., and Fülöp, F. (2006) Enantioselective addition of diethylzinc to aldehydes catalyzed by gamma-amino alcohols derived from (+)- and (−)-alpha-pinene. *Tetrahedron: Asymmetry*, **17**, 199–204.

27 Gyónfalvi, S., Szakonyi, Z., and Fülöp, F. (2003) Synthesis and transformation of novel cyclic β-amino acid derivatives from (+)-3-carene. *Tetrahedron: Asymmetry*, **14**, 3965–3972.

28 Szakonyi, Z., Martinek, T.A., Sillanpää, R., and Fülöp, F. (2007) Regio- and stereoselective synthesis of the enantiomers of monoterpene-based β-amino acid derivatives. *Tetrahedron: Asymmetry*, **18**, 2442–2447.

29 Mittendorf, J., Benet-Buchholz, J., Fey, P., and Mohrs, K.-H. (2003) Efficient asymmetric synthesis of β-amino acid BAY 10-8888/PLD-118, a novel antifungal for the treatment of yeast infections. *Synthesis*, 136–140.

30 (a) Bolm, C., Schiffers, I., Atodiresei, I., and Hackenberger, C.P.R. (2003) An alkaloid-mediated desymmetrization of *meso*-anhydrides via a nucleophilic ring opening with benzyl alcohol and its application in the synthesis of highly enantiomerically enriched β-amino acids. *Tetrahedron: Asymmetry*, **14**, 3455–3467; (b) see also a review: Atodiresei, I., Schiffers, I., and Bolm, C., (2007) Stereoselective anhydride openings. *Chemical Reviews*, **107**, 5683–5712.

31 Hamersak, Z., Roje, M., Avdagic, A., and Sunjic, V. (2007) Quinine-mediated parallel kinetic resolution of racemic cyclic anhydride: stereoselective synthesis, relative and absolute configuration of novel alicyclic β-amino acids. *Tetrahedron: Asymmetry*, **18**, 635–644.

32 Pai Fondekar, K.P., Volk, F.-J., Khaliq-uz-Zaman, S.M., Bisel, P., and Frahm, A.W. (2002) Carbocyclic α,β-diamino acids: asymmetric Strecker synthesis of stereomeric 1,2-diaminocyclohexanecarboxylic acids. *Tetrahedron: Asymmetry*, **13**, 2241–2249.

33 Priego, J., Flores, P., Ortiz-Nava, C., and Escalante, J. (2004) Synthesis of enantiopure *cis*- and *trans*-2-aminocyclohexane-1-carboxylic acids from octahydroquinazolin-4-ones. *Tetrahedron: Asymmetry*, **15**, 3545–3549.

34 Miyata, O., Muroya, K., Kobayashi, T., Yamanaka, R., Kajisa, S., Koide, J., and Naito, T. (2002) Radical cyclization in heterocycle synthesis. Part 13: sulfanyl radical addition-cyclization of oxime ethers and hydrazones connected with alkenes for synthesis of cyclic β-amino acids. *Tetrahedron*, **58**, 4459–4479.

35 Kiss, L., Forró, E., Bernáth, G., and Fülöp, F. (2005) Synthesis of alicyclic *trans*-β-amino acids from *cis*-β-hydroxycycloalkanecarboxylates. *Synthesis*, 1265–1268.

36 Gnad, F. and Reiser, O. (2003) Synthesis and applications of β-aminocarboxylic acids containing a cyclopropane ring. *Chemical Reviews*, **103**, 1603–1623.

37 Mangelinckx, S. and De Kimpe, N. (2003) Synthesis of racemic *cis*-1-alkyl- and 1-aryl-2-aminocyclopropanecarboxylic esters. *Tetrahedron Letters*, **44**, 1771–1774.

38 Ortuno, R.M., Moglioni, A.G., and Moltrasio, G.Y. (2005) Cyclobutane biomolecules: synthetic approaches to amino acids, peptides and nucleosides. *Current Organic Chemistry*, **9**, 237–259.

39 Gauzy, C., Pereira, E., Faure, S., and Aitken, D.J. (2004) Synthesis of (+)-(1*S*,2*R*) and (−)-(1*R*,2*S*)-2-aminocyclobutane-1-carboxylic acids. *Tetrahedron Letters*, **45**, 7095–7097.

40 Fernandes, C., Gauzy, C., Yang, Y., Roy, O., Pereira, E., Faure, S., and Aitken, D.J. (2007) [2 + 2] photocycloadditions with chiral uracil derivatives: Access to all four stereoisomers of 2-aminocyclobutanecarboxylic acid. *Synthesis*, 2222–2232.

41 Roy, O., Faure, S., and Aitken, D.J. (2006) A solution to the component instability problem in the preparation of peptides containing C2-substituted *cis*-cyclobutane β-amino acids: synthesis of a stable rhodopeptin analogue. *Tetrahedron Letters*, **47**, 5981–5984.

42 Mondiere, A., Peng, R., Remuson, R., and Aitken, D.J. (2007) Efficient synthesis of 3-hydroxymethylated *cis*- and *trans*-cyclobutane β-amino acids using an intramolecular photocycloaddition strategy. *Tetrahedron*, **64**, 1088–1093.

43 Izquierdo, S., Rua, F., Sbai, A., Parella, T., Alvarez-Larena, A., Branchadell, V., and Ortuno, R.M. (2005) (+)- and (−)-2-Aminocyclobutane-1-carboxylic acids and their incorporation into highly rigid β-peptides: stereoselective synthesis and a structural study. *The Journal of Organic Chemistry*, **70**, 7963–7971.

44 Bunnage, M.E., Ganesh, T., Masesane, I.B., Orton, D., and Steel, P.G. (2003) Asymmetric synthesis of the putative structure of (−)-oryzoxymycin. *Organic Letters*, **5**, 239–242.

45 Masesane, I.B. and Steel, P.G. (2004) Stereoselective routes to 3-hydroxy and 3,4-dihydroxy derivatives of 2-aminocyclohexanecarboxylic acid. *Tetrahedron Letters*, **45**, 5007–5009.

46 Fülöp, F., Palkó, M., Forró, E., Dervarics, M., Martinek, T.A., and Sillanpää, R. (2005) Facile regio- and diastereoselective syntheses of hydroxylated 2-aminocyclohexanecarboxylic acids. *European Journal of Organic Chemistry*, 3214–3220.

47 Szakonyi, Z., Gyónfalvi, S., Forró, E., Hetényi, A., De Kimpe, N., and Fülöp, F. (2005) Synthesis of 3- and 4-hydroxy-2-aminocyclohexanecarboxylic acids by iodolactonization. *European Journal of Organic Chemistry*, 4017–4023.

48 Kiss, L., Forró, E., and Fülöp, F. (2006) A new strategy for the regio- and stereoselective hydroxylation of *trans*-2-aminocyclohexenecarboxylic acid. *Tetrahedron Letters*, **47**, 2855–2858.

49 Kiss, L., Forró, E., Martinek, T.A., Bernáth, G., De Kimpe, N., and Fülöp, F. (2008) Stereoselective synthesis of hydroxylated β-aminocyclohexanecarboxylic acids. *Tetrahedron*, **64**, 5036–5043.

50 Soengas, R.G., Estevez, J.C., and Estevez, R.J. (2003) Stereocontrolled transformation of nitrohexofuranoses into cyclopentylamines via 2-oxabicyclo[2.2.1] heptanes: incorporation of polyhydroxylated carbocyclic β-amino acids into peptides. *Organic Letters*, **5**, 1423–1425.

51 Soengas, R.G., Pampin, M.B., Estevez, J.C., and Estevez, R.J. (2005) Stereocontrolled transformation of nitrohexofuranoses into cyclopentylamines via 2-oxabicyclo[2.2.1] heptanes. Part 2: synthesis of (1*S*,2*R*,3*S*,4*S*,5*R*)-3,4,5-trihydroxy-2-aminocyclopentanecarboxylic acid. *Tetrahedron: Asymmetry*, **16**, 205–211.

52 Kiss, L., Szatmári, I., and Fülöp, F. (2006) A simple synthesis of orthogonally protected 2,5-diaminocyclohexanecarboxylic acids. *Letters in Organic Chemistry*, **3**, 463–467.

53 Kiss, L., Forró, E., Sillanpää, R., and Fülöp, F. (2007) Diastereo- and enantioselective synthesis of orthogonally protected 2,4-diaminocyclopentanecarboxylates: a flip from β-amino- to β,γ-diaminocarboxylates. *The Journal of Organic Chemistry*, **72**, 8786–8790.

54 Kanerva, L.T., Csomós, P., Sundholm, O., Bernáth, G., and Fülöp, F. (1996) Approach to highly enantiopure β-amino acid esters by using lipase catalysis ín organic media. *Tetrahedron: Asymmetry*, **7**, 1705–1716.

55 Gyarmati, Z.C., Liljeblad, A., Rintola, M., Bernáth, G., and Kanerva, L.T. (2003) Lipase-catalyzed kinetic resolution of 7-, 8- and 12-membered alicyclic β-amino esters and *N*-hydroxymethyl-β-lactam enantiomers. *Tetrahedron: Asymmetry*, **14**, 3805–3814.

56 Csomós, P., Kanerva, L.T., Bernáth, G., and Fülöp, F. (1996) Biocatalysis for the preparation of optically active β-lactam precursors of amino acids. *Tetrahedron: Asymmetry*, **7**, 1789–1796.

57 Kámán, J., Forró, E., and Fülöp, F. (2000) Enzymatic resolution of alicyclic β-lactams. *Tetrahedron: Asymmetry*, **11**, 1593–1600.

58 Gyarmati, Z.C., Liljeblad, A., Argay, G., Kálmán, A., Bernáth, G., and Kanerva, L.T. (2004) Chemoenzymatic preparation of enantiopure homoadamantyl β-amino acid and β-lactam derivatives. *Advanced Synthesis and Catalysis*, **346**, 566–572.

59 Forró, E., Árva, J., and Fülöp, F. (2001) Preparation of (1*R*,8*S*)- and (1*S*,8*R*)-9-azabicyclo[6.2.0]dec-4-en-10-one: potential starting compounds for the synthesis of anatoxin-a. *Tetrahedron: Asymmetry*, **12**, 643–649.

60 Fülöp, F., Palkó, M., Kámán, J., Lázár, L., and Sillanpää, R. (2000) Synthesis of all four enantiomers of 1-aminoindane-2-carboxylic acid, a new cispentacin benzologue. *Tetrahedron: Asymmetry*, **11**, 4179–4187.

61 Forró, E. and Fülöp, F. (2003) Lipase-catalyzed enantioselective ring opening of unactivated alicyclic-fused β-lactams in an organic solvent. *Organic Letters*, **5**, 1209–1212.

62 Forró, E. and Fülöp, F. (2004) Synthesis of enantiopure 1,4-ethyl- and 1,4-ethylene-bridged cispentacin by lipase-catalyzed enantioselective ring opening of β-lactams. *Tetrahedron: Asymmetry*, **15**, 573–575.

63 Forró, E. and Fülöp, F. (2004) Advanced procedure for the enzymatic ring opening of unsaturated alicyclic β-lactams. *Tetrahedron: Asymmetry*, **15**, 2875–2880.

64 Lloyd, R.C., Lloyd, M.C., Smith, M.E.B., Holt, K.E., Swift, J.P., Keene, P.A., Taylor, S.J.C., and McCague, R. (2004) Use of hydrolases for the synthesis of cyclic amino acids. *Tetrahedron*, **60**, 717–728.

65 Forró, E. and Fülöp, F. (2006) An efficient enzymatic synthesis of benzocispentacin and its new six- and seven-membered homologues. *Chemistry – A European Journal*, **12**, 2587–2592.

66 Forró, E. and Fülöp, F. (2006) Do lipases also catalyse the ring cleavage of inactivated cyclic *trans* β-lactams? *Tetrahedron: Asymmetry*, **17**, 3193–3196.

67 Winkler, M., Martinková, L., Knall, A.C., Krahulec, S., and Klempier, N. (2005) Synthesis and microbial transformation of β-amino nitriles. *Tetrahedron*, **61**, 4249–4260.

68 Forró, E. and Fülöp, F. (2007) The first direct enzymatic hydrolysis of alicyclic β-amino esters: a route to enantiopure *cis* and *trans* β-amino acids. *Chemistry – A European Journal*, **13**, 6397–6401.

69 Forró, E. and Fülöp, F. (2006) PCT Int. Appl. WO 2006/055528 A2. Enzymatic process for the preparation of cyclic beta-amino acid and ester enantiomers.

70 Péter, A., Török, G., and Fülöp, F. (1998) Effect of temperature on retention of cyclic β-amino acid enantiomers on a chiral crown ether stationary phase. *Journal of Chromatographic Science*, **36**, 311–317.

71 Török, G., Péter, A., and Fülöp, F. (1998) High-performance liquid chromatographic separation of enantiomers of unusual amino acids possessing cycloalkane or cycloalkene skeletons on a chiral crown ether-containing stationary phase. *Chromatographia*, **48**, 20–26.

72 Péter, A. and Fülöp, F. (1995) High-performance liquid chromatographic method for the separation of isomers of *cis*- and *trans*-2-amino-1-cyclopentanecarboxylic acid. *Journal of Chromatography. A*, **715**, 219–226.

73 Péter, A., Török, G., Csomós, P., Péter, M., Bernáth, G., and Fülöp, F. (1997) High-performance liquid chromatographic methods for the separation of enantiomers of alicyclic amino acids. *Journal of Chromatography. A*, **761**, 103–113.

74 Péter, A., Péter, M., Fülöp, F., Török, G., Tóth, G., Tourwé, D., and Sápi, J. (2000) High-performance liquid chromatographic separation of unusual amino acid enantiomers derivatized with (1*S*,2*S*)-1,3-diacetoxy-1-(4-nitrophenyl)-2-propyl-isothiocyanate. *Chromatographia*, **51**, S148–S154.

75 Péter, M., Péter, A., and Fülöp, F. (1999) Development of a new isothiocyanate-based chiral derivatizing agent. *Chromatographia*, **50**, 373–375.

76 Török, G., Péter, A., Csomós, P., Kanerva, L.T., and Fülöp, F. (1998) High-performance liquid chromatographic methods for separation of isomers of β-amino acids possessing bicyclo[2.2.1] heptane skeletons. *Journal of Chromatography. A*, **797**, 177–186.

77 D'Acquarica, I., Gasparrini, F., Misiti, D., Zappia, G., Cimarelli, C., Palmieri, G., Carotti, A., Cellamare, S., and Villani, C. (2000) Application of a new chiral stationary phase containing the glycopeptide antibiotic A-40,926 in the direct chromatographic resolution of β-amino acids. *Tetrahedron: Asymmetry*, **11**, 2375–2385.

78 Péter, A., Árki, A., Tourwé, D., Forró, E., Fülöp, F., and Armstrong, D.W. (2004) Comparison of the separation efficiencies of Chirobiotic T and TAG columns in the separation of unusual amino acids. *Journal of Chromatography. A*, **1031**, 159–170.

79 Berkecz, R., Török, R., Ilisz, I., Forró, E., Fülöp, F., Armstrong, D.W., and Péter, A. (2006) LC enantioseparation of β-lactam and β-amino acid stereoisomers and a comparison of the performances of macrocyclic glycopeptide based columns. *Chromatographia*, **63**, S37–S43.

9
Synthetic Approaches to α, β-Diamino Acids

Alma Viso and Roberto Fernández de la Pradilla

9.1
Introduction

The discovery of α,β-diamino acids among natural products, either in the native state or as fragments of complex molecules, has stimulated interest in these molecules. Their structural complexity, having two vicinal chiral centers, represents a current challenge for synthetic organic chemists, especially the synthesis of enantiopure materials. Therefore, this chapter will provide a general overview of the existing methodology for the synthesis of aliphatic α,β-diamino acids and their simple derivatives, esters or amides [1]. Within this context, in recent years a number of synthetic routes of variable length, yield, and complexity have been reported. Figure 9.1 gathers some representative routes that can be classified into two main categories: methods that require construction of the carbon backbone, and methods that start from the basic carbon skeleton and introduce the C−N bonds modifying the nature of the functional groups.

9.2
Construction of the Carbon Backbone

The search for new methodology to form C−C bonds is critical for the development of organic synthesis. The established methods to build the carbon skeleton of α,β-diamino acids can be classified with respect to the C−C bond formed in the key step and therefore the following section has been organized accordingly (Figure 9.1).

9.2.1
Methods for the Formation of the C_b−C_c Bond

9.2.1.1 Reaction of Glycinates and Related Nucleophiles with Electrophiles

The Mannich reaction of glycinates and imines is a fundamental solution within the plethora of available methods. This strategy can be applied to the synthesis

Amino Acids, Peptides and Proteins in Organic Chemistry. Vol.1 – Origins and Synthesis of Amino Acids.
Edited by Andrew B. Hughes

ISBN: 978-3-527-32096-7

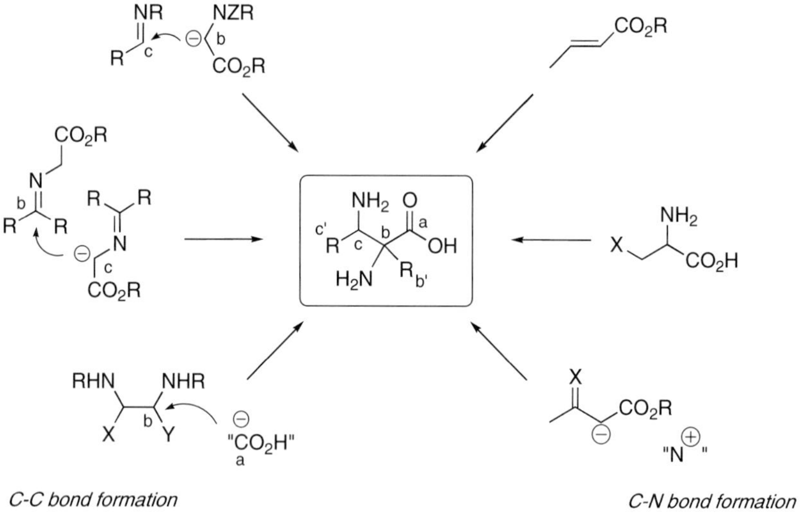

Figure 9.1 Representative routes towards α,β-diamino acids.

of 3-amino-β-lactams [2]; nevertheless the use of suitably functionalized precursors can circumvent the cyclization [3]. In a recent example high *syn* diastereoselectivities (>90 : 10) are obtained when the glycinate **2** is generated *in situ* under reductive conditions from an α-iminoester **1** and TiI_4. In these examples the relative stereochemistry of the final diamino esters **3** depends on the nature of the substituents of the imine counterpart with a complete reversal to the *anti* isomer when the imine contains a triple bond [4]. *N*-(*p*-Toluenesulfonyl)-α-chloroaldimines **5** can also act as electrophiles affording racemic γ-chloro-α,β-diamino esters **6** upon reaction with benzophenone imine glycinate **4** with low *syn*/*anti* diastereoselectivity. After separation, both diastereoisomers were efficiently cyclized by treatment with K_2CO_3 to β,γ-aziridino α-amino ester derivatives **7a,b** [5]. Substoichiometric amounts of Lewis acids such as $Zn(OTf)_2$ allow for the smooth reaction of benzophenone imine glycinates and enamines, but again low diastereomeric ratios are produced [6] (Scheme 9.1).

Asymmetric versions of this strategy involving chiral glycinates are found in the literature. The enantiopure Ni(II) complex of benzophenone α-imino glycinate and the aluminum enolate of a chiral oxazinone derived from glycine have been used for the synthesis of fluorinated 2,3-diamino acids (98% d.e.) [7] and capreomycidine [8], respectively. Chiral imines can participate as inductors of asymmetry in the Mannich process. Recently, Yadav *et al.* have reported the one-pot synthesis of 4-amino perhydropyrimidines **10** upon a Biginelli reaction. Upon microwave irradiation and in the presence of a Ce(III) salt, D-xylose and D-glucose react with ureas **8** and thioureas generating aldose-derived C=N species that are attacked *in situ* by an 1,3-oxazol-5-one **9** and finally evolve to the perhydropyrimidine **10** diastereoselectively (Scheme 9.2) [9]. *p*-Toluenesulfinimines **11** can participate as chiral inductors in the $BF_3{\cdot}Et_2O$-mediated addition of lithiated α-imino glycinates to afford enantiopure

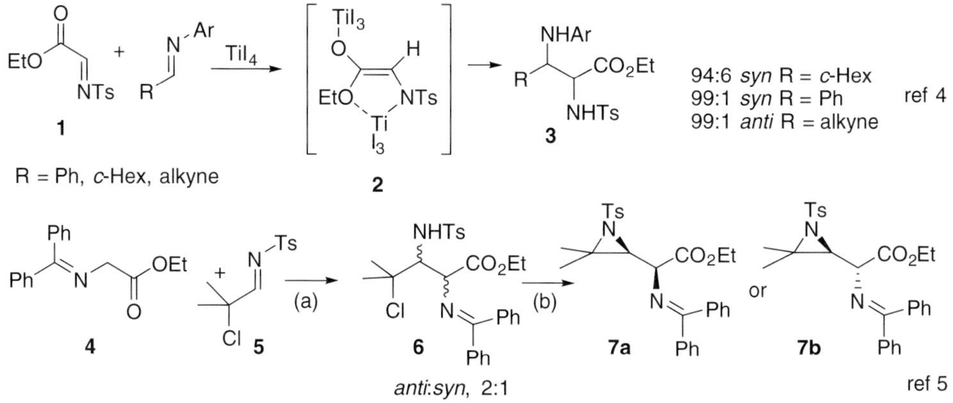

Reagents and conditions: (a) LDA, THF. (b) separation and then K_2CO_3, acetone

Scheme 9.1 Mannich reaction of imines and glycinates.

4,5-*trans* 4-carbomethoxy *N*-sulfinylimidazolidines **12** (Scheme 9.2). Interestingly in the absence of $BF_3{\cdot}Et_2O$, enolates derived from phenylalanine, alanine, and leucine treated with lithium diisopropylamide (LDA) provide imidazolidines **12a** containing quaternary centers with *endo* stereocontrol and excellent diastereofacial discrimination

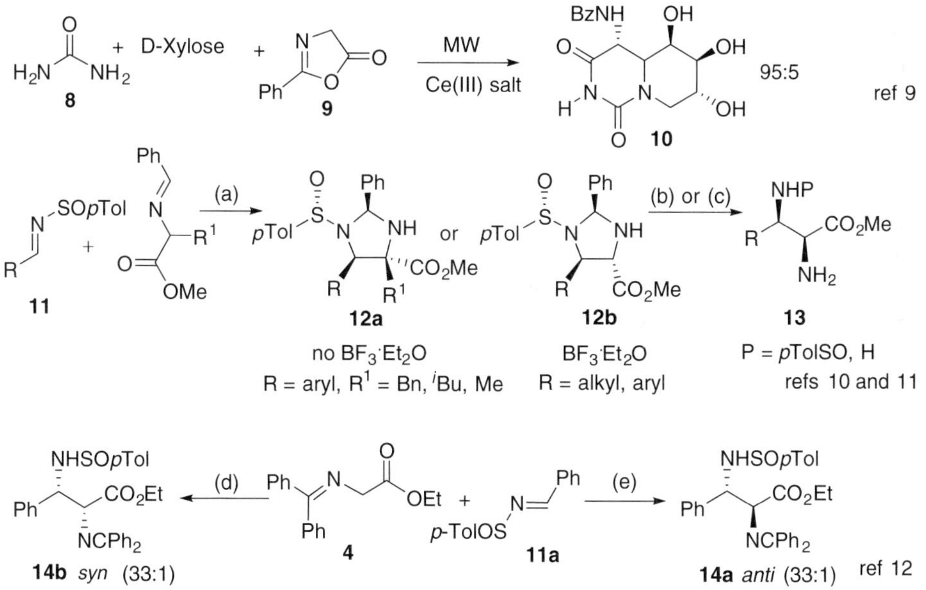

Reagents and conditions: (a) LDA, –78 °C then $BF_3{\cdot}Et_2O$, –78 °C to rt. (b) H_3PO_4, THF-H_2O, rt, 68-84 % (c) H_3PO_4, MeOH-H_2O, rt, 59-65 %. (d) LDA, THF –78 °C. (e) H_2O-LDA, THF–78 °C.

Scheme 9.2 Mannich reaction of chiral imines and glycinates.

of the chiral sulfiminine [10]. Subsequently, enantiopure *syn* *N*-sulfinyl-α,β-diamino esters **13** can be selectively produced under acidic conditions modulated by choosing a non-nucleophilic solvent [tetrahydrofuran (THF) : H_2O versus MeOH] to preserve the sulfinamide moiety and thus achieve different degrees of functionalization at the amino groups [11]. A parallel strategy has been addressed by the group of Davis in the synthesis of *syn* and *anti* α,β-diamino esters recently used in the total synthesis of (−)-agelastatin A and (+)-CP-99 994. Diastereoselectivity in the addition to sulfinimine **11a** was completely controlled by the nitrogen protecting groups of the glycinate that provide different geometries, (*E*) versus (*Z*), respectively, for the enolate transition state. In addition, a thorough study has demonstrated that for the *N*-benzhydrylglycinate **4** a small amount of H_2O is crucial in preventing the retro-Mannich reaction of the kinetic *anti* isomer **14a** thus yielding complete *anti* diastereoselectivity, while in the absence of H_2O the addition provides *syn* isomers **14b** exclusively (Scheme 9.2) [12]. Double diastereoselection has also been applied for the asymmetric synthesis of β-carboline-containing α,β-diamino esters with fairly good results [13].

Asymmetric catalysis also constitutes a useful tool in the synthesis of α,β-diamino esters (Scheme 9.3). In this context, chiral Lewis acid-stabilized benzophenone imino glycinate enolates (from **15**), generated in the presence of $CuClO_4$ and a chiral oxazoline **17**, react with a group of *N*-sulfonyl imines **16** to afford, as major products, *syn* α,β-diamino esters **18** with good enantiomeric excesses (88–97%). Interestingly, the aliphatic nature of *N*-sulfonyl imines is crucial to reach a good *syn* : *anti* diastereoselectivity (>95 : 5) [14]. A parallel approach has been reported by Willis to provide enantioenriched *anti* α,β-diamino acid derivatives **21** as major products by using an oxazolidinone **19** as nucleophilic partner and a C-2 symmetric bis(oxazoline) **20** as ligand in the presence of $Mg(ClO_4)_2$ [15]. In contrast, the system CuOTf/(*R*,*R*)-Me-DuPhos has been used for the catalytic direct-type Mannich reaction of aliphatic aldehydes, secondary amines, and benzophenone imine glycinates, providing low diastereo- and enantioselectivities [16]. Similarly, phase-transfer-catalyzed Mannich reaction has been examined using chiral quaternary ammonium salts. For this purpose, catalytic amounts of an *N*-spiro C-2-symmetric chiral quaternary ammonium salt **24** have been used to produce enantiomerically enriched *syn*-diamino succinates **25** [64% d.e.; 91% e.e. (*syn*)] from benzophenone imine glycinate **22** and α-iminoester **23**, in the context of the synthesis of a precursor of streptolidine [17]. Alternatively, *syn* α,β-diamino ester derivatives **28** have been generated from *N*-Boc imines **26** using a tartrate-derived diammonium salt [(*S*,*S*)-TaDiAS] **27** with excellent *syn* : *anti* ratio and moderate enantiomeric excesses [18].

Imines are also susceptible to nucleophilic addition of nitroalkanes (Scheme 9.4). Recently, the group of Jørgensen has developed one of the most efficient protocols for the synthesis of α,β-diamino esters **32**, the catalytic enantioselective aza-Henry reaction between imino ester **23** and nitro compounds **29**. The reaction proceeds at room temperature with high *anti* : *syn* diastereoselectivity (from 77 : 23 to 95 : 5) and enantiomeric excess values above 95% for the major *anti* β-nitro-α-amino esters **31**, in the presence of base and a copper complex based on a chiral bisoxazoline

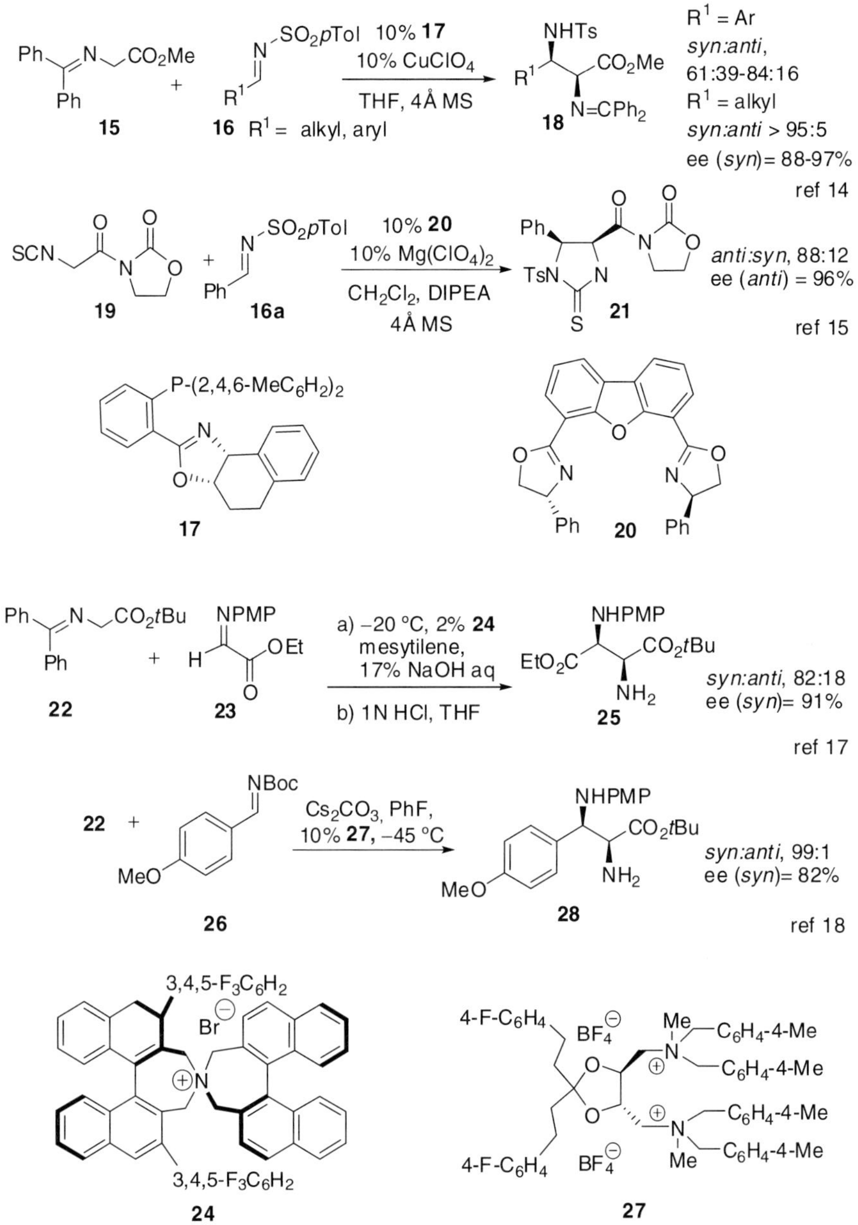

Scheme 9.3 Chiral catalysts for the enantioselective Mannich reaction.

ligand [19]. Furthermore, the use of base can be avoided using silyl nitronates **30** instead of nitro compounds [20]. Recently, Johnston *et al.* have reported the enantioselective Brønsted acid catalyzed addition of nitroacetates **33** to aromatic *N*-Boc imines **34** using chiral *trans*-1,2-cyclohexane diamine derived quinolinium

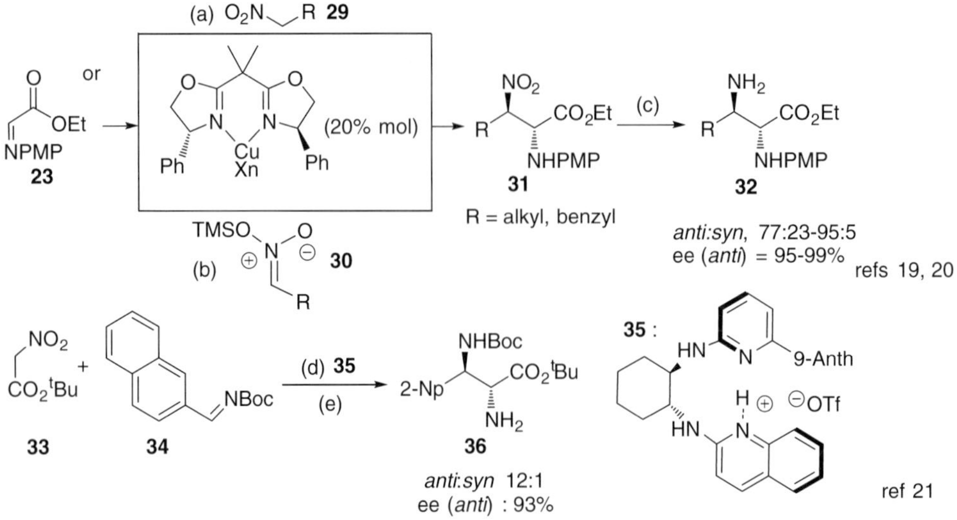

Reagents and conditions: (a) Et_3N, $Cu(OTf)_2$, rt or 0 °C. (b) $CuPF_6$, −100 °C, THF. (c) Raney Ni, H_2, 80%. (d) Tol, −78 °C, **35** (5% mol). (e) $NaBH_4$, $CoCl_2$.

Scheme 9.4 Enantioselective aza-Henry reactions.

salts (**35**). After reduction of the resulting nitro compound the best results are 12 : 1 *anti* : *syn* ratio and 93% e.e. for the major *anti* diaminoester **36** [21].

The reactions between glycinates and electrophiles other than imines have also been employed to build the C_b–C_c bond; however, these protocols need additional steps to transform the primary adduct into the final α,β-diamino acid. For this purpose, halogenated electrophiles like dibromomethane [22] and *N*-bromomethylphthalimide [23] have been successfully employed. Esters such as ethyl formate have been submitted to the addition of ethyl hippurate to give (±)-quisqualic acid, further submitted to enzymatic resolution [24], and aldehydes have been demonstrated to be valuable intermediates in the multistep synthesis of L-capreomycidine [25]. Amide acetals are also useful electrophiles in these reactions [26]. In particular, an enantioselective synthesis of α,β-diaminopropionic acid derivatives **40** has been achieved by condensation of hippuric acid **37** with dimethylformamide dimethyl acetal, **38** followed by treatment with NH_4OAc in MeOH and then acylation of the resulting enamide. The α,β-dehydro α,β-diamino ester **39** was submitted to a remarkably efficient asymmetric hydrogenation using (*R*,*R*)-EtDuPhos-Rh(I) complex that provides α,β-diamino esters in yields above 99% and enantiomeric excesses of 99% (Scheme 9.5) [27].

9.2.1.2 Dimerization of Glycinates

In some instances dimerizations of glycinates and related compounds [28] have been used for the synthesis of diamino succinates. Early attempts entailing photodimerization or treatment with sodium hydride of ethyl *N*-acetyl malonate and 2-acetoxy glycinate [29] led to equimolecular mixtures of diastereoisomers. Alternatively,

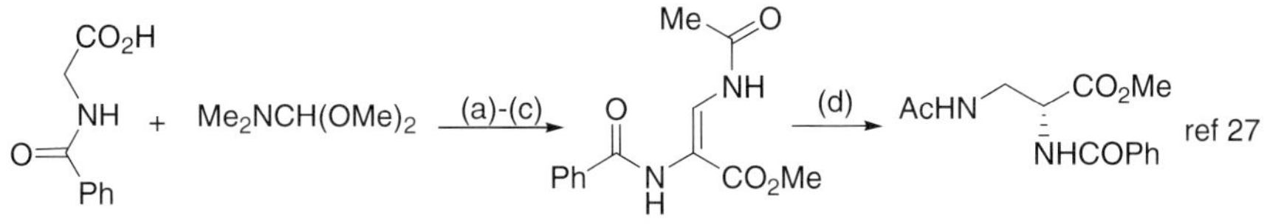

Reagents and conditions: (a) Tol, reflux. (b) NH_4OAc, MeOH, rt. (c) MeCOCl, base, CH_2Cl_2-Et_2O. (d) (*R,R*)-EtDuPhos-Rh (I), H_2, 60-90 psi, 15-60 h.

Scheme 9.5 Enantioselective hydrogenation of an α,β-dehydro α,β-diamino ester.

the oxidative dimerization of a chiral Ni(II) complex derived from α-imino alaninate occurs by treatment with *n*BuLi followed by addition of pentyl iodide or MnO_2 to afford after acidic hydrolysis (−)-(2*S*,3*S*)-2,3-dimethyl-2,3-diaminosuccinic acid [30]. In another example, a highly diastereoselective oxidative dimerization of glycinates has been reported by enolization (*t*BuLi, LDA, or *s*BuLi) and treatment with iodine to afford racemic *syn* (*threo*) diamino succinates (*syn* : *anti* >98 : 2) in high yield [31]. Recently, a highly diastereoselective route that relies on the coupling of α-ethylthio-glycinate **41** and α-acyliminoglycinate **42** mediated by PPh_3 to provide selectively a good yield of (*Z*)-dehydrodiamino succinate **43** that can be readily converted to the (*E*) isomer **44** under basic conditions has been described. The *cis*-selective catalytic hydrogenation of both (*Z*)- and (*E*)-dehydrodiamino acids allows for the efficient and diastereoselective synthesis of *anti* and *syn* orthogonally protected diamino succinates **45** and **46** [32]. Finally, enantioselective catalysis has also been applied to the synthesis of diamino succinates **48**. Indeed, a 2-acetoxy imino glycinate **47** was submitted to a palladium-mediated π-azaallylic substitution using the benzophenone α-iminoglycinate **15** derived sodium enolate as nucleophile and chiral phosphine ligands. Unfortunately, although the yields are fairly good, diastereomeric ratios (*dl* : *meso*) are moderate and enantiomeric excess values for the *dl* pair are low (Scheme 9.6) [33].

9.2.1.3 **Through Cyclic Intermediates**

Cycloaddition reactions, one of the most useful tools for the construction of C−C bonds, have also been used in the synthesis of α,β-diamino acids. In this context, imines and related compounds containing C=N bonds can undergo stereocontrolled cycloaddition with a number of species that render cyclic precursors of α,β-diamino acids.

Among these cyclic precursors are 2-alkoxycarbonyl aziridines, readily available by Lewis acid-mediated [2 + 1] cycloaddition of imines and alkyl diazoacetates [34]. Studies on this process comprise the use of Lewis acids [35], metal complexes based on Cu(I) and chiral oxazolines [36], Rh(II) and sulfur ylides [37], boron-based catalysts [38], and chiral phase-transfer catalysts [39]. Alternatively, 1,3,5-triazines, *N*-methoxymethylanilines, and 2-amino nitriles can be used as source of imines in the presence of the Lewis acid [40]. In particular, $SnCl_4$ and 2-amino nitriles derived from 1-(*R* or *S*)-phenylethylamine **50**, provide *in situ* formation of the

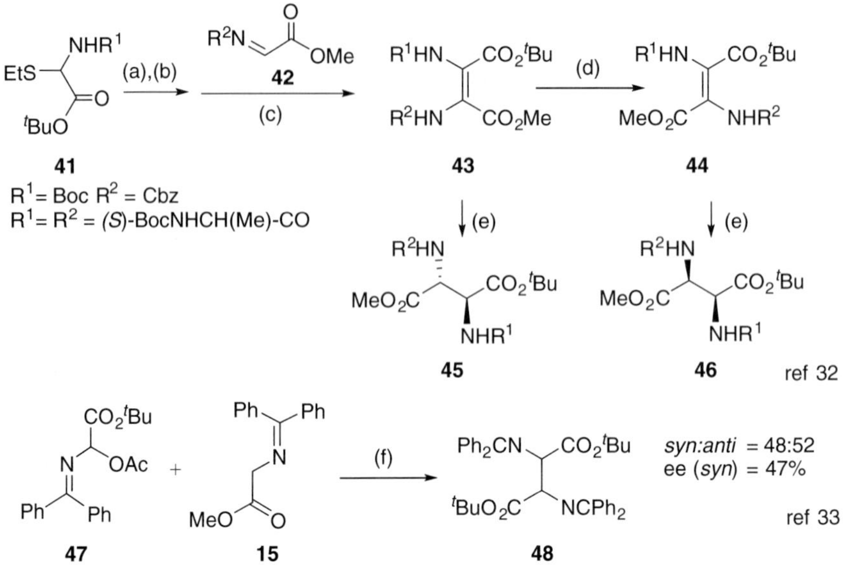

NHR¹ EtS ᵗBuO O (a),(b) R²N OMe O 42 (c) R¹HN $CO_2{}^tBu$ R²HN CO_2Me (d) R¹HN $CO_2{}^tBu$ MeO_2C NHR²

41 43 44

R¹= Boc R² = Cbz
R¹= R² = (*S*)-BocNHCH(Me)-CO

(e) R²HN MeO_2C $CO_2{}^tBu$ NHR¹ 45 (e) R²HN MeO_2C $CO_2{}^tBu$ NHR¹ 46 ref 32

$CO_2{}^tBu$ N OAc Ph Ph 47 + Ph Ph N MeO O 15 (f) Ph_2CN $CO_2{}^tBu$ $^tBuO_2C$ $NCPh_2$ 48 *syn:anti* = 48:52 ee (*syn*) = 47% ref 33

Reagents and conditions: (a) SO_2Cl_2, CH_2Cl_2, 0 °C. (b) PPh_3. (c) Et_3N, THF, −78 °C, 61%. (d) Et_2NH, MeOH, 70 °C, 52%. (e) $[Rh(COD)Cl]_2$dppf, 90 bar, H_2, toluene, 80 °C. (f) $Pd(OAc)_2$, MeCN, (*R*)-BINAP, 24 h, NaH, 77%.

Scheme 9.6 Diaminosuccinates through dimerization of glycinates.

iminium ion in the reaction with diazoacetate **49**. In this example, aziridine **51** is obtained with complete *cis* selectivity and with moderate diastereoselectivity relative to the chiral auxiliary. A further three-step sequence transformed the 2-ethoxycarbonyl aziridine **51** into the corresponding enantiopure free α,β-diamino acid [*syn*-(2*R*,3*S*)- diaminobutanoic acid (Dab)] (Scheme 9.7) [[40]b].

3-Amino-β-lactams are valuable precursors to α,β-diamino acid derivatives. These intermediates are available by the [2 + 2] ketene–imine cycloaddition, also referred to as the Staudinger reaction. Using a suitable amino ketene equivalent or another ketene equivalent such as 2-acetoxyacetyl chloride allows for straightforward introduction of an amino group at C-3 after β-lactam formation [41]. The preparation of amino taxol side-chain precursors has been addressed using this strategy [42].

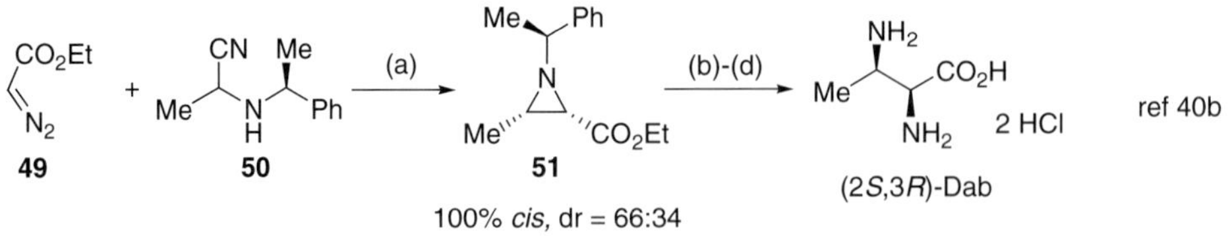

CO_2Et N_2 49 + CN Me Me N H Ph 50 (a) Me Ph N Me CO_2Et 51 100% *cis*, dr = 66:34 (b)-(d) NH_2 Me CO_2H NH_2 2 HCl (2*S*,3*R*)-Dab ref 40b

Reagents and conditions: (a) $SnCl_4$, CH_2Cl_2, rt. (b) $TMSN_3$, $BF_3{\cdot}OEt_2$. (c) H_2, Pd-C, $(Boc)_2O$. (d) 6 N HCl.

Scheme 9.7 α,β-diamino acids through aziridines.

Alternatively, 3-azido β-lactams are available from azido ketenes and this approach was employed in the total synthesis of dimeric diketopiperazine antibiotic (±)-593A [43]. A more straightforward route to enantiopure α,β-diamino acids has been reported using chiral Evans–Sjögren acid chloride **52** and imine **53**. The resulting *cis*-3-amido-4-styryl-β-lactam **54** can be then alkylated diastereoselectively at C-3 (**55**), furnishing after complete removal of protecting groups and β-lactam cleavage (3 steps), free (*S*,*R*)-α-methyl-α,β-diamino acids **56** (Scheme 9.8) [44]. Alternatively, the *cis* β-lactam **54** can undergo epimerization to the *trans* isomer and further acidic hydrolysis afforded *anti* (*R*,*R*)-α,β-diamino acids in good yield [45]. The highly efficient transformation of enantiopure *N*-Boc-protected 3-amino-β-lactams **57** into α-alkyl-α,β-diamino acid residues **58** incorporated as part of a peptide backbone has also been developed by the group of Palomo. Indeed, upon treatment with (*S*)-phenylalanine or (*S*)-valine methyl esters and sodium azide or potassium cyanide, β-lactams afforded dipeptides (Scheme 9.8) [46].

Carboxyimidazolines and imidazolidines are cyclic analogs of α,β-diamino acids, easily accessible via [3 + 2] cycloaddition of imines and azomethine ylides [47]. The palladium-mediated synthesis of imidazolines reported by Arndtsen can be held in this category [48]. Within this context, the group of Harwood reported that chiral azomethine ylides **61** generated *in situ* from (5*S*)-phenylmorpholin-2-one **59** in the presence of aromatic aldimines **60** and pyridinium *p*-toluenesulfonate can undergo cycloaddition with excess of imine to give imidazolidines **62** as single products. Finally, hydrogenolysis under acidic conditions released the corresponding enantiopure *syn* α,β-diamino acids in excellent yields **63** (Scheme 9.9) [49]. Another remarkable example of this approach is the reaction of methyl isocyanoacetate **64** with *N*-sulfonyl imines **16** [R (aryl, (*E*)-styryl] catalyzed by transition metal complexes such as AuCl(NC*c*-hex) to give racemic *cis*-imidazolines **65** with high

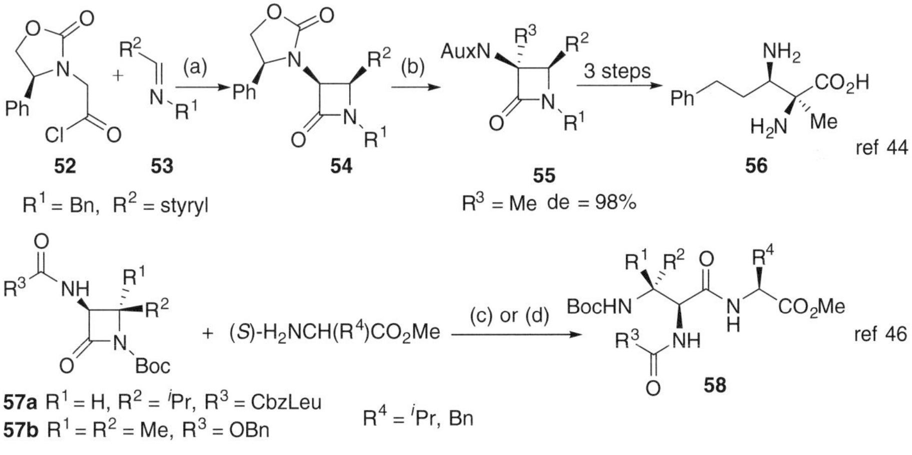

52 **53** **54** **55** **56** ref 44

R^1 = Bn, R^2 = styryl

R^3 = Me de = 98%

57a R^1 = H, R^2 = ^{i}Pr, R^3 = CbzLeu
57b R^1 = R^2 = Me, R^3 = OBn

+ (*S*)-$H_2NCH(R^4)CO_2Me$ → (c) or (d) → **58** ref 46

R^4 = ^{i}Pr, Bn

Reagents and conditions: (a) Et_3N, CH_2Cl_2, –78 °C to rt. (b) LHMDS, MeI. (c) DMF, NaN_3, rt 10-14 h. (d) KCN , DMF, rt, 10-14 h.

Scheme 9.8 α,β-Diamino acids through β-lactams.

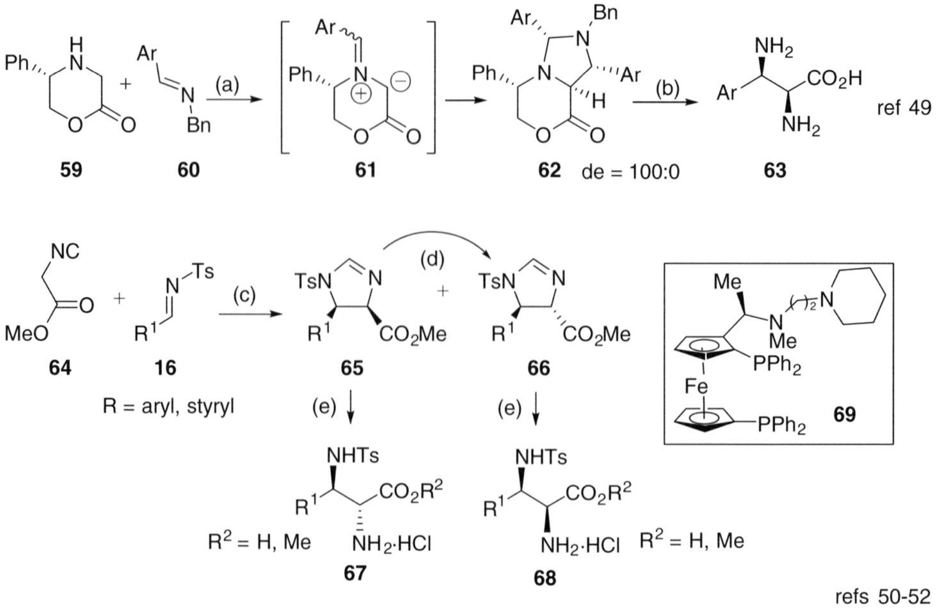

Reagents and conditions: (a) PPTS, toluene, reflux. (b) TFA, MeOH-H_2O (10:1), $Pd(OH)_2$, H_2, 5 bar. (c) AuCl (*c*-hexNC); 89:11-95:5 cis: trans. $RuH_2(PPh_3)_4$; 13:87-5:95 trans:cis. AuCl·SMe_2, ferrocenyl phosphine; (4*R*,5*S*) cis, ee = 96%-99%. (d) Et_3N, CH_2Cl_2, reflux. (e) 6 N HCl for R = H or HCl, MeOH for R = Me.

Scheme 9.9 α,β-Diamino acids through imidazolidines and imidazolines.

diastereoselectivity. *cis*-Imidazolines were readily converted into *anti* α,β-diamino acids **67** or their methyl esters and can be isomerized (R^1 = Ph) into the thermodynamically more stable *trans* isomer **66** by treatment with Et_3N leading to the corresponding *syn* α,β-diamino acids or esters **68** (Scheme 9.9) [50]. Interestingly, $RuH_2(PPh_3)_4$ as catalyst and CH_2Cl_2/MeOH 1 : 3 as solvent, provide opposite diastereoselectivity (*trans*) for the cycloaddition [51]. Furthermore, in the presence of Au(I) salts and a chiral ferrocenylphosphine ligand **69**, *cis* imidazolines (R^1 = aryl) were obtained as optically pure materials (96–99% e.e.) and were then converted to *anti* diamino esters (2*R*,3*R*) using a parallel procedure [52].

9.2.2
Methods in Which the C_a-C_b Bond is Formed

9.2.2.1 Nucleophilic Synthetic Equivalents of CO_2R

Construction of the C_a-C_b bond is another key approach in the synthesis of α,β-diamino acids. In fact, the carboxylic group is introduced into the molecule by means of a synthetic equivalent as a nucleophile or electrophile and then additional manipulation to give the carboxylic group is usually needed. Within this context, nitrocompounds have been employed as nucleophiles to form the C_a-C_b bond [53].

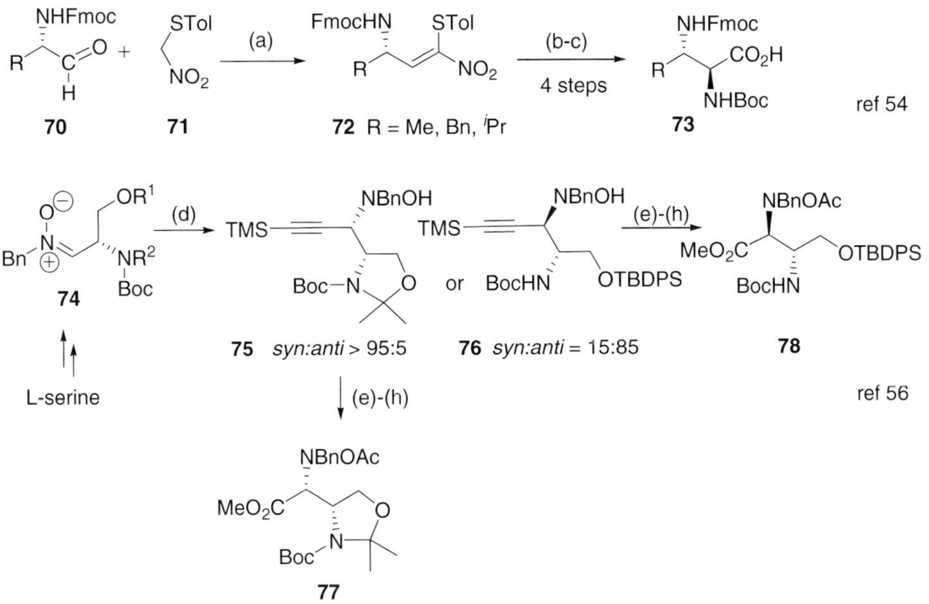

Reagents and conditions: (a) i. ^{t}BuOK, ^{t}BuOH-THF; ii. $MeSO_2Cl$, $^i Pr_2EtN$,CH_2Cl_2, 69-80%. (b) i. $LiOO^tBu$, toluene; ii. NH_3, 35-57%. (c) i. Boc_2O, THF, 97-99%; ii. H_2O_2, NaOH, 63-91%. (d) LiCCTMS, THF, –80 °C, 68-97%. (e) Bu_4NF, THF, rt, 70-80 %. (f) Ac_2O, pyr, 1 h, rt. (g) $RuCl_3$-$NaIO_4$, MeCN, rt. (h) CH_2N_2, Et_2O, 72-76% (3 steps).

Scheme 9.10 Introduction of the carboxylate as nucleophile.

Jackson *et al.* have developed a new route to enantiopure *anti* α,β-diamino acids based on the stepwise condensation of (*p*-tolylthio)nitromethane **71** and α-amino aldehydes **70** (Scheme 9.10). The resulting nitroalkenes **72** were submitted to nucleophilic epoxidation and diastereoselective epoxide cleavage using NH_3 to render α,β-diamino thioesters that were transformed in two steps into differentially protected *anti* α,β-diamino acids **73** [54]. The use of terminal alkynes as synthetic equivalents of the carboxylic acid moiety is well documented in organic synthesis and has been applied to the synthesis of α,β-diamino acids (Scheme 9.10). The group of Merino has demonstrated that addition of lithium trimethylsilylacetylide to α-amino nitrones **74**, takes place with a high degree of diastereocontrol (>95 : 5) to give *syn* propargylhydroxylamines **75**. In contrast, changing protecting groups in the starting nitrone allows for an inversion of the diastereoselectivity (**76**, *anti* : *syn*, 85 : 15). Then, removal of the trimethylsilyl group, *O*-acylation, alkyne oxidation with $RuCl_3/NaIO_4$, and esterification provide *syn* and *anti* α,β-diamino esters **77** and **78**, respectively. Pyrrolidinyl glycinates have been produced with this approach [55]. A close strategy can be carried out using 2-lithiofuran instead of lithium trimethylsilylacetylide in the addition to the nitrone [56].

The Bucherer–Bergs reaction of α-amido ketones, potassium cyanide, and ammonium carbonate produced hydantoins in good yields that, after cleavage of the protecting carbamate and saponification, rendered α-alkyl α,β-diamino acids [57], as well as α,β-diamino acids derived from proline [58] and pipecolic acid [59] that have served as monomers for assembling, by solid-phase synthesis, larger molecular entities with a well-defined secondary structure in solution. On the other hand, the Strecker reaction, involving addition of cyanide to α-amino imines provides efficient access to α,β-diamino nitriles that have been transformed into different α,β-diamino acid derivatives. This strategy was employed to synthesize peptidic enalapril analogs containing an α,β-diamino acid residue, as inhibitors of the angiotensin-converting enzyme [60], as well as for the synthesis of octapeptide angiotensin II analogs containing an α,β-diamino acid unit [61]. Asymmetric Strecker reactions have been carried out using chiral imines generated from enantiopure (*R*)-α-phenylethylamine with moderate diastereoselectivity [62]. In contrast, α,β-aziridinosulfinimines [63] and glucopyranose-derived aldimines [64] provide, with high diastereocontrol, amino nitriles that have been used in the synthesis of α,β-diamino acids. The Strecker methodology has been used in the total synthesis of enantiopure dysibetaine and related compounds (Scheme 9.11) [65]. In fact, an enantiopure bicyclic δ-lactam **79** was used as precursor of an *N*-acyliminium ion for the highly diastereoselective addition of trimethylsilyl cyanide (**80**). After diastereoselective lactam hydroxylation at C-4 (**81**, diastereomeric ratio > 80 : 1), hydrogenolysis, acidic hydrolysis and esterification, hydroxymethyl lactam **82** was produced. Substitution of the side-chain hydroxyl group was effected by means of azide displacement onto a mesylate derivative and then suitable manipulation of the functional groups afforded natural (2*S*,4*S*)-dysibetaine. Within the context of the Strecker protocol, the modular synthesis of *syn* α,β-diamino esters has been reported via sequential addition of cyanocuprates and cyanide. The key intermediate

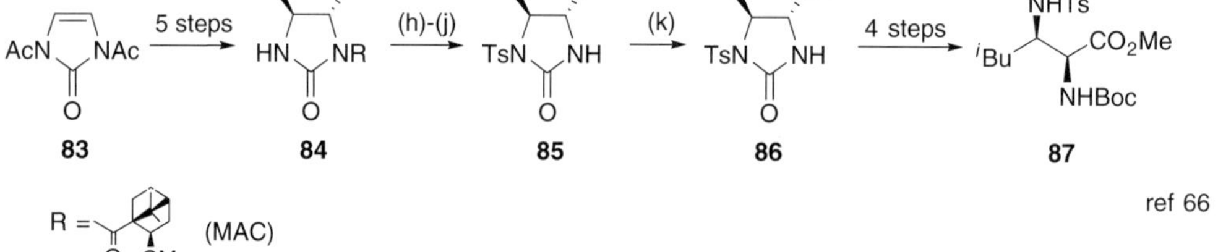

79 → (a) → 80, dr = 100 :0 → (b) → 81, dr = 80:1 → (c-d) → 82 → (e-g) → (*S*)-dysibetaine (ref 65)

83 → 5 steps → 84 → (h)-(j) → 85 → (k) → 86 → 4 steps → 87 (ref 66)

R = (MAC)

Reagents and conditions: (a) TMSCN, $SnCl_4$, 65%. (b) KHMDS, MoOPH, toluene, THF, 81%. (c) 6 N HCl. (d) CH_2N_2, Et_2O, 59% over 2 steps. (e) i. MsCl-Pyr; ii. NaN_3, DMF, 28% over 2 steps. (f) H_2, Pd-C, MeOH. (g) i. MeI, $^iPr_2EtN$, THF; ii. Dowex 550A, 61%. (h) $^iBuCuCNMgBr$, LiCl, $BF_3{\cdot}Et_2O$, THF. (i) TsCl, BuLi, THF. (j) BnSH, BuLi, THF. (k) TMSCN, $BF_3{\cdot}Et_2O$, CH_2Cl_2.

Scheme 9.11 Diastereoselective Strecker reaction in the synthesis of α,β-diamino acids.

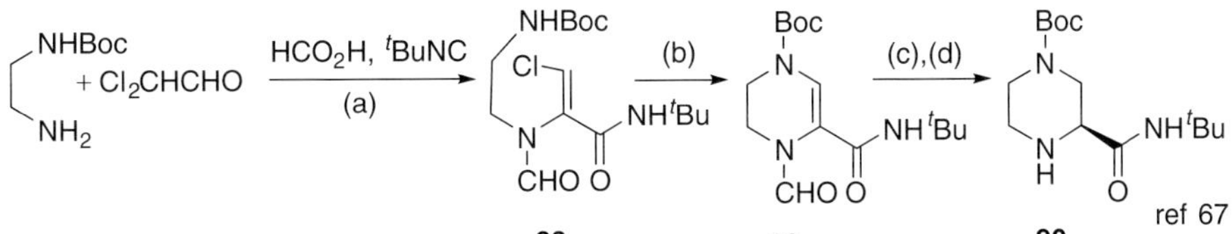

Reagents and conditions: (a) toluene, 50 °C, MeOH, 2 days, rt, then Et_3N, 3 h, rt, (~100%). (b) KO^tBu, THF, 3 h, 60%. (c) MeOH, [(*R*)-BINAP(COD)Rh]OTf, H_2, 100 atm. (d) aqueous NH_2NH_2, 100 °C, 91%.

Scheme 9.12 Multicomponent Ugi reaction for piperazine-2-carboxamides.

enantiopure dimethoxyimidazolidinone **84**, generated in five steps from a nonchiral imidazolone **83**, underwent nucleophilic addition of isobutylcyanocuprate to an *in situ* generated acyliminium ion followed by introduction of a tosyl group and removal of the chiral auxiliary (MAC) to afford an *N*-tosyl methoxyimidazolidinone **85** that was submitted to diastereoselective Lewis acid mediated addition of cyanotrimethylsilane (**86**). Finally, four steps, including acidic hydrolysis of the cyanide group and imidazolidinone cleavage, rendered the enantiopure *syn* α,β-diamino ester **87** in good yield (Scheme 9.11) [66].

Finally, the multicomponent Ugi condensation has been used as the key step for the assembly of piperazine-2-carboxamides (Scheme 9.12). This one-pot procedure consists of reaction between a mono-*N*-alkylethylenediamine, an α-chloroacetaldehyde, an isocyanide, and a carboxylic acid to give piperazine carboxamides in a very efficient manner. Furthermore, access to enantiopure piperazines is also possible using α,α-dichloroacetaldehyde that allows for the isolation of **88** that cyclizes to tetrahydropyrazine **89**. Asymmetric hydrogenation of the latter in the presence of Rh-2,2′-Bis(diphenylphosphino)1,1′-binaphthyl (BINAP) catalyst provides the enantiopure piperazine **90**, a fragment of the HIV protease inhibitor indinavir [67].

9.2.2.2 Electrophilic Synthetic Equivalents of CO_2R and Other Approaches

In contrast to the approaches above, the introduction of the carboxylic group as an electrophile has been scarcely explored. An example developed by the group of Seebach starts from a glycine-derived imidazolidinone (*R*)-[2-*tert*-butyl-3-methyl-4-imidazolidinone (BMI)] (Scheme 9.13). Deoxygenation using lithium triethylborohydride in combination with lithium borohydride rendered an acylimidazolidine **91** that upon directed metallation with *t*BuLi followed by trapping with CO_2 diastereoselectively afforded carboxyimidazolidines **92** (diastereomeric ratio > 98 : 2). Subsequent diastereocontrolled alkylation using alkyl halides **93** and then acidic hydrolysis allowed for the isolation of a number of enantiopure free α-alkyl α,β-diamino acids **94**. In addition, since (*S*)-(BMI) is equally accessible the enantiomeric α,β-diamino acids are also available. A parallel approach has been used in the synthesis of 2-carboxypiperazines by the group of Quirion [68]. Finally, [2 + 2] ketene–imine cycloadditions have been applied to form the C_a-C_b bond of

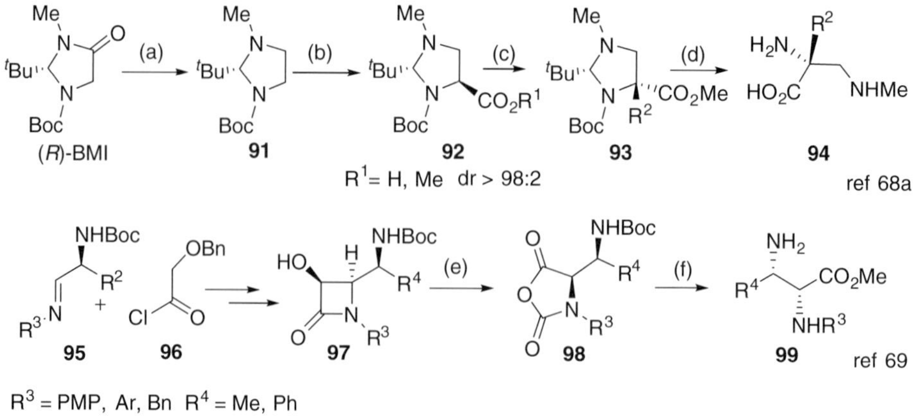

R^1 = H, Me dr > 98:2

ref 68a

ref 69

R^3 = PMP, Ar, Bn R^4 = Me, Ph

Reagents and conditions: (a) $LiBH_4$ cat., $LiBHEt_3$, THF, reflux, 90 min, 90%. (b) TMEDA, ^tBuLi, Et_2O, −70 °C, 30 min then CO_2, 3 h then CH_2N_2 84%. (c) $^iPr_2NH$, THF, ^nBuLi, −60 °C to rt, 20 min, R^2X, 12 h, 60-86%. (d) i. TFA, CH_2Cl_2, 14 h; ii. 2 N NaOH, rt, 1 h; iii. Dowex 50x8, 60-75%. (e) NaOCl, TEMPO cat., $NaHCO_3$, KH_2PO_4-K_2HPO_4, KBr, CH_2Cl_2, 0 °C, 92%. (f) MeOH, TMSCl, rt, 24 h or MeOH, reflux, 80%.

Scheme 9.13 Introduction of C_a through electrophilic reagents.

α,β-diamino acids with the oxidative treatment of β-lactams as a crucial step of this strategy (Scheme 9.13). In particular, enantiopure 3-hydroxy-4-(aminoalkyl) β-lactams **97**, readily available from acid chlorides **96** and optically pure α-amino imines **95**, were efficiently submitted to an oxidative cycloexpansion protocol using TEMPO (2,2,6,6-tetramethylpiperidinooxy) that provides *N*-carboxy anhydrides **98** without epimerization. Ensuing esterification rendered enantiopure differentially protected *syn* α,β-diamino esters **99** [69].

9.2.3
Methods in Which the $C_bC_{b'}$ or $C_cC_{c'}$ Bonds are Formed

The number of routes to introduce carbon substituents on the diamino acid skeleton is considerably smaller than the number of approaches found for the above disconnections. Some of these methods are based on alkylation of asparagine or aspartic derivatives [70]. Within this family of methods, the group of Merino has reported an effective route for the diastereoselective synthesis of *syn* or *anti* α,β-diamino esters based on nucleophilic additions of Grignard reagents to nitrones prepared from L-serine. The authors have observed different diastereoselectivities depending on the protecting groups of the substrates [71]. Similarly, addition of allylmagnesium bromide to Garner's aldehyde derived imine **100** has recently been applied to the synthesis of 2-piperidinyl glycinate **101** (Scheme 9.14) [72]. Finally, in a report from the group of Snider, the total synthesis of (−)- and (()-dysibetaine has been addressed using as a key step the intramolecular cleavage of an oxirane to build a pyrrolidinone ring (Scheme 9.14). Initial *N*-acylation of ethyl amino(cyano)acetate

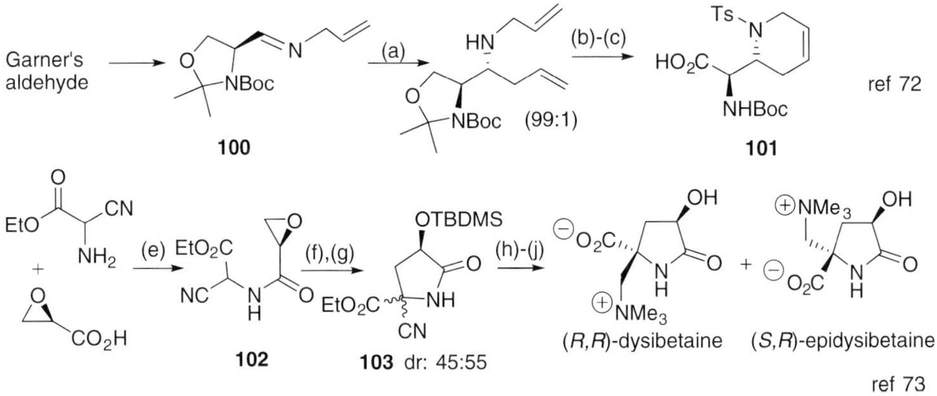

Reagents and conditions: (a) $CH_2{=}CHCH_2MgBr$, THF:Et_2O, −30 °C. (b) TsCl, NEt_3, CH_2Cl_2. (c) Grubbs' cat 5%, rt. (d) Jones' reagent, acetone, 0 °C. (e) DCC, EtOAc. (f) NaOEt, THF. (g) TBSOTf, 2,6-lutidine, CH_2Cl_2. (h) separation; PtO_2, H_2, EtOH, HCl. (i) i. H_2CO, Pd-C, H_2; ii. $NaHCO_3$, MeI, THF. (j) Dowex 550A, MeOH, 55 °C.

Scheme 9.14 Examples of formation of $C_cC_{c'}$ and $C_bC_{b'}$ bonds.

with (*R*)-glycidic acid gave a glycinamide **102** that underwent intramolecular alkylation upon treatment with NaOEt to give a 45 : 55 mixture of diastereomeric hydroxy pyrrolidinones **103** that was readily separated after silylation. Subsequent hydrogenation of the cyano group, followed by permethylation and saponification rendered (*R,R*)-dysibetaine and (*S,R*)-epidysibetaine, respectively [73].

9.3 Introduction of the Nitrogen Atoms in the Carbon Backbone

9.3.1 From Readily Available α-Amino Acids

Natural α-amino acids are key starting materials for the synthesis of α,β-diamino acids. Indeed, a plethora of synthetic approaches can be found through the literature consisting of the manipulation of functional groups that already exist in available α-amino acids with the additional advantage of using enantiopure starting materials. Most of these methods are already reviewed [1] and only illustrative examples are gathered herein.

Serine, threonine, and *allo*-threonine have been frequently submitted to conversion of the alcohol into an amine, and by far the most studied approach is the Mitsunobu reaction (for a recent example, see [74]). However, a suitable combination of protecting groups at the acid, the α-nitrogen as well as the newly introduced β-nitrogen should be chosen to prevent undesired side-reactions such as β-elimination, aziridine formation, and so on, and to ensure a reasonable yield of the α,β-diamino acid derivative [75]. The synthesis of β-pyrazol-1-yl-L-alanine [76] analogs of willardine [77] as well as enantiopure 2-carboxypiperazine derivatives [78] have been addressed using Mitsunobu protocols. In contrast, the direct nucleophilic substitution on suitably protected β-tosyl serine [79] and threonine [80] derivatives or even on β-chloroalanine [81] is less documented. A recent and simple example describes *N,N*-dibenzyl-*O*-methylsulfonyl serine methyl ester **104** as a suitable precursor of α,β-diamino esters **105ab** by reaction with nitrogen nucleophiles via regioselective (C-2) opening of a transient aziridinium ion (Scheme 9.15) [82]. Finally, there are a few reports in which the hydroxymethyl group of L-serine is converted into the new carboxylic group in the target α,β-diamino acid [83]. In many of these examples the Garner's aldehyde obtained from L-serine is an essential synthetic intermediate [84].

Aside from using β-hydroxy amino acids, a common approach to L-2,3-diaminopropionic acid involves either a Curtius rearrangement of free L-aspartic acid or a Hofmann reaction of an L-asparagine derivative [85,86]. The main side-products in these reactions are imidazolidinones [87]; however, a clever use of the reagents can provide orthogonally protected L-2,3-diaminopropionic acid [88]. Within this general approach, L-aspartic acid has been used as starting material for the preparation of a key intermediate for the synthesis of (+)-biotin [89]. Indeed, an enantiopure β-hydroxymethyl asparagine derivative **106**, prepared from *N*-Cbz-L-aspartic acid in five steps, was submitted to Hofmann rearrangement to afford cyclic urea **107** with no epimerization. Removal of the protecting group (Bom) by hydrogenation led to bicyclic γ-lactone **108** that was submitted to *N*-benzylation followed by treatment with KSAc to furnish thiolactone intermediate **109** (Scheme 9.15).

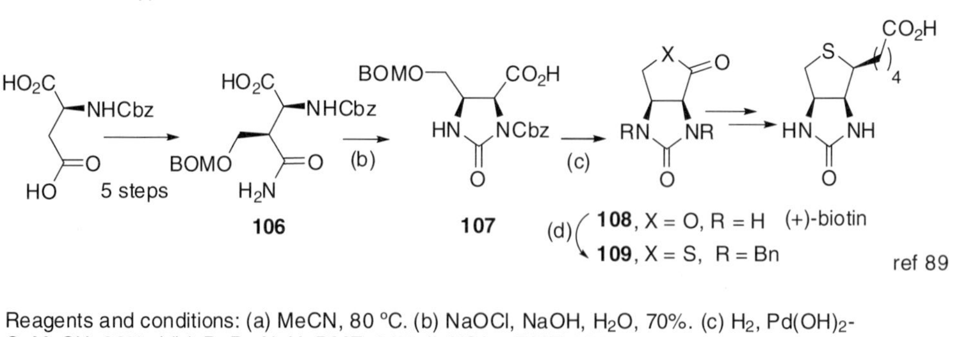

Reagents and conditions: (a) MeCN, 80 °C. (b) NaOCl, NaOH, H_2O, 70%. (c) H_2, $Pd(OH)_2$-C, MeOH, 80%. (d) i. BnBr, NaH, DMF, 84%; ii. KSAc, DMF, 92%.

Scheme 9.15 Serine and aspartic derivatives as starting materials.

9.3.2
From Allylic Alcohols and Amines

Optically pure allylic alcohols and amines are valuable precursors to α,β-diamino acids. The key step in these approaches is the internal delivery of a nitrogen atom from the group attached to the hydroxy or amino group to the unsaturated C–C bond [90]. Consequently, the degree of diastereoselectivity in the intramolecular transfer is crucial for efficient access to the target molecules. An approach to a taxol side-chain analog illustrates this methodology (Scheme 9.16) [91]. Enantiopure epoxy carbamate **110** was prepared from cinnamyl alcohol via Sharpless asymmetric epoxidation and was then treated with sodium hydride to produce an oxazolidinone **111** with complete inversion of stereochemistry at the epoxide carbon. Activation of the hydroxyl group as a methanesulfonate ester **112** and further displacement using potassium phthalimide with undesired retention of configuration, gave the *anti* oxazolidinone **113**. In contrast, when the mesylate was treated with sodium azide and chlorotrimethylsilane (TMSCl), the *syn* oxazolidinone **114** was produced exclusively that was transformed into the target *syn* α,β-diamino acid derivative in five steps. Alternatively, other approaches rely on sigmatropic rearrangements for the construction of the C–N bond [92]. Recently, a stereocontrolled route to *anti*- and *syn*-2,3-Dabs

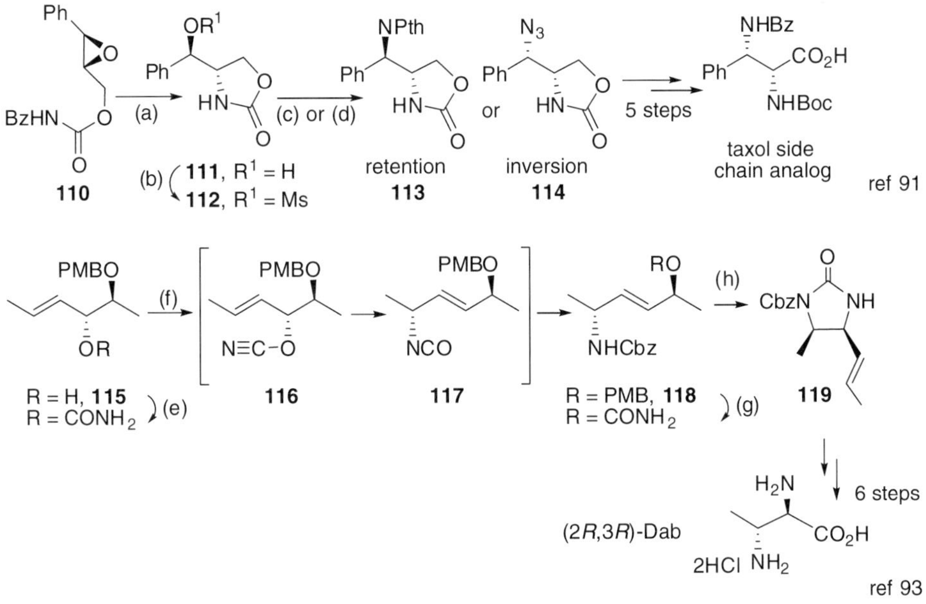

Reagents and conditions: (a) i. NaH, THF, reflux, 88%; ii. LiOH, H_2O, THF, rt, 85 %. (b) MsCl, NEt_3, 89 %. (c) NaN_3, TMSCl, DMF, 64%. (d) KPhthalimide, DMF, 79 %. (e) i. CCl_3CONCO, CH_2Cl_2. ii. K_2CO_3, MeOH. (f) i. PPh_3, CBr_4, NEt_3, CH_2Cl_2. ii. BnOH, Bu_3SnOBn, 93%. (g) i. DDQ, CH_2Cl_2, H_2O. ii. CCl_3CONCO, CH_2Cl_2. iii. K_2CO_3, MeOH. (h) PPh_3, CBr_4, Et_3N, CH_2Cl_2, 70-77 %.

Scheme 9.16 α,β-Diamino acids from allylic alcohols.

has been reported that entails two consecutive [3,3]-sigmatropic rearrangements with complete transfer of chirality from the C–O bonds in the starting material to the new C–N bonds. Thus, the starting *anti* allyl diol **115**, available in four steps from L-lactic acid methyl ester, was transformed into an allyl cyanate **116** that underwent rearrangement to an allyl isocyanate **117** captured with benzylic alcohol to provide a benzyloxy carbamate **118**. The remaining allyl alcohol is again transformed into a cyanate that suffered the second [3,3]-sigmatropic rearrangement to yield *cis*-4-propenyl imidazolidinone **119**. Finally, a six-step sequence that includes oxidative cleavage of the propenyl chain rendered *anti*-(2*R*,3*R*)-Dab. Through a similar sequence, the *syn* allyl diol provides enantiopure *syn*-(2*R*,3*S*)-Dab (Scheme 9.16) [93].

9.3.3
From Halo Alkanoates

One of the most efficient routes found among the earlier endeavors to synthesize these compounds consists of treating α,β-dibromo propionates **120** or succinates **121** with amines (Scheme 9.17). Following this approach *meso, dl*-2,3-diaminosuccinic acid derivatives [94] and 2-carboxypiperazines can be readily obtained [95]. However, when ammonia is used as the nucleophile, racemic methoxycarbonyl aziridines, readily resolved by enzymatic means, were produced. This protocol has been used for the synthesis of enantiopure 2,3-diaminopropionic acid [96] and of VLA-4 antagonists containing an α,β-diamino acid residue [97]. Besides, sequential introduction of the nitrogen nucleophiles has also been reported using ethyl β-bromo α-hydroxy propanoate as starting material [98]. Alternatively, a facile route for the synthesis of α,β-diamino acids has been reported by reaction of 2-methoxycarbonyl aziridine **122**, with chiral (+)-α-methylbenzylisocyanate **123** in the presence of sodium iodide to produce a mixture of diastereomeric imidazolidinones **124** (65 : 35), by means of initial iodide-mediated aziridine opening. After separation and acidic cleavage both enantiomers of 2,3-diaminopropionic acid are available. α-Alkyl-α,β-diamino acids are also available by this approach

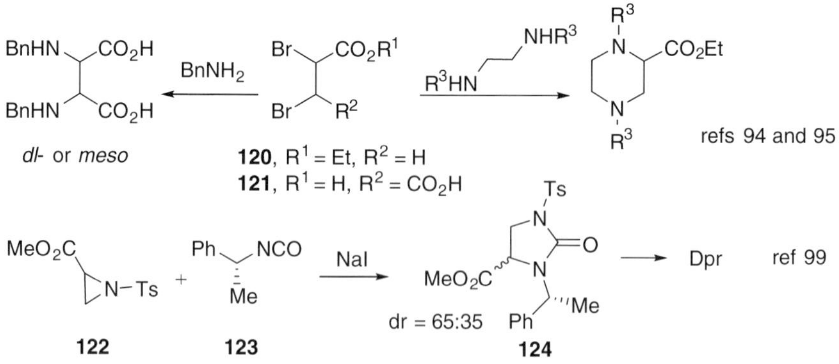

Scheme 9.17 α,β-Diamino acids from halo and aziridino alkanoates.

(Scheme 9.17) [99]. Similarly, the group of Ha has examined the TMSCl-mediated ring expansion of enantiopure aziridine-2-carboxylates with achiral isocyanates. In these reactions, imidazolidin-2-one-4-carboxylates are obtained regio- and stereospecifically in good yields [100].

9.3.4
From Alkenoates

Aside from the above examples, it could be easily envisioned that enantioselective diamination of α,β-unsaturated acids would be the shorter route to α,β-diamino acids; however, the direct asymmetric diamination of olefins is a reaction comparatively less studied than the related dihydroxylation process. In fact, the stoichiometric osmium mediated diamination of a chiral cinnamate **125** with bis(*tert*-butylimido) dioxoosmium(VIII) provides an osmaimidazolidinone **126** in excellent diastereomeric ratio (94 : 6). Subsequent reaction with *tert*-butylamine and $ZnCl_2$ provides an amide that was submitted to osmium removal with sodium borohydride furnishing an α,β-diamino amide **127** in good yield (Scheme 9.18). Furthermore, the synthesis of enantioenriched osmaimidazolidinones **129** (enatiomeric ratio > 90 : 10) from achiral *N*-alkenoyl oxazolidinones **128** has been accomplished using a 10% TiTADDOLate (TADDOL = 4,5-bis[hydroxy(diphenyl)methyl]-2,2-dimethyl-1,3-dioxolane)

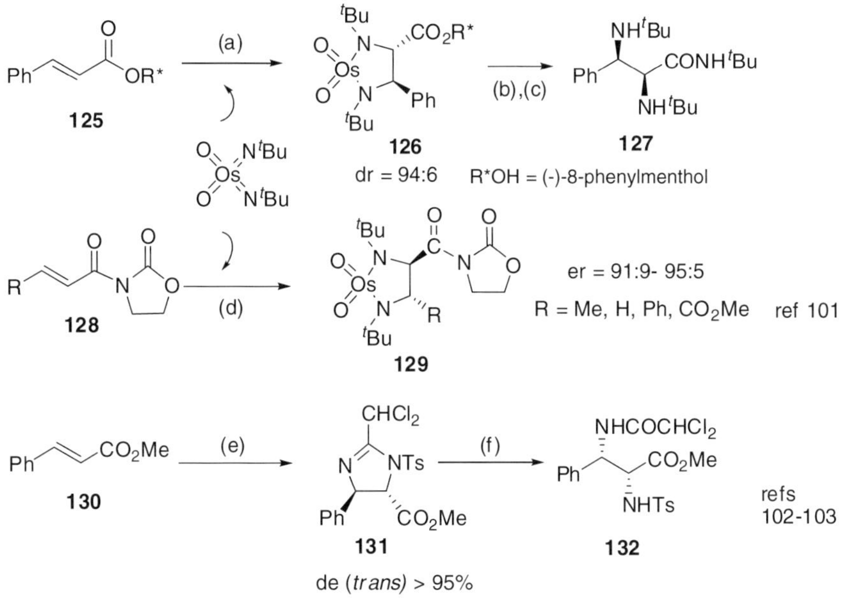

Reagents and conditions: (a) THF,-15 °C. (b) $^tBuNH_2$, $ZnCl_2$. (c) $NaBH_4$, EtOH, rt, 95%. (d) Ti-TADDOLate, Tol, 5 °C. (e) $TsNCl_2$, MeCN, $FeCl_3{\cdot}PPh_3$ (cat.), 63%. (f) 6 N HCl, 70 °C, 92%.

Scheme 9.18 Direct diamination of α,β-unsaturated esters and amides.

as enantioselective catalyst [101]. Within the same context, the group of Li has reported the first direct electrophilic diamination of α,β-unsaturated esters **130** using *N,N*-dichloro 2-nitrobenzenosulfonamide (2-$NsNCl_2$) in acetonitrile [102]. The process takes place with complete *anti* diastereoselectivity and the use of acetonitrile is crucial since a molecule of solvent is apparently responsible for delivering the second nitrogen atom to the final racemic *anti* α,β-diamino ester. Alternatively, in the presence of Rh(II), Fe(III), or Mn(IV) catalysts *trans* imidazolines **131** are produced that can be readily transformed into *syn* α,β-diamino esters **132** upon acidic hydrolysis (Scheme 9.18) [103].

Different groups have envisioned that α,β-diamino acid derivatives could be prepared efficiently by aminohydroxylation of α,β-unsaturated esters [104,105]. The group of Sharpless has outlined a short and highly versatile route to a series of α,β-diamino esters from an 87 : 13 mixture of regioisomeric *anti* amino hydroxy propanoates **133**, prepared from a *trans* glycidic ester and a secondary amine (Scheme 9.19). Subsequent mesylation of the mixture provided exclusively the β-chloroamino ester **134** in quantitative yield via *in situ* chloride attack to an aziridinium ion. Regeneration of the aziridinium ion followed by *in situ* reaction with a wide range of nitrogen nucleophiles took place with excellent yields and regioselectivities, furnishing racemic *anti* α,β-diamino esters **135** as major products. Furthermore, the starting mixture of *anti* amino hydroxy propanoates underwent a one-pot procedure, mesylation and nucleophilic attack of the amine, to give diamino esters **135** through the aziridinium ion in an extremely efficient manner [106].

A different approach takes advantage of the Michael acceptor character of α,β-unsaturated acid derivatives to nitrogen nucleophiles. In particular, α,β-dehydroalanine derivatives have resulted extremely useful starting materials containing the α-nitrogen [107,108]. In addition, the reaction of vinyl triflates derived from α-keto esters with secondary amines has been recently examined as an efficient route to α,β-diamino acid derivatives. However, the low stereoselectivies

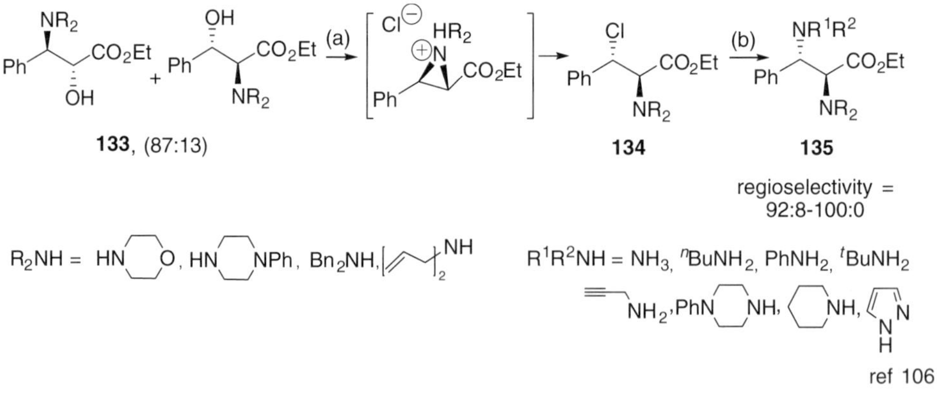

133, (87:13) **134** **135**

regioselectivity = 92:8-100:0

R_2NH = morpholine, N-phenylpiperazine, Bn_2NH, diallylamine

R^1R^2NH = NH_3, $^nBuNH_2$, $PhNH_2$, $^tBuNH_2$, propargylamine, N-phenylpiperazine, piperidine, pyrazole

ref 106

Reagents and conditions: (a) MsCl, Et_3N, CH_2Cl_2, 0 °C, 94-99%. (b) R^1R^2NH, K_2CO_3, MeCN, 72-99%.

Scheme 9.19 α,β-Diamino esters from aminohydroxy alkanoates.

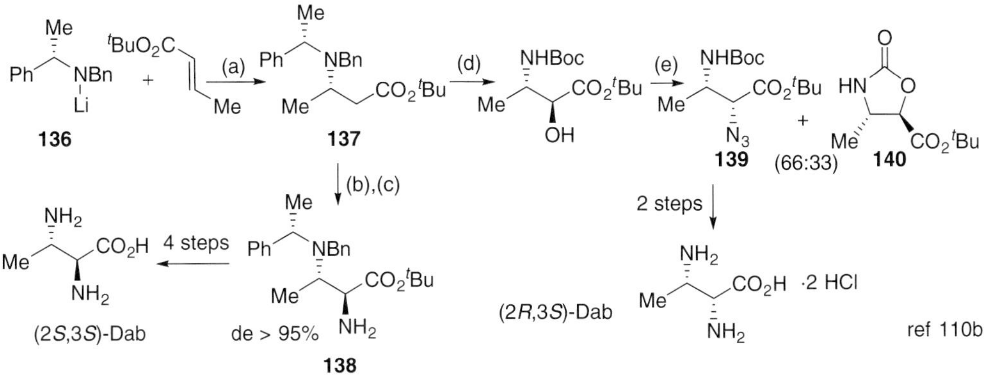

Reagents and conditons: (a) THF, −78 °C. (b) i. LDA, THF, −78 °C; ii. trisylazide, THF, −78 °C; iii. HOAc, THF, −78 °C to rt, 32%. (c) PPh_3, THF, H_2O, rt, 91%. (d) i. LDA, THF, −78 °C; ii. (−)-(camphorsulfonyl)oxaziridine. (e) i. MsCl, Et_3N, CH_2Cl_2, rt. ii. NaN_3, DMF, 55%.

Scheme 9.20 α,β-Diamino acids via Michael addition of chiral amines.

represent a severe drawback for this route [109]. Alternatively, when the starting α,β-unsaturated ester does not have a pre-existing nitrogen atom in the molecule, the synthesis of the α,β-diamino acid requires two steps, Michael addition of an amine to introduce the β-nitrogen and subsequent amination of an enolate to introduce the α-nitrogen. The groups of Seebach and Davies have independently examined the above methodology [110]. In particular, Davies showed that addition of chiral secondary lithium amide **136** was crucial to obtain β-amino esters **137** with complete diastereocontrol. Further diastereoselective trapping of the enolate with trisylazide (**138**, *anti*, >95% d.e.) followed by Staudinger reaction and hydrolysis of the iminophosphorane intermediate led after four more steps to enantiopure *anti* α,β-diamino acid, (2*S*,3*S*)-Dab. In contrast, after *anti* α-hydroxylation of the enolate using chiral (−)-(camphorsulfonyl)oxaziridine and displacement of the mesylate with sodium azide a 66 : 33 mixture of *syn* azide **139** and *trans* oxazolidinone **140** was obtained. After separation, hydrogenation and acidic deprotection, enantiopure *syn* α,β-diamino acid (2*R*,3*S*)-Dab was also available (Scheme 9.20) [110b].

9.3.5
Electrophilic Amination of Enolates and Related Processes

The diastereoselective electrophilic amination of β-amino enolates provides an efficient access to α,β-diamino acid derivatives of diverse structures. *N*-Sulfonyl azides have been used as electrophiles for the introduction of a nitrogen atom at C-3 of β-lactams within the synthesis of an analog of rhodopeptin B5 and also for the amination of piperidinyl acetic acid derivatives, in a study focused on the synthesis of analogs of streptolutin as enantiopure materials [111]. Furthermore, this approach has been successfully employed for the diastereocontrolled synthesis of 2,3-diamino

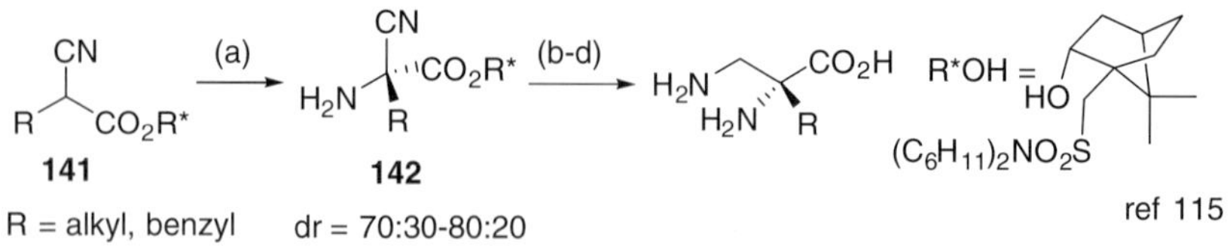

Reagents and conditions: (a) LiHMDS, $Ph_2P(O)ONH_2$, THF, –78 °C to rt, 60-70%. (b) separation. (c) H_2, Rh-Al_2O_3, MeOH-NH_3, 83%-quantitative. (d) KOH, MeOH, 90-95%.

Scheme 9.21 α,β-Diamino acids through electrophilic amination.

succinates [112] and for the synthesis of orthogonally protected units in the preparation of somatostatin analogs [113]. Alternatively, the electrophilic amination of β-amino acids can be performed with di-*tert*-butylazodicarboxylate as electrophile [114]. In addition, the group of Cativiela found that 2-cyano propanoates **141** containing an isobornyloxy group as a chiral auxiliary (R*) can be converted into 2-amino-2-cyano propanoates **142** with moderate to good diastereoselectivites upon treatment with lithium hexamethyldisilazane and *O*-(diphenylphosphinyl)hydroxylamine as the source of electrophilic nitrogen (Scheme 9.21). Separation and suitable manipulation of the functional groups resulted in α-substituted α,β-diamino acids [115].

Finally, a recent and interesting example was brought to light by the group of Wardrop in the total synthesis of (−)-dysibetaine (Scheme 9.22) [116]. The synthetic route commenced from an α,β-unsaturated ester **143** readily converted into an enantiopure α-silyloxy methoxylamide **144** in four steps that entail Sharpless asymmetric dihydroxylation and regioselective reduction of the β-hydroxyl group. Subsequent generation of an *N*-acylnitrenium ion using phenyliodine(III) bis(trifluoroacetate) promoted spirocyclization to afford an spirodienone **145** as an inseparable 9 : 1 mixture of C-5 epimers. The mixture was converted into an azido γ-lactam **146** through a sequence that entails ozonolysis and methanosulfonate displacement with azide. Further removal of the silyl group, hydrogenation of the azide, permethylation and ester hydrolysis finally furnished (−)-dysibetaine.

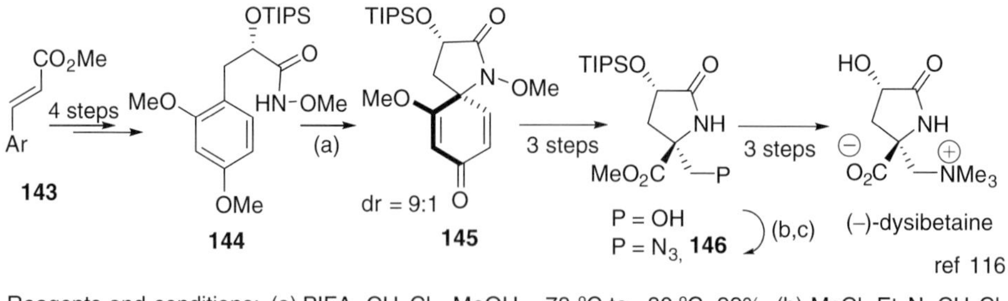

Reagents and conditions: (a) PIFA, CH_2Cl_2, MeOH, −78 °C to −30 °C, 99%. (b) MsCl, Et_3N, CH_2Cl_2, 0 °C, 87%. (c) NaN_3, DMF, 80 °C, 63%.

Scheme 9.22 Total synthesis of (−)-dysibetaine.

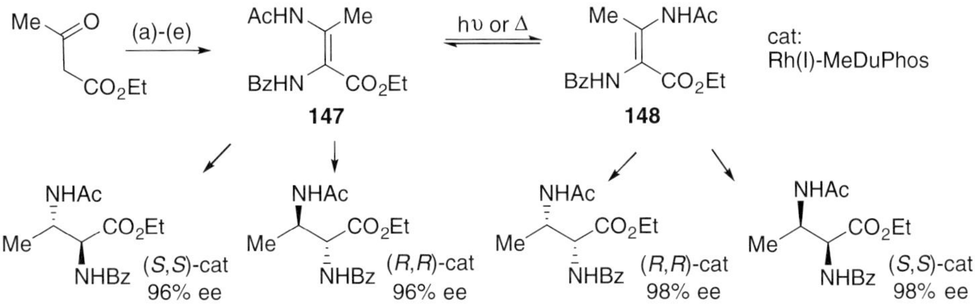

Reagents and conditions: (a) $NaNO_2$, aqueous HOAc, 0 °C, 81%. (b) Al(Hg), THF. (c) BzCl, 63% (2 steps). (d) NH_4OAc, MeOH, 81%. (e) AcCl, pyr, CH_2Cl_2-Et_2O.

ref 118

Scheme 9.23 α,β-Diamino esters by catalytic enantioselective hydrogenation.

9.3.6
From β-Keto Esters and Related Compounds

The use of β-keto esters as precursors to α,β-diamino acid derivatives is also documented in the literature. For instance, enantioenriched alkoxycarbonyl aziridines were prepared by an asymmetric Neber reaction to generate disubstituted azirines followed by reduction with sodium borohydride [117]. In another recent report, ethyl acetoacetate was transformed in five steps into (*E*) and (*Z*) α,β-unsaturated α,β-diamino esters (**147** and **148**) that can be thermally or photochemically interconverted. The key process in this report is the highly enantioselective hydrogenation using (*R,R*)- or (*S,S*)-MeDuPhos-Rh(I) triflate as chiral catalyst that provides independently all four isomers of α,β-diaminobutanoic acid derivatives in greater than 95% e.e. (Scheme 9.23) [118]. A bisferrocenyl phosphine has also been used as chiral ligand in the asymmetric hydrogenation of α,β-dehydro-α,β-diamino esters catalyzed by a Rh(I) complex [119]. Similarly, the catalytic enantioselective hydrogenation of 2-carboxamide tetrahydropyrazines using [(*R*)-BINAP(COD)Rh]OTf (COD = cyclooctadiene) afforded orthogonally protected carboxamide piperazines (99% e.e.), which are valuable intermediates in the synthesis of indinavir [120].

9.4
Conclusions

The development of short, general, and efficient synthetic routes to α,β-diamino acids is a focus of current chemical research. α,β-Diamino acids are becoming useful tools in many areas and the need for these molecules demands not only application of existing methodology, but discovery of new strategies that contribute to the advance of organic synthesis. This overview brings to light that in spite of the increasing number of approaches to these diamino acids, there are still many challenges. Future endeavors in this area will provide new routes and applications of these fascinating molecules.

9.5
Experimental Procedures

9.5.1
(S_S,2*R*,3*S*)-(+)-Ethyl-2-*N*-(diphenylmethyleneamino)-3-*N*-(*p*-toluenesulfinyl)-amino-3-phenylpropanoate (14b) [12a]

1. LDA
2. 11a
4 → 14b

In a 50-ml, one-necked, round-bottomed flask equipped with a magnetic stirring bar, rubber septum and argon balloon was placed *N*-(diphenylmethylene)glycine ethyl ester (**4**) (0.302 g, 1.13 mmol), in water-free THF (8 ml, purified through columns packed with activated alumina and supported copper catalyst). The solution was cooled to −78 °C and LDA (0.56 ml, 2.0 M in heptane/THF/ethylbenzene from Aldrich) was added dropwise. The red-brown solution was stirred at −78 °C; the color slowly changed to yellow within 40 min and the solution became slightly turbid. After stirring for 60 min, sulfinimine (*S*)-(+)-**11a** [121] (0.055 g, 0.23 mmol) in anhydrous THF (1.6 ml) at −78 °C was added dropwise via a cannula. The reaction mixture was stirred at −78 °C for 30 min, quenched by addition of saturated aqueous NH_4Cl (3 ml), stirred for 5 min, and warmed to room temperature. The phases were separated and the aqueous phase was extracted with EtOAc (3 × 5 ml). The combined organic phases were washed with brine (10 ml), dried (Na_2SO_4), and concentrated. Chromatography (20% EtOAc in hexanes) afforded 0.099 g (86%) of a clear oil (**14b**). $[\alpha]_D^{20} = +182$ (*c* 0.4, $CHCl_3$); infrared (IR) (neat): 3281, 3058, 1738, 1093 cm^{-1}; ^{1}H-nuclear magnetic resonance (NMR) ($CDCl_3$) δ 1.18 (t, *J* = 7.2 Hz, 3 H), 2.42 (s, 3 H), 4.13 (m, 2 H), 4.18 (d, *J* = 3.6 Hz, 1 H), 5.08 (m, 1 H), 5.75 (d, *J* = 7.2 Hz, 1 H), 6.50 (d, *J* = 7.2 Hz, 2 H), 7.20–7.60 (m, 17 H); ^{13}C-NMR δ 14.4, 21.7, 60.9, 61.8, 71.3, 126.2, 127.5, 127.8, 127.9, 128.4, 128.6, 128.6, 128.9, 129.3, 129.8, 131.1, 136.2, 139.1, 140.8, 141.5, 142.7, 169.8, 173.1. High-resolution mass spectroscopy (MS) calculated for $C_{31}H_{30}N_2O_3SLi$ (M + Li): 517.2137; found: 517.2146.

9.5.2
Synthesis of Ethyl (2*R*,3*R*)-3-amino-2-(4-methoxyphenyl)aminopentanoate 32a via Asymmetric aza-Henry Reaction [20]

30a + 23 → (cat (20 mol%), THF, −100 °C) → 31a → (H_2, RaNi, EtOH) → 32a

cat

Ethyl (2*R*,3*R*)-2-(4-methoxyphenyl)amino-3-nitropentanoate 31a To a flame-dried Schlenk tube was added $CuPF_6 \cdot 4MeCN$ (11.8 mg, 0.040 mmol, 20% mol) and 2,2′-methylenebis[(4*R*,5*S*)-4,5-diphenyl-2-oxazoline] (20.2 mg, 0.044 mmol, 22 mol %) under N_2. The mixture was stirred in vacuum for 1 h, then anhydrous THF (2.0 ml) was added and the resulting solution was stirred for 1 h. To this solution was added ethyl α-(4-methoxyphenyl)iminoacetate **23** (39 mg, 0.2 mmol, available from ethyl glyoxylate and *p*-methoxyaniline in CH_2Cl_2/4 Å MS/room temperature) [19]. The solution was cooled to −100 °C using an ether/liquid N_2 cooling bath. A solution of trimethylsilyl propanenitronate [122] **30a** (42 mg, 0.3 mmol) in THF (1.0 ml) was added over 1 h using a syringe pump. The reaction was left to warm to −20 °C overnight and quenched with EtOH (0.5 ml). The crude material was purified by flash chromatography using CH_2Cl_2/pentane 1 : 1 as eluent. The product was isolated in 68% yield with a diastereomeric ratio of 10 : 1 favoring the title compound. The enantiomeric excess was 97% for the major *erythro* isomer and 88% for the minor *threo* isomer detected by high-performance liquid chromatography using an OD column (hexane/*i*PrOH, 97 : 3). **31a**: $[\alpha]_D^{20} = -25.7$ (*c* 0.017, $CHCl_3$); ^{1}H-NMR ($CDCl_3$) δ 0.97 (dd, *J* = 7.2, 7.2 Hz, 3 H), 1.20 (dd, *J* = 7.2, 7.2 Hz, 3 H), 1.9–1.3 (bs, 1 H), 1.93 (ddq, *J* = 14.8, 7.2, 4.0 Hz, 1 H), 2.16 (ddq, *J* = 14.8, 10.0, 7.2 Hz, 1 H), 4.13 (dq, *J* = 11.2, 7.2 Hz, 1 H), 3.68 (s, 3 H), 4.18 (dq, *J* = 11.2, 7.2 Hz, 1 H), 4.41 (d, *J* = 5.6 Hz, 1 H), 4.65 (ddd, *J* = 10.0, 5.6, 4.0 Hz, 1 H), 6.58 (d, *J* = 8.8 Hz, 2 H), 6.72 (d, *J* = 8.8 Hz, 2 H); ^{13}C-NMR δ 10.6, 14.0, 23.1, 55.6, 60.7, 62.4, 90.3, 114.9, 116.2, 139.6, 153.7, 170.0; mass (time-of-flight electrospray +ve): *m*/*z* 319; high-resolution MS calculated for $C_{14}H_{20}N_2O_5Na$: 319.1270; found: 319.1263.

Ethyl (2*R*,3*R*)-3-amino-2-(4-methoxyphenyl)aminopentanoate 32a Ethyl (2*R*,3*R*)-2-(4-methoxyphenyl)amino-3-nitropentanoate **31a** (74 mg, 0.25 mmol) was dissolved in EtOH (3.6 ml) and Raney nickel (100 mg) was added. The reaction was treated with H_2 at 1 atm and left for 48 h. The catalyst was filtered off and the crude was purified by flash chromatography in 20% EtOAc/CH_2Cl_2 to yield the title compound 53.0 mg, 80% yield. **32a**: $[\alpha]_D^{20} = -23.0$ (*c* 0.026, $CHCl_3$); ^{1}H-NMR ($CDCl_3$) δ 1.02 (dd, *J* = 7.2, 7.2 Hz, 3 H), 1.23 (t, *J* = 7.2 Hz, 3 H), 1.35 (ddq, *J* = 14.0, 8.8, 7.2 Hz, 1 H); 1.55–1.35 (br, 3 H), 1.62 (ddq, *J* = 14.0, 7.2, 4.4 Hz, 1 H), 2.97 (ddd, *J* = 8.8, 4.4, 4.4 Hz, 1 H), 3.73 (s, 3 H), 3.97 (d, *J* = 4.4 Hz, 1 H), 4.16 (q, *J* = 7.2 Hz, 2 H); 6.65 (d, *J* = 8.8 Hz, 2 H), 676 (d, *J* = 8.8 Hz, 2 H); ^{13}C-NMR δ 11.2, 14.5, 28.1, 55.5, 55.9, 61.1, 62.6, 115.0, 115.6, 141.4, 153.0, 173.3; mass (time-of-flight electrospray, +ve): *m*/*z* 289; high-resolution MS calculated for $C_{14}H_{22}N_2O_3Na$: 289.1528; found: 289.1528.

References

1 Viso, A., Fernández de la Pradilla, R., García, A., and Flores, A. (2005) *Chemical Reviews*, **105**, 3167.

2 van Maanen, H.L., Kleijn, H., Jastrzebski, J.T.B.H., Verweij, J., Kieboom, A.P.G., and van Koten, G. (1995) *The Journal of Organic Chemistry*, **60**, 4331.

3 Kavrakova, I.K. and Lyapova, M.J. (2000) *Collection of Czechoslovak Chemical Communications*, **65**, 1580.

4 Shimizu, M., Inayoshi, K., and Sahara, T. (2005) *Organic and Biomolecular Chemistry*, **3**, 2237.

5 (a) Kiss, L., Mangelinckx, S., Sillanpää, R., Fülöp, F., and De Kimpe, N. (2007) *The Journal of Organic Chemistry*, **72**, 7199; (b) Kiss, L., Mangelinckx, S., Fülöp, F., and De Kimpe, N. (2007) *Organic Letters*, **9**, 4399.

6 Kobayashi, J., Yamashita, Y., and Kobayashi, S. (2005) *Chemistry Letters*, **34**, 268.

7 Soloshonok, V.A., Avilov, D.V., Kukhar, V.P., Van Meervelt, L., and Mischenko, N. (1997) *Tetrahedron Letters*, **38**, 4671.

8 DeMong, D.E. and Williams, R.M. (2003) *Journal of the American Chemical Society*, **125**, 8561.

9 Yadav, L.D.S., Rai, A., Rai, V.K., and Awasthi, C. (2007) *Synlett*, 1905.

10 Viso, A., Fernández de la Pradilla, R., García, A., Guerrero-Strachan, C., Alonso, M., Tortosa, M., Flores, A., Martínez-Ripoll, M., Fonseca, I., André, I., and Rodríguez, A. (2003) *Chemistry – A European Journal*, **9**, 2867.

11 Viso, A., Fernández de la Pradilla, R., López-Rodríguez, M.L., García, A., Flores, A., and Alonso, M. (2004) *The Journal of Organic Chemistry*, **69**, 1542.

12 (a) Davis, F.A., Zhang, Y., and Qiu, H. (2007) *Organic Letters*, **9**, 833; (b) Davis, F.A. and Deng, J. (2005) *Organic Letters*, **7**, 621; (c) Davis, F.A., Zhang, Y., and Li, D. (2007) *Tetrahedron Letters*, **48**, 7838.

13 Polniaszek, R.P. and Bell, S.J. (1996) *Tetrahedron Letters*, **37**, 575.

14 Bernardi, L., Gothelf, A.S., Hazell, R.G., and Jørgensen, K.A. (2003) *The Journal of Organic Chemistry*, **68**, 2583.

15 Cutting, G.A., Stainforth, N.E., John, M.P., Kociok-Köhn, G., and Willis, M.C. (2007) *Journal of the American Chemical Society*, **129**, 10632.

16 Salter, M.M., Kobayashi, J., Shimizu, Y., and Kobayashi, S. (2006) *Organic Letters*, **8**, 3533.

17 Ooi, T., Kameda, M., Fujii, J.-I., and Maruoka, K. (2004) *Organic Letters*, **6**, 2397.

18 Okada, A., Shibuguchi, T., Ohshima, T., Masu, H., Yamaguchi, K., and Shibasaki, M. (2005) *Angewandte Chemie (International Edition in English)*, **44**, 4564.

19 Nishiwaki, N., Knudsen, K.R., Gothelf, K.V., and Jørgensen, K.A. (2001) *Angewandte Chemie (International Edition in English)*, **40**, 2992.

20 Knudsen, K.R., Risgaard, T., Nishiwaki, N., Gothelf, K.V., and Jørgensen, K.A. (2001) *Journal of the American Chemical Society*, **123**, 5843.

21 Singh, A., Yoder, R.A., Shen, B., and Johnston, J.N. (2007) *Journal of the American Chemical Society*, **129**, 3466.

22 Hartwig, W. and Mittendorf, J. (1991) *Synthesis*, 939.

23 Gilbert, I.H., Rees, D.C., Crockett, A.K., and Jones, R.C.F. (1995) *Tetrahedron*, **51**, 6315.

24 Bycroft, B.W., Chhabra, S.R., Grout, R.J., and Crowley, P.J. (1984) *Journal of the Chemical Society, Chemical Communications*, 1156.

25 Shiba, T., Ukita, T., Mizuno, K., Teshima, T., and Wakamiya, T. (1977) *Tetrahedron Letters*, 2681.

26 Herdeis, C. and Nagel, U. (1983) *Heterocycles*, **20**, 2163.

27 Robinson, A.J., Lim, C.Y., He, L., Ma, P., and Li, H.-Y. (2001) *The Journal of Organic Chemistry*, **66**, 4141.

28 Vartak, A.P., Young, V.G., and Johnson, R.L. (2005) *Organic Letters*, **7**, 35.

29 Ozaki, Y., Iwasaki, T., Miyoshi, M., and Matsumoto, K. (1979) *The Journal of Organic Chemistry*, **44**, 1714.

30 Belokon, Y.N., Chernoglazova, N.I., Batsanov, A.S., Garbalinskaya, N.S., Bakhmutov, V.I., Struchkov, Y.T., and Belikov, V.M. (1987) *Bulletin of the Academy of Sciences of the USSR, Division of Chemical Sciences*, **36**, 779.

31 Álvarez-Ibarra, C., Csákÿ, A.G., Colmenero, B., and Quiroga, M.L. (1997) *The Journal of Organic Chemistry*, **62**, 2478.

32 Zeitler, K. and Steglich, W. (2004) *The Journal of Organic Chemistry*, **69**, 6134.

33 Chen, Y. and Yudin, A.K. (2003) *Tetrahedron Letters*, **44**, 4865.

34 Rinaudo, G., Narizuka, S., Askari, N., Crousse, B., and Bonnet-Delpon, D. (2006) *Tetrahedron Letters*, **47**, 2065.

35 Casarrubios, L., Pérez, J.A., Brookhart, M., and Templeton, J.L. (1996) *The Journal of Organic Chemistry*, **61**, 8358.

36 Hansen, K.B., Finney, N.S., and Jacobsen, E.N. (1995) *Angewandte Chemie (International Edition in English)*, **34**, 676.

37 Aggarwal, V.K., Thompson, A., Jones, R.V.H., and Standen, M.C.H. (1996) *The Journal of Organic Chemistry*, **61**, 8368.

38 Antilla, J.C. and Wulff, W.D. (1999) *Journal of the American Chemical Society*, **121**, 5099.

39 Aires-de-Sousa, J., Lobo, A.M., and Prabhakar, S. (1996) *Tetrahedron Letters*, **37**, 3183.

40 (a) Ha, H.-J., Suh, J.-M., Kang, K.-H., Ahn, Y.-G., and Han, O. (1998) *Tetrahedron*, **54**, 851; (b) Lee, K.-D., Suh, J.-M., Park, J.-H., Ha, H.-J., Choi, H.-G., Park, C.S., Chang, J.W., Lee, W.K., Dong, Y., and Yun, H. (2001) *Tetrahedron*, **57**, 8267.

41 (a) Ojima, I. and Delaloge, F. (1997) *Chemical Society Reviews*, **26**, 377; (b) Palomo, C., Aizpurua, J.M., Ganboa, I., and Oiarbide, M. (2001) *Synlett*, 1813.

42 (a) Endo, M. and Droghini, R. (1993) *Bioorganic & Medicinal Chemistry Letters*, **3**, 2483; (b) Moyna, G., Williams, H.J., and Scott, A.I. (1997) *Synthetic Communications*, **27**, 1561.

43 Fukuyama, T., Frank, R.K., and Jewell, C.F. Jr. (1980) *Journal of the American Chemical Society*, **102**, 2122.

44 (a) Evans, D.A. and Sjogren, E.B. (1985) *Tetrahedron Letters*, **26**, 3783; (b) Ojima, I. and Pei, Y. (1990) *Tetrahedron Letters*, **31**, 977.

45 Ojima, I. (1995) *Accounts of Chemical Research*, **28**, 383.

46 Palomo, C., Aizpurua, J.M., Galarza, R., and Mielgo, A. (1996) *Journal of the Chemical Society, Chemical Communications*, 633.

47 Brown, D., Brown, G.A., Andrews, M., Large, J.M., Urban, D., Butts, C.P., Hales, N.J., and Gallagher, T. (2002) *Journal of the Chemical Society, Perkin Transactions 1*, 2014.

48 Dghaym, R.D., Dhawan, R., and Arndtsen, B.A. (2001) *Angewandte Chemie (International Edition in English)*, **40**, 3228.

49 Alker, D., Harwood, L.M., and Williams, C.E. (1998) *Tetrahedron Letters*, **39**, 475.

50 Hayashi, T., Kishi, E., Soloshonok, V.A., and Uozumi, Y. (1996) *Tetrahedron Letters*, **37**, 4969.

51 Lin, Y.-R., Zhou, X.-T., Dai, L.-X., and Sun, J. (1997) *The Journal of Organic Chemistry*, **62**, 1799.

52 Zhou, X.-T., Lin, Y.-R., and Dai, L.-X. (1999) *Tetrahedron: Asymmetry*, **10**, 855.

53 Foresti, E., Palmieri, G., Petrini, M., and Profeta, R. (2003) *Organic and Biomolecular Chemistry*, **1**, 4275–4281.

54 Ambroise, L., Dumez, E., Szeki, A., and Jackson, R.F.W. (2002) *Synthesis*, 2296.

55 Merino, P., Pádár, P., Delso, I., Thirumalaikumar, M., Tejero, T., and Kovács, L. (2006) *Tetrahedron Letters*, **47**, 5013.

56 Merino, P., Franco, S., Merchan, F.L., and Tejero, T. (1998) *The Journal of Organic Chemistry*, **63**, 5627.

57 Obrecht, D., Karajiannis, H., Lehmann, C., Schönholzer, P., Spiegler, C., and Müller, K. (1995) *Helvetica Chimica Acta*, **78**, 703.

58 Levins, C.G. and Schafmeister, C.E. (2003) *Journal of the American Chemical Society*, **125**, 4702.

59 Gupta, S., Das, B.C., and Schafmeister, C.E. (2005) *Organic Letters*, **7**, 2861.

60 Greenlee, W.J., Allibone, P.L., Perlow, D.S., Patchett, A.A., Ulm, E.H., and Vassil, T.C. (1985) *Journal of Medicinal Chemistry*, **28**, 434.

61 Mohan, R., Chou, Y.L., Bihovsky, R., Lumma, W.C., Erhardt, P.W., and Shaw, K.J. (1991) *Journal of Medicinal Chemistry*, **34**, 2402.

62 Fondekar, K.P.P., Volk, F.-J., Khaliq-uz-Zaman, S.M., Bisel, P., and Frahm, A.W. (2002) *Tetrahedron: Asymmetry*, **13**, 2241.

63 Li, B.-F., Yuan, K., Zhang, M.-J., Wu, H., Dai, L.-X., Wang, Q.-R., and Hou, X.-L. (2003) *The Journal of Organic Chemistry*, **68**, 6264.

64 Wang, D., Zhang, P.-F., and Yu, B. (2007) *Helvetica Chimica Acta*, **90**, 938.

65 Langlois, N. and Le Nguyen, B.K. (2004) *The Journal of Organic Chemistry*, **69**, 7558.

66 Seo, R., Ishizuka, T., Abdel-Aziz, A.A.-M., and Kunieda, T. (2001) *Tetrahedron Letters*, **42**, 6353.

67 Rossen, K., Pye, P.J., DiMichele, L.M., Volante, R.P., and Reider, P.J. (1998) *Tetrahedron Letters*, **39**, 6823.

68 (a) Pfammatter, E. and Seebach, D. (1991) *Liebigs Annalen der Chemie*, 1323; (b) Schanen, V., Cherrier, M.-P., de Melo, S.J., Quirion, J.-C., and Husson, H.-P. (1996) *Synthesis*, 833.

69 Palomo, C., Aizpurua, T.M., Ganboa, I., Carreaux, F., Cuevas, C., Maneiro, E., and Ontoria, J.M. (1994) *The Journal of Organic Chemistry*, **59**, 3123.

70 (a) Castellanos, E., Reyes-Rangel, G., and Juaristi, E. (2004) *Helvetica Chimica Acta*, **87**, 1016; (b) Dunn, P.J., Häner, R., and Rapoport, H. (1990) *The Journal of Organic Chemistry*, **55**, 5017.

71 Merino, P., Lanaspa, A., Merchán, F.L., and Tejero, T. (1998) *Tetrahedron: Asymmetry*, **9**, 629.

72 Chattopadhyay, S.K., Sarkar, K., Thander, L., and Roy, S.P. (2007) *Tetrahedron Letters*, **48**, 6113.

73 Snider, B.B. and Gu, Y. (2001) *Organic Letters*, **3**, 1761.

74 Kelleher, F. and Proinsias, K.O. (2007) *Tetrahedron Letters*, **48**, 4879.

75 (a) Luo, Y., Blaskovich, M.A., and Lajoie, G.A. (1999) *The Journal of Organic Chemistry*, **64**, 6106; (b) Cherney, R.J. and Wang, L. (1996) *The Journal of Organic Chemistry*, **61**, 2544.

76 Arnold, L.D., May, R.G., and Vederas, J.C. (1988) *Journal of the American Chemical Society*, **110**, 2237.

77 Jane, D.E., Hoo, K., Kamboj, R., Deverill, M., Bleakman, D., and Mandelzys, A. (1997) *Journal of Medicinal Chemistry*, **40**, 3645.

78 Warshawsky, A.M., Patel, M.V., and Chen, T.-M. (1997) *The Journal of Organic Chemistry*, **62**, 6439.

79 Atherton, E. and Meienhofer, J. (1972) *Journal of Antibiotics*, **25**, 539.

80 Nakamura, Y., Hirai, M., Tamotsu, K., Yonezawa, Y., and Shin, C.-G. (1995) *Bulletin of the Chemical Society of Japan*, **68**, 1369.

81 Eaton, C.N., Denny, G.H. Jr., Ryder, M.A., Ly, M.G., and Babson, R.D. (1973) *Journal of Medicinal Chemistry*, **16**, 289.

82 Couturier, C., Blanchet, J., Schlama, T., and Zhu, J. (2006) *Organic Letters*, **8**, 2183.

83 Falorni, M., Porcheddu, A., and Giacomelli, G. (1997) *Tetrahedron: Asymmetry*, **8**, 1633.

84 (a) Guibourdenche, C., Roumestant, M.L., and Viallefont, P. (1993) *Tetrahedron: Asymmetry*, **4**, 2041; (b) Chhabra, S.R., Mahajan, A., and Chan, W.C. (2002) *The Journal of Organic Chemistry*, **67**, 4017.

85 (a) Rao, S.L.N. (1975) *Biochemistry*, **14**, 5218; (b) Belov, V.N., Brands, M., Raddatz, S., Krüger, J., Nikolskaya, S., Sokolov, V., and de Meijere, A. (2004) *Tetrahedron*, **60**, 7579; (c) Otsuka, M., Kittaka, A., Iimori, T., Yamashita, H., Kobayashi, S., and Ohno, M. (1985) *Chemical & Pharmaceutical Bulletin*, **33**, 509.

86 Zhang, L.-H., Kauffman, G.S., Pesti, J.A., and Yin, J. (1997) *The Journal of Organic Chemistry*, **62**, 6918.

87 Olsen, R.K., Hennen, W.J., and Wardle, R.B. (1982) *The Journal of Organic Chemistry*, **47**, 4605.

88 (a) Englund, E.A., Gopi, H.N., and Appella, D.H. (2004) *Organic Letters*, **6**, 213; (b) Schirlin, D. and Altenburger,

J.-M. (1995) *Synthesis*, 1351; (c) Rao, R.V.R., Tantry, S.J., and Suresh Babu, V.V. (2006) *Synthetic Communications*, **26**, 2901.

89 Seki, M., Shimizu, T., and Inubushi, K. (2002) *Synthesis*, 361.

90 Cardillo, G., Orena, M., Penna, M., Sandri, S., and Tomasini, C. (1991) *Tetrahedron*, **47**, 2263.

91 (a) Rossi, F.M., Powers, E.T., Yoon, R., Rosenberg, L., and Meinwald, J. (1996) *Tetrahedron*, **52**, 10279; (b) Knapp, S., Kukkola, P.J., Sharma, S., Dhar, T.G.M., and Naughton, A.B.J. (1990) *The Journal of Organic Chemistry*, **55**, 5700.

92 Gonda, J., Helland, A.-C., Ernst, B., and Bellus, D. (1993) *Synthesis*, 729.

93 Ichikawa, Y., Egawa, H., Ito, T., Isobe, M., Nakano, K., and Kotsuki, H. (2006) *Organic Letters*, **8**, 5737.

94 McKennis, H. Jr. and Yard, A.S. (1958) *The Journal of Organic Chemistry*, **23**, 980.

95 Demaine, D.A., Smith, S., and Barraclough, P. (1992) *Synthesis*, 1065.

96 Davoli, P., Forni, A., Moretti, I., and Prati, F. (1995) *Tetrahedron: Asymmetry*, **6**, 2011.

97 Astles, P.C., Harris, N.V., and Morley, A.D. (2001) *Bioorganic and Medicinal Chemistry*, **9**, 2195.

98 McCort, G.A. and Pascal, J.C. (1992) *Tetrahedron Letters*, **33**, 4443.

99 Nadir, U.K., Krishna, R.V., and Singh, A. (2005) *Tetrahedron Letters*, **46**, 479.

100 Kim, M.S., Kim, Y.-W., Hahm, H.S., Jang, J.W., Lee, W.K., and Ha, H.-J. (2005) *Journal of the Chemical Society, Chemical Communications*, 3062.

101 (a) Muñiz, K. and Nieger, M. (2003) *Synlett*, 211; (b) Muñiz, K. and Nieger, M. (2005) *Journal of the Chemical Society, Chemical Communications*, 2729.

102 Li, G., Kim, S.H., and Wei, H.-X. (2000) *Tetrahedron Letters*, **41**, 8699.

103 (a) Wei, H.-X., Kim, S.H., and Li, G. (2002) *The Journal of Organic Chemistry*, **67**, 4777; (b) Timmons, C., Mcpherson, L.M., Chen, D., Wei, H.-X., and Li, G. (2005) *Journal of Peptide Research*, **66**, 249.

104 (a) Han, H., Yoon, J., and Janda, K.D. (1998) *The Journal of Organic Chemistry*, **63**, 2045; (b) Hennings, D.D. and Williams, R.M. (2000) *Synthesis*, 1310.

105 Lee, S.-H., Yoon, J., Chung, S.-H., and Lee, Y.-S. (2001) *Tetrahedron*, **57**, 2139.

106 Chuang, T.-H. and Sharpless, K.B. (2000) *Organic Letters*, **2**, 3555.

107 Ferreira, P.M.T., Maia, H.L.S., and Monteiro, L.S. (1999) *Tetrahedron Letters*, **40**, 4099.

108 (a) Sagiyan, A.S., Avetisyan, A.E., Djamgaryan, S.M., Djilavyan, L.R., Gyulumyan, E.A., Grigoryan, S.K., Kuz'mina, N.A., Orlova, S.A., Ikonnikov, N.S., Larichev, V.S., Tararov, V.I., and Belokon, Y.N. (1997) *Russian Chemical Bulletin*, **46**, 483; (b) Belov, V.N., Funke, C., Labahn, T., Es-Sayed, M., and Meijere, A. (1999) *European Journal of Organic Chemistry*, 1345.

109 Tranchant, M.-J. and Dalla, V. (2006) *Tetrahedron*, **62**, 10255.

110 (a) Estermann, H. and Seebach, D. (1988) *Helvetica Chimica Acta*, **71**, 1824; (b) Bunnage, M.E., Burke, A.J., Davies, S.G., Millican, N.L., Nicholson, R.L., Roberts, P.M., and Smith, A.D. (2003) *Organic and Biomolecular Chemistry*, **1**, 3708.

111 Durham, T.B. and Miller, M.J. (2003) *The Journal of Organic Chemistry*, **68**, 35.

112 Fernández-Megía, E. and Sardina, F.J. (1997) *Tetrahedron Letters*, **38**, 673.

113 Riemer, C., Bayer, T., Schmitt, H., and Kessler, H. (2004) *Journal of Peptide Research*, **63**, 196.

114 Capone, S., Guaragna, A., Palumbo, G., and Pedatella, S. (2005) *Tetrahedron*, **61**, 6575.

115 Badorrey, R., Cativiela, C., Díaz-de-Villegas, M.D., and Gálvez, J.A. (1995) *Tetrahedron: Asymmetry*, **6**, 2787.

116 Wardrop, D.J. and Burge, M.S. (2004) *Journal of the Chemical Society, Chemical Communications*, 1230.

117 Verstappen, M.M.H., Ariaans, G.J.A., and Zwanenburg, B. (1996) *Journal of the American Chemical Society*, **118**, 8491.

118 Robinson, A.J., Stanislawski, P., Mulholland, D., He, L., and Li, H.-Y. (2001) *The Journal of Organic Chemistry*, **66**, 4148.

119 Kuwano, R., Okuda, S., and Ito, Y. (1998) *Tetrahedron: Asymmetry*, **9**, 2773.

120 (a) Kukula, P. and Prins, R. (2002) *Journal of Catalysis*, **208**, 404; (b) Rossen, K., Weissman, S.A., Sager, J., Reamer, R.A., Askin, D., Volante, R.P., and Reider, P.J. (1995) *Tetrahedron Letters*, **36**, 6419.

121 Davis, F.A., Zhang, Y., Andemichael, Y., Fang, T., Fanelli, D.L., and Zhang, H. (1999) *The Journal of Organic Chemistry*, **64**, 1403.

122 (a) Torssell, K.B.G. and Zeuthen, O. (1978) *Acta Chemica Scandinavica*, **B32**, 118; (b) Colvin, E.W., Beck, A.K., Bastani, B., Seebach, D., Kaj, Y., and Dunitz, J.D. (1980) *Helvetica Chimica Acta*, **63**, 697.

10
Synthesis of Halogenated α-Amino Acids

Madeleine Strickland and Christine L. Willis

10.1
Introduction

Biological halogenation occurs on a wide range of organic scaffolds including polyketides, terpenes, and nonribosomal peptides, and more than 4500 halogenated natural products have been reported [1]. Furthermore, halogenated analogs of natural products often exhibit significant differences in their bioactivity and bioavailability compared with the parent compound [2]. Thus, it is unsurprising that halogenated amino acids have found various applications in medicine, as intermediates in organic synthesis, and are of value as probes to investigate biological mechanisms. For example, not only is ^{19}F active in nuclear magnetic resonance, the strong carbon–fluorine bond is particularly resistant to metabolic transformations and hence fluorinated compounds are widely used as tracers, and for studies on the structure and properties of enzymes. Furthermore, positron emission topography is proving of particular value in the diagnosis of disease and there is a burgeoning interest in the synthesis of compounds incorporating the short-lived isotope ^{18}F such as [^{18}F]L-3,4-dihydroxyphenylalanine (L-DOPA) [3]. Hence, fluorinated amino acids have been a focus of intense synthetic activity and a number of comprehensive reviews have been published [4]. It is not possible to describe the entire gamut of synthetic approaches to halogenated amino acids in this chapter and, since fluorinated amino acids have been reviewed, we have described only recently published methods for their synthesis, and mainly focused on the preparation and uses of chlorinated, brominated and iodinated α-amino acids. These targets have a variety of applications (e.g., 3,5-dibromotyrosine has shown activity against brain ischemia and hence is useful in the fight against schizophrenia [5], 5-bromotryptophan is active against the gelation of sickle cells [6], and 2-fluoro- and 2-iodohistidines inhibit the growth of chloroquine-resistant *Plasmodium falciparum* – a cause of malaria [7]). We have arranged the amino acids into three groups based on their side-chains: (i) aliphatic hydrocarbon, (ii) aromatic, and (iii) aliphatic side-chains that include heteroatoms. A review on the resolution of amino acids using biotransformation methods is a further valuable source of background information [8].

Amino Acids, Peptides and Proteins in Organic Chemistry. Vol.1 – Origins and Synthesis of Amino Acids.
Edited by Andrew B. Hughes

ISBN: 978-3-527-32096-7

10.2
Halogenated Amino Acids with a Hydrocarbon Side-Chain

10.2.1
Halogenated Alanines and Prolines

Haloalanines are readily prepared and are important building blocks in organic synthesis, and indeed L-chloroalanine is commercially available. The use of biotransformations for its conversion into further α-amino acids has been reviewed [9]. If a synthesis of L-3-chloroalanine is required, a valuable method involves reductive amination of 3-chloropyruvate **1** using the commercially available enzymes alanine, leucine, or phenylalanine dehydrogenase (Scheme 10.1). A second enzyme, formate dehydrogenase is used to recycle the cofactor NADH giving L-3-chloroalanine in greater than 90% yield [10].

A general method for the preparation of 3-chloroalanine derivatives is from the corresponding serine analog using, for example, PCl_5, Ph_3P/CCl_4, or chlorotrimethylsilane (TMSCl) ([11] and references cited therein). Lowpetch and Young have described the synthesis of D-3-chloroalanine using an approach which enables the selective incorporation of deuterium at C-3 (Scheme 10.2) [12]. Amino hydroxylation of methyl acrylate using 1,3-phthalazinediyl-bisdihydroquinidine [$(DHQD)_2PHAL$] as the ligand, gave (2*S*)-*N*-benzyloxycarbonylisoserinate **2** in 60% yield and 86% e.e. Hydroxy ester **2** was then converted in to the D-*N*-tritylaziridine **3** via hydrogenolysis, protection of the resultant amino ester as the *N*-trityl derivative followed by cyclization using mesyl chloride and triethylamine. The trityl group was replaced by the more electron-withdrawing Cbz group prior to reaction with $TiCl_4$ and finally an acid-mediated deprotection gave D-3-chloroalanine. A similar approach but using the ligand 1,3-phthalazinediyl-bisdihydroquinine [$(DHQ)_2PHAL$] in the amino hydroxylation step gave L-3-chloroalanine. A further approach to the synthesis of D-3-chloroalanine

Amino acid dehydrogenases
1
NADH
NAD^+
L-3-chloroalanine > 90% yield
CO_2
HCO_3^-
FDH

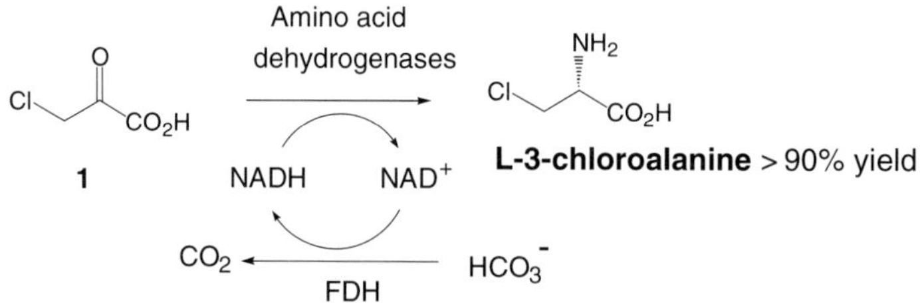

Scheme 10.1 Synthesis of L-3-chloroalanine.

$CBzNH_2$, NaOH, tBuOCl
$(DHQD)_2PHAL$, $K_2OsO_2(OH)_4$, rt
2
60% yield
3
D-3-chloroalanine

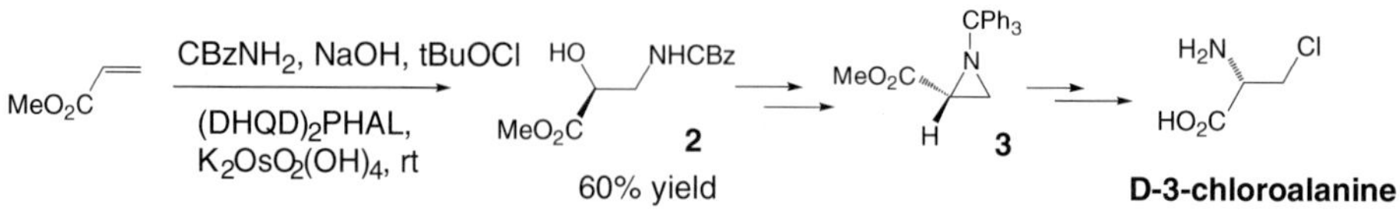

Scheme 10.2 Synthesis of D-3-chloroalanine.

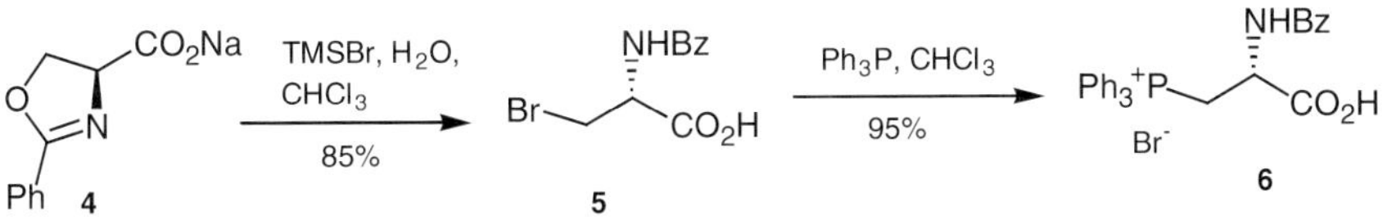

Scheme 10.3 Synthesis of protected L-3-bromoalanine.

selectively labeled with deuterium utilized [3-^{2}H]malic acid in the synthesis of an intermediate protected aziridine which was ring opened with $TiCl_4$ in $CHCl_3$/CH_2Cl_2 [13].

Turning to 3-bromoalanine derivatives, these have proved valuable in synthesis as electrophilic intermediates, radical precursors and in the preparation of phosphonium salts. For example, oxazoline **4** was prepared via condensation of L-serine with phenyl imino ether, PhC(=NH)OMe. Ring opening of **4** with bromotrimethylsilane yielded 3-bromo acid **5**, which in turn was readily converted to phosphonium salt **6** (Scheme 10.3). Using TMSCl or iodotrimethylsilane in the ring-opening step gave the analogous 3-chloro- and 3-iodo-protected amino acids in 75 and 90% yield, respectively [14].

Protected 3-bromoalanine has been used as an intermediate in the synthesis of (2*S*)-4-exomethylene proline **10** (Scheme 10.4) [15]. First, protected serine **7** was *N*-alkylated with propargyl bromide prior to conversion of tetrahydopyranyl (THP) ether **8** directly to bromide **9** with Ph_3P/Br_2. Radical cyclization followed by deprotection gave the target **10**. More recently protected 3-bromoalanine derivative **12** was prepared in excellent yield from the parent alcohol **11** using PPh_3/CBr_4 and radical cyclization gave a series of protected 4-substituted prolines **13** in good yield and variable stereocontrol [16].

Barton reported the conversion of protected aspartate **14** to *N*-Cbz L-3-bromoalanine benzyl ester **16** via halodecarboxylation of ester **15** derived from *N*-hydroxythiopyridine (Scheme 10.5) [17]. Irradiation of **15** using bromotrichloromethane as the source of bromide cleanly gave protected 3-bromoalanine **16**.

Variously protected 3-iodoalanines are readily prepared from serine via displacement of a 3-tosylate with sodium iodide [18] and Jackson *et al.* have utilized these products to good effect in organic synthesis via conversion of the iodide to an organozinc reagent ([19] and references cited therein). For example, Fmoc-L-3-chloroalanine *tert*-butyl ester **17** was treated with sodium iodide in acetone to give protected 3-iodoalanine **18** in excellent yield (Scheme 10.6). Formation of the organozinc reagent **19** followed by a palladium-mediated cross-coupling with aryl iodides gave a series of functionalized L-phenylalanines including bromide **20**. The conversion of the *tert*-butyl esters to the corresponding acids was achieved in 63–95% yield using TFA/Et_3SiH.

In 2007, Li and van der Donk described the synthesis of *N*-Fmoc-β,β-difluoroalanine **25** required for studies on lantibiotic synthetases [20]. First, (diethylamino) sulfur trifluoride (DAST) was used to effect difluorination of protected glyceraldehyde **21** (Scheme 10.7). Protecting group manipulation of **22** gave alcohol **23** which was activated for nucleophilic displacement with azide then reduction and Fmoc

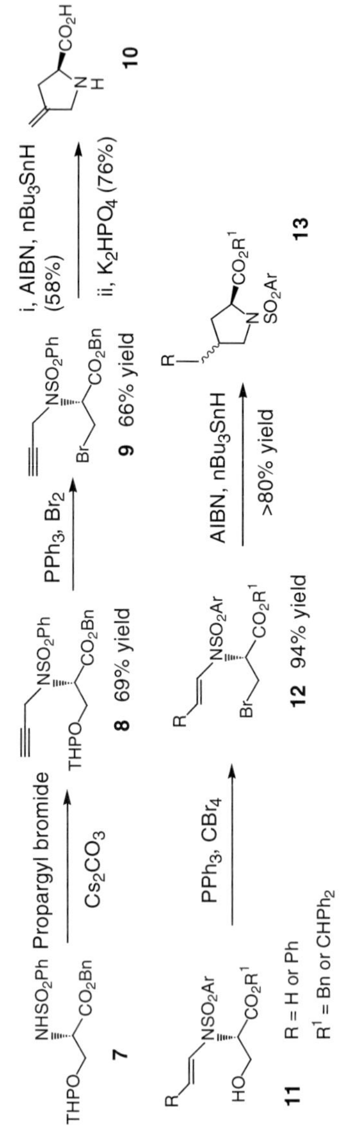

Scheme 10.4 Synthesis of 4-substituted prolines via protected bromoalanines.

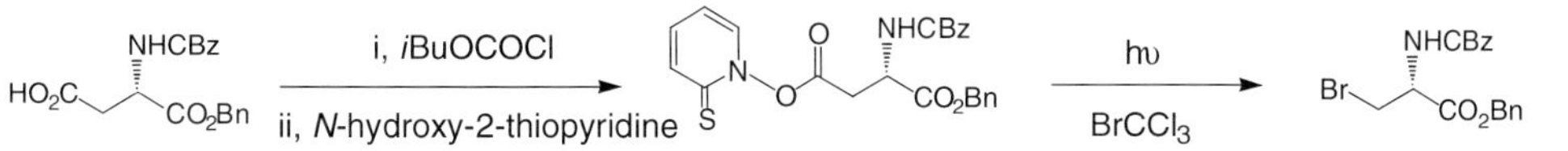

Scheme 10.5 Synthesis of protected 3-bromoalanine.

protection gave difluoride **24**. Finally, deprotection of the primary alcohol and oxidation with Jones reagent gave the target, protected L-3,3-difluoroalanine **25**, in 81% yield.

The most common method for the synthesis of halogenated prolines is from commercially available 4-hydroxyproline. For example, treatment of *N*-Boc-L-4α-hydroxyproline ethyl ester with Ph_3P and either CBr_4 or CCl_4 gave the analogous 4-bromo- and chloro-L-proline derivatives with inversion of configuration in 53% and 76% yields, respectively [21]. Similarly reaction of both *cis*- and *trans*-*N*-Boc-L-4-hydroxyproline methyl esters with DAST gave the analogous protected 4-fluoroprolines and further studies on the alkylation of these products at the α-position have been reported by Filosa *et al.* [22]. Three members of the astin family of cyclopentapeptides, astins A–C, contain the unusual amino acid (–)-*cis*-(3*S*,4*R*)-dichloro-L-proline; *trans*-4-hydroxy-L-proline **26** was used as the starting material for the synthesis of the analogous methyl ester **29** (Scheme 10.8). Hydroxy acid **26** was converted to alkene **27** in four steps and 60% overall yield. Osmylation gave a mixture of diastereomeric *syn* diols and the major diastereomer **28** was transformed to dichloride **29** under Appel conditions and deprotection [23].

Amino acids possessing two fluorine atoms at the β-carbon may act as potent inactivators of certain enzymes and hence have the potential to block metabolic pathways. This biological activity inspired Uneyama *et al.* to explore the synthesis of such compounds. In 2004, they reported the preparation of β,β-difluoroglutamic acid (see Section 10.4.1) and β,β-difluoroproline derivatives (Scheme 10.9) [24]. The approach involved a reductive defluorination of trifluoromethyl iminoester **30** using magnesium and TMSCl to give **31** in 50% yield. Treatment of enaminoester **31** with *N*-bromosuccinimide (NBS) followed by asymmetric reduction gave protected amino ester **32** in 88% e.e. Radical substitution of the bromide by an allyl group using 2,2′-azobisisobutyronitrile as the initiator gave **33** in 54% yield from **31**. Further manipulation of the protecting groups and oxidative cleavage of the terminal olefin gave hemiaminal **34**. Finally, dehydration and a Rh(I)-catalyzed hydrogenation of the resultant alkene gave protected L-3,3-difluoroproline **35** in high yield.

10.2.2
Halogenated α-Amino Acids with Branched Hydrocarbon Side-Chains

10.2.2.1 Halogenated Valines and Isoleucines

Easton *et al.* have conducted pioneering studies into the mechanism and synthetic applications of radical halogenation of α-amino acids with particular emphasis on those with branched hydrocarbon side-chains [25]. For example, reaction of

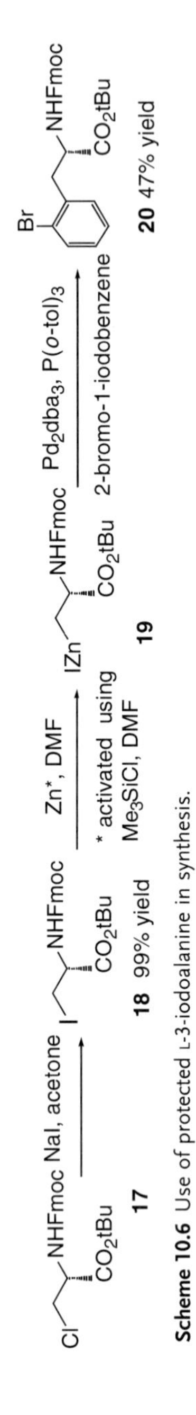

Scheme 10.6 Use of protected L-3-iodoalanine in synthesis.

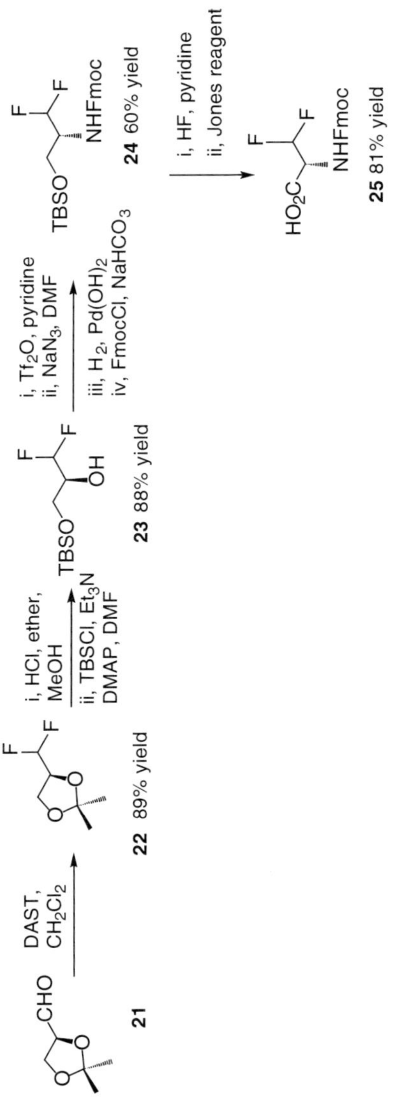

21 → (DAST, CH_2Cl_2) → **22** 89% yield → (i, HCl, ether, MeOH; ii, TBSCl, Et_3N, DMAP, DMF) → **23** 88% yield → (i, Tf_2O, pyridine; ii, NaN_3, DMF; iii, H_2, $Pd(OH)_2$; iv, FmocCl, $NaHCO_3$) → **24** 60% yield → (i, HF, pyridine; ii, Jones reagent) → **25** 81% yield

Scheme 10.7 Preparation of L-3,3-difluoroalanine.

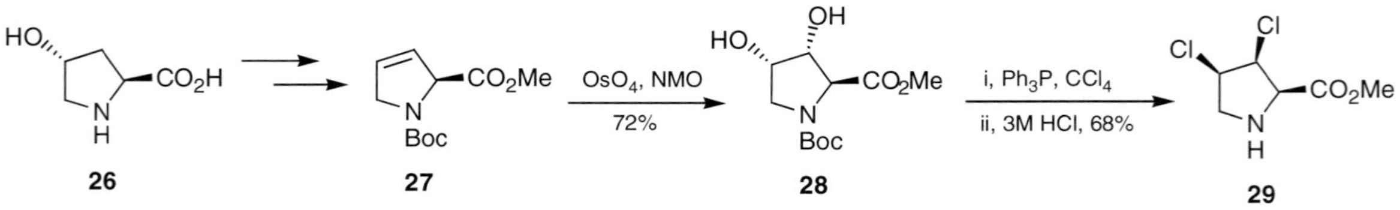

HO, N H, CO_2H **26** → → N Boc, CO_2Me **27** — OsO_4, NMO / 72% → HO, OH, N Boc, CO_2Me **28** — i, Ph_3P, CCl_4 / ii, 3M HCl, 68% → Cl, Cl, N H, CO_2Me **29**

Scheme 10.8 Preparation of (3*S*,4*R*)-dichloroproline methyl ester.

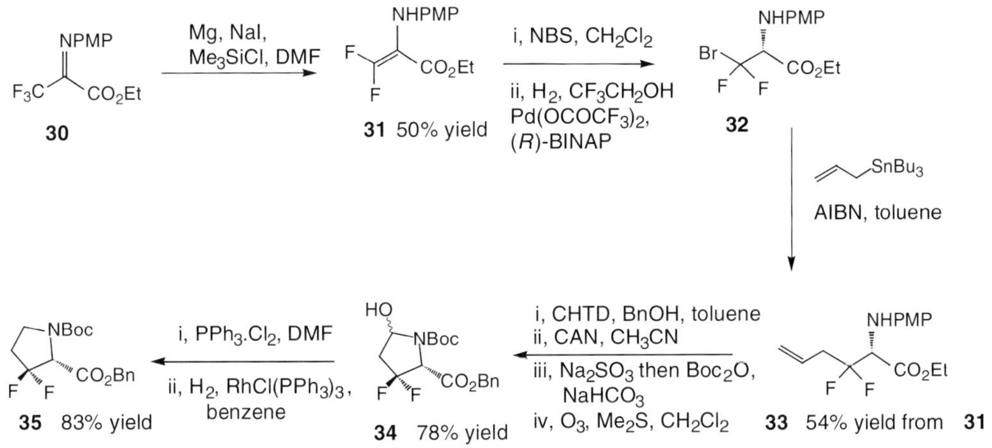

NPMP F_3C CO_2Et **30** — Mg, NaI, Me_3SiCl, DMF → **31** 50% yield — i, NBS, CH_2Cl_2; ii, H_2, CF_3CH_2OH, $Pd(OCOCF_3)_2$, (*R*)-BINAP → **32** — allyl-$SnBu_3$, AIBN, toluene → **33** 54% yield from **31** — i, CHTD, BnOH, toluene; ii, CAN, CH_3CN; iii, Na_2SO_3 then Boc_2O, $NaHCO_3$; iv, O_3, Me_2S, CH_2Cl_2 → **34** 78% yield — i, $PPh_3.Cl_2$, DMF; ii, H_2, $RhCl(PPh_3)_3$, benzene → **35** 83% yield

Scheme 10.9 Preparation of *N*-Boc-L-3,3-difluoroproline benzyl ester.

N-benzoyl valine methyl ester **36** with sulfuryl chloride gave a mixture of chlorinated products and, following high-performance liquid chromatography, the major product, protected 3-chlorovaline **37**, was isolated in 40% yield (Scheme 10.10); in addition the crystalline diastereomeric 4-chlorovaline derivatives **38** and **39** were

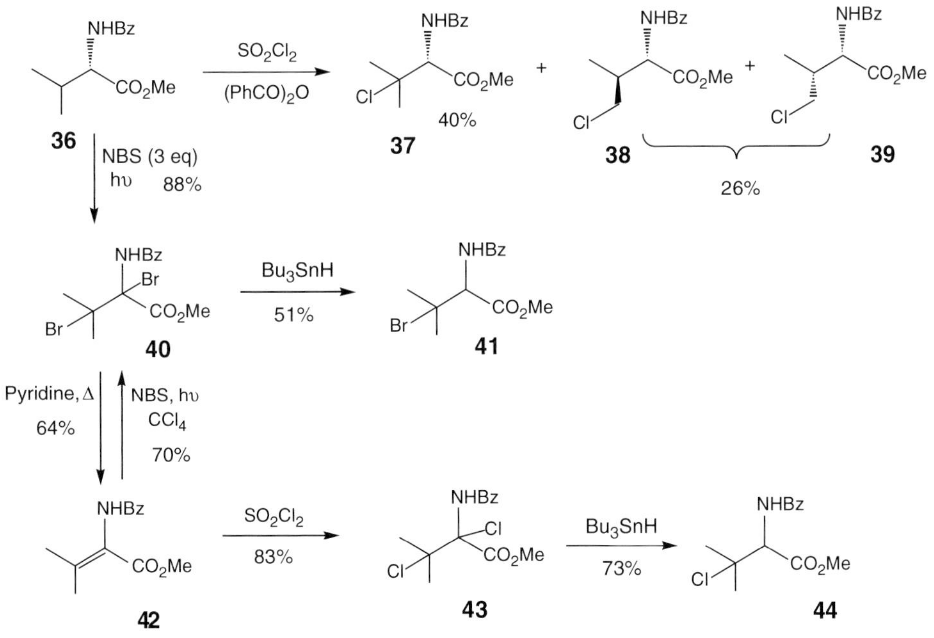

36 — SO_2Cl_2, $(PhCO)_2O$ → **37** 40% + **38** + **39** (26%)

36 — NBS (3 eq), hυ, 88% → **40** — Bu_3SnH, 51% → **41**

40 — Pyridine, Δ, 64% → **42**; **42** — NBS, hυ, CCl_4, 70% → **40**

42 — SO_2Cl_2, 83% → **43** — Bu_3SnH, 73% → **44**

Scheme 10.10 Radical reactions of *N*-benzoyl valine methyl ester derivatives.

obtained (about 1 : 1 mixture, 26% yield) [26]. On photolysis of the analogous *N*-chloroamide (obtained from reaction of **36** with *tert*-butyl hypochlorite) a similar ratio of products was obtained along with the parent protected valine **36** and it was proposed that the reactions proceed by intermolecular β-hydrogen atom transfer. In contrast, it was found that on reaction of amino ester **36** with 3 equiv. NBS under radical conditions hydrogen-atom abstraction occurs at the α-position giving dibromide **40** in 88% yield [27]. Heating **40** in pyridine gave acylenamine **42**, which on irradiation with NBS returned dibromide **40**. Selective radical reduction of **40** led to protected 3-bromovaline **41**. In addition, treatment of unsaturated ester **42** with sulfuryl chloride gave 2,3-dichloride **43**, which in turn was reduced with tributyltin hydride to protected 3-chlorovaline **44**. Kaushik *et al.* have reported a general method for the *N*-chlorination of protected amino esters and peptides using *N,N'*-dichlorobis (2,4,6-trichlorophenyl)urea giving products in high yields (>85%) and with short reaction times [28].

Further studies on radical reactions with a series of *N*-phthaloylamino methyl esters with NBS led Easton *et al.* to conclude that with this protecting group the α-position is much less reactive than in the corresponding *N*-acyl amino esters [29]. Indeed, treatment of *N*-phthaloyl α-amino methyl ester **45** with NBS followed by reaction with silver nitrate in aqueous acetone afforded homochiral β-hydroxy-α-amino ester derivative **46** (Scheme 10.11).

Furthermore, it was reported that treatment of protected isoleucine **47** with NBS gave β-bromoisoleucine derivative **48** that, following elimination of HBr, resulted in a mixture of alkenes that were separated by chromatography (Scheme 10.12) [30]. Reaction of the alkenes with chlorine gave allyl chlorides **49**, **50**, and **51** in modest yields (28–37%).

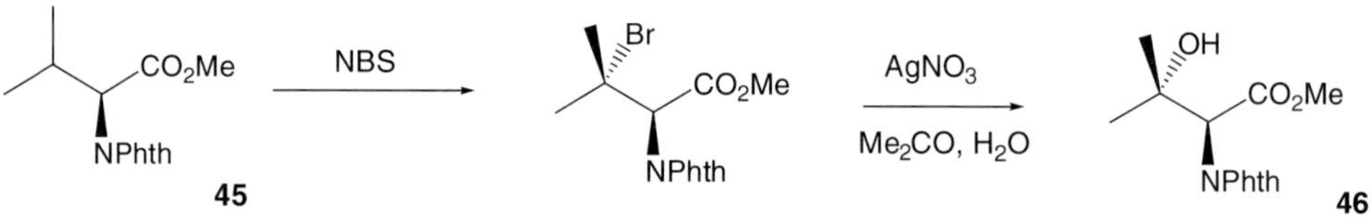

Scheme 10.11 Preparation of *N*-phthaloyl 3-hydroxy-L-valine methyl ester.

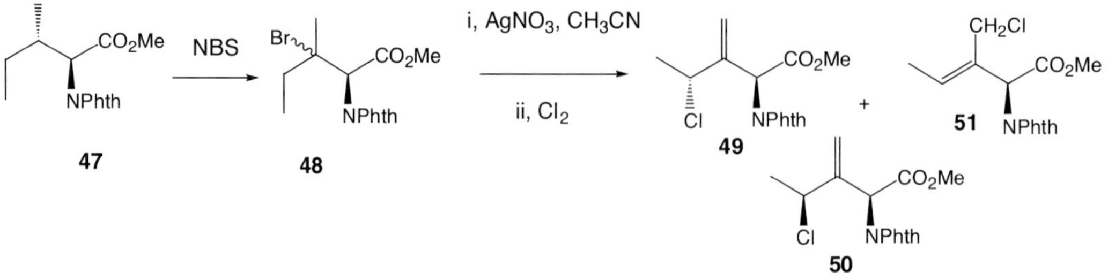

Scheme 10.12 Synthesis and reactions of *N*-phthaloyl L-3-bromoisoleucine methyl ester **48**.

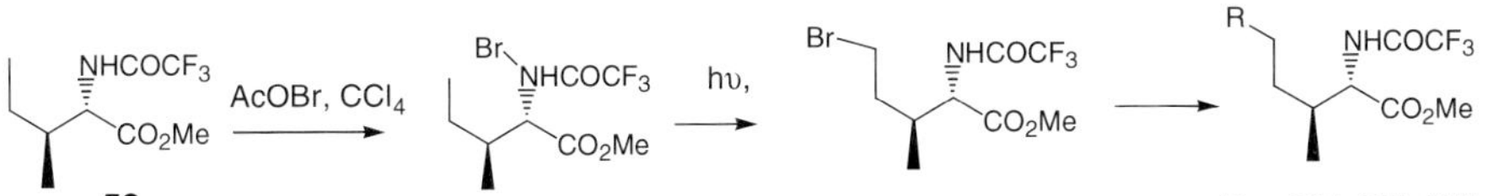

Scheme 10.13 Synthesis of N-trifluoroacetyl-L-5-bromoisoleucine methyl ester.

In the 1960s, Kollonitsch *et al.* described the radical chlorination of a series of α-amino acids under acidic conditions giving β- and γ-chlorinated products [31]. It was shown that irradiation of L-isoleucine in either trifluoroacetic acid (TFA) or 50% sulfuric acid in the presence of chlorine gave the L-5-chloroisoleucine which was not purified but treated directly with base to give a concise synthesis of *trans*-L-3-methylproline in 33% yield [32]. Whilst this chapter is focused on the synthesis of bromo-, chloro-, and iodo-α-amino acids, it is pertinent to note that in 2004 Young *et al.* reported the stereoselective synthesis of (2*S*,3*S*)-3′-fluoroisoleucine from (*S*)-pyroglutamic acid for incorporation into ubiquitin to study protein interactions ([33] and references cited therein).

Corey *et al.* have described an efficient use of a Hofmann–Löffler–Freytag-type reaction for the conversion of *N*-trifluoroacetylisoleucine methyl ester **52** to the 5-bromo derivative **54** in 90% yield (Scheme 10.13). It was proposed that the intermediate amide radical is especially reactive in hydrogen-atom abstraction leading to formation of the radical at C-5 which then propagates the chain reaction by intermolecular bromine abstraction from the *N*-bromo substrate **53**. Bromide **54** proved to be a valuable synthetic intermediate, and, for example, was readily converted to the analogous alcohol, amine, and thiol [34].

10.2.2.2 **Halogenated Leucines**

Following the theme of manipulation of amino acids using radical chemistry, interestingly it has been shown that the regioselectivity of bromination of *N*-trifluoroacetyl-L-leucine methyl ester **55** can be controlled by use of an intra- or intermolecular pathway (Scheme 10.14) [34]. Thus when **55** was *N*-brominated and then exposed to light at room temperature it was transformed into a 1.5 : 1 mixture of diastereomeric 5-bromides **56**. In contrast, reaction of **55** with NBS gave the 4-bromo derivative **57** as the major product (80% yield) along with 10% of a mixture of the (*E*/*Z*)-α,β-unsaturated esters.

Easton *et al.* isolated γ-bromide **59** in a similar yield (82%) when *N*-phthaloyl leucine methyl ester **58** was treated with NBS in CCl_4 and used it as an intermediate in the synthesis of further amino acids (Scheme 10.15) [35]. For example, treatment

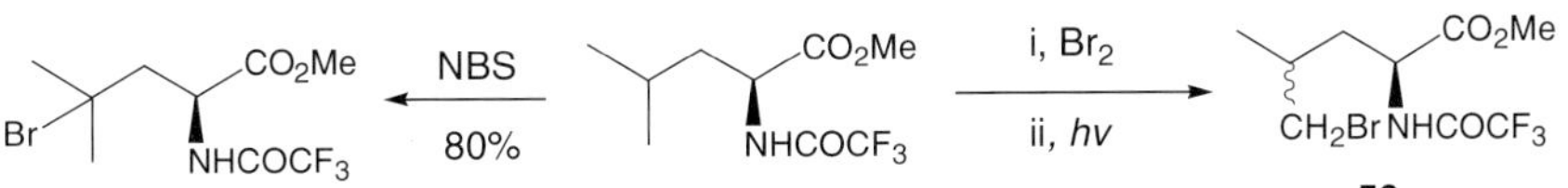

Scheme 10.14 Regioselective bromination of N-trifluoroacetyl-L-leucine methyl ester.

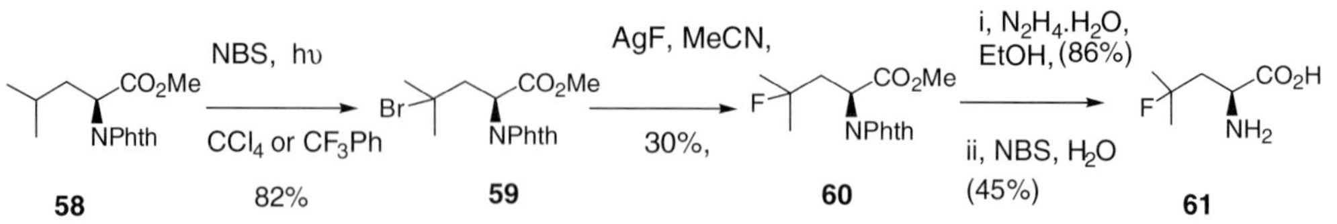

58 → (NBS, hυ; CCl_4 or CF_3Ph; 82%) → 59 → (AgF, MeCN, 30%) → 60 → (i, $N_2H_4.H_2O$, EtOH, (86%); ii, NBS, H_2O (45%)) → 61

Scheme 10.15 Regioselective bromination of *N*-phthaloyl-L-leucine methyl ester.

of bromide **59** with silver fluoride gave the 3-fluoro analog **60** and two byproducts (a γ-lactone and olefin) then a two-stage deprotection strategy gave (*S*)-γ-fluoroleucine **61** [36].

Marine organisms are a rich source of halogenated secondary metabolites that exhibit a range of biological activities (see [37] and earlier reviews in the same series). The majority of the halogens are located in positions that indicate their biochemical incorporation as electrophilic species (catalyzed by haloperoxidases) [38]. However, there are some cyanobacterial and sponge-cyanobacterial metabolites that possess halogenated functional groups (e.g., the trichloromethyl group in barbamide and dichloromethyl group in dechlorobarbamide – secondary metabolites isolated from cultures of the cyanobacterium *Lyngbya majuscula*) for which the electronic nature of the halogenating species was until recently uncertain [39]. To gain an understanding of the mechanism of biochlorination, synthetic routes to L-5-chloroleucines were developed that could be readily adapted for incorporation of ^{13}C.

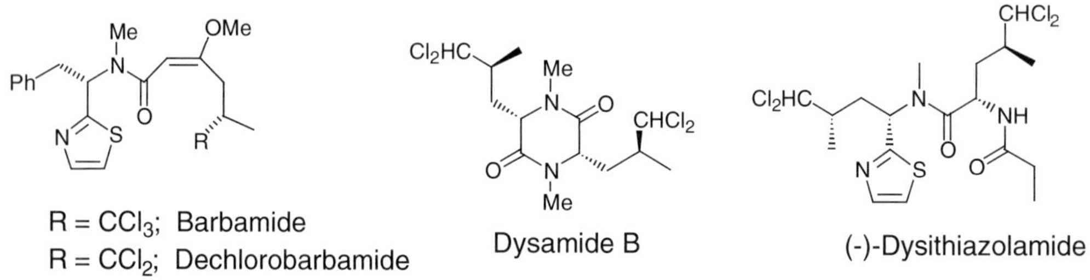

R = CCl_3; Barbamide
R = CCl_2; Dechlorobarbamide

Dysamide B

(-)-Dysithiazolamide

To this end, Willis *et al.* prepared (2*S*,4*S*)-5-chloroleucine from L-glutamic acid (Scheme 10.16) [40]. A key step was alkylation of the lithium enolate of protected glutamate **62** giving **63** in greater than 60% yield and with excellent diastereoselection [41]. Use of $^{13}CH_3I$ as the electrophile enabled the incorporation of an isotopic label for biosynthetic studies. The α-amino functionality was further protected as the di-Boc derivative prior to a stepwise reduction of the γ-ester, via aldehyde **64**, to alcohol **65**. However, the selective reduction of the terminal ester was somewhat capricious and good yields could only be obtained reliably when fresh diisobutylaluminum hydride (DIBALH) was employed. Hence, the preferred approach is to use a selective reduction of a terminal acid rather an ester as described below (Scheme 10.17). Treatment of alcohol **65** with Ph_3P/CCl_4 gave, after acid-mediated deprotection, the target (2*S*,4*S*)-5-chloroleucine **66** in 40% overall yield from **63**. Furthermore, the unusual amino acid **67** containing a vinyl dichloride was readily

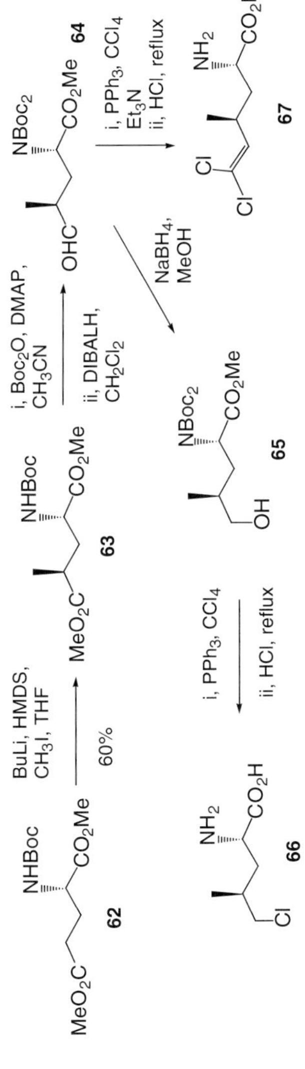

Scheme 10.16 Preparation of (2*S*,4*S*)-5-chloroleucine.

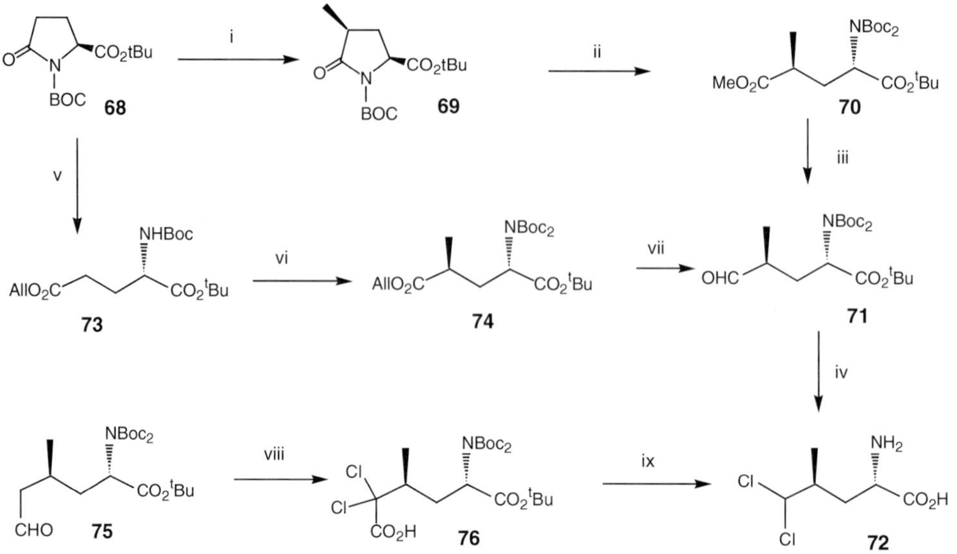

i,(a) LiHMDS, Eschenmoser salt, THF (75%); (b) MeI, MeOH; (c) $NaHCO_3$, H_2O; (d) H_2, Pd/C, EtOAc, (95%); ii, (a) LiOH , THF; (b) $ClCO_2Me$, Et_3N, DMAP, (70%); iii, (a) DIBALH, Et_2O, (b) Dess-Martin, (66%); iv, (a) $NH_2NH_2.H_2O$, mol. sieves; (b) $CuCl_2$, Et_3N, (40%); (c) 4N HCl, reflux (75%); v, (a) LiOH, (b) Cs_2CO_3, allylbromide (82%); (vi) (a) LiHMDS, MeI, THF (92%), (b) Boc_2O, DMAP, (92%); vii, (a) $Pd(PPh_3)_4$, $NaBH_4$ (77%); (b) $BH_3.SMe_2$ (96%); (c) $(COCl)_2$, DMSO, Et_3N (87%); viii, (a) $tBuNH_2$, NCS (65%); (b) $NaClO_2$, H_2O_2, KH_2PO_4 (72%); ix, (a) $Pb(OAc)_4$, 1,4-cyclohexadiene (58%); (b) TFA (100%).

Scheme 10.17 Preparation of (2*S*,4*S*)-5,5-dichloroleucine.

prepared via treatment of aldehyde **64** with Ph_3P/CCl_4 in the presence of Et_3N to prevent epimerization at C-4.

The first synthesis of (2*S*,4*S*)-5,5-dichloroleucine **72** was reported by Rodríguez *et al.* via aldehyde **71** [42]. First, protected pyroglutamate **68** was converted to protected 4-methylene pyroglutamate and then catalytic hydrogenation gave (4*S*)-methylglutamate **69** as a single diastereomer (Scheme 10.17). Hydrolytic cleavage of the ring followed by protection of the resultant acid gave **70**, which was reduced to aldehyde **71** for the key chlorination reaction. Using a modification of the Takeda conditions [43], aldehyde **71** was transformed to its hydrazone with hydrazine monohydrate in anhydrous MeOH in the presence of molecular sieves then treated with $Et_3N/CuCl_2$ giving, after acid-mediated deprotection, the required dichloride **72** in 30% yield from aldehyde **71**. A similar yield (35%) for the *gem*-dichlorination of the (2*S*,4*R*)-diastereomer of **71** was obtained during the synthesis of (2*S*,4*R*)-5,5-dichloroleucine and was accompanied by a minor product (7%) from elimination of HCl [44].

An alternative approach to the key intermediate aldehyde **71** began with hydrolytic cleavage of protected pyroglutamate using lithium hydroxide and the resultant γ-acid was temporarily protected as an allyl ester **73** [45]. An efficient stereoselective alkylation of **73** was followed by introduction of a second Boc protecting to give **74** in 85% yield over the two steps. The allyl ester of **74** was selectively cleaved via a

Scheme 10.18 Proposed biosynthesis of barbamide in *L. majuscula*.

π-allyl complex using $Pd(PPh_3)_4/NaBH_4$ and the resultant acid was reduced using borane dimethylsulfide to a primary alcohol, which was oxidized to the required aldehyde **71** in seven steps and 42% overall yield from protected pyroglutamate **68**.

The use of the Takeda conditions for the *gem*-dichlorination of aldehyde **71** is somewhat capricious giving yields in the range 30–40%, hence an alternative approach for the synthesis of (2*S*,5*S*)-5,5-dichloroleucine **72** has been investigated (Scheme 10.17). A one-carbon homologation of aldehyde **71** to **75** was achieved using a series of straightforward transformations [45]. Treatment of aldehyde **75** with *tert*-butylamine and *N*-chlorosuccinimide (NCS) in the presence of molecular sieves led to the required α,α-dichlorination and oxidation of the intermediate dichloroaldehyde gave acid **76**. The pivotal step required decarboxylation of **76** in the presence of a hydride source which would not give concomitant reduction of the dichloride. Hence, acid **76** was treated with $Pb(OAc)_4$ using 1,4-cyclohexadiene as the hydride source that, after deprotection, gave dichloroleucine **72** in 58% yield from **76**. Whilst all these transformations are simple to perform in good yield and give access to a series of novel protected amino acids, clearly the *gem*-dichlorination of aldehyde **71** under the Takeda conditions is more direct.

(2*S*,4*S*)-5,5-Dichloroleucine **72** has proved to be a valuable intermediate in the total syntheses of the marine natural products dysamide B and (−)-dysithiazolamide [45, 46]. Furthermore, access to chlorinated leucines **66** and **72** labeled with ^{13}C has been important in the elucidation of the biosynthesis of barbamide in the cyanobacterium *L. majuscula* by Gerwick *et al.* [47]. It has been shown that selective chlorination of the *pro*-(*R*) methyl group of leucine occurs leading via dichloroleucine **72** to trichloroleucine **77** prior to incorporation into the marine natural product (Scheme 10.18). In 2006, a novel iron-dependent mechanism of biochlorination at unactivated positions was described by the groups of Walsh [48] and Gerwick [49], giving an insight into these fascinating transformations.

5,5,5-Trichloroleucine **77** is not only an intermediate in the biosynthesis of barbamide, it is a common scaffold in a number of marine natural products including herbacic acid and dysidenamide [50].

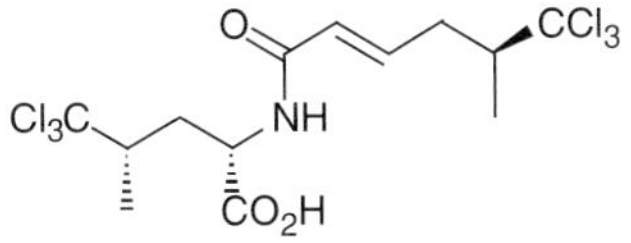

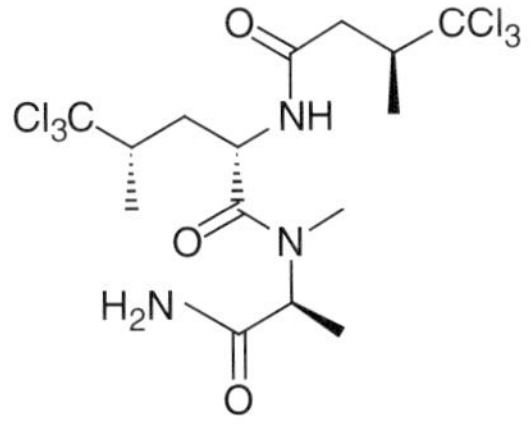

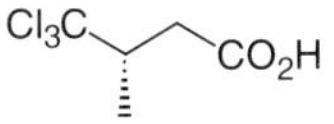

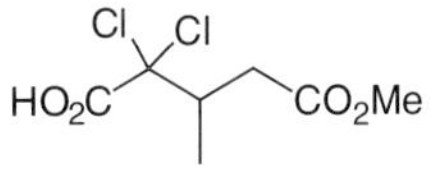

3-Trichloromethylbutanoic acid **78** has been used in the synthesis of such compounds and may be prepared in a number of ways. For example, a halodecarboxylation strategy was pioneered by de Laszlo and Williard in which dichloro acid **79** was prepared via α,α-dichlorination of an imine (*cf.* Scheme 10.17) followed by oxidation with $KMnO_4$ [51]. A modified Hunsdiecker reaction with $Pb(OAc)_4$ and LiCl installed the required trichloromethyl group. Further approaches to the enantioselective synthesis of trichloro acid **78** involve either conjugate addition of a trichloromethyl group to the crotonate derivative of a chiral auxiliary [52] or resolution of 3-trichloromethylbutanoic acid prepared via radical addition of bromotrichloromethane to crotonic acid followed by reductive cleavage of the α-bromide with zinc and acetic acid [53].

The final site of halogenation of leucine to be considered is at C-3. A challenge to be addressed in the synthesis of β-halo-α-amino acids is their propensity to undergo β-elimination. In 2006, Hanessian *et al.* described the total synthesis of the marine metabolite chlorodysinosin A, which contains an unusual 3-chloroleucine residue [54]. It is a potent inhibitor of the serine protease thrombin, a crucial enzyme in the process leading to platelet aggregation. Epoxy alcohol **80** was converted to *tert*-butylsulfonamide **81** and then a regioselective ring opening of the aziridine was achieved in 80% yield using cerium chloride as the halide source giving, after deprotection of the amine, chloroamino alcohol **82** (Scheme 10.19). Coupling the amine with (*R*)-3-*tert*-butyldiphenylsilyloxy-2-methoxypropanoic acid **83** and oxidation of the primary alcohol gave acid **84**, which was used in the assembly of chlorodysinosin A.

Bonjoch *et al.* used the selective ring opening of *N*-protected leucine β-lactones in the synthesis of (2*S*,3*R*)- and (2*S*,3*S*)-3-chloroleucine (Scheme 10.20) [55]. *O*-Benzothiazol-1-yl-*N*,*N*,*N*′,*N*′-tetramethyluronium hexafluorophosphate (HBTU) was used to effect the lactonization of Cbz-(2*R*,3*S*)-hydroxyleucine **85**, giving lactone

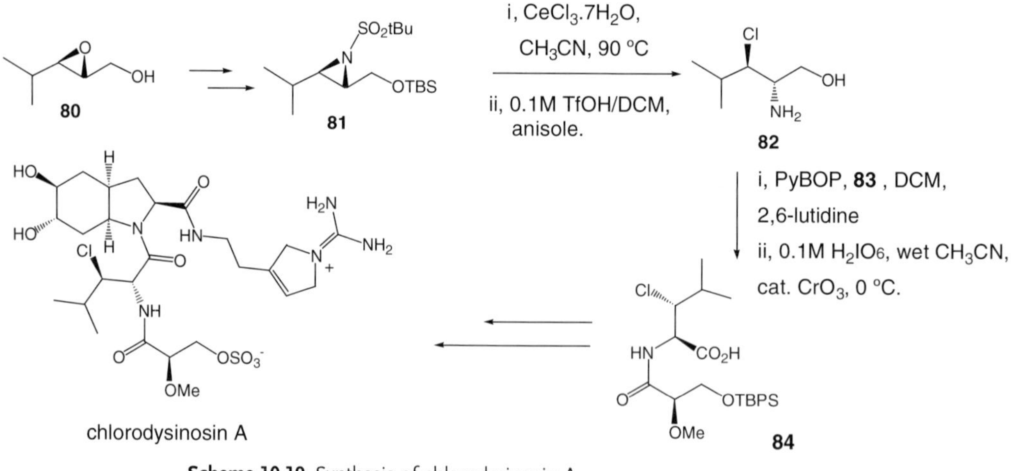

Scheme 10.19 Synthesis of chlorodysinosin A.

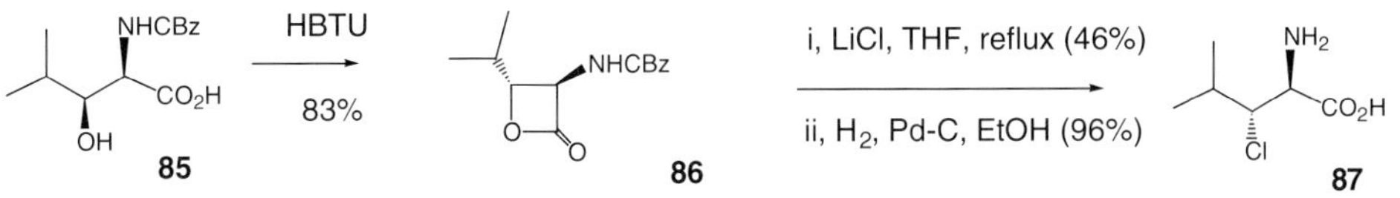

Scheme 10.20 Synthesis of (2*S*,3*R*)-3-chloroleucine.

86 in 83% yield. Treatment of **86** with LiCl in THF under reflux followed by deprotection gave (2*S*,3*R*)-3-chloroleucine **87**. The (2*S*,3*S*) diastereomer was prepared by an analogous procedure using (2*R*,3*R*)-3-hydroxyleucine as the starting material.

10.3 Halogenated Amino Acids with an Aromatic Side-Chain

10.3.1 Halogenated Phenylalanines and Tyrosines

Palladium cross-couplings of aryl halides are widely used for the creation of C–C and hence aromatic amino acids halogenated in the ring are versatile building blocks in synthesis (*cf.* Scheme 10.6). The most common method for the preparation of phenylalanine and tyrosine derivatives halogenated in the ring is to use electrophilic aromatic substitution processes. For example, nitration of L-phenylalanine gave the *p*-nitro derivative in 40% yield following recrystallization. Reduction of the nitro group using H_2 and Pd/$BaSO_4$ then diazotization and a Sandmeyer reaction with CuCl gave *p*-chlorophenylalanine in excellent yield and with no loss of enantiopurity [56]. It has been reported that chlorine, bromine, and iodine substituents in aromatic rings strongly increase the lipophilicity, and thus Escher and coworkers were inspired to prepare a number of L-phenylalanine derivatives brominated in the aromatic ring to incorporate them into angiotensin II [57]. For example, (3′,4′,5′)-tribromo-L-phenylalanine was prepared from *p*-amino-L-phenylalanine via an *ortho*-directed dibromination using Br_2 in CH_3CO_2H, followed by a Sandmeyer reaction mediated by $CuSO_4$, NaBr, and Na_2SO_4. Direct monobromination of L-phenylalanine can be achieved using 1.1 equiv. bromoisocyanuric acid sodium salt (BICA-Na) in 60% sulfuric acid giving a mixture of *o*- and *p*-bromo-L-phenylalanines **88** and **89**, which were separated by reverse-phase column chromatography (Scheme 10.21) [58]. In contrast, reaction of the more activated ring of L-tyrosine under similar conditions led to the mono-bromo derivative (29% yield) accompanied by dibromotyrosine **90** (19%) and an improved yield (71%) of the dibromide was achieved using 2.0 equiv. BICA-Na. An alternative method for the synthesis of 3′,5′-dichloro- and dibromotyrosine methyl esters **91** and **92** simply involves treatment of L-tyrosine in a 3 : 1 mixture of methanol and glacial acetic acid with either chlorine gas or bromine under reflux giving the products in good yield [59].

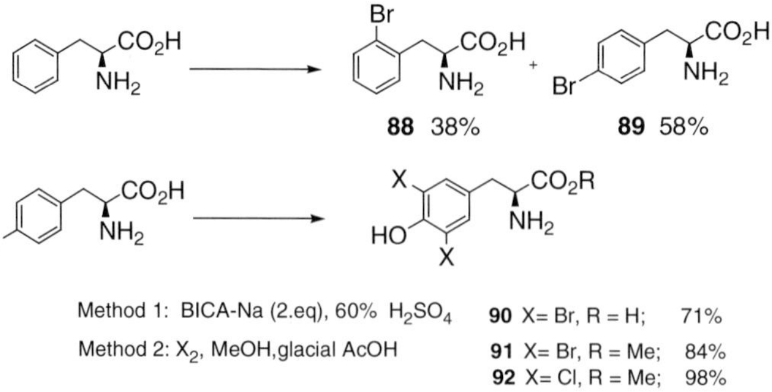
CO2H
NH2
Br
88 38%
89 58%
HO
X
CO2R
Method 1: BICA-Na (2.eq), 60% H2SO4
Method 2: X2, MeOH,glacial AcOH
90 X= Br, R = H; 71%
91 X= Br, R = Me; 84%
92 X= Cl, R = Me; 98%

Scheme 10.21 Halogenation of L-phenylalanine and L-tyrosine.

In certain bacteria it has been shown that 2′-chloro- and 2′-bromotyrosines have more potent growth inhibitory effects than their 3-halotyrosine counterparts [60]. In 1975, the racemic halogenated amino acids were prepared from the appropriate 2-halo-4-methoxybenzyl bromide **93** via the base mediated reaction with acetamidomalonic ester (EtAAM) to give **94** followed by ester hydrolysis and decarboxylation to afford **95** (Scheme 10.22). There are now many chiral glycinate equivalents and chiral catalysts available which have the potential to enable modification of this approach for the synthesis of enantioenriched products. For example, in 2004, Krasikova *et al.* used the achiral glycine derivative NiPBPgly and (*S*)-NOBIN as the catalyst to achieve a stereoselective alkylation of **96** in the synthesis of [^{18}F]2′-fluorotyrosine **97** [61].

Radiolabeled iodophenylalanine has been used for pancreatic imaging studies and, in particular, to detect tumors [62]. Many methods are known for introducing iodine

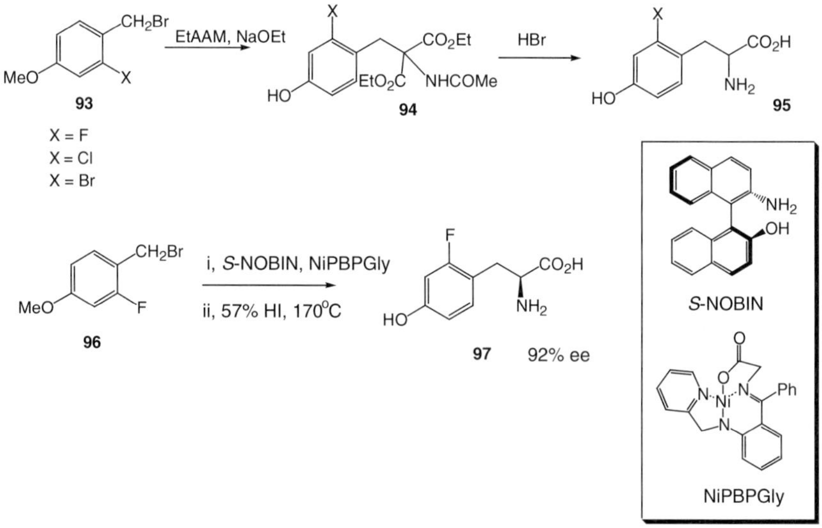
CH2Br
EtAAM, NaOEt
MeO
X
93
X = F
X = Cl
X = Br
CO2Et
EtO2C
NHCOMe
HO
94
HBr
CO2H
NH2
95
NH2
OH
S-NOBIN
CH2Br
i, S-NOBIN, NiPBPGly
ii, 57% HI, 170°C
F
96
97 92% ee
Ph
NiPBPGly

Scheme 10.22 Synthesis of 2′-halotyrosine.

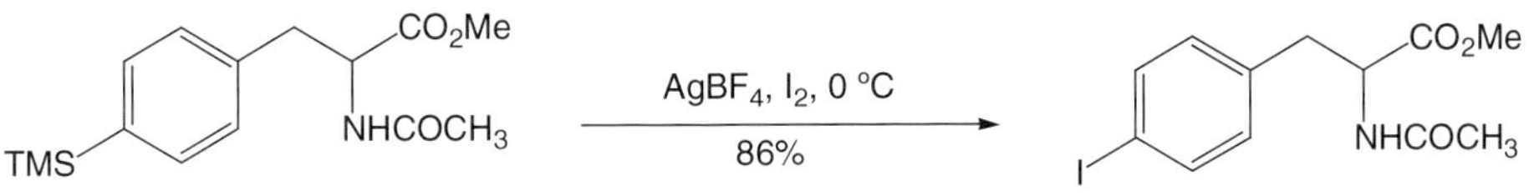

Scheme 10.23 Synthesis of *N*-acetyl *p*-iodophenylalanine methyl ester.

into an aromatic ring and an efficient synthesis of *N*-acetyl *p*-iodophenylalanine methyl ester **99** from the *p*-trimethylsilyl precursor **98** used iodine in the presence of silver tetrafluoroborate giving **99** in 86% yield (Scheme 10.23) [63].

As well as halogenation of the aromatic ring, there have been a number of reports of the introduction of halogens into the side-chain of protected aromatic α-amino acids. For example, epoxy ester **100** was transformed into the chloroamine **104** via the intermediate amino alcohols **101** and **102** in excellent (>90%) yield (Scheme 10.24). Chloride **104** was used as a substrate in the introduction of a variety of amines at C-3 with good regio- and stereocontrol via the intermediate aziridine **103** [64].

β-Hydroxyphenylalanine and β-hydroxytyrosine are components of several biologically active compounds including vancomycin which contains both (2*S*,3*R*)- and (2*R*,3*R*)-β-hydroxytyrosine residues whilst the structure of lysobactin includes a (2*S*,3*R*)-β-hydroxyphenylalanine ([65, 66] and references cited therein). Protected β-hydroxyphenylalanine and β-hydroxytyrosine have been prepared by side-chain bromination of *N*-phthaloyl-amino esters followed by hydrolysis of the resultant bromides [67]. For example, on treatment of *N*-phthaloyl-L-phenylalanine methyl ester **105** with NBS a 1 : 1 mixture of diastereomeric bromides **106** was obtained, which was separated by fractional crystallization (Scheme 10.25). Treatment of the (2*R*,3*R*) ester with silver nitrate in acetone and water gave a 2 : 1 mixture of the (2*S*,3*R*)

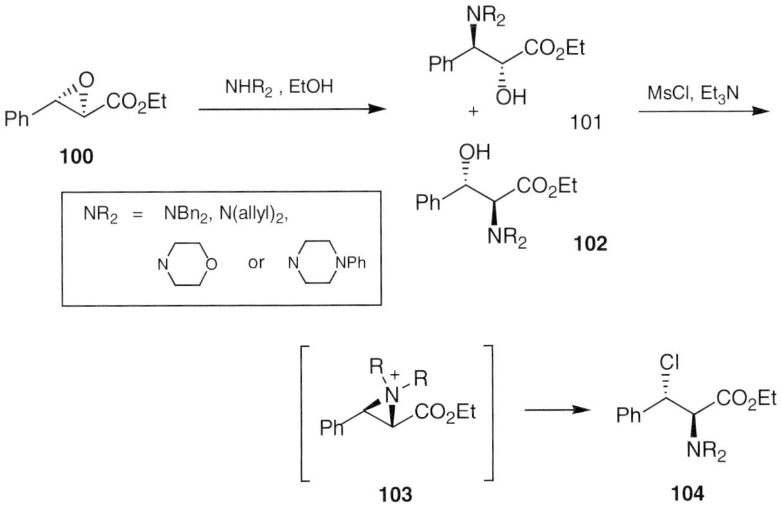

Scheme 10.24 Synthesis of protected β-chlorophenylalanine.

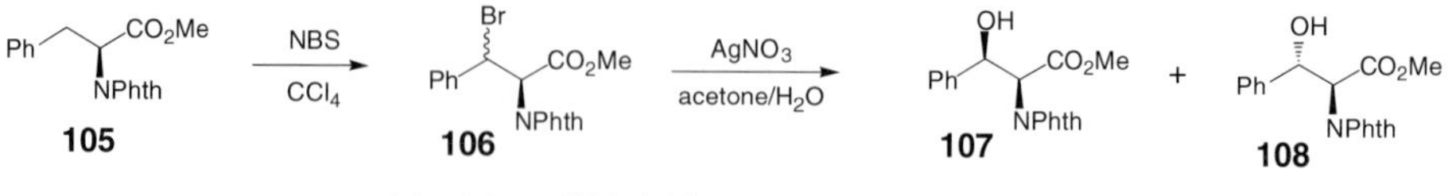

1:1 mixture, 82% yield

Scheme 10.25 Synthesis of *N*-phthaloyl L-3-hydroxyphenylalanine methyl esters.

β-hydroxy-α-amino ester derivative **107** and the (2*S*,4*S*) diastereomer **108**, whilst under the same conditions the (2*R*,3*S*) bromide afforded only the (2*S*,3*R*) diastereomer **108** in 93% yield (note the change in Cahn–Prelog–Ingold priorities).

Crich and Banerjee have extended the utility of side-chain bromination chemistry and reported the synthesis of a series of aromatic β-hydroxy-α-amino acids (Phe, Tyr, His, and Trp) [68]. For example, the NBS-mediated radical bromination of *N*,*N*-di-Boc phenylalanine methyl ester **109** gave **110** as a mixture of diastereomers and subsequent treatment with silver nitrate in acetone provided predominantly the *trans* oxazolidinones **111** (Scheme 10.26). The epimers **111** and **112** were readily separated by column chromatography, and cesium carbonate-catalyzed hydrolysis then generated the β-hydroxy amino esters **113** and **114** in good overall yield.

10.3.2
Halogenated Histidines

Crich and Banerjee [68] also described the synthesis of L-3-hydroxyhistidine. Ring halogenated histidines are components of biologically active peptides, and 2-iodo- and 2-fluoro-L-histidines display significant antimalarial activity against *Plasmodium falciparum* [7]. Halogens have been introduced into the ring of *N*-Boc-L-histidine methyl ester **115** (e.g., by reaction with Br_2/Et_3N giving dibromohistidine **119** in 25% yield) [69]. A more versatile and efficient approach to ring halogenation involves treatment of protected histidine **115** with one equivalent of either NCS, NBS, or

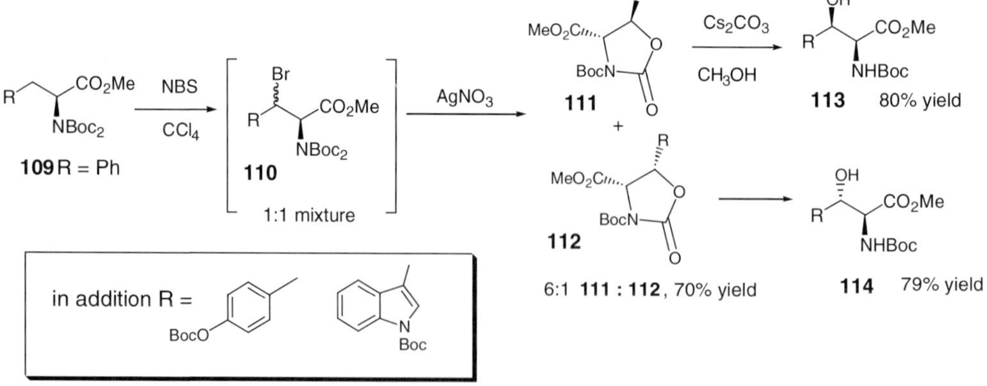

Scheme 10.26 Synthesis of *N*-phthaloyl L-3-hydroxyphenylalanine methyl esters.

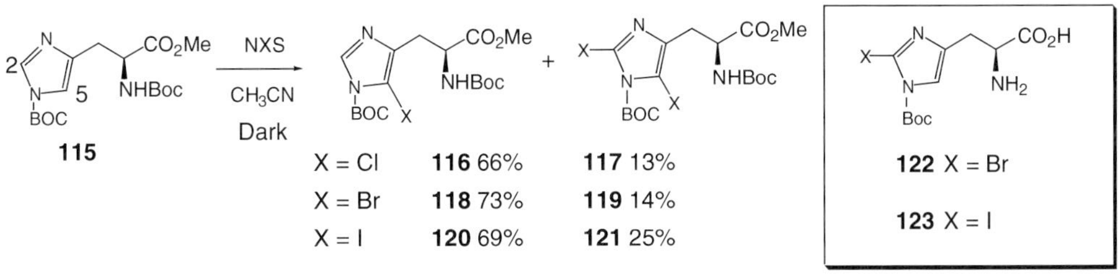

X		
X = Cl	**116** 66%	**117** 13%
X = Br	**118** 73%	**119** 14%
X = I	**120** 69%	**121** 25%

122 X = Br

123 X = I

Scheme 10.27 Synthesis of histidine derivatives selectively halogenated in the ring.

N-iodosuccinimide in acetonitrile at room temperature in the dark yielding the protected 5-halohistidines **116**, **118**, and **120**, respectively, as the major product in each case accompanied by minor amounts of the dihalogenated derivatives **117**, **119**, and **121** (Scheme 10.27) [70]. The protecting groups were removed under acidic conditions at room temperature giving the analogous halogenated L-histidines in greater than 60% yield. Monobromide **118** and monochloride **116** could be further halogenated at C-2 with *N*-halosuccinimides to give mixed dihalogenated products in 60–90% yield. Interestingly when the dihalogenated compounds with either bromine or iodine at C-5 were treated with refluxing HCl, regioselective dehalogenation occurred giving the 2-bromo- and 2-iodohistidines **122** and **123**. Attempts to form mixed haloderivatives from iodide **120** using NCS or NBS proved unsuccessful as the iodide at C-5 of the imidazole ring was replaced by either bromide or chloride, leading to the 2′,5′-dichlorides or dibromides **117** and **119**.

Aryl histidines occur naturally in the active site of heme–copper oxidases as well as in cytotoxic and antifungal marine peptides. Protected L-5-bromohistidine **125** is a valuable intermediate for the synthesis of these compounds (Scheme 10.28) ([71] and references cited therein). Following 2-(trimethylsilyl)ethoxymethyl chloride

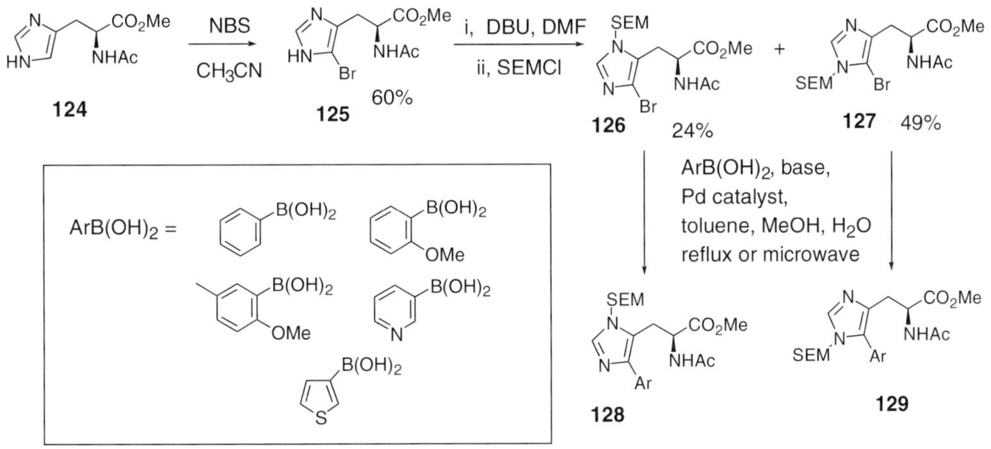

Scheme 10.28 Protected 5-bromohistidine derivatives in Suzuki–Miyaura reactions.

protection of the imidazole ring of **125**, microwave irradiation efficiently promoted the Suzuki–Miyaura reaction of protected bromides **126** and **127** with various boronic acids in the presence of a palladium catalyst, giving protected 5-arylhistidines **128** and **129**.

10.3.3
Halogenated Tryptophans

A comprehensive review on the synthesis and applications of tryptophans brominated in the aromatic ring was published in 2007, and so these topics are not further discussed herein [72]. Chlorotryptophan residues are found in several medicinally important compounds, including the anticancer agents rebeccamycin and diazonamide A [73]. The first example of the synthesis of a chlorotryptophan was by Rydon and Tweddle in 1955 [74] and since then a number of approaches have been published [75]. For example, D-5-chlorotryptophan was isolated following hydrolytic cleavage of the antibiotic longicatenamycin, and in the same paper a preparation of D,L-5-chlorotryptophan was reported from 3-cyanopropanal and *p*-chlorophenylhydrazine via a Fischer indole synthesis [76]. Taniguchi described a key ring closing and opening procedure in the synthesis of D,L-5-chlorotryptophan [77]. Treatment of protected tryptophan **130** with phosphoric acid leads to cyclization, thus protecting the amino ester **131** during chlorination with NCS in AcOH to give the 5-chloro derivative **132** in 89% yield. Ring opening with NaOH returned 5-chlorotryptophan **133** (Scheme 10.29).

D-6-Chlorotryptophan has stimulated the interest of the scientific community as it is a non-nutritive sweetener with a sucrose-like taste. It is readily prepared, for example, via nitration (60%) of D-tryptophan followed by reduction to the corresponding amine (91%) and finally a Sandmeyer reaction using CuCl as the source of halide (51%) ([78] and references cited therein). Goss and Newill have described a versatile approach to the synthesis of a series of L-halotryptophan derivatives using a readily prepared bacterial lysate containing tryptophan synthase (Scheme 10.30) [79]. The reaction proceeds via condensation of L-serine and pyridoxal 5′-phosphate (PLP) to form a Schiff's base, then dehydration and nucleophilic attack by the haloindole **134** and hydrolysis returns PLP and the halotryptophan **135**. This simple one-pot reaction gives variable yields (e.g., 5-fluoro-, 5-chloro-, and 5-bromoindoles are converted to the corresponding halotryptophans in 63, 54, and 16% yield, respectively).

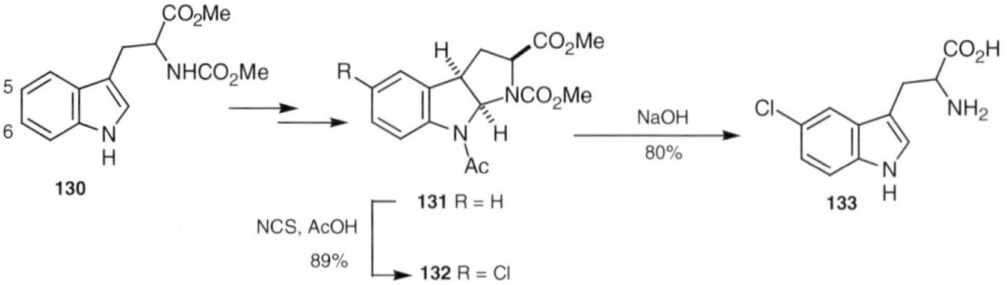

CO_2Me $NHCO_2Me$ 5 6 N H **130** → R H CO_2Me NCO_2Me N H Ac **131** R = H; NCS, AcOH 89% → **132** R = Cl; NaOH 80% → Cl CO_2H NH_2 N H **133**

Scheme 10.29 Synthesis of 5-chlorotryptophan.

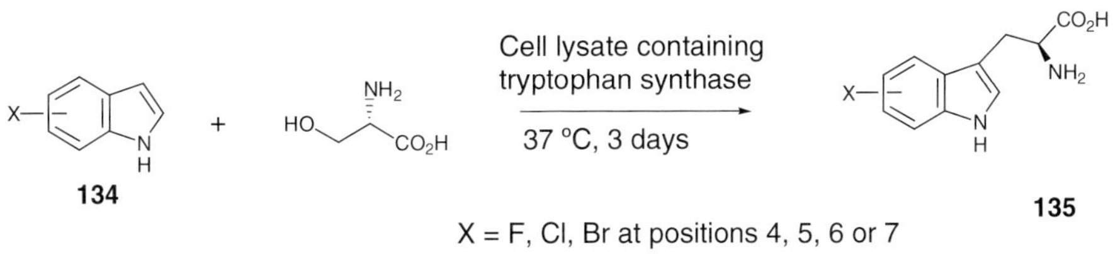

Cell lysate containing tryptophan synthase

37 °C, 3 days

134 + HO–CH₂–CH(NH₂)–CO₂H → 135

X = F, Cl, Br at positions 4, 5, 6 or 7

Scheme 10.30 Synthesis of L-halotryptophans.

10.4
Halogenated Amino Acids with Heteroatoms in the Aliphatic Side-Chain

10.4.1
Halogenated Aspartic and Glutamic Acids

A direct approach to the stereoselective synthesis of 3-chloro-L-aspartic acids **137** was reported by Akhtar *et al.* using biotransformations (Scheme 10.31) [80]. The enzyme 3-methylaspartate ammonia lyase catalyzes the reversible α,β-elimination of ammonia from (2*S*,3*S*)-3-methylaspartic acid to give mesaconic acid **138**. This enzyme from *Clostridiun tetnomorphum* was used efficiently in a retro-physiological reaction direction to prepare (2*R*,3*S*)-3-chloroaspartic acid **137** in 60% yield from chloro acid **136**. The analogous bromide **139** was a substrate for the enzyme, but the product was unstable to the incubation conditions and cyclized to form the 2,3-aziridine dicarboxylic acid and also alkylated the enzymes. Unsaturated iodide **140** was not a substrate for 3-methylaspartase and the reaction rate with the fluoride **141** was extremely slow.

From previous sections of this chapter, it is evident the important role played by radical chemistry and it has found further application in the synthesis of halogenated glutamic acids. For example, in 1981, Silverman *et al.* showed that photochlorination of L-glutamic acid in sulfuric acid gave approximately a 1 : 1 mixture of L-*threo* and L-*erythro*-β-chloroglutamic acids **142** and **143** in 33% yield based on recovered starting material (Scheme 10.32) [81]. The acids were separated by ion-exchange chromatography using Dowex 50(H^+), 0.1 N HCl, then chloride **142** was used as an intermediate in the synthesis of the antitumor agent **144**.

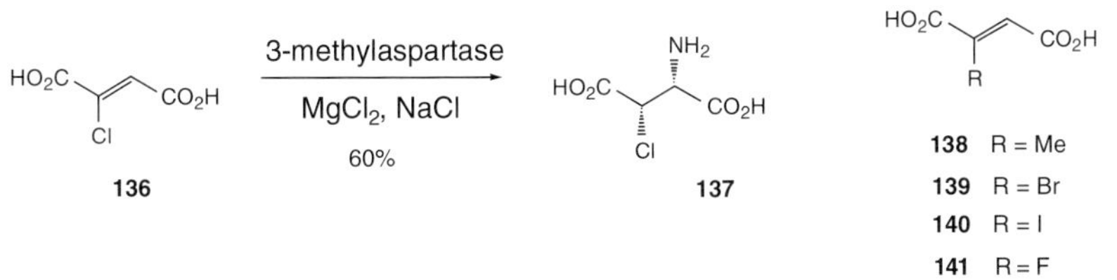

3-methylaspartase

$MgCl_2$, NaCl

60%

136 → 137

138 R = Me

139 R = Br

140 R = I

141 R = F

Scheme 10.31 Synthesis of (3*S*)-chloro-L-aspartic acid.

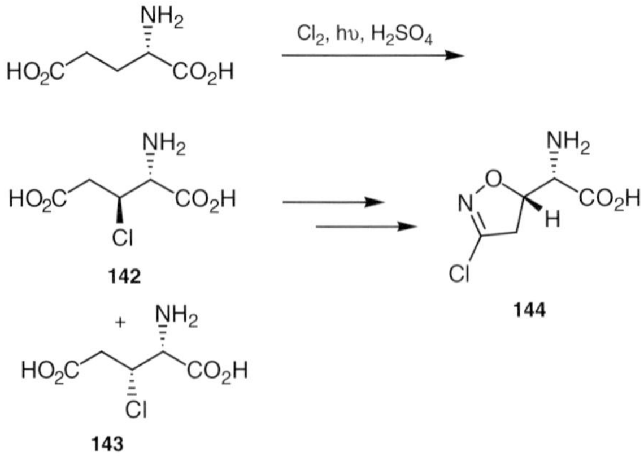

1:1 mixture, 33% yield (brsm)

Scheme 10.32 Chlorination of L-glutamic acid.

Interestingly, Kollonitsch discovered that by varying the reaction conditions, chlorination could be directed predominantly to C-3 or C-4. For the synthesis of 3-chloroglutamic acid, the parent amino acid was irradiated with chlorosulfuric acid (6 M), whilst starting with glutamic anhydride in 3 M chlorosulfuric acid, 4-chloroglutamic acid was the major product [31]. An alternative approach to introduce a chloride at C-3 of glutamate was reported by Iwasaki *et al.* in which 3-hydroxy-D,L-glutamic acid diethyl ester was treated with PCl_5 in $CHCl_3$ [82].

The synthesis of protected 3,3-difluoroglutamic acid **146** has been reported from amino ester **33** (Scheme 10.33), which was also used as an intermediate in the synthesis of protected β,β-difluoroproline (Scheme 10.9) [24]. Following a protecting group exchange on the amine from the paramethoxyphenyl **33** to the Boc derivative **145**, oxidative cleavage of the olefin afforded an acid that was protected as the benzyl ester **146**.

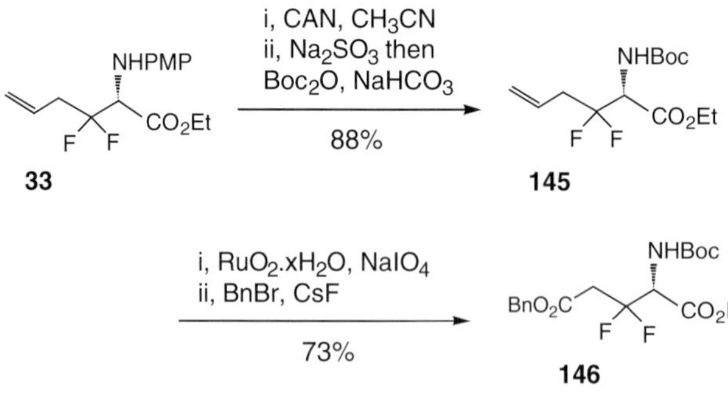

Scheme 10.33 Synthesis of protected 3,3-difluoroglutamic acid.

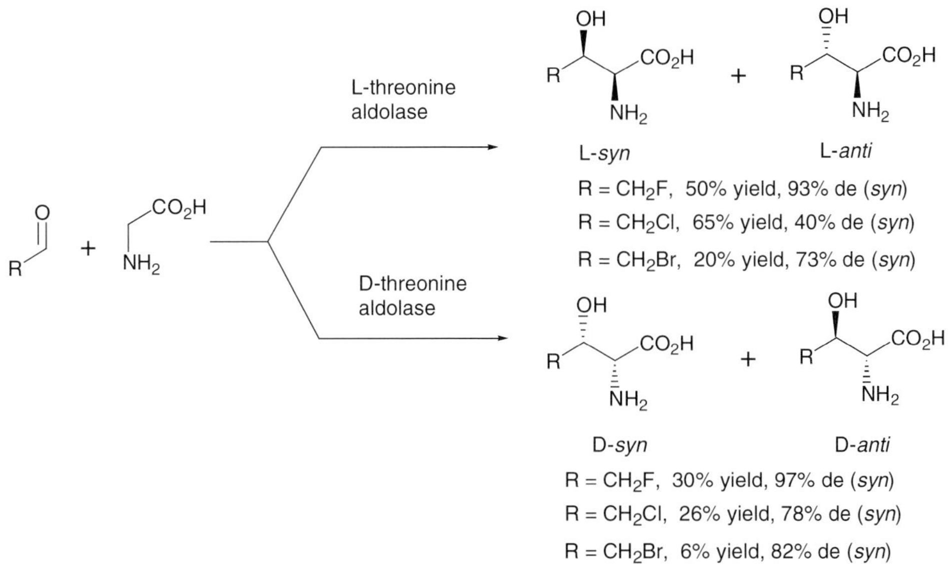

L-threonine aldolase

L-*syn* L-*anti*

R = CH_2F, 50% yield, 93% de (*syn*)
R = CH_2Cl, 65% yield, 40% de (*syn*)
R = CH_2Br, 20% yield, 73% de (*syn*)

D-threonine aldolase

D-*syn* D-*anti*

R = CH_2F, 30% yield, 97% de (*syn*)
R = CH_2Cl, 26% yield, 78% de (*syn*)
R = CH_2Br, 6% yield, 82% de (*syn*)

Scheme 10.34 Synthesis of halogenated β-hydroxy-α-amino acids.

10.4.2 Halogenated Threonine and Lysine

Biotransformations have proved valuable for the synthesis of a series of 4-halogenated threonines [83]. The reaction requires PLP and may be catalyzed either by L-threonine aldolase from *Pseudomonas putida* leading to the analogous halogenated L-amino acids or by D-threonine aldolase from *Alcaligenes xylosoxidans* giving the D series. As shown in Scheme 10.34, a stereocontrolled addition of glycine to a series of halogenated α-haloacetaldehydes gives the γ-halo-β-hydroxy-α-amino acids in favor of the *syn* diastereomers.

As well as the synthesis of Fmoc-3,3-difluoroalanine **25** (Scheme 10.7), in 2007, Li *et al.* described the preparation of protected 4,4-difluorothreonine **149** from α-hydroxy ester **147** (prepared in three steps and 76% yield from L-ascorbic acid) [20]. Following protection of the secondary alcohol of **147**, the ester was reduced to an aldehyde for reaction with DAST to introduce the difluoromethyl group **148**

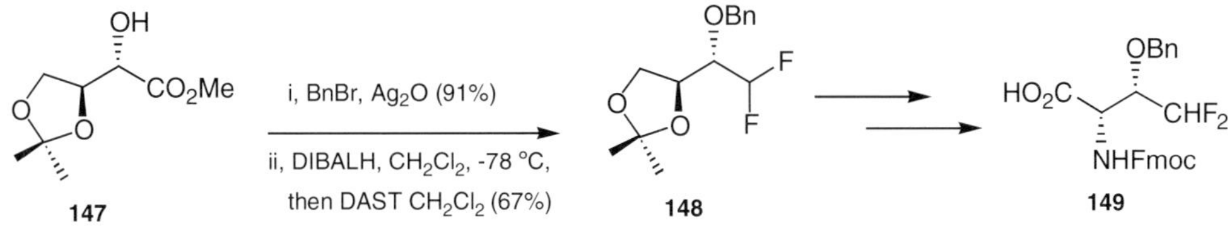

i, BnBr, Ag_2O (91%)
ii, DIBALH, CH_2Cl_2, -78 °C, then DAST CH_2Cl_2 (67%)

147 **148** **149**

Scheme 10.35 Synthesis of protected 4,4-difluorothreonine.

(Scheme 10.35). Using a similar series of reactions as for the synthesis of protected difluoroalanine **25** (Scheme 10.7), acetal **148** was converted in nine steps and 33% overall yield to protected 4,4-difluorothreonine **149**.

There are few reports of the synthesis of halogenated amino acids with the nitrogen-containing aliphatic side-chains of lysine and arginine, and most involve fluorination [84]. However, photochlorination of L-lysine in concentrated HCl was found to be both regio- and stereoselective, giving the *erythro* product, (2*S*,4*S*)-4-chlorolysine in 74% yield, which was then converted to γ-hydroxy-L-lysine using silver nitrate and used in the synthesis of *threo*-γ-hydroxyhomo-L-arginine [31, 85].

In conclusion, it is evident from this chapter that halogenated amino acids are common components of natural products and have a number of applications. Whilst a wide variety of methods exist for their synthesis, challenges remain in the development of further efficient, regio- and stereoselective syntheses of halogenated amino acids.

References

1 Gribble, G.W. (2004) Natural organohalogens: a new frontier for medicinal agents? *Journal of Chemical Education*, **81**, 1441–1449.

2 Niedleman, N.L. and Geigert, J. (1986) *Biohalogenation: Principles, Basic Roles and Applications*, Ellis Horwood, Chichester.

3 Lasne, M.-C., Perrio, C., Rouden, J., Barre, L., Roeda, D., Dolle, F., and Crouzel, C. (2002) Chemistry of β^+-emitting compounds based on fluorine-18. *Topics in Current Chemistry*, **222**, 201–258.

4 (a) Sutherland, A. and Willis, C.L. (2000) Synthesis of fluorinated amino acids. *Natural Product Reports*, **17**, 621–631; (b) Qiu, X.-L., Meng, W.-D., and Qing, F.-L. (2004) Synthesis of fluorinated amino acids. *Tetrahedron*, **60**, 6711–6745.

5 Kagiyama, T., Glushakov, A.V., Sumners, C., Roose, B., Dennis, D.M., Phillips, M.I., Ozcan, M.S., Seubert, C.N., and Martynyuk, A.E. (2004) Neuroprotective action of halogenated derivatives of L-phenylalanine. *Stroke*, **35**, 1192–1196.

6 Poillon, W.N. (1982) Noncovalent inhibitors of sickle haemoglobin gelation: effects of aryl-substituted alanines. *Biochemistry*, **21**, 1400–1406.

7 Panton, L.J., Rossan, R.N., Escajadillo, A., Matsumoto, Y., Lee, A.T., Labroo, V.M., Kirk, K.L., Cohen, L.A., Aikawa, M., and Howard, R.J. (1988) *In vitro* and *in vivo* studies of the effects of halogenated histidine analogs on *Plasmodium falciparum*. *Antimicrobial Agents and Chemotherapy*, **32**, 1655–1659.

8 Miyazawa, T. (1999) Enzymic resolution of amino acids via ester hydrolysis. *Amino Acids*, **16**, 191–213.

9 Nagasawa, T. and Yamada, H. (1986) Enzymic transformations of 3-chloroalanine into useful amino acids. *Applied Biochemistry and Biotechnology*, **13**, 147–165.

10 Kato, Y., Fukumoto, K., and Asano, Y. (1993) Enzymatic synthesis of L-β-chloroalanine using amino acid dehydrogenase. *Applied Microbiology and Biotechnology*, **39**, 301–304.

11 Choi, D. and Kohn, H. (1995) Trimethylsilyl halides: effective reagents for the synthesis of β-halo amino acid derivatives. *Tetrahedron Letters*, **36**, 7011–7014.

12 Lowpetch, K. and Young, D.W. (2005) A short, versatile chemical synthesis of L- and D-amino acids stereoselectively labelled solely in the *beta* position. *Organic and Biomolecular Chemistry*, **3**, 3348–3356.

13 Axelsson, B.S., O'Toole, K.J., Spencer, P.A., and Young, D.W. (1991) A versatile synthesis of stereospecifically labelled D-amino acids and related enzyme inhibitors. *Journal of the Chemical Society, Chemical Communications*, 1085–1086.

14 Meyer, F., Laaziri, A., Papini, A.M., Uziel, J., and Jugé, S. (2003) Efficient synthesis of β-halogeno protected L-alanines and their β-phosphonium derivatives. *Tetrahedron: Asymmetry*, **14**, 2229–2238.

15 Adlington, R.M. and Mantell, S.J. (1992) A radical route to 2(*S*)-4-exomethylene proline. *Tetrahedron*, **48**, 6529–6536.

16 Basak, A., Bag, S.S., Rudra, K.R., Barman, J., and Dutta, S. (2002) Diastereoselective synthesis of 4-substituted L-prolines by intramolecular radical cyclization of *N*-aryl sulphonyl-*N*-allyl 3-bromoalanines: interesting dependence of selectivity on the nature of the sulphonamido groups. *Chemistry Letters*, 710–711.

17 Barton, D.H.R., Hervé, Y., Potier, P., and Thierry, J. (1988) Manipulation of the carboxyl groups of α-amino acids and peptides using radical chemistry based on esters of *N*-hydroxy-2-thiopyridine. *Tetrahedron*, **44**, 5479–5486.

18 Itaya, T. and Mizutani, A. (1985) Synthesis of (*S*)-(−)-wybutine, the fluorescent minor base from yeast phenylalanine transfer ribonucleic acids. *Tetrahedron Letters*, **26**, 347–350.

19 Deboves, H.J.C., Montalbetti, C.A.G.N., and Jackson, R.W.F. (2001) Direct synthesis of Fmoc-protected amino acids using organozinc chemistry: application to polymethoxylated phenylalanines and 4-oxoamino acids. *Journal of the Chemical Society, Perkin Transactions 1*, 1876–1884.

20 Li, G.L. and van der Donk, W.A. (2007) Efficient synthesis of suitably protected β-difluoroalanine and γ-difluorothreonine from L-ascorbic acid. *Organic Letters*, **9**, 41–44.

21 Webb, T.R. and Eigenbrot, C. (1991) Conformationally restricted arginine analogs. *The Journal of Organic Chemistry*, **56**, 3009–3016.

22 Filosa, R., Holder, C., and Auberson, Y.P. (2006) Diastereoselectivity in the alkylation of 4-fluoroproline methyl esters. *Tetrahedron Letters*, **47**, 8929–8932.

23 Schumacher, K.K., Jiang, J., and Joullié, M.M. (1998) Synthetic studies toward astins A, B and C. Efficient syntheses of *cis*-3,4-dihydroxyprolines and (−)-(3*S*,4*R*)-dichloroproline esters. *Tetrahedron: Asymmetry*, **9**, 47–53.

24 Suzuki, A., Mae, M., Amii, H., and Uneyama, K. (2004) Catalytic route to the synthesis of optically active β,β-difluoroglutamic acid and β,β-difluoroproline derivatives. *The Journal of Organic Chemistry*, **69**, 5132–5134.

25 Easton, C.J. (1997) Free-radical reactions in the synthesis of α-amino acids and derivatives. *Chemical Reviews*, **97**, 53–82.

26 Bowman, N.J., Hay, M.P., Love, S.G., and Easton, C.J. (1988) Regioselective chlorination of valine derivatives. *Journal of the Chemical Society, Perkin Transactions 1*, 259–264.

27 Easton, C.J., Hay, M.P., and Love, S.G. (1988) Regioselective formation of amidocarboxy-substituted free radicals. *Journal of the Chemical Society, Perkin Transactions 1*, 265–268.

28 Sathe, M., Karade, H., and Kaushik, M.P. (2007) *N,N'*-Dichlorobis(2,4,6-trichlorophenyl)urea (CC-2): An efficient reagent for *N*-chlorination of amino esters, amide and peptides. *Chemistry Letters*, **36**, 996–997.

29 (a) Easton, C.J., Hutton, C.A., Tan, E.W., and Tiekink, E.R.T. (1990) Synthesis of homochiral hydroxy-α-amino acid derivatives. *Tetrahedron Letters*, **31**, 7059–7062; (b) Easton, C.J., Hutton, C.A., Merrett, M.C., and Tiekink, E.R.T. (1996) Neighbouring group effects promote substitution reactions over elimination and provide a stereocontrolled route to chloramphenicol. *Tetrahedron*, **52**, 7025–7036; (c) Easton, C.J. and Merrett, M.C. (1997) Stereoselective synthesis of (2*S*,3*S*)-γ-hydroxyvaline utilising an asymmetric radical hydrogen bromide addition. *Tetrahedron*, **53**, 1151–1156.

30 Easton, C.J., Edwards, A.J., McNabb, S.B., Merrett, M.C., O'Connell, J.L., Simpson, G.W., Simpson, J.S., and Willis, A.C. (2003) Allylic halogenation of unsaturated amino acids. *Organic and Biomolecular Chemistry*, **1**, 2492–2498.

31 Kollonitsch, J., Rosegay, A., and Doldouras, G. (1964) Reactions in strong acids. II. New concept in amino acid chemistry: C-derivatization of amino acids. *Journal of the American Chemical Society*, **86**, 1857–1858.

32 Kollonitsch, J., Scott, A.N., and Doldouras, G.A. (1966) C-Derivatization of amino acids. Synthesis and absolute configuration of 3-methyl proline. *cis–trans* Isomers with unusual vicinal proton coupling. *Journal of the American Chemical Society*, **88**, 3624–3626.

33 Charrier, J.-D., Hadfield, D.S., Hitchcock, P.B., and Young, D.W. (2004) Synthesis of (2*S*,3*S*)-3′-fluoroisoleucine. *Organic and Biomolecular Chemistry*, **2**, 797–802.

34 Reddy, L.R., Reddy, B.V.S., and Corey, E.J. (2006) Efficient method for selective introduction of substituents as C(5) of isoleucine and other α-amino acids. *Organic Letters*, **8**, 2819–2821.

35 Easton, C.J., Hutton, C.A., Rositano, G., and Tan, E.W. (1991) Regioselective functionalization of *N*-phthaloyl-substituted amino acid and peptide derivatives. *The Journal of Organic Chemistry*, **56**, 5614–5618.

36 Padmakshan, D., Bennett, S.A., Otting, G., and Easton, C.J. (2007) Stereocontrolled synthesis of (*S*)-γ-fluoroleucine. *Synlett*, 1083–1084.

37 Blunt, J.W., Copp, B.R., Hu, W.-P., Munro, M.H.G., Northcote, P.T., and Prinsep, M.R. (2007) Marine natural products. *Natural Product Reports*, **24**, 31–86.

38 Butler, A. (1998) Vanadium haloperoxidases. *Current Opinion in Chemical Biology*, **2**, 279–285.

39 Orjala, J. and Gerwick, W.H. (1996) Barbamide – a chlorinated metabolite with molluscicidal activity from the Caribbean cyanobacterium *Lyngbya majuscula*. *Journal of Natural Products*, **59**, 427–430.

40 Gerwick, W.H., Leslie, P., Long, G.C., Marquez, B.L., and Willis, C.L. (2003) [6-^{13}C]-(2*S*,4*S*)-5-Chloroleucine: synthesis and incubation studies with cultures of the cyanobacterium, *Lyngbya majuscula*. *Tetrahedron Letters*, **44**, 285–288.

41 Hanessian, S. and Margarita, R. (1998) 1,3-Asymmetric induction in dianionic allylation reactions of amino acid derivatives – synthesis of functionally useful enantiopure glutamates, pipecolates and pyroglutamates. *Tetrahedron Letters*, **39**, 5887–5890.

42 Ardá, A., Jiménez, C., and Rodríguez, J. (2004) A study of polychlorinated leucine derivatives: synthesis of (2*S*,4*S*)-5,5-dichloroleucine. *Tetrahedron Letters*, **45**, 3241–3243.

43 Takeda, T., Sasaki, R., Yamauchi, S., and Fujiwara, T. (1997) Transformation of ketones and aldehydes to *gem*-dihalides via hydrazones using copper(II) halides. *Tetrahedron*, **53**, 557–566.

44 Ardá, A., Jiménez, C., and Rodríguez, J. (2006) Polychlorinated leucine derivatives: synthesis of (2*S*,4*R*)-5,5-dichloroleucine and its *J*-based analysis. *European Journal of Organic Chemistry*, 3645–3651.

45 Durow, A.C., Long, G.C., O'Connell, S.J. and Willis, C.L. (2006) Total synthesis of the chlorinated marine natural product dysamide B. *Organic Letters*, **8**, 5401–5404.

46 Ardá, A., Soengas, R.G., Nieto, M.I., Jiménez, C., and Rodríguez, J. (2008) Total synthesis of (−)-dysthiazolamide. *Organic Letters*, **10**, 2175–2178.

47 Sitachitta, N., Márquez, B.L., Williamson, R.T., Rossi, J., Roberts, M.A., Gerwick, W.H., Nguyen, V.-A., and Willis, C.L. (2000) Biosynthetic pathway and origin of the chlorinated methyl group in barbamide and dechlorobarbamide, metabolites from the marine cyanobacterium *Lyngbya majuscula*. *Tetrahedron*, **56**, 9103–9113.

48 Galonić, D.P., Vaillancourt, F.H., and Walsh, C.T. (2006) Halogenation of unactivated carbon centers in natural product biosynthesis: trichlorination of leucine during barbamide biosynthesis. *Journal of the American Chemical Society*, **128**, 3900–3901.

49 Flatt, P.M., O'Connell, S.J., McPhail, K.L., Zeller, G., Willis, C.L., Sherman, D.H., and Gerwick, W.H. (2006) Characterization of the initial enzymatic steps of barbamide biosynthesis. *Journal of Natural Products*, **69**, 938–944.

50 (a) MacMillan, J.B. and Molinski, T.F. (2000) Herbacic acid, a simple prototype of 5,5,5-trichloroleucine metabolites from the sponge *Dysidea herbacea*. *Journal of Natural Products*, **63**, 155–157; (b) Jiménez, J.I. and Scheuer, P.J. (2001) New lipopeptides from the Caribbean cyanobacterium *Lyngbya majuscula*. *Journal of Natural Products*, **64**, 200–203.

51 de Laszlo, S.E. and Williard, P.G. (1985) Total synthesis of (+)-demethyldysidenin and (–)-demethylisodysidenin, hexachlorinated amino acids from the marine sponge *Dysidea herbacea*. Assignment of absolute stereochemistry. *Journal of the American Chemical Society*, **107**, 199–203.

52 (a) Helmchen, G. and Wegner, G. (1985) Synthesis of enantiomerically pure (*S*)-3-trichloromethylbutyric acid via asymmetric conjugate addition of trichloromethyl metal compounds to a chiral enoate. Activation effect of a sulfonylamino group. *Tetrahedron Letters*, **26**, 6047–6050; (b) Brantley, S.E. and Molinski, T.F. (1999) Synthetic studies of trichloroleucine marine natural products. Michael addition of $LiCCl_3$ to *N*-crotylcamphor sultam. *Organic Letters*, **1**, 2165–2167.

53 Nguyen, V.-A., Willis, C.L., and Gerwick, W.H. (2001) Synthesis of the marine natural product barbamide. *Journal of the Chemical Society, Chemical Communications*, 1934–1935.

54 Hanessian, S., Del Valle, J.R., Xue, Y., and Blomberg, N. (2006) Total synthesis and structural confirmation of chlorodysinosin A. *Journal of the American Chemical Society*, **128**, 10491–10495.

55 Valls, N., Borregán, M., and Bonjoch, J. (2006) Synthesis of β-chloro α-amino acids: (2*S*,3*R*)- and (2*S*, 3S)-3-chloroleucine. *Tetrahedron Letters*, **47**, 3701–3705.

56 Houghten, R.A. and Rapoport, H. (1974) Synthesis of pure *p*-chlorophenyl-L-alanine from L-phenylalanine. *Journal of Medicinal Chemistry*, **17**, 556–558.

57 Leduc, R., Bernier, M., and Escher, E. (1983) Angiotensin-II analogues. I: synthesis and incorporation of the halogenated amino acids 3-(4′-iodophenyl) alanine, 3-(3′,5′-dibromo-4′-chlorophenyl) alanine, 3-(3′,4′,5′-tribromophenyl)alanine and 3-(2′,3′,4′,5′,6′-pentabromophenyl) alanine. *Helvetica Chimica Acta*, **66**, 960–970.

58 Yokoyama, Y., Yamaguchi, T., Sato, M., Kobayashi, E., Murakami, Y., and Okuno, H. (2006) Chemistry of unprotected amino acids in aqueous solution: direct bromination of aromatic amino acids with bromoisocyanuric acid sodium salt under strong acid condition. *Chemical & Pharmaceutical Bulletin*, **54**, 1715–1719.

59 Pajpanova, T.I. (2000) A simple and convenient procedure for the preparation of 3,5-dihalogenated tyrosine derivatives useful in peptide synthesis. *Dokladi na Bulgarskata Akademiya na Naukite*, **53**, 53–56.

60 McCord, T.J., Smith, D.R., Winters, D.W., Grimes, J.F., Hulme, K.L., Robinson, L.Q., Gage, L.D., and Davis, A.L. (1975) Synthesis and microbiological activities of some monohalogenated analogs of tyrosine. *Journal of Medicinal Chemistry*, **18**, 26–29.

61 Krasikova, R.N., Zaitsev, V.V., Ametamey, S.M., Kuznetsova, O.F., Fedorova, O.S.,

Mosevich, I.K., Belokon, Y.N., Vyskočil, S., Shatik, S.V., Nader, M., and Schubiger, P.A. (2004) Catalytic enantioselective synthesis of ^{18}F-fluorinated α-amino acids under phase-transfer conditions using (*S*)-NOBIN. *Nuclear Medicine and Biology*, **31**, 597–603.

62 El-Wetery, A.S., El-Mohty, A.A., Ayyoub, S., and Raieh, M. (1997) Catalytic effect of copper(II) chloride on the radioiodination of L-*p*-iodophenylalanine. *Journal of Labelled Compounds & Radiopharmaceuticals*, **39**, 631–644.

63 Wilson, S.R. and Jacob, L.A. (1986) Iodination of aryltrimethylsilanes: a mild approach to iodophenylalanine. *The Journal of Organic Chemistry*, **51**, 4833–4836.

64 Chuang, T.-H. and Sharpless, K.B. (2000) Applications of arizidinium ions. Selective syntheses of α,β-diamino esters, α-sulfanyl-β-amino esters, β-lactams and 1,5-benzodiazepin-2-one. *Organic Letters*, **2**, 3555–3557.

65 Williams, D.H. (1984) Structural studies on some antibiotics of the vancomycin group, and on the antibiotic–receptor complexes, by proton NMR. *Accounts of Chemical Research*, **17**, 364–369.

66 Tymiak, A.A., McCormick, T.J., and Unger, S.E. (1989) Structure determination of lysobactin, a macrocyclic peptide lactone antibiotic. *The Journal of Organic Chemistry*, **54**, 1149–1157.

67 Easton, C.J., Hutton, C.A., Roselt, P.D., and Tiekink, E.R.T. (1994) Stereocontrolled synthesis of β-hydroxyphenylalanine and β-hydroxytyrosine derivatives. *Tetrahedron*, **50**, 7327–7340.

68 Crich, D. and Banerjee, A. (2006) Expedient synthesis of *threo*-β-hydroxy-α-amino acid derivatives: phenylalanine, tyrosine, histidine and tryptophan. *The Journal of Organic Chemistry*, **71**, 7106–7109.

69 Brown, T., Jones, J.H., and Richards, J.D. (1992) Further studies on the protection of histidine side chains in peptide synthesis: the use of the π-benzyloxymethyl group. *Journal of the Chemical Society, Perkin Transactions 1*, 1553–1561.

70 Jain, R., Avramovitch, B., and Cohen, L.A. (1998) Synthesis of ring-halogenated histidines and histamines. *Tetrahedron*, **54**, 3235–3242.

71 Cerezo, V., Afonso, A., Planas, M., and Feliu, L. (2007) Synthesis of 5-arylhistidines via a Suzuki–Miyaura cross-coupling. *Tetrahedron*, **63**, 10445–10453.

72 Bittner, S., Scherzer, R., and Harlev, E. (2007) The five bromotryptophans. *Amino Acids*, **33**, 19–42.

73 (a) Yeh, E., Garneau, S., and Walsh, C.T. (2005) Robust *in vitro* activity of RebF and RebH, a two-component reductase/halogenase, generating 7-chlorotryptophan during rebeccamycin biosynthesis. *Proceedings of the National Academy of Sciences of the United States of America*, **102**, 3960–3965; (b) Lindquist, N., Fenical, W., Van Duyne, G.D., and Clardy, J. (1991) Isolation and structure determination of diazonamides A and B, unusual cytotoxic metabolites from the marine ascidian *Diazona chinensis*. *Journal of the American Chemical Society*, **113**, 2303–2304.

74 Rydon, H.N. and Tweddle, J.C. (1955) Experiments on the synthesis of Bz-substituted indoles and tryptophans. III. The synthesis of four Bz-chloroindoles and -tryptophans. *Journal of the Chemical Society*, 3499–3503.

75 (a) Ma, C., Liu, X., Li, X., Flippen-Anderson, J., Yu, S., and Cook, J.M. (2001) Efficient asymmetric synthesis of biologically important tryptophan analogues via a palladium-mediated heteroannulation reaction. *The Journal of Organic Chemistry*, **66**, 4525–4542; (b) Perry, C.W., Brossi, A., Deitcher, K.H., Tautz, W., and Teitel, S. (1977) Preparation of (*R*)-(+)- and (*S*)-(−)-α-methyl-*p*-nitrobenzylamines and their use as resolving agents. *Synthesis*, 492–494; (c) Konda-Yamada, Y., Okada, C., Yoshida, K., Umeda, Y., Arima, S., Sato, N., Kai, T.,

Takayanagi, H., and Harigaya, Y. (2002) Convenient synthesis of 7′- and 6′-bromo-D-tryptophan and their derivatives by enzymatic optical resolution using D-aminoacylase. *Tetrahedron*, **58**, 7851–7861.

76 Shiba, T., Mukunoki, Y., and Akiyama, H. (1975) Component amino acids of the antibiotic longicatenamycin. Isolation of 5-chloro-D-tryptophan. *Bulletin of the Chemical Society of Japan*, **48**, 1902–1906.

77 Taniguchi, M., Gonsho, A., Nakagawa, M., and Hino, T. (1983) Cyclic tautomers of tryptophans and tryptamines. VI. Preparation of N_a-alkyl-, 5-chloro-, and 5-nitrotryptophan derivatives. *Chemical & Pharmaceutical Bulletin*, **31**, 1856–1865.

78 Moriya, T., Hagio, K., and Yoneda, N. (1975) Facile synthesis of 6-chloro-D-tryptophan. *Bulletin of the Chemical Society of Japan*, **48**, 2217–2218.

79 Goss, R.J.M. and Newill, P.L.A. (2006) A convenient enzymatic synthesis of L-halotryptophans. *Journal of the Chemical Society, Chemical Communications*, 4924–4925.

80 (a) Akhtar, M., Botting, N.P., Cohen, M.A., and Gani, D. (1987) Enantiospecific synthesis of 3-substituted aspartic acids via enzymic amination of substituted fumaric acids. *Tetrahedron*, **43**, 5899–5908; (b) Akhtar, M., Cohen, M.A., and Gani, D. (1987) Stereochemical course of the enzymic amination of chloro- and bromofumaric acid by 3-methylaspartate ammonia-lyase. *Tetrahedron Letters*, **28**, 2413–2416.

81 Silverman, R.B. and Holladay, M. (1981) Stereospecific total syntheses of the natural antitumor agent (α*S*,5*S*)-α-amino-3-chloro-4,5-dihydro-5-isoxazoleacetic acid and its unnatural C-5 epimer. *Journal of the American Chemical Society*, **103**, 7357–7358.

82 Iwasaki, H., Kamiya, T., Oka, O., and Ueyanagi, J. (1965) Synthesis of tricholomic acid, a flycidal amino acid. I. *Chemical & Pharmaceutical Bulletin*, **13**, 753–758.

83 Steinreiber, J., Fesko, K., Mayer, C., Resinger, C., Schürmann, M., and Griengl, H. (2007) Synthesis of γ-halogenated and long-chain β-hydroxy-α-amino acids and 2-amino-1,3-diols using threonine aldolases. *Tetrahedron*, **63**, 8088–8093.

84 (a) Tolman, V. and Benes, J. (1976) Monofluorinated analogues of some aliphatic basic amino acids. *Journal of Fluorine Chemistry*, **7**, 397–407; (b) Kitagawa, O., Hashimoto, A., Kobayashi, Y., and Taguchi, T. (1990) Michael addition of 2,2-difluoroketene silyl acetal. Preparation of 4,4-difluoroglutamic acids and 5,5-difluorolysine. *Chemistry Letters*, 1307–1310.

85 Fujita, Y., Kollonitsch, J., and Witkop, B. (1965) The stereospecific synthesis of *threo*-γ-hydroxyhomo-L-arginine from *Lathyrus* species. *Journal of the American Chemical Society*, **87**, 2030–2033.

11
Synthesis of Isotopically Labeled α-Amino Acids

Caroline M. Reid and Andrew Sutherland

11.1
Introduction

α-Amino acids specifically labeled with stable (^2H, ^{13}C, ^{15}N, ^{17}O, and ^{18}O) and radioactive (^3H, ^{11}C, ^{14}C, ^{13}N, ^{15}O, and ^{35}S) isotopes are valuable tools for a range of bioorganic chemical studies. These include the elucidation of biosynthetic pathways, the investigation of peptide and protein conformations using nuclear magnetic resonance (NMR) spectroscopy as well as a wide range of clinical applications. The importance of these studies is reflected in the large number of methods that have been developed for the synthesis of α-amino acids incorporating isotopic labels [1]. The challenge in this area is the efficient synthesis of the α-amino acid while incorporating the sometimes expensive and/or unstable isotope at a late stage of the synthesis, thereby maximizing the utilization of the isotopic label. This chapter describes the main synthetic approaches that have been developed to achieve this goal, such as, enzyme-catalyzed procedures, the use of the chiral pool as a readily available source of chiral starting materials, and asymmetric synthesis including the use of chiral auxiliaries.

11.2
Enzyme-Catalyzed Methods

One of the most common methods for the stereoselective synthesis of α-amino acids involves the use of enzyme-catalyzed reactions. The advantages of using enzymes include fast, mild reactions that give high yields and enantioselectivities.

The class of enzyme that has been most commonly used for the synthesis of isotopically labeled α-amino acids are the amino acid dehydrogenases (AADHs) [1]. These are a superfamily of enzymes that catalyze the reductive amination of α-keto acids using ammonia as the nitrogen source and either NADH or NADPH as the reductant (Scheme 11.1). Thus, this process allows the introduction of the stereogenic center at C-2 of the α-amino acid.

Amino Acids, Peptides and Proteins in Organic Chemistry. Vol.1 – Origins and Synthesis of Amino Acids.
Edited by Andrew B. Hughes

ISBN: 978-3-527-32096-7

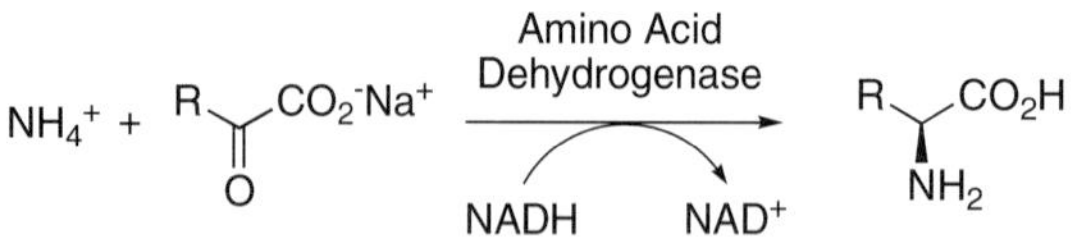

$$NH_4^+ + RC(=O)CO_2^-Na^+ \xrightarrow[\text{NADH} \rightarrow \text{NAD}^+]{\text{Amino Acid Dehydrogenase}} RCH(NH_2)CO_2H$$

Scheme 11.1

When first developed as a synthetic process, stoichiometric amounts of the expensive cofactor, NAD(P)H were used. However, a number of different recycling strategies have been developed resulting in an efficient process that uses only catalytic amounts of the AADH and the cofactor [2].

The use of the [^{15}N]ammonium ion or isotopically labeled α-keto acids for this reaction has led to a general method for the synthesis of a range of isotopically labeled α-amino acids [3–6]. An early example of this approach developed by Ducrocq *et al.*, involved the preparation of 4-[$^2H_2$]L-glutamic acid **1** using glutamate dehydrogenase to carry out the reductive amination and alcohol dehydrogenase to recycle the NAD$^+$ [3]. Using ammonium chloride as the source of nitrogen gave 4-[$^2H_2$]L-glutamic acid **1** in quantitative yield (Scheme 11.2).

More recently, the formate dehydrogenase (FDH) recycling system for the cofactor, NADH has been used to good effect by Willis and coworkers for the general synthesis of isotopically labeled α-amino acids [7–11]. The use of [^{15}N]ammonium formate not only provides the source of ammonium ion and isotopic label, but the formate counterion is also utilized to recycle the NADH (Table 11.1). Moreover, this recycling system leads to the generation of the isotopically labeled α-amino acid very cleanly, as the byproduct from the FDH-catalyzed reaction is carbon dioxide [7].

AADHs have a narrow scope of substrate specificity as the pockets that bind the side-chain of the α-keto acid tend only to accept groups similar in steric and electronic character to that of the natural substrate. However, as a family of AADH enzymes exist, an enzyme can normally be found that catalyzes the reductive amination of a particular α-keto acid. This was the case for the synthesis of [^{15}N]L-*allo*-threonine **2** that was prepared using leucine dehydrogenase while [^{15}N]L-threonine **3** was prepared using phenylalanine dehydrogenase [12]. Both substrates for these enzyme-catalyzed reactions were prepared from the corresponding lactic acid derivative. The resulting α-keto ester was then transformed into the desired [^{15}N]α-amino acid using a one-pot, dual-enzyme procedure involving the hydrolysis of the ester using lipase followed by the reductive amination of the ketone using the

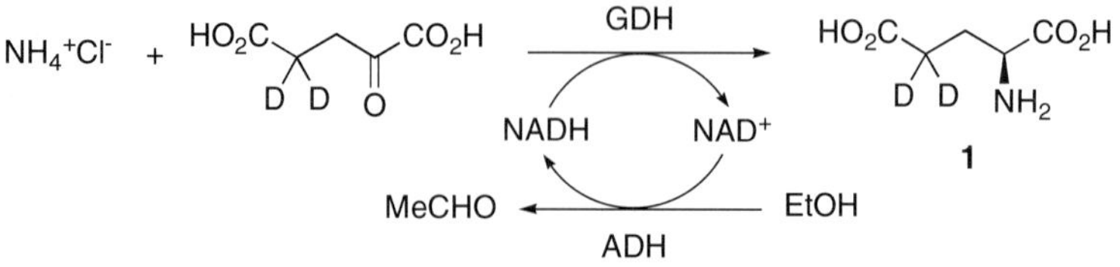

$$NH_4^+Cl^- + HO_2C\text{-}CD_2\text{-}CH_2\text{-}C(=O)CO_2H \xrightarrow[\text{NADH} \rightarrow \text{NAD}^+]{\text{GDH}} HO_2C\text{-}CD_2\text{-}CH_2\text{-}CH(NH_2)CO_2H \;(\mathbf{1})$$

(NAD$^+$ recycled to NADH by ADH: EtOH → MeCHO)

Scheme 11.2

Table 11.1 AADH-catalyzed formation of [^{15}N]L-amino acids.

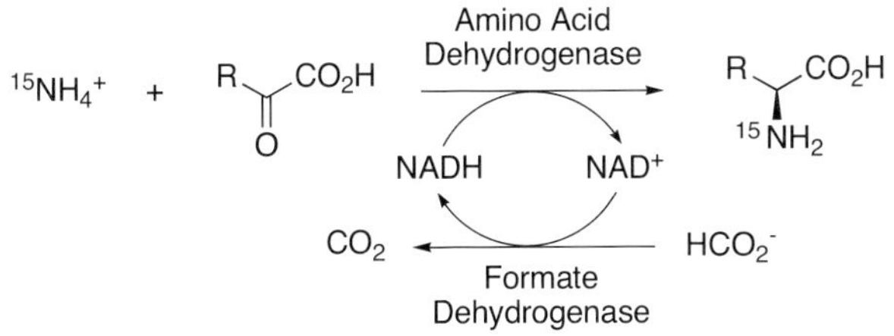

$^{15}NH_4^+$ + R–CO–CO_2H → (Amino Acid Dehydrogenase; NADH → NAD^+) R–CH($^{15}NH_2$)–CO_2H; HCO_2^- → CO_2 (Formate Dehydrogenase)

R	Enzyme	Yield (%)
CH_3	AlaDH	93
$HOCH_2$	AlaDH	66
$(CH_3)_2CH$	LeuDH	84
$(CH_3)_2CHCH_2$	LeuDH	95
$PhCH_2$	PheDH	92

appropriate AADH (Scheme 11.3). The advantage of this dual-enzyme procedure is that mild lipase hydrolysis of the ester prevents any racemization of the C-3 stereogenic center. Acid-mediated hydrolysis of the methoxymethyl (MOM) protecting group then completed the preparation of the [^{15}N]β-hydroxy-α-amino acids.

One of the limitations of using AADHs, especially during large-scale reactions, is substrate inhibition of the enzyme. One method of overcoming this problem is to add the substrate in portions to the reaction mixture [7]. Such an approach can produce labeled α-amino acids in quantities up to 50 mmol. Alternative techniques include a repetitive batch method where the product of the enzyme reaction is filtered from the reaction vessel leaving behind the enzyme for a further batch of substrate [13]. Using this technique with glutamate dehydrogenase and the formate/FDH recycling system has produced 37 g of [^{15}N]L-glutamic acid after seven batches in 85% overall yield.

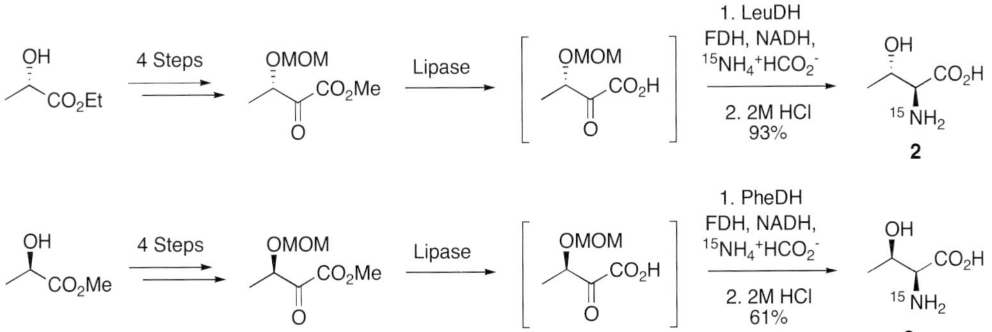

OH, CO_2Et —4 Steps→ OMOM, CO_2Me, O —Lipase→ [OMOM, CO_2H, O] —1. LeuDH, FDH, NADH, $^{15}NH_4^+HCO_2^-$; 2. 2M HCl, 93%→ OH, CO_2H, $^{15}NH_2$ **2**

OH, CO_2Me —4 Steps→ OMOM, CO_2Me, O —Lipase→ [OMOM, CO_2H, O] —1. PheDH, FDH, NADH, $^{15}NH_4^+HCO_2^-$; 2. 2M HCl, 61%→ OH, CO_2H, $^{15}NH_2$ **3**

Scheme 11.3

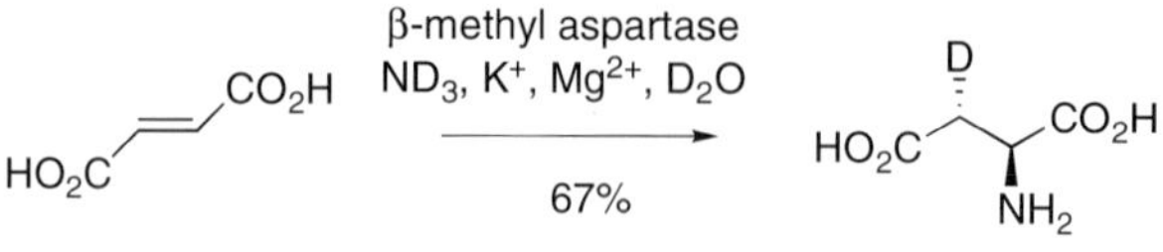

Scheme 11.4

A second class of biotransformation commonly used for the stereoselective synthesis of isotopically labeled α-amino acids involves synthase enzymes. Like AADHs, a number of synthase enzymes are commercially available, and thus these reactions can thus be routinely performed without having to grow organisms and isolate and purify the enzymes. This general class of enzyme carry out key bond-forming reactions during the biosynthesis of α-amino acids. Thus, the use of organisms that overexpress these enzymes in combination with isotopically labeled substrates or solvent is an effective method of preparing isotopically labeled α-amino acids. Gani and coworkers used such a strategy for the preparation of (2*S*,3*R*)-[3-$^2H_1$] aspartic acid **5** where β-methylaspartase catalyzed the stereoselective addition of ammonia across the double bond of fumaric acid **4** [14]. The use of deuteriated ammonia and D_2O gave (2*S*,3*R*)-[3-$^2H_1$]aspartic acid **5** in 67% yield (Scheme 11.4). In a similar manner, β-methylaspartase has been utilized for the synthesis of (2*S*,3*S*)-[4-^{13}C]valine and (2*S*,4*R*)-[5-^{13}C]leucine using mesaconic acid and 2-ethylfumaric acid as substrates, respectively [15].

Tryptophan synthases and β-tyrosinases are enzymes that typically catalyze C–C bond-forming reactions allowing the introduction of an aromatic side-chain. The use of labeled substrates for these reactions has led to the synthesis of a wide range of isotopically labeled analogs of tyrosine, phenylalanine, and tryptophan [16–21]. For example, reaction of L-serine **7** with [2,6-$^{13}C_2$]phenol **6** in the presence of a β-tyrosinase enzyme from *Erwinia herbicola* gave [3′,5′-$^{13}C_2$]L-tyrosine **8** in 91% yield (Scheme 11.5) [20]. Deoxygenation of the aromatic ring was also carried out via an activated ether leading to the preparation of [3′,5′-$^{13}C_2$]L-phenylalanine **9**.

One of the major applications of enzymes in synthetic organic chemistry is the kinetic resolution of racemic mixtures to produce enantiomerically enriched products. This approach has also been used for the preparation of isotopically labeled L-amino acids and generally involves enzyme-catalyzed hydrolysis of *N*-acyl derivatives [22–25]. This methodology was used for the synthesis of [$^{13}C_3$]L-leucine **12** (Scheme 11.6) [22]. Racemic [$^{13}C_3$]leucine **11** was initially prepared in nine steps from sodium acetate **10** using a modified Strecker reaction to install the amino acid functional groups. Acetylation with trifluoroacetic anhydride followed by resolution

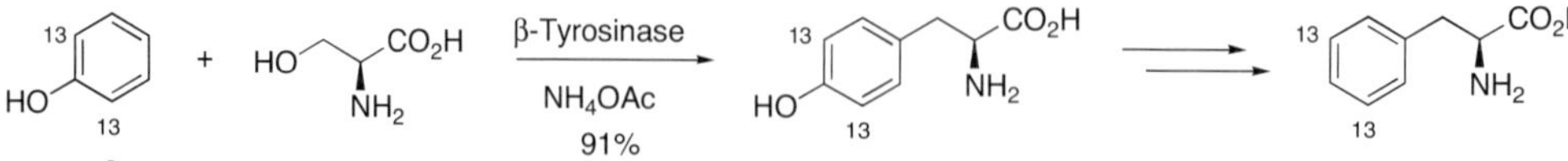

Scheme 11.5

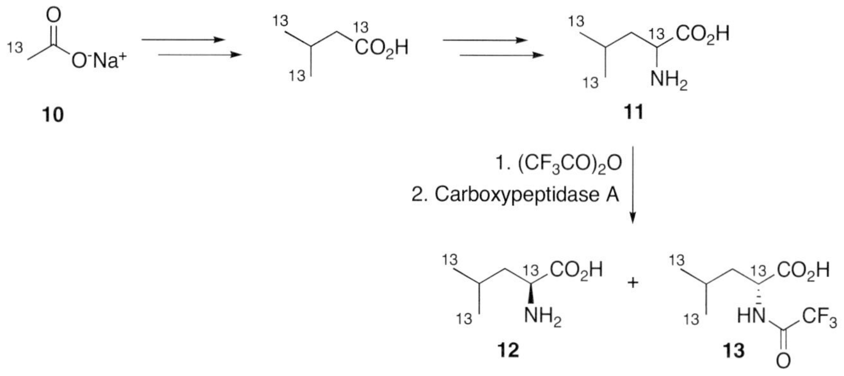

Scheme 11.6

with carboxypeptidase A gave [$^{13}C_3$]L-leucine **12** in 30% yield and the corresponding D-trifluoroacetate derivative **13** in 40% yield. Although hydrolytic enzymes are widespread and readily available, as highlighted by this example, the major drawback with this approach is that at least 50% of the isotopically labeled material is lost during this key step.

The methods described above are the most commonly used enzyme reactions for the synthesis of α-amino acids incorporating isotopes. However, to a lesser extent enzyme-catalyzed transferase [26–29], decarboxylation [30], and reduction reactions [31] have also been utilized.

11.3
Chiral Pool Approach

A simple, direct and cost-efficient approach for the synthesis of enantiomerically pure α-amino acids is the use of the chiral pool strategy. In particular, the use of readily available proteinogenic α-amino acids has led to a large number of methods for the synthesis of α-amino acids incorporating isotopic labels. Ragnarsson and coworkers produced a general method for the preparation of ^{15}N-labeled Boc-protected L-amino acids from the corresponding D-amino acid [32]. Diazotization of the D-amino acid **14** and ester formation gave the corresponding α-hydroxy ester **15** with retention of configuration (Scheme 11.7). Activation of the hydroxyl group as

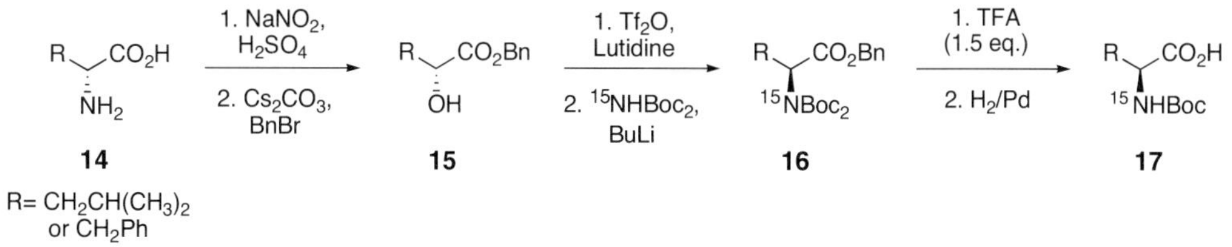

Scheme 11.7

the triflate and displacement with the lithium salt of $Boc_2{}^{15}NH$ gave the di-Boc-protected L-amino ester **16** with inversion of configuration. Selective deprotection then yielded the ^{15}N-labeled Boc-protected amino acids **17** with enantiomeric excesses greater than 98%.

The approach above uses the amino group to incorporate the isotopic label. However, the most common strategy normally involves the elaboration of the side-chain of a protected α-amino acid while simultaneously incorporating the isotopic label. Two α-amino acids that are highly amenable to this strategy are the carboxylic acid derivatives, L-aspartic acid and L-glutamic acid. The challenge in using these compounds is the differentiation between the α- and side-chain carboxylic acid groups. This is normally achieved using orthogonal carboxylic acid protecting groups or bulky amino protecting groups that shield the α-carboxylic acid position. Chamberlin and coworkers utilized this type of approach for the stereoselective synthesis of (2*S*,3*S*)-[4-^{13}C]valine **21** (Scheme 11.8) [33]. The first stage involved a stereoselective alkylation of the protected aspartate analog **18** using [^{13}C]methyl iodide to introduce the isotopic label. Regioselective reduction of the less sterically hindered β-ester **19** was then carried out to give the corresponding alcohol **20** and this was followed by a two-step deoxygenation protocol that produced the desired carbon skeleton. Deprotection using standard conditions then gave (2*S*,3*S*)-[4-^{13}C] valine **21**. This synthesis is significant as it allows the specific labeling of one of the diastereotopic methyl groups. The ability to do this with either valine or leucine

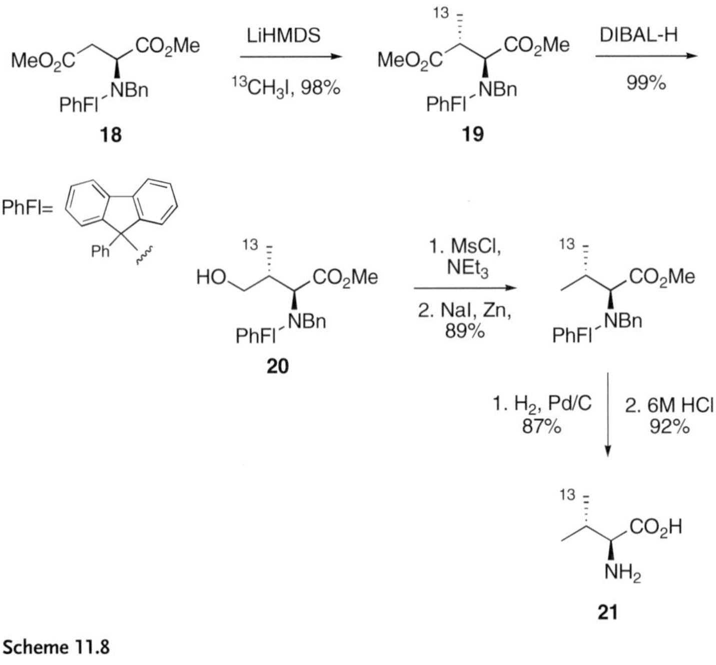

MeO2C CO2Me PhFl NBn **18** — LiHMDS, $^{13}CH_3I$, 98% → MeO2C CO2Me 13 PhFl NBn **19** — DIBAL-H, 99% → HO CO2Me 13 PhFl NBn **20** — 1. MsCl, NEt3; 2. NaI, Zn, 89% → CO2Me 13 PhFl NBn — 1. H2, Pd/C 87%; 2. 6M HCl 92% → 13 CO2H NH2 **21**

PhFl= Ph

Scheme 11.8

and the subsequent incorporation of these amino acids within protein structures leads to more precisely defined NMR spectra and the tools to probe hydrophobic interactions.

Reported syntheses of specifically labeled L-arginine are relatively rare in the literature. However, [5-^{13}C]L-arginine **29** was prepared by Hamilton and Sutherland using a flexible approach from L-aspartic acid **22** (Scheme 11.9) [34]. In this example, L-aspartic acid **22** was converted in two steps to *N,N*-di-*tert*-butoxycarbonyl L-aspartic acid dimethyl ester **23**. Like the previous example, di-protection of the amino group sterically hinders the α-ester and allows the selective manipulation of the β-ester [35]. Thus, treatment with diisobutylaluminum hydride (DIBAL-H) gave only the β-aldehyde **24**. Further reduction with sodium borohydride gave alcohol **25** that was converted in two steps to the corresponding iodide **26**. Reaction of **26** with sodium [^{13}C]cyanide then gave the nitrile **27** while also incorporating the isotopic label. Selective deprotection of one of the Boc protecting groups and the methyl ester followed by hydrogenation of the nitrile functional group gave the ornithine derivative **28**. The last stage involved coupling of the ornithine side-chain with the commercially available carboxamidine followed by deprotection of the Boc groups, which gave [5-^{13}C]L-arginine **29**. These two examples show the different types of isotopically labeled amino acids that can be easily prepared from carboxylic acid-derived amino acids. Similar strategies have been used for the synthesis of isotopically labeled isoleucine, asparagine, proline, lysine, histidine, and glutamic acid [33, 36–42].

Selective transformation of the hydroxyl group of L-serine has also resulted in the preparation of a range of isotopically labeled amino acids [43–45]. Lodwig and Unkefer prepared [4-^{13}C]L-aspartic acid **33** in three steps from Boc-protected L-serine **30** (Scheme 11.10) [46]. Formation of the serine β-lactone **31** was done using the Vederas protocol involving triphenylphosphine and dimethyl azodicarboxylate (DMAD) [47]. Ring opening using potassium [^{13}C]cyanide gave the corresponding β-cyanoalanine **32** with incorporation of the ^{13}C label. Acid-mediated hydrolysis of the nitrile and Boc deprotection of the amino group gave [4-^{13}C] L-aspartic acid **33**.

Oba and coworkers used D-serine for the stereoselective synthesis of (2*S*,3*R*)-[3-^{2}H] serine **38** (Scheme 11.11) [48]. A protected derivative of D-serine **34** was converted to the pyrazolide **35** by reaction with 3,5-dimethylpyrazole. Reduction with $LiAlD_4$ gave the 1-deuterio aldehyde **36** which was then stereoselectively reduced using (*S*)-Alpine-Borane. Acylation of the resulting hydroxyl **37**, followed by deprotection of the benzyl ether and ruthenium-catalyzed oxidation gave the protected L-serine analog. Acid hydrolysis of the remaining protecting groups gave (2*S*,3*R*)-[3-^{2}H]serine **38** in excellent overall yield (66%). A similar approach was used for the preparation of (2*R*,2′*R*,3*S*,3′*S*)-[3,3′-$^2H_2$]cysteine.

The research groups of Young and Nishiyama have used pyroglutamate analogs as precursors for the synthesis of a range of isotopically labeled α-amino acids such as leucine, isoleucine, valine, and proline [49–57]. Their strategy utilizes the pyroglutamate ring to introduce additional functionality, followed by facile ring-opening

Scheme 11.9

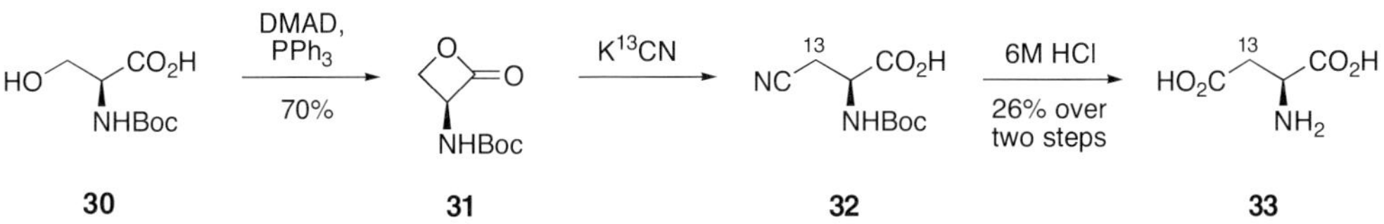

Scheme 11.10

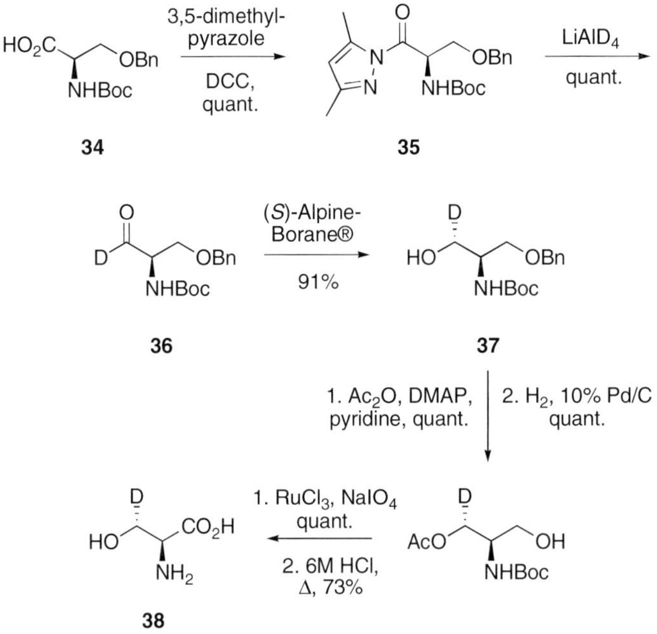

Scheme 11.11

reactions for the synthesis of linear amino acids. For example, Young and coworkers used pyroglutamate analog **39** to good effect for the preparation of (2*S*,3*R*)-[3′,3′,3′-$^2H_3$] valine **43** (Scheme 11.12) [51]. Reaction of **39** with a copper Grignard reagent allowed the stereoselective incorporation of a CD_3 group on the least hindered face of the ring. Hydrolysis of the lactam **40** (100% diastereomeric selectivity) was then achieved using lithium hydroxide and the resulting carboxylic acid **41** was converted to the valine side-chain using chemistry developed by Barton and coworkers [58]. Deprotection of the silyl protecting group and subsequent oxidation using ruthenium chloride and sodium periodate gave the α-carboxylic acid **42**. Finally, deprotection of the amino group gave (2*S*,3*R*)-[3′,3′,3′-$^2H_3$]valine **43** in excellent overall yield. Again, this synthesis is significant as it allows the diastereotopic labeling of one of the methyl groups.

The other major source of material from the chiral pool for the synthesis of isotopically labeled amino acids has been the use of carbohydrates [59, 60].

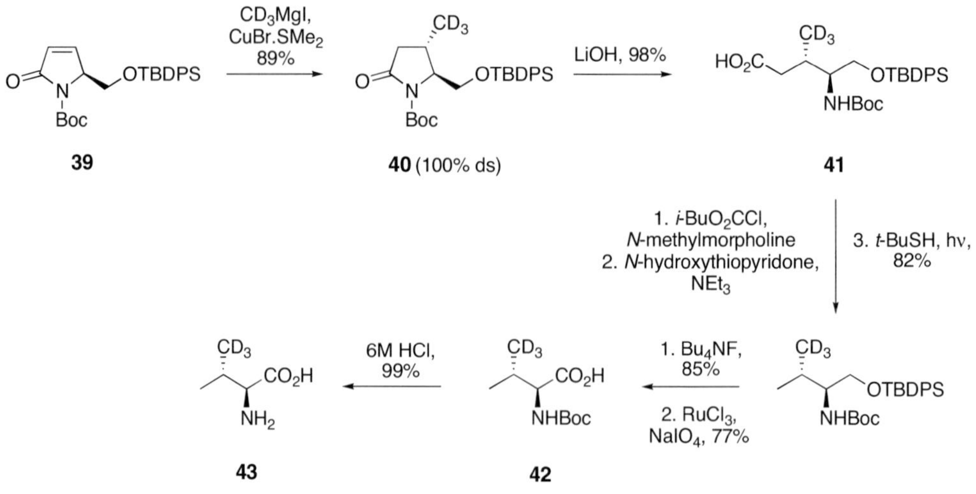

Scheme 11.12

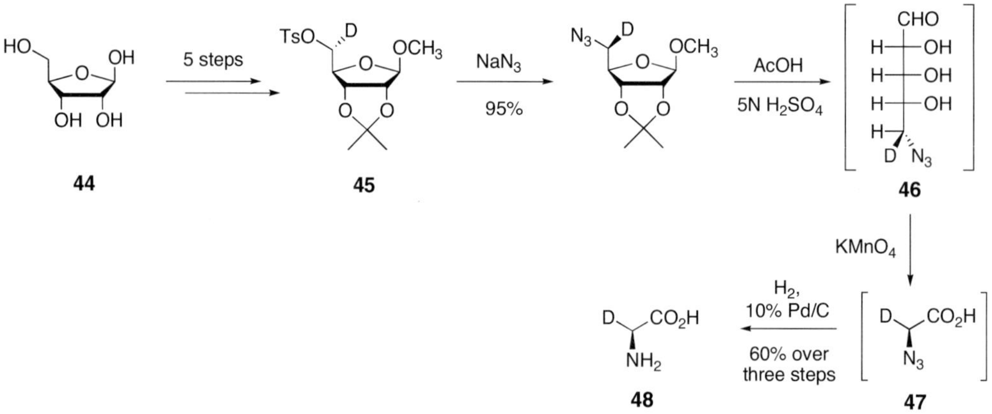

Scheme 11.13

Both glucose and ribose have been used primarily for the synthesis of isotopically labeled chiral glycine. Ohrui *et al.* synthesized (*S*)-[2-$^2H_1$]glycine **48** in nine steps from D-ribose **44** (Scheme 11.13) [61]. A chiral tosylate analog **45** of D-ribofuranoside was initially prepared in five steps from D-ribose **44**. Displacement of the tosylate group using sodium azide was then achieved in excellent yield. Treatment with a mixture of acetic acid and sulfuric acid gave 5-azido-5-deoxypentose **46** that was directly oxidized with potassium permanganate, yielding (*S*)-[2-$^2H_1$]azidoacetic acid **47**. Finally, reduction of the azide functional group gave (*S*)-[2-$^2H_1$] glycine **48** in 60% yield.

11.4 Chemical Asymmetric Methods

As well as using biotransformations for the asymmetric synthesis of isotopically labeled α-amino acids, a large number of chemical methods have also been developed. In particular, the chiral auxiliaries and chiral glycine equivalents, including Oppolzer's camphorsultams **49** [62], Schöllkopf's bislactim ethers **50** [63], Williams' oxazinones **51** [64], Evans' oxazolidinones **52** [65], and Seebach's imidazolidinones **53** [66], have all been used to good effect.

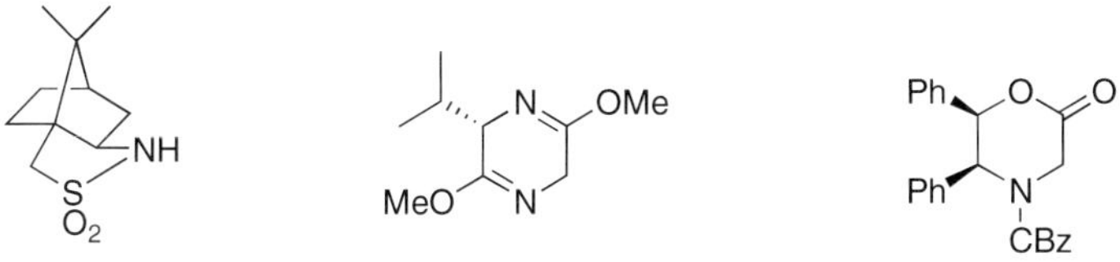

The general strategy with chiral glycine equivalents involves a stereoselective alkylation or acylation to introduce the side-chain of the amino acid followed by cleavage of the auxiliary to give the amino acid. For the preparation of isotopically labeled α-amino acids, the isotopic label can be incorporated during the synthesis of the chiral glycine equivalent or, more efficiently, during the alkylation or acylation step. Both of these approaches have been utilized with the Schöllkopf bislactim for the synthesis of a range of isotopically labeled amino acids. For example, Ragnarsson and coworkers used an isotopically labeled analog of the bislactim **54** for the preparation of Boc-protected $[1,2\text{-}^{13}C_2,{}^{15}N]$D-leucine **57** [67] (Scheme 11.14).

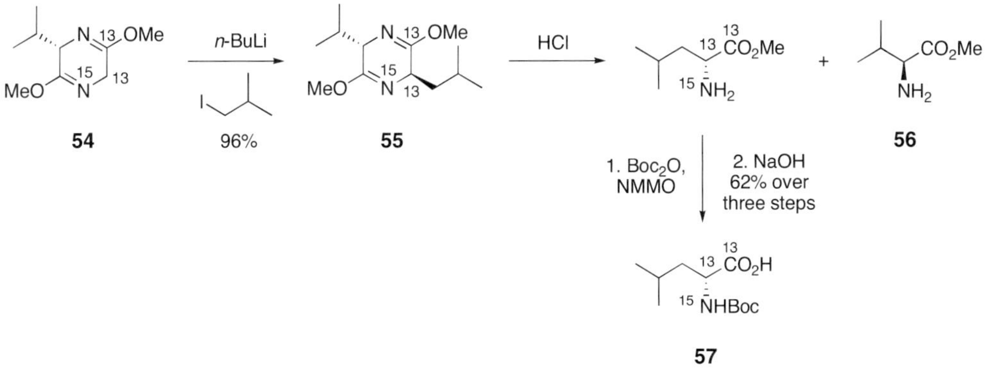

Scheme 11.14

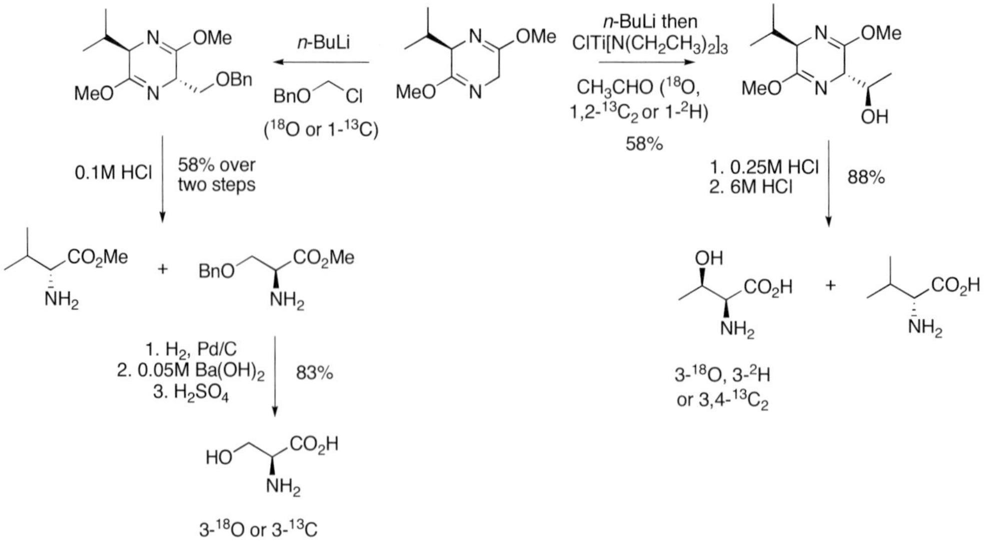

Scheme 11.15

Alkylation with *n*-butyl lithium and isobutyl iodide proceeded in excellent yield. Hydrolysis of **55**, separation from the byproduct L-valine methyl ester **56** followed by manipulation of the protecting groups gave Boc-[1, 2-$^{13}C_2$, ^{15}N]D-leucine **57** in 97% e.e.

Raap and coworkers have utilized the latter strategy for the preparation of ^{13}C-, ^{18}O-, and ^{2}H-labeled L-serines and L-threonines [68]. The general approach involved alkylation with labeled benzylchloromethyl ether or acylation with labeled acetaldehyde to give the serine and threonine analogs, respectively (Scheme 11.15). For the synthesis of the labeled serines, the alkylated products were hydrolyzed to the corresponding methyl esters and separated from the valine byproduct using silica gel chromatography. Hydrogenation of the benzyl ether and hydrolysis of the ester then gave either [3-^{18}O]L- or [3-^{13}C]L-serine. The acylation products were formed using a titanium enolate that favors formation of the (*R*)-center at the 2′-position (15 : 1). These were then hydrolyzed to the amino acids using a two-step approach. The corresponding labeled threonines were then separated from the valine byproduct using cationic-exchange chromatography.

While the Schöllkopf bislactim ether has been used for the synthesis of a number of labeled amino acids [69, 70], purification of these products from the valine byproduct and the level of diastereoselectivity during the alkylation or acylation step can sometimes be problematic [67]. Using oxazines, in particular the Williams' oxazines supersedes some of these issues as the oxazine ring is cleaved under hydrogenation conditions and the isotopically labeled amino acid is isolated by ion-exchange chromatography. Oxazines have been used extensively for the synthesis of labeled amino acids including aspartic acid, glutamic acid, phenylalanine, tyrosine, lysine, glycine as well as alanine [71–77]. For example, Williams and coworkers prepared (5*S*,6*R*)-[2, 3-$^{13}C_2$, ^{15}N]4-benzyloxy-5,6-diphenyl-2,3,5,6-tetrahydro-4*H*-oxazine-2-one

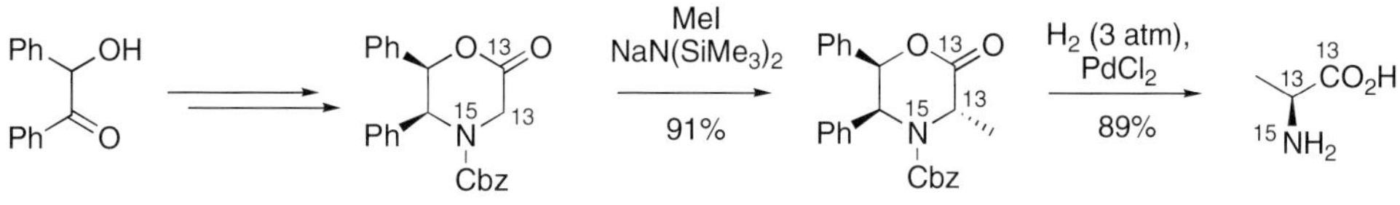

Scheme 11.16

59 from benzoin 58 using a lipase-mediated kinetic resolution as the key step (Scheme 11.16) [77]. Stereoselective alkylation with methyl iodide and sodium bis (trimethylsilylamide) gave a single diastereomer in 91% yield. Catalytic hydrogenation then gave $[^{15}N, ^{13}C_2]$L-alanine **60** in 89% yield and in 98% e.e.

Asymmetric reactions involving Oppolzer's camphorsultam also generally proceed with excellent diastereoselectivity and hydrolysis of the auxiliary results in easy isolation of the amino acid [78]. The most common approach for the synthesis of isotopically labeled amino acids involving the sultam uses a bis(methylsulfanyl) glycine equivalent that when coupled to the auxiliary undergoes facile alkylation reactions [67, 79, 80]. Such an approach was used for the preparation of Boc-$[1,2\text{-}^{13}C_2, ^{15}N]$L-proline **64** [80]. Initially, the camphorsultam was coupled with the bis(methylsulfanyl)glycinate from $[^{13}C_2, ^{15}N]$glycine. The resulting glycine equivalent **61** was alkylated with 1-chloro-3-iodopropane in 78% yield (Scheme 11.17).

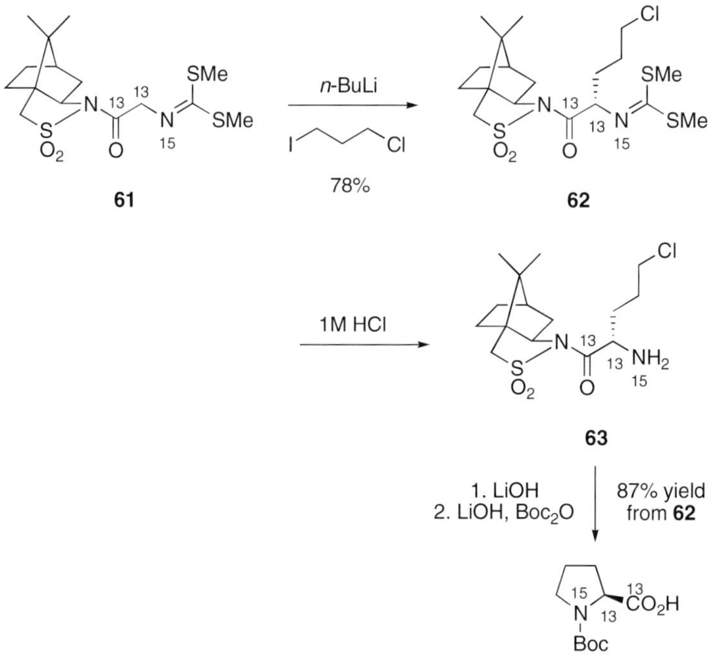

Scheme 11.17

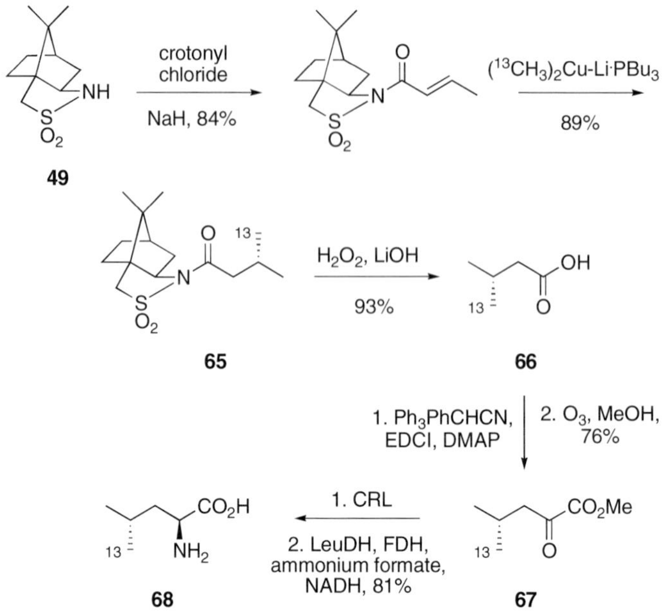

Scheme 11.18

Reaction of **62** with dilute hydrochloric acid gave the corresponding amine **63**, which was subsequently treated with lithium hydroxide resulting in cyclization to form the proline ring and cleavage of the sultam. Protection of the amino group then gave Boc-[1, 2-$^{13}C_2$, ^{15}N]L-proline **64** in 87% yield over the three steps.

Oppolzer's camphorsultams have also been used in combination with conjugate addition reactions for the synthesis of isotopically labeled amino acids [10, 81]. This strategy was utilized by Willis and coworkers for the diastereotopic labeling of one of the methyl groups of leucine [10]. The (1*S*)-1,10-camphorsultam **49** was acylated using crotonyl chloride (Scheme 11.18). Conjugate addition with a phosphine stabilized Gilman reagent gave **65** in 89% yield with good regio- and stereoselectivity (10 : 1 ratio of diastereomers). The auxiliary was hydrolyzed using lithium hydroperoxide and the resulting acid **66** was subjected to a one-carbon homologation. Hydrolysis of the α-keto ester **67** and subsequent reductive amination using a one-pot, dual-enzyme procedure involving lipase and then leucine dehydrogenase gave (2*S*,4*S*)-[5-^{13}C]leucine **68** in 81% yield.

While the auxiliaries and chiral glycine equivalents outlined above are the most common used for the asymmetric synthesis of isotopically labeled α-amino acids, it should be noted that other systems including imidazole-, piperazine-, and menthol-based moieties have also been used [82–87].

The synthesis of isotopically labeled amino acids has also been achieved using catalytic asymmetric methods. While less common than the use of chiral auxiliaries

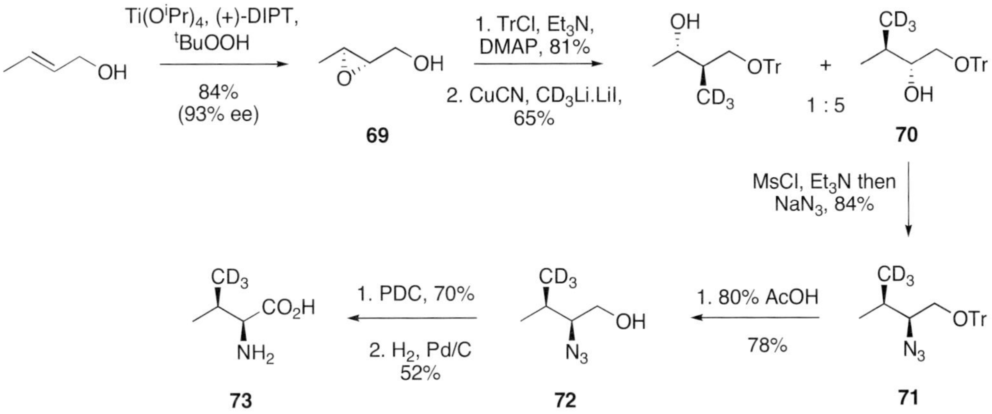

Scheme 11.19

or chiral glycine equivalents, several elegant approaches have been developed for the efficient and highly stereoselective synthesis of isotopically labeled α-amino acids. Shattuck and Meinwald synthesized (2*S*,3*S*)-[4-$^{2}H_{3}$]valine **73** using a Sharpless asymmetric epoxidation to effect the key step (Scheme 11.19) [88]. The epoxide **69** was then ring-opened using a higher-order mixed organocuprate introducing the CD_3 group in good yield and with modest regioselectivity. Alcohol **70** was then converted to the azide **71** with inversion of configuration using a two-step protocol via a mesylate. Deprotection and subsequent oxidation of **72**, followed by reduction of the azide functional group gave (2*S*,3*S*)-[4-$^{2}H_{3}$]valine **73**.

Asymmetric hydrogenations using chiral rhodium catalysts have also been used for the synthesis of isotopically labeled amino acids [89, 90]. For example, Kendall synthesized a multilabeled dehydrotyrosine derivative **75** in five steps from D_6-phenol **74**, which was then hydrogenated in the presence of a chiral rhodium catalyst (Scheme 11.20) [90]. This gave **76** in 98% e.e. after recrystallization. Acid-mediated hydrolysis in 5 N DCl of the acetyl protecting groups gave [2,3,3,2′,3′,5′,6′-$^{2}H_{7}$] L-tyrosine **77** in 97% overall yield from **75**.

The Lygo research group have used chiral phase transfer catalyzed alkylation of achiral glycine equivalents for the synthesis of α-carbon deuterium-labeled L-α-amino acids while Lugtenburg and coworkers have used a similar approach for the preparation of all possible ^{13}C and ^{15}N isotopomers of L-lysine, L-ornithine, and L-proline [91, 92]. Phase transfer catalysts such as **79** have been used in these procedures for the alkylation of imine substrates **78** (Scheme 11.21). For example, in the presence of a mixture of 11.7 M KOD/D_2O, the α-deuterated alkylated products **80** were obtained in enantiomeric excesses of 86–96%. These products were then hydrolyzed under mild conditions to give the α-carbon deuterium-labeled L-α-amino esters **81** in excellent yield and with greater than 90% deuterium content.

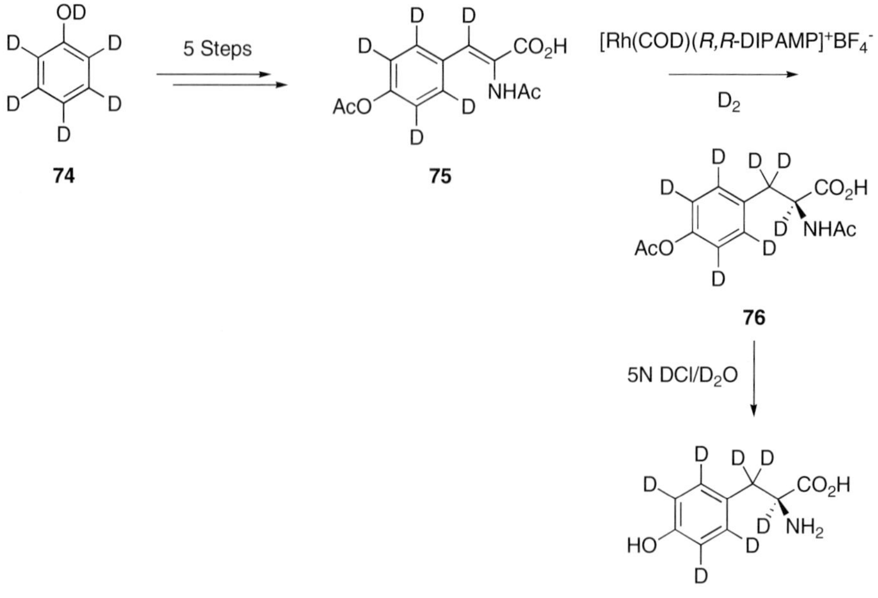

Scheme 11.20

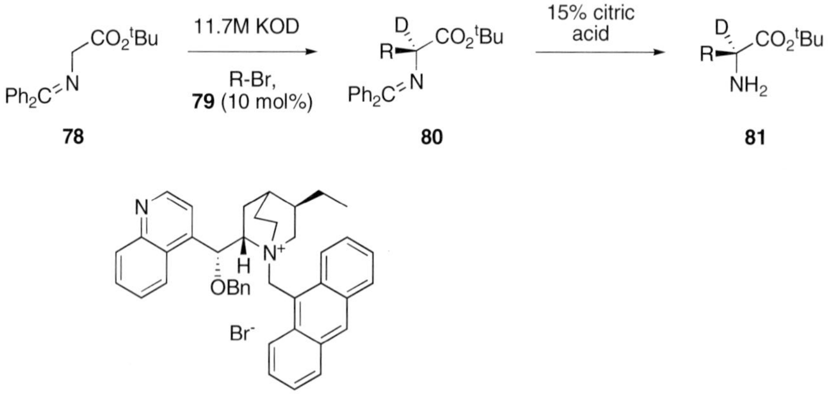

Scheme 11.21

11.5
Conclusions

As this chapter illustrates, many methods exist for the synthesis of L-α-amino acids that allow the efficient incorporation of isotopic labels at specific positions. Using either labeled substrates, reagents, or solvents, any of the proteinogenic amino acids can be easily accessed in excellent enantiopurity. The best strategy generally involves introduction of the label at a late stage of the synthetic route, thus maximizing utilization of the isotope.

11.6 Experimental Procedures

11.6.1 Biocatalysis: Synthesis of [^{15}N]L-amino Acids from α-Keto Esters using a One-Pot Lipase-Catalyzed Hydrolysis and Amino Acid Dehydrogenase-Catalyzed Reductive Amination

$RC(O)CO_2R'$ → $RCH(NH_2)CO_2H$

1. *C. rugosa* lipase, pH 7-7.5
2. DTT, AADH, FDH, NADH, $^{15}NH_4^+HCO_3^-$
3. NH_4OH

The α-keto ester (1 mmol) was dissolved in 5 mM phosphate buffer (40 ml). *Candida rugosa* lipase (10 000 eU) was added and the pH was adjusted to 7.5 by the addition of 1.0 M hydrochloric acid. The pH was maintained at a value between 7.0 and 7.5 by the addition of 0.1 M sodium hydroxide until the pH had stopped changing or until 1 equiv. of base had been added. The solution was deoxygenated by bubbling a stream of nitrogen through it for 1 h. Dithiothreitol (20 μl) was then added followed by AADH (1 eU), FDH (10 mg), [^{15}N]ammonium formate (1.2 mmol), and NADH (10 mg). The reaction was left stirring under a nitrogen atmosphere with the pH being kept constant at approximately 7.1 by the addition of 1.0 M hydrochloric acid. After the pH had stopped changing or 1 equiv. of acid had been added, the reaction mixture was concentrated. The resulting residue was redissolved in water (5 ml) and loaded onto a Dowex 50W ion-exchange column. The column was washed with water (3 × 30 ml) followed by concentrated ammonium hydroxide (3 × 30 ml). The combined ammonia washings were concentrated to give the desired amino acid.

11.6.2 Chiral Pool: Preparation of Aspartic Acid Semi-Aldehydes as Key Synthetic Intermediates; Synthesis of Methyl (2*S*)-*N*,*N*-di-tert-butoxycarbonyl-2-amino-4-oxobutanoate from L-aspartic Acid

(Note: the same approach can be used for the glutamate analog from L-glutamic acid.)

$HO_2CCH_2CH(NH_2)CO_2H$ → $HC(O)CH_2CH(NBoc_2)CO_2Me$

1. TMSCl, MeOH then Boc_2O, Et_3N
2. Boc_2O, DMAP
3. DIBAL-H, -78 °C

L-Aspartic acid (5.0 g, 38 mmol) was suspended in methanol (100 ml) under an argon atmosphere and cooled to 0° C. Chlorotrimethylsilane (21.1 ml, 167.2 mmol) was added dropwise. The reaction mixture was warmed to room temperature after 1 h and left stirring for 12 h. Triethylamine (23.3 ml, 167.2 mmol) was added along with di-*tert*-butyl dicarbonate (9.1 g, 42 mmol). After 2 h, the reaction mixture was concentrated *in vacuo*. The resulting residue was dissolved in diethyl ether (150 ml) which was filtered to remove the resulting white precipitate. The filtrate was concentrated *in vacuo*. Purification by flash column chromatography, eluting with 35% ethyl acetate in hexane gave dimethyl (2*S*)-*N*-*tert*-butoxycarbonyl-2-aminobutanodioate as a colorless oil (7.35 g, 75%). $[\alpha]_D$ +43.9 (*c* 1.5, $CHCl_3$); ^1H-NMR ($CDCl_3$, 300 MHz) δ 1.42 (s, 9H, O*t*Bu), 2.79 (dd, 1H, J = 16.9, 4.7 Hz, 3-C*H*H), 2.97 (dd, 1H, J = 16.9, 4.7 Hz, 3-CH*H*), 3.67 (s, 3H, OMe), 3.73 (s, 3H, OMe), 4.55 (m, 1H, 2-H), 5.48 (br d, 1H, J = 7.5 Hz, NH); electrospray mass spectroscopy (ES-MS): m/z 284.1 (MNa^+, 100%). Dimethyl (2*S*)-*N*-*tert*-butoxycarbonyl-2-aminobutanodioate (7.35 g, 28 mmol) was dissolved in acetonitrile (70 ml) under an argon atmosphere. 4-dimethylamino-pyridine (DMAP) (0.7 g, 5.6 mmol) was added along with a solution of di-*tert*-butyl dicarbonate (6.72 g, 28 mmol) in acetonitrile (40 ml). After 3 h, di-*tert*-butyl dicarbonate (3.36 g, 14 mmol) in acetonitrile (20 ml) was added and the reaction was left stirring for 12 h. The reaction mixture was concentrated *in vacuo*. Purification by flash column chromatography, eluting with 20% ethyl acetate in hexane gave the dimethyl (2*S*)-*N*,*N*-di-*tert*-butoxycarbonyl-2-aminobutanedioate as a white solid (9.61 g, 95%). $[\alpha]_D$ −69.3 (*c* 1.5, $CHCl_3$); ^1H-NMR ($CDCl_3$, 300 MHz) δ 1.49 (s, 18H, 2 × O*t*Bu), 2.72 (dd, 1H, J = 16.4, 6.6 Hz, 3-C*H*H), 3.23 (dd, 1H, J = 16.4, 7.1 Hz, 3-CH*H*), 3.69 (s, 3H, OMe), 3.72 (s, 3H, OMe), 5.43 (t, 1H, J = 6.9 Hz, 2-H); ES MS m/z 384.2 (MNa^+, 100%). Dimethyl (2*S*)-*N*,*N*-di-*tert*-butoxycarbonyl-2-aminobutanedioate (5.0 g, 14 mmol) was dissolved in diethyl ether (60 ml) under an argon atmosphere and cooled to −78 °C. A solution of DIBAL-H in hexane (17 ml, 16.8 mmol, 1.0 M) was added dropwise. After 1 h, a saturated solution of ammonium chloride (10 ml) was added and the reaction mixture was warmed to room temperature. The reaction mixture was filtered through a pad of Celite. The filtrate was concentrated *in vacuo* to give a viscous oil. Purification by flash column chromatography, eluting with 25% ethyl acetate in hexane gave methyl (2*S*)-*N*,*N*-di-*tert*-butoxycarbonyl-2-amino-4-oxobutanoate as a viscous colorless oil (4.00 g, 87%). $[\alpha]_D$ −55.5 (*c* 1.0, $CHCl_3$); ^1H-NMR ($CDCl_3$, 300 MHz) δ 1.48 (s, 18H, 2 × O*t*Bu), 2.82 (dd, 1H, J = 17.0, 6.6 Hz, 3-C*H*H), 3.38 (dd, 1H, J = 17.0, 7.0 Hz, 3-CH*H*), 3.71 (s, 3H, OMe), 5.51 (t, 1H, J = 6.4 Hz, 2-H), 9.76 (s, 1H, C*H*O); ES-MS: m/z 354.1 (MNa^+, 100%).

11.6.3
Asymmetric Methods: Asymmetric Alkylation Using the Williams' Oxazine and Subsequent Hydrogenation to Give the α-Amino Acid

Ph, Ph, O, O, N, Cbz

1. KHMDS, R-X

2. $PdCl_2$, H_2, 50 psi

X= Halide

R, CO_2H, NH_2

The Williams' oxazine (~10 mmol) was dissolved in THF (30 ml) under an argon atmosphere and cooled to −78 °C. A solution of potassium bis(trimethylsilyl) amide (1.2 equiv.) in THF (15 ml) was added dropwise over a 15-min period. After 40 minutes, the alkyl halide (1.5 equiv.) was added and the reaction mixture was kept at −78 °C for 2 h. The reaction mixture was warmed to room temperature and concentrated *in vacuo*. The resulting residue was dissolved in ethyl acetate (40 ml) and washed with 2 M hydrochloric acid (2 × 20 ml). The organic layer was dried over $MgSO_4$ and concentrated at reduced pressure. Purification by flash column chromatography eluting with ethyl acetate in hexane gave the alkylated oxazine. The alkylated oxazine was dissolved in ethanol (20 ml) and palladium chloride (1.0 g) was added. The reaction mixture was allowed to stir under a hydrogen atmosphere at 50 psi for 48 h. The reaction mixture was filtered through a Celite pad that was then washed with ethanol (100 ml). The combined ethanol washings were concentrated *in vacuo* to give a yellow solid. Purification by ion-exchange chromatography eluting with dilute ammonia gave the α-amino acid.

References

1 For general reviews of this area, see: (a) Vogues, R. (1995) In *Synthesis and Applications of Isotopically Labelled Compounds 1994* (eds J. Allen and R. Vogues), John Wiley & Sons, Inc., New York, p. 1;(b) Young, D.W. (1991) In *Isotopes in the Physical and Biomedical Sciences, Vol. 1: Labelled Compounds (Part B)* (eds E. Buncel and J.R. Jones), Elsevier, Amsterdam, p. 341;(c) Adriaens, P. and Vanderhaeghe, H. (1991) In *Isotopes in the Physical and Biomedical Sciences, Vol 1: Labelled Compounds (Part A)* (eds E. Buncel and J.R. Jones), Elsevier, Amsterdam, p. 330; (d) Winkler, F.J., Kuhnl, K., Medina, R., Schwarz-Kaske, R., and Schmidt, H.L. (1995) *Isotopes in Environmental and Health Studies*, **31**, 161; (e) Kelly, N.M., Sutherland, A., and Willis, C.L. (1997) *Natural Product Reports*, **14**, 205.

2 (a) Hanson, R.L., Schwinden, M.D., Banerjee, A., Brzozowski, D.B., Chen, B.-C., Patel, B.P., McNamee, C.G., Kodersha, G.A., Kronenthal, D.R., Patel, R.N., and Szarka, L.J. (1999) *Bioorganic & Medicinal Chemistry Letters*, **7**, 2247; (b) Shaked, Z. and Whitesides, G. (1980) *Journal of the American Chemical Society*, **102**, 7104; (c) Wong, C.-H. and Whitesides, G. (1981) *Journal of the American Chemical Society*, **103**, 4890; (d) Liese, A., Karutz, M., Kamphuis, J., Wandrey, C., and Kragl, U. (1996) *Biotechnology and Bioengineering*, **51**, 544; (e) Lamed, R., Keinan, E., and Zeikus, J.G. (1981) *Enzyme and Microbial Technology*, **3**, 144.

3 Ducrocq, C. Decottignies-Le Maréchal, P. and Azerad, R. (1985) *Journal of Labelled Compounds & Radiopharmaceuticals*, **22**, 61.

4 Bojan, O., Bologa, M., Niac, G., Palibroda, N., Vargha, E., and Bârzu, O. (1979) *Analytical Biochemistry*, **101**, 23.

5 Goux, W.J., Rench, L., and Weber, D.S. (1993) *Journal of Labelled Compounds & Radiopharmaceuticals*, **33**, 181.

6 Raap, J., Nieuwenhuis, S., Creemers, A., Hexspoor, S., Kragl, U., and Lugtenburg, J. (1999) *European Journal of Organic Chemistry*, 2609.

7 Kelly, N.M., O'Neill, B.C., Probert, J., Reid, G., Stephen, R., Wang, T., Willis, C.L., and Winton, P. (1994) *Tetrahedron Letters*, **35**, 6533.

8 Kelly, N.M., Reid, R.G., Willis, C.L., and Winton, P.L. (1995) *Tetrahedron Letters*, **36**, 8315.
9 Kelly, N.M., Reid, R.G., Willis, C.L., and Winton, P.L. (1996) *Tetrahedron Letters*, **37**, 1517.
10 Fletcher, M.D., Harding, J.R., Hughes, R.A., Kelly, N.M., Schmalz, H., Sutherland, A., and Willis, C.L. (2000) *Journal of the Chemical Society, Perkin Transactions 1*, 43.
11 Harding, J.R., Hughes, R.A., Kelly, N.M., Sutherland, A., and Willis, C.L. (2000) *Journal of the Chemical Society, Perkin Transactions 1*, 3406.
12 Sutherland, A. and Willis, C.L. (1997) *Tetrahedron Letters*, **38**, 1837.
13 Kragl, U., Gödde, A., Wandrey, C., Kinzy, W., Cappon, J.J., and Lugtenburg, J. (1993) *Tetrahedron: Asymmetry*, **4**, 1193.
14 Barclay, F., Chrystal, E., and Gani, D. (1996) *Journal of the Chemical Society, Perkin Transactions 1*, 683.
15 Siebum, A.H.G., Woo, W.S., and Lugtenburg, J. (2003) *European Journal of Organic Chemistry*, 4664.
16 Bjurling, P., Watanabe, Y., Tokushige, M., Oda, T., and Langstrom, B. (1989) *Journal of the Chemical Society, Perkin Transactions 1*, 1331.
17 Malthouse, J.P.G., Fitzpatrick, T.B., Milne, J.J., Grehn, L., and Ragnarsson, U. (1997) *Journal of Peptide Science*, **3**, 361.
18 Unkefer, C.J., Lodwig, S.N., Silks, L.A. III, Hanners, J.L., Ehler, D.S., and Gibson, R. (1991) *Journal of Labelled Compounds & Radiopharmaceuticals*, **29**, 1247.
19 Boroda, E., Rakowska, S., Kanski, R., and Kanska, M. (2003) *Journal of Labelled Compounds & Radiopharmaceuticals*, **46**, 691.
20 Walker, T.E., Matheny, C., Storm, C.B., and Hayden, H. (1986) *The Journal of Organic Chemistry*, **51**, 1175.
21 Faleev, N.G., Ruvinov, S.B., Saporovskaya, M.B., Belikov, V.M., Zakomyrdina, L.N., Sakharova, I.S., and Torchinsky, Y.M. (1990) *Tetrahedron Letters*, **31**, 7051.
22 Yuan, S.-S. and Foos, J. (1981) *Journal of Labelled Compounds & Radiopharmaceuticals*, **18**, 563.
23 Uchida, K. and Kainosho, M. (1991) *Journal of Labelled Compounds & Radiopharmaceuticals*, **29**, 867.
24 Hasegawa, H., Shinohara, Y., Tagoku, K., and Hashimoto, T. (2001) *Journal of Labelled Compounds & Radiopharmaceuticals*, **44**, 21.
25 Kerr, V.N. and Ott, D.G. (1978) *Journal of Labelled Compounds & Radiopharmaceuticals*, **15**, 503.
26 Baldwin, J.E., Ng, S.C., Pratt, A.J., Russell, M.A., and Dyer, R.L. (1987) *Tetrahedron Letters*, **28**, 2303.
27 Maeda, H., Takata, K., Toyoda, A., Niitsu, T., Iwakura, M., and Shibata, K. (1997) *Journal of Fermentation and Bioengineering*, **83**, 113.
28 Hanners, J., Gibson, R., Velarde, K., Hammer, J., Alvarez, M., Griego, J., and Unkefer, C.J. (1991) *Journal of Labelled Compounds & Radiopharmaceuticals*, **29**, 781.
29 Polach, K.J., Shah, S.A., LaIuppa, J.C., and LeMaster, D.M. (1993) *Journal of Labelled Compounds & Radiopharmaceuticals*, **33**, 809.
30 Santaniello, E., Casati, R., and Manzocchi, A. (1985) *Journal of the Chemical Society, Perkin Transactions 1*, 2389.
31 Yamada, H., Kurumaya, K., Eguchi, T., and Kajiwara, M. (1987) *Journal of Labelled Compounds & Radiopharmaceuticals*, **24**, 561.
32 Degerbeck, F., Fransson, B., Grehn, L., and Ragnarsson, U. (1993) *Journal of the Chemical Society, Perkin Transactions 1*, 11.
33 Humphrey, J.M., Hart, J.A., and Chamberlin, A.R. (1995) *Bioorganic & Medicinal Chemistry Letters*, **5**, 1315.
34 Hamilton, D.J. and Sutherland, A. (2004) *Tetrahedron Letters*, **45**, 5739.
35 (a) Padrón, J.M., Kokotos, G., Martín, T., Markidis, T., Gibbons, W.A., and Martín, V.S. (1998) *Tetrahedron: Asymmetry*, **9**, 3381; (b) Kokotos, G., Padrón, J.M., Martín, T., Gibbons, W.A., and Martín, V.S.

(1998) *The Journal of Organic Chemistry*, **63**, 3741.

36 Baldwin, J.E., Adlington, R.M., Crouch, N.P., Mellor, L.C., Morgan, N., Smith, A.M., and Sutherland, J.D. (1995) *Tetrahedron*, **51**, 4089.

37 Lam-Thanh, H., Mermet-Bouvier, R., Fermandjian, S., and Fromageot, P. (1983) *Journal of Labelled Compounds & Radiopharmaceuticals*, **20**, 143.

38 Dieterich, P. and Young, D.W. (1993) *Tetrahedron Letters*, **34**, 5455.

39 Sutherland, A. and Willis, C.L. (1996) *Journal of Labelled Compounds & Radiopharmaceuticals*, **38**, 95.

40 Furuta, T., Katayama, M., Shibasaki, H., and Kasuya, Y. (1992) *Journal of the Chemical Society, Perkin Transactions 1*, 1643.

41 Stapon, A., Li, R., and Townsend, C.A. (2003) *Journal of the American Chemical Society*, **125**, 8486.

42 Field, S.J. and Young, D.W. (1983) *Journal of the Chemical Society, Perkin Transactions 1*, 2387.

43 Kovacs, J., Jham, G., and Hui, K.Y. (1982) *Journal of Labelled Compounds & Radiopharmaceuticals*, **19**, 83.

44 Blaskovich, M.A., Evindar, G., Rose, N.G.W., Wilkinson, S., Luo, Y., and Lajoie, G.A. (1998) *The Journal of Organic Chemistry*, **63**, 3631.

45 (a) Kenworthy, M.N., Kilburn, J.P., and Taylor, R.J.K. (2004) *Organic Letters*, **6**, 19; (b) Barfoot, C.W., Harvey, J.E., Kenworthy, M.N., Kilburn, J.P., Ahmed, M., and Taylor, R.J.K. (2005) *Tetrahedron*, **61**, 3403.

46 Lodwig, S.N. and Unkefer, C.J. (1992) *Journal of Labelled Compounds & Radiopharmaceuticals*, **31**, 95.

47 (a) Arnold, L.D., Kalantar, T.H., and Vederas, J.C. (1985) *Journal of the American Chemical Society*, **107**, 7105; (b) Arnold, L.D., Drover, J.C.G., and Vederas, J.C. (1987) *Journal of the American Chemical Society*, **109**, 4649.

48 Oba, M., Iwasaki, A., Hitokawa, H., Ikegame, T., Banba, H., Ura, K., Takamura, T., and Nishiyama, K. (2006) *Tetrahedron: Asymmetry*, **17**, 1890.

49 August, R.A., Khan, J.A., Moody, C.M., and Young, D.W. (1992) *Tetrahedron Letters*, **33**, 4617.

50 August, R.A., Khan, J.A., Moody, C.M., and Young, D.W. (1996) *Journal of the Chemical Society, Perkin Transactions 1*, 507.

51 Charrier, J.-D., Hitchcock, P.B., and Young, D.W. (2004) *Organic and Biomolecular Chemistry*, **2**, 1310.

52 Barraclough, P., Spray, C.A., and Young, D.W. (2005) *Tetrahedron Letters*, **46**, 4653.

53 Oba, M., Terauchi, T., Hashimoto, J., Tanaka, T., and Nishiyama, K. (1997) *Tetrahedron Letters*, **38**, 5515.

54 Oba, M., Terauchi, T., Miyakawa, A., Kamo, H., and Nishiyama, K. (1998) *Tetrahedron Letters*, **39**, 1595.

55 Oba, M., Terauchi, T., Miyakawa, A., and Nishiyama, K. (1999) *Tetrahedron: Asymmetry*, **10**, 937.

56 Oba, M., Miyakawa, A., Shionoya, M., and Nishiyama, K. (2001) *Journal of Labelled Compounds & Radiopharmaceuticals*, **44**, 141.

57 Oba, M., Ohkuma, K., Hitokawa, H., Shirai, A., and Nishiyama, K. (2006) *Journal of Labelled Compounds & Radiopharmaceuticals*, **49**, 229.

58 Barton, D.H.R., Hervé, Y., Potier, P., and Thierry, J. (1988) *Tetrahedron*, **44**, 5479.

59 Kakinuma, K., Imamura, N., and Saba, Y. (1982) *Tetrahedron Letters*, **23**, 1697.

60 Kakinuma, K., Koudate, T., Li, H.-Y., and Eguchi, T. (1991) *Tetrahedron Letters*, **32**, 5801.

61 Ohrui, H., Misawa, T., and Meguro, H. (1985) *The Journal of Organic Chemistry*, **50**, 3007.

62 (a) Oppolzer, W., Pedrosa, R., and Moretti, R. (1986) *Tetrahedron Letters*, **27**, 831; (b) Oppolzer, W., Moretti, R., and Thomi, S. (1989) *Tetrahedron Letters*, **30**, 6009.

63 Schöllkopf, U., Hartwig, W., and Groth, U. (1979) *Angewandte Chemie (International Edition in English)*, **18**, 863.

64 Sinclair, P.J., Zhai, D., Reibenspies, J., and Williams, R.M. (1986) *Journal of the American Chemical Society*, **108**, 1103.

65 Evans, D.A., Ellman, J.A., and Dorow, R.L. (1987) *Tetrahedron Letters*, **28**, 1123.

66 Seebach, D., Aebi, J.D., Naef, R., and Weber, T. (1985) *Helvetica Chimica Acta*, **68**, 144.

67 Lankiewicz, L., Nyassé, B., Fransson, B., Grehn, L., and Ragnarsson, U. (1994) *Journal of the Chemical Society, Perkin Transactions 1*, 2503.

68 Karstens, W.F.J., Berger, H.J.F.F., van Haren F E.R., Lugtenburg, J., and Raap, J. (1995) *Journal of Labelled Compounds & Radiopharmaceuticals*, **36**, 1077.

69 Raap, J., Wolthuis, W.N.E., Hehenkamp, J.J.J., and Lugtenburg, J. (1995) *Amino Acids*, **8**, 171.

70 Rose, J.E., Leeson, P.D., and Gani, D. (1995) *Journal of the Chemical Society, Perkin Transactions 1*, 157.

71 Takatori, K., Nishihara, M., and Kajiwara, M. (1999) *Journal of Labelled Compounds & Radiopharmaceuticals*, **42**, 701.

72 Takatori, K., Sakamoto, T., and Kajiwara, M. (2006) *Journal of Labelled Compounds & Radiopharmaceuticals*, **49**, 445.

73 Takatori, K., Nishihara, M., Nishiyama, Y., and Kajiwara, M. (1998) *Tetrahedron*, **54**, 15861.

74 Takatori, K., Hayashi, A., and Kajiwara, M. (2004) *Journal of Labelled Compounds & Radiopharmaceuticals*, **47**, 787.

75 Williams, R.M., Zhai, D., and Sinclair, P.J. (1986) *The Journal of Organic Chemistry*, **51**, 5021.

76 Takatori, K., Toyama, S., Narumi, S., Fujii, S., and Kajiwara, M. (2004) *Journal of Labelled Compounds & Radiopharmaceuticals*, **47**, 91.

77 Aoyagi, Y., Iijima, A., and Williams, R.M. (2001) *The Journal of Organic Chemistry*, **66**, 8010.

78 Lodwig, S.N. and Unkefer, C.J. (1996) *Journal of Labelled Compounds & Radiopharmaceuticals*, **38**, 239.

79 Lodwig, S.N. and Unkefer, C.J. (1998) *Journal of Labelled Compounds & Radiopharmaceuticals*, **41**, 983.

80 Budesinsky, M., Ragnarsson, U., Lankiewicz, L., Grehn, L., Slaninova, J., and Hlavacek, J. (2005) *Amino Acids*, **29**, 151.

81 Stocking, E.M., Martinez, R.A., Silks, L.A., Sanz-Cervera F J.F., and Williams, R.M. (2001) *Journal of the American Chemical Society*, **123**, 3391.

82 Fasth, K.-J., Hörnfeldt, K., and Långström, B. (1995) *Acta Chemica Scandinavica*, **49**, 301.

83 Strauss, E. and Begley, T.P. (2003) *Bioorganic & Medicinal Chemistry Letters*, **13**, 339.

84 Koch, C.-J., Simonyiová, S., Pabel, J., Kärtner, A., Polborn, K., and Wanner, K.T. (2003) *European Journal of Organic Chemistry*, 1244.

85 Davies, S.G., Rodríguez-Solla, H., Tamayo, J.A., Cowley, A.R., Concellón, C., Garner, A.C., Parkes, A.L., and Smith, A.D. (2005) *Organic and Biomolecular Chemistry*, **3**, 1435.

86 Hamon, D.P.G., Razzino, P., and Massy-Westropp R.A. (1991) *Journal of the Chemical Society, Chemical Communications*, 332.

87 Barnett, D.W., Panigot, M.J., and Curley, R.W. Jr. (2002) *Tetrahedron: Asymmetry*, **13**, 1893.

88 Shattuck, J.C. and Meinwald, J. (1997) *Tetrahedron Letters*, **38**, 8461.

89 Torizawa, T., Ono, A.M., Terauchi, T., and Kainosho, M. (2005) *Journal of the American Chemical Society*, **127**, 12620.

90 Kendall, J.T. (2000) *Journal of Labelled Compounds & Radiopharmaceuticals*, **43**, 917.

91 Lygo, B. and Humphreys, L.D. (2002) *Tetrahedron Letters*, **43**, 6677.

92 Siebum, A.H.G., Tsang, R.K.F., van der Steen F R., Raap, J., and Lugtenburg, J. (2004) *European Journal of Organic Chemistry*, 4391.

12
Synthesis of Unnatural/Nonproteinogenic α-Amino Acids

David J. Ager

12.1
Introduction

Nonproteinogenic and unnatural amino acids are finding widespread application especially in the pharmaceutical arena to modify structure. The incorporation of an unnatural amino acid can reduce the susceptibility of a compounds to enzymatic degradation, act as a mechanistic probe, or be used as a synthetic handle. Methods to incorporate unnatural amino acids into peptides and proteins are not covered in this chapter. Of course, nature can perform transformations on proteinogenic amino acids in proteins as part of post-translational modifications to make a functional enzyme or protein; these transformations are also not covered in this chapter.

Unnatural amino acids may be required in small (milligram) quantities for incorporation into a polypeptide to investigate three-dimensional structure to large amounts (megatonnes) as a starting material for a commercial drug. The optimal scale where a methodology can be used is addressed in this chapter. For the purposes of this chapter, any nonproteinogenic amino acid has been included even though it may occur in nature. An example is the D-amino acid, D-alanine. However, preparations of simple derivatives of proteinogenic amino acids, such as *O*-methylserine, have been omitted.

As many synthetic uses of unnatural α-amino acids require protection of the amino or carboxylate moieties, many of the methods described will be about derivatives. Conditions for the removal of these protecting groups are usually routine and methods can be found in standard texts [1]. The majority of methods provide monosubstituted α-amino acid derivatives, which still have an acidic hydrogen that can lead to epimerization. Methods for α,α-disubstituted amino acids are also covered where epimerization cannot occur and approaches are more limited.

There are many methods to prepare α-amino acids [2]. It has been noted that there is no single methodology that has yet emerged for the asymmetric synthesis of all unnatural amino acids [3]. Figure 12.1 summarizes the major approaches that are discussed further in this chapter. The transformations may involve one or more steps. The methods provide enantiomerically enriched product.

Amino Acids, Peptides and Proteins in Organic Chemistry. Vol.1 – Origins and Synthesis of Amino Acids.
Edited by Andrew B. Hughes

ISBN: 978-3-527-32096-7

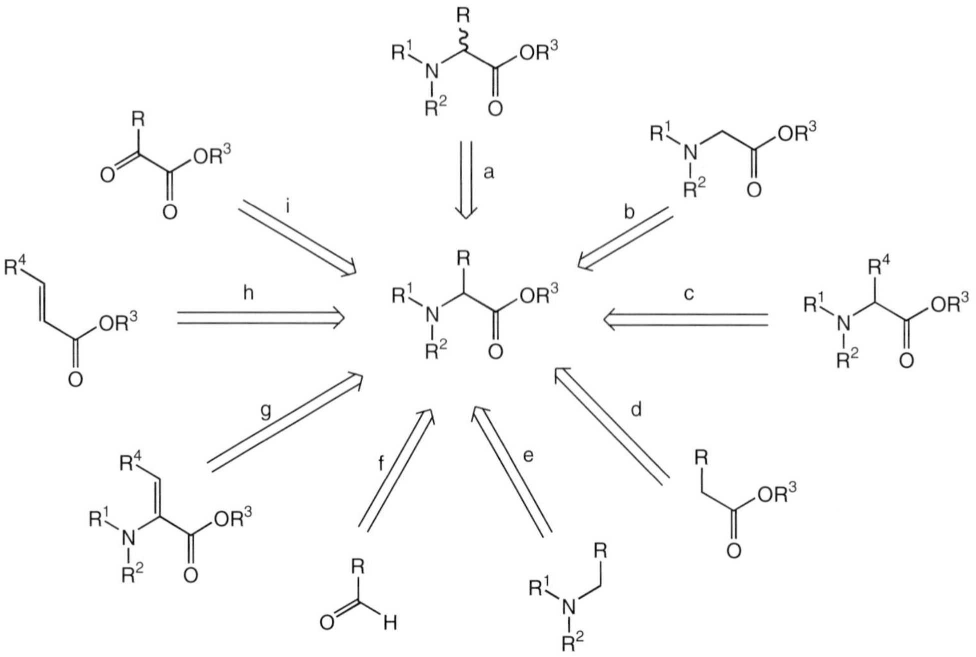

Figure 12.1 Approaches to α-amino acids and derivatives. For method *a*, see Sections 12.2.1, 12.3.1, 12.3.2, and 12.3.6; method *b*, Section 12.2.2.1; method *c*, Section 12.2.2.2; method *d*, Section 12.2.3.1; method *e*, Section 12.2.3.2; method *f*, Section 12.2.3.2.1; method g, Section 12.2.4; method *h*, Section 12.3.3; and method *i*, Sections 12.3.4 and 12.3.5.

Method *a* is a resolution approach where the racemic mixture is separated into single enantiomers. The approach can also include dynamic resolutions where the undesired antipode is isomerized to allow a single product enantiomer to be formed; this increases the theoretical yield from 50 to 100%. Approach *b* involves the introduction of the side-chain; this could be through an alkylation reaction. Method *c* denotes that the side-chain of a substrate amino acid is modified to produce the desired product.

Functionality can be introduced to produce the amino acid as depicted by method *d*, where the nitrogen moiety is added, and method *e*, where the carboxylic acid group is introduced. Method *f* combines these last two approaches and is a Strecker reaction. Hydrogenation is denoted by method *g*. Method *h* introduces the nitrogen group to unsaturation and this can be performed by an enzyme. Method *i* denotes either a transamination or reductive amination approach, usually achieved by enzymes.

This chapter only discusses general approaches with illustrative examples. In many instances, large-scale methods have been exemplified, as these are usually better understood and can be easier to apply to analogous systems.

12.2 Chemical Methods

12.2.1 Resolution Approaches

As α-amino acids contain an epimerizable center, D-amino acids are accessible through isomerization of the L-isomer, followed by a resolution. Of course, the racemic mixture can also be obtained by synthesis. As resolution can be either wasteful, if the undesired isomer is discarded, or clumsy, when the other isomer is recycled through an epimerization protocol, many large-scale methods now rely upon a dynamic resolution, where all of the starting material is converted to the desired isomer. With the advent of asymmetric reactions that can be performed at large scale, a substrate can now be converted to the required stereoisomer without the need for any extra steps associated with a resolution approach. Some enzymatic methods rely on an enzyme acting on only one antipode while the unwanted isomer is isomerized *in situ*; these approaches are discussed in the enzymatic approaches section (Section 12.3).

An example of a chemical method is provided by an approach to D-proline where a dynamic resolution is performed with L-proline (**1**). The presence of butyraldehyde allows racemization of the L-proline in solution, while the desired D-isomer (**2**) is removed as a salt with D-tartaric acid (Scheme 12.1) [4, 5].

An example of a very efficient asymmetric transformation is the preparation of (*R*)-phenylglycine amide (**3**) where the synthesis of the amino acids, by a Strecker reaction, and racemization are all performed at the same time (Scheme 12.2) [6]. This offers a good alternative to the enzymatic resolution of phenylglycine amide with an (*S*)-specific acylase (see Section 12.3.1) [7].

Another example is provided by D-*tert*-leucine (**4**) (Scheme 12.3) where an asymmetric hydrogenation approach cannot be used due to the lack of a hydrogen atom at the β-position. In this example, quinine was used as the resolving agent [8, 9].

12.2.2 Side-Chain Methods

12.2.2.1 Introduction of the Side-Chain

A number of methods have been developed using chiral auxiliaries or templates where a glycine surrogate is alkylated to give the unnatural amino acid. The alternatives

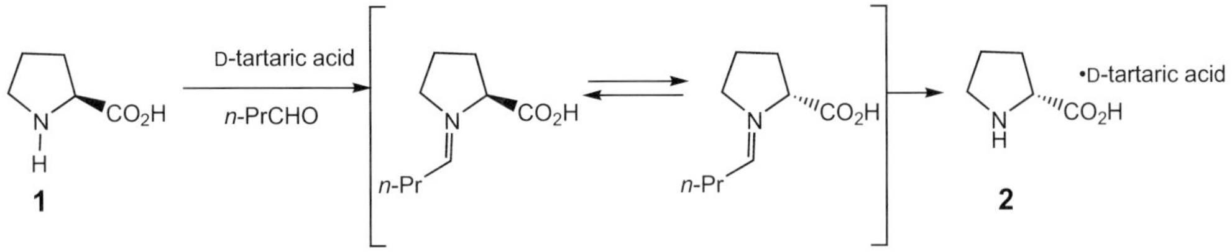

D-tartaric acid
n-PrCHO
CO_2H
N
H
1
n-Pr
•D-tartaric acid
2

Scheme 12.1

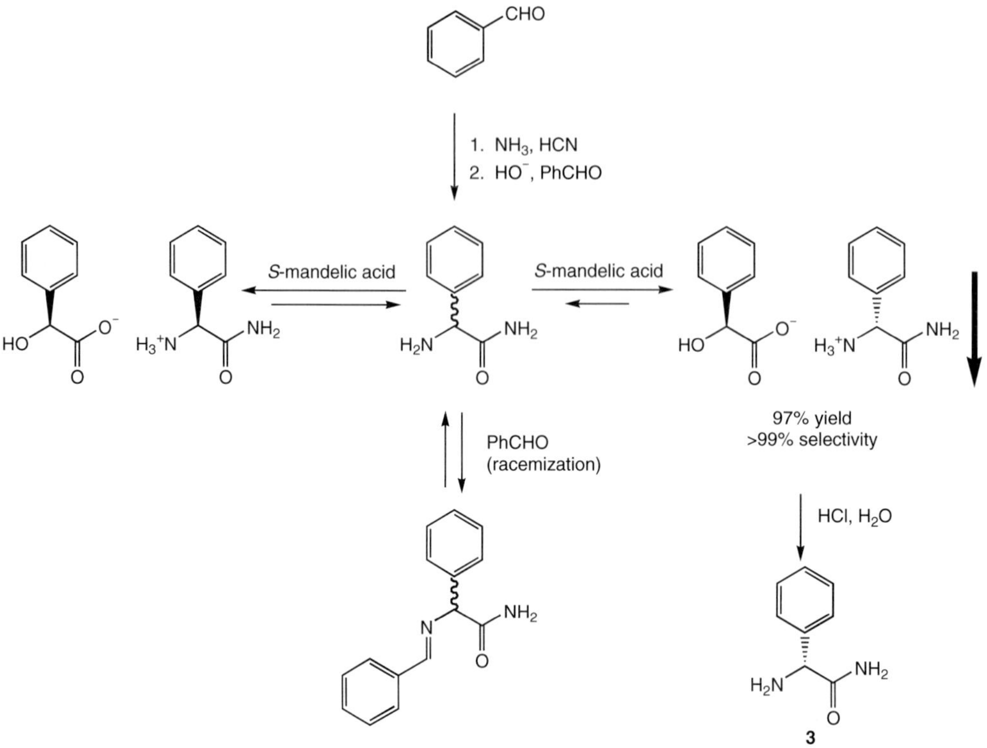

Scheme 12.2

are to use chiral catalysts and bases. To date, phase-transfer catalysts have provided the most general approach and highest asymmetric induction.

There are numerous examples of the Schöllkopf [10, 11] and Evans [12–14] auxiliary methods, and there are many examples in the literature that there is a high probability of success with an analogous system [13]. The methodology is excellent for smaller-scale applications. Benzyl halides tend to be good alkylating agents and this is illustrated with the lactim (**5**) developed by Schöllkopf (Scheme 12.4) [15].

To overcome the limitations of needing a reactive halide so that competing elimination is not a problem, an anion can be reacted with a carbonyl compound and the product either reduced or a conjugate addition performed. In some cases,

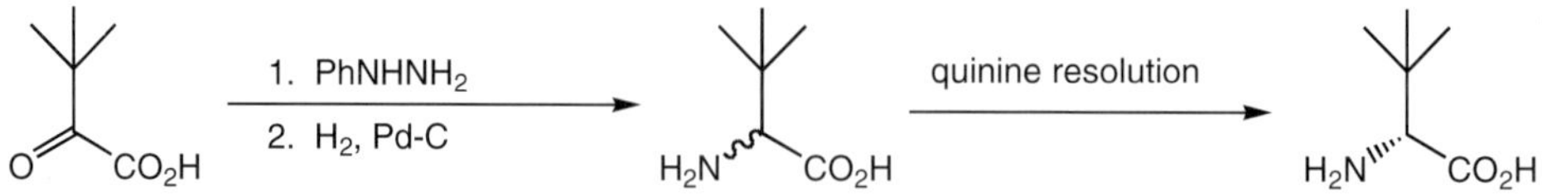

Scheme 12.3

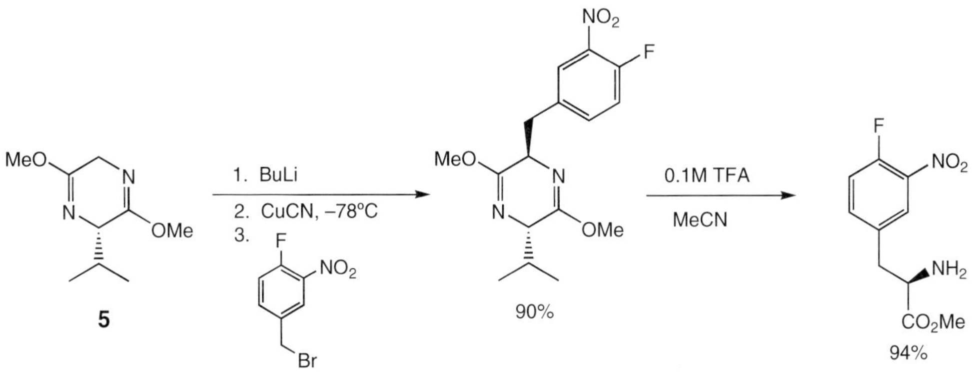

Scheme 12.4

the additional functionality in the side-chain may be desired, as in the example with an Evans' oxazolidinone (Scheme 12.5) [16, 17].

An amino acid can be used as the origin of stereochemistry and still reacted as in the concept of the regeneration of stereogenic centers (Scheme 12.6) [18–20]. The side-chain, R, can also be an alkene, which allows for conjugate additions or reductions to provide an unnatural amino acid; the stereochemical outcome is controlled by the *tert*-butyl group [19, 21]. The concept can be applied to glycine derivatives, although the enantiomers of the imidazidinone **6** (R = H) need to be separated to achieve an enantioselective synthesis [18].

Additions to imines can also be used to introduce the side-chain of an amino acid. These methods are usually based on the use of a chiral control group. An example is the addition of an alkene to a sulfinyl imine (Scheme 12.7). The sulfur group can be removed by treatment with HCl and the alkene is available for further elaboration, such as by a cross-metathesis reaction [22].

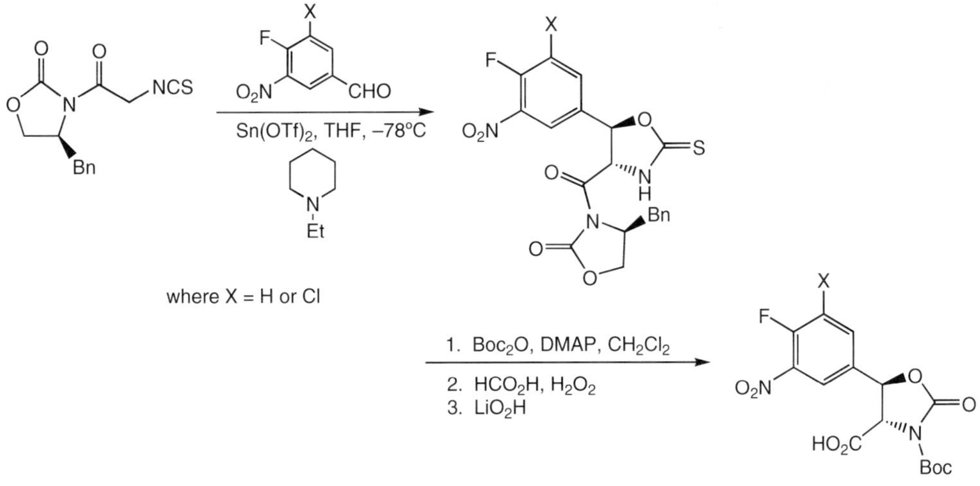

Scheme 12.5

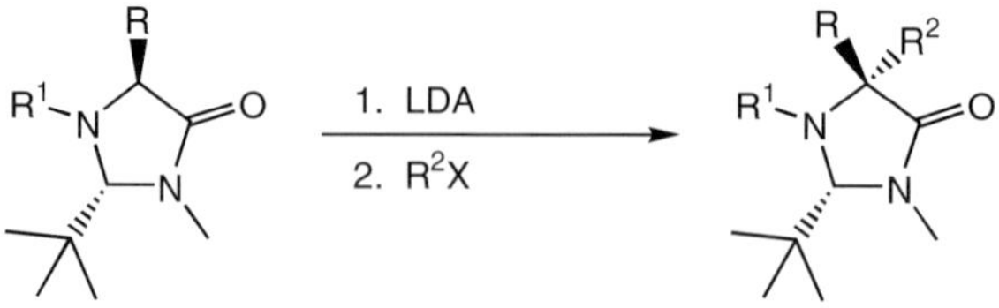

R
R¹
N
O
N
1. LDA
2. R²X
R
R²
R¹
N
O
N

6

where R^1 = Boc or Bz
R = alkyl or aryl
R^2 = alkyl, benzyl or allyl

Scheme 12.6

The introduction of the side-chain can also be achieved with a chiral catalyst, which is much more amenable to large-scale reactions.

Many phase-transfer catalysts have been developed to achieve an asymmetric alkylation of a glycine derivative. The use of an ester, particularly the *tert*-butyl ester, of the benzophenone Schiff's base of glycine tends to provide the highest asymmetric induction (Scheme 12.8) (see Section 12.5.1). The most useful catalytic systems are based on the cinchona alkaloids [23–25], which have evolved to the 9-anthraceneylmethyl derivatives [26–28]. Catalysts based on biaryl systems also provide excellent chiral induction [29–33]. The approach can also be used to prepare α,α-disubstituted amino acids.

12.2.2.2 **Modifications of the Side-Chain**

Standard chemical modifications to amino acids with functionality in the side-chains, such as serine, cysteine, aspartic, and glutamic acids, allow for a wide variety of unnatural amino acids to be accessed [34–37]. Although the D-series of these functionalized amino acids are not as readily available, the use of the approach can still be extremely powerful, concise and cost effective.

One example is provided by the synthesis of L-homoserine (7) from methionine (Scheme 12.9) [38]. The problems associated with the formation of 1 equiv. of dimethylsulfide have to be overcome!

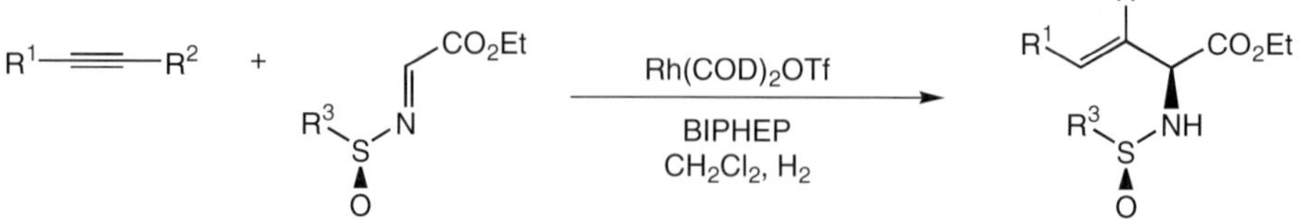

R¹
R²
+
CO₂Et
R³
S
N
O
Rh(COD)₂OTf
BIPHEP
CH₂Cl₂, H₂
R²
R¹
CO₂Et
R³
S
NH
O

where R^1 = vinyl or alkynyl
R^2 = alkyl or aryl
R^3 = *t*-Bu or 2,4,6-(iPr)$_3C_6H_2$

66-99%
dr >95:5

Scheme 12.7

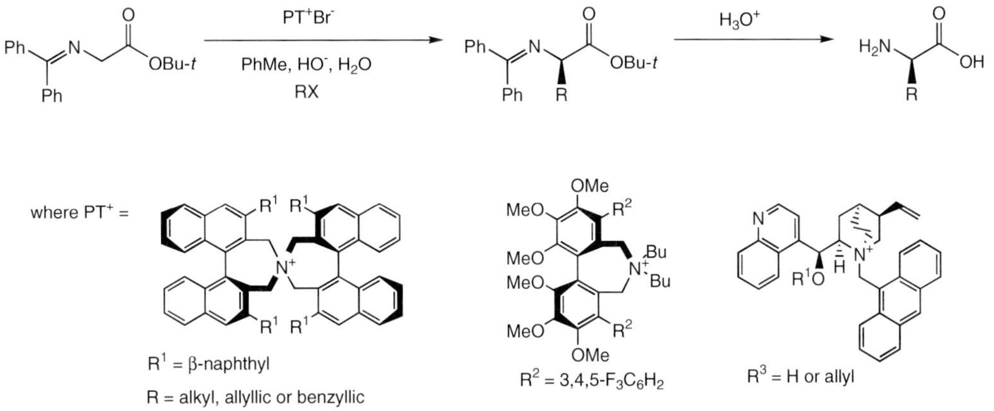

Scheme 12.8

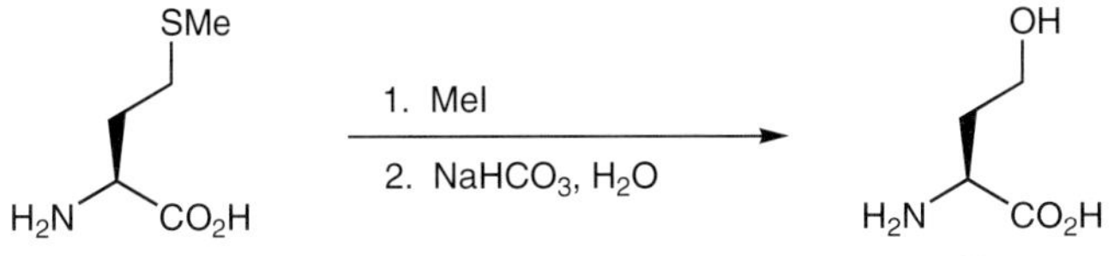

Scheme 12.9

A wide variety of simple transformations on chiral pool materials can lead to unnatural amino acid derivatives. This is illustrated by the Pictet–Spengler reaction of L-phenylalanine, followed by amide formation and reduction of the aromatic ring (Scheme 12.10) [8]. The resultant amide **8** is an intermediate in a number of commercial HIV protease inhibitors.

A wide variety of side-chains can be introduced through the use of chain extension reactions, such as metathesis and cross coupling reactions [39, 40]. The Suzuki reaction has proven useful for the latter, although the amino acid does need to be protected (Scheme 12.11) [41], and sometimes the protected amino alcohol, rather than acid, has been employed [42–44].

In addition to acyclic transformations, glutamic acid can be cyclized to pyroglutamic whose subsequent chemistry can benefit from the stereocontrol of the ring system [45]. Some unnatural amino acids have drawn considerable attention and many syntheses involving side-chain manipulations have been used as in approaches to kainic acid (**9**) and it analogs [46].

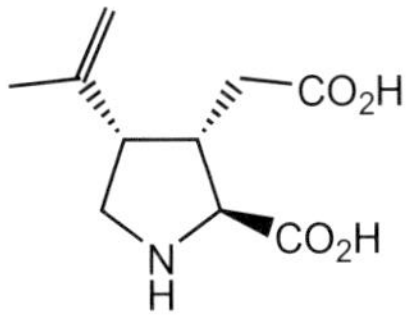

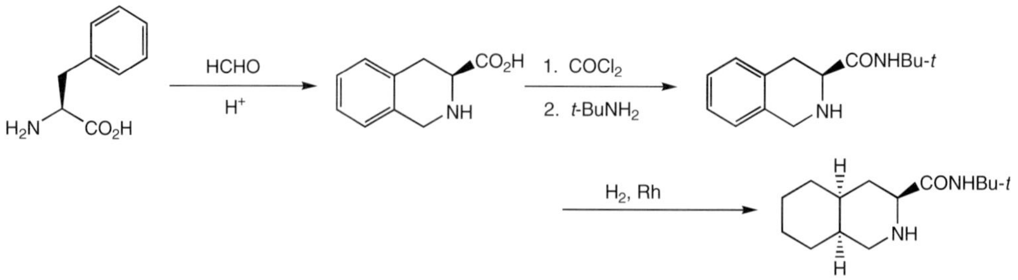

Scheme 12.10

12.2.3
Introduction of Functionality

12.2.3.1 Nitrogen Introduction

This approach usually relies on the use of a chiral auxiliary or template to control the stereoselectivity. As a consequence, the method is more useful for smaller-scale preparations. The nitrogen can be introduced as an electrophile or nucleophile.

The stereogenic center of a chiral template controls the stereochemistry at a newly formed center, but it is destroyed during removal of the control group. As an example, dehydromorpholinones **10** can be used to prepare unnatural amino acids and α-substituted amino acids from α-keto esters (Scheme 12.12) [47, 48]. The phenylglycinol unit is cleaved by hydrogenolysis, effectively transferring the nitrogen from the amino alcohol control moiety to the product.

N-Acyl-oxazolidinones are readily available and the use of this class of compounds for the preparation of unnatural amino acids has been well documented [13, 14]. Di-*tert*-butyl azodicarboxylate reacts readily with the lithium enolates of *N*-acyloxazolidinones to provide hydrazides **11** in excellent yield and high diastereomeric ratio (Scheme 12.13) (see Section 12.5.2) [49, 50]. The hydrazides can then be converted to the amino acid.

The electrophilic introduction of azide with chiral imide enolates has also been used to prepare α-amino acids with high diastereoselection (Scheme 12.14). The reaction can be performed with either the enolate directly [51–55] or through a halo intermediate [56]. The resultant azide can be reduced to an amine [16].

Although general, these approaches suffer from low atom economy with the related high costs.

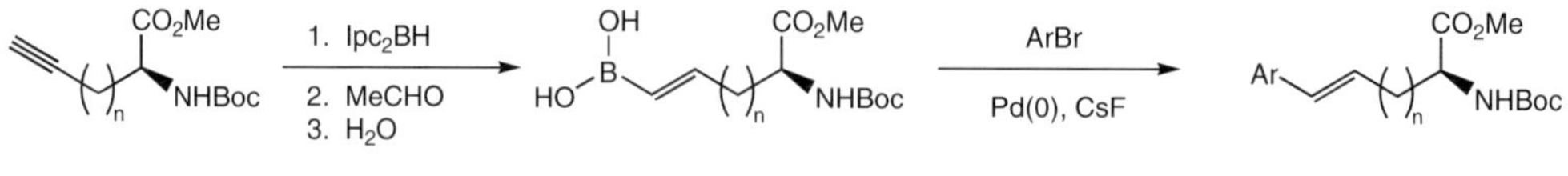

where n = 1 or 2
Ar = 3-Br-pyridine or thiophenene

Scheme 12.11

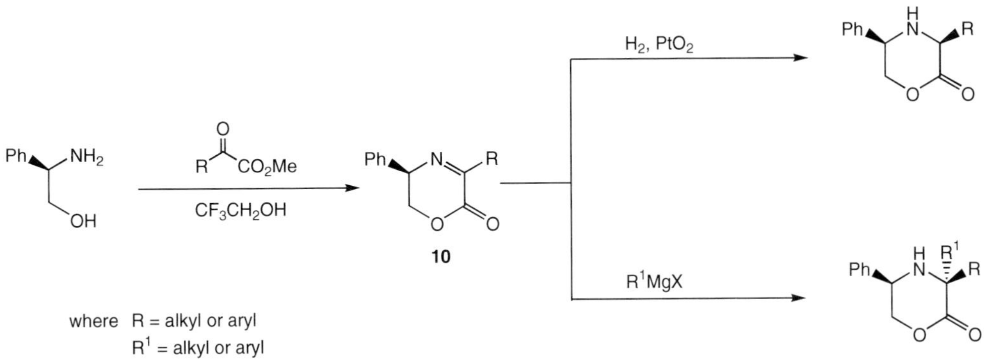

where R = alkyl or aryl
R^1 = alkyl or aryl

Scheme 12.12

12.2.3.2 **Carboxylic Acid Introduction**

Most of these methods use the Strecker reaction where a nitrile is used as a masked carboxylic acid. In effect, the nitrile adds to an imine.

12.2.3.2.1 **Strecker Reaction** The original Strecker reaction leads to a racemic product, which can be resolved by a variety of methods but results in a maximum 50% yield for a single enantiomer.

The reaction can be performed asymmetrically by the use of chiral induction from an auxiliary or template as well as by the use of a chiral catalyst [57, 58]. The latter methods are more amenable to use at larger scales.

The chiral template (*R*)-phenylglycine amide (**3**) gave the aminonitrile **12** into greater than 98% d.e. (Scheme 12.15), as it precipitates from the reaction medium [59]. The HCN was generated *in situ* from NaCN/AcOH for convenience in the laboratory. The reaction was performed in water at 70 °C for 24 h, as the initial experiment in methanol gave a much lower diastereomeric excess. This diastereoselectivity can be explained by the reversible nature of the conversion of the amino nitriles into the intermediate imine and by the difference in solubility between both diastereoisomers under the conditions applied. The aminonitrile **12** can be converted to (*S*)-*tert*-leucine (**13**) [59].

Phenylglycinol (**14**) itself can be used as a chiral template transferring the nitrogen in an asymmetric Strecker reaction (Scheme 12.16). However, in some examples of this approach, caution must be exercised as epimerization has been observed during hydrolysis of the nitrile group [60–63].

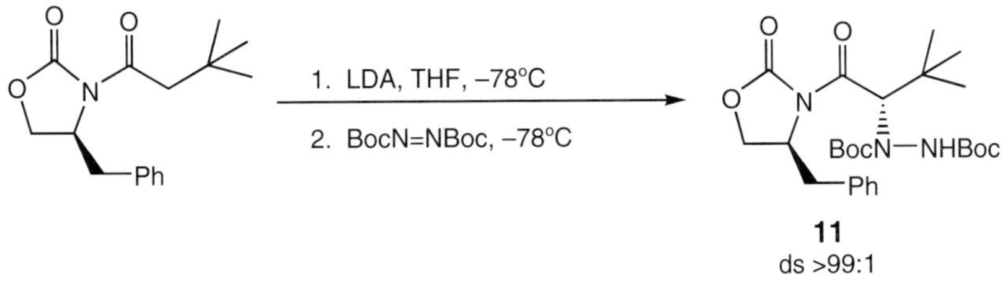

Scheme 12.13

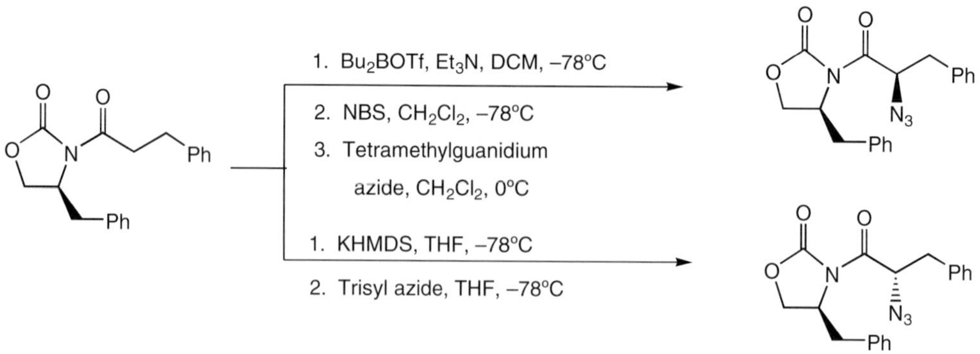

1. Bu_2BOTf, Et_3N, DCM, –78°C
2. NBS, CH_2Cl_2, –78°C
3. Tetramethylguanidium azide, CH_2Cl_2, 0°C

1. KHMDS, THF, –78°C
2. Trisyl azide, THF, –78°C

Scheme 12.14

12.2.4
Hydrogenation

There are literally thousands of catalysts and ligand systems available for the preparation of unnatural amino acids and some can be used at scale [64, 65]. In many cases, the ligands were developed to allow a company freedom to operate. Despite the plethora of ligands in the literature, few are available at scale and, thus, most of these have arisen from a company applying its own technology.

Perhaps the most famous example of the use of asymmetric hydrogenation at scale for the product of an "unnatural" amino acid is the Monsanto synthesis of L-3,4-dihydroxyphenylalanine (DOPA), a drug used for the treatment of Parkinson's disease (Scheme 12.17) [66, 67]. The methodology with the Knowles' catalyst system has been extended to a number of other unnatural amino acids [68, 69].

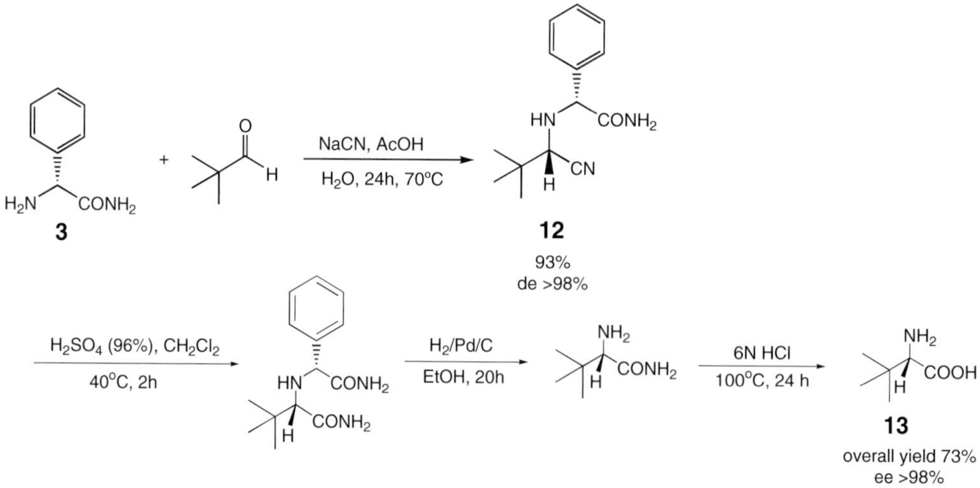

NaCN, AcOH, H_2O, 24h, 70°C

3 → 12 (93%, de >98%)

H_2SO_4 (96%), CH_2Cl_2, 40°C, 2h

H_2/Pd/C, EtOH, 20h

6N HCl, 100°C, 24 h

13 (overall yield 73%, ee >98%)

Scheme 12.15

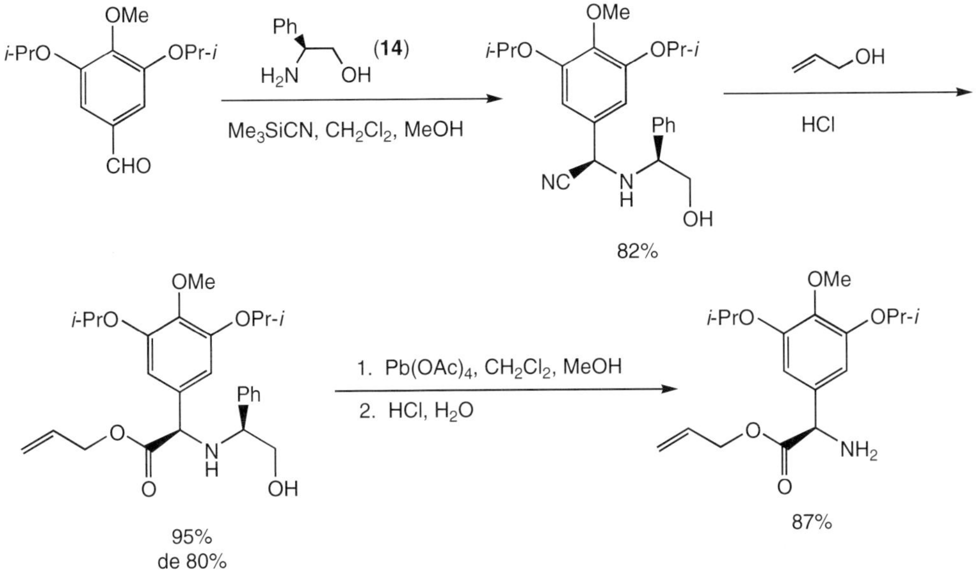

Scheme 12.16

The asymmetric catalyst is based on the chiral bisphosphine, (*R,R*)-DIPAMP (**15**) (DIPAMP = 1,2-bis[(*o*-anisyl)(phenyl)phosphino]ethane), that has chirality at the phosphorus atoms and can form a five-membered chelate ring with rhodium. The asymmetric reduction of the (*Z*) enamide proceeds in 96% e.e. [67]. The pure isomer of the protected amino acid intermediate **16** can be obtained upon crystallization from the reaction mixture as it is a conglomerate [66]. Although the catalyst system is amenable to the preparation of a wide variety of amino acids, especially substituted

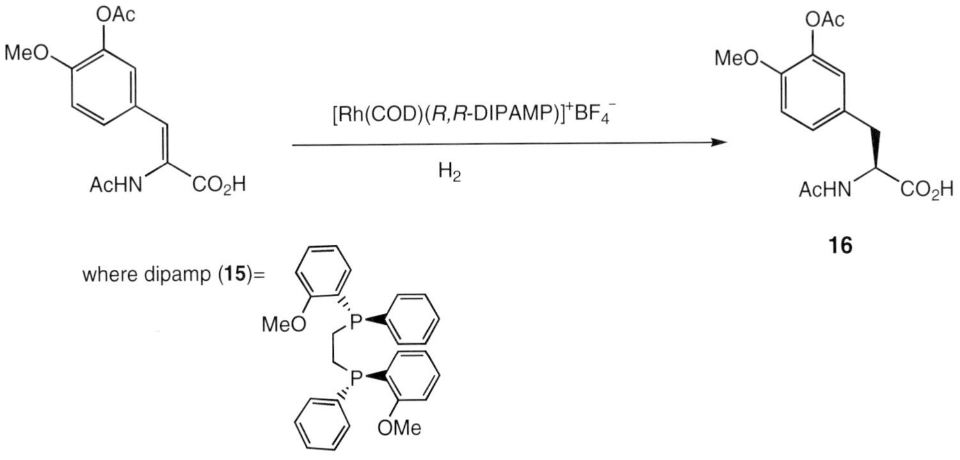

Scheme 12.17

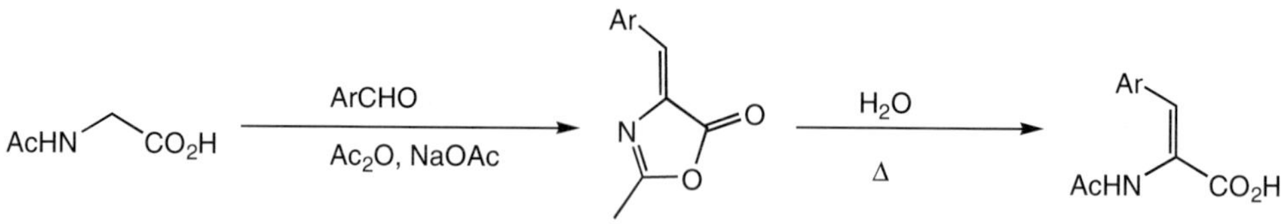

Scheme 12.18

phenylalanine derivatives, a major shortcoming of the approach is the need to have just the (*Z*) enamide isomer as the substrate (see Section 12.6.3).

Fortunately, the general method for the preparation of the enamides is stereoselective to the (*Z*) isomer when an aryl aldehyde is used (Scheme 12.18) [70]. Other approaches to the alkene are also available such as the Heck or Suzuki reactions [71].

In addition to DIPAMP, which is expensive to synthesize, DuPhos and MonoPhos ligands have been employed to prepare unnatural amino acids at scale.

The MonoPhos family of ligands for the reduction of C=C double bonds, including enamides, are based on the 3,3′-bisnaphthol (BINOL) backbone. These phosphoramidite ligands are comparatively inexpensive to prepare compared to bisphosphine ligands. For enamide reductions with MonoPhos (**17a**) as ligand, it was found that the reaction is strongly solvent dependent [72]. Very good enantioselectivities were obtained in nonprotic solvents [73–75]. A large range of substituted *N*-acetyl phenylalanines and their esters can be prepared by asymmetric hydrogenation with Rh/MonoPhos with enantioselectivities ranging from 93 to 99% (Scheme 12.19). The configuration of the α-amino acid products always is the opposite of that of the BINOL in the ligand used in the hydrogenation.

Due to the relative ease of synthesis of the phosphoramidites, it is simple to introduce variations within the ligand, especially substitution on the nitrogen atom. In turn, this allows for rapid screening of ligands to find the best one for a specific substrate. The screen can be used to not only increase enantioselectivities but also

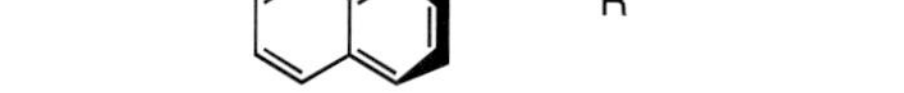

where R = alkyl, aryl or heterocyclic
R^1 = alkyl or H

17 a $R^2 = R^3$ = Me (MonoPhos)
b $R^2, R^3 = -(CH_2)_5-$ (PipPhos)
c $R^2 = R^3$ = Et
d R^2 = (*R*)-CHMePh, R^3 = H

Scheme 12.19

turnover numbers and frequencies. In a number of cases the piperidine analog (**17b**) has been found to be superior for α-amino acid and ester production over MonoPhos (**17a**) itself [73, 76, 77]. The best ligands in terms of enantioselectivity are **17c** [77] and **17b** [76]. The rate of the hydrogenation is retarded by bulky groups on the nitrogen atom and by electron withdrawing substituents on the BINOL.

Phosphites (**18**), also based on BINOL, are excellent ligands for the asymmetric hydrogenation of dehydroamino acids and esters. As with the MonoPhos ligands, synthesis is short and simple and screening may be necessary to find the optimal ligand. The methodology has also been applied to the synthesis of β,β-disubstituted amino acids [78, 79].

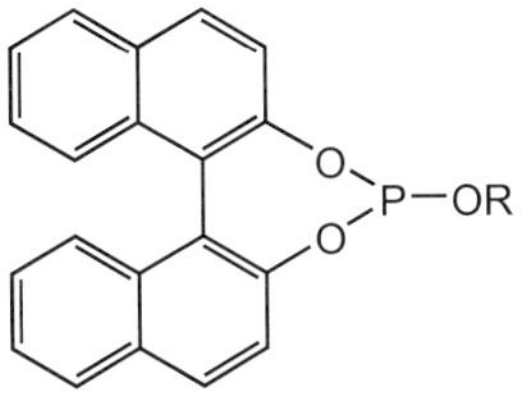

18

Rhodium complexes of the type $[(COD)Rh(DuPhos)]^+X^-$ (X = weakly or non-coordinating anion; COD = cyclooctadiene) have been developed as one of the most general classes of catalyst precursors for efficient, enantioselective low-pressure hydrogenation of enamides (**19**) (Scheme 12.20) [80, 81]. The DuPhos approach overcomes some of the limitations of the DIPAMP system as the substrates may be present as mixtures of (*E*) and (*Z*) geometric isomers. For substrates that possess a single β-substituent (e.g., $R^3 = H$), the Me-DuPhos-Rh and Et-DuPhos-Rh catalysts were found to give enantioselectivities of 95–99% for a wide range of amino acid derivatives [81, 82].

As with the DIPAMP and MonoPhos ligand systems, a variety of *N*-acyl protecting groups, such as acetyl, Cbz, or Boc, may be on the substrate nitrogen, and enamides **19** may be used either as carboxylic esters or acids. The high catalyst activities with the DuPhos ligands and catalyst productivities are an advantage for use in large-scale reactions. In addition, unlike other catalyst systems, there is little difference between the use of (*E*) or (*Z*) dehydroamino acids or esters as substrates [82, 83].

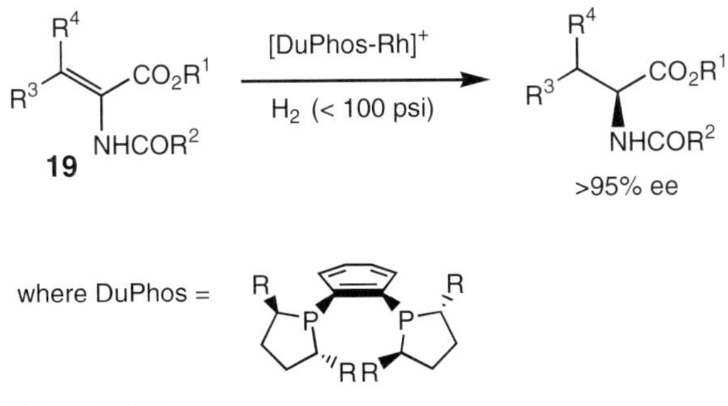

19 → $[DuPhos\text{-}Rh]^+$, H_2 (< 100 psi) → >95% ee

where DuPhos =

Scheme 12.20

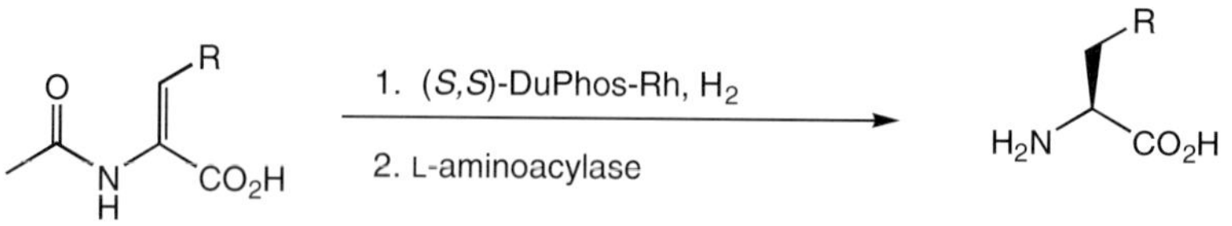

1. (*S*,*S*)-DuPhos-Rh, H_2

2. L-aminoacylase

Scheme 12.21

The rhodium catalysts derived from Me-DuPhos and the more electron-rich Me-BPE [1,2-bis(2*R*,5*R*)-2,5-dimethylphospholano)ethane; **20**, R = Me) catalysts can be used to access β-branched amino acids (Scheme 12.20, R^2, $R^3 \neq H$) [84, 85]. Hydrogenation of just the (*E*) or (*Z*) enamide isomers allows the generation of the β-stereogenic center with high selectivity.

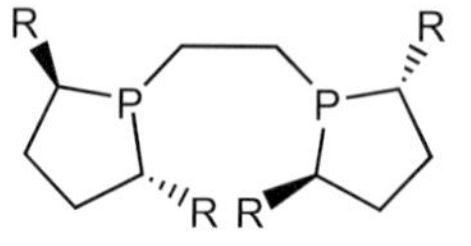

20

The DuPhos reduction has been used to prepare a wide range of amino acids with substituted aromatic, heteroaromatic, alkyl, fluoroalkyl, and other functionalized organic groups (see Section 12.5.4) [82, 83, 86–90]. Polyamino acids can also be accessed [91–94].

Although asymmetric hydrogenations can give high enantioselectivity, invariably an acetyl group is present on the nitrogen. In some circumstances, cleavage of this amide bond to reveal the free amino group can be troublesome, as harsh conditions such as hot, concentrated hydrochloric acid may be required; this can lead to racemization. A milder method, which can also enhance the stereoselection is to use an enzyme to perform the unmasking reaction (Scheme 12.21) [95]. This approach allowed the upgrade of enantioselectivity from 86% for the hydrogenation of the 3-fluorophenylalanine derivative to greater than 99% e.e. by enzymatic hydrolysis to the parent amino acid, which was converted to the *N*-Boc derivative prior to isolation.

12.2.5
Other Chemical Methods

A wide variety of functionality can be incorporated into an amino acid by preparing them with a pericyclic reaction. The approach also allows for α,α-disubstituted amino acids to be accessed [96]. The use of a Claisen rearrangement illustrates the power of this reaction class (Scheme 12.22) [97, 98].

12.3
Enzymatic Methods

Most of the enzymatic and biological methods rely on the ability of an enzyme to differentiate between enantiomers and, thus, perform resolutions. For larger-scale applications, organisms that contain two are more enzymes are being developed.

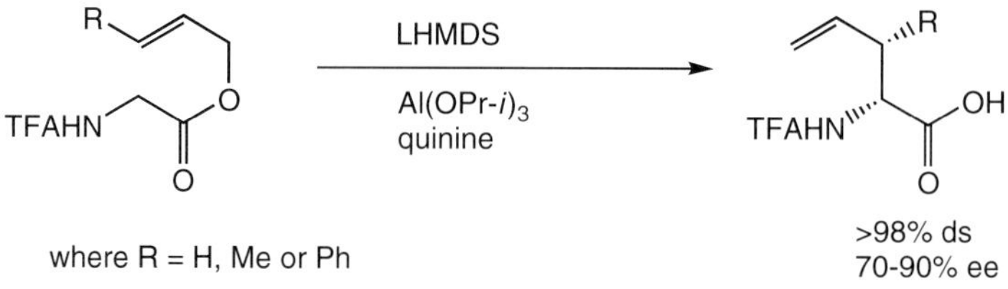

where R = H, Me or Ph

>98% ds
70-90% ee

Scheme 12.22

The side-chain of an amino acid can also be manipulated by biological catalysts [99]. One example takes advantage of the naturally broad substrate specificity of *O*-acetylserine sulfhydrylase, the final enzyme in L-cysteine biosynthesis. This enzyme accepts a variety of alternative nucleophiles, in addition to the usual sulfide required for cysteine biosynthesis [100], to provide *S*-phenyl-L-cysteine (**21**), *S*-hydroxyethyl-L-cysteine (**22**), or phenylseleno-L-cysteine (**23**).

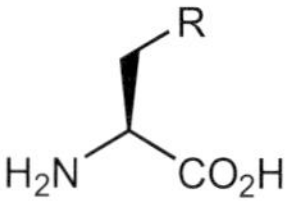

21, R = SPh
22, R = $S(CH_2)_2OH$
23, R = SePh

12.3.1 Acylases

Although some of the methods that use enzymes were developed to prepare proteinogenic amino acids, the wide substrate specificity of some of the enzymes has allowed the same methodology to be applied for the preparation of unnatural amino acids. Such an example is the use of acylases [101], where the aminoacylase I from *Aspergillus oryzae* is used for the production of L-amino acids (Scheme 12.23) [102]. The unwanted D-*N*-acyl compound is recycled into the synthesis stream of the racemate, which causes racemization. The enzyme can be used in a membrane reactor or immobilized to allow multiple catalytic cycles and simplify separation [99, 102, 103]. In addition, the racemization can be performed with an enzyme or by imine formation. This approach has been used to provide L-α-methyl-DOPA [104, 105], as well as a variety of β-heterocyclic α-amino acids [106]. The methodology provides a general,

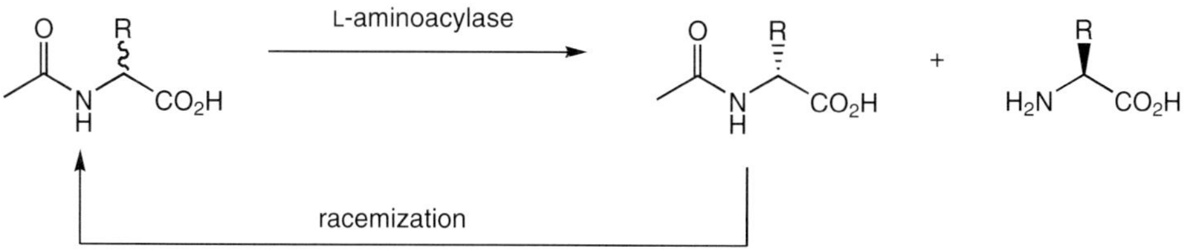

Scheme 12.23

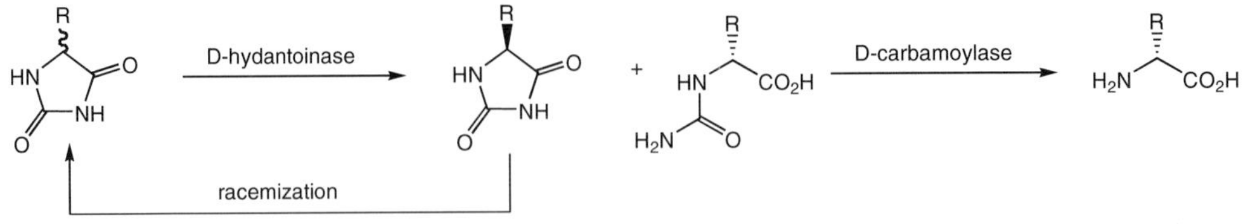

D-hydantoinase

D-carbamoylase

racemization

Scheme 12.24

industrial-scale process for the production of either L- or D-amino acids through the hydrolysis of the amide [107–110]. However, isolation of the D-aminoacylase from the L-selective enzymes invariably present in an organism is not always trivial; the versatility of alternative approaches, therefore, detracts from the use of the methodology with D-aminoacylases.

12.3.2 Hydantoinases

Hydantoins can offer a simple racemization approach, although the side-chain does have an influence on the acidity of the α-hydrogen. The use of a carbamoylase removes the desired product from the equilibria (Scheme 12.24) [111, 112]. The racemization can be accomplished through pH or an additional enzyme. The methodology has been used for the preparation of, amongst other D-amino acids, D-phenylglycine and D-*p*-hydroxyphenylglycine, side-chains in semisynthetic β-lactam antibiotics [3, 103, 113–118]. Rather than use an enzyme, it is possible to convert the urea to the amine with chemical reagents such as HNO_2. The enzymatic method, however, uses milder conditions and tolerates a wide range of functional groups in the side-chain.

The same approach can also be used to prepare L-amino acids, by use of the analogous L-specific enzymes [119–122]. All the enzymes can be expressed in a single whole cell [117], but productivity still remains low.

12.3.3 Ammonia Lyases

Lyases are an attractive group of enzymes from a commercial perspective as demonstrated by their use in many industrial processes [123]. Two well-studied ammonia lyases, aspartate ammonia lyase (aspartase) and phenylalanine ammonia lyase (PAL) catalyze the *trans*-elimination of ammonia from the amino acids, L-aspartate and L-phenylalanine, respectively. The reverse reaction has been utilized to prepare the L-amino acids at the ton scale (Scheme 12.25) [123–128]. These reactions are conducted at very high substrate concentrations such that the equilibrium is shifted resulting in very high conversion to the amino acid products, although this may be offset by substrate inhibition [129, 130]. Immobilization has been used to aid product isolation [102, 131]. Aspartase exhibits incredibly strict substrate specificity and, thus, is of little use in the preparation of L-aspartic acid analogs.

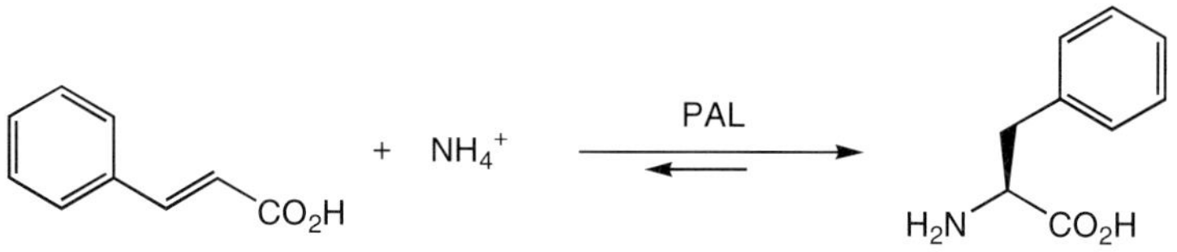

+ NH_4^+ PAL CO_2H H_2N CO_2H

Scheme 12.25

However, a number of L-phenylalanine analogs have been prepared with various PAL enzymes from the yeast strains *Rhodotorula graminis, Rhodotorula rubra,* and *Rhodoturula glutinis,* and several other sources, which have been cloned into *Escherichia coli* [127, 128]. Use of tyrosine phenol lyase has been applied to the preparation of L-DOPA [103], while molecular biology is providing more enzymes capable of producing unnatural amino acids.

12.3.4 Transaminases

The aminotransferase class of enzymes, also known as transaminases, are ubiquitous enzymes that have been utilized to prepare natural L-amino acids and other chiral compounds [132–137]. The L-aminotransferases catalyze the general reaction where an amino group from one L-amino acid is transferred to an α-keto acid to produce a new L-amino acid and the respective α-keto acid (Scheme 12.26). The enzymes most commonly used as industrial biocatalysts have been cloned, overexpressed, and are generally used as whole-cell or immobilized preparations.

An ω-aminotransferase, ornithine (lysine) aminotransferase, has been described for the preparation of chiral intermediates (Figure 12.2) used in the synthesis of omapatrilat (**24**), a vasopeptidase inhibitor from Bristol-Myers Squibb, as well as the unnatural amino acid Δ^1-piperidine-6-carboxylic acid (**25**), which is the byproduct from the amino donor [138–140]. The approach can be used to access L-2-aminobutyrate (**26**), a component of the drug levetiracetam (**27**). Another large-scale application of the technology is for the synthesis of L-phosphinothricin (**28**), which is used in the herbicide Basta [136, 141]. Immobilization of the enzyme allows for the implementation of a continuous process.

Ornithine can also be used as the amino donor with the enzyme ornithine γ-aminotransferase because the byproduct is glutamate semialdehyde, which spontaneously cyclizes to the dihydropyrrole (Scheme 12.27) [142].

O R CO_2^- + CO_2^- n ^+H_3N CO_2^- aminotransferase R ^+H_3N CO_2^- + CO_2^- n O CO_2^-

where n = 1 or 2

Scheme 12.26

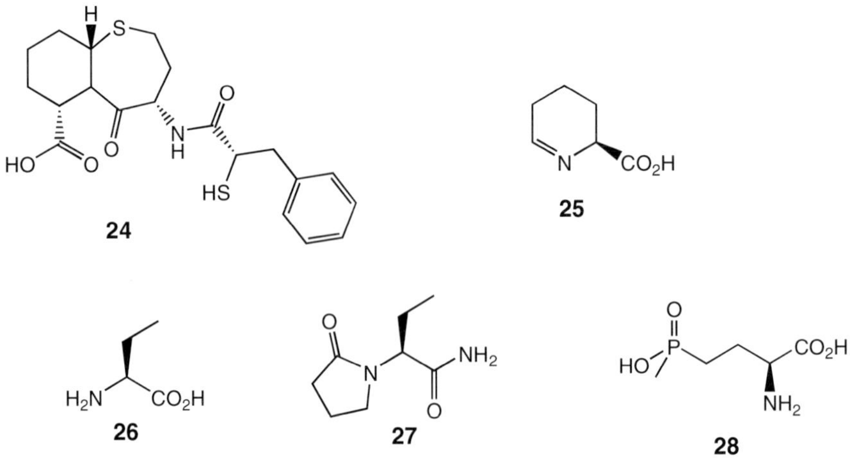

24 25 26 27 28

Figure 12.2 Products from the use of lysine aminotransferase.

Another useful transaminase, D-amino acid transaminase (EC 2.6.1.21), has been the subject of much study [143–145]. This enzyme catalyzes the reaction utilizing a D-amino acid donor, alanine, aspartate, or glutamate, to produce another D-amino acid (Scheme 12.28).

Although the utility of transaminases has been widely examined, one limitation is the equilibrium constant for the reaction, which is near unity. Therefore, a shift in this equilibrium is necessary for the reaction to be synthetically useful. A number of approaches to address this problem can be found in the literature [133, 143]. An elegant solution is to use aspartate as the amino donor, as this is converted into oxaloacetate (**29**) (Scheme 12.21). Since **29** is unstable, it decomposes to pyruvate (**30**) and, thus, favors product formation. However, since pyruvate is itself an α-keto acid and can be converted into alanine, which could potentially cause downstream processing problems, it must be removed. This is accomplished by including the *alsS* gene encoding for the enzyme acetolactate synthase, which condenses 2 mol of pyruvate to form (*S*)-acetolactate (**31**). The (*S*)-acetolactate then undergoes decarboxylation either spontaneously or by the enzyme acetolactate decarboxylase to the final byproduct, (*R*)-acetoin (**32**), which is metabolically inert. This process can be applied to the production of L- or D-2-aminobutyrate (**26** and **33**, respectively) (Scheme 12.29) [133, 134, 139, 140, 146–149].

For D-amino acids, the aminotransferases also have broad substrate specificity and can be used in an analogous manner to L-aminotransferases. For the amino donor, the D-isomer of the amino acid is required and this can be obtained by

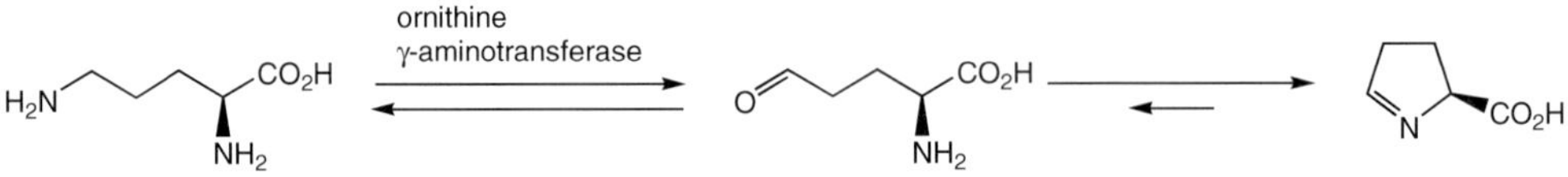

ornithine γ-aminotransferase

Scheme 12.27

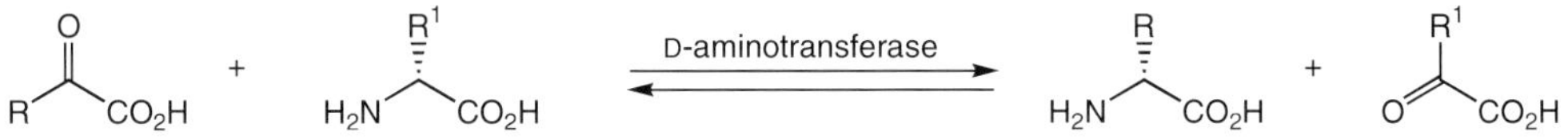

where R^1 = Me, CH_2CO_2H or $(CH_2)_nCO_2H$

Scheme 12.28

racemization [148, 150]. Fortunately, the racemases for the amino acids, alanine, aspartic acid, and glutamic acid are specific, and will not degrade the enantiopurity of the desired product [143, 146, 148, 151, 152].

12.3.5 Dehydrogenases

The alternative to transamination is a reductive amination. A dehydrogenase, such as leucine dehydrogenase for the formation of L-*tert*-leucine (**13**), uses ammonia as the nitrogen source and performs a reduction to give the amino acid (Scheme 12.30) [153]. To date, leucine and phenylalanine dehydrogenases have been the most successful in commercial applications [154, 155]. To make the process commercially viable,

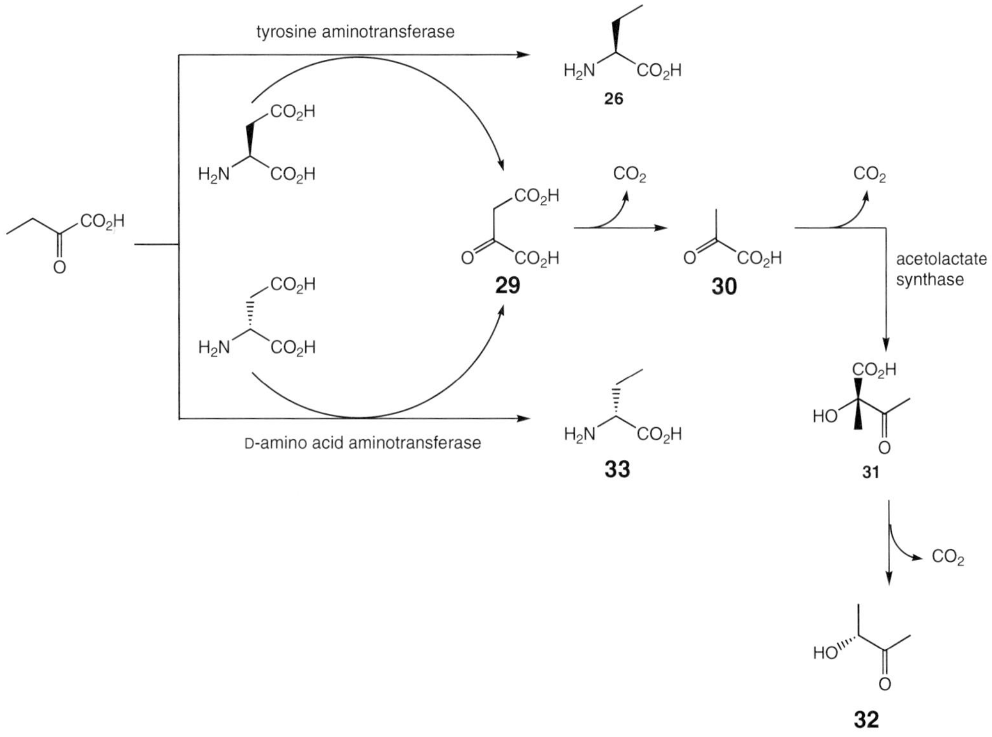

Scheme 12.29

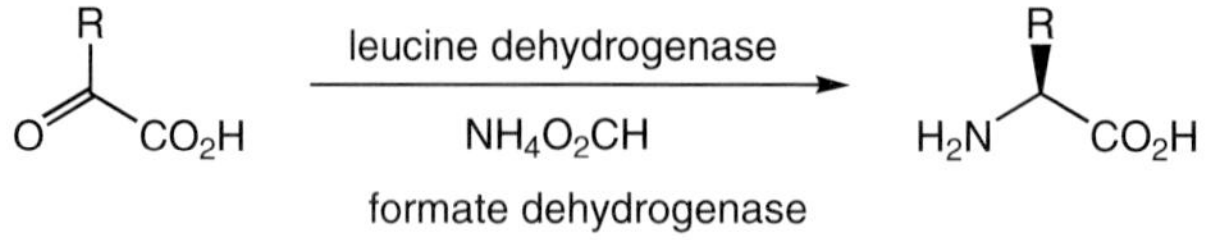

leucine dehydrogenase
NH_4O_2CH
formate dehydrogenase

Scheme 12.30

the NAD^+ cofactor is recycled by coupling the reaction to formate dehydrogenase where formate is converted to CO_2, thus driving the equilibria in the desired direction [156–162]. Other reductants, such as glucose, can also be used [163]. The processes can be run in membrane reactors [164–166].

Although D-hydrogenases are known, significant work still needs to be done with regard to selectivity before they can be viable at scale.

12.3.6
Amino Acid Oxidases

A further variation on the manipulation of the amino group is to perform a dynamic kinetic resolution with an enzyme and a chemical reagent.

One resolution approach that is showing promise is the use of an enzyme, such as an amino acid oxidase to convert one isomer of an amino acid to the corresponding achiral imine. A chemical reducing agent then returns the amino acid as a racemic mixture (Scheme 12.31) [167]. As the enzyme only acts on one enantiomer of the amino acid, the desired product builds up. After only seven cycles, the product at greater than 99% e.e. [168]. The key success factor is to have the reducing agent compatible with the enzyme [169, 170]. Rather than use the antipode of the desired product, the racemic mixture can be used as the starting material. The enzymes are available to prepare either enantiomeric series of amino acid.

The approach is amenable to access β-substituted α-amino acids [171]. The methodology has culminated in a way to prepare all four possible isomers of β-aryl α-amino acids by a combination of asymmetric hydrogenation (See Section 12.2.4) and the use of the deracemization process to invert the α-center (Scheme 12.32) (see Section 12.5.4) [71].

Of course, a racemic mixture of an amino acid can be resolved by the action of an amino acid oxidase in the absence of a chemical reducing agent, but then this is

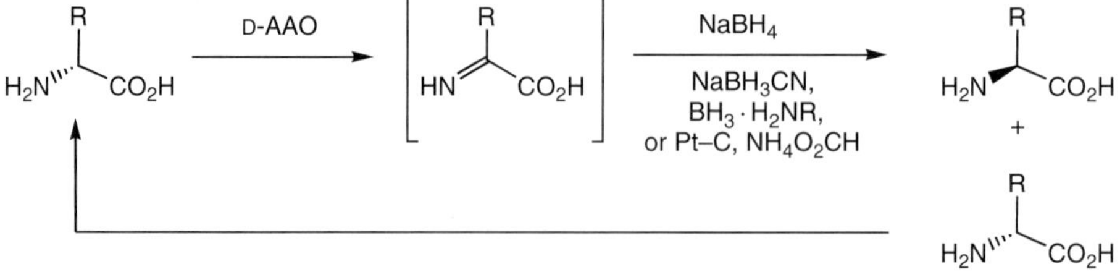

D-AAO
$NaBH_4$
$NaBH_3CN$,
$BH_3 \cdot H_2NR$,
or Pt–C, NH_4O_2CH

Scheme 12.31

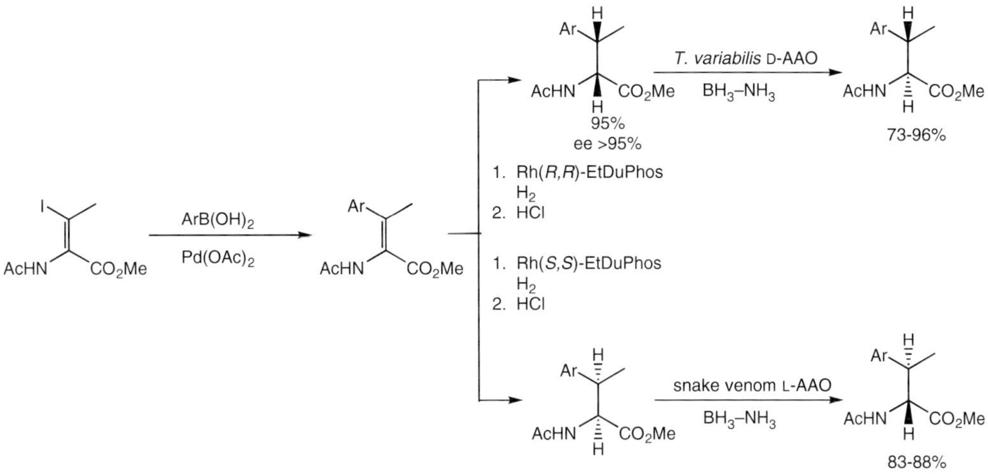

Scheme 12.32

a standard resolution and the maximum yield is 50% and one of the amino acid antipodes is destroyed [172].

12.3.7 Decarboxylases

A number of decarboxylase enzymes have been described as catalysts for the preparation of chiral synthons, which are difficult to access chemically [173]. The amino acid decarboxylases catalyze the pyridoxal 5′-phosphate-dependent removal of CO_2 from their respective substrates. This reaction has found great industrial utility with one specific enzyme in particular, L-aspartate-β-decarboxylase from *Pseudomonas dacunhae*. This biocatalyst, as immobilized whole cells, has been utilized for the industrial-scale synthesis of L-alanine [102]. Another use for this biocatalyst is the resolution of racemic aspartic acid to produce L-alanine and D-aspartic acid (Scheme 12.33) [174, 175]. The two amino acids can be separated by crystallization. The cloning of the L-aspartate-β-decarboxylase from *Alcaligenes faecalis* into *E. coli* offers additional potential to produce both of these amino acids [176].

Numerous other amino acid decarboxylases have been isolated and characterized, and much interest has been shown due to the irreversible nature of the reaction with the release of CO_2 as the thermodynamic driving force. Although these enzymes have narrow substrate specificity profiles, their utility has been widely demonstrated.

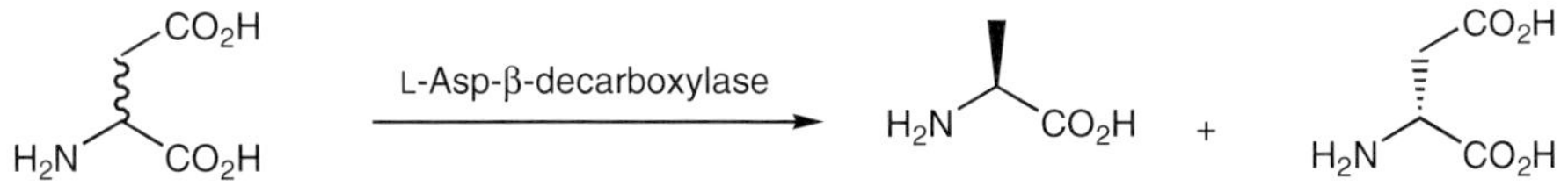

Scheme 12.33

12.4
Conclusions

Today, there are a multitude of methods available to prepare unnatural amino acids either in the L- or D-series. At laboratory scale, the use of chiral auxiliaries or side-chain manipulations provide the quickest and most tested approaches. At larger scale, asymmetric hydrogenation and the use of chiral pool starting materials offer methods for fast implementation. For very large production, biological methods offer the cheapest solutions and with our understanding of these processes increasing, implementation timelines have been drastically reduced [177].

12.5
Experimental Procedures

This section contains experimental procedures for reactions described in the previous sections. The emphasis is on chemical transformations as these are faster to implement in a laboratory setting. In addition, many of the biological approaches do not have experimental details for the actual conversion of the substrate to the unnatural amino acid. The examples are illustrative only.

12.5.1
Side-Chain Introduction with a Phase-Transfer Catalyst

Preparation of D-Phenylalanine *tert*-Butyl Ester (Scheme 12.8; R = $PhCH_2$, R^3 = H) [28]

Ph
Ph N CO_2tBu
H_2N CO_2tBu

A solution of the benzophenone imine of glycine *tert*-butyl ester (0.5 mmol) in toluene (5 ml) is treated sequentially with the catalyst (0.05 mmol), benzyl bromide (0.6 mmol), and 50% aqueous KOH (1 ml). The resulting mixture is stirred vigorously (about 1000 rpm) at room temperature for 18 h. The aqueous layer is then extracted with EtOAc (3 × 5 ml), and the combined organics dried (Na_2SO_4) and concentrated under reduced pressure to give the cruse imine product. The material is dissolved in tetrahydrofuran (THF) (3 ml) and 15% aqueous citric acid (1.5 ml) added. The mixture is stirred vigorously at room temperature for 3 h, then diluted with water (5 ml). The mixture is extracted with diethyl ether (2 × 5 ml) to remove any excess alkylating agent and benzophenone, then the aqueous layer is basified (K_2CO_3). Extraction with EtOAc (3 × 5 ml) followed by drying of the extracts (Na_2SO_4) and concentration under reduced pressure gives the crude amino acid *tert*-butyl ester which can be purified by passing down a plug of silica. Yield of the ester was 63% (89% e.e.).

12.5.2
Introduction of Nitrogen Through an Oxazolidinone Enolate with a Nitrogen Electrophile

Preparation of Di-*tert*-butyl 1-((*S*)-1-((*S*)-4-benzyl-2-oxooxazolidin-3-yl)-3,3-dimethyl-1-oxobutan-2-yl) hydrazine-1,2-dicarboxylate (11) (Scheme 12.13) [49]

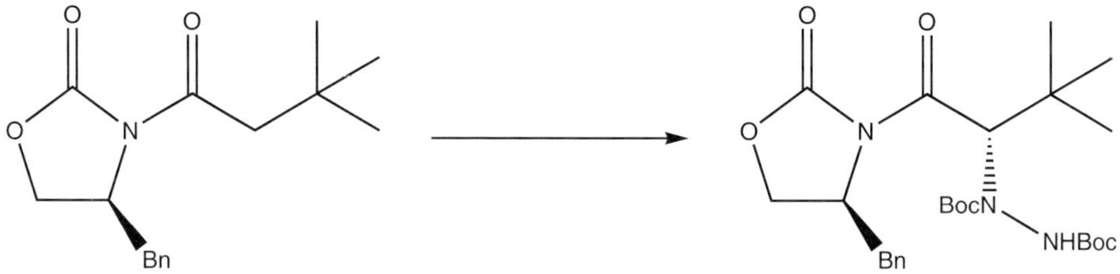

To a freshly solution of lithium diisopropylamide (10.0 mmol) in THF (30 ml), stirred at −78 °C under N_2, was added via a cannula a precooled (−78 °C) solution of the imide (2.75 g, 10.0 mmol) in THF (30 ml). The residual imine was rinsed in with THF (2 × 10 ml) and stirring was continued at −78 °C for 40 min. A precooled solution of di-*tert*-butyl azodicaboxylate (2.65 g, 11.5 mmol) in dry CH_2Cl_2 (60 ml) was added via cannula to the enolate solution and after an additional 3 min the reaction was quenched with AcOH (26 mmol). The mixture was partitioned between CH_2Cl_2 and pH 7 phosphate buffer. The aqueous phase was washed with three portions of CH_2Cl_2. The CH_2Cl_2 phases were combined, washed with saturated $NaHCO_3$, dried ($MgSO_4$), and purified by medium-pressure liquid chromatography (MPLC) (CH_2Cl_2/hexane/MeCN 70 : 30 : 5) to give the hydrazine **11** (4.83 g, 96%) as a colorless glass foam with a diastereomeric ratio of 99.7 : 0.3.

Preparation of 2-(*N*,*N*′-bis(*tert*-butoxycarbonyl) hydrazino)-3,3-dimethylbutanoic Acid, Benzyl Ester

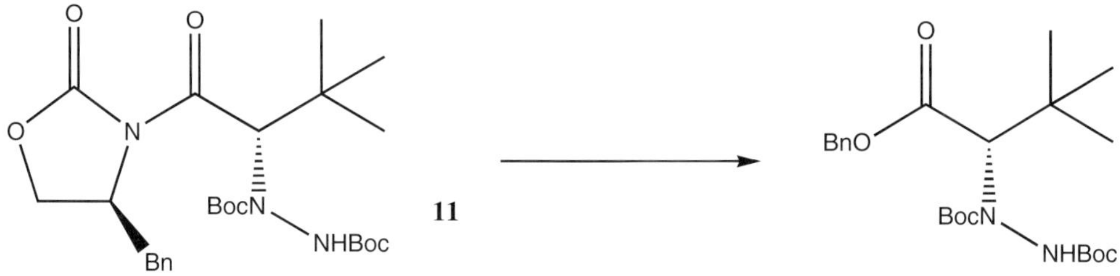

To a cold (−78 °C) solution of LiOBn [10.88 mmol, prepare from benzyl alcohol (16.32 mmol) and *n*-butyl lithium (10.88 mmol as a 1.59 M solution in hexane)] in dry THF (33.4 ml) was added via cannula to the imide **11** (2.75 g, 5.44 mmol) in THF (21.7 ml). The residual imine was rinsed with THF (2 × 11 ml) and the resultant solution stirred at −50 °C for 50 h prior to quenching with aqueous pH 7 phosphate buffer (34 ml). The mixture was partitioned between H_2O and CH_2Cl_2. The aqueous phase was extracted with three portions of CH_2Cl_2, and the organic phases were combined, dried ($MgSO_4$), and evaporated *in vacuo* to give the crude ester, which was purified by MPLC (two portions on 165 g silica; hexane/EtOAc 93 : 7) to give 1.21 g (51%). The enantiomeric excess was determined by conversion to

2-[(+)-α-methoxy-α-trifluoromethylphenylacetylamino]-3,3-dimethylbutanoic acid, methyl ester. The benzyl ester (43.6 mg, 0.100 mmol) was converted to the corresponding methyl ester by debenzylation (H_2, 1 atm, 5% Pd–C, 7 mg) in EtOAc (1.5 ml), 3 h, 25 °C) and subsequent diazomethane treatment (CH_2Cl_2, 0 °C). Without purification, the methyl ester in CH_2Cl_2 (2.0 ml) was subjected to treatment with trifluoroacetic acid (TFA) (2.0 ml) at 25 °C under N_2. After 30 min, Raney nickel (~300 mg wet weight) was added and the stirred mixture hydrogenated under 550 psig H_2 for 16 h. The mixture was purged with N_2 and then filtered through Celite. The filter cake was washed with MeOH (4 times) and the filtrate evaporated *in vacuo*, residual TFA and MeOH being removed by azeotroping with toluene (3 times). The residue was suspended in CH_2Cl_2 (3 ml) and split into two equal portions. One was acylated with (+)-MTPA chloride [MTPA = (*S*)-(+)-α-methoxy-α-trifluoromethylphenylacetyl; 19.4 μl, 0.10 mmol, 2.0 equiv.] and triethylamine (28 μl, 0.20 mmol, 4.0 equiv.). The crude product was purified by flash chromatography (5 g of silica; hexane/EtOAc 9 : 1) to give the product (16.0 mg, 89%) as a colorless, viscous oil, greater than 99% e.e. (capillary gas chromatography).

12.5.3
Asymmetric Hydrogenation with Knowles' Catalyst

Preparation of (*S*)-2-Amino-3-(naphthalen-2-yl) propanoic Acid (Scheme 12.18, Ar = 2-Naphthyl; See also Scheme 12.17) [69]

CHO
+
Ac N H CO₂H
N O O
H₂N OH O

A 2-l round-bottom flask is charged with 2-naphthaldehyde (600 g, 3.85 mol), NaOAc (420.5 g, 5.13 mol), *N*-acetylglycine (600 g, 5.13 mol), and acetic anhydride (1.27 l, 13.5 mol), and heated to 95–105 °C for 2 h. The reaction mixture is cooled to 75 °C, diluted with HOAc (240 ml), and cooled to 45 °C. H_2O (80 ml) is added to the slurry and the reaction mixture is cooled to room temperature. The solids are filtered, rinsed with HOAc (250 ml, 2 times) and H_2O (600 ml, 3 times), and dried in a vacuum oven (25 °C, 28 mmHg) overnight to give (*Z*)-2-methyl-4-(naphthalen-2-ylmethylene)oxazol-5(4*H*)-one as a yellow solid (658 g, 72%).

A 2-l round-bottom flask is charged with the oxazolone (640 g, 2.76 mol) just prepared, NaOAc (1.16 g, 14.2 mmol), HOAc (1.1 l), and H_2O (1.1 l), and the mixture is heated to 85 °C for 2 h. The reaction mixture is cooled to room temperature and the solids filtered, rinsed with H_2O (600 ml, 2 times), and dried in a vacuum oven (50 °C, 28 mmHg) overnight to give the dehydroamino acid as a yellow solid (657 g, 95%).

A 2-l Parr bomb is charged with the dehydroamino acid (150 g, 0.6 mol), [Rh(COD)(*R*,*R*-DIPAMP)]BF_4 (0.09 g, 0.12 mmol) and 75% *i*PrOH in H_2O (1.65 l). The vessel is sealed, sparged with N_2 for 15 min, heated to 60 °C, purged with H_2 (4 times), and pressurized to 50 psig H_2. Hydrogen pressure is maintained at 50 psig with a regulated reservoir. The reaction mixture is cooled to room temperature and depressurized after 18 h, and the volatiles are removed *in vacuo* to give the *N*-acetyl amino acid as a solid (149 g, 98%). The enantioselectivity of the product has been determined by chiral high-performance chromatography (HPLC) to be 92% e.e.

A round bottom flask is charged with the *N*-acetyl amino acid (578 g, 2.23 mol) and 3 M HCl (3.7 ml), and heated to 95 °C for 6 h. The reaction mixture is cooled to room temperature and pH is adjusted to around 5 with 50% NaOH. The solids are filtered, rinsed with H_2O (500 ml, 2 times), and dried overnight under vacuum (50 °C) to give (*S*)-2-amino-3-(naphthalen-2-yl)propanoic acid (427 g, 89%). Chiral analysis by HPLC indicated 94.7% e.e.

12.5.4 Asymmetric Hydrogenation with Rh(DuPhos) Followed by Enzyme-Catalyzed Inversion of the α-Center

Preparation of all Four Isomers of 2-Amino-3-naphthalen-2-yl-butyric Acid (Scheme 12.32, Ar = 2-Naphthyl) [71]

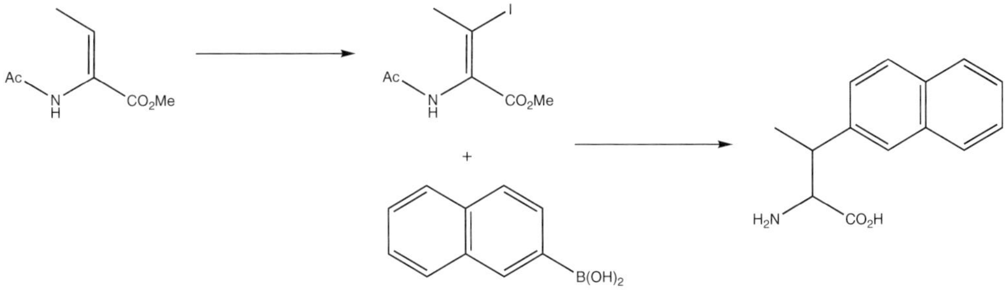

N-Iodosuccinimide (12.9 g, 0.057 mol) was added to a stirring solution of methyl 2-(*N*-acetylamino)-(*Z*)-butenoate (7.50 g, 0.048 mol) in 2% TFA in CH_2Cl_2 (250 ml with 5 ml TFA). The mixture was stirred at room temperature overnight. The mixture was cooled in an ice bath before triethylamine (20 ml) was added slowly. The mixture was allowed to stir for a further 1 h. The mixture was concentrated and taken up in CH_2Cl_2, washed with 1 M $KHSO_4$, water, and brine before concentrating and running down a column of silica (CH_2Cl_2/EtOAc 1 : 1) to obtain (*Z*)-2-acetamino-3-iodobut-2-enoic acid methyl ester as a yellow/brown solid (5.41 g, 40%).

A mixture of the vinyl iodide (800 mg, 2.82 mmol), prepared as described in the previous paragraph, 2-naphthaleneboronic acid (729 mg, 4.24 mmol), $Pd(OAc)_2$ (64 mg, 0.242 mmol), Na_2CO_3 (600 mg, 5.66 mmol), and EtOH (25 ml) was heated at 55 °C for 4 h. The mixture was concentrated and taken up in CH_2Cl_2/H_2O (1 : 1). The CH_2Cl_2 layer was washed once with water, once with brine, and dried over $MgSO_4$ before concentrating to the crude product. Wet flash column chromatography

(CH_2Cl_2/EtOAc 1 : 1) gave (*Z*)-2-acetamino-3-naphthalen-2-yl-but-2-enoic acid methyl ester as a pale yellow solid (575 mg, 72%).

A mixture of the dehydroamino acid ester (300 mg, 0.99 mmol) and [Rh(R,R)-Et-DuPhos(COD)]BF_4 (6.5 mg, 0.010 mmol) and degassed MeOH (5 ml) was stirred under 100 psi H_2 for 72 h. The mixture was concentrated and run down a short column (EtOAc) of silica to yield (2*R*,3*S*)-2-acetamino-3-naphthalen-2-ylbutyric acid methyl ester as a pale yellow solid (300 mg, 99%). HPLC showed 99.4% e.e.

HCl (4 N, 20 ml) was added to a stirred solution of the N-acetyl amino acid ester (250 mg, 0.83 mmol) and acetone (2.5 ml), and the mixture was heated under reflux for 3 h. The mixture was treated with decolorizing charcoal before filtering and concentrating. Trituration with acetone gave (2*R*,3*S*)-2-amino-3-naphthalen-2-ylbutyric acid, hydrochloride salt as a white solid (168 mg, 71%, 100% e.e.).

A mixture of the amino acid hydrochloride (2 ml of a 5 mM solution), *Trigonopsis variabilis*D-amino acid oxidase (5 drops of immobilized slurry, 222.7 U/g dry weight), and ammonia/borane complex (5 mg, 0.16 mmol) were shaken on a blood rotator at 30 °C. HPLC showed after 20 h complete conversion to (2*S*,3*S*)-2-amino-3-naphthalen-2-ylbutyric acid (99% d.e.).

A mixture of the dehydroamino acid ester (300 mg, 0.99 mmol) and [Rh(S,S)-Et-DuPhos(COD)]BF_4 (7.9 mg, 0.012 mmol) and degassed MeOH (5 ml) were stirred under 100 psi H_2 for 72 h. The mixture was concentrated and run down a short column (EtOAc) of silica to yield (2S,3R)-2-acetamino-3-naphthalen-2-ylbutyric acid methyl ester as a pale yellow solid (300 mg, 99%). HPLC showed 98.2% e.e.

HCl (4 N, 20 ml) was added to a stirred solution of the *N*-acetyl amino acid ester (250 mg, 0.83 mmol) and acetone (2.5 ml), and the mixture was heated under reflux for 3 h. The mixture was treated with decolorizing charcoal before filtering and concentrating. Trituration with acetone gave (2*R*,3*R*)-2-amino-3-naphthalen-2-ylbutyric acid, hydrochloride salt as an off-white solid (203 mg, 87%, 100% e.e.).

A mixture of the salt from the previous paragraph (4 ml of a 5 mM solution), snale venom L-amino acid oxidase (20 mg of a 0.5 U/mg enzyme), and ammonia/borane complex (10 mg, 0.32 mmol) were shaken on a blood rotator at 30 °C. HPLC analysis after 40 h showed an 80% yield (>99% d.e.) of (2*R*,3*R*)-2-amino-3-naphthalen-2-yl-butyric acid.

References

1 Wuts, P.G.M. and Greene, T.W. (2006) *Greene's Protective Groups in Organic Synthesis*, 4th edn, John Wiley & Sons, Inc., New York.

2 Williams, R.M. (1989) *Synthesis of Optically Active α-Amino Acids*, Pergamon Press, Oxford.

3 Breuer, M., Ditrich, K., Habicher, T., Hauer, B., Keßeler, M., Stürmer, R., and Zelinski, T. (2004) *Angewandte Chemie (International Edition in English)*, **43**, 788.

4 Shiraiwa, T., Shinjo, K., and Kurokawa, H. (1989) *Chemistry Letters*, 1413.

5 Shiraiwa, T., Shinjo, K., and Kurokawa, H. (1991) *Bulletin of the Chemical Society of Japan*, **64**, 3251.

6 Kaptein, B., Vries, T.R., Nieuwenhuijzen, J.W., Kellogg, R.M., Grimbergen, R.F.P.,

and Broxterman, Q.B. (2005) In *Handbook of Chiral Chemicals*, 2nd edn (ed. D.J. Ager), CRC, Boca Raton, FL, p. 97.

7 Sonke, T., Kaptein, B., Boesten, W.H.J., Broxterman, Q.B., Kamphuis, J., Formaggio, F., Toniolo, C., Rutjes, F.P.J.T., and Schoemaker, H.E. (1993) In *Comprehensive Asymmetric Catalysis* (eds E.N. Jacobsen, A.H. Pfaltz, and H. Yamamoto), Springer, Berlin, p. 23.

8 Ager, D.J. (2005) In *Handbook of Chiral Chemicals*, 2nd edn (ed. D.J. Ager), CRC, Boca Raton, FL, p. 11.

9 Miyazawa, T., Takashima, K., Mitsuda, Y., Yamada, T., Kuwata, S., and Watanabe, H. (1979) *Bulletin of the Chemical Society of Japan*, **52**, 1539.

10 Schöllkopf, U., Busse, U., Lonsky, R., and Hinrichs, R. (1986) *Annalen Der Chemie – Justus Liebig*, 2150.

11 Schöllkopf, U. (1983) *Topics in Current Chemistry*, **109**, 65.

12 Evans, D.A. (1982) *Aldrichimica Acta*, **15**, 23.

13 Ager, D.J., Prakash, I., and Schaad, D.R. (1996) *Chemical Reviews*, **96**, 835.

14 Ager, D.J., Prakash, I., and Schaad, D.R. (1997) *Aldrichimica Acta*, **30**, 3.

15 Bois-Choussy, M., Neuville, L., Beugelmans, R., and Zhu, J. (1996) *The Journal of Organic Chemistry*, **61**, 9309.

16 Evans, D.A., Evrard, D.A., Rychnovsky, S.D., Fruh, T., Whittingham, W.G., and DeVries, K.M. (1992) *Tetrahedron Letters*, **33**, 1189.

17 Evans, D.A., Wood, M.R., Trotter, B.W., Richardson, T.I., Barrow, J.C., and Katz, J.L. (1998) *Angewandte Chemie (International Edition in English)*, **37**, 2700.

18 Seebach, D., Juaristi, E., Miller, D.D., Schickli, C., and Weber, T. (1987) *Helvetica Chimica Acta*, **70**, 237.

19 Seebach, D., Sting, A.R., and Hoffmann, M. (1996) *Angewandte Chemie (International Edition in English)*, **35**, 2708.

20 Seebach, D., Bürger, H.M., and Schickli, C.P. (1991) *Annalen Der Chemie – Justus Liebig*, 669.

21 Seebach, D., Dziadulewicz, E., Behrendt, L., Cantoreggi, S., and Fitzi, R. (1989) *Annalen Der Chemie – Justus Liebig*, 1215.

22 Kong, J.-R., Cho, C.-W., and Krische, M.J. (2005) *Journal of the American Chemical Society*, **127**, 11269.

23 O'Donnell, M.J. (2004) *Accounts of Chemical Research*, **37**, 506.

24 O'Donnell, M.J. (2001) *Aldrichimica Acta*, **34**, 3.

25 O'Donnell, M.J., Wu, S.D., and Huffman, J.C. (1994) *Tetrahedron*, **50**, 4507.

26 Lygo, B. and Andrews, B.I. (2004) *Accounts of Chemical Research*, **37**, 518.

27 Corey, E.J., Xu, F., and Noe, M.C. (1997) *Journal of the American Chemical Society*, **119**, 12414.

28 Lygo, B. and Wainwright, P.G. (1997) *Tetrahedron Letters*, **38**, 8595.

29 Ooi, T., Kameda, K., and Maruoka, K. (1999) *Journal of the American Chemical Society*, **121**, 6519.

30 Ooi, T., Takeuchi, M., Kameda, M., and Maruoka, K. (2000) *Journal of the American Chemical Society*, **122**, 5228.

31 Ooi, T., Uematsu, Y., Kameda, M., and Maruoka, K. (2002) *Angewandte Chemie (International Edition in English)*, **41**, 1551.

32 Han, Z., Yamaguchi, Y., Kitamura, M., and Maruoka, K. (2005) *Tetrahedron Letters*, **46**, 8555.

33 Wang, Y.-G., Ueda, M., Wang, X., Han, Z., and Maruoka, K. (2007) *Tetrahedron*, **63**, 6042.

34 Ager, D.J. and Fotheringham, I.G. (2001) *Current Opinion in Drug Discovery & Development*, **4**, 800.

35 Kenworthy, M.N., Kilburn, J.P., and Taylor, R.J.K. (2004) *Organic Letters*, **6**, 19.

36 Koskinen, A.M.P., Hassila, H., Myllmäki, V.T., and Rissanen, K. (1995) *Tetrahedron Letters*, **36**, 5619.

37 Dunn, M.J. and Jackson, R.F.W. (1997) *Tetrahedron*, **53**, 13905.

38 Boyle, P.H., Davis, A.P., Dempsey, K.J., and Hosken, G.D. (1995) *Tetrahedron: Asymmetry*, **6**, 2819.

39 Rutjes, F.P.J.T., Wolf, L.B., and Schoemaker, H.E. (2000) *Journal of the Chemical Society, Perkin Transactions 1*, 4197.

40 Kotha, S., Lahiri, K., and Kashinath, D. (2002) *Tetrahedron*, **58**, 9633.

41 Collet, S., Danion-Bougot, R., and Danion, D. (2001) *Synthetic Communications*, **31**, 249.

42 Sabat, M. and Johnson, C.R. (2000) *Organic Letters*, **2**, 1089.

43 Collier, P.N., Campbell, A.D., Patel, I., Raynham, T.M., and Taylor, R.J.K. (2002) *The Journal of Organic Chemistry*, **67**, 1802.

44 Harvey, J.E., Kenworthy, M.N., and Taylor, R.J.K. (2004) *Tetrahedron Letters*, **45**, 2467.

45 Acevedo, C.M., Kogut, E.F., and Lipton, M.A. (2001) *Tetrahedron*, **57**, 6353.

46 Parsons, A.F. (1996) *Tetrahedron*, **52**, 4149.

47 Cox, G.G. and Harwood, L.M. (1994) *Tetrahedron: Asymmetry*, **5**, 1669.

48 Iyer, M.S., Yan, C., Kowalczyk, R., Shone, R., Ager, D., and Schaad, D.R. (1997) *Synthetic Communications*, **27**, 4355.

49 Evans, D.A., Britton, T.C., Dorow, R.L., and Dellaria, J.F. (1986) *Journal of the American Chemical Society*, **108**, 6395.

50 Trimble, L.A. and Vederas, J.C. (1986) *Journal of the American Chemical Society*, **108**, 6397.

51 Doyle, M.P., Dorow, R.L., Terpstra, J.W., and Rodenhouse, R.A. (1985) *The Journal of Organic Chemistry*, **50**, 1663.

52 Evans, D.A. and Britton, T.C. (1987) *Journal of the American Chemical Society*, **109**, 6881.

53 Evans, D.A., Britton, T.C., Ellman, J.A., and Dorow, R.L. (1990) *Journal of the American Chemical Society*, **112**, 4011.

54 Evans, D.A., Clark, J.S., Metternich, R., Novack, V.J., and Sheppard, G.S. (1990) *Journal of the American Chemical Society*, **112**, 866.

55 Evans, D.A. and Lundy, K.M. (1992) *Journal of the American Chemical Society*, **114**, 1495.

56 Evans, D.A., Ellman, J.A., and Dorow, R.L. (1987) *Tetrahedron Letters*, **28**, 1123.

57 Gröger, H. (2003) *Chemical Reviews*, **103**, 2795.

58 Yet, L. (2001) *Angewandte Chemie (International Edition in English)*, **40**, 875.

59 de Lange, B., Boesten, W.H.J., van der Sluis, M., Uiterweerd, P.G.H., Elsenberg, H.L.M., Kellogg, R.M., and Broxterman, Q.B. (2005) In *Handbook of Chiral Chemicals*, 2nd edn (ed. D.J. Ager), CRC, Boca Raton, FL, p. 487.

60 Vergne, C., Bouillon, J.-P., Chastanet, J., Bois-Choussy, M., and Zhu, J. (1998) *Tetrahedron: Asymmetry*, **9**, 3095.

61 Zhu, J., Bouillon, J.-P., Singh, G.P., Chastanet, J., and Bengelmans, R. (1995) *Tetrahedron Letters*, **36**, 7081.

62 Chakraborty, T.K., Reddy, G.V., and Hussain, K.A. (1991) *Tetrahedron Letters*, **32**, 7597.

63 Rao, A.V.R., Chakraborty, T.K., and Joshi, S.P. (1992) *Tetrahedron Letters*, **33**, 4045.

64 Blaser, H.-U. and Schmidt, E. (2004) In *Asymmetric Catalysis on Industrial Scale* (eds H-.U. Blaser and E. Schmidt), Wiley-VCH Verlag GmbH, Weinheim, p. 1.

65 Tang, W. and Zhang, X. (2003) *Chemical Reviews*, **103**, 3029.

66 Knowles, W.S. (1983) *Accounts of Chemical Research*, **16**, 106.

67 Knowles, W.S. (2002) *Angewandte Chemie (International Edition in English)*, **41**, 1998.

68 Ager, D.J. and Laneman, S.A. (2004) In *Asymmetric Catalysis on Industrial Scale* (eds H-.U. Blaser and E. Schmidt), Wiley-VCH Verlag GmbH, Weinheim, p. 259.

69 Laneman, S.A., Froen, D.E., and Ager, D.J. (1998) In *Catalysis of Organic Reactions* (ed. F.E. Herkes), Marcel Dekker, New York, p. 525.

70 Herbst, R.M. and Shemin, D. (1943) *Organic Syntheses*, **75**, Col. vol. II, 1.

71 Roff, G., Lloyd, R.C., and Turner, N.J. (2004) *Journal of the American Chemical Society*, **126**, 4098.

72 de Vries, J.G. (2005) In *Handbook of Chiral Chemistry*, 2nd edn (ed. D.J. Ager), CRC, Boca Raton, FL, p. 269.

73 van den Berg, M., Minnaard, A.J., Haak, R.M., Leeman, M., Schudde, E.P., Meetsma, A., Feringa, B.L., de Vries, A.H.M., Maljaars, C.E.P., Willans, C.E., Hyett, D., Boogers, J.A.F., Henderick, H.J.W., and de Vries, J.G. (2003) *Advanced Synthesis and Catalysis*, **345**, 308.

74 van den Berg, M., Minnaard, A.J., Schudde, E.P., van Esch, J., de Vries, A.H.M., de Vries, J.G., and Feringa, B.L. (2000) *Journal of the American Chemical Society*, **122**, 11539.

75 Minnaard, A.J., Feringa, B.L., Lefort, L., and de Vries, J.G. (2007) *Accounts of Chemical Research*, **40**, 1267.

76 Bernsmann, H., van den Berg, M., Hoen, R., Minnaard, A.J., Mehler, G., Reetz, M.T., de Vries, J.G., and Feringa, B.L. (2005) *The Journal of Organic Chemistry*, **70**, 943.

77 Jia, X., Li, X., Xu, L., Shi, Q., Yao, X., and Chan, A.S.C. (2003) *The Journal of Organic Chemistry*, **68**, 4539.

78 Reetz, M.T. and Mehler, G. (2000) *Angewandte Chemie (International Edition in English)*, **39**, 3889.

79 Reetz, M.T. and Li, X. (2005) *Synthesis*, 3183.

80 Burk, M.J. (1991) *Journal of the American Chemical Society*, **113**, 8518.

81 Burk, M.J. and Ramsden, J.A. (2005) In *Handbook of Chiral Chemicals*, 2nd edn (ed. D.J. Ager), CRC, Boca Raton, FL, p. 249.

82 Burk, M.J. and Bienewald, F. (1998) In *Transition Metals for Organic Synthesis and Fine Chemicals* (eds C. Bolm and M. Beller), VCH, Weinheim, p. 13.

83 Burk, M.J., Feaster, J.E., Nugent, W.A., and Harlow, R.L. (1993) *Journal of the American Chemical Society*, **115**, 10125.

84 Burk, M.J., Gross, M.F., and Martinez, J.P. (1995) *Journal of the American Chemical Society*, **117**, 9375.

85 Hoerrner, R.S., Askin, D., Volante, R.P., and Reider, P.J. (1998) *Tetrahedron Letters*, **39**, 3455.

86 Stammers, T.A. and Burk, M.J. (1999) *Tetrahedron Letters*, **40**, 3325.

87 Masquelin, T., Broger, E., Mueller, K., Schmid, R., and Obrecht, D. (1994) *Helvetica Chimica Acta*, **77**, 1395.

88 Jones, S.W., Palmer, C.F., Paul, J.M., and Tiffin, P.D. (1999) *Tetrahedron Letters*, **40**, 1211.

89 Adamczyk, M., Akireddy, S.R., and Reddy, R.E. (2001) *Tetrahedron: Asymmetry*, **12**, 2385.

90 Adamczyk, M., Akireddy, S.R., and Reddy, R.E. (2001) *Organic Letters*, **3**, 3157.

91 Rizen, A., Basu, B., Chattopadhyay, S.K., Dossa, F., and Frejd, T. (1998) *Tetrahedron: Asymmetry*, **9**, 503.

92 Hiebl, J., Kollmann, H., Rovenszky, F., and Winkler, K. (1999) *The Journal of Organic Chemistry*, **64**, 1947.

93 Maricic, S., Ritzén, A., Berg, U., and Frejd, T. (2001) *Tetrahedron*, **57**, 6523.

94 Shieh, W.-C., Xue, S., Reel, N., Wu, R., Fitt, J., and Repic, O. (2001) *Tetrahedron: Asymmetry*, **12**, 2421.

95 Cobley, C.J., Johnson, N.B., Lennon, I.C., McCague, R., Ramsden, J.A., and Zanotti-Gerosa, A. (2004) In *Asymmetric Catalysts on Industrial Scale* (eds H.-U. Blaser and E. Schmidt), Wiley-VCH Verlag GmbH, Weinheim, p. 269.

96 Ohfune, Y. and Shinada, T. (2005) *European Journal of Organic Chemistry*, 5127.

97 Kazmaier, U. and Krebs, A. (1995) *Angewandte Chemie (International Edition in English)*, **34**, 2012.

98 Mues, H. and Kazmaier, U. (2001) *Synthesis*, 487.

99 Tosa, T., Mori, T., Fuse, N., and Chibata, I. (1969) *Agricultural and Biological Chemistry*, **33**, 1047.

100 Maier, T.H.P. (2003) *Nature Biotechnology*, **21**, 422.

101 Chenault, H.K., Dahmer, J., and Whitesides, G.M. (1989) *Journal of the American Chemical Society*, **111**, 6354.

102 Chibata, I., Tosa, T., and Shibatani, T. (1992) In *Chirality in Industry* (eds A.N. Collins, G.N. Sheldrake, and J. Crosby), John Wiley & Sons, Ltd, Chichester, p. 351.

103 Gröger, H. and Drauz, K. (2004) In *Asymmetric Catalysis on Industrial Scale* (eds H-.U. Blaser and E. Schmidt), Wiley-VCH Verlag GmbH, Weinheim, p. 131.

104 Schulze, B. and de Vroom, E. (2002) In *Enzyme Catalysis in Organic Synthesis*, 2nd edn (eds K. Drauz and H. Waldmann), Wiley-VCH Verlag GmbH, Weinheim, Chapter 12.2.3.

105 Kamphuis, J., Hermes, H.F.M., van Balken, J.A.M., Schoemaker, H.E., Boesten, W.H.J., and Meijer, E.M. (1990) In *Amino Acids: Chemistry, Biology, Medicine* (eds G. Lubec and G.A. Rosenthal), ESCOM Science, Leiden, p. 119.

106 Rolland-Fulcrand, V., Haroune, N., Roumestant, M.-L., and Martinez, J. (2000) *Tetrahedron: Asymmetry*, **11**, 4719.

107 Kamphuis, J., Boesten, W.H.J., Broxterman, Q.B., Hermes, H.F.M., van Balken, J.A.M., Meijer, E.M., and Schoemaker, H.E. (1990) *Advances in Biochemical Engineering/Biotechnology*, **42**, 133.

108 Rutjes, F.P.J.T. and Schoemaker, H.E. (1997) *Tetrahedron Letters*, **38**, 677.

109 Schoemaker, H.E., Boesten, W.H.J., Broxterman, Q.B., Roos, E.C., Kaptein, B., van den Tweel, W.J.J., Kamphuis, J., Meijer, E.M., and Rutjes, F.P.J.T. (1997) *Chimia*, **51**, 308.

110 Kamphuis, J., Boesten, W.H.J., Kaptein, B., Hermes, H.F.M., Sonke, T., Broxterman, Q.B., van den Tweel, W.J.J., and Schoemaker, H.E. (1992) In *Chirality in Industry* (eds A.N. Collins, G.N. Sheldrake, and J. Crosby), John Wiley & Sons, Ltd, Chichester, p. 187.

111 Syldatk, C., Müller, R., Siemann, M., and Wagner, F. (1992) In *Biocatalytic Production of Amino Acids and Derivatives* (eds D. Rozzell and F. Wagner), Hansen, Munich, p. 75.

112 Altenbuchner, J., Siemann-Herberg, M., and Syldatk, C. (2001) *Current Opinion in Biotechnology*, **12**, 559.

113 Drauz, K., Kottenhahn, M., Makryaleas, K., Klenk, H., and Bernd, M. (1991) *Angewandte Chemie (International Edition in English)*, **30**, 712.

114 Olivieri, R.E., Fascetti, E., Angelini, L., and Degen, L. (1979) *Enzyme and Microbial Technology*, **1**, 201.

115 Olivieri, R.E., Fascetti, E., Angelini, L., and Degen, L. (1981) *Biotechnology and Bioengineering*, **23**, 2173.

116 Pietzsch, M. and Syldatk, C. (2002) In *Enzyme Catalysis in Organic Synthesis* (eds K. Drauz and H. Waldmann), Wiley-VCH Verlag GmbH, Weinheim, p. 761.

117 May, O., Nguyen, P.T., and Arnold, F.H. (2000) *Nature Biotechnology*, **18**, 317.

118 Wegman, M.A., Janssen, M.H.A., van Rantwijk, F., and Sheldon, R.A. (2001) *Advanced Synthesis and Catalysis*, **343**, 559.

119 May, O., Verseck, S., Bommarius, A., and Drauz, K. (2002) *Organic Process Research & Development*, **6**, 452.

120 Ishikawa, T., Mukohara, Y., Watabe, K., Kobayashi, S., and Nakamura, H. (1994) *Bioscience, Biotechnology, and Biochemistry*, **58**, 265.

121 Ness, J.E., Welch, M., Giver, L., Bueno, M., Cherry, J.R., Borchert, T.V., Stemmer, W.P.C., and Minshull, J. (1999) *Nature Biotechnology*, **17**, 893.

122 Watabe, K., Ishikawa, T., Mukohara, Y., and Nakamura, H. (1992) *Journal of Bacteriology*, **174**, 7989.

123 van den Werf, M.J., van den Tweel, W.J.J., Kamphuis, J., Hartmans, S., and de Bont, J.A.M. (1994) *Trends in Biotechnology*, **12**, 95.

124 Chibata, I., Tosa, T., and Sato, T. (1974) *Applied Microbiology*, **27**, 878.

125 Tosa, T. and Shibatani, T. (1995) *Annals of the New York Academy of Sciences*, **750**, 364.

126 Terasawa, M., Yukawa, H., and Takayama, Y. (1985) *Process Biochemistry*, **20**, 124.

127 Gloge, A., Zoń J., Kövári, Á., Poppe, L., and Rétey, J. (2000) *Chemistry – A European Journal*, **6**, 3386.

128 Renard, G., Guilleux, J.-C., Bore, C., Malta-Valette, V., and Lerner, D.A. (1992) *Biotechnology Letters*, **14**, 673.

129 Takaç S., Akay, B., and Özdamar, T.H. (1995) *Enzyme and Microbial Technology*, **17**, 445.

130 Yamada, S., Nabe, K., Izuo, N., Nakamichi, K., and Chibata, I. (1981) *Applied and Environmental Microbiology*, **42**, 773.

131 Tosa, T., Sato, T., Mori, T., and Chibata, I. (1974) *Applied Microbiology*, **27**, 886.

132 Stewart, J.D. (2001) *Current Opinion in Chemical Biology*, **5**, 120.

133 Crump, S.P. and Rozzell, J.D. (1992) In *Biocatalytic Production of Amino Acids and Derivatives* (eds J.D. Rozzell and F. Wagner), Hanser, Munich, p. 43.

134 Rozzell, J.D. and Bommarius, A.S. (2002) In *Enzyme Catalysis in Organic Synthesis* (eds K. Drauz and H. Waldmann), Wiley-VCH Verlag GmbH, Weinheim, p. 873.

135 Cho, B.-K., Seo, J.-H., Kang, T.-W., and Kim, B.-G. (2000) *Biotechnology and Bioengineering*, **83**, 226.

136 Schulz, A., Taggeselle, P., Tripier, D., and Bartsch, K. (1990) *Applied and Environmental Microbiology*, **56**, 1.

137 Hanzawa, S., Oe, S., Tokuhisa, K., Kawano, K., Kobayashi, T., Kudo, T., and Kakidani, H. (2001) *Biotechnology Letters*, **23**, 589.

138 Patel, R.N., Banerjee, A., Nanduri, V.B., Goldberg, S.L., Johnston, R.M., Hanson, R.L., McNamee, C.G., Brzozowski, D.B., Tully, T.P., Ko, R.Y., LaPorte, T.L., Cazzulino, D.L., Swaminathan, S., Chen, C.-K., Parker, L.W., and Venit, J.J. (2000) *Enzyme and Microbial Technology*, **27**, 376.

139 Ager, D.J., Li, T., Pantaleone, D.P., Senkpeil, R.F., Taylor, P.P., and Fotheringham, I.G. (2001) *Journal of Molecular Catalysis B, Enzymatic*, **11**, 199.

140 Ager, D.J., Fotheringham, I.G., Pantaleone, D.P., and Senkpeil, R.F. (2000) *Enantiomer*, **5**, 235.

141 Bartsch, K., Schneider, R., and Schulz, A. (1996) *Applied and Environmental Microbiology*, **62**, 3794.

142 Li, T., Kootstra, A.B., and Fotheringham, I.G. (2002) *Organic Process Research & Development*, **6**, 533.

143 Galkin, A., Kulakova, L., Yamamoto, H., Tanizawa, K., Tanaka, H., Esaki, N., and Soda, K. (1997) *Journal of Fermentation and Bioengineering*, **83**, 299.

144 Tanizawa, K., Asano, S., Masu, Y., Kuramitsu, S., Kagamiyama, H., Tanaka, H., and Soda, K. (1989) *The Journal of Biological Chemistry*, **264**, 2450.

145 Taylor, P.P. and Fotheringham, I.G. (1997) *Biochimica et Biophysica Acta*, **1350**, 38.

146 Taylor, P.P., Pantaleone, D.P., Senkpeil, R.F., and Fotheringham, I.G. (1998) *Trends in Biotechnology*, **16**, 412.

147 Bessman, S.P. and Layne, E.C. Jr. (1950) *Archives of Biochemistry*, **26**, 25.

148 Fotheringham, I. (2000) *Current Opinion in Chemical Biology*, **4**, 120.

149 Fotheringham, I. and Taylor, P.P. (2005) In *Handbook of Chiral Chemicals*, 2nd edn (ed. D.J. Ager), CRC, Boca Raton, FL, p. 31.

150 Fotheringham, I.G., Bledig, S.A., and Taylor, P.P. (1998) *Journal of Bacteriology*, **180**, 4319.

151 Galkin, A., Kulakova, L., Yoshimura, T., Soda, K., and Esaki, N. (1997) *Applied and Environmental Microbiology*, **63**, 4651.

152 Yohda, M., Okada, H., and Kumagai, H. (1089) *Biochimica et Biophysica Acta – Gene Structure and Expression*, **1991**, 234.

153 Leuchtenberger, W., Huthmacher, K., and Drauz, K. (2005) *Applied Microbiology and Biotechnology*, **69**, 1.

154 Brunhuber, N.M.W. and Blanchard, J.S. (1994) *Critical Reviews in Biochemistry and Molecular Biology*, **29**, 415.

155 Brunhuber, N.M.W., Thoden, J.B., Blanchard, J.S., and Vanhooke, J.L. (2000) *Biochemistry*, **39**, 9174.

156 Drauz, K., Jahn, W., and Schwarm, M. (1995) *Chemistry – A European Journal*, 538.

157 Bommarius, A.S., Schwarm, M., Stingl, K., Kottenhahn, M., Huthmacher, K., and Drauz, K. (1995) *Tetrahedron: Asymmetry*, **6**, 2851.

158 Bommarius, A.S., Schwarm, M., and Drauz, K. (1998) *Journal of Molecular Catalysis B, Enzymatic*, **5**, 1.

159 Bommarius, A.S., Drauz, K., Hummel, W., Kula, M.-R., and Wandrey, C. (1994) *Biocatalysis*, **10**, 37.

160 Hummel, W. and Kula, M.-R. (1989) *European Journal of Biochemistry*, **184**, 1.

161 Shaked, Z. and Whitesides, G.M. (1980) *Journal of the American Chemical Society*, **102**, 7104.

162 Schütte, H., Flossdorf, J., Sahm, H., and Kula, M.-R. (1976) *European Journal of Biochemistry*, **62**, 151.

163 Hanson, R.L., Schwinden, M.D., Banerjee, A., Brzozowski, D.B., Chen, B.-C., Patel, B.P., McNamee, C.G., Kodersha, G.A., Kronenthal, D.R., Patel, R.N., and Szarka, L.J. (1999) *Bioorganic and Medicinal Chemistry*, **7**, 2247.

164 Wöltinger, J., Drauz, K., and Bommarius, A.S. (2001) *Applied Catalysis A – General*, **221**, 171.

165 Krix, G., Bommarius, A.S., Drauz, K., Kottenhahn, M., Schwarm, M., and Kula, M.-R. (1997) *Journal of Biotechnology*, **53**, 29.

166 Kula, M.-R. and Wandrey, C. (1987) *Methods in Enzymology*, **136**, 9.

167 Soda, K., Oikawa, T., and Yokoigawa, K. (2001) *Journal of Molecular Catalysis B, Enzymatic*, **11**, 149.

168 Turner, N.J. (2003) *Trends in Biotechnology*, **21**, 474.

169 Alexandre, F.-R., Pantaleone, D.P., Taylor, P.P., Fotheringham, I.G., Ager, D.J., and Turner, N.J. (2002) *Tetrahedron Letters*, **43**, 707.

170 Beard, T.M. and Turner, N.J. (2002) *Journal of the Chemical Society, Chemical Communications*, 246.

171 Enright, A., Alexandre, F.-R., Roff, G., Fotheringham, I.G., Dawson, M.J., and Turner, N.J. (2003) *Journal of the Chemical Society, Chemical Communications*, 2636.

172 Benz, P. and Wohlgemuth, R. (2007) *Journal of Chemical Technology and Biotechnology*, **82**, 1082.

173 Ward, O.P. and Baev, M.V. (2000) In *Stereoselective Biocatalysis* (ed. R.N. Patel), Marcel Dekker, New York, p. 267.

174 Bommarius, A.S., Schwarm, M., and Drauz, K. (2001) *Chimia*, **55**, 50.

175 Tosa, T., Takamatsu, S., Furui, M., and Chibata, I. (1984) *Annals of the New York Academy of Sciences*, **434**, 450.

176 Chen, C.-C., Chou, T.-L., and Lee, C.-Y. (2000) *Journal of Industrial Microbiology & Biotechnology*, **25**, 132.

177 Schoemaker, H.E., Mink, D., and Wubbolts, M.G. (2003) *Science*, **299**, 1694.

13
Synthesis of γ- and δ-Amino Acids

Andrea Trabocchi, Gloria Menchi, and Antonio Guarna

13.1
Introduction

An area of significant importance that provides new dimensions to the field of molecular diversity and drug discovery is the area of peptidomimetics [1]. Biomedical research has reoriented towards the development of new drugs based on peptides and proteins, by introducing both structural and functional specific modifications, and maintaining the features responsible for biological activity. As a part of this research area, unnatural amino acids are of valuable interest in drug discovery and their use as new building blocks for the development of peptidomimetics with a high structural diversity level is of key interest. In particular, medicinal chemistry has taken advantage of the use of amino acid homologs to introduce elements of diversity for the generation of new molecules as drug candidates in the so-called "peptidomimetic approach", where a peptide lead is processed into a new nonpeptidic molecule. The additional methylenic unit between the N- and C-termini in β-amino acids results in an increase of the molecular diversity in terms of the higher number of stereoisomers and functional group variety, and many synthetic approaches to the creation of β-amino acids have been published. There is great interest in these as a tool for medicinal chemistry [2] and in the field of β-peptides generally [3]. The γ- and δ-amino acids have garnered similar interest. In particular, the folding properties of γ- [4] and δ-peptides [5] have been investigated, as they have been proved to generate stable secondary structures. The homologation of α-amino acids into γ- and δ-units allows an enormous increase of the chemical diversity within the three and four atoms between the amino and carboxyl groups. Thus, additional substituents and stereocenters expand the number of compounds belonging to the class of γ- and δ-amino acids. γ-Amino acids have been reported both in linear form, and with the amino and carboxyl groups separated or incorporated in a cyclic structure. In the case of δ-amino acids, five- and six-membered rings have been mainly reported having two substituents variously tethered between the amino and carboxyl groups. Moreover, bicyclic and spiro compounds have been reported as constrained δ-amino acids, especially in the field of peptidomimetics. The synthetic approaches for the preparation of

Amino Acids, Peptides and Proteins in Organic Chemistry. Vol.1 – Origins and Synthesis of Amino Acids.
Edited by Andrew B. Hughes

ISBN: 978-3-527-32096-7

γ- and δ-amino acids take into account a wide array of tools, including asymmetric synthesis, enzymatic processes, resolution procedures, and the use of building blocks from the chiral pool. Among the use of chiral auxiliaries, pantolactone [6], Evans' oxazolidinone [7], and natural product derivatives [8] have been commonly applied for stereoselective processes, such as alkylations, double-bond additions, and aldol condensations. The building blocks from the chiral pool are usually natural amino acids as starting precursors for homologation to γ- and δ-amino acid structures, and carbohydrates particularly suited to the preparation of sugar amino acids (SAAs). Modern stereoselective syntheses include the use of chiral metal complexes as catalysts in hydrogenations or double-bond additions, and more recently the application of proline- and cinchona-based organocatalysts. Also, enzymatic syntheses have been reported as a tool for desymmetrization or selective transformation of ester functions. The amino group is typically obtained directly from reduction of azides, cyano, or nitro groups, or from carboxamides and carboxylic acids through Hoffmann and Curtius degradations, respectively. The carboxylic group is achieved by hydrolysis of esters, cyanides, or oxazolines as protecting groups, or by oxidations of hydroxymethyl moieties.

13.2 γ-Amino Acids

One of the major goals in the development of γ-amino acids is the generation of γ-aminobutyric acid (GABA) analogs, as this molecule is a neurotransmitter in the central nervous system of mammals and its deficiency is associated with several neurological disorders, such as Parkinson's and Huntington's diseases [9]. Thus, a great deal of interest in the synthesis of GABA analogs is well documented, with the aim to modulate the pharmacokinetic properties and the selectivity of various GABA receptors. Also, β-hydroxy-γ-amino acids constitute the family of statines, which have been developed as inhibitors of aspartic acid proteases, thus finding important applications in the therapy of many infectious diseases, including HIV and malaria, and also Alzheimer's disease and hypertension. Finally, γ-amino acids have been developed so as to constrain peptide sequences into β- and γ-turns, and to produce new carbohydrate-based amino acids and oligomers. The asymmetric synthesis of many structurally diverse γ-amino acids has been extensively reviewed, comprising linear hydroxy-functionalized molecules and cyclic and azacyclic systems [10]. The most representative examples of the various synthetic approaches are described herein so as to cover the huge area of synthetic methodologies leading to γ-amino acids.

13.2.1 GABA Analogs

The inhibitor neurotransmitter GABA has served as a structural template for a number of substances that have effects on the central nervous system [11]. A common

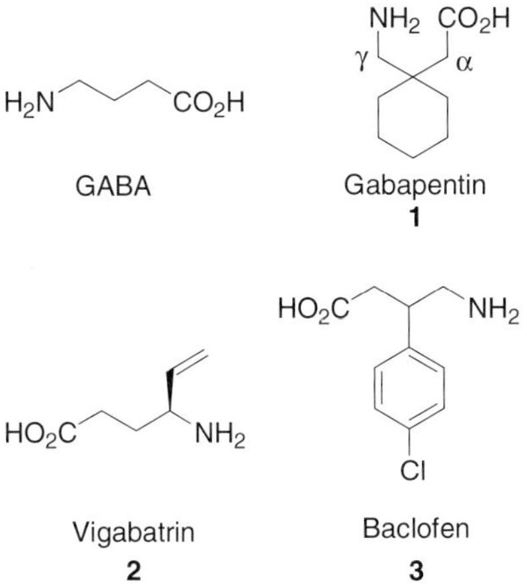
NH2 CO2H
γ
α
H2N
CO2H
GABA
Gabapentin
1
HO2C
NH2
HO2C
NH2
Cl
Vigabatrin
2
Baclofen
3

Scheme 13.1

strategy to design many of these compounds is to manipulate the GABA molecule in order to increase its lipophilicity, thus allowing it to gain access to the central nervous system [12]. In particular, the incorporation of the third carbon atom of GABA into a cyclohexane ring produced the anticonvulsant agent gabapentin (Neurontin) (**1**). Other GABA analogs such as vigabatrin (Sabril) (**2**) [13] and baclofen (Kemstro and Lioresal) (**3**) [14] have been reported as enzyme inhibitors of the GABA metabolic pathway, while gabapentin activity results from a different biological pathway (Scheme 13.1).

A number of alkylated analogs have been synthesized and evaluated *in vitro* for binding to the gabapentin binding site [15]. The synthetic approach, shown in Scheme 13.2, is exemplified by 4-methyl cyclohexanone (**4**), which is allowed to react with a cyanoacetate to give the α,β-unsaturated ester (**5**), followed by insertion of a second cyano group, and final conversion into spiro lactam (**6**). The insertion of the

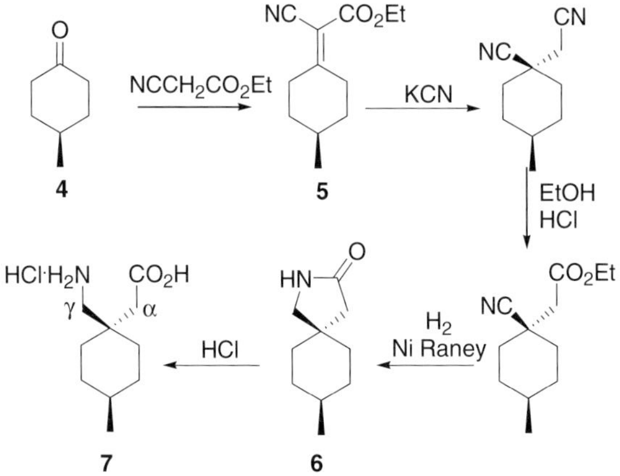
O
NCCH2CO2Et
NC
CO2Et
KCN
CN
NC
4
5
EtOH
HCl
HCl·H2N
CO2H
γ
α
HN
O
CO2Et
NC
H2
Ni Raney
HCl
7
6

Scheme 13.2

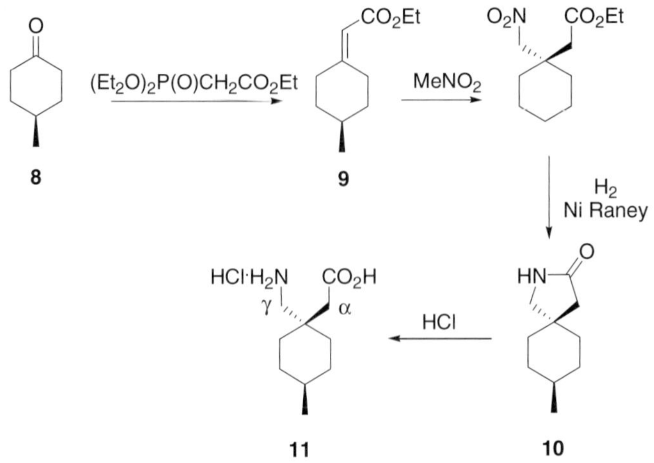

Scheme 13.3

second CN occurs so as to minimize the unfavorable diaxial interactions with C-3 and C-5 protons, thus placing both the second CN and the 4-methyl group in axial positions. Acid-mediated ring opening of (**6**) produces γ-amino acids of general formula (**7**) as gabapentin analogs.

Compound (**11**), stereoisomeric to (**7**), is obtained by Wittig olefination of 4-methyl cyclohexanone (**8**), to give cyclohexylidenes (**9**). Insertion of nitromethyl function at the β-position from the less hindered equatorial direction, followed by catalytic hydrogenation, gives the spiro lactam (**10**) (Scheme 13.3).

The synthesis of (*S*)- and (*R*)-2MeGABA (**17a** and **17b**, respectively) has been reported by Duke *et al.* starting from tiglic acid (**12**), through insertion of the amino group via the *N*-bromosuccinimide/potassium phthalimide procedure [6a]. The racemic carboxylic acid (**13**) was then coupled with (*R*)-pantolactone (**14**) to allow the resolution, via chromatographic separation of the resulting diastereoisomers (**15** and **16**) (Scheme 13.4), and the achievement of the title enantiomeric γ-amino acids after ester hydrolysis and phthalimide deprotection.

Another example of synthesis employing chiral auxiliaries is the preparation of β-aryl-GABAs (**22**) starting from benzylidenemalonates (**18**) (Scheme 13.5) [8d]. γ-Lactam (**19**) resulting from addition of KCN to the double bond, followed by reduction to the corresponding amine and subsequent alkaline cyclization, was coupled with (*R*)-phenylglycinol to allow the chromatographic separation of the resulting diastereomeric amides (**20**). Final amide hydrolysis to remove the chiral auxiliary, followed by decarboxylation produced the lactam (**21**), which was converted to enantiopure γ-amino acid (**22**) by alkaline hydrolysis.

The synthesis of pregabalin (*S*)-(**25**) (Scheme 13.6) was achieved through asymmetric hydrogenation in the presence of (*R*,*R*)-(Me-DuPHOS)Rh(COD)BF_4 (COD = cyclooctadiene) (**23**) as chiral catalyst [16]. The intermediate nitrile compound (**24**) was then reduced by hydrogenation over nickel to achieve the corresponding γ-amino acid (**25**).

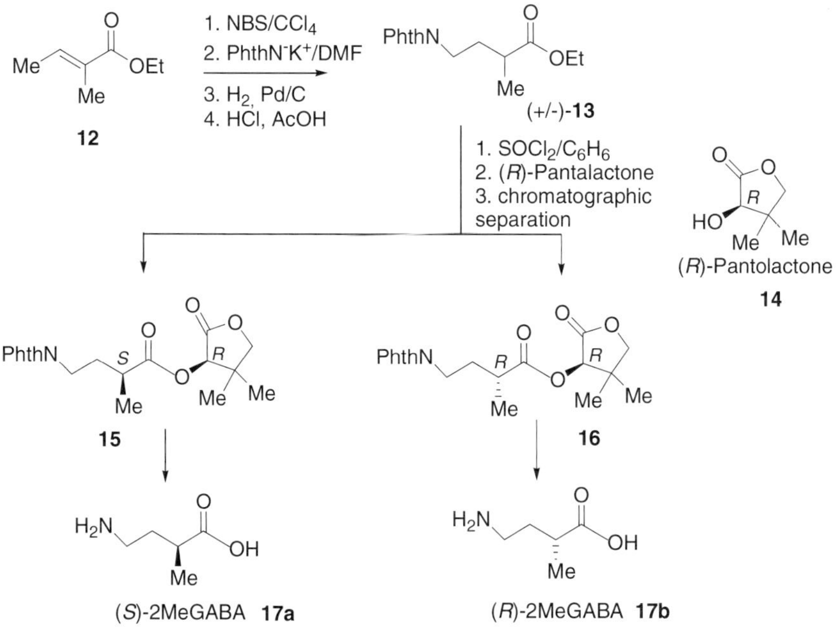

Scheme 13.4

The Ru(II)-(S)-BINAP [BINAP = 2,2′-bis(diphenylphosphino)1,1′-binaphthyl] complex has been employed as catalyst for the asymmetric reduction of the keto group of keto ester (**26**) in 96% e.e. Subsequent treatment of the OH group of (*R*)-(**27**) with PBr_3 and NaCN allowed introduction of the cyano group as an amine precursor. Finally, reductive cyclization to the corresponding γ-lactam (*R*)-(**28**) was achieved by $NaBH_4$ and $NiCl_2$, followed by acid hydrolysis to obtain the title γ-amino acid (*R*)-(**29**) (Scheme 13.7) [17].

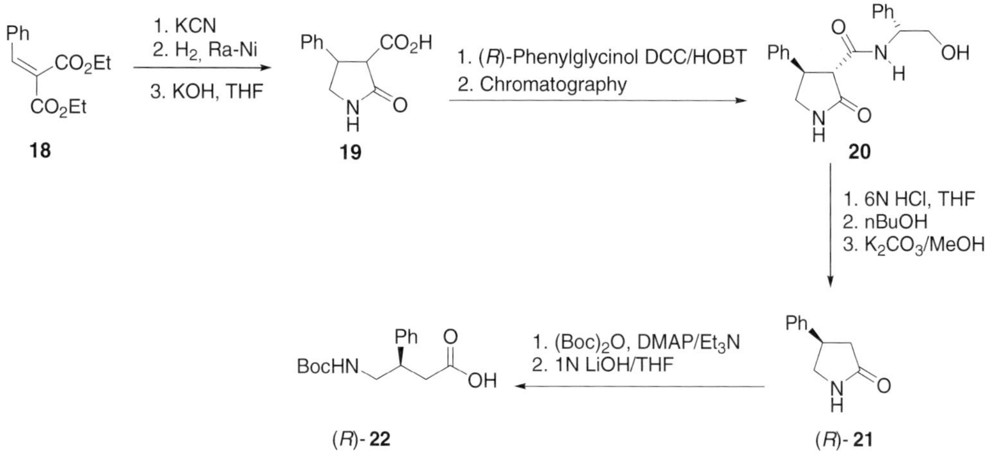

Scheme 13.5

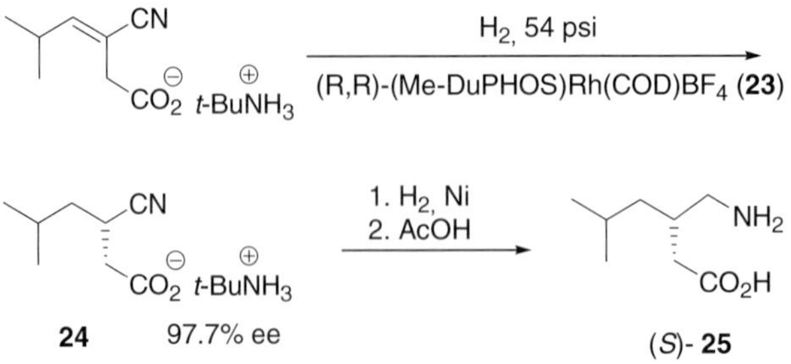

Scheme 13.6

As an example of enzymatic asymmetric synthesis, the preparation of (*R*)-baclofen (**3**) using microbiological mediated Baeyer–Villiger oxidation has been reported (Scheme 13.8) [18]. Specifically, oxidation of cyclobutanone (**30**) produced the corresponding γ-lactone (*R*)-**31** with high enantioselectivity. Subsequent manipulation of (*R*)-**31** consisted of regioselective ring opening using iodotrimethylsilane, followed by treatment with sodium azide, and finally catalytic hydrogenation to afford the amino group of the title compound (*R*)-**3**.

The use of sugars as building blocks from the chiral pool has been applied to *N*-D-mannose substituted nitrones (**32**) in the reaction with acrylates using SmI_2, giving the corresponding adduct (**33**) in 90% e.e. with the major (*R*) diastereoisomer (Scheme 13.9) [19]. The same reaction when applied to *N*-D-ribose substituted nitrones (**34**) afforded the corresponding γ-substituted-α-amino acid precursor (**35**) with opposite (*S*) configuration, indicating the choice of sugar nitrone as a tool for obtaining either (*R*) or (*S*) enantiomers.

One example of the application of α-amino acids as building blocks from the chiral pool to obtain γ-substituted γ-amino acids was achieved through double Arndt–Eistert

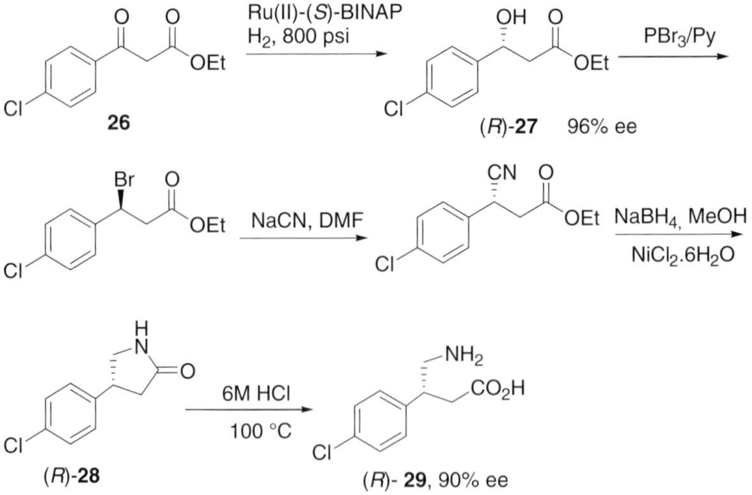

Scheme 13.7

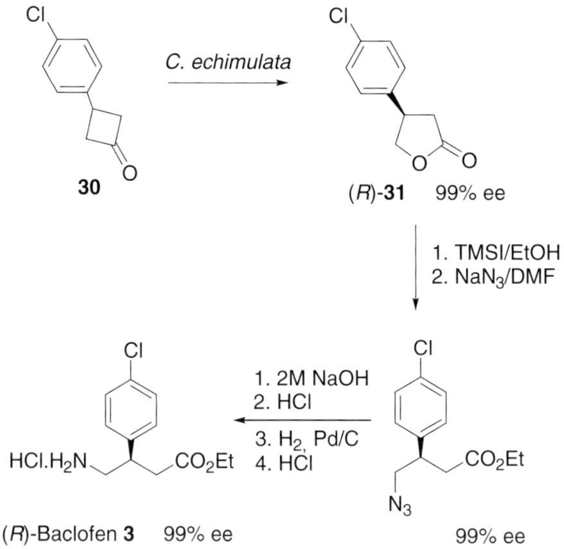

Cl

C. echimulata

30

(*R*)-**31** 99% ee

1. TMSI/EtOH
2. NaN_3/DMF

1. 2M NaOH
2. HCl
3. H_2, Pd/C
4. HCl

$HCl.H_2N$ CO_2Et

CO_2Et

N_3

(*R*)-Baclofen **3** 99% ee

99% ee

Scheme 13.8

homologation (Scheme 13.10) [20]. The β-amino acid was obtained from the corresponding Cbz-α-amino acid (**36**) by reaction with oxalyl chloride and diazomethane to give the diazoketone (**37**). Subsequent Wolff rearrangement of (**37**) using AgOBz and Et_3N in MeOH afforded the fully protected Cbz-γ-amino acid (**38**).

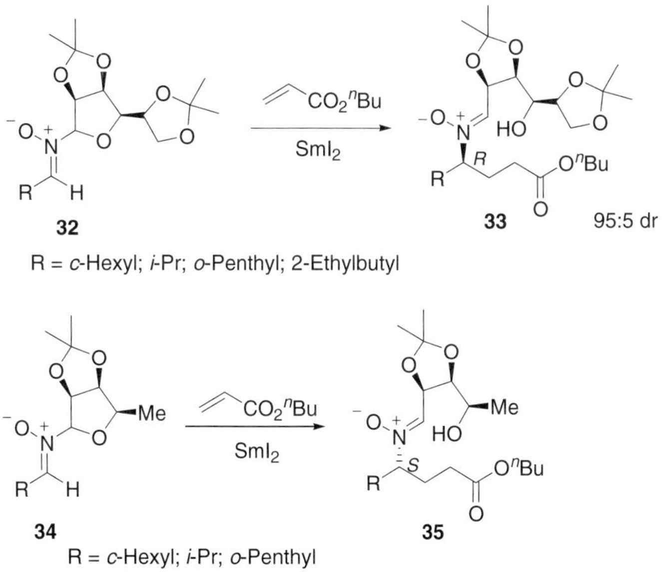

$CO_2{}^nBu$

SmI_2

32

R = *c*-Hexyl; *i*-Pr; *o*-Penthyl; 2-Ethylbutyl

HO *R* O^nBu

33 95:5 dr

Me

$CO_2{}^nBu$

SmI_2

34

R = *c*-Hexyl; *i*-Pr; *o*-Penthyl

Me HO *S* O^nBu

35

Scheme 13.9

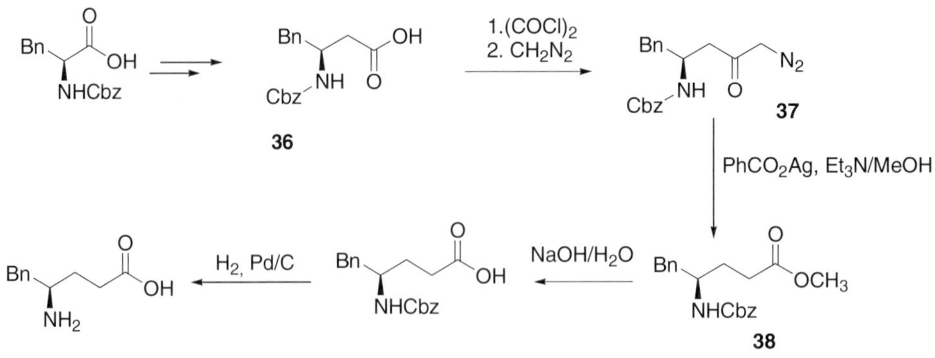

Scheme 13.10

13.2.2
α- and β-Hydroxy-γ-Amino Acids

Ley *et al.* recently reported an attractive approach to chiral α-hydroxy-γ-amino acids using Michael addition of a chiral glycolic acid-derived enolate to the corresponding α,β-unsaturated carbonyl compound or nitro olefin [21]. Specifically, lithium enolate Michael addition of butane-2,3-diacetal desymmetrized-glycolic acid (**39**) to nitro olefins affords the corresponding Michael adducts (**40**) with high diastereoselectivity. Subsequent hydrogenation of the nitro group lead to γ-lactams (**41**), which can be converted to α-hydroxy-γ-amino acids (**42**) (Scheme 13.11).

Reetz *et al.* have shown a route to prepare α-hydroxy-γ-amino acids starting from L-amino acids, by stereoselective [2,3]-Wittig rearrangement [22]. The use of tetramethylethylenediamine in the formation of the lithium enolate led to the preferential formation of one of the four possible diastereoisomers (**43**), as shown in Scheme 13.12. Compound (**44**) displays intramolecular hydrogen bonds arising from the amide group and the alcohol function as a conformational feature of the γ-turn mimetic.

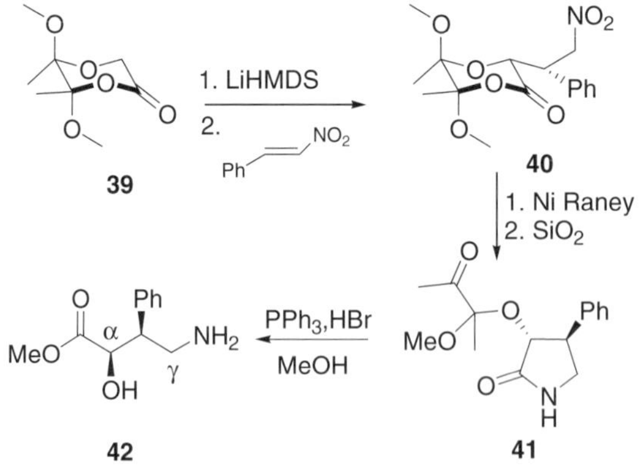

Scheme 13.11

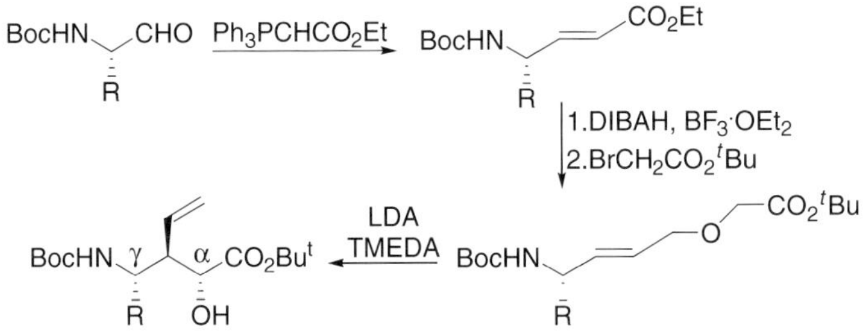

BocHN CHO Ph_3PCHCO_2Et BocHN CO_2Et R R 1.DIBAH, $BF_3 \cdot OEt_2$ 2.$BrCH_2CO_2{}^tBu$ LDA TMEDA BocHN γ α CO_2Bu^t R OH **44** BocHN O $CO_2{}^tBu$ R **43**

R = Me, CH_2Ph, *i*-Pr, *i*-Bu

Scheme 13.12

The structural class of β-hydroxy-γ-amino acids in recent years has been the object of much attention, especially in connection with the development of new pharmaceuticals based on protease inhibitors [1b,23]. Statine, (3*S*,4*S*)-4-amino-3-hydroxy-6-methylheptanoic acid (**45**) (Scheme 13.13), is an essential component of pepstatine, a natural hexapeptide antibiotic, which acts as an inhibitor of renin, pepsin and cathepsin D aspartyl proteases [24]. The α- and β-hydroxy-γ-amino acids derivatives are among the most synthesized γ-amino acids, as these key structural units are found in molecules showing several types of pharmacological activity, including antibacterials and anticancer drugs [25]. The low selectivity of pepstatine has induced the development of more specific synthetic analogs. In particular, the substitution of the isobutyl moiety of statine by the more lipophilic cyclohexylmethyl substituent has led to the widely used analog cyclohexylstatine (**46**), which is a key component of renin inhibitors (Scheme 13.13).

Castejón *et al.* reported stereodivergent and enantioselective approaches to statine analogs, which allows the synthesis of any of the four possible stereoisomers (**47–50**) of a given β-hydroxy-γ-amino acid in fully protected form, ultimately arising from a single allylic alcohol (**51**) of (*E*) configuration (Scheme 13.14). This approach relies on the ready availability of *anti N*-protected-3-amino-1,2-diols of general structure (**52**) and (**53**) in high enantiomeric excess [26].

The *anti* β-hydroxy-γ-amino acids (**48**) and (**50**) have been obtained starting from 3-amino-1,2-alkanediols (**52**) and (**53**) by introduction of a carboxyl-synthetic equivalent at C-1, followed by selective activation of this position (Scheme 13.15). This methodology is also applicable to the preparation of *syn* amino acids (**47**) and (**49**), provided that a configuration inversion at the C-2 position is included in the sequence.

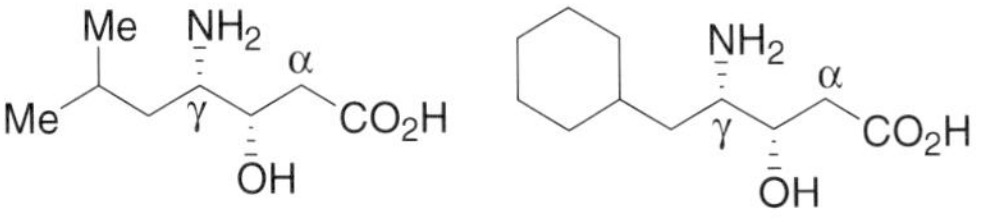

Me NH_2 α Me γ CO_2H OH Statine **45** NH_2 α γ CO_2H OH Cyclohexylstatine **46**

Scheme 13.13

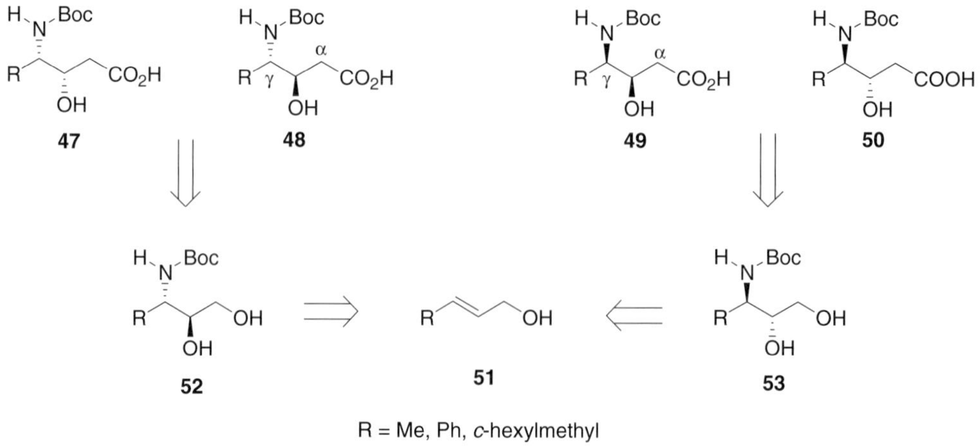

R = Me, Ph, c-hexylmethyl

Scheme 13.14

A general method for the synthesis of α- and γ-alkyl-functionalized β-hydroxy-γ-amino acids was described by Kambourakis *et al.* [27]. Some methods for the synthesis of such compounds are based on low yielding resolutions of ester derivatives using lipases [28]. The synthesis of both classes of amino acids according to Kambourakis *et al.* proceeds through a common chiral alcohol intermediate (**55**), which is generated from ketone diester (**54**) via the action of a nicotinamide-dependent ketoreductase. Regioselective chemical or enzymatic hydrolysis, followed by rearrangement under Hofmann or Curtius conditions, gives the final amino acid γ- and α-alkyl functionalized products (**56**) and (**57**), respectively. The chemo-enzymatic method for the synthesis of individual diastereomers of β-hydroxy-γ-amino acids, using nonchiral starting materials, has a key step in the form of a diastereoselective enzymatic reduction that generates two stereocenters in a single step: the keto diester is reduced diastereoselectively to form the hydroxyl diester

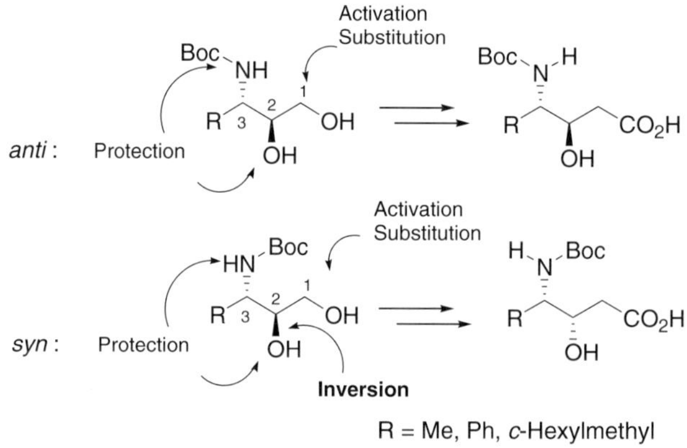

R = Me, Ph, c-Hexylmethyl

Scheme 13.15

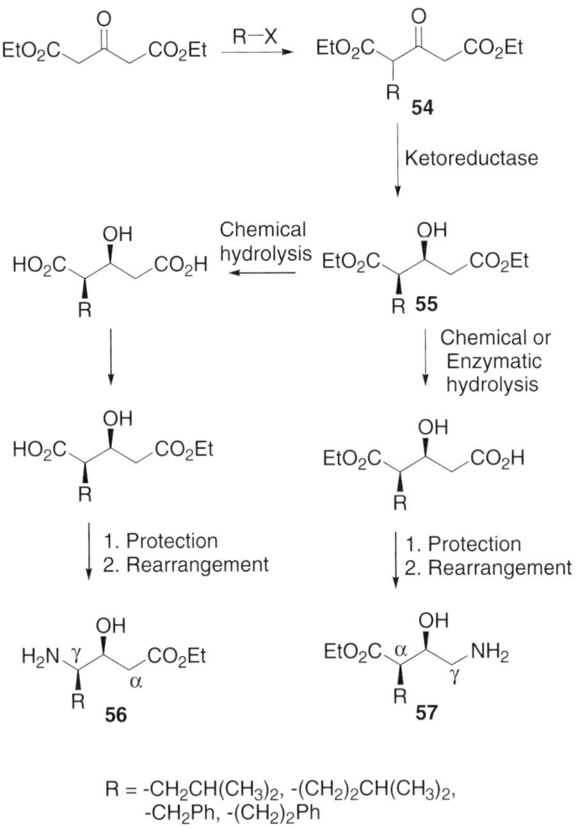

R = $-CH_2CH(CH_3)_2$, $-(CH_2)_2CH(CH_3)_2$, $-CH_2Ph$, $-(CH_2)_2Ph$

Scheme 13.16

intermediate (**55**) in high yield by the action of a ketoreductase enzyme which sets the absolute stereochemistry in the molecule (Scheme 13.16).

Selective enzymatic transesterification of racemic *O*-acetyl cyanohydrin (**58**) using the yeast *Candida cylindracea* lipase (CCL) afforded enantiopure (*R*)-cyanohydrin (**59a**), and the enriched (*S*)-*O*-acetyl-cyanohydrin (**59b**) (Scheme 13.17). Subsequent treatment of (**59b**) with porcine pancreas lipase (PPL) gave enantiopure (*S*)-cyanohydrin (**60**). Reduction of the cyano group in (*R*)-**61a** and (*S*)-**60** using borane–tetrahydrofuran (THF) complex and $NiCl_2$ produced the corresponding enantiomers of (*S*)-4-amino-3-hydroxybutanoic acid (GABOB; **61a** and **61b**, respectively) [29].

The chemoselective reduction of (*R*)-malic acid dimethyl ester with borane–Me_2S complex in the presence of $NaBH_4$ gave the corresponding diol (*R*)-**62**. Subsequent manipulation of the primary hydroxylic group to introduce an azide consisted of S_N2 reaction on the preformed tosylate (*R*)-**63** to give (*R*)-**64**, which by hydrogenation in the presence of Boc_2O generated the corresponding protected β-hydroxy-γ-amino ester (*R*)-**65** in one pot (Scheme 13.18) [30].

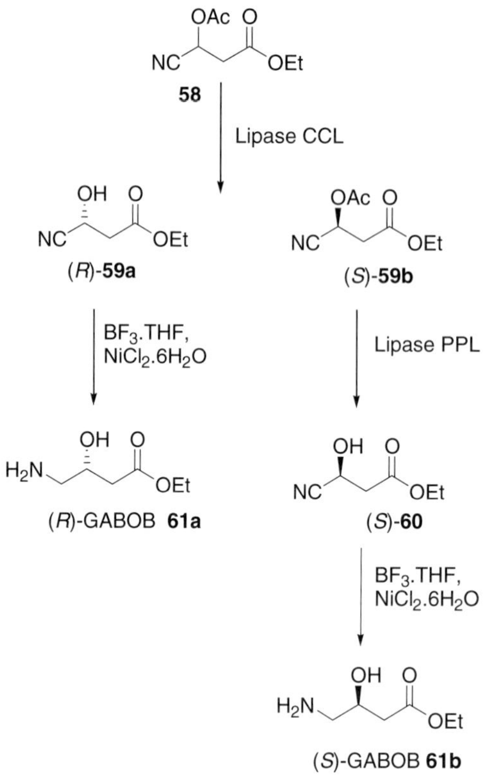

Scheme 13.17

A stereoselective method for the synthesis of γ-substituted-γ-amino acids from the corresponding α-amino acids using the α-amino acyl Meldrum's acid (**66**) as precursor was described by Smrcina (Scheme 13.19) [31]. Changing the order of $NaBH_4$ reduction–lactam ring formation allows the introduction of a hydroxyl function at β-position. Final basic ring opening of the five-membered lactam gives the corresponding γ-amino acids (**67**) and (**68**).

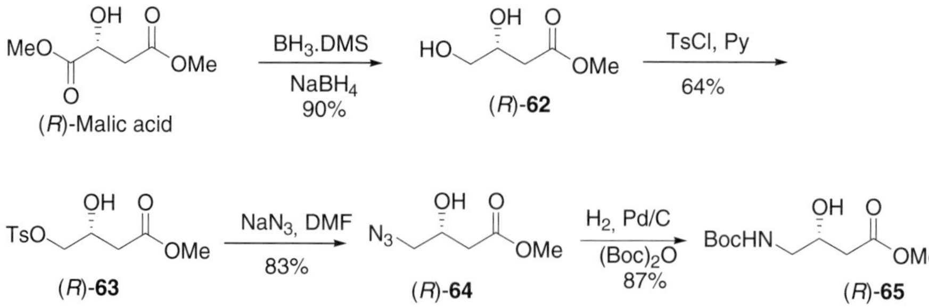

Scheme 13.18

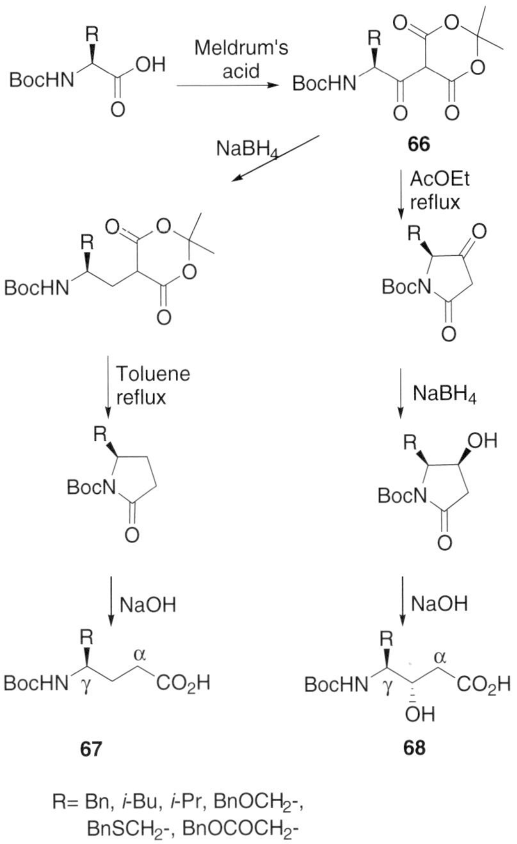

R= Bn, *i*-Bu, *i*-Pr, $BnOCH_2$-, $BnSCH_2$-, $BnOCOCH_2$-

Scheme 13.19

13.2.3
Alkene-Derived γ-Amino Acids

Several nitro-olefin derivatives have been reported as good intermediates for the synthesis of γ-amino acids. In particular, variously functionalized nitrooxazolines have been described as versatile intermediates for the synthesis of γ-amino acids by converting the nitro and oxazoline groups of (**69**) into the corresponding amino and carboxylic functions as in (**70**) (Scheme 13.20) [32].

Another example of the preparation of enantiomerically enriched γ-amino acids from unsaturated compounds involves a palladium-catalyzed allylic nucleophile substitution [33]. In particular, an oxazoline-based phosphine ligand catalyst is used to effect an efficient enantioselective reaction, such as the conversion of the standard allyl acetate (**71**) into the substitution product (**72**) (Scheme 13.21).

Subsequent conversion of (**72**) into the corresponding γ-amino acid (**75**) is achieved with a Krapcho decarboxylation reaction on the cyano ester (**72**), giving the corresponding nitrile (**73**), that is successively reduced to give (**74**). Finally, an

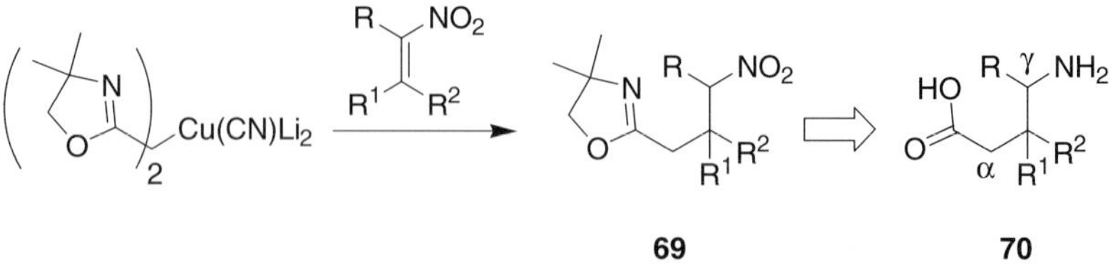

R = H or alkyl; R^1 = H or alkyl; R^2 = H or alkyl

Scheme 13.20

oxidative cleavage of the alkene (**74**) affords the carboxylic acid (**75**). Otaka *et al.* describe facile access to functionalized γ-amino acid derivatives for the design of foldamers via SmI_2-mediated reductive coupling between γ-acetoxy-α,β-enoate (**76**) and *N*-Boc-α-amino aldehydes (**77**) (Scheme 13.22). The reaction is reported to give diastereomeric mixtures of γ-amino acids (**78**) in good yields, although without any diastereoselection [34]. The additional functional groups present on the resulting γ-amino acid derivatives may allow further chemical transformations to increase the chemical diversity.

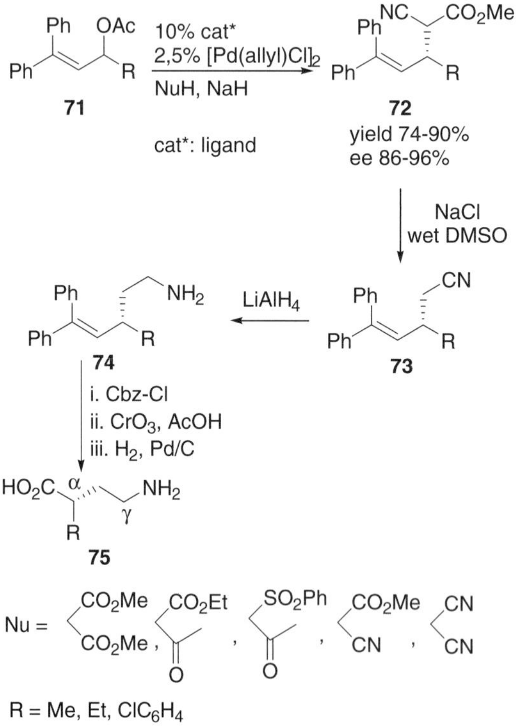

R = Me, Et, ClC_6H_4

Scheme 13.21

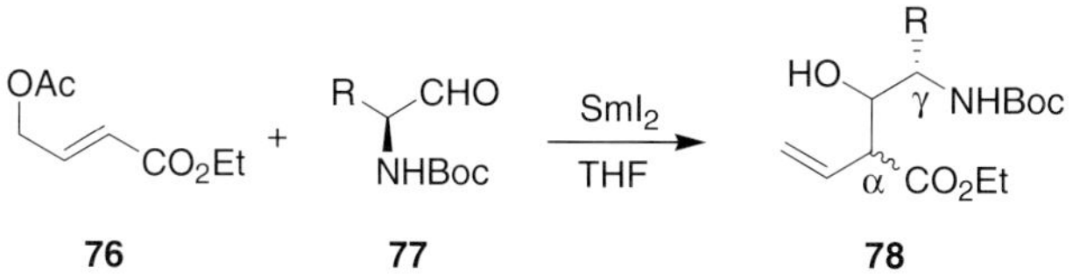

R = Bn, *i*-Pr, 3-Butenyl

Scheme 13.22

13.2.4
SAAs

SAAs are defined as carbohydrates bearing at least one amino and one carboxyl functional group directly attached to the sugar frame. Thus, such compounds represent a new class of building blocks for the generation of peptide scaffolds and constrained peptidomimetics, due to the presence of a relatively rigid furan or pyran ring decorated with space-oriented substituents. Recently, Kessler *et al.* reported a detailed survey about the synthesis of SAAs and their applications in oligomer, peptidomimetic and carbohydrate synthesis [35]. Generally, the amino group is introduced by azidolysis of a hydroxyl group followed by reduction and protection of the resulting amine, although cyanide and nitro equivalents have been also reported. The carboxylic group is usually obtained by oxidation of a primary alcohol function. Hydrolysis of cyanide or direct insertion of CO_2 have been also described. Fleet *et al.* reported the generation of THF-templated γ- and δ-amino acids starting from sugar-derived lactones (Scheme 13.23) [36]. The 2-triflate of the carbohydrate δ-lactones (**79**) when treated with methanol in the presence of either an acid or base catalyst undergo efficient ring contraction to highly substituted THF-2-carboxylates (**80**). Also, azide introduction to the six-membered ring lactone, followed by subsequent S_N2-type ring

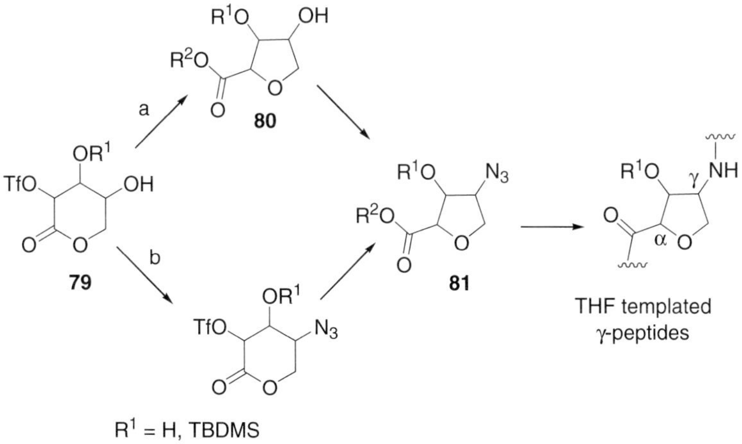

THF templated γ-peptides

R^1 = H, TBDMS
R^2 = Me, *i*-Pr

Scheme 13.23

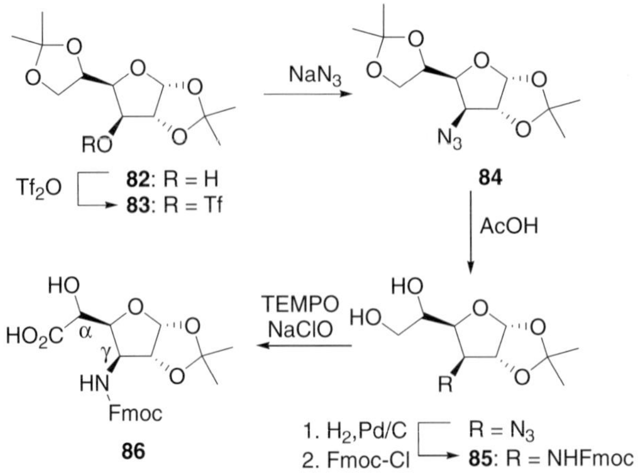

Scheme 13.24

closure of an intermediate hydroxy triflate forms the THF ring, with inversion of configuration at the C-2 position. Thus, synthesis of THF-γ-azido esters (**81**) using this strategy allows the introduction of the C-4 azido group either after (route *a*) or before (route *b*) (Scheme 13.23) formation of the THF ring.

Access to bicyclic furanoid γ-SAA has been reported by Kessler *et al.* starting from diacetone glucose (**82**) and an application for solid-phase oligomer synthesis has also been disclosed [37]. Specifically, azidolysis (**83**) gives azide (**84**), which, after deprotection of the exocyclic hydroxyl groups, is subjected to azide conversion to Fmoc-protected amine (**85**) in a one-pot process. Final oxidation of the primary hydroxyl group furnishes the corresponding furanoid α-hydroxy-γ-amino acid (**86**) (Scheme 13.24).

The first effective solid-phase chemical method for the preparation of carbohydrate-based universal pharmacophore mapping libraries was reported by Sofia *et al.* [38]. The sugar scaffold (**89**) has three sites of diversification, with an amino and a carboxyl group of the γ-amino acid scaffold, and an additional hydroxyl group. The synthesis starts from D-glucose derivative (**87**) (Scheme 13.25), which is treated with $NaIO_4$ and nitromethane to introduce a nitro group at C-3. Subsequent orthogonal protections, and conversion of the nitro group to the corresponding protected amino group, gives (**88**), which is further oxidized to γ-amino acid (**89**) by the TEMPO–NaClO (TEMPO = 2,2,6,6-tetramethylpiperidinooxy) system. By anchoring the carboxyl group on a solid-phase, libraries of 1648 members have been prepared using eight amino acids as acylating agents of the amino group and six isocyanates for hydroxyl functionalization.

13.2.5
Miscellaneous Approaches

An enantiopure cyclobutane-based γ-amino acid was obtained from (*S*)-verbenone, readily available from α-pinene as a building block (Scheme 13.26) [39]. Oxidative

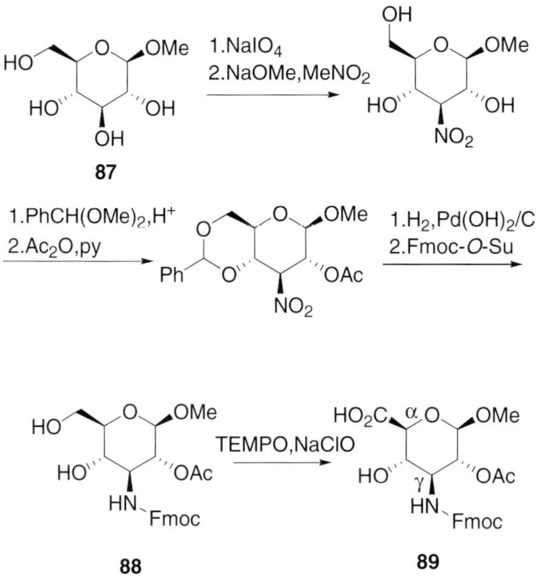

Scheme 13.25

cleavage of (*S*)-verbenone with $NaIO_4$ in the presence of catalytic $RuCl_3$ gave the keto acid (**90**), which was esterified with benzyl chloride. The resulting ester (**91**) was subjected to a haloform reaction with hypobromite to convert the methyl ketone into the corresponding carboxylic acid, which in turn was transformed in the amino group via Curtius rearrangement, giving the protected γ-amino acid (**92**).

Synthesis of polysubstituted γ-amino acids of general formula (**95**) by rearrangement of α-cyanocyclopropanone hydrates (**93**) to the corresponding β-cyano acid (**94**) was reported by Doris *et al.* [40]. Further reduction of the nitrile moiety of β-cyano acid affords γ-amino acid (**95**) with substituents in α- and β-positions (Scheme 13.27).

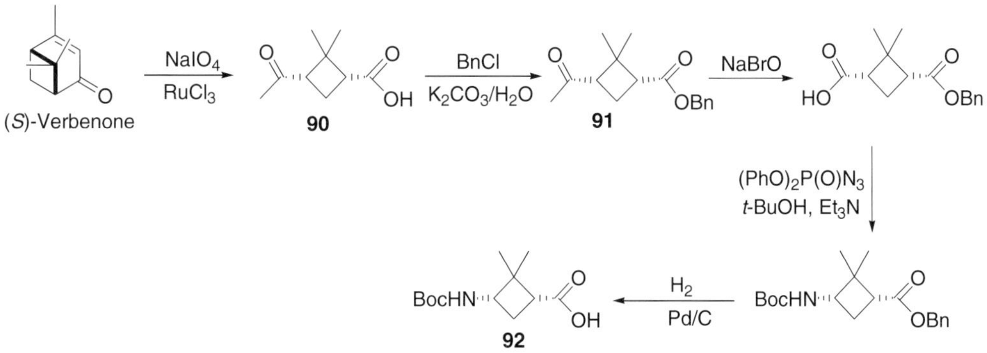

Scheme 13.26

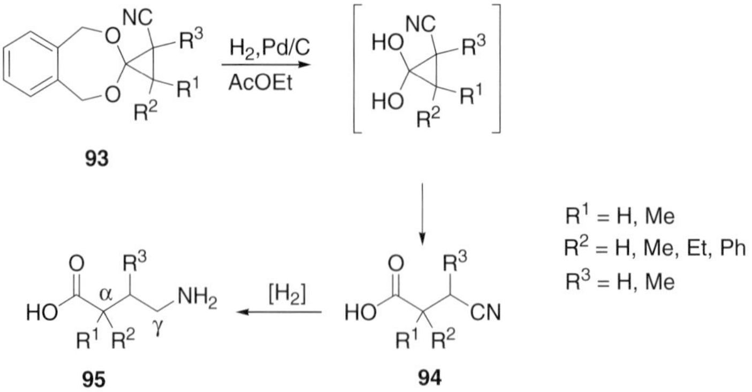

Scheme 13.27

Sato *et al.* reported an efficient and practical access to enantiopure γ-amino acids using allyl titanium complexes (**97**) [41]. These species are prepared by the reaction of allylic acetal (**96**) with a divalent titanium reagent (η^2-propene)Ti(O-*i*Pr)$_2$, readily generated *in situ* (Scheme 13.28), and reacts as an efficient chiral homoenolate equivalent with aldehydes, ketones and imines at the γ-position [42].

γ-Amino acids have also been prepared by modification of glutamic acid [43] and by homologation of α-amino acids. Use of a zinc reagent in straight asymmetric synthesis of aryl γ-amino acids (**98**) is described by Jackson *et al.* (Scheme 13.29) [44].

Entry to bicyclic γ-amino acids was proposed by Tillequin *et al.*, who described the synthesis of constrained scaffolds from the naturally occurring iridoid glycoside aucubin (**99**). The two glycosylated hydroxyl-γ-amino acids (**101**) and (**102**) are prepared by chiral pool synthesis in eight steps. The amino function was introduced via a phthalimido group, and the carboxylic function was introduced on the double bond by formyl insertion using the Vilsmeier reaction, followed by carbonyl oxidation and amine deprotection with hydrazine [45]. Bicyclic γ-amino acid (**102**) was obtained with the same procedure as for (**101**) after oxidation of the double bond of (**100**) to an

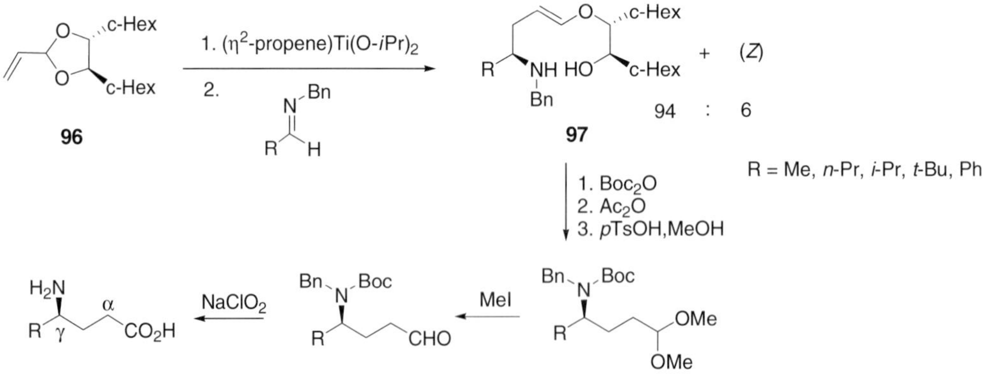

Scheme 13.28

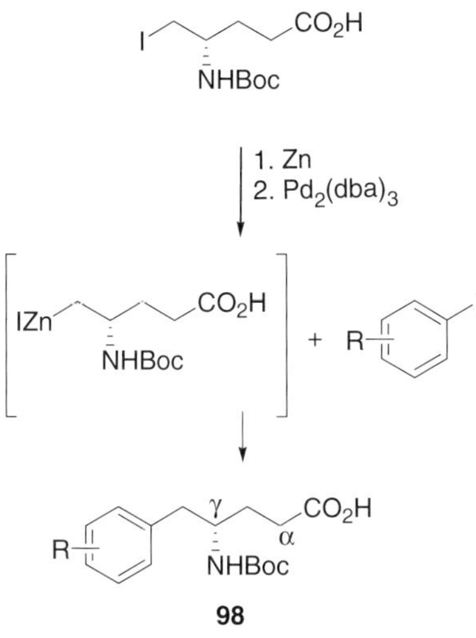

I–CH2–CH(NHBoc)–CH2–CH2–CO_2H

1. Zn
2. $Pd_2(dba)_3$

[IZn–CH2–CH(NHBoc)–CH2–CH2–CO_2H] + R–C6H4–I

98

R = H, 4-Me, 4-OMe, 2-OMe, 4-Br, 2-F, 4-F, 2-NO_2, 3-NO_2, 4-NO_2, 2-NH_2

Scheme 13.29

iodolactol derivative and alkaline rearrangement to the corresponding THF (Scheme 13.30).

Early investigations by Seebach [4a], revealed that γ-amino acid oligopeptides can also adopt helical conformations in solution. *N*-Boc-protected monomers (**105**) have been prepared by homologation of L-alanine and L-valine, followed by stereoselective

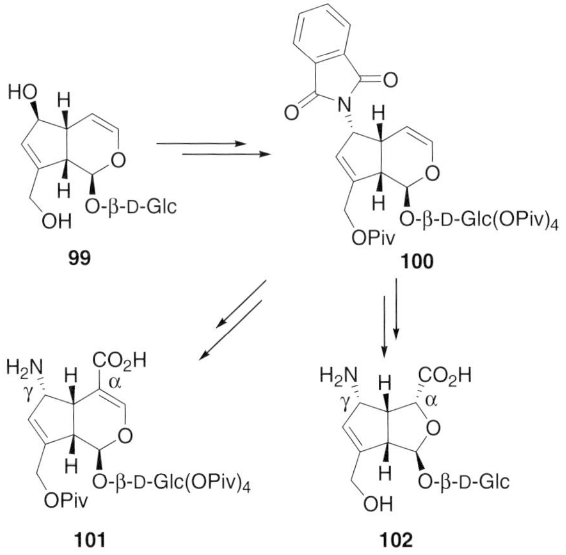

99 (O-β-D-Glc) → 100 (O-β-D-Glc(OPiv)$_4$, OPiv) → 101 (O-β-D-Glc(OPiv)$_4$, OPiv) ; 100 → 102 (O-β-D-Glc, OH)

Scheme 13.30

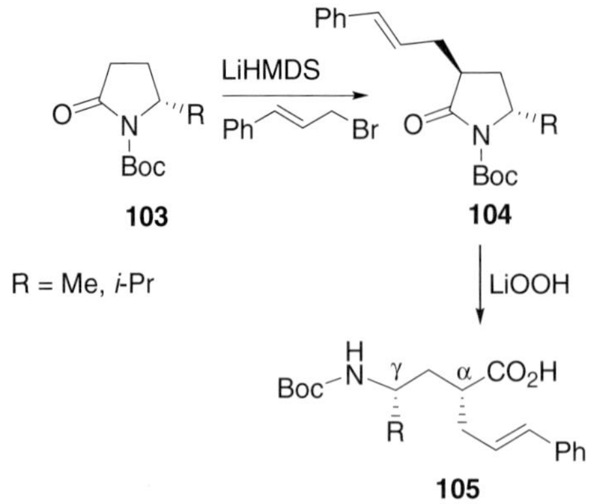

Scheme 13.31

alkylation. Specifically, pyrrolidinones (**103**) are alkylated with cinnamyl bromide in the presence of lithium bis(trimethylsilyl)amide at −78 °C, giving (**104**) in good isolated yields and with high diastereoselectivities (*anti* : *syn* from 18 : 1 to 40 : 1, depending on R). Mild hydrolysis of the lactam moiety affords (**105**) without epimerization at the newly generated stereocenters (Scheme 13.31).

The bicyclic pyrrolizidinone skeleton of **112** and **110** (Scheme 13.32) was obtained by 1,3-dipolar cycloaddition of the cyclic nitrone **106** and acrylamide **107** [46]. Two main products were obtained as racemic mixtures (**108** and **109**) and, after separation, they underwent the same synthetic route, here depicted for (±)-**108** only. The reductive cleavage/cyclization step was then followed by alcohol group transformation into the target γ-amino acid **112**.

The synthesis of enantiopure compound **112** was achieved by separation of diastereoisomeric intermediates obtained from (±)-**111** containing (1*R*)-1-phenylethylamine as a chiral auxiliary.

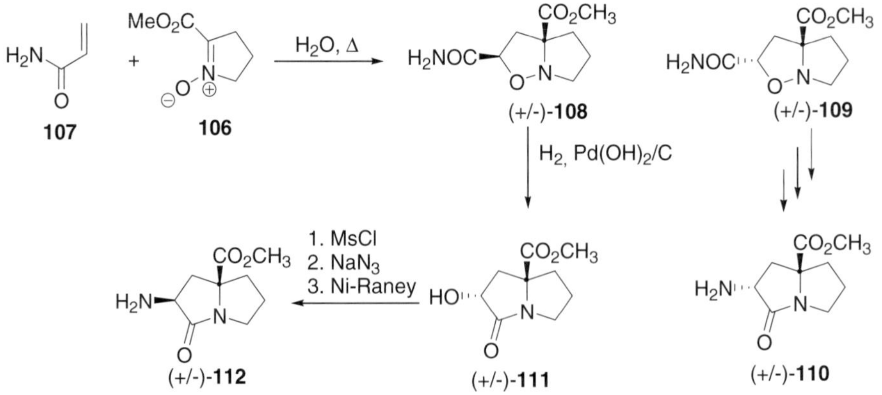

Scheme 13.32

13.3 δ-Amino Acids

δ-Amino acids are isosteric replacements of dipeptide units and their application in the field of peptidomimetics has been extensively reported. In particular, since the β-turn is a common structural feature of proteins associated with the dipeptide unit, much research about δ-amino acids has been concentrated on the creation of reverse turn mimetics, where the central amide bond is replaced by a rigid moiety. Moreover, δ-amino acids have been involved in the generation of new peptide nucleic acid (PNA) monomers, since the six-atom length of these amino acid homologs corresponds to the optimal distance to mimic the ribose unit found in RNA and DNA polymers. Relevant examples of δ-amino acids include linear, cyclic, bicyclic and spiro compounds, and different templates have been applied for the generation of such amino acids.

13.3.1 SAAs

δ-Amino acids of the SAA class have been synthesized as furanoid or pyranoid compounds, and both cyclic and bicyclic scaffolds have been reported (Scheme 13.33).

13.3.1.1 Furanoid δ-SAA

Among furanoid δ-SAA, either monocyclic compounds or oxabicyclo[3.3.0]octane and oxabicyclo[3.2.0]heptane structures have been synthesized by several authors according to different synthetic routes shown in Scheme 13.34.

Furanoid δ-SAA (**115**) and (**117**) have been obtained using different strategical approaches by three different authors: Le Merrer *et al.*, Chakraborty *et al.*, and Fleet *et al.*, as reviewed by Kessler *et al.* [35]. Le Merrer *et al.* used mannose as starting material to generate the enantiomerically pure double epoxide (**113**) [47], which was treated with NaN_3 and silica gel to generate the corresponding azidomethyl-furanoid sugar (**114**). Oxidation of primary hydroxyl group, and conversion of the azido group to Boc-protected amine produces the furanoid δ-SAA (**115**). Starting from the enantiomeric epoxide (**116**), also obtained from D-mannitol in six steps, it is possible to achieve the synthesis of the δ-amino acid (**117**) having the same orientation of functional groups relative to the ring, but inverted configurations of the amino and carboxylic functions at the C-1 and C-5 positions, respectively (Scheme 13.35).

Chakraborty *et al.*'s approach consists of an intramolecular 5-*exo* ring-opening of a terminal *N*-Boc-aziridine [48], derived from α-glucopyranose, during alcohol to

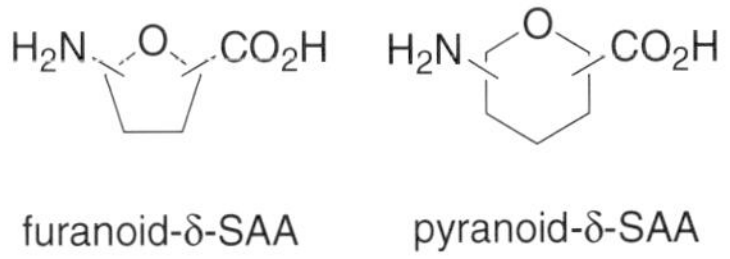

H_2N–(O)–CO_2H furanoid-δ-SAA; H_2N–(O)–CO_2H pyranoid-δ-SAA

Scheme 13.33

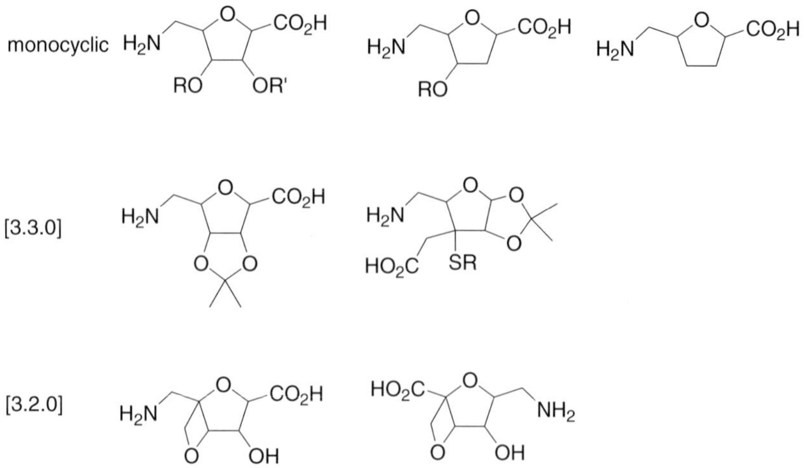

Scheme 13.34

acid oxidation, resulting in the protected furanoid δ-SAA (**119**) similar to (**115**) with complete stereocontrol (Scheme 13.36). The stereoisomerically pure aziridine (**118**) obtained from the same treatment from the D-mannose precursor generates the corresponding isomeric δ-amino acid (**120**), having the configuration at C-1 inverted.

Fleet *et al.* reported a range of stereoisomeric furanoid δ-SAA starting from sugar-derived lactones [49]. For example, the previously described compound (**119**) was obtained as the azido ester (**123**) from D-mannono-γ-lactone (**121**) by acid-catalyzed ring rearrangement of the corresponding triflate (**122**) (Scheme 13.37).

More recently, the same authors reported the synthesis of all diastereomeric precursors to THF-templated δ-amino acids lacking the hydroxyl at C-2, starting from mannono- and gulono-lactones [50], in analogy with the corresponding THF-templated γ-amino acids (Scheme 13.38) [36]. Two different strategic approaches

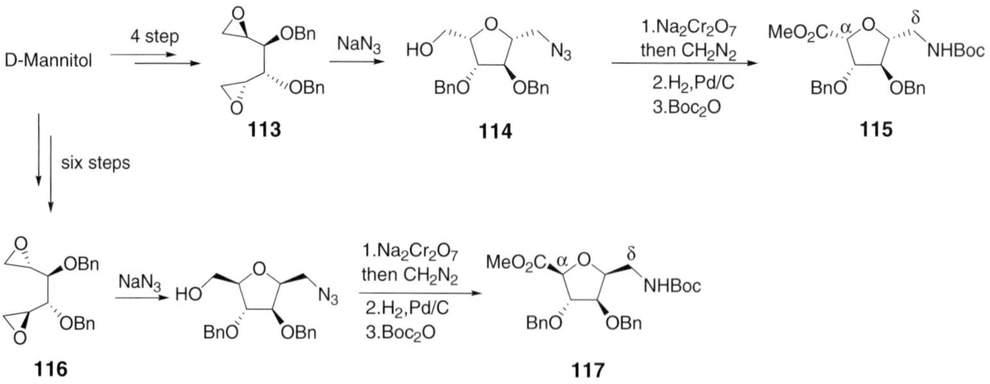

Scheme 13.35

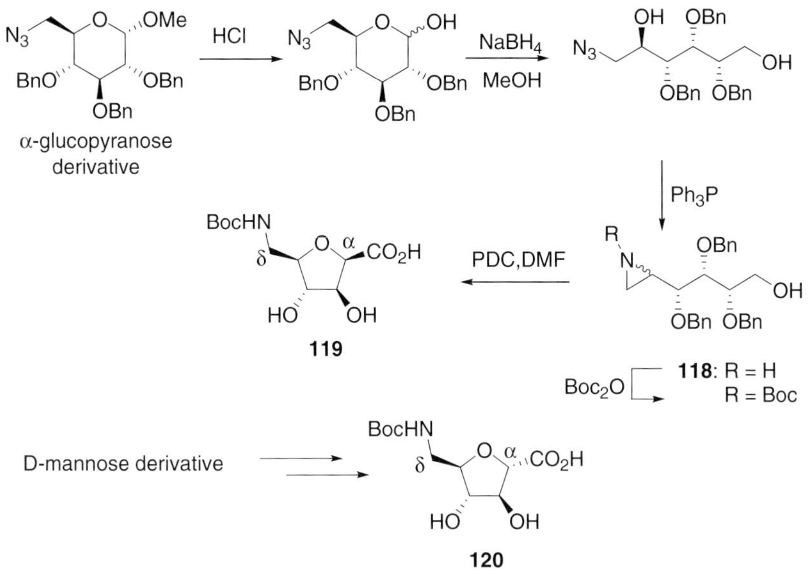

Scheme 13.36

have been proposed, by changing the order of deoxygenation and THF formation reactions.

The hydroxylated THF-carboxylic acid derivatives have been further manipulated to obtain the azido esters as δ-amino acid precursors: selective activation of primary hydroxyl group with tosyl chloride was followed by azide insertion at the δ-position.

Recently, a set of conformationally locked δ-amino acids have been proposed, based on furan rings [51]. In particular, a bicyclic furano-oxetane core has been proposed as scaffold for a constrained δ-amino acid. The synthetic strategy is based upon CO insertion on fully protected β-D-ribofuranoside (**124**), followed by conversion of a primary alcohol function at C-1 to azide to give (**125**). After hydroxyl group

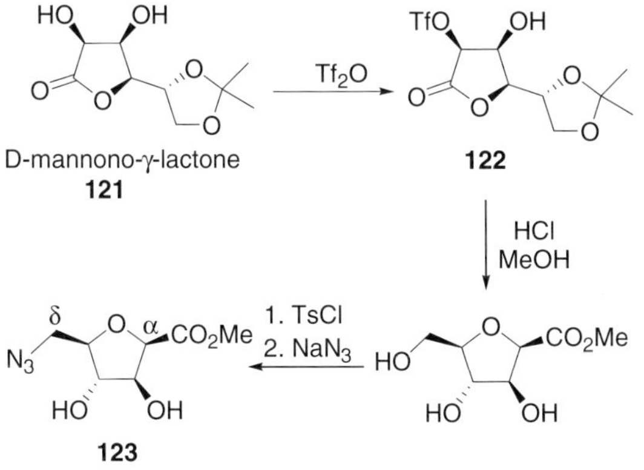

Scheme 13.37

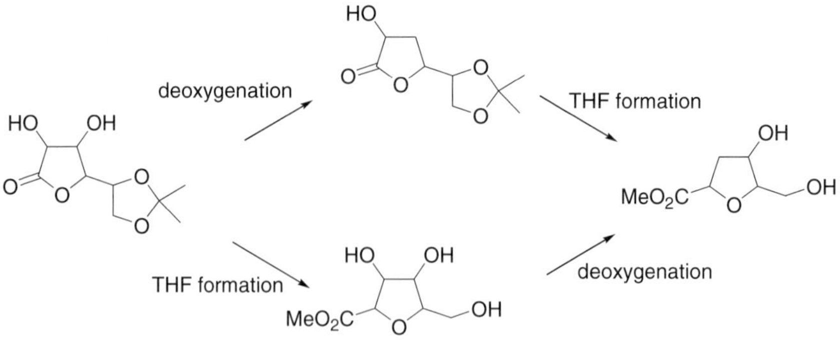

Scheme 13.38

protection/deprotection steps, oxidation of primary alcohol group, followed by aldol condensation with formaldehyde and oxetane cyclization step produces (**126**), which is treated with Boc-on and Me_3P to convert the azido group into the corresponding Boc-protected δ-amino acid (**127**) (Scheme 13.39).

Inversion of functional groups to obtain the isomeric δ-amino acid (**129**) was accomplished by protection of the alcohol function at C-1 of (**128**) as the *tert*-

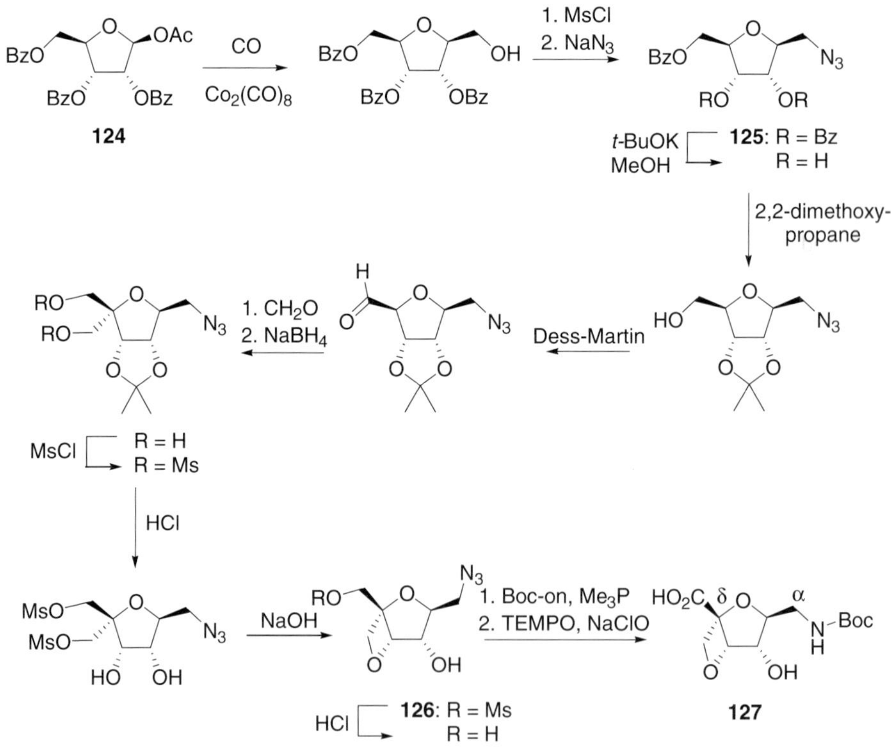

Scheme 13.39

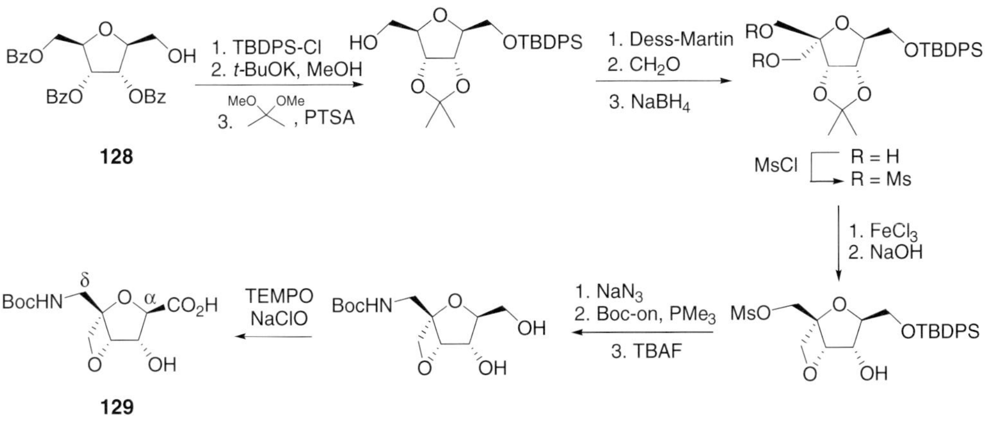

Scheme 13.40

butyldiphenylsilyl ether, followed by conversion of C-5 to a Boc-protected amino group via azide formation (Scheme 13.40).

The conformational rigidity of the pyran and furan rings makes carbohydrate-derived amino acids interesting building blocks for the introduction of specific secondary structures in peptides. For example, compound **132** (Scheme 13.41) was incorporated into the cyclic peptide containing the RGD (Arg–Gly–Asp) loop sequence by solid-phase peptide synthesis using Fmoc chemistry [52]. Reduction of the azide to amine group and coupling with the desired amino acid was realized in one pot in the presence of Bu_3P and carboxylic acid activating agents. Allyl compounds **130**, derived from allylation of 2,3,5-tri-*O*-benzyl-D-arabinofuranose, underwent iodocyclization to **131** as a diastereomeric mixture, easily separated by chromatography. This step, crucial for the formation of bicyclic scaffolds, consisted of an intermediate iodonium ion opening by attack of the γ-benzyloxy group and formation of a cyclic iodoether with simultaneous debenzylation. The final azido acid **132** was obtained by reaction with Bu_4NN_3, followed by selective deprotection steps and primary alcohol Jones oxidation.

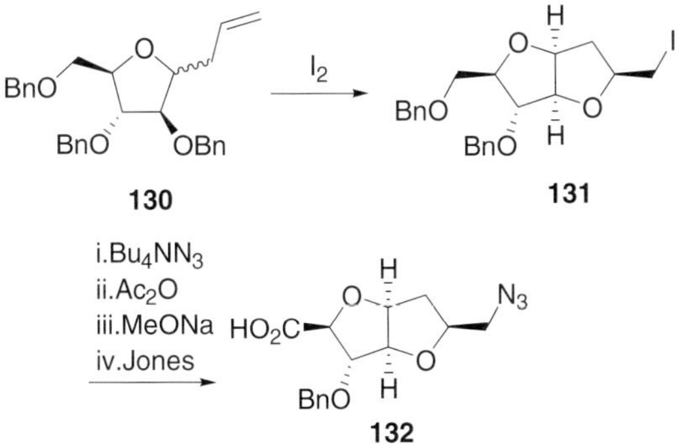

Scheme 13.41

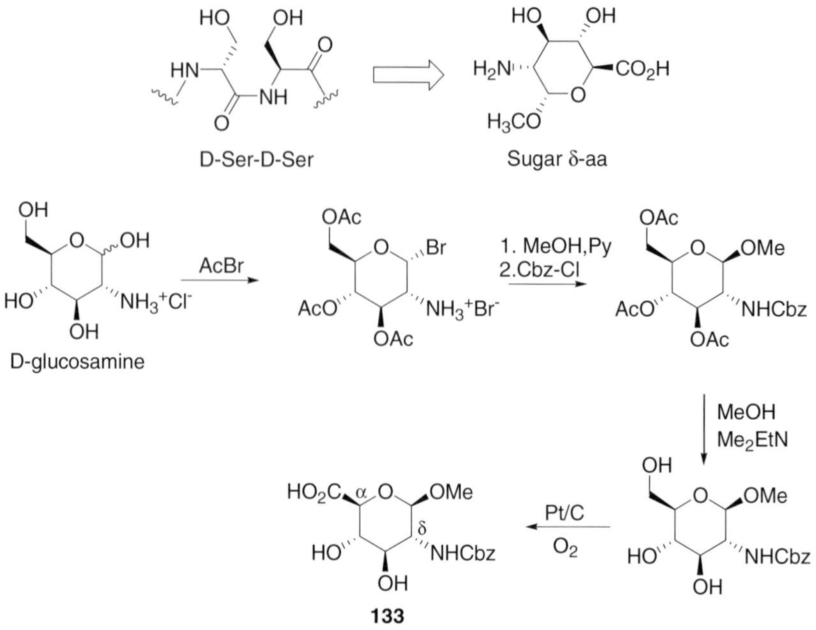

Scheme 13.42

13.3.1.2 **Pyranoid δ-SAA**

δ-Amino acids belonging to the SAA class constrain a linear peptide chain when the NH_2 and COOH groups are in the 1,4-positions. In particular, such δ-SAAs have been thought of as rigidified D-Ser–D-Ser dipeptide isosteres, as shown in Scheme 13.42 [53]. The synthesis of β-anomer **133** has been reported starting from glucosamine.

Ichikawa *et al.* reported the synthesis and incorporation in oligomeric structures of a series of glycamino acids, a family of SAAs that possesses a carboxyl group at the C-1 position and the amino group at C-2, -3, -4, or -6 position [54]. In particular, δ-amino acids with the amino group in position 4 or 6 have been reported, and the syntheses are shown in Schemes 13.43 and 13.44. Benzylidene-protected ester (**135**) is obtained starting from methyl β-D-galactopyranosyl-*C*-carboxylate (**134**) by treatment with benzaldehyde and formic acid. *O*-Benzylation and reductive opening of the benzylidene group are followed by treatment with Tf_2O and NaN_3, to give the Boc-protected amino group at C-4 position after reduction and protection of the azido derivative (**136**). Oligomerization has been carried out using deprotected hydroxyl functions of Boc-protected δ-amino acid (**137**).

The synthesis of δ-amino acid (**139**) was accomplished starting from D-glucose, using an aldol reaction to introduce a nitromethylene group at the anomeric position of (**138**) as an aminomethylene equivalent, followed by selective oxidation of the primary hydroxyl group to a carboxylic acid (Scheme 13.44) [55].

As an alternative approach to pyranosidic glucose-derived δ-amino acids, Xie *et al.* proposed a synthetic route from β-C-1-vinyl glucose to generate δ-SAA

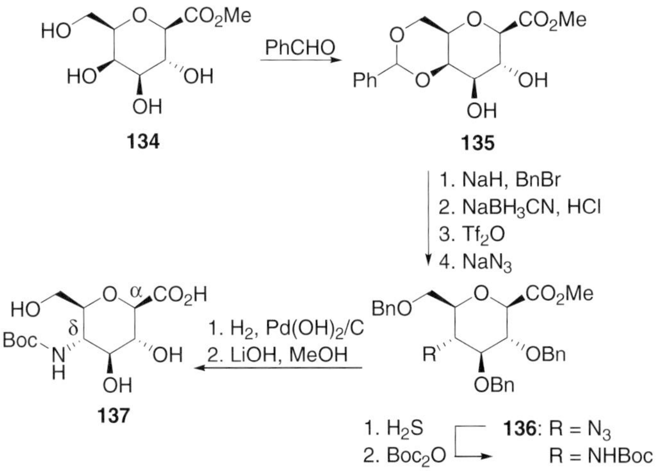

Scheme 13.43 δ-Amino acid monomer for β-1–4-linked oligomers.

(Scheme 13.45) [56]. Starting from vinyl-glucoside (**140**), selective deprotection of 6-benzyloxy group affords (**141**). Primary alcohol oxidation with pyridinium chlorochromate leads to (**143**), which is treated with $O_3/NaBH_4$ as an oxidation–reduction step to insert a hydroxymethyl function at C-1. Final activation and azide insertion produces the δ-amino acid precursor (**144**). Alternatively, azide insertion at C-6 via mesylation of hydroxymethyl group followed by treatment with NaN_3 allows the synthesis of the δ-amino acid precursor (**142**), thus inverting the order of reactions. Oligomerization has been carried out in solution using the corresponding azido ester and protected amino acid for subsequent coupling reactions.

Pyranoid δ-amino acids have also been described by Sofia *et al.* for solid-phase generation of carbohydrate-based universal pharmacophore mapping libraries [38]. The sugar scaffold is provided with three sites of diversification, using an amino and a

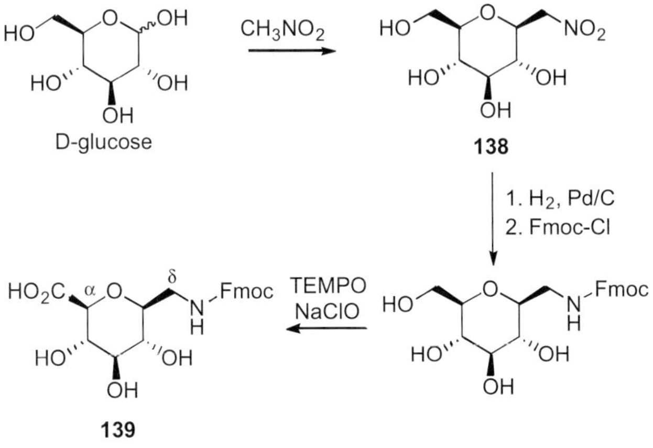

Scheme 13.44

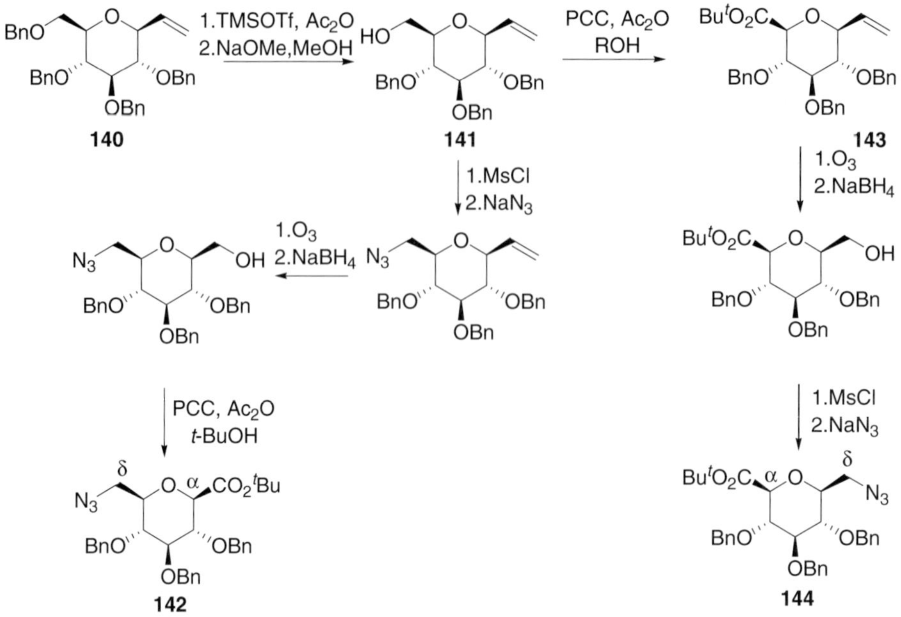

Scheme 13.45

carboxyl group of the δ-amino acid scaffold, and an additional hydroxyl group. The synthesis of selected δ-amino acid scaffolds was achieved starting from glucosamine by amine protection as Cbz-urethane, followed by treatment with 2,2-dimethoxypropane to protect hydroxyl groups at C-5 and C-6. Methylation of the hydroxyl at C-4 afforded the orthogonally protected (**145**). Subsequent oxidation with TEMPO–NaClO of the deprotected primary hydroxylic function, hydrogenation and Fmoc protection, and gave the δ-amino acid (**146**) with a free hydroxyl at C-4 (Scheme 13.46).

13.3.2
δ-Amino Acids as Reverse Turn Mimetics

Among δ-amino acids as reverse turn inducers, a classification can be made depending upon their structural features. In particular, linear, cyclic, and bicyclic compounds have been developed as dipeptide isosteres and reverse turn inducers. Recently, an overview of all the possible folding alternatives of δ-amino acids in oligomeric δ-peptides with respect to the parent α-peptides was reported using various methods of an *ab initio* molecular orbital theory [57]. In particular, cyclic δ-amino acids have been investigated as possible reverse turn mimetics, showing the strict relationship between stereochemistry on the ring system and secondary structure of the sequence containing the δ-amino acid. An early example of a linear δ-amino acid as a reverse turn inducer was first reported by Gellman *et al.* [58], in which a *trans* C=C double bond was introduced to replace the amide bond of the

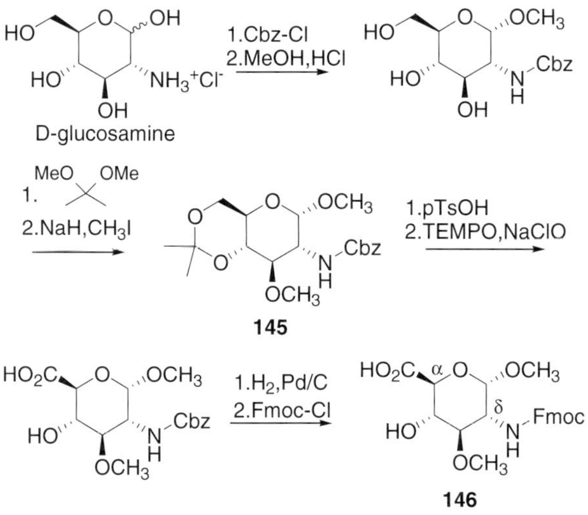

Scheme 13.46

central dipeptide of the β-turn. Thus, such alkene-based β-turn mimetics have been thought of as rigidified mimetics of a Gly–Gly dipeptide (Scheme 13.47).

Other authors reported the synthesis and conformational analysis of alkene-based dipeptide isosteres in which the δ-amino acid is substituted in the α- and δ-positions with methyl groups [59]. The synthetic strategies allow one to obtain δ-amino acid (**147**) in diastereomerically and enantiomerically pure form as a D-Ala–L-Ala isostere via S_N2' addition of cuprate reagents to the alkenyl aziridine (**148**) (Scheme 13.48).

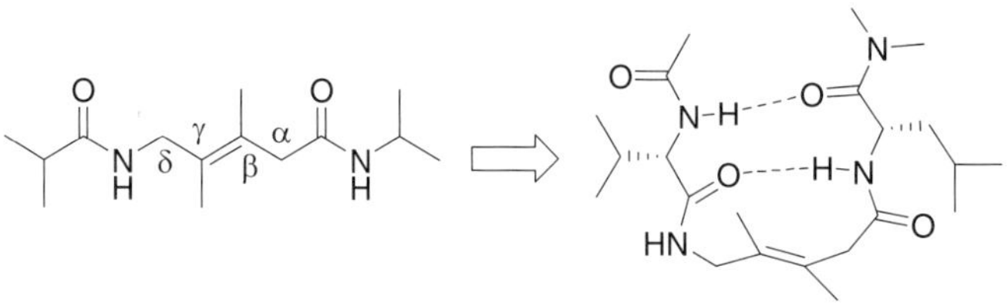

Scheme 13.47

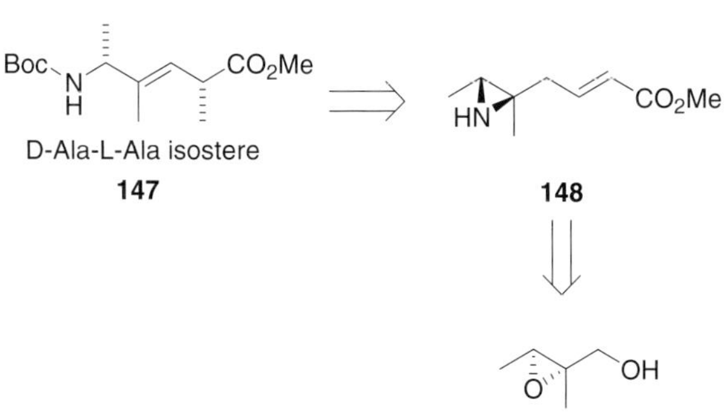

Scheme 13.48

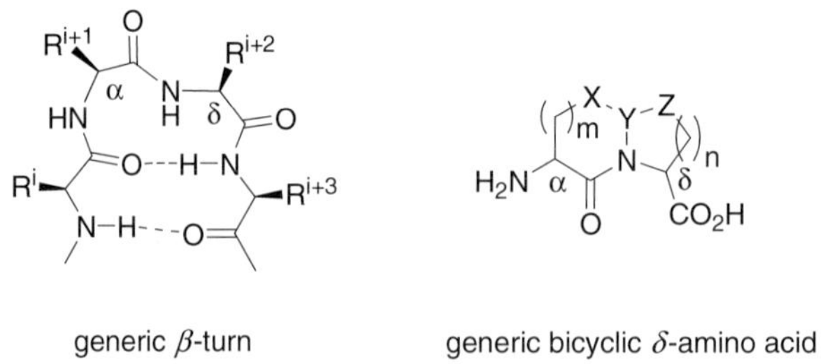

generic β-turn

generic bicyclic δ-amino acid

Scheme 13.49

Many of the bicyclic δ-amino acids have been designed as dipeptide mimetic replacements of the $i + 1$ and $i + 2$ residues in β-turn systems. For this reason, attempts to correctly positioning the N- and C-termini of the dipeptidomimetic unit resulted in an obligatory δ-amino acid, often incorporating an amide bond in the bicyclic backbone (Scheme 13.49).

The most common approach to substitute the central dipeptidic sequence of β-turns with peptidomimetics has been to use tethered prolines within bicyclic scaffolds, since such amino acid are often found in β-turns, especially in types I and II in the $i + 1$ position, and in type VI in the $i + 2$ position, the last having the amide bond in a *cis* configuration. Most syntheses focused on proline-based bicyclic δ-amino acids, can be divided into numerous azabicyclo[*x*.3.0]alkane subclasses (see Scheme 13.50 for a few examples).

In most cases the key step of the synthesis is the lactam ring formation that can be obtained by different approaches (Scheme 13.51) [60, 61]: the radical addition to an olefinic double bond, alkylation of malonate enolate, ring-closing metathesis, intramolecular alkylation followed by amidation and Hoffman rearrangement, and aldol condensation [62]. Further functional groups at the α-position could be inserted on the final scaffold by alkylation under basic conditions [63].

More generally, Lubell *et al.* reported an extensive overview on tethered prolines to generate bicyclic compounds of general formula as shown in Scheme 13.52. In particular, a systematic description of synthetic procedures leading to different sized bicycles having additional heteroatoms is reported [64]. The introduction of substituents on both cycles to mimic the side-chain of the central dipeptide of β-turns is also described.

Other examples include the generation of 3-aminooxazolidino-2-piperidones as Ala–Pro dipeptide surrogates, and conformational analysis to demonstrate the type

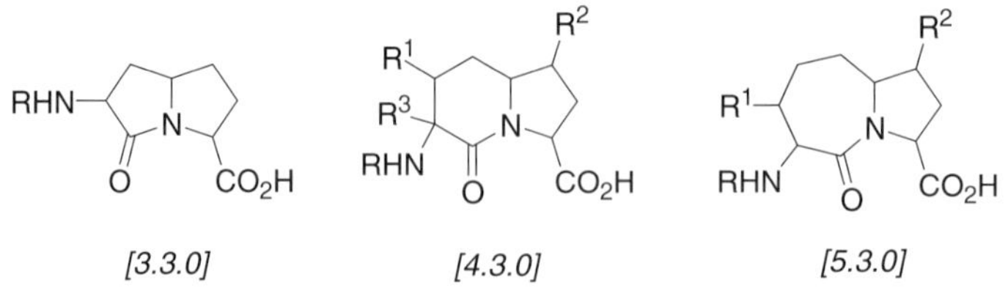

[3.3.0] [4.3.0] [5.3.0]

Scheme 13.50

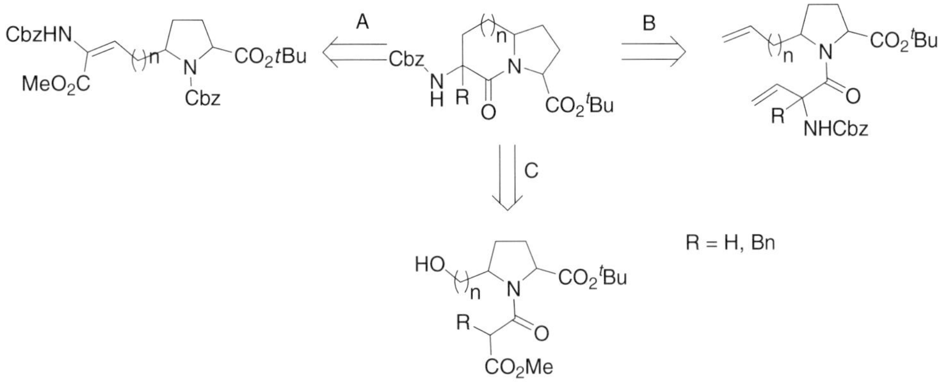

Scheme 13.51

II′ β-turn inducing properties of such δ-amino acids (Scheme 13.53) [65]. Compound (**150**) was obtained in a one-pot process from the glutamic acid-derived aldehyde (**149**) and Ser-OMe. Structural analysis of the end-protected compound (**151**) revealed the all-(*S*) stereoisomer to act as the best reverse turn mimetic.

Bicycles constituted by six- and five-membered rings have been used as type VI β-turn mimetic and antiparallel β-ladder nucleators [66]. Bicyclic scaffold (**152**) derives from proline and carries the carboxylic function at the bridgehead position (Scheme 13.54). Detailed conformational analysis revealed its peculiar reverse turn properties.

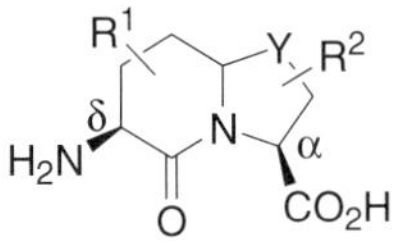

Scheme 13.52

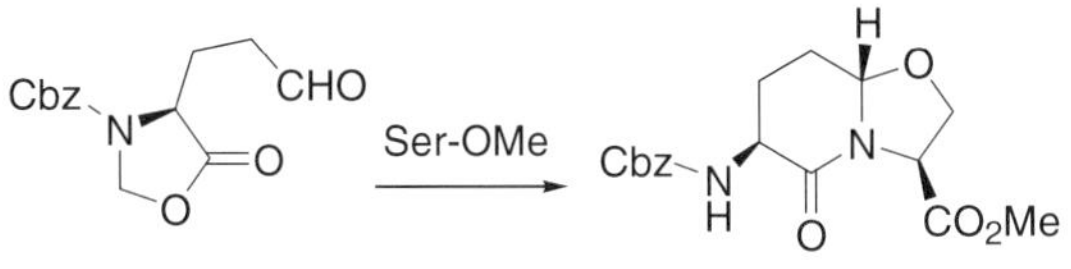

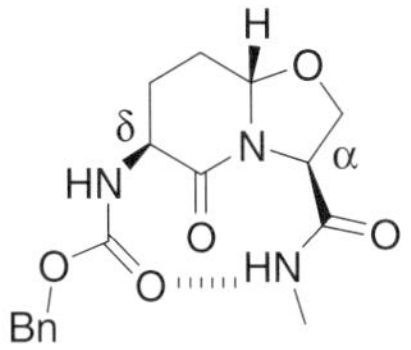

Scheme 13.53

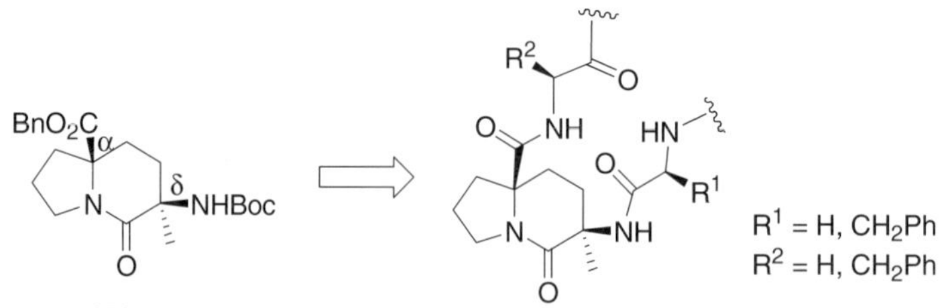

R^1 = H, CH_2Ph
R^2 = H, CH_2Ph

152

Scheme 13.54

Two approaches (i.e., radical and nonradical) have been investigated by Scolastico *et al.* to accomplish lactam ring formation [60]. Radical cyclization involves radical addition to an olefinic double bond, and, among nonradical methods, alkylation of malonate enolate, ring-closing metathesis, and lactam bond formation have been taken into account (Scheme 13.55).

Detailed conformational analysis by nuclear magnetic resonance, infrared, and molecular modeling techniques [67], as well as applications as thrombin inhibitors [68], and in RGD-based cyclopeptides such as $\alpha_v\beta_3$-integrin ligands [69], proved this class of compounds to act as δ-amino acid reverse turn mimetics.

A recent example of a spiro-β-lactam system was obtained via Staudinger reaction of the disubstituted ketene (**153**) deriving from Cbz-proline, giving a reverse turn mimetic by reaction with a glycine-derived imine [70]. Compound (**154**) as depicted in Scheme 13.56 was obtained exclusively as a consequence of *cis* stereoselectivity of the process. Such a compound is designed using high-level *ab initio* methods, and conformational analysis via nuclear magnetic resonance demonstrated the scaffold stabilized a β-turn conformation with a geometry close to an ideal type II β-turn, by restricting the ϕ and φ torsion angles in the $i + 1$ position.

Extremely constrained spiro-bicyclic lactams of 5.6.5 and 5.5.6 size, as shown in Scheme 13.57, have been investigated as PLG (Pro–Leu–Gly) peptidomimetics for the modulation of dopamine receptor activity [71]. An earlier report by the same authors described the generation of the corresponding 5.5.5-sized scaffold [72].

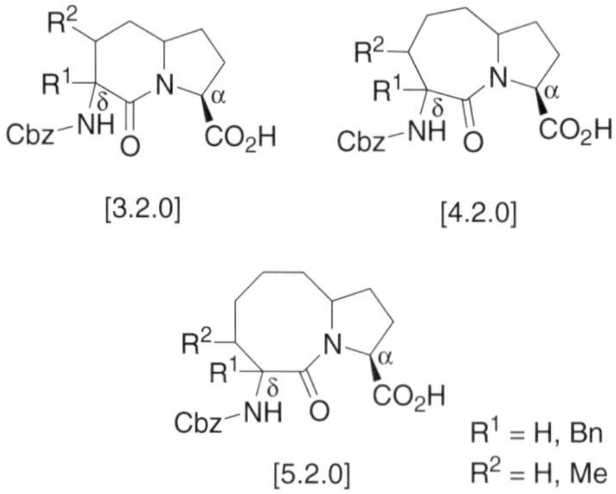

[3.2.0]

[4.2.0]

[5.2.0]

R^1 = H, Bn
R^2 = H, Me

Scheme 13.55

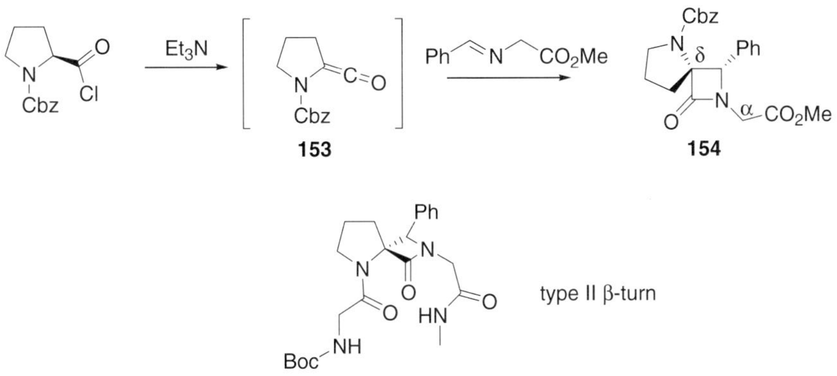

Scheme 13.56

Structural analysis by X-ray and computational methods revealed these compounds to mimic a type II β-turn, and a strong dependence of torsion angles with ring size has been evidenced.

The synthetic strategy to produce the 5.5.6 scaffold consists of stereoselective allyl alkylation of proline followed by protection of the amino and carboxylic functions. Double-bond oxidative cleavage of (**155**) with $OsO_4/NaIO_4$ followed by condensation with homocysteine produces the thiazinane (**156**). Direct intramolecular amide bond formation generates the title fully protected δ-amino acid (**157**) after esterification of the carboxylic function with diazomethane (Scheme 13.58).

The synthesis of spiro bicyclic δ-amino acid (**161**) starts with alkylation of proline-derived oxazolidinone (**158**) with 4-bromo-1-butene (Scheme 13.59). Subsequent lactone ring opening, protection of amino and carboxylic functions, and double-bond oxidative cleavage produces aldehyde (**159**), which was condensed with cysteine, and the resulting adduct (**160**) was allowed to cyclize to furnish the 5.6.5 δ-amino acid (**161**).

Another example of a bicyclic structure bearing a pyrrolidine ring was reported by Johnson *et al.*, which used 2-allylproline to generate a 4,4-spirolactam scaffold as a type II β-turn mimetic [73]. δ-Valerolactam has been used to generate a lactam-based δ-amino acid as reverse turn mimetic [74]. In particular, diversity in the six-membered ring of compound (**162**) is obtained by conjugate addition on the double bond, followed by Curtius rearrangement to introduce the amino group (Scheme 13.60).

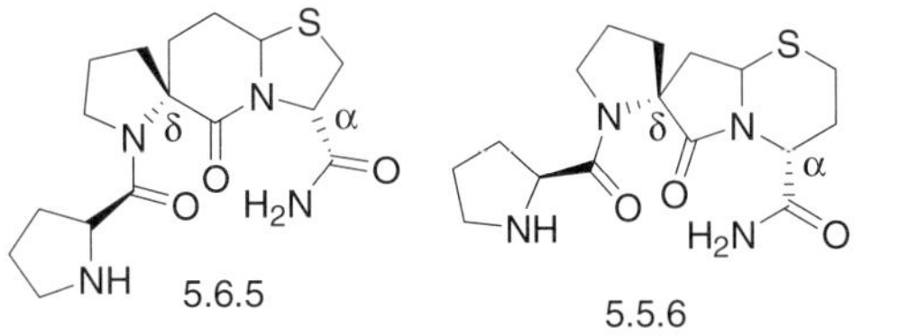

Scheme 13.57

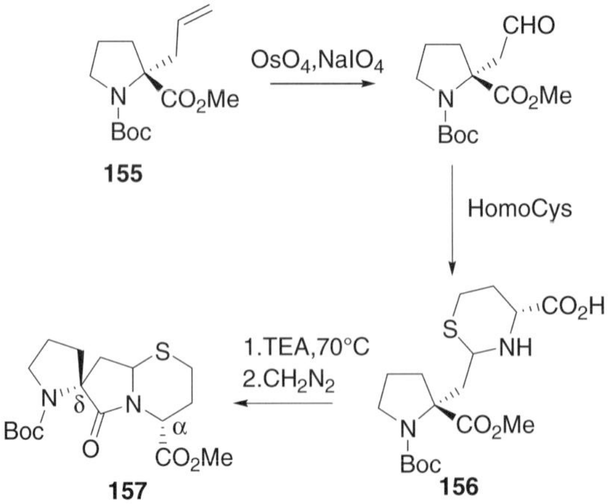

Scheme 13.58

A unique example of a macrocyclic δ-amino acid as a reverse turn mimetic was reported in 1998 (Scheme 13.61). Katzenellenbogen *et al.* described the synthesis of a 10-membered ring as a type I β-turn mimetic by dimerization of the α-amino acid 2-amino-hexenoic acid (**163**), followed by ring-closing metathesis of the resulting adduct (**164**) [75].

The class of bicyclo[*x.y.*1] γ/δ-amino acid scaffolds is mainly represented by 6,8-dioxa-3-azabicyclo[3.2.1]octane-7-carboxylic acids (**165**) named BTAa (Bicycles from Tartaric acid and Amino acid; Scheme 13.62) [76], easily synthesized from the condensation of α-amino aldehyde derivatives (**166**) with tartaric acid (**167**) or sugar derivatives in a stereoselective fashion [77]. The insertion of 6,8-dioxa-3-azabicyclo[3.2.1]octane-7-carboxylic acids in cyclic and linear peptidic sequences demonstrates the ability of these scaffolds to revert a peptide chain.

Phenylalanine-derived amino alcohol (**168**) was coupled with L-tartaric acid monoester derivative (**169**) to give the corresponding amide (**170**), which was successively oxidized at the primary hydroxyl group and cyclized in refluxing toluene in the presence

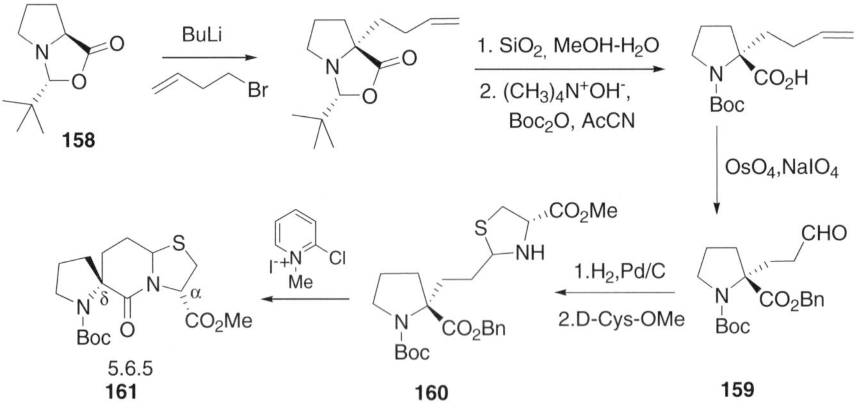

Scheme 13.59

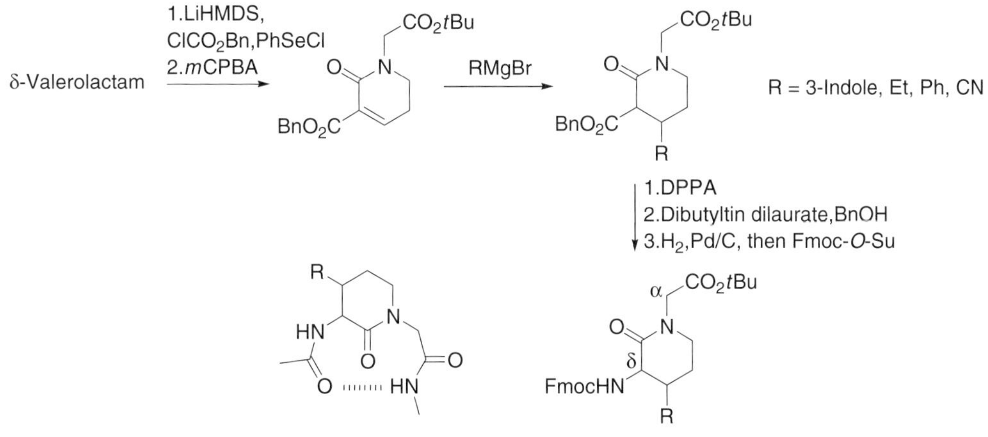

Scheme 13.60

of acid silica gel. Further manipulations of (**171**) to the corresponding δ-amino acid (**172**) consists of complete $LiAlH_4$ reduction, *N*-debenzylation, Fmoc protection, and final oxidation using Jones' methodology with CrO_3–H_2SO_4 (Scheme 13.63).

In particular, if C-7 has an *endo* configuration, a dipeptide isostere reverse turn mimetic can be obtained, which can mimic the central portion $i + 1 - i + 2$ of a common β-turn (Scheme 13.64).

The first example was the introduction of a 7-*endo*-BTAa in a cyclic Bowman–Birk inhibitor peptide as an Ile–Pro mimetic, showing that the scaffold is able to maintain the existing turn [78]. Successively, a detailed conformational analysis on linear model peptides containing leucine-derived BTAa demonstrated the reverse turn inducing properties of these bicyclic γ/δ-amino acids and the effect of substituent δ-position [79].

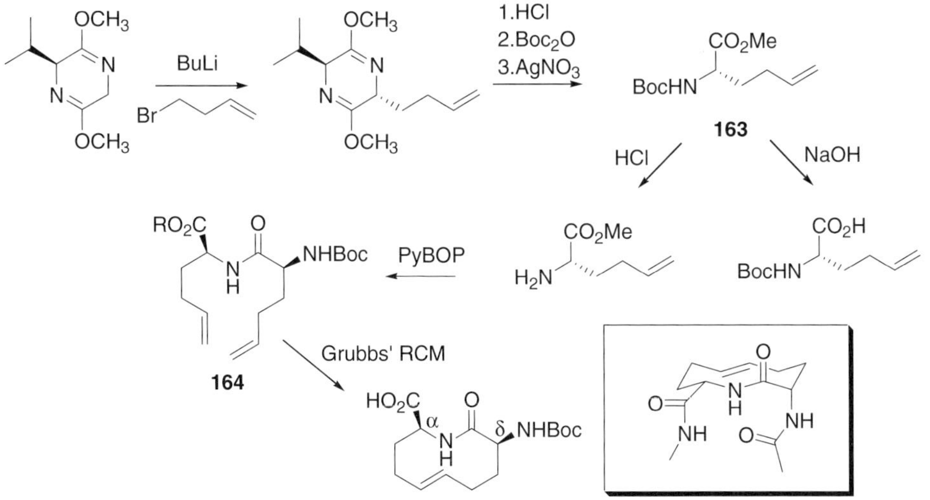

Scheme 13.61

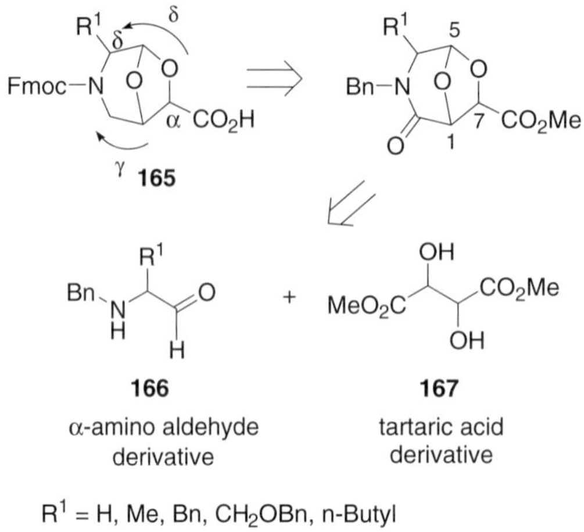

R¹ δ δ O Fmoc–N O α CO₂H γ **165** ⇒ R¹ 5 O Bn–N O 7 CO₂Me O 1

Bn N H R¹ O H + OH MeO₂C CO₂Me OH

166 α-amino aldehyde derivative

167 tartaric acid derivative

R^1 = H, Me, Bn, CH_2OBn, n-Butyl

Scheme 13.62

13.3.3
δ-Amino Acids for PNA Design

In 1991, Nielsen *et al.* reported the synthesis and properties of new DNA analogs based on complete substitution of the DNA backbone with a polyamidic chain covalently linked to the DNA nucleobases [80]. Most of PNA monomers have been designed according to a "golden rule": the number of atoms of the repeating unit in

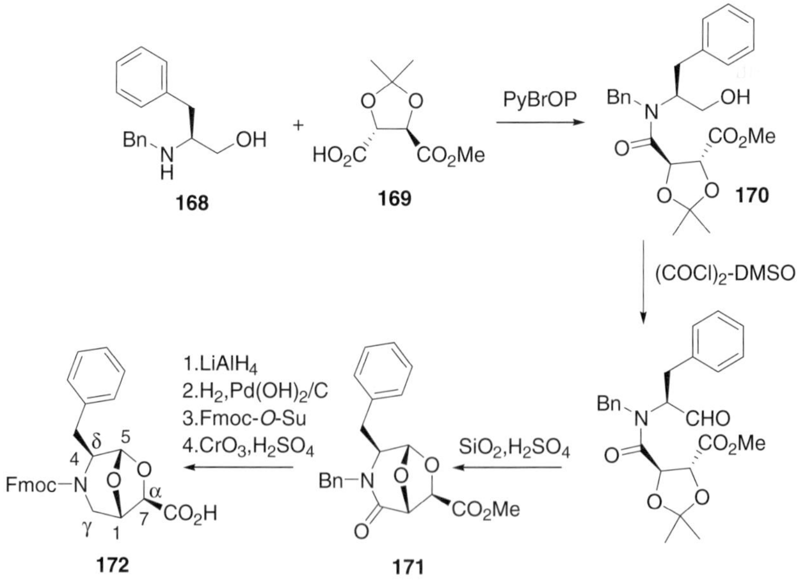

Bn N H OH **168** + O O HO_2C CO_2Me **169** PyBrOP → Bn N OH O CO_2Me O O **170**

$(COCl)_2$-DMSO

Bn N CHO O CO_2Me O O → SiO_2,H_2SO_4 → Bn–N O O O CO_2Me **171**

1.$LiAlH_4$
2.H_2,$Pd(OH)_2$/C
3.Fmoc-*O*-Su
4.CrO_3,H_2SO_4

δ 5 4 O Fmoc–N O α 7 CO_2H γ 1 **172**

Scheme 13.63

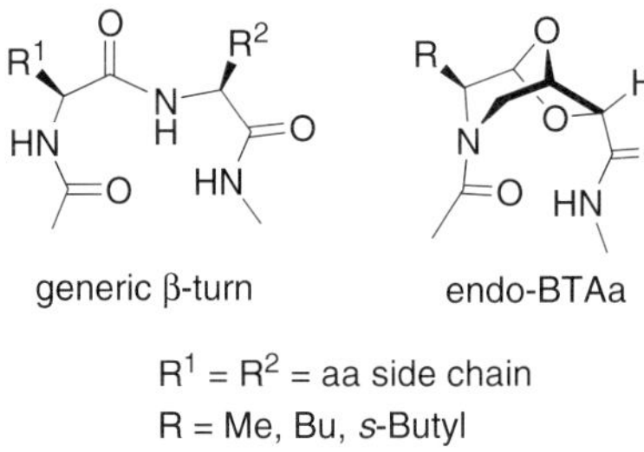

generic β-turn   endo-BTAa

$R^1 = R^2$ = aa side chain
R = Me, Bu, *s*-Butyl

Scheme 13.64

the polyamide chain should be six and the number of atoms between the backbone and the nucleobase (B) should be two, as shown in Scheme 13.65 [81].

Various examples of PNA monomers have been proposed to satisfy the golden rule, and to provide a switch towards selectivity in the complexation with DNA and RNA. Introduction of constraints by means of five- or six-membered ring structures in the aminoethylglycyl-PNA contributes to reducing the entropic loss during complex formation, and maintains a balance between rigidity and flexibility in the backbone. Pyrrolidine and pyrrolidone rings have been selected as scaffolds for cyclic PNA monomers since the nitrogen atom of the ring serves as the amino group for backbone or side-chain conjugation with purinyl- or pyrimidinyl-acetic acid derivatives. In particular, 4-OH-proline has become the most popular starting material to generate such PNA monomers (Scheme 13.66).

Compounds (**173**), (**174**), (**175**), and (**176**) derive from 4-OH-proline, while compound (**177**) shows a thiazolidine ring. In compound (**173**), the hydroxy group at position 4 of the starting amino acid is used to insert the base via Mitsunobu reaction. Thus, the pyrrolidine ring of the resulting PNA monomer is thought of as a constraint of the β,γ-positions, with the side-chain containing the nucleobase [82]. The synthesis is shown in Scheme 13.67 starting from Boc-4-OH-proline. The amino group is inserted in position 5 of the ring via ester reduction of (**178**), followed by azide introduction in (**179**). The carboxylic group is inserted in (**180**) by reaction of pyrrolidine nitrogen atom with methyl α-bromoacetate, and after azide reduction-amine protection, the resulting Boc-δ-amino acid (**181**) was obtained, and succes-

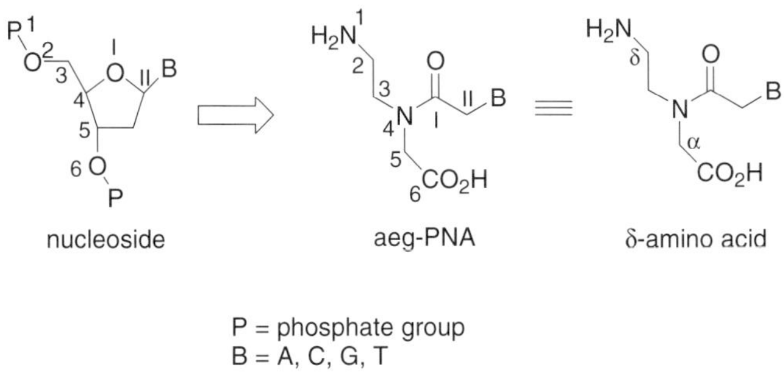

nucleoside   aeg-PNA   δ-amino acid

P = phosphate group
B = A, C, G, T

Scheme 13.65

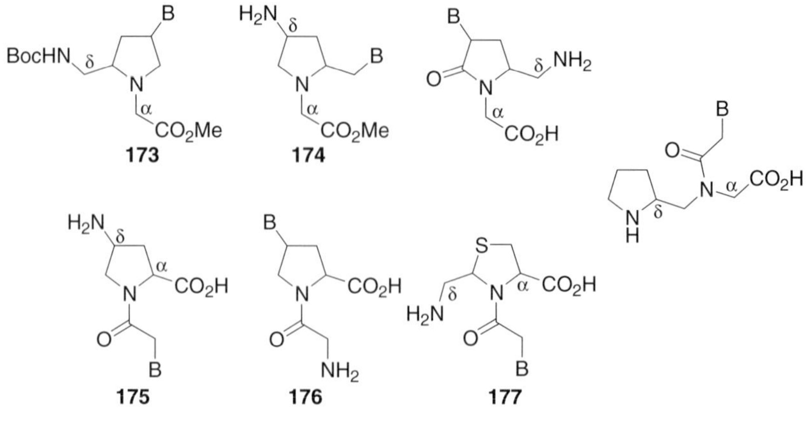

B = Nucleo base

Scheme 13.66

sively transformed in the PNA monomer (**182**) by means of 4-hydroxyl deprotection-activation and nucleobase insertion.

Six-membered ring PNA monomers have been developed using mainly cyclohexane or pyranosidic sugars as scaffolds. Cyclohexane decorated with the amino, carboxyl, and base groups on the cycle has been synthesized starting from butadiene via enantioselective cycloaddition with acryloyl-oxazolidinone (**183**) in the presence of TADDOL (4,5-bis[hydroxy(diphenyl)methyl]-2,2-dimethyl-1,3-dioxolane) as chiral catalyst (Scheme 13.68) [83]. Nucleobase insertion was accomplished via nucleophilic substitution reaction on the bicyclic species (**184**), which gives the title δ-amino acid (**185**) after final ring opening.

13.3.4 Miscellaneous Examples

Similar to the work by Smrcina *et al.* [31], β-amino acids have been used to generate functionalized δ-amino acids. In particular, it is reported by Guichard *et al.* at synthesis of α,δ-disubstituted δ-amino acids (**189**) starting from β-amino acids (**186**), through the generation of δ-valerolactam (**187**), which are successively alkylated and opened to give the corresponding linear δ-amino acid (**189**) [84]. As shown in Scheme 13.69, the valerolactam ring is generated from the β-amino acid by reaction with Meldrum's acid, followed by carbonyl reduction and cyclization. Enolate alkylation at C-3 of (**187**) to give (**188**) is *trans* stereoselective with respect to the side-chain functional group of the starting β-amino acid.

The morpholine nucleus has been recently used as a scaffold to generate δ- and ε-amino acids starting from carbohydrates (Scheme 13.70) [85]. The synthetic strategy relies on a two-step glycol cleavage/reductive amination process, to give the morpholine-based amino acid. In order to obtain such δ-amino acids, ribose was chosen as the starting carbohydrate and transformed into the corresponding unprotected methyl azido ester derivative (**190**) [86].

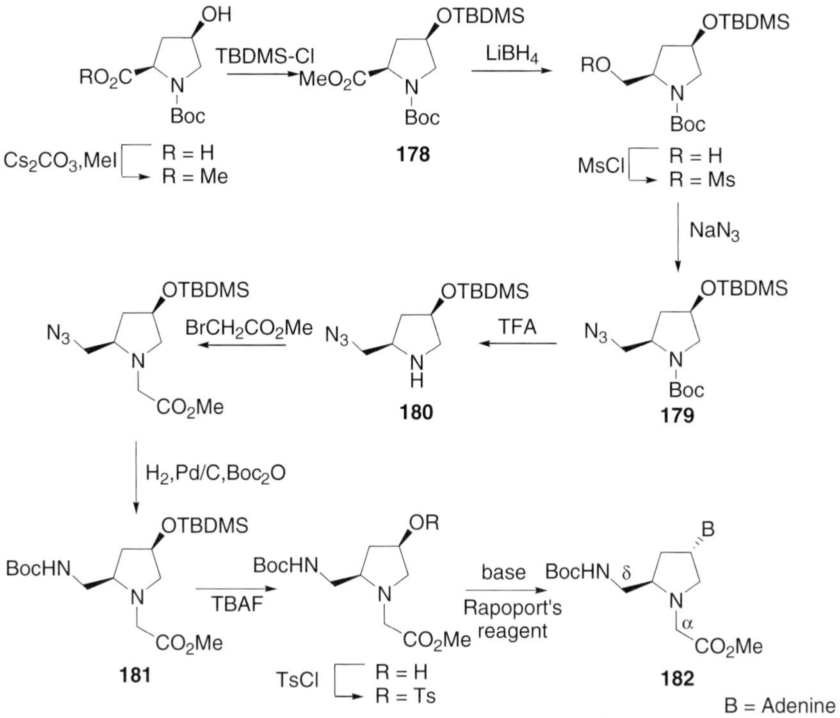

OH
RO_2C
N
Boc
TBDMS-Cl
OTBDMS
MeO_2C
N
Boc
$LiBH_4$
OTBDMS
RO
N
Boc
Cs_2CO_3,MeI
R = H
R = Me
178
MsCl
R = H
R = Ms
NaN_3
OTBDMS
N_3
N
Boc
179
TFA
OTBDMS
N_3
N
H
180
$BrCH_2CO_2Me$
OTBDMS
N_3
N
CO_2Me
H_2,Pd/C,Boc_2O
OTBDMS
BocHN
N
CO_2Me
181
TBAF
OR
BocHN
N
CO_2Me
TsCl
R = H
R = Ts
base
Rapoport's reagent
B
BocHN
δ
N
α
CO_2Me
182
B = Adenine

Scheme 13.67

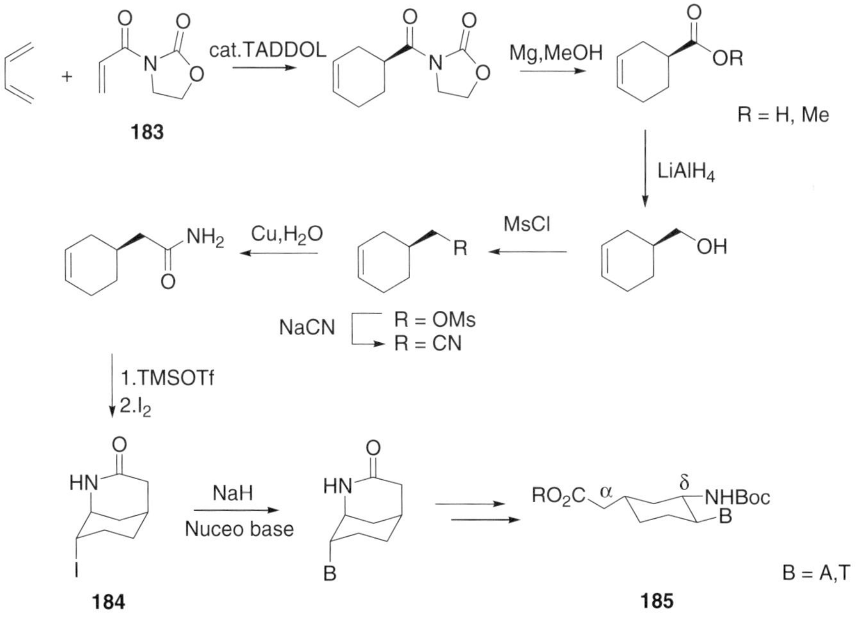

O
O
N
O
+
183
cat.TADDOL
O
O
N
O
Mg,MeOH
O
OR
R = H, Me
$LiAlH_4$
OH
MsCl
R
NaCN
R = OMs
R = CN
Cu,H_2O
NH_2
O
1.TMSOTf
2.I_2
O
HN
I
184
NaH
Nuceo base
O
HN
B
RO_2C
α
δ
NHBoc
B
185
B = A,T

Scheme 13.68

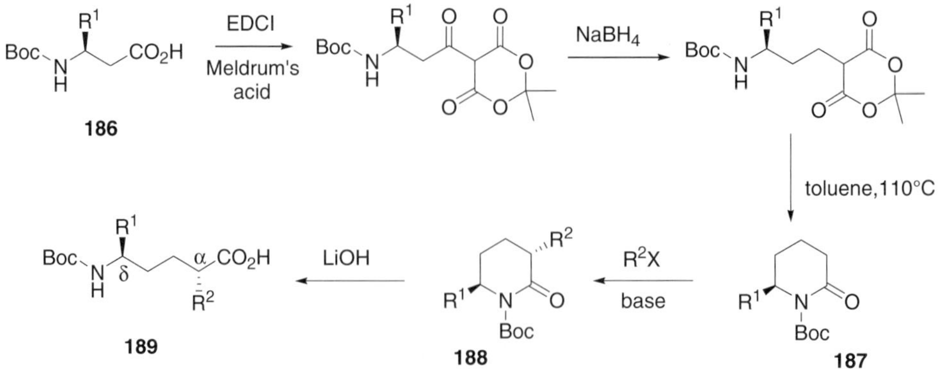

R^1 = Me, *n*-Bu, *s*-Bu
R^2 = Me, *n*-Bu, CH_2=$CHCH_2$, CH_2=$C(CH_3)CH_2$, $(CH_3)_2CHCH_2$
X = Br, I

Scheme 13.69

Unavoidable epimerization during glycol cleavage forced the authors to devise a new strategy to avoid an undesired β-keto ester. Thus, starting from either (**193**) or (**196**), respectively, stereoisomeric δ-azido esters (**194**) and (**197**) have been achieved as amino acid precursors of the corresponding (**195**) and (**198**) (Scheme 13.71).

13.4
Conclusions

The possibility of having a broad spectrum of γ- and δ-amino acids, either linear or cyclic, allows the generation of different species able to interact with biological

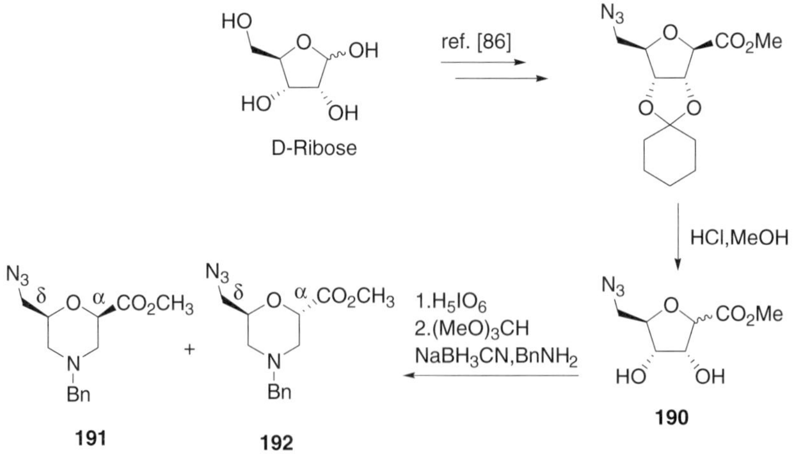

Scheme 13.70

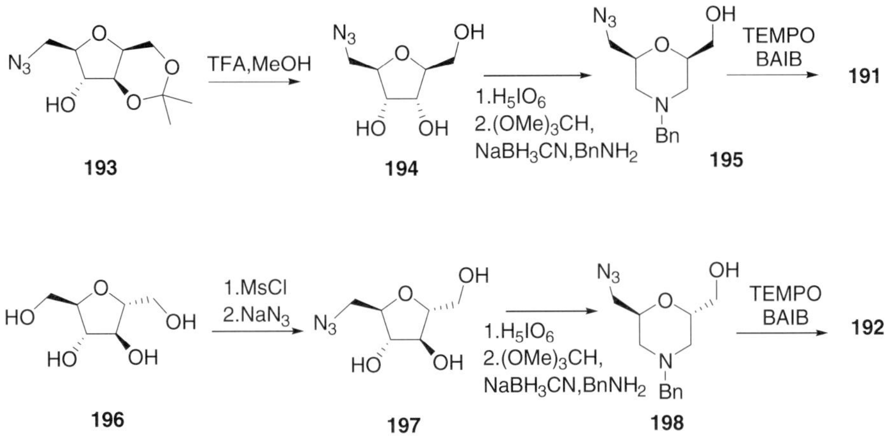

Scheme 13.71

systems. In particular, the high tendency of such new amino acids to fold into stable secondary structures enables their application in oligomer synthesis of more complex structures. γ-Amino acids are particularly relevant in the context of lead development, since they allow presentation of different functional groups in a relatively compact structure, either linear or cyclic in nature. Thus, much effort has been dedicated to the development of asymmetric synthetic strategies to obtain γ-amino acids with high stereocontrol. Moreover, interest in the field of medicinal chemistry related to the design of GABA and statine analogs is ever increasing. δ-Amino acids have been reported to play a central role in the design of peptidomimetics, especially in the field of β-turn analogs. Sugar amino acid-based scaffolds gathered interest in both oligomer synthesis and peptidomimetics due to the attractive feature of presenting many functional groups anchored in a rigid scaffold in well-defined positions. Moreover, research in the field of PNAs turned much attention towards both linear and cyclic enantiopure δ-amino acids for developing new PNA strands with increased selectivity with respect to complex formation with RNA and DNA. Thus, research in the field of γ- and δ-amino acids is ever active in diverse fields of medicinal chemistry, spanning from peptide-based drug design to oligomer synthesis, and also in the development of new efficient synthetic methods for γ- and δ-amino acid-based compounds.

References

1 (a) Giannis, A. and Kolter, T. (1993) *Angewandte Chemie (International Edition in English)*, **32**, 1244; (b) Gante, J. (1994) *Angewandte Chemie (International Edition in English)*, **33**, 1699.

2 (a) Abdel-Magid, A.F., Cohen, J.H., and Maryanoff, C.A. (1999) *Current Medicinal Chemistry*, **6**, 955; (b) Juaristi, E. and Lopez-Ruiz, H. (1999) *Current Medicinal Chemistry*, **6**, 983; (c) Sewald, N. (1996)

Amino Acids, **11**, 397; (d) Cole, D.C. (1994) *Tetrahedron*, **50**, 9517.

3 (a) Gellman, S.H. (1998) *Accounts of Chemical Research*, **31**, 173; (b) Iverson, B.L. (1997) *Nature*, **385**, 113; (c) Seebach, D. and Matthews, J.L. (1997) *Journal of the Chemical Society, Chemical Communications*, 2015.

4 (a) Hanessian, S., Luo, X., Schaum, R., and Michnick, S. (1998) *Journal of the American Chemical Society*, **120**, 8569; (b) Hintermann, T., Gademann, K., Jaun, B., and Seebach, D. (1998) *Helvetica Chimica Acta*, **81**, 983.

5 Szabo, L., Smith, B.L., McReynolds, K.D., Parrill, A.L., Morris, E.R., and Gervay, J. (1998) *The Journal of Organic Chemistry*, **63**, 1074.

6 (a) Duke, R.K., Chebib, M., Hibbs, D.E., Mewett, K.N., and Johnston, G.A.R. (2004) *Tetrahedron: Asymmetry*, **15**, 1745; (b) Allan, R.D., Bates, M.C., Drew, C.A., Duke, R.K., Hambley, T.W., Johnston, G.A.R., Mewett, K.N., and Spence, I. (1990) *Tetrahedron*, **46**, 2511.

7 (a) Hoveyda, A.H., Evans, D.A., and Fu, G.C. (1993) *Chemical Reviews*, **93**, 1307; (b) Evans, D.A., Gage, J.R., Leighton, J.L., and Kim, A.S. (1992) *The Journal of Organic Chemistry*, **57**, 1961; (c) Belliotti, T.R., Capiris, T., Ekhato, I.V., Kinsora, J.J., Field, M.J., Heffner, T.G., Meltzer, L.T., Schwarz, J.B., Taylor, C.P., Thorpe, A.J., Vartanian, M.G., Wise, L.D., Ti, Z.-S., Weber, M.L., and Wustrow, D.J. (2005) *Journal of Medicinal Chemistry*, **48**, 2294; (d) Brenner, M. and Seebach, D. (1999) *Helvetica Chimica Acta*, **82**, 2365; (e) Yuen, P.-W., Kanter, G.D., Taylor, C.P., and Vartanian, M.G. (1994) *Bioorganic & Medicinal Chemistry Letters*, **4**, 823.

8 (a) Palomo, C., Aizpurua, J.M., Oiarbide, M., García, J.M., Gonzaléz, A., Odriozola, I., and Linden, A. (2001) *Tetrahedron Letters*, **42**, 4829; (b) Palomo, C., Gonzaléz, A., García, J.M., Landa, C., Oiarbide, M., Rodríguez, S., and Linden, A. (1998) *Angewandte Chemie (International Edition in English)*, **37**, 180; (c) Camps, P., Munóz-Torrero, D., and Sánchez, L. (2004) *Tetrahedron: Asymmetry*, **15**, 2039; (d) Zelle, R.E. (1991) *Synthesis*, 1023.

9 Wall, G.M. and Baker, J.K. (1989) *Journal of Medicinal Chemistry*, **32**, 1340.

10 Ordóñez, M. and Cativiela, C. (2007) *Tetrahedron: Asymmetry*, **18**, 3.

11 Bryans, J.S. and Wustrow, D.J. (1999) *Medicinal Research Reviews*, **19**, 149.

12 (a) Pardridge, W.M. (1985) *Annual Reports in Medicinal Chemistry*, **20**, 305; (b) Shashoua, V.E., Jacob, J.N., Ridge, R., Campbell, A., and Baldessarini, R.J. (1984) *Journal of Medicinal Chemistry*, **27**, 659.

13 Lippert, B., Metcalf, B.W., Jung, M.J., and Casara, P. (1977) *European Journal of Biochemistry*, **74**, 441.

14 (a) Allan, R.D., Bates, M.C., Drew, C.A., Duke, R.K., Hambley, T.W., Johnston, G.A.R., Mewett, K.N., and Spence, I. (1990) *Tetrahedron*, **46**, 2511; (b) Olpe, H.-R., Demieville, H., Baltzer, V., Bencze, W.L., Koella, W.P., Wolf, P., and Haas, H.L. (1978) *European Journal of Pharmacology*, **52**, 133; (c) Ong, J., Kerr, D.I.S., Doolette, D.J., Duke, R.K., Mewett, K.N., Allen, R.D., and Johnston, G.A.R. (1993) *European Journal of Pharmacology*, **233**, 169.

15 Bryans, J.S., Davies, N., Gee, N.S., Dissanayake, V.U.K., Ratcliffe, G.S., Horwell, D.C., Kneen, C.O., Morrell, A.I., Oles, R.J., O'Toole, J.C., Perkins, G.M., Singh, L., Suman-Chauhan, N., and ONeill, J.A. (1998) *Journal of Medicinal Chemistry*, **41**, 1838.

16 Burk, M.J., De Koning, P.D., Grote, T.M., Hoekstra, M.S., Hoge, G., Jennings, R.A., Kissel, W.S., Le, T.V., Lennon, I.C., Mulhern, T.A., Ramsden, J.A., and Wade, R.A. (2003) *The Journal of Organic Chemistry*, **68**, 5731.

17 Thakur, V.V., Nikalje, M.D., and Sudalai, A. (2003) *Tetrahedron: Asymmetry*, **14**, 581.

18 Mazzini, C., Lebreton, J., Alphand, V., and Furstoss, R. (1997) *Tetrahedron Letters*, **38**, 1195.

19 Johannesen, S.A., Albu, S., Hazell, R.G., and Skrydstrup, T. (1962) *Journal of the Chemical Society, Chemical Communications*, 2004.

20 (a) Tseng, C.C., Terashima, S., and Yamada, S.-I. (1977) *Chemical & Pharmaceutical Bulletin*, **25**, 29; (b) Matthews, J.L., Braun, C., Guibourdenche, C., Overhand, M., and Seebach, D. (1997) In *Enantioselective Synthesis of β-Amino Acids* (ed. E. Juaristi), Wiley-VCH Verlag GmbH, Weinheim, pp. 105–126; (c) Buchschacher, P., Cassal, J.-M., Fürst, A., and Meier, W. (1977) *Helvetica Chimica Acta*, **60**, 2747.

21 Dixon, D.J., Ley, S.V., and Rodríguez, F. (2001) *Organic Letters*, **3**, 3753.

22 Reetz, M.T., Griebenow, N., and Goddard, R. (1995) *Journal of the Chemical Society, Chemical Communications*, 1605.

23 Adang, A.E.P., Hermkens, P.H.H., Linders, J.T.M., Ottenheijm, H.C.J., and van Staveren, C.J. (1994) *Recueil des Travaux Chimiques des Pays-Bas*, **113**, 63.

24 Umezawa, H., Aoyagi, T., Morishima, H., Matsuzaki, M., Hamada, H., and Takeuchi, T. (1970) *Journal of Antibiotics*, **23**, 259.

25 (a) Hom, R.K., Fang, L.Y., Mamo, S., tung, J.S., Guinn, A.C., Walker, D.E., Davis, D.L., Gailunas, A.F., Thorsett, E. D., Leanna, M.R., DeMattei, J.A., Li, W., Nichols, P.J., Rasmussen, M., and Morton, E. (2000) *Organic Letters*, **2**, 3627; (b) Sun, C.I., Chen, C.H., Kashiwada, Y., Wu, J.H., Wang, H.K., and Lee, K.H. (2002) *Journal of Medicinal Chemistry*, **45**, 4271; (c) Thorsett, E.D., Sinha, S., Knops, J.E., Jewett, N.E. Anderson, J.P. and Varghese, J. (2003) *Journal of Medicinal Chemistry*, **46**, 1799; (d) Liang, B., Richard, D.J., Portonovo, P.S., and Joullie, M.M. (2001) *Journal of the American Chemical Society*, **123**, 4469.

26 Castejón, P., Moyano, A., Pericàs, M.A., and Riera, A. (1996) *Tetrahedron*, **52**, 7063.

27 Kambourakis, S. and Rozzell, J.D. (2004) *Tetrahedron*, **60**, 663.

28 (a) Patel, R.N., Banerjee, A., Howell, J.M., McNamee, C.G., Brozozowski, D., Mirfakhrae, D., Nanduri, V., Thottathil, J.K., and Szarka, L.J. (1993) *Tetrahedron: Asymmetry*, **4**, 2069; (b) Gou, D.M., Liu, Y.C., and Chen, C.S. (1993) *The Journal of Organic Chemistry*, **58**, 1287.

29 (a) Lu, Y., Miet, C., Kunesch, N., and Poisson, J.E. (1993) *Tetrahedron: Asymmetry*, **4**, 893; (b) Lu, Y., Miet, C., Kunesch, N., and Poisson, J.E. (1990) *Tetrahedron: Asymmetry*, **1**, 703.

30 (a) Shioiri, T., Sasaki, S., and Hamada, Y. (2003) *ARKIVOC*, 103; (b) Sasaki, S., Hamada, Y., and Shioiri, T. (1999) *Synlett*, 453; (c) for the synthesis of diprotected (*R*)-GABOB from (*S*)-malic acid, see: Rajashekhar, B. and Kaiser, E.T. (1985) *The Journal of Organic Chemistry*, **50**, 5480.

31 Smrcina, M., Majer, P., Majerová, E., Guerassina, T.A., and Eissenstat, M.A. (1997) *Tetrahedron*, **53**, 12867.

32 Simonelli, F., Clososki, G.C., dos Santos, A.A., de Oliveira, A.R.M., Marques, F. de A., and Zarbin, P.H.G. (2001) *Tetrahedron Letters*, **42**, 7375.

33 Martin, C.J., Rawson, D.J., and Williams, J.M.J. (1998) *Tetrahedron: Asymmetry*, **9**, 3723.

34 Otaka, A., Yukimasa, A., Watanabe, J., Sasaki, Y., Oishi, S., Tamamura, H., and Fujii, N. (2003) *Journal of the Chemical Society, Chemical Communications*, 1834.

35 Gruner, S.A.W., Locardi, E., Lohof, E., and Kessler, H. (2002) *Chemical Reviews*, **102**, 491.

36 Sanjayan, G.J., Stewart, A., Hachisu, S., Gonzalez, R., Watterson, M.P., and Fleet, G.W.J. (2003) *Tetrahedron Letters*, **44**, 5847.

37 Gruner, S.A.W., Truffault, V., Voll, G., Locardi, E., Stöckle, M., and Kessler, H. (2002) *Chemistry – A European Journal*, **8**, 4365.

38 Sofia, M.J., Hunter, R., Chan, T.Y., Vaughan, A., Dulina, R., Wang, H., and Gange, D. (1998) *The Journal of Organic Chemistry*, **63**, 2802.

39 Burgess, K., Li, S., and Rebenspies, J. (1997) *Tetrahedron Letters*, **38**, 1681.

40 Royer, F., Felpin, F.-X., and Doris, E. (2001) *The Journal of Organic Chemistry*, **66**, 6487.

41 Okamoto, S., Teng, X., Fujii, S., Takayama, Y., and Sato, F. (2001) *Journal of the American Chemical Society*, **123**, 3462.

42 Teng, X., Takayama, Y., Okamoto, S., and Sato, F. (1999) *Journal of the American Chemical Society*, **121**, 11916.

43 El Marini, A., Roumestant, M.L., Viallefont, P., Razafindramboa, D., Bonato, M., and Follet, M. (1992) *Synthesis*, 1104.

44 Dexter, C.S. and Jackson, R.W.F. (1998) *Journal of the Chemical Society, Chemical Communications*, 75.

45 Mouriès, C., Deguin, B., Lamari, F., Foglietti, M.-J., Tillequin, F., and Koch, M. (2003) *Tetrahedron: Asymmetry*, **14**, 1083.

46 Cordero, F.M., Pisaneschi, F., Batista, K.M., Valenza, S., Machetti, F., and Brandi, A. (2005) *The Journal of Organic Chemistry*, **70**, 856.

47 Poitout, L., Le Merrer, Y., and Depezay, J.-C. (1995) *Tetrahedron Letters*, **36**, 6887.

48 (a) Chakraborty, T.K., Jayaprakash, S., Diwan, P.V., Nagaraj, R., Jampani, S.R.B., and Kunwar, A.C. (1998) *Journal of the American Chemical Society*, **120**, 12962; (b) Chakraborty, T.K. and Ghosh, S. (2001) *Journal of the Indian Institute of Science*, **81**, 117.

49 (a) Long, D.D., Smith, M.D., Marquess, D.G., Claridge, T.D.W., and Fleet, G.W.J. (1998) *Tetrahedron Letters*, **39**, 9293; (b) Smith, M.D., Long, D.D., Claridge, T.D.W., Fleet, G.W.J., and Marquess, D.G. (1998) *Journal of the Chemical Society, Chemical Communications*, 2039; (c) Smith, M.D., Claridge, T.D.W., Fleet, G.W.J., Tranter, G.E., and Sansom, M.S.P. (1998) *Journal of the Chemical Society, Chemical Communications*, 2041.

50 Watterson, M.P., Edwards, A.A., Leach, J.A., Smith, M.D., Ichihara, O., and Fleet, G.W.J. (2003) *Tetrahedron Letters*, **44**, 5853.

51 Van Well, R.M., Meijer, M.E.A., Overkleeft, H.S., van Boom, J.H., van der Marel, G.A., and Overhand, M. (2003) *Tetrahedron*, **59**, 2423.

52 Peri, F., Cipolla, L., La Ferla, B., and Nicotra, F. (2000) *Journal of the Chemical Society, Chemical Communications*, 2303.

53 Graf von Roedern, E., Lohof, E., Hessler, G., Hoffmann, M., and Kessler, H. (1996) *Journal of the American Chemical Society*, **108**, 10156.

54 Suhara, Y., Yamaguchi, Y., Collins, B., Schnaar, R.L., Yanagishita, M., Hildreth, J.E.K., Shimada, I., and Ichikawa, Y. (2002) *Bioorganic and Medicinal Chemistry*, **10**, 1999.

55 Graf von Roedern, E. and Kessler, H. (1994) *Angewandte Chemie (International Edition in English)*, **33**, 687.

56 Durrat, F., Xie, J., and Valéry, J.-M. (2004) *Tetrahedron Letters*, **45**, 1477.

57 Baldauf, C., Günther, R., and Hofmann, H.-J. (2004) *The Journal of Organic Chemistry*, **69**, 6214.

58 Gardner, R.R., Liang, G.-B., and Gellman, S.H. (1995) *Journal of the American Chemical Society*, **117**, 3280.

59 Wipf, P., Henninger, T.C., and Geib, S.J. (1998) *The Journal of Organic Chemistry*, **63**, 6088.

60 Belvisi, L., Colombo, L., Manzoni, L., Potenza, D., and Scolastico, C. (2004) *Synlett*, 1449.

61 Wang, W., Yang, J., Ying, J., Xiong, C., Zhang, J., Cai, C., and Hruby, V.J. (2002) *The Journal of Organic Chemistry*, **67**, 6353.

62 Bravin, F.M., Busnelli, G., Colombo, M., Gatti, F., Manzoni, L., and Scolastico, C. (2004) *Synthesis*, 353.

63 Manzoni, L., Belvisi, L., Di Carlo, E., Forni, A., Invernizzi, D., and Scolastico, C. (2006) *Synthesis*, 1133.

64 Hanessian, S., McNaughton-Smith, G., Lombart, H.-G., and Lubell, W.D. (1997) *Tetrahedron*, **53**, 12789.

65 Estiarte, M.A., Rubiralta, M., Diez, A., Thormann, M., and Giralt, E. (2000) *The Journal of Organic Chemistry*, **65**, 6992.

66 (a) Kim, K. and Germanas, J.P. (1997) *The Journal of Organic Chemistry*, **62**, 2847; (b) Kim, K. and Germanas, J.P. (1997) *The Journal of Organic Chemistry*, **62**, 2853.

67 Belvisi, L., Gennari, C., Mielgo, A., Potenza, D., and Scolastico, C. (1999) *European Journal of Organic Chemistry*, 389.

68 Salimbeni, A., Paleari, F., Canevotti, R., Criscuoli, M., Lippi, A., Angiolini, M., Belvisi, L., Scolastico, C., and Colombo, L.

(1997) *Bioorganic & Medicinal Chemistry Letters*, **7**, 2205.

69 Belvisi, L., Bernardi, A., Checchia, A., Manzoni, L., Potenza, D., Scolastico, C., Castorina, M., Cupelli, A., Giannini, G., Carminati, P., and Pisano F C. (2001) *Organic Letters*, **3**, 1001.

70 Alonso, E., López-Ortiz, F., del Pozo, C., Peralta, E., Macías, A., and González, J. (2001) *The Journal of Organic Chemistry*, **66**, 6333.

71 Khalil, E.M., Ojala, W.H., Pradhan, A., Nair, V.D., Gleason, W.B., Mishra, R.K., and Johnson, R.L. (1999) *Journal of Medicinal Chemistry*, **42**, 628.

72 Genin, M.J. and Johnson, R.L. (1992) *Journal of the American Chemical Society*, **114**, 8778.

73 Genin, M.J., Ojala, W.H., Gleason, W.B., and Johnson, R.L. (1993) *The Journal of Organic Chemistry*, **58**, 2334.

74 Ecija, M., Diez, A., Rubiralta, M., and Casamitjana, N. (2003) *The Journal of Organic Chemistry*, **68**, 9541.

75 Fink, B.E., Kym, P.R., and Katzenellenbogen, J.A. (1998) *Journal of the American Chemical Society*, **120**, 4334.

76 Trabocchi, A., Menchi, G., Guarna, F., Machetti, F., Scarpi, D., and Guarna, A. (2006) *Synlett*, 331.

77 Trabocchi, A., Cini, N., Menchi, G., and Guarna, A. (2003) *Tetrahedron Letters*, **44**, 3489.

78 Scarpi, D., Occhiato, E.G., Trabocchi, A., Leatherbarrow, R.J., Brauer, A.B.E., Nievo, M., and Guarna, A. (2001) *Bioorganic and Medicinal Chemistry*, **9**, 1625.

79 Trabocchi, A., Occhiato, E.G., Potenza, D., and Guarna, A. (2002) *The Journal of Organic Chemistry*, **67**, 7483.

80 Nielsen, P.E., Egholm, M., Berg, R.H., and Buchardt, O. (1991) *Science*, **254**, 1497.

81 (a) Hyrup, B. and Nielsen, P.E. (1996) *Bioorganic and Medicinal Chemistry*, **4**, 5; (b) Uhlmann, E., Peyman, A., Breipohl, G., and Will, D.W. (1998) *Angewandte Chemie (International Edition in English)*, **37**, 2796.

82 Püschl, A., Tedeschi, T., and Nielsen, P.E. (2000) *Organic Letters*, **2**, 4161.

83 (a) Karig, G., Fuchs, A., Büsing, A., Brandstetter, T., Scherer, S., Bats, J.W., Eschenmoser, A., and Quinkert, G. (2000) *Helvetica Chimica Acta*, **83**, 1049; (b) Schwalbe, H., Wermuth, J., Richter, C., Szalma, S., Eschenmoser, A., and Quinkert, G. (2000) *Helvetica Chimica Acta*, **83**, 1079.

84 Casimir, J.R., Didierjean, C., Aubry, A., Rodriguez, M., Briand, J.-P., and Guichard, G. (2000) *Organic Letters*, **2**, 895.

85 Grotenbreg, G.M., Christina, A.E., Buizert, A.E.M., van der Marel, G.A., Overkleeft, H.S., and Overhand, M. (2004) *The Journal of Organic Chemistry*, **69**, 8331.

86 Brittain, D.E.A., Watterson, M.P., Claridge, T.D.W., Smith, M.D., and Fleet, G.W.J. (2000) *Journal of the Chemical Society, Perkin Transactions 1*, 3655.

14
Synthesis of γ-Aminobutyric Acid Analogs

Jane R. Hanrahan and Graham A.R. Johnston

14.1
Introduction

Given the likely varied roles of γ-aminobutyric acid (GABA; 4-aminobutanoic acid) (**1**) as a ubiquitous signaling molecule in most living organisms, new GABA analogs are continuously being sought in order to probe metabolic roles and to develop new biologically active agents. The considerable interest in synthesizing analogs of GABA derives from the vital role of this nonprotein amino acid in brain function. GABA is essential to maintaining a functioning brain by acting as the major inhibitory neurotransmitter by balancing the effects of the major excitatory neurotransmitter L-glutamate [1]. Insufficiency of synaptic inhibition may lead to anxiety, epilepsy, insomnia, or memory problems. Overactive synaptic inhibition may be associated with anesthesia, coma, or sedation. Recent reviews on aspects of the central nervous system (CNS) pharmacology of GABA include the ionotropic GABA receptors ($GABA_A$ and $GABA_C$) as drug targets [2–4], especially receptor subtypes [5], the G-protein-coupled $GABA_B$ receptors [6], GABA transporters [7], and patents on novel GABA analogs [8].

Several GABA analogs are used clinically (Figure 14.1), including the antiepileptic drug vigabatrin (Sabril) (**2**), which was designed as an irreversible inhibitor of the enzyme that degrades GABA, resulting in increases in the level of GABA in the brain [9]. GABA analogs substituted in the 3-position have been of particular interest [10]. The selective $GABA_B$ agonist, baclofen (Lioresal) (**3**) has been used for many years to treat spinal spasticity. Gabapentin (Neurotenin) (**4**) and pregabalin (Lyrica) (**5**) are the newest GABA analogs on the CNS drug market for the treatment of epilepsy and neuropathic pain, with sales of the recently marketed Lyrica reaching US$1.8 billion in 2007. While they are 3-substituted GABA analogs, they do not appear act directly on any known GABA neurotransmitter system in the brain. They act as selective inhibitors of voltage-gated calcium channels containing the $\alpha_2\delta_1$ subunit [11]. They are not the first therapeutically useful drugs to be designed as GABA analogs that have turned out to act predominately on systems apparently unrelated to GABA – the prototypical antipsychotic drug haloperidol acting

Amino Acids, Peptides and Proteins in Organic Chemistry. Vol.1 – Origins and Synthesis of Amino Acids.
Edited by Andrew B. Hughes

ISBN: 978-3-527-32096-7

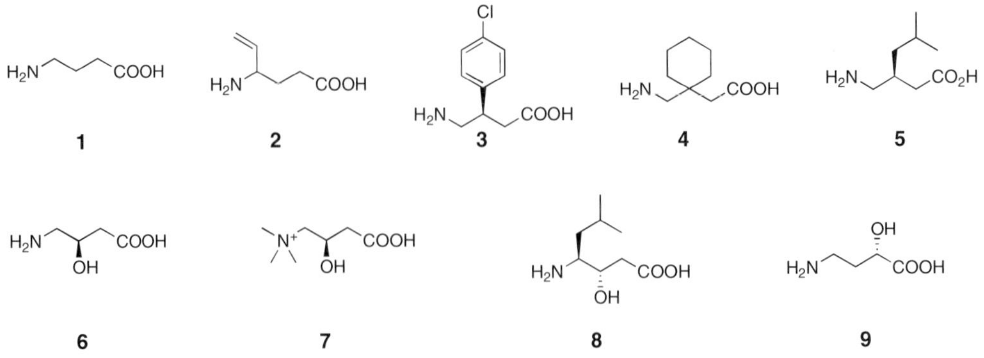

Figure 14.1

at dopamine receptors was designed as a GABA analog able to penetrate the blood–brain barrier [12].

GABA occurs extensively in plants, fungi, bacteria, insects, invertebrates, and vertebrates. It has been described as a conserved and ubiquitous signaling system [13]. Recently, it has been suggested that GABA mediates communication between plants and other organisms [14]. GABA is thought to facilitate communication via diverse mechanisms, including activation of neuronal receptors and induction of enzymes and transporters.

Several biologically active GABA analogs occur naturally (Figure 14.1). These include (*S*)-4-amino-3-hydroxybutanoic acid (GABOB) (**6**), also found in the cyclic depsipeptide hapalosin, and L-carnitine (**7**) (used in the treatment of myopathies) found in mammalian tissues [15]. Statine [(3*S*,4*S*)-4-amino-3-hydroxy-6-methylheptanoic acid] (**8**) is regarded as the key pharmacophore in the rennin inhibitor pepstatin isolated from bacteria [16]. Statine (**8**) and its analogs are useful as building blocks for peptidomimetics [17]. Homoisoserine (2-hydroxy-4-aminobutyric acid) (**9**) acts as an inhibitor for GABA uptake, exhibits antitumor activity, is found in many antibiotics, and is also used to construct peptidomimetics [18].

In mammals, GABA is found in many organs outside of the CNS where it serves various functions [19]. GABA is involved in cell proliferation and migration, and may play a role in cancer [20]. Recent evidence implicates GABA receptors in mucus overproduction in asthma acting on airway epithelial cells [21]. GABA regulates insulin secretion from pancreatic β cells in concert with changes in glucose concentration [22] and may be involved with type 1 diabetes inhibiting the development of proinflammatory T cell responses acting via GABA receptors [23]. Functional $GABA_A$ receptors have also been described in T cells [24] and macrophages [25]. Thus, asthma, cancer, diabetes, and the immune system may also be targets for GABA analogs.

The development of synthetic methodologies for the synthesis of substituted alkyl and conformationally restricted GABA analogs has been the focus of much endeavor [17, 26, 27]. The availability of a range of GABA analogs has played a major role in elucidating the physiological roles of GABA receptors. More recently, substituted GABA analogs have become of interest for their presence in novel peptides and potential application in constraining peptide sequences in to β- and γ-turns. This chapter focuses on the synthetic procedures for the preparation of linear and alicyclic analogs of GABA.

14.2 α-Substituted γ-Amino Acids

(*R*)- and (*S*)-α-Methyl-GABAs (**13**) were originally prepared by reaction of 2-methyl-butyrolactone with potassium phthalimide and resolution by fractional crystallization of the phthalimido derivatives as quinine salts. Hydrolysis and cleavage of the phthalimido group afforded the enantiomerically pure amino acids in low overall yields [28]. Duke *et al.* described the synthesis of (**13**) starting from tiglic acid (**10**), which was esterified, allylically brominated, and coupled with phthalimide to introduce the nitrogen atom. Hydrogenation and ester hydrolysis afforded (±)-(**11**) which was resolved by derivatization as the (*R*)-pantolactone ester (**12**) and chromatographic separation and acid hydrolysis afforded enantiomerically pure (*R*)-(**13**) and (*S*)-(**13**) in good overall yield (Scheme 14.1) [29].

Esterification of racemic α-aryl-β-cyanopropionic acid chlorides (**15a–c**) (Scheme 14.2), from methyl 2-arylacetates (**14a–c**), with (*R*)-*N*-phenylpantolactam (*R*)-(**16**) as the chiral auxiliary results in the predominant formation of (α*R*,3′*R*)-(**17a–c**) in nearly quantitative yields with diastereomeric ratios of up to 93 : 7 achieved through a dynamic kinetic resolution process. The use of (*S*)-(**16**) similarly affords (α*S*,3′*S*)-(**17a–c**). Chromatographic purification of the diastereoenriched cyano esters, followed by hydrolysis of the resulting diastereopure cyano esters under nonracemizing conditions, gave enantiopure α-aryl-β-cyanopropionic acids (**18a–c**), which were readily converted in high yields into enantiopure *N*-Boc-α-aryl-GABAs (*R*)-(**19a–c**) or (*S*)-(**19a–c**) [30].

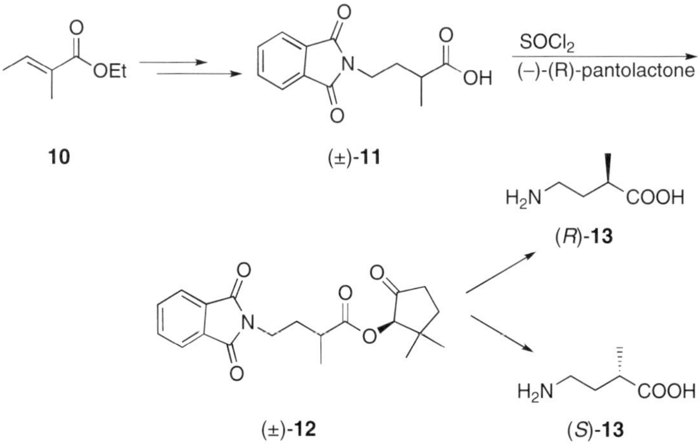

OEt

10

OH

(±)-**11**

$SOCl_2$

(–)-(R)-pantolactone

(±)-**12**

H_2N COOH

(*R*)-**13**

H_2N COOH

(*S*)-**13**

Scheme 14.1

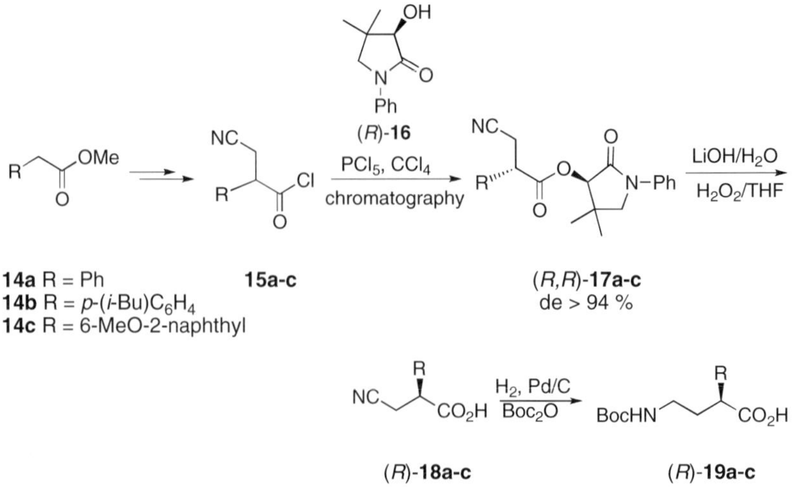

(R)-16
PCl_5, CCl_4
chromatography
LiOH/H_2O
H_2O_2/THF

14a R = Ph
14b R = p-(i-Bu)C_6H_4
14c R = 6-MeO-2-naphthyl
15a-c
(R,R)-17a-c
de > 94 %

H_2, Pd/C
Boc_2O
(R)-18a-c
(R)-19a-c

Scheme 14.2

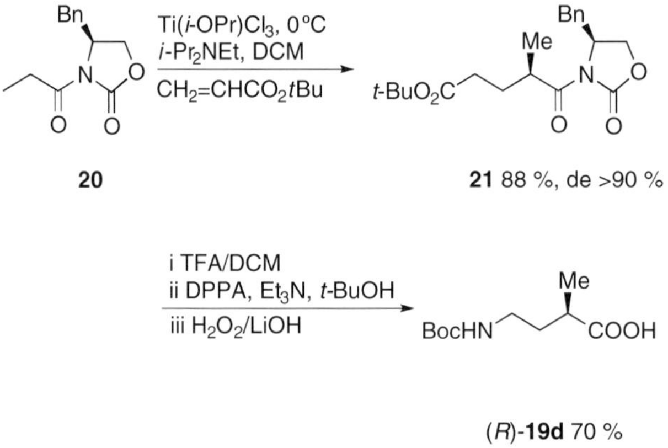

Ti(i-OPr)Cl_3, 0°C
i-Pr_2NEt, DCM
CH_2=$CHCO_2$tBu

20
21 88 %, de >90 %

i TFA/DCM
ii DPPA, Et_3N, t-BuOH
iii H_2O_2/LiOH

(R)-19d 70 %

Scheme 14.3

Evans *et al.* have reported a diastereoselective Michael reaction of *tert*-butyl acrylate with the titanium enolate derived from *N*-propionyloxazolidinone (**20**) (Scheme 14.3) that affords enantiomerically pure ester (**21**). Hydrolysis of the ester, subsequent Curtius rearrangement, and removal of the chiral auxiliary afforded the desired amino acid (*R*)-(**19d**) [31].

Similarly, Seebach *et al.* have used the titanium enolates of valine-derived auxiliaries (**22a–c**) (Scheme 14.4) to give a range of 4-nitro derivatives (**23a–c**) in high diastereoselectivities (80 to >95%) and good yields (50–75% of purified samples of diastereoselctivity > 98%) Reduction of the nitro group, recovery of the insoluble auxiliary and ring opening of the *N*-Boc-lactams affords the (*R*)-*N*-Boc-amino acids (**19d–f**) [32].

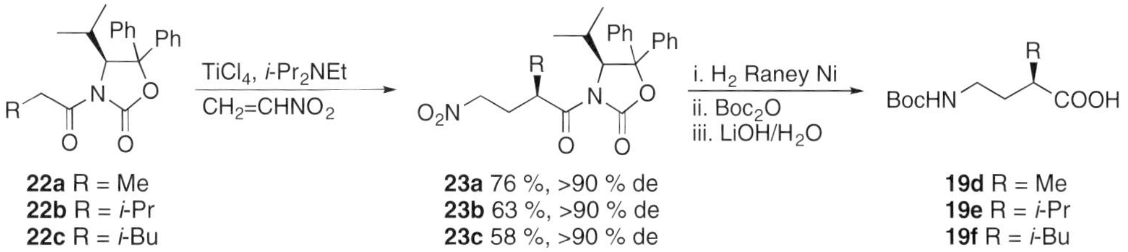

22a R = Me
22b R = i-Pr
22c R = i-Bu

$TiCl_4$, i-Pr_2NEt
$CH_2=CHNO_2$

23a 76 %, >90 % de
23b 63 %, >90 % de
23c 58 %, >90 % de

i. H_2 Raney Ni
ii. Boc_2O
iii. $LiOH/H_2O$

19d R = Me
19e R = i-Pr
19f R = i-Bu

Scheme 14.4

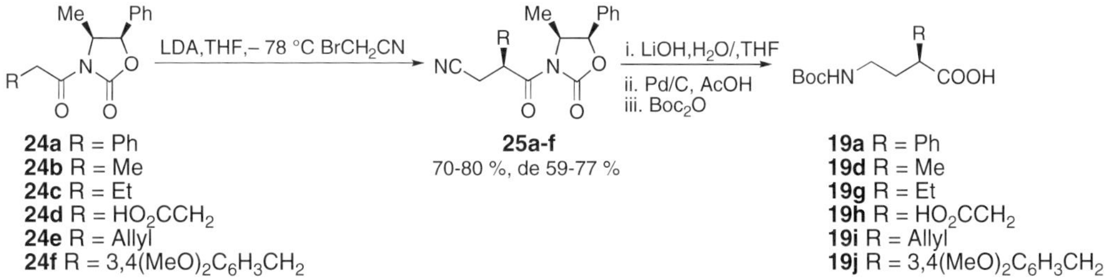

24a R = Ph
24b R = Me
24c R = Et
24d R = HO_2CCH_2
24e R = Allyl
24f R = $3,4(MeO)_2C_6H_3CH_2$

LDA,THF,– 78 °C $BrCH_2CN$

25a-f
70-80 %, de 59-77 %

i. $LiOH,H_2O/,THF$
ii. Pd/C, AcOH
iii. Boc_2O

19a R = Ph
19d R = Me
19g R = Et
19h R = HO_2CCH_2
19i R = Allyl
19j R = $3,4(MeO)_2C_6H_3CH_2$

Scheme 14.5

Cyanomethylation of sodium or lithium enolates from 3-acyloxazolidinones (**24a–f**) (Scheme 14.5) with bromoacetonitrile affords the nitriles (**25a–f**) providing a general approach to the enantioselective synthesis of 2-substituted GABA derivatives (**19a**, **d** and **g–j**). Relative poor diastereomeric excesses (60–77%) compared to similar methodology are attributed to the relatively small size of bromoacetonitrile, combined with its high reactivity as an electrophile [33].

Enantioselective palladium-catalyzed allylic substitution of readily available allylic acetates (**26**) (Scheme 14.6) using methyl cyanoacetate provides substitution

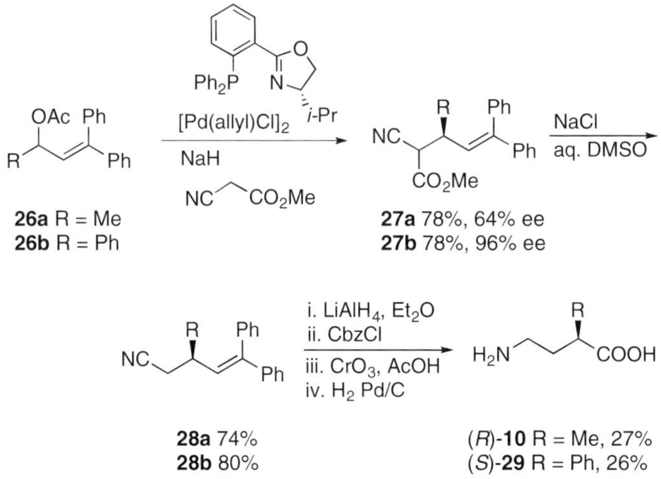

26a R = Me
26b R = Ph

$[Pd(allyl)Cl]_2$
NaH

27a 78%, 64% ee
27b 78%, 96% ee

NaCl
aq. DMSO

i. $LiAlH_4$, Et_2O
ii. CbzCl
iii. CrO_3, AcOH
iv. H_2 Pd/C

28a 74%
28b 80%

(R)-**10** R = Me, 27%
(S)-**29** R = Ph, 26%

Scheme 14.6

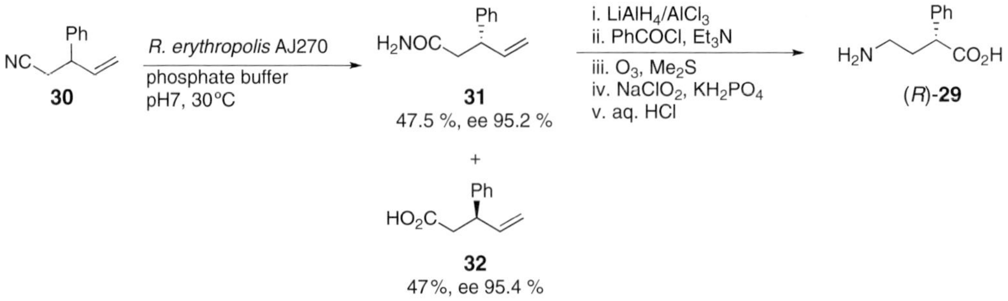

NC–CH(Ph)–CH=CH₂ **30** → *R. erythropolis* AJ270, phosphate buffer, pH7, 30°C → H_2NOC (Ph) **31** 47.5 %, ee 95.2 % + HO_2C (Ph) **32** 47%, ee 95.4 %

31 → i. $LiAlH_4/AlCl_3$; ii. PhCOCl, Et_3N; iii. O_3, Me_2S; iv. $NaClO_2$, KH_2PO_4; v. aq. HCl → H_2N–CH₂CH₂–CH(Ph)–CO_2H (*R*)-**29**

Scheme 14.7

products (**27**) that can be rapidly converted in to α-substituted γ-amino acids. Krapcho decarboxylation, reduction of the nitrile (**28**) to the amine which was protected as a carbamate, allows oxidative cleavage of the alkene affording (*R*)-(**10**) or (*S*)-(**29**) in good yields and moderate to high enantiomeric excess [34].

(±)-α-Substituted GABA analogs have been prepared by a range of synthetic methodologies, including the palladium-catalyzed rearrangement of α-cyanocyclopropanone hydrates, which occurs with internal proton transfer, providing access to selectively deuterated α-amino acids [35] and zirconium-mediated intramolecular ester transfer reaction of *N*-alkenyl-*N*-tosyl carbamates derivatives [36].

Enantioselective biotransformation of racemic 3-arylpent-4-enenitrile (**30**) (Scheme 14.7) with *Rhodococcus erythropolis* AJ270 provides a straightforward and scaleable route to highly enantiopure (*S*)-3-phenylpent-4-enoic acid amide (**31**) and (*R*)-3-phenylpent-4-enoic acid (**32**). The amide (**31**) can be converted to (*R*)-(**29**) via reduction of the carboxamido group and oxidation of the vinyl group [37].

Two groups have reported organocatalyzed asymmetric conjugate addition reactions of aldehydes (**33**) to nitroethylene (**35**). The reactions, catalyzed by either proline containing peptides [38], or (*S*)-diphenylprolinol silyl ether (**34**) (Scheme 14.8) [39], afford (**36**) in high yield (95%) and enantioselectivity (>95% e.e.), after reduction with $NaBH_4$. The use of *m*-nitrobenzoic acid with (*S*)-(**34**) results in greater catalyst efficiency. Jones oxidation of the aldehyde, reduction of the nitro group, and protection of the resulting amine affords the protected amino acids (**19d**, **e**, **g** and

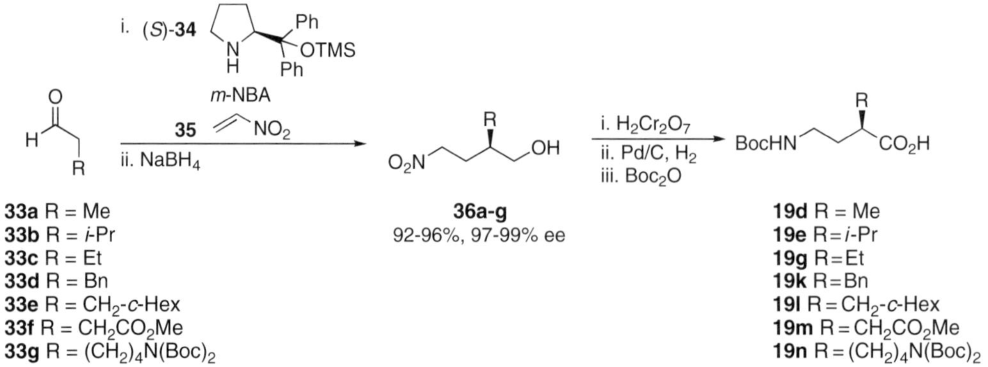

33 (H–CO–CH₂R) → i. (*S*)-**34** (diphenylprolinol OTMS ether), *m*-NBA, **35** (CH₂=CH–NO_2); ii. $NaBH_4$ → O_2N–CH₂–CH(R)–CH₂OH **36a-g** 92-96%, 97-99% ee → i. $H_2Cr_2O_7$; ii. Pd/C, H_2; iii. Boc_2O → BocHN–CH₂–CH(R)–CO_2H

33a R = Me
33b R = *i*-Pr
33c R = Et
33d R = Bn
33e R = CH_2-*c*-Hex
33f R = CH_2CO_2Me
33g R = $(CH_2)_4N(Boc)_2$

19d R = Me
19e R = *i*-Pr
19g R = Et
19k R = Bn
19l R = CH_2-*c*-Hex
19m R = CH_2CO_2Me
19n R = $(CH_2)_4N(Boc)_2$

Scheme 14.8

k–n) in high overall yield [39]. The use of the peptide catalyst produces compounds with the opposite stereochemistry to those produced by (*S*)-(**34**) [38].

The addition of bromoacetonitrile to the dianions of carboxylic acids has been reported as a two-step process to afford α-substituted amino acids. Although high yielding, the use of chiral bases resulted in only minimal chiral induction [40].

14.3
β-Substituted γ-Amino Acids

Many of the methods employed for the synthesis of α-substituted γ-amino acids are also applicable to the synthesis of β-substituted γ-amino acids, including reaction of titanium enolates of valine-derived auxiliaries [32] and the rearrangement of α-cyanocyclopropanone hydrates [35].

(*R*)- and (*S*)-4-Amino-3-methylbutanoic acids (**41**) (Scheme 14.9) were synthesized in a stereodivergent route via an initial enantioselective hydrolysis of dimethyl 3-methylglutarate (**37**) to give methyl (*R*)-3-methylglutarate (**38**) with pig liver esterase (PLE). Conversion of the ester group to an amide (**39**) by ammonolysis and subsequent Hoffmann rearrangement gives (*R*)-4-amino-3-methylbutanoic acid (*R*)-(**41**). Conversion of the carboxylic acid to the carbamate (**40**) via a modified Curtius rearrangement affords (*S*)-4-amino-3-methylbutanoic acid (*S*)-(**41**) [41].

A general route to a range of 3-vinyl-substituted GABA derivatives (**43** and **45**) from substituted 4-phthalimido-2-buten-l-ols (**42**) has been described (Schemes 14.10

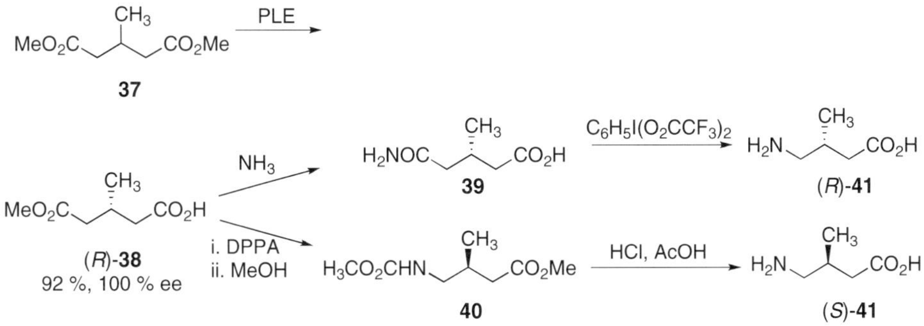

Scheme 14.9

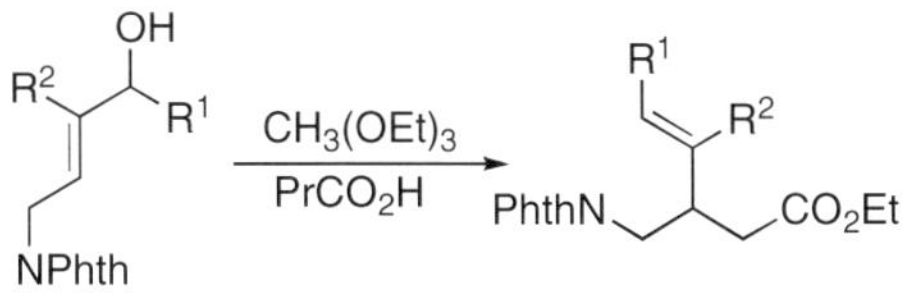

42a $R^1 = R^2 = H$
42b $R^1 = Me$, $R^2 = H$
42c $R^1 = Et$, $R^2 = H$

Scheme 14.10

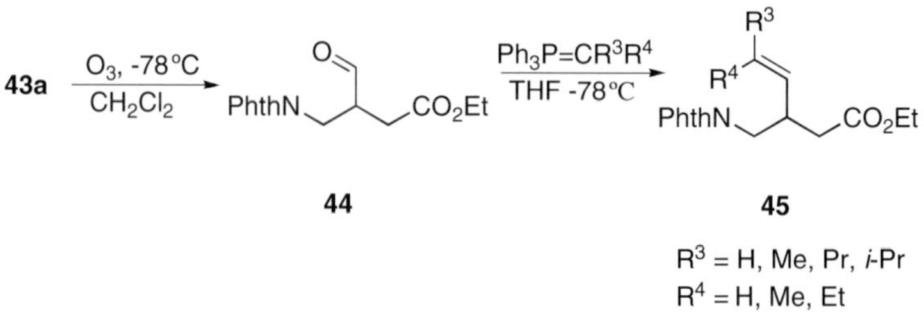
43a
O3, -78°C
CH2Cl2
PhthN
CO2Et
44
Ph3P=CR3R4
THF -78°C
R3
R4
PhthN
CO2Et
45
R3 = H, Me, Pr, i-Pr
R4 = H, Me, Et

Scheme 14.11

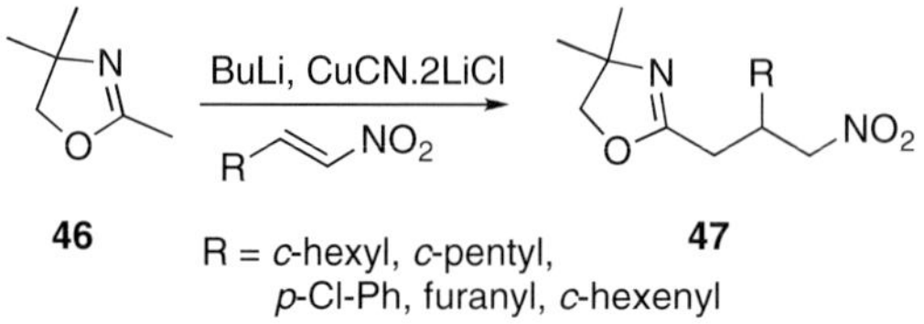
BuLi, CuCN.2LiCl
NO2
46
47
R = c-hexyl, c-pentyl,
p-Cl-Ph, furanyl, c-hexenyl

Scheme 14.12

and 14.11). Acid-catalyzed Claisen rearrangement in ethyl orthoacetate formed the (*E*)-3-substituted phthalimido esters (**43**), which can be easily deprotected to yield the amino acids (Scheme 14.10). Alternatively, the synthesis of ω-disubstituted or *cis*-monosubstituted derivatives (**45**) can be obtained by a Wittig reaction on the aldehyde (**44**) obtained by ozonolysis of ethyl 4-phthalimidomethylpent-4-enoate (**43a**) (Scheme 14.11) [42].

The addition of cyanocuprates of oxazolines (**46**) to conjugated nitroalkenes has been reported as an efficient method for the synthesis of a variety of β-substituted γ-nitrooxazolines (**47**) (Scheme 14.12), which are precursors of a variety of β-substituted γ-amino acids [43].

Enantiomerically pure 3-methyl- (**41**) and 3-mercaptomethyl-γ-amino acids (**50**) have been prepared by a novel route for the synthesis of optically pure β-substituted GABA derivatives (Scheme 14.13). (*S*)-Phenylethylamine was used as a chiral auxiliary of the radical precursor (*S*)-(**48**) for a 5-*exo*-trig radical cyclization as the strategy for the construction of the 4-substituted pyrrolidinones (**49**). Chromatographic separation of

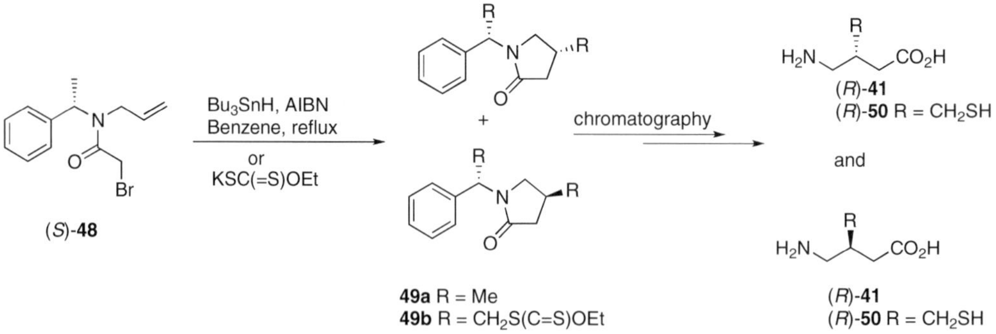
Bu3SnH, AIBN
Benzene, reflux
or
KSC(=S)OEt
Br
(S)-48
chromatography
49a R = Me
49b R = CH2S(C=S)OEt
H2N
CO2H
(R)-41
(R)-50 R = CH2SH
and
(R)-41
(R)-50 R = CH2SH

Scheme 14.13

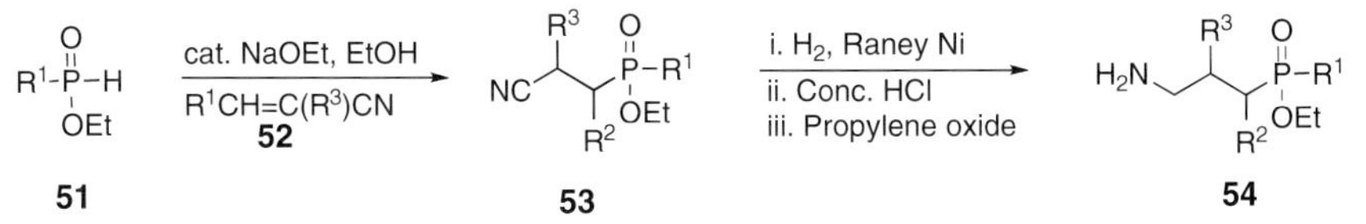

R^1-P(O)(H)-OEt **51** → cat. NaOEt, EtOH; $R^1CH{=}C(R^3)CN$ **52** → **53** → i. H_2, Raney Ni; ii. Conc. HCl; iii. Propylene oxide → **54**

R^1 = $CH(OEt)_2$, $CCH_3(OEt)_2$ Me, CHF_2, *n*-Bu, $CH_2C_6H_5$
R^2 = H, Me, Et, C_6H_5, *p*-ClC_6H_4
R^3 = H, CH_3, *p*-ClC_6H_4

R^1 = H, Me, CHF_2, *n*-Bu, $CH_2C_6H_5$
R^2 = H, Me, Et, C_6H_5, *p*-ClC_6H_4
R^3 = H, CH_3, *p*-ClC_6H_4

Scheme 14.14

the diastereoisomers, removal of the chiral auxiliary followed by an aqueous hydrolysis provided the corresponding optically pure β-substituted GABA derivatives (**41** and **50**) [44].

Pregabalin (**5**) has also been prepared via a stereoselective Bu_3SnH-mediated radical cyclization. In the presence of the Lewis acid $BF_3{\cdot}OEt_2$ a diastereoisomeric ratio of 88 : 12 was achieved [45].

A wide range of α-, β-, and α,β-substituted phosphinic and alkylphosphinic GABA analogs have been reported [46, 47]. Conjugate addition of the phosphinate (**51**) to acrylonitrile (**52**) gave the cyano esters (**53**) (Scheme 14.14), which upon hydrogenation (or reductive amination) yielded amino ester. After acidic hydrolysis, phosphinic or alkyl phosphinic acids (**54**) were obtained [46, 47].

14.3.1 Pregabalin

The multistep discovery synthesis of pregabalin (**5**) utilized an asymmetric Evans alkylation with alkyl bromoacetate of the chiral auxiliary prepared from (+)-norephedrine. Cleavage of the chiral auxiliary to give the acid, reduction to the alcohol, conversion to the amine via the tosylate, and deprotection afforded pregabalin in high enantiomeric purity, but relatively low overall yield. This synthesis was not suitable for the large-scale preparation of the anticonvulsant and the significant efforts to develop a low-cost manufacturing process have been fully reviewed [48]; however, pregabalin (**5**) has remained an interesting synthetic target.

A Baylis–Hillman reaction between isobutyrylaldehyde (**55**) and acrylonitrile (**56**) yields 3-hydroxy-4-methyl-2-methylenepentanenitrile (**57**) (Scheme 14.15) [49]. Conversion to the *N*-ethyl carbonate and palladium-catalyzed carbonylation in ethanol afforded the ester (**58**), which was hydrolyzed. Hydrogenation of the ethyl ester proceeded with poor enantioselectivity, however hydrogenation of the *tert*-butyl ammonium salt and hydrogenation in the presence of a chiral rhodium catalyst resulted in high enantioselectivity. Hydrogenation of the nitrile with a heterogeneous nickel catalyst afforded (*S*)-(**5**) [50].

Conjugate addition of cyanide, generated *in situ* from trimethylsilyl cyanide to unsaturated imides using commercially available (salen)Al(III) catalysts has been reported to provide a facile synthesis of β-substituted γ-amino acids. Hydrolysis of the resulting imide and nitrile reduction affords the amino acids such as pregabalin

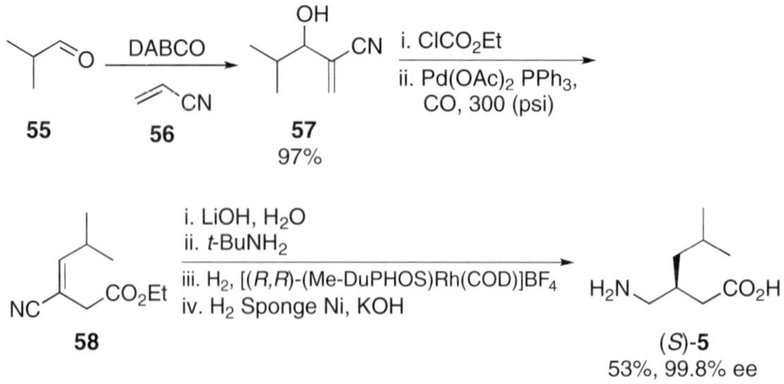

Scheme 14.15

(5) [51]. Alternatively, (*R*)-pregabalin (*R*)-(**5**) and baclofen (**3**) have been prepared via conjugate addition of cyanide catalyzed by Sm(III) isopropoxide to an appropriately substituted Evan's chiral auxiliary [52].

A catalytic enantioselective Michael reaction of nitroalkanes with α,β-unsaturated aldehydes using (*R*)-(**34**), as an asymmetric organocatalyst has been used to synthesize both baclofen (**3**) and pregabalin (**5**) (Scheme 14.16). Conjugate addition of nitromethane (**60**) to 5-(*p*-chlorophenyl)-enal (**59a**) or 5-isobutyl-hex-2-enal (**59b**) afforded enantiomerically pure 3-substituted nitrobutanals (**61**), which were converted to (**3**) or (**5**), respectively, in two steps via the nitro acid (**62**) [53].

3-Isobutylglutaric anhydride (**63**) (Scheme 14.17) prepared from cyanoacetamide and isovaleraldehyde undergoes desymmetrization via a quinine-mediated ring opening with cinnamyl alcohol. The crude monoacid (**64**) prepared in 72% e.e. underwent purification by crystallization with (*S*)-phenylethylamine, affording the acid (**64**) in 73% yield and 97% e.e. A Curtius rearrangement and subsequent deprotection provided pregabalin (*S*)-(**5**) in 45% overall yield [54].

A general, convenient and scalable synthetic method for enantiomerically pure β-substituted γ-butyrolactones, of either configuration, has been reported. The desired chiral bicyclic lactone (**67**) (Scheme 14.18) is prepared in one pot from chiral

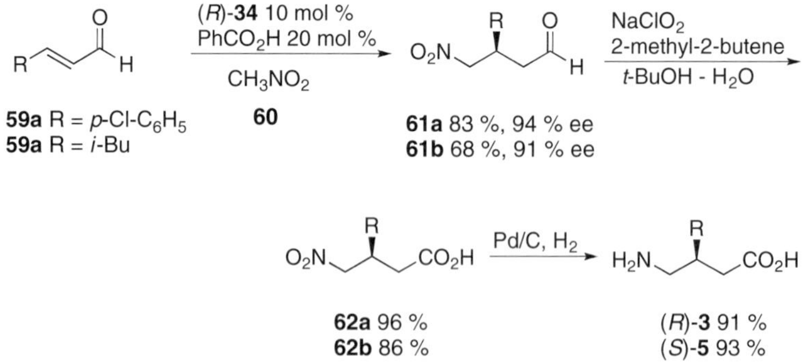

Scheme 14.16

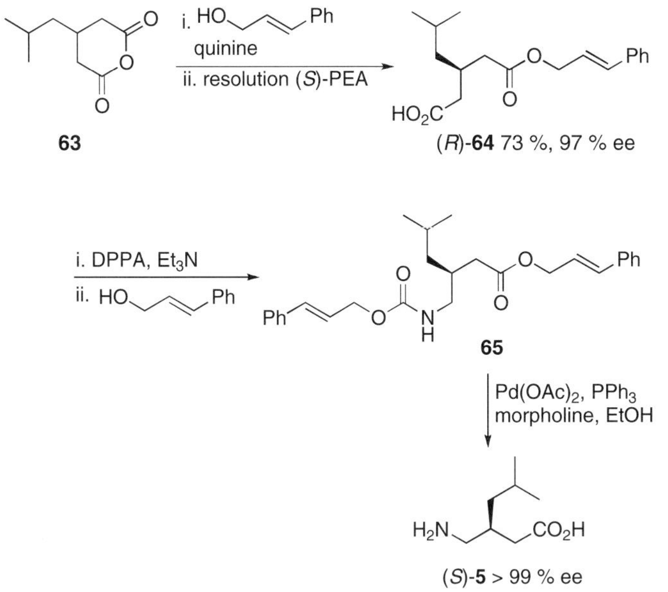

Scheme 14.17

epichlorohydrin (**66**) by intramolecular double displacement of the alkyl malonate anion followed by lactonization [55]. Ring opening of the cyclopropane ring with isopropyl Grignard reagent and decarbethoxylation provides access to the isopropyl substituted lactone (**68**), which can be converted to pregabalin (**2**) via the azide (**69**) in

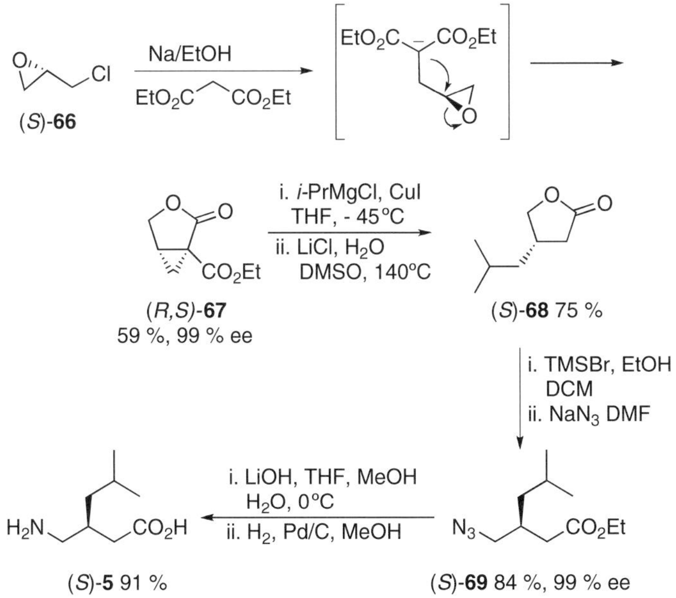

Scheme 14.18

four steps. The use of other Grignard reagents affords access to a range of β-substituted γ-amino acids [56].

(±)-*N*-Boc-pregabalin has been synthesized in low overall yields via a hetero-Diels–Alder addition of 3-nitrosoacrylate to ethyl vinyl ether [57]. D-Mannitol has also been used as the primary source of chirality in a high-yielding (28% overall yield), multistep synthesis of (*S*)-pregabalin (**5**) [58]. A series of racemic heteroaromatic analogs of pregabalin have been synthesized via reaction of the anion of heteroaryl β-esters with *tert*-butyl bromoacetate [59].

14.3.2 Gabapentin

The original syntheses of gabapentin (**4**) starting from cyclohexanone, based on formation of the Guareshi salt and its conversion to a spiro anhydride, or a Knoevenagel condensation with ethyl cyanoacetate, have been reviewed [10]. Since the discovery of the anticonvulsant activity of gabapentin, a variety of analogs have been prepared, including alkyl substituted cyclohexanes [60–62], carboxylate bioisosteres including tetrazole and phosphonic acid [63], and conformationally restricted spiro and fused bicyclic compounds [64].

Cu(I)-mediated radical cyclization of halo-amides (**70**) has been utilized to afford functionalized pyrrolidinones via a 5-*exo*-trig radical cyclization pathway (Scheme 14.19). The chloro-substituted cyclohexyl pyrrolidinones (**71**) can be dechlorinated to the pyrrolidinone (**72**) and converted to gabapentin (**4**). The chlorine atoms potentially allow for the selective introduction of a variety of substituents at the 2- and 4-positions of gabapentin, providing a useful route to synthesizing substituted analogs [65]. The use of the benzoylamino group as the radical cyclization auxiliary facilitates concomitant dehalogenation and deprotection of the chlorinated *N*-substituted pyrrolidin-2-one (**72b**) using Raney nickel [66].

14.3.3 Baclofen and Analogs

Enantiomerically pure β-phenyl-GABA (**76**) was originally prepared via separation of (*R*)-(−)-pantolactone esters. Ethyl (*E*)-3-phenylbut-2-enoic acid (**73**) (Scheme 14.20) was synthesized by a modified Wittig–Horner reaction on acetophenone and

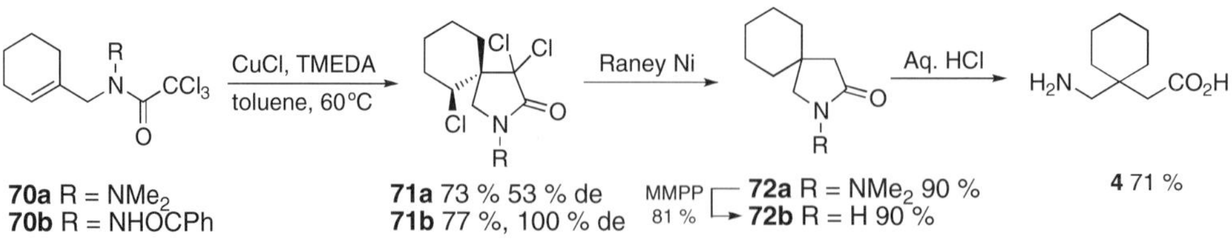

CuCl, TMEDA, toluene, 60°C; Raney Ni; Aq. HCl; MMPP 81 %

70a R = NMe_2
70b R = NHOCPh

71a 73 % 53 % de
71b 77 %, 100 % de

72a R = NMe_2 90 %
72b R = H 90 %

4 71 %

Scheme 14.19

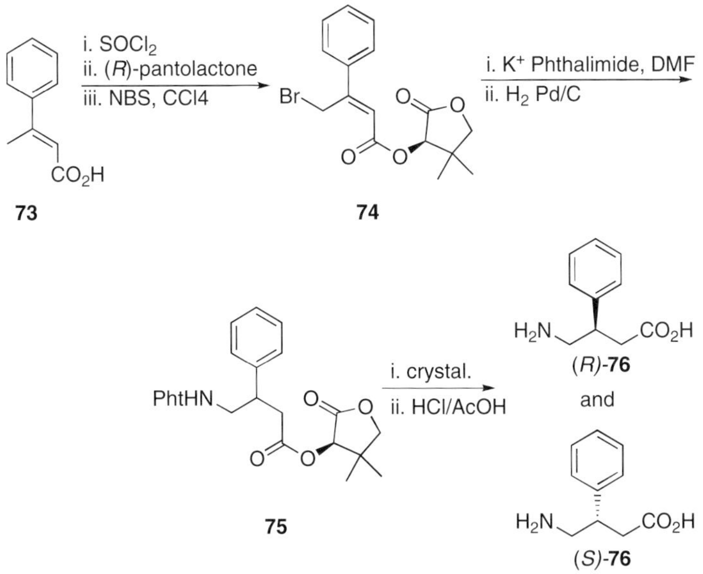

Scheme 14.20

subsequent alkaline hydrolysis of the ester product. Conversion to the (*R*)-(−)-pantolactone ester (**74**), allylic bromination, reaction with potassium phthalimide, and catalytic hydrogenation afforded a mixture of diastereoisomeric esters (**75**) that could be separated by crystallization. Acid hydrolysis of the resolved esters afforded the enantiomerically pure (*R*)- and (*S*)-β-phenyl-GABA (**76**) [67]. This methodology has also been applied to the synthesis of [^{3}H]β-phenyl-GABA [68].

Resolution of racemic 4-nitro-3-(*p*-chlorophenyl)butanoic acid with (*R*)- or (*S*)-*R*-phenylpantolactam yields the corresponding (3*R*,3′*R*)- or (3*S*,3′*S*)-nitro esters with greater than 96% d.e. after column chromatography. Hydrolysis of the resulting diastereopure nitro esters yields the enantiopure nitro acids, which afford (*R*)- or (*S*)-baclofen after reduction of the nitro group to the amine [69]. (*R*)- and (*S*)-Baclofen have also been prepared in high enantiomeric purity by resolution of the (*S*)-(2)-phenylethylamine salt of 3-(*p*-chlorophenyl)glutaramic acid [70] and β-phenyl-GABA by chromatographic separation of (*R*)-phenylglycinol carboxamides of 4-phenylpyrrolidinone [71].

A variety of chemo-enzymatic methods have been used in the synthesis of baclofen and related analogs. Stereoselective enzymatic hydrolysis of dimethyl 3-(*p*-chlorophenyl)glutarate (**77**) by chymotrypsin afforded the chiral half-ester (**78**) in 85% yield and greater than 98% optical purity (Scheme 14.21). The carboxyl group of the product from the chymotrypsin reaction was converted to an amine with retention of configuration through a Curtius rearrangement and hydrolysis gave (*R*)-baclofen (*R*)-(**3**). Alternatively, ammonolysis of the ester produced the corresponding amide that was submitted to a Hofmann rearrangement and hydrolysis giving (*S*)-baclofen (*S*)-(**3**) [72, 73]. Similarly, chymotrypsin-mediated kinetic

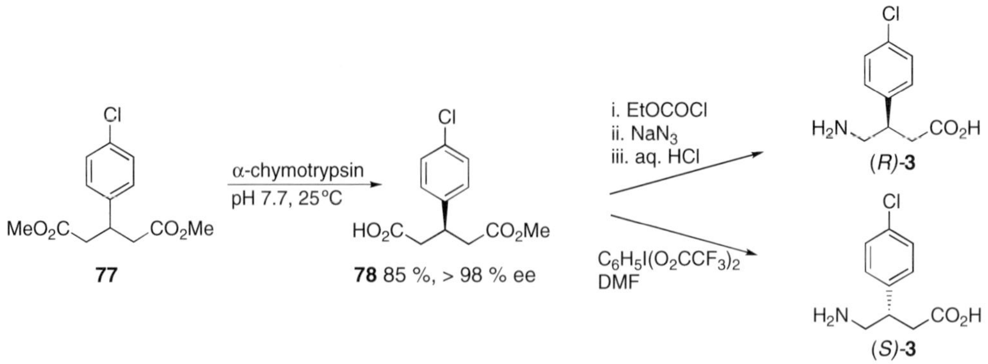

Scheme 14.21

resolutions of racemic of 3-phenyl- and 3-(*p*-chlorophenyl)-4-nitrobutyric acid methyl esters have also been reported to afford both enantiomers of β-phenyl-GABA and baclofen [74].

Another approach involves the lipase-mediated asymmetric acetylation of prochiral 2-(aryl)-1,3-propanediols (e.g., **79**) (Scheme 14.22). The enantiomerically enriched monoacetate (**80**) is prepared in good chemical yield and high optical purity [73]. (*R*)-3-Acetoxy-2-(*p*-chlorophenyl)-1-propanol (**80**) was converted in a six-step sequence via the hydroxy nitrile (**81**) and an azido nitrile into (*R*)-baclofen (*R*)-(**3**) [75].

The allylic alcohol (±)-(**82**) undergoes porcine pancreatic lipase (PPL)-catalyzed kinetic resolution with vinyl acetate affording enantiomerically pure allylic alcohol (*R*)-(**82**) and acetate (*S*)-(**83**) (Scheme 14.23). An orthoester Claisen rearrangement of (*R*)-(**82**) affords (*S,E*)-γ,δ-unsaturated esters (*R*)-(**84**) with high stereoselectivity; (*R*)-(**84**) was converted into (*R*)-(**3**) through a one-pot reduction of an ozonolysis mixture followed by alkyl ester hydrolysis [76].

(±)-2-(*p*-Chlorophenyl)-4-pentenenitrile undergoes enantioselective hydrolyses catalyzed by *Rhodococcus* sp. AJ270 microbial cells to afford excellent yields of enantiomerically pure (*R*)-(−)-2-(4-chlorophenyl)-4-pentenamides and (*S*)-(+)-2-(*p*-chlorophenyl)-4-pentenoic acids. Reduction of the amide to the amine and direct oxidation of the alkene afforded (*R*)-(−)-baclofen [77] in a manner analogous to that described for α-phenyl-GABA (**29**) (Scheme 14.7).

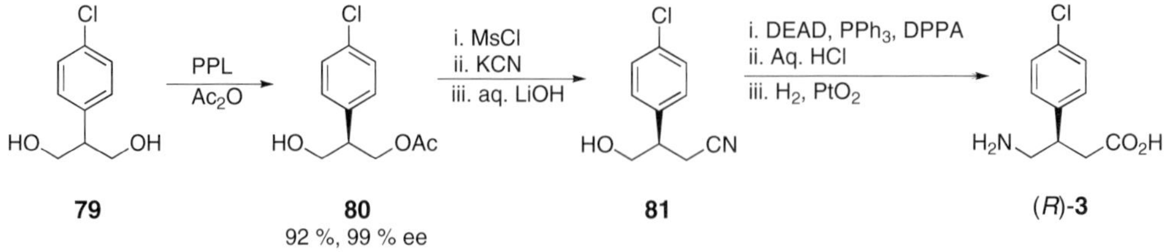

Scheme 14.22

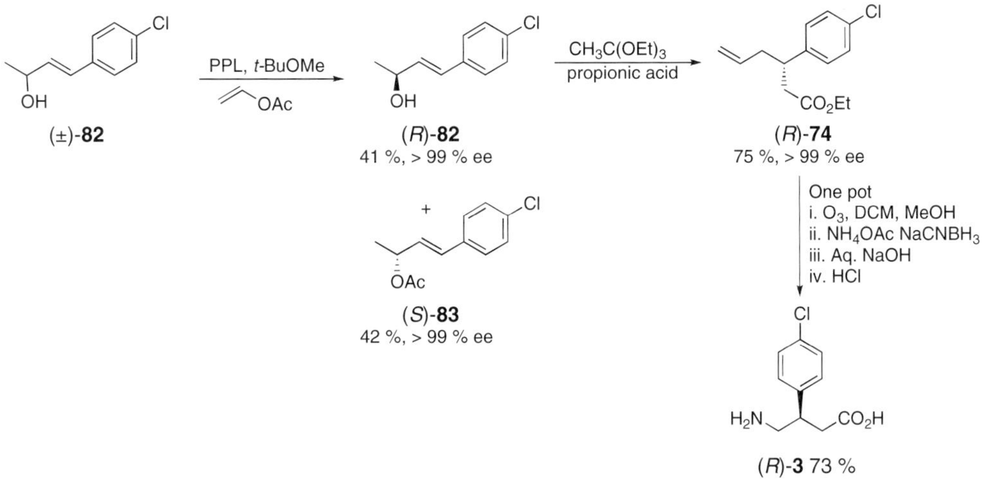

Scheme 14.23

A microbiological Baeyer–Villiger oxidation of the prochiral 3-(*p*-chlorobenzyl)-cyclobutanone (**85**) mediated by the fungus *Cunninghamella echinulata* yielded the optically pure lactone (**86**) in very high enantiomeric purity, although modest yield (30%) (Scheme 14.24). The lactone was further transformed to the target molecule (*R*)-baclofen (*R*)-(**3**) via the azide (**87**) [78].

Rhodium-catalyzed site-selective intramolecular C–H insertion of a variety of α-diazoacetamides has been used to prepare the corresponding γ-lactams (**89**) that are hydrolyzed to the desired β-aryl GABA analogs (**90**). These include α-diazoacetamides (**88a**) prepared from *N-tert*-butyl amine with $Rh_2(cap)_4$ catalyst (Scheme 14.25) [79], α-methoxycarbonyl-α-diazoacetamides (**88b**) [80], *N*-arylalkyl, *N*-bis(trimethylsilyl) methyl α-diazoacetamides, and α-carboalkoxy-α-diazoacetamides (**88c**) [81]. Enantioselectivities of up to 69% were achieved by the use of the di-Rh(II) tetrakis[*N*-phthaloyl-(*S*)-*tert*-leucinate] catalyst $Rh_2[(S)\text{-PTTL}]_4$ [80, 81].

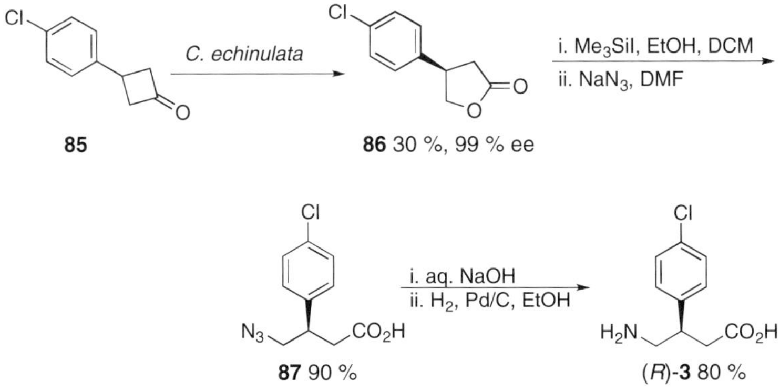

Scheme 14.24

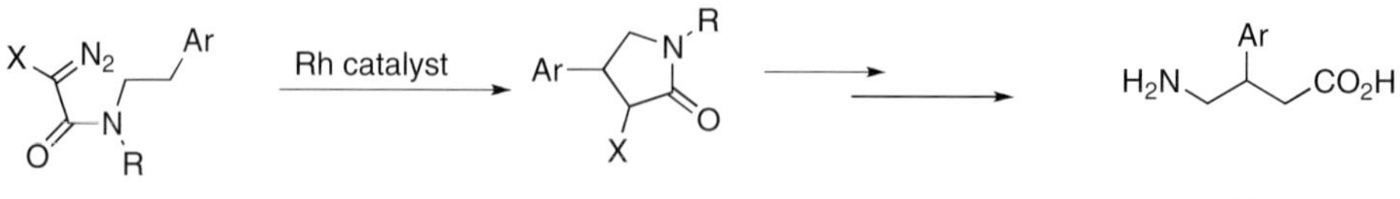

Rh catalyst
Ar
H_2N
CO_2H

88a R = *t*-Bu, X = H
Ar = Ph, *p*-ClPh, *m*-ClPh, *p*-MeOPh, *p*-NO_2Ph, *o*-MePh, *m*-MePh
88b R = *p*-NO_2Ph, X = H, CO_2Me
Ar = Ph, *p*-ClPh, *p*-MeOPh, *p*-NO_2Ph
88c R = BTMSM, X = H, CO_2Et
Ar = Ph, CH_2Ph, *p*-MeOPh

89

90

Scheme 14.25

Michael reactions using a variety of substrates as both acceptors and donors have been thoroughly explored as a methodology for the synthesis of β-aryl-GABA derivatives. These include the addition of anions of a chiral carbene to nitroolefins which proceed in low to moderate diastereomeric excess [82], catalytic conjugate addition of cyanide to β-substituted α,β-unsaturated *N*-acylpyrroles using a chiral gadolinium complex [83], addition of the Grignard cuprate $(p\text{-ClPh})_2CuMgCl$ to the (*S*)-pyroglutamic acid-derived γ-lactam [84], addition of aryl cuprates to (+)-4-cumyloxycyclopent-2-en-1-one [85], and conjugate addition of cyanide to α,β-unsaturated oxazolines derived from (*R*)-phenyl glycinols [86].

Under phase-transfer conditions, Michael addition of nitromethane to enantiomerically pure tricarbonyl(ethyl-4-chloro-2-trimethylsilylcinnamate)Cr(0) (*S*)-(**91**) yielded the tricarbonylchromium nitroester (*S*,*R*)-(**92**) intermediate in good yield (88%) and high stereoselectivity (96% d.e.) (Scheme 14.26). Desilylation and decomplexation, afforded the nitrobutanoate (*R*)-(**93**). Hydrogenation over Raney nickel and subsequent hydrolysis then provided (*R*)-(**3**) [87].

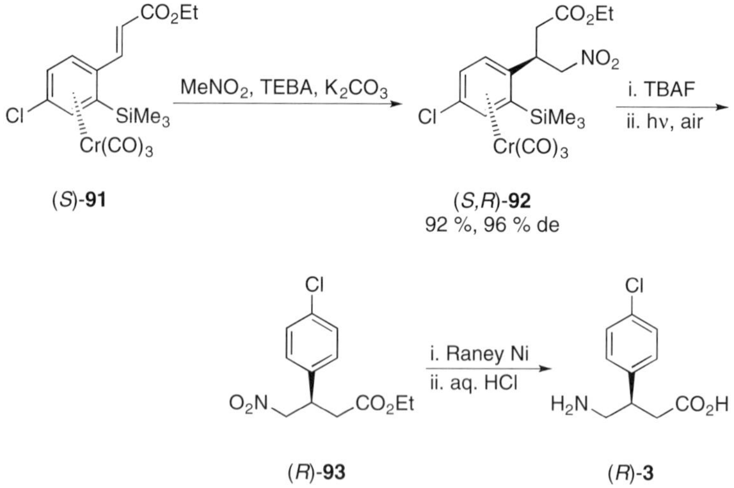

CO_2Et
Cl
$SiMe_3$
$Cr(CO)_3$
$MeNO_2$, TEBA, K_2CO_3
NO_2
i. TBAF
ii. hν, air
(*S*)-**91**
(*S*,*R*)-**92**
92 %, 96 % de
O_2N
i. Raney Ni
ii. aq. HCl
H_2N
CO_2H
(*R*)-**93**
(*R*)-**3**

Scheme 14.26

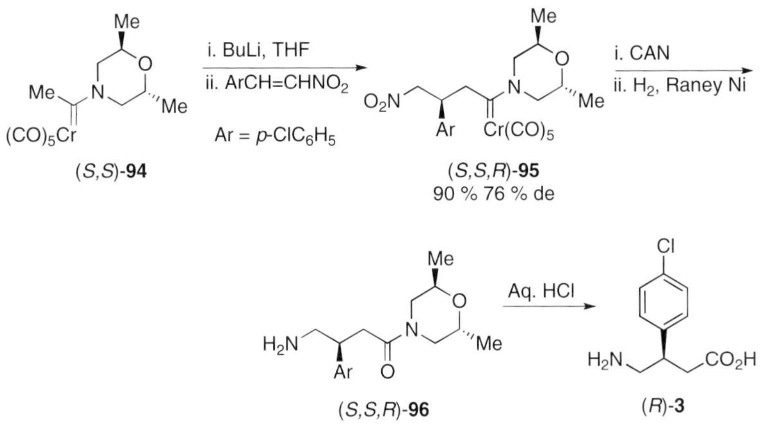

Scheme 14.27

Diastereoselective Michael addition of the conjugate base of the enantiopure chromium complex (*S*,*S*)-(**94**) to (*E*)-*p*-chloro-nitrostyrene provided enantiopure (*S*,*S*,*R*)-(**95**) (Scheme 14.27) The carbene complex was oxidized to the corresponding amide and the nitro group was converted to the amine providing (*S*,*S*,*R*)-(**96**). Deprotection of the carboxylic function by acid hydrolysis afforded (*R*)-(**3**) and the chiral auxiliary as hydrochloride salts which were separated by reverse-phase chromatography [88].

Rhodium-catalyzed asymmetric 1,4-additions of arylboronic acids to 4-amino-2,3-butenoic acid derivatives (**99**) provides a short efficient synthesis of baclofen and related compounds (Scheme 14.28). Substrates were prepared in two steps from 2-aminoethanol derivatives (**97**). The conjugate addition of the arylboronic acid (**98**) was carried out in the presence of Rh(acac)(C_2H_4)$_2$, and (*S*)-2,2′-bis(diphenylphosphino)1,1′-binaphthyl (BINAP), affording the protected amino acid (*R*)-(**100**) in good yield and enantiopurity. Compound (**100**) can be converted to (*R*)-**3** [89, 90].

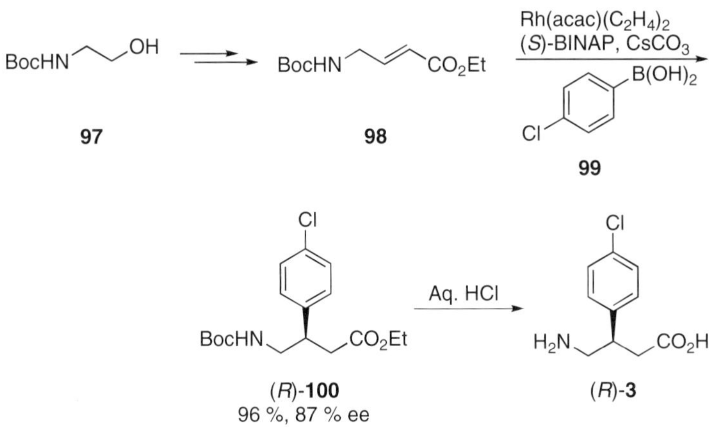

Scheme 14.28

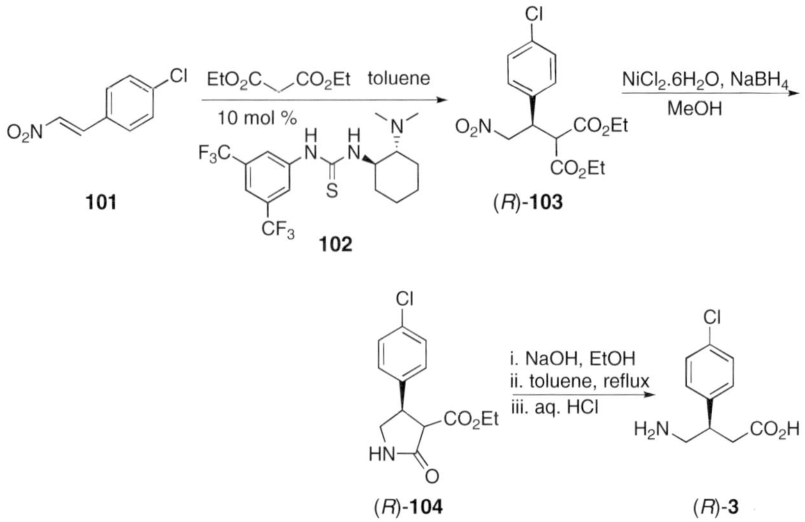

Scheme 14.29

1,4-Addition of diethyl malonate to the nitroolefin (**101**) in the presence of 10 mol% of the novel thiourea catalyst (**102**) affords the adduct (**103**) in 80% yield with 94% e.e. (>99% e.e. after a single recrystallization) (Scheme 14.29). Reduction of the nitro group with nickel borite and *in situ* lactamization gave the lactam (**104**) in 94% yield. Ester and lactam hydrolysis and decarboxylation afforded (*R*)-(**5**) [91].

Chiral rhodium complexes have been employed to effect enantioselective reductions of both ketones and alkenes in the synthesis of baclofen. Enantioselective hydrogenation of ethyl 3-(4-chlorophenyl)-3-oxopropanoate using (*S*)-BINAP-Ru(II) complex (800 psi) affords the corresponding (*R*)-hydroxy ester in 95% yield and 96% e.e., the hydroxy ester was converted to (*R*)-(**3**) in four steps, 26% overall yield and 90% e.e. Alternatively, (*S*)-BINAP-Ru(II) catalyzed hydrogenation (200 psi) of ethyl 4-azido-3-(4-chlorophenyl)but-2-enoate yields the alkyl azide (200 psi), which was hydrolyzed to (*R*)-**3** [92]. Similarly, (*Z*)-3-(4-chlorophenyl)-4-phthalimidobut-2-enoate ester has been reduced to the protected baclofen derivative using a range of modular chiral BoPhoz-type phosphine-aminophosphine ligands with high yields and enantioselectivities (800 psi) [93]. However, the high pressures required for ruthenium catalyzed reductions limit their utility.

A [2 + 2] cycloaddition of 4-chlorostyrene (**105**) and dichloroketene yields (±)-2,2-dichloro-3-(*p*-chlorophenyl)cyclobutanone (**106**) (Scheme 14.30), which provides an efficient and novel route to (*R*)-(**3**). Reductive dechlorination and subsequent enantioselective deprotonation with the lithium (*S,S*′)-di(α-methylbenzyl)amide (*S,S*′)-(**107**) and trapping of the resulting enolate with triethylsilyl chloride provided the silylenol ether (*R*)-(**108**). A one-pot ozonolysis and reductive amination afforded (*R*)-(**3**) [94].

α-Alkylation of (*S*)-1-amino-2-methoxymethylpyrrolidine (SAMP)-hydrazone with methyl bromoacetate has been used to provide a range of β-substituted GABA analogs

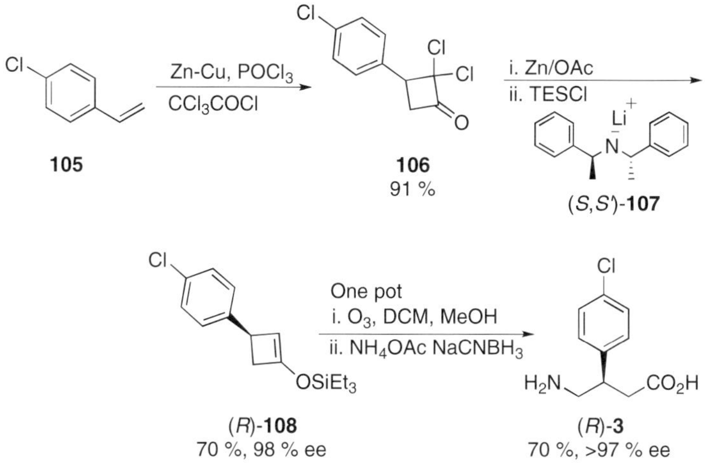

Scheme 14.30

in excellent enantiomeric purity (Scheme 14.31). Alkylation of *p*-chlorophenyl aldehyde SAMP-hydrazone (**109**) affords the α-substituted aldehyde hydrazone (*S*,*R*)-(**110**) in excellent yield and diastereoselectivity. Oxidative removal of the chiral auxiliary yields the nitrile (*R*)-(**111**) which was reduced with nickel borite and hydrolyzed to (*R*)-(**3**) [95].

Baclofen has recently been prepared using Sharpless asymmetric dihydroxylation methodology to introduce the chiral centre. Asymmetric dihydroxylation of (*E*)-ethyl

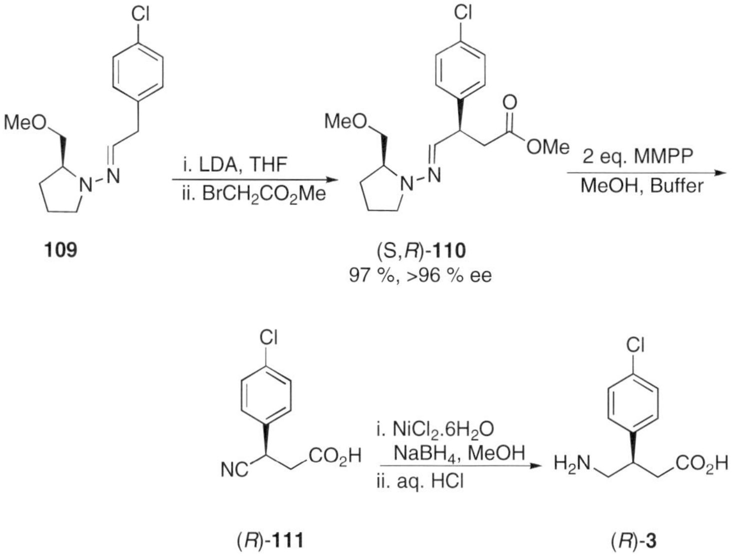

Scheme 14.31

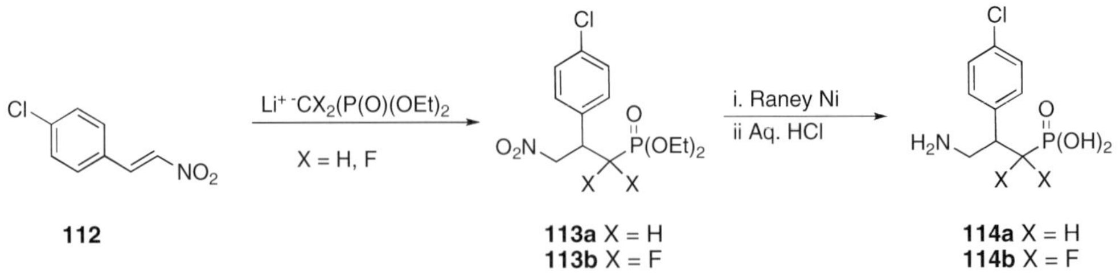

Scheme 14.32

p-chlorocinnamate affords the dihydroxy ester which was converted to (*R*)-(**3**) in 14% overall yield and 85% e.e. [96].

The synthesis of a wide range of baclofen analogs has also been described including the phosphonic analog, phaclofen [97], iodobaclofen [98], heterocyclic analogs [99, 100], hydroxysaclofen [101, 102], homologs [103–106], pyrolidinone analogs [107], and conformationally restricted derivatives [108, 109]. A highly efficient three-step synthesis of the phosphonic analog of baclofen, (±)-phaclofen (**114a**) [110] and its α,α-difluoro analog (**114b**) has been reported (Scheme 14.32) [111]. The key step is a Michael addition of the lithium anion of diethyldifluoromethanephosphonate to a *p*-chloro-β-nitro styrene (**112**). Reduction of the nitro functionality and ester hydrolysis of (**113**) affords the desired phosphonic amino acids (**114**).

14.4 γ-Substituted γ-Amino Acids

γ-Substituted γ-amino acids have been prepared via a two-carbon homologation of α-amino acids using a variety of methodologies. The traditional method being a double Arndt–Eistert homologation in which the multistep sequence can result in low yields [103]. Alternatively, a Wittig reaction with an alkyl (triphenylphosphoranylidene)acetate on α-amino aldehydes such as phenylalaninal (**115**) and subsequent reduction of the resulting alkene (**116**) has been employed (Scheme 14.33). Hydrolysis of the protecting group affords the γ-substituted γ-amino acid (**117**) [112].

A facile procedure for a two-carbon homologation involves coupling of an *N*-Boc-α-amino acid (**118**) with Meldrum's acid (**119**) and complete reduction of the keto

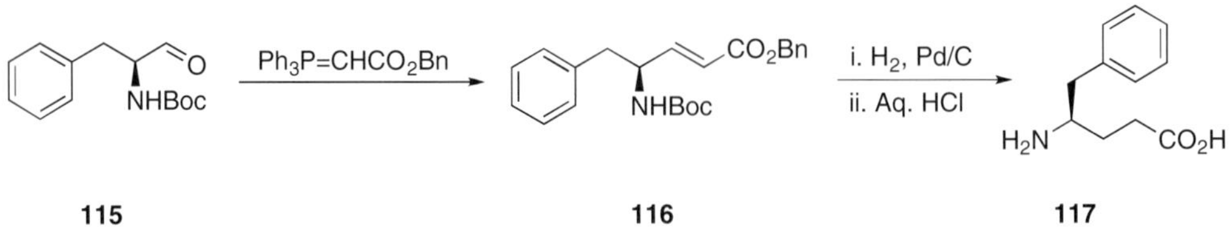

Scheme 14.33

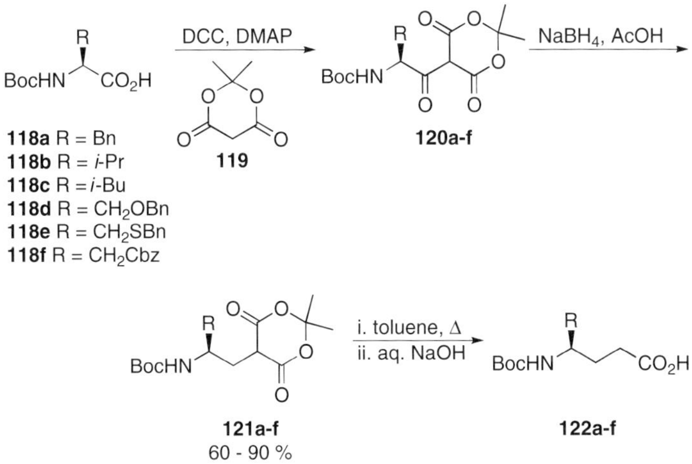

BocHN–CH(R)–CO_2H (**118a** R = Bn; **118b** R = *i*-Pr; **118c** R = *i*-Bu; **118d** R = CH_2OBn; **118e** R = CH_2SBn; **118f** R = CH_2Cbz) —DCC, DMAP, **119**→ **120a-f** —$NaBH_4$, AcOH→ **121a-f** (60 - 90 %) —i. toluene, Δ; ii. aq. NaOH→ **122a-f**

Scheme 14.34

functionality of the α-amino acyl compound (**120**) (Scheme 14.34). The resulting amino alkyl Meldrum's acid (**121**) undergoes thermal decarboxylative ring closure to a 5-substituted pyrrolidinone that yields the corresponding γ-amino acid (**122**) after hydrolysis. The overall yields of the procedure ranges from 40 to 65% [113].

Alternatively, *N*-Boc-α-amino acids (**118**) have been converted to the corresponding Weinreb amides and then reduced to the α-amino aldehydes (**123**) (Scheme 14.35). Olefination yielded the α,β-unsaturated *N*-Boc-protected γ-amino acid methyl esters (**124**) as *trans/cis* mixtures. Hydrogenation with Pd/C and saponification produced the desired Boc-protected γ-amino acids (**122**) in 55–72% yield [114].

The use of glutamic acid derivatives provides access to γ-substituted γ-amino acids of the opposite chirality to those derived from α-amino acids. The iodozinc derivative of (**126**) prepared in three steps from the protected L-glutamic acid (**125**)

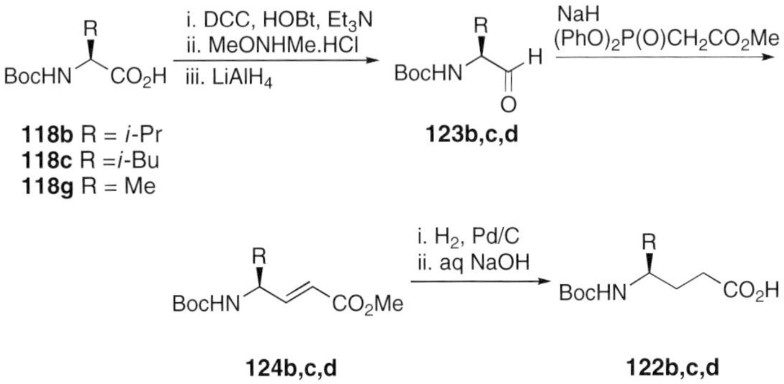

BocHN–CH(R)–CO_2H (**118b** R = *i*-Pr; **118c** R = *i*-Bu; **118g** R = Me) —i. DCC, HOBt, Et_3N; ii. MeONHMe.HCl; iii. $LiAlH_4$→ **123b,c,d** —NaH, $(PhO)_2P(O)CH_2CO_2Me$→ **124b,c,d** —i. H_2, Pd/C; ii. aq NaOH→ **122b,c,d**

Scheme 14.35

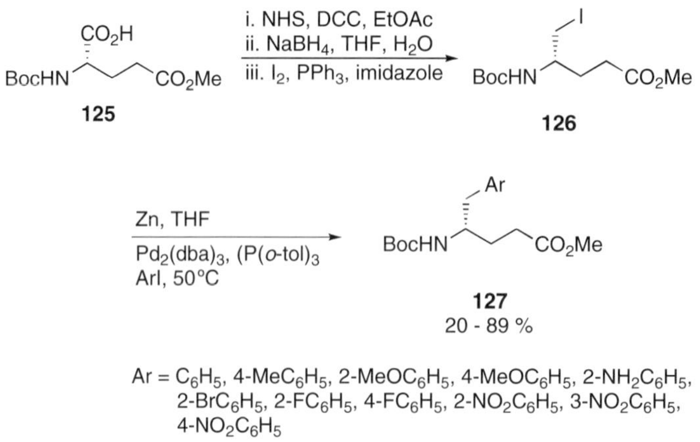

CO_2H BocHN CO_2Me **125** — i. NHS, DCC, EtOAc; ii. $NaBH_4$, THF, H_2O; iii. I_2, PPh_3, imidazole → I, BocHN CO_2Me **126**

Zn, THF; $Pd_2(dba)_3$, (P(*o*-tol)$_3$; ArI, 50°C → Ar, BocHN CO_2Me **127** 20 - 89 %

Ar = C_6H_5, 4-MeC_6H_5, 2-MeOC_6H_5, 4-MeOC_6H_5, 2-$NH_2C_6H_5$, 2-BrC_6H_5, 2-FC_6H_5, 4-FC_6H_5, 2-$NO_2C_6H_5$, 3-$NO_2C_6H_5$, 4-$NO_2C_6H_5$

Scheme 14.36

has been reported to undergo coupling with a range of substituted aryl iodides to afford enantiopure γ-aryl substituted GABA (**127**) derivatives in moderate to good yields (Scheme 14.36) [115, 116].

A range of hetero- and benzenoid-aromatics undergo Friedel–Crafts reactions with chiral *N*-protected (α-aminoacyl)benzotriazoles (**129**), prepared in three steps from L-glutamic acid derivatives (**128**) (Scheme 14.37). The resulting α-amino ketones (**130**) are reduced to yield the corresponding γ-amino acid derivatives (**131**), in moderate to good overall yields with preservation (>99%) of chirality [117].

A range of *N*-protected γ-aryl GABA (**136**) derivatives have been synthesized in good yields without loss of optical activity in a three-step, one-pot process [118]. Sequential reduction of benzonitriles (**132**) with trialkylborane (**133**) and methanolysis, allylboration and hydroboration of the intermediate (**134**) yields the δ-amino alcohols (**135**) (Scheme 14.38). The aldimine–borane adducts were prepared by room

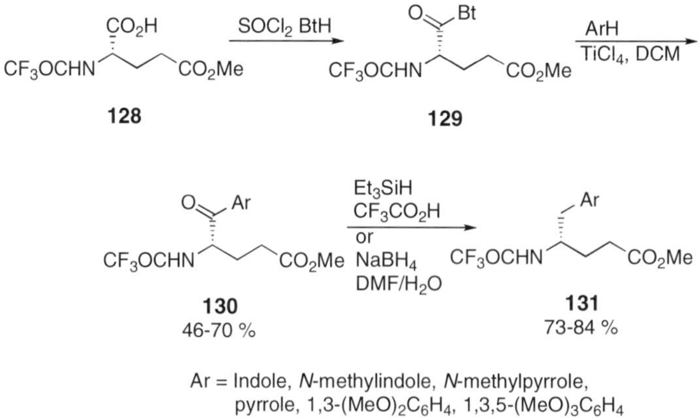

CO_2H CF_3OCHN CO_2Me **128** — $SOCl_2$ BtH → O, Bt, CF_3OCHN CO_2Me **129** — ArH, $TiCl_4$, DCM →

O, Ar, CF_3OCHN CO_2Me **130** 46-70 % — Et_3SiH, CF_3CO_2H or $NaBH_4$ DMF/H_2O → Ar, CF_3OCHN CO_2Me **131** 73-84 %

Ar = Indole, *N*-methylindole, *N*-methylpyrrole, pyrrole, 1,3-$(MeO)_2C_6H_4$, 1,3,5-$(MeO)_3C_6H_4$

Scheme 14.37

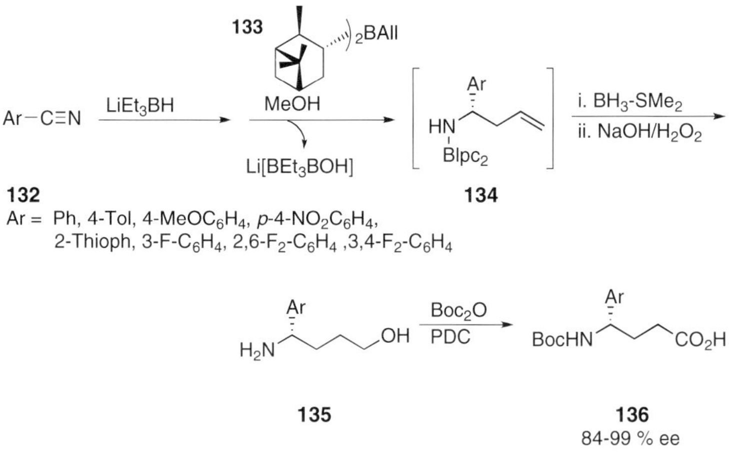

133 $)_2$BAll

Ar−C≡N LiEt$_3$BH MeOH Li[BEt$_3$BOH] Ar HN BIpc$_2$ i. BH$_3$-SMe$_2$ ii. NaOH/H$_2$O$_2$

132 134

Ar = Ph, 4-Tol, 4-MeOC$_6$H$_4$, p-4-NO$_2$C$_6$H$_4$, 2-Thioph, 3-F-C$_6$H$_4$, 2,6-F$_2$-C$_6$H$_4$,3,4-F$_2$-C$_6$H$_4$

Ar H$_2$N OH Boc$_2$O PDC Ar BocHN CO$_2$H

135 136 84-99 % ee

Scheme 14.38

temperature reduction of the nitrile (**132**) with Super-Hydride. Addition of one equivalent of methanol results in the formation of the aldimine–triethylborane complex (**134**), which undergoes allylboration and hydroboration. Direct oxidation of the hydroboration product (**135**) directly to the GABA derivatives with Pyridinium dicarbonate (PDC) resulted in a mixture of products. Therefore, the δ-amino alcohols were protected as the *N*-Boc derivatives and then oxidized to the corresponding GABA derivatives (**135**) [118].

The use of *n*BuLi/(−)-sparteine (−)-(**138**) as a chiral base in an asymmetric deprotonation and electrophilic substitution sequence has also been used in the synthesis of enantiomerically pure γ-phenyl-GABA (Scheme 14.39). Alkylation of *N*-Boc-*N*-(*p*-methoxyphenyl)benzylamine (**137**) with acrolein and Jones oxidation of the aldehyde (**139**) yields (*S*)-*N*-protected γ-phenyl-GABA (*S*)-(**140**) in 77% overall yield. Alternatively, a lithiation–stannylation–transmetalation protocol yields (*R*)-*N*-protected γ-phenyl-GABA (*R*)-(**140**) after oxidation in 46% overall yield [119].

Chiral *N*-benzylidene-*p*-toluenesulfinamides (**141**) prepared by the reaction of benzonitrile with an alkyllithium followed by the addition (−)-menthyl (*S*)-*p*-tolylsulfinate have been used to induce chirality in the synthesis of γ-amino acids (Scheme 14.40). Reaction with allylmagnesium bromide gave adducts (**142**) with excellent stereoselectivity. Transformation to the *N*-acetyl derivative (**143**), oxidation and deprotection afforded enantiomerically pure aryl, alkyl-disubstituted γ-amino acid derivatives (**144**) [120].

γ-Substituted γ-amino acids have also been prepared by olefination of β-enamino phosphorus compounds with alkyl glyoxylates and selective reduction of the resulting 1-azadienes with NaBH$_4$ to yield (*E*)-γ-amino-α,β-unsaturated esters, which can be further reduced to the corresponding saturated γ-amino esters [121]; 1,4-addition of carbon radicals, generated from a Barton ester derivative of α-amino acids, to form acrylic derivatives [122], and samarium iodide promoted addition of alkyl nitrones to α,β-unsaturated amides and esters [123]. In the presence of *N*-substituted sugars as

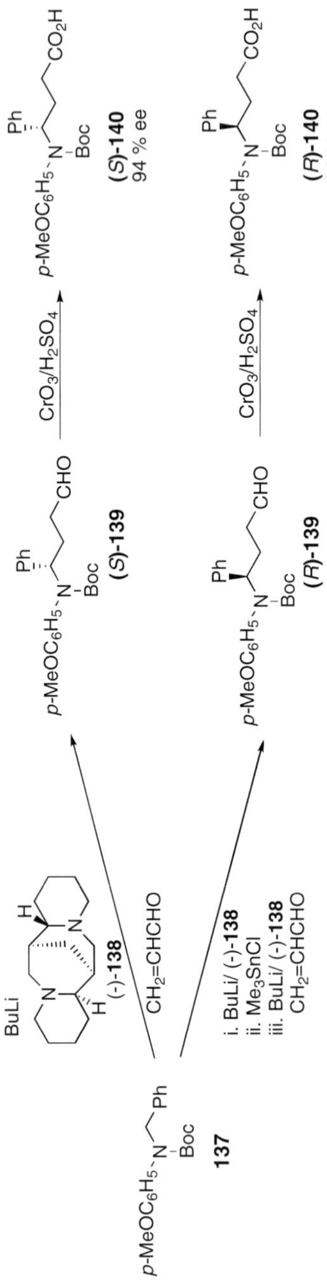

Scheme 14.39

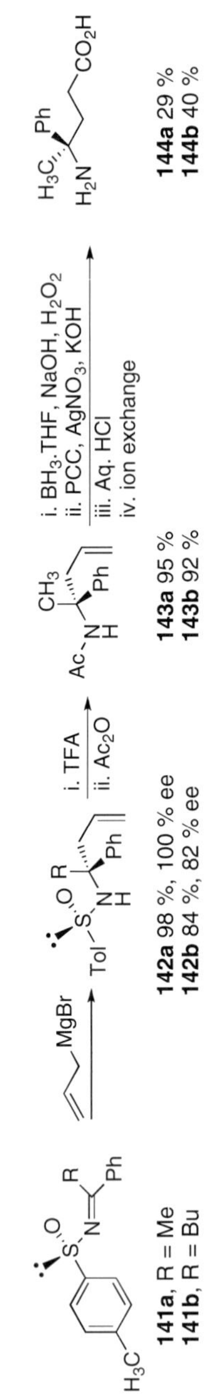

Scheme 14.40

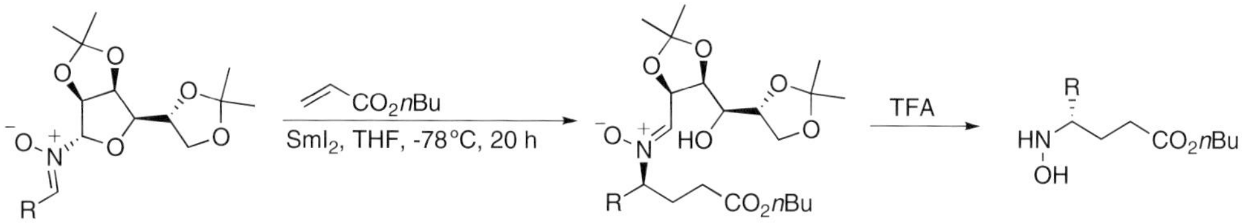

145a R = Cyclohexyl
145b R = *i*-Pr
145c R = Cyclopentyl
145d R = 2-Ethylbutyl
145e R = *i*-Bu
145f R = $Ph(CH_2)_2$

CO_2nBu / SmI_2, THF, -78°C, 20 h → 146a-f 54-80 %, > 90 % de → TFA → 147a-f

Scheme 14.41

chiral auxiliaries (**145**) this reaction provides access to variously substituted nitrones (**146**) with high diastereoselectivity (diastereomeric ratio > 95 : 5) (Scheme 14.41). Acid hydrolysis affords the *N*-hydroxyl amino acid derivatives (**147**) [124].

14.4.1 Vigabatrin

4-Amino-5-hexenoic acid (vigabatrin) (**2**) was originally synthesized via Birch reduction of racemic 4-amino-5-hexynoic acid or semihydrogenation of methyl 4-amino-5-hexynoate acid using Lindlar's catalyst and subsequent acid hydrolysis [9]. (±)-Vigabatrin has also been prepared by reaction of ethyl hexenoate (**148**) with *N*-sulfinyl benzenesulfonamide (Scheme 14.42). The adduct (**149**) undergoes a [2,3] sigmatropic rearrangement on silylation. The resulting allylamine derivative (**150**) can then be converted to vigabatrin (**2**) by deprotection in 26.5% overall yield [125].

Ethyl 6-hydroxyhex-4-enoate (**152**), prepared in three steps from erythritol (**151**), undergoes an Overman rearrangement on heating in trichloromethyl acetimidate, introducing the amine functionality (Scheme 14.43). Deprotection of (**153**) afforded (±)-vigabatrin (**2**) in 25% overall yield [126].

[^{14}C]Vigabatrin was synthesized in five steps from the tosylate (**154**) and [^{14}C] NaCN. After displacement of the tosylate by cyanide, reduction of the resulting nitrile (**155**) in the presence of excess dimethylamine gave the amine, which was

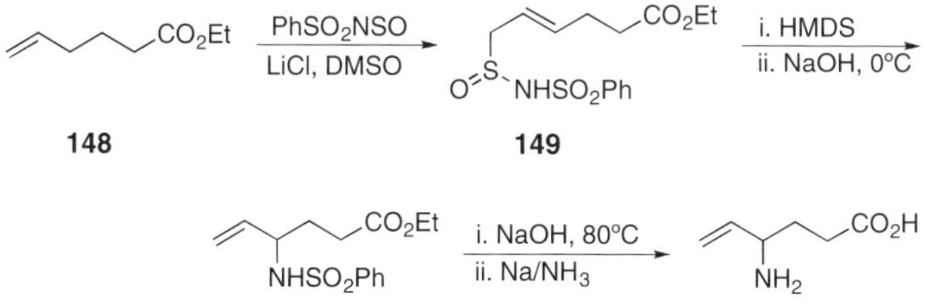

148 → $PhSO_2NSO$ / LiCl, DMSO → 149 → i. HMDS ii. NaOH, 0°C → 150 → i. NaOH, 80°C ii. Na/NH_3 → 2

Scheme 14.42

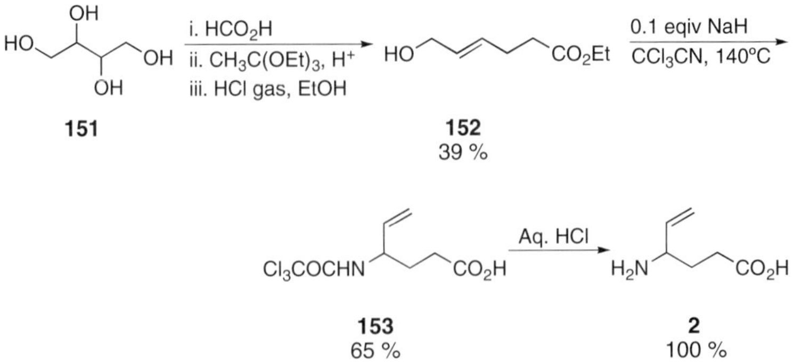

Scheme 14.43

oxidized to the *N*-oxide (**156**) (Scheme 14.44). Elimination to the alkene (**157**) and acid hydrolysis afforded an vigabatrin (**2**) with an overall radiochemical yield of 22% and radiochemical purity greater than 98% [127].

(*S*)-Vigabatrin (*S*)-(**2**) has been prepared in 38% overall yield and 90% enantiomeric purity via a Wittig reaction of the ω-aldehyde of glutamic acid and methyl triphenylphosphonium bromide [128]. Alternatively, both enantiomers of vigabatrin have been prepared in five steps from (*R*)- or (*S*)-methionine, respectively, via a two-carbon homologation. Formation of the γ-lactam, oxidation of the thiol to the sulfoxide and thermal elimination afforded the 5-vinyl-γ-lactam, which was hydrolyzed to vigabatrin [129].

Catalytic deracemization of butadiene monoepoxide (**158**) using a palladium catalyzed asymmetric allylic alkylation yields optically pure 2-phthalimido-but-3-en-ol (**159**) (Scheme 14.45). Conversion to the triflate and substitution with the anion of diethyl malonate yielded the diester (**160**). Deprotection and concomitant decarboxylation affords (*R*)-vigabatrin (*R*)-(**2**) in 59% overall yield and high enantiopurity. Use of the other enantiomer of the catalyst would afford (*S*)-vigabatrin (*S*)-(**2**) [130, 131].

Starting from inexpensive pyrrole, a four-step synthesis of both isomers of *N*-Boc-protected 3,4-didehydropyrohomoglutamate (**161**) in 91% e.e., involving resolution by simulated moving bead chromatography, has been reported (Scheme 14.46). Conjugate reduction of the enone, deprotection of the amine, and subsequent reduction of the methyl ester gave the alcohol (**162**). The final transformation to (*S*)-vigabatrin (*S*)-(**2**) was carried out by a three-step protocol, forming first the

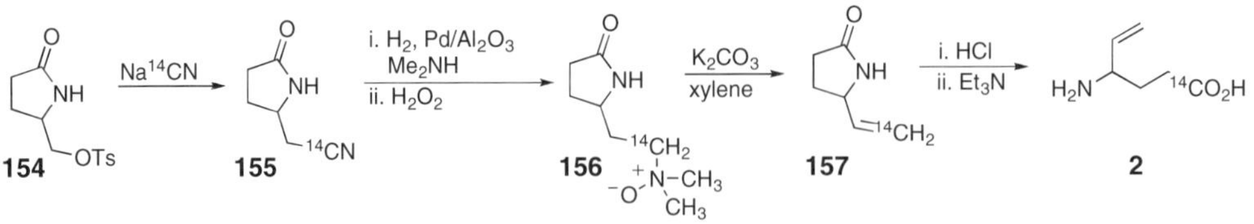

Scheme 14.44

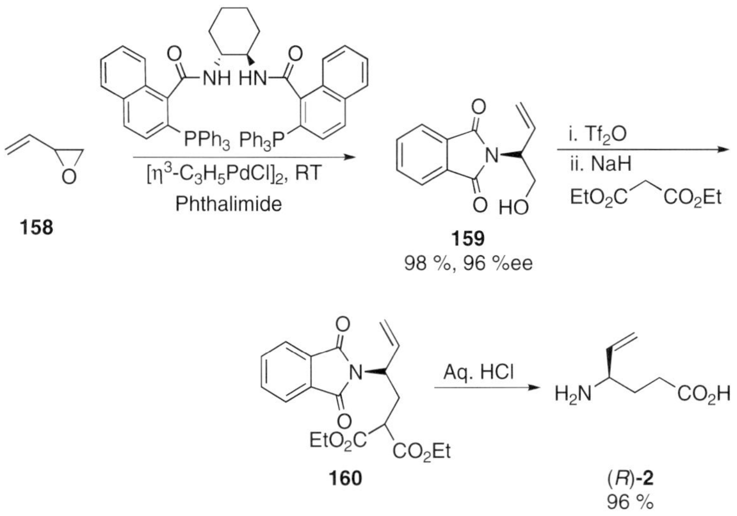

Scheme 14.45

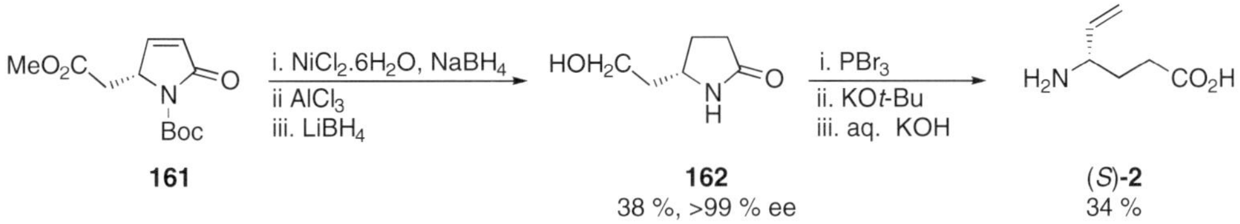

Scheme 14.46

bromide, followed by dehydrobromination, and finally hydrolysis of the vinylpyrrolidinone to yield (*S*)-(**2**) [132].

(*S*)-Vigabatrin (*S*)-(**2**) has been prepared via multistep syntheses employing Sharpless epoxidation of 5-phenylpent-2-en-1-ol [133] and Sharpless asymmetric aminohydroxylation of ethyl 6-hydroxyhex-2-enoate as the sole source of chirality [134], and the addition of alkyl 3-lithiopropiolates [135] or SmI_2-catalyzed addition of methyl acrylate [136] to the nitrone of D-glyceraldehyde. The cyclopropane analog of vigabatrin, (±)-γ-amino-γ-cyclopropylbutanoic acid, has been prepared in six steps via a Mannich reaction of malonic acid, cyclopropanecarbaldehyde, and ammonium acetate, and subsequent Arndt–Eistert homologation [137].

14.5 Halogenated γ-Amino Acids

γ-Amino-α-fluorobutanoic acid [(±)-α-fluoro-GABA] has been prepared by the addition of the sodium salt of diethylfluoromalonate to *N*-(*p*-nitrobenzoyl)aziridine yielding diethyl 2-(*p*-nitrobenzamido)ethylfluoromalonate, which undergoes acid hydrolysis to afford (±)-α-fluoro-GABA [138].

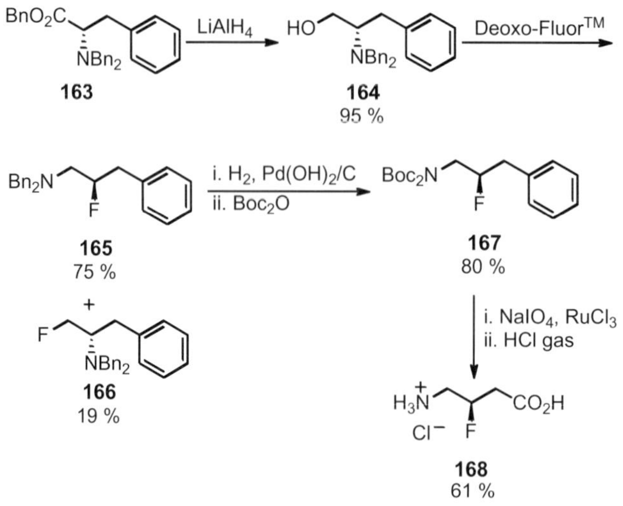

Scheme 14.47

The synthesis of (*R*)-β- and (*S*)-β-fluoro-GABA (**168**) was initiated from (*R*)- or (*S*)-phenylalanine, respectively (Scheme 14.47). Tribenzyl protection of phenylalanine gave the *N*,*N*-dibenzyl benzyl ester (**163**), which was followed by reduction of the ester with $LiAlH_4$ to generate the *N*,*N*-dibenzyl alcohol (**164**). Treatment of the alcohol with Deoxo-Fluor gave rise to a 4 : 1 mixture of the separable regioisomeric products (**165**) and (**166**). Conversion of the dibenzyl amine moiety to the di-Boc-protected amine (**167**) allowed successful oxidation of the phenyl group and deprotection to afford the respective isomer as the hydrochloride salt (**168**) [139].

(±)-β-Chloro-GABA has been prepared by treatment of methyl β-hydroxy-γ-aminobutyrate with excess PCl_5 [140], and photochlorination of GABA using chlorine gas in concentrated HCl and photolysis with a mercury arc lamp [141].

14.6 Disubstituted γ-Amino Acids

14.6.1 α,β-Disubstituted γ-Amino Acids

Only limited reports of α,β-substituted GABA analogs exist. A range of α,β-dialkyl substituted GABA derivatives were initially prepared by Raney nickel reduction of the corresponding γ-nitro acids [142]. The α-methyl analog of pregabalin has been reported [143]. Reduction of the carboxylic acid of (*S*)-(**169**) and cyclization yields the γ-lactone (*S*)-(**170**) (Scheme 14.48). Reaction of the lithium enolate of (*S*)-(**170**) with methyl iodide results in a 4 : 1 mixture of *trans* : *cis* isomers (**171**). Separation of the *trans* isomer and conversion to the azide in two steps yields the azidoester (*S*,*S*)-(**172**). Reduction of the azide and hydrolysis of the lactam affords (*S*,*S*)-α-methyl pregabalin (*S*,*S*)-(**173**) [143].

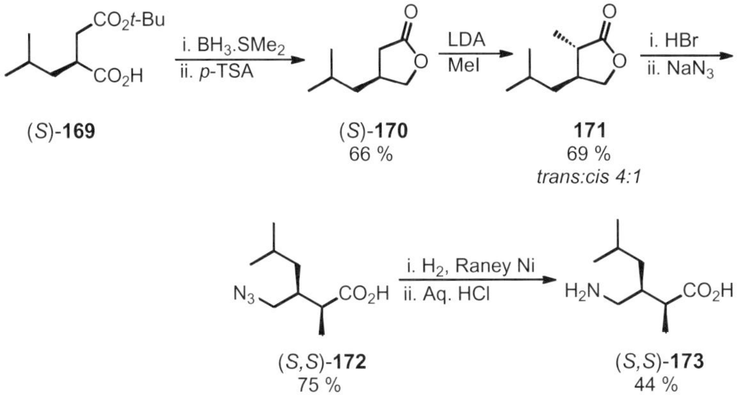

(*S*)-**169** → i. $BH_3.SMe_2$ ii. *p*-TSA → (*S*)-**170** 66 % → LDA, MeI → **171** 69 % *trans:cis 4:1* → i. HBr ii. NaN_3 → (*S,S*)-**172** 75 % → i. H_2, Raney Ni ii. Aq. HCl → (*S,S*)-**173** 44 %

Scheme 14.48

14.6.2
α,γ-Disubstituted γ-Amino Acids

Alkylation, under a range of conditions, of the enolates of methyl *N*-protected γ-amino esters (**174**) affords the *syn*-2,4-disubstituted products (**175**), through 1,3-asymmetric induction (Scheme 14.49) [144, 145].

Alternatively, reaction of pyrrolidine-derived lithium enolates (**176**) with cinnamyl bromide yields the *anti*-disubstituted γ-lactams (**177**) with high diastereoselectivities (Scheme 14.50). Basic hydrolysis affords the *anti*-2,4-disubstituted *N*-Boc-γ-amino acids (**178**) [146].

(*R,S*)-Dimethyl-GABA has been prepared from *N*-Boc-L-alaninal (**179**) via a Horner–Emmons–Wadsworth reaction, yielding the unsaturated (*E*)-ester (**180**) as the major product (Scheme 14.51). Hydrogenation resulted in a 2 : 1 mixture of *anti* : *syn* products. However, conversion to the γ-lactam (**181**), subsequent hydrogenation, and hydrolysis yielded the *syn* compound (**182**) as the major product [147]. Coupling to L-serine methyl ester gave dipeptides that were readily separable by chromatography [148].

A highly diastereoselective manganese-mediated photolytic addition of alkyl iodides to the C=N bond of chiral hydrazones (**183**), provides enantiomerically pure

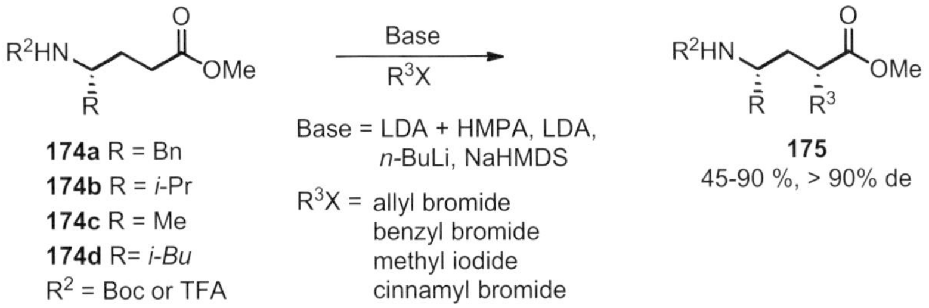

174a R = Bn
174b R = *i*-Pr
174c R = Me
174d R= *i-Bu*
R^2 = Boc or TFA

→ Base, R^3X →

Base = LDA + HMPA, LDA, *n*-BuLi, NaHMDS

R^3X = allyl bromide, benzyl bromide, methyl iodide, cinnamyl bromide

175 45-90 %, > 90% de

Scheme 14.49

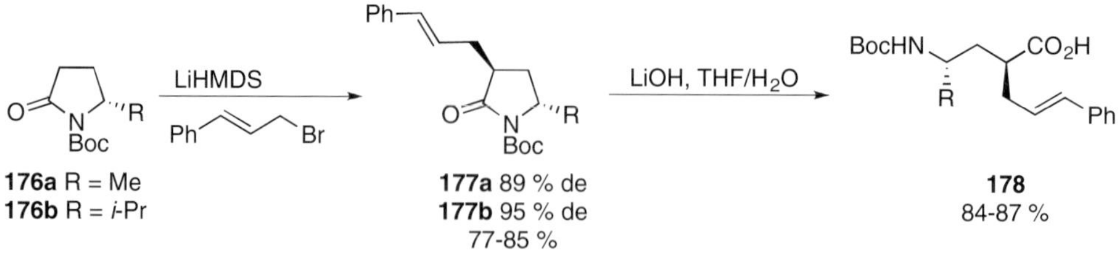

176a R = Me
176b R = *i*-Pr

LiHMDS, Ph/Br; LiOH, THF/H_2O

177a 89 % de
177b 95 % de
77-85 %

178
84-87 %

Scheme 14.50

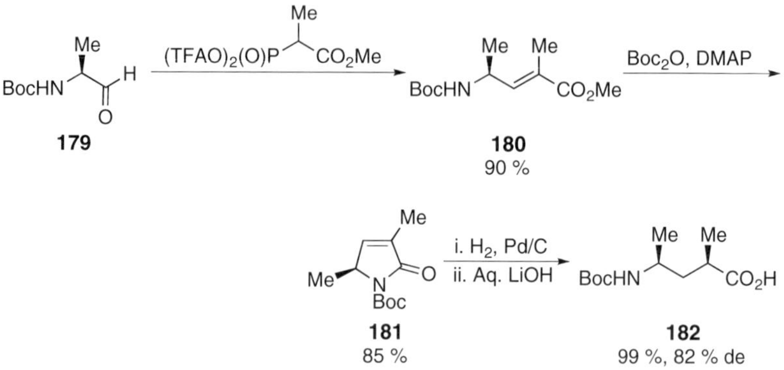

$(TFAO)_2(O)P$–CH(Me)–CO_2Me; Boc_2O, DMAP; i. H_2, Pd/C ii. Aq. LiOH

179

180
90 %

181
85 %

182
99 %, 82 % de

Scheme 14.51

adducts (**184**) (Scheme 14.52). Reductive N–N bond cleavage yields the amine (**185**). Deprotection of the alcohol and then oxidation affords *anti*-2,4-disubstituted γ-amino acids such as tubuphenylalanine (**186**) [149].

Zirconium-mediated diastereoselective alkene–carbonyl coupling reactions using chiral *tert*-butyl 3-butenylcarbamates (**187**) having a substituent at the homoallylic position proceed in a highly diastereoselective manner to give α-methyl-γ-substituted γ-amino acid derivatives (**188**) (Scheme 14.53) [150].

Stereocontrolled aziridine ring-opening reactions with chiral enolates derived from (*S*,*S*)-(+)-pseudoephedrine amides (**189**) lead to γ-amino amides in good yields (Scheme 14.54). The diastereoselectivity of the reaction is controlled by the presence of the chiral auxiliary on the enolate, although the stereogenic center contained in the structure of the aziridine has a strong influence on the stereochemical course of the reaction which results in matched and mismatched combinations. The enolates derived from (*S*,*S*)-(+)-pseudoephedrine propionamide (*S*,*S*)-(**189**) and (*S*)-aziridines (*S*)-(**190**) form a matched combination leading to γ-aminoamides with a 1,3-*syn* configuration (**191**) in good diastereomeric excess. On the contrary, the same enolate and (*R*)-aziridines (*R*)-(**190**) form a mismatched combination leading to the corresponding adducts (**191**) with a relative 1,3-*anti* configuration with moderate to poor diastereoselectivity. Hydrolysis and esterification affords the amino acid derivatives (**192**) [151].

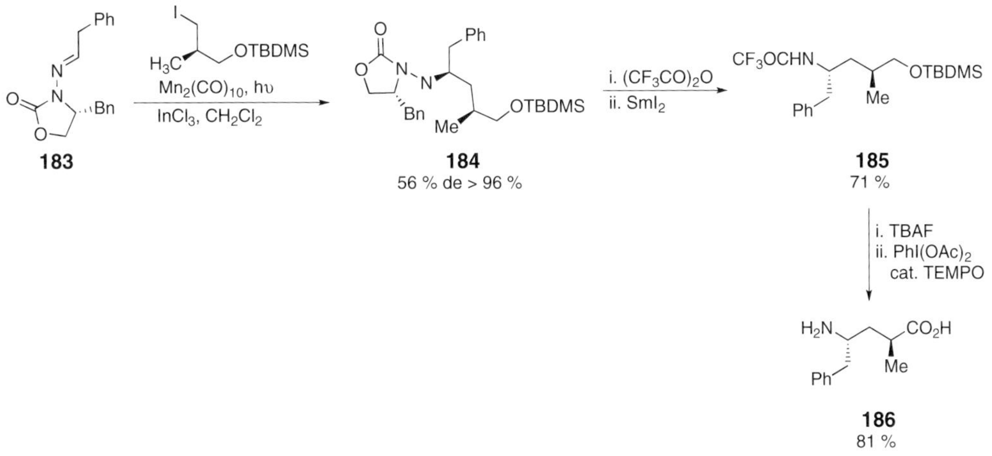

$Mn_2(CO)_{10}$, hυ
$InCl_3$, CH_2Cl_2
i. $(CF_3CO)_2O$
ii. SmI_2
i. TBAF
ii. $PhI(OAc)_2$ cat. TEMPO

183

184
56 % de > 96 %

185
71 %

186
81 %

Scheme 14.52

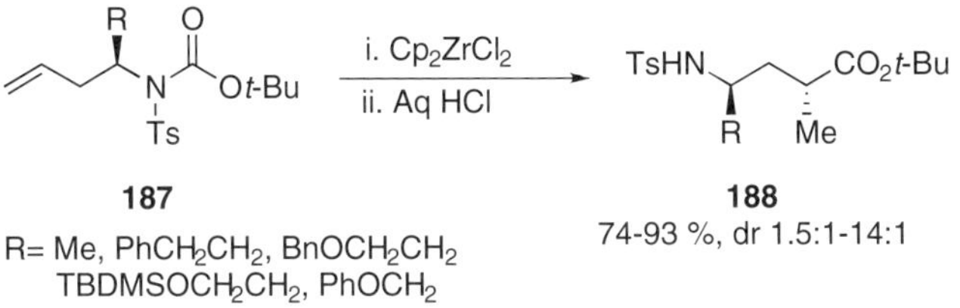

i. Cp_2ZrCl_2
ii. Aq HCl

187
R= Me, $PhCH_2CH_2$, $BnOCH_2CH_2$ $TBDMSOCH_2CH_2$, $PhOCH_2$

188
74-93 %, dr 1.5:1-14:1

Scheme 14.53

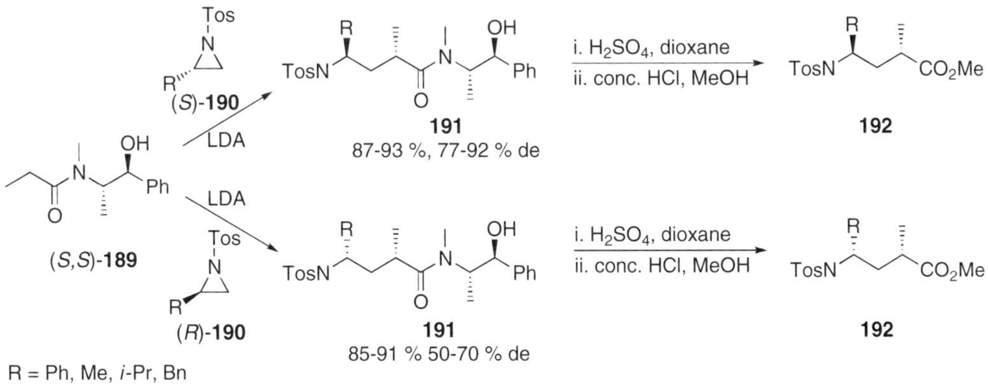

(S)-**190** LDA
(R)-**190** LDA
i. H_2SO_4, dioxane
ii. conc. HCl, MeOH

(S,S)-**189**

191
87-93 %, 77-92 % de

192

191
85-91 % 50-70 % de

192

R = Ph, Me, i-Pr, Bn

Scheme 14.54

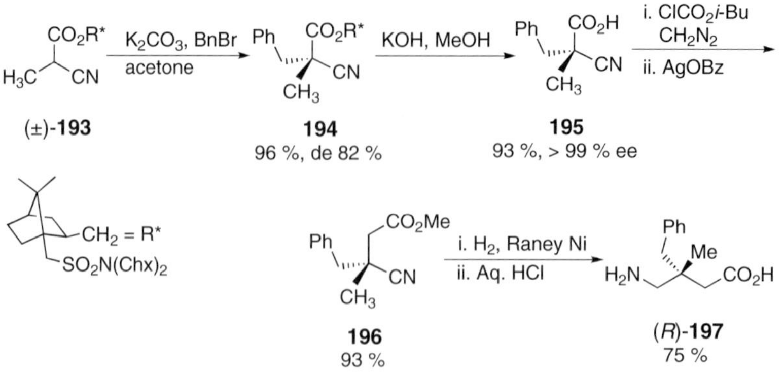

CO_2R^* H_3C CN (±)-**193** — K_2CO_3, BnBr, acetone → Ph CO_2R^* CN CH_3 **194** 96 %, de 82 % — KOH, MeOH → Ph CO_2H CN CH_3 **195** 93 %, > 99 % ee — i. $ClCO_2i$-Bu, CH_2N_2; ii. AgOBz →

$CH_2 = R^*$ $SO_2N(Chx)_2$

Ph CO_2Me CN CH_3 **196** 93 % — i. H_2, Raney Ni; ii. Aq. HCl → Ph Me H_2N CO_2H (*R*)-**197** 75 %

Scheme 14.55

Ph Ph N Ph **198** — TEBA, MeCN, Aq. NaOH, Ph CO_2R **199** R = Me, Et, CH_2Ph, $CH(CH_3)_2$, $C(CH_3)_3$ → Ph Ph Ph N CO_2R Ph **200** 48-79%, de 100 % — Aq. HCl → Ph H_2N CO_2H Ph **201**

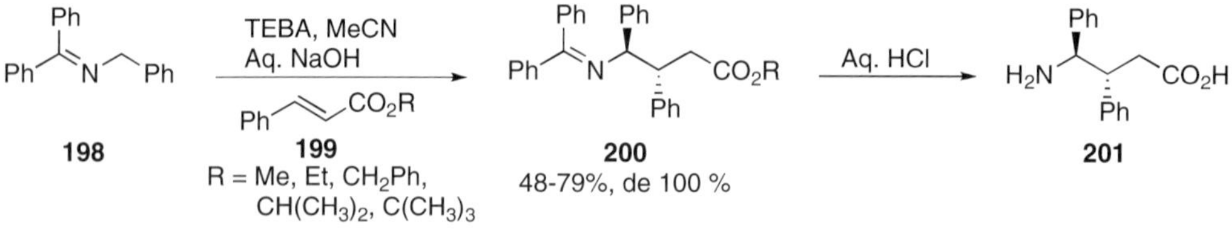

Scheme 14.56

14.6.3 β,β-Disubstituted γ-Amino Acids

Enantiomerically pure (*R*)-γ-amino-β-benzyl-β-methylbutyric acid (*R*)-(**197**) has been efficiently synthesized using an Oppolzer chiral auxiliary derivative (**193**) (Scheme 14.55). Diastereoselective benzylation of the chiral enolate gives the product (**194**) in excellent yield and good diastereomeric excess. Isolation of the major isomer by recrystallization and cleavage of the chiral auxiliary yields the enantiomerically pure acid (**195**). Arndt–Eistert homologation yields the methyl ester (**196**). Reduction of the nitrile and acid hydrolysis afforded (*R*)-(**197**) in 65% overall yield [152].

14.6.4 β,γ-Disubstituted γ-Amino Acids

Phase-transfer catalyzed reaction of the Schiff's base *N*-(diphenylmethylene)benzylamine (**198**) with cinnamates (**199**) and hydrolysis of the resulting adduct (**200**) affords *threo*-4-amino-3,4-diphenylbutanoic acid (**201**) (Scheme 14.56) [153].

14.7 Trisubstituted γ-Amino Acids

(*R,R,R*)-Trialkyl-γ-amino acids have been prepared by the stereoselective Michael addition of modified Evans acyloxazolidinones with nitroolefins (Scheme 14.57).

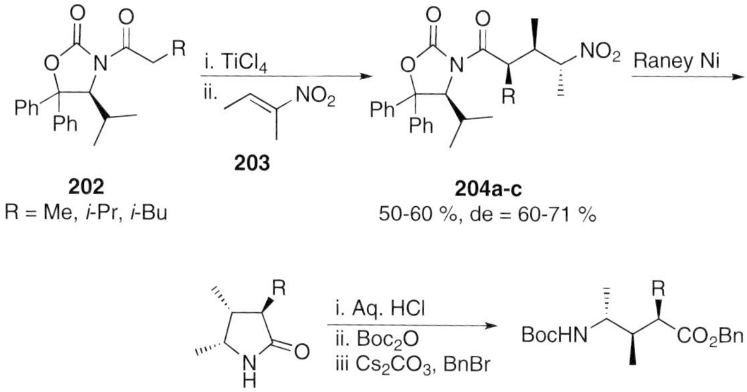

Scheme 14.57

Reaction of the titanium enolate of the acyloxazolidinones (**202**) with (*E*)-2-nitro-2-butene (**203**) yielded the nitro derivatives (**204a–c**) in moderate diastereomeric excess, which were separable by chromatography. Raney nickel reduction which was accompanied by some epimerization, afforded the pyrrolidinones (**205a–c**), which were hydrolyzed and converted to the protected amino acid derivatives (**206a–c**) [154, 155].

14.8
Hydroxy-γ-Amino Acids

14.8.1
α-Hydroxy-γ-Amino Acids

(±)-α-Hydroxy-γ-amino acid (homoisoserine, α-hydroxy-GABA) (**208**) was originally prepared by dehydration of L-2-hydroxysuccinamic acid and catalytic reduction of the ω-cyano acid [156] or by hydrogenolysis of a racemic isoxazolidine carboxylic acid [157]. A number of chemo-enzymatic syntheses of α-hydroxy-GABA have also been reported including reduction of *N*-(benzyloxycarbonyl) methyl 4-amino-2-oxobutanoate with baker's yeast [158] and various *Saccharomyces* sp. [159]. These reductions proceeded with poor yields and only moderate enantioselectivity. However, lactate dehydrogenase from *Bacillus stearothermophilus* and *Staphylococcus epidermidis* catalyzed reductions of the sodium 4-benzyloxycarbonylamino-2-oxobutanoate (**207**) have been used to prepare both (*S*)- and (*R*)-benzyloxycarbonylamino-2-hydroxybutanoic acids, respectively (Scheme 14.58). Deprotection affords the corresponding (*S*)-α-hydroxy-GABA (*S*)-(**208**) and (*R*)-α-hydroxy-GABA (*S*)-(**208**) [160].

Fully protected L-α-methyl-homoisoserine (α-hydroxy-α-methyl-GABA) (**213**) has been prepared in seven steps from L-citramalic acid (**209**), via protection of the α-hydroxyl and adjacent carboxylic acid as the 2,2-bis(trifluoromethyl)-1,3-dioxalanone (**210**) (Scheme 14.59). Conversion to the diazoketone (**211**) and Wolff

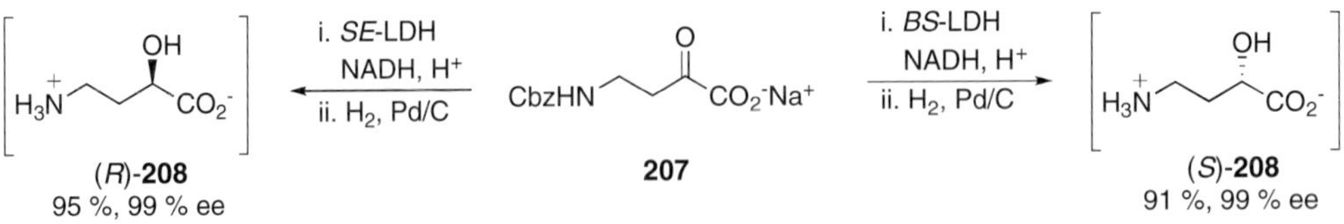

Scheme 14.58

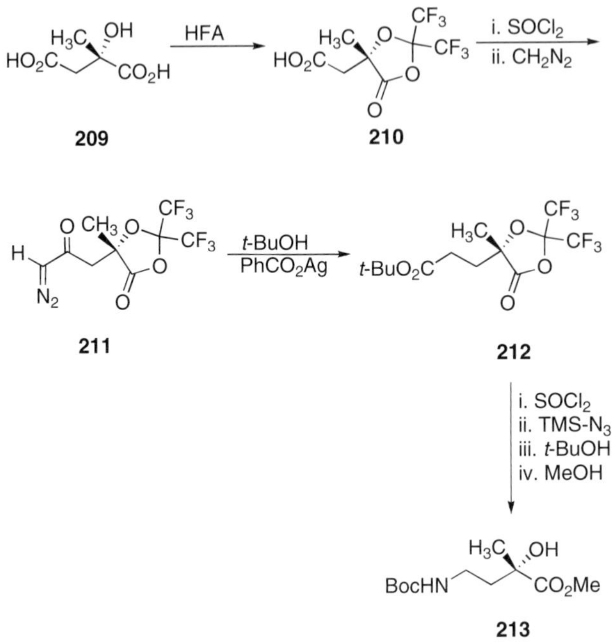

Scheme 14.59

rearrangement yields the homolog (**207**). Curtius rearrangement and protection affords (**213**) [161].

α-Hydroxy and α-mercapto acids (**214**) react with hexafluoroacetone to give 2,2-bis(trifluoromethyl)-1,3-dioxolan-4-ones (**215a**) and 2,2-bis(trifluoromethyl)-1,3-oxathiolan-5-ones (**215b**), respectively, in excellent yields (Scheme 14.60). Arndt–Eistert homologation yields the methyl esters (**216**). Treatment with thionyl chloride, reaction with trimethylsilyl azide, and Curtius rearrangement affords protected homoisoserine (**217a**) and homoisocysteine (**217b**) [18].

14.8.2 β-Hydroxy-γ-Amino Acids

Enantiomerically pure (*R*)- and (*S*)-GABOB (**6**) have been prepared via chromatographic separation of the oxazolidin-2-ones formed from (*S*)-ethyl 4-(1-phenylethylamino)but-2-enoate [162]. Alternatively, reaction of trimethylamine and

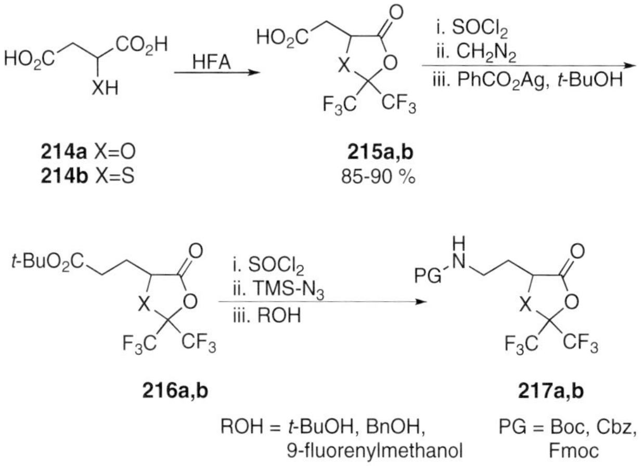

Scheme 14.60

L-tartaric acid with epichlorohydrin yields the (3-chloro-2-hydroxypropyl)trimethylammonium tartrate salts which were resolved by crystallization and converted to enantiomerically pure (*R*)-carnitine in two steps [163].

GABOB and carnitine have been prepared from a wide variety of chiral starting materials. Both isomers of glycerol acetonide have been used in the syntheses of (*R*)-GABOB. (*R*)-Glycerol acetonide, prepared in two steps from ascorbic acid, was converted to (*R*)-GABOB in seven steps and 10% overall yield [164]. However, later reports suggest that a hydrolysis using 98% H_2SO_4 results in significant racemization [165]. Alternatively, (*S*)-glycerol acetonide (**218**) prepared from D-mannitol was transformed to (*R*)-GABOB (*R*)-(**6**) via conversion to the phthalimide (**219**) and subsequent formation of the cyclic sulfite (**220**) (Scheme 14.61). Reaction with potassium cyanide yielded the hydroxy nitrile (**221**) which was hydrolyzed to (*R*)-(**6**) in 37.5% overall yield [166].

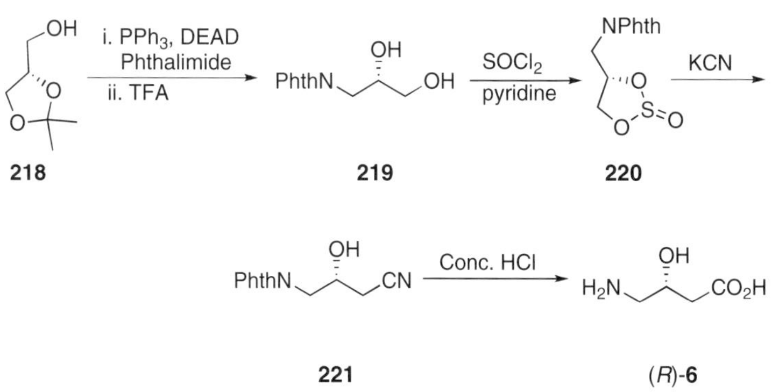

Scheme 14.61

Bols *et al.* have described the use of 3-deoxy-D-galactono-1,4-lactone (**222**) in two simple syntheses of (*R*)-carnitine (*R*)-(**7**) (Scheme 14.62). In the first route, treatment of the lactone (**222**) with dimethylamine, reduction of the amide (**223**) with borane-dimethylsulfide and methylation yielded the trimethylammonium iodide (**224**) that underwent oxidative cleavage of the polyol chain with $KMnO_4$ to afford (*R*)-(**7**) in 58% overall yield. In the alternative approach, the C-5–C-6 bond of the diol (**222**) was cleaved with $NaIO_4$, giving a quantitative yield of the aldehyde (**225**) that was hydrogenated in the presence of methylamine, hydrolyzed, and quarternized to the trimethylammonium salt (**226**), which then underwent oxidative cleavage to give (*R*)-(**7**) [167].

Chemoselective opening of (*R*)-3-hydroxy-γ-butyrolactone (*R*)-(**227**) with trimethylsilyl iodide, treatment of the resulting iodide (**228**) with sodium azide, ester hydrolysis, and reduction of the azide (**229**) afforded (*R*)-GABOB (*R*)-(**6**) in 56% overall yield and 99% e.e. (Scheme 14.63) [168]. Conversely, (*S*)-3-hydroxy-γ-butyrolactone (*S*)-(**227**) has been transformed into (*R*)-(**6**) via opening of (*S*)-(**227**) with HBr, conversion of the bromide (**230**) to the epoxide (**231**), and opening of the epoxide with cyanide to yield the cyano ester (**215**). Subsequently, (**215**) can be converted to (*R*)-(**6**) via a Curtius reaction of the ester and hydrolysis of the nitrile, resulting in a net inversion of stereochemistry to afford (*R*)-(**6**) [169].

Commercially available (*R*)- and (*S*)-ethyl 4-chloro-3-hydroxybutyrate (**233**) are reacted with phenyl selenide anions to give the corresponding β-hydroxyalkyl phenylselenides (Scheme 14.64). Reaction with benzoyl isocyanate yields the *N*-benzoyl-carbamate derivative (**234**) that undergoes cyclization to the oxazolidin-2-one (**235**) on treatment with *m*-chloroperoxybenzoic acid. Hydrolysis affords optically pure (*R*)- or (*S*)-GABOB (**6**) [170].

A number of syntheses of enantiomerically pure GABOB (**6**) and carnitine (**7**) from malic acid (**236**) have been reported [171–174]. These include a seven-step synthesis via a cyclic anhydride [171], and a 13-step [172] and later six-step synthesis [173] via an oxazolidin-2-one. Both (*R*)- or (*S*)-carnitine (**7**) have been synthesized from the respective enantiomer of malic acid (**236**) (Scheme 14.65). Conversion to the dibenzyl ester (**237**) and chemoselective reduction of the α-hydroxyester group yielded a diol, which was converted to the monotosylate (**238**). Substitution with trimethylamine gave a carnitine derivative that was deprotected to afford the corresponding isomer of carnitine (**7**) in good overall yield [174].

Epichlorohydrin has also proved a versatile starting material for the synthesis of β-hydroxy-GABA analogs. Ring opening of (*R*)-epichlorohydrin (*R*)-(**66**) with phenyl lithium/cuprous cyanide yields the alcohol (**239**) that was converted to the azide (**240**) and Sharpless oxidation yielded the acid (**241**) (Scheme 14.66). Reduction of the azide and alcohol deprotection afforded (*R*)-GABOB (*R*)-(**6**) in 57% overall yield [175]. Similarly, opening of the epoxide (*R*)-(**66**) with vinyl Grignard/cuprate and reaction of the resulting chloride (**242**) with trimethylamine yields the ammonium salt (**243**). Ozonolysis and oxidation produces enantiomerically pure (*R*)-(**7**) in 67% overall yield [176]. (*R*)-Glycidyl tosylate has also been converted in eight steps via the corresponding oxazoline to enantiopure (*R*)-GABOB (*R*)-(**6**) in good overall yield [177].

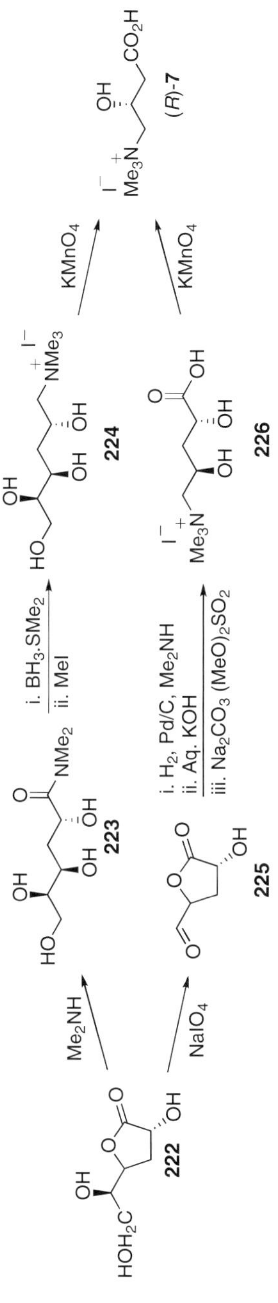

Scheme 14.62

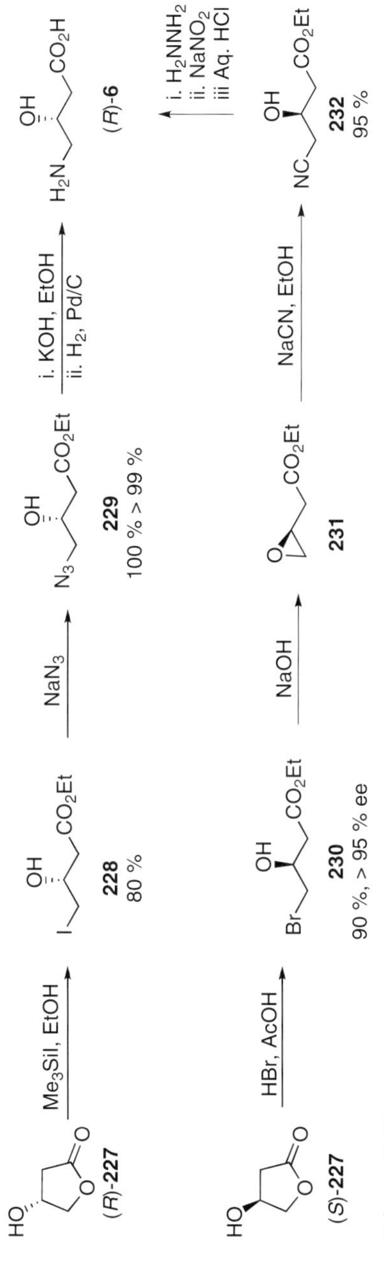

Scheme 14.63

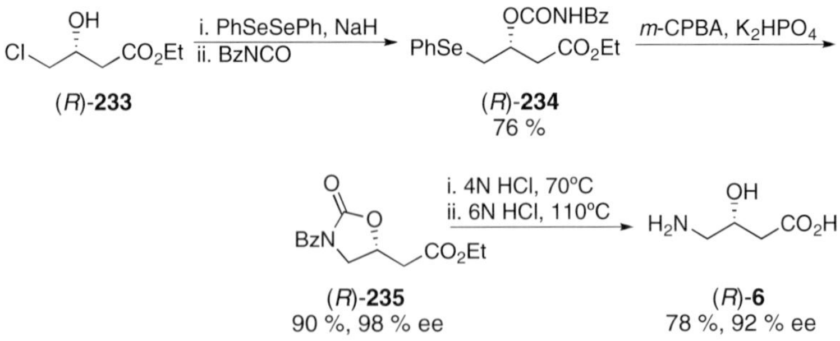

OH
Cl / CO$_2$Et — i. PhSeSePh, NaH; ii. BzNCO → PhSe / OCONHBz / CO$_2$Et — m-CPBA, K_2HPO_4 →
(R)-**233** — (R)-**234** 76 %

BzN / O / O / CO$_2$Et — i. 4N HCl, 70°C; ii. 6N HCl, 110°C → H_2N / OH / CO_2H
(R)-**235** 90 %, 98 % ee — (R)-**6** 78 %, 92 % ee

Scheme 14.64

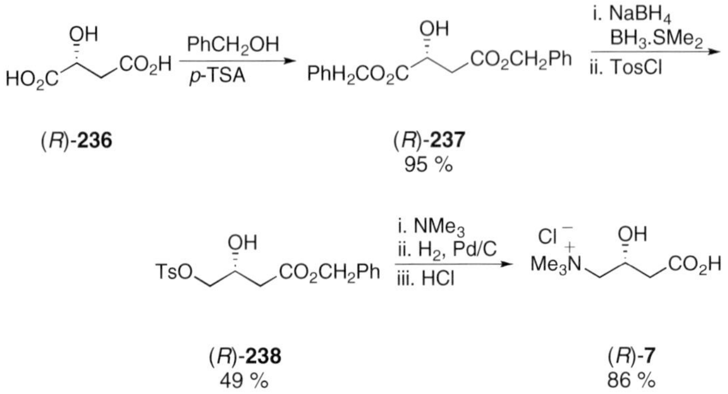

OH
HO_2C / CO_2H — $PhCH_2OH$, p-TSA → PhH_2CO_2C / OH / CO_2CH_2Ph — i. $NaBH_4$, $BH_3.SMe_2$; ii. TosCl →
(R)-**236** — (R)-**237** 95 %

TsO / OH / CO_2CH_2Ph — i. NMe_3; ii. H_2, Pd/C; iii. HCl → Cl^- Me_3N^+ / OH / CO_2H
(R)-**238** 49 % — (R)-**7** 86 %

Scheme 14.65

Amino acids have also been used as sources of chirality in the synthesis of (**6**) and (**7**). Enantiomerically pure (R)-2-*tert*-butyl-1,3-oxazoline prepared from serine has been converted to (S)-GABOB in six steps and 23% overall yield [178]. A novel procedure involving the electrochemical oxidation of (S,R)-N-acetyl-4-hydroxyproline (**244**) in methanol that yields the 2-methoxy derivative (**240**) as a mixture of diastereoisomers has been reported (Scheme 14.67). Oxidation of the aminal and opening of the resulting lactam (**246**) affords optically pure (R)-(**6**) in good overall yield that can be converted to (R)-(**7**) [179].

The utility of the commercially available (R,S)-4-(benzyloxycarbonyl)-5,6-diphenyl-2,3,5,6-tetrahydro-4H-1,4-oxazin-2-one (**247**) in the synthesis of (R)- and (S)-(**6**) and (R)- and (S)-(**7**) and β-alkyl analogs has been demonstrated. The acetoxy hemiacetal (**248**) undergoes a stereoselective substitution reaction with allyltrimethylsilane (Scheme 14.68). Oxidative cleavage of (**249**) yields the acid (**250**) which was deprotected to afford (S)-(**6**) which can be converted to (S)-(**7**) in two steps [180]. A second route involves an asymmetric Mukaiyama-type aldol reaction on (**247**) providing the *tert*-butyldimethylsilyl (TBDMS)-protected hemiketal (**251**) which undergoes elimination to the alkene (**252**) and subsequent reduction to give the all *syn*-substituted oxazine (**253**). Deprotection affords (R)-(**6**) which can undergo conversion to (R)-(**7**) [181].

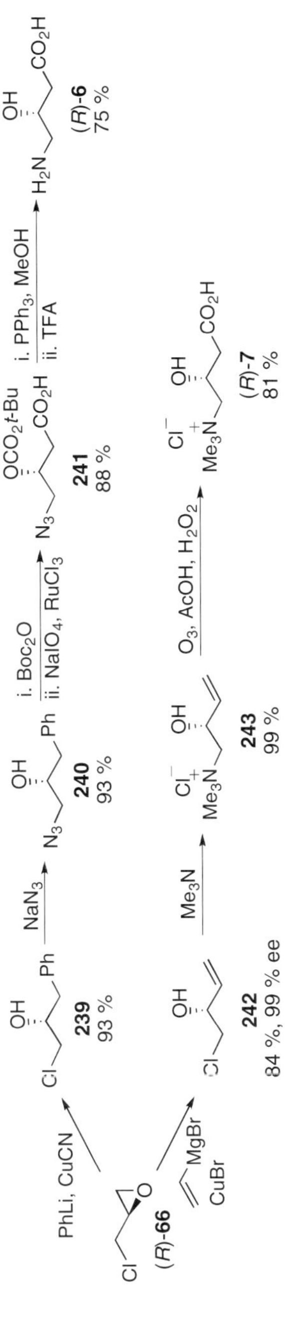

Scheme 14.66

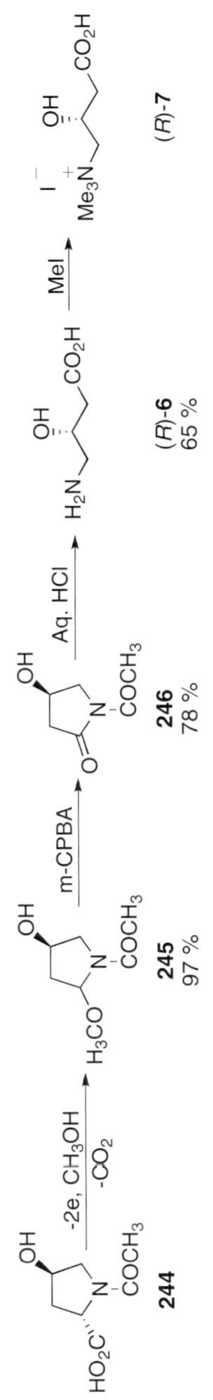

Scheme 14.67

Scheme 14.68

Scheme 14.69

Only a limited number of GABOB and carnitine syntheses based on catalytic asymmetric reactions have been described, and these suffer from relatively low yields and/or low enantiomeric purities. Sharpless epoxidation of but-3-enol yields (*S*)-2-epoxyethanol that can undergo oxidation to the acid. Opening of the epoxide with ammonium hydroxide affords GABOB in 66% overall yield and 49% e.e., which can be improved by repeated crystallizations (95–95% e.e., 7.3% overall yield) [165]. Catalytic asymmetric dihydroxylation of allyl bromide provides access to (*S*)-3-bromopropane-1,2-diol in 74% yield and 72% e.e., which can be converted via a γ-chloro-β-hydroxy nitrile, hydrolysis, and recrystallization to afford (*R*)-GABOB in 90% e.e. Alternatively, treatment with methylamine, recrystallization, and hydrolysis affords (*R*)-carnitine in 95% e.e. [182]. Sharpless asymmetric aminohydroxylation of the 4-nitrophenyl ether of but-3-en-1-ol provides a mixture of 2-hydroxy and 4-hydroxy regioisomeric products in a 10 : 1 ratio. Separation, recrystallization, deprotection, and oxidation to the acid affords the (*R*)-GABOB precursor in 23% overall yield and 96% e.e. [183].

Jacobsens' hydrolytic kinetic resolution technique using the cobalt chiral salen complex (*R,R*)-(**254**) has been used to prepare enantiomerically pure (*R*)- and (*S*)-[2-benzyloxy)ethyl]oxirane (**255**) (Scheme 14.69). Debenzylation of the (*S*)-isomer (*S*)-(**255**) yields (*S*)-(**257**), alcohol oxidation and opening of the epoxide with ammonia yields (*R*)-(**6**) and *N*-methylation affords (*R*)-(**7**) in high overall yield [184].

Enantiomerically pure (*R*)-4-(trichloromethyl)-oxetan-2-one (*R*)-(**260**) was obtained from the poly(acryloyl quinidine) catalyzed [2 + 2] cycloaddition of ketene (**258**) and chloral (**259**) (Scheme 14.70). Ethanolysis of (*R*)-(**260**) in the presence of catalytic amounts of *p*-toluenesulfonic acid and treatment with tributyltinhydride [185] or hydrogenation [186] yields ethyl (*R*)-3-hydroxy-4-chlorobutyrate (*R*)-(**233**) that can be transformed (*R*)-(**6**) and (*R*)-(**7**).

Condensation of 3-benzyloxycyclobutanone (**261**) with α-methylbenzylamine and oxidation yields the oxaziridine (**262**) as a mixture of stereoisomers that undergo a photochemical rearrangement to afford readily separable diastereoisomeric lactams (**263**) in 43 and 40% yields (Scheme 14.71). After chromatographic separation, removal of protecting groups gave (*R*)-4-hydroxypyrrolidin-2-one (**264a**) in 51% yield, which was converted to (*R*)-(**6**) and (*R*)-(**7**) [187].

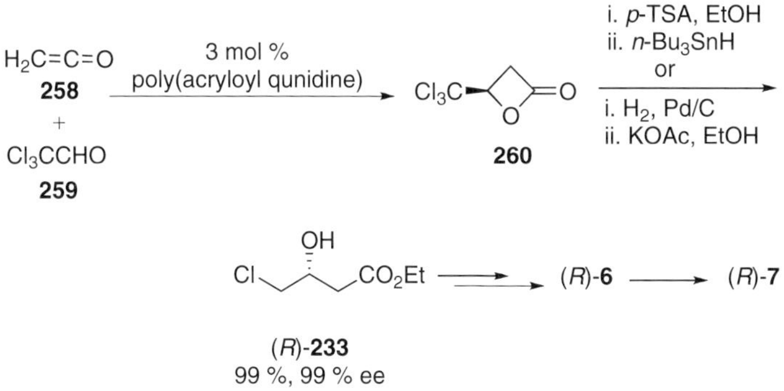

$H_2C{=}C{=}O$ **258** + Cl_3CCHO **259** —(3 mol % poly(acryloyl qunidine))→ Cl_3C–oxetanone **260** —(i. *p*-TSA, EtOH; ii. *n*-Bu_3SnH; or i. H_2, Pd/C; ii. KOAc, EtOH)→ Cl–CH(OH)–CO_2Et (*R*)-**233** 99 %, 99 % ee → (*R*)-**6** → (*R*)-**7**

Scheme 14.70

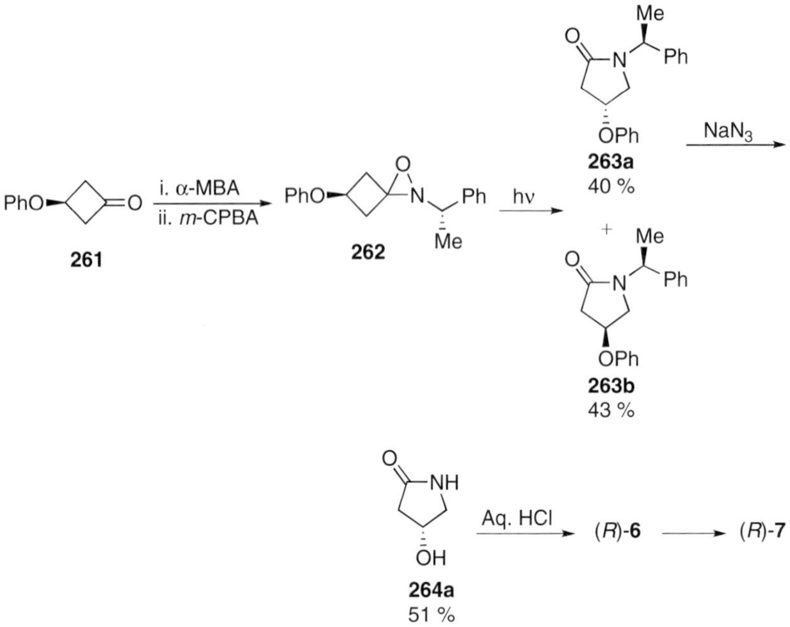

Scheme 14.71

Asymmetric hydrogenation of both ethyl 4-chloro-3-oxobutanoate (**265a**) with Ru-BINAP [188] and ethyl 4-(dimethylamino)-3-oxobutanoate (**265b**) with a range of chiral pyrrolidine-based rhodium catalysts [189, 190] affords the corresponding alcohol (**266**) that can be converted to (*R*)-(**7**) [188–190] and (*R*)-norcarnitine (*R*)-(**267**) (Scheme 14.72) [189, 190]. Highest yields (97%) and enantiomeric excess of 97% were obtained with Ru-BINAP catalyzed reactions, carried out on 100-g scale.

A low-cost, high-yielding seven-step synthesis of (*R*)-(**6**) from glycerol (**268**) has been achieved through the use of the "Oppolzer" (1*R*)-(−)-10-camphorsulfonamide chiral auxiliary (**269**) to desymmetrize glycerol (Scheme 14.73). Reaction of (**268**) with the camphorsulfonamide (**269**) resulted in only one of the four possible spiro-acetals

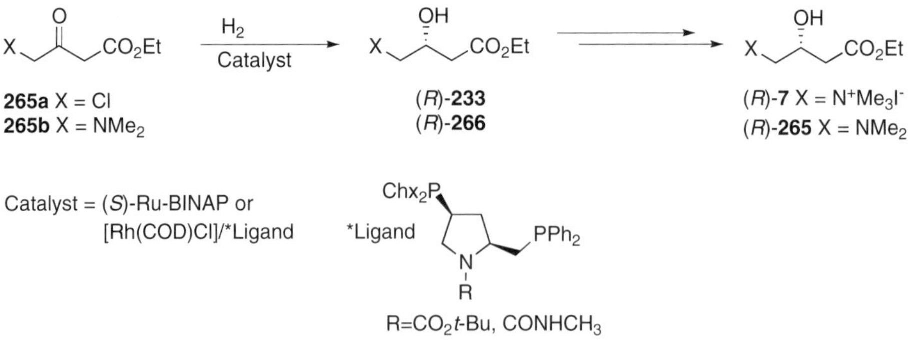

Scheme 14.72

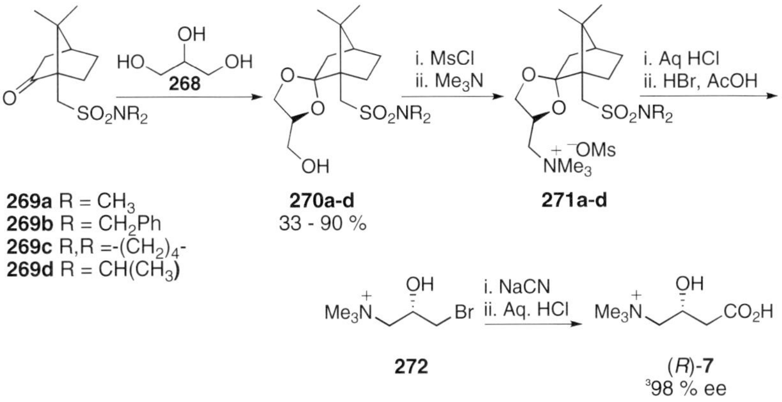

Scheme 14.73

forming. Conversion of the alcohol (**270**) to the mesylate and reaction with trimethylamine yielded trimethylamonium salts (**271**). Cleavage of the chiral auxiliary and treatment with HBr gave exclusively the primary bromide (**272**), which was converted to the nitrile and hydrolyzed to afford (*R*)-(**7**) in 56% overall yield and 98% e.e. or higher [191].

A wide range of chemo-enzymatic syntheses of GABOB and carnitine have been reported, with many of these based on the production of enantiomerically pure epoxides that can be further transformed into GABOB and carnitine. Enantioselective hydrolysis by PLE of (±)-methyl 3,4-epoxybutanoate (**274**), prepared from commercially available 3-butenoic acid (**273**) in two steps, yielded 40% (*R*)-epoxy acid (*R*)-(**275**) which was treated with ammonia and hydrolyzed to afford (*S*)-GABOB (*S*)-(**6**) in 97% e.e. (Scheme 14.74) [192]. The effect of ester length and enzyme on the hydrolysis of a series of alkyl 3,4-epoxybutanoates has also been investigated, with good results being obtained with steapsin 700 hydrolysis of isopropyl and *n*-octyl esters (30 and 40% yield, respectively, and 95% e.e.) [193].

2,2,2-Trichloroethyl 3,4-epoxybutanoate has been resolved by enantioselective transesterification with polyethylene glycol using PPL in diisopropyl ether at 45 °C. The unchanged (*R*) enantiomer, isolated from the reaction mixture by cooling and filtration, was converted to (*R*)-(**7**) (>96% e.e.) in two steps [194]. Optically pure (*S*)-epichlorohydrin (*S*)-(**66**) (>99% e.e.) has also been obtained via microbial

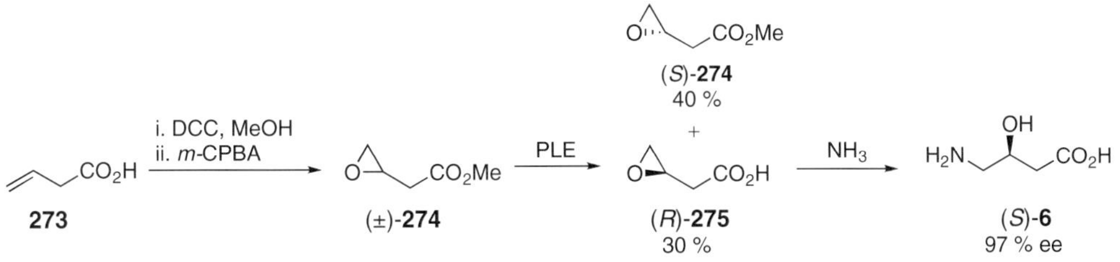

Scheme 14.74

resolution of (±)-2,3-dichloro-l-propanol using *Alcaligenes* sp. DS-K-S389 and converted to (*R*)-(**7**) [195].

Lipases have been used in both enantioselective hydrolyzes and transesterification reactions of cyanohydrins. *Candida cylindracea* lipase (CCL) hydrolysis of *O*-acetyl cyanohydrin (±)-(**277**), prepared in three steps from (**276**), yields the (*R*)-cyanohydrin (*R*)-(**278**) (Scheme 14.75). Treatment of the residual *O*-acetyl cyanohydrin (*S*)-(**277**) with PPL gives the (*S*)-cyanohydrin (*S*)-(**278**). Reduction of the cyanohydrins affords enantiomerically pure (*R*)-(**6**) and (*S*)-(**6**) in good yields [196].

Alternatively, lipase-catalyzed esterification of (±)-*N*-(3-cyano-2-hydroxypropan-1-yl)phthalimide with *Pseudonomas cepacia* lipase (PS) supported on ceramic particles (PS-C) affords (*R*)-*N*-(3-cyano-2-acetoxypropan-1-yl)phthalimide (46%, 99% e.e.), which was converted to (*R*)-(**6**) and (*R*)-(**7**) in high yields and enantioselectivity [197]. Similarly, *P. cepacia* lipase supported on diatomite (PS-D)-catalyzed enatioselective esterification of (±)-3-hydroxy-4-(tosyloxy)butanenitrile provides optically pure (*R*)-3-(acetoxy)-4-(tosyloxy)-butanenitrile which was converted to (*R*)-(**6**) and (*R*)-(**7**) [198].

Candida antarctica lipase-catalyzed aminolysis of the prochiral diester dimethyl 3-ydroxyglutarate (**279**) affords the enantiopure monoamide (*S*)-(**280**) in high yield (Scheme 14.76). Conversion to the acetate and Hofmann rearrangement produced the protected amino acid (*R*)-(**281**), which was deprotected to afford a high yield of enantiopure (*R*)-(**6**) [199].

An alternative approach involves the enantioselective microbial hydrolysis of diethyl-3-hydroxyglutarate by *Corynebacterium equi* (IFO-3730). The resulting (*S*)-monoacid (97% e.e.) was transformed into (*R*)-GABOB and (*S*)-carnitine, via Curtius and Hunsdiecker rearrangements, respectively [200].

Bakers' yeast reductions have been employed to prepare a number of enantiomerically pure intermediates in the synthesis of GABOB and carnitine. Reduction of methyl 4-(*N*-Boc)-3-oxobutanoate (**282**) affords the (*R*)-hydroxy ester (*R*)-(**283**) in high yield and enantiomeric excess (Scheme 14.77). Deprotection affords (*R*)-(**6**) [201]. Ethyl 4-azido- and 4-bromo-3-oxobutanoate [202] and octyl 4-chloro-3-oxobutanoate [203] undergo similar reductions in good yields and enantiomeric excesses.

Both isomers of GABOB (**6**) and carnitine (**7**) have also been prepared by methods which have previously been described for the synthesis of β-substituted γ-amino acids such as the addition of chiral alkyl acetates to α-amino acids [204] and enantioselective di-Rh(II) catalyzed intramolecular C–H insertion of α-diazoacetamides [79, 205].

A number of syntheses of the phosphonic acid analogs of GABOB (GABOB^P; **290**) and carnitine (phosphocaritine; **287**), including resolution of dimethyl (±)-3-(*N*,*N*-dibenzylamino)-2-hydroxypropylphosphonate with (*S*)-*O*-methylmandelic acid [206] and the conversion of (*R*)-epichlorohydrin (*R*)-(**66**) to (*R*)-phosphocarnitine (*R*)-(**287**), have been reported [207]. Compounds (*R*)-and (*S*)-(**287**) have been prepared by bakers' yeast reduction of diethyl 3-azido-2-oxopropanephosphonate in a similar method to that described above for the carboxylic compounds (Scheme 14.77) [208, 209]. Alternatively, resolution of diethyl 3-chloro-2-chloroacetyloxypropanephosphonate (**284**) with *Mucor miehei* lipase provides access to both isomers of 3-chloro-2-

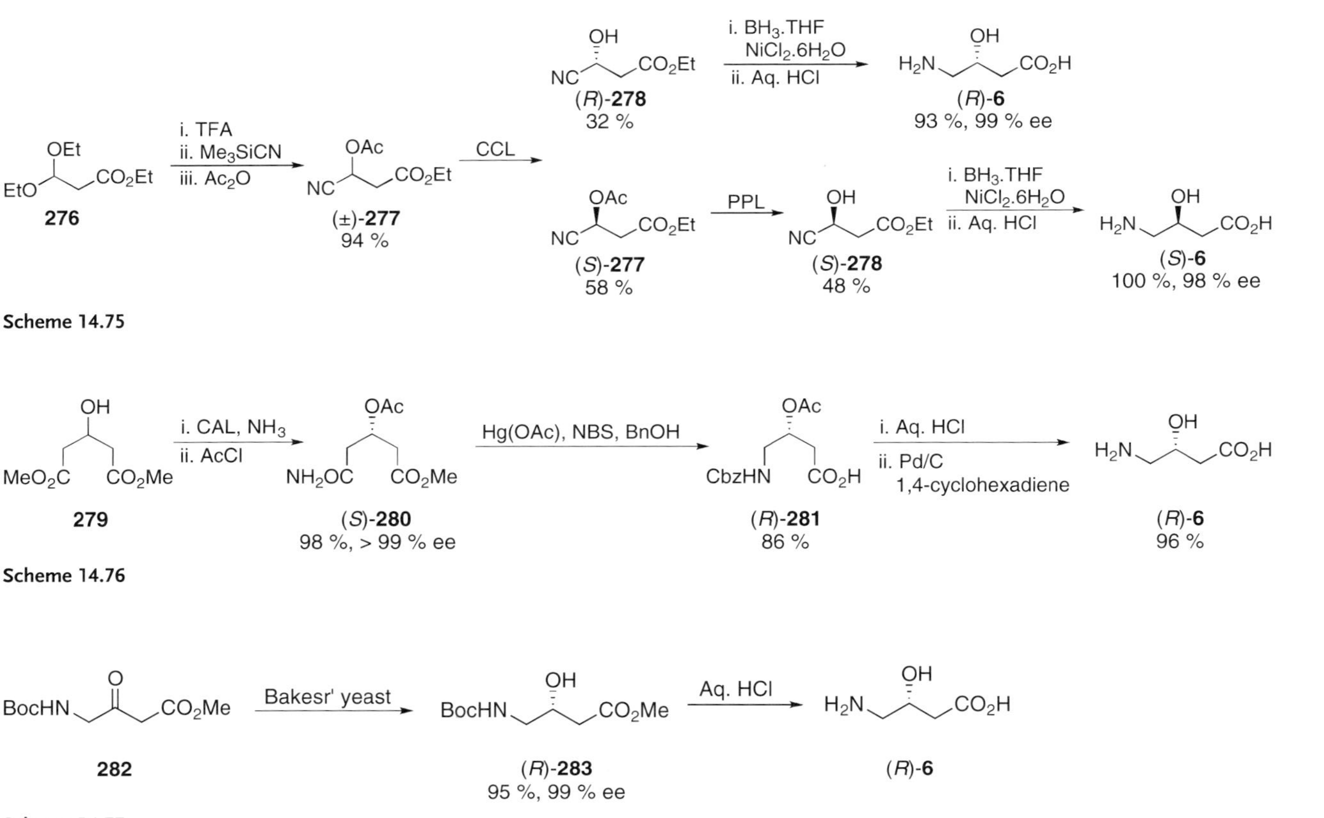

Scheme 14.75

Scheme 14.76

Scheme 14.77

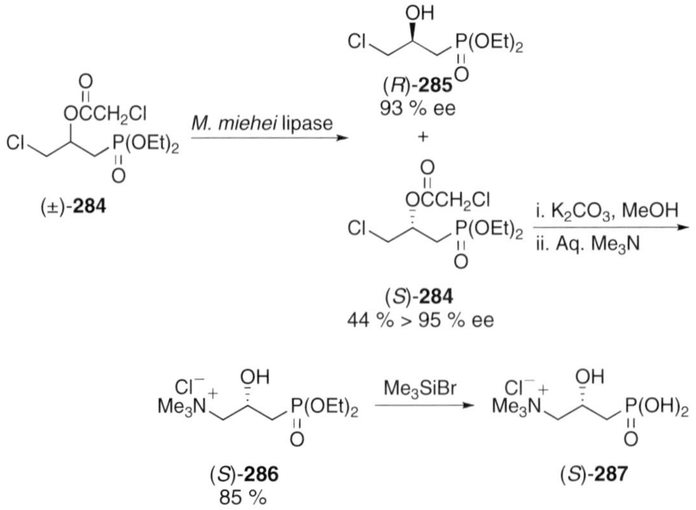

Scheme 14.78

hydroxypropanephosphonate (**285**) that can be converted to phosphocarnitine (**287**) in two steps via the ester (*S*)-(**286**) in 45% overall yield (Scheme 14.78) [209].

In an example of 1,4-induction, the diastereoselective reduction of (*R*,*R*)-dimethyl 3-*N*,*N*-di(α-methylbenzyl)amino-2-ketophosphonate (*R*,*R*)-(**288**) or (*S*,*S*)-(**288**) with catecholborane (CCB) at −78 °C in the presence of $LiClO_4$, yields and (*S*,*R*,*R*)-γ-amino-β-hydroxyphosphonate (*S*,*R*,*R*)-(**289**) and (*R*,*S*,*S*)-(**289**), respectively (Scheme 14.79). Solvolysis and hydrogenolysis afforded the corresponding isomers of (**290**) [210].

(*S*)-Dimenthyl 3-chloro-2-hydroxyphosphonate (**291**) has been used to prepare the (*R*)-epoxide (*R*)-(**292**) without loss of enantiomeric purity (Scheme 14.80). Opening of the epoxide with dibenzylamine yields (*R*)-(**293**), which was deprotected to afford enantiomerically pure $GABOB^P$ (*R*)-(**290**) [211].

Both enantiomers of β-trifluoromethyl- and β-difluoromethyl-GABOB (**297**) have been prepared by the addition of trimethylsilyl cyanide to the corresponding β-alkoxyvinyl polyfluoromethyl ketones (**294**) (Scheme 14.81). Reduction of the cyanohydrins (**295**) and treatment with phthalic anhydride yields the protected amine (**296**). Resolution with (*S*)-phenylethyl amine, deprotection of the aldehyde and oxidation to the carboxylic acid afforded the fluorinated GABOB derivatives (**297**) [212].

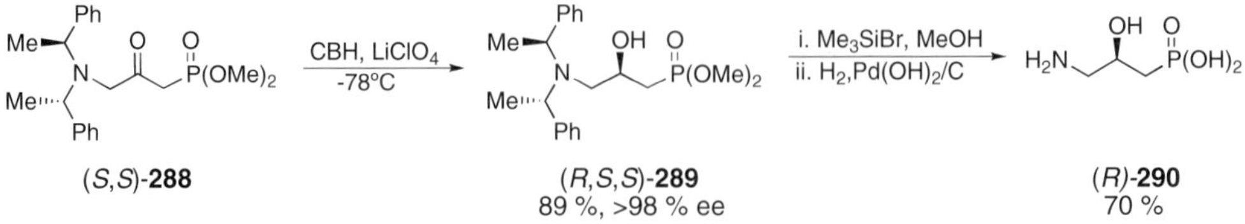

Scheme 14.79

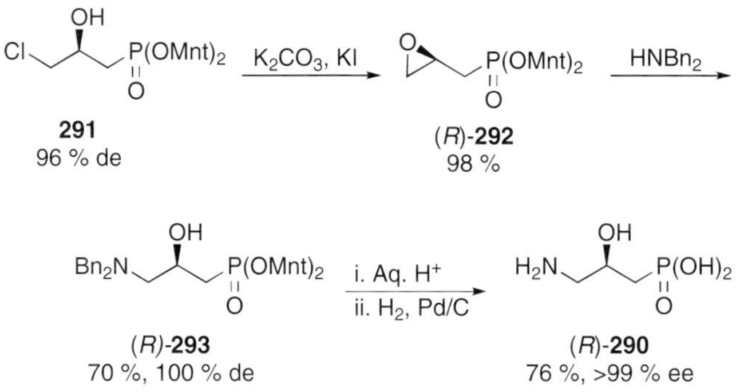

Scheme 14.80

14.8.3
α-Hydroxy-γ-Substituted γ-Amino Acids

The α-hydroxy γ-substituted amino acid (**301**) has been prepared as an intermediate in the synthesis of tubuvaline from the peptide tubulysin. Oxidation of *N*-Cbz-(*S*)-valinol (**298**) with 2,2,6,6-tetramethylpiperidinooxy (TEMPO) and homologation using a Wittig condensation gave the enoate (**299**) (Scheme 14.82). To prevent lactamization, the reduction was carried out with *rac*-BINAP, *t*BuONa, CuCl, and polymethylhydrosiloxane. Treatment of the resulting *N*-Cbz-protected γ-amino acid (**300**) with sodium hexamethyldisilazane in tetrahydrofuran (THF) at −78 °C, followed by the achiral Davis reagent, gave the α-hydroxy derivative (**301**) as a single diastereoisomer. Finally, the hydroxyl group was protected as the *tert*-butyldiphenylsilyl (TBDPS) ether, affording fully protected tubuvaline (**302**) in five steps and 31% yield from valinol [213].

14.8.4
β-Hydroxy-γ-Substituted γ-Amino Acids

Elongation of α-amino acids or α-amino aldehydes by a C-2 synthon has proved a versatile synthetic methodology in the preparation of γ-substituted β-hydroxy-γ-amino acids. The addition of achiral ester enolates to protected α-amino aldehydes and chromatographic isolation of the major product has routinely been used to

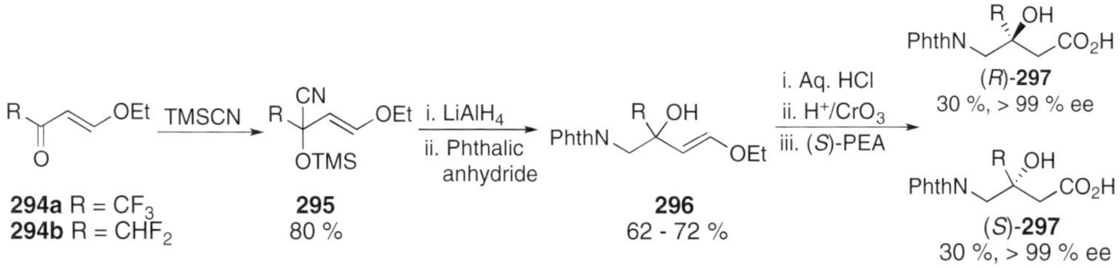

Scheme 14.81

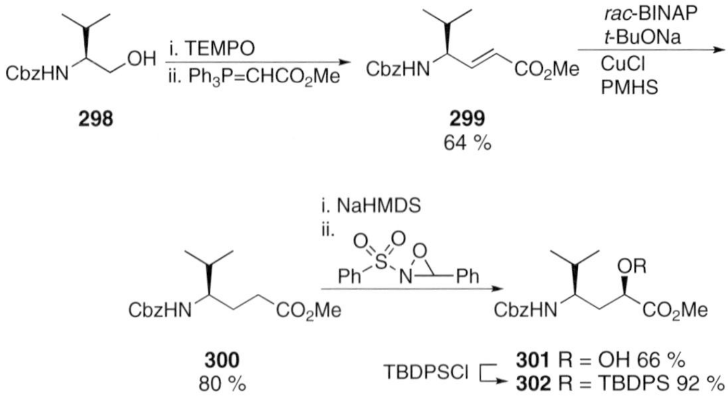

Scheme 14.82

prepare (*S,S*)-4-amino-3-hydroxy-6-methylheptanoic (statine) (*S,S*)-(**8**) [214, 215]. (*R,S*)-Statine [216], and related analogs 4-amino-3-hydroxy-5-phenylpentanoate (AHPPA) [217] and (*S,S*)-4-amino-5-cyclohexyl-3-hydroxypentanoic acid (ACHPA) [218], heterocyclic analogs [219], isostatine [220], all eight possible stereoisomers of isostatine from isoleucinal and *allo*-isoleucinal [221], (*R,S,S*)-dolaisoleuine [222], and phosphostatines [223, 224]. Treatment of isoleucinal (**303**) with ethyl lithioacetate yields the β-hydroxy acid as a 3 : 2 mixture of diastereoisomers (**304**) (Scheme 14.83). Chromatographic separation and hydrolysis affords (*S,S*)-(**8**) [214].

The lithium enolates of alkyl acetates add to Boc- and Cbz-protected amino aldehydes with *syn* selectivity [214], whereas the addition of the same enolates to *N,N*-dibenzyl aminoaldehydes afford the *anti* products [225]. Similarly, *N,N*-dibenzyl aminoaldehydes (**305**) treated with Reformatsky's reagent have also been reported to yield *anti*-γ-dibenzylamino-β-hydroxy esters (**306**) as the major diastereoisomer with moderate diastereoselectivity (3 : 2–5 : 1) in yields of 30–87% (Scheme 14.84). The diastereoisomers of most products can be easily separated by chromatography.

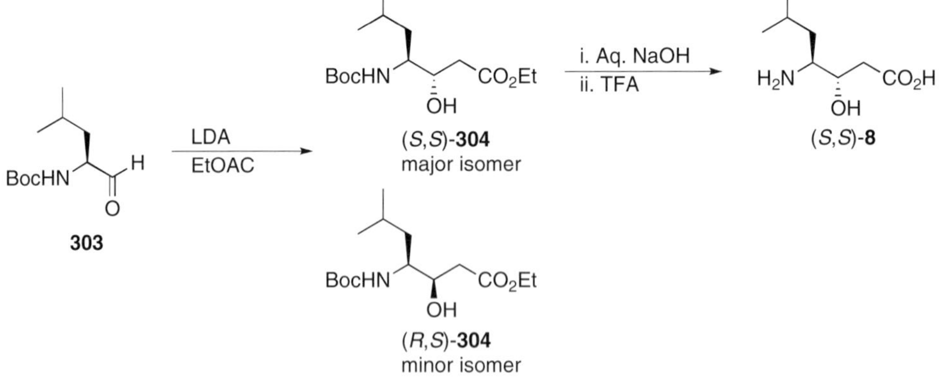

Scheme 14.83

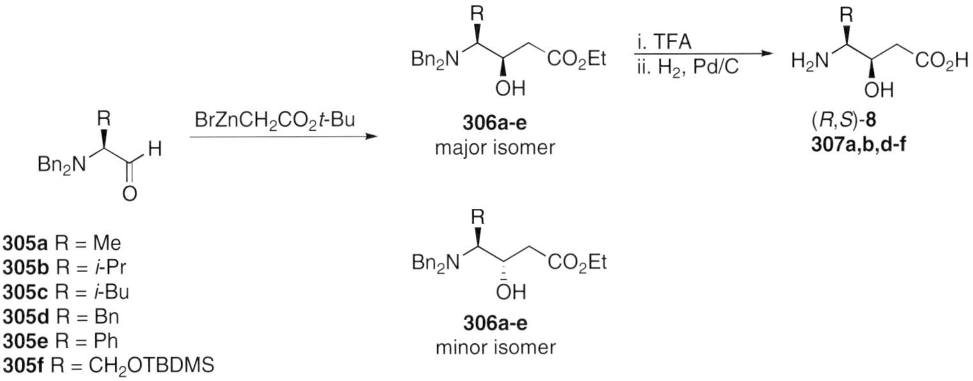

305a R = Me
305b R = *i*-Pr
305c R = *i*-Bu
305d R = Bn
305e R = Ph
305f R = $CH_2OTBDMS$

Scheme 14.84

Deprotection with trifluoroacetic acid (TFA) and hydrogenolysis afford the corresponding γ-amino-β-hydroxy acids (**8**, **307**) [226].

Alternatively, fully protected (*S,S*)-AHPPA has been prepared from (*S*)-*N*-Boc-phenylglycinal via Grignard reaction with allylmagnesium bromide that proceeds in moderate diastereoselectivity. Isolation of the major *syn* product by chromatography, protection of the amine and alcohol as the oxazolidine, and oxidation of the terminal olefin afforded the desired protected amino acid (*S,S*)-(**307d**) [227].

Another approach has been the highly diastereoselective addition of ketene silyl acetals to α-amino aldehydes in an aldol-type reaction catalyzed by Lewis acids [228–230]. Reaction of ketene silyl acetal (**308**) with L-leucinal (**305c**) in the presence of $EtAlCl_2$ affords the *anti* products (**306**, **309**) as the major diastereoisomer (Scheme 14.85). The amino protecting group and the Lewis acid employed is of importance with addition to the *N*-Boc-protected compound in the presence of $SnCl_2$ giving the *syn* adduct [229].

An alternative amino acid based synthesis of statines and related compounds involves the addition of alkyl lithioacetates to activated carboxylic acids and subsequent reduction of the resulting β-keto ester. (*S,R*)-Statines and (*S,R,S*)-isostatine have been prepared by addition of alkyl lithioacetates to the imidazolide (**311**) (Scheme 14.86) [231, 232] or pentafluorophenyl esters [233–235] of *N*-protected D-leucine (*R*)-(**310**) and D-*allo*-isoleucine derivatives, respectively. Reduction of the β-keto ester (**312**) with KBH_4 or $NaBH_4$ affords the corresponding β-hydroxy esters (**313**) in good yield and high diastereomeric excess (up to 91%).

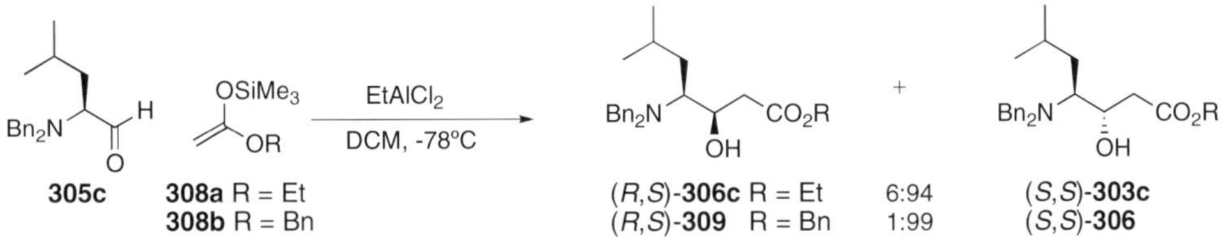

Scheme 14.85

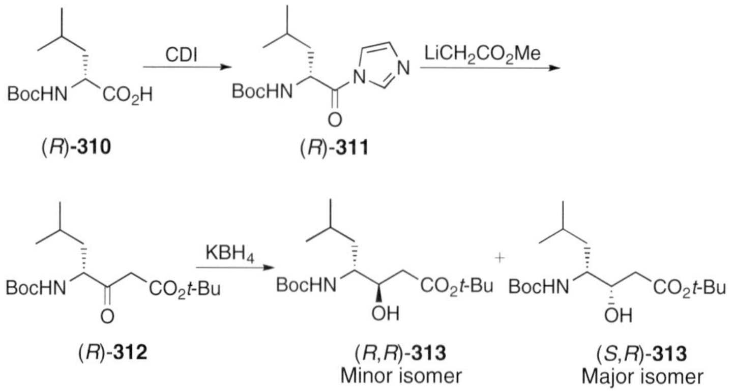

Scheme 14.86

Similarly, protected norstatine has been prepared from valine [236], (*R*,*S*,*S*)-dolaisoleuine from L-isoleucine [237], (3*R*,4*S*)-AHPPA from L-phenylalanine [238–242], and (*S*,*S*)-ACHPA from the cyclohexyl amino acid prepared by PtO_2-catalyzed hydrogenation of L-phenylalanine [243]. Acid chloride-activated *N*-Boc-protected [244] and *N*-Fmoc-protected [245] amino acids also yield β-hydroxy esters in high enantiomeric purity. The Fmoc protecting group allows for a final purification by crystallization to afford diastereomerically pure products [245]. Using this methodology, a variety of statine analogs, with both natural and unnatural configurations, and with branched and unbranched R groups, have been prepared in four steps from activated α-amino esters in excellent overall yields and diastereoselectivity [246].

Pd(0) coupling of *N*-phthaloyl L-leucine acid chloride (**314**) with benzyl zincbromide has also been used to prepare the ketone (**315**) that undergoes a *syn*-selective reduction with the bulky and chemoselective $LiAlH(OtBu)_3$ (Scheme 14.87). Protection of the alcohol (**316**), Sharpless oxidation of the phenyl ring and esterification afforded the fully protected statine (**317**) [247].

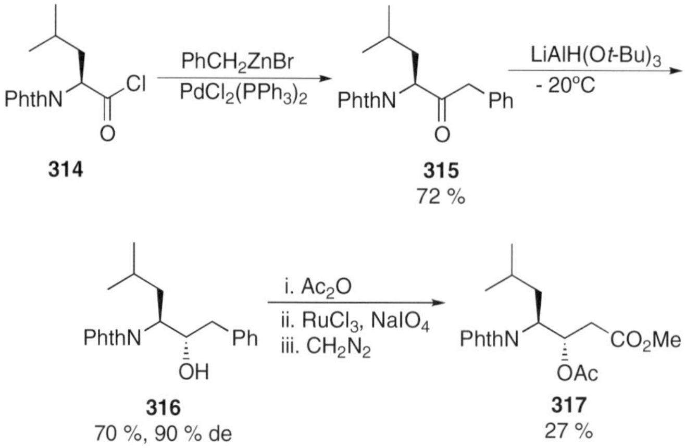

Scheme 14.87

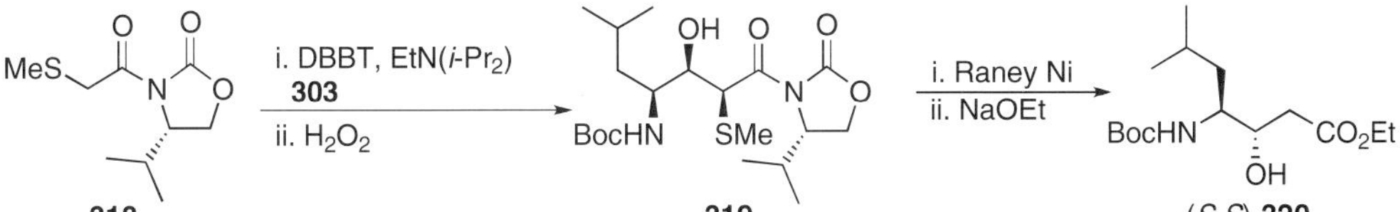

MeS 318 — i. DBBT, EtN(*i*-Pr$_2$) 303; ii. H_2O_2 → BocHN, OH, SMe 319 94 % ee — i. Raney Ni; ii. NaOEt → BocHN, OH, CO_2Et (*S,S*)-320 > 99 % ee

Scheme 14.88

The use of an Evan's chiral auxiliary allows the two stereocenters of the target statines to be established by a single adol reaction. Treatment of the imide of (**315**) with dibutylboron triflate, reaction of the resulting enolate with *N*-Boc-leucinal (**303**), and oxidative decomposition of the boron complex yielded the trisubstituted product (**319**), which was purified chromatographically (Scheme 14.88). Desulfurization and cleavage of the oxazolidine afforded the product (*S,S*)-(**320**) in 24% overall yield [248].

(*S,S*)-4-Amino-3-hydroxy-2-methylbutanoic acid [249], AHPPA (**307e**), *N*-MeAHPPA [250], and a range of statine analogs [251] have also been prepared using Evan's aldol methodology in multistep syntheses. Interestingly, increasing the amount of reagents has been reported to reverse the stereoselectivity of the aldol reaction albeit with lower yields and selectivity [252].

The highly *syn* diastereo- and enantioselective alkylation of the (*S*)-1-amino-2-(1-methoxy-1-methylethyl)pyrrolidine hydrazone of protected glycol aldehydes (**321**) yields alkylated hydrazone (**322**) which has been used in the preparation of (*R,R*)-statine (*R,R*)-(**8**) and analogs (Scheme 14.89). The hydrazone was prepared by condensation of the glycol aldehyde with the chiral auxiliary. α-Alkylation and

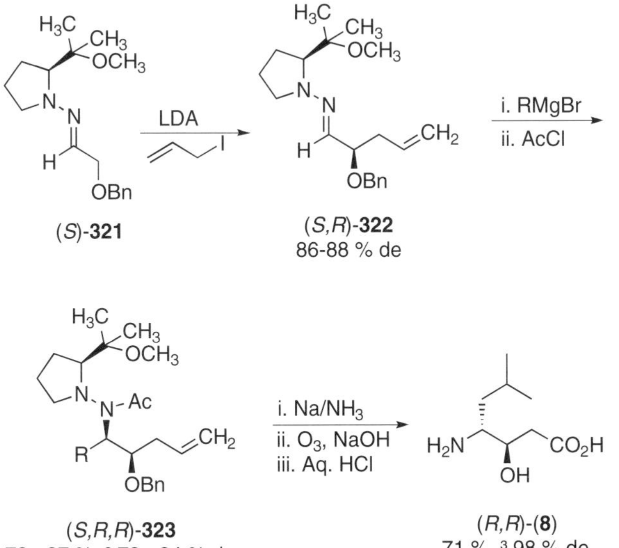

H_3C, CH_3, OCH_3, N, N, H, OBn (*S*)-321 — LDA → H_3C, CH_3, OCH_3, N, N, H, CH_2, OBn (*S,R*)-322 86-88 % de — i. RMgBr; ii. AcCl →

H_3C, CH_3, OCH_3, N, N-Ac, R, CH_2, OBn (*S,R,R*)-323 78 - 87 %, [3] 78 - 84 % de — i. Na/NH_3; ii. O_3, NaOH; iii. Aq. HCl → H_2N, CO_2H, OH (*R,R*)-(**8**) 71 %, [3] 98 % de

Scheme 14.89

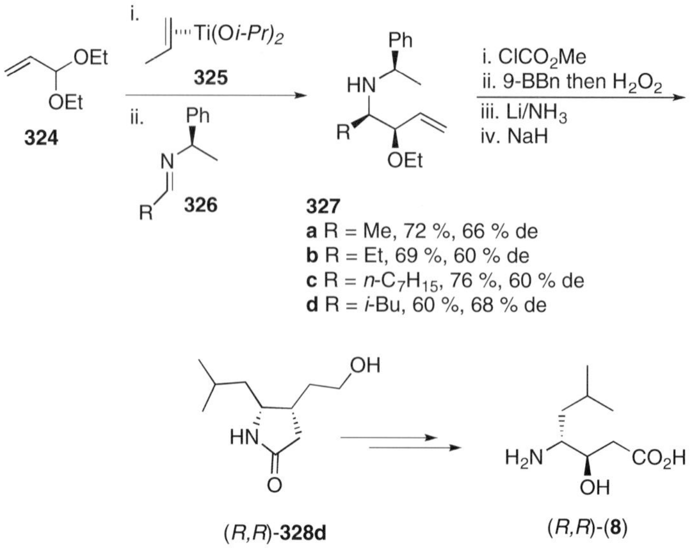

i. $Ti(O\textit{i}\text{-}Pr)_2$ **325**
ii. Ph N R **326**

OEt OEt **324**

Ph HN R OEt

i. $ClCO_2Me$
ii. 9-BBn then H_2O_2
iii. Li/NH_3
iv. NaH

327
a R = Me, 72 %, 66 % de
b R = Et, 69 %, 60 % de
c R = *n*-C_7H_{15}, 76 %, 60 % de
d R = *i*-Bu, 60 %, 68 % de

OH HN O

(*R*,*R*)-**328d**

H_2N OH CO_2H

(*R*,*R*)-(**8**)

Scheme 14.90

subsequent reaction with excess Grignard reagent and the acetyl chloride gives the *N*-acetyl hydrazides (**323**). Removal of the chiral auxiliary with concomitant deprotection of the alcohol, ozonolysis of the terminal alkene, and hydrolysis of the acetamide affords (*R*,*R*)-(**8**) in 38% overall yield [253, 254].

(γ-Alkoxyallyl)titaniums, generated by the reaction of acrolein diethyl acetals (**324**) and a divalent titanium reagent (η-propene)Ti(O*i*Pr)$_2$ (**325**), react readily with chiral imines (**326**), in a regiospecific manner to give optically active *syn*-1-vinyl-2-amino alcohol derivatives (**327**) with moderate diastereoselectivity and yield (Scheme 14.90). Protection of the amine, hydroboration, debenzylation, and subsequent cyclization affords the oxazolidine (**328d**), which was converted to (*R*,*R*)-(**8**) in three steps [255].

The addition of menthone-derived chiral boron enolates of *tert*-butyl thioacetate (**329**) to amino acid-derived chiral α-amino aldehydes (**305**) yields either the 3,4-*anti* or the 3,4-*syn* adduct (**330**) with very high diastereoselectivity depending on the configuration of the chiral boron ligand (Scheme 14.91). This methodology has been used in the synthesis of (*S*,*S*)-statine (*S*,*S*)-(**8**) and related analogs (**307**). The use of the other stereoisomer of the boron ligand yields the *anti* adducts as the major product with a ratio *syn* : *anti* > 1.8 : 98.2. The higher diastereoselectivities a result of "matched" stereochemistry between the aldehyde and the boron ligand [256, 257].

(*S*)- and (*R*)-2-Acetoxy-1,1,2-triphenylethanol has also been used as a chiral auxiliary in the addition of lithium enolates to various aldehydes in the synthesis of statine and its C-3 epimer [258], and a range of statine analogs [259], with good diastereoselectivity. Reaction of lithium enolate (**332**) with the protected aldehyde (**331**) affords the adduct (**333**) in moderate diastereomeric excess (Scheme 14.92). Chromatographic separation and removal of the chiral auxiliary by transesterification affords enantiomerically pure protected statine analogs (**334**) [259].

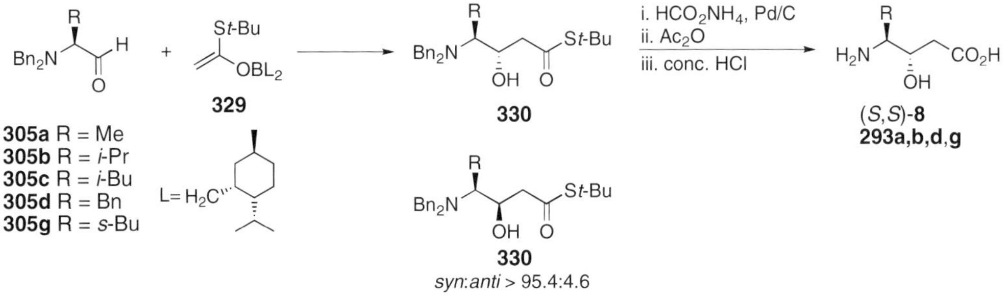

Scheme 14.91

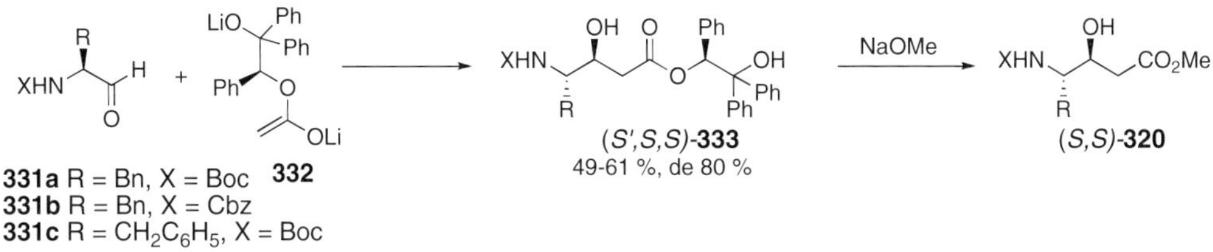

Scheme 14.92

A stereoselective synthesis of statine utilizing an iron acetyl complex as a chiral acetate enolate equivalent has been reported [260]. Diethylaluminium enolates derived from the iron acetyl complex (**335**) undergo highly diastereoselective aldol reactions with *N,N*-dibenzyl α-amino aldehydes (Scheme 14.93). Reaction of complex (**335**) with *N,N*-dibenzyl leucinal (*S*)-(**305c**) yields the adduct (*S,R,S*)-(**336**). Decomplexation and deprotection affords (*R,S*)-statine (*R,S*)-(**8**) [260].

An efficient synthesis of enantiomerically pure (*S,S*)-statine was achieved with the stereoselective intramolecular conjugate addition of a hydroxyl group tethered to the

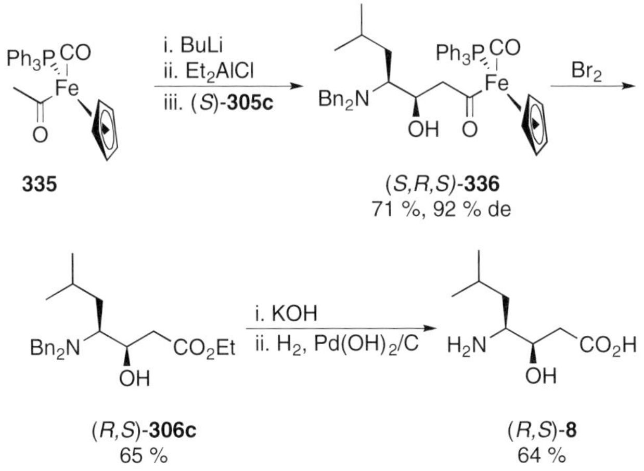

Scheme 14.93

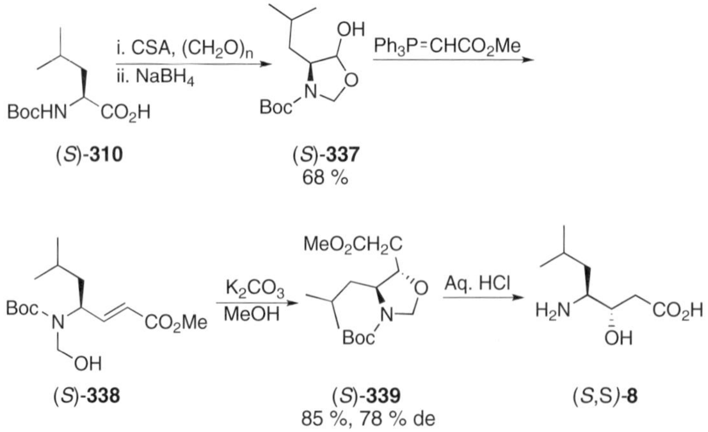

Scheme 14.94

amine of a configurationally stable *N*-Boc-L-leucinal derivative. Conversion of *N*-Boc-L-leucine (**310**) to the *N*-Boc-L-leucinal derivative (**337**) is achieved in two steps and Wittig olefination yields the α,β-unsaturated ester (*S*)-(**338**) (Scheme 14.94). Intramolecular conjugate addition of the hydroxyl group was successful to give the expected oxazolidine (*S*)-(**339**). Deprotection and recrystallization afforded enantiomerically pure (*S*,*S*)-(**8**) [261].

The development of flexible non-amino acid-based approaches to the synthesis of *anti*-γ-amino-β-hydroxy carboxylic acids has been explored. Malimides (**340**) derived from malic acid undergo regioselective alkylation at the C-2 position with good yields as a mixture of diastereoisomers (Scheme 14.95). Catalytic hydrogenation of the diastereomeric mixture gives the lactam as a 3:1 mixture of diastereoisomers [262], however Lewis acid mediated ionic hydrogenation yields only the *trans* isomer (**341**)

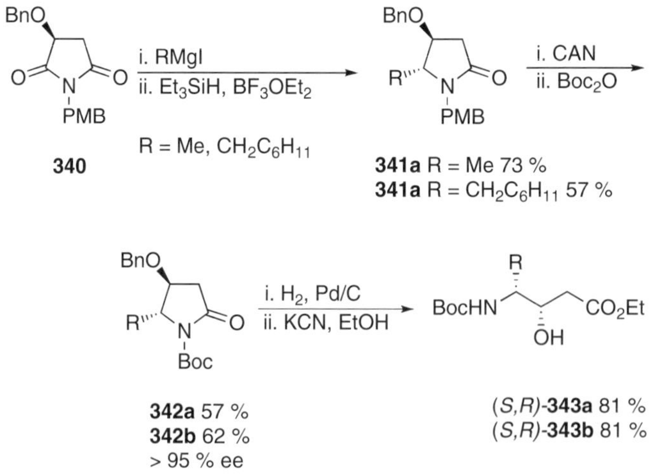

Scheme 14.95

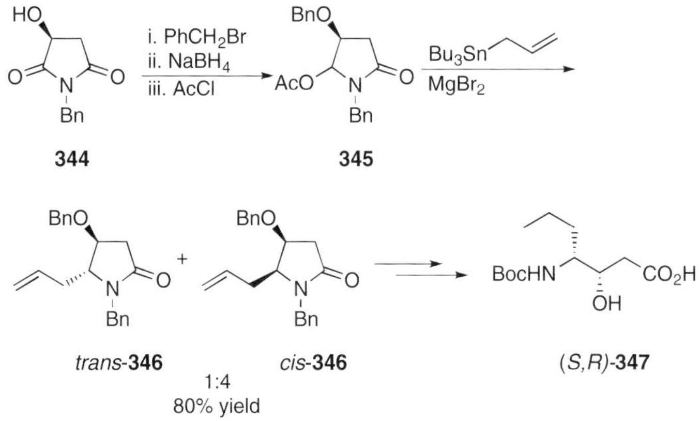

Scheme 14.96

indicating that the reaction proceeds through an *N*-acyliminium intermediate [263]. The pyrrolidin-2-ones (**342**) have been converted to the corresponding protected statine analogs (*S*,*R*)-(**343a,b**) in four steps [262, 263]. Similar methodology has been used to generate small libraries of β-hydroxy-α-amino acids used in a combinatorial synthesis of hapalosin analogs [264].

Statine analogs have also been synthesized from the malimide derivative (**344**) with diastereomeric ratios up to 4 : 11 via a *cis*-selective allylation of the α-alkoxy N-acyliminium intermediate of (**345**) (Scheme 14.96). Alkylation with allyltin in the presence of $MgBr_2$ proceeds in high yield affording the lactam (**346**) as mixture of diasteroisomers which can be converted to the statine analog (**347**) in three steps [265]. A similar procedure has also been reported using methallyltrimethylsilane in the presence of $BF_3{\cdot}Et_2O$ to facilitate the *trans*-selective alkylation (*cis* : *trans* 1 : 11) [266].

Several syntheses of statine or *syn* analogs starting from carbohydrates have been described. These include a multistep syntheses of statine from 3-deoxy-furanose derivatives [267, 268], AHPPA and statine from glucosamine [269, 270], and *N*-methyl AHPPA from (*R*)-cyclohexylidene glyceraldehydes [271]. One of the most versatile and efficient carbohydrate-based syntheses of statines is via the addition of Grignard reagents to the glyceraldehyde-derived chiral imine (**348**) (Scheme 14.97) [272]. The amine (**349**) is produced in high yields and diastereoselectivity, and converted to the amidosulfate (**350**). Displacement of the sulfate with cyanide, methanolysis, hydrogenolysis, and carbamoylation of the *N*-benzyl afforded the statine derivatives (**351**) [273].

The stereocontrolled elaboration of various chiral auxiliaries has provided access to a range of γ-substituted β-hydroxy-γ-amino acids and related analogs. Chiral sulfoxides have been used as a chiral auxiliary in the synthesis of (*S*,*R*)-4-amino-5,5,5-trifluoro-3-hydroxypentanoic acid [(*S*,*R*)-γ-Tfm-GABOB] (*S*,*R*)-(**357**) (Scheme 14.98). Addition of lithiated (*R*)-*p*-tolyl-γ-butenyl sulfoxide (*S*)-(**352**) to *N*-*p*-methoxyphenyl imine (**353**) affords the adduct (**354**) with poor diastereomeric excess. After cleavage of *N*- *p*-methoxyphenyl group the major (*S*,*R*,R_S) isomer (**355**) was isolated by chromatography. Conversion of the amine, to the *N*-Cbz derivative,

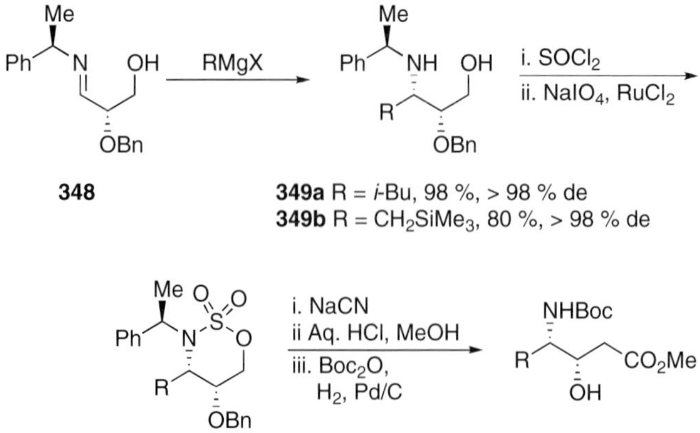

Me
Ph N OH
OBn
RMgX
Me
Ph NH OH
R
OBn
i. $SOCl_2$
ii. $NaIO_4$, $RuCl_2$

348

349a R = *i*-Bu, 98 %, > 98 % de
349b R = CH_2SiMe_3, 80 %, > 98 % de

Me O O
Ph N S O
R
OBn
i. NaCN
ii Aq. HCl, MeOH
iii. Boc_2O,
H_2, Pd/C
NHBoc
R CO_2Me
OH

350a 72 %
350b 71 %

(*S,S*)-**351a** 58 %
(*S,S*)-**351b** 65 %

Scheme 14.97

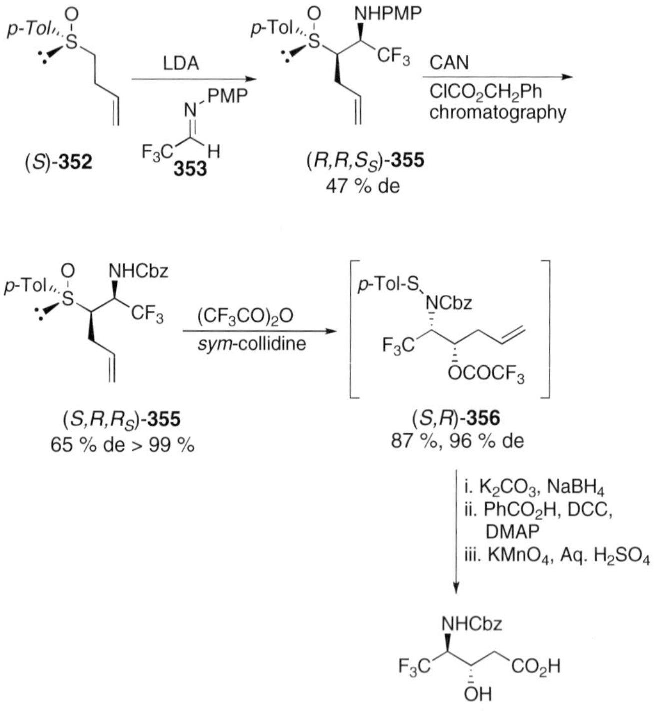

p-Tol S O
LDA
N PMP
F_3C H
p-Tol S O NHPMP
CF_3
CAN
$ClCO_2CH_2Ph$
chromatography

(*S*)-**352**

353

(*R,R,S_S*)-**355**
47 % de

p-Tol S O NHCbz
CF_3
$(CF_3CO)_2O$
sym-collidine
p-Tol-S NCbz
F_3C
$OCOCF_3$

(*S,R,R_S*)-**355**
65 % de > 99 %

(*S,R*)-**356**
87 %, 96 % de

i. K_2CO_3, $NaBH_4$
ii. $PhCO_2H$, DCC, DMAP
iii. $KMnO_4$, Aq. H_2SO_4

NHCbz
F_3C CO_2H
OH

(*S,R*)-**357**
77 %

Scheme 14.98

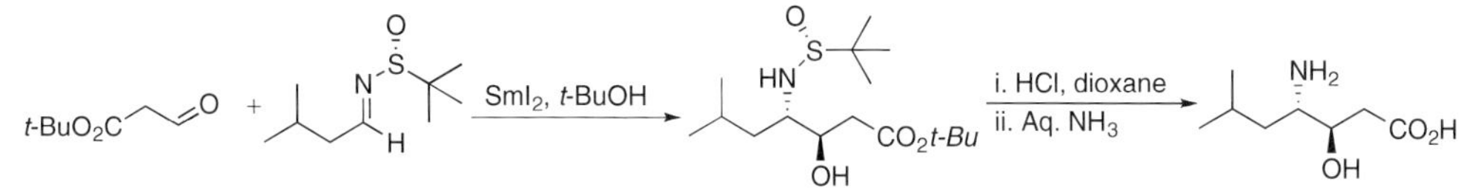

358 (S_S)-359 (R,S,S_S)-360 58 %, 99 % de (R,S)-8 80 %

Scheme 14.99

a nonoxidative Pummerer reaction, and reduction of the sulfenamide intermediate affords (*S*,*R*)-(**356**), which was converted to orthogonally protected enantiomerically pure (*S*,*R*)-γ-Tfm-GABOB (*S*,*R*)-(**357**). The synthesis of (*R*,*S*)-(**357**) from the *N*,*S*-thioaminal of (*R*)-trifluoropyruvaldehyde has also been reported [274].

Alternatively, chiral *N*-*tert*-butanesulfinyl imine of 3-methylbutanal (**359**) undergoes a samarium diiodide catalyzed cross-coupling reaction with *tert*-butyl 3-oxopropanoate (Scheme 14.99). The corresponding β-amino alcohol (**360**) is attained in 58% yield with 99% d.e. Cleavage of the *tert*-butyl ester and *N*-sulfinyl group in one step by acid hydrolysis affords optically pure (*R*,*S*)-statine (*R*,*S*)-(**8**) in high yield [275].

Stereospecific synthesis of *N*-Boc-(*S*,*S*)-statine (*S*,*S*)-(**364a**) and *N*-Boc-(*S*,*S*)-AHPPA (*S*,*S*)-(**364b**) was achieved via a novel Wittig reaction of oxazolidinones (**361**) derived from *N*-Cbz-α-amino acids (Scheme 14.100). Treatment of the oxazolidines (**362**) with HCl yields the tetramic acids (**363**) that can be converted to (*S*,*S*)-(**364a**) and (*S*,*S*)-(**364b**) in four steps with high enantiopurity and good yields [276, 277].

A range of optically pure tetramic acid (**367**) derivatives have also been prepared in high yields by reaction of urethane-*N*-carboxyanhydrides (**362**) from protected α-amino acids, with Meldrum's acid (**119**) and subsequent cyclization (Scheme 14.101). Reduction to the alcohol (**368**) and hydrolysis affords the statine derivatives (**369**) in high overall yields and diastereomeric purity [278].

Reaction of (*E*)-methyl 4-chloro-3-methoxybut-2-enoate (**370**) with ammonia, direct condensation with the appropriate aldehyde, and subsequent hydrolysis yields

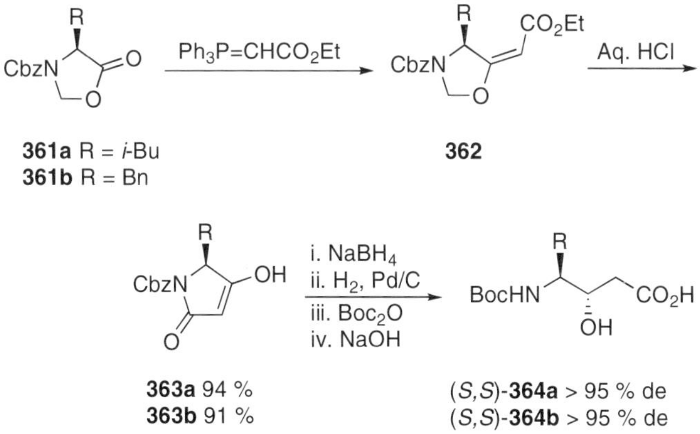

361a R = i-Bu
361b R = Bn
362
363a 94 %
363b 91 %
(S,S)-364a > 95 % de
(S,S)-364b > 95 % de

Scheme 14.100

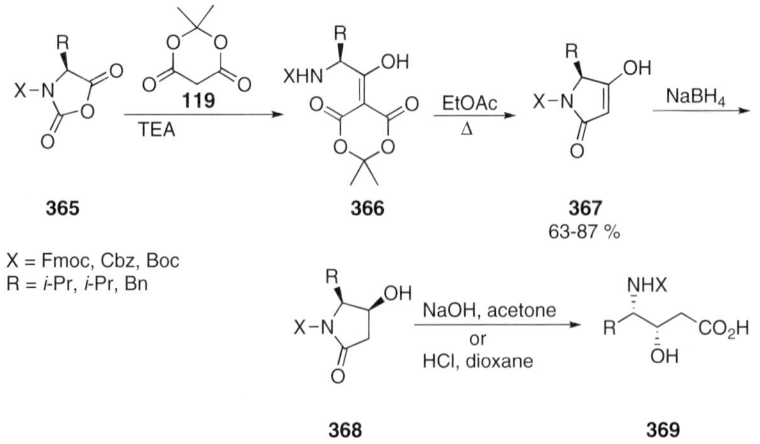

Scheme 14.101

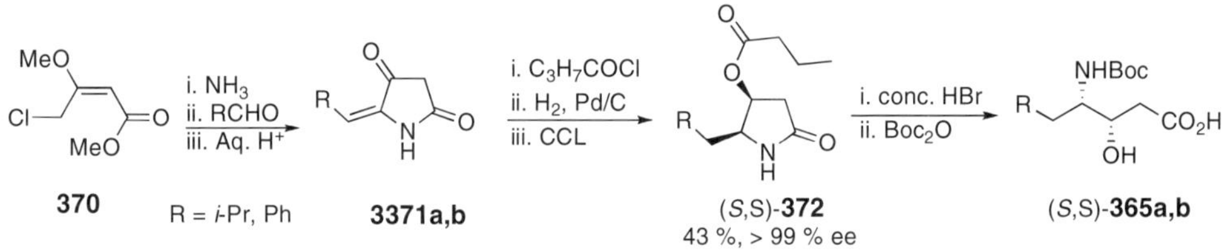

Scheme 14.102

the 5-substituted tetramic acid (**371**) (Scheme 14.102). Conversion to the enol butyrate, hydrogenation, and crystallization afforded the racemic *cis* butyrates which were resolved via a kinetic resolution catalyzed by CCL to provide (*S*,*S*)-(**372**). Acid hydrolysis and protection of the amine affords the *N*-Boc derivatives (*S*,*S*)-(**365**) [279].

Intramolecular Pd(0)-catalyzed reaction of the *O*-acetyl 4-amino-2-alken-1-ols (**373**), prepared in four steps from α-amino acids, affords the *trans* oxazolines (**374**) in good yield and high diastereoselectivity (Scheme 14.103). A hydroboration sequence yields the alcohol (**375**), which was oxidized and esterified to afford the oxazoline protected γ-amino-β-hydroxy esters (**376**). Deprotection to the corresponding γ-amino-β-hydroxy acids can be achieved using known methodology [280].

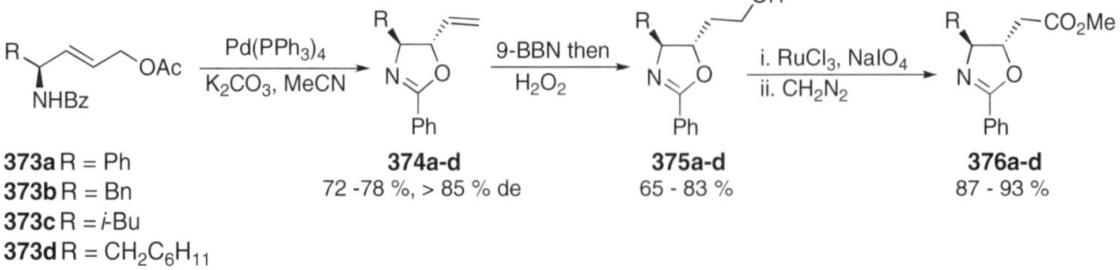

Scheme 14.103

R
Cl
NHCO2Si(t-Bu)2Me
AgF-Pd(II)
R
HN
O
O
i. 9-BBN
ii. Ba(OH)2
iii. (Boc)2O
iv. O2, PtO2
BocHN
CO2H
OH

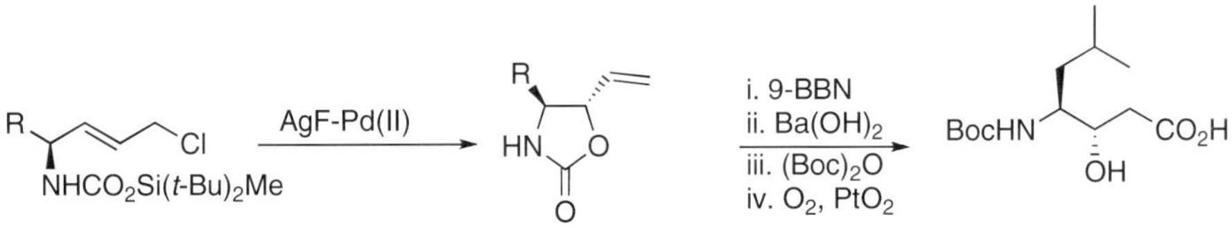

377a R = *i*-Bu
377b R = Bn

(*S,S*)-**378a** 78 % de 87.5 %
(*S,S*)-**378b** 70 % de 81.8 %

(*S*,S)-**379a**
(*S,S*)-**379b**

Scheme 14.104

Stereoselective S_N2' cyclic carbamate formation [initiated by AgF or AgF–Pd(II)] of the *tert*-butyldimethyl silyl carbamate of 4-amino-6-methylhept-2-enyl chloride (**377a**) yields the oxazolidone (**378a**) (Scheme 14.104). Hydroboration of the terminal alkene, cleavage of the oxazolidone, protection of the amine, and oxidation affords the *N*-Boc-protected statine (*S,R*)-(**379a**). AHPPA (*S,R*)-(**379a**) was also prepared by this method [281, 282].

The stereodefined functionalization of oxazolones (**380**) results in the highly diastereoselective formation of substituted oxazolidinone derivatives, which are versatile chiral synthons for vic-amino hydroxy compounds (Scheme 14.105). Allylation at the 5-position and oxidative cleavage of the allyl group yields the ester (**381**). Stereospecific displacement of the 4-methoxy group with a methallyl group afforded the oxazolidinone (**382**), which was converted to (*R,R*)-statine (*R,R*)-(**8**) [283]. In a similar synthesis, all four stereoisomers of statine have been prepared from oxazolidinones derived from (*R*)- and (*S*)-methyl α-hydroxyphenylpropanoate via a highly diastereoselective isobutenylation [284].

A novel intramolecular Ru(II)-catalyzed cyclization of the chiral oxazolones (**383**) results in the exclusive formation of the 12-membered cycloadducts (**384**) with complete diastereoselectivity (Scheme 14.106). Reductive cleavage of the adducts

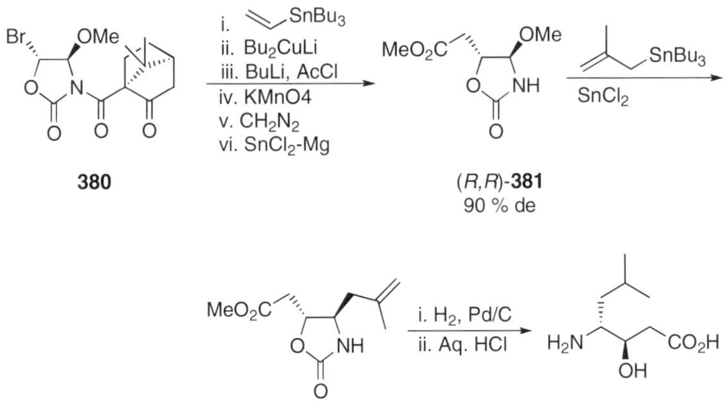

Br
OMe
O
N
O
O
O
i.
SnBu3
ii. Bu2CuLi
iii. BuLi, AcCl
iv. KMnO4
v. CH2N2
vi. SnCl2-Mg
MeO2C
OMe
O
NH
O
SnBu3
SnCl2
MeO2C
O
NH
O
i. H2, Pd/C
ii. Aq. HCl
H2N
CO2H
OH

380

(*R,R*)-**381**
90 % de

(*R,R*)-**382**

(*R,R*)-**8**

Scheme 14.105

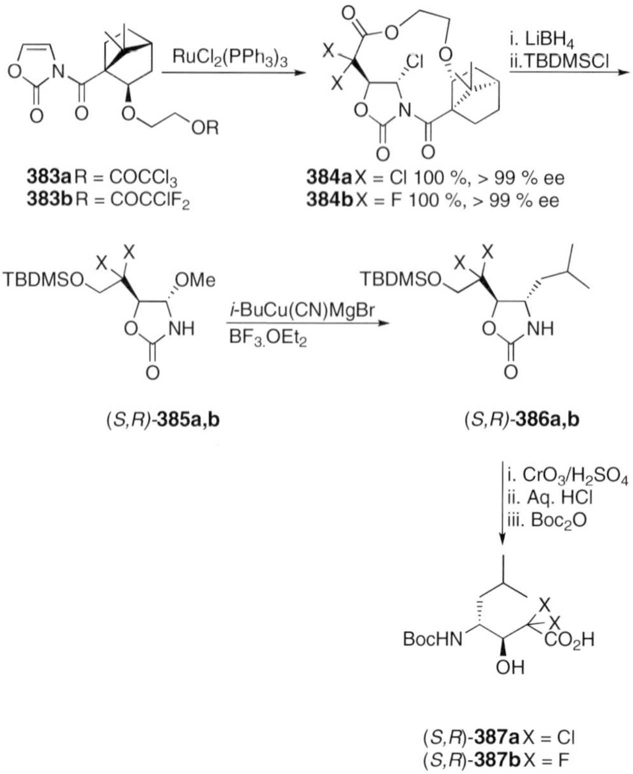

Scheme 14.106

affords the oxazolidinones (*S*,*R*)-(**385**) which can be alkylated and converted to enantiomerically pure dichloro and difluoro statine analogs (*S*,*R*)-(**387**) [285].

All four isomers of statine [286] and a range of analogs [287] have been prepared via a tandem stereoselective reduction/hydroboration strategy. 1-Trialkylsilyl acetylenic ketones (**389**) were derived from the appropriate α-amino acid by reaction of the Weinreb amide (**388**) with the lithium acetylide (Scheme 14.107). Reduction of the ketones with the chiral oxazaborolidine (**390**) afforded the corresponding alcohol (**391**) in good yields and high diastereoselectivity. Oxidative hydroboration yields the *N*-Boc-statine (**365a**) [286, 288].

The reduction of *N*-protected γ-amino-β-ketophosphonates derived from L-amino acids has been thoroughly investigated as a route to phosphostatines. Reaction of protected L-α-amino acids yields the protected γ-amino-β-ketophosphonates (**392**) (Scheme 14.108). Reduction of (*S*)-(**393**) with CCB affords the *syn* product (*R*,*S*)-(**394**) [289, 290], whereas the reduction of (*S*)-*N*-benzylamino-β-ketophosphonates with $Zn(BH_4)_2$ yields the (*S*,*S*) or *anti* product through chelation control [290]. In both cases the reduction proceeds with good chemical yields and high diastereoselectivity. Hydrolysis and hydrogenolysis affords the corresponding aminohydroxyphosphonic acids (**395**). However, protection as the *N*-*p*-toluenesulfonamide is reported to result

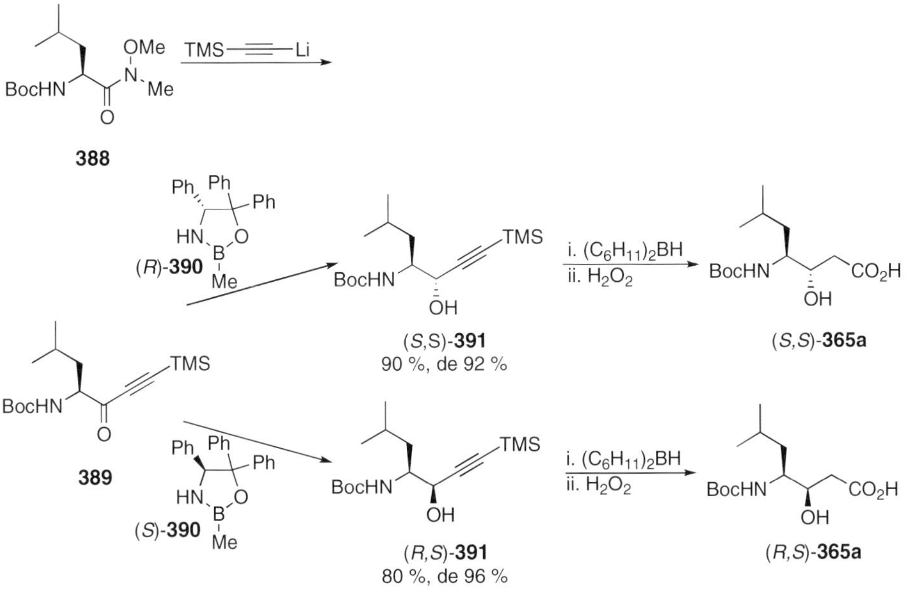

Scheme 14.107

in good yields but poor diastereoselectivity on reduction with a range of hydrides [291].

Both (*R*,*S*)- and (*S*,*S*)-*N*-Boc-statine have also been synthesized in high diastereomeric purity from the readily available β-keto sulfoxide by a stereodivergent sequence involving reduction of the keto sulfoxide with diisobutylaluminum hydride (DIBAH) or DIBAH/$ZnBr_2$, respectively [292].

Statine and statine analogs with natural and unnatural side chains have been prepared via a diastereoselective reduction of a 2-alkyl-substituted 3-ketoglutarate (**396**) by an NADPH-dependent ketoreductase (Scheme 14.109). Various enzymes

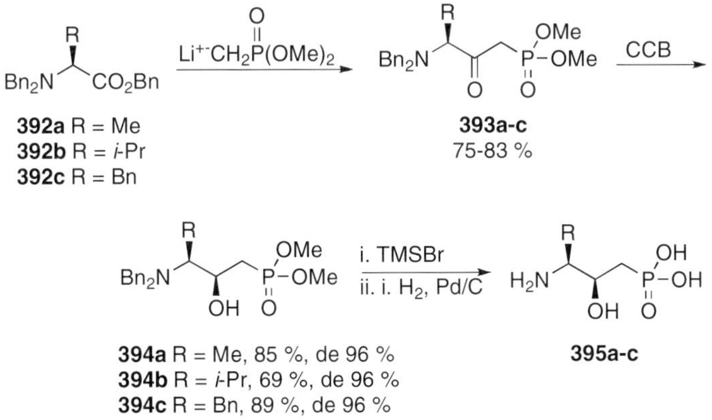

Scheme 14.108

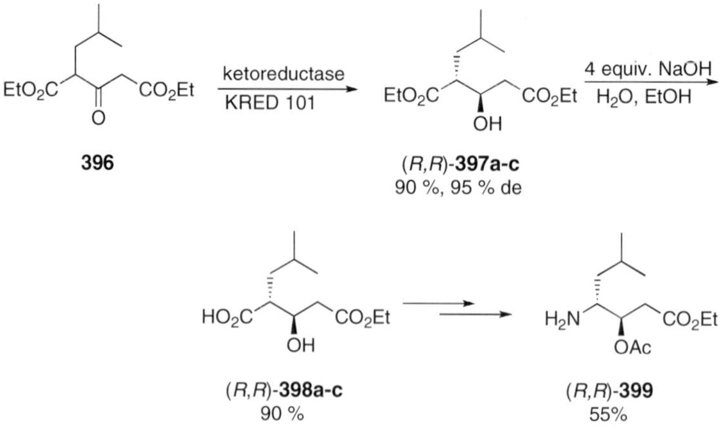

396 → (ketoreductase, KRED 101) → (R,R)-397a-c 90 %, 95 % de → (4 equiv. NaOH, H_2O, EtOH) → (R,R)-398a-c 90 % → (R,R)-399 55%

Scheme 14.109

were evaluated for stereoselectivity, with the enzyme KRED 101 found to have the highest stereoselectivity in the reduction of (**396**). Due to the rapid isomerization of (**396**) the two chiral centers of (**397**) are generated in one step with high yields of a single stereoisomer obtained. Subsequent chemical or enzymatic regioselective hydrolysis to the mono-acid followed by rearrangement under Curtius or Hofmann conditions generates the final statine protected (*R*,*R*)-(**399**) [293, 294].

The diastereoselective hydrogenation of *N*-protected γ-amino β-keto esters catalyzed by BINAP-Ru(II) complexes has also been reported to provide an efficient entry to the statine series with high enantiomeric purities [295]. The protected (*R*,*R*)-AHPPA derivative (*R*,*R*)-(**401**) has been prepared in high yields and enantiomeric purity by a one-pot, two-catalyst sequential reduction of γ-(acylamino)-γ,δ-unsaturated-β-keto esters (**400**) using a combination of Rh(I) and Ru(II) catalysts (Scheme 14.110) [296]. However, these reactions also require high pressures (90 atm) over extended periods of time.

Sharpless catalytic asymmetric functionalization of allylic alcohols affords a convenient entry to enantiopure *syn* or *anti* β-hydroxy-γ-amino acids. Sharpless epoxidation of 4-substituted (*E*)-but-2-en-1-ols (**402**) provides the epoxide (**403**) that is converted to the enantiomerically enriched *anti*-3-amino-1,2-diols (**404**) and subsequently transformed through a stereodivergent sequence to both *N*-Boc-aminoalkyl epoxides (**406**) (Scheme 14.111) [297].

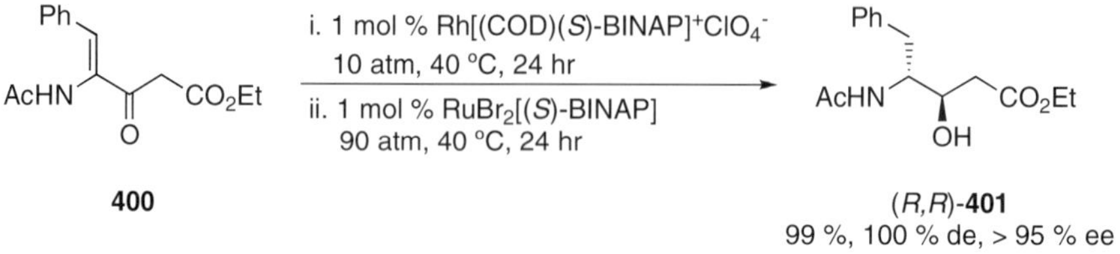

400 → (i. 1 mol % Rh[(COD)(*S*)-BINAP]$^+ClO_4^-$, 10 atm, 40 °C, 24 hr; ii. 1 mol % $RuBr_2$[(*S*)-BINAP], 90 atm, 40 °C, 24 hr) → (R,R)-401 99 %, 100 % de, > 95 % ee

Scheme 14.110

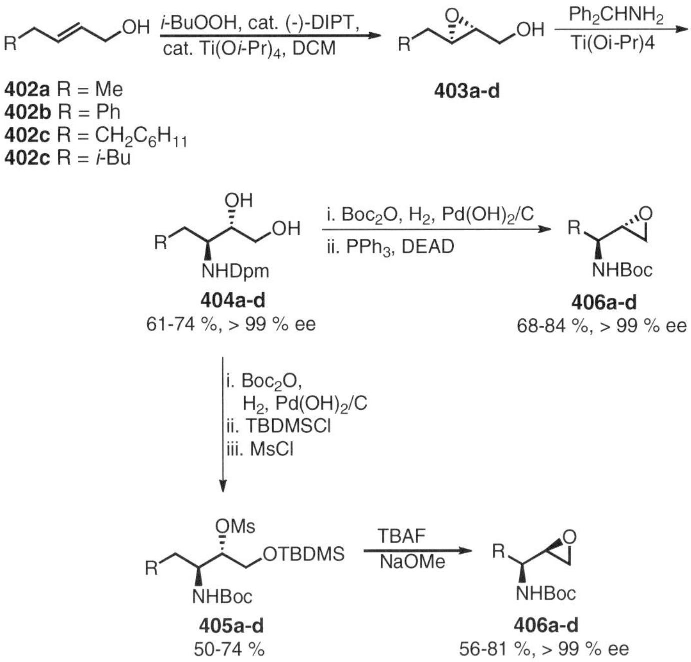

Scheme 14.111

Subsequent regioselective ring-opening of the epoxide (**406b**) with cyanide yields the nitrile (**407**), formation of the oxazolidine (**408**) and nitrile to carboxyl conversion afford, in good yields, protected γ-hydroxy-β-amino acids (**409**) belonging to either the *anti* or *syn* series, depending on the stereochemistry of the epoxide (**406b**) (Scheme 14.112) [298]. Similar methodology has been used in the synthesis of (*R*,*S*)-MeAHPPA [299]. Statine and its 3-epimer have also been prepared by the regioselective epoxide opening of the 2,3-epoxy-1-alkanol (**400d**) by azide [300].

Asymmetric dihydroxylations have also been employed as the sole source of chirality in the synthesis of statines. *N*-MeAHPPA has been prepared in an 11-step synthesis from (*E*)-5-(benzyloxy)pent-2-en-1-ol [301]. Reaction of (*E*)-ethyl 5-methylhex-2-enoate (**410**) under Sharpless asymmetric dihydroxylation conditions yields the *syn*-2,3-dihydroxy ester (**411**) which was converted to the γ-azide (**412**) either directly or via the tosylate (**413**), providing access to both diastereoisomers (Scheme 14.113). Arndt–Eistert homologation and deprotection yielded the *anti*-statine (*S*,*R*)-(**8**) and

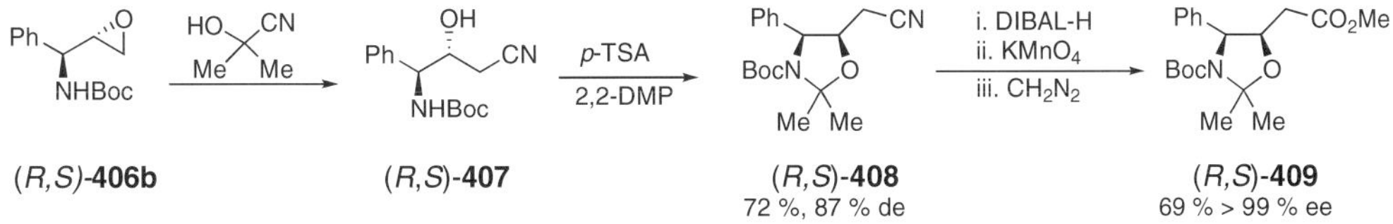

Scheme 14.112

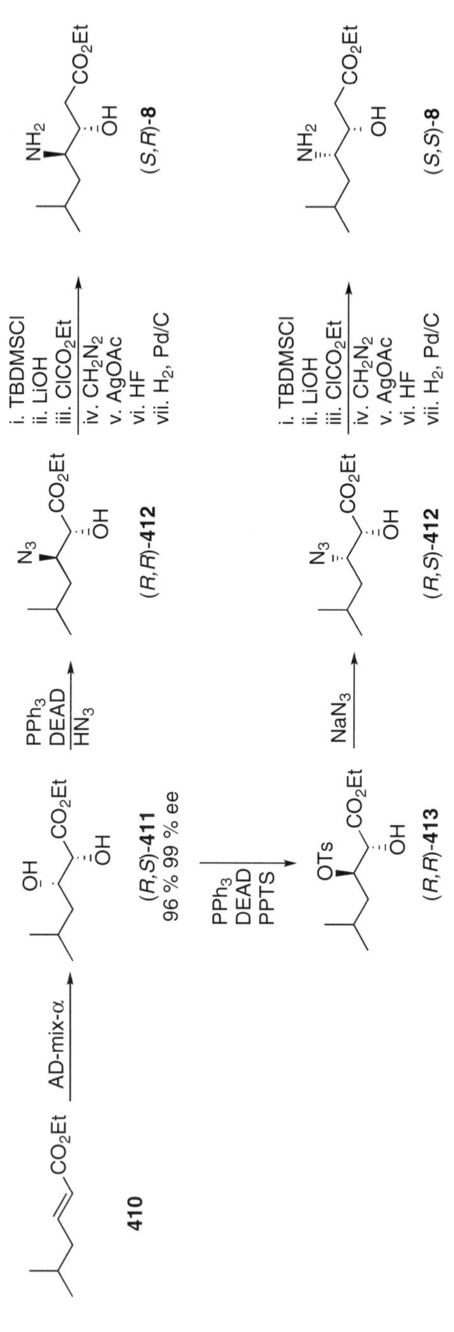

Scheme 14.113

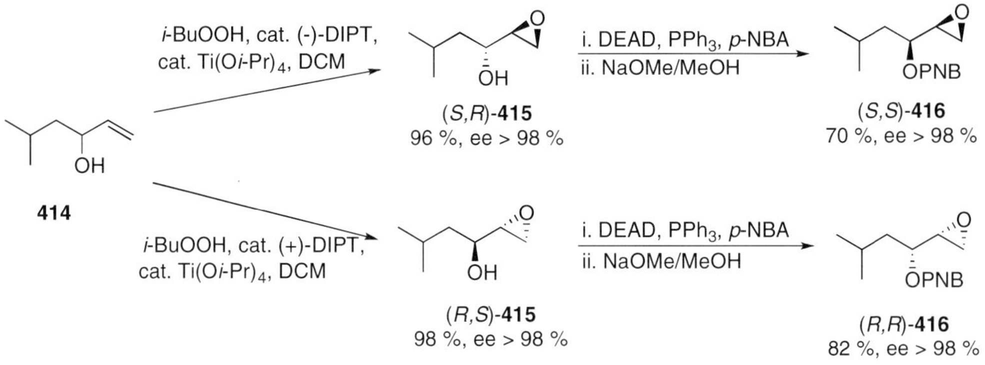

i-BuOOH, cat. (-)-DIPT, cat. Ti(O*i*-Pr)$_4$, DCM
(*S*,*R*)-**415** 96 %, ee > 98 %
i. DEAD, PPh$_3$, *p*-NBA ii. NaOMe/MeOH
(*S*,*S*)-**416** 70 %, ee > 98 %
414
i-BuOOH, cat. (+)-DIPT, cat. Ti(O*i*-Pr)$_4$, DCM
(*R*,*S*)-**415** 98 %, ee > 98 %
i. DEAD, PPh$_3$, *p*-NBA ii. NaOMe/MeOH
(*R*,*R*)-**416** 82 %, ee > 98 %

Scheme 14.114

the natural *syn*-statine (*S*,*S*)-(**8**), respectively. The other two stereoisomers are attainable by the use of AD-mix-β [302].

All four stereoisomers of statine amide (*S*,*S*)-4-amino-3-hydroxy-6-methylheptanamide have been synthesized via a kinetic resolution using the Sharpless asymmetric epoxidation of a racemic 3-hydroxy-5-methyl-l-hexene (**414**) (Scheme 14.114). Epoxidation with diisopropyl D-tartrate gave (*R*,*S*)-(**415**) and diisopropyl L-tartrate gave (*S*,*R*)-(**415**). The *anti* compounds (*S*,*S*)- and (*R*,*R*)-(**416**) were prepared through inversion of the C-2 hydroxyl under Mitsunobu conditions. The epoxides (**416**) were converted to the corresponding statine amides via conversion to the azide and opening of the epoxide with cyanide [303].

AHPPA has also been prepared via Sharpless asymmetric aminohydroxylation of ethyl cinnamate with *N*-bromoacetamide as the nitrogen source. The *N*,*O*-protected aminohydroxyl ester was afforded in moderate yield as a 10 : 1 mixture of regioisomers and 89% e.e. [304].

The phenyl glycine-derived alkylated oxazinone (**417**) was alkylated to yield (**418**) and reduced to the lactol acetate (**419**), which undergoes coupling with the ketene silyl acetals or allyl silanes [305] in the presence of a Lewis acid to afford the corresponding product (**420**) (Scheme 14.115). In the case of the allyl silanes, the reaction proceeds with good to excellent stereoselectivity and yields. The smaller methyl substituent results in the reverse stereoselectivity. However, a substantial amount of a rearrangement product, resulting from a 1,2-alkyl migration, was formed in a number of cases. These coupling products are easily converted to the β-hydroxy-γ-amino acids (**307**) by oxidation to the carboxylic acid and deprotection [306].

Other methods have also been used to prepare statine and related analogs, such as chromatographic resolution of α-methoxyphenylacetates [307], the cycloaddition of chiral enol ethers with dichloroketene coupled with the Beckmann ring expansion [308, 309], and the addition of cyanide to 4-alkoxy-trichloro-but-3-en-2-ones which occurs with poor diastereoselectivity [310].

A novel protected α-methylene-statine (**424**) has been prepared as an intermediate in the synthesis of epopromycin B (Scheme 14.116). Baylis–Hillman reaction of (*S*)-*N*-Fmoc-leucinal (**422**) with 1,1,1,3,3,3-hexafluoroisopropyl acrylate (**423**)

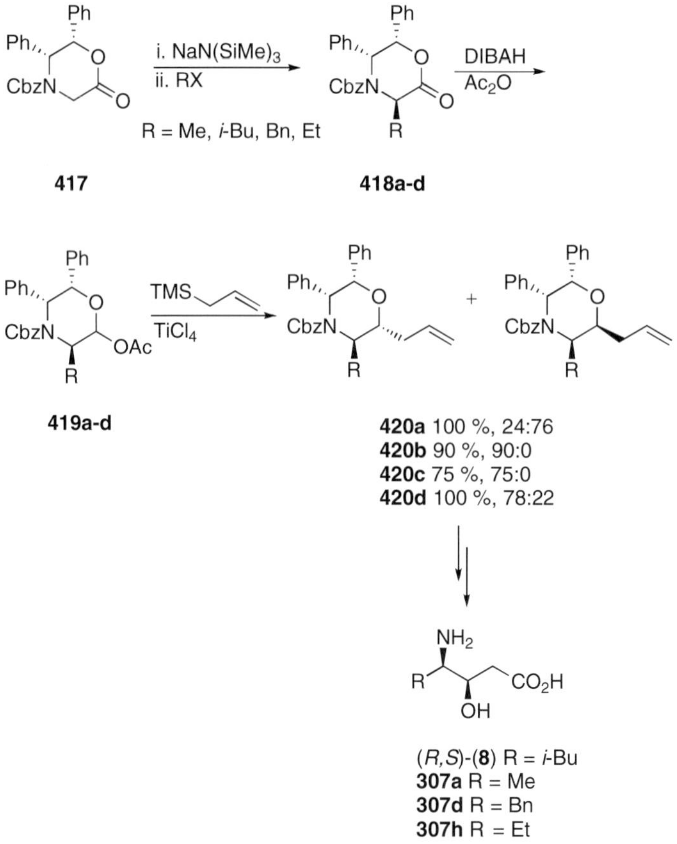

Scheme 14.115

catalyzed by a stoichiometric amount of the *Cinchona* alkaloid proceeded smoothly to yield a mixture of the ester (**424**) and the dioxanone (**425**). Methanolysis of the mixture afforded the *syn* diastereoisomer (**426**) as the major product [311].

14.8.5
β-Hydroxy-Disubstituted γ-Amino Acids

Many of the standard synthetic methods used for the synthesis of β-hydroxy-substituted γ-amino acids have also been used in the synthesis of β-hydroxy-disubstituted derivatives. The 2,3-*anti*-2-isobutyl statines were prepared by the addition of achiral ester enolates to *N*-Boc-leucinal and the 2,3-*syn* isomers via a β-keto ester reduction [312]. Evan's aldol methodology has been employed in the synthesis of β-hydroxy-disubstituted γ-amino acids (2*S*,3*S*,4*R*)-4-amino-3-hydroxy-2-methylpentanoic acid (AHMPA) subunit of bleomycin A_2 [313, 314] and a 2-substituted analog of ACHPA [315]. (2*S*,3*S*,4*R*)-AHMPA has been prepared via a nonstereoselective $NaBH_4$ reduction of the corresponding 3-keto derivative [316],

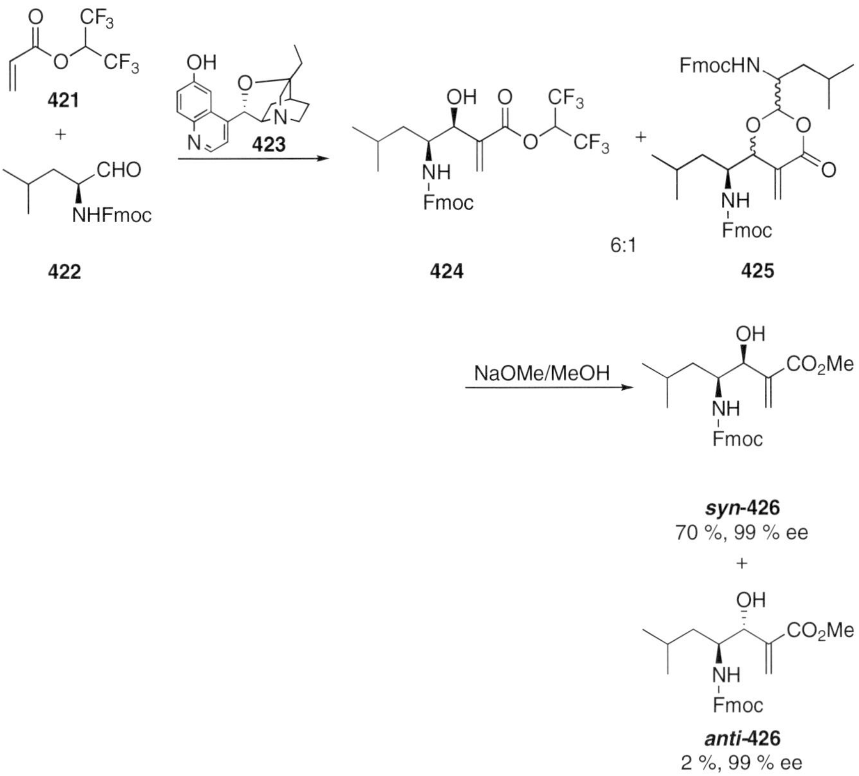

Scheme 14.116

and from L-rhamnose via a multistep synthesis resulting in a low overall yield [317]. A more facile synthesis of (2*S*,3*S*,4*R*)-AHMPA, was achieved through a stereoselective aldol condensation of protected D-alaninal and chiral (*E*)-vinyloxyboranes. Highest diastereoselectivity (35 : 1) was achieved by the addition of phenylthio (*E*)-vinyloxyborane to the protected alaninal [318]. The addition of ethyl lithioacetate to the α-amino acid-derived methyl ketones affords 3-methyl-statine and AHPPA derivatives. Coupling of the diastereoisomers to alanine isoamylamide facilitated chromatographic separation [319].

A range of α-methyl-β-substituted β-hydroxy-γ-amino acids (**430**) have been prepared using a two-step Evans chiral auxiliary methodology. The addition of Et_2AlCN in the presence of $ZnBr_2$ or Et_2AlCl to 1,3-dicarbonyl compounds (**428**) derived from (*S*)-4-isopropyl-2-oxazolidinone (**427**), proceeds with good to excellent diastereoselectivity and good yields (Scheme 14.117). This type of addition to chiral-dicarbonyl substrates represents a new synthetic methodology leading to the formation of enantiomerically pure cyanohydrins (**429**) which can be converted in a three-step procedure to the α-methyl-β-methyl-β-hydroxy-γ-amino acid (**430**) [320].

The achiral TFA-protected glycine crotyl ester (**431**) has been converted to *N*-protected isostatine in four steps via an ester enolate Claisen rearrangement

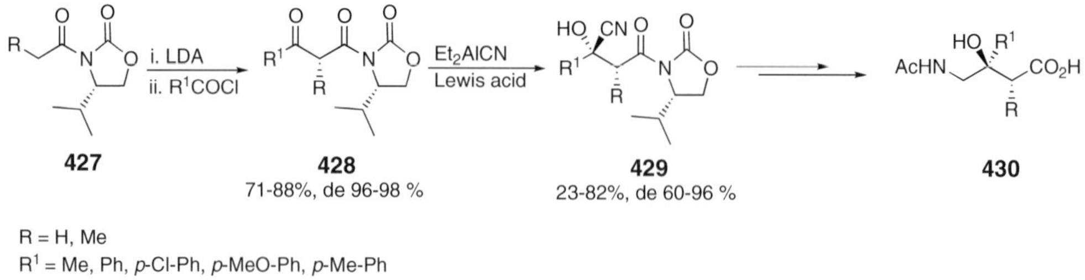

R = H, Me
R^1 = Me, Ph, *p*-Cl-Ph, *p*-MeO-Ph, *p*-Me-Ph

Scheme 14.117

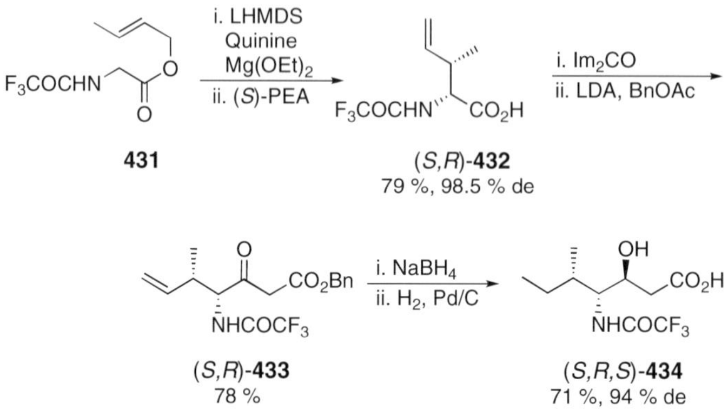

Scheme 14.118

(Scheme 14.118). Deprotonation of the achiral glycine crotyl ester (**431**) in the presence of quinine affords (*S*,*R*)-(**432**) in near quantitative yield and 88% e.e. A single recrystallization with (*S*)-phenylethylamine provides (*S*,*R*)-(**432**) in high enantiomeric purity. Claisen condensation with the imadazolide and conversion to the benzyl ester yields the β-keto ester (*S*,*R*)-(**433**) which is reduced and deprotected to afford (*S*,*R*,*S*)-(**434**) [321].

Enantiomerically pure 4-amino-allyloxyacetates (**436**), prepared from L-α-amino aldehydes (**300**, **435**), undergo a stereoselective Wittig rearrangement in the presence of *N*,*N*,*N*′,*N*′-tetramethylethylenediamine (Scheme 14.119). The *anti*-α-hydroxy-β,γ-substituted γ-amino acid esters (**437**) are produced as the major diastereoisomer in good yield [322].

14.9
Unsaturated γ-Amino Acids

A range of 4-aminobut-2-ynoic acids (4-aminotetrolic acid) (**441**) (Scheme 14.120), the alkyne analog of GABA, have been prepared by direct nucleophillic attack of the appropriate amine on 4-chlorotetrolic acid (**440**), which was prepared via oxidation of 4-chlorobut-2-yn-1-ol (**439**) [323]. More recently, γ-substituted α,β-acetylenic γ-amino

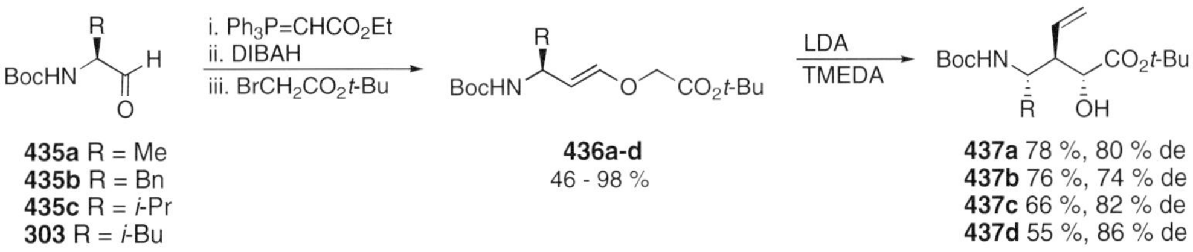

BocHN–CH(R)–CHO → i. $Ph_3P=CHCO_2Et$ ii. DIBAH iii. $BrCH_2CO_2t$-Bu → BocHN–CH(R)–CH=CH–CH₂–O–CH₂–CO_2t-Bu → LDA, TMEDA → BocHN–CH(R)–CH(CH=CH₂)–CH(OH)–CO_2t-Bu

435a R = Me
435b R = Bn
435c R = *i*-Pr
303 R = *i*-Bu

436a-d 46 - 98 %

437a 78 %, 80 % de
437b 76 %, 74 % de
437c 66 %, 82 % de
437d 55 %, 86 % de

Scheme 14.119

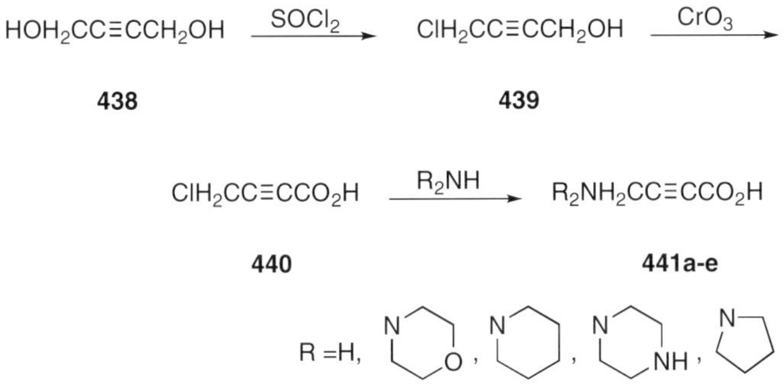

$HOH_2CC{\equiv}CCH_2OH$ (**438**) —$SOCl_2$→ $ClH_2CC{\equiv}CCH_2OH$ (**439**) —CrO_3→ $ClH_2CC{\equiv}CCO_2H$ (**440**) —R_2NH→ $R_2NH_2CC{\equiv}CCO_2H$ (**441a-e**)

R = H, morpholine (N, O), piperidine (N), piperazine (N, NH), pyrrolidine (N)

Scheme 14.120

esters were obtained in moderate yields (30–40%) by a two-step procedure involving formation and flash vacuum pyrolysis of chiral aminoacyl phosphorus ylides [324, 325].

Hydrogenation of 4-aminotetrolic acid (**441a**) over 10% palladium-on-barium sulfate catalyst in the presence of quinoline afforded a mixture of GABA (**1**) and *cis*-4-aminocrotonic acid (**442**) that was purified by ion-exchange chromatography and recrystallization (Scheme 14.121). The *trans* isomer of 4-aminocrotonic acid has been prepared by dehydration of 3-hydroxy-4-aminobutyric acid (**6**) [326].

14.9.1 Unsaturated Substituted γ-Amino Acids

Allan and Twitchin have prepared a range of substituted *trans*-4-aminocrotonic acids (**445**) via amination of the allylic bromides (**444**) (Scheme 14.122). The configuration of (**443**) was found not to be important as the allylic bromination resulted in

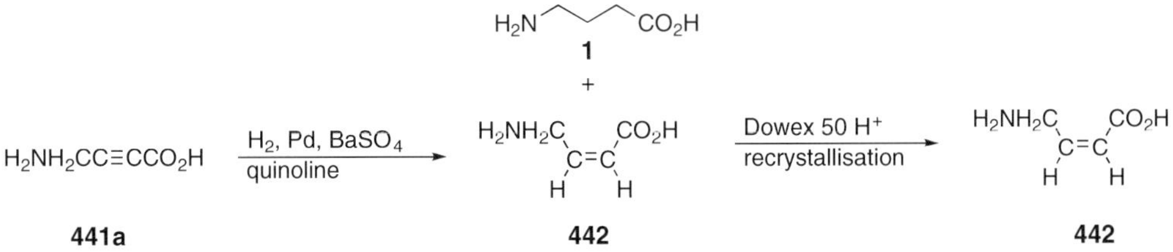

$H_2NH_2CC{\equiv}CCO_2H$ (**441a**) —H_2, Pd, $BaSO_4$, quinoline→ H_2N–CH₂CH₂CH₂–CO_2H (**1**) + H_2NH_2C(H)C=C(H)CO_2H (**442**) —Dowex 50 H^+, recrystallisation→ H_2NH_2C(H)C=C(H)CO_2H (**442**)

Scheme 14.121

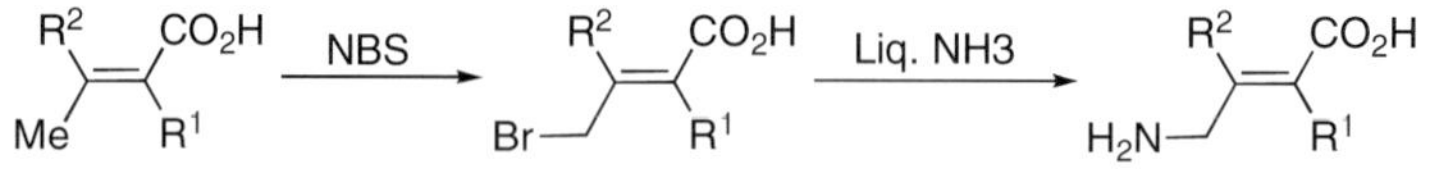

NBS Liq. NH3

443 444 445

R^1 = H, Me, Cl, Br
R^2 = H, Me, Br

Scheme 14.122

isomerization to equilibrium mixtures. The reactions proceed in low to moderate yields and also resulted in the formation of vinyl glycine analogs [327, 328]. 4-Amino-3-halogenobut-2-enoic acids were prepared by *trans* addition of HX to 4-chlorotetrolic acid (**440**) and subsequent amination [329].

The preparation of the *cis* isomer of the unsaturated baclofen analog (**449**) from 4-chloroacetophenone (**446**) via a Reformatsky reaction has been reported (Scheme 14.123) [330]. The α,β-unsaturated acid (**447**) which was isolated by crystallization from a mixture with the alternative α,β-unsaturated acid. Allylic bromination gave a 12 : 1 mixture of monobrominated derivatives (**448**) with the (*Z*) product predominating. Treatment of the (*Z*) isomer with liquid ammonia gave (*Z*)- and (*E*)-4-amino-3-(4-chlorophenyl)but-2-enoic acids (**449**) as a 1 : 1 mixture from which the (*Z*) isomer could be isolated in low yield [330].

γ-Amino-α,β-unsaturated esters have been prepared by Horner–Wadsworth–Emmons olefination of a range of aldehydes. *N*-Boc-protected α-amino aldehydes (**303**, **435b**, **e** and **f**) react smoothly with a variety of ylides and in general without racemization to afford 2,4-disubstitued α,β-unsaturated γ-amino acids (**450**) (Scheme 14.124). Racemization was reported to occur with (**435e**) [331].

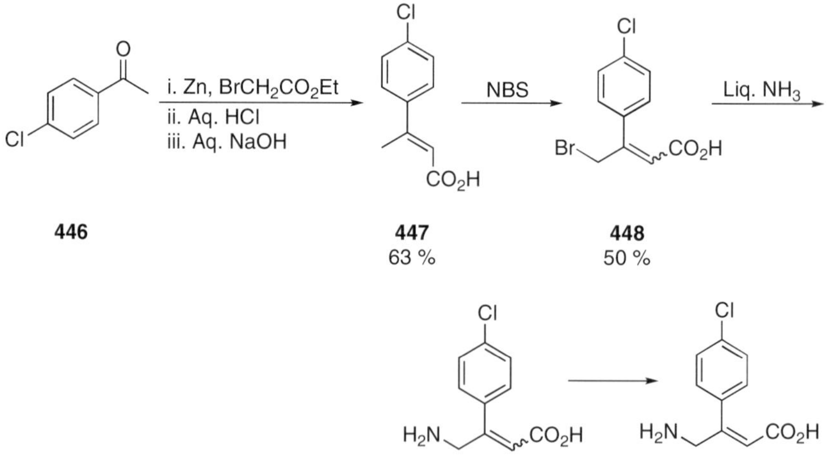

i. Zn, $BrCH_2CO_2Et$
ii. Aq. HCl
iii. Aq. NaOH

NBS Liq. NH_3

446 447 63 % 448 50 %

449 *E*:*Z* 1:1 16 % *Z*-449 4 %

Scheme 14.123

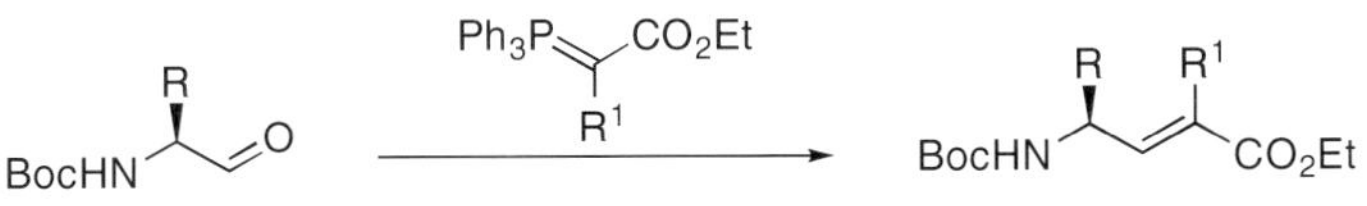

303 R = i-Bu
435b R = Bn
435e R = CH_2OBn
435f R = pyridine

$Ph_3P=C(R^1)CO_2Et$, R^1 = alkyl, aryl

450
48-80 %

Scheme 14.124

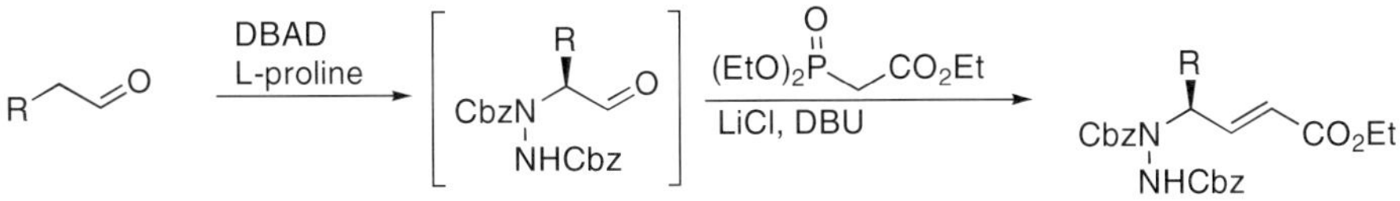

DBAD, L-proline; $(EtO)_2P(O)CH_2CO_2Et$, LiCl, DBU

451
R = Me, Et, i-Pr, Bn

452

453
85 -90%, 92-99% ee

Scheme 14.125

Similar methodology has been used in the solid-phase preparation of olefin-containing protease inhibitors [332]. A one-pot tandem proline-catalyzed direct α-amination/Horner–Wadsworth–Emmons olefination of aldehydes has also been described. Reaction of the aldehydes (**451**) and trapping of the intermediate (**452**) with diethyl phosphonacetate affords the γ-amino-α,β-unsaturated esters (**453**) (Scheme 14.125) [333].

Protected chiral γ-amino acetylenic esters have been synthesized using naturally occurring amino acids as the chiral source. Enantiomerically enriched propargylamines (**454**) [334, 335] or vinyldibromides (**455**) [334] were treated with BuLi at low temperature affording, after alkoxy carboxylation and carbamoylation, enantiomerically enriched derivatives of alkynologous amino esters (**456**) (Scheme 14.126). Cyclopentadienylruthenium (1,4-cyclooctadiene [COD]) chloride-catalyzed reaction

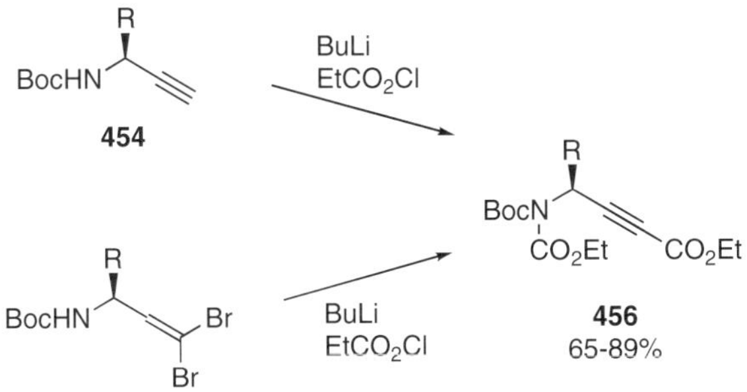

454

BuLi, $EtCO_2Cl$

456
65-89%

455

R = i-Bu, Bn, i-Pr

Scheme 14.126

of the γ-amino acetylenic esters with alkenes affords a convenient synthesis of α-alkylated-γ-amino-Z-alkenoates [335].

γ-Substituted γ-amino α,β-unsaturated esters have also been prepared by the nucleophilic reaction of a planar chiral allyl η3-allyldicarbonylnitrosyliron complex with benzylamine [336], the flash vacuum pyrolysis of α-aminoacyl-stabilized phosphorus ylides [324, 325], and the reaction of nitrones with the alkyl lithiopropiolates and subsequent reduction [337].

14.10 Cyclic γ-Amino Acids

14.10.1 Cyclopropyl γ-Amino Acids

Ethyl 2-cyanocyclopropanecarboxylate (**458**) was initially prepared by cyclopropanation of acrylonitrile (**457**) (Scheme 14.127). The *trans* and *cis* isomers of the cyanoester were separated by fractional distillation. Hydrogenation of the *trans* cyanoester (±)-*trans*-(**458**) in acetic acid, hydrolysis of the resulting amide, and recrystallization yielded *trans*-2-(aminomethyl)cyclopropanecarboxylic acid (±)-*trans*-(**459**) [338]. Hydrogenation of the less-stable *cis* isomer (±)-*cis*-(**458**) was accompanied by isomerism to the *trans* isomer; however, pure *cis*-2-(aminomethyl)cyclopropanecarboxylic acid (±)-*cis*-(**459**) could be isolated after hydrolysis of the amide by repeated slow recrystallizations [339]. All four enantiomers of 2-(aminomethyl)-1-carboxycyclopropane prepared as described above, have also been resolved by chromatographic separation of the diastereomeric (*R*)-pantolactone esters [340].

The (±)-*cis*-(**459**) has also been achieved by reaction of the cyclopropyl lactone (**460**) with potassium phthalimide to give the *cis* acid (**461**), overcoming the problems of racemization (Scheme 14.128). Dephthaloylation was accomplished in ethanolic methylamine solution and the *cis* amino acid was obtained as a crystalline solid. The *trans*-(**459**) was also prepared by Gabriel synthesis of *trans*-ethyl 2-(bromomethyl) cyclopropanecarboxylate and subsequent hydrolysis [341].

Polymer-supported PLE has been used for the resolution of the *meso* diester *cis*-(**462**) to yield the enantiopure monoacid (*R*,*S*)-(**463**) (Scheme 14.129). Borane

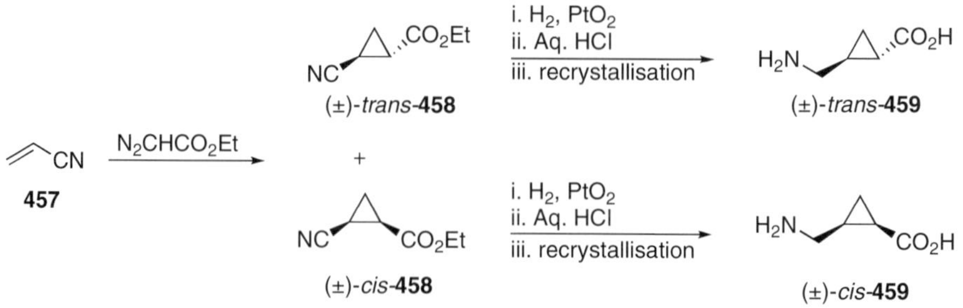

CN **457** —N_2CHCO_2Et→ NC–cyclopropane–CO_2Et (±)-*trans*-**458** + NC–cyclopropane–CO_2Et (±)-*cis*-**458**

i. H_2, PtO_2; ii. Aq. HCl; iii. recrystallisation → H_2N–cyclopropane–CO_2H (±)-*trans*-**459**; H_2N–cyclopropane–CO_2H (±)-*cis*-**459**

Scheme 14.127

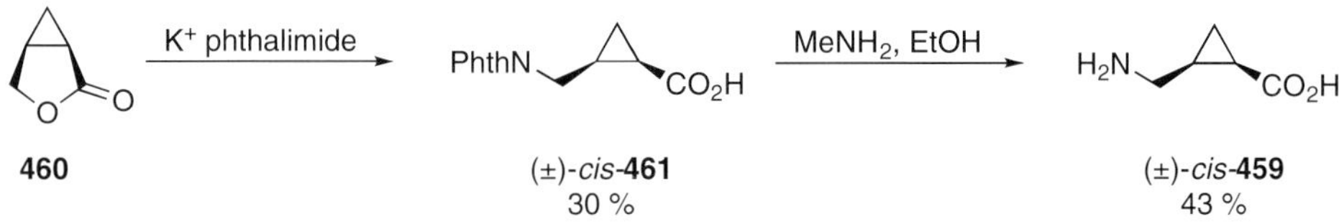

K+ phthalimide; PhthN; CO_2H; MeNH$_2$, EtOH; H_2N; CO_2H

460 (±)-*cis*-461 30 % (±)-*cis*-459 43 %

Scheme 14.128

reduction of the carboxylic acid and removal of boric acid with a borane-specific resin yielded a mixture of alcohol (*R,S*)-(**464**) and lactone (*R,S*)-(**460**), which were both transformed to the bromide (*R,S*)-(**465**). Conversion of (*R,S*)-(**465**) to (*R,S*)-(**459**) was carried out, either via reaction with azide or phthalimide [342].

(*R,S*)-(**459**) has also been prepared by a novel tin-free chemo- and stereoselective radical protocol (Scheme 14.130) [343]. The 4-alkyl-pyrrolin-2-ones (**467**) were synthesized from chiral *N*-allyl-α-bromoacetamides (**466**), via a sequential 5-*exo*-trig radical cyclization-hydrogen or bromine atom-transfer process, and the major isomer isolated by chromatography. Formation of the cyclopropane (**468**) and deprotection afforded (*R,S*)-(**459**) in high yield and enantiopurity [343].

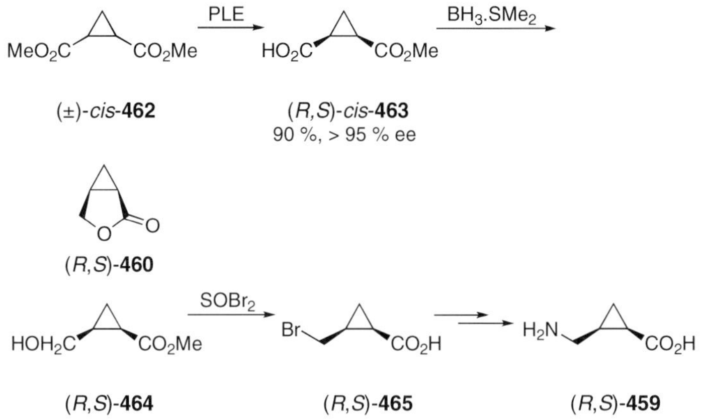

MeO_2C CO_2Me PLE HO_2C CO_2Me $BH_3.SMe_2$

(±)-*cis*-**462** (*R,S*)-*cis*-**463** 90 %, > 95 % ee

(*R,S*)-**460**

HOH_2C CO_2Me $SOBr_2$ Br CO_2H H_2N CO_2H

(*R,S*)-**464** (*R,S*)-**465** (*R,S*)-**459**

Scheme 14.129

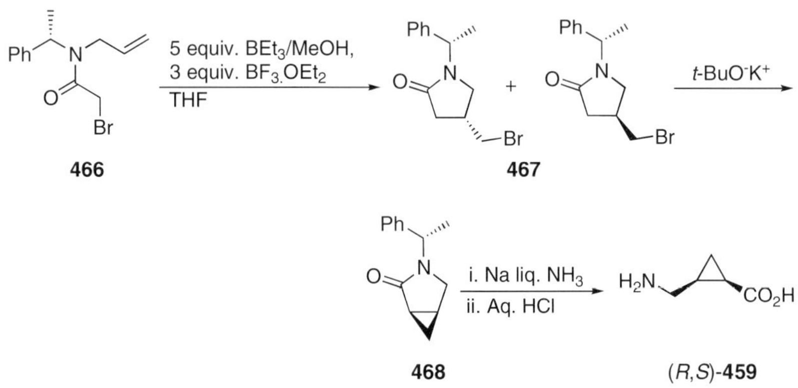

Ph N Br O; 5 equiv. BEt_3/MeOH, 3 equiv. $BF_3.OEt_2$, THF; Ph N O Br + Ph N O Br; *t*-BuO⁻K⁺

466 **467**

Ph N O; i. Na liq. NH_3 ii. Aq. HCl; H_2N CO_2H

468 (*R,S*)-**459**

Scheme 14.130

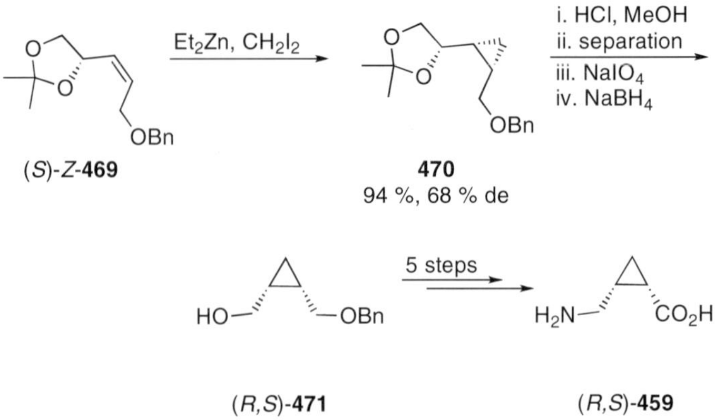

Et$_2$Zn, CH$_2$I$_2$
OBn
OBn
i. HCl, MeOH
ii. separation
iii. NaIO$_4$
iv. NaBH$_4$
(S)-Z-469
470
94 %, 68 % de
5 steps
HO
OBn
H$_2$N
CO$_2$H
(R,S)-471
(R,S)-459

Scheme 14.131

All four enantiomers of (**459**) are available from Simmons–Smith reactions of (*Z*)- and (*E*)-allyl alcohol derivatives (**469**), respectively, obtained from (*R*)-2,3-*O*-isopropylideneglyceraldehyde (Scheme 14.131). The cyclopropane (**470**) was afforded in good yields. Cleavage of the chiral auxiliary, chromatographic separation, and oxidation provided access to the desymmetrized diol (**471**), which could be converted to the corresponding isomer of (**459**) in five steps. The terminal allylic hydroxyl protecting group was found to greatly influence the diastereoselectivity of the cyclopropanation, with the TBDMS ether affording a single diastereoisomer [344].

An efficient synthesis of (*S,S*)-(**459**) has been achieved via asymmetric cyclopropanation of *trans*-cinnamyl alcohol (**472**) in the presence of the (+)-tartaric acid-derived chiral dioxaborolane chiral ligand (*R,R*)-(**473**) (Scheme 14.132). The (*S,S*)-cyclopropyl alcohol (**474**) was obtained in high enantiomeric excess and good yield. Conversion of the alcohol to the azide via the mesylate and reduction with Sn(II) chloride followed by Boc protection yields the *N*-protected amine (**475**). Oxidative degradation of the phenyl moiety to a carboxylic acid and deprotection completed the synthesis of (*S,S*)-(**459**) [345].

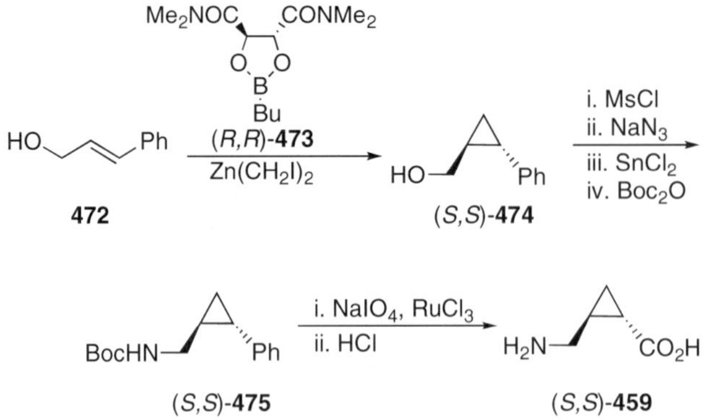

Me$_2$NOC
CONMe$_2$
B
Bu
HO
Ph
(R,R)-473
Zn(CH$_2$I)$_2$
HO
Ph
i. MsCl
ii. NaN$_3$
iii. SnCl$_2$
iv. Boc$_2$O
472
(S,S)-474
BocHN
Ph
i. NaIO$_4$, RuCl$_3$
ii. HCl
H$_2$N
CO$_2$H
(S,S)-475
(S,S)-459

Scheme 14.132

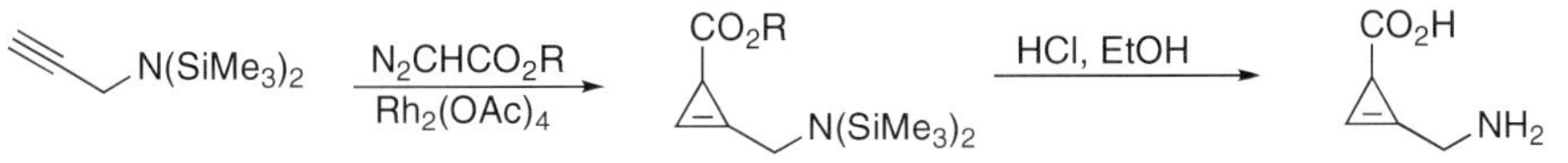

$N(SiMe_3)_2$ $\xrightarrow{N_2CHCO_2R,\ Rh_2(OAc)_4}$ CO_2R $N(SiMe_3)_2$ $\xrightarrow{HCl,\ EtOH}$ CO_2H NH_2

476 **477a** R = Me, 71 % **477b** R = *t*-Bu, 56 % **478**

Scheme 14.133

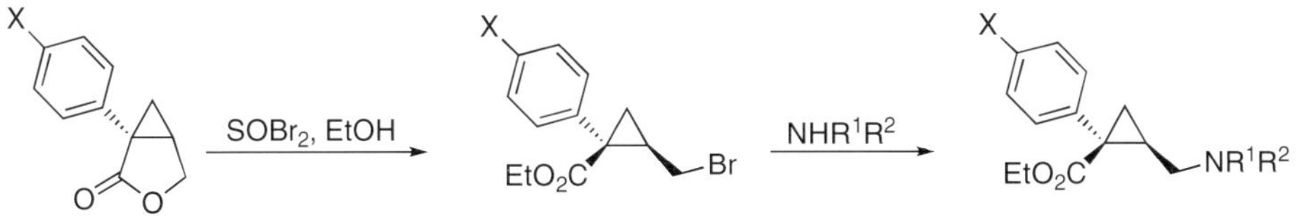

X $\xrightarrow{SOBr_2,\ EtOH}$ X EtO_2C Br $\xrightarrow{NHR^1R^2}$ X EtO_2C NR^1R^2

479 **480** **481**

X = H, Cl, F, CH_3, OCH_3,

R^1 = H, Me, CH_2CH_2OH, *i*-Pr
R^2 = H, CH3, Bn, CH_2CH_2OH, *i*-Pr, *n*-Bu

Scheme 14.134

Treatment of *N*-silylated propargylamine (**476**) with alkyl diazoacetates in the presence of rhodium acetate affords the cyclopropene (**477**) in good yields (Scheme 14.133). Hydrolysis yields racemic 2-(aminomethyl)cyclopropenecarboxylic acid (**478**) [346]. Similarly, β,γ-unsaturated, *N*-silylated amines undergo reaction with diazoacetates to afford 1-substituted 2-(aminomethyl)cyclopropanecarboxylic acids [347] and *N*-silylated allylamines react with substituted methyl diazoacetate to yield 2-substituted 2-(aminomethyl)cyclopropanecarboxylic acids [348].

(*S,R*)-(2-(Aminomethyl)cyclopropanecarboxylic acid [349] and a series of (*Z*)-2-substituted 2-(aminomethyl)cyclopropanecarboxylic acids have also been prepared from the appropriately substituted lactones [350]. Reaction of (**479**) with thionyl bromide affords the bromo ester (**480**) which can be converted to the disubstituted amine (**481**) (Scheme 14.134). Alternatively, reaction with phthalimide and hydrolysis with methylamine affords the primary amine ($R^1 = R^2 = H$) [350, 351].

The multicomponent condensation of organozirconocene, an aldimine, and a zinc carbenoid has been applied to the stereoselective synthesis of γ-substituted α,β–cyclopropane amino acid derivatives. Reaction of the organozirconocene with propargylic ethers (**482**) or homopropargylic ethers, followed by sequential transmetallation to dimethylzinc, addition to *N*-diphenylphosphinylimine, and treatment with bis(iodomethyl)-zinc/dimethoxy ether (DME) complex afforded the desired amide (±)-(**483**) (Scheme 14.135). Removal of the TBDPS group and oxidation of the resulting alcohol afforded the carboxylic acid, which could be converted into the methyl ester. Hydrolysis of the amide and resolution as the tartrate salts afforded diastereomerically pure amino acids (**484**) [352]. Homopropargylic ethers have also been converted to γ-substituted α,β-cyclopropane amino acids [353].

Cyclopropanation of optically active nitroalkenes (**485**) with sulfur ylides or dibromocarbene affords nitrocyclopropanes (**586**) in a diastereoselective manner (Scheme 14.136). Reduction of the nitro group, protection as the *N*-trifluoroacetyl

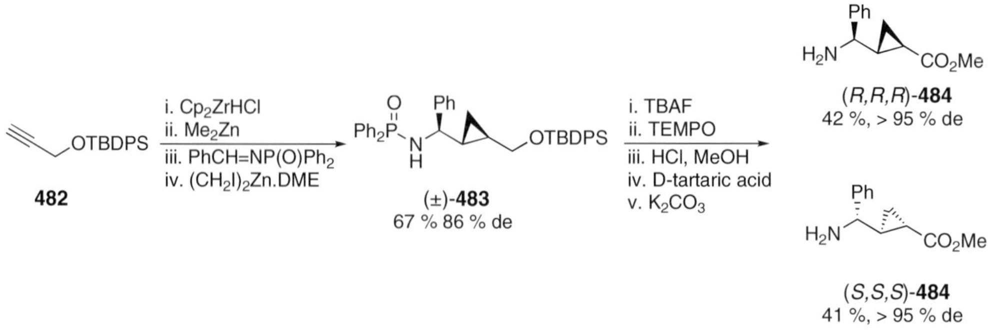

OTBDPS
482
i. Cp_2ZrHCl
ii. Me_2Zn
iii. PhCH=NP(O)Ph_2
iv. $(CH_2I)_2Zn$.DME
Ph_2P
Ph
N
H
OTBDPS
(±)-483
67 % 86 % de
i. TBAF
ii. TEMPO
iii. HCl, MeOH
iv. D-tartaric acid
v. K_2CO_3
Ph
H_2N
CO_2Me
(R,R,R)-484
42 %, > 95 % de
Ph
H_2N
CO_2Me
(S,S,S)-484
41 %, > 95 % de

Scheme 14.135

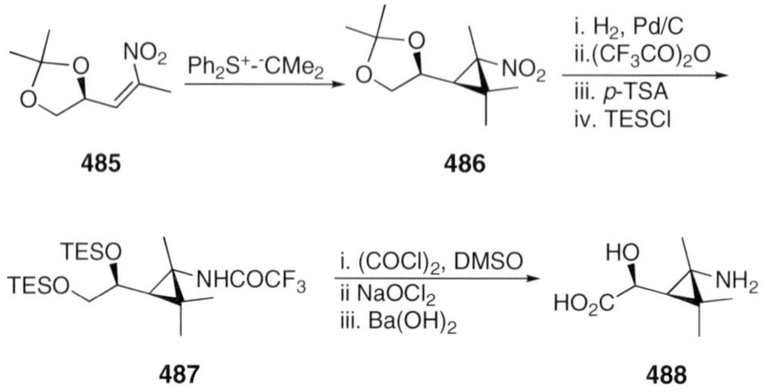

NO_2
485
Ph_2S^+-$^-CMe_2$
NO_2
486
i. H_2, Pd/C
ii.$(CF_3CO)_2O$
iii. p-TSA
iv. TESCl
TESO
TESO
$NHCOCF_3$
487
i. $(COCl)_2$, DMSO
ii $NaOCl_2$
iii. $Ba(OH)_2$
HO
HO_2C
NH_2
488

Scheme 14.136

derivative, and replacement of the dioxolane protecting group yields the bistriethylsilyloxy ether (**487**). Selective oxidation of the primary triethylsilyl ether and subsequent deprotection afforded 2,3,3,-trisubstituted (2-aminocyclopropyl)-2-hydroxyacetic acid (**488**) [354].

The *cis*- and *trans*-2-aminocyclopropylacetic acids have been prepared in seven steps starting from the corresponding diethyl cyclopropane-l,2-dicarboxylic ester (**489**) (Scheme 14.137). Partial hydrolysis yielded the monoacid, which was reduced to the alcohol (**490**). Conversion to the bromide and treatment with sodium cyanide yielded the nitrile (**491**). Curtius rearrangement of the acid hydrazide and hydrolysis afforded the desired *trans* amino acid (**492**). In the case of the *cis* isomer, conversion to the acid hydrazide was carried out at 0 °C to control racemization and required purification by high-performance chromatography (HPLC) to remove the *trans* isomer was required [355].

14.10.2
Cyclobutyl γ-Amino Acids

Ethyl 3-azidoocyclobutane-l-carboxylate (**493**) was synthesized from epibromohydrin and diethyl malonate in seven steps. Catalytic reduction to the amine and ester hydrolysis gave an approximately 1 : 1 mixture of *cis*- and *trans*-3-aminocyclobutane-l-carboxylic acids (**494**) that on careful crystallization yielded the pure *cis*-(**494**)

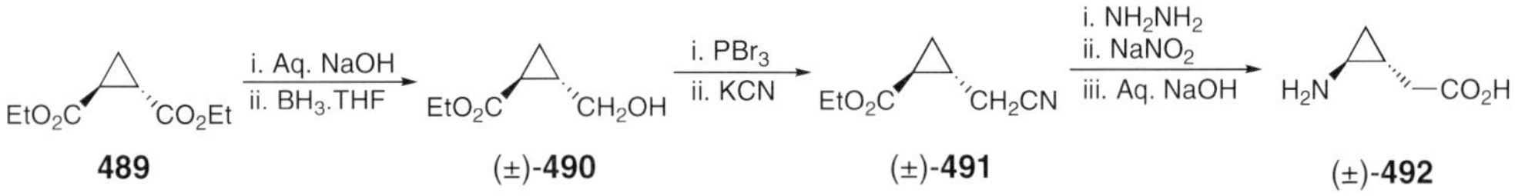

Scheme 14.137

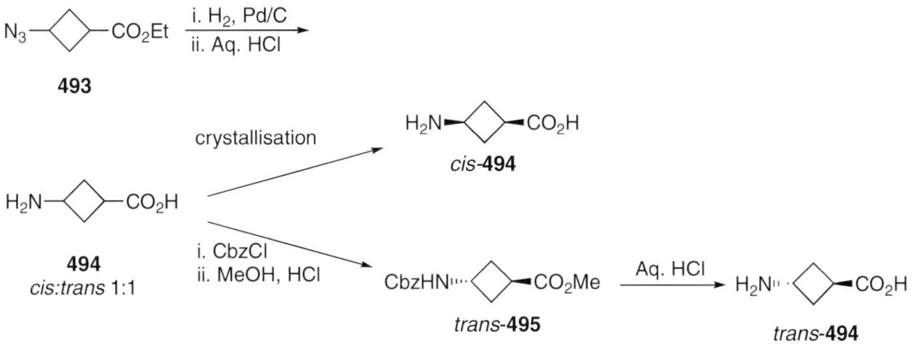

Scheme 14.138

(Scheme 14.138). The *trans* isomer *trans*-(**494**) was obtained by chromatography of the *N*-benzyloxycarbonyl methyl esters *trans*-(**495**) of the residual mixture and subsequent hydrolysis and crystallization. The *cis* isomer *cis*-(**494**) was also prepared by reaction of cyclobutane cyclic anhydride with an equivalent of trimethylsilyl azide followed by hydrolysis and Curtius rearrangement of the resulting isocyanate [356].

The (±)-*trans*-2-(aminomethyl)cyclobutanecarboxylic acid (±)-*trans*-(**498**) has been prepared by Gabriel synthesis of (±)-*trans*-2-(bromomethyl)cyclobutanecarboxylic acid (±)-*trans*-(**496**) and hydrolysis of the protected intermediate *trans*-(**497**) (Scheme 14.139). The *cis*-(**498**) was prepared by opening of the cyclobutane lactone with phthalimide as described above for the cyclopropyl analog (Scheme 14.128) [341].

Both diastereoisomers of *cis*-(**498**) have been prepared by a stereodivergent synthesis that is dependent on the conditions used for the intramolecular cyclization of (*R*)-(**499**) (Scheme 14.140). The use of NaH in THF leads to (*R,S,S*)-(**500**) and NaOEt in EtOH (*R,R,R*)-(**500**). Demethoxycarboxylation of (**500**), reduction of the alcohol, and conversion to the iodide affords (**501**). Reaction with lithium hexamethyldisilazane provides the β-lactam in good yield as a sole diastereoisomer and deprotection affords the corresponding isomer of (**498**) in 30% overall yield [357].

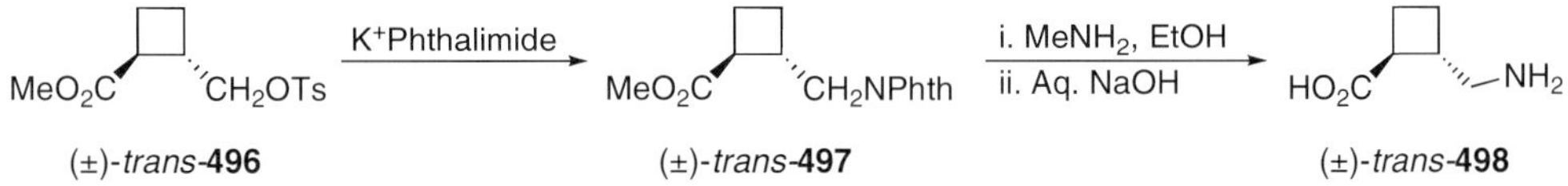

Scheme 14.139

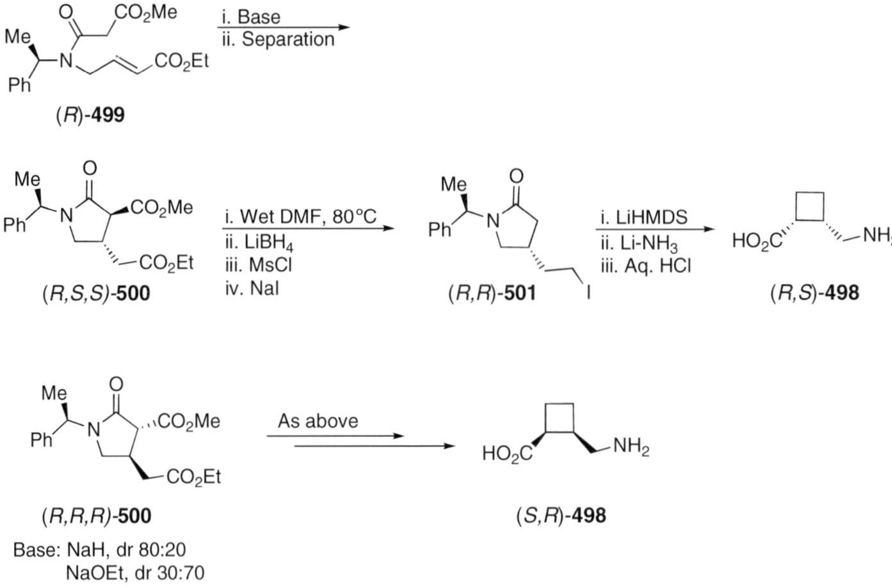

(R)-499 → i. Base; ii. Separation

(R,S,S)-500 → i. Wet DMF, 80°C; ii. $LiBH_4$; iii. MsCl; iv. NaI → (R,R)-501 → i. LiHMDS; ii. Li-NH_3; iii. Aq. HCl → (R,S)-498

(R,R,R)-500 → As above → (S,R)-498

Base: NaH, dr 80:20
NaOEt, dr 30:70

Scheme 14.140

A number of methodologies used for the synthesis of the cyclopropane compounds are also applicable to the synthesis of cyclobutane γ-amino acids. Both (±)-*cis*- and (±)-*trans*-(**498**) have also been prepared by opening of the cyclobutyl lactone with potassium phthalimide and a Gabriel synthesis on the tosylate, respectively, in a manner identical to that described for the corresponding cyclopropyl compounds [341]. The (±)-*trans*-(**498**) has also been prepared by conversion of the tosylate to the azide and subsequent reduction to the amine [358]. The (*S*,*R*)-(**498**) has also been obtained by a PPL resolution of the *meso* diester as described above for the cyclopropyl analog [342].

Likewise, the synthesis of (±)-*trans*-2-aminocyclobutylacetic acid (±)-*trans*-(**505**) in seven steps from the corresponding cyclobutyl-l,2-dicarboxylic acid, in a method analogous to that described above for the 2-aminocyclopropylacetic acids, has been reported [355]. The (±)-*cis*-2-aminocyclobutylacetic acid (±)-*cis*-(**505**) has been prepared from the readily available (±)-*cis*-(**502**) (Scheme 14.141) [359]. Conversion to the methyl carbamate (±)-*cis*-(**503**) prior to reaction with cyanide prevents the ring-opening side-reaction from occurring, yielding (±)-*cis*-(**504**) in good yield. Hydrolysis affords (±)-*cis*-(**505**) (Scheme 14.141) [355].

A cyclobutane analog of GABOB has been obtained via reaction of 2-(dibenzylamino)cyclobutanone (**506**) with *tert*-butylbromoacetate under Reformatsky conditions (Scheme 14.142). Both (±)-*cis*- and (±)-*trans*-(**507**) were readily separated by flash chromatography on silica gel. Debenzylation of each product gave the racemic amino alcohols, which upon treatment with TFA/anisole and subsequent purification by ion-exchange chromatography and HPLC gave the (±)-l-hydroxy-2-

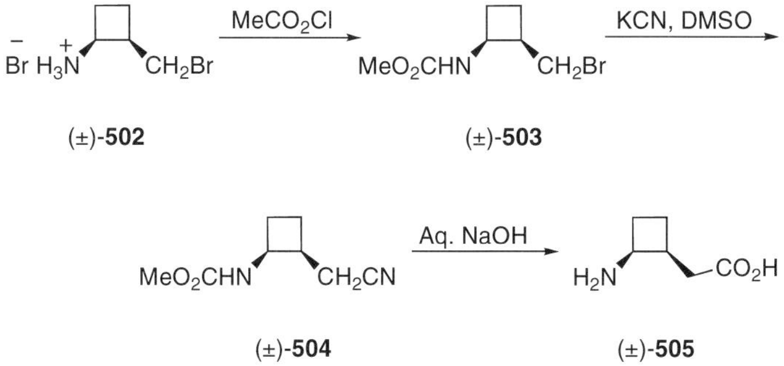

Scheme 14.141

aminocyclobutaneacetic acids (**508**). The *trans*-(*R*,*R*)- and (*S*,*S*)-l-hydroxy-2-amino-cyclobutane-l-acetic acid were resolved by HPLC or by coupling to *N*-*tert*-butoxycarbonyl-(*S*)-valine and recrystallization [360].

Pinene has proved to be a versatile substrate in the synthesis of 2,2-dimethyl-substituted cyclobutane γ-amino acid derivatives. Oxidative cleavage of (*R*)-verbenone (*R*)-(**509**), available from the allylic oxidation of (+)-α-pinene, produced (+)-pinonic acid (*S*,*R*)-(**510**) with concomitant loss of CO_2 (Scheme 14.143). The (*S*,*R*)-(**510**) was converted to both isomers of *cis*-3-amino-2,2-dimethylcyclobutanecarboxylic acid (**512**) by a stereodivergent synthesis. Benzylation of (*S*,*R*)-(**510**) and subsequent haloform reaction yields the acid (**511**) which undergoes a Curtius rearrangement in *tert*-butanol to give the protected amino ester. Cleavage of the benzyl ester by hydrogenolysis affords (*R*,*S*)-3-(Boc-amino)-2,2-dimethylcyclobutanecarboxylic acid (*R*,*S*)-(**512**) in good yield. The (*S*,*R*)-(**512**) is available by Curtius rearrangement to give the keto-acid (**513**) and subsequent haloform reaction [361]. Similar syntheses of (**512**) have also been reported from (*S*)-verbenone [362] and (+)-(*R*)-α-pinene [363]. The highly stereoselective conjugate addition of nitromethane to α,β-unsaturated cyclobutyl esters derived from (−)-(*S*)-verbenone furnishes 3-substituted cyclobutyl gabapentin analogs [364].

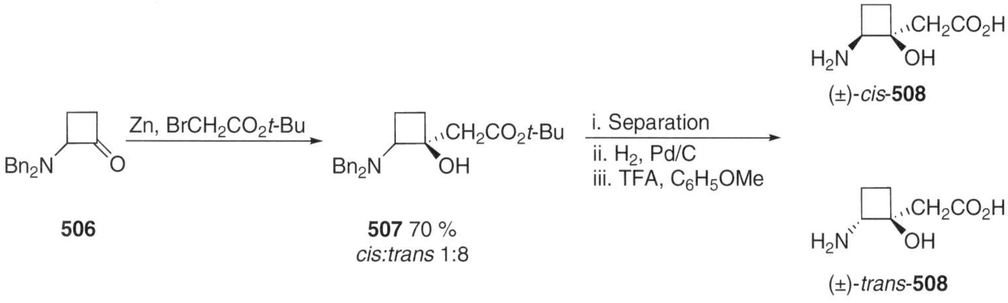

Scheme 14.142

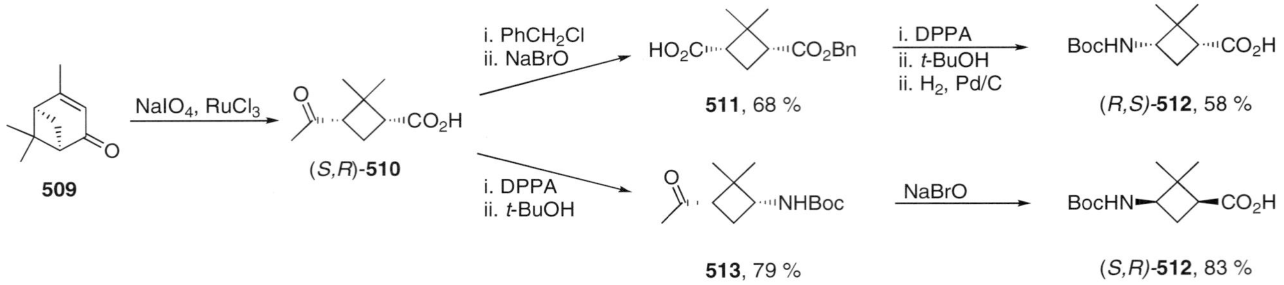

Scheme 14.143

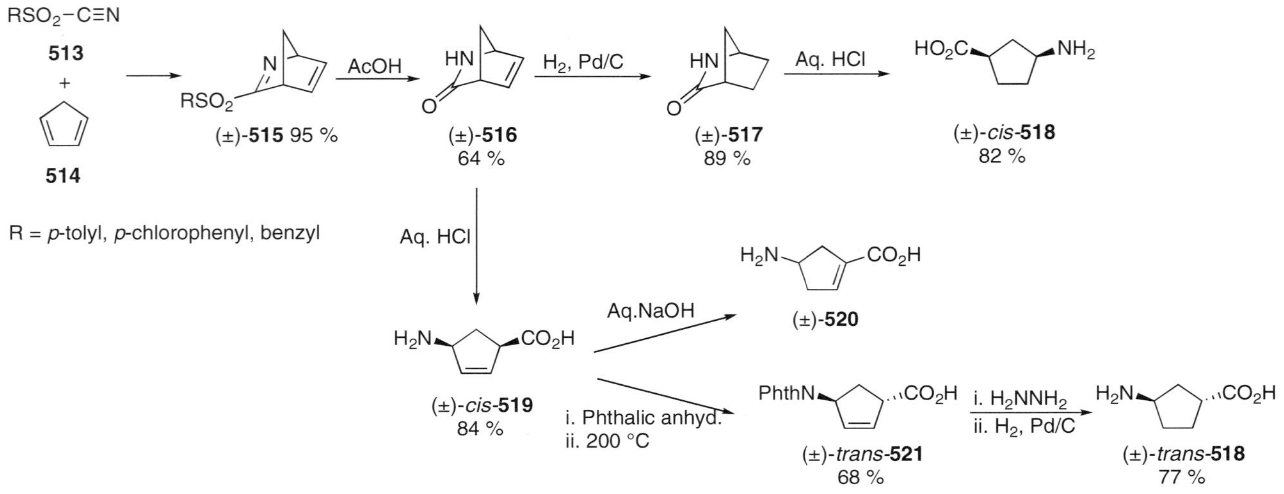

R = *p*-tolyl, *p*-chlorophenyl, benzyl

Scheme 14.144

14.10.3
Cyclopentyl γ-Amino Acids

Diels-Alder cycloaddition of tosyl cyanide (**513**) to cyclopentadiene (**514**) at room temperature produces 3-tosyl-2-azabicyclo[2.2.l]hepta-2,5-diene (±)-(**515**) (Scheme 14.144). Hydrolysis yields the unsaturated lactam (±)-(**516**). Catalytic reduction to the saturated lactam (±)-(**517**) and acid-catalyzed hydrolysis affords (±)-*cis*-3-aminocyclopentanecarboxylic acid (±)-*cis*-(**518**) [365]. Hydrolysis of the unsaturated lactam affords 4-aminocyclopent-2-enecarboxylic acid (±)-*cis*-(**519**) [366], which undergoes isomerization to gave 4-aminocyclopent-1-ene-1-carboxylic acid (±)-(**520**) on treatment with 2 M NaOH [367]. Thermal *cis–trans* isomerization of (±)-*cis*-(**519**) yields (±)-*trans*-4-aminocyclopent-2-ene-1-carboxylic acid (±)-*trans*-(**521**) [368]. Resolution of (±)-(**521**) was achieved by crystallization of isopropylideneribonolactone esters or pantolactone esters of the phthalimido-protected intermediates to afford (+)-(*S*)-(**521**) and (−)-(*R*)-(**521**) [368].

Reduction of 3-(hydroxyamino)cyclopentanecarboxylic acid or its ethyl ester has also been used in the preparation of (**518**). Hydrogenation over platinum oxide in ethanolic HCl provided a mixture of isomeric amino esters, which on distillation afforded the *cis* lactam and the *trans* amino ester. Hydrolysis afforded the (±)-*cis*-(**518**) and (±)-*trans*-(**518**), respectively [355]. All four stereoisomers of 3-aminocyclopentanecarboxylic acid have also been prepared via reduction of the oximes prepared from optically pure (*R*)- and (*S*)-ethyl 3-oxocyclopentanecarboxylates. Reduction using sodium/ammonia/methanol yielded a separable mixture (*cis*: *trans* 55 : 45) of the desired amino acids. However, reduction using zinc in HCl proceeded stereoselectively to the *cis* amino acids in low yield to afford after crystallization (*R*,*S*)-*cis*-(**518**) and (*S*,*R*)-*cis*-(**518**) [369].

Allylic bromination of the cyclopentene nitrile (**522**) gave a 3 : 1 mixture of the desired product (±)-(**523**) and 5-bromo regioisomer (±)-(**524**) (Scheme 14.145). Amination of the crude mixture in liquid ammonia and immediate hydrolysis

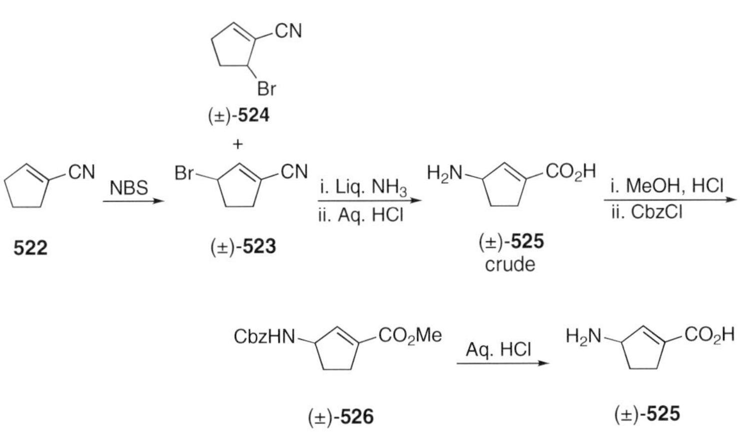

Scheme 14.145

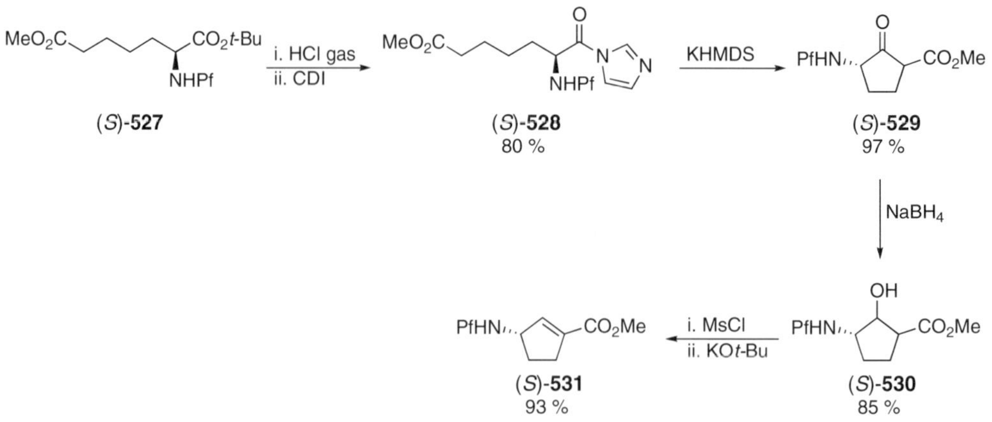

Scheme 14.146

yielded the desired amino acid (±)-(**525**); however, the 5-bromo byproduct hindered purification. Derivatization as the *N*-benzyloxycarbonyl methyl ester (±)-(**526**) facilitated purification by chromatography and crystallization. Regeneration of the amino acid by hydrolysis afforded pure (±)-3-aminocyclopent-1-enecarboxylate derivative (±)-(**525**) [367].

Dieckmann cyclization of the *N*-(9-phenylfluoren-9-yl)-protected (S)-2-aminoadipic acid derivative (*S*)-(**527**), activated by conversion to the imidazolide (*S*)-(**528**) yields the cyclopentane keto ester (*S*)-(**529**) (Scheme 14.146). Reduction of the keto ester to the hydroxy ester (*S*)-(**530**) and subsequent elimination affords enantiomerically pure 3-aminocyclopent-1-enecarboxylate derivative (*S*)-(**531**) in high overall yield. Hydrogenation of the double bond under a variety of conditions gave *cis* : *trans* ratios of 1 : 1 to 1 : 5.[370]

Enzymatic resolution of the racemic diester (±)-(**533**), prepared via ozonolysis of norbornylene (**532**), with cholesterol esterase yields the mono-ester (*S*,*R*)-(**534**) in high yield and enantiopurity (Scheme 14.147). Conversion of the isocyanate (*R*,*S*)-(**535**) and subsequent Curtius rearrangement afforded (*R*,*S*)-(**518**). Alternatively, ammonolysis of the monoester and subsequent Hofmann rearrangement of the amide (*S*,*R*)-(**536**) using bis(trifluoroacetoxy)iodobenzene afforded the amino acid (*S*,*R*)-(**518**) [371].

Nitrilase-mediated hydrolysis of *N*-protected *cis*- or *trans*-3-aminocyclopentane carbonitriles (**538**), prepared via Michael addition of cyanide to α,β-unsaturated cyclic ketone (**537**), has been reported as an efficient method for the enantioselective synthesis of *cis*-(*R*,*S*)-3-aminocyclopentanecarboxylic acid in high optical purity (Scheme 14.148). In contrast, the nitrilase-mediated hydrolysis of the *trans* isomer resulted in much lower optical purity (55% e.e.). A range of enzymes and *N*-protecting groups were investigated, with the *N*-tosyl moiety found to be optimal for both yield and stereoselectivity [372].

Enzymatic hydrolysis of (±)-4-acetamido-cyclopent-2-enecarboxylates (±)-(**540**) has been investigated as a route to enantiomerically pure cyclopentene/ane γ-amino

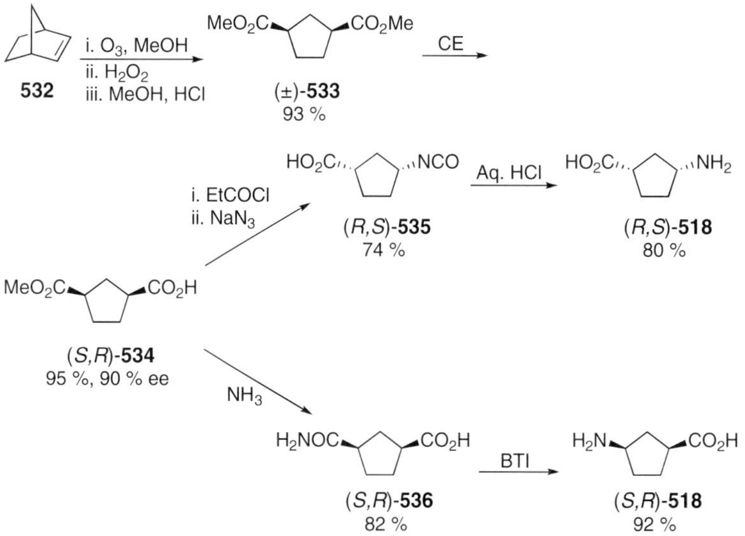

Scheme 14.147

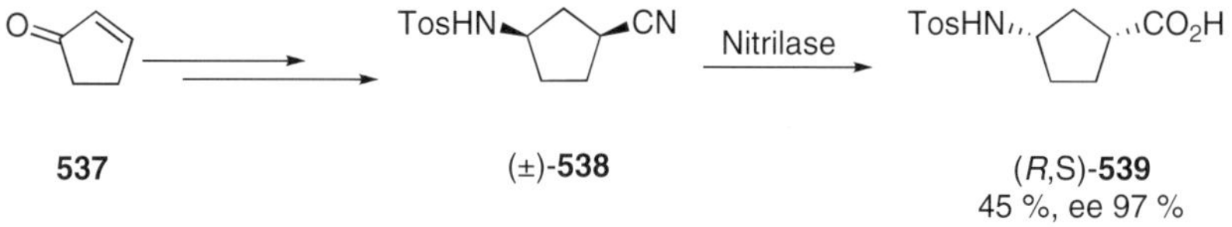

Scheme 14.148

acid analogs. Hydrolysis of (±)-(**540a**) by PLE was originally reported to afford high yields and moderate to high enaniopurities of both (+)-(**540**) and (−)-(**541**) (Scheme 14.149) [373]. However, a later study reported much lower yields and enantioselectivity for this enzymatic hydrolysis [374]. Alternatively, CCL-catalyzed hydrolysis of (±)-(**540b,c**) afforded enantiopure (+)-(**540b,c**) and (−)-(**541**) [374].

The lactam (±)-(**516**), undergoes enantiospecific and enantiocomplementary hydrolyses using whole cell catalysts or immobilized enzyme to give enantiopure lactams and *cis*-4-aminocyclopent-2-enecarboxylic acid (Scheme 14.150). The use of *Rhodococcus equi* NCIB 40 213 (ENZA 1) or *Aureobacterium* sp. (ENZA 25) affords (+)-lactam (+)-(*R,S*)-(**516**) and the (−)-acid (−)-(*S,R*)-(**519**), while the use of *Pseudomonas solanacearum* NCIB 40 249 (ENZA 20) and *Pseudomonas fluorescens* (ENZA 22) affords the (−)-lactam (−)-(*S,R*)-(**516**) and the (+)-acid (+)-(*R,S*)-(**519**) [375, 376]. Treatment of the lactam (−)-(*S,R*)-(**516**) with bromine gave the adduct (−)-(**542**) that, through rearrangement of the bromonium ion intermediate, undergoes a net "inversion" of the carbocyclic skeleton. Reduction of the dibromo compound with tributyltin hydride gave the saturated lactam (+)-(*R,S*)-(**517**), which on hydrolysis of gave the (+)-(*S,R*)-(**518**) [377].

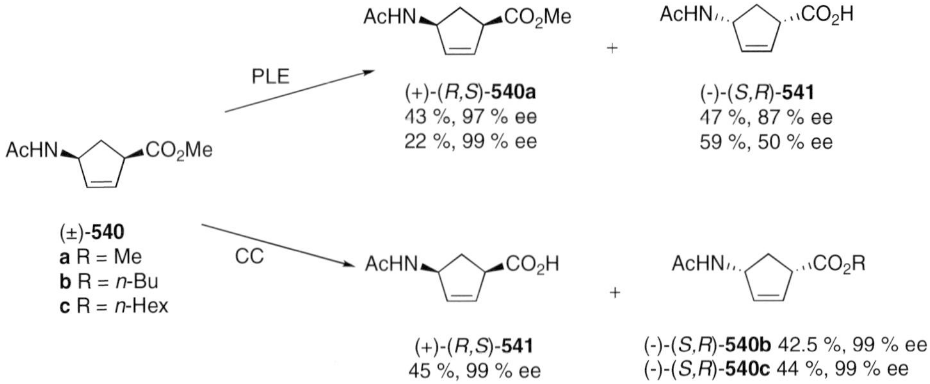

AcHN CO$_2$Me

(±)-**540**
a R = Me
b R = *n*-Bu
c R = *n*-Hex

PLE

AcHN CO$_2$Me + AcHN CO$_2$H

(+)-(*R*,*S*)-**540a**
43 %, 97 % ee
22 %, 99 % ee

(-)-(*S*,*R*)-**541**
47 %, 87 % ee
59 %, 50 % ee

CC

AcHN CO$_2$H + AcHN CO$_2$R

(+)-(*R*,*S*)-**541**
45 %, 99 % ee

(-)-(*S*,*R*)-**540b** 42.5 %, 99 % ee
(-)-(*S*,*R*)-**540c** 44 %, 99 % ee

Scheme 14.149

The enantiomerically pure lactams (−)-(*S*,*R*)-(**516**) and (+)-(*R*,*S*)-(**516**), which can be prepared on a multitonne scale, have been used in the synthesis of all eight stereoisomers of 3-(*tert*-butoxycarbonylamino)-4-hydroxycyclopentanecarboxylic acid methyl ester (**548**) (Scheme 14.151). Treatment of (−)-(*S*,*R*)-(**516**) with thionyl chloride in methanol and subsequent reaction with Boc-anhydride yielded (−)-(*S*,*R*)-(**543**). The protected cyclopentene (−)-(*S*,*R*)-(**543**) was subjected to an NBS-promoted bromocyclization, forming the cyclic carbamate (**544**) in high yield and introducing an oxygen atom with defined stereochemistry. Elimination of HBr, hydrolysis, and reprotection of the amine as the *N*-Boc-carbamate yielded (**545**). Hydrogenation

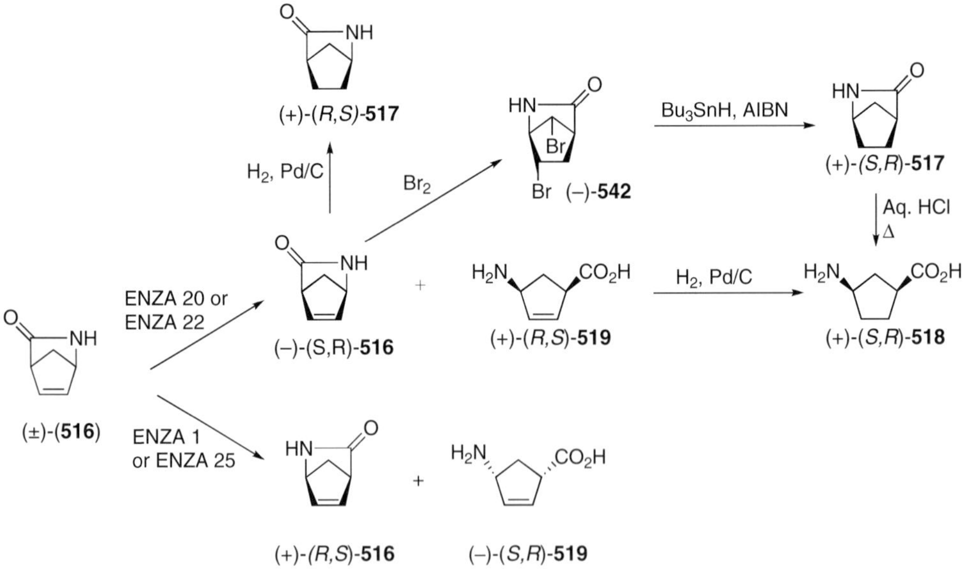

(±)-(**516**)

ENZA 20 or ENZA 22

(−)-(S,R)-**516** + H$_2$N CO$_2$H (+)-(*R*,*S*)-**519**

H$_2$, Pd/C

(+)-(*R*,*S*)-**517**

Br$_2$

HN Br Br (−)-**542**

Bu$_3$SnH, AIBN

(+)-(*S*,*R*)-**517**

Aq. HCl Δ

H$_2$, Pd/C

H$_2$N CO$_2$H (+)-(*S*,*R*)-**518**

ENZA 1 or ENZA 25

(+)-(*R*,*S*)-**516** + H$_2$N CO$_2$H (−)-(*S*,*R*)-**519**

Scheme 14.150

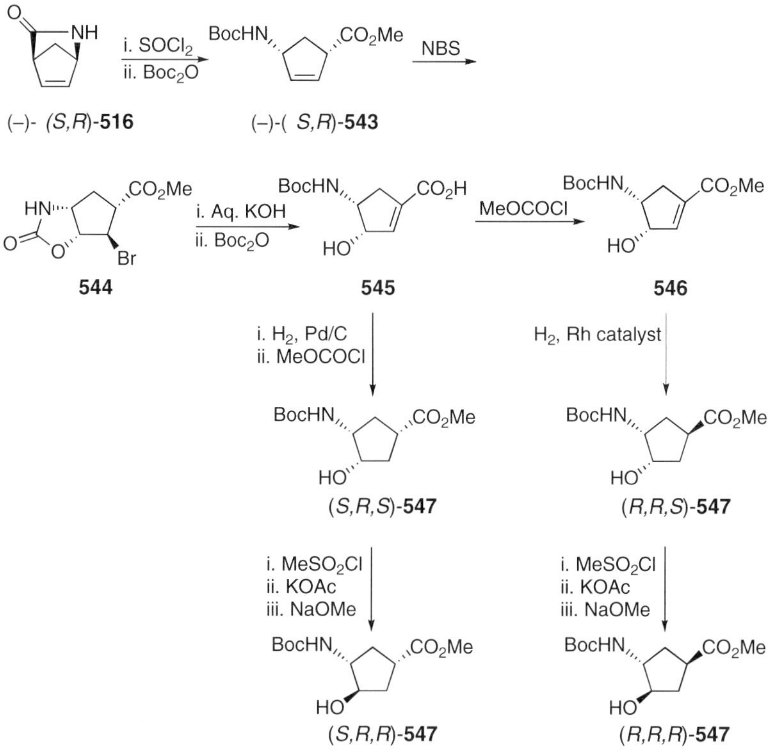

Scheme 14.151

provided the all-*cis* stereoisomer, which was esterified to yield (*S*,*R*,*S*)-(**541**). Alternatively, esterification to afford (**540**) and homogeneous hydrogenation in the presence of (*R*,*R*)-{[MeDuPhos]-Rh(COD)}BF_4 [378] gave (*R*,*R*,*S*)-(**547**). Inversion of the hydroxyl of both (*S*,*R*,*S*)-(**547**) and (*R*,*R*,*S*)-(**547**) afforded (*S*,*R*,*R*)-(**547**) and (*R*,*R*,*R*)-(**547**), respectively. An analogous process starting from the (+)-(*R*,*S*)-(**516**) afforded the enantiomeric products [379]; (**547**) has also been formed via epoxidation of (**543**) and treatment with base [378].

The lactam (−)-(*R*,*S*)-(**516**) has also been used in the preparation of the dihydroxy-γ-amino esters (**550**) (Scheme 14.152). Treatment of (−)-(*R*,*S*)-(**516**) with thionyl chloride and subsequent reaction with benzyl chloroformate provides the fully protected (−)-(*R*,*S*)-(**548**). Dihydroxylation gave two diastereomeric *cis* diols in a 1 : 1 ratio that could be separated by column chromatography after conversion to the acetonides (**549**). Deprotection afforded the dihydroxy-γ-amino esters (*R*,*S*,*R*,*S*)-(**550**) and (*R*,*R*,*S*,*S*)-(**550**). Epimerization of (*R*,*S*,*R*,*S*)-(**549**) and (*R*,*R*,*S*,*S*)-(**549**) and deprotection afford (*S*,*S*,*R*,*S*)-(**550**) and (*S*,*R*,*S*,*S*)-(**550**), respectively [380].

The (*S*,*R*,*S*,*R*)- and (*S*,*R*,*R*,*R*)-4-amino-2,3-dihydroxycyclopentanecarboxylic acids are available by hydrolysis of (*S*,*R*,*S*,*R*)- and (*S*,*R*,*R*,*R*)-(**555**) (Scheme 14.153). The

Scheme 14.152

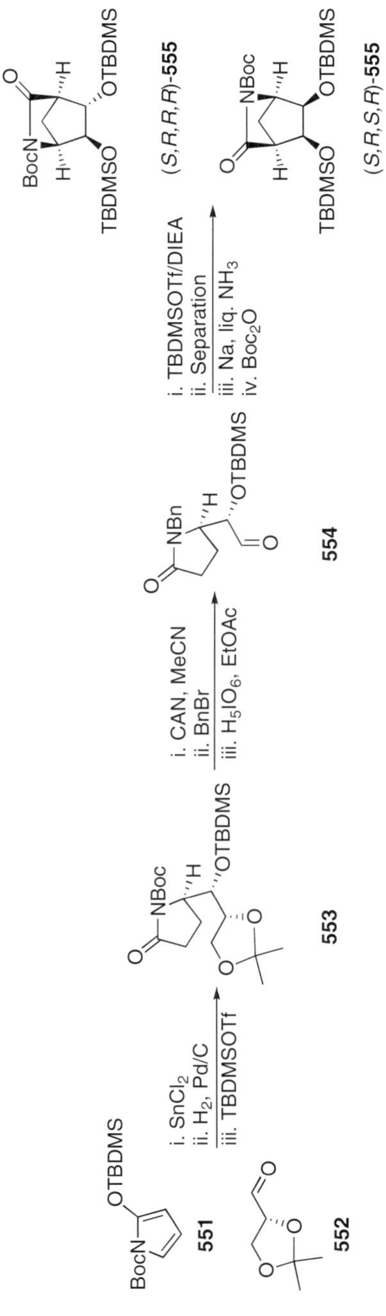

551 (BocN, OTBDMS) + **552**

i. $SnCl_2$
ii. H_2, Pd/C
iii. TBDMSOTf

553 (NBoc, H, OTBDMS)

i. CAN, MeCN
ii. BnBr
iii. H_5IO_6, EtOAc

554 (NBn, H, OTBDMS)

i. TBDMSOTf/DIEA
ii. Separation
iii. Na, liq. NH_3
iv. Boc_2O

(*S*,*R*,*R*,*R*)-**555** (BocN, H, TBDMSO, OTBDMS)

(*S*,*R*,*S*,*R*)-**555** (NBoc, H, TBDMSO, OTBDMS)

Scheme 14.153

dihydroxy lactams are prepared as a separable mixture of *trans* : *cis* (8 : 2) isomers in a multistep synthesis via a diastereoselective crossed vinylogous Mukaiyama aldol coupling of the pyrrole (**551**) and 2,3-*O*-isopropylidene-D-glyceraldehyde (**552**). Reduction and *O*-protection yields (**553**), which can be converted to the aldehyde (**554**) in four steps. A high-yielding silylative cycloaldolization of (**554**), chromatographic separation, and conversion of the *N*-benzyl to the *N*-Boc-carbamate affords enantiomerically pure (*S*,*R*,*S*,*R*)- and (*S*,*R*,*R*,*R*)-(**555**) [381, 382].

Modifications of the rearrangement of (−)-(*S*,*R*)-(**516**) to (−)-(**542**) (Scheme 14.150) or derivatives have been used to prepare versatile 5,(7)-substituted 2-azabicycloheptan-3-ones, which can be converted to a range of 3-substituted cyclopentene/ane γ-amino acids. Reaction of *N*-PMB protected lactam (**556**) with 1,3-dibromo-5,5-dimethylhydantoin in glacial acetic acid afforded the bromo acetate (**557**) (Scheme 14.154). Hydrolysis of the acetate furnished alcohol (**558**) [383]. Reaction of (**558**) with (diethylamino)sulfur trifluoride (DAST) and debromination yields the monofluoro compound (**561**), which can be deprotected and hydrolyzed to afford the monofluoro amino acid (**562**). Alternatively, debromination and oxidation of (**558**) to the ketone (**559**) and treatment with DAST affords the difluoro derivative (**564**) and the corresponding difluoro amino acid (**564**) [384]. Conversion of (**559**) to the iodo derivative (**560**) facilitates the synthesis of 3-fluoro-cyclopentene and 3-fluoroalkyl-cyclopentene amino acids (**568a–c**) [385]. Debromination and hydrolysis of (**557**) furnishes the 3-hydroxy derivative (**565**) which can be converted to the 3-bromo amino acid (**566**) [384]. Finally the ketone (**559**) can be converted to alkenes by reaction with fluoromethyl phenylsulfone to yield (**567a**) [386] or via Horner–Wadsworth–Emmons reactions to yield (**567b** and **c**) [387].

The cycloaddition of cyclopentadiene (**514**) and the *N*-sulfinyl carbamate (**569**) afforded the bicyclic 3,6-dihydrothiazine oxide (**570**) (Scheme 14.155). Base hydrolysis afforded the carbamate protected sulfinic acid (**571**). Oxidation and hydrolysis gave (±)-*cis*-4-aminocyclopent-2-ene-1-sulfonic acid (±)-(**572**) [388].

A stereodivergent synthesis of the alkyl phosphinate cyclopentane γ-amino acid analogs (±)-*cis*- and (±)-*trans*-(**576**) has been reported. Both (±)-*cis*- and (±)-*trans*-(**576**) were obtained in five steps from the key (±)-(3-hydroxycyclopent-1-ene)alkylphosphinate esters (**574**) which are prepared via a Pd(0)-catalyzed C–P bond-forming reaction between trimethylsilyl-protected 3-iodocyclopent-2-enol (**573**) and alkylphosphinate esters (Scheme 14.156). Protection of (**574**) as the sterically demanding TBDMS ether and hydrogenation yielded the *cis* alcohols (±)-*cis*-(**577**) that were converted to (±)-*trans*-(**576**) via a Staudinger–Mitsunobu reaction and subsequent hydrolysis. Alternatively, conversion of (**574**) to the amine and protection as the *N*-Boc-carbamate yields (**575**). Hydrogenation and subsequent hydrolysis affords (±)-*cis*-(**576**). Both reaction pathways proceed in moderate to good overall yield and high diastereomeric excess [389]. Similar methodology has been used for the enantioselective synthesis of the 4-amino-cyclopent-1-enyl phosphinic acid analogs from (*R*)- and (*S*)-4-(*tert*-butyldimethylsilyloxy)cyclopent-1-enyl trifluoromethanesulfonate [390].

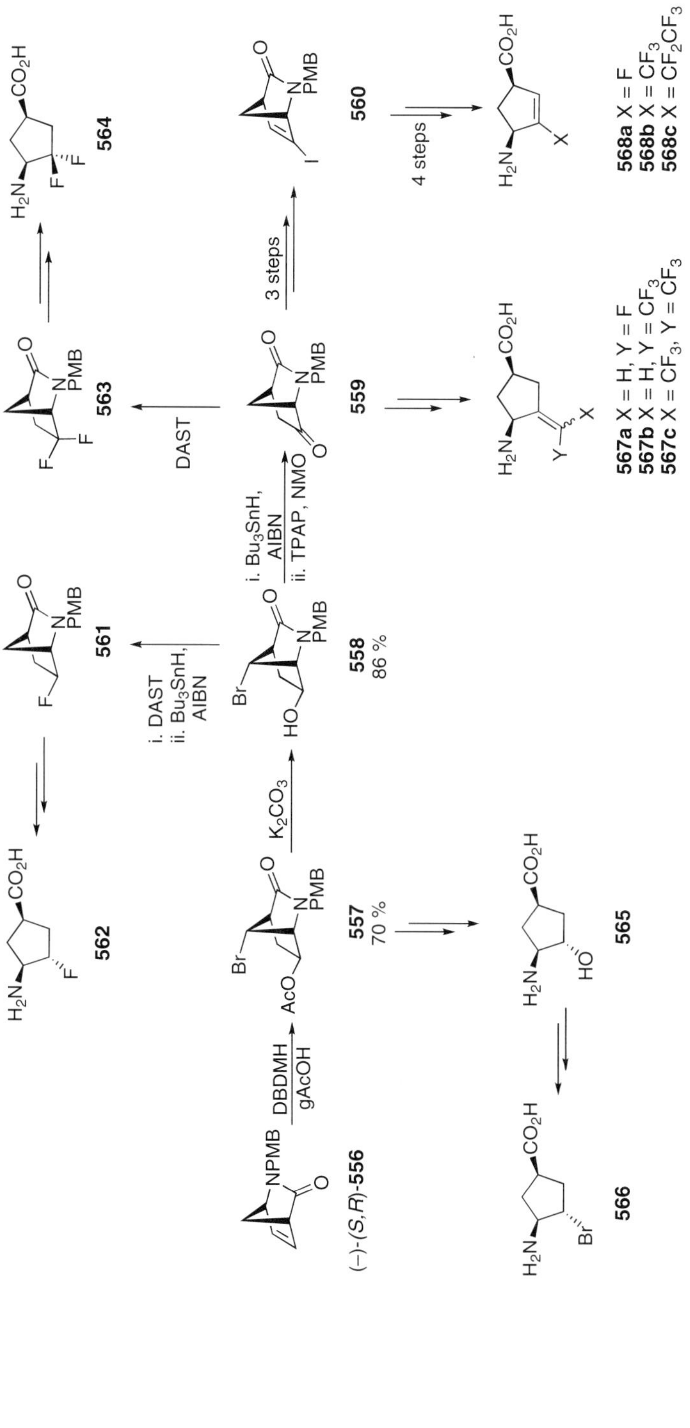

(−)-(*S*,*R*)-**556**

DBDMH, gAcOH

557 70 %

K_2CO_3

558 86 %

i. DAST ii. Bu_3SnH, AIBN

561

562

i. Bu_3SnH, AIBN ii. TPAP, NMO

559

DAST

563

564

3 steps

560

4 steps

565

566

567a X = H, Y = F
567b X = H, Y = CF_3
567c X = CF_3, Y = CF_3

568a X = F
568b X = CF_3
568c X = CF_2CF_3

Scheme 14.154

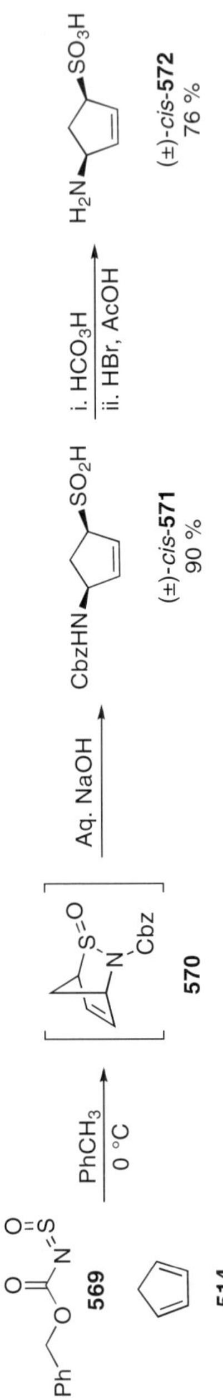

569

514

$PhCH_3$, 0 °C

570

Aq. NaOH

(±)-*cis*-**571** 90 %

i. HCO_3H
ii. HBr, AcOH

(±)-*cis*-**572** 76 %

Scheme 14.155

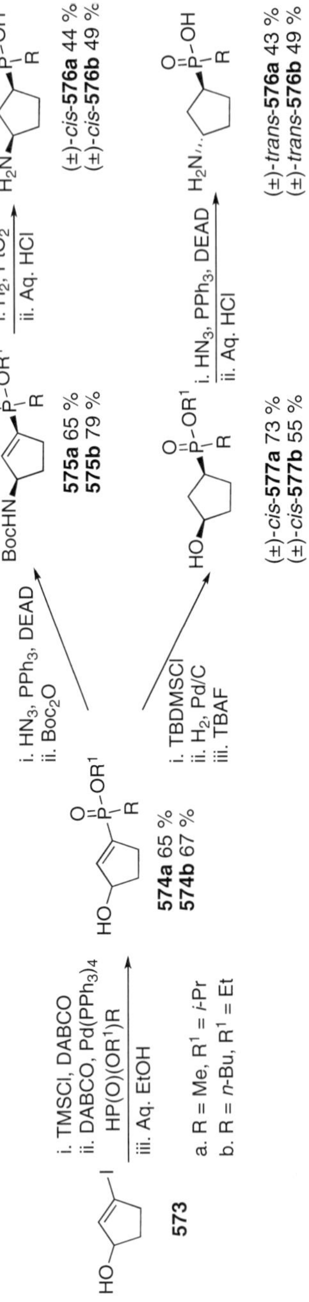

573

i. TMSCl, DABCO
ii. DABCO, $Pd(PPh_3)_4$, $HP(O)(OR^1)R$
iii. Aq. EtOH

a. R = Me, R^1 = *i*-Pr
b. R = *n*-Bu, R^1 = Et

574a 65 %
574b 67 %

i. HN_3, PPh_3, DEAD
ii. Boc_2O

575a 65 %
575b 79 %

i. H_2, PtO_2
ii. Aq. HCl

(±)-*cis*-**576a** 44 %
(±)-*cis*-**576b** 49 %

i. TBDMSCl
ii. H_2, Pd/C
iii. TBAF

(±)-*cis*-**577a** 73 %
(±)-*cis*-**577b** 55 %

i. HN_3, PPh_3, DEAD
ii. Aq. HCl

(±)-*trans*-**576a** 43 %
(±)-*trans*-**576b** 49 %

Scheme 14.156

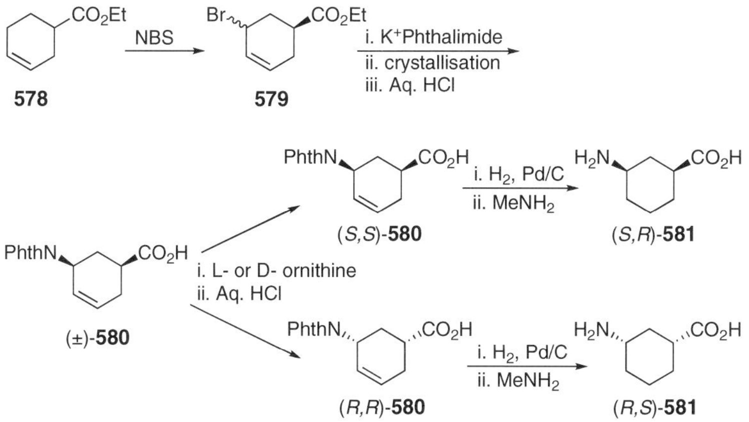

Scheme 14.157

14.10.4 Cyclohexyl γ-Amino Acids

Allylic bromination of ethyl cyclohex-3-ene-1-carboxylate (**578**) gave the unsaturated bromo ester (**579**) (Scheme 14.157). Reaction of this crude product with potassium phthalimide gave ethyl 5-phthalimidocyclohex-3-ene-1-carboxylate as a mixture of isomers from which the major product, (±)-*cis*-(**580**), was purified by crystallization. Hydrolysis of the ester and resolution via crystallization with L- or D-ornithine yielded enantiomerically pure (*S*,*S*,)- and (*R*,*R*)-(**580**). Catalytic reduction and removal of the phthaloyl protecting group produced (*S*,*R*)- and (*R*,*S*)-(**581**), respectively, in high optical purity [391].

Birch reduction of (*R*)-(**583**), prepared from (*R*)-(**582**), yields a mixture from which enantiomerically pure (*R*)-(**584**) is obtained, after recrystallization (Scheme 14.158) [391].

Oxime reduction has also been used in the synthesis of (±)-*trans*- and (±)-*cis*-2-(2-aminocyclohexyl)acetic acids. Hydrogenation of (**585**) and distillation of the crude product gave a mixture of the *cis* and *trans* lactams (**586**) that were separated by HPLC (Scheme 14.159). Hydrolysis afforded pure (±)-*cis*- and (±)-*trans*-(**587**) [355].

The preparation of (±)-*trans*-(**592**) was carried out by S_N2 azide substitution of the lactam (±)-(**590**) which had been prepared from (**588**) via the iodolactam (±)-(**589**) (Scheme 14.160). Reduction of the azide afforded (±)-(**592**) [392].

The (±)-*cis*- and (±)-*trans*-5-amino-2-methylenecyclohexanecarboxylic acids (±)-*cis*- and (±)-*trans*-(**597**) have both been prepared from ethyl 4-oxocyclohexanecarboxylate

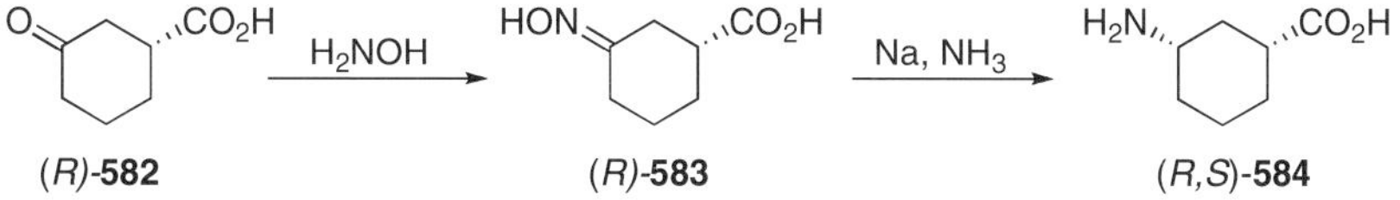

Scheme 14.158

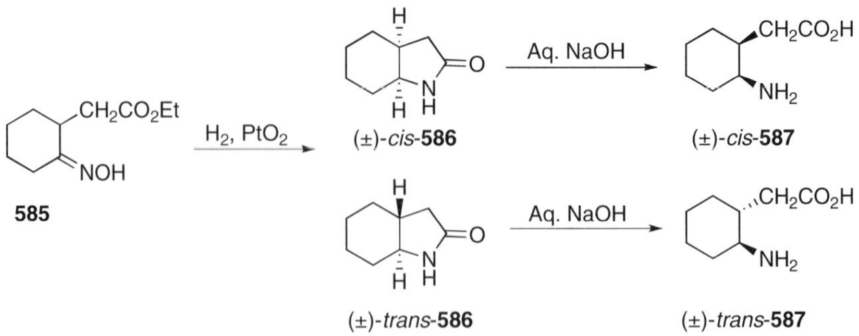

Scheme 14.159

(**593**) in a divergent synthesis (Scheme 14.161). Conversion of (**593**) to the alkene (**594**) by Wittig reaction and oxidation yielded the *trans* allylic alcohol (±)-*trans*-(**595**) as the major product. Mitsunobu reaction with *N*-Boc-ethyl oxamate and hydrolysis furnished the protected *cis* amino acid (±)-*cis*-(**596**) that can be deprotected to (±)-*cis*-(**597**). Mitsunobu reaction of the alcohol (±)-*trans*-(**595**), followed by methanolysis in the presence of sodium azide gave the *cis* alcohol (±)-*cis*-(**595**). This alcohol was converted to (±)-*trans*-(**596**) and hydrolyzed to (±)-*trans*-(**597**) as described above [386].

Hydrogenation of *m*-substituted benzoic acid derivatives affords a convenient synthesis of cyclohexane γ-amino acid analogs. Hydrogenation of the sodium salt of *m*-aminobenzoic acid in the presence of alkaline Raney nickel at 90–100 atm and 150 °C followed by treatment with Boc-anhydride provides (±)-*cis*-cyclohexane amino acid. Resolution by successive recrystallizations from chloroform containing one equivalent of (+)-1-phenylethanamine gave (*R*,*S*)-*N*-Boc-3-aminocyclohexanecarboxylic acid with an enantiomeric purity greater than 95% [393]. Similarly, substituted cyclohexane γ-amino acid analogs have been prepared by hydrogenation of 3-amino-4-hydroxybenzoic acid over Rh/Al_2O_3 and treatment with DAST to yield 4-fluorocyclohexane amino acid in low yield or hydrogenation of (±)-3-amino-5-hydroxybenzoic acid and subsequent Dess–Martin oxidation and reaction with DAST to afford (±)-*cis*-3-amino-2,2-difluorocyclohexanecarboxylic acid [394]. Nitration of 5-chlorosalicylic acid, esterification, and then hydrogenation over PtO_2 affords the

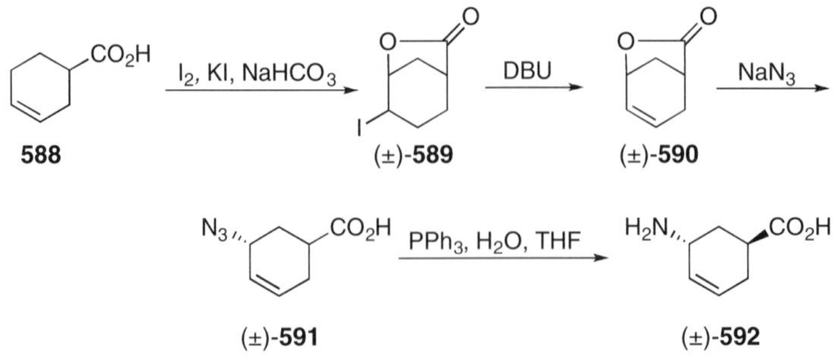

Scheme 14.160

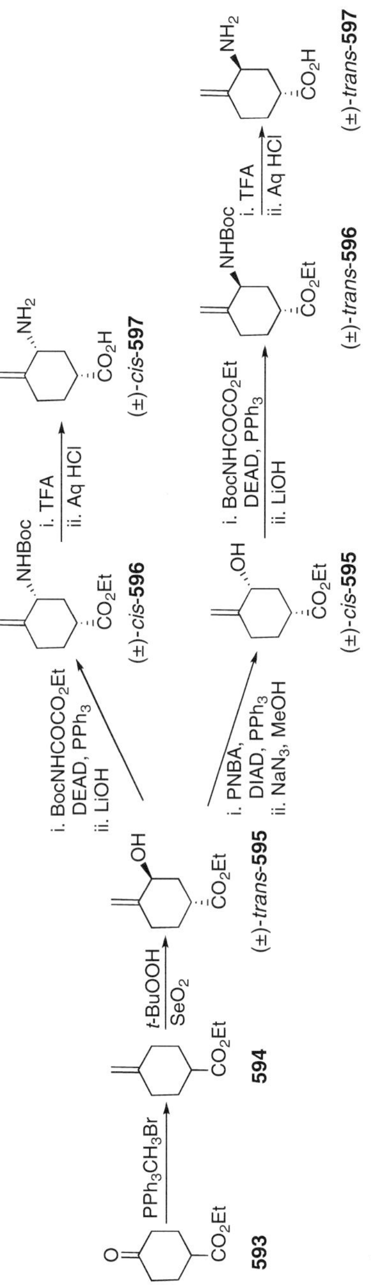

Scheme 14.161

cyclohexane analog of GABOB as a mixture of diastereoisomers that can be separated by chromatography of the benzyl carbamate derivates [395].

The camphorsultam-derived dienophile (**598**) was reacted with 1-acetoxy-1,3-diene (**599**) to yield two products in a ratio of 96 : 4, in which the desired *endo* product (*S*,*R*)-(**600**) predominated (Scheme 14.162). The (*S*,*R*)-(**600**) was hydrolyzed to give the intermediate hydroxy acid (*S*,*R*)-(**601**) in nearly quantitative yield. Iodination gave the iodolactone and removal of the iodine under radical conditions yielded the lactone (*S*,*S*,*R*)-(**602**). Transesterification of the lactone to the methyl ester and conversion of the hydroxyl to the triflate provided (*S*,*S*,*R*)-(**603**). Displacement of the triflate using buffered tetramethylguanidinium azide and subsequent deprotection afforded the enantiomerically pure GABOB analog (*S*,*S*,*R*)-3-amino-2-hydroxycyclohexane-carboxylic acid (*S*,*S*,*S*)-(**604**) [396, 397].

A number of substituted cyclohexane γ-amino acid analogs have been recently prepared via a [3 + 3] ring-forming strategy between the nitro phosphonate (**605**) and unsaturated aldehydes (**606**) (Scheme 14.163). Reaction in the presence of 2 equiv. 1,8-diazabicyclo[5.4.0]undec-7-ene produces the cyclic nitro ester (**607**) in moderate yields predominantly as a single diastereoisomer. Reduction of the nitro group with Raney nickel in ethanol afforded the cyclohexene amino esters (**608**) [398].

14.11 Conclusions

Interest in the synthesis of new analogs of GABA will continue to develop in view of the increasing range of targets associated with the GABA structural motif. These targets include not only subtypes of ionotropic ($GABA_A$ and $GABA_C$) and metabotropic ($GABA_B$) receptors, GABA transporters (GAT1–4), and metabolizing enzymes (glutamic acid decarboxylase and GABA-T), but also voltage-gated calcium channels and non-neuronal sites in the immune system. In addition to their use in the treatment of epilepsy and other neurological disorders, GABA analogs may have future roles in the treatment of disorders such as asthma, cancer and diabetes. The GABA motif is also found as a component of many peptide antibiotics and may be used to construct peptidomimetics. The continued development of new synthetic routes to GABA analogs is important in order to increase the availability of new agents for evaluation and to provide procedures for the larger-scale synthesis of analogs of commercial interest.

14.12 Experimental Procedures

14.12.1 (*R*)-2-Ethyl-4-nitrobutan-1-ol (36c) (Scheme 14.8 [39])

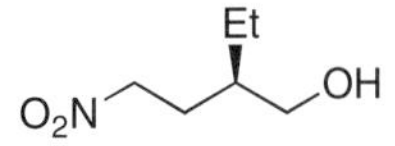

Et

O_2N OH

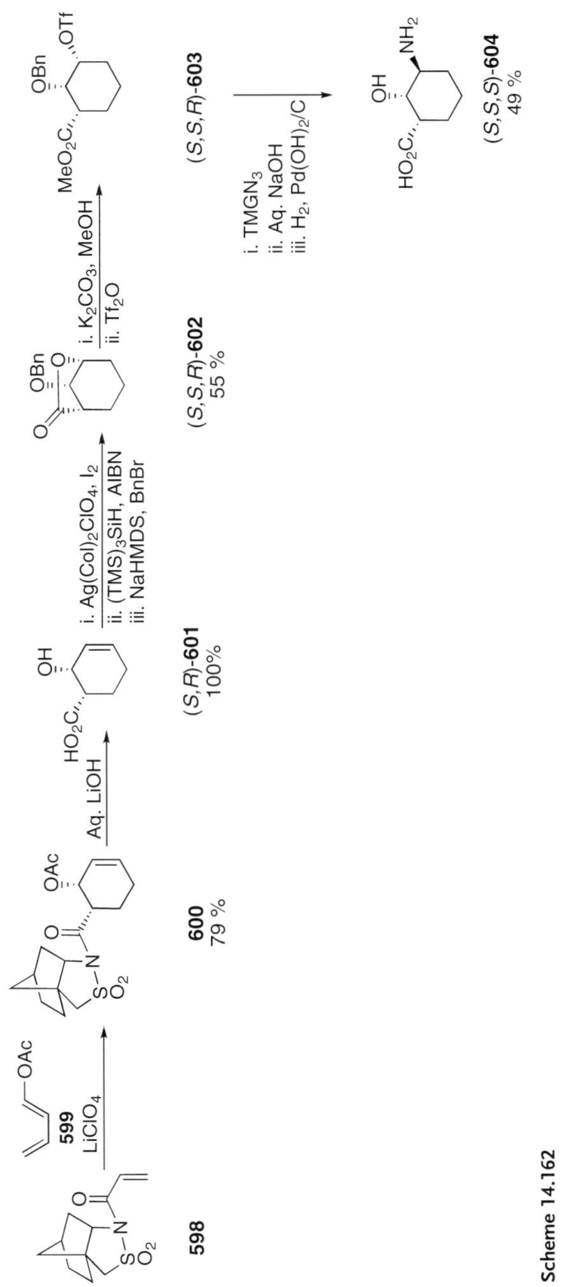

598

599, $LiClO_4$

600 79 %

Aq. LiOH

(S,R)-**601** 100%

i. $Ag(Col)_2ClO_4$, I_2
ii. $(TMS)_3SiH$, AIBN
iii. NaHMDS, BnBr

(S,S,R)-**602** 55 %

i. K_2CO_3, MeOH
ii. Tf_2O

(S,S,R)-**603**

i. $TMGN_3$
ii. Aq. NaOH
iii. H_2, $Pd(OH)_2/C$

(S,S,S)-**604** 49 %

Scheme 14.162

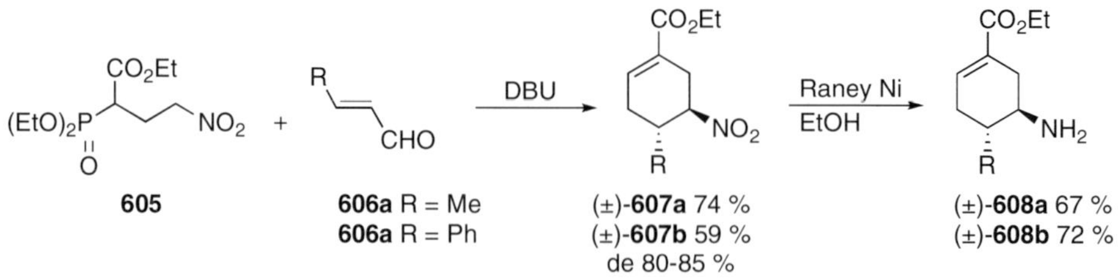

605 + 606a R = Me; 606a R = Ph → (DBU) → (±)-607a 74 %; (±)-607b 59 %; de 80-85 % → (Raney Ni, EtOH) → (±)-608a 67 %; (±)-608b 72 %

Scheme 14.163

To an 8-ml vial equipped with a small magnetic stir bar was added dry toluene (1.2 ml), (*S*)-diphenylprolinol silyl ether (2 mol%, 0.4 ml stock solution in toluene, 0.05 M), butyraldehyde (2.0 mmol, neat), and 3-nitrobenzoic acid (33.4 mg, 0.2 mmol). The mixture was stirred in an ice bath for 5 min and then nitroethylene (1 mmol, 0.2 ml stock solution in toluene) was added. The mixture was stirred at 3 °C and the reaction progress was monitored by ^{1}H-nuclear magnetic resonance analysis of the crude reaction mixture. After the reaction was complete, excess $NaBH_4$ (3.4 mmol, 128.5 mg) was added, followed by MeOH (10 ml), and the mixture was stirred for a few minutes. The mixture was then slowly poured into a 100-ml beaker containing aqueous NH_4Cl (15 ml, 1 M) at 0 °C and the resulting mixture was extracted with EtOAc (3 × 10 ml). The EtOAc layers were collected, washed with brine (20 ml), dried over $MgSO_4$, and filtered. The filtrate was concentrated to give the crude alcohol product, which was purified via silica gel column chromatography (EtOAc/hexane) to yield (*R*)-2-ethyl-4-nitrobutan-1-ol (96%, 99% e.e.).

(*R*)-2-Ethyl-4-nitrobutanoic Acid

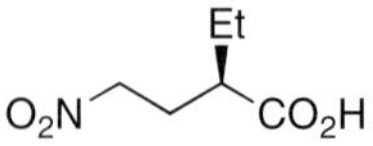

To the alcohol (1.0 mmol) dissolved in acetone (10 ml) at 0 °C was added $H_2Cr_2O_7$ (3 ml, 1.5 mmol). The mixture was stirred for 5 h, during which time the mixture warmed to room temperature. Excess isopropanol was added and the mixture was stirred for 10 min. The mixture was filtered, and the solution was diluted with aqueous HCl (2 ml, 2 N) and extracted with Et_2O. Complete extraction of the product into the Et_2O phase was monitored by thin-layer chromatography (TLC). The organic layers were washed with saturated NaCl solution, dried over $MgSO_4$, filtered, and concentrated to give a viscous oil, from which the desired nitrobutanoic acid was purified via column chromatography eluting with EtOAc/hexane (1 : 10 to 1 : 3; v/v) to give pure product (92%, >95% e.e.).

(*R*)-4-(*tert*-Butoxycarbonylamino)-2-ethylbutanoic Acid

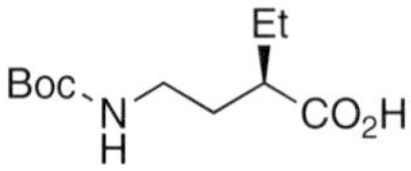

A mixture of (*R*)-2-ethyl-4-nitrobutanoic acid (0.36 g, 2.2 mmol), ammonium formate (0.70 g, 11 mmol), and 10% Pd/C (dry, 0.21 g) in anhydrous MeOH (10 ml) was refluxed overnight under an atmosphere of N_2 until the starting material disappeared as indicated by TLC analysis. The mixture was cooled to room temperature and filtered through a pad of Celite. The filtrate was collected and concentrated to give the crude amine, which was dissolved in CH_2Cl_2 (10 ml) containing *N,N*-diisopropylethylamine (0.6 ml, 3.3 mmol) and Boc_2O (0.7 g, 3.3 mmol) was added. The mixture was stirred at room temperature for 2 h and then it was concentrated to give the crude product, which was purified by silica gel chromatography EtOAc/hexane to yield pure (*R*)-4-(*tert*-butoxycarbonylamino)-2-ethylbutanoic acid (71%).

14.12.2
N-*tert*-Butyl-N-(*p*-chlorophenylethyl) α-Diazoacetamide (Scheme 14.25 [79])

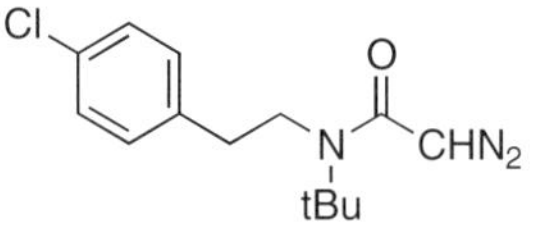

To *N-tert*-butyl-*N*-(*p*-chlorophenylethyl)amine (5.29 g, 25 mmol) in THF (30 ml) was added diketene (2 ml, 25 mmol). The mixture was stirred at 0 °C for 0.5 h, then allowed to warm to room temperature and stirred overnight. To the resulting reaction mixture was added 4-acetamidobenzene sulfonyl azide (7.3 g, 29 mmol), followed by the addition of 1,8-diazabicyclo[5.4.0]undec-7-ene (4.4 ml, 29 mmol). The resulting solution was stirred at room temperature for 8 h. An aqueous solution of lithium hydroxide (50 ml, 95 mmol) was added and the resulting orange–brown mixture was stirred vigorously for 8 h. The reaction mixture was diluted with ethyl acetate (60 ml), and the organic layer was washed with water (2 × 30 ml) and dried over anhydrous $MgSO_4$. Solvent was removed under reduced pressure to yield a red–brown mixture. Silica gel column chromatography petroleum ether/ethyl acetate (5 : 1 v/v), afforded the product as a yellow solid in 84% yield.

N-*tert*-Butyl-β-*p*-chlorophenyl-γ-lactam

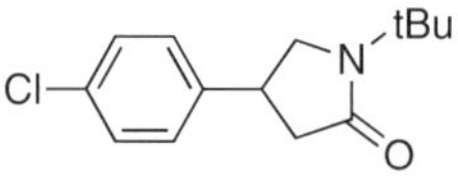

To a solution of $Rh_2(cap)_4$ (1 mol%) in the CH_2Cl_2 (5 ml) at reflux was added the α-diazoacetamide (1.0 mmol) in the CH_2Cl_2 (5 ml) via a syringe pump over a 2 h period. After the addition was complete, the reaction mixture was stirred for an additional 30 min and then the solvent was removed under reduced pressure. Silica gel column chromatography purification (petroleum ether/ethyl acetate, 3 : 1 v/v) yielded the γ-lactam as a pale yellow (or colorless) oil (77% yield).

(±)-Baclofen Hydrochloride

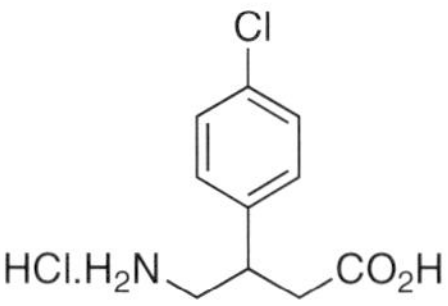

A mixture of the γ-lactam (0.85 mmol) in aqueous HCl (6 ml, 25–28%) was heated at 90–120 °C for 18 h. After cooling to room temperature, the solution was extracted with diethyl ether (3 × 20 ml). The water was removed under reduced pressure to give the desired product (as the corresponding hydrochloride salt) in 95% yield.

14.12.3
(*R*)-5-[(1-Oxo-2-(*tert*-butoxycarbonylamino)-3-phenyl)-propyl]-2,2-dimethyl-1,3-dioxane-4,6-dione (Scheme 14.34 [113])

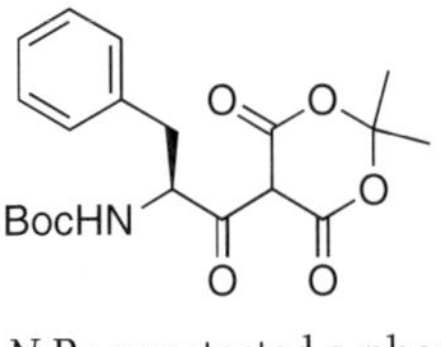

BocHN O O O O O

N-Boc-protected L-phenylalanine (5.3 g, 20 mmol) was dissolved with 2,2-dimethyl-1,3-dioxane-4,6-dione (3.02 g, 22 mmol) and 4-(dimethylamino)pyridine (3.85 g, 31 mmol) in CH_2Cl_2 (100 ml). The reaction mixture was cooled to −5 °C and a solution of dicyclohexylcarbodiimide (4.74 g, 22 mmol) in CH_2Cl_2 (50 ml) was added dropwise over 1 h. The mixture was left at below 0 °C overnight, during which time dicyclohexylurea precipitated. After filtration the reaction mixture was washed 4 times with aqueous 5% $KHSO_4$ and once with brine, and dried in the refrigerator with $MgSO_4$ for 5 h. This solution was used in the second step without further purification.

(*R*)-5-[(2-*tert*-Butoxycarbonylamino-3-phenyl)-propyl]-2,2-dimethyl-1,3-dioxane-4,6-dione

BocHN O O O O

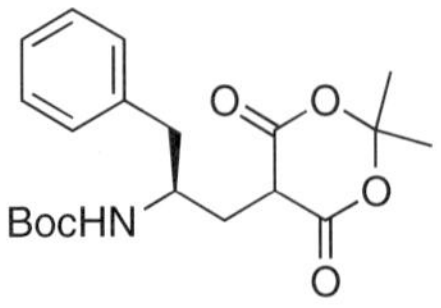

The solution of (*R*)-5-[(1-oxo-2-(*tert*-butoxycarbonylamino)-3-phenyl)-propyl]-2,2-dimethyl-1,3-dioxane-4,6-dione from the previous reaction was cooled to −5 °C and 98% AcOH (13.3 ml, 220 mmol) was added. Then, $NaBH_4$ (1.85 g, 50 mmol) was added in small portions while stirring over 1 h. The reaction mixture was left in the refrigerator overnight, and then washed 3 times with brine and 2 times with water. The organic phase was dried with $MgSO_4$, filtered, evaporated to dryness, and purified by column chromatography with hexane/ethyl acetate (1 : 1) to afford the desired product (76% yield).

(*R*)-(*tert*-Butoxycarbonyl)-5-benzyl-2-pyrrolidinone

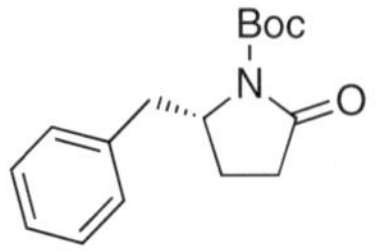

Boc N O

Pure (*R*)-5-[(2-*tert*-butoxycarbonylamino-3-phenyl)-propyl]-2,2-dimethyl-1,3-dioxane-4,6-dione was refluxed in toluene (50 ml). TLC in ethyl acetate/hexane 2 : 1 indicated complete conversion after 3 h. After evaporation of solvent, the compound was isolated in 96% yield and used without further purification.

(*R*)-4-(*tert*-Butoxycarbonylamino)-5-phenyl-pentanoic Acid

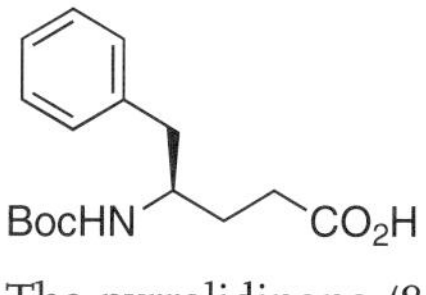

The pyrrolidinone (8 mmol) was dissolved in acetone (15 ml) and aqueous NaOH (1 M, 24 ml) was added. The reaction mixture was stirred at 22 °C for 30 min. Acetone was removed under reduced pressure and the reaction mixture was acidified to pH 2 (6 M HCl). The desired product precipitated and was filtered, washed twice with water, and recrystallized from a mixture of ethyl acetate/hexane to afford (*R*)-4-(*N*-*tert*-butoxycarbonylamino)-5-phenyl-pentanoic acid in 72% yield.

14.12.4
(*R*)- and (*S*)-[2-(Benzyloxy) ethyl]oxirane (Scheme 14.69 [184])

A stirred solution of (±)-[2-(benzyloxy)ethyl]oxirane (8.0 g, 44.9 mmol) and (*R*,*R*)-(salen)Co(III)-OAc catalyst (145 mg, 0.22 mmol) was cooled to 0 °C, then H_2O (0.45 ml, 25.0 mmol) was added dropwise over a period of 45 min. The reaction mixture was stirred at room temperature for 5 h, diluted with EtOAc, dried over anhydrous Na_2SO_4, and concentrated at reduced pressure. The residue was purified by silica gel chromatography (EtOAc/hexane, 1 : 9). The first fraction contained (*R*)-(benzyloxy)ethyl]oxirane and the second fraction afforded (*S*)-4-(benzyloxy)butane-1,2-diol (3.8 g, 43%). The diol (3.5 g, 17.8 mmol), Ph_3P (7.0 g, 26.7 mmol), and diisopropylazadicarboxylate (5.16 ml, 26.8 mmol) in benzene (50 ml) were heated under reflux for 20 h. The solvent was removed under reduced pressure and Et_2O (80 ml) added precipitating the Ph_3PO which was removed by filtration. The filtrate was concentrated and the resulting residue was purified by column silica gel chromatography (EtOAc/hexane, 1 : 9) to afford (*S*)-(benzyloxy)ethyl]oxirane (2.55 g, 80%).

(3*R*)-GABOB

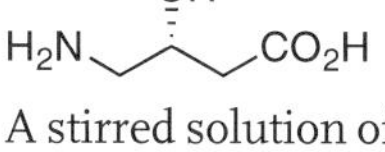

A stirred solution of (*R*)-(benzyloxy)ethyl]oxirane (1.5 g, 8.4 mmol) in ethanol (10 ml) was hydrogenated over 5% Pd/C (150 mg) under an atmosphere of H_2 (1 atm). After 5 h the reaction mixture was filtered through Celite and the filtrate was concentrated to afford (*S*)-2-(2-(hydroxy)ethyl)oxirane (630 mg, 85%), which was used without further purification. The alcohol (0.5 g, 5.7 mmol) was added slowly to a vigorously stirred solution of sodium periodate (3.33 g 15.5 mmol) in $CH_3CN/H_2O/CCl_4$ (1 : 1 : 1, 10 ml) and $RuCl_3{\cdot}5H_2O$ (8.0 mg). The reaction mixture was stirred for 1 h, filtered, and the heterogeneous mixture partitioned. The aqueous phase was extracted with THF (3 × 5 ml). The combined organic phases were treated with excess concentrated NH_4OH (2 ml) and the reaction mixture was warmed on a steam bath for 24 h. The solution was evaporated under reduced pressure to give a tan solid which was purified on Dowex 50W-X8 (H^+ form) to yield GABOB as a white crystalline solid (400 mg, 60%).

14.12.5
Dimethyl (*S*,*S*)-(−)-3-*N*,*N*-bis(α-Methylbenzyl)-amino-2-oxopropylphosphonate (Scheme 14.79 [210])

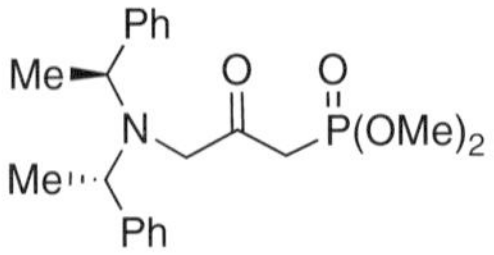

A solution of dimethyl methylphosphonate (1.43 g, 11.6 mmol) in anhydrous THF (20 ml) was cooled to −78 °C before the slow addition of *n*BuLi (0.56 g, 3.61 ml of 2.4 M solution in hexanes, 8.7 mmol). The resulting solution was stirred at −50 °C for 1.5 h, and then the solution cooled to −78 °C and slowly added to a solution of ethyl glycinate (0.9 g, 2.9 mmol) in anhydrous THF (20 ml). The reaction mixture was stirred at −78 °C for 4 h, quenched with aqueous NH_4Cl solution (10 ml), and extracted with EtOAc (3 × 20 ml). The combined organic extracts were washed with brine solution (2 × 10 ml), dried over anhydrous Na_2SO_4, filtered, and concentrated under reduced pressure. The crude product was purified by column chromatography (ethyl acetate/hexane 70 : 30) to afford 0.98 g, 87% yield of the desired product as a colorless oil.

Dimethyl (*S*,*S*)-(−)-3-*N*,*N*-bis(α-Methylbenzyl)-amino-2-hydroxypropylphosphonate

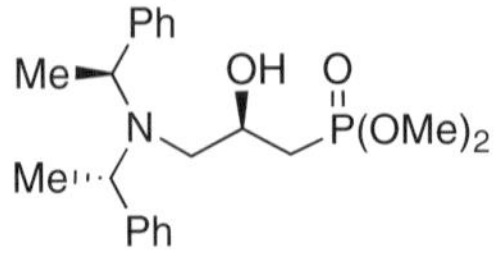

To a solution of the β-ketophosphonate (100 mg, 0.26 mmol) and $LiClO_4$ (69 mg, 0.65 mmol) in dry THF (20 ml) cooled at −78 °C was added 124 mg, 1.1 ml, 1.04 mmol of CCB 1 M in THF. The reaction mixture was stirred at −78 °C for 4 h, and at room temperature for 10 h. The reaction was quenched with aqueous

NH_4Cl solution. The solvent was evaporated in vacuum and the residue dissolved in water (10 ml) and extracted with ethyl acetate (3 × 20 ml). The combined organic extracts were dried over anhydrous Na_2SO_4, filtered, and concentrated under reduced pressure. The crude β-hydroxyphosphonates were purified by flash chromatography to afford dimethyl 3-[(*R*,*R*)-*N*,*N*-bis(α-methylbenzylamino)]-(2*S*)-hydroxypropylphosphonate as a white solid (88 mg, 87%).

(*R*)-3-Amino-2-hydroxypropylphosphonic Acid

Dimethyl 3-[(*S*,*S*)-*N*,*N*-bis(α-methylbenzylamino)]-(2*R*)-hydroxypropylphosphonate (228 mg, 0.58 mmol) in CH_2Cl_2 (5 ml) was treated under a nitrogen atmosphere with bromotrimethylsilane (178 mg, 0.15 ml, 1.16 mmol). The reaction mixture was stirred at room temperature for 6 h, and after this period of time the volatiles materials were removed under reduced pressure and water was added. After 4 h under stirring, the solvents were removed *in vacuo* to give 3-[(*S*,*S*)-*N*,*N*-bis(α-methylbenzylamino)]-(2*R*)-hydroxypropylphosphonic acid, which without isolation was treated with PdOH/C (22 mg, 20 wt%) in methanol (20 ml) and five drops of HCl/*i*PrOH (20%), and stirred for 48 h under hydrogen gas at 60 °C and 60 psi. The mixture was filtered through a pad of Celite and the solvent removed under reduced pressure. The residue was treated with propylene oxide (3 ml) to afford (*R*)-3-amino-2-hydroxypropylphosphonic acid as a white solid (63 mg, 70%).

14.12.6
Boc-L-leucinal (Scheme 14.83 [214])

To a stirred solution of Boc-L-leucine methyl ester (4.0 g, 16.3 mmol) in dry toluene (70 ml) was added a hexane solution of DIBAH (40.8 mmol) at −78 °C under a nitrogen atmosphere. After 6 min, the reaction was quenched with methanol (4 ml) and Rochelle salt solution was added immediately. The mixture was allowed to warm to 25 °C and ether (100 ml) was added. The etheral layer was separated and combined with ether extracts of the aqueous layer. The combined layers were dried ($MgSO_4$) and concentrated under reduced pressure. The crude oil was passed through a short pad of silica gel, eluting with 4% ethyl acetate in benzene to remove the alcohol side-product to yield Boc-leucinal (2.98 g, 85%), which was used without further purification.

(3*RS*,4*S*)-Ethyl *N*-Boc-4-amino-3-hydroxy-6-methylheptanoate

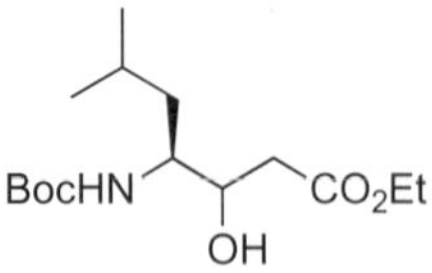

To dry THF (5 ml) cooled in dry ice/CCl_4 was added diisopropylamine (15 mmol) under a nitrogen atmosphere, followed by a solution of *n*BuLi in hexane (15 mmol). After 1 h, the bath temperature was lowered to −78 °C and dry EtOAc (15 mmol) was added via syringe and stirred for 15 min. Boc-leucinal (2.15 g, 10 mmol) in THF (10 ml) was added and the reaction mixture was stirred for 5 min before 1 N HCl was added. The flask was warmed to room temperature and the reaction mixture acidified with cold 1 N HCl to pH 2–3, and then extracted with EtOAc 3 times. The organic layer was washed with saturated NaCl and dried ($MgSO_4$). Evaporation under reduced pressure gave an oil which after silica gel column chromatography gave Boc-Sta-OEt (2.42 g, 80%) as a mixture of diastereomers. Chromatography of the mixture on silica gel eluting with a gradient of 10% ethyl acetate in benzene to 50% ethyl acetate in benzene afforded pure (3*S*,4*S*)-ethyl *N*-Boc-4-amino-3-hydroxy-6-methylheptanoate (38–40%) and (3*R*,4*S*)-ethyl *N*-Boc-4-amino-3-hydroxy-6-methylheptanoate.

(3*S*,4*S*)-*N*-Boc-4-amino-3-hydroxy-6-methylheptanoic Acid

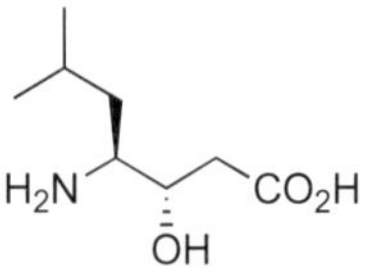

A solution of the ester (548 mg, 1.8 mmol) in aqueous dioxane was maintained at pH 10 for 30 min. The solution was acidified (pH 2.5) with cold HCl (1 N) and the aqueous layer was washed with EtOAc. The organic layer was dried over $MgSO_4$ and evaporated to give the desired acid (428 mg, 86%).

14.12.7
Cross-Coupling of *N-tert*-Butylsulfinyl Imine and *tert*-Butyl 3-oxopropanoate (Scheme 14.99 [275])

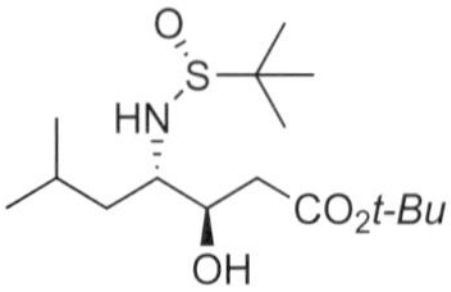

A freshly prepared solution of samarium diiodide (1.0 mmol) in THF (5 ml) was cooled to −78 °C under an atmosphere of nitrogen. A mixture of *tert*-butyl alcohol (1.0 mmol), *tert*-butyl 3-oxopropanoate (1.0 mmol), and chiral *N-tert*-butylsulfinyl imine (0.5 mmol) in THF (6 ml) was added dropwise. The reaction was monitored by TLC and quenched with saturated aqueous $Na_2S_2O_3$ (5 ml) when the reaction had gone to completion. Extraction with ethyl acetate and purification by flash column chromatography afforded the cross-coupling product in 58% yield.

(*R*,*S*)-Statine

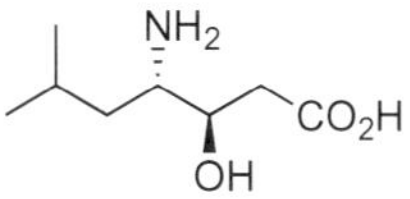

To the solution of HCl in dioxane (5.8 N, 0.5 ml) was added the β-amino alcohol (42 mg, 0.125 mmol) and stirred at room temperature for 5 h. The reaction solution was washed with ether; the separated aqueous layer was neutralized with ammonia. After concentration under reduced pressure, the resulting residue was purified by preparative TLC (silica gel) to give (*R*,*S*)-statine (18 mg, 80%).

14.12.8
(1′*S*,4*S*)-4-Bromomethyl-1-(1-phenyleth-1-yl)-pyrrolidin-2-one (Scheme 14.130 [343])

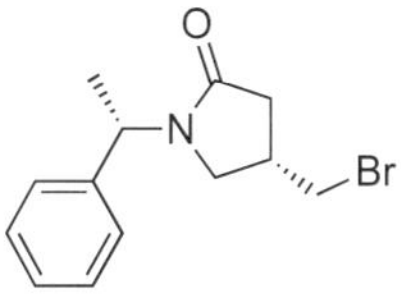

To a solution of (*S*)-*N*-allyl-2-bromo-*N*-(1-phenylethyl)-acetamide (0.27 g, 0.9 mmol) in dry toluene (50 ml) at −78 °C were added triethylborane (0.4 g, 2.8 mmol of a 1 M solution in hexane) and $BF_3{\cdot}OEt_2$ (0.5 g, 5.6 mmol). The resulting solution was stirred for 15 min before adding methanol (0.8 ml, 2.8 mmol). Then, an air balloon was placed on the reaction flask and the reaction was stirred for 6 h to ensure completion. Finally, the reaction mixture was concentrated under reduced pressure and the residue was purified by column chromatography (silica gel) eluting with hexane/EtOAc (45 : 1) giving the (1′*S*,4*S*) product as the major diastereoisomer (0.23 g, 86%) and the (1′*S*,4*R*) product as the minor diastereoisomer (0.04 g, 4%), both as colorless oils.

(1′*S*,1*R*,5*S*)-3-(1-Phenyleth-1-yl)-3-aza-bicyclo[3.1.0] hexane-2-one

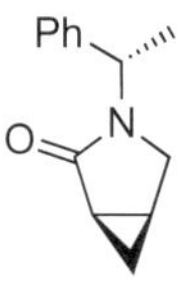

To a solution of the (1′*S*,4*S*) product (0.34 g, 1.2 mmol) in dry THF (30 ml) at 0 °C was added a solution of potassium *tert*-butoxide (2.4 mmol, 1 M solution in THF). The resulting solution was stirred for 1 h. The reaction mixture was quenched with water (30 ml) and extracted with EtOAc (3 × 30 ml). The organic layer was dried with $NaSO_4$, evaporated under reduced pressure, and the residue was purified by column chromatography (silica gel) eluting with hexane/EtOAc (8 : 1) to give the desired product as a colorless oil (0.24 g, 98%).

(1*R*,5*S*)-3-Aza-bicyclo[3.1.0] hexan-2-one

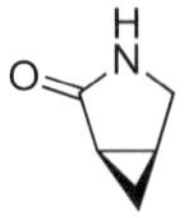

A solution of (1′*S*,1*R*,5*S*)-3-(1-phenyleth-1-yl)-3-aza-bicyclo[3.1.0]hexane-2-one (0.38 g, 1.90 mmol) in THF (30 ml) was added dropwise to a deep blue solution of Li (0.1 g) in liquid NH_3 (20 ml) at −78 °C. The reaction mixture was allowed to stir for 1 h at −78 °C before the addition of an aqueous solution of NH_4Cl (40 ml), then the reaction mixture was neutralized with dilute aqueous HCl, extracted with EtOAc, dried over Na_2SO_4, and evaporated under reduced pressure. The reaction was purified by column chromatography through silica gel eluting with hexane/EtOAc (1 : 1) to give (1*R*,5*S*)-3-aza-bicyclo[3.1.0]hexan-2-one as a white solid (0.2 g, 96%).

(1*R*,2*S*)-2-(Aminomethyl) cyclopropanecarboxylic Acid

A solution of (1*R*,5*S*)-3-aza-bicyclo[3.1.0]hexan-2-one (0.15 g, 2.1 mmol) in of HCl (20 ml, 1 M) was stirred at 70 °C for 8 h. The resulting solution was evaporated under reduced pressure and the residue was crystallized from EtOAc/methanol (1 : 5) to give (1*R*,2*S*)-2-(aminomethyl)cyclopropanecarboxylic acid as a white solid (0.12 g, 80%).

14.12.9
(+)-(*R*,*S*)-4-Amino-2-cyclopentene-1-carboxylic Acid (Scheme 14.150 [376])

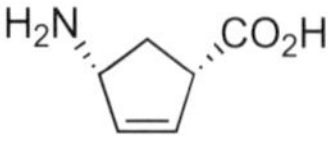

(±)-2-Azabicyclo[2.2.1] hept-5-en-3-one was dissolved to 10 g/l in a whole-cell suspension of ENZA 20. The cultures were incubated on orbital shakers (350 rpm) at 30 °C in conical flasks. Biotransformations were monitored by HPLC, following the disappearance of 2-azabicyclo[2.2.1]hept-5-en-3-one and appearance of 4-amino-2-cyclopentenecarboxylic acid. The cells were first removed by centrifugation then (−)-(*S*,*R*)-2-azabicyclo[2.2.1]hept-5-en-3-one continuously extracted from the biotransformation culture into CH_2Cl_2 for 48 h. The dichloromethane solutions were concentrated under reduced pressure, (−)-(*S*,*R*)-2-Azabicyclo[2.2.1]hept-5-en-3-one was recrystallized by addition of *n*-hexane. The aqueous solution after extraction of the residual lactam was concentrated to 150 g/l and diluted with an equal volume of acetone. (+)-(*R*,*S*)-4-Amino-2-cyclopentene-1-carboxylic acid crystallized.

References

1 Chebib, M. and Johnston, G.A.R. (2000) *Journal of Medicinal Chemistry*, **43**, 1427.
2 Whiting, P.J. (2003) *Current Opinion in Drug Discovery & Development*, **6**, 648.
3 Johnston, G.A.R., Chebib, M., Hanrahan, J.R., and Mewett, K.N. (2003) *Current Drug Targets*, **2**, 260.
4 Johnston, G.A.R., Hanrahan, J.R., Chebib, M., Duke, R.K., and Mewett, K.N. (2006) *Advances in Pharmacology*, **54**, 285.
5 Rudolph, U. and Möhler, H. (2006) *Current Opinion in Pharmacology*, **6**, 18.
6 Bowery, N.G. (2006) *Current Opinion in Pharmacology*, **6**, 37.
7 Clausen, R.P., Madsen, K., Larsson, O.M., Frolund, B., Krogsgaard-Larsen, P., and Schousboe, A. (2006) *Advances in Pharmacology*, **54**, 265.
8 Yogeeswari, P., Ragavendran, J.V., and Sriram, D. (2006) *Recent Patents on CNS Drug Discovery*, **1**, 113.
9 Lippert, B., Metcalf, B.W., Jung, M.J., and Casara, P. (1977) *European Journal of Biochemistry*, **74**, 441.
10 Bryans, J.S. and Wustrow, D.J. (1999) *Medicinal Research Reviews*, **19**, 149.
11 Sills, G.J. (2006) *Current Opinion in Pharmacology*, **6**, 108.
12 Allan, R.D. and Johnston, G.A.R. (1983) *Medicinal Research Reviews*, **3**, 91.
13 Bouché, N., Lacombe, B., and Fromm, H. (2003) *Trends Cell Biology*, **13**, 607.
14 Shelp, B.J., Bown, A.W., and Faure, D. (2006) *Plant Physiology*, **142**, 1350.
15 Borum, P. (1981) *Nutrition Reviews*, **39**, 385.
16 Rich, D.H. (1985) *Journal of Medicinal Chemistry*, **28**, 263.
17 Trabocchi, A., Guarna, F., and Guarna, A. (2005) *Current Organic Chemistry*, **9**, 1127.
18 Radics, G., Pires, R., Koksch, B., El-Kousy, S.M., and Burger, K. (2003) *Tetrahedron Letters*, **44**, 1059.
19 Erdö, S.L. and Wolff, J.R. (1990) *Journal of Neurochemistry*, **54**, 363.
20 Watanabe, M., Maemura, K., Oki, K., Shiraishi, N., Shibayama, Y., and Katsu, K. (2006) *Histology and Histopathology*, **21**, 1135.
21 Xiang, Y.-Y., Wang, S., Liu, M., Hirota, J.A., Li, J., Ju, W., Fan, Y., Kelly, M.M., Ye, B., Orser, B., O'Byrne, P.M., Inman, M.D., Yang, X., and Lu, W.-Y. (2007) *Nature Medicine*, **13**, 862.
22 Dong, H., Kumar, M., Zhang, Y., Gyulkhandanyan, A., Xiang, Y.Y., Ye, B., Perrella, J., Hyder, A., Zhang, N., Wheeler, M., Lu, W.Y., and Wang, Q. (2006) *Diabetologia*, **49**, 697.
23 Tian, J., Lu, Y., Zhang, H., Chau, C.H., Dang, H.N., and Kaufman, D.L. (2004) *Journal of Immunology*, **173**, 5298.
24 Alam, S., Laughton, D.L., Walding, A., and Wolstenholme, A.J. (2006) *Molecular Immunology*, **43**, 1432.
25 Reyes-Garcia, M.G., Hernández-Hernández, F., Hernández-Téllez, B., and Garcia-Tamayo, F. (2007) *Journal of Neuroimmunology*, **188**, 64.
26 Galeazzi, R., Martelli, G., Mobbili, G., Orena, M., and Rinaldi, S. (2003) *Tetrahedron: Asymmetry*, **14**, 3353.
27 Ordonez, M. and Cativiela, C. (2007) *Tetrahedron: Asymmetry*, **18**, 3.
28 Adams, R. and Fles, D. (1959) *Journal of the American Chemical Society*, **81**, 4946.
29 Duke, R.K., Chebib, M., Hibbs, D.E., Mewett, K.N., and Johnston, G.A.R. (2004) *Tetrahedron: Asymmetry*, **15**, 1745.
30 Camps, P., Munoz-Torrero, D., and Sanchez, L. (2004) *Tetrahedron: Asymmetry*, **15**, 311.
31 Evans, D.A., Gage, J.R., Leighton, J.L., and Kim, A.S. (1992) *The Journal of Organic Chemistry*, **57**, 1961.
32 Brenner, M. and Seebach, D. (1999) *Helvetica Chimica Acta*, **82**, 2365.
33 Azam, S., D'Souza, A.A. and Wyatt, P.B. (1996) *Journal of the Chemical Society, Perkin Transactions 1*, 621.

34 Martin, C.J., Rawson, D.J., and Williams, J.M.J. (1998) *Tetrahedron: Asymmetry*, **9**, 3723.

35 Royer, F., Felpin, F.-X., and Doris, E. (2001) *The Journal of Organic Chemistry*, **66**, 6487.

36 Ito, H., Omodera, K., Takigawa, Y., and Taguchi, T. (2002) *Organic Letters*, **4**, 1499.

37 Gao, M., Wang, D.-X., Zheng, Q.-Y., and Wang, M.-X. (2006) *The Journal of Organic Chemistry*, **71**, 9532.

38 Wiesner, M., Revell, J.D., Tonazzi, S., and Wennemers, H. (2008) *Journal of the American Chemical Society*, **130**, 5610.

39 Chi, Y., Guo, L., Kopf, N.A., and Gellman, S.H. (2008) *Journal of the American Chemical Society*, **130**, 5608.

40 Gil, S., Parra, M., and Rodríguez, P. (2008) *Molecules*, **13**, 716.

41 Andruszkiewicz, R., Barrett, A.G.M., and Silverman, R.B. (1990) *Synthetic Communications*, **20**, 159.

42 Serfass, L. and Casara, P.J. (1998) *Bioorganic & Medicinal Chemistry Letters*, **8**, 2599.

43 Simonelli, F., Clososki, G.C., dos Santos, A.A., de Oliveira, A.R.M., de, F., Marques, A., and Zarbin, P.H.G. (2001) *Tetrahedron Letters*, **42**, 7375.

44 Rodriguez, V., Sanchez, M., Quintero, L., and Sartillo-Piscil, F. (2004) *Tetrahedron*, **60**, 10809.

45 Rodriguez, V., Quintero, L., and Sartillo-Piscil, F. (2007) *Tetrahedron Letters*, **48**, 4305.

46 Froestl, W., Mickel, S.J., Hall, R.G., Von Sprecher, G., Strub, D., Baumann, P.A., Brugger, F., Gentsch, C., Jaekel, J., Olpe, H.R., Rihs, G., Vassout, A., Waldmeier, P.C., and Bittiger, H. (1995) *Journal of Medicinal Chemistry*, **38**, 3297.

47 Froestl, W., Mickel, S.J., Von Sprecher, G., Diel, P.J., Hall, B.J., Maier, L., Strub, D., Melillo, V., Baumann, P.A., Bernasconi, R., Gentsch, C., Hauser, K., Jaekel, J., Karlsson, G., Klebs, K., Maitre, L., Marescaux, C., Pozza, M.F., Schmutz, M., Steinmann, M.W., Van Riezen, H., Vassout, A., Monadori, C., Olpe, H.R., Waldmeier, P.C., and Bittiger, H. (1995) *Journal of Medicinal Chemistry*, **38**, 3313.

48 Hoekstra, M.S., Sobieray, D.M., Schwindt, M.A., Mulhern, T.A., Grote, T.M., Huckabee, B.K., Hendrickson, V.S., Franklin, L.C., Granger, E.J., and Karrick, G.L. (1997) *Organic Process Research & Development*, **1**, 26.

49 Basavaiah, D., Rao, A.J., and Satyanarayana, T. (2003) *Chemical Reviews*, **103**, 811.

50 Burk, M.J., De Koning, P.D., Grote, T.M., Hoekstra, M.S., Hoge, G., Jennings, R.A., Kissel, W.S., Le, T.V., Lennon, I.C., Mulhern, T.A., Ramsden, J.A., and Wade, R.A. (2003) *The Journal of Organic Chemistry*, **68**, 5731.

51 Sammis, G.M. and Jacobsen, E.N. (2003) *Journal of the American Chemical Society*, **125**, 4442.

52 Armstrong, A., Convine, N.J., and Popkin, M.E. (2006) *Synlett*, 1589.

53 Gotoh, H., Ishikawa, H., and Hayashi, Y. (2007) *Organic Letters*, **9**, 5307.

54 Hamersak, Z., Stipetic, I., and Avdagic, A. (2007) *Tetrahedron: Asymmetry*, **18**, 1481.

55 Pirrung, M.C., Dunlap, S.E., and Trinks, U.P. (1989) *Helvetica Chimica Acta*, **72**, 1301.

56 Ok, T., Jeon, A., Lee, J., Lim, J.H., Hong, C.S., and Lee, H.-S. (2007) *The Journal of Organic Chemistry*, **72**, 7390.

57 Gallos, J.K., Alexandraki, E.S., and Stathakis, C.I. (2007) *Heterocycles*, **71**, 1127.

58 Izquierdo, S., Aguilera, J., Buschmann, H.H., García, M., Torrens, A., and Ortuño, R.M. (2008) *Tetrahedron: Asymmetry*, **19**, 651.

59 Schelkun, R.M., Yuen, P.-W., Wustrow, D.J., Kinsora, J., Su, T.-Z., and Vartanian, M.G. (2006) *Bioorganic & Medicinal Chemistry Letters*, **16**, 2329.

60 Bryans, J.S., Davies, N., Gee, N.S., Dissanayake, V.U.K., Ratcliffe, G.S., Horwell, D.C., Kreen, C.O., Morrell, A.I.,

Oles, R.J., O'Toole, J.C., Perkins, G.M., Singh, L., Suman-Chauhan, N., and O'Neill, J.A. (1998) *Journal of Medicinal Chemistry*, **41**, 1838.

61 Bryans, J.S., Davies, N., Gee, N.S., Horwell, D.C., Kneen, C.O., Morrell, A.I., O'Neill, J.A., and Ratcliffe, G.S. (1997) *Bioorganic & Medicinal Chemistry Letters*, **7**, 2481.

62 Bryans, J.S., Horwell, D.C., Ratcliffe, G.S., Receveur, J.-M., and Rubin, J.R. (1999) *Bioorganic & Medicinal Chemistry*, **7**, 715.

63 Burgos-Lepley, C.E., Thompson, L.R., Kneen, C.O., Osborne, S.A., Bryans, J.S., Capiris, T., Suman-Chauhan, N., Dooley, D.J., Donovan, C.M., Field, M.J., Vartanian, M.G., Kinsora, J.J., Lotarski, S.M., El-Kattan, A., Walters, K., Cherukury, M., Taylor, C.P., Wustrow, D.J., and Schwarz, J.B. (2006) *Bioorganic & Medicinal Chemistry Letters*, **16**, 2333.

64 Receveur, J.-M., Bryans, J.S., Field, M.J., Singh, L., and Horwell, D.C. (1999) *Bioorganic & Medicinal Chemistry Letters*, **9**, 2329.

65 Bryans, J.S., Chessum, N.E.A., Huther, N., Parsons, A.F., and Ghelfi, F. (2003) *Tetrahedron*, **59**, 6221.

66 Cagnoli, R., Ghelfi, F., Pagnoni, U.M., Parsons, A.F., and Schenetti, L. (2003) *Tetrahedron*, **59**, 9951.

67 Allan, R.D., Bates, M.C., Drew, C.A., Duke, R.K., Hambley, T.W., Johnston, G.A.R., Mewett, K.N., and Spence, I. (1990) *Tetrahedron*, **46**, 2511.

68 Duke, R.K., Allan, R.D., Drew, C.A., Johnston, G.A.R., Mewett, K.N., Long, M.A., and Than, C. (1993) *Journal of Labelled Compounds & Radiopharmaceuticals*, **33**, 767.

69 Camps, P., Munoz-Torrero, D., and Sanchez, L. (2004) *Tetrahedron: Asymmetry*, **15**, 2039.

70 Caira, M.R., Clauss, R., Nassimbeni, L.R., Scott, J.L., and Wildervanck, A.F. (1997) *Journal of the Chemical Society, Perkin Transactions 2*, 763.

71 Zelle, R.E. (1991) *Synthesis*, 1023.

72 Chenevert, R. and Desjardins, M. (1991) *Tetrahedron Letters*, **32**, 4249.

73 Chenevert, R. and Desjardins, M. (1994) *Canadian Journal of Chemistry*, **72**, 2312.

74 Felluga, F., Gombac, V., Pitacco, G., and Valentin, E. (2005) *Tetrahedron: Asymmetry*, **16**, 1341.

75 Shishido, K. and Bando, T. (1998) *Journal of Molecular Catalysis B – Enzymatic*, **5**, 183.

76 Brenna, E., Caraccia, N., Fuganti, C., Fuganti, D., and Grasselli, P. (1997) *Tetrahedron: Asymmetry*, **8**, 3801.

77 Wang, M.-X. and Zhao, S.-M. (2002) *Tetrahedron Letters*, **43**, 6617.

78 Mazzini, C., Lebreton, J., Alphand, V., and Furstoss, R. (1997) *Tetrahedron Letters*, **38**, 1195.

79 Chen, Z., Chen, Z., Jiang, Y., and Hu, W. (2005) *Tetrahedron*, **61**, 1579.

80 Anada, M. and Hashimoto, S-i. (1998) *Tetrahedron Letters*, **39**, 79.

81 Wee, A.G.H., Duncan, S.C., and Fan, G.-J. (2006) *Tetrahedron: Asymmetry*, **17**, 297.

82 Maiorana, S., Licandro, E., Capella, L., Perdicchia, D., and Papagni, A. (1999) *Pure and Applied Chemistry*, **71**, 1453.

83 Mita, T., Sasaki, K., Kanai, M., and Shibasaki, M. (2005) *Journal of the American Chemical Society*, **127**, 514.

84 Herdeis, C. and Hubmann, H.P. (1992) *Tetrahedron: Asymmetry*, **3**, 1213.

85 Hayashi, M. and Ogasawara, K. (2003) *Heterocycles*, **59**, 785.

86 Langlois, N., Dahuron, N., and Wang, H.-S. (1996) *Tetrahedron*, **52**, 15117.

87 Baldoli, C., Maiorana, S., Licandro, E., Perdicchia, D., and Vandoni, B. (2000) *Tetrahedron: Asymmetry*, **11**, 2007.

88 Licandro, E., Maiorana, S., Baldoli, C., Capella, L., and Perdicchia, D. (2000) *Tetrahedron: Asymmetry*, **11**, 975.

89 Becht, J.-M., Meyer, O., and Helmchen, G. (2003) *Synthesis*, 2805.

90 Meyer, O., Becht, J.-M., and Helmchen, G. (2003) *Synlett*, 1539.

91 Okino, T., Hoashi, Y., Furukawa, T., Xu, X., and Takemoto, Y. (2005) *Journal of the American Chemical Society*, **127**, 119.

92 Thakur, V.V., Nikalje, M.D., and Sudalai, A. (2003) *Tetrahedron: Asymmetry*, **14**, 581.

93 Deng, J., Duan, Z.-C., Huang, J.-D., Hu, X.-P., Wang, D.-Y., Yu, S.-B., Xu, X.-F., and Zheng, Z. (2007) *Organic Letters*, **9**, 4825.

94 Resende, P., Almeida, W.P., and Coelho, F. (1999) *Tetrahedron: Asymmetry*, **10**, 2113.

95 Enders, D. and Niemeier, O. (2005) *Heterocycles*, **66**, 385.

96 Thakur, V.V., Paraskar, A.S., and Sudalai, A. (2007) *Indian Journal of Chemistry, Section B: Organic Chemistry Including Medicinal Chemistry*, **46**, 326.

97 Chiefari, J., Galanopoulos, S., Janowski, W.K., Kerr, D.I.B., and Prager, R.H. (1987) *Australian Journal of Chemistry*, **40**, 1511.

98 Wakita, Y., Kojima, M., Schwendner, S.W., McConnell, D., and Counsell, R.E. (1990) *Journal of Labelled Compounds & Radiopharmaceuticals*, **28**, 123.

99 Berthelot, P., Vaccher, C., Flouquet, N., Luyckx, M., Brunet, C., Boulanger, T., Frippiat, J.P., Vercauteren, D.P., Debaert, M., Evrard, G., and Durant, F. (1991) *European Journal of Medicinal Chemistry*, **26**, 395.

100 Garcia, A.L.L., Carpes, M.J.S., de Oca, A.C.B.M., dos Santos, M.A.G., Santana, C.C., and Correia, C.R.D. (2005) *The Journal of Organic Chemistry*, **70**, 1050.

101 Abbenante, G. and Prager, R.H. (1992) *Australian Journal of Chemistry*, **45**, 1791.

102 Desideri, N., Galli, A., Sestili, I., and Stein, M.L. (1992) *Archiv der Pharmazie*, **325**, 29.

103 Buchschacher, P., Cassal, J.-M., Fuerst, A., and Meier, W. (1977) *Helvetica Chimica Acta*, **60**, 2747.

104 Abbenante, G., Hughes, R., and Prager, R.H. (1997) *Australian Journal of Chemistry*, **50**, 523.

105 Prager, R.H. and Schafer, K. (1997) *Australian Journal of Chemistry*, **50**, 813.

106 Karla, R., Ebert, B., Thorkildsen, C., Herdeis, C., Johansen, T.N., Nielsen, B., and Krogsgaard-Larsen, P. (1999) *Journal of Medicinal Chemistry*, **42**, 2053.

107 Ansar, M., Ebrik, S., Mouhoub, R., Vaccher, C., Vaccher, M.P., Flouquet, N., Bakkas, S., Taoufik, J., Debaert, M., Caignard, D.H., Renard, P., and Berthelot, P. (1999) *Therapie*, **54**, 651.

108 Shuto, S., Shibuya, N., Yamada, S., Ohkura, T., Kimura, R., and Matsuda, A. (1999) *Chemical & Pharmaceutical Bulletin*, **47**, 1188.

109 Hanessian, S., Cantin, L.-D., Roy, S., Andreotti, D., and Gomtsyan, A. (1997) *Tetrahedron Letters*, **38**, 1103.

110 Hall, R.G. (1989) *Synthesis*, 442.

111 Howson, W., Hills, J.M., Blackburn, G.M., and Broekman, M. (1991) *Bioorganic & Medicinal Chemistry Letters*, **1**, 501.

112 Loukas, V., Noula, C., and Kokotos, G. (2003) *Journal of Peptide Science*, **9**, 312.

113 Smrcina, M., Majer, P., Majerova, E., Guerassina, T.A., and Eissenstat, M.A. (1997) *Tetrahedron*, **53**, 12867.

114 Hintermann, T., Gademann, K., Jaun, B., and Seebach, D. (1998) *Helvetica Chimica Acta*, **81**, 983.

115 Dexter, C.S. and Jackson, R.F.W. (1998) *Journal of the Chemical Society, Chemical Communications*, 75.

116 Dexter, C.S., Jackson, R.F.W., and Elliott, J. (1999) *The Journal of Organic Chemistry*, **64**, 7579.

117 Katritzky, A.R., Tao, H., Jiang, R., Suzuki, K., and Kirichenko, K. (2007) *The Journal of Organic Chemistry*, **72**, 407.

118 Ramachandran, P.V. and Biswas, D. (2007) *Organic Letters*, **9**, 3025.

119 Park, Y.S. and Beak, P. (1997) *The Journal of Organic Chemistry*, **62**, 1574.

120 Hua, D.H., Miao, S.W., Chen, J.S., and Iguchi, S. (1991) *The Journal of Organic Chemistry*, **56**, 4.

121 Palacios, F., Aparicio, D., Garcia, J., Rodriguez, E., and Fernandez-Acebes, A. (2001) *Tetrahedron*, **57**, 3131.

122 Corvo, M.C. and Pereira, M.M.A. (2007) *Amino Acids*, **32**, 243.

123 Riber, D. and Skrydstrup, T. (2003) *Organic Letters*, **5**, 229.

124 Johannesen, S.A., Albu, S., Hazell, R.G., and Skrydstrup, T. (2004) *Journal of the Chemical Society, Chemical Communications*, 1962.

125 Deleris, G., Dunogues, J., and Gadras, A. (1988) *Tetrahedron*, **44**, 4243.

126 Casara, P. (1994) *Tetrahedron Letters*, **35**, 3049.

127 Schuster, A.J. and Wagner, E.R. (1993) *Journal of Labelled Compounds & Radiopharmaceuticals*, **33**, 213.

128 Wei, Z.Y. and Knaus, E.E. (1993) *The Journal of Organic Chemistry*, **58**, 1586.

129 Wei, Z.-Y. and Knaus, E.E. (1994) *Tetrahedron*, **50**, 5569.

130 Trost, B.M. and Lemonine, R.C. (1996) *Tetrahedron Letters*, **37**, 9161.

131 Trost, B.M., Bunt, R.C., Lemoine, R.C., and Calkins, T.L. (2000) *Journal of the American Chemical Society*, **122**, 5968.

132 Gheorghe, A., Schulte, M., and Reiser, O. (2006) *The Journal of Organic Chemistry*, **71**, 2173.

133 Alcon, M., Poch, M., Moyano, A., Pericas, M.A., and Riera, A. (1997) *Tetrahedron: Asymmetry*, **8**, 2967.

134 Chandrasekhar, S. and Mohapatra, S. (1998) *Tetrahedron Letters*, **39**, 6415.

135 Dagoneau, C., Tomassini, A., Denis, J.-N., and Vallee, Y. (2001) *Synthesis*, 150.

136 Masson, G., Zeghida, W., Cividino, P., Py, S., and Vallee, Y. (2003) *Synlett*, 1527.

137 Wang, Z. and Silverman, R.B. (2004) *Journal of Enzyme Inhibition and Medicinal Chemistry*, **19**, 293.

138 Buchanan, R.L. and Pattison, F.L.M. (1965) *Canadian Journal of Chemistry*, **43**, 3466.

139 Deniau, G., Slawin, A.M.Z., Lebl, T., Chorki, F., Issberner, J.P., van Mourik, T., Heygate, J.M., Lambert, J.J., Etherington, L.-A., Sillar, K.T., and O'Hagan, D. (2007) *ChemBioChem*, **8**, 2265.

140 Buu, N.T. and Van Gelder, N.M. (1974) *British Journal of Pharmacology*, **52**, 401.

141 Silverman, R.B. and Levy, M.A. (1981) *The Journal of Biological Chemistry*, **256**, 11565.

142 Cologne, J. and Pouchol, J.M. (1962) *Bulletin de la Societe Chimique de France*, 598.

143 Belliotti, T.R., Capiris, T., Ekhato, I.V., Kinsora, J.J., Field, M.J., Heffner, T.G., Meltzer, L.T., Schwars, J.B., Taylor, C.P., Thorpe, A.J., Vartanian, M.G., Wise, L.D., Zhi-Su, T., Weber, M.L., and Wustrow, D.J. (2005) *Journal of Medicinal Chemistry*, **48**, 2294.

144 Hanessian, S. and Schaum, R. (1997) *Tetrahedron Letters*, **38**, 163.

145 Brenner, M. and Seebach, D. (2001) *Helvetica Chimica Acta*, **84**, 2155.

146 Hanessian, S., Luo, X., and Schaum, R. (1999) *Tetrahedron Letters*, **40**, 4925.

147 Koskinen, A.M.P. and Pihko, P.M. (1994) *Tetrahedron Letters*, **35**, 7417.

148 Pihko, P.M. and Koskinen, A.M.P. (1998) *The Journal of Organic Chemistry*, **63**, 92.

149 Friestad, G.K., Marie, J.-C., and Deveau, A.M. (2004) *Organic Letters*, **6**, 3249.

150 Takigawa, Y., Ito, H., Omodera, K., Koura, M., Kai, Y., Yoshida, E., and Taguchi, T. (2005) *Synthesis*, 2046.

151 Vicario, J.L., Badia, D., and Carrillo, L. (2001) *The Journal of Organic Chemistry*, **66**, 5801.

152 Aguirre, D., Cativiela, C., Diaz-de-Villegas, M.D., and Galvez, J.A. (2006) *Tetrahedron*, **62**, 8142.

153 Dryanska, V., Pashkuleva, I., and Angelov, V. (2003) *Journal of Chemical Research Synopses*, **2**, 89.

154 Seebach, D., Brenner, M., Rueping, M., Schweizer, B., and Jaun, B. (2001) *Journal of the Chemical Society, Chemical Communications*, 207.

155 Seebach, D., Brenner, M., Rueping, M., and Jaun, B. (2002) *Chemistry – A European Journal*, **8**, 573.

156 Yoneta, T., Shibahara, S., Fukatsu, S., and Seki, S. (1978) *Bulletin of the Chemical Society of Japan*, **51**, 3296.

157 Brehm, L. and Krogsgaard-Larsen, P. (1979) *Acta Chemica Scandinavica – Series B*, **B33**, 52.
158 Iriuchijima, S. and Ogawa, M. (1982) *Synthesis*, 41.
159 Harris, K.J. and Sih, C.J. (1992) *Biocatalysis*, **5**, 195.
160 Bentley, J.M., Wadsworth, H.J., and Willis, C.L. (1995) *Journal of the Chemical Society*, 231.
161 Radics, G., Koksch, B., El-Kousy, S.M., Spengler, J., and Burger, K. (2003) *Synlett*, 1826.
162 Bongini, A., Cardillo, G., Orena, M., Porzi, G., and Sandri, S. (1987) *Tetrahedron*, **43**, 4377.
163 Voeffray, R., Perlberger, J.C., Tenud, L., and Gosteli, J. (1987) *Helvetica Chimica Acta*, **70**, 2058.
164 Jung, M.E. and Shaw, T.J. (1980) *Journal of the American Chemical Society*, **102**, 6304.
165 Rossiter, B.E. and Sharpless, K.B. (1984) *The Journal of Organic Chemistry*, **49**, 3707.
166 Lohray, B.B., Reddy, A.S., and Bhushan, V. (1996) *Tetrahedron: Asymmetry*, **7**, 2411.
167 Bols, M., Lundt, I., and Pedersen, C. (1992) *Tetrahedron*, **48**, 319.
168 Larcheveque, M. and Henrot, S. (1990) *Tetrahedron*, **46**, 4277.
169 Wang, G. and Hollingsworth, R.I. (1999) *Tetrahedron: Asymmetry*, **10**, 1895.
170 Tiecco, M., Testaferri, L., Temperini, A., Terlizzi, R., Bagnoli, L., Marini, F., and Santi, C. (2005) *Synthesis*, 579.
171 Rajashekhar, B. and Kaiser, E.T. (1985) *The Journal of Organic Chemistry*, **50**, 5480.
172 Misiti, D., Zappia, G., and Delle Monache, G. (1995) *Gazzetta Chimica Italiana*, **125**, 219.
173 Misiti, D. and Zappia, G. (1995) *Synthetic Communications*, **25**, 2285.
174 Bellamy, F.D., Bondoux, M., and Dodey, P. (1990) *Tetrahedron Letters*, **31**, 7323.
175 Takano, S., Yanase, M., Sekiguchi, Y., and Ogasawara, K. (1987) *Tetrahedron Letters*, **28**, 1783.
176 Kabat, M.M., Daniewski, A.R., and Burger, W. (1997) *Tetrahedron: Asymmetry*, **8**, 2663.
177 Kazi, A.B., Shidmand, S., and Hajdu, J. (1999) *The Journal of Organic Chemistry*, **64**, 9337.
178 Stucky, G. and Seebach, D. (1989) *Chemische Berichte*, **122**, 2365.
179 Renaud, P. and Seebach, D. (1986) *Synthesis*, 424.
180 Jain, R.P. and Williams, R.M. (2001) *Tetrahedron*, **57**, 6505.
181 Jain, R.P. and Williams, R.M. (2001) *Tetrahedron Letters*, **42**, 4437.
182 Kolb, H.C., Bennani, Y.L., and Sharpless, K.B. (1993) *Tetrahedron: Asymmetry*, **4**, 133.
183 Harding, M., Bodkin, J.A., Hutton, C.A., and McLeod, M.D. (2005) *Synlett*, 2829.
184 Bose, D.S., Fatima, L., and Rajender, S. (2006) *Synthesis*, 1863.
185 Song, C.E., Lee, J.K., Lee, S.H., and Lee, S-g. (1995) *Tetrahedron: Asymmetry*, **6**, 1063.
186 Song, C.E., Lee, J.K., Kim, I.O., and Choi, J.H. (1997) *Synthetic Communications*, **27**, 1009.
187 Aube, J., Wang, Y., Ghosh, S., and Langhans, K.L. (1991) *Synthetic Communications*, **21**, 693.
188 Kitamura, M., Ohkuma, T., Takaya, H., and Noyori, R. (1988) *Tetrahedron Letters*, **29**, 1555.
189 Takeda, H., Hosokawa, S., Aburatani, M., and Achiwa, K. (1991) *Synlett*, 193.
190 Sakuraba, S., Takahashi, H., Takeda, H., and Achiwa, K. (1995) *Chemical & Pharmaceutical Bulletin*, **43**, 738.
191 Marzi, M., Minetti, P., Moretti, G., Tinti, M.O., and De Angelis, F. (2000) *The Journal of Organic Chemistry*, **65**, 6766.
192 Mohr, P., Roesslein, L., and Tamm, C. (1987) *Helvetica Chimica Acta*, **70**, 142.
193 Bianchi, D., Cabri, W., Cesti, P., Francalanci, F., and Ricci, M. (1988) *The Journal of Organic Chemistry*, **53**, 104.
194 Wallace, J.S., Reda, K.B., Williams, M.E., and Morrow, C.J. (1990) *The Journal of Organic Chemistry*, **55**, 3544.
195 Kasai, N. and Sakaguchi, K. (1992) *Tetrahedron Letters*, **33**, 1211.

196 Lu, Y., Miet, C., Kunesch, N., and Poisson, J.E. (1993) *Tetrahedron: Asymmetry*, **4**, 893.

197 Kamal, A., Krishnaji, T., and Khan, M.N.A. (2007) *Journal of Molecular Catalysis B – Enzymatic*, **47**, 1.

198 Kamal, A., Khanna, G.B.R., and Krishnaji, T. (2007) *Helvetica Chimica Acta*, **90**, 1723.

199 Puertas, S., Rebolledo, F., and Gotor, V. (1996) *The Journal of Organic Chemistry*, **61**, 6024.

200 Gopalan, A.S. and Sih, C.J. (1984) *Tetrahedron Letters*, **25**, 5235.

201 Hashiguchi, S., Kawada, A., and Natsugari, H. (1992) *Synthesis*, 403.

202 Fuganti, C., Grasselli, P., Casati, P., and Carmeno, M. (1985) *Tetrahedron Letters*, **26**, 101.

203 Zhou, B.N., Gopalan, A.S., VanMiddlesworth, F., Shieh, W.R., and Sih, C.J. (1983) *Journal of the American Chemical Society*, **105**, 5925.

204 Braun, M. and Waldmueller, D. (1989) *Synthesis*, 856.

205 Du, Z., Chen, Z., Chen, Z., Yu, X., and Hu, W. (2004) *Chirality*, **16**, 516.

206 Ordonez, M., Gonzalez-Morales, A., Ruiz, C., De La Cruz-Cordero, R., and Fernandez-Zertuche, M. (2003) *Tetrahedron: Asymmetry*, **14**, 1775.

207 Tadeusiak, E., Krawiecka, B., and Michalski, J. (1999) *Tetrahedron Letters*, **40**, 1791.

208 Yuan, C-y., Wang, K., Li, J., and Li, Z-y. (2002) *Phosphorus, Sulfur, Silicon, and Related Elements*, **177**, 2391.

209 Wang, K., Zhang, Y., and Yuan, C. (2003) *Organic and Biomolecular Chemistry*, **1**, 3564.

210 Ordonez, M., Gonzalez-Morales, A., and Salazar-Fernandez, H. (2004) *Tetrahedron: Asymmetry*, **15**, 2719.

211 Nesterov, V.V. and Kolodyazhnyi, O.I. (2006) *Russian Journal of General Chemistry*, **76**, 1677.

212 Shaitanova, E.N., Gerus, I.I., Belik, M.Y., and Kukhar, V.P. (2007) *Tetrahedron: Asymmetry*, **18**, 192.

213 Wipf, P., Takada, T., and Rishel, M.J. (2004) *Organic Letters*, **6**, 4057.

214 Rich, D.H., Sun, E.T., and Boparai, A.S. (1978) *The Journal of Organic Chemistry*, **43**, 3624.

215 Rittle, K.E., Homnick, C.F., Ponticello, G.S., and Evans, B.E. (1982) *The Journal of Organic Chemistry*, **47**, 3016.

216 Chattopadhaya, S., Chan, E.W.S., and Yao, S.Q. (2005) *Tetrahedron Letters*, **46**, 4053.

217 Rich, D.H., Sun, E.T.O., and Ulm, E. (1980) *Jounal of Medicinal Chemistry*, **23**, 27.

218 Boger, J., Payne, L.S., Perlow, D.S., Lohr, N.S., Poe, M., Blaine, E.H., Ulm, E.H., Schorn, T.W., LaMont, B.I., Lin, T.-Y., Kawai, M., Rich, D.H., and Veber, D.F. (1985) *Journal of Medicinal Chemistry*, **28**, 1779.

219 Sham, H.L., Rempel, C.A., Stein, H., and Cohen, J. (1987) *Journal of the Chemical Society, Chemical Communications*, 683.

220 Lloyd-Williams, P., Monerris, P., Gonzalez, I., Jou, G., and Giralt, E. (1994) *Journal of the Chemical Society, Perkin Transactions 1*, 1969.

221 Rinehart, K.L., Sakai, R., Kishore, V., Sullins, D.W., and Li, K.M. (1992) *The Journal of Organic Chemistry*, **57**, 3007.

222 Shioiri, T., Hayashi, K., and Hamada, Y. (1993) *Tetrahedron*, **49**, 1913.

223 Dellaria, J.F. Jr. and Maki, R.G. (1986) *Tetrahedron Letters*, **27**, 2337.

224 Dellaria, J.F. Jr., Maki, R.G., Stein, H.H., Cohen, J., Whittern, D., Marsh, K., Hoffman, D.J., Plattner, J.J., and Perun, T.J. (1990) *Journal of Medicinal Chemistry*, **33**, 534.

225 Reetz, M.T., Drewes, M.W., and Schmitz, A. (1987) *Angewandte Chemie (International Edition in English)*, **26**, 1141.

226 Andres, J.M., Pedrosa, R., Perez, A., and Perez-Encabo, A. (2001) *Tetrahedron*, **57**, 8521.

227 Jayasinghe, L.R., Datta, A., Ali, S.M., Zygmunt, J., Vander Velde, D.G., and Georg, G.I. (1994) *Journal of Medicinal Chemistry*, **37**, 2981.

228 Takemoto, Y., Matsumoto, T., Ito, Y., and Terashima, S. (1990) *Tetrahedron Letters*, **31**, 217.

229 Mikami, K., Kaneko, M., Loh, T.-P., Terada, M., and Nakai, T. (1990) *Tetrahedron Letters*, **31**, 3909.

230 Kiyooka, S., Suzuki, K., Shirouchi, M., Kaneko, Y., and Tanimori, S. (1993) *Tetrahedron Letters*, **34**, 5729.

231 Harris, B.D., Bhat, K.L., and Joullie, M.M. (1987) *Tetrahedron Letters*, **28**, 2837.

232 Harris, B.D. and Joullie, M.M. (1988) *Tetrahedron*, **44**, 3489.

233 Liang, B., Portonovo, P., Vera, M.D., Xiao, D., and Joullie, M.M. (1999) *Organic Letters*, **1**, 1319.

234 Joullie, M.M., Portonovo, P., Liang, B., and Richard, D.J. (2000) *Tetrahedron Letters*, **41**, 9373.

235 Liang, B., Richard, D.J., Portonovo, P.S., and Joullie, M.M. (2001) *Journal of the American Chemical Society*, **123**, 4469.

236 Adrio, J., Cuevas, C., Manzanares, I., and Joullie, M.M. (2006) *Organic Letters*, **8**, 511.

237 Hamada, Y., Hayashi, K., and Shioiri, T. (1991) *Tetrahedron Letters*, **32**, 931.

238 Parkes, K.E.B., Bushnell, D.J., Crackett, P.H., Dunsdon, S.J., Freeman, A.C., Gunn, M.P., Hopkins, R.A., Lambert, R.W., Martin, J.A., Merret, J.H., Redshaw, S., Spurden, W.C., and Thomas, G.J. (1994) *The Journal of Organic Chemistry*, **59**, 3656.

239 Ohmori, K., Okuno, T., Nishiyama, S., and Yamamura, S. (1996) *Tetrahedron Letters*, **37**, 3467.

240 Okuno, T., Ohmori, K., Nishiyama, S., Yamamura, S., Nakamura, K., Houk, K.N., and Okamoto, K. (1996) *Tetrahedron*, **52**, 14723.

241 Wagner, B., Beugelmans, R., and Zhu, J. (1996) *Tetrahedron Letters*, **37**, 6557.

242 Wagner, B., Gonzalez, G.I., Dau, M.E.T.H., and Zhu, J. (1999) *Bioorganic & Medicinal Chemistry*, **7**, 737.

243 Schuda, P.F., Greenlee, W.J., Chakravarty, P.K., and Eskola, P. (1988) *The Journal of Organic Chemistry*, **53**, 873.

244 Maibaum, J. and Rich, D.H. (1988) *The Journal of Organic Chemistry*, **53**, 869.

245 Kessler, H. and Schudok, M. (1990) *Synthesis*, 457.

246 Hoffman, R.V. and Tao, J. (1997) *The Journal of Organic Chemistry*, **62**, 2292.

247 Sengupta, S. and Sarma, D.S. (1999) *Tetrahedron: Asymmetry*, **10**, 4633.

248 Woo, P.W.K. (1985) *Tetrahedron Letters*, **26**, 2973.

249 Gallagher, P.O., McErlean, C.S.P., Jacobs, M.F., Watters, D.J., and Kitching, W. (2002) *Tetrahedron Letters*, **43**, 531.

250 Pais, G.C.G. and Maier, M.E. (1999) *The Journal of Organic Chemistry*, **64**, 4551.

251 Hermann, C., Giammasi, C., Geyer, A., and Maier, M.E. (2001) *Tetrahedron*, **57**, 8999.

252 Hayashi, K., Hamada, Y., and Shioiri, T. (1991) *Tetrahedron Letters*, **32**, 7287.

253 Enders, D. and Reinhold, U. (1995) *Angewandte Chemie (International Edition in English)*, **34**, 1219.

254 Enders, D. and Reinhold, U. (1996) *Liebigs Annalen*, 11.

255 Okamoto, S., Fukuhara, K., and Sato, F. (2000) *Tetrahedron Letters*, **41**, 5561.

256 Gennari, C., Pain, G., and Moresca, D. (1995) *The Journal of Organic Chemistry*, **60**, 6248.

257 Gennari, C., Moresca, D., Vulpetti, A., and Pain, G. (1997) *Tetrahedron*, **53**, 5593.

258 Wuts, P.G.M. and Putt, S.R. (1989) *Synthesis*, 951.

259 Devant, R.M. and Radunz, H.-E. (1988) *Tetrahedron Letters*, **29**, 2307.

260 Cooke, J.W.B., Davies, S.G., and Naylor, A. (1993) *Tetrahedron*, **49**, 7955.

261 Yoo, D., Oh, J.S., and Kim, Y.G. (2002) *Organic Letters*, **4**, 1213.

262 Ohta, T., Shiokawa, S., Sakamoto, R., and Nozoe, S. (1990) *Tetrahedron Letters*, **31**, 7329.

263 He, B.-Y., Wu, T.-J., Yu, X.-Y., and Huang, P.-Q. (2003) *Tetrahedron: Asymmetry*, **14**, 2101.

264 Dai, C.-F., Cheng, F., Xu, H.-C., Ruan, Y.-P., and Huang, P.-Q. (2007) *Journal of Combinatorial Chemistry*, **9**, 386.

265 Bernardi, A., Micheli, F., Potenza, D., Scolastico, C., and Villa, R. (1990) *Tetrahedron Letters*, **31**, 4949.

266 Koot, W.J., Van Ginkel, R., Kranenburg, M., Hiemstra, H., Louwrier, S., Moolenaar, M.J., and Speckamp, W.N. (1991) *Tetrahedron Letters*, **32**, 401.

267 Kinoshita, M., Hagiwara, A., and Aburaki, S. (1975) *Bulletin of the Chemical Society of Japan*, **48**, 570.

268 Yanagisawa, H., Kanazaki, T., and Nishi, T. (1989) *Chemistry Letters*, 687.

269 Shinozaki, K., Mizuno, K., Oda, H., and Masaki, Y. (1992) *Chemistry Letters*, 2265.

270 Shinozaki, K., Mizuno, K., Oda, H., and Masaki, Y. (1996) *Bulletin of the Chemical Society of Japan*, **69**, 1737.

271 Mula, S. and Chattopadhyay, S. (2006) *Letters in Organic Chemistry*, **3**, 54.

272 Veith, U., Leurs, S., and Jaeger, V. (1996) *Journal of the Chemical Society, Chemical Communications*, 329.

273 Meunier, N., Veith, U., and Jaeger, V. (1996) *Journal of the Chemical Society, Chemical Communications*, 331.

274 Bravo, P., Corradi, E., Pesenti, C., Vergani, B., Viani, F., Volonterio, A., and Zanda, M. (1998) *Tetrahedron: Asymmetry*, **9**, 3731.

275 Zhong, Y.-W., Dong, Y.-Z., Fang, K., Izumi, K., Xu, M.-H., and Lin, G.-Q. (2005) *Journal of the American Chemical Society*, **127**, 11956.

276 Reddy, G.V., Rao, G.V., and Iyengar, D.S. (1999) *Tetrahedron Letters*, **40**, 775.

277 Reddy, G.V., Rao, G.V., and Iyengar, D.S. (1999) *Journal of the Chemical Society, Chemical Communications*, 317.

278 Fehrentz, J.-A., Bourdel, E., Califano, J.C., Chaloin, O., Devin, C., Garrouste, P., Lima-Leite, A.-C., Llinares, M., Rieunier, F., Vizavonna, J., Winternitz, F., Loffet, A., and Martinez, J. (1994) *Tetrahedron Letters*, **35**, 1557.

279 Bänziger, M., McGarrity, J.F., and Meul, T. (1993) *The Journal of Organic Chemistry*, **58**, 4010.

280 Lee, K.-Y., Kim, Y.-H., Park, M.-S., Oh, C.-Y., and Ham, W.-H. (1999) *The Journal of Organic Chemistry*, **64**, 9450.

281 Sakaitani, M. and Ofune, Y. (1987) *Tetrahedron Letters*, **28**, 3987.

282 Sakaitani, M. and Ohfune, Y. (1990) *Journal of the American Chemical Society*, **112**, 1150.

283 Kunieda, T., Ishizuka, T., Higuchi, T., and Hirobe, M. (1988) *The Journal of Organic Chemistry*, **53**, 3381.

284 Kano, S., Yuasa, Y., Yokomatsu, T., and Shibuya, S. (1988) *The Journal of Organic Chemistry*, **53**, 3865.

285 Yamamoto, T., Ishibuchi, S., Ishizuka, T., Haratake, M., and Kunieda, T. (1993) *The Journal of Organic Chemistry*, **58**, 1997.

286 Alemany, C., Bach, J., Farras, J., and Garcia, J. (1999) *Organic Letters*, **1**, 1831.

287 Bonini, B.F., Comes-Franchini, M., Fochi, M., Laboroi, F., Mazzanti, G., Ricci, A., and Varchi, G. (1999) *The Journal of Organic Chemistry*, **64**, 8008.

288 Alemany, C., Bach, J., Garcia, J., Lopez, M., and Rodriguez, A.B. (2000) *Tetrahedron*, **56**, 9305.

289 Ordonez, M., de la Cruz, R., Fernandez-Zertuche, M., and Munoz-Hernandez, M.-A. (2002) *Tetrahedron: Asymmetry*, **13**, 559.

290 de la Cruz-Cordero, R., Labastida-Galvan, V., Fernandez-Zertuche, M., and Ordonez, M. (2005) *Journal of the Mexican Chemical Society*, **49**, 312.

291 Ordonez, M., De la Cruz-Cordero, R., Fernandez-Zertuche, M., Munoz-Hernandez, M.A., and Garcia-Barradas, O. (2004) *Tetrahedron: Asymmetry*, **15**, 3035.

292 Yuste, F., Diaz, A., Ortiz, B., Sanchez-Obregon, R., Walls, F., and Ruano, J.L.G. (2003) *Tetrahedron: Asymmetry*, **14**, 549.

293 Kambourakis, S. and Rozzell, J.D. (2003) *Advanced Synthesis and Catalysis*, **345**, 699.

294 Kambourakis, S. and Rozzell, J.D. (2004) *Tetrahedron*, **60**, 663.

295 Nishi, T., Kitamura, M., Ohkuma, T., and Noyori, R. (1988) *Tetrahedron Letters*, **29**, 6327.

296 Doi, T., Kokubo, M., Yamamoto, K., and Takahashi, T. (1998) *The Journal of Organic Chemistry*, **1998**, 428.

297 Castejon, P., Pasto, M., Moyano, A., Pericas, M.A., and Riera, A. (1995) *Tetrahedron Letters*, **36**, 3019.

298 Castejon, P., Moyano, A., Pericas, M.A., and Riera, A. (1996) *Tetrahedron*, **52**, 7063.

299 Catasus, M., Moyano, A., Pericas, M.A., and Riera, A. (1999) *Tetrahedron Letters*, **40**, 9309.

300 Bertelli, L., Fiaschi, R., and Napolitano, E. (1993) *Gazzetta Chimica Italiana*, **123**, 521.

301 Maier, M.E. and Hermann, C. (2000) *Tetrahedron*, **56**, 557.

302 Ko, S.Y. (2002) *The Journal of Organic Chemistry*, **67**, 2689.

303 Bessodes, M., Saiah, M., and Antonakis, K. (1992) *The Journal of Organic Chemistry*, **57**, 4441.

304 Kondekar, N.B., Kandula, S.R.V., and Kumar, P. (2004) *Tetrahedron Letters*, **45**, 5477.

305 Williams, R.M., Colson, P.-J., and Zhai, W. (1994) *Tetrahedron Letters*, **35**, 9371.

306 Aoyagi, Y. and Williams, R.M. (1998) *Tetrahedron*, **54**, 10419.

307 Comber, R.N. and Brouillette, W.J. (1987) *The Journal of Organic Chemistry*, **52**, 2311.

308 Nebois, P. and Greene, A.E. (1996) *The Journal of Organic Chemistry*, **61**, 5210.

309 Kanazawa, A., Gillet, S., Delair, P., and Greene, A.E. (1998) *The Journal of Organic Chemistry*, **63**, 4660.

310 Zanatta, N., da Silva, F.M., da Rosa, L.S., Jank, L., Bonacorso, H.G., and Martins, M.A.P. (2007) *Tetrahedron Letters*, **48**, 6531.

311 Iwabuchi, Y., Sugihara, T., Esumi, T., and Hatakeyama, S. (2001) *Tetrahedron Letters*, **42**, 7867.

312 Travins, J.M., Bursavich, M.G., Veber, D.F., and Rich, D.H. (2001) *Organic Letters*, **3**, 2725.

313 DiPardo, R.M. and Bock, M.G. (1983) *Tetrahedron Letters*, **24**, 4805.

314 Boger, D.L., Colletti, S.L., Honda, T., and Menezes, R.F. (1994) *Journal of the American Chemical Society*, **116**, 5607.

315 Rivero, R.A. and Greenlee, W.J. (1991) *Tetrahedron Letters*, **32**, 2453.

316 Yoshioka, T., Hara, T., Takita, T., and Umezawa, H. (1974) *The Journal of Antibiotics*, **27**, 356.

317 Ohgi, T. and Hecht, S.M. (1981) *The Journal of Organic Chemistry*, **46**, 1232.

318 Narita, M., Otsuka, M., Kobayashi, S., Ohno, M., Umezawa, Y., Morishima, H., Saito, S., Takita, T., and Umezawa, H. (1982) *Tetrahedron Letters*, **23**, 525.

319 Kawai, M., Boparai, A.S., Bernatowicz, M.S., and Rich, D.H. (1983) *The Journal of Organic Chemistry*, **48**, 1876.

320 Flores-Morales, V., Fernandez-Zertuche, M., and Ordonez, M. (2003) *Tetrahedron: Asymmetry*, **14**, 2693.

321 Kazmaier, U. and Krebs, A. (1999) *Tetrahedron Letters*, **40**, 479.

322 Reetz, M.T., Griebenow, N., and Goddard, R. (1995) *Journal of the Chemical Society, Chemical Communications*, 1605.

323 Beart, P.M. and Johnston, G.A.R. (1972) *Australian Journal of Chemistry*, **25**, 1359.

324 Aitken, R.A. and Karodia, N. (1996) *Journal of the Chemical Society, Chemical Communications*, 2079.

325 Aitken, R.A., Karodia, N., Massil, T., and Young, R.J. (2002) *Journal of the Chemical Society, Perkin Transactions 1*, 533.

326 Johnston, G.A.R., Curtis, D.R., Beart, P.M., Game, C.J.A., McCulloch, R.M., and Twitchin, B. (1975) *Journal of Neurochemistry*, **24**, 157.

327 Allan, R.D. and Twitchin, B. (1978) *Australian Journal of Chemistry*, **31**, 2283.

328 Allan, R.D. (1979) *Australian Journal of Chemistry*, **32**, 2507.

329 Allan, R.D., Johnston, G.A.R., and Twitchin, B. (1980) *Australian Journal of Chemistry*, **33**, 1115.

330 Allan, R.D. and Tran, H. (1981) *Australian Journal of Chemistry*, **34**, 2641.

331 Scholz, D., Weber-Roth, S., Macoratti, E., and Francotte, E. (1999) *Synthetic Communications*, **29**, 1143.

332 Bang, J.K., Naka, H., Teruya, K., Aimoto, S., Konno, H., Nosaka, K., Tatsumi, T., and Akaji, K. (2005) *The Journal of Organic Chemistry*, **70**, 10596.

333 Kotkar, S.P., Chavan, V.B., and Sudalai, A. (2007) *Organic Letters*, **9**, 1001.

334 Reginato, G., Mordini, A., Capperucci, A., Degl'Innocenti, A. and Manganiello, S. (1998) *Tetrahedron*, **54**, 10217.

335 Trost, B.M. and Roth, G.J. (1999) *Organic Letters*, **1**, 67.

336 Yamaguchi, H., Nakanishi, S., Okamoto, K., and Takata, T. (1997) *Synlett*, 722.

337 Denis, J.-N., Tchertchian, S., Tomassini, A., and Vallee, Y. (1997) *Tetrahedron Letters*, **38**, 5503.

338 Ivanskii, V.I. and Maksimov, V.N. (1972) *Journal of Organic Chemistry of the USSR*, 8.

339 Allan, R.D., Curtis, D.R., Headley, P.M., Johnston, G.A.R., Lodge, D., and Twitchin, B. (1980) *Journal of Neurochemistry*, **34**, 652.

340 Duke, R.K., Allan, R.D., Chebib, M., Greenwood, J.R., and Johnston, G.A.R. (1998) *Tetrahedron: Asymmetry*, **9**, 2533.

341 Kennewell, P.D., Matharu, S.S., Taylor, J.B., Westwood, R., and Sammes, P.G. (1982) *Journal of the Chemical Society, Perkin Transactions 1*, 2563.

342 Baxendale, I.R., Ernst, M., Krahnert, W.-R., and Ley, S.V. (2002) *Synlett*, 1641.

343 Rodríguez-Soria, V., Quintero, L., and Sartillo-Piscil, F. (2008) *Tetrahedron*, **64**, 2750.

344 Morikawa, T., Sasaki, H., Hanai, R., Shibuya, A., and Taguchi, T. (1994) *The Journal of Organic Chemistry*, **59**, 97.

345 Mohapatra, D.K. (1999) *Synthetic Communications*, **29**, 4261.

346 Paulini, K. and Reissig, H.U. (1992) *Synlett*, 505.

347 Paulini, K. and Reissig, H.U. (1991) *Liebigs Annalen der Chemie*, 455.

348 Paulini, K. and Reissig, H.-U. (1995) *Journal für Praktische Chemie*, **337**, 55.

349 Galeazzi, R., Mobbili, G., and Orena, M. (1997) *Tetrahedron: Asymmetry*, **8**, 133.

350 Bonnaud, B., Cousse, H., Mouzin, G., Briley, M., Stenger, A., Fauran, F., and Couzinier, J.P. (1987) *Journal of Medicinal Chemistry*, **30**, 318.

351 Shuto, S., Takada, H., Mochizuki, D., Tsujita, R., Hase, Y., Ono, S., Shibuya, N., and Matsuda, A. (1995) *Journal of Medicinal Chemistry*, **38**, 2964.

352 Wipf, P., Werner, S., Woo, G.H.C., Stephenson, C.R.J., Walczak, M.A.A., Coleman, C.M., and Twining, L.A. (2005) *Tetrahedron*, **61**, 11488.

353 Wipf, P. and Stephenson, C.R.J. (2005) *Organic Letters*, **7**, 1137.

354 Hubner, J., Liebscher, J., and Patzel, M. (2002) *Tetrahedron*, **58**, 10485.

355 Kennewell, P.D., Matharu, S.S., Taylor, J.B., Westwood, R., and Sammes, P.G. (1982) *Journal of the Chemical Society, Perkin Transactions 1*, 2553.

356 Allan, R.D., Curtis, D.R., Headley, P.M., Johnston, G.A.R., Kennedy, S.M.E., Lodge, D., and Twitchin, B. (1980) *Neurochemical Research*, **5**, 393.

357 Galeazzi, R., Mobbili, G., and Orena, M. (1999) *Tetrahedron*, **55**, 261.

358 O'Donnell, J.P., Johnson, D.A., and Azzaro, A.J. (1980) *Journal of Medicinal Chemistry*, **23**, 1142.

359 Wheeler, J.W., Shroff, C.C., Stewart, W.S., and Uhm, S.J. (1971) *The Journal of Organic Chemistry*, **36**, 3356.

360 Baldwin, J.E., Adlington, R.M., Parisi, M.F., and Ting, H.H. (1986) *Tetrahedron*, **42**, 2575.

361 Burgess, K., Li, S., and Rebenspies, J. (1997) *Tetrahedron Letters*, **38**, 1681.

362 Rouge, P.D., Moglioni, A.G., Moltrasio, G.Y., and Ortuno, R.M. (2003) *Tetrahedron: Asymmetry*, **14**, 193.

363 Balo, C., Caamano, O., Fernandez, F., and Lopez, C. (2005) *Tetrahedron: Asymmetry*, **16**, 2593.

364 Moglioni, A.G., Brousse, B.N., Alvarez-Larena, A., Moltrasio, G.Y., and Ortuno, R.M. (2002) *Tetrahedron: Asymmetry*, **13**, 451.

365 Jagt, J.C. and Van Leusen, A.M. (1974) *The Journal of Organic Chemistry*, **39**, 564.

366 Daluge, S. and Vince, R. (1978) *The Journal of Organic Chemistry*, **43**, 2311.

367 Allan, R.D. and Twitchin, B. (1980) *Australian Journal of Chemistry*, **33**, 599.

368 Allan, R.D. and Fong, J. (1986) *Australian Journal of Chemistry*, **39**, 855.

369 Allan, R.D., Johnston, G.A.R., and Twitchin, B. (1979) *Australian Journal of Chemistry*, **32**, 2517.

370 Bergmeier, S.C., Cobas, A.A., and Rapoport, H. (1993) *The Journal of Organic Chemistry*, **58**, 2369.

371 Chenevert, R. and Martin, R. (1992) *Tetrahedron: Asymmetry*, **3**, 199.

372 Winkler, M., Knall, A.C., Kulterer, M.R., and Klempier, N. (2007) *The Journal of Organic Chemistry*, **72**, 7423.

373 Sicsic, S., Ikbal, M., and Le Goffic, F. (1987) *Tetrahedron Letters*, **28**, 1887.

374 Csuk, R. and Dörr, P. (1994) *Tetrahedron: Asymmetry*, **5**, 269.

375 Taylor, S.J.C., Sutherland, A.G., Lee, C., Wisdom, R., Thomas, S., Roberts, S.M., and Evans, C. (1990) *Journal of the Chemical Society, Chemical Communications*, 1120.

376 Taylor, S.J.C., McCague, R., Wisdom, R., Lee, C., Dickson, K., Ruecroft, G., O'Brien, F., Littlechild, J., Bevan, J., Roberts, S.M., and Evans, C.T. (1993) *Tetrahedron: Asymmetry*, **4**, 1117.

377 Evans, C., McCague, R., Roberts, S.M., and Sutherland, A.G. (1991) *Journal of the Chemical Society, Perkin Transactions 1*, 656.

378 Smith, M.E.B., Derrien, N., Lloyd, M.C., Taylor, S.J.C., Chaplin, D.A., and McCague, R. (2001) *Tetrahedron Letters*, **42**, 1347.

379 Smith, M.E.B., Lloyd, M.C., Derrien, N., Lloyd, R.C., Taylor, S.J.C., Chaplin, D.A., Casy, G., and McCague, R. (2001) *Tetrahedron: Asymmetry*, **12**, 703.

380 Rommel, M., Ernst, A., and Koert, U. (2007) *European Journal of Organic Chemistry*, 4408.

381 Rassu, G., Auzzas, L., Pinna, L., Zambrano, V., Zanardi, F., Battistini, L., Marzocchi, L., Acquotti, D., and Casiraghi, G. (2002) *The Journal of Organic Chemistry*, **67**, 5338.

382 Casiraghi, G., Rassu, G., Auzzas, L., Burreddu, P., Gaetani, E., Battistini, L., Zanardi, F., Curti, C., Nicastro, G., Belvisi, L., Motto, I., Castorina, M., Giannini, G., and Pisano, C. (2005) *Journal of Medicinal Chemistry*, **48**, 7675.

383 Palmer, C.F., Parry, K.P., Roberts, S.M., and Sik, V. (1992) *Journal of the Chemical Society, Perkin Transactions 1*, 1021.

384 Qiu, J. and Silverman, R.B. (2000) *Journal of Medicinal Chemistry*, **43**, 706.

385 Lu, H. and Silverman, R.B. (2006) *Journal of Medicinal Chemistry*, **49**, 7404.

386 Pan, Y., Calvert, K., and Silverman, R.B. (2004) *Bioorganic & Medicinal Chemistry*, **12**, 5719.

387 Pan, Y., Qiu, J., and Silverman, R.B. (2003) *Journal of Medicinal Chemistry*, **46**, 5292.

388 Fusi, S., Papandrea, G., and Ponticelli, F. (2006) *Tetrahedron Letters*, **47**, 1749.

389 Hanrahan, J.R., Mewett, K.N., Chebib, M., Matos, S., Eliopoulos, C.T., Crean, C., Kumar, R.J., Burden, P., and Johnston, G.A.R. (2006) *Organic and Biomolecular Chemistry*, **4**, 2642.

390 Kumar, R.J., Chebib, M., Hibbs, D.E., Kim, H.-L., Johnston, G.A.R., Salam, N.K., and Hanrahan, J.R. (2008) *Journal of Medicinal Chemistry*, **51**, 3825.

391 Allan, R.D., Johnston, G.A.R., and Twitchin, B. (1981) *Australian Journal of Chemistry*, **34**, 2231.

392 Choi, S. and Silverman, R.B. (2002) *Journal of Medicinal Chemistry*, **45**, 4531.

393 Amorin, M., Castedo, L., and Granja, J.R. (2005) *Chemistry – A European Journal*, **11**, 6543.

394 Wang, Z. and Silverman, R.B. (2006) *Bioorganic & Medicinal Chemistry*, **14**, 2242.

395 Brouillette, W.J., Saeed, A., Abuelyaman, A., Hutchison, T.L., Wolkowicz, P.E., and McMillin, J.B. (1994) *The Journal of Organic Chemistry*, **59**, 4297.

396 Xiao, D., Carroll, P.J., Mayer, S.C., Pfizenmayer, A.J., and Joullie, M.M. (1997) *Tetrahedron: Asymmetry*, **8**, 3043.

397 Xiao, D., Vera, M.D., Liang, B., and Joullie, M.M. (2001) *The Journal of Organic Chemistry*, **66**, 2734.

398 Kraus, G.A. and Goronga, T. (2007) *Synthesis*, 1765.

Index

Amino Acids, Peptides and Proteins in Organic Chemistry. Vol.1 – Origins and Synthesis of Amino Acids.
Edited by Andrew B. Hughes

ISBN: 978-3-527-32096-7

d

e